Digital VLSI Systems

OTHER IEEE PRESS BOOKS

Digital VLSI Systems

Edited by

Mohamed I. Elmasry
Professor of Electrical Engineering
University of Waterloo

A volume in the IEEE PRESS Selected Reprint
Series, prepared under the sponsorship of the
IEEE Solid-State Circuits Council.

IEEE PRESS

The Institute of Electrical and Electronics Engineers, Inc., New York

IEEE Order Number: PC01842

Library of Congress Cataloging in Publication Data
Main entry under title:

Digital VLSI systems.

(IEEE Press selected reprint series)
"Prepared under the sponsorship of the IEEE
Solid-State Circuits Council."
Bibliography: p.
Includes indexes.
1. Integrated circuits—Very large scale integration—
Addresses, essays, lectures. 2. Digital electronics—
Addresses, essays, lectures. I. Elmasry, Mohamed I.,
1943- . II. IEEE Solid-State Circuits Council.
TK7874.D538 1985   621.381'73   85-10724
ISBN 0-87942-190-8

This book is dedicated to those young men and women who are determined to become fluent in VLSI and who are contributing to the area of digital VLSI systems, now and for years to come.

ACKNOWLEDGMENT

The editor would like to thank members of the VLSI Group, members of the Departments of Electrical Engineering, Systems Design, and Computer Science, and members of the Institute for Computer Research at the University of Waterloo for their enthusiastic support and encouragement. The futuristic approach for designing digital VLSI systems taken by colleagues in industry and universities has provided the inspiration for this project. The administration at the University of Waterloo, the National Sciences and Engineering Research Council of Canada (NSERCC), and my students have provided an excellent research, scientific, and teaching atmosphere in which I was able to complete this project.

Contents

Preface

THE spectacular growth in the scale of integration, measured in device count, will continue in the years to come. The use of small size and three-dimensional devices will contribute to this growth, and by the year 1990 chips and/or wafers capable of accommodating a few million devices will be a reality. At that time, we will be designing digital systems into these chips in a way completely different from the way we were designing chips five years ago. This book represents an effort to bridge the gap and is offered at a crossroad in time: 1985.

VLSI design methodologies, design, and simulation tools are going through a substantial change to handle the complexity imposed by VLSI. In order to use VLSI economically, the design phase should not be too expensive and should not take too long. The key is to use structured hierarchical design methods with efficient design and simulation tools. Moreover, the fabrication phase should not be too expensive and should be reliable. The question of time is of little concern here since processing steps are controlled by physical and chemical mechanisms. Similar to the design phase, the testing phase should not be too expensive and should not take too long. The key is to incorporate a testing methodology which is an integral part of the design. If these requirements for design, fabrication, and testing phases are met simultaneously, we will witness a growth in the application of VLSI to our life like never before.

This book is a collection of some of the key research work related to two phases, namely design and testing. The book attempts to bring to the reader the area of digital VLSI systems of today. It is hoped that this will inspire him/her to contribute to this area, so we can all be ready for the few million device digital VLSI chips of the 1990's.

The book is divided into five parts. Part I is an overview of the design and the analysis of digital systems. It includes three tutorial papers covering digital VLSI systems, MOS, and bipolar digital circuits. In Part II, the different topics related to the design automation are covered. Part III examines simulation of digital VLSI systems at different levels of the design. In Part IV, the topics of testability and fault tolerance design are examined. Finally, in Part V design examples are given. The book ends with a bibliography which contains close to 800 references.

I hope this book will be of value to digital VLSI system designers of today and those of tomorrow.

M. I. ELMASRY
Waterloo, Ont., Canada
January 1985

Part I
Design and Analysis of Digital Systems

THIS part consists of three sections. The first section deals with an overview of the topic: design and analysis of digital VLSI systems. Three tutorial papers written by the editor review the design of digital systems, digital MOS circuits, and digital bipolar circuits. This material is followed by a paper by Séquin discussing how to manage VLSI complexity. Digital system design in general, and digital VLSI system design in particular, are the topics of a number of books [1]–[7] and the reader is referred to them for background and extra information. The reader can find more information on digital circuit design in [8]–[10]. The area of VLSI design is also covered in more recent books [11]–[15].

Section I-B covers design methodologies, including hierarchical and structure design and the knowledge-based approach to VLSI design. Section I-C gives an overview of design automation and simulation.

REFERENCES

[1] T. R. Blakeslee, *Digital Design with Standard MSI and LSI.* New York: Wiley Interscience, 1975.
[2] D. J. Hamilton and W. G. Howard, *Basic Integrated Circuit Engineering.* New York: McGraw-Hill, 1975.
[3] A. B. Glaser and G. E. Subak-Sharpe, *Integrated Circuit Engineering.* Reading, MA: Addison-Wesley, 1977.
[4] S. Muroga, *VLSI System Design.* New York: Wiley, 1982.
[5] J. D. Uliman, *Computational Aspects of VLSI.* Rockville, MD: Computer Science Press, 1984.
[6] C. Mead and L. Conway, *Introduction to VLSI Systems.* Reading, MA: Addison-Wesley, 1980.
[7] B. A. Bowen and W. R. Brown, *VLSI Systems Design for Digital Signal Processing*, vol. I. Englewood Cliffs, NJ: Prentice-Hall, 1982.
[8] D. A. Hodges and H. G. Jackson, *Analysis and Design of Digital Integrated Circuits.* New York: McGraw-Hill, 1983.
[9] M. I. Elmasry, Ed., *Digital MOS Integrated Circuits.* New York: IEEE PRESS, 1981.
[10] M. I. Elmasry, *Digital Bipolar Integrated Circuits.* New York: Wiley, 1983.
[11] P. Antonetti, D. O. Pederson, and H. DeMan, Eds., *Computer Design Aids for VLSI Circuits.* Rockville, MD: Sijthoff & Noordhoff, 1981.
[12] P. G. Jespers, C. H. Sequin, and F. Van De Wiele, *Design Methodologies for VLSI Circuits.* Rockville, MD: Sijthoff & Noordhoff, 1982.
[13] R. F. Ayres, *VLSI Silicon Compilation and the Art of Automatic Microchip Design.* Englewood Cliffs, NJ: Prentice-Hall, 1983.
[14] Weste and Eshraghian, *Principles of CMOS VLSI Design.* Reading, MA: Addison-Wesley, 1985.
[15] Glasser and Dobberpuhl, *Design and Analysis of VLSI Circuits.* Reading-MA: Addison-Wesley, 1985.

Digital VLSI Systems: A Tutorial

M. I. ELMASRY, SENIOR MEMBER, IEEE

Abstract—Basic approaches to the design of digital VLSI systems are presented. Design methodologies, simulation aids, and testing are discussed.

1. INTRODUCTION

THE purpose of this tutorial paper is to review the main methodologies used to design digital VLSI systems, to offer a brief treatment of circuit design approaches, and to discuss design and simulation aids used in the different phases of the design.

Since the invention of the transistor in the late forties, we have seen continuous progress in the area of electronics, specifically digital electronics, and today we are in the era of VLSI. Table I shows the progress made in the number of devices integrated on a single silicon chip every five years. The increase in the number of devices was, and still is, coupled to an increase in the complexity of the function implemented on the chip and an increase in performance. Today, the VLSI equation could read

$$\text{VLSI} \longleftrightarrow (\text{more functions} + \text{higher performance} + \text{shorter design time}). \tag{1}$$

Indeed, it is the last entry in this equation that separates VLSI from LSI, MSI, and SSI. Shorter design time is today a necessary condition for a successful implementation of a VLSI system, and its weighting factor in (1) would increase as the level of integration increases.

In turn, shorter design time (SDT) implies the following:

$$\text{SDT} \longleftrightarrow (\text{Hierarchical structured design} + \text{use of design and simulation aids}), \tag{2}$$

i.e., in order to achieve shorter design time, the design methodology has to be structured (top-down, bottom-up, and/or middle-out, more on this later), and the design has to be carried out in a number of well-defined hierarchical levels and at each level extensive use is made of design and simulation aids. The detail of the design methodology is a function of many technical and economical factors; for example, it is a function of the number of hierarchical levels and the sophistication of the design and the simulation aids. The design aid itself could be a high-level silicon compiler or a simple special-purpose circuit simulator.

2. DESIGNING A DIGITAL SYSTEM

Two basic factors affect the design of a digital system: the *function* of the system and the *technology* used to implement

Manuscript received December 21, 1984. This work was supported in part by grants from NSERC and the General Electric Company.

The author is with the Department of Electrical Engineering, University of Waterloo, Waterloo, Ont., Canada, N2L 3G1.

it. As the design progresses from the high-level specification of the system to the lower level of implementation, the technology starts to influence many of the design decisions. In the VLSI implementation of a system, it is preferable to make as many of the design levels technology-independent as possible. There is a tradeoff: if the design is made technology-independent in most of the design levels, the system performance will suffer since VLSI chip performance (chip area, speed, power dissipation) is a strong function of technology. Making full use of the technology features in the early stages of the design will reflect on the system performance. Alternatively, allowing the technology to influence the design in the early stages improves the system performance, but makes the design dependent on that particular technology and increases the time required to switch from one technology to another and the time required to adopt the system design to new improvements in the same technology.

Fig. 1 is a flow diagram which shows the different phases of designing a digital VLSI system. At the beginning, there is a need, which is technically and economically defined. This is followed by Phase 2, which includes a preliminary system specification defining key system requirements and factors related to optimization, criteria, and constraints related to performance. This is followed by detailed system specification which allows the designer to decide on which part of the system will be implemented in software and which in hardware. This followed by Phases 3 and 4, which deal with hardware and software development. Although VLSI technology could have an affect as early as Phase 1, it is Phase 3 where it starts to play a major role. Here the designer deals with the choice of the technology: e.g., NMOS, CMOS, or bipolar; the design methodology, e.g., gate arrays, standard cells, or custom design. Moreover, the designer in Phase 3 could mix between technologies and between methodologies to satisfy the three components of (1), i.e., to increase the number of functions integrated, to improve the performance, and to shorten the design time.

The hardware and the software development phases are combined to yield a VLSI system which meets the original system specifications. In the implementation phase (Phase 7) there are two results: mask information necessary for fabrication and testing information necessary to test the final product.

If we look closer at Phase 7 we find that it is closely coupled to the fabrication technology as shown in Fig. 2. The chip fabrication and processing information is used to define device and circuit parameters which, in turn, are used in the device and circuit design and layout. This is followed by device and circuit simulation which characterizes the basic devices and circuits. This is used to build the parameterized design tool associated with the methodology under consideration as indi-

This paper was specially prepared for this volume.

The Progress of IC State-of-the-Art. $\mu P \equiv$ Microprocessor, $\mu P^+ \equiv$ Enhanced Microprocessor, $\mu C \equiv$ Microcomputer, $\mu C^+ \equiv$ Enhanced Microcomputer, mC $\equiv$ Minicomputer, VLSI$^+ \equiv$ Higher Level of Integration Than VLSI, VLSI$^{++} \equiv$ Higher Level Than VLSI$^+$

YEAR	INTEGRATION LEVEL	NUMBER OF DEVICES	FUNCTION INTEGRATED	TECHNOLOGY	PRODUCTION DESIGN RULES	WAFER SIZE
1960	SSI	10's	FLIP—FLOP	BIPOLAR	10 μm	$1\frac{1}{2}$"
1965	MSI	100's	COUNTER	BI, MOS	8 μm	2"
1970	LSI	1,000's	μP	MOS, BI	6 μm	2"
1975	LSI$^+$	10,000's	μP$^+$	MOS, BI	5 μm	3"
1980	VLSI	100,000's	μC	MOS, BI	3 μm	$\frac{4}{5}$"
1985	VLSI$^+$	IM	μC$^+$	MOS, BI	1.5 μm	6"
1990	VLSI^{++}	10 M	mC	MOS, BI	1-0.5 μm	?

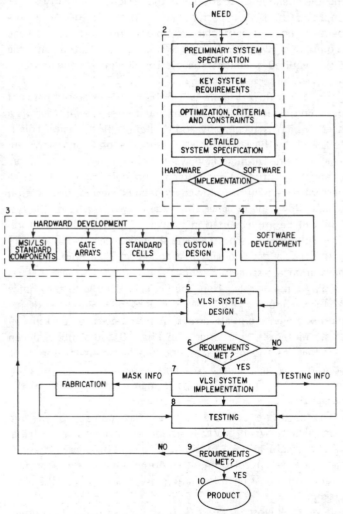

Fig. 1. Typical design cycle of digital systems.

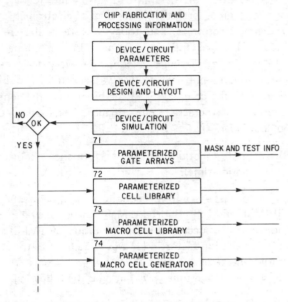

Fig. 2. A typical VLSI system implementation: the bottom-up approach.

practical to assume technology-independent models. This is followed by an architecture/structure design phase. At this point the design can be transferred to a silicon compiler 7A, a computer program which takes the architecture/structure design and compiles it into different lower levels of details until it delivers mask and test information of the silicon chip. The compiler does not need to be universal, i.e., for all possible architectures and structures. Indeed, 7A is a set of targetable compilers for different architectures and structures, e.g., one for pipelined digital signal processing, one for parallel data processing, etc.

Alternatively, 7d could be followed by a logic design phase where the designer has more control on the detailed implementation of the different functions, e.g., PLA's (programmable logic arrays), random logic, or structured logic arrays. After a simulation phase the design is handled by logic compilers which compile logic presentations, e.g., logic equations, tables, state diagrams, etc., into mask and test information data. However, the designer might like to carry out the circuit and the layout phase because of the critical delays involved.

cated by 71, 72, ..., etc. These methodologies are discussed in detail in Section III of this paper.

Another way of handling Phase 7 is given in Fig. 3. Here we start from the subsystem specification to obtain a behavioral description. This is a followed by a simulation of the behavior of the subsystem including timing information. Simple technology-dependent models are used in Phase 7c since it is im-

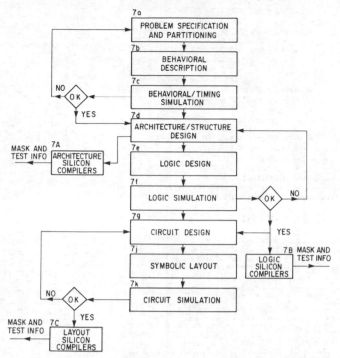

Fig. 3. A typical VLSI system implementation: the top-down approach.

In this case, 7f is followed by 7g to 7k and a compiler is used to compile from the symbolic layout 7j, where the layout is abstracted using simple symbols, to mask and test information.

It should be emphasized here that the designer can use the different approaches indicated—7A, 7B, 7C—in a single design where the system is partitioned into subsystems. Moreover, a lower mask level could be added (not shown in Fig. 3) for controlling the details of the layout. This level could be used, for example, in designing memory cells, analog circuits, sense amplifiers, and A/D and D/A circuits, where fine details of the layout affect performance. It should be noted that in Fig. 3 the simulation phase could be part of the compilers, e.g., 7f could be part of the logic silicon compiler.

3. DESIGN METHODOLOGY

The term design methodology has been associated with VLSI circuits where the design of these circuits has, in fact, become a design of an integrated *system*. A team of designers from various disciplines cooperate in the design of such a system and each, in addition to being fluent in VLSI, should know the different aspects of the design methodology which will be followed. Computer-aided design tools are used to help manage the large amount of data at the different levels of the design. The design itself as we indicated should be hierarchical in order to cope with the complexity of the design. This means that the design should be decomposed into several levels with clear interfaces defined between the levels above and below. During the design process, the designer has to cycle the design top-down and bottom-up to optimize performance. The *top-down* approach implies that one starts with the desired function and decomposes it into functional building blocks with more details at the lower level of description including geometric silicon floor planning. The *bottom-up* approach implies that one starts with given available technol-

ogy and design rules and builds basic cells and macrocells. Cycling between the top-down and the bottom-up approaches could be referred to as *middle-out*. At each level of the hierarchy a language is used to express the data at that level. A universal language used at all levels would of course make the designer's job a lot easier.

To implement a digital system into silicon the designer has different methodologies to follow. Among the widely used are
1) gate arrays (masterslice, uncommitted logic arrays, programmable arrays);
2) standard cells (polycells, macrocells);
3) handcrafted design;
4) hierarchical structured design (with or without the use of silicon compilers).

The first methodology is usually referred to as the *semicustom* approach, the second as the *custom* approach, and the last two as the *full custom* approaches. Most digital system designers at lower levels of integration (LSI or MSI) are using handcrafted design. Even today VLSI digital systems with high volume markets are in part designed using the handcrafted approach where the designer has full control over the details of the layout. However, as design tools become more powerful, the designer will be less dependent on the handcrafted approach.

Designing VLSI systems using gate arrays is similar to using PC boards and SSI/MSI logic. A floor plan of a gate array chip is as shown in Fig. 4(a). Identical cells of logic gates (e.g., two-input NAND gates), several transistor pairs, or transistor/resistor combinations are laid out in rows and back to back with routing channels in between to allow for interconnecting the different gates to implement the system under design. The array size could vary from 50 to 20 000 equivalent logic gates.

A closely related approach is using standard predesigned and pretested cells. The designer would place and route these cells in a typical floor plan as shown in Fig. 4(b). Usually the cells are of the same height and variable width.

In both approaches design aids, which include computer programs for logic partitioning, testability analysis, logic simulation, placement and routing, are used. In the gate array approach, few mask levels in the chip fabrication are customized, whereas in the standard cell approach all the mask levels are customized. This leads to the conclusion that gate arrays cost less and take less time to develop than standard cells, while standard cells are more flexible and yield denser chips. However, this conclusion is influenced by how sophisticated are the design tools used for each methodology. Moreover, each methodology is evolving; gate array could be more flexible by incorporating special macros (a bigger block than a simple cell, e.g., a sequencer, a RAM block, . . . , etc.) and the standard cell approach can give up some customization to reduce turnaround time.

A general feature of the gate array approach is that chip density has been traded for short design time. The lower chip density stems from the fact that the final chip contains unused cells because the designer cannot route to them and connect them to other cells. The higher the ratio of unused cells to those used, the lower the chip density. However, this situation can be rectified by using a channel-less layout, in which the basic cell accommodates the first level of wiring in its interior instead of in the routing channels. An increase in density of

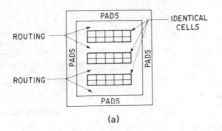
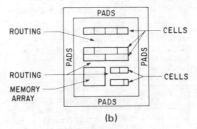

Fig. 4. Floor plan of gate arrays (a) and standard cell chips (b).

TABLE II
A Relative Comparison of VLSI Methodologies

	1. Gate Arrays	2. Standard Cells	3. Handcrafted	4. Silicon Compilers
1. Cost and time to develop methodology	Low / Medium	Medium	Low	High
2. Methodology	• Top-down chip design, starting from the logic level. • Cycling top-down/ bottom-up array design.	• Top-down chip design, starting from the logic level. • Bottom-up cell design.	• Cycling top-down/ bottom-up, starting from the system level.	• Top-down starting from system, logic or layout levels.
3. Chip density	Lowest	Medium	Highest	High
4. Chip performance	Lowest	Medium	Highest	High
5. Design time	Shortest	Medium	Long	Short
6. Chip development cost	Lowest	Medium	High	Low
7. Testability	Easy	Less Difficult	Difficult	Easy
8. Suitability for high volume production	Low	Medium	High	High

25 percent has been reported using this technique. Moreover, the array layout could be complemented by several RAM and PLA macros that reside in the software design system which supports that methodology.

The short turnaround time for gate arrays stems from the fact that gate array wafers are processed up to a given processing step, stockpiled, and then personalized to implement a given design. This personalization can be done in a number of ways: single layer of metal, single layer of metal and contacts, or double layer of metal and contacts.

Before customization, the design of gate array chips includes basic cell selection, transistor sizing, selection of the area and the width of the routing channels, the placement and nature of wiring, the directional control of the wiring (e.g., metal horizontal, polyvertical), and if macros are used, what type and what is the ratio of macros to the gate array.

The third methodology mentioned above is the handcrafted design. This is the oldest method of designing integrated circuits and today it is used in part or in whole to design and lay out high volume chips, e.g., standard memory and microprocessor/microcomputer chips.

The last methodology of hierarchical structured design (with or without the use of silicon compilers) is gaining widespread use, especially the use of silicon compilers at the system, logic, and layout levels. The silicon compiler approach to designing VLSI chips is based on writing a computer program to do the design in part. These compilers will take an input design specification at the system, logic, or layout levels and compile this specification into detailed design at a lower level *or* at the final mask/test information level. As in writing application programs, silicon compilers should be *targetable for a specific technology and a specific subset of applications*. This is similar to the situation where a computer program is written to

solve a subset of differential equations with specified boundary conditions. The method of solving these differential equations is a very important decision to the efficiency of running this program, accuracy of the solution and its convergence, and eventually to the success of this program. Similarly, in silicon compilers, the *designer* of the compiler determines all the compilation steps followed by the compiler to convert from one design level to a lower one. The choice of compiling levels, computer language, computer system, input and output interface, and where the user-compiler feedback is established in the design cycle are other important factors which will affect the time required to develop the compiler, its appeal to users, and its maintainability. In the near future, all VLSI digital system designs will be done at least in part by silicon compilers. This is because the use of silicon compilers in VLSI digital system design makes perfect sense both economically and technically.

Table II compares the above four methodologies based on eight different criteria. The three basic ones mentioned in (1) are given in items 3, 4, and 5. In all three, the silicon compiler approach is superior. The only drawback is the relatively high cost and time to develop the methodology. However, the payback is handsome. The main feature of each methodology from the structured/hierarchical viewpoint is mentioned in item 2 of the table. Here we see that the gate array methodology is based on top-down structured design as far as the chip designer (the user) is concerned. The design cycle usually starts at the logic level, although higher levels are also used. As for the array design itself, the design is cycled between top-down and bottom-up. A similar situation arises for the standard cell approach. As for the handcrafted approach, the design cycles between top-down and bottom-up starting at the system level. This leads to a long design time and the highest density and performance. The compiler methodology, as far as the user is concerned, is a top-down approach which starts at the system, logic, or layout levels according to the compiler under consideration. However, the compiler designer can take any approach: a top-down, a bottom-up, or cycling between the two, in the design of the compiler. The chosen methodology by the compiler designer is not transparent to the user.

4. SIMULATION, DESIGN VERIFICATION, AND MODELING

As shown in Figs. 2 and 3, simulation is used at the different phases of the design. The purpose of simulation is to obtain information about the expected performance of a given design before fabrication, to verify the design from the mask data of the chip, and to verify the production test.

For a given design, simulation is used at different hierarchical levels to obtain the expected performance. Alternatively, after the design is complete and mask data are available, a simulation can be used, again at different hierarchical levels, to predict the performance and verify the design. In this case, an extractor is used to extract the building blocks, their parameters, and their interconnections. In both simulation stages suitable models of the building blocks are used.

Simulation, and hence modeling, can be done at different levels:

1) function/behavior
2) structure/architecture/RTL
3) gate/logic
4) switch
5) timing
6) circuit
7) mixed-mode
8) device
9) process
10) mixed-mode.

Let us examine these simulation levels starting with circuit simulation (level 6). This is the heart of simulation levels; it is the most widely used, well known, and will continue to be of critical importance to the design of VLSI chips. Circuit simulation is an electrical simulation of the circuit blocks of a chip giving accurate waveform analysis of electrical signals, voltages, and currents. However, as the chip size increases, the time and the memory requirements of such simulation become prohibitive [1]. Improved simulation algorithms and simplified models can be used to improve on the time and the memory requirements. Moreover, if the circuit simulator is targeted to a specific technology (e.g., MOS) and a specific circuit operation (e.g., dynamic), substantial speed improvements can be achieved. Moreover, the use of macromodeling, where a model is generated for a large subcircuit from the model of a smaller one, can also improve the memory requirement.

The simulation speed can be improved if one is to use a timing simulator (level 5). Timing simulators are electrical simulators which are characterized by an improvement in simulation speed while maintaining an acceptable waveform accuracy. The modeling for this type of simulator is done with either delay equations for each block of circuit or by using simplified lookup tables for the nonlinear devices. Usually the delay equations are synthesized from conventional circuit simulation

of the circuit block under consideration. The delay equations can be expressed in terms of parameters which have a given statistical distribution. This requires an increase in computer power. However, the speed of simulation can be improved by considering only the activated circuits. In both circuit and timing simulators, the circuit elements are resistors, capacitors, and transistors.

If the speed of the timing simulator is insufficient for the analysis of the circuit blocks under consideration, one can use gate/logic level simulators (level 3). In these types of simulators, we deal with logic gates rather than resistors, capacitors, and transistors. The signals are discrete logic states rather than voltage and current waveforms. Each logic state consists of a level and strength pair. The number of the logic states determine the accuracy and the speed of the simulator. The modeling and the simulation are based on the evaluation of a new logic state using Boolean operations; hence the improvement in the simulation speed over electrical (circuit and timing) simulators.

The techniques used in timing and logic simulators can be combined in one simulator which is referred to as the switch level simulator (level 4). In this case, a good compromise can result between the logic simulators, where the simulation is fast but modeling and accuracy are a problem, and timing simulators, where modeling is more precise but the simulation, although accurate, is time consuming.

In VLSI design, the simulation levels 3–6 are widely used, while levels 1 and 2 have recently been introduced to VLSI design, although they have been used extensively by system designers. The purpose of the first level is to be able to simulate the performance of the system to make global design descriptions. For example, this level would aid the designer to maximize throughput, to choose what function should go on each chip, to develop an instruction set to optimize software/hardware tradeoffs, to develop algorithms to execute the instruction set, to specify the main blocks and buses for the chip, and to proceed to the next level of the design. The simulation and the modeling at this level involve the use of design languages, delay, and timing equations and should use technology-dependent parameters to enable the designer to evaluate the design at this early stage.

As the designer firms up the design and specifies logic blocks and architecture structures, the use of level 2 becomes necessary. Here the connectivity plays a more dominant role and leads to logic implementation. The simulation and the modeling at this level involve, as in level 1, the use of design (hardware description) languages, delay, and timing equations.

In many cases, it may be necessary to obtain the waveform accuracy of a circuit simulation for parts of a chip while a more efficient logic simulation is sufficient for other parts. In this case, mixed-mode analysis can be used. In a mixed-mode simulator, different levels of simulation are combined: circuit, timing, logic [2], and functional.

Other levels of simulation are related to the semiconductor device used in circuit design and to the process used to fabricate that device. These two levels differ in nature from each other and also differ from the other levels.

Device level simulation is the numerical simulation of the dc and transient characteristics of a given device via solution of partial differential equations which describe the electrical mechanisms governing the device behavior. The differential equations include carrier transport, carrier concentrations, and distribution of electric potentials. Device simulations provide information regarding the influence of the geometry on device performance, effect of scaling the size of the device, device characteristics, and limitations. Moreover, device simulation is used to formulate device models for circuit simulations. One-, two- and three-dimensional device simulations may be used depending on the required accuracy. Accurate simulation runs are computer-bound and more research is directed towards simulation techniques to improve computation time [3].

Process simulation involves the solution of the partial differential equations defining the fabrication processes, e.g., diffusion, ion implantation, oxidization, and epitaxial growth. Moreover, process simulation includes the simulation of the lithographic process, e.g., mask making and pattern etching. This becomes more critical as the dimensions of the device become smaller. Process simulation offers necessary information as to how the process affects the device and what are the limitations.

Finally, the last level of simulation is a mixed-mode simulator which simulates a given process via a process simulator and offers a device simulator the necessary data for calculating device characteristics, and hence generating necessary data for a device model incorporated in a circuit simulator. The process of generating a model for a higher level of simulation proceeds until the highest level is reached. Such a mixed-mode simulator, although not in existence today, would offer a very powerful tool to the design and simulation of VLSI chips.

At the beginning of this section we mentioned that one purpose of simulating the VLSI chip before fabrication is to verify the design directly from the mask data, hence predicting the chip performance. This design verification might prove to be expensive, since the designer has already simulated the chip performance throughout the design phase. In this case, the designer might like to verify the design in whole or in part through some checker programs, such as a design rule checker (DRC) or an electric rule checker (ERC). DRC and ERC are used to check if any of the design rules (minimum dimensions, spacings, etc.), or any of the electric design rules (the three terminals of a transistor are connected, etc.) are violated, respectively [1]. Their use becomes unnecessary if the chip design methodology relies on the correct-by-design principle, e.g., in a silicon compiler, where the designer is not allowed to commit a design error at any level of the design cycle.

5. TESTING, FAULT MODELING, AND DESIGNING FOR TESTABILITY

Testing can be divided into two phases. One phase is before the chips are fabricated to determine *how* a good chip is distinguished from a faulty one, and the other phase is after the chips are fabricated to determine *which* one is a good one. Both testing phases should be considered in the early design stages. The first phase is closely related to the design methodology and includes fault modeling, fault simulation, test generation, and designing the chip for ease of testing.

Fault modeling deals with modeling the behavior of a circuit (a single transistor, a logic gate, or a functional block) under a faulty condition (e.g., an open or a short connection). One typical example is the model of stuck-at-0 or stuck-at-1 for logic gates. Another is the modeling of a functional block which produces the wrong sequence of 1's and 0's.

Fault simulation offers information regarding how a given test will discover if a given fault has occurred and, if possible, its location. Fault simulation is one of three basic types [2]: parallel, deductive, and concurrent. In the parallel approach, all faulty chips and the good chips are simulated in parallel. Although this approach has the advantage of making use of the parallelism offered by today's computer hardware when performing the Boolean word operations required in fault simulation, it has the disadvantage of wasting computing time in simulating all the faulty chips. In deductive simulation, logic simulation is performed only on the good chip and selective trace procedures are used to determine which of the possible stuck-at faults on the inputs would propagate to the output. Concurrent fault simulation is similar to deductive fault simulation except that logic simulation of the faulty chips is performed.

Test generation deals with generating test patterns that provide an adequate covering of all possible faults, hence distinguishing between a good or a faulty chip. These patterns are used in testing the chip after fabrication. Generating test patterns for combinational circuits is easier compared to generating ones for sequential circuits. In order to ease the problem of generating test patterns, especially for sequential circuits, extra circuits are added to the original design for that purpose. For example, in the level sensitive scan (LSSD) and the scan path approaches, the problem of test generation for sequential circuits is reduced to test generation for combinational circuits. Adding extra circuitry for testing off course will increase chip area, hence increase fabrication cost and may reduce performance. However, a major reduction in testing cost results.

A useful information for the designer regarding testing is a "testability measure" which shows how difficult it may be to test a given circuit block [1]. This testability measure would need a calibration based on fabrication data and fault simulation data.

Another phase of testing is to deal with the chip after its fabrication. In the prototyping stage, one would be interested in locating the fault and identifying it, if it is a process or a design fault. While in the production stage, one would be interested if the chip operates according to specifications. Although the time for testing in the prototyping stage could be long and the equipment could be expensive (e.g., microprobing, electron beam, etc.), the time for testing in the production stage must be short and the equipment less sophisticated. Another stage of testing for the fabricated chip takes place when the chip is in the field as part of an overall system. In this case, self-testing chips could be important.

6. DISCUSSION AND COMMENTS

In the early 1960's, the first digital IC's were available commercially. Now the technology is capable of integrating a complete digital VLSI system on a single chip. This happened over a period of 25 years, when device counts went from 100's to 100 000's. However, design and simulation methods and tools have not improved 1000 times as did the device count. This situation has to change, before we are faced with a processing capability and a design incapability.

Substantial effort towards the goal of improving design and simulation methods and tools started only ten years ago and more work is needed. Steady evolution of CAD design tools has occurred and looks certain to continue. Both chip designers and the developers of CAD tools have to collaborate to produce the most effective CAD tool at a reasonable cost.

REFERENCES

[1] A. R. Newton, "CAD of VLSI circuits," *Proc. IEEE*, pp. 1189–1199, Oct. 1981.
[2] R. K. Brayton and G. D. Hachtel, "A taxonomy of CAD for VLSI," in *Proc. ICCC Conf.*, 1980, pp. 34–57.
[3] H. E. Ruehli and G. S. Ditlow, "Circuit analysis, logic simulation and design verification for VLSI," *Proc. IEEE*, pp. 34–48, Jan. 1983.

Digital MOS Integrated Circuits: A Tutorial

MOHAMED I. ELMASRY, SENIOR MEMBER, IEEE

Abstract—Basic digital NMOS and CMOS integrated circuits are analyzed. Dc and transient performance is studied using first-order design equations. Static and dynamic circuits are discussed. The effects of device parameters on circuit performance are explained.

1. INTRODUCTION

THE dual purpose of this tutorial paper is to review the main characteristics of MOS transistors as they are related to digital circuit design and to analyze basic static and dynamic NMOS digital circuits using first-order dc and transient design equations. The study of digital NMOS circuits is followed by a similar study of digital CMOS circuits, and the most basic digital circuit, the static inverter, is analyzed in detail. This is followed with an explanation of how the design of logic gates and flip-flops, both static and dynamic, is related to the design of the simple static inverter.

2. MOS DEVICE CHARACTERISTICS

Figure 1 shows a diagrammatic cross-section in a NMOS transistor [1]. It consists of two n^+ regions, introduced in a P substrate by diffusion or ion implantation. In circuit operation, the more positive region is called the drain while the other region is called a source. The surface region between the source and drain is called the channel. The conduction through this channel is controlled by the voltage on the gate, which is either metal or polysilicon. The gate is separated from the channel by a thin layer of a dielectric, usually silicon oxide.

If the voltage of the source terminal is taken as a reference, then V_{DS}, V_{GS}, and V_{BB} are the voltages of the drain, gate, and substrate respectively. V_{BB} is referred to as the back-gate bias. For NMOS digital circuits V_{BB} is an applied negative or zero voltage. The negative V_{BB} is either supplied from off-chip, or is generated on-chip from the available positive power supply V_{DD} (see Appendix A).

An applied positive V_{DS} allows electrons, when present in the channel, to drift from the source to the drain causing I_{DS} to flow from the drain to the source. In depletion-type NMOS devices, electrons are present in the channel even at $V_{GS} = 0$. This is achieved by ion implanting the surface channel with n-type material. An increase in V_{GS} increases I_{DS}. However, if V_{GS} is negative and larger than $|V_T|$, where V_T is the threshold voltage of the depletion-type device, the channel is depleted and I_{DS} is reduced to zero. In enhancement-type NMOS, electrons are only present at the surface if V_{GS} is positive and

Manuscript received December 21, 1984. It is based on a tutorial paper by the author published in the IEEE PRESS Book, *Digital MOS Integrated Circuits*, M. I. Elmasry, Ed., 1981. This work was supported in part by NSERCC and by the University of Waterloo.

The author is with the Department of Electrical Engineering, University of Waterloo, Waterloo, Ont., Canada, N2L 3G1.

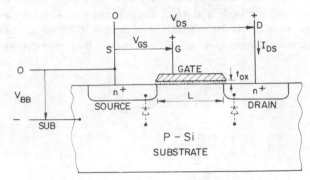

Fig. 1. Diagrammatic cross-section in an NMOS transistor showing parasitic diodes and terminal voltages.

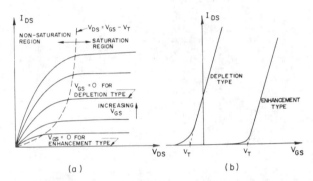

Fig. 2. (a) I_{DS} vs. V_{DS} and (b) I_{DS} vs. V_{GS} for a given V_{BB}. Note that V_T is positive for enhancement-type NMOS devices and negative for depletion-type.

larger than V_T where V_T is the threshold voltage of the enhancement-type device. An increase in V_{GS} increases I_{DS}.

2.1 DC Characteristics

Two important dc characteristics of the NMOS transistor are shown in Fig. 2:

- the drain current I_{DS} vs. the drain voltage V_{DS} for different values of V_{GS} at a given substrate bias V_{BB}; and
- the drain current I_{DS} vs. the gate voltage V_{GS} at a given V_{BB}.

Three regions of operation can be distinguished on the I_{DS} vs. V_{DS} characteristic:

1. The off region, where

$$V_{GS} < V_T$$

$$I_{DS} \simeq 0$$

This region is also referred to as the subthreshold region where I_{DS} increases exponentially with V_{DS} and V_{GS} [1]. The value of I_{DS} in this region is much smaller than its value when $V_{GS} > V_T$. Thus, in many NMOS circuits for $V_{GS} < V_T$, the transis-

This paper was specially prepared for this volume.

tor is considered off and $I_{DS} \simeq 0$. However, the small value of I_{DS} in this region could affect the circuit performance as in MOS dynamic circuits.

2. The nonsaturation region, where

$$V_{DS} < V_{GS} - V_T$$

$$I_{DS} = \beta[(V_{GS} - V_T) V_{DS} - \tfrac{1}{2} V_{DS}^2] \qquad (1)$$

3. The saturation region, where

$$V_{DS} \geqslant V_{GS} - V_T$$

$$I_{DS} = \frac{\beta}{2} [V_{GS} - V_T]^2 \qquad (2)$$

where

$$\beta = \frac{W}{L} \frac{\epsilon_{ox} \mu}{t_{ox}} = \frac{W}{L} K' \qquad (3)$$

W = width of the MOS channel

L = length of the MOS channel (in the direction of current flow)

ϵ_{ox} = permittivity of the gate oxide

t_{ox} = thickness of the gate oxide

μ = average surface mobility of carriers (μ_n in the case of electrons in NMOS, and μ_p in the case of holes in PMOS)

V_T = threshold voltage

C_{ox} = gate capacitance per unit area = ϵ_{ox}/t_{ox}

2.1.1 The Conduction Factor K'

The conduction factor K' ($\equiv \epsilon_{ox} \mu / t_{ox}$) is technology dependent and is specified for a given MOS process. Thus it is not a circuit design variable. For a t_{ox} range of 1000–500°A, it has a typical value of 12–25 $\mu A/V^2$ for NMOS devices and 5–10 $\mu A/V^2$ for PMOS. As the value of t_{ox} decreases with advances in technology, K' increases. The difference between K'_n and K'_p results from the fact that $\mu_n \sim 2.5 \, \mu_p$.

K' is a function of temperature because of its dependency on μ:

$$\frac{K'}{K'_0} = \left(\frac{T}{T_0}\right)^{-3/2} \qquad (4)$$

where K'_0 is the value of K' at room temperature ($T_0 = 298°K$) and T is the absolute temperature (°K).

2.1.2 The Geometrical Ratio (W/L)

This ratio is a circuit design parameter. The minimal value of L (L_{min}) is determined by the MOS fabrication process. L_{min} is determined mainly by the mask channel length, the tolerances on that length and the lateral diffusions of both the source and the drain regions. The minimum value of W is usually in the order of the minimum value of L. Increasing (W/L) will increase the drain current for a given set of operating voltages. However, increasing W increases the gate area and the source and the drain diffusion areas and hence increases the value of the capacitances associated with the gate and with the source-substrate and the drain-substrate junctions.

2.1.3 The Threshold Voltage V_T

V_T is a function of the MOS processing parameters and the substrate bias V_{BB}. In general, it is also a function of V_{DS}. In order to highlight these different functional dependencies, V_T of an enhancement-type NMOS can be written as:

$$V_T = V_{TO} + \Delta V_T(V_{BB}) - \Delta V_T(V_{DS}) \qquad (5)$$

where

$$V_{TO} = \left(\phi_{GS} - \frac{Q_{SS}}{C_{ox}}\right) + \gamma(2\phi_F)^{1/2} + 2\phi_F \qquad (5a)$$

$$= V_{FB} + \gamma(2\phi_F)^{1/2} + 2\phi_F$$

$$\Delta V_T(V_{BB}) = \gamma[(|V_{BB}| + 2\psi_F)^{1/2} - (2\psi_F)^{1/?}] \qquad (5b)$$

$$\Delta V_T(V_{DS}) = z(V_{DS} + 2|V_{BB}| + 2V_{Bi}) \qquad (5c)$$

$$\gamma = \frac{(2\epsilon_S q N_B)^{1/2}}{C_{ox}} \qquad (5d)$$

$$z = \frac{\eta_0(x_j, N_B)}{C_{ox} L^n} \qquad (5e)$$

ϕ_{GS} = gate voltage necessary to counter balance the gate-to-silicon work function difference.

Q_{SS}/C_{ox} = gate voltage necessary to counter balance the effect of the oxide surface charge Q_{SS}.

$V_{FB} = \phi_{GS} - (Q_{SS}/C_{ox})$ = gate voltage necessary to cause the flat band condition at the silicon surface, hence the name "flat band voltage." The flat band condition occurs when the energy bands in the substrate are flat at the surface [1]. In this condition, there is zero electric field at the silicon surface. If the gate voltage is more positive than V_{FB}, for P-type substrates, the silicon surface is in depletion, i.e., there are no mobile carriers at the surface. If the gate voltage is further increased the surface starts to "invert," i.e., electrons are attracted to the surface forming a conductive channel.

$\gamma(2\phi_F)^{1/2}$ = gate voltage necessary to counter balance the effect of the charge created by the exposed dopants at the surface. ϕ_F is the substrate Fermi potential at the surface, and is equal to (KT/q) $\ln(N_B/n_i)$, where (KT/q) is the thermal voltage, n_i is the intrinsic concentration for silicon, $n_i^2 = 1.5 \times 10^{33} \, T^3 \, e^{-1.2q/KT} \, cm^{-6}$, N_B is the average substrate concentration at the silicon surface. Note that a surface implant at the MOS channel is usually used to influence N_B [2].

$2\phi_F$ = additional gate voltage, by definition necessary to produce a "strong inversion" condition at the silicon surface.

$\Delta V_T(V_{BB})$ = increase in the threshold voltage due to the reverse bias on the back-gate (substrate). If $V_{BB} = 0$ then $\Delta V_T(V_{BB}) = 0$.

$\Delta V_T(V_{DS})$ = decrease in the threshold voltage due to the short channel effect. For large L, $\Delta V_T(V_{DS})$ tends to zero.

11

$\eta_0(x_j, N_B)$ = factor which is a function of the source and drain junction depth x_j and N_B [3].

n = factor which ranges between 2.6 and 3.2 for 10^{15} cm$^{-3} \leqslant N_B \leqslant 10^{16}$ cm^{-3}, 1.5 μm $\leqslant x_j \leqslant$ 0.41 μm [3].

V_{Bi} = source (or drain)-substrate built-in voltage = (KT/q) ℓn $(N_B N/n_i^2)$ where N is the average impurity concentration of the source and drain diffusions.

Examining equation (5) reveals the following:

1. V_{TO} is not a function of the operating voltages and is a function of temperature.
2. The sensitivity of the threshold voltage V_T to V_{BB}: ΔV_T (V_{BB}) is determined by γ which is a function of $N_B^{1/2} C_{ox}^{-1}$. Increasing N_B or decreasing C_{ox}, i.e., increasing t_{ox}, will increase that sensitivity.
3. The sensitivity of the threshold voltage V_T to V_{DS}: ΔV_T (V_{DS}) is determined by z which tends to zero for long channel devices. z is an empirical factor [3] which is proportional to x_j, $1/N_B$, $1/C_{ox}$ in addition to $1/L$. Decreasing x_j, increasing N_B or decreasing t_{ox} (hence increasing C_{ox}) reduces the sensitivity of V_T to V_{DS} in short channel devices.
4. In short channel and narrow channel devices. γ is a function of both L and W; a decrease in L decreases γ and a decrease in W increases γ. This dependency of γ and L and W affects V_{TO} and $\Delta V_T(V_{BB})$ and could be incorporated empirically in equation (5d). This effect should be considered in the design of digital circuits using MOS devices of small dimensions because a transistor having $W = W_{min}$, $L = L_{min}$ may have a different dc characteristic from one having $W = mW_{min}$ and $L = mL_{min}$, although (W/L) of the two transistors are the same.

Note that temperature affects both K' and V_T in such a way that the effects on I_{DS} could cancel each other. Thus, MOS transistors can be operated so that they exhibit positive, negative, or zero temperature coefficient [1].

2.2 Transient Characteristics

The transient performance of an MOS integrated circuit is a function of the total capacitance at the output node. This capacitance C_{out} is the summation of the parasitic output capacitance C_0 and the input gate capacitance(s) of the loading stage(s) C_{IN}.

The parasitic output capacitance C_0 is the summation of two capacitances:

• C_J—the junction capacitance of the output diffusion(s). This capacitance varies with the junction voltage V_j:

$$\frac{C_J}{C_{JO}} = \left(1 - \frac{V_j}{V_{Bi}}\right)^n$$

where

C_J = junction capacitance at voltage V
C_{JO} = junction capacitance at zero voltage
V_j = junction voltage (negative for reverse-bias)

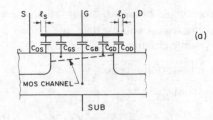

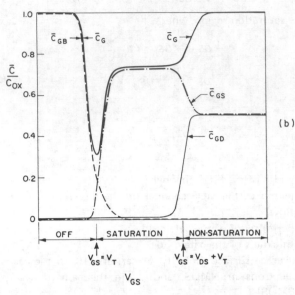

Fig. 3. (a) The different components of C_{IN}; (b) the voltage dependency of the capacitances per unit area: $\bar{C}_{GS}$, $\bar{C}_{GD}$, $\bar{C}_{GB}$, and $\bar{C}_G$ vs. V_{GS}.

V_{Bi} = built-in junction potential
n = factor between (−0.5) and (−0.3) depending upon junction abruptness

• C_{INT}—the interconnector capacitance associated with metal, polysilicon, or diffusion interconnection lines. C_{INT} is voltage independent and usually contributes to C_{out} in LSI circuits where complex interconnection patterns exist.

The input capacitance C_{IN} of an MOS transistor consists of the following components [1, 4, 5, 6] as shown in Fig. 3(a).

C_{OS}, C_{OD}—the source and the drain overlap capacitances resulting from the overlap of the gate on the source and the drain diffusions: $C_{OS} = C_{ox}\ell_S W$, $C_{OD} = C_{ox}\ell_D W$ where ℓ_S and ℓ_D are the overlap lengths.

C_{GS}, C_{GD}—represent gate to channel capacitances lumped at the source and drain regions of the channel respectively:

$$C_{GS} = C_{ox}WLf_S(V), \quad C_{GD} = C_{ox}WLf_D(V)$$

C_{GB}—the gate-substrate capacitance:

$$C_{GB} = C_{ox}WLf_B(V)$$

where $f_S(V)$, $f_D(V)$ and $f_B(V)$ are voltage dependent functions. Figure 3(b) demonstrates the nature of the voltage dependency of the capacitances per unit areas: $\bar{C}_{GS}$, $\bar{C}_{GD}$, $\bar{C}_{GB}$, and $\bar{C}_G$ of an NMOS transistor vs. V_{GS} where $\bar{C}_G = \bar{C}_{GS} + \bar{C}_{GD} + \bar{C}_{GB}$. When the transistor is off, the only nonzero component is $\bar{C}_{GB}$. This component is due to the series combination of the surface depletion layer and gate oxide capacitances. As the transistor turns on, $\bar{C}_{GB}$ reduces to zero because of the shielding effect of the inversion layer. In the nonsaturation region, the source and the drain regions of the MOS channel are inverted and $\bar{C}_{GS} \simeq \bar{C}_{GD} \simeq C_{ox}/2$. In the saturation region, where the drain region of the channel is pinched-off, $\bar{C}_{GD}$ reduces to zero from $C_{ox}/2$ and $\bar{C}_{GS}$ increases from that value to approximately $\frac{2}{3} C_{ox}$. Figure 3(b) shows that $\bar{C}_G$ vs. V_{GS} has a minimum just below $V_{GS} = V_T$.

2.3 Leakage and Breakdown

It is important to consider leakage currents in an MOS chip, especially if the circuit is operating in a dynamic mode and particularly in dynamic memories. The leakage currents are associated with the source and drain p-n junctions. The absolute values of these currents should be minimized, and their variation with temperature should be considered [1].

A potential source of parasitic current in MOS circuits is the current associated with the thick field oxide MOS transistors which have a thick (field) oxide. This current increases exponentially as the gate voltage approaches the threshold voltage of the field-oxide MOSFETs. The threshold voltage of the field-oxide MOSFETs can be increased by increasing the surface doping density under the field oxide.

In MOS circuits, breakdown can occur by different mechanisms. Avalanche breakdown can occur in the reverse-biased drain-substrate junction. Punch-through breakdown can also occur if the depletion region of the drain-substrate junction reaches the source-substrate junction. Either type of breakdown may predominate for a given MOS structure, e.g., punch-through breakdown could be the predominate for short channel MOSFETs. Although both types of breakdowns are related to the p-n junction characteristics, they are strongly affected by the presence of a gate oxide and a conducting gate [4].

3. THE STATIC NMOS INVERTER

The NMOS inverter, as shown in Fig. 4(a), consists of an enhancement type driver transistor and a load. The load is one of the following: (a) a saturated enhancement-type NMOS device, (b) a nonsaturated enhancement-type NMOS device, (c) a depletion-type NMOS device, or (d) a polysilicon resistor.

Figure 4(b) shows the load lines of the above four loads superimposed on the I_{DS} vs. V_{DS} of the driver. The intersection of the load line with the driver characteristic for $V_{GS} = V_{IN} = V_0$, gives $V_{DS} = V_{OUT} = V_1$ where V_0 is the low voltage level representing logical '0' (see Appendix B), $V_0 < V_{TD}$, V_{TD} is the threshold voltage of the driver and V_1 is the high voltage level representing logical '1.' Similarly, the intersection of the load line with the driver characteristics for $V_{GS} = V_{IN} = V_1$, gives $V_{DS} = V_{OUT} = V_0$.

3.1 DC Analysis

3.1.1 Saturated Enhancement-Type Load

The saturated enhancement-type load was used in the early digital MOS integrated circuits. Figure 5 shows an inverter using NMOS devices. The load device has its gate connected to its drain (i.e., $V_{DS} = V_{GS}$), and operates in the saturation region when it conducts, since $V_{DS} > (V_{GS} - V_{TL})$, where V_{TL} is

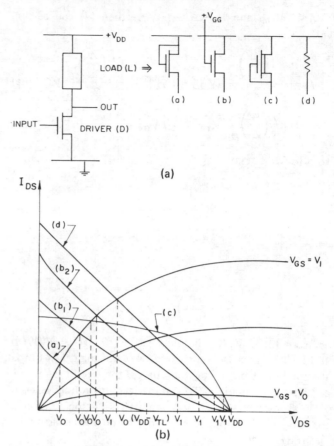

(a)

(b)

Fig. 4. (a) NMOS inverter with different loads; (b) load lines superimposed on the driver transistor I_{DS} vs. V_{DS} characteristic. V_{GG} for case $(b_2) > V_{GG}$ for case (b_1).

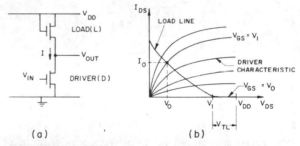

(a)

(b)

Fig. 5. (a) NMOS inverter with saturated enhancement-type load; (b) load line superimposed on the driver I_{DS} vs. V_{DS} characteristic, showing the operating current I_0 when $V_{OUT} = V_0$.

the threshold voltage of the load device. The dc circuit operation is explained as follows:

(a) when $V_{IN} = V_0$, $V_0 < V_{TD}$, the driver is off, $I \approx 0$ and

$$V_{OUT} = V_1 = V_{DD} - V_{TL} \qquad (6)$$

(b) when $V_{IN} = V_1$, $V_1 > V_{TD}$, the driver is operating in the nonsaturation region, the load device is in saturation and $I = I_0$ where:

$$I_0 = \frac{K' \left(\frac{W}{L}\right)_L}{2} [(V_{DD} - V_0) - V_{TL}]^2 \qquad (7a)$$

$$= K' \left(\frac{W}{L}\right)_D \left[(V_1 - V_{TD}) V_0 - \frac{1}{2} V_0^2\right] \qquad (7b)$$

13

If $V_0 \ll V_{DD}$ and $V_0 \ll V_1 - V_{TD}$ then (7) can be simplified:

$$I_0 \simeq \frac{K' \left(\dfrac{W}{L}\right)_L}{2} (V_{DD} - V_{TL})^2 = \frac{K' \left(\dfrac{W}{L}\right)_L}{2} V_1^2 \quad (8a)$$

$$\simeq K' \left(\frac{W}{L}\right)_D (V_1 - V_{TD}) V_0 \quad (8b)$$

From (8) we obtain:

$$\frac{\left(\dfrac{W}{L}\right)_D}{\left(\dfrac{W}{L}\right)_L} \geqslant \frac{V_1}{2V_0} \quad (9)$$

and

$$V_0 \simeq \frac{I_0}{K' \left(\dfrac{W}{L}\right)_D (V_1 - V_{TD})} \quad (10)$$

Thus from (9) if V_1/V_0 is taken to be 10, $(W/L)_D/(W/L)_L$ should be $\geqslant 5$. Because the ratio of (W/L) of the driver to that of the load must be greater than unity, the circuit is referred to as "a ratioed circuit." To summarize: the voltage levels of the NMOS inverter V_1, V_0, V_t, where V_t is the logic threshold of the inverter (see Appendix B) and the dc power dissipation are given by:

$$V_1 = V_{DD} - V_{TL}$$

$$V_0 \cong \frac{I_0}{K' \left(\dfrac{W}{L}\right)_D (V_1 - V_{TD})}$$

$$V_t \geqslant V_{TD}$$

$$P_{DC}(@ V_{OUT} = V_1) \simeq 0$$

$$P_{DC}(@ V_{OUT} = V_0) = I_0 V_{DD} \quad (11)$$

where I_0 is given by (8a).

In (11) I_0 is determined by transient considerations or by the allowable power dissipation. V_{DD} is usually fixed by subsystem design considerations and today it is typically $= 5$ V.

It is clear that V_1 increases as the power supply voltage V_{DD} increases. The maximum allowable V_{DD} must be less than the junction breakdown voltage of the drain-substrate junction. It should also be less than the voltage at which the parasitic field-oxide MOS transistors start to conduct.

Because the substrate-source junction of the load device is more reverse biased than the driver, V_{TL} is larger than V_{TD} due to the back-gate bias effect. The higher the V_{BB} value of the load, the higher the value of V_{TL}, and the lower the value of V_1. Because $V_1 = V_{DD} - V_{TL}$, the saturated enhancement-type MOS load is said to introduce "threshold losses" in the value of V_1. As we explain in the following sections, depletion-type nonsaturated enhancement type and bootstrapped MOS loads do not introduce threshold losses.

The low logical voltage level, V_0, is a function of the operating current I_0. As I_0 increases, the dc power dissipation increases and the speed of operation also increases as we explain in 3.2. V_0 can be reduced (hence increasing the logic swing and the NM_o noise margin) by increasing the geometrical ratio $(W/L)_D$ of the driver transistor, hence increasing its size, with respect to $(W/L)_L$.

The threshold voltage V_t of the inverter (see Appendix B) is $\geqslant V_{TD}$. The higher the ratio $[(W/L)_D/(W/L)_L]$ is, the closer V_t to V_{TD} becomes. If the driver is replaced with a number of stacked transistors connected in a series, as in the case of an MOS NAND gate (see 4), then V_{TD} and hence V_T of the upper input transistors are higher than that of the lower transistors because of the back-gate bias effect. This is a drawback for using NAND gates in NMOS logic design.

3.1.2 Nonsaturated Enhancement-Type Load

One drawback of using a saturated enhancement-type load is the threshold voltage losses caused by the load, resulting in $V_1 = V_{DD} - V_{TL}$.

This situation can be rectified if the load is operating in the nonsaturation region by connecting its gate to V_{GG} where $V_{GG} > (V_{DD} + V_{TL})$. In this case V_1 can be made close to V_{DD} by increasing V_{GG} as shown in Fig. 4(b). The inverter dc circuit operation is explained as follows:

(a) when $V_{IN} = V_0$, $V_0 < V_{TD}$ the driver is off, the load is nonsaturated and $V_{OUT} = V_1 = V_{DD} - V_{DS}|_{load}$ where $V_D|_{load}$ is the voltage drop across the load device. If the inverter is driving only MOS gates, which is usually the case with high input impedances, then $V_{DS}|_{load} \simeq 0$ and $V_1 \simeq V_{DD}$.

It should be noted that $V_{DS}|_{load}$ can be reduced for a given load current by increasing $(W/L)_L$ or $(V_{GG} - V_{TL})$ at the expense of increasing the gate power dissipation.

(b) when $V_{IN} = V_1$, $V_1 > V_{TD}$, the driver is operating in the nonsaturation region, the load is nonsaturated and $I = I_0$ where

$$I_0 = K' \left(\frac{W}{L}\right)_L \left[(V_{GG} - V_{TL} - V_0)(V_{DD} - V_0) \right.$$
$$\left. - \frac{1}{2}(V_{DD} - V_0)^2 \right] \quad (12a)$$

$$= K' \left(\frac{W}{L}\right)_D \left[(V_1 - V_{TD}) V_0 - \frac{1}{2} V_0^2 \right] \quad (12b)$$

If $V_0 \ll V_{DD}$, $V_0 \ll V_1 - V_{TD}$, $V_{GG} - V_{TL} \gg V_{DD}$ then (12) can be simplified:

$$I_0 \simeq K' \left(\frac{W}{L}\right)_L (V_{GG} - V_{TL}) V_{DD} \quad (13a)$$

$$\simeq K' \left(\frac{W}{L}\right)_D (V_1 - V_{TD}) V_0 \quad (13b)$$

From (13) we obtain

$$\frac{\left(\dfrac{W}{L}\right)_D}{\left(\dfrac{W}{L}\right)_L} \geqslant \frac{V_{GG} - V_{TL}}{V_0} \quad (14)$$

14

and

$$V_0 \simeq \frac{I_0}{K'\left(\frac{W}{L}\right)_D (V_1 - V_{TD})} \tag{15}$$

where I_0 is given by (13a).

In summary, using a nonsaturated enhancement-type load offers a higher V_1, and hence higher logic swing and higher noise margins. Moreover, it offers an improvement in transient performance, as shown in 3.2. This improved performance is obtained at the expense of adding the extra power supply V_{GG}.

3.1.3 Depletion-Type Load

Depletion-type devices are widely used as loads in NMOS logic circuits. They offer a voltage logic level $V_1 \simeq V_{DD}$ as shown in Fig. 4(b). In addition, the constant current characteristic improves the transient performance of the circuit [7]. The inverter dc circuit operation is explained as follows:

(a) when $V_{IN} = V_0$, $V_0 < V_{TD}$, the driver is off, the load is nonsaturated and similar to the nonsaturated enhancement-type case:

$$V_{OUT} = V_1 = V_{DD} - V_{DS}|_{load} \simeq V_{DD}$$

It should be noted that $V_{DS}|_{load}$ can be reduced for a given load current by increasing $(W/L)_L$ or $|V_{TL}|$.

(b) when $V_{IN} = V_1$, $V_1 > V_{TD}$, the driver is operating in the nonsaturation region, the load is saturated and $I = I_0$ where

$$I_0 = \frac{K'\left(\frac{W}{L}\right)_L}{2} V_{TL}^2 \tag{16a}$$

$$= K'\left(\frac{W}{L}\right)_D \left[(V_1 - V_{TD})V_0 - \frac{1}{2}V_0^2\right] \tag{16b}$$

If $V_0 \ll V_1 - V_{TD}$, then (16) can be simplified:

$$I_0 = \frac{K'\left(\frac{W}{L}\right)_L}{2} V_{TL}^2 \tag{17a}$$

$$\simeq K'\left(\frac{W}{L}\right)_D (V_1 - V_{TD})V_0 \tag{17b}$$

From (17) we obtain:

$$\frac{\left(\frac{W}{L}\right)_D}{\left(\frac{W}{L}\right)_L} \geqslant \frac{V_{TL}^2}{2V_0(V_1 - V_{TD})} \tag{18}$$

$$V_0 \simeq \frac{I_0}{K'\left(\frac{W}{L}\right)_D (V_1 - V_{TD})} \tag{19}$$

where I_0 is given by (17a). It should be noted here that because of the back-gate bias effect on the depletion load, V_{TL},

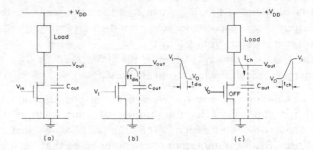

Fig. 6. (a) NMOS inverter showing output node capacitance; (b) simple circuit for discharging; (c) simple circuit for charging.

and hence I_0, should be either taken as an average or a worse case value.

3.1.4 Resistive Load

Polyresistors have recently been used in the design of NMOS digital integrated circuits, especially static memories. As shown in Fig. 4(b), a resistive load provides $V_1 \simeq V_{DD}$. In fact, the nonsaturated enhancement-type MOS load dc characteristic approaches that of a resistance when $V_{GG} \gg V_{DD} + V_{TL}$ as shown in Fig. 4(b), case (b_2). The inverter dc circuit operation is explained as follows:

(a) when $V_{IN} = V_0$, $V_0 < V_{TD}$, the driver is off and similar to the nonsaturated enhancement-type case:

$$V_{OUT} = V_1 = V_{DD} - V|_{load} \simeq V_{DD}$$

Similarly $V|_{load}$ can be reduced for a given load current by decreasing the value of the load resistance R.

(b) when $V_{IN} = V_1$, $V_1 > V_{TD}$, the driver is operating in the nonsaturation and $I = I_0$ where

$$I_0 = \frac{V_{DD} - V_0}{R} \tag{20a}$$

$$= K'\left(\frac{W}{L}\right)_D \left[(V_1 - V_{TD})V_0 - \frac{1}{2}V_0^2\right] \tag{20b}$$

If $V_0 \ll V_{DD}$, then (20) can be simplified

$$I_0 \simeq \frac{V_{DD}}{R} \tag{21a}$$

$$\simeq K'\left(\frac{W}{L}\right)_D (V_1 - V_{TD})V_0 \tag{21b}$$

From (21) we obtain

$$RK'\left(\frac{W}{L}\right)_D \geqslant \frac{V_{DD}}{V_0(V_1 - V_{TD})} \tag{22}$$

$$V_0 \simeq \frac{I_0}{K'\left(\frac{W}{L}\right)_D (V_1 - V_{TD})} \tag{23}$$

where I_0 is given by (21a).

3.2 Transient Analysis

The transient performance of an MOS inverter is a function of two times (see Fig. 6):

15

1. The discharging time t_{dis}, which is the time taken by the output node capacitance C_{out} to discharge through the driver transistor, from V_1 to V_0.
2. The charging time t_{ch}, which is the time taken by C_{out} to charge, through the load from V_0 to V_1.

In calculating t_{dis} and t_{ch} we will direct our attention to simplified expressions for the purpose of using them in the first phase of circuit design. The approach follows that of Crawford [8]. It is important to remember here that the time (t) taken by a capacitance (C) to change its voltage by ΔV, using a constant current (I), is given by the simple relation $t = C\Delta V/I$. In the following analysis we shall assume the input voltage to the MOS inverter is an ideal voltage step.

3.2.1 The Discharging Time t_{dis}

As shown in the dc analysis of the MOS inverter, the ratio of

$$\left(\frac{W}{L}\right)_D \quad \text{to} \quad \left(\frac{W}{L}\right)_L$$

is large, hence the on resistance of the driver transistor is much less than the on resistance of the load transistor. As a result, during the discharging of the output node capacitance, when $V_{IN} > V_{TD}$ and the driver is on, it is reasonable to neglect the current through the load [8]. Hence, the discharging circuit can be simplified, as shown in Fig. 6(b), to C_{out} with initial output voltage $= V_1$ and an on driver transistor with $V_{IN} = V_1$. The final output voltage $= V_0$.

During the discharging time, the driver is initially in the saturation region and the constant drain current:

$$I_D = \frac{\beta_D}{2}(V_1 - V_{TD})^2$$

will partially discharge C_{out} in a time given by

$$(t_{dis})_1 = \frac{C_{out}\Delta V}{I_D} \tag{24}$$

where

$$\Delta V = V_1 - V''_{OUT}$$

and V''_{OUT} is the output voltage at which the driver begins operating in the nonsaturation region (i.e., $V''_{OUT} = V_1 - V_{TD}$).

Equation (24) can be written in the form

$$(t_{dis})_1 = 2\frac{C_{out}}{g_{mD}}\frac{V_{TD}}{V_1 - V_{TD}} \tag{25}$$

where

$$g_{mD} = \beta_D(V_1 - V_{TD})$$

When the driver transistor begins to operate in the nonsaturation region at $V_{OUT} = V''_{OUT}$, the available current to discharge the output capacitance is not a constant and the discharging time $(t_{dis})_2$ for C_{out} to discharge from V''_{OUT} to V_0 can be obtained from the solution of the differential equation which results from equating the capacitance current through C_{out} to the current through the driver [8]:

$$C_{out}\frac{dV_{OUT}}{dt} = -\beta_D(V_1 - V_{TD})V_{OUT} + \frac{\beta_D}{2}V_{OUT}^2$$

which gives

$$V_{OUT} = (V_1 - V_{TD})\frac{2e^{-t/\tau_D}}{1 + e^{-t/\tau_D}} \tag{26}$$

where

$$\tau_D = C_{out}/g_{mD}$$

$$g_{mD} = \beta_D(V_1 - V_{TD})$$

Solving (26) for $(t_{dis})_2$ as a difference between the time at which $V_{OUT} = V''_{OUT}$ and the time at which $V_{OUT} = V_0$ gives

$$(t_{dis})_2 = \tau_D \ln\left[\frac{2(V_1 - V_{TD})}{V_0} - 1\right] \tag{27}$$

From (25) and (27):

$$t_{dis} = \tau_D\left(\frac{2V_{TD}}{V_1 - V_{TD}} + \ln\left[\frac{2(V_1 - V_{TD})}{V_0} - 1\right]\right) \tag{28}$$

As shown from (28), t_{dis} is reduced if (V_1/V_{TD}) is increased, hence the higher the logical '1' voltage level, the lower the discharging time.

Because $(t_{dis})_2 \gg (t_{dis})_1$, $V_1 \gg V_{TD}$, and $V_1 \gg V_0$, (28) can be simplified:

$$t_{dis} = \tau_D \ln\frac{2V_1}{V_0} \simeq \tau_D \ln\frac{V_1}{V_0} = \frac{C_{out}}{\beta_D(V_1 - V_{TD})}\ln\frac{V_1}{V_0} \tag{28a}$$

i.e., t_{dis} is linearly proportional to C_{out}, β_D^{-1} and $(V_1 - V_{TD})^{-1}$, and is logarithmically proportional to V_1/V_0. Thus, t_{dis} can be reduced, for a given C_{out}, by increasing $(W/L)_D$ or the driving voltage factor $(V_1 - V_{TD})$.

3.2.2 The Charging Time t_{ch}

During the charging time the driver is off and C_{out} will charge through the load as shown in Fig. 6(c).

Case 1: Saturated Enhancement-Type Load

Equating the capacitive current through C_{out} to that through the load we obtain the differential equation for the output voltage V_{OUT}:

$$C_{out}\frac{dV_{OUT}}{dt} = \frac{\beta_L}{2}[V_{DD} - V_{OUT} - V_{TL}]^2$$

With the initial condition:

$$\text{at} \quad t = 0, \quad V_{OUT} = V_0$$

the solution is:

$$V_{OUT} = (V_1 - V_0)\frac{t/\tau_L}{2 + \dfrac{t}{\tau_L}} + V_0$$

hence

$$t_{ch} = \tau_L\left(\frac{V_1 - V_0}{V_0}\right) = \tau_L\left(\frac{V_1}{V_0} - 1\right) = \tau_L\frac{V_\ell}{V_0} \tag{29}$$

16

where

$$\tau_L = \frac{C_{out}}{g_{mL}}$$

$$g_{mL} = \beta_L (V_{DD} - V_{TL})$$

From (29) it is clear that the charging time is proportional to the time constant τ_L. Reducing C_{out}, increasing β_L (i.e., the size of the load device), or the voltage factor $(V_{DD} - V_{TL})$ will decrease t_{ch}. Moreover, reducing the logic swing V_ℓ (or reducing V_1/V_0), at the expense of reducing the noise margins, will also reduce t_{ch}. As the output voltage rises during t_{ch}, the substrate-source bias of the load increases and hence V_{TL} increases. Thus, in calculating g_{ml}, V_{TL} may be taken as an average value in the voltage range $V_0 \leqslant V_{OUT} \leqslant V_1$. However, a pessimistic value of t_{ch} can be obtained if V_{TL} is taken as the maximum value where $V_{OUT} = V_1$.

Case 2: Nonsaturated Enhancement-Type Load

Similarly, the differential equation describing the output voltage V_{OUT} is given by:

$$C_{out} \frac{dV_{OUT}}{dt} = \frac{\beta_L V_{DD}^2}{2m} \left(1 - \frac{V_{OUT}}{V_{DD}}\right) \left(1 - m\frac{V_{OUT}}{V_{DD}}\right)$$

Calculating t_{ch} requires calculating

$$t_{ch} = \int_{t@V_{OUT} = V_0}^{t@V_{OUT} = V_1} dt$$

$$\frac{V_{OUT}}{V_{DD}} = \frac{V_1}{V_{DD}}$$

$$= C_{out} \frac{2m}{\beta_L V_{DD}} \int \frac{d(V_{OUT}/V_{DD})}{\left(1 - \frac{V_{OUT}}{V_{DD}}\right)\left(1 - m\frac{V_{OUT}}{V_{DD}}\right)}$$

$$\frac{V_{OUT}}{V_{DD}} = \frac{V_0}{V_{DD}}$$

which gives

$$t_{ch} = \tau_c \frac{1}{1-m} \ln \left[\frac{\left(1 - m\frac{V_1}{V_{DD}}\right)\left(1 - \frac{V_0}{V_{DD}}\right)}{\left(1 - m\frac{V_0}{V_{DD}}\right)\left(1 - \frac{V_1}{V_{DD}}\right)} \right]$$

$$= 2 \frac{C_{out}}{\beta_L V_{DD}} \frac{m}{1-m} \ln \left[\frac{\left(1 - m\frac{V_1}{V_{DD}}\right)\left(1 - \frac{V_0}{V_{DD}}\right)}{\left(1 - m\frac{V_0}{V_{DD}}\right)\left(1 - \frac{V_1}{V_{DD}}\right)} \right]$$

$$0 \leqslant m < 1 \qquad (30)$$

where

$$m = \frac{V_{DD}}{2(V_{GG} - V_{TL}) - V_{DD}}$$

The factor m is a measure of the biasing conditions on the load [8]. If $V_{GG} \gg V_{DD}$ and $V_{GG} \gg V_{TL}$, m tends to

zero and the MOS load transistor acts like a resistance. As m approaches unity, i.e., $(V_{GG} - V_{TL})$ approaches V_{DD}, the nonsaturated load approaches the saturated load. Note that (30) is not valid for $m = 1$.

It is clear that as m approaches zero, t_{ch} approaches zero. Reducing C_{out}, increasing β_L or increasing V_{DD} will decrease t_{ch}. Because of the back-gate bias effect during charging, in calculating m, V_{TL} may be taken as a pessimistic value as is done in Case 1 or can be taken as an average value for the voltage range $V_0 \leqslant V_{OUT} \leqslant V_1$. For example, it can be taken at:

$$V_{OUT} = \frac{V_1 + V_0}{2}.$$

Case 3: Depletion-Type Load

Assuming that the load current is constant during the charging time as the output voltage rises from V_0 to V_1 then

$$t_{ch} = \frac{C_{out} V_\ell}{I_0} \qquad (31)$$

where I_0 is given by (16a) and an average value of V_{TL} is used. The charging time can be reduced by decreasing V_ℓ C_{out} or increasing I_0, where I_0 can be increased by increasing V_{TL} or β_L.

Case 4: Resistive Load

From the analysis of a simple RC circuit:

$$t_{ch} = R_L C_{out} \ln \frac{V_{DD} - V_0}{V_{DD} - V_1} \qquad (32)$$

NMOS Transient Performance—In 3.2, the transient performance of the basic NMOS inverter has been analyzed. Let us now calculate from the simple relations (28a), (29), (30), (31), and (32), typical values for t_{ch} and t_{dis}. Assume $V_{DD} = 5$ V, $V_{TD} = 0.5$ V and $K' = 25$ $\mu A/V^2$:

1. For $V_1 = 5$ V, $V_0 = 0.1$ V, (28a) gives: $t_{dis} \simeq 30$ nsec for $C_{out} = 1$ pF and $(W/L)_D = 1$
2. For $V_1 = V_{DD} - V_{TL} = 5 - 0.7 = 4.3$ V, $V_0 = 0.1$ V, (29) gives: $t_{ch} \simeq 400$ nsec for $C_{out} = 1$ pF and $(W/L)_L = 1$
3. For $V_1 = 4.9$ V, $V_0 = 0.1$ V, (3) gives: $t_{ch} \simeq 6.7, 25.6, 259$ nsec for $m = 0.1, 0.5, 0.9$ respectively, and $C_{out} = 1$ pF and $(W/L)_L = 1$
4. For $V_1 = 5$ V, $V_0 = 0.1$ V, $V_{TL} = -2.5$ V, (31) gives: $t_{ch} = 62.7$ nsec for $C_{out} = 1$ pF and $(W/L)_L = 1$
5. For $V_1 = 0.9 V_{DD}$, $V_0 = 0.1 V_{DD}$ (32) gives:
   $t_{ch} = 2.2 R_L C_{out}$
   $= 2.2$ nsec for $C_{out} = 1$ pF and $R_L = 1$ kΩ.

Examining the above typical values shows that the charging time in the case of saturated enhancement-type load is relatively high. The nonsaturated enhancement-type load offers improvement; the lower the value of m is (i.e., $V_{GG} \gg V_{DD}$) the lower t_{ch}. The depletion-type load also offers improvement. Because $(W/L)_D$ must be greater than $(W/L)_L$, $t_{ch} \gg t_{dis}$. The resistive case with $R_L = 1$ kΩ offers a reference time for t_{ch} for the different cases. For example, a non-saturated enhancement-type load, for $m = 0.1$, 0.5 and 0.9 resembles a resistive load with $R_L = 1$, 10, and 100 kΩ respectively.

Examining the above typical values shows also that increasing $(W/L)_L$ (i.e., the device size) of the load improves t_{ch}. Similarly, increasing the size of the driving transistor improves t_{dis}. However, it should be noted that increasing the size of these transistors increases their input capacitances, i.e., their loading on the previous stages, and the overall delay may not improve. Moreover, increasing the size of these transistors adds more capacitance to C_{out} and again the overall delay may not improve. Thus, increasing the size of the driving and the load transistors should be considered in association with their effect on the overall delay of a digital MOS chip [14].

NMOS Delay-Power Tradeoffs—It has been shown previously that the NMOS static inverter consumes dc power only when the output voltage is low. Thus

$$P_{DC}|_{av} = \tfrac{1}{2} I_0 V_{DD} \tag{33}$$

where I_0 is the inverter dc current when the output voltage is low. In addition to this component of power dissipation, there is a transient power component due to the switching of the output node capacitance C_{out}. This is given by:

$$P_t = C_{out} V_{\ell}^2 f \tag{34}$$

where V_{ℓ} is the logic swing and f is the switching frequency. The maximum switching frequency is given by: $f_{max} = 1/(t_{ch} + t_{dis})$.

In order to *highlight* the tradeoff that exists in the design of MOS digital circuits between delay and power dissipation, let us consider the case of a static inverter with a depletion-type load and let us assume that $t_{dis} \ll t_{ch}$ and the inverter is on half of the time. Thus, the total power dissipation at the maximum switching frequency is given by:

$$P = \tfrac{1}{2} I_0 V_{DD} + (C_{out} V_{\ell}^2 / t_{ch})$$

and the inverter average delay time is given by:

$$\tau_D = \frac{t_{ch} + t_{dis}}{2} \simeq \frac{t_{ch}}{2} = \frac{C_{out} V_{\ell}}{2 I_0}$$

Thus, the delay-power product is given by:

$$\tau_D P = \tfrac{1}{4} C_{out} V_{\ell} V_{DD} + \tfrac{1}{2} C_{out} V_{\ell}^2$$
$$\simeq \tfrac{3}{4} C_{out} V_{DD}^2 \tag{35}$$

This shows that the delay-power product is proportional to $C_{out} V_{DD}^2$.

NMOS Source-Followers and Push-Pull Drivers—In addition to the basic NMOS inverter discussed in 3.1 and 3.2, other circuit configurations are also used in the design of NMOS digital circuits. Two of these are the source-follower and the push-pull driver. Source-followers are used to reduce the capacitive loading on a given node while push-pull drivers are used to drive high capacitive loads, e.g., off-chip loads. Because of the close relationship between the two circuits, they are discussed here with the aid of Fig. 7.

The static inverter in Fig. 7 consists of a driving transistor Q and a load. The output node of that inverter (node 2) is driving the source-follower transistor Q_1. The load of that source-follower is the switchable transistor Q_2. The capacitive loading of node 2 is the gate-source and the gate-drain capacitances of

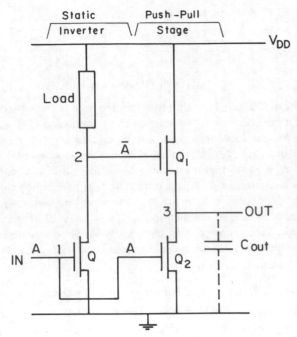

Fig. 7. NMOS push-pull driver.

Q_1, rather than C_{out}. Thus, Q_1 isolates node 2 from the loading capacitance C_{out}.

Q_1 and Q_2 are operating in the push-pull mode to charge and discharge C_{out}, thus the name "push-pull driver." As Q_1 is turned on, Q_2 is turned off and vice versa. It can be assumed that the circuit consists of a source-follower transistor Q_1 with a switchable load Q_2, or a driving transistor Q_2 with a switchable load Q_1. The push-pull operation allows high charging and discharging currents resulting in low switching times. The transistor Q_1 could be an enhancement or a depletion type. If it is an enhancement-type, the push-pull driver does not consume dc power since either Q_1 or Q_2 is on during the steady state. Thus, there is no dc requirement on the ratio $[(W/L)_2/(W/L)_1]$ and the circuit configuration is "ratioless." However, the voltage at node 3 suffers from threshold losses. In the case of a depletion-type load, threshold losses are eliminated at the expense of dc power consumption. The circuit in this case must be ratioed.

Bootstrapped Loads—It has been shown in 3.1 and 3.2 that when operating the NMOS load transistor in the nonsaturation region by tying its gate to V_{GG}, $V_{GG} > V_{DD} + V_{TL}$, threshold losses are eliminated and the transient performance is improved. These two results can be obtained without the need for the extra power supply, V_{GG}, by bootstrapping the load.

Bootstrapping allows the gate of the NMOS load transistor to rise above V_{DD} *during switching*, thus allowing the load to operate in the nonsaturation region. The basic circuit is shown in Fig. 8(a) and consists of the load transistor Q_1, a biasing transistor Q_2, and an MOS capacitor C. The biasing transistor Q_2 allows node 1 to float at $(V_{DD} - V_{T2})$, where V_{T2} is the threshold voltage of Q_2. As the input voltage drops, the output voltage rises by an amount $= \Delta V$. Because of the capacitive coupling, node 1 rises by an amount $= \Delta V'$: $\Delta V' = \Delta V C / C_1 + C$, i.e., if $C \gg C_1$, where C_1 is the parasitic capacitance of node 1, then $\Delta V' \approx \Delta V$. In the limit $\Delta V = V_{DD}$, the voltage

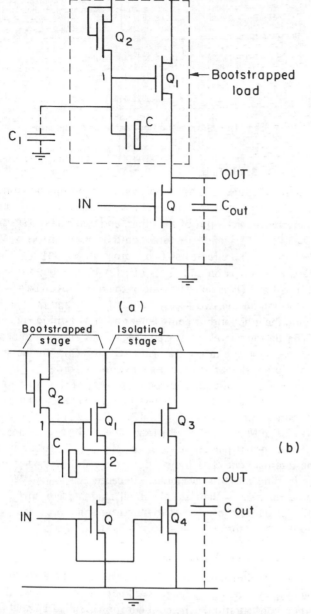

(a)

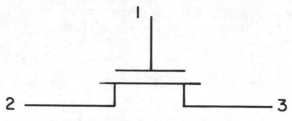

Fig. 9. NMOS transmission gate.

Fig. 8. (a) Bootstrapped load; (b) isolated bootstrapped load.

It consists of a single transistor as shown in Fig. 9. The gate is simply a symmetrical switch controlled by the voltage on node 1. If that voltage is greater than V_T, then nodes 2 and 3 are connected through the on resistance of the NMOS device. If that voltage is less than V_T then the NMOS device is off and nodes 2 and 3 are separated by an open circuit. Thus, a transmission gate can be used as a two-input AND gate where the inputs are applied to node 1 and node 2 (or 3) and the output is at node 3 (or 2).

The back-gate bias effect should be considered in the design of NMOS transmission gates. Thus V_T is a function of the voltage on nodes 2 and 3. Nodes 2 and 3 take the role of a source or drain depending on the values of their respective voltages. The on resistance of the NMOS device can be reduced by increasing (W/L) or the voltage driving factor $[(V$ at node 1$) - V_T)]$.

Because of the capacitive coupling between node 1 and nodes 2 and 3, a feed-through voltage appears at nodes 2 and 3 as the voltage at node 1 changes. The effect of the feed-through voltages on the overall circuit performance where the transmission gate is used should be evaluated.

4. STATIC LOGIC GATES AND FLIP-FLOPS

To realize static MOS logic gates, the driver transistor of the inverter circuit is replaced with a number of MOS transistors. If they are connected in series, as shown in Fig. 10(a), a NAND gate results, since the output is low only if all the inputs are high, i.e., all the transistors are on. If the transistors are connected in parallel, as shown in Fig. 10(b), a NOR gate results, since the output is low if any of the inputs are high, i.e., any one transistor is on. The transistors can also be connected in series-parallel arrangements, such as shown in Fig. 10(c), to realize more complex logic functions.

The design of an m-input MOS logic gate usually follows that of an inverter. First, an inverter is designed to meet a given dc and transient performance and $(W/L)_D$ and $(W/L)_L$ are determined. Then $(W/L)_D$ of the parallel transistors in the case of the NOR gate and the series transistors in the case of the NAND are determined. In the case of a NOR gate, since a worse situation results if only one of the m parallel transistors is on at a given time, $(W/L)_D$ of the NOR gate drivers is taken to be $= (W/L)_D$ of the inverter driver. However, in the case of a NAND gate, since the driver transistors are connected in series, the $(W/L)_D$ of the NAND gate drivers is taken to be m times $(W/L)_D$ of the inverter driver. As a result NAND MOS logic gates take more area than NOR gates.

In designing NOR gates, with $(W/L)_L$ and $(W/L)_D$ the same as that of an inverter, the logic levels V_0 and V_1 would be the

at node 1 can reach up to $(2V_{DD} - V_{T2})$ allowing Q_1 to operate in the nonsaturation region.

It should be noted that as node 1 rises above V_{DD}, Q_2 shuts off and isolates node 1 from the rest of the circuit [9].

It should also be noted that bootstrapping is a dynamic operation occurring when the output voltage switches. If the output node capacitance C_{out} is large and the output voltage rises slowly, then the bootstrap coupling between the output node and node 1 occurs at a slower rate. This can be resolved by isolating the output node capacitance C_{out} from the bootstrapped load using a source-follower as shown in Fig. 8(b). The isolating stage consists of a bootstrapped transistor Q_3 and a driver transistor Q_4. The output node is now isolated from the bootstrapped load.

NMOS Transmission Gates—The NMOS transmission gate is used in the design of static and dynamic NMOS digital circuits.

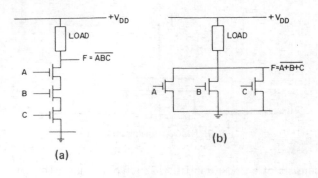

(a)

(b)

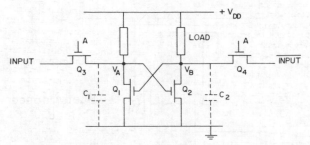

Fig. 11. NMOS static flip-flop with access transistors.

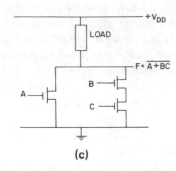

(c)

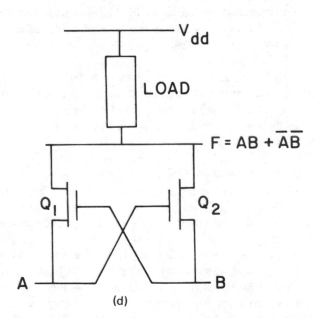

(d)

Fig. 10. (a) NMOS NAND gate; (b) NMOS NOR gate; (c) NMOS logic gate with series-parallel arrangement of driver transistors; (d) realization of an Exclusive-NOR using two NMOS transistors and a load.

same as that of an inverter. However, because the m parallel drivers contribute more capacitance to the output node than a single driver, the (W/L) of the load and the driver transistors of the NOR should be increased to obtain the same transient performance as that of the inverter.

4.1 Special Circuit Configurations

In addition to the NMOS logic gates shown in Fig. 10(a–c), innovative NMOS circuit configurations are sometimes used to efficiently realize specific complex logic functions. An example is shown in Fig. 10(d), where the circuit realizes the Exclusive-OR function F, $F = A\bar{B} + \bar{A}B$ by using both the gate and the drain of Q_1 and Q_2 as inputs.

4.2 NMOS Static Flip-Flops

Figure 11 shows a cross-coupled static flip-flop consisting of two inverters connected back to back, i.e., the output of one is connected to the input of the other, and two access transistors Q_3 and Q_4. The load structures could be polyresistors or MOS transistors. The circuit has two storage states: (1) V_A is low, V_B is high, Q_1 is on and Q_2 is off, (2) V_A is high, V_B is low, Q_1 is off and Q_2 is on. The state of the flip-flop can be changed by turning on the two access transistors (Q_3 and Q_4), and applying the input and its complement as shown in Fig. 11.

The dc logic levels at V_A and V_B nodes are determined by the loads and the driver transistors Q_1 and Q_2. The dc design procedure is similar to that of an inverter. In order to minimize the dc power dissipation (stand-by power) of the flip-flop, the load structures are chosen to have high value resistances, i.e., in the case of polyresistors high value resistors are chosen and in the case of MOS transistors, a low value of $(W/L)_L$ is chosen.

The transient performance of the flip-flop is determined by the internal node capacitances C_1 and C_2 and the charging and discharging currents available. Because the loads have high value resistances they supply small charging currents to the internal nodes and most of these currents are provided by the input circuits through the access transistors.

4.3 MOS Current Mode Logic

Since the introduction of MOS devices in the early 1960's, the devices have mainly been used in a pull-up pull-down (PUD) circuit configuration. The configuration in single channel static NMOS logic has serious shortcomings [7], [17]: high delay-power product, high sensitivity to interconnections, unsymmetrical static and dynamic operation (unequal noise margins and unequal charging and discharging times) and a logic swing which is highly dependent on the value of the power supply and difficult to change by design to meet different circuit requirements. The most important shortcoming is the configuration's inability to deliver bipolar-like performance with delays in the nanosecond range even with high-performance VLSI NMOS devices.

One circuit configuration which has been recently considered for MOS logic is the current-mode logic (CML) configuration. CML offers symmetrical high-speed operation, equal noise margins, and equal rise and fall times. The symmetrical operation allows operation at lower logic swing, thus enhancing the speed of operation even further. Moreover, the availability of the output logic function and its complementary (the NOR and the OR functions) offers flexibility in logic design and

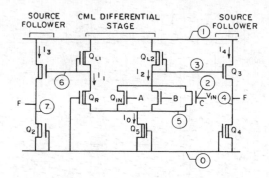

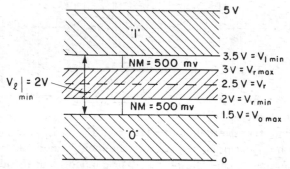

Fig. 12. Basic NMOS CML gate and typical V_{TO} and W/L for the different transistors.

TRANSISTOR	Q_R	Q_{IN}	Q_{L1}	Q_{L2}	Q_5	Q_1	Q_2	Q_3	Q_4
V_{TO} volts	-2.6	0.36	-2.6	-2.6	-2.6	0	-2.6	0	-2.6
$W_{\mu m}/L_{\mu m}$	25/2.5	25/2.5	2.5/2.5	2.5/2.5	2.5/2.5	25/2.5	10/2.5	25/2.5	10/2.5

Fig. 13. Logic levels V_1, V_0, noise-margins (NM) and logic swing (V_l).

provides two complementary output signals to drive long interconnections.

The basic MOS CML gate is shown in Fig. 12 [18]. It consists of a differential stage and two source-follower stages. The reference voltage V_r is established inherently by Q_R and Q_{IN} similar to that given in [19]. V_r is chosen to be $V_{DD}/2$, i.e., in a 5 V design $V_r = 2.5$ V where

$$V_r \simeq |V_T|Q_R + V_T|Q_{IN}$$

where $V_T|Q_R$ and $V_T|Q_{IN}$ are the threshold voltages of Q_R and Q_{IN}, respectively, at the operating values of the substrate bias.

The design of an MOS CML gate starts with choosing a value for V_r and determining the allowable tolerances on this value. The tighter these tolerances are, the lower the logic swing of the gate, the higher the speed, and the lower the dynamic power of the gate. We assume here that reasonable tolerances on V_r, which is assumed to be 2.5 V, are ±500 mV. If we further assume noise margins of 500 mV, then the logic levels of the gate are $V_1 \geqslant 3.5$ V and $V_0 \leqslant 1.5$ V. Thus, the minimum logic swing is 2 V, as shown in Fig. 13.

Because the MOS CML gate relies for its operation on the gate-source voltage drops V_{GS} of the different transistors, it is important to examine analytically the value of V_{GS} of an MOS device.

This is given by [7]:

$$V_{GS} = V_T + \sqrt{\frac{2I}{n(2-n) K'(W/L)}} \qquad (36)$$

where I is the operating drain–source current

$$n = \frac{V_{DS}}{V_{GS} - V_T}$$

$$= 1, \text{ if the transistor is saturated}$$

$$< 1, \text{ if the transistor is nonsaturated}$$

$$V_T = V_{T0} + \Delta V_T(V_{BB}) - \Delta V_T(V_{DS}).$$

V_{T0} is that part of the threshold voltage which is only a function of processing parameters and not a function of operating voltages. $\Delta V_T(V_{BB})$ is the shift in the threshold voltage due to the substrate source voltage bias V_{BB}. $\Delta V_T(V_{DS}) = 0$ for MOS devices which have no short-channel effects, i.e., devices with proper scaled processing parameters.

Equation (36) shows that V_{GS} depends on the operating current and this function dependency is stronger if the transistor is operating in the nonsaturation region. The ratio $[I/(W/L)]$ controls this current-dependent term.

It has been shown [18] that NMOS CML gates offer nanosecond performance at few picojoules delay-power products. The gates offer two logic output functions and for any added logic input only one transistor is added. This is in contrast with CMOS where only one logic output function is available and for any added logic input two transistors are needed, one of them being a bulky PMOS device.

The NMOS CML gate reported here uses a simple realization for the reference voltage, the current source, the loads, and the source followers, e.g., no biasing chain for V_r or the current source is used.

The NMOS CML circuit configuration could offer the answer to high-performance nanosecond and subnanosecond MOS VLSI, avoiding the complicated clock circuitry required for dynamic circuits and the high area and the complex processing associated with CMOS. NMOS CML bipolar-like performance is obtained at approximately half the area and half the delay-power product of bipolar CML.

5. DYNAMIC MOS CIRCUITS

The delays of static MOS circuits can be reduced by using advances in MOS technologies. However, it is possible to increase the throughput (the number of processed computations per unit time) of an MOS logic system by pipelining the circuit operation, where the delays of the system building blocks do not accumulate. The key feature of this approach is that the delay contribution of each block is with respect to a master clock, and that delay does not transfer from one block to another. Thus, the maximum frequency of operation of a pipelined system is determined by the delay of a *single*, although the slowest, block. Dynamic MOS logic circuits lead inherently to a pipeline operation, and can realize logic and arithmetic functions with a high throughput rate.

In dynamic MOS circuits, a master clock is used to generate different timing clocks, which are used to control the dynamic

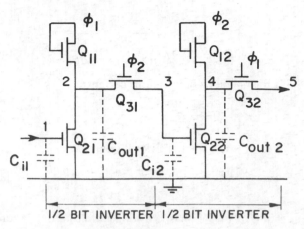

Fig. 14. Two 2ϕ inverters connected in cascade.

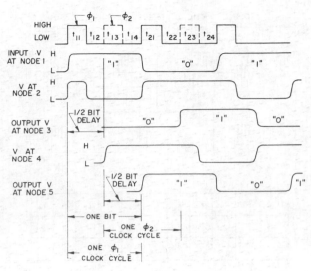

Fig. 15. Timing diagram of Fig. 14.

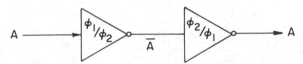

Fig. 16. Logic symbol of 2ϕ inverters.

operation of the circuit. These clocks are referred to as multiphase clocks [2 phase (2ϕ), 3 phase (3ϕ), four phase (4ϕ), etc.]. Although in principle the number of clocks can be increased to any number, four clocks is a practical compromise [5, 10], and 4ϕ dynamic circuits are commonly used. However, first we shall study 2ϕ dynamic circuits since they are the simplest.

5.1 Two-Phase (2ϕ) Circuits

The basic operations in 2ϕ dynamic circuits are:

1. *Charging* a capacitance through an MOS transistor during a first time slot (precharge time).
2. *Logically discharging* that capacitance (discharging it or not depending on the logical state(s) of the input(s)) through input transistor(s) during a second time slot (evaluation time).
3. *Transferring* the logical state (the voltage level on the capacitance) to the input of the next gate during the proper time slot (sampling time).

Figure 14 shows two 2ϕ inverters connected in cascade. Each inverter consists of three transistors Q_1, Q_2 and Q_3. The charging transistor Q_1 permits the output capacitance C_{out} to charge. The input transistor Q_2 provides a discharging path for C_{out}. The transferring transistor Q_3 allows a charge transfer between C_{out} and the input capacitance C_i of the next stage.

The above three basic operations are performed under the control of two nonoverlapping clocks ϕ_1 and ϕ_2 as shown in the timing diagram of Fig. 15. We assume that the input sequence is logical '1' – '0' – '1.' The logical inputs are presented by voltage signals which maintain their levels during the ϕ_1 clock cycles. For the case when the input is '1' the following sequence of operations occurs.

During the time interval t_{11} (precharge time), C_{out1} is precharged through Q_{11}. During t_{12} (evaluation time), because the input is high, C_{out1} is discharged through Q_{21}. During t_{13} (sampling time), Q_{31} conducts and a charge sharing occurs between C_{out1} and C_{i2}. If $C_{out1} \gg C_{i2}$ the voltage level at node 3 will be approximately the same as that of node 2: (V at node 3) = (V at node 2) $(C_{out1}/C_{out1} + C_{i2})$. Note that the input information which was available at node 1 during t_{11} has been inverted at node 3 and is available during t_{13}, i.e., the

input imformation has been inverted with a half-bit delay (a half clock cycle).

During the same time interval t_{13}, C_{out2} is precharged through Q_{12}. During t_{14}, because the input at node 3 is low, node 4 stays high. During t_{21}, Q_{32} conducts and the input capacitance of the next stage shares the charge of C_{out2} and a high voltage results at node 5, i.e., the input information at node 3 has been inverted with another half-bit delay.

The sequence of operations for a logical '0' input at node 1 is shown in Fig. 15 and can be similarly explained.

Figure 16 shows a logic symbol of the two inverters of Fig. 14. The first inverter type (ϕ_1/ϕ_2) accepts input at ϕ_1 clock cycles and provides output at ϕ_2 clock cycles. The second inverter type (ϕ_2/ϕ_1) accepts input at ϕ_2 clock cycles and provides output at ϕ_1 clock cycles. As a result, two inverters of the same type cannot be successively connected.

Logic gates can be realized by replacing the input transistor Q_2 of a two-phase inverter with a combination of transistors. As before, parallel transistors provide the NOR function, serial transistors provide the NAND function and mixed serial/parallel arrangements provide complex logic functions. As in the case of static gates, NOR gates take less area than NAND gates. Flip-flops can be realized using such 2ϕ gates. Figure 17 shows some examples where 2ϕ AND, OR, NOR and noninverting gates are used.

With reference to Fig. 14 fanning-out in 2ϕ circuits can be done at node 3. In this case the basic gate consists of Q_1 and Q_2 followed by the transferring transistor Q_3. The interconnection capacitances in this case are added to that of node 3. Alternatively, fanning-out can be done at node 2. In this case the basic gate consists of a transferring transistor Q_3 followed by Q_1 and Q_2. Thus, for each input, a Q_3 and a Q_2 are required. The interconnection capacitances in this case are added to that

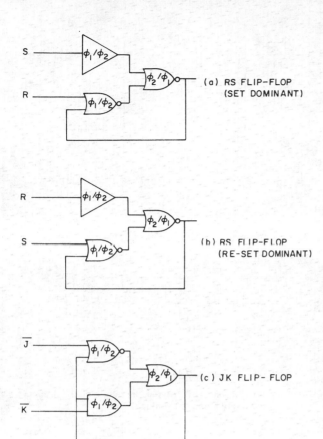

Fig. 17. Dynamic 2ϕ flip-flops.

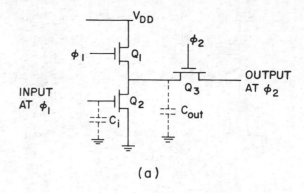

(a)

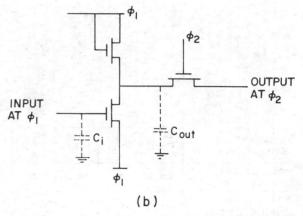

(b)

Fig. 18. Two other 2ϕ configurations: (a) ratioed configuration and (b) ratioless configuration with overlapping clocks.

of node 2. The advantage of this method of fanning out is having $C_{out} > C_i$, thus efficient charge sharing results during sampling times. However, the first method uses fewer transistors and fewer diffusion regions than the second method [11].

The basic 2ϕ circuit can be modified to meet different requirements. For example, Fig. 18(a) shows one configuration where Q_1 is operating from V_{DD}, while its gate is clocked with ϕ_1. The transferring transistor Q_3 is clocked with ϕ_2. The inverter (Q_1 and Q_2) consumes power when ϕ_1 is high. Its design is based on the static inverter design given in 3. Note that in the configuration of Fig. 14 both the gate and the drain of the load transistor are clocked with ϕ_1. This results in saving the power line V_{DD}. Again the design of the inverter is based on the static inverter design. The configuration is also a ratioed configuration.

In Fig. 18(b), both load and driver transistors are clocked using ϕ_1 and in this case no dc path exists. The power dissipation of the inverter is only a transient power given by $C_{out} V_\ell^2 f$. Moreover, because of the absence of a dc path, the load and the driver transistors do not have to satisfy any dc requirements. As a result, there is no need for a ratio between $(W/L)_D$ and $(W/L)_L$. In this case the (W/L) of both transistors are determined from transient considerations: the charging time is determined by $(W/L)_L$ and the discharging time is determined by $(W/L)_D$. It has the advantage of not using a V_{DD} or a ground line, no dc power dissipation, and using relatively small size devices for the load and the driver. However, it has the disadvantage of contributing large load capacitance to the clock lines.

5.2 Four-Phase (4ϕ) Circuits

One disadvantage of 2ϕ MOS dynamic circuits is the reduction of logic levels during the transfer operation through the transmission gate. This can be overcome by using more than two clocks to control the dynamic operation of the circuits. In this section we will discuss the operation of 4ϕ circuits using a given circuit configuration. The discussion can be easily extended to other circuit configurations.

Figure 19(a) shows four 4ϕ inverters connected in cascade. They are of four types and use 4ϕ overlapping clocks as shown in Fig. 19(b). Type 2 inverters operate under the control of the ϕ_1 and $\phi_{1\&2}$ clocks. With reference to the timing diagram of Fig. 19(b), the output capacitance of type 2 gates precharge during the time slot 1 (i.e., when both ϕ_1 and $\phi_{1\&2}$ are at high level). During time slot 2 the logic voltage level on the output capacitance is logically discharged through Q_{23} depending upon the logic input signal. In the same time slot, the output voltage logic level is transferred to the input capacitance of the next stage (Q_{22} is on). Thus, an inverting logic operation has been realized with a fourth-bit time delay. During time slots 3 and 4, the input node capacitance of the next stage will hold that logic level. This operation is summarized in the table of Fig. 19(d). It should be noted that in Fig. 19(a) each gate could be of the ratioed type or the ratioless type as we have explained in dealing with 2ϕ circuits. For example, for type 2 gates, the drain of Q_{21} can be connected to V_{DD} or ϕ_1 and the source of Q_{23} can be connected to ground or ϕ_1.

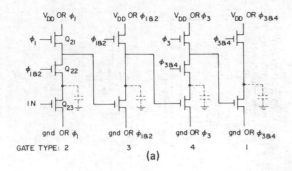

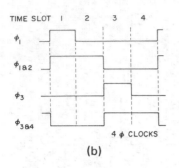

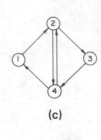

GATE TYPE: 2   3   4   1

(a)

TIME SLOT 1   2   3   4

ϕ_1

$\phi_{1\&2}$

ϕ_3

$\phi_{3\&4}$

4 ϕ CLOCKS

(b)

(c)

SUMMARY OF OPERATION	GATE TYPE 1	GATE TYPE 2	GATE TYPE 3	GATE TYPE 4
PRECHARGE DURING TIME SLOT	3 & 4	1	1 & 2	3
EVALUATION DURING TIME SLOT	1	2	3	4
HOLD DURING TIME SLOT	2	3 & 4	4	1 & 2

(d)

Fig. 19. Four 4ϕ inverters with overlapping clocks (a) circuit diagram; (b) timing diagram; (c) connection diagram; (d) summary of operation.

The operation of the other 4ϕ inverters of Fig. 19 can be similarly explained. The connection diagram of Fig. 19(c) shows, for example, that type 1 inverters can feed into type 2, while type 2 can feed into type 3 or type 4. The connection diagram can be easily obtained from a detailed timing diagram of the system.

Figure 20 shows another example of 4ϕ logic inverters along with the nonoverlapping clocks used. A connection diagram and a table for the summary of operations are left as an exercise for the reader.

Four-phase NOR gates are obtained by replacing the single input transistor of the inverter (e.g., Q_{23} in Fig. 19) with transistors connected in parallel. Four-phase flip-flops are realized using 4ϕ logic gates in a manner similar to the realization of 2ϕ flip-flops using 2ϕ logic gates.

Maximum Frequency of Operation—One important parameter in dynamic MOS circuits is the maximum clock frequency f_{max} at which the circuit can operate. If τ_1 and τ_2 are the worst case charging and discharging times of the output node capacitance then

$$f_{max} = \frac{1}{n(\tau_1 + \tau_2)}$$

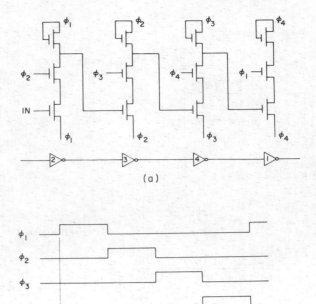

(a)

ϕ_1

ϕ_2

ϕ_3

ϕ_4

|← 1 CLOCK CYCLE →|

(b)

Fig. 20. Four 4ϕ inverters with nonoverlapping clocks (a) circuit diagram and (b) timing diagram.

where n is a factor which is a function of the clocking scheme used, e.g., in the two-phase nonoverlapping clocks of Fig. 14, $n = 2$.

Minimum Frequency of Operation—Because in dynamic MOS logic circuits the logic binary information is represented by a voltage level on a capacitance, the clock frequency has to be higher than a certain minimum f_{min} at which the stored charge of the capacitance starts to leak and the voltage level starts to deteriorate. The evaluation of f_{min} involves knowing the leakage currents of the MOS transistors under consideration. In a 2ϕ nonoverlapping logic system:

$$\frac{1}{2f_{min}} = \frac{\Delta V.C}{I_\ell}$$

where ΔV is the allowable deterioration in the high logic level, C is the total node capacitance, and I_ℓ is the leakage current associated with that node.

Clock Noise in Dynamic Circuits—Because of the capacitive coupling between the clock lines and the various nodes in a dynamic circuit, the nodes are subjected to changes in their voltages (noise) during the switching of the clocks. If the magnitude of these voltage changes is high enough, faulty operations could result. Computer simulation is used in these situations to assure that the circuit can tolerate this clock noise.

Clock Power Dissipation—In dynamic MOS circuits, the different clocks required by the circuits are generated on-chip and contribute to the total power dissipation. This clock power dissipation is a transient (dynamic) power given by $\Sigma C_\phi V_\phi^2 f_\phi$ where C_ϕ is the capacitive loading on the individual clock lines, V_ϕ is the clock voltage swing, and f_ϕ is the clock frequency.

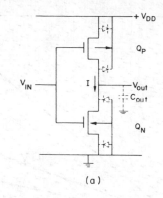

(a)

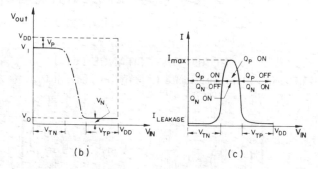

(b)　　　　　　　(c)

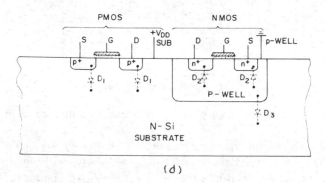

(d)

Fig. 21. CMOS inverter (a) circuit diagram; (b) transfer characteristic; (c) I vs. V_{IN}; (d) cross-section in a P-well CMOS technology.

6. COMPLEMENTARY MOS (CMOS) CIRCUITS

The use of a p-channel MOS as a load for an n-channel MOS driver provides the basic complementary CMOS inverter, as shown in Fig. 21(a). Logic gates and flip-flops can be realized using the complementary pair. Today, CMOS is the only MOS logic family available realizing standard SSI, MSI, and LSI functions. The family offers many advantages, including high noise immunity, operation at a wide range of power supply voltages, low power dissipation, relatively high speed, and compatibility to other logic families. The main disadvantage of the family is the relatively high silicon area consumed because of the need of complementary MOS pairs for each added logic input as shown in 6.2.

CMOS circuits can be fabricated using P-well, N-well or twin-tub CMOS technologies [12]. In a P-well CMOS technology the starting material is N-silicon substrate and is used to fabricate the PMOS devices, while a diffused or ion implanted P-well is used to fabricate the NMOS devices. More recently an N-well CMOS technology has been introduced which is compatible with NMOS technologies where the starting material is a P-sili-

con substrate and is used to fabricate NMOS devices and an ion implanted N-well is used to fabricate the PMOS devices. In a twin-tub CMOS technology, the starting material is a high resistivity silicon and two ion implanted N and P tubs are used to fabricate the PMOS and NMOS devices respectively.

Figure 21(d) shows a cross-section in NMOS and PMOS devices in a P-well CMOS technology. The following parasitic diodes are shown: D_1 is the substrate-source/drain junction diode, D_2 is the well-source/drain junction diode, and D_3 is the well-substrate junction diode. The three diodes make up a three junction thyrister (SCR) and during circuit operation it may latch-up. Techniques are used to minimize this hazardous circuit operation [12].

6.1 The CMOS Static Inverter

The basic inverter operation can be best explained with the aid of the transfer characteristic of Fig. 21(b).

1. If V_{IN} is a logic '0' (i.e., $V_{IN} = V_0$, $V_0 < V_{TN}$, where V_{TN} is the threshold voltage of Q_N), Q_N is off. If $|V_0 - V_{DD}| > |V_{TP}|$, where V_{TP} is the threshold voltage of Q_P, then Q_P is on. The output voltage in this case is given by

$$V_{OUT} = V_1 = V_{DD} - V_P \qquad (37)$$

where V_P is the voltage drop across Q_P and is given by

$$V_P = I_P R_P \simeq I_P \frac{1}{\beta_P [(V_{DD} - V_0) - |V_{TP}|]}$$

where I_P is the current supplied by Q_P to Q_N and to the loads. If the inverter is loaded with MOS gates, as is often the case, then I_P is negligible in the order of the leakage current of the substrate-source and drain junctions and $V_1 \simeq V_{DD}$. If I_P is not negligible, then by increasing β_P (i.e., the size of Q_P) it is possible to achieve $V_P \ll V_{DD}$ and, hence, $V_1 \approx V_{DD}$.

2. If V_{IN} is a logical '1' (i.e., $V_{IN} = V_1$, $V_1 > V_{TN}$) then Q_N is on. If $|V_1 - V_{DD}| < |V_{TP}|$ then Q_P is off. The output voltage in this case is given by

$$V_{OUT} = V_0 = V_N \qquad (38)$$

where V_N is the voltage drop across Q_N and is given by

$$V_N = I_N R_N \simeq I_N \frac{1}{\beta_N (V_1 - V_{TN})}$$

where I_N is the current sunk by Q_N from Q_P and from the loads. If the loads are MOS gates then I_N is a leakage current and $V_N \approx 0$. However, if the loads are bipolar circuits (e.g., $T^2 L$ gates) then I_N could be considerable. In this case V_0 can be reduced by increasing β_N.

Thus, the logic levels of a CMOS inverter can be made close to V_{DD} and ground; and a logic swing V_ℓ of the order of V_{DD} results. This is a main feature of CMOS gates. The threshold voltage of CMOS logic gate V_t (see Appendix B for the definition of V_t) is close to $(V_{DD}/2)$ if symmetrical devices are used, i.e., $\beta_N = \beta_P$. The $\beta_N = \beta_P$ condition can be achieved if

$$\left[\left(\frac{W}{L} \right)_P \Big/ \left(\frac{W}{L} \right)_N \right] = \frac{\mu_N}{\mu_P} \approx 2.5.$$

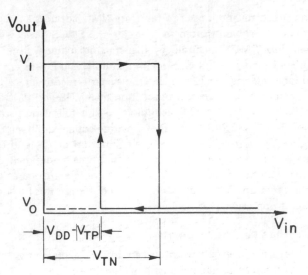

Fig. 22. Hysterises transfer characteristic of a CMOS inverter, $V_{DD} <$ $(V_{TN} + |V_{TP}|)$.

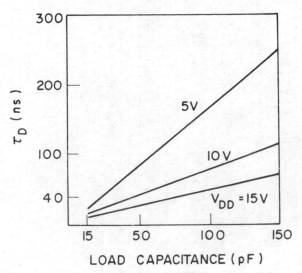

Fig. 23. Delay vs. load capacitance for different values of V_{DD} for a given CMOS inverter.

Thus, equal noise margins result which are $\approx (V_{DD}/2)$. This is another feature of CMOS gates.

As shown in Fig. 21(c), the gate current I passes between V_{DD} and ground only when both Q_P and Q_N are on, i.e., during the transition region. During the two steady states, $I \approx I_{leakage}$ and the dc power dissipation of the gate is given by

$$P_{DC} = V_{DD} I_{leakage} \qquad (39)$$

This power dissipation is a negligible component of the total power dissipation of the gate as will be explained.

From Fig. 21(b) it is clear that the minimum value of V_{DD} is given by:

$$V_{DD}|min = V_{TN} + |V_{TP}| \qquad (40)$$

If V_{DD} is lower than that value, the gate demonstrates a hysteresis transfer characteristic as shown in Fig. 22, and it can not be used as a logic gate.

Transient Analysis—If we consider a CMOS inverter with an ideal step input voltage from 0 to V_{DD}, the charging and discharging times can be calculated following the approach outlined in 3.2. The discharging time, t_{dis}, which is the time taken for the output capacitance C_{out} to discharge through Q_N from an output logic level of $V_1 \simeq V_{DD}$ to an output logic level of V_0, is given by:

$$t_{dis} = \tau_N \left[\frac{2}{\dfrac{V_{DD}}{V_{TN}} - 1} + \ell n \left(\frac{2(V_{DD} - V_{TN})}{V_0} - 1 \right) \right] \qquad (41)$$

where

$$\tau_N = \frac{C_{out}}{g_{mN}} = \frac{C_{out}}{\beta_N (V_{DD} - V_{TN})}$$

Equation (41) is similar to (28). Because of the symmetry of the problem, the charging time t_{ch} which is the time taken for C_{out} to charge through Q_P from V_0 to V_1 is given by:

$$t_{ch} = \tau_P \left[\frac{2}{\left(\dfrac{V_{DD}}{|V_{TP}|}\right) - 1} + \ell n \left(\frac{2(V_{DD} - |V_{TP}|)}{V_{DD} - V_1} - 1 \right) \right] \qquad (42)$$

where

$$\tau_P = \frac{C_{out}}{g_{mp}} = \frac{C_{out}}{\beta_p (V_{DD} - |V_{TP}|)}$$

It should be noted that (41) and (42) can be simplified as in (28a).

Another useful transient parameter is the delay through a CMOS inverter τ_D, defined as the delay time between the input and the output waveforms measured at the $(V_{DD}/2)$ points for a chain of CMOS inverters or a ring oscillator. This is given by [13]:

$$\tau_D \simeq \frac{0.9 \, C_{out}}{V_{DD} \beta_N} \left[\frac{1}{\left(1 - \dfrac{V_{TN}}{V_{DD}}\right)^2} + \frac{1}{\dfrac{\beta_P}{\beta_N} \left(1 - \dfrac{|V_{TP}|}{V_{DD}}\right)^2} \right] \qquad (43)$$

If $(V_{TN}/V_{DD}) \ll 1$ and $|V_{TP}|/V_{DD} \ll 1$ then:

$$\tau_D \simeq \frac{0.9 \, C_{out}}{V_{DD} \beta_N} \left[1 + \frac{\beta_N}{\beta_P} \right] \qquad (43a)$$

i.e., under these conditions V_{TN} and V_{TP} have only a small effect on the delay time and τ_D is linearly proportional to (C_{out}/V_{DD}) and β_N^{-1} ($\beta_N = \beta_p$ by design). This result is demonstrated in Fig. 23.

For $V_{DD} = 5$ V, $K' = 25$ $\mu A/V^2$, $\beta_N = \beta_p$, the simplification of (41) and (42) gives $t_{ch} = t_{dis} = 30$ nsec while (43a) gives $\tau_D = 14.4$ nsec for a $C_{out} = 1$ pF and $(W/L)_N = 1$. This compares favorably with the corresponding transient performance of NMOS circuits.

Transient Power Dissipation—The transient power dissipation of a CMOS inverter has two components. The first results

26

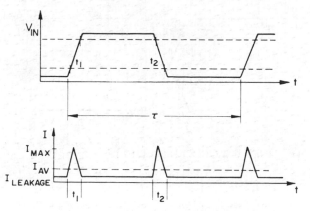

Fig. 24. Input voltage and current waveforms for a CMOS inverter.

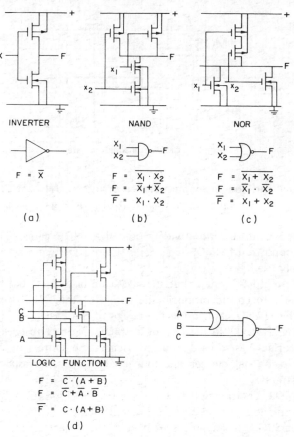

Fig. 25. CMOS logic gates (a) inverter; (b) two-input NAND; (c) two-input NOR; (d) a two-input complex logic gate.

from charging and discharging the output capacitance through the finite on resistance of Q_N and Q_P. This component is given by:

$$P_{t1} = C_{out} V_\ell^2 f \qquad (44)$$

where C_{out} is the total output-node capacitance including the capacitive effect of the interconnections, V_ℓ is the logic swing of the gate ($V_\ell = V_1 - V_0 \simeq V_{DD}$), and f is the frequency of switching. The second transient power dissipation component arises from the fact that the input (and hence the output) voltage waveforms have finite rise and fall times. The waveform of the current supplied by V_{DD} to a CMOS gate is shown in Fig. 24, and as a result a power dissipation P_{t2} occurs:

$$P_{t2} = V_{DD} I_{av},$$

$$I_{av} = \frac{1}{2} I_{max} \left[\frac{V_{DD} - (V_{TN} + |V_{TP}|)}{V_{DD}} \right] \left(\frac{t_1 + t_2}{\tau} \right) \qquad (45)$$

where

$$I_{max} = \frac{V_{DD}}{R_N + R_p}$$

If t_1 and $t_2 \ll \tau$ then $P_{t2} \ll P_{t1}$.

CMOS Delay-Power Trade-Offs—In CMOS circuits the main component of power dissipation is P_{t1}. Thus, the total power dissipation at the maximum frequency of operation is given by:

$$P \approx C_{out} V_\ell^2 / 2\tau_D \qquad (46)$$

and the delay-power product is given by:

$$\tau_D P = 0.5 C_{out} V_\ell^2 = 0.5 C_{out} V_{DD}^2. \qquad (47)$$

This shows that the delay-power product in CMOS digital circuits, as in the case of NMOS circuits, is proportional to $C_{out} V_{DD}^2$.

6.2 CMOS Static Gates

Logic gates (and hence flip-flops) can be realized using the complementary pairs. Figures 25(b) and (c) show a two-input NAND and a two-input NOR. Each input requires a complementary pair. In the case of the NAND gate the p-channel devices are connected in parallel while the n-channel devices are connected in series. But in the case of the NOR gate the

n-channel devices are connected in parallel while the p-channel devices are connected in series.

In CMOS logic gates the standby dc current of the gate is ≈ 0 because there is no dc path between V_{DD} and ground for any logic combination of the input. For example, for the two-input NOR and NAND gates shown in Fig. 25 for any input logic combination ($x_1 x_2 = 00, 01, 11$ or 10) the dc current of the gate is zero.

The design of a CMOS logic gate follows that of an inverter. First, an inverter is designed to meet a given dc and transient performance, and $(W/L)_N$ and $(W/L)_P$ are determined. Then $(W/L)_N$ and $(W/L)_P$ of the devices of a logic gate are determined as follows. If a CMOS m-input NAND gate is to be designed to have the same dc and transient performance as that of the inverter, then for the same values of C_{out}: $(W/L)_P$ of the NAND gate devices should be $\geqslant (W/L)_P$ of the inverter while $(W/L)_N$ should be $\geqslant m(W/L)_N$ of the inverter. On the other hand, if a CMOS m-input NOR gate is to be designed to have the same dc and transient performance as that of the inverter, then for the same values of C_{out}, $(W/L)_P$ of the NOR gates should be $\geqslant m(W/L)_P$ of the inverter, while $(W/L)_N$ should be $\geqslant (W/L)_N$ of the inverter. Note that the increase in the size of the NMOS devices in the case of the NAND gate and the increase in the size of the PMOS devices in the case of the NOR gates allows the logic levels V_0 and V_1 to be the same as that of the inverter if the dc currents are nonzero. The size of these transistors should be further increased because C_{out} is larger than that of the inverter. Note also that for the same perfor-

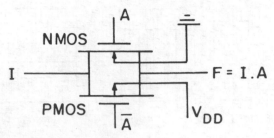

Fig. 26. CMOS transmission gate.

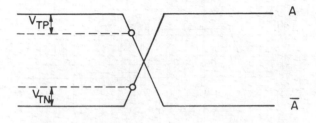

(a) INPUT LOGIC LEVEL (I) $\approx \dfrac{V_{DD}}{2}$

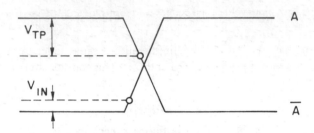

(b) INPUT LOGIC LEVEL (I) $\approx$ 0

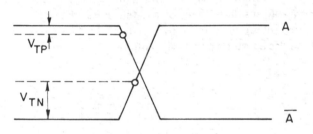

(c) INPUT LOGIC LEVEL (I) $\approx V_{DD}$

Fig. 27. CMOS transmission gate waveforms.

mance and for the same number of inputs the NAND gate consumes less silicon area than that of a NOR gate because of the smaller area taken by the NMOS devices. Hence, CMOS NAND gates are more widely used than NOR gates. This is the opposite of NMOS logic gates where NOR gates are more widely used.

It should be noted that the back-gate bias effect has to be taken into consideration in the design of the NMOS devices of the NAND gate and in the design of the PMOS devices of the NOR gate. In the case of a CMOS inverter, no back-gate bias effect has to be considered because the sources of the n- and p-channel devices are connected to their corresponding substrates.

CMOS Transmission Gates—As shown in Fig. 26, a CMOS transmission gate consists of a complementary pair connected in parallel. It acts as a switch, with the logical variable A as the control input. Let us assume that A is connected to the n-channel device and $\overline{A}$ is connected to the p-channel device. If A is high, the gate is on and acts as a switch with an on resistance of R_N and R_P in parallel. If A is low, the gate is off and presents a high resistance between the terminals. It should be noted that in CMOS transmission gates the back-gate bias effect on both the NMOS and PMOS devices must be taken into account in designing CMOS transmission gates to meet specific dc and transient requirements.

The advantage of using a complementary pair, rather than a single NMOS or PMOS device, to realize a transmission gate in CMOS can be explained with the aid of Fig. 27. Although in Fig. 27 we assume that there is no time delay between A and $\overline{A}$, the conclusion of the following discussion is general. Figure 27 shows A and $\overline{A}$ waveforms and also shows the times at which the NMOS and PMOS devices turn on. Figure 27(a) shows these times for a reference case where the input logic level is $\approx (V_{DD}/2)$, and as a result $V_{TN} \simeq |V_{TP}|$. In this case the two devices turn on at the same time. Fig. 27(b) shows these times for the case where the input logic level $\approx$ 0, and $|V_{TP}| > V_{TN}$ because of the back-gate bias effect. In this case the NMOS device turns on before the PMOS and the transmission gate delay time between the input and the output is almost the same as in case (a). Similarly, if the input logic level $\approx V_{DD}$ and as a result $V_{TN} > |V_{TP}|$, as shown in Fig. 27(c), the delay time of the gate will be unaffected. In conclusion, independent of the voltage level of the input variable of the CMOS transmission gate, the gate delay time is approximately the same. It is easy to see that this is *not* the case when single-channel type is used in designing transmission gates.

A drawback of the CMOS transmission gate is that it con-

sumes more area than a single-channel transmission gate. Thus, if the area is of prime concern, non-complementary n-channel transmission gates are used.

6.3 CMOS Static Flip-Flops

The basic building blocks of CMOS flip-flops are the CMOS inverter, logic gates, and transmission gates. Figure 28 shows a cross-coupled CMOS static flip-flop. In the storage mode where V_A is high, V_B is low, Q_1 and Q_4 are on while Q_2 and Q_3 are off. Similarly, in the storage mode where V_A is low, V_B is high, Q_1 and Q_4 are off while Q_2 and Q_3 are on. The standby power dissipation of the cell is very small. The state of the flip-flop is changed by using two CMOS transmission gates connected to V_A and V_B nodes as shown in Fig. 28.

Figure 29 shows a D-type CMOS master-slave static flip-flop. Each of the master and the slave requires the presence of the clock to latch up and store the information. Thus they differ from the basic cross-coupled configuration of Fig. 28. The transmission gates are represented with ideal switches: $(TG)_1$, $(TG)_2$, $(TG)_3$, and $(TG)_4$. The transmission gate $(TG)_1$ con-

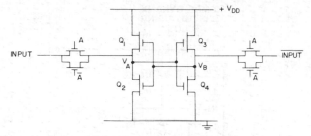

Fig. 28. CMOS cross-coupled static flip-flop.

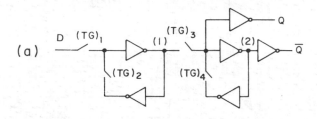

(a)

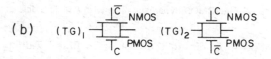

(b)

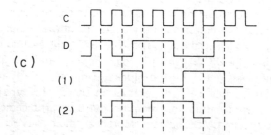

(c)

Fig. 29. CMOS Master-Slave static D-type flip-flop (a) circuit diagram; (b) transmission gate types; (c) timing diagram.

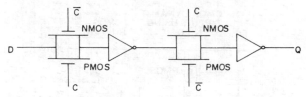

Fig. 30. CMOS Master-Slave dynamic D-type flip-flop.

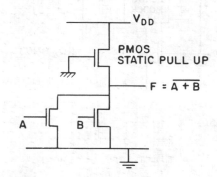

Fig. 31. Pseudo-NMOS NOR-circuit.

nects the master section to the input data (D) when the clock input (C) is low, while $(TG)_3$ connects the slave section to the output of the master section (node 1) when the clock input (C) is high. The gates $(TG)_2$ and $(TG)_4$ are identical to $(TG)_3$ and $(TG)_1$ respectively, and are used to latch the master and the slave sections during C and $\bar{C}$ time slots respectively. Figure 29(c) shows the flip-flop timing diagram.

6.4 CMOS Dynamic Flip-Flops

The D-type static flip-flop can be modified to obtain the dynamic one shown in Fig. 30. Although the term dynamic flip-flop is commonly used to refer to Fig. 30, each stage is basically a dynamic shift-register stage. Each stage consumes less area than its corresponding stage, Fig. 29, because of the absence of the latching paths. The information is stored on the output node capacitances of the two inverters. As in dynamic circuits, the minimum clock frequency is limited by the leakage paths at these output nodes.

The D-type CMOS flip-flop is the simplest to realize and it is advantageous to use it in LSI subsystem designs as a storage or shift register element. However, if other flip-flop types are required, e.g., JK-flip-flops, the D-type flip-flop can be modified to realize these types.

6.5 Dynamic CMOS Circuits

Static CMOS gates tend to become inefficient in terms of chip area because an n-input logic gate requires $2n$ transistors while the equivalent NMOS gate requires $(n+1)$ transistors. The gate area can be reduced if CMOS circuits are designed in a similar way to NMOS circuits; a PMOS device is used as a load to replace the depletion-type device in NMOS circuits. These types of circuits are referred to as pseudo-NMOS, and an NOR gate is shown in Fig. 31. The design of this class of circuit follows that of static NMOS circuits with no real advantage of using this circuit over NMOS except that it is compatible with CMOS technologies.

To reduce the area taken by CMOS circuits, the dynamic mode of operation can be used. Both PMOS and NMOS devices are used to an advantage and the principles of dynamic logic outlined in Section 5 are applied. One such circuit is shown in Fig. 32, and is referred to as a domino CMOS circuit [20]. The domino gate shown in Fig. 32 consists of a dynamic CMOS circuit followed by a static CMOS buffer. The dynamic circuit consists of a PMOS precharge transistor Q_1, an evaluation NMOS transistor Q_2, a storage capacitor C, and an N logic block which is a series-parallel combination of NMOS devices activated by the inputs and implementing the required logic.

During the precharge time T_1, the storage capacitance is charged through the PMOS device Q_1, the inputs have no effect during T_1, and the output of the buffer is low; hence the input transistors of the next stage are off. Also during T_1, the next stage is evaluating the logic state since it operates from $\bar{\phi}$.

During the evaluation time T_2, Q_2 is on, and depending on the logic performed by the N logic block, the voltage at node (a) is either discharged through the logic block and Q_2, or the voltage will be maintained. The output voltage at the OUT node will be the inversion of the voltage at node (a).

The circuit has the following features. Domino gate opera-

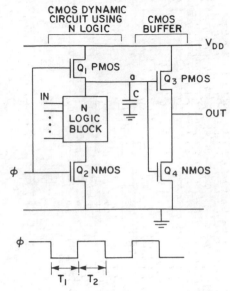

Fig. 32. Domino CMOS dynamic circuit with N logic block.

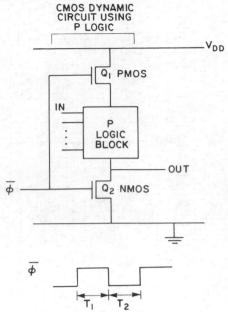

Fig. 34. Domino CMOS dynamic circuit with P logic block.

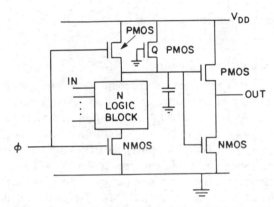

Fig. 33. The circuit of Fig. 32 with a pull-up PMOS transistor at the storage node.

tion requires a single clock, ϕ to operate one stage and $\bar{\phi}$ to operate the loading stage. Since the dynamic circuits are buffered by the CMOS static inverters, the dynamic nodes [as node (a)] are isolated from the loading capacitances. Thus, the gate is suitable for high fan-out operation. Because the gate is efficient in area for high fan-in (it requires $(n + 4)$ transistors compared to $2n$ for CMOS static gate), it is suitable for realizing complex logic functions. Because the output of the gate is always low during precharge time, the output node can only go high if switching occurs; hence the gate operation is glitch free if the inputs to the gates are set up during the precharge time.

One limitation of the gate is that it implements noninverting logic functions. However, this is not a serious limitation and can be overcome if the need arises by using CMOS static inverters. The designer can mix both static and dynamic CMOS logic circuits in a given design to optimize the overall performance.

In some applications, it is desirable to have dynamic logic operation at low frequency. This can be obtained in a domino circuit by the addition of a low-current (low W/L) pull-up PMOS transistor Q as shown in Fig. 33. This transistor is chosen small enough so there is no significant impact on the

pull-down operation during the evaluation time. Adding Q will allow a dc path during evaluation time and the gate power dissipation will increase.

In Figure 32 the domino gate uses a dynamic circuit with an N logic block. The complementary circuit is shown in Fig. 34 where the percentage transistor is an NMOS Q_2, the evaluation transistor is a PMOS Q_1 and the logic implementation uses PMOS transistors in the P logic block. Time slot T_1 is the precharge time, while T_2 is the evaluation time. This type of circuit can also be used in a domino dynamic arrangement separated by static CMOS buffers. Moreover, the two types of dynamic CMOS circuits, one with N logic and the other with P logic, can directly drive each other without the use of buffers [21].

The circuit design of CMOS dynamic circuits follows that of NMOS dynamic circuits, and overall performance optimization of these circuits, including the buffering stages, can be performed [22].

6.6 Digital Design with NMOS and CMOS

NMOS circuits can be considered to be a subset of CMOS. The circuit designer can use a CMOS technology to design both type of circuits operating in either a static or a dynamic mode. CMOS offers lower power dissipation at the expense of higher silicon area. At low frequency of operation, the CMOS power dissipation is a few nanowatts. In static NMOS, because of the dc path, dc power dissipation may be of the order of tens of microwatts and dominate the low-frequency operation. When the dynamic power dissipation $CV_1^2 f$ becomes dominant, the total power dissipation of the two types of circuits approach each other, as shown in Fig. 35.

The dc power dissipation of static NMOS circuits can be eliminated in dynamic circuits at the expense of higher overhead circuitry. Similarly, the high silicon area of static CMOS can be reduced in dynamic circuits at the expense of higher overhead circuitry. The NMOS current-mode logic configuration

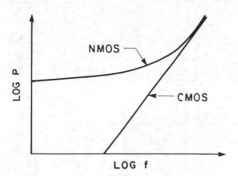

Fig. 35. Power versus frequency of operation for NMOS and CMOS.

can be used for high fan-out high driving capability circuit operation.

Memory design has tended to remain largely NMOS dominated [23], mainly due to the area efficiency and because clocking circuitry can be shared among memory arrays. However, CMOS can be used in peripheral circuit design to reduce the standby power.

In design automation (DA), it is easier to deal with static NMOS and CMOS circuits. Dynamic circuits are sensitive to loading effects and this puts constraints on the design automation tool. CMOS topologies are more complex than NMOS and do not lend themselves easily to design automation layout tools. Thus, one has to be innovative in designing the layout structures and floor plans for CMOS DA tools to avoid slow speed of operation and high silicon area. Dynamic CMOS circuits can be candidates for such tools because of their simpler layout topology.

The high noise immunity of CMOS circuits which can approach half the value of the power supply makes CMOS circuits very attractive in noisy system environments such as in automobile electronics. The low-power supply requirements make CMOS circuits attractive for portable electronic environments.

7. COMMENTS AND DISCUSSION

In this paper a simple analysis of some basic NMOS and CMOS digital circuits is given. This analysis is useful in the first phase of circuit design. At a later phase, circuit analysis computer programs with adequate models are used to analyze and to aid in the design of MOS integrated circuits. It is suggested that the reader use these programs to compare the computed performance to that predicted by the simple analysis. We like to stress here that a simple circuit analysis, with reasonable approximations, is *always* possible even for the most complicated circuit configurations. It is hoped that the reader would develop such analysis for the proposed projects of Appendix C and for the circuit configurations covered in this book.

The following is a summary of some basic guidelines, based on the material covered in this paper, to be used in the design of digital MOS integrated circuits:

1. A simple digital model for an MOS transistor is that it is a switch, either on or off, which can be controlled by the gate voltage. The off resistance of this switch, between the drain and the source, is very high and is close to an open circuit. The on resistance can be reduced by increasing the (W/L) of the transistor or by increasing the gate voltage.

The gate voltage required to turn-on or turn-off the transistor is V_T and is is a function of process parameters, the substrate bias V_{BB}, and, in general, the drain-source potential $V_{DS} \cdot V_T$ can be controlled by ion implanting the MOS channel. Thus, the MOS circuit designer can choose from MOS transistors with different values of V_T. For example, an NMOS transistor could be enhancement-type, e.g., $V_T = 0.7$ V, weak-enhancement, $V_T = 0.3$ V, depletion-type, $V_T = -2$ V, weak-depletion, $V_T = -1$ or zero-V_T type, $V_T \approx 0$.

2. The drain current I_{DS} of an MOS transistor is a function of the operating voltages.

Below V_T, the MOS transistor allows a "small" I_{DS} to flow and in this region I_{DS} increases exponentially with V_{DS} and V_{GS}. This region of operation is called the subthreshold region, and the value of I_{DS} is much smaller than when V_{GS} is greater than V_T.

Above V_T, I_{DS} is a function of $(V_{GS} - V_T)$ and V_{DS}. At very low V_{DS} where $V_{DS} \ll (V_{GS} - V_T)$, I_{DS} is approximately linearly proportional to the product $(V_{GS} - V_T) V_{DS}$. As V_{DS} increases, I_{DS} becomes less dependent on V_{DS}, and I_{DS} approximately "saturates" for $V_{DS} \geq (V_{GS} - V_T)$, i.e., the rate of change of I_{DS} with respect to V_{DS} reduces. This is called the saturation region.

For the same V_T and the same operating voltages, I_{DS} is linearly proportional to (W/L).

3. It is useful to restate guideline 2 from another viewpoint—by considering the voltage drop across the gate-source terminals—V_{GS} required for a given I_{DS} (see Appendix C, project 5).

Above V_T, the required V_{GS} is the summation of V_T plus *current dependent* terms. These terms can be made small by increasing (W/L) or by reducing the operating current. For example in the saturation region V_{GS} is given by:

$$V_{GS} = V_T + \left(\frac{2I_{DS}}{\beta}\right)^{1/2}$$

4. If an enhancement-type NMOS transistor has its gate connected to its drain, the resulting two terminal device conducts in one direction in a diode-like manner with a square-law I-V characteristic as shown in Fig. 36(a):

$$I = \frac{K'}{2} \left(\frac{W}{L}\right) (V - V_T)^2$$

Alternatively, if a depletion-type NMOS transistor has its gate connected to its source, the resulting two terminal device conducts in both directions as shown in Fig. 36(b). For positive small values of V, it acts as a modulated nonlinear resistance and for positive large values of V it acts as a current source. For negative values of V, it conducts in a diode-like manner with a square-law I-V characteristic similar to the enhancement-type case.

5. Although PMOS are currently used to fabricate a number of MOS products, in the state-of-the-art MOS chips, PMOS transistors are generally limited to CMOS circuits. Single-channel state-of-the-art chips use NMOS circuits.

6. During the operation of static digital MOS circuits, the output logic node is either pulled "up" towards V_{DD} through

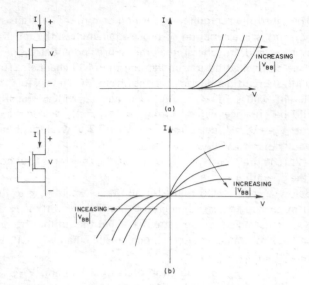

Fig. 36. (a) Two terminal enhancement-type NMOS; (b) two terminal depletion-type NMOS.

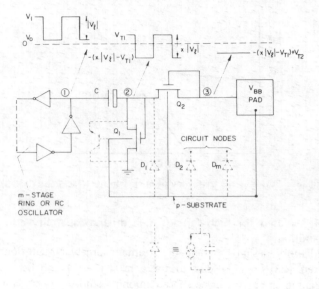

Fig. A1. Substrate bias generator: circuit diagram, waveforms, and substrate diode equivalent circuit.

pull-up polyresistors or MOS load transistors, or pulled "down" towards ground through pull-down MOS drivers. In NMOS circuits, the pull-up MOS load transistors are either enhancement or depletion type while in CMOS circuits they are enhancement PMOS transistors. This MOS circuit configuration is referred to as the pull-up pull-down (PUD) configuration.

In NMOS circuits with depletion-type loads, the pull-down circuit path is "stronger" than the pull-up path, thus V_0 would have the range 0.1 to 0.05 V_{DD} while V_1 would be $\simeq V_{DD}$. However, in CMOS circuits the pull-down and the pull-up circuit "strengths" are usually the same and $V_0 \simeq 0$ and $V_1 \simeq V_{DD}$.

7. Not only do the pull-up and pull-down circuit paths in static MOS gates affect V_0 and V_1, but also they affect t_{ch} and t_{dis} by affecting the average available currents, $\bar{I}_{ch}$ and $\bar{I}_{dis}$, to charge and discharge the output node capacitance C_{out}. Because MOS transistors do not suffer from charge storage effects, as in bipolar transistors, t_{ch} and t_{dis} can be expressed simply by $V_\ell C_{out}/I_{ch}$ and $V_\ell C_{out}/I_{dis}$ respectively.

8. In addition to polyresistors, NMOS and PMOS transistors with various values of V_T, other circuit elements which may be available to MOS digital circuit designers in a given technology are MOS capacitances which are formed between gates and MOS channels and poly capacitances which are formed between two layers of poly or between poly and metal.

9. MOS transistors and capacitances can be arranged to realize NMOS or CMOS dynamic circuits which are controlled by multiphase clocks where information is stored on capacitances.

10. In MOS circuits, especially dynamic and CMOS circuits, a transient power dissipation, $CV_\ell^2 f$ can be an important component of the total power dissipation.

ACKNOWLEDGMENT

The author would like to thank Dr. L. M. Terman of IBM for his critical review of the original manuscript and for his many valuable comments; and Dr. Y. A. El-Mansy of Intel Corporation and H. Tuan, L. Peterson, and D. Pearson of the Microcomponents Organization, Burroughs Corporation, for useful discussions. Feedback from students attending my integrated circuit classes at Waterloo has been extremely valuable.

APPENDIX A

Substrate Bias Generators

In many circuit applications it is desirable to reverse bias the substrate of an NMOS chip. Reverse biasing the substrate-source junctions, rather than zero biasing them, reduces the parasitic capacitances of these junctions and reduces the injected carriers from the source regions into the substrate. Reducing the parasitic capacitances enhances the performance of both static and dynamic NMOS digital circuits. Reducing the injected carriers is important in NMOS dynamic memories because these carriers contribute to the discharging of the storage nodes. Moreover, a substrate bias reduces the sensitivity of V_T on V_{BB} and increases the field-oxide threshold voltage.

From the system viewpoint, it is convenient to have the reverse bias of the substrate generated on-chip, saving a V_{BB} pin and eliminating the need for generating a negative power supply.

An on-chip substrate bias generator, as shown in Fig. A1, consists of an oscillator and a charge pumping circuit. The oscillator is either a ring or an RC type which consists of an odd number of inverters, usually 5 or 7. An advantage of the RC oscillator is that its frequency of oscillation is less sensitive to processing parameters. The R and C are integrated on chip, using a depletion-type NMOS as an R and an MOS capacitor as a C as shown in Fig. A2. The charge pumping circuit consists of an MOS capacitance C and two rectifying (diode-like connected) enhancement NMOS transistors Q_1 and Q_2. The output of the circuit, node 3, is connected to a pad which is connected, on the package, to the back of the chip (the substrate). Figure A1 shows the substrate parasitic diodes: D_1 is associated with node 2 while D_2 to D_m are associated with the different circuit nodes. These nodes are at different positive potentials and some of them are at $+V_{DD}$. These diodes represent leakage currents and parasitic capacitances as shown in Fig. A1.

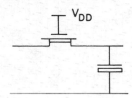

Fig. A2. RC network for RC oscillators.

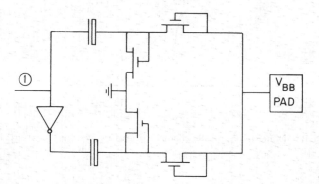

Fig. A3. Double pumping circuit.

Also shown in Fig. A1 is a parasitic bipolar transistor associated with Q_1.

The circuit operates as follows. The oscillator delivers a waveform at node 1, roughly a square wave with logic swing V_ℓ, from V_0 to V_1. The frequency of oscillation is typically 10 MHz, and V_0 and V_1 are typically 0.05 V_{DD} and 0.95 V_{DD} respectively. Neglecting leakage, C and Q_1 clamp the input wave form to a V_{T1} above ground, as shown at node 2, with a voltage swing of χV_ℓ where χ is determined by C and C_1 where C_1 is the parasitic capacitance at node 2: $\chi \simeq C/C_1 + C$ and approaches unity for $C \gg C_1$. Q_2 and the large capacitance at node 3 (the summation of the capacitances of D_2 to D_m) stabilize the voltage at node 3 to $[-(\chi|V_\ell| - V_{T1}) + V_{T2}]$.

Because Q_2 has less substrate bias than Q_1, $V_{T2} < V_{T1}$. Q_2 can be eliminated from the circuit and in this case D_1 and the large capacitance at node 3 stabilizes the voltage at node 3 to $[-(\chi|V_\ell| - V_{T1}) + V_j]$ where V_j is the turn-on voltage of D_1. Because V_{T2} can, by design, be made smaller than V_j, the value of V_{BB} is more negative in the case where Q_2 is used.

In the case of dc loading conditions when leakage currents pass through the substrate diodes, the output voltage V_{BB} at node 3 would be less negative than its open circuit value. The value of V_{BB} under loading conditions can be enhanced by the double pumping circuit shown in Fig. A3.

Because the substrate-drain junction of Q_1 is slightly forward biased by an amount $= V_{T2}$, a parasitic bipolar transistor results which makes the voltage at node 2 less negative. This bipolar action can be reduced by increasing the channel length of Q_1.

A drawback of using an on-chip substrate bias generator is the fact that the substrate terminal is not dc grounded, i.e., it floats. As the circuit nodes switch, because of the capacitive coupling between these nodes and the substrate, the substrate potential is pulsed around its dc value. The value of this pulse is proportional to $C_a/C_a + C_b$ where C_a is the capacitance associated with the switching nodes and C_b is the capacitance asso-

ciated with the nonswitching nodes including V_{DD} (ac ground). Although usually $C_a \ll C_b$, in some chips where simultaneous switching is occurring at the same time, as in the case of dynamic memories where signal lines are precharged simultaneously, the effect of pulsing the substrate on the circuit performance should be evaluated.

APPENDIX B

Logic Levels and Noise Margins

The operation of a digital binary circuit is quite distinct from that of a linear circuit in that devices are intended to be either "on" or "off." The input and output voltages are at two levels, either "high" representing a logic '1,' or "low" representing a logic '0.' These two levels are actually a band of voltages separated by a buffer voltage zone to ensure that the state of a device can be determined even in the presence of noise.

In a digital system, two basic operations are performed. The first operation is to realize a logic function, that is to determine a logic output variable in terms of input logic variables (e.g., NAND gate) or to determine a logic output variable in terms of the input logic variables and the previous state of the output (e.g., a flip-flop). The second operation performed in a digital system is transferring a logic signal from one logic block (logic source) to the input of another logic block(s) (logic detector(s)) through a transmission system (Fig. B1). This transmission can be either on-chip (metal lines), or between chips (on a printed-circuit board). A logic block has usually more than one input and each output is used as an input for a number of other logic blocks. The number of inputs is called the *fan-in* and the number of blocks driven by one output is called the *fan-out*.

Ideally, at the input of a logic detector, the voltage signal is detected with respect to a *threshold voltage* V_t. Although in NMOS digital circuits V_t is related to the MOS device threshold voltage V_T, V_t is a parameter defined for any digital logic block: in ideal inverting logic blocks, if the input voltage is greater than V_t the output voltage will be at logical '0' and if the input voltage is less than V_t the output will be at logical '1.'

Signal distortion, due to attenuation or system noise, causes a reduction of the system reliability. Since distortion is primarily introduced by the noise in the system, it is conventional to refer to the margin allowed for distortion of the logic levels through the transmission system as the noise margins. Two *voltage noise margins* are defined (Fig. B1):

$$VNM_0 \equiv \text{voltage noise margin for a logical '0'}$$
$$\equiv V_{t\min} - V_{0\max}$$
$$VNM_1 \equiv \text{voltage noise margin for a logical '1'} \qquad \text{(B1)}$$
$$\equiv V_{1\min} - V_{t\max}$$

The difference between the two levels V_1 and V_0 is called the *logic swing* V_ℓ.

DC Transfer Characteristics

Noise margin definitions have been stated by considering the transmission system. Alternatively, the dc transfer character-

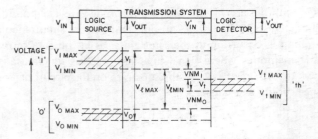

Fig. B1. System model for digital circuits, logic levels, and noise margins.

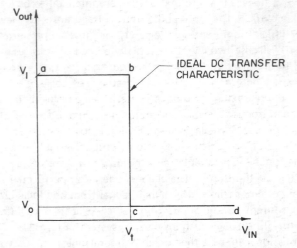

Fig. B2. Ideal dc transfer characteristic of an inverting digital circuit.

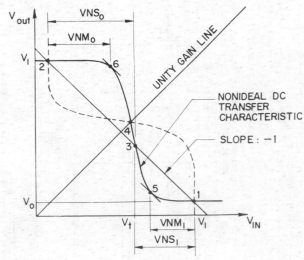

Fig. B3. Nonideal dc transfer characteristic of an inverting digital circuit.

istic (V_{OUT} vs. V_{IN}) of a logic gate also provides information regarding logic levels and noise margins. Figure B2 shows an ideal dc transfer characteristic of an inverting logic gate. It consists of three sections: two zero-gain (for steady-state operation) sections (ab and cd), where the output voltages are V_1 and V_0 respectively, and an infinite-gain transition section (bc) where the input voltage level is V_t. The noise margins of the circuit are the same as the ones given in Fig. B1.

Nonideal dc transfer characteristics, as shown in Fig. B3, have a finite gain in the transition region. Because the input and the output levels must be compatible, i.e., V_0 and V_1 must be the same for both input and output, in order to determine V_0 and V_1, V_{OUT} vs. V_{IN} is replotted (shown dotted) using the V_{OUT} axis as V_{IN} and the V_{IN} axis as V_{OUT}. This intersects the transfer characteristic at point (1) and (2), giving V_0 and V_1, and at the reference point (4). The line joining points (1) and (2) has a slope of -1 and intersects the transfer characteristic at point (3) where by definition $V_{IN} = V_t$. Two extra points on the transfer characteristic are of importance: (5) and (6) where the slopes are -1. Between these two points the gain is higher than unity.

In Fig. B3 the voltage noise margins (VNM) to the circuit are defined as the input voltage change required to cause the output to change from an operating point to the nearest unity gain point. Moreover, voltage noise sensitivities (VNS) are defined as the input voltage change required to cause the output voltage to change from an operating point to the threshold point. Note that if the dc transfer characteristic of Fig. B3 is approximated by an ideal one, then points 3, 4, 5, 6 coincide at $V_{IN} \equiv V_t$

and $VNM \equiv VNS$. The deviation of a transfer characteristic from an ideal one is measured by the voltage transition width: $VTW \equiv [V_{IN}@5 - V_{IN}@6]$. If VTW is small, the transfer characteristic can be approximated by an ideal one.

DC Transfer Characteristics: Examples

In this section we shall examine the transfer characteristics of an NMOS inverter with an enhancement-type NMOS driver transistor and one of the following types of loads:

(a) a resistive load
(b) a saturated enhancement-type NMOS load
(c) a nonsaturated enhancement-type NMOS load

The dc transfer characteristics are obtained analytically using two relations:

(i) the relationship between the driver current I_D and the input voltage V_{IN}:
if $V_{IN} - V_{TD} \leqslant V_{OUT}$, the driver is in saturation and

$$I_D = \frac{\beta_D}{2}(V_{IN} - V_{TD})^2 \qquad (B2)$$

if $V_{IN} - V_{TD} > V_{OUT}$, the driver is in the nonsaturation region and

$$I_D = \beta_D[(V_{IN} - V_{TD})V_{OUT} - \tfrac{1}{2}V_{OUT}^2] \qquad (B3)$$

(ii) the relationship between the load current I_L and the output voltage V_{OUT}:
(a) a resistive load:

$$I_L = \frac{V_{DD} - V_{OUT}}{R_L} \qquad (B4)$$

(b) a saturated enhancement-type load:

$$I_L = \frac{\beta_L}{2}(V_{DD} - V_{OUT} - V_{TL})^2 \qquad (B5)$$

(c) a nonsaturated enhancement-type load:

34

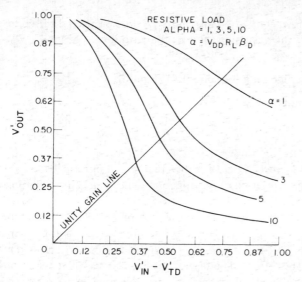

Fig. B4. DC transfer characteristic, resistive load.

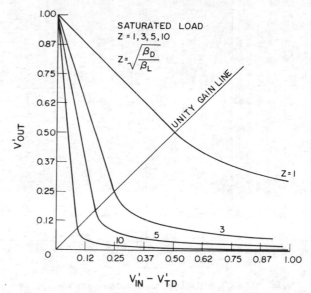

Fig. B5. DC transfer characteristic, saturated load.

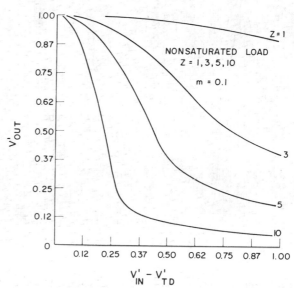

Fig. B6. (i) DC transfer characteristic, nonsaturated load, $m = 0.1$.

$$I_L = \beta_L \left[(V_{GG} - V_{OUT} - V_{TL})(V_{DD} - V_{OUT}) \right.$$

$$\left. - \frac{1}{2}(V_{DD} - V_{OUT})^2 \right]$$

$$= \frac{\beta_L V_{DD}^2}{2m} \left(1 - \frac{V_{OUT}}{V_{DD}}\right)\left(1 - m\frac{V_{OUT}}{V_{DD}}\right) \quad \text{(B6)}$$

where

$$m = \frac{V_{DD}}{2(V_{GG} - V_{TL}) - V_{DD}} \quad \text{for } V_{GG} - V_{TL} > V_{DD} \quad \text{(B7)}$$

The parameter m [8] is a measure of the biasing conditions on the load. As $V_{GG} \gg V_{DD}$ and $V_{GG} \gg V_{TL}$, m tends to zero and the load approaches a resistance load characteristic. The inequality shown in (B7) demonstrates the fact that m and hence (B6) are only valid if the load is operating in the nonsaturation region.

In computing the dc transfer characteristics we shall assume that there is no dc load current at the output of the inverter, i.e., $I_D = I_L$ [8].

A Resistive Load

(i) $V_{IN} - V_{TD} \leqslant V_{OUT}$

Equating (B2) and (B4) we obtain (Fig. B4):

$$1 - V'_{OUT} = V_{DD}\frac{R_L \beta_D}{2}(V'_{IN} - V'_{TD})^2$$

where V'_{OUT}, V'_{IN}, V'_{TD} are normalized voltages with respect to the maximum value of the output voltage V_{DD}.

(ii) $V_{IN} - V_{TD} > V_{OUT}$

Equating (B3) and (B4) we obtain (Fig. B4):

$$1 - V'_{OUT} = V_{DD}R_L\beta_D \left[(V'_{IN} - V'_{TD})(V'_{OUT}) - \tfrac{1}{2}V'^2_{OUT}\right]$$

A Saturated Enhancement-Type Load

(i) $V_{IN} - V_{TD} \leqslant V_{OUT}$

Equating (B2) and (B5) (Fig. B5):

$$(1 - V'_{OUT}) = \sqrt{\frac{\beta_D}{\beta_L}}(V'_{IN} - V'_{TD})$$

where V'_{OUT}, V'_{IN} and V'_{TD} are voltages normalized with respect to the maximum value of the output voltage $(V_{DD} - V_{TL})$.

(ii) $V_{IN} - V_{TD} > V_{OUT}$

Equating (B3) and (B5) (Fig. B5):

$$(1 - V'_{OUT}) = \sqrt{\frac{\beta_D}{\beta_L}}\left[2(V'_{IN} - V'_{TD})(V'_{OUT}) - V'^2_{OUT}\right]^{1/2}$$

A Nonsaturated Enhancement-Type Load

(i) $V_{IN} - V_{TD} \leqslant V_{OUT}$

Equating (B2) and (B6) (Fig. B6):

$$(1 - V'_{OUT})^{1/2}(1 - mV'_{OUT})^{1/2} = \sqrt{m}\sqrt{\frac{\beta_D}{\beta_L}}(V'_{IN} - V'_{TD})$$

35

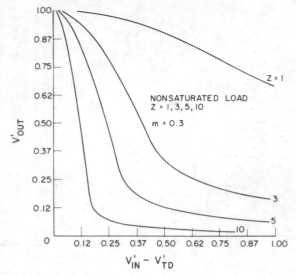

Fig. B6. (ii) DC transfer characteristic, nonsaturated load, $m = 0.3$.

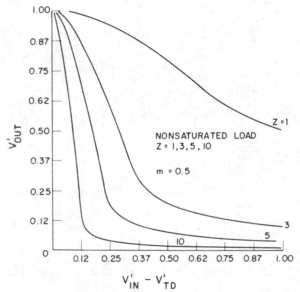

Fig. B6. (iii) DC transfer characteristic, nonsaturated load, $m = 0.5$.

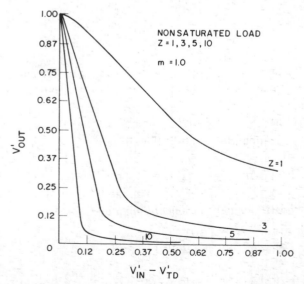

Fig. B6. (iv) DC transfer characteristic, nonsaturated load, $m = 1.0$.

where V'_{OUT}, V'_{IN} and V'_{TD} are normalized with respect to the maximum value of the output voltage V_{DD}.

(ii) $V_{IN} - V_{TD} > V_{OUT}$

Equating (B3) and (B6) (Fig. B6):

$$(1 - V'_{OUT})^{1/2}(1 - mV'_{OUT})^{1/2}$$

$$= \sqrt{m} \sqrt{\frac{\beta_D}{\beta_L}} \{2(V'_{IN} - V'_{TD})V'_{OUT} - V'^2_{OUT}\}^{1/2}$$

Figures B4 to B6 show that the transition width V_{TW} of the transfer characteristic, and hence the gain of the inverter, is a function of the circuit parameters: V_{TW} can be reduced by increasing $R_L \beta_D$ in the case of resistive loads, by increasing $\sqrt{\beta_D/\beta_L}$ in the case of saturated loads and by increasing $\sqrt{\beta_D/\beta_L}$, or by increasing m in the case of nonsaturated loads.

The threshold voltage of the inverter V_t is a function of V_{TD} and the circuit parameters. For example, in the case of a resistive load for $\alpha = 10$, $V_t \simeq V_{TD} + 0.4 V_{DD}$, in the case of a saturated load for $(\beta_D/\beta_L) = 9$, $V_t \simeq V_{TD} + 0.25 (V_{DD} - V_{TL})$, in the case of a nonsaturated load for $(\beta_D/\beta_L) = 9$, $m = 0.5$, $V_t \simeq V_{TD} + 0.4 V_{DD}$ and in all cases the higher the value of (β_D/β_L), the lower the value of V_t.

APPENDIX C

Digital MOS Integrated Circuits Projects

In the following projects make any necessary, but reasonable assumptions, e.g., value of available power supply; layout design rules [14]; process parameters: V_{TO}, γ, z . . . , etc.; and system/circuit constraints: speed of operation, allowable power dissipation, operating temperatures . . . , etc. Use any level of analysis, first-order analytical equations and/or computer simulations. If integrated circuit fabrication facilities are available, layout, fabricate, test, and evaluate your design.

1. Compare static NMOS integrated inverters with different loads regarding V_1, V_0, V_t, VNMs, P_{DC}, t_{ch}, t_{dis}, delay-power product and area.

2. Compare static CMOS logic circuits to static E/D NMOS (NMOS circuits with depletion-type load) regarding area, delay times and speed-power product. Use different logic circuits with different complexity in the comparison, e.g., inverter, four-input NAND (or NOR), full-adder, and a multiplier cell.

3. Design multiphase dynamic CMOS logic circuits [15]. Compare this design to dynamic 2ϕ and 4ϕ NMOS logic circuits regarding area, maximum frequency of operation, and total power dissipation.

4. In CMOS digital circuits $V_{DD}|_{min} = V_{TN} + |V_{TP}|$. What is $V_D|_{min}$ in static NMOS digital circuits? Compare the operations of the two circuit types at $V_{DD}|_{min}$ regarding V_0, V_1, V_t, VNM, t_{dis} and t_{ch}.

5. In bipolar transistors, the base-emitter voltage V_{BE} is given by:

$$V_{BE} \simeq \frac{KT}{q} \ln \frac{I_c}{J_S h_{fe} A_c}$$

where I_c is the operating collector current, J_S is the saturation current density, a process parameter, h_{fe} is the transistor current gain, a process parameter, and A_c is the collector-base junction area, a layout design parameter. Obtain a similar expression for the V_{GS} of an NMOS transistor. Compare V_{BE} to V_{GS} regarding sensitivity to processing parameters, circuit design parameters, operating currents and voltages, and temperature.

6. Design CMOS circuits for CMOS/TTL and TTL/CMOS interfaces.

7. Design NMOS circuits for NMOS/TTL and TTL/NMOS interfaces [16]. Compare with the circuits of project 6 regarding area.

8. Design cross-coupled static flip-flops with minimum standby power dissipation using (a) poly resistor loads, (b) depletion loads, and (c) complementary loads. Compare regarding area, power dissipation and speed of operation.

9. Design JK CMOS static and dynamic flip-flops using the D-type flip-flop to operate at 10 MHz clock. Compute static and dynamic power dissipations and area.

10. Design a 4ϕ JK NMOS flip-flop to operate at 10 MHz. Compare to that of project 9 regarding area, static, and dynamic power dissipations.

11. Design an 8-bit shift register to operate at 10 MHz and to consume the minimum area. Choose the technology and the circuit configuration: static or dynamic NMOS, static or dynamic CMOS.

12. Use a multiplier cell to compare the area taken by static NMOS and CMOS circuits with the following interconnection systems:

    (1) metal and poly; (2) metal, poly 1, and poly 2; and (3) poly 1, poly 2, and two layers of metal.

13. Design a substrate bias generator to have an open circuit voltage of −3 volts. Examine the design for loading conditions, capacitive coupling to the substrate, and power dissipation.

14. Design a static half-adder using (1) NMOS NOR gates, (2) NMOS single gate with series/parallel arrangement, (3) CMOS NAND gates, and (4) CMOS single gate with series/parallel arrangement to operate from $V_{DD} = 5$ V and to have a total $t_D = 20$ nsec, $V_0 \leqslant 0.1$ V and $V_1 \geqslant 4$ V. Compare the design regarding area, power dissipation, and VNMs.

15. Design a clock driver for dynamic circuits using (1) NMOS E/D and (2) CMOS. $V_{DD} = 5$ V, capacitive load = $C_\phi = 20$ pF. Calculate and compare V_ϕ, f_{max}, and power dissipation. Redesign the circuits for $V_\phi = 7$ V with $V_{DD} = 5$ V.

REFERENCES

[1] R. S. C. Cobbold, *Theory and Applications of Field-Effect Transistors*, Wiley Interscience: 1970.

[2] T. W. Sigman and R. Swanson, "MOS threshold shifting by ion implantation," *Solid State Electronics*, vol. 16, pp. 1217–1232, 1973.

[3] H. Masuda *et al.*, "Characteristics and limitation of scaled-down MOSFET's due to two-dimensional field effect," *IEEE Trans. on Electron Devices*, vol. ED-26, no. 5, pp. 980–986, June 1979.

[4] A. S. Grove, *Physics and Technology of Semiconductor Devices*, John Wiley: 1976.

[5] W. M. Penny and L. Lau, *MOS Integrated Circuits*, Van Nostrand Reinhold Company: 1972.

[6] Y. A. El-Mansy and A. R. Boothroyd, "A new approach to the theory and modeling of insulated-gate field-effect transistors," *IEEE Trans. on Electron Devices*, pp. 241–253, March 1977.

[7] M. I. Elmasry, *Digital MOS Integrated Circuits*, IEEE Press: 1981, Section 1.2.

[8] R. H. Crawford, *MOSFET in Circuit Design*, McGraw-Hill: 1967.

[9] M. I. Elmasry, *Digital MOS Integrated Circuits*, IEEE Press: 1981, Section 1.3.

[10] B. G. Watkins, "A low-power multiphase circuit technique," *IEEE JSSC*, pp. 213–220, December 1967.

[11] M. I. Elmasry, "An optimal two-phase MOS-LSI logic system," *International EE Conference—Toronto*, pp. 136–137, October 1973.

[12] M. I. Elmasry, *Digital MOS Integrated Circuits*, IEEE Press: 1981, Section 2.2.

[13] J. R. Burns, "Switching response of complementary symmetry MOS transistor logic circuits," *RCA Review*, pp. 627–661, December 1964.

[14] C. Mead and L. Conway, *Introduction to VLSI Systems*, Addison Wesley: 1980.

[15] Y. Suzuji *et al.*, "Clocked CMOS calculator circuitry," *IEEE JSSC*, pp. 462–469, December 1973.

[16] W. N. Carr and J. P. Mize, *MOS/LSI Design and Application*, McGraw-Hill: 1972.

[17] M. I. Elmasry, "Interconnection delays in MOSFET VLSI," *IEEE JSSC*, vol. SC-16, pp. 585–591, October 1981.

[18] M. I. Elmasry, "Nanosecond NMOS VLSI current mode logic," *IEEE Trans. on Electron Devices*, pp. 781–784, April 1982.

[19] R. A. Blauschild *et al.*, "An NMOS voltage reference," in *ISSCC Tech. Digest*, pp. 50–51, 1978.

[20] R. H. Krambeck *et al.*, "High speed compact circuits with CMOS," *IEEE JSSC*, vol. SC-17, pp. 614–619, June 1982.

[21] V. Friedman and S. Liu, "Dynamic logic CMOS circuits," *IEEE JSSC*, vol. SC-19, pp. 263–266, April 1984.

[22] J. A. Pretorius *et al.*, "Optimization of domino CMOS logic and its applications to standard cells," *Dig. of the Custom Integrated Circuits Conf.*, pp. 150–153, 1984.

[23] M. I. Elmasry, Ed., *Special Issue on VLSI*, Proc. IEE, vol. 130, Part I, June 1983.

Digital Bipolar Integrated Circuits: A Tutorial

MOHAMED I. ELMASRY, SENIOR MEMBER, IEEE

Abstract—This paper offers an overview of today's digital bipolar integrated circuits, their characteristics, and applications to the design of logic families. T^2L, ST^2L, I^2L, ECL, and EFL are discussed. General circuit design guidelines are included.

1. INTRODUCTION

THE bipolar transistor has been used in digital integrated circuit design for the last 25 years. Today, although bipolar digital IC's make up a small percentage of the total digital integrated circuits market, they offer unique characteristics which make them candidates for very high-speed ($\leqslant 1$ ns) and very low-voltage ($\leqslant 1$ V) applications. In general, it is more difficult to design a bipolar circuit than an MOS one. This fact mainly stems from the following. The MOS transistor is an open circuit at the input, between the gate and the source, hence only the input voltage controls its characteristics. In the bipolar transistor, both the base voltage *and* current control the device characteristics. Moreover, the collector current of the bipolar transistor is an exponential function of the base voltage while the drain current of the MOS transistor is dependent on the gate voltage through a binomial relation which can be simplified to a square law. In addition, the bipolar transistor suffers from charge storage in the different regions of the device (base, collector, and emitter) while the MOS device does not have a charge storage problem, a fact that simplifies MOS circuit design and analysis.

This paper reviews the basic bipolar characteristics and discusses some of the commonly used digital families. The reader is referred to [1] for further analysis of digital bipolar integrated circuits.

2. BIPOLAR DEVICE CHARACTERISTICS

Fig. 1 shows a diagrammatic cross section in an npn bipolar transistor. It consists of two nested surface regions of n^+ and p introduced by ion implantation into an n epitaxial region which has been grown on top of mask-selected n^+ regions. These selected n^+ regions are referred to as underlayers and are ion implanted into a p substrate. Two pn junctions can be identified: an n^+p junction near the surface and a p n(epi) junction. The two junctions have a common *thin* layer of p type semiconductor which greatly affects the performance of the bipolar transistor and is referred to as the base B of the transistor. The transistor can operate in two modes of operation: a downward d and an upward u mode. In the downward mode of operation, the n^+p junction acts as an emitting junction E, while the p n(epi) junction acts as a collecting junction C. In the upward

The author is with the Department of Electrical Engineering, University of Waterloo, Waterloo, Ont., Canada, N2L 3G1.

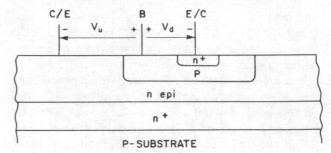

Fig. 1. Diagrammatic cross section in an npn bipolar transistor.

mode of operation, the role of the two junctions is reversed. In either case, an emitting junction emits electrons (carriers with negative charge) into the base; some will be transported and collected by the collecting junction. The collected current can be controlled by the base current I_B. The emitted electrons are transported through the base by a diffusion and a drift mechanism. The diffusion results because the electron's concentration near the emitter is higher than that near the collector. The drift results from an electric field which develops across the base. Another mechanism which affects the electrons as they travel through the base is recombination with holes (carriers with positive charge) provided by the base region.

2.1 DC Characteristics

Two important dc characteristics of the npn transistor are shown in Fig. 2:
- the collector current I_C versus the collector voltage V_{CE} for different values of I_B;
- the base current I_B versus the base voltage V_{BE}.

Three regions of operation relevant to digital circuit design can be distinguished on the I_C versus V_{CE} characteristic:

1) Cutoff Region: In this region both the E and the C junctions are reverse biased or forward biased with a voltage less than their turn on voltage V_{on}. The collector current is only due to the leakage of the collector junction, and in the ideal case both input and output equivalent circuits are presented by open circuits as shown in Fig. 3:

$$V_{BE} > V_{on}, \quad V_{BC} < V_{on}, \quad I_C \simeq 0, \quad I_B \simeq 0.$$

2) Active Region: In this region, $V_{BE} \geqslant V_{on}$ and $V_{BC} < V_{on}$. The output circuit is presented by an equivalent current source: $\beta_F I_B = I_C$. As V_{CE} approaches the soft saturation voltage ($\simeq 300$ mV) the transistor enters into a soft saturation region where β_F starts falling:

$$V_{BE} \geqslant V_{on}, \quad V_{BC} < V_{on}, \quad I_C = \beta_F I_B = I_{Se}V_{BE}/V_T$$

where

I_S = transport saturation current of the transistor; an area and a technology dependent parameter.

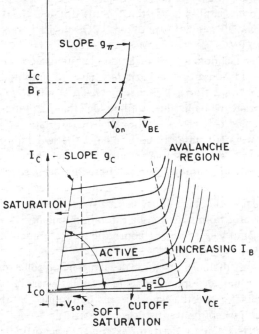

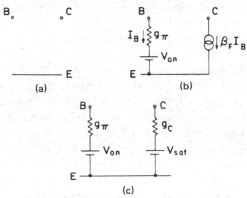

Fig. 2. DC characteristics of an npn transistor.

Fig. 3. Ideal switching equivalent circuits of the bipolar transistor: (a) cutoff; (b) active; (c) saturation.

V_T = thermal voltage; a temperature dependent parameter.

β_F = forward current gain of the transistor, a technology and operating-current dependent parameter.

3) Saturation Region: In this region, both V_{BE} and V_{BC} are $\geqslant V_{on}$. The output circuit is presented by a saturation voltage source V_{sat} in series with a saturation resistance $(1/g_c)$:

$$V_{BE} \geqslant V_{on}, \ V_{BC} \geqslant V_{on}, \ I_B \gg I_C/\beta_F.$$

2.2 Transient Characteristics

The transient characteristics of the bipolar transistor are a function of charge storage within the device itself, in addition to node capacitances contributed by interconnections and other loading devices. The charge storage within the device can be divided into two components: q_{BE}, q_{BC} (one associated with the E junction and the other associated with the C junction). These two components in turn can be divided into two types: one due to minority carrier charge storage (q_F and q_R for the emitter and collector junctions, respectively) and the other is due to space charge effects (q_{VE} and q_{VC}). Thus,

$$q_{BE} = q_F + q_{VE} \tag{1}$$

$$q_{BC} = q_R + q_{VE}. \tag{2}$$

The components q_F and q_R are proportional to the current of their corresponding junctions, while q_{VE} and q_{VC} are proportional to the logic swing (V_ℓ) across their corresponding junction. These proportionality functions are technology and operating-condition dependent [2]. To reduce the delay times of a circuit, q_{BE} and q_{BC} should be reduced by operating the circuit at lower current and/or reducing the logic swing at the two junctions.

3. THE BIPOLAR INVERTER

The analysis of a simple bipolar transistor is given in this section to demonstrate how the calculations of different gate parameters are done. The simple inverter used to be the basic building block in the RTL (resistor transistor logic) logic family. However, today it is used as part of other logic families, e.g., $T^2 L$ and $I^2 L$.

3.1 DC Analysis

Fig. 4(a) shows the basic inverter with a fan-out n. An input resistor R_b is added to the input to reduce the dependence of the input characteristic of the gate on the diode-like characteristic of the base-emitter junction. This makes the circuit less susceptible to "current hogging," which results from the unmatched base-emitter junction characteristics.

According to the piecewise linear approximation of the input characteristic, the threshold voltage of the inverter (V_t) is V_{on} of the base-emitter junction, because if $V_{in} \simeq V_{on}$, Q_1 is on and operates in the saturation region, and if $V_{in} < V_{on}$, Q_1 is off. Fig. 4(b) shows the dc equivalent circuit of the gate when $V_{in} = V_1$, and Fig. 4(c) shows the dc equivalent circuit when $V_{in} = V_0$.

With reference to Fig. 4, the output voltages can be calculated:

$$V_0 \simeq V_{sat} + V_{CC} \frac{r_c}{R_l} \tag{3}$$

$$V_1 = V_{on} \frac{R_L}{R_L + (R_b + r_\pi)/n} + V_{CC} \frac{(R_b + r_\pi)/n}{R_L + (R_b + r_\pi)/n} \tag{4}$$

For typical values of $V_{CC} = 5$ V, $R_L = 1$ kΩ, $R_b = 3$ kΩ, $V_{on} = 700$ mV, $V_{sat} = 50$ mV, $r_c = 50$ Ω, $r_\pi \ll R_b$, and $n = 1$:

$V_0 = 300$ mV

$V_1 = 3925$ mV

$V_\ell = 3625$ mV

$V_t = 700$ mV

$NM_1 = 3225$ mV

$NM_0 = 400$ mV

$I_{on} = 4.7$ mA

$I_{off} = 1.075$ mA, $I_{on}/I_{off} = 4.4$

$P_{av} = 14.4$ mW. $\tag{5}$

Equation (5) shows that the gate has a high logic swing, unsymmetrical noise margins, a large ratio between I_{on}/I_{off} and a relatively high power dissipation. The high ratio between I_{on} and I_{off} causes current transient spikes at the power supply line, which could cause faulty logic operation.

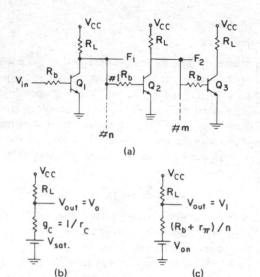

Fig. 4. (a) Bipolar inverter circuit; (b) equivalent circuit when $V_{in} = V_1$; (c) equivalent circuit when $V_{in} = V_0$.

3.2 Transient Analysis

We deal with three times in the transient analysis of the circuit of Fig. 3:

1) t_r, the rise time of the voltage waveform at node F_1 when the Q_2 transistors are off and the Q_3 transistors are saturated.

2) t_d, the delay time of the voltage waveform at node F_2 when the Q_2 transistors are on, the Q_3 transistors are on, and are turning off.

3) t_f, the fall time of the voltage waveform at node F_2 when the Q_3 transistors are off.

Neglecting the effect of R_b, these times are given by [2]:

$$t_r \simeq \frac{V_{on}}{V_{CC}} \tau \qquad (6)$$

$$\tau \simeq nR_L (C_S + 2C_{JE} + 2.23C_{JC}) \qquad (7)$$

$$t_d \simeq m \frac{3\tau_F + \tau_S}{\beta_F} \qquad (8)$$

$$t_f \simeq m \frac{2C_{JC}R_L}{[(V_{CC}/V_{on}) - 1]} \qquad (9)$$

where

C_s = substrate–collector junction capacitance

C_{JE} = emitter junction capacitance

C_{JC} = collector junction capacitance

τ_F = forward time constant of the device

τ_S = saturation time constant of the device

β_F = forward collector current gain

V_{on} = the turn on of the base–emitter junction.

For $C_{JE} = C_{JC} = C_S = 0.5$ pf, $V_{on} = 0.6$ V, $V_{CC} = 3$ V, $R_L = 240$ kΩ, $n = m = 1$, $\beta_F = 50$, $\tau_F = 0.1$ ns, and $\tau_S = 1$ ns: $t_r = 0.124$ μs, $t_d = 0.1$ ns, and $t_f = 60$ ns.

4. Bipolar Logic Families

In MOS technology, there are two basic logic families: single-channel transistor families (NMOS or PMOS) and a comple-mentary transistor family (CMOS). Both families use the pull-up pull-down (PUD) circuit configuration. The pull-down device in both families is an NMOS transistor, and the pull-up device in NMOS families is another NMOS device or a resistor while it is a PMOS device in CMOS.

In bipolar technology, the logic families are categorized into saturated and nonsaturated logic families. In saturated logic families the transistor is switched between an off state and a saturated state, while in nonsaturated logic families the transistor is switched between an off (or slightly on) state and an active (nonsaturated) state. In both families the switching device is an npn transistor. Complementary pnp transistors are only used as loads or current sources. Resistors are also used as loads or to implement current sources or bias networks.

Two main circuit techniques can be used to prevent the bipolar transistor from entering the saturation region:

1) To clamp V_{CE} to a value above the soft saturation voltage ($\simeq 300$ mV).

2) To limit the current (the base, emitter, or collector current) to a value corresponding to $I_B = I_C/\beta_F$, where I_B and I_C are the collector and the base currents determined by the external circuit when the transistor is conducting.

The first technique is used, for example, in Schottky transistor-transistor logic (STTL) where a Schottky diode, with a turn-on voltage of V_{on}(SBD)($\simeq 400$ mV), is connected between the collector and the base. As a result the minimum V_{CE} when the transistor is conducting is $[V_{on}(BE) - V_{on}(SBD)] \simeq 300$ mV.

The second technique is used in current-mode logic families such as emitter-coupled logic (ECL), where a current source I_0 is used to limit the emitter current (and hence the collector current) to I_0. As a result, the base current will be limited to $I_0/(\beta + 1)$, provided that $V_{CE} \geqslant 300$ mV.

When the transistor enters into saturation, excess charge is stored in the device. This charge causes extra delay when the transistor is switched from the on state to the off state. As a result, nonsaturated logic families are faster than saturated logic families. Saturated logic families include direct-coupled and resistor-transistor logic (DCTL and RTL), diode-transistor logic (DTL), transistor-transistor logic (T²L), and integrated injection logic (I²L). Nonsaturated logic families include emitter-coupled logic (ECL) and emitter-function logic (EFL).

In the early days of integrated circuits, the RTL circuit was used as the basic building block. The circuit was chosen because it was a simple discrete component circuit that was already in use and because it was possible to realize it using the state-of-the-art integrated circuits of the day. The basic elements for a NOR gate were a set of npn transistors connected in parallel and fabricated on the same isolation land sharing a common collector region and diffused resistors for the base and the collectors.

The shortcomings of RTL led to diode-transistor logic (DTL)—another discrete component circuit adopted for the integrated form. The DTL has led to the use of a truly integrated element—the multi-emitter npn transistor connected to provide multiple BE diodes—and eventually to the transistor-transistor logic (T²L) family. A totem-pole output circuit was added to that basic T²L circuit if a high driving capability was

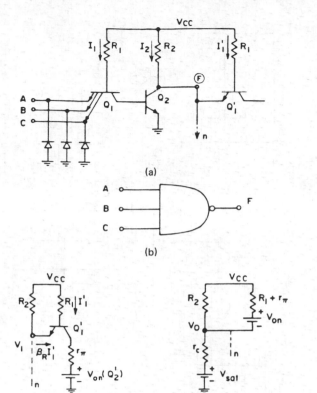

Fig. 5. Basic T^2L circuit: (a) circuit diagram; (b) logic equivalent; (c) and (d) equivalent circuits.

needed. Hence, the family has been widely used in MSI. However, with Schottky T^2L (ST^2L), the family has been adopted for LSI and VLSI applications.

4.1 Transistor-Transistor Logic (T^2L)

The basic T^2L gate is shown in Fig. 5 with a fan-out of n. The multi-emitter input transistor Q_1 controls the conduction state of the output transistor Q_2. The gate realizes the NAND function since the output F is at logic "0" (low) only if all the inputs are at logical "1" (high).

The logical "1" output voltage V_1 and the logical "0" output voltage V_0 are given by [3]:

$$V_1 = V_{CC} - n\beta_R R_2 \left(\frac{V_{CC} - 2V_{on}}{R_1 + 2r_\pi} \right) \tag{10}$$

$$V_0 \simeq V_{sat} + V_{CC}\frac{r_C}{R_2} + (V_{CC} - V_{on})\frac{nr_c}{R_1} \tag{11}$$

where V_{CC} is the power supply voltage, n the fan-out, β_R is the reverse current gain of the transistor, V_{on} is the on base–emitter voltage drop of the transistor, r_π is the resistor of the base–emitter diode, V_{sat} is the saturation voltage of the transistor, and r_C is the collector equivalent resistor in saturation.

From (10) we can conclude that it is advantageous to minimize β_R because the lower β_R is, the higher the maximum allowable fan-out number n. Moreover, the unmatched values of β_R from one loading gate to another result in a current-hogging problem, where more output current is supplied to gates with higher values of β_R. It should be noted that in deriving (10) we have assumed that all the emitters of the input transistors

of the loading gates are connected to a high-voltage level, that is, all the input transistors of the loading gates are operating in the inverse active mode. This assumption of course gives the lowest possible value for V_1.

From (11), it is clear that V_0 is always higher than the saturation voltage of the output transistor Q_2. As the fan-out increases, V_0 increases and it is advantageous to reduce r_C.

For typical values of V_{CC} = 5 V, R_1 = 4 kΩ, R_2 = 1 kΩ, V_{on} = 700 mV, V_{sat} = 50 mV, r_C = 50 Ω, $r_\pi \ll R_2$, $n = 1$, β_R = 3:

$$V_1 = 2300 \text{ mV}$$

$$V_0 \simeq 353 \text{ mV}.$$

Examining the basic T^2L gate from the transient performance viewpoint, we find that Q_1 provides a low impedance path from the base of Q_2 to ground during turn-off transients, and the stored charge flows out through Q_1. This provides a faster path for the stored charge than if a simple inverter is used without Q_1. This is one of the key features of T^2L. The speed of operation can be further improved if Schottky diodes are added in parallel with the base–collector junctions of Q_1 and Q_2, resulting in a ST^2L gate.

Another key feature of T^2L is the fact that adding inputs to a single gate requires adding extra emitters to the multi-emitter input transistor Q_1. This is a very important feature because the silicon area consumed by a single gate will increase slightly as the number of inputs increases. This explains the suitability of the family for LSI/VLSI.

The basic T^2L of Fig. 5 can be modified in a number of ways to meet specific performance requirements [3]. For example, to increase the driving capability of the circuit a totem-pole output circuit is added. Also, the gate can be modified to allow wired-OR logic to be performed at the output, hence a saving of silicon area results. The circuit can also be modified to operate with lower power dissipation at slower speed, to offer tristate output, or to clamp the output transistor to prevent it from operating in saturation. It is the last feature which led to ST^2L.

4.2 Schottky T^2L

One method of preventing the transistor from going into saturation is to clamp the base–collector junction with a Schottky diode, which has a lower turn-on voltage than a junction silicon diode. This reduces the storage charge in the base of the transistor, hence reducing the storage turn-off time. The Schottky clamped transistor (Schottky transistor) is fabricated as a single structure. The diode is built within the same collector isolation region as the transistor and is formed by the interconnection metallization. However, additional metal deposition is sometimes used to reduce the diode leakage.

Fig. 6 shows a basic ST^2L gate. The high-voltage level is the same as T^2L, while the low-voltage level V_0 is given by (V_{on} – V_{SBD}), which is on the order of 300–400 mV. Thus, the ST^2L logic swing (V_ℓ) is lower, and this results in even higher speed (delay is proportional to V_ℓ) and lower power dissipation (dynamic power dissipation = $CV_\ell^2 f$).

It should be noted that ST^2L has a lower logic swing (due to

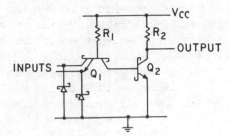

Fig. 6. Basic ST^2L circuit.

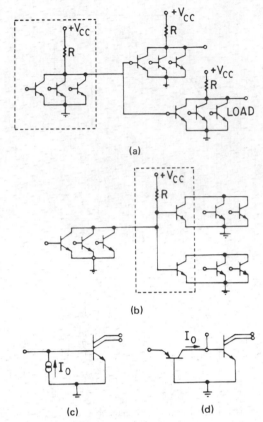

Fig. 7. The development of I^2L from DCTL. (a) DCTL structure. (b) I^2L transitional structure. (d) I^2L.

the higher V_0) and lower voltage noise margins than T^2L. Moreover, the base–collector breakdown voltage is lower because of the Schottky diodes. However, these disadvantages do not affect the performance of the circuit at the chip level, where voltage levels are low and the noise generated is not excessive. Although ST^2L circuits are based on their T^2L counterparts, it is possible to extract the best possible performance from ST^2L by adapting new circuit configurations. Some of these are discussed in [1].

4.3 Integrated Injection Logic (I^2L)

Integrated injection logic (I^2L), also known as merged transistor logic (MTL), is a low-power, high-density bipolar logic family suitable for LSI. It was introduced in 1972 by Berger and Wiedman [4] and Hart and Slob [5]. It is based on repartitioning the direct-coupled transistor logic (DCTL) and merging of a vertical npn transistor with a lateral pnp transistor. The vertical transistor acts as a switching transistor, while the lateral one is used as a current source.

Fig. 7 shows the development of I^2L from DCTL. If the load resistance of DCTL is considered to be associated with the input circuit as shown in Fig. 7(b), rather than the output, the input circuit becomes a current source I_0 steered by the input level, and the output circuit becomes a multicollector transistor as shown in Fig. 7(c). In I^2L this current source I_0 is realized using a lateral pnp transistor as shown in Fig. 7(d). The npn transistor is allowed to operate in the active upward mode with the epitaxial layer and n^+ buried layer acting as an emitter and the surface n^+ diffusions acting as collectors. This is basically the same structure as used for the T^2L input device, but the collector–emitter terminals are functionally reversed. Inspection of the basic gate shown in Fig. 7(d) reveals that the pnp and npn devices share a common ground n^+ region, and the collector of the pnp is interconnected to another p region, which is the base of the npn. As a result, it is possible to merge the pnp and npn transistors to save silicon area. In this case, the pnp collector and npn base is one region and the pnp base and the npn emitter is also one region. The emitter of the pnp is called the injector of the I^2L structure. The pnp is forward-biased using a power supply ($+V_{CC}$) and a biasing resistance R_B. The supply voltage ($+V_{CC}$) could be as low as 1 V, since all that is required is for the p^+-n injector to be forward-biased enough for the desired injector current to flow. Hence, the structure can operate at low-power levels with only microamperes as bias.

On the other hand, the supply voltage could be as high as 10–15 V, depending on the allowed power dissipation in R_B.

The higher the operating injector current is, the larger the power dissipation of the structure and the faster the operating speed. This represents an advantage for I^2L because of the flexibility in operating the circuit at different current levels, and hence with different speed and power dissipation levels.

The isolation between structures is made using either shallow or deep n^+ diffusions or using a dielectric isolation. These isolation regions are usually referred to as collars and can touch the p diffusion walls since only low voltages appear across these junctions. The dielectric isolation provides the best approach since it reduces the leakage currents and parasitic capacitors typical of junction isolation. The second best solution is the deep n^+ isolation collar followed by the shallow n^+ collar. It is appropriate to note that the relation between stage delays and power dissipation becomes nonlinear at increased power levels. In fact, a delay minima occurs so that diminished returns in stage delays are expected as one approaches the bottom of the curve. This minimum delay point occurs when the saturated charge storage in the n regions dominates over space-charge-storage effects. Owing to the device topology with the npn emitter having the largest area, it is difficult to reduce the stored charge and hence the minimum delay of the basic gate.

4.4 I^2L Related Families

Schottky I^2L (SI^2L): In I^2L, because a logic swing of about 700 mV is still larger than required by the transconductance of the switching transistor (≥ 150 mV), power-delay improvement is expected if the logic swing is reduced. This is achieved by using Schottky diodes for the output coupling as shown in Fig.

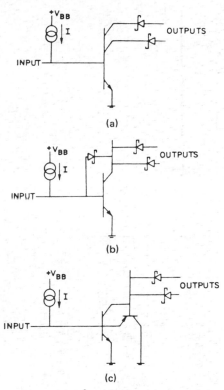

Fig. 8. (a) SI^{2}L; (b) STL; (c) ISL gates.

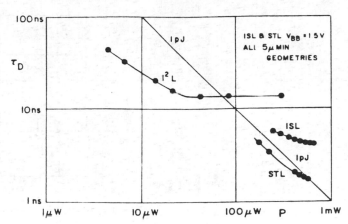

Fig. 9. Power-speed comparison of I^{2}L, ISL, and STL.

8(a). This reduces the logic swing from about 700 mV to about 200–300 mV, resulting in a power-delay improvement by a factor of 5 [6].

Schottky Transistor Logic (STL): If the I^{2}L collectors are replaced by Schottky collectors as shown schematically in Fig. 8(b), a similar improvement in power-delay performance is expected because of the reduction in the logic swing and in the total stored charge [7]. STL also employs a Schottky clamp on the npn transistor. The clamp and collector Schottky diodes are made with different barrier-metal systems so that the difference in their forward voltages at any current level is about 200 mV. This low logic swing contributes to the high speed of the family. Moreover, the clamp Schottky diode, which is made of Pt–Si, keeps the npn transistor out of saturation and hence reduces the stored charge.

Integrated Schottky Logic (ISL): In SI^{2}L and STL, the npn transistor is operating in the upward mode of operation. In ISL the npn transistor operates in the downward mode of operation. The basic gates are shown in Fig. 8(c), which consists of an npn transistor and a set of Schottky diodes used as outputs. To control the saturation of the npn, a merged lateral pnp transistor is added. The base–current source I is realized using an integrated resistor or a nonsaturated pnp transistor [8]. We note here that this pnp transistor cannot be merged into the npn transistor as in I^{2}L because of the fact that the npn transistor is operating in the downward mode. This requires an extra isolated n island for the current source where a pnp or a resistor is integrated. This results in lowering the packing density compared to I^{2}L. However, mainly because the logic swing of the family is low (about 200 mV) and controlling the saturation of the npn, the family speed of operation is higher than conventional I^{2}L.

Fig. 9 shows a power-speed comparison of I^{2}L, ISL, and STL using the same layout design rules [9]. I^{2}L offers a lower speed-power product, while ISL and STL offer lower delay. For the same basic dielectric-isolated double-metal ion-implanted bipolar process, I^{2}L requires 9 masks, ISL 10 masks, and STL 12 masks. Thus, in general, I^{2}L is the simplest technology, STL offers higher speed performance, and ISL is between both in terms of complexity and speed.

4.5 Emitter-Coupled Logic (ECL)

As mentioned in Section 2, in order to increase the frequency of operation of a bipolar logic family, it is necessary to prevent the transistor from operating in the saturation region. One method to do this is to make sure that the operating V_{CE} is greater than the soft saturation voltage and that the base current is just enough to maintain the transistor operating in the active region. Because it is difficult to control the base current at a level corresponding to the active region of operation ($I_B = I_C/\beta$) due to the wide tolerance in β, it is more practical to control the emitter current (or the collector current for that matter) at a constant value that is determined by the bias circuit.

This leads us to the basic emitter-coupled logic gate of Fig. 10. It consists of a fixed bias transistor Q_2 whose base is connected to a reference voltage V_r at node 4 and a number of input transistors connected in parallel ($Q_{1A} \cdots Q_{1B}$).

The input transistors are able to steer the current I_0 between Q_2 and (Q_{1A} to Q_{1B}). As the current is steered, the logic output voltage at 2 and 5 also will change, giving an NOR and an OR logical output, respectively. Because the fact that the steering of the current changes the output voltage, the name current-mode logic (CML) is sometimes used.

In ECL circuits the logic output is usually referred to as a true ground potential, that is, V_{CC} is taken to be ground and the circuit will operate from a negative supply voltage. This provides a less noisy signal at the output.

In order to increase the driving capability of the basic ECL gate, an emitter-follower stage is added. The addition of emitter followers lowers the logic levels at the outputs by a V_{on}. Thus, the reference V_r must also be lowered to a value of $1\frac{1}{2}$ V_{on}.

If the ECL input voltage changes by a diode drop V_{on}, the collector current will change by approximately I_0. The logic

43

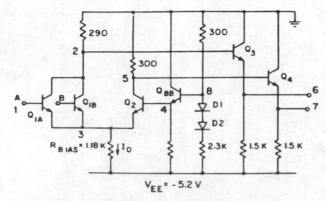

Fig. 10. ECL gate with emitter-follower outputs.

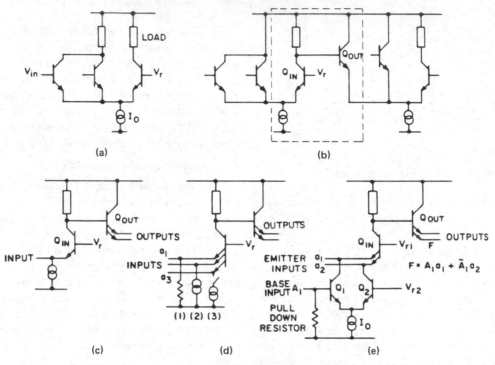

Fig. 11. The development of EFL from ECL. (a) ECL gate. (b) ECL with single load. (c) Noninverting gate. (d) Multi-emitter EFL. (e) Two-level EFL cell.

swing of the circuit is $I_0 R$ (where $R = R_1$ or R_2 depending on the output node). Usually, the output logic swing is limited to a diode drop and in many ECL circuits this is obtained by shunting the load resistance with a diode.

The current source is usually realized using a transistor working in the active region rather than a resistor.

4.6 Emitter-Function Logic (EFL)

The emitter-function logic (EFL) cell can be viewed as a repartitioned form of ECL [10]. The main feature of EFL is the increase in logic function complexity at minimum increment in the device and circuit complexity. This is achieved without additional power dissipation in comparison to the single-function current switch gate. This increase in functional complexity results in significant improvements in relative density, power per gate, and speed-power product. It should also be noted that all of these gains are realized only if this high functional complexity can be fully utilized. Fig. 11 shows the similarities

with ECL. Fig. 11(a) shows the conventional ECL gate with two collector loads, shown as boxes for generality. The time delay between the input and the OR output is less than that between the input and the NOR output. This is because of the Miller effect at the NOR output. At the NOR side the input and output voltages swing simultaneously in opposite directions. On the other hand, for the OR side the base is fixed so that only the collector-voltage swing affects speed. The time delay of the circuit would be that of the OR output if the NOR output is not used. Thus, if only a single load, the one connected to the fixed-biased transistor, is left in the circuit, two ECL gates connected in cascade will appear as shown in Fig. 11(b). Furthermore, if a simple repartition of the circuit is done, the input now is at the emitter of the fixed-biased transistor Q_{IN} and the output(s) is (are) at the output transistor Q_{OUT}. This is shown in Fig. 11(c). The structure now realizes a noninverting gate, that is, the logical output is the same at the logic input, which is of limited use in logic design. However, if Q_{IN} is made

44

as a number of multi-emitter transistors as shown in Fig. 11(d), the structure now realizes an AND gate, in that a_1, a_2, and a_3 must be high for the outputs to be high. The input pull-downs can be implemented as 1) a resistor, 2) active current source, or 3) logically controlled switched current source. The logical capability of the structure is further enhanced if a two-level EFL cell, as shown in Fig. 11(e), is used as a building block for logic design. Fig. 11(e) shows Q_{IN} with two emitters (the number of emitter inputs can be more than two), and one current source is used which is controlled by a second level of current steering transistors Q_1 and Q_2, where Q_2 is a fixed bias transistor with reference voltage V_{r2}. As a result, the output logic function F is given by

$$F = A_1 a_1 + \bar{A}_1 a_2 \qquad (12)$$

where A_1, a_1, and a_2 are, in general, wired-OR functions. That is, multiple logic outputs can be directly attached (ORed) to a_1, for example.

Equation (12) can be explained more physically as follows. If A_1 is logical "1," that is, its voltage level is greater than V_{r2}, the current I_0 is steered through Q_1 and, hence, the output logical level is controlled by a_1: $F = a_1$. Alternatively, if A_1 is logical "0," that is, its voltage level is less than V_{r2}, the current I_0 is steered through Q_2 and, hence, the output logical level is controlled by a_2: $F = a_2$.

In order to allow the outputs to be directly coupled either to an input emitter or an input base(s), V_{r1} must be one V_{on} higher than V_{r2}. Hence, the threshold voltage of the EFL gate is V_{r2}. It should be mentioned that each output emitter can fan-out to only one emitter input. The reason for this is that when the current I_0 is steered through Q_1 or Q_2 to the emitter input, it will be further steered through Q_{IN} or through Q_{OUT} of the previous stage. Allowing output emitters to fan-out to only one emitter input prevents current sharing problem between inputs. On the other hand, Q_{OUT} can fan-out to any number of bases since these inputs have a negligible current input requirement.

In EFL only one lower level input is allowed to be "1" at a given time to prevent the current source I_0 from being shared between Q_{OUT} of the previous stages. Multiple outputs will result in a nonspecific variation in delay. Although this restriction of allowing only one control input to be "1" at a given time presents a limitation on logic designs, in practice it does not present a serious drawback.

6. COMMENTS AND DISCUSSION

In this paper the bipolar transistor and its digital logic families have been discussed. Some simple design equations have been given. The reader is referred to detailed treatments in [1].

The following is a summary of some basic guidelines, based on the material covered in this paper, to be used in the design of digital bipolar integrated circuits.

1) A simple digital model for a bipolar transistor is that it is a switch, either on or off, and can be controlled by the base current and/or base voltage. The off resistance of this switch, between the collector and the emitter, is high and can be considered as an open circuit. When the switch is closed, it is represented by a current source if the transistor is operating in the active region, and it is represented by a dc source in series with a resistor if the transistor is operating in the saturation region.

2) The base voltage required to turn on or off the bipolar transistor is not sensitive to processing and operating parameters and it has a typical range of 600–800 mV.

3) The distinction between active and saturation regions is very important in operating the bipolar transistor. At the base input, the active region is characterized by the fact that the input current is just enough to supply the base current necessary to maintain a collector current (i.e., $I_B = I_C/\beta$). When the transistor is in saturation, the input current is more than the necessary base current needed to maintain the collector current ($I_B = (I_C/\beta) + I_X$). The extra component I_X results in an extra charge storage and slows down the switching speed.

4) The bipolar circuit designer has an option to use any of the available circuit structures referred to as logic families. He/she can mix families in a given chip design to make use of the different unique characteristics of each family, e.g., I^2L for low power and EFL for higher speed.

5) The basic switching device is the npn bipolar transistor. The complementary pnp transistor is only used for the design of current sources, bias networks, and load implementation. It is an inferior switching device and it is not recommended to be used as a switch.

6) The npn transistor can be used as a transfer gate (transmission gate) as in the case of MOS transistors. However, the transmission characteristic is highly unsymmetrical. This mode of operating the npn transistor as a transfer gate is used somewhat in implementing the input transistor of a T^2L gate.

7) The npn transistor can be used in an upward or downward mode of operation. The upward mode is when the surface n^+ region is acting as a collector (e.g., as in I^2L) while the downward mode is when the surface n^+ region is acting as an emitter (e.g., as in T^2L and ECL).

8) Because of the geometrical structure of the npn transistor, it is recommended that the logic capability of a device can be increased by increasing the number of surface n^+ regions; thus the total planar area of the transistor does not substantially increase. This feature is used in I^2L, T^2L, and EFL.

9) Because the minimum required V_{CE} to operate the transistor is in the order of the turn-on-voltage of the base–emitter junction ($\simeq 600$–800 mV), bipolar logic families can easily be designed to operate from a 1 V battery. This is a unique characteristic of bipolar logic families and is not easily matched by NMOS or CMOS digital circuits.

10) Although bipolar logic families can operate in a dynamic mode of operation, similar to MOS logic families, since it is not easy to fabricate a capacitor on a bipolar chip, dynamic bipolar circuits are seldom used.

11) The static power dissipation ($V_{CC} I_{av}$, $I_a = (I_{ON} + I_{OFF})/2$) of a bipolar circuit is usually dominant over the dynamic power dissipation ($\Sigma C V_{\ell}^2 f$, C is the node capacitance, V_{ℓ} is the logic swing, and f is the switching frequency).

12) Although both bipolar and MOS digital circuits can approach each other in terms of power dissipation and speed, bipolar circuits are used when speed is the main requirement and MOS circuits are used when power dissipation is of main concern.

REFERENCES

[1] M. I. Elmasry, *Digital Bipolar Integrated Circuits.* New York: Wiley Interscience, 1983.

[2] ——, *Digital Bipolar Integrated Circuits.* New York: Wiley Interscience, 1983, ch. 3.

[3] ——, *Digital Bipolar Integrated Circuits.* New York: Wiley Interscience, 1983, ch. 4.

[4] H. H. Berger and S. K. Wiedmann, "Merged transistor logic (MTL)— A low cost bipolar logic concept," *IEEE J. Solid-State Circuits,* vol. SC-7, pp. 340–346, 1972.

[5] K. Hart and A. Slob, "Integrated injection logic: A new approach to LSI," *IEEE J. Solid-State Circuits,* vol. SC-7, pp. 346–351, 1972.

[6] F. W. Hewlett, Jr., "Schottky I^{2}L," *IEEE J. Solid-State Circuits,* vol. SC-10, pp. 343–348, 1975.

[7] H. H. Berger and S. K. Wiedmann, "Schottky transistor logic," in *Dig. Tech. Papers, IEEE Int. Solid-State Circuits Conf.,* Feb. 22, 1975, pp. 172–173.

[8] J. Lohstroh, "ISL: A fast and dense low power logic, made in a standard Schottky process," *IEEE J. Solid-State Circuits,* vol. SC-14, pp. 585–590, 1979.

[9] T. J. Sanders and B. R. Doyle, "Bipolar gate array technology: I^{2}L versus ISL," in *Proc. Custom Integrated Circuit Conf.,* May 1980, pp. 87–91.

[10] M. I. Elmasry, *Digital Bipolar Integrated Circuits.* New York: Wiley Interscience, 1983, ch. 6.

Managing VLSI Complexity: An Outlook

CARLO H. SÉQUIN, FELLOW, IEEE

Invited Paper

Abstract—The nature of complexity in the context of VLSI circuits is examined, and similarities with the complexity problem in large software systems are discussed. Lessons learned in software engineering are reviewed, and the applicability to VLSI systems design is investigated. Additional difficulties arising in integrated circuits such as those resulting from their two-dimensionality and from the required interconnections are discussed. The positive aspects of VLSI complexity as a way to increase performance and reduce chip size are reviewed.

With this discussion as a basis, the evolution of VLSI system design environments is outlined for the near-term, medium-term, and long-term future. The changing role of the designer is discussed. Recommendations are made for enhancements to our engineering curriculums which would provide the next generation of designers with skills relevant to managing VLSI complexity.

I. Introduction

VERY-LARGE-SCALE-INTEGRATION (VLSI) will soon make it economically viable to place 1 000 000 devices on a single chip, and the technological evolution will continue to double this number every 1–2 years for at least another decade [1]. According to G. Moore [2], the major hurdle faced in the construction of ever larger integrated systems is a *complexity barrier*. In order to exploit fully the technological potential of VLSI, new ways of managing the information associated with the design of a VLSI chip must be developed.

Why is it that complexity and its management have suddenly become such popular concerns? Mankind is routinely building systems with more than a million components: skyscrapers, air planes, telephone systems. . . . System complexity and the associated engineering issues do not seem to differ markedly for alternative implementations, e.g., whether a circuit is contained within a cabinet, on a printed-circuit board, or on a single silicon chip.

Packaging *does*, however, have an impact because of the partitioning that it enforces. Nobody would dare to insert a million discrete devices into a large chassis using discrete point-to-point wiring. Large systems built from discrete devices are broken down into subchassis, mother boards, and module boards carrying the actual components. This physical partitioning encourages careful consideration of the logical partitioning and of the interfaces between the modules at all levels of the hierarchy. Since such systems are typically designed by large teams, early top-down decisions concerning the partitioning and the interfaces must be made and enforced rather rigidly—for better or for worse. This keeps the total complexity in the scope of each individual designer limited in magnitude, and thus manageable.

VLSI permits the whole system to be concentrated in the basically unstructured domain of a single silicon chip which does not *a priori* force any partitioning or compartmentalization. On the positive side, this freedom may be exploited for significant performance advantages. On the negative side, it may result in a dangerous situation where the complexity within a large, unstructured domain simply overwhelms the designer. A similar crisis was faced by software engineers when unstructured programs started to grow to lengths in excess of 10 000 lines of code. The crisis was alleviated by the development of suitable design methodologies, structuring techniques, and documentation styles. Many of the lessons learned in the software domain are also applicable to the design of VLSI systems.

Much as programming was done throughout the 1950's by small groups of highly qualified people solving problems in their own style, integrated circuit design in the 1970's was still done by small clusters of layout wizards. This often led to intricate, if not mysterious, circuits. This "cottage industry" style of creating systems [3] starts to fail once programs approach tens of thousands of lines of code or once circuits end up with 100 000 devices. More formal and organized methods are required to create larger systems.

Sections II and III explore the nature of complexity in general and analyze methods to deal with complexity that were developed in the framework of software engineering. In Sections IV and V the point is made that the problems associated with VLSI designs are even harder than those encountered in large programs.

Section VI presents general developments in the emerging VLSI system design tools that address the discussed problems, and Section VII tries to predict the evolution of such tools for three different time frames. In Sections VIII and IX the role of the designer is re-evaluated in the face of this changing design environment, and some recommendations are made for the education of the next generation of designers.

Most sections are relatively self-contained and can be read as a sequence of minipapers. On the whole, they should not be viewed as the final word on the subject of VLSI complexity, but rather as food for thought and a nucleus for further discussion.

II. The Nature of Complexity

VLSI is more than just "a lot of LSI"—in the same sense that a city is not just a large village and the human brain is not just a large collection of ganglions. The 256-kbit RAM's

Manuscript received April 23, 1982; revised November 8, 1982. This work was sponsored by the Defense Advance Research Projects Agency (DOD) under ARPA Order 3803; monitored by the Naval Electronic System Command under Contract N00039-81-K-0251.

The author is with the Computer Science Division, Electrical Engineering and Computer Sciences, University of California, Berkeley, CA 94720.

Reprinted from *Proc. IEEE*, vol. 71, pp. 149–166, Jan. 1983.

(random-access memories) of 1982 are quite different from the first 1-kbit RAM's that appeared on the market a decade ago. As the scale of any system is increased by orders of magnitude, its organization typically changes because different issues become relevant. Internal structuring starts to appear, specialized subparts emerge, communication between the different parts gains importance, and, at some level, the *complexity* of the system becomes an explicitly stated concern.

The introduction of new design methodologies to deal with this concern requires an understanding of complexity and its special demands. Webster's Dictionary gives the following definition:

> complex: referring to that which is made up of many elaborately interrelated or interconnected parts, so that much study or knowledge is needed to understand or operate it.

This section tries to provide some insight into the notion of system complexity. It will emerge that it is useful to distinguish between explicit (apparent) complexity and implicit (hidden) complexity.

A. Explicit Complexity

Unlike the information content of a transmitted message, there is no formal absolute measure for complexity. In general one tends to call those systems more complex that need more words or more bytes to be adequately described or specified. Thus a comparison of the lengths of the corresponding descriptions can be used as a measure for the *relative* complexities of two or more *similar* systems. In order for such a comparison to make sense, the systems to be compared must be described in the same language and in terms of the same primitive elements.

This *apparent* or *explicit* complexity of a system depends strongly on the language or notation used for its description. By using suitable abstractions, the details of the subcomponents of the system can be hidden, and the apparent complexity can be reduced dramatically. It is equally important to exploit any structure or regularity to reduce the length of the description and to express more succinctly the "essence" of a particular system. This can be achieved with the use of suitable notations designed to express repetition, symmetry, and other regularities in an efficient and compact manner. In this light, an early microprocessor such as the 8008, with only a few thousand transistors, is more complex than one of the large memory chips with, say, 16 384 bits. Because of its regularity, the latter can be grasped and described more easily.

B. Implicit Complexity

However, complexity is not simply proportional to the number of different parts or the length of an optimally encoded layout description. The interactions between the parts play an important role. Steward [4] defines complexity in the following way:

> Given the parts and their behaviors, complexity is the difficulty involved in using the relations among the parts to infer the behavior of the whole.
> Or phrased another way:
> Complexity is how much more the whole is than just the sum of its parts.

This *hidden* or *implicit* complexity of a system involves issues such as the behavior of the system and the way in which it achieves a particular function. A collection of gates that im-

plement the well-defined Boolean function of an adder is far less complex than some historical radio receiver stage with a single vacuum tube which simultaneously amplifies the incoming HF signal, demodulates the LF component, and preamplifies the sound signal.

These intricate aspects of a system's behavior could, in principle, be reduced to the previously defined measure of explicit complexity. One would have to take into account the complete systems documentation, including: all its functional specifications, a description of its operation, the plans for its physical realization, and the set of repair instructions. Since it is normally impractical to sum up all this diverse information, one needs some other guidelines to evaluate this more elusive notion of implicit complexity. Looking at the way in which systems are composed from smaller components is a suitable approach.

C. Complexity in a Structured System

Few systems are described or documented in a hierarchically flat manner. Large systems are viewed as being composed of subsystems, which, in turn, consist of components on a lower hierarchical level. *Cohesion* and *coupling* are two useful concepts for a more detailed evaluation of the complexity of a module at a particular level of the system hierarchy. They address the nature by which components are combined into modules at the next higher level of the description.

Cohesion: Cohesion is a measure of how closely the internal parts that make up a module belong together when seen from different perspectives. In general, a high level of cohesion is desirable, since it leads to a simpler description at the next higher level and thus reduces its explicit complexity. High cohesion can be derived from functional similarity, from logical or physical grouping, or from the fact that all subparts work on the same data. Yourdon and Constantine [5] define several different criteria by which the cohesion of a compound software module can be judged. They can readily be extended to the domain of integrated circuits as demonstrated by the examples in Table I.

Spatial cohesion is something that is unique to the hardware domain and plays a big role in analog VLSI components such as capacitor arrays or optical sensor systems. Temporal, logical, or coincidental cohesion are very weak and are primarily used to keep things ordered from an organizational point of view.

Coupling: Coupling measures how much the submodules interact. Good structuring and clever systems partitioning aim at reducing implicit complexity by minimizing the amount of interaction between subparts. The interaction patterns should be simple and regular. This leads to independence of design, simple layout, high testability, and easy modifiability. Table II gives some measures by which the degree of coupling can be judged.

TABLE I
MEASURES OF COHESION IN COMPOUND MODULES

Name	Explanation	Example
functional	all parts contribute to function.	operational amplifier
sequential	portion of a data-flow diagram	filter cascade
communicational	data abstraction, same data used	LIFO stack
spatial	need to be physically close	sensor array
procedural	same procedural block	bus controller
temporal	used at a similar time	start-up circuitry
logical	a group of similar functions	I/O-pad library
coincidental	random collection	miscellaneous wiring

TABLE II
MEASURES OF COUPLING BETWEEN SUBMODULES

Perspective	Degree of Coupling	Example
Topology:	linear flow-through	filter chain
	hierarchical tree	carry look-ahead
	lattice modularity	memory array
	irregular mosaic	processor control
Interaction:	continuous	light sensor
	periodic	timer
	occasional	reset
Bandwidth:	high	video signal
	bursty	disk head
	low	voice signal

Minimal coupling, using explicit communications between submodules rather than parasitic interactions, supports modularity and makes possible plug-compatible replacements. The topology of the interactions should be kept as simple as possible. Clean modularization reduces implicit complexity since self-contained modules are easier to specify than a collection of highly interacting components.

III. DEALING WITH COMPLEXITY

Complexity is an integral part of large, "interesting" systems. In this section, general techniques for managing this complexity will be discussed. Most of them have been developed or refined in the context of large programs and form the basis of good software engineering. They are also applicable to the design of complex VLSI systems.

A. Design Equals Documentation

A complex system often can be understood more easily when viewed from the proper perspective. This is the role of good documentation. *Design* and *documentation* should become synonymous.

Software maintenance costs usually exceed initial production costs by a significant factor [6]. Software systems are constantly being modified, upgraded, and expanded [7]. This means that other people must be able to thoroughly understand what the original developer(s) had in mind. A similar issue is emerging in VLSI design as the development cycle of many products starts to exceed the average time an employee spends with the same company, and thus designs have to be passed from one designer to the next one.

Among good software designers, it is now common practice to make the documentation an integral part of the program. It has been realized that a separately kept documentation folder is never up-to-date and, before long, describes something quite different from the actual code. Documentation for a typical integrated circuit, however, is still normally scattered over several different media and/or several organizational groups, including the systems architects, the logic designers, and the layout crew. Often some pieces of the documentation are lacking entirely, making the circuit a mystery to all but the original design group. Modifying or even just debugging of such a system by anybody but the original designer(s) tends to be a real nightmare.

Notations and Representations: An important step in minimizing the *apparent* complexity of a system at any level of its description is to use suitable languages and/or notations, tailored to the particular issue that needs to be documented. Hoare emphasizes the importance of the proper notation by pointing out that the history of mathematics is, to a large degree, the story of improvements of notation [8]. The use of bad, hard-to-decipher notation is not only unproductive, it breaks the thought process aimed at finding problem solutions.

One should use all possibilities of the representation medium to enhance the clarity of the constructs that are being used: suitable mnemonic labels, symbolic shapes, differentiating colors, sound.... To simply ask for more extended English descriptions for each item is not the right way to go. Abstract, terse symbols are crucial to making the "grand picture" visible.

B. Use of Abstraction

Humans can effectively deal with only one problem at a time and can concern themselves with only a few constraints simultaneously [9]. This necessitates judicious use of hierarchy and abstraction. If the representation of a system can be rendered more succinct, one has taken a big step towards managing its complexity. Abstraction in the form of *user-defined macros* or *parameterized subroutine or cell calls* permits one to focus on the few qualities that are essential at a particular phase of the design process. It hides internal interactions and irrelevant details that contribute nothing to the solution but would clutter the picture and thus hamper the relevant thought processes.

Examples from the world of VLSI are: lumped devices hiding the complicated physical processes inside a contiguous piece of silicon; logic gates abstracting from a particular realization; register-transfer modules showing the basic data operations while hiding details of timing; and instruction-set descriptions for computer architectures hiding all but the primitives that the programmer has to deal with.

In addition, abstraction permits the creation of *generic modules*, usable with different data types or implementable with different technologies. Abstract encapsulation of modules in well-defined interfaces is the basis for efficient design of large systems. Well-defined modules, properly documented and made available through module libraries, can be used repeatedly by many designers. Design systems for integrated circuits often provide several plug-compatible versions of such modules which vary in a few parameters such as power and speed, or the availability of an explicit testing interface. Efficient construction of large systems is not possible without a library of well-designed modular components.

C. High-Level Descriptions

High-level languages lead to shorter and more readable programs. They make it easier to understand the organization and behavior of programs or systems and result in designs that are more likely to be free of errors, easier to debug and modify.

Programmer Productivity: Long-term average productivity in a commercial software production environment is about 10 lines of code per day. One arrives at this number by dividing the total number of lines in the final program by the amount of programmer time used for its development. The number is surprisingly independent of the language used—but the resulting net programmer effectiveness is not. A high-level language statement such as CASE or a complicated algebraic expression assignment is more powerful than an assembler-level LOAD instruction. Thus the net programmer productivity in terms of the functional effect produced by his/her code scales strongly with the power of the primitives employed. Similarly, in integrated circuit design, designers can place on average only five to ten *items* per day [10]. If ten *processors* rather than ten *transistors* are inserted into the layout per day, the effective productivity is increased dramatically.

Compilers versus Code Efficiency: High-level descriptions must be mapped down to the level of the actual implementation. Traditionally, coding in lower level languages has resulted in more compact and faster-executing code than what can be produced by a compiler. However, compiler technology is improving. Present-day compilers for high-level programming languages will do an adequate job for most applications and will do it far more quickly and reliably than any programmer. Of course, skilled programmers, willing to devote a large amount of time, can still outperform them by a factor 2–3 in small program segments, but this effort is only justified if there are hard constraints such as real-time requirements or the need to fit a program into a limited amount of on-chip memory.

High-level languages are now being introduced to the development of VLSI systems. There are several long-term research efforts in VLSI design aimed at the construction of a *silicon compiler* [11]–[13]. Driven by high-level specification of the desired circuit, these automated design systems are expected to translate high-level systems descriptions into correct circuit implementations in a particular technology. These layout compilers are still in their infancy and produce relatively inefficient results for all but narrowly defined, highly structured subsystems.

However, when the rising development costs for VLSI systems become comparable to the fabrication costs of ten thousands of devices, a faster development cycle at the price of a less efficient implementation becomes an attractive alternative for many systems. The increasing market share of gate arrays [14] and macro cells [15] must be seen in this light.

D. Partitioning

Problems that are too large to be handled as a whole must be partitioned into smaller, more manageable parts. Section III-E reviews some formal methods for finding suitable partitionings. In any such subdivision, the interactions between the subparts must be given particular attention.

Intermediate Decisions and Specifications: When a module is partitioned into submodules, the latter must be specified in enough details so that one can verify that the partitioning step does not violate any of the original specifications. In carrying out the successive refinement steps, one must also assure that the implementation of every module agrees with the abstract view used at the higher level. The properties of the composition of the new submodules must be assured for *all* possible concerns, e.g., functional correctness, timing constraints, signal levels, and even testability.

If *formal* verification is not possible—and it may be a long time before this becomes practical in the domain of circuit design—another solution has to be found. Intermediate system simulation, using proper models for the submodules, is a usable approach. VLSI system design tools must have the means to create, retain, and make use of such intermediate models. They should readily permit partial and mixed-mode simulations [16], [17] of a system in a top-down manner, making use of the abstractions for the subparts. Such a simulation can cut off all branches of the design hierarchy for which suitable simulation models exist. The resulting savings in simulation efforts are considerable.

Separation of Functionality and Implementation: The specifications of a system are normally separated into a description of what the system is supposed to do plus a list of performance objectives. This provides another boundary along which a problem can be partitioned.

During the design of a system it is also useful to separate the concern for functional correctness from the details of the actual implementation, which must often include an optimization step; the number of packages is minimized or a set of packages is distributed evenly on a few printed circuit boards. Constantine refers to this optimization and packaging step as clustering pieces of a problem solution into physical constraints without unduly compromising the integrity of the original design [5]. Such packaging optimizations should be made as the final step and only after the algorithmic optimizations have been exploited to the fullest. Once the design has been compromised by low-level packaging optimizations, it loses a lot of its generality, portability, and durability in the face of emerging technologies. Such optimization tricks are difficult to comprehend and to reverse at a later time.

Conceptually, the design process should thus contain the following three concerns:

1) functional design: guaranteeing proper behavior;
2) implementation: finding a suitable structure;
3) optimization: fine tuning the physical arrangement.

These basic steps should also be taken in the design of a VLSI system. However, several iterations through these three steps may be required to make sure that the nature of the implementation medium has been properly taken into account in the high-level design decisions.

E. Structuring Methods

Structuring is a key approach to large programs, systems, or problems in general. Good structuring implies partitioning. Many of the concepts suitable for judging the structure of programs [18] are also applicable to VLSI systems. As a start, one can demand that a well-designed subsystem implement a single, independent function, have only a few "low-bandwidth" entry or exit points, be separately testable from the module boundaries, and be itself constructed of a limited number of submodules that obey the same criteria.

Unfortunately it is often not clear *how* one should partition a particular problem or system; it can involve the most important and most difficult decisions a designer must face. This section reviews some program structuring strategies and discusses their applicability for VLSI systems design.

Functional Decomposition: Functional Decomposition is an application of the very general and popular *divide-and-conquer technique*. It consists of a recursive subdivision into parts, so that the joint operation of the properly interconnected parts performs the specified operation of the whole.

It can be used in a top-down manner, splitting a large system into cohesive blocks [19]. The recursive refinement step to be executed in this design strategy is as follows. Start from the definition of the function to be performed and its desired interface. Identify logical subparts, define the function and interfaces of these parts, and specify their interactions. Then verify that when the subparts are interconnected in the specified manner, their overall behavior corresponds to the original specification at the previous level. Unfortunately, this last step is nontrivial in programming and even harder in VLSI systems design.

Any significant design effort will have to iterate between top-down and bottom-up techniques because the characteristics of the medium used in the final implementation might not be visible at the top. Functional decomposition also plays a role in conjunction with the bottom-up phase. The natural building blocks of a particular technology and effective implementations thereof are derived in a bottom-up manner. The overall problem specification is then scanned for the usage of such blocks and partitioned accordingly.

The advantage as well as the problem with functional decomposition lie in its generality. If there is a large gap between the top-level problem specification and the natural building blocks dictated by the technology, the methodology will not provide guidelines for reproducible implementations and an excessive variety of possible decompositions results. The quality of the final result will then depend strongly on the intuitive insights of the designer. This is equally true for VLSI systems and for large programs.

Flow Graph Design: A more specific way to decompose a problem is to follow its *flow of data or of control.* In the first approach, the *data* flowing through the system are viewed as the primary ingredients, and all functional blocks are viewed as filters transforming the data as they pass through [20]–[22]. This approach suggests a further decomposition of each module into input circuitry, transformation circuitry, and output circuitry. Particular concern has to be given to synchronization, flow control, and possibly to buffering of intermediate data at the interfaces.

This is a particularly important and suitable approach for integrated circuits since it deals explicitly with the communication between blocks and puts proper emphasis on spatial ordering as dictated by the flow of data. It is being used in signal-processing devices and is particularly well exemplified in systolic arrays [23]. The formal data-flow graph of a system is a good starting point for a layout. Even in laying out the datapath of a microprocessor, the flow of data and of control often form the basis for the floor plan. Typically, they are assigned to the directions of the two coordinate axes so that the two flows cross at right angles [24], [25], [13].

The price to formally partition each "filtering module" into input circuitry, transformation circuitry, and output circuitry appears to be too high for present-day VLSI systems. In regular structures, such as systolic arrays [23] implemented as an iteration of identical cells, it is preferable to design outputs and inputs as matched pairs. Significant savings in power and layout area result from this optimization step. For modules that are used more randomly and are combined with a variety of other modules, the extra overhead of buffering provides abstraction and modifiability. So far, the control flow of a system has been used less directly to produce a structured circuit layout, except possibly for the case of a microcoded control section.

Data-Structure Design: Another successful structuring method in software engineering is to start with the *data structures.* The idea is to describe a problem through its associated objects, and to specify these objects through suitable abstractions and their interactions [20], [26]. This technique applies particularly well to small subsystems performing a specific function. Because there are fewer arbitrary and different ways to find data structures that correspond to a particular model of a subsystem, the resulting system structuring is more reproducible than with other methods. However, it is not so clear how this approach can be used for large systems.

The application of this method to VLSI systems design is also difficult. Direct mappings into hardware have been realized for only a rather limited number of data structures. The parallelism desired in VLSI circuits is a further difficulty. The interactions between objects and their spatial ordering are crucial. There is a high price on interactions between submodules, which is normally not considered in the setup of data structures.

There is plenty of room for innovation. VLSI will provide ever more sophisticated and efficient implementations for such data structures as trees, queues, and content-addressable memories. Such structures can either be implemented directly with special hardware primitives, e.g., an array of associative memory cells, or they can be built using standard RAM's with suitable control structures. In either case, these VLSI data structures should be properly encapsulated so that the user sees only the desired external behavior, regardless of implementation.

Programming Calculus: Perhaps the most intellectually satisfying approach to the construction of correct and reliable software is to prove correctness during construction of the program [27], [28]. The design freedom is deliberately limited so that program and proof can be constructed hand-in-hand. This design strategy requires that the system to be built has solid, unambiguous specifications. One way to do this formally is to state the required result as an assertion in predicate calculus. Each top-down refinement step is then carried out with rigorous strength to guarantee that the specifications are still met and to produce the specifications for the next lower level of modules. Till now this approach has been limited to fairly small program segments.

There is some expectation that in the long run the circuits resulting from silicon compilers (Section III-C) can be shown to be correct by construction. However, the day is still far in the future when such generators will produce correct and economically viable solutions for arbitrary systems for all sensible combinations of input parameters, and formal proofs of correctness should not be expected in this century.

F. Restriction to a Limited Set of Constructs

Böhm and Jacopini [29] have shown that the logic of any program can be represented as a hierarchical arrangement of three basic constructs: *sequence, iteration,* and *selection.* Others [30], [31], [20] have also recommended that the number and type of control flow constructs in programs be limited to these few well-defined constructs and that unstructured control flow, as resulting from indiscriminate use of the controversial GOTO statement, should be minimized. For software systems, Jackson [20], Brooks [32], and Dijkstra [33] recommend the use of a rather limited set of allowable structuring topologies and even suggest restriction to pure *tree structures.*

For the structuring of integrated circuits, such a formal restriction to tree structures leads to implementations that are too inefficient. However, in general, restriction to few robust, well-understood constructs has many advantages. In particular, the use of standard modules that have been proven to work previously can make a system more modular and easier to understand, and reduces the possibility for errors. This comes at the price of some inefficiency and associated performance loss; thus the degree to which one can adhere to that philosophy depends on the application. Here is a sample of possible restrictions in VLSI design:

1) At the geometrical layout level, one might restrict oneself to only "Manhattan" geometry, i.e., only rectilinear features. This simplifies the required design tools and results in substantial speedup of such operations as layout rule checking for the price of some loss in layout density.

2) At the logic level, one could rely exclusively on pretested standard cells interconnected in a standardized automatic or

semiautomatic way. This leads quickly to relative risk-free implementations; but the chips are large and dominated by wiring area.

3) At the register level, one can avoid problems with hazards and race conditions by using strictly synchronous sequential circuits with no unclocked feedback loops.

4) At the large system level, hung flip-flops and similar synchronization problems can be avoided with a self-timed sequencing approach.

At all levels, the improvement in testability and understandability gained from such restrictions in the types of constructs used will make the introduced inefficiencies well worthwhile, if overall systems constraints can still be met.

G. Testing the Design

Testing is an integral part of the design process. Every design decision must be checked for its appropriateness. Any large system designed in an "open-loop" manner has a very small chance to work correctly.

There are many different ways in which people can convince themselves that a part or a design "works." Two extreme approaches are:

1) by running the part in its intended application environment;
2) by formally proving the correctness of the design from its specifications.

Both approaches have obvious drawbacks. The first one corresponds to driving a car around the block to see whether it works. In a complicated system this can never give full confidence that there are no undiscovered flaws; there is nothing to tell the designer what has *not* been tested yet. The second approach is like taking a lawyer's affidavit based on an inspection of the car assembly line; unless the inspection is complete and the conclusions are derived in an impeccable manner, the result is just as questionable.

Any test requires exact knowledge of what the system is supposed to do. However, at the onset of the design of a system, its specifications are seldom completely defined. As the design is refined and the constraints become better understood, specifications get added and fine tuned. In the end, specifications must become an exact description of what the designer believes that the system will do so that this information can then become the basis for tests or design verifications. The formulation of exact systems specifications is thus another important part of the *design* effort.

Testing should be kept in mind during the design of any system, program, or VLSI part. The designer should constantly ask questions of the kind: What is this part supposed to do? How will it be used? What are the extremes of the input parameters or operating conditions it will have to withstand? How can it be asserted that the completed part will be doing what it is supposed to do? Asking these tough questions and trying to provide answers to them will lead to cleaner and safer designs, which are also more amenable to debugging and testing.

H. Tools and Design Methodologies

Of crucial importance in the construction of large and complex systems is a good set of tools and a suitable design method.

Design Tools: The user interface is an important part of any design tool, be it for program development or the design of VLSI circuits. As computer time becomes relatively less expensive than the designer's time, the tools become more interactive in nature. Batch jobs should be used only for large

program runs where the interaction of the designer is not required. Program development, on the other hand, should be done in an environment where the programmers get all the help they need to be most productive, such as syntax directed editors [34] and powerful high-level debuggers [35]. An equivalent set of tools in the domain of integrated circuit design would include design stations with built-in design rule checking, automatic layout compaction, and tight coupling to a set of simulators.

Most currently available tools fall short of the designer's expectations in many ways. Here is a list of frequently heard complaints about today's commercially available design tools for integrated circuits. Similar criticism has been voiced for software development tools [36].

1) *Cumbersome commands.* Too many actions (key strokes or cursor moves) are required to invoke frequently used operations such as placing a rectangle in a layout.

2) *Not responsive enough.* Manipulations on a graphics screen should be at least as fast as the corresponding manipulations with pencil and paper.

3) *Lack of uniformity.* Design tools reside on several machines with quite different accessing methods and interfacing protocols; this discourages the effective and frequent use of these tools.

4) *Poor integration with other tools.* Tools use different data formats or different hierarchical structuring; this makes it cumbersome to use different tools on the same design.

5) *Capabilities too narrow.* Some tools, efficiently tailored to a specific approach such as Gate Arrays [14], [37] do not permit the integration of other useful devices or functional blocks, even when these devices are compatible with the fabrication technology.

Support for Good Methodologies: An even more serious, general shortcoming is that most tools do not support a clean design methodology. Layouts done on a graphics editor can be rather ad hoc, since there is nothing to enforce high-level structuring disciplines. Normally there is no place where the designer can capture semantic information, either for the benefit of other designers re-using a particular cell or for other tools such as circuit extractors or logic simulators. Graphics editors and design tools centered around layout information alone are not sufficient for the design of complex VLSI systems.

The analysis of tradeoffs, resolution of conflicting demands, and finding the best structure are at the heart of the design process. According to Parnas [38], this leads to three key criteria by which different design methodologies should be compared:

1) In what order are decisions made?
2) How are these decisions recorded?
3) When is their correctness verified?

The search continues for good formal structuring methods giving consistent results. The search for the right design methodology for VLSI systems and for tools that properly support the latter has only just begun.

Consistency of Style: A good design methodology is one that reproducibly generates a particular system implementation, independent of whoever is applying it. So far, no design method, even in the field of software engineering, automatically produces unique solutions. As a result, one can find widely differing programming or layout styles. Even if the advantages of one style over another are hard to quantify, adherence to a consistent style in the design of a large system is important from the point of view of clarity and maintainability.

TABLE III
DIFFERENT REPRESENTATIONS OF A SUBSYSTEM

Particular Aspect	Description Format
Semantic behavior:	English text
Test specification:	text, tables
Logic function:	equations, state diagram
Timing behavior:	waveforms
Geometrical area:	graphics display
Connection diagram:	display, wire list
Power consumption:	text, tables

Representations, too, should use a consistent notation to make understanding easy. The use of familiar constructs greatly enhances understandability. The pattern-matching processes in the human brain are very powerful and can be trained to quickly find relevant patterns. This should be exploited in the selection of the notations for the various representations of a problem.

For similar reasons, interactive tools also should have consistent user interfaces. A particular command such as *delete* should have a corresponding effect in all tools in a design environment, regardless whether it is used in a graphics layout tool, in a text editor, or in a file manager. A uniform design environment will greatly enhance designer productivity.

IV. ADDITIONAL DIFFICULTIES IN VLSI SYSTEMS DESIGN

The general problems associated with complex systems that were discussed in the previous sections also occur in VLSI systems design. The related techniques developed in the field of software engineering can thus readily be applied to VLSI design. However, VLSI systems cause additional problems based primarily on the two-dimensional nature of integrated circuits and the need for physical interconnections between modules, i.e., the inability to "jump" to a "submodule." This section will outline problems that go beyond those that are normally faced by the software engineer.

A. Multiple Views and Representations

In VLSI systems, the number of different concerns is quite large and diverse. This necessitates a much richer set of views and representations than is normally used to document a software system. First, there is a larger spectrum of hierarchical abstractions, ranging from functional specifications to machine-level layout descriptions. Secondly, for a particular level of detail, several different representations may be used to optimally capture a particular concern of the designer. For instance, at the module level of a digital system, a pipelined piece of logic might be described by:

1) its logic function and test patterns for switch-level functional simulation;
2) suitable delay models to be used in signal-independent worst case timing verification;
3) separate logic gates with timing parameters useful for timing simulation and checking for glitches.

The different representations cannot always be fit into a single hierarchical ordering. Table III gives a sampling of various concerns in VLSI systems and possible notations to capture these concerns. This set of representations is richer than the one normally encountered in software engineering.

Unfortunately, these many different representations do not easily follow from one another. In a VLSI chip, most information can be computed from a final layout, but this is a very time-consuming and computation-intensive task. In order to provide an interactive design environment, most derived information should be stored explicitly in a database once it has been computed, so that it is readily available to the designer. Other information, such as functional specification, is provided top-down before any layout exists. This must be stored and later compared against the information derived from the actual implementation.

Keeping the various representations up-to-date and consistent in the face of daily changes is a major task. The search continues for a centralized representation from which most of the other representations can be derived quickly and inexpensively. Conversions between the different representations are particularly difficult if the designers do not relinquish part of their freedom and do not restrict themselves to a subset of well-defined modules and constructs.

B. Partitioning a VLSI System

Section III-E dealt with the difficulties in top-down design methods related to the fact that it is not clear *how* a large system should be partitioned into smaller blocks. In highly integrated systems, such as a VLSI chip, there are additional difficulties. Because of the lack of explicit interfaces, the partitioning step is often done in an *ad hoc* manner, improperly documented, and never formally verified to meet the original specifications of the overall system. Specifications that might have been precise at the higher level become more fuzzy or may even be violated in the intermediate partitioning steps. It is often not until the whole partitioning hierarchy has been instantiated, and a global systems simulation is being performed, that such errors get noticed. At this late stage it is often no longer clear *who* was responsible for the original violation, and an innocent group of people may have to sweat to correct the problem.

Clean and precise specifications are needed *at all levels* of the design hierarchy and in all transformations of representations. This is tedious and difficult, but unless it is done properly, the original problem has not *really* been partitioned; at best, a more terse and economical notation has been found for something that is still functionally unstructured. As an example, consider a switched-capacitor filter comprising some switches and capacitors and five identical operational amplifiers, each with, say, 30 transistors. In any reasonable description, the amplifier will be factored out as some kind of macro cell. But unless a simpler and more abstract model is used to represent it in the systems simulation, i.e., as long as the filter circuit is presented to the simulator with all 200-odd components visible, the circuit has not been partitioned functionally. Proper functional partitioning is accomplished only if each amplifier is viewed as a new lumped element with its own higher level simulation model.

There is a lot more work involved in creating satisfactory models in *all* relevant representations of a physical component than what it takes to give an abstract description of a cleanly designed program subroutine.

C. Optimization and Packaging

Because of constraints on implementation, physical and logical hierarchy need not have a direct one-to-one correspondence. The logical design hierarchy results from a suitable partitioning into significant functional blocks. The physical hierarchy may be determined by packaging constraints or by geometrical placement restrictions.

In present-day integrated circuits, functional solution and technical implementation are rather narrowly intertwined.

The performance loss resulting from the separation of the two concerns and from the use of automated implementation has been considered too high by most companies producing high-volume integrated circuits. The emergence of integrated, customized systems of VLSI complexity will force us to re-think this issue. The intertwining of functional solution and implementation should be considered an optimization step that has to be justified, rather than being the default approach.

To make these tradeoffs wisely, a better understanding needs to be developed of where the boundary between functional solutions and implementation can be drawn. The special nature of VLSI technology and its physical realities must be considered already at the system level. Algorithmic changes may be called for when one switches from a software solution to a printed-circuit board implementation and later to a VLSI chip. Yet, at the same time, one cannot afford to tie a design exclusively to a single fabrication process.

Many of the high-level partitioning steps are relatively implementation independent; e.g., the decomposition of a large processor into register–transfer level modules is independent of whether the system will be implemented in NMOS or in CMOS technology. Thus the top end of such a design system can be common to many different technologies. In principle, one could decompose a computer down to the gate level and then simply use either the NMOS or CMOS gate implementations. However, that would lead to highly inefficient realizations of such functions as an operand shifter for which compact, technology-specific realizations exist. The more one wants to push performance and optimize an implementation, the higher up the technology-specific concerns have to enter the selection of the algorithm and the structuring process. This optimization step is crucial for the fabrication of present-day VLSI chips and will result in a certain loss of clarity in the design.

D. Layout

A particularly important set of constraints stems from the two-dimensional nature of integrated circuits.

Electrical Connectivity: Whereas high-level language programmers normally need not concern themselves with the overhead of jump instructions, communication between modules in a physical machine is expensive, both in terms of space and time. For components interacting with high frequency, physical proximity is crucial since the penalties for long, high-bandwidth interconnections are severe. In large integrated circuits, the partitioning into blocks and their location on the chip must be chosen carefully since the physical distances between components and the total bandwidth between them remains frozen once the chip is made; there is no equivalent of a memory hierarchy managed by an operating system which can move frequently used information closer to the point where it is used, e.g., into a cache memory.

The communications overhead is particularly severe if the signals must travel from chip to chip or from one printed-circuit board to another. This results in a substantial expense in power and in loss of bandwidth. In addition, the number of pins is often strictly limited. (In the software domain this would correspond to the situation where the number of parameters that can be passed to a subroutine is limited to, say, two integers or one floating-point variable!) The physical packaging hierarchy must thus be chosen with particular care.

Two-Dimensionality: The concern with interconnectivity and geometrical placement is particularly intensive in VLSI systems since the implementation is restricted to the two-dimen-

TABLE IV
Degrees of Concurrency

Type	Example
Bit concurrency:	n-bit parallel adder
Vector operations:	matrix multiplication
Pipelining:	overlapped instruction execution
Set concurrency:	evaluation of alternatives
Specialist functions:	co-processors
Task concurrency:	communicating processes
Random concurrency:	everything else ...

sional space of the surface of a silicon wafer. The signal paths to other modules compete with the computational elements themselves for the rather limited chip surface. This places severe topological restrictions on implementation. A particular module can have only a very small set of close neighbors, and it is often hard to decide which of several contenders should get the preferred spot. By comparison, the sequence and ordering of the definitions of subroutines is rarely a concern to a programmer (except when forward references are disallowed).

Furthermore, changing the placement of a module in a layout is a much harder task than rearranging software modules. Very few design tools reroute the attached wires automatically when a circuit block is moved to a different corner of the chip floor plan; i.e., the equivalent of a link editor for two-dimensional layouts is still in the research stage. Moreover, no really good language exists for expressing topology.

E. Exploiting Concurrency

To realize their true potential, VLSI systems should be designed for as much concurrent action as possible. Ideally, one would like to see all the gates on a chip do useful work most of the time. But so far, the concurrency exploited on VLSI chips has been of a relatively simple and straightforward nature. That is not surprising since the problem of exploiting concurrency in a general manner is not even solved in the software domain.

A first goal must be to find adequate high-level descriptions for highly parallel but irregular operations. Timing sheets with multiple traces, a method typically employed by integrated circuit designers, does not scale effectively to the VLSI level. Table IV gives an overview over various levels of concurrency, ordered by their generality and, alas, by their difficulty of implementation.

The simplest kind is bit concurrency. Parallel operation on the bits of fixed-sized operands, such as parallel addition, is routinely employed in microprocessors and normally implemented in a straightforward manner by iteration of the proper 1-bit cells.

The next level, the lock-step concurrency, as employed in single-instruction multiple-data vector machines, is conceptually not much harder. Problems arise when the number of data components exceeds the available hardware resources, and the job has to be done partly parallel, partly serial.

Control for the third level, pipelining, is more complicated. The execution times in the various stages have to be matched carefully in order to efficiently process different data elements through the same basic functional blocks. Particular problems arise at discontinuities in the data stream where the pipeline has to be refilled with relevant information.

Really difficult problems arise with the higher levels of concurrency, such as set, task, or random concurrency. At this level, one tries to extract concurrency from a problem in such a manner that different operations may be performed simultaneously on different data. The potential improvements in functional throughput resulting from an exploitation of parallelism in this most general form are tremendous. This problem is still unsolved even at the algorithmic level and remains one of the major challenges for this decade.

F. Technology Changes

VLSI designs are aimed at a moving target. Because of the length of the development cycle, a new design is often aimed at a presently emerging, still rather speculative implementation process, with the hope that the process will be mature by the time that the design is finished. As the envisioned process changes, the emerging layout has to be adjusted, demanding modular and modifiable designs.

Because the technology changes so fast, the adaptation to a new process must occur in a reasonably short time span. If the design takes longer than this period, it will have missed a fast moving target. This also speaks for the usage of higher level, and thus more technology-independent descriptions. If the system design has been carried out at too low a level of representation, it will not be usable with new fabrication processes. High-level language descriptions with suitable layout compilers will be more adaptable.

G. Debugging and Testing a VLSI Chip

Debugging the prototype design of a VLSI system involves all the problems that the software engineer faces and then some additional ones. Because a VLSI chip is a rather monolithic system, it cannot be easily tested in bits and pieces, i.e., one subroutine at a time—unless explicit measures have been taken to partition the system for debugging or testing purposes. Because of the small physical features on the chip, it is very tedious and costly to gain access to the inner parts of a VLSI circuit. Only signals that are brought out to the terminals of the chip are readily observable. In other words, the equivalent of a debugger with breakpoints and monitoring of variables has not yet been invented for VLSI chips.

Checking a monolithic system becomes exponentially harder with increasing size [39] and the amount of complexity that can be packed into a single VLSI chip is far beyond the maximum reasonable size of a testable block. Special measures are thus necessary to break the overall system into smaller, testable parts, such as providing internal access points or scan-in scan-out registers [40].

Imperfect Implementation: In addition to debugging, i.e., verifying that the *design* of a chip is correct, there is also the problem of testing each copy to make sure that it is acceptable. This problem can usually be ignored in software, since a simple redundancy check is normally sufficient to guarantee that the copy is a perfect replica of the original.

When writing a program, one typically assumes that there will be a system that guarantees, often by complicated means, that all instructions are properly executed in the sequence specified. The VLSI designer faces a less ideal world. Signal voltages are only defined in certain *ranges;* in addition, they are subject to *noise* that may well exceed the stated tolerances. Alpha particles can change the contents of a memory cell. Temporary voltage surges occur on the supply lines when large numbers of signal lines change state simultaneously.

The overall system has to be immune against such erratic behavior. Overdesign, redundancy, checking, and recovery mechanisms have to be built into larger functional blocks and subsystems so that they can present a more nearly ideal picture to the observer at the next higher level of abstraction. In debugging the design as well as in testing the final product, a particular operation has to be checked over a range of "environmental" conditions, such as supply voltage variations, temperature ranges, and processing parameter deviations. This adds an extra dimension to the debugging/testing problem.

V. Tradeoffs in VLSI Chip Complexity

In software systems there is a clear trend to higher level languages. With ever faster processors and cheaper memories, the inefficiencies in code density and execution speed in a compiled program become insignificant in comparison to the advantages associated with the usage of high-level languages (Section III-C). Hand optimization at the assembly language level is reserved for critical inner loops, often-used routines, or for applications with real-time constraints.

The issue of layout density is more important for VLSI chips than code density is for programs because of the severe technological restrictions on maximum chip size. Efficiency of implementation cannot be ignored. It will be necessary to sacrifice a certain amount of modularity and modifiability in order to fit a system on an economically viable chip. This section considers these tradeoffs.

A. Hardware versus Software

In designing a complete system based primarily on some VLSI chips, a new tradeoff has to be considered: what part of the function should be implemented in silicon and what part can be done with software? This tradeoff actually spans a whole spectrum of possibilities. Consider the example of a microprocessor: should a complicated instruction be wired into the decoding logic, should it be programmed into the microcode memory, or should the compiler compose this function from a collection of more primitive instructions? These decisions are of particular importance in VLSI system design, since the individual chip is an entity that must work as a whole. If the chip design is too ambitious, the chip cannot be fabricated economically. Thus at any given point in time, the number of active devices of a certain type that can be used economically on a single chip represents a rather rigid resource limitation. One must decide very carefully how to best use this resource.

B. Simplicity to Get the Job Done

A very direct approach to managing complexity is to question whether the complexity asked for in a set of specifications is indeed reasonable and necessary. Some recent studies into instruction sets and organizations of microprocessors showed that evolving products need not necessarily follow a trend to ever increasing complexity [41]. Complexity introduced to provide rarely used instructions may not be worth the extra costs in terms of increased chip area, a slower overall machine cycle due to the delays through more complicated decoders or micro stores, and delayed market entry because of a prolonged design cycle. Worse yet, such additional circuitry may run against the hard limit of resources on the chip and thus take away devices from other functions that might have made a larger contribution to the overall performance of the system. In the domain of integrated circuits, unnecessary functionality as well as irregularity [42] comes at a much higher price than in software engineering.

C. Complexity for Higher Yield

Beyond a certain point, the yield of good integrated circuits emerging from the fabrication process drops off exponentially with increasing active chip area [43]. Due to these hard constraints on chip size, a reduction in the number of devices on the chip is well worth some increase in complexity.

In large regular arrays such as memory blocks, groups of two or four adjacent cells are often designed jointly so that proper rotation and mirroring can be used to share contacts or power bus lines. Such a cluster of two or four cells is then repeated to produce the overall array. Similarly, large amounts of random logic are typically not implemented as a single, large, and sparsely populated programmed logic array (PLA), but are broken down into several smaller PLA's [44]. In both cases, while apparent complexity is slightly increased, considerable area savings may result.

As integrated circuit technology evolves, chip functionality will increase. But the limit of how many fully functional devices can be placed on a single chip cannot rise indefinitely. To make significant improvements in the total functionality of a chip, new methods have to be exploited. The emerging 64-kbit memory parts already employ redundant memory cells and circuit elements which take over for the devices that have fabrication flaws [45]. So far, the parts have to be tested individually, and the proper rewiring or reprogramming has to be done from the outside. It is reasonable to expect that at some point in the future there will be more complex, self-testing circuits that reconfigure themselves internally to portray to the outside a flawless function of a given specification.

D. Complexity for Better Performance

Among the reasons for casting systems in silicon is the aim of compactness and performance. This implies a certain degree of system optimization. Simply taking the most regular and straightforward implementation might result in an intolerably bulky or slow chip.

Even RISC [41] is not just a simple Turing machine. Instruction fetch and execution are overlapped. This pipelining increases the system complexity, but no microprocessor designer would give up the resulting performance gain at the present time. New commercial microprocessors go even further and use more sophisticated pipelining schemes. Increased complexity and increased performance need to be carefully evaluated to find the optimum approach. Similarly, increasing the richness of the instruction set leads to a point of diminishing return. A general guideline on this issue has been given by Wulf [46]. He demands that instruction sets offer *primitives, not solutions.*

A look into the future sometimes justifies cramming more complexity and functionality onto a particular chip than is justified by present-day resource limitations. First, one expects that these limits will expand, and that the decision to add extra features will be right in the light of future developments. Secondly, computer architectures tend to be around for a long time. A company that plans to launch a whole family of processors of various performance ranges needs to accommodate the basic instruction set even in the lowest performer; this might distort the proper balance of resource allocation for this particular member of the family.

VI. General Trends in VLSI Systems Design

Many of the techniques for complexity management reviewed in Section III are slowly being integrated into presently emerging VLSI circuit design systems. This section summarizes the main trends and explores how the discussed methods are adapted to the special problems of VLSI systems design.

TABLE V
COMPARISON OF LANGUAGES

IC DESIGN and	PROGRAMMING
Functional specifications	Specification languages
Self-generating layouts	APL, ...
Protected, abstract modules	Ada, ...
Subcircuits, explicit connections	Pascal, ...
Symbolic description	Basic, ...
λ-based conceptual features	Intermediate code
Mask geometry	Assembler language
Mann or MEBES format	Machine language

A. Towards High-Level Languages for VLSI Systems

There is a rich spectrum of possible views which can be used to discuss the design of a VLSI system. Table V shows the most important hierarchical levels in a high-to-low order and attempts to put them in perspective by pairing these levels with representations from the world of programming languages.

In the 1970's, the practice of laying out integrated circuits corresponded to assembly language programming. Even today, most designers still have an overriding concern for density in order to end up with a chip that is small enough to be manufactured economically. Traditional designers feel that the compactness and performance achievable with good hand layout are worth the tediousness of the approach.

In the domain of software engineering, high-level languages are now used routinely for system developments of substantial size. Assembly language code is reserved for small, high-volume systems with limited memory, such as games and controllers, or for real-time systems with absolute timing constraints. The same trends towards high-level languages seen in software engineering are also apparent in the domain of integrated circuit design. As integrated circuit technology progresses, it will soon be possible to place more than a million devices on a single chip. For random control circuits, this is an inordinate amount—far more than any designer can reasonably handle manually. In such complicated systems, efficiency is a lesser issue than achieving correct operation.

High-level languages also reduce development time. This is even more important in the fast-changing field of integrated circuit fabrication. Earlier market entry will translate into large economical advantages in this highly competitive field. The evolution of single-chip VLSI systems creates a demand for more personalized systems that fit the customer's needs exactly; multiyear development times are no longer tolerable in this context. Minimizing the size of integrated circuits will only pay off for high-volume products (more then 100 000 samples). For customized VLSI applications, some layout inefficiencies can normally be tolerated; the most important consideration is to get the job done correctly and on time.

B. Partitioning of Concerns and Problem-Specific Representation

The development of a VLSI system involves even more diverse considerations than writing a large software system. If possible, the various concerns should be addressed independently. Jackson recommends the separation of problem-oriented concerns from machine-oriented ones [20]. The use of two complementary languages is suggested: a programming language to solve the problem, and an execution language to specify how to compile and execute with efficiency. The notation for the two languages should be optimized for the task at hand.

Similarly, in VLSI design, different languages and notations should be used at different points of the abstraction hierarchy.

Representations and notations must be matched to the problems that one tries to solve. In the design of VLSI systems there is a large variety of different concerns; a rich spectrum of notations is thus necessary. The proper definition of these representations is important because notations not only determine the form of the final solution, but also shape the way we think about a problem [47].

This issue has recently been addressed by Stefik et al. [48]. They define several representations and develop appropriate notations and composition rules. Each one deals with only a few concerns and is thus most effective in avoiding a particular class of bugs:

"*Linked Module Abstraction*" deals with event sequencing at the systems or subsystem level. Modules get started by accepting a token; they do exactly one task at a time; when finished they emit a token to the next module. This representation helps avoid deadlocks and the sampling of data that are not ready.

"*Clocked Registers and Logic*" are concerned with the details of timing. This abstraction tries to eliminate errors such as race conditions in unclocked feedback loops, e.g., by using strictly alternating two-phase clocking.

"*Clocked Primitive Switches*" try to assure proper digital behavior by addressing the ratios of digital NMOS inverters and avoiding passive charge sharing on intermediate nonrestoring nodes.

"*Layout Geometry*" is concerned with geometrical layout rules and tries to prevent spacing errors or unrealizable mask features.

Stefik et al. [48] point out that such problem-specific representations need not fall into a strict hierarchical ordering. The relative ranking of the abstractions along different axes of concern might well be permuted.

C. Emphasis on Structure and Abstraction

Without the introduction of some structure, i.e., hierarchy and regularity, the problem of VLSI system design would be unmanageable. As the organizational problems start to overshadow the fabricational difficulties, the use of a *constrained hierarchy* [49] becomes more attractive. Additional restrictions are imposed on the structuring process:

1) The hierarchical partitioning is truly functional with proper abstractions at each node.

2) The hierarchies in the various representations (geometrical, logical) must correspond to one another.

These restrictions on the designer's freedom result in a higher degree of clarity and in more efficient operation of most design tools, at the price of some loss of layout density.

Regularity, i.e., reusing the same submodules as often as possible, also increases effectiveness. Lattin has introduced the *regularity factor*, derived by dividing the total number of features (transistors, rectangles) by the number of features actually drawn by the designer. Present-day microprocessor chips have regularity factors ranging around 5 for commercial products [10] and reaching about 20 for experimental devices [50].

Traditional Logic Cell Macros: A traditional approach to structuring large digital circuits relies on the logical abstractions of the well-known small- and medium-scale components, such as gates, flip-flops, and registers. This approach is used in the *standard cell* and *gate array* systems. The *standard cell* or *polycell* approach [15] relies on a library of layouts of such logic components and on some automatic or semiautomatic procedure to place the cells and to route the interconnections. The *macro cell* or *gate array* approach [14], [37] uses a combination of preprocessed wafers and libraries of predefined

wiring options to generate the same standard logic functions. The design systems normally contain powerful routing algorithms that wire most of the chip automatically. The designer has to help only when there are special timing or area constraints. The increasing market share of such products indicates that these are indeed practical approaches. In both cases, engineers can approach the VLSI design in the traditional manner and map their logic designs onto a silicon chip with relatively little effort.

The drawback with these approaches is that they use the wrong weighting factors. They encourage the designer to minimize the number of logic gates, assuming that wires are free, whereas on VLSI chips, they dominate the layout area as well as the timing delays. Suitable representations for VLSI must properly address the interconnections at all levels of the partitioning hierarchy.

Advanced systems permit the integration of custom-designed macro cells into the final layout like any other library cell. This approach has the advantage that higher level functional blocks, such as the datapath of a microprocessor or a control PLA, can be designed by alternative design methods, leading to more compact implementations.

Hierarchical Analysis Tools: As regularity and hierarchy is empasized in the synthesis of our designs, the design tools also need to be restructured to exploit this new methodology in the analysis phase. Design rule checkers or circuit extractors that walk rectangle by rectangle through a regular array are no longer tolerable for 64-kbit memory chips. The generating cells and their generic constellations in the array should be checked *once*; repeated instances of checked constellations can then be skipped. This requires some sophisticated book-keeping, but the performance advantages of hierarchical circuit extractors and design rule checkers are well worth it [51]–[54].

A well thought-out abstraction hierarchy must also be exploited in the simulation tools. Once a low-level module, such as a 1-bit adder, has been constructed and checked to behave correctly, that information must be tightly linked to the definition of that cell. System simulations then no longer need to work all the way to the level of the physical differential equations but can stop at the module level and use a more abstract model describing the behavior of the module at that level.

The use of hierarchical analysis tools can only succeed if it can be guaranteed that the introduced abstractions remain valid. In particular, the cell has to be protected so that its behavior cannot be changed accidentally. In the layout domain, some kind of protection frame can be employed so that accidental interference into inner parts of a cell can be detected. There are some inefficiencies associated with this approach; cells cannot be packed as tightly as they could be in an unrestricted environment. However, this mechanism should only be invoked for cells of a reasonable size, where introducing a new level of abstraction is appropriate. The protection frames serve the same roles as the *module* construct in some high-level languages; across module boundaries all interactions have to be declared explicitly. At that level, the loss in packing density can be tolerated.

D. Separation of Design and Implementation

Another application of abstraction and partitioning of concerns occurs in the transition from design to implementation. Traditionally, the layout engineer had to stay in close contact with the fabrication line and the mask makers in order to understand what exact geometrical patterns (what polarity, suitably grown or shrunk) had to be submitted in order to receive the desired physical features on the fabricated silicon wafer. Certain process variations force the designer to rework one or more mask levels. In addition, the selection of the proper processing test structures and alignment marks was also a responsibility of the designer.

The experiments with multiproject chips shared among many designers at several different universities [55] demonstrated that design and implementation can indeed be separated. The designer is responsible only for his or her own design and need not worry about mask polarity, alignment marks, process compensation, or monitors for critical dimensions. The designs are submitted in a standardized low-level geometrical descriptive form such as CIF2.0, the Caltech Intermediate Form [56]. A centralized service organization will then merge the different designs onto several reasonably sized chips together with a *starting frame* containing the alignment marks and some process control monitors. These chips, in turn, are assembled into a suitable tiling pattern, which also contains a few slots for gross registration marks and more extensive test structures, to cover a whole wafer. The same service organization will arrange and coordinate mask making and the actual fabrication of the silicon wafer and will subsequently redistribute the fabricated and possibly packaged chips to the original designers. This approach distributes the costs of mask making and wafer fabrication among all the designers and brings the cost per design down to an affordable range for small volume silicon systems; such services have been called "silicon foundries" [57].

This approach shields the occasional designer from the time-consuming and confusing details that one has to consider when submitting a design for fabrication. The silicon foundry approach abstracts the implementation process to a single "module" with a well-defined interface. The designer specifies the physical features that should appear on the final chip. All the expand/shrink operations necessary to compensate for the actual processing are performed by the implementation service. Only chips originating from wafers that pass the process control tests will be returned to the customers. This approach can only be successful with well-established, stable fabrication technologies. Again, the additional abstractions and the corresponding reduction in complexity in the design process comes at the price of somewhat lower performance.

E. Procedural Generation of Modules and Systems

While a silicon compiler for complete systems is still some time off in the future, procedural generators for special, frequently used, functional blocks become commonplace. The most important ones produce PLA's and read-only memories (ROM's). These module generators take a set of parameters or, in the case of ROM's and PLA's, the Boolean equations of the desired logic function, and then compose the complete layout from predefined low-level cells. Because these blocks are so important and used so frequently, a lot of effort has been spent to optimize the basic cells from which the final layout is composed.

Other, more complicated functional blocks, are special data processing elements, such as the datapath of a typical microprocessor forming a linear arrangement of registers, shifter elements, incrementers, or complete arithmetic/logic units [24], [13]. Hewlett–Packard claims [58] that about 90 percent of its data processing chips could be cast in the format of such a generalized datapath.

The first set of these module generators is aimed at a specific process and a specific set of layout rules. In later versions, more fabrication independence will be gained by performing the cell composition at the logic level and passing the result through a fabrication-rule-dependent circuit compactor [59]–[61].

Eventually there will be module generators that handle many *diverse* fabrication technologies (NMOS, CMOS, SOS, . . .). However, because the most suitable topology for a PLA is different for these different implementation technologies, only the higher levels of these generators, e.g., the algebraic minimization and the optimizations through folding and splitting, can be shared. To be most effective, these module generators must be fully integrated with the layout system. Their output must have the complete documentation of a typical macro module. This requires that the module generator produce all necessary representations, ideally including test specifications.

F. Enhanced Use of Graphics and Interactive Tools

VLSI system design is strongly tied to the two-dimensional nature of the implementation medium. Because of the lack of suitable languages to express the relative geometrical arrangements of individual modules and the wiring between them, the use of graphics plays an important role. Presently graphics display hardware is becoming inexpensive enough, so that there is no reason not to provide every tool that can profit from it with a graphics interface. A highly responsive, interactive graphics display is going to be the core of every integrated circuit design station.

The success of the module generators discussed in the previous section has led many people to postulate that *all* design work should be done in a procedural manner. This approach would have two major drawbacks:

1) The description of the geometry of low-level cells is rather tedious.

2) Without a two-dimensional representation the designer lacks an important element in the feedback from the design system.

Graphics is an invaluable aid at many levels of the design. For the geometrical layout of complicated leaf cells, graphics is clearly the preferred medium of interaction. At the higher levels, where a procedural description of the modules is more appropriate, graphics still plays an important role for checking the results of the procedural constructions. A good interactive graphics system can also play an important role in the strategic planning of the chip layout. It permits the presentation of the problem to a human in a format where the designer's intuition can be tapped most effectively.

On the other hand, it seems more appropriate to use a procedural approach to specify the placement of the "1"-cells in a large PLA. Because of these different tradeoffs, there is an ongoing dispute of the use of "Pictures versus Parentheses" [62]. Both approaches are optimal for certain tasks, and a good design system thus cannot ignore either one of them.

A major disadvantage with present-day procedural design tools stems from their "batch-processing" nature. It is often necessary to make small changes in a nearly finished design. If placement or routing is specified in a procedural manner, the introduction of a small change in the specifications may result in a completely different layout. This will invalidate the simulation and performance verification efforts expended on the 95 percent of the chip that were correct beforehand. Restating the constraints in such a manner that only the *desired* changes take place is normally not possible.

In the future, these procedural tools need to be integrated into a design system with interactive interfaces on which desired changes in placement or routing can be specified in a natural manner. New algorithms are required that make only *incremental* changes and derive the next solution as a *minimal change* from the previous solution. Many problems need to be solved to create such an integrated system in which the designer gets the best benefits from both approaches.

VII. Evolution of Design Tools

To a large degree, the evolution of design environments for systems on silicon will follow the evolution of programming environments [63]–[66]. However, future tools need to take into account the special demands of VLSI design discussed in Section IV. This section attempts to project how this will be achieved in three different time frames. Tomorrow's design tools already exist in prototype form in many research laboratories, and so one can have a fairly clear idea what they will look like. The remarks concerning future design environments are based on work that is currently in active pursuit in several research institutions with the goal of creating prototypes within the next couple of years. The last subsection on ultimate ways of creating solid-state systems is a rather speculative extrapolation of some ideas currently being discussed in research papers and workshops.

A. Tomorrow's Design Tools

The trends discussed in Section VI will gradually make their way into the emerging design tools. Based on the work currently in progress in many research and development organizations, one can have the following expectations about tomorrow's design systems: they will have strongly improved human interfaces, addressing most of the issues discussed in Section III-H. The tools for layout planning, module placement, and routing will make use of ever more powerful algorithms to do a large fraction of the task, but retain a strong interactive nature to permit the designers to make full use of their intuition and judgment.

There will be an increased use of higher level symbolic descriptions coupled with more and more automatic generation of frequently needed modules. Explicit design methodologies, such as the use of a constrained hierarchy, will gain preference, and the analysis tools will start to exploit the hierarchy used in the design process.

The two main components of tomorrow's design system will be a modular set of tools and a database.

A Modular Set of Tools: The same considerations that apply to large programs or VLSI systems also apply to the development of the design environment itself: the latter should also be highly structured and composed of modular components with simple, well-defined interfaces.

The art of VLSI design is still not fully understood, and new methodologies are still evolving. It is thus too early to specify a rigid design system that performs the complete design task; quite likely, such a system would be obsolete by the time it becomes available to the user. It is more desirable to create a framework that permits the usage of many common tools in different approaches and that supports a variety of different design methods and styles. In short, the environment should provide *primitives* rather than *solutions* or, in other words, *mechanisms* rather than *policies*.

Intricate interaction between the various tools must be avoided. Every tool should do one task well and with reasonable efficiency [36]. Compatible data formats are needed to make possible ready exchange of information among the tools or direct piping of data through a whole chain of tools. These formats should be comprehensive enough to carry all relevant information for various tools working on a particular design.

At Berkeley such a collection of tools [67] is being developed, embedded in the UNIX [64] operating system. UNIX already provides many of the facilities needed for our planned environment; a suitable hierarchical *file structure*, a powerful monitor program in the form of the UNIX *shell* [68] and convenient mechanisms for *piping* the output of one program directly into the input of a successor program.

Database and Data Management: All information concerning a particular design should be kept in one place, ideally in some *database* readily accessible to designers and design tools. In an emerging system at Berkeley [67], a VLSI design is mapped onto the structure of the UNIX file and directory system. Each module of the system under development is represented as a directory of files on the computer's storage system. Submodules are represented as immediate subdirectories or as files in special library directories. Each such directory may contain a varying number of files corresponding to different representations of this module. There may be files describing layouts, circuit diagrams, symbolic representations, suitable models for high-level simulation, sets of test vectors for this module, and possibly some plain English documentation. The number of files at each node is potentially unlimited.

Currently, many design systems for custom circuitry use the geometrical layout information to "glue" everything together. It is from this low-level description that other representations are derived; many of the analysis tools work from that level, e.g., circuit extraction and design rule checking. This is an unsatisfactory approach. Too much of the designer's intent has been lost in that low-level representation and has to be pieced together again by the analysis tools. If there is to be a "core" description from which other representations are derived, it has to be at a higher level. The trend is to move to a symbolic description [59] that is still close enough to the actual geometry so that ambiguities in the layout specification can be avoided. Yet at the same time, this description must have provisions to specify symbolically the electrical connections and functional models of subcircuits [69], [70].

Regardless of the exact structure of the database, the various different representations of a design should be at the fingertips of the designer so that one can readily go back and forth to the one representation that best captures the problem formulation with which the designer is grappling at the moment.

B. Future Design Environments

Based on the improved tools discussed in the previous section, five to ten years from now one will see the emergence of integrated yet flexible development *environments* for solid-state systems. Major parts of such future environments are currently in the active research stage at several institutions.

Design Management: An ever larger fraction of VLSI systems will be generated in a semiautomatic manner in which the tedious low-level operations are performed automatically. Just as future programming environments will tend to deemphasize programming by end users [71], the system design environments of the 1990's will encourage most users to work at a higher level. These environments will contain parameterized components in a variety of the technologies available at that time and the necessary tools to place and interconnect them semiautomatically. The designs will be described at much higher levels and significant parts of them will be compiled into the layouts for the chosen technology. "Design *management* tools" rather than design tools will play an increasingly important role.

Tools to help with the management of the overall design effort, including task distribution, activity scheduling, and design reviews, will gain importance with respect to the tools that just support the technical aspects of the design. Some utilities that have proven beneficial in the domain of software engineering have started to make inroads into solid-state systems design. One such utility, the Source Code Control System (SCCS) [72], plays a role when a system is too big for a single designer. It maintains a record of versions of a file or a system. This is useful when several people work together on a project and thus need to make changes to a document that is jointly used by all of them. SCCS keeps a record with each set of changes of what the changes are, why they were made, by whom, and when. Old versions of the document can be recovered, and different versions can be maintained simultaneously. It also assures that two persons are not editing the same file at the same time, which could lead to the loss of one person's modifications.

In the realm of VLSI system design, a more formal *configuration management* [73] discipline should be adopted to control modification of the specification files, the emerging microcode, or the lists of the intermodule connections. These are typical examples of documents that change continuously throughout the design phase and which may have to receive input from several people.

Design Data Base and Module Libraries: The file system containing the information about a design will evolve into a full database. This database containing the emerging custom design will be complemented by a library of cells that will be shared by many designers working on different designs. For custom designs using hand layout, these libraries have only been moderately successful. There is a lot of reinventing the wheel. Weinberg [74] noted that the problem with program libraries is that "everyone wants to put something in, but no one wants to take anything out." Most of this is caused by cumbersome access methods, inadequate documentation, poor adaptability of the modules, and a "not invented here" mentality [75], [76].

The libraries of the next decade must evolve into database systems that include convenient, interactive search procedures, employing hierarchical menus or the possibility of querying in a subset of a natural language to find the desired function. For the selected cells, the system must produce layers of more and more detailed documentation so that the designer can unambiguously determine whether a particular cell will suit his needs or will have to be modified. All relevant design information must also be available on-line so that modifications can be done easily.

Such a library should contain all regularly used, generic, functional units, such as memory blocks, stacks, ALU's, analog-to-digital converters—to name just a few. These components should be suitably parameterized to span a wide range of possible applications and should be available in a range of different technologies. In addition, they should be properly encapsulated and provided with unambiguous, easy-to-use, standardized interfaces by which they can be fit together without a need for the designer to pay attention to the device-level details of this interface. The effort needed to create, document, maintain, and adapt all these modules to an evolving

technology can be prohibitive for small companies. It is conceivable that such libraries will be sold or leased on a commercial basis by special companies.

Automatic Consistency Checking: As outlined in the previous section, the various pieces that make up the whole VLSI system will be designed at a higher level and stored in that form to increase the designer's productivity and to maintain some technology independence. A rich set of representations is thus associated with each module. Special software is needed to maintain the consistency of the various views of each module. While such consistency checks could be built into the database, it is preferable to put the know-how of this task into special tools. As more sophisticated tools become available, they can readily be introduced without affecting the structure of the database.

By 1990 silicon compilers will have made a lot of progress. They will then hopefully be able to handle fairly complex subsystems. However, they will still not be mature enough to produce guaranteed correct results from a set of functional specifications. It may not even be possible to prove that they produce proper results for all sensible parameter combinations. A substantial amount of checking will, therefore, still be needed.

A mechanism to alleviate this situation and to catch many possible errors at compile time is to rely on built-in assertions. These assertions verify that certain restrictions hold; they can perform bounds checking on parameters or combination of parameters, monitor current densities, or control module dimensions. Today, assertions are frequently used to verify timing constraints in large systems [77], [78]. They will play an ever increasing role in future systems.

In the design of a large system, individual modules are separately constructed. For a specific implementation, these descriptions have to be compiled into a representation that unambiguously describes the information to be passed to the fabrication line. Some simulation and performance checks may have to be done after this transformation. Rather than doing these steps jointly for a whole VLSI system, it will become more practical to do it on a per-module basis. This corresponds to the notion of separate compilation of the modules of large programs; the required facilities for a suitable environment for VLSI design [79] would be rather similar. A good model is the *make* facility [80] in UNIX. A *makefile* at each node in the file system describes the dependencies between the various data descriptions which may exist at that node, as well as the rules by which a desired description can be obtained from the information on which it depends. In addition, all files carry a time stamp. When the *make* facility is invoked, all files which have predecessors with more recent time stamps will be regenerated. In this manner, incremental updates propagate through the hierarchy.

At all levels, submodules can be tested or verified independently and then protected against further accidental changes. If a deliberate change is made inside such a module, then the flags indicating that the module has been verified are reset to false. An automatic sequence of test programs will recheck the new version and assert the corresponding flags if the specified tests are passed. Thus recompilation or rechecking of modules is required only if something changes inside. Small corrections in a large layout may thus be contained to a small fraction of the overall design.

Testing and Debugging: The areas that are most in need of

advancement are debugging and testing of VLSI systems. Because of the monolithic structure of integrated circuits, more than ever before, debugging and testing must be carefully planned during the design of the system [81]. The key issue is to subdivide the VLSI system into smaller parts that are easier to test. The controllability and observability of internal nodes can be greatly improved with such scan-in scan-out techniques as LSSD [40]. This methodology can be integrated into the design systems that rely on logic gates and register-transfer components. In such a system, a special type of latch is used for every flip-flop, unless explicitly told otherwise, which can then be strung together into scan-in scan-out registers. This approach is particularly powerful if for every encapsulated module the design environment automatically generates a set of test vectors, or, if that is not possible, prompts the designer for the delivery of such a set to encourage him or her to address the issue.

In the same context, better and more technology-specific fault models will have to be developed. It is too simplistic just to deal with stuck-at faults in a solid-state system that is dominated by interconnects. *Shorts* between different wires need to be considered. The simulation of the effect of such faults *per se* is no problem. However, it is impractical to test all possible pairs of nodes for the effect of a short between them. With circuit extractors operating from the layout or from a suitable symbolic description that defines the topology unambiguously, all adjacencies can be determined, and short simulation can thus be limited to relevant pairs. Broken wires, creating *open circuits* can be reduced to corresponding stuck-at faults.

C. Ultimate Ways for Creating Systems

This section is rather speculative; it is not expected that the concepts discussed herein will become reality in this century.

Specification-Driven System Generation: There is a possibility that ultimately routine system design will be driven directly by very-high-level language system specifications. This goal has been pursued in the software domain for the last two decades; but automatic, specification-driven software generation has yet to be demonstrated on a *general* program of substantial size.

In the VLSI domain, such specification-driven designs might emerge as an outgrowth of ever more sophisticated module generators. PLA generators currently require only a few specifications concerning their intended logic functions, constraints on the ordering of inputs and outputs, and some indication to what degree minimization, splitting, and folding should be performed. As these generators evolve, one may soon be able to select from several different structural variants to cover a wide range of delay/power tradeoffs, and to map the resulting design into most common integrated circuit fabrication processes. Once such facilities have been created, a specification-driven system is not too far away. The provision of such facilities for *arbitrary* VLSI systems, however, will lag many years behind. Here, as in software engineering, the difficulty of providing exact, unambiguous specifications is a major part of the problem.

Formal verification for programs [82]–[84] has been discussed for more than a decade. However, it is still impractical, too cumbersome, and mathematically too difficult for most programmers [85]. Application to the domain of VSLI design lies even further in the future because of the more complicated

medium and its "analog" behavior. However, most inventions that have revolutionized our lives, have required 20 to 50 years from first demonstration to the point where they started to have a significant impact [71]. This is true for the telegraph, the steam engine, photography, as well as for television. More readable specification languages will be developed, and most of the tedium of grinding through the formal verification steps will be automated to the point where the computer is taking care of the details under high-level directions by the designer. So, while verification methods have not yet become a smashing success in their first 15 years, it may be well before the turn of the century that they will affect the way in which large designs will be approached.

Knowledge Engineering and Expert Systems: Techniques developed in the field of Artificial Intelligence are expected to have a major influence on our development environments for software engineering as well as for VLSI design. The goal is to capture the essence of the design process and establish a formal design methodology. Once VLSI design is well enough understood that we can start to express explicitly some of the plans followed by good designers, some of that reasoning and background knowledge can be incorporated into *Expert Systems* [86]. Such expert systems combined with rich databases of suitable software tools or electronic subsystems will then be able to piece together reasonable solutions to routine problems. The specially constrained, high-performance systems will be left to the human experts for a long time to come.

Self-Testing and Self-Reorganizing Systems: Even on that day far in the future, when the emerging designs will be provably correct because of the methodology that produced them, fabrication yields will still be less than 100 percent. Testing of the fabricated circuits will thus still be a necessity. When done in the traditional way, it can become a major, if not the dominant, part of their costs. More of the testing chore has to be off-loaded to the VLSI system itself.

Here is a possible scenario. The wafer carries, in addition to the desired VLSI chips, a grid of power and control lines that connects all chips through some suitable circuitry to a few external connectors. In the testing phase, the whole array is powered up and some control signal will initiate a self-test in all the chips on the wafer. Bad chips that draw an excessive amount of power will automatically trip a circuit that disconnects them from the power grid. The remaining circuits will start a small internal state machine that tests a local ROM. If the first test is successful, a second phase will be initiated, in which these registers and the tested ROM from a more sophisticated testing machine that now can test systematically all major parts of the associated chip. If necessary, more layers of this testing hierarchy can be introduced. The final result of all those tests, a single bit of information, will be written nondestructively into an easily readable spot; this might be done by selectively blowing a small fuse, so that good and bad chips can be identified by visible inspection.

This general idea can be extended to systems that coninue to perform self-checks while they are in operation. A collection of multiple redundant blocks, constantly monitoring themselves and each other, may be a way to move into wafer scale integration. Failing blocks will be disconnected from the power supplies and spare parts will be turned on instead. This approach is only viable if the blocks themselves are large enough (more than a 100 000 devices) so that the control and monitoring overhead can be tolerated. The size of these blocks, in turn, requires a more mature technology so that blocks of a few hundred thousand devices can be built with reasonable yield.

VIII. THE ROLE OF THE DESIGNER

A. The Design Tool Manager

Just as evolving programming environments show a trend away from programming, designers of solid-state systems will spend an ever smaller fraction of their time designing at the solid-state level. The general trend has already started. The low-level layout functions get off-loaded to tools of increasing sophistication. The "tall, thin designer" [87] who carries the design from the top-level down to the layout will be a short-lived phenomenon. Solid-state systems engineers who work too much at the layout level neglect the more important design decisions that need to be made at the higher levels of abstraction. A lot of the low-level technical tasks can soon be left to computer-based design tools. The systems designer will be able to rely on tools and on prototype modules generated by expert designers. He or she will thus change from being a technical designer to being a *manager of design tools*. At the same time, *computer-aided design* will evolve into synergistic *designer-directed semiautomatic design*.

A good manager, rather than doing the job himself, will concentrate on creating an environment in which the job can get done most efficiently. This requires an attitude change on the part of the typical engineer. The most leverage out of human ingenuity can be obtained if the latter is used to build new and better tools, which then can do the job more or less automatically. In this mode, the impact of the work of individual engineers can be compounded.

B. The Expert Designer

In all the excitement about the potential of these sophisticated design environments and the power of the proposed tools, one should not overlook the role of the technical expert designer. As technology scales down, the medium gets more difficult, devices are dominated by edge effects and parasitics, and more design parameters have to be considered. It takes an expert to know whether the constrained design space provided for the average designer should be left, where it pays to cross abstraction boundaries, and when large advantages can be gained by violating the rules. Feedback from various tools can be used to check the intuitive insights of the expert.

Experts, too, will rely heavily on design tools. Even a less sophisticated tool can be used to survey the design space in order to decide which approaches are worth studying in detail. The layout expert who plans to hand-pack a certain array structure to the limit possible in a given technology, will be well advised to use a less skillful, automatic compaction program first to quickly find promising topologies that can then be optimized further by hand. Without such an exploratory search of the design space, the expert designer might get "trapped in a local minimum," overlooking a deeper global minimum resulting from a different topology. Seeing a succession of many possible solutions also leads to a better understanding of the tradeoffs, and develops heuristics that permit homing-in on desirable solutions more quickly.

Designers most interested and skilled in low-level design will make their contributions by improving the basic cells on which the solid-state systems compilers must rely. They will formalize their knowledge and incorporate it into the systems that drive the design tools. The support by the expert designer will be vital in keeping the automatic design environment from becoming obsolete.

C. Multi-Designer Teams

Even with a very powerful set of almost fully automatic design tools, large VLSI systems will very likely still require the help of more than one designer. A big problem, well known in the area of software engineering, lies in the management of a design team. Too large a team can become completely ineffective [32] if so much time is spent in meetings discussing *what should be done*, that there is no time left for *actually doing* something. Computers can alleviate this situation in two ways. First, by providing more sophisticated tools, computers will allow the individual designer to achieve more, and this will reduce the size of design teams. Secondly, properly set up electronic communication can also increase the efficiency of the interactions between designers.

The foundation is provided by a sophisticated electronic mail system. A lot of time is wasted when busy people try to get in touch with each other in real time. Forwarding messages to an electronic mailbox that is read when the other party next uses the computer permits more frequent interaction. In addition, electronic messages tend to be relatively terse and devoid of a lot of the additional overhead encountered in many telephone conversations. Most people can deal with an order of magnitude more computer messages than phone conversations per unit time.

Such a system can be expanded readily to contain interest-oriented mailing lists for semipublic announcements and electronic bulletin boards that store relevant information about an emerging system design or a family of design tools. By browsing through these bulletin boards periodically, every member of a large design department can keep informed about issues of general concern.

Electronic mailboxes and bulletin boards can be integrated with the design system itself. Analysis tools, such as geometrical design rule checkers, or update tools such as the described *make* facility, may do some of their tedious work during the night and then send the results and suitable diagnostics to the mailbox of the designer. Changes to cells in a common library may be announced on the bulletin board, and reported to every designer employing these cells in a system that is currently in the active design stage.

Computers can also help formalize the tools used in the management of the design team. They can keep records of the formal distribution of tasks and responsibilities, and of planned and actual project schedules. They can record design decisions, preserve the specifications of the various subsystems and their interfaces, and use them to verify the design in functional simulations. Such a system could even initiate group meetings for design reviews and structured walk-throughs.

IX. RECOMMENDATIONS FOR ENGINEERING CURRICULA

Currently the more innovative design systems are being developed by people with a strong background in computer science, software engineering, or artificial intelligence. The tools required as well as the VSLI systems to be built require an explicit methodology to cope with the complexity involved; ad hoc methods are no longer sufficient.

Hands-on Experience with Complexity Management: It is thus recommended that courses be introduced into the engineering curriculum that explicitly and implicitly teach some of the techniques already well established in software engineering. Courses that concentrate on one big project, demonstrating how the techniques can be used to subdivide a large task into manageable parts, will prove highly beneficial. At the University of California at Berkeley several courses that give the students experience in complexity management exist in the area of Computer Science, even at the undergraduate level. These are courses in which the students build major parts of a compiler, modify significant pieces of an operating system, or design complete subsystems in a VLSI layout course. In an advanced course sequence, a small group of students has recently built a complete microprocessor [41]. On the other hand, such courses are much sparser in the traditional engineering schools, where the emphasis is typically more on design at the "component" level. There is, for instance, rarely a course on how to design a complete television set or a frequency spectrum analyzer.

Emphasis on Specification and Testing: Functional partitioning goes hand in hand with properly specifying the resulting submodules. To emphasize this point, all designers should be forced to specify precisely what it is that they are going to design or implement, be it hardware or software. Before they write a single line of code or draw a single circuit schematic, they should write down exactly how they are going to check whether their product fits certain specifications.

They should also have prepared a complete debugging plan and hand in their test files with their project proposal. Such a debugging plan starts with the simplest possible test, e.g., measuring the power consumption, for the case when chips come back in which not a single recognizable signal sequence appears at the output. Then, in case that this simple test works, there is a sequence of incrementally more sophisticated experiments building on the results of earlier tests. Thinking about testing early in the design process and doing the thorough preparations outlined above will strongly influence the design. The changes that such discipline enforces will be most appreciated when the prototype chips come back from fabrication.

X. Conclusion

Many of the techniques that are needed to manage the complexity of VLSI system design are already known. During the last decade, several of them have reached a stage of maturity in the domain of software engineering. However, they are only slowly entering the domain of integrated circuit design; they have not yet appeared as integral parts of commercial VLSI design tools.

The introduction of these new design techniques and their emergence in VLSI design tools will change the style in which VLSI systems will be produced in the future. The bulk of the products will no longer be designed by circuit wizards who, by ad hoc methods based on their experience and intuition, will come up with chip designs that do miraculous and mysterious things. More circuits will be specified by high-level descriptions and by functional specifications from which the lower level descriptions and ultimately the layout will be compiled in an automatic or semiautomatic manner. The time will soon come where a systems engineer can draw on a set of sophisticated semiautomatic tools that will produce a properly structured, understandable, and testable chip that will adhere to specifications and has a good chance to work the first time.

To make this happen sooner rather than later requires primarily a change in attitude. Such a change is taking place in many research laboratories. It can gain momentum if the engineering schools in their curricula put proper emphasis on explicit and practical instruction in complexity management.

Acknowledgment

The thoughts expressed in this paper have clearly been influenced by my environment: the many colleagues at UC Berkeley and other academic and industrial institutions with whom I interact during my work, at conferences, and indirectly through publications and correspondence. I am sure that many of my friends will recognize their models and metaphors appearing implicitly or explicitly in this text. I would like to express to all of them my sincere gratitude for the stimulating environment and the fruitful personal interactions that make it so exciting to work in this rapidly evolving field. Special thanks go to S. C. Johnson, H. T. Kung, J. K. Ousterhout, R. L. Russo, S. Trimberger, P. W. Verhofstadt, and A. I. Wasserman who have given me a lot of constructive criticism on earlier drafts.

References

[1] G. E. Moore, "Progress in digital integrated electronics," presented at the IEEE Int. Electron Devices Meet., Talk 1.3 (Washington, DC, Dec. 1975).

[2] G. E. Moore, Quote at the First Caltech Conference on VLSI, Pasadena, CA, Jan. 1979.

[3] J. N. Buxton, "Software engineering," in *Programming Methodology*, D. Gries, Ed. New York: Springer, 1978, pp. 23–28.

[4] D. V. Steward, "Analysis and complexity," in *Systems Analysis and Management*. New York: Petrocelli Books, 1981, p. 2.

[5] E. Yourdon and L. L. Constantine, *Structured Design*. Englewood Cliffs, NJ: Prentice-Hall, 1979.

[6] P. Freeman and A. I. Wasserman, *Tutorial: Software Design Techniques*, 3rd ed. Los Alamitos, CA: IEEE Computer Society, 1980.

[7] L. A. Belady and M. M. Lehman, "Characteristics of large systems," in *Research Directions in Software Technology*, P. Wegner, Ed. Cambridge, MA: MIT Press, 1979, pp. 106–142.

[8] C.A.R. Hoare, "Hints on programming language design," Memo AIM 224, Stanford Artificial Intelligence Laboratory (Oct. 1973). Reprinted in *Tutorial: Programming Language Design*, A. I. Wasserman, Ed. Los Alamitos, CA: IEEE Computer Society, 1980.

[9] G. A. Miller, "The magical number seven, plus or minus two: Some limitations on our capacity for processing information," *Psychol. Rev.*, vol. 63, pp. 81–97, 1956.

[10] W. W. Lattin, J. A. Bayliss, D. L. Budde, J. R. Rattner, and W. S. Richardson, "A methodology for VLSI chip design," *Lambda*, vol. 2, no. 2, pp. 34–44, 2nd quarter, 1981.

[11] D. L. Johannsen, "Bristle blocks: A silicon compiler," in *Proc. Caltech Conf. VLSI* (Pasadena, CA), pp. 303–310, Jan. 1979.

[12] J. M. Siskind, J. R. Southard, and K. W. Couch, "Generating custom high performance VLSI designs from succinct algorithmic descriptions," in *Proc. Conf. on Adv. Research in VLSI* (MIT, Cambridge, MA), pp. 28–40, Jan. 1982.

[13] H. E. Shrobe, "The data path generator," in *Proc. Conf. on Adv. Research in VLSI* (MIT, Cambridge, MA), pp. 175–181, Jan. 1982.

[14] R. J. Blumberg and S. Brenner, "A 1500 gate, random logic, large-scale integrated masterslice," *IEEE J. Solid-State Circuits*, vol. SC-14, pp. 818–823, 1979.

[15] B. W. Kernigham, D. G. Schweikert, and G. Persky, "An optimum channel routing algorithm for polycell layouts of integrated circuits," in *Proc. 10th Design Automation Workshop* (Portland, OR), pp. 50–59, June 1973.

[16] V. D. Agrawal *et al.*, "The mixed mode simulator," in *Proc. 17th Design Automation Conf.*, pp. 618–625, June 1980.

[17] A. R. Newton, "Timing, logic, and mixed mode simulation for large MOS integrated circuits," NATO Advanced Study Institute on Computer Design Aids for VLSI Circuits, Sogesta-Urbino, Italy, 1980.

[18] G. D. Bergland, "A guided tour of program design methodologies," *Computer*, vol. 14, no. 10, pp. 13–36, Oct. 1981.

[19] N. Wirth, "Program development by stepwise refinement," *Commun. ACM*, vol. 14, no. 4, pp. 221–227, Apr. 1971.

[20] M. A. Jackson, *Principles of Program Design*. New York: Academic Press, 1975.

[21] T. DeMarco, *Structured Analysis and System Specification*. Englewood Cliffs, NJ: Prentice-Hall, 1979.

[22] C. Gane and T. Sarson, in *Structured Systems Analysis: Tools and Techniques*. Englewood Cliffs, NJ: Prentice-Hall, 1979.

[23] H. T. Kung and C. E. Leiserson, "Algorithms for VLSI processor arrays," in *Introduction to VLSI Systems*. C. A. Mead and L. A. Conway, Eds. Reading, MA: Addison-Wesley, 1980.

[24] D. L. Johannsen, "Our machine: A microcoded LSI processor," Display File 1826, Dept. Comp. Science, Caltech, Pasadena, CA, July 1978.

[25] R. W. Sherburne, M.G.H. Katevenis, D. A. Patterson, and C. H. Séquin, "Datapath design for RISC," in *Proc. Conf. on Adv. Research in VLSI* (MIT, Cambridge, MA), pp. 53–62, Jan. 1982.

[26] J. D. Warner, *Logical Construction of Programs*. New York: Van Nostrand, 1974.

[27] E. W. Dijkstra, *A Discipline of Programming*. Englewood Cliffs,

NJ: Prentice-Hall, 1976.

[28] D. Gries, "An illustration of current ideas on the derivation of correctness proofs and correct programs," *IEEE Trans. Software Eng.*, vol. SE-2, no. 4, pp. 238–244, Dec. 1976.

[29] C. Böhm and G. Jacopini, "Flow diagrams, turing machines and languages with only two formation rules," *Commun. ACM*, vol. 9, no. 5, pp. 366–371, May 1966.

[30] H. D. Mills, "Mathematical foundations for structured programming," IBM Tech. Rep. FSC 72-6012, Federal Syst. Div., Gaithersburg, MD, 1972.

[31] D. E. Knuth, "Structured programming with goto statements," *Computing Surveys*, vol. 6, no. 4, pp. 261–301, Dec. 1974.

[32] F. P. Brooks, Jr., *The Mythical Man-Month.* Reading, MA: Addison-Wesley, 1975.

[33] E. W. Dijkstra, "The humble programmer," *Commun. ACM*, vol. 15, no. 10, pp. 859–866, Oct. 1972.

[34] T. Teitelbaum and T. Reps, "The Cornell program synthesizer: A syntax-directed programming environment," *Commun. ACM*, vol. 24, no. 9, pp. 563–573, Sept. 1981.

[35] M. Linton, "A debugger for the Berkeley Pascal system," Master's Report, U.C. Berkeley, June 1981.

[36] S. Gutz, A. I. Wasserman, and M. J. Spier, "Personal development systems for the professional programmer," *Computer*, vol. 14, no. 4, pp. 45–53, Apr. 1981.

[37] D. Hightower and F. Alexander, "A mature I^2L/STL gate array layout system," in *Dig. Papers, COMPCON*, pp. 149–155, Feb. 1980.

[38] D. L. Parnas, "The use of precise specifications in the development of software," in *Information Processing 77 (Proc. IFIP Congress).* Amsterdam, The Netherlands: North Holland, 1977, pp. 861–867.

[39] G. J. Myers, *The Art of Software Testing.* New York: Wiley, 1980.

[40] E. B. Eichelberger and T. W. Williams, "A logic design structure for LSI testability," *J. Des. Automat. Fault-Tolerant Comput.*, vol. 2, pp. 165–178, May 1978.

[41] D. A. Patterson and C. H. Séquin, "RISC I: A reduced instruction set VLSI computer," in *Proc. 8th Int. Symp. on Computer Architecture* (Minneapolis, MN), pp. 443–457, May 1981.

[42] W. A. Wulf, "Compilers and computer architecture," *Computer*, vol. 14, no. 7, pp. 41–47, July 1981.

[43] A. B. Glaser and G. E. Subak-Sharpe, "Failure, reliability and yield of integrated circuits," in *Integrated Circuit Engineering.* Reading, MA: Addison-Wesley, 1978, pp. 746–799.

[44] R. Ayres, "Silicon compilation—Hierarchical use of PLAs," in *Proc. Caltech Conf. on VLSI* (Pasadena, CA), pp. 311–326, Jan. 1979.

[45] S. S. Eaton, D. Wooton, W. Slemmer, and J. Brady, "Circuit advances propel 64-k RAM across the 100 ns barrier," *Electronics*, vol. 55, no. 6, pp. 132–136, Mar. 24, 1982.

[46] W. A. Wulf, "Keynote address," presented at the Symp. on Arch. Supp. for Prog. Lang. and Operat. Syst., Palo Alto, CA, Mar. 1982.

[47] ——, "Trends in the design and implementation of programming languages," *Computer*, vol. 13, no. 1, pp. 14–25, Jan. 1980.

[48] M. Stefik, D. G. Bobrow, A. Bell, H. Brown, L. Conway, and C. Tong, "The partitioning of concerns in digital systems design," in *Proc. Conf. on Adv. Research in VLSI* (MIT, Cambridge, MA), pp. 43–52, Jan. 1982.

[49] M. Tucker and L. Scheffer, "A constrained design methodology for VLSI," *VLSI Des.*, vol. 3, no. 3, pp. 60–65, May 1982.

[50] D. T. Fitzpatrick, J. K. Foderaro, M.G.H. Katevenis, H. A. Landman, D. A. Patterson, J. B. Peek, Z. Peshkess, C. H. Séquin, R. W. Sherburne, and K. S. VanDyke, "VLSI implementation of a reduced instruction set computer," in *Proc. CMU Conf. on VLSI Systems and Computations* (Pittsburgh, PA), pp. 327–336, Oct. 1981.

[51] M. E. Newell and D. T. Fitzpatrick, "Exploiting structure in integrated circuit design analysis," in *Proc. Conf. on Adv. Research in VLSI* (MIT, Cambridge, MA), pp. 84–92, Jan. 1982.

[52] L. K. Scheffer, "A methodology for improved verification on VLSI designs without loss of area," in *Proc. 2nd Caltech Conf. on VLSI* (Jan. 19–21, 1981).

[53] T. Whitney, "A hierarchical design analysis front end," in *Proc. VLSI81, Int. Conf. on Very Large Scale Integration* (Edinburgh, Scotland), pp. 217–225, 1981.

[54] S. C. Johnson, "Hierarchical design validation based on rectangles," in *Proc. Conf. on Adv. Research in VLSI* (MIT, Cambridge, MA), pp. 97–100, Jan. 1982.

[55] L. A. Conway, A. Bell, and M. E. Newell, "MPC79: A large-scale demonstration of a new way to create systems in silicon," *Lambda*, vol. 1, no. 2, pp. 10–19, 2nd quarter, 1980.

[56] B. Hon and C. H. Séquin, *Guide to LSI Implementation* (2nd revised and extended edition), Xerox PARC, Palo Alto, CA, Jan. 1980.

[57] W. D. Jansen and D. G. Fairbairn, "The silicon foundry: Concepts and reality," *Lambda*, vol. 2, no. 1, pp. 16–26, 1st quarter, 1981.

[58] W. J. Haydamack, Public Lecture, University of California, Berkeley, Fall 1981.

[59] J. D. Williams, "STICKS—A graphical compiler for high level LSI design," in *AFIPS Conf. Proc., NCC*, vol. 47, pp. 289–295, 1978.

[60] A. Dunlop, "SLIP—Symbolic layout of integrated circuits with compaction," *Computer-Aided Des.*, vol. 10, pp. 387–391, Nov. 1978.

[61] M. Y. Hsueh and D. O. Pederson, "Computer-aided layout of LSI circuit building blocks," in *Proc. IEEE Int. Solid-State Circuits Conf.* (Tokyo, Japan), pp. 474–477, 1979.

[62] C. H. Séquin, "Pictures versus parentheses: Design methodologies of the 1980's," Panel Discussion at COMPCON, San Francisco, CA, Feb. 1982.

[63] A. I. Wasserman, "Automated development environments," *Computer*, vol. 14, pp. 7–10, Apr. 1981.

[64] B. W. Kernighan and J. R. Mashey, "The UNIX programming environment," *Computer*, vol. 14, no. 4, pp. 12–22, Apr. 1981.

[65] W. Teitelman and L. Masinter, "The INTERLISP programming environment," *Computer*, vol. 14, no. 4, pp. 25–33, Apr. 1981.

[66] A. I. Wasserman and S. Gutz, "The future of programming," *Comm. ACM*, vol. 25, no. 3, pp. 196–206, Mar. 1982.

[67] A. R. Newton, D. O. Pederson, A. L. Sangiovanni-Vincentelli, and C. H. Séquin, "Design aids for VLSI: The Berkeley perspective," *IEEE Trans. Circuits Syst.*, vol. CAS-28, no. 7, pp. 666–680, July 1981.

[68] S. R. Bourne, "UNIX time-sharing system: The UNIX shell," *Bell Syst. Tech. J.*, vol. 57, no. 6, pp. 1971–1990, Jul.–Aug. 1978.

[69] C. H. Séquin and A. R. Newton, "Description of STIF 1.0," in *Design Methodologies for VLSI.* Groningen, The Netherlands: Noordhoff, Jan. 1982, pp. 147–171.

[70] S. A. Ellis, K. H. Keller, A. R. Newton, D. O. Pederson, A. L. Sangiovanni-Vincentelli, and C. H. Séquin, "A symbolic layout design system," presented at Int. Symp. on Circuits and Systems, Rome, Italy, May 1982.

[71] A. K. Graham, "Software design: Breaking the bottleneck," *IEEE Spectrum*, vol. 19, no. 3, pp. 43–50, Mar. 1982.

[72] M. J. Rochkind, "The source code control system," *IEEE Trans. Software Eng.*, vol. SE-1, no. 4, pp. 364–370, Dec. 1975.

[73] E. H. Bersoff, V. D. Henderson, and S. G. Siegel, "Software configuration management: A tutorial," *Computer*, vol. 12, no. 1, pp. 6–14, Jan. 1979.

[74] G. M. Weinberg, *The Psychology of Computer Programming.* New York: Van Nostrand-Reinhold, 1971.

[75] L. A. Belady, "Evolved software for the 80's," *Computer*, vol. 12, no. 2, pp. 79–82, Feb. 1979.

[76] A. I. Wasserman and L. A. Belady *et al.*, "Software engineering: The turning point," *Computer*, vol. 11, no. 9, pp. 30–41, Sept. 1978.

[77] T. M. McWilliams, "Verification of timing constraints on large digital systems," in *Proc. 17th Design Automation Conf.* (Minneapolis, MN), pp. 139–147, June 1980.

[78] W. E. Cory and W. M. VanCleemput, "Development in verification and design correctness," in *Proc. 17th Design Automation Conf.* (Minneapolis, MN), pp. 156–164, June 1980.

[79] A. R. Newton, "The VLSI design challenge of the 80's," in *Proc. 17th Design Autom. Conf.*, pp. 343–344, June 1980.

[80] S. I. Feldman, "Make—A program for maintaining computer programs," *Software—Practice and Experience*, vol. 9, pp. 255–265, Apr. 1979.

[81] J. Grason and A. W. Nagle, "Digital test generation and design for testability," in *Proc. 17th Design Automation Conf.* (Minneapolis, MN), pp. 175–189, June 1980.

[82] P. Naur, "Proof of algorithms by general snapshot," *BIT*, vol. 6, pp. 310–316, 1966.

[83] R. Floyd, "Assigning meaning to programs," in *Proc. Symp. in Applied Mathematics*, vol. 19, pp. 19–32, American Math. Soc., Providence, RI, 1967.

[84] C.A.R. Hoare, "An axiomatic basis for computer programming," *Commun. ACM*, vol. 12, no. 10, pp. 576–583, Oct. 1969.

[85] R. DeMillo, R. Lipton, and A. J. Perlis, "Social processes and proofs of computer programs," *Commun. ACM*, vol. 22, no. 5, pp. 271–280, May 1979.

[86] M. Stefik, J. Aikins, R. Balzer, J. Benoit, L. Birnbaum, F. Hayes-Roth, and E. Sacerdoti, "The organization of expert systems: A perspective tutorial," Tech. Rep., Xerox PARC, Palo Alto, CA, Jan. 1982.

[87] C. A. Mead, "VLSI and technological innovation," Keynote Address, presented at the Caltech Conf. on VLSI, Pasadena, CA, Jan. 1979.

Hierarchical Design Methodologies and Tools for VLSI Chips

C. NIESSEN

Invited Paper

Abstract—Hierarchical design methods are considered to be a means of managing the VLSI design problem. This paper will consider why this problem exists and discuss alternative means that can be used to arrive at a solution. The merits of design methodologies, with emphasis on hierarchical techniques, will be compared with those of automated design approaches. The discussion of hierarchy will lead to the conclusion that the method requires formal abstraction facilities in order to be effective. Hierarchical design methods permit the creation of a new generation of CAD programs that can both give a designer better support and can be much more efficient than the present generation of tools. An example of such a tool, VOILA, will be given.

I. Introduction

HIERARCHICAL design methods are considered as a means of managing the VLSI design problem. Remarkably, the discussion on such a methodology mainly happens in the CAD environment and less among chip designers. In discussing the method of hierarchical design it is necessary to examine the nature of the IC design problem and to compare this method with alternative ones. As we do so it becomes more obvious that the discussion of design methods originated in the CAD domain, because the underlying complexity of the problem can be perceived more readily there.

Today's design problem is caused by the ever-increasing ability of technologists to integrate more components in a single integrated circuit. The maximum number of components in a chip has on average doubled annually for almost two decades [1]. The maximum number of components is at present in excess of 100 000. An analysis of technological possibilities and limitations [2] leads to predictions of future densities that may well be two orders of magnitude greater. Despite this technological potential, present circuits, with the exception of memories and other very regular circuits, contain a significantly smaller number of components. This has been caused by, *inter alia*, the increased difficulty of designing such complex circuits.

The exponential growth of IC complexity has had a dramatic impact on the time needed to design a circuit. It has increased exponentially from a few weeks for small-scale integrated circuits (SSI) to tens of man-years for large-scale circuits. Simple extrapolations [3] would yield future design times in the order of thousands of man-years. Obviously this cannot take place, it demonstrates, however, that continued use of current design procedures will be a barrier to growth of IC complexity in the future. The progress in circuit integration, again with the exception of memories and very regular structures, will no

longer be limited only by technology, but also, and to a much greater extent, by our design capability [1].

Today's design methods, which emerged in the SSI and MSI period, apparently need adaptation. In the early days of IC design, almost unlimited freedom was required in order to achieve the best possible results. The limited technological possibilities were the prime justification for this attitude, while the restricted design problem allowed for this. That freedom is now the main cause of the increase in design time. The justification for that freedom has, however, disappeared. The sheer fact that our design capability is going to impede progress, means that we have to explore methods to enhance designer productivity. The purpose of such methods is a drastic reduction of the number of detailed design decisions to be taken by a designer. The products made using these methods will inevitably be less optimal than those designed with all possible freedom. It appears that many designers are reluctant to realize that some trade-in of design efficiency may well yield a decrease of the sum of design plus production cost, thus opposition has to be anticipated. Yet one has to head for that opposition. The former controversy between synchronous and asynchronous logic may serve as an example. Asynchronous logic, because of its freedom in timing, utilizes the time domain inherently better than synchronous logic. For some time synchronous logic has been banned from IC's. Nevertheless, the difficulty of designing and debugging large asynchronous circuits has finally led to acceptance of the concept of synchronous logic.

For a long time now automation has been explored as a means to enhance design productivity. Although a massive amount of work has been and is still being done on the subject and although some of the results are impressive, the degree of acceptance is on the whole disappointing. Automated design produces results that are inferior in most cases to manual designs [4]. Still more unfortunately, the quality of the result deteriorates as the size of the problem increases. This might be a "contra indication" for its use in VLSI. Be that as it may, one of the aims of this paper is to indicate a feasible position for automation in a superimposed design architecture.

Another alternative for improving designer productivity is the exploration of various design architectures. Although it has to be admitted that this subject is still an area of research from which significant results are to be expected, it is already widely accepted that hierarchical design techniques must be a cornerstone for a VLSI design methodology.

In this paper we shall try to find the most important deficiencies of current design methods and the associated CAD

Manuscript received May 17, 1982; revised October 18, 1982.
The author is with the Philips Research Laboratories, Eindhoven, The Netherlands.

Reprinted from *Proc. IEEE*, vol. 71, pp. 66–75, Jan. 1983.

tools. On that basis we shall formulate concepts that should be included in a VLSI design environment. We shall then depict a VLSI design method that adheres to such concepts and sketch a hierarchical CAD system that supports that method. Finally, we shall present some details of VOILA, a hierarchical CAD system, which we are currently implementing.

II. PRESENT-DAY IC DESIGN METHODS

A designer who has to create a function by means of discrete components on a printed-circuit board is confronted with many constraints. As a general rule, he cannot define the components to be used, but can merely select them from a standard catalog. Only in rare cases will he find components with precisely the specifications he needs; normally, he has to be content with components that approach without exactly satisfying his specifications. Similarly, he is confronted with limitations regarding the printed-circuit board. All such boundary conditions lead to less perfect results than those obtained with components specially designed for the needs of that application. These limitations are, of course, not created to tease the designer. They are necessary when a catalog of standard components is compiled, with a manufacturing infrastructure for general use in mind. Against this disadvantage there is an obvious advantage; namely, that it is almost inconceivable how time-consuming printed-circuit board design would be without such an infrastructure.

Quite in contrast, the pioneers of IC design were equipped with a blank sheet of paper, a ruler, a pencil, and an eraser. The potentials and the unknowns of the new technology demanded unlimited freedom. The results are familiar, astonishing innovations have come about. Currently designers are not standing with empty hands, but their primary tools are still paper, ruler, pencil, and eraser. Design styles have, of course, changed in the meantime, but is that enough?

In the pioneering phase IC design meant making a circuit diagram; a breadboard model was used to predict the ultimate behavior on the chip as accurately as possible. The layout was drawn by hand, CAD was an unknown term. The first breakthrough was the introduction of computer aids for layout design. Layout design was cumbersome and error-prone, even for small circuits. The laborious tasks of the designer were taken over by programs. Important contributions to design methodology date back to that period. The use of computers meant that precise circuit descriptions had to be made available. The ability of the computer to take over much of the boring repetitive work required the development of concepts for the description of regular and repetitive structures. Formal layout description languages satisfying these requirements were evolved. Precise textual design descriptions rather than drawings became the design documents.

Tools for electrical analysis and logical simulation rapidly succeeded each other. Languages were again designed to describe analog and digital circuits accurately. This led to a design procedure which is still typical of most present-day design activities (Fig. 1).

A typical feature of this figure is that the synthesis activity from left to right is performed manually. The task of transforming a required function into a chip by successively adding detail is inherently difficult. Up till now this has best been dealt with by a creative designer. By contrast, the time-consuming and boring task of checking whether a subsequent step in a design is in effect the implementation of the preceding specification cannot be done effectively by a human being.

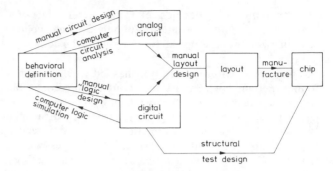

Fig. 1. Traditional IC design procedure.

There have been many examples of design errors overlooked by the designer, while the computer excels in performing the verification task with great accuracy.

As has been mentioned, description languages exist for analog circuits, digital circuits, and layout, but hardly any exist for the behavioral definition. In many cases, such definitions reside only in the designer's mind, or at best are written down in natural language. As a result, the analog and/or digital circuit description is the first document that accurately defines the intended functions. The disadvantages are many. The feedback of accurate design data to customers is delayed considerably. Hence it is difficult to do electrical and logic design as a team and if it is so done, misunderstanding and different interpretations can result.

Despite its disadvantages, the method described has been applied successfully for many circuits. Increasing chip sizes have created a need for additional methodology.

The cumbersome task of designing and verifying asynchronous designs has resulted in the acceptance of synchronism as a means to master complexity, notwithstanding the related disadvantages. The time-consuming nature of layout design activity has led to methods that reduce its complexity. Standard cell approaches and cells in row organizations are examples. These methods have now gained wide acceptance despite some loss of efficiency.

The possibility of shifting parts of the design work to CAD by exploiting the capability of CAD to handle regularity and repetition resulted in a design style which is referred to as "informal hierarchy" [5]. The macro facilities available in many of the IC description languages were used to describe circuits as a collection of modules rather than as a monolithic entity. The underlying thought is, of course, that the repeated design of modules is simpler than the design of one complex circuit. The notion underlying the word "informal" will be explained more thoroughly later. Roughly, it means that designs are broken down into modules, but that rules enabling modules to be designed independently of each other do not exist. The simplification that can be obtained by the use of hierarchical methods is severely diminished by the absence of such rules.

In addition to these approaches aimed at reducing design effort by the introduction of methodological concepts, another approach has been tried whereby design tasks are shifted from the designer to the computer, design automation proper. The main topics of design automation have been automation of the layout task and automatic test pattern generation. Although automatic layout tools have gained some acceptance, there is a disproportionate ratio between the effort spent in creating such tools and their utilization. The resulting products of the automatic programs are large compared to their manually designed counterparts. The difference between the two in fact increases with increasing design size. In the present state of affairs, automatic layout tools are useful only for small volumes and circuit sizes that are not at the far end of the complexity range.

Automatic test pattern generation has led to an even more profound trauma. In contrast to layout automation, which was to provide an alternative for manual layout, automatic test pattern generation had to replace a manual task which had already become impossible. Automatic test pattern generation itself proved to be an unrealistic tool. With excessive computer run times and incomplete sets of tests, IC testing threatened to become an unsolvable problem. It was design methodology that came to the rescue. Scan testing [6], [7] proved to be the method of transforming an unsolvable problem into an easy one. Provided that a designer fulfills a simple set of requirements that makes a circuit scan testable, test pattern generation can be done automatically. In fact, the impact of scan testing is so dramatic that IC testing can almost be considered a mechanistic issue that needs no further attention from the designer.

In summary, present-day design methods have emerged from an initial situation with almost total freedom. The difficulty of designing ever larger circuits has been dealt with by the introduction of some methodological concepts and the application of passive analytical tools. Automated design has until now played a less important role. The best prospects for the future seem to lie in devising methodological concepts which further simplify the design activity.

III. PRESENT POSITION OF IC DESIGN TOOLS

The extremes in IC design are total manual design and fully automated design. Computer aids for design have been developed on the basis of these extremes. One line being followed is the assisting role, the intention of which is to aid the designer in the evaluation of his work and to take over from him the more laborious and dreary routine tasks. This approach is usually referred to as CAD. The other approach attempts to replace human design tasks by computerized counterparts. This latter situation is called design automation, DA. However, quite often the term DA is used when actually CAD is meant. The term DA has been in existence since at least 1964 [8].

There is a significant difference between the two approaches. Manual design assisted by CAD can yield highly optimized results at the cost of long design time. Because of the restrictions imposed by the automatic tools, automated design leads to much less efficient solutions, which are, however, obtained with considerably reduced design effort. In fact, these approaches represent two extremes in a much larger design space. This can be represented graphically as in Fig. 2, in which design cost and circuit cost are related. One can arrange the drawing in a number of areas. There is an area, below the indicated design space, that is inaccessible because a certain minimum effort is required to obtain a desired result. The area above the design space is unattractive because it yields results that are inferior to the state of the art. Within the design space, a variety of design methods can be located. The better methods are, of course, located near the lower left boundary. The design space symbolizes the design limitations, inventions are needed to shift it towards the origin.

In appraising CAD and DA, their purpose and their location in the design space have to be considered. The search for alternatives has also to take into account the design space for variants meeting the requirements of the application. In this

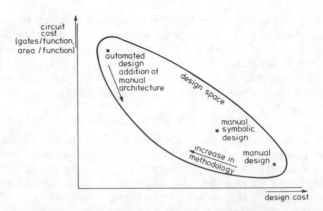

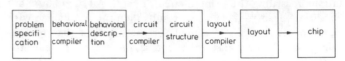

Fig. 2. Design space: acceptable design methods are located within the design space. Cost of design can be substituted for circuit cost.

Fig. 3. The ultimate in design automation.

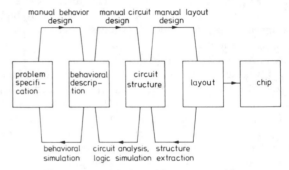

Fig. 4. Manual design with computer aids.

process, it becomes clear that the popular proposal to replace manual design by automated design ignores properties of the problems in hand.

The overall reach of the design process is from problem specification to final realization. Synthesis is both the step-wise refinement of the initial specification towards its realization via a number of intermediate specifications (behavioral definition, circuit structure, layout) and the addition of design detail at each of the intermediate specification stages. Analysis is the verification of a newly created intermediate specification against the preceding one.

If this view is combined with the view of design methods, the resulting extremes are a totally automated design (Fig. 3) and a manual computer-aided design (Fig. 4). The view with regard to design space shows that many intermediate situations are possible.

Similarly to the previously mentioned lack of description languages for problem specification and behavioral description, the set of tools to transform a problem specification into a circuit structure is also meagre. Automation of the transformation from behavioral description to circuit structure is in its infancy and a research topic [9]. Analytical tools for verifying the correct operation of a circuit structure with respect to the original specification are the main achievements. Circuit analysis and gate level logic simulation tools are used extensively. They may be considered as two different tools, but it is much better to regard them simply as two items in a hierarchy of

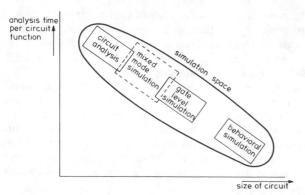

Fig. 5. Simulation space. Simulation tools must be located within the area shown. The area above it is unsuitable because simulation cost becomes excessive. The area below it is unattractive because it gives lower quality result than are essential.

simulation tools. Circuit analysis examines in detail circuits composed of transistors, resistors, etc., limiting it to circuits of modest size. Logic simulation can deal with much larger circuits, because of the abstraction from analog-to-digital circuits. *As will be discussed later, correct abstraction techniques are the key towards hierarchical design techniques.* Abstraction can only be properly applied if certain rigid abstraction rules are given, as in the case of analog-to-digital abstraction. An immediate consequence of the use of abstraction techniques is the need to have facilities to verify that the abstraction rules have been adhered to, for analog-to-digital abstraction simulators of this kind are being made [10].

Although gate-level logic simulators can handle several thousands of gates, the corresponding computer runtime then becomes large. Considerable increases in the number of gates will result in prohibitive run times. Further abstraction is thus unavoidable. This gives rise to the notion of simulation space (Fig. 5), according to which circuit granularity must grow with increasing circuit size in order to keep cost under control.

The majority of tools presently available are for the transformation of a circuit structure into a layout. The tools for assisting manual design are especially well known. Some examples are interactive graphic programs for assisting the designer to enter a layout, design rule checkers, and layout-to-structure checkers. Among the tools for automation are automatic placers and automatic routers. There is much concern at present about the applicability of these programs in the future. The increase in circuit complexity has caused a dramatic increase in the run time of the layout-analysis programs. Automated tools are equally affected in that the compactness of their results, which is not too good for small designs in the first place, degrades further with the increase of problem size. A solution to these problems is not simple. The problems are caused by the intrinsic computational complexity of the tool [11], which is roughly polynominal for analysis and exponential for synthesis tools. Attempts to perform program optimizations which yield decrease in run times by a constant factor have no long-term effect. Such improvements will be consumed rapidly by the growth of the circuits to be designed.

The real problem with design aids is that they deal with layouts as monolithic entities. Solutions are required which support abstraction techniques, similar to the ones mentioned for circuit structure design. Investigations are now in progress which examine whether present design methods can be further exploited to improve CAD performance. The informal hierarchy in particular is receiving much attention, the idea being

to do as much work as possible on lower level modules in a hierarchy of this kind, e.g., by using filtering techniques [12]. The thesis referred to indeed shows that given good organization in a design, significant savings can be obtained. It also shows that the method is very dependent on and sensitive to the quality of the structure presented. With badly structured designs the intended savings aimed at can turn into dramatic increases in computer run times.

It must be concluded that present-day informal design methods are like quicksand, they are too weak a foundation on which to construct a set of VLSI design tools.

IV. CONCEPTS FOR A VLSI DESIGN METHODOLOGY

The goals to be met by a VLSI design methodology are in order of importance:

a) It must provide complexity control such that a reasonable confidence in the correctness of designed circuits is made possible.

b) The method must comprise the whole design traject.

c) An efficient utilization of technological possibilities should be provided.

d) A considerable increase in design productivity should result.

e) It must enable the creation of efficient CAD tools.

From the preceding section on IC design tools it may be concluded that a further automation of the design activity is in conflict with requirements b) and c). Layout design, the main subject of design automation, is just a fairly small portion of the design activity as a whole. The replacement of human creativity by automatic algorithms is an impoverishment of the design process, of which increase of chip area is just one symptom in a larger spectrum of resulting drawbacks. In short, it is questionable whether automated design on VLSI scale is at all feasible.

This brings us back to methodology. Hierarchical design has been already referred to several times in this connection. It is, however, just one method though probably the most important for dealing with complexity. Design methodology is not an established speciality, but has to be subjected to further investigations. At present, the primary concepts are abstraction, repetition, and, perhaps somewhat trivially, the use of past experience.

Abstraction: Abstraction is a method to replace an object by a simplified one that only defines the interactions of the object with its environment, while deleting the internal organization of the object.

The abstraction of analog-to-digital circuits is an example. Such an abstraction is often only valid within certain constraints. A logic gate, for instance, only behaves as a gate when certain load conditions are not exceeded.

The virtue of abstraction is data reduction, sometimes by one or more orders of magnitude. For very large systems, one level of abstraction may not suffice, it has to be applied a number of times in succession, "hierarchical abstraction." A system can thus be considered as being broken down into a number of subsystems, each of which is further broken down into smaller entities, and so on. The building block in such a breakdown is often called a module. Fig. 6 shows a possible representation of a system composed of modules.

The hierarchy, when properly implemented, is the self-evident implementation of the divide and conquer principle. Dividing and conquering itself is a principle that can be applied hierarchically, as is the case in many human organizations. To

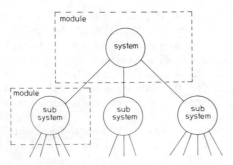

Fig. 6. A hierarchical representation of a system.

do so, concepts are consequently needed for the implementation of both "division" and of "conquering." Division necessitates means to delegate responsibilities, i.e., principles for *communication*. Conquering implies a need to express *local autonomy*. A so-called hierarchical system that does not embody these principles is not a hierarchical system at all. Such a hierarchy is informal, it can provide some intellectual help, but it does not simplify the problem in hand. The main drawback of the informal hierarchy is its lack of rigid abstractions. A hierarchy that does contain these principles will be called a formal hierarchy instead.

Locality of concern is an extremely important attribute of a hierarchical design system. Simplification of the design problem is its predominant virtue. However, it introduces a disadvantage for which a price has to be paid. A module at a higher level in the hierarchy expresses requirements applying to a lower level module in global terms (e.g., as a transfer function $Y = f(x)$). The detailed implementation is a matter for the lower level module. Neighboring modules thus possess only limited knowledge about each other's implementation. It may well happen that certain similar subproblems are solved in adjacent modules. It can also happen that modules are given requirements that cannot be fulfilled within the constraints of the local autonomy. This can be partially overcome by supplying the local module with an oversized set of resources. The disadvantage of the clean hierarchy concept introduces the necessity to make iterations through the hierarchical levels possible. In principle, it is possible to perform so many iterations through a hierarchy that its disadvantage is eliminated; unfortunately, this is achieved at the cost of a complete loss of the simplification provided by the hierarchy. The other extreme is the total absence of iterations. This is pure top-down or pure bottom-up design. The realistic approaches are somewhere in between these extremes.

As such, this reasoning tells us that hierarchical design methods are not restricted to certain problem classes, whether low or high volume, whether custom circuits or whatever. *It does, however, also tell us that a hierarchical CAD tool which prohibits such iterations restricts itself to a limited set of problem classes.*

Repetition: Repetition is an easy and often applied method of simplifying design. It can be used in a regular array type manner (RAM's, ROM's) but also irregularly. Ratios such as the total device count versus the drawn device count give a measure of the use of repetition. This ratio [13] has increased with more recent designs. The concept of hierarchy is of advantage to that of repetition because it provides the module as the atom for repetition.

Use of Past Experience: It is a waste of effort not to use experience gained in previous activities. Stated more concretely

in the context of IC design, it is wasteful to design each chip from scratch rather than to use modules that have been done previously. The advantages are many. The time taken to design a chip is reduced. The risk of errors is reduced because modules copied from previous designs are more likely to be error-free. The disadvantages, too, are worth mentioning. It is difficult to define standard circuits other than the familiar ones. Moreover, such standard circuits probably do not precisely match the requirements of the new application, causing a certain degree of inefficiency. The most important reason not to use standardized modules for IC design is the absence of facilities for precisely defining the interface requirements of the standard module.

Perhaps the concept of formal hierarchy will come to the rescue. It provides the necessary grips for formulating the interface requirements of the standard module. Not improbably, it will have a similar effect on IC design as the dual-in-line package standard has had on printed-circuit board design.

To sum up, the primary concept for a VLSI design methodology must be one that permits design to be done modularly, the formal hierarchy. This concept leads to well-defined interactions between the various modules in a design, yielding simplified design as an advantage and some inefficiency as a disadvantage. It may also be the carrier for a design style which makes use of standardized modules.

Hierarchical design may even provide the infrastructure on which design automation will flourish. Instead of design automation being regarded as a replacement for human design, it might be seen as subordinate to human design. In a hierarchical environment the human being can assume responsibility for design at a high level, i.e., the architecturing tasks, while the computer can undertake design at the lower levels, i.e., the implementation tasks. Experiments in silicon compilation [14] are, in fact, supportive of this idea.

V. The Architecture of a Hierarchical VLSI Design Environment

Scope of Hierarchy

Hierarchical design is often associated with layout design. From the preceding it may be clear that hierarchical concepts also have to be extended to the stage of circuit structure and behavioral definition (Fig. 4), such that every design specification can be created with the aid of abstraction facilities. But even that is not enough. The creation of a new design specification refers back to a previous one, the circuit structure is created with reference to the behavioral definition, and the layout is created with reference to the circuit structure. If during such an activity reference has to be made to hierarchical structures which are different, the complexity issue returns in all its magnitude. The transformation from one hierarchy to another requires an intermediate stage in which the structure is eliminated by expansion, which is undesirable. This drastically broadens the scope and the impact of the hierarchy concept. The same hierarchical breakdown should be present, in principle, throughout the whole design trajectory, although compromises may sometimes be possible and justified. These compromises, however, ought to be of limited scope.

A good example of such a compromise is the organization of standard cells in rows. The distribution of power and clock lines over standard cells makes any other organization very inefficient (Fig. 7).

This means that the layout structure has to be different

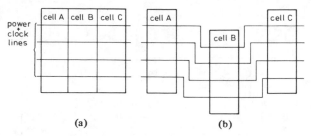

(a)   (b)

Fig. 7. Power and clock distribution may force a layout organization in rows (a), as other arrangements (b) are very inefficient.

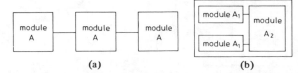

(a)   (b)

Fig. 8. Example of aliasing: a module *A* used three times in a circuit structure may need different implementations in layout owing to constraints present here. (a) Circuit structure. (b) Layout.

from that present in the circuit structure. However, the applicability of this layout technique has shown to be limited. Beyond a certain complexity, the transformation of a circuit structure into a row-type layout becomes cumbersome, while the area occupied for signal distribution tends to increase disproportionately. In addition, the inefficiency of an irregular power distribution, though significant at the standard cell level, is of less importance at the level of larger modules.

Finally, because modules tend to be distributed over several rows in a layout, alterations that are local at the circuit stage have a more global impact on the layout. This layout technique consequently does not lend itself to redesigns.

Given these limitations of the standard cell approach, it is not a threat to the ambition to have similar hierarchy in circuit structure and layout. A limited discrepancy in hierarchical organization can be easily eliminated via selective expansion.

Similarity of hierarchical structure at the various design specification stages does not come about spontaneously. If the constraints of the other stages are not considered, detailed design at, for example, the behavioral stage may not be implementable at those stage. Since rejection of the considerable work done up to this point seems wasteful, there is only one conclusion which can be drawn; namely, that the hierarchy at the next design stage must be different.

If the constraints at the various design stages had been taken into account simultaneously, however, a uniform hierarchical organization might have been possible. This leads to the conclusion that a hierarchical design environment requires decisions taken at the high level of design to be taken simultaneously at all the specification stages. Floorplanning [15] is becoming the accepted term for denoting this kind of activity. This word, with its geometrical undertones, was first used in the context of layout planning but its meaning is now extending to the circuit and behavioral design stages as well.

The uniformity of hierarchy in the various design phases was so far not used in the proposed rigid form. Although it seems that it is being used more and more as circuit size increases, general acceptance cannot yet be said to exist. The provision of flexibility within the given constraints may well help confidence in the method to become established.

One such provision may be the aliasing concept. When a module has been used several times at a certain design stage, its instances at a subsequent stage may need to be different owing to conditions at that stage. A good example can again be found in layout, where geometrical boundary conditions may oblige one or more module instances to be deformed (Fig. 8).

The design of alternative implementations for a particular module certainly does increase the design effort. A decision to do so thus must, therefore, be taken carefully. It does not affect the similarity of hierarchy concept as such, its impact being limited to a single module.

The Implementation of "Divide and Conquer"

In the section headed "Concepts for a VLSI Design Methodology" it was stated that a formal hierarchy requires means, so far as implementation of communication is concerned, delegating responsibilities as well as means to express local autonomy. Globally, these concepts must be the same at all design stages in order to allow hierarchies to be mapped. More detailed examination of these aspects will reveal that there are also differences in order to cope with specific aspects.

An important architectural decision is whether these concepts must be the same from top to bottom in a hierarchy or whether there is a need for them to be different. Identity of concept is a useful property as it enables design to be tackled recursively with the same approach.

The recursive concept has an important consequence, namely that the "grammar" for the atom of recursion, in our case the module, can only be defined once. This means that the grammar has to incorporate all the primitives needed for composition.

It is not at all clear yet whether this recursion concept is applicable throughout the IC design trajectory.

We will treat the "divide-and-conquer" subject in the reverse order as design normally proceeds, viz., starting from layout because the concepts there can be more easily identified and visualized. We will not discuss all the primitives needed for design but restrict ourselves to the discussion of such aspects as are of concern to the hierarchical structure.

a) Layout: A layout composed in a hierarchical manner consists of a number of modules which are mutually interconnected with interconnection tracks, often called nets. Every module can, in turn, be broken down in a similar fashion.

The concept of locality must clearly be expressed in terms of area. We will call the area occupied by a module its domain. Rules for the definition of domains can be given with various degrees of refinement. The roughest method is to allocate every module a single rectangular area as its domain. A more subtle method is to associate each mask in the layout with one or more polygons as its domain, permitting more efficient management of silicon area. This approach makes it possible for a module to provide (partially) unused masks for use at the next design level.

The layout of a module has to be designed inside its domain in the usual manner, making use of design rules prescribed by the relevant technology. The concept of locality means that the designer need not bother about details outside a module while the details inside a module are being designed. For that purpose, design rules applying between details in a layout must be replaced by new design rules expressing relations between details and domains. There have to be rules to express rela-

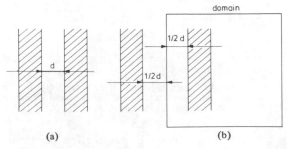

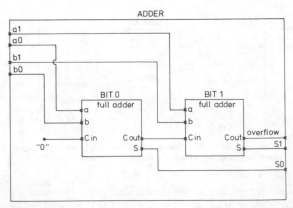

Fig. 11. A circuit structure of a 2-bit adder composed of full adders. Each of the modules, a 2-bit adder and a full adder, has its own domain and signal names.

Fig. 9. An example showing how a design rule (minimum distance *d*) between layout details (a) can be expressed as a design rule with respect to domains (b) in hierarchical environment. The latter rules eliminate the need to varify the relations inside and outside a module.

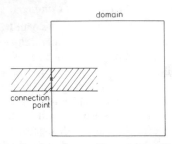

Fig. 10. Communication in a layout hierarchy is effected with the aid of connection points. Details outside and inside a domain are enabled to connect at a connection point.

tionships between details inside a domain and the domain and rules for those between details outside the domain and the domain. The situation obviously becomes simplest when the inside and outside rules are symmetrical (Fig. 9).

A good set of design rules relating to domains makes the application of the normal design rules on details inside and outside the domain superfluous. Hence it also makes any communication from the inside to the outside and vice versa impossible.

In addition to the domain as a means expressing local autonomy, a primitive for communication is also needed; namely, the connection point. The connection point is the means for controlled intervention in details inside and outside a module. Just as rules are needed for domains, also rules are needed regarding connection points.

Connection points must be located at domain boundaries. Each connection point is available for connection from both the outside and the inside of a domain (Fig. 10). Apart from these rules for connection points—rules concerned with geometrical aspects—connection points can also be given certain attributes. One example is electrical loading. For instance, the supply current drain can be passed from the lowest level modules up to the bonding pads via the connection points. Derivation of the width required for supply lines in every module is then a simple and local problem.

The domain and the connection point, in combination with rules on how they are to be used, are the necessary and sufficient primitives needed to make hierarchically structured layout design possible.

It may be appropriate at this point to discuss the effect which hierarchical concepts have on layout (in)efficiency. There are two sources affecting efficiency, the first being the local treatment of layout modules without detailed knowledge of the environment. As has been explained, this source can be eliminated to any degree by multiple top-down and bottom-up iterations. The other source is the desired introduction of design rules specifically for hierarchical modules. At domain boundaries, these rules result in inefficiencies which would not otherwise have occurred. It will be clear that the penalty is more severe for small than for large modules.

b) Circuit Structure: The locality concept of a circuit module is more complex than it is for layout. A circuit structure not only expresses connectivity of circuit modules but it also expresses timing relationships, adding a new dimension to the problem [16].

The connectivity aspects can be arranged in similar fashion to program structures. For a circuit module one can define a domain, in this case expressed in terms of signals. The set of signals available to a module may be composed of three subsets. First, there is the set of signals which are created and used locally; their scope is limited to the particular module. Second, there are signals which are passed from outside the module via a mechanism still to be described. Finally, there are global signals of which clocks and system resets are examples.

The counterpart of a global signal, the global net, does not exist in layout. A global net might perhaps be implemented as a solid layer of metal, completely dedicated to that net. From a design point of view the absence of such layers is a deficiency of the technology.

Communicaton between modules can be established by specification of the set of signals to be passed. By analogy with terminology used in programming languages, that set may be called a formal signal list. A higher level module using the particular module has to define the signals it passes. Again by analogy, this is called an actual signal list. This mechanism binds actual and formal signals while retaining local names. Fig. 11 shows how this works out for a simple example.

Unlike the domain and connection-point concepts for layout which are relatively new and not yet in general use, the corresponding concepts for circuit modules are not new at all and widely used. They can be used recursively at any level in a design. This seems not to be the case with regard to timing aspects. Timing concepts are expressed in terms such as asynchronous, synchronous, handshakes, self-timed [16], etc. They cannot be used interchangeably. An elegant solution appears to be to associate attributes with a module, e.g., asynchronous module, self-timed module, etc.

A quick look into the possibility of mapping a circuit hierarchy on to a layout counterpart reveals that it must be relatively easy to accomplish. Circuit domains and layout domains map easily on to each other. In outline this also applies to the mapping of signals in a circuit module onto nets in layout, except for two problems. Difficulties may be encountered when the signals are to be mapped on to two-dimensional space, a limitation not present in the circuit module. A more difficult problem, however, is that global signals at the circuit stage have no counterpart in layout but must be embedded in the hierarchy of intermodular interconnections.

c) Behavioral Description: The use of behavioral descriptions for IC design is as yet very limited [17]. Expertise and experience are still lacking in this field. A discussion of the subject cannot present solutions but is rather restricted to indicate directions that may be taken.

A circuit exhibits both sequential and parallel behavior. Apart from measures of scale there is thus similarity to operating systems in computers that also exhibit sequentialism and parallelism. In the computer, the sequential behavior is expressed in terms of programs. Parallelism is obtained by having multiple programs running concurrently. Each such program then is called a process. The possibility to create and control a number of processes is obtained by having facilities in the programming language to create, terminate, and communicate with processes. There is recursion in this scheme, a process can launch and control a number of processes, while each of the created processes can do the same. All the processes running at a given point in time are the ultimate result of a single start-up process.

The behavioral description of an integrated circuit can be modeled along similar lines. One needs the concept of process, one needs to be able to describe sequential behavior, and one needs constructs to create and interact with other processes. Such a scheme seems to contain the necessary primitives to describe circuit behavior. The manner how behavior has to be mapped on to circuit structure seems as yet to be unresolved. An interesting question in view of uniformity of hierarchies is whether there can be an association between a process in a behavioral description and a circuit module in the circuit structure.

VI. THE ROLE OF CAD IN A HIERARCHICAL DESIGN ENVIRONMENT

CAD for hierarchical design has to satisfy three conditions. First, it must support the hierarchical design methods. Second, it must provide improvements in capacity and efficiency that can be expected from hierarchical tools. Finally, it should open up new possibilities that become feasible in this environment.

Support Given by CAD Tools

Precise hierarchical languages in which structured designs can be described are the chief sources of support of the design method. For example, a description language for layout must, besides permitting the description of components, facilitate the description of domains and connection points.

CAD programs must ensure fast excursions through the hierarchical structures, both through the levels at a certain design stage and also between design stages. A typical use of CAD could be

a top level behavioral description is created;
simulation is done at this level;
a floorplan for the circuit structure is created, using a hierarchical graphic editor;
the top level behavior is made more detailed, the top level itself is translated to circuit structure;

a mixed mode behavioral, circuit simulation is done;
a floorplan for layout is created;
etc.

This picture shows that there will be a need for high-level simulators, mixed-level simulators, hierarchical graphic editors, etc.

Capacity and Efficiency of Hierarchical CAD Tools

As explained, most of the classical CAD tools do not use structure in a design, it is even eliminated when present. This has resulted in inefficient programs with limited capacity, considering present requirements. New CAD tools have to and can do away with the disadvantages of the classical tools.

By retaining the hierarchy of a design in internal data structures, the problem of data storage is completely eliminated. Where a data structure for a VLSI design may consume hundreds of megabytes of storage in the classical approach, a hierarchical approach only needs at the most a few megabytes. What is more, any increase in circuit complexity that is obtained by better utilization of structure and repetition does not increase the size of a hierarchical data structure, in complete contrast to the classical. Hierarchical tools can solve the capacity problem of design data storage once and for all.

The run time of CAD programs is affected in a similar way. The execution of CAD programs on modules instead of on a complete circuit in itself constitutes an advantage because program run time increases faster than linearly with problem complexity. But the main advantage derives from the fact that in a hierarchy a CAD task needs only be performed once per module, regardless of the number of times it occurs in the design. The classical CAD program does the task as many times as there are occurrences. Hence the run time of a hierarchical CAD program is hardly affected by an increase in circuit complexity as long as it is obtained by a better utilization of structure and repetition.

New Possibilities in CAD

Hierarchical design may do much to encourage the silicon compilation approach. Especially layout automation of the design of small modules is within reach of present-day design automation tools. A recursive application of the tool on modules at various levels in the design hierarchy may result in a situation where considerable parts of an IC are done automatically: This will then be due to the symbiosis of man and the computer, the hierarchy, i.e., the architecture, being ascribable to man and the implementation to the computer.

The efficiency improvements to be expected from new CAD tools promise that certain former batch programs will become interactive. Graphic editors containing interactive design rule checkers are within reach of hierarchical CAD.

VII. VOILA—A HIERARCHICAL VLSI LAYOUT DESIGN SYSTEM

At our laboratories we are currently implementing a hierarchical CAD system for VLSI layout which is based upon previously described principles. It is aimed at circuit complexities to be expected in the next decade (10^7 components on a chip as a guideline). The approach adopted is a combination of manual design of chip architecture and automatic implementation of modules in a design. Consequently, emphasis is placed on high-quality and high-speed man–machine interaction, while using state-of-the-art interactive color graphics. Integrity of design will be achieved by having check programs which complement manual design activities in such a way that layouts developed with VOILA are proven to be correct prior to manufacturing. VOILA takes as a starting point a circuit structure described in NDL (a proprietary network description language). From there, a symbolic layout having the same hierarchy is produced, which is an abstracted form of the detailed geometrical layout that will be implemented on the chip. The abstraction employed in the symbolic layout agree with methods which have been described in literature [18]; namely, simplification of geometries and the replacement of actual mask layers by symbolic mask layers. For the symbolic layout, we have developed a hierarchical description language SLDL (symbolic layout description language). The detailed geometrical layout is described in GLDL (geometric layout description language). The detailed layout is compiled from the symbolic layout supplemented with blocks designed at this level.

In this presentation we will first concentrate on the design languages and then provide some information about CAD programs that have been developed in VOILA.

Design Languages

a) Network Description Language, NDL: The network language used to describe circuit structures has now been in use for several years, during which it has proved its capability. Basically, a network is a module, called a block, which is broken down into a number of smaller blocks, etc. Every block consists of an interface to its environment, described in a block header, and a block body which defines its contents. The body contains references to lower level blocks. In outline the adder block in Fig. 11 would be described as

```
ADDER:  2BIT ADDER  I(A0, A1, B0, B1)
                    O(S0, S1, OVERFLOW)
BEGIN
        BIT0:  FULLADDER  I(A0, B0, 0)
                          O(S0, C1)
        BIT1:  FULLADDER  I(A1, B1, C1)
                          O(S1, OVERFLOW)
END
```

In this example all signals except C1 are passed from the outside, C1 is an internal signal with a scope limited to the 2-bit adder.

This network language, though simple in concept, is powerful enough to express any degree of hierarchical structure.

b) Symbolic Layout Description Language, SLDL: This language has been newly developed to express hierarchically structured layouts. In symbolic layout, the module has been called a cell, a layout being a cell which can be broken down into a number of smaller cells, etc.

The major subdivision of a cell is into a cell header, describing the interface to its environment, and the cell body, describing its contents. The cell header declares the name of the cell, its domain in a domain part, and its connection points in a pin part. The cell body consists of a contents part describing its detailed contents and a drawing part in which a symbolic drawing can be specified. The drawing part can be used as a replacement of the cell contents if one is not interested in that degree of detail. Before examining some of these parts more closely we will look at some other aspects of the symbolic environment which are relevant to cell description.

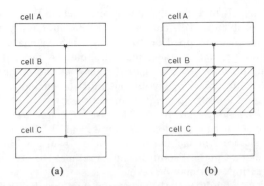

Fig. 12. Alternative crossover approaches. The hatched area denotes the cell domain. An unused area in cell *B* can be made available for use by others using a disjunct domain ((a), VOILA approach) or filled with a crossover track (b).

Cells have to be described in symbolic masks. Many actual masks, e.g., masks for contact holes, are omitted. Their function is replaced by a symbol, for example, the contact hole providing vertical connection is replaced by a VIA symbol.

The *x, y* coordinate system is selectable. The unit of size can be specified (micrometer, mil, etc.), by default it is micrometer. In addition to this unit, a grid to which all coordinates are rounded off can also be specified.

The selectable grid concept makes design rule violations possible. To permit checking, the design rules must be stated.

Cell geometries are restricted: they must be orthogonal at symbolic level.

Careful consideration was devoted to the domain part and the pin part. Two objectives had to be met, clear demarcation of a cell in relation to its environment had to be possible with a minimal loss of chip area. We adopted the following domain concept: A cell can have in each of the symbolic masks any number (including zero) of orthogonal polygons defining its domain. Polygons in a mask need not be tied together. Unused and partially unused masks can then be made available to other cells. This approach is the consequence of a fundamental decision on how to deal with "crossovers" by other cells over an area of a cell that is unoccupied. The other approach would be to incorporate the crossover as part of the cell (Fig. 12). We rejected the latter alternative because it is less flexible and presents a threat to equality between layout and circuit structure. Design rules versus domains are as already explained (Fig. 9).

The pin part defines the locations of the connection points (pins) on the perimeter of the domain and associates signals to pins. This provides a reference to the NDL description. More than one pin can be associated with a signal, providing alternative connection points. Nets inside a domain can be connected to a pin as can nets outside the domain. We ran into a difficulty with regard to design rules for unconnected pins; namely, that the symmetry in design rules cannot be maintained near such pins (Fig. 13) It was for this reason that we created the "automatic pin." An automatic pin is also located at the domain boundary, but a net inside the domain may approach the pin to within a distance of $1/2d$. A CAD program will automatically close the "gap" when a connection is done from the outside.

In the cell contents part one can describe nets and cell references as could be expected.

c) Geometrical Layout Description Language, GLDL: This language has a great similarity to SLDL. The constructs for

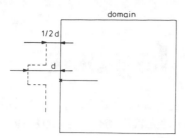

Fig. 13. The symmetrical design rule concept for domains as shown in Fig. 9 becomes more complex in the vicinity of a pin.

the description of domain and connection points are the same. The main difference is the larger degree of freedom to describe geometry, plus the fact that actual layout dimensions are used.

CAD Programs

We intend to create the VOILA CAD programs in a number of phases. In the first phase we are implementing the more traditional programs, interactive graphics, design rule checkers, checkers verifying layout to circuit, mask manipulators, post processors for pattern generation, etc., but with a drastic improvement of program capacity and efficiency by exploiting the advantages of the hierarchy. In later phases we will provide more advanced man–machine symbiosis by creating programs that automatically implement detailed design of modules.

Our aim is to have CAD programs work directly from a hierarchical design description. Since hierarchical descriptions are compact they can consequently be held in main memory, dramatically improving program performance. We will not discuss all the VOILA programs exhaustively, but take a selection that illustrates the benefits most clearly.

The VOILA graphics editor works directly from the hierarchical description. It can provide a detailed view of a part of a layout as well as overviews of larger portions. For overviews it uses the SLDL drawing part to replace layout details by a simplified presentation. It creates a picture using a selective trace through the hierarchy with the aid of bounding boxes. The efficiency of this approach is startling, a picture of a portion of a layout can be selected in a short time (1 s on a VAX 780) irrespective of the overall complexity of the layout. The simplicity of the algorithm is astonishing. Edit operations can be performed on any module of the layout, which automatically affect all its instances.

Rule checkers in VOILA, both design rule checkers and checkers verifying that a layout is in conformity with the required function, also benefit from the hierarchy concept. The batch rule checker will do its job in a fraction of the time needed by a conventional checker. Moreover, incremental checks, i.e., checks verifying that a component added to a good design still yields a good design, can be executed in real time. Incremental checks are an integral part of VOILA graphics.

We have also incorporated in VOILA a program named LOCAL, which was developed in our group but outside the VOILA project. Its function is to extract an electrical schematic from a layout, providing layout to circuit verification. Although circuit extraction is technology dependent, LOCAL has been designed as a technology-independent and circuit-description-language-independent program. The dependencies are concentrated in so-called application files which describe the design context. Another technology only requires a re-

write of a compact application file. Initially developed for the extraction of limited sized circuits, its efficiency allows it to be used for complex circuits as well. Many thousands of transistors can be extracted from a layout in a reasonable computer run time. It turns out that VOILA and LOCAL can be combined nicely. LOCAL provides a verification of modules designed at a detailed level, making use of the module's domain, VOILA is used for system integration.

The programs referred to are in the domain of aids for manual design. Further tools are envisaged beyond this level. If the VOILA coordinate space is interpreted as mouldable, compaction programs can be incorporated. Module design can be automated with placers and routers, e.g., for irregularly shaped blocks. As already stated, such programs will be economically viable because they can work on a module after module basis.

VIII. Conclusion

The design of complex circuits is becoming difficult and time consuming. Hierarchical design is gaining in popularity as a remedy for that problem. However, the introduction of hierarchical design methods is not a simple matter. It requires a change of attitude to design, calls for new concepts to support hierarchical design, and a new generation of CAD tools. But once the capital expenditure on these innovations has been made, the reward will be proportionately great. They will enable us to go on designing increasingly complex circuits.

References

[1] G. E. Moore, "Are we really ready for VLSI?" presented at the CALTECH Conf. on VLSI, Pasadena, CA, Jan. 1979.
[2] J. D. Meindl, K. N. Ratnakumar, L. Geizberg, and K. C. Saraswat, "Circuit scaling limits for ultra large-scale integration," presented at the Int. Solid State Circuits Conf., 1981.
[3] D. P. Siewiorek, D. E. Thomas, and D. L. Scharfetter, "The use of LSI modules in computer structures: Trends and limitations," *Computer*, July 1978.
[4] W.G.J. Kreuwels, "CAD at Philips-Europe reviewed in a historical perspective," presented at the Custom Integrated Circuits Conf., May 1982.
[5] W. M. van Cleemput, "Hierarchical design for VLSI: Problems and advantages," presented at the CALTECH Conf. on VLSI, Pasadena, CA, Jan. 1979.
[6] E. B. Eichelberger and T. W. Williams, "A logic design structure for LSI testability," presented at the 14th Design Automation Conf., June 1977.
[7] C. Mulder, C. Niessen, and R.M.G. Wijnhoven, "Layout and test design of dynamic LSI structures," presented at the Int. Solid State Circuits Conf., 1979.
[8] The First Share Design Automation Conf., 1964.
[9] J. A. Darringer, W. H. Joyner, L. Berman, and L. Trevillyan, "Experiments in logic synthesis," presented at the IEEE Int. Conf. on Circuits and Computers, ICCC, 1980.
[10] H. de Man, J. Rabaey, G. Arnout, and J. Vandewalle, "DIANA as a mixed-mode simulator for MOSLSI sampled-data circuits," presented at the IEEE 1980 Int. Symp. on Circuits and Systems.
[11] S. Sahni and A. Bhatt, "The complexity of design automation problems," presented at the Design Automation Conf., 1980.
[12] T. Whitney, "A hierarchical design-rule checking algorithm," *LAMBDA*, first quarter 1981.
[13] B. Lattin, "VLSI design methodology, the problem of the 80's for microprocessor design," presented at the CALTECH Conf. on VLSI, Pasadena, CA, Jan. 1979.
[14] D. L. Johannsen, "Bristle blocks: A silicon compiler," in *Proc. 16th Design Automation Conf.*, 1979.
[15] S. Trimberger, J. A. Rowson, C. R. Lang, and J. P. Gray, "A structured design methodology and associated software tools," *IEEE Trans. Circuits Syst.*, vol. CAS-28, pp. 618–634, July 1981.
[16] C. L. Seitz, "System timing," in C. Mead, L. Conway, *Introduction to VLSI systems*. Reading, MA: Addison-Wesley, ch. 7.
[17] M. R. Barbacci, "Instruction set processor specifications (ISPS): The notation and its application," *IEEE Trans. Comput.*, vol. C-30, pp. 24–40, Jan. 1981.
[18] N.H.E. Weste, "MULGA, an interactive symbolic layout system for the design of integrated circuits," *Bell Syst. Tech. J.*, vol. 60, no. 6, July–Aug. 1981.

Automating chip layout

New computer-based layout systems are faster than their human counterparts and produce designs that almost match those created manually

Only a small portion of new large-scale and very large-scale integrated-circuit designs are laid out automatically today. But that picture is changing. IC fabrication costs are now so low that the design process offers the only major avenue for reducing IC costs further. IC producers are certain to adopt automated layout tools—sophisticated programs that translate functional specifications into physical descriptions of chips—increasingly over the next few years, and by 1990 layout will probably be predominantly automatic.

Automating layout not only eliminates the need for highly skilled layout people, but it also virtually eliminates mistakes and greatly reduces the turnaround time for new IC designs. These features all reduce costs, of course.

Automated layout tools have been criticized because they use silicon chip area less efficiently than a human layout designer would and produce ICs that operate more slowly and consume more power. To some extent, these charges are true; some automated layout tools do waste chip real estate and use excessively long internal connections, resulting in speed and dissipation that are not optimum. Some automated designs use half again as much area as a manual design for the same function. But such faults often fade in comparison with the benefits of guaranteed correctness and fast turnaround. Moreover, improvements are continually being made in automated tools.

Indeed, a new automated layout tool known as the standard floor-plan system comes close to human proficiency and occasionally exceeds it. Although the standard floor-plan approach is currently limited in the types of ICs it can accommodate, the repertory is being expanded.

Industry giants automate

Today two industry giants—IBM Corp. and the Bell System—use the automated layout for virtually all their LSI and VLSI designs. Their IC chips are designed for in-house use. To a lesser extent, large commercial IC suppliers, such as Texas Instruments Inc., Dallas, and National Semiconductor Corp., Sunnyvale, Calif., also use automated tools for laying out random-logic chips and others. The smaller custom design houses use automated layout hardly at all, except for memories, which require only simple software because of their repetitive structures. But under pressure to stay competitive in cost and delivery time, large and small manufacturers inevitably will turn to automation more and more.

How will automated layout tools affect the people who lay out ICs? Automated layout specification requires different talents from those necessary to hand layout. Automated layout is more like programming; designers who use it do more textual specification and less graphic specification than those who use

Stephen Trimberger
California Institute of Technology

hand layout. They are much less concerned with the precise geometrical form of patterns on silicon and more concerned with the function of the circuit. The effects of job displacement have not yet been felt in the IC industry because of the acute shortage of qualified designers and the relatively small number of designs being made automatically.

Four basic ways to automate

IC layout—the process of translating a description, or specification, of an integrated circuit into photolithographic masks for fabricating the circuit—is being totally automated with four basic methods [Fig. 1]:

1. **Standard cell.** A large library of predefined cells (small logic elements, such as three-input NAND gates or 4-bit counters) is stored in the layout system. The designer of a circuit tells the system which cells are needed and the kinds of connections that will be required between the cells. The standard-cell-system software then assigns the cells to positions on the silicon chip and determines metallization routes, or "wiring," for connecting the cells.

2. **Gate array.** A prefabricated chip contains hundreds of thousands of identical cells, such as NAND-gates, arranged in arrays. The designer tells the layout system the logic functions the chip is to perform and the system selects the cells needed and establishes wiring routes among them.

3. **Programmed logic array (PLA).** A chip contains two arrays—consisting of NAND-gates or NOR-gates—that in series perform any Boolean logic operation. The designer supplies the PLA layout system with the general logic equations. The system selects the signals to be included in the array gates to implement the equations.

4. **Standard floor plan.** The system contains a generic structure for the chip that is much like a housing developer's basic floor plan for a house, which gives the relative positions of rooms and hallways and the means for connecting utilities to the rooms, but leaves unspecified room sizes and contents. Similarly, the chip floor plan specifies the positions of functional units, bus orientation, and wiring strategy, without predetermining the kinds of elements that will be connected.

The floor plan generating the most interest is known as the microprocessor data path. Here a designer tells the layout system the number of bits in the data path, the kinds of processing elements, and the input conditions that cause each element to perform its function. For each element, the system calls a subprogram to lay out the subcircuit with requisite parameters derived from the floor plan.

Each of these layout tools accepts a functional description of the circuit and may also accept some clues to the placement of logic on the chip. The functional description varies widely among automated layout tools and is tailored to the particular tool and the designer's needs [see table]. Each tool embodies a characteristic layout strategy that gives the additional information needed to

Reprinted from *IEEE Spectrum*, vol. 19, pp. 38–45, June 1982.

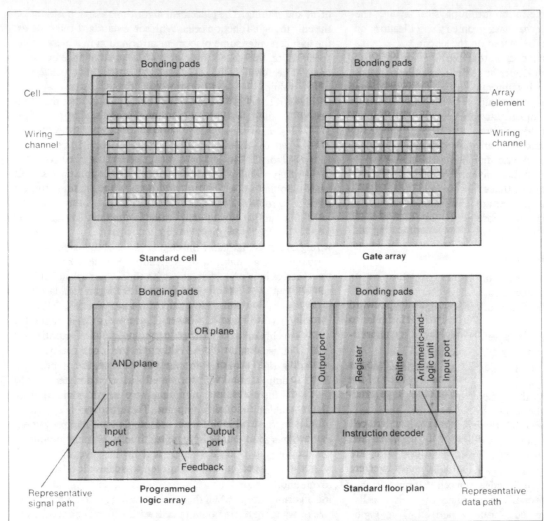

Standard cell

Cell

Wiring channel

Bonding pads

Gate array

Bonding pads

Array element

Wiring channel

Programmed logic array

Bonding pads

OR plane

AND plane

Input port

Output port

Feedback

Representative signal path

Standard floor plan

Bonding pads

Output port

Register

Shifter

Arithmetic-and-logic unit

Input port

Instruction decoder

Representative data path

[1] Among the four major automated layout concepts for IC chips are standard cell and gate-array layout, which are place-and-route schemes; circuit elements are assigned positions, and connections ("wiring") are routed among them as necessary. Unlike standard cells, the gate-array chip image is predefined, so wiring space is limited. The programmed logic array and standard floor plan emphasize wiring rather than placement; regular wiring paths are determined first, and elements are positioned to accommodate them. PLAs perform combinational logic, whereas the standard floor plan here is restricted to microprocessor data paths.

[2] A standard-cell chip made by ZyMOS Corp., Sunnyvale, Calif., contains wiring in channels of various widths between rows of cells. A relatively modest effort for a telephony application, this metal-gate CMOS chip contains 1110 transistors. The company has applied its automated layout system to design chips containing 10 000 random-logic devices in addition to memory on a single chip.

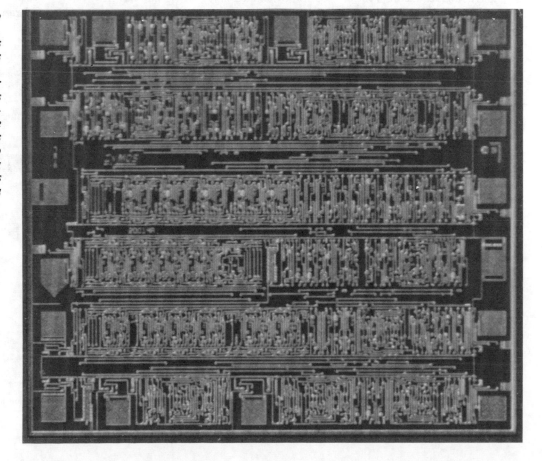

translate the functional description into a physical layout. The output of the layout tool is the mask geometry specification, or the geometrical patterns to be placed on the silicon.

An advantage of human over automated techniques is the ability to select different strategies in different situations. Most engineering organizations using automated layout have only one tool with one strategy. A good match between problem and strategy yields an efficient implementation. Standard cell layout, for example, does well on shift registers, PLA layout produces excellent control circuits, and standard floor-plan layout yields efficient processor circuits. A bad match, on the other hand, yields a messy implementation that may require extensive human intervention to complete, or one that cannot be completed at all. Some semiconductor manufacturers are now experimenting with hybrid layout systems that incorporate two or more layout strategies.

The straightforward approach

The standard cell approach is one of the most straightforward for automating the layout process. To the designer, it is similar to PC-board layout; the tools and methods to solve the layout problem are the same.

Standard cell technology was developed at Bell Laboratories and is now used there for virtually all LSI designs. A few of the more recent circuits from Bell include PLA or other large structures on the chips.

The logic cells selected by the designer from the standard cell library are positioned by the layout software in rows, between which are wiring channels [Fig. 2]. Power and ground wiring are laid out along the sides of the chip so that it does not interfere with the data signals. To ensure that all connections to a cell can be made, the cells are designed with a few simple rules, such as: "All inputs and outputs lie on the edge of a cell." Input to standard cell systems is in the form of a so-called netlist—a list of cells and connections between cells—usually derived from logic diagrams or logic equations. (The netlist can also be used as input to a simulator, to verify that the logic performs the required function before the chips are made.) The standard cell system examines the netlist, determines the positions of each of the cells on the chip, and creates the wiring pattern to connect the cells into a working system. Thus, both the input language and the design constraints are much the same for standard cell design as they are for PC-board layout.

The two steps to the standard-cell layout process are place-ment and routing. The placement algorithms assign positions on the chip to logic-function cells. With some standard cell systems, the user can enter some placement information in a hierarchical specification that determines which cells will be placed close to one another. Ordinarily, however, an algorithm has much leeway in determining the precise placement of cells. Those that communicate directly are placed close together, and those that connect to bonding pads are placed near the edge of the chip.

Routing algorithms select wiring paths to make the connections given by a designer in a functional specification. Most routing algorithms route one wire at a time, remembering the used wiring channels and avoiding them with future wires. This causes problems when one wire blocks another, so some routers attempt to route all wires simultaneously.

The amount of logic that can fit on a chip is limited by the number of wiring channels used. Long wires require additional wiring channels that make the chip larger, thereby increasing the probability of a fatal flaw in fabrication. Signals take longer to travel across long wires, so a chip with short wires will not only be smaller, but will also run faster. Therefore, most standard cell-optimization routines attempt to minimize total wire length, wiring channel use, or both. Although this does not guarantee minimum chip area or optimum performance, the quantities are easy to measure and well correlated with area and speed.

Finding the absolute minimum wire length is not practical from a computational standpoint. Therefore, no systems try to find it but instead look for an acceptable local minimum, using an acceptable amount of computer time. A clustering algorithm is used in the placement step, and an iterative improvement algorithm is used in the routing step to achieve local minima.

Clustering algorithms attempt to group cells that have many connections among themselves, so those cells may be placed close to one another to make their connections short. A typical clustering algorithm stops when the clusters have reached a maximum size or when there are no more cells with enough connections to existing clusters.

If the designer provides a hierarchical specification—that is, describes the chip in terms of cells containing smaller cells—the user's specification is used as the initial clustering. Since cells defined hierarchically vary greatly in size, systems that allow hierarchical descriptions often relax the restriction that the cells must lie in rows, and the largest blocks of cells are positioned anywhere on the chip.

Iterative improvement algorithms come into play after the first

Layout tools at a glance

Tool	Applications	Percent of space for wiring	Input	Major users
Manual	All	Low	Cells and connections	Virtually universal
Standard cell	All	High (up to 80)	Cells and connectons	Bell Labs, others
Gate array	All	High	Logic specification	IBM, Motorola, others
Programmed logic array	Combinational logic, state machines	Low	Logic equations	IBM
Standard floor plan	Frequently designed large structures (such as processors)	Low	Functional parameters	Caltech, IBM, Bell Labs (experimental)

Note: Place-and-route systems—that is, standard cell and gate-array layout systems—are versatile but use chip area extravagantly for wiring. Programmed logic arrays and standard floor plans, by contrast, conserve area almost as much as manual layout, but they have limited applications.

IEEE spectrum JUNE 1982

cell placement: they exchange the positions of pairs of cells if this reduces the total wire length or channel use. The iterative improvement proceeds until no further improvement can be found by any exchange, yielding a local minimum for the total wire length or channel use.

Some standard cell layout systems employ Karnaugh minimization to find similar gating configurations. The system can then implement the same logic with fewer cells.

In addition, many layout systems recognize that there usually are several possible physical forms of a cell for a given function. The forms differ in the orientation of their inputs and outputs and the driving power of their output. The system chooses the form that minimizes the wiring and can drive the fanout at a reasonable speed. Further, some systems perform critical path analysis and optimize critical paths so the chip will run faster. For example, an algorithm checks all logic stages from input to output to identify the signal paths having the most gates and hence the greatest delay. The system gives these critical paths special attention, selecting particularly fast gates from the library and positioning them for minimum wiring length.

Standard cell layout is fast because the positioning of the cells and routing of the wiring are done entirely by computer. A relatively modest LSI circuit—one, say, with about 10 000 gates—can be laid out and fabricated in perhaps two months, whereas the lead time for hand layout of a similar circuit is well over a year.

Designers need not know details of the IC technology that will be used to execute the design; the layout rules for CMOS, TTL, and other technologies are built into the algorithms. A design may be reimplemented when more advanced fabrication technologies become available.

On the other hand, standard cell layout undeniably wastes chip real estate, with as much as 80 percent of the area taken up by wiring. Moreover the automatic placement and routing system may not be able to produce a chip within the desired size, performance, and power-consumption limits. When this occurs, a human designer must modify the placement and wiring, and much of the advantage of automated layout is lost.

Exploiting mass production with gate arrays

Gate-array layout—also known as master-slice layout at IBM and uncommitted logic-array layout by Japanese and European manufacturers—is a variant of standard cell layout that takes advantage of the manufacturing economics of integrated circuitry. A gate-array chip consists of hundreds or thousands of identical structures placed at regular points on the surface of the chip [Fig. 3]. The structures can be sets of transistors and resistors, complete NAND-gates (the usual form), or other functional elements. The gates in an array are separated by wiring channels, which contain the user-specified interconnections and predefined power routing to ensure good power connections to all parts of the chip.

A manufacturer produces identical gate-array chips by the thousands—with all gates in position but without connections between the gates—and stores them. Then when a user orders a few or a few hundred chips to perform a given function, the manufacturer customizes the gate-array chips merely by adding the interconnections needed for the function.

IBM is the pioneer in gate-array technology. Layout techniques have matured there over the last 15 years or so and gate-array technology is now the mainstay of the company's IC design effort. Virtually every IC in IBM's computers is a gate-array circuit. Gate arrays are also catching on elsewhere; many custom IC fabricators are using the technology, including Silicon Systems

Inc., Tustin, Calif.; Interdesign Inc., Sunnyvale, Calif.; International Microcircuits Inc., Santa Clara, Calif.; Dionics Inc., Westbury, N.Y.; Motorola Inc., Phoenix, Ariz.; and Exar Integrated Systems Inc., Sunnyvale, Calif.

Gate-array layout algorithms map the logic specification onto the logic gates by first defining the correspondence between logic in the functional description and physical structures on the chip, then specifying the wiring between the gates to implement the circuit. These two tasks in a gate-array system are essentially the same as the placement and routing tasks in a standard cell system. However, gate-array layout is more constrained than standard cell layout because all cells are identical and the amount of wiring space available is defined when the structure of the gate array is defined. As with standard cell layout, the input specification may contain information about the placement of the gates on the chip, but the system will function without it.

A gate-array layout system requires the same input as a standard cell system: a list of components and the connections to be made between them. The components are either broken down into the array elements—for example, NAND-gates—or mapped into "macros," structures made up of several array elements with predefined wiring within those elements—for example, a two-bit adder. Macros speed up layout and use chip space more efficiently.

Advantages of gate arrays

Gate arrays offer all the advantages of standard cells: the designers need not know details of the fabrication technology, and the designs can be reimplemented in many different technologies with a minimum of human intervention. The same functional description used to generate the layout can be used for simulation; therefore, the simulation is likely to model the implementation accurately.

But gate arrays have a great advantage not found in standard cells: the identical gate pattern is used by the manufacturer for all gate-array chips, regardless of the end use. Most of the processing steps are identical for all gate-array chips; only the final wiring is unique. The cost of new masks and separate production runs is avoided. Gate-array chips can be processed in advance, shortening the delay from product specification to delivery to a few weeks instead of several months.

One challenge to manufacturers, of course, is estimating the best number of wiring channels for the number of gates and I/O pads on the chip. Although the amount of wiring needed is not known until the customer's needs are known, probability calculations are fairly reliable guides in setting aside chip area for wiring. Naturally, there are some chips for which the wiring capacity of the gate array is exceeded, leading to a condition akin to overbooking on airlines. At such times, lack of wiring channel space makes it difficult or impossible to wire a particular function on a gate array, so that a human must often complete the wire routing after the layout system's initial guess.

With some layout systems, the wiring can trespass over areas occupied by gates when more wiring space is needed. Then there may not be enough gates left on the chip to implement the function and the design may have to be partitioned into several chips. Partitioning creates its own problems, since there may not be enough pins on the chips to accommodate the input and output connections among chips. If a human finally must solve such problems, the fast turnaround and other automation advantages of the gate array may be lost.

Layout problems do not ordinarily crop up with gate arrays. But there is a fundamental assumption in place-and-route systems that, on an integrated circuit, as on a printed-circuit

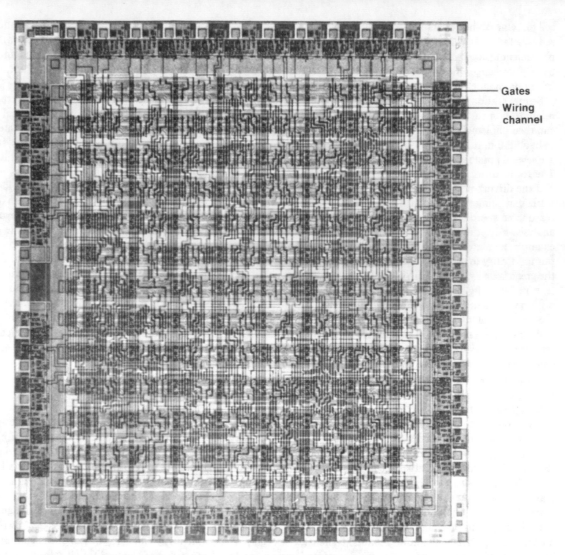

Gates

Wiring channel

[3] Gate-array chip manufactured by Digital Equipment Corp., Maynard, Mass., for its computers consists of a custom wiring pattern on prefabricated gates.

board, wiring is cheap and logic expensive. This is a false assumption: since logic function and wiring are made by the same processes, there is no additional expense for additional functions. In fact, as ever larger chips are made, the space required for wiring grows faster than the space required for devices, and the priorities are reversed: wiring is more expensive than logic in terms of chip real estate. Programmed logic-array and standard floor-plan layout systems recognize the new priorities with layouts in which the wiring is predetermined and logic is placed where it is needed on the wiring.

A programmed logic array, sometimes called programmed array logic, brings all inputs along parallel connections into the AND plane, where they are used selectively as inputs to several gates. The outputs from the gates in the AND-plane, called minterms, run perpendicularly into the OR-plane, where the minterms become the inputs to the OR-plane gates. The outputs from the OR-plane are the outputs of the PLA and run parallel to the inputs. Outputs may be run back into the inputs, in which case they are feedback signals [Fig. 4].

The AND-plane and the OR-plane may both contain NAND gates, giving the AND-OR combination of inputs (hence the terms AND-plane and OR-plane); they may both be made of NOR-gates, giving a NOT-AND-OR-NOT function; or the AND-plane may be NAND-gates, the OR-plane may be NOR gates, and the minterms may be inverted with extra logic between the planes to give an AND-OR-NOT result. If necessary, inputs and outputs are inverted at the respective ports. The choice of

gate types in the arrays depends on the preferred structures in a particular fabrication technology—for example, NMOS makes better NOR gates, while a NAND-NOR combination is preferred in some CMOS schemes.

Gates in the AND-plane are spread horizontally across the entire array, so a minterm is the AND of any set of inputs. The PLA system simply lays a transistor on top of the wiring where the input line meets the minterm line. The same strategy is applied to build the OR-plane.

The regular grid pattern of wiring in PLAs allows them to perform complex logic functions without long wiring connections. PLA layout is similar to read-only memory programming: locations in the PLA are connected if the Boolean operation in that part of the PLA includes the input signal.

Some algorithms generate PLA layout directly from state machine equations, synchronous logic equations (equations with clocking information), or microcode specifications. Indeed, the PLA technique excels at circuits based on state machines—for example, microcode controllers. With the PLA scheme, inputs can be latched and held and outputs can be recycled from the output drivers to the inputs on the next clock cycle, where they provide the state needed by a state machine. But the PLA approach is less successful at implementing such chips as shift registers, which contain little or no logic. Forcing such functions into the PLA mold means very large chips.

PLAs were originally developed by IBM, although they were never widely used in that company. More recently there has been a resurgence of interest in them throughout the industry, because of their regular form and because they can be used to build new types of controllers quickly and economically.

PLA layout systems cannot optimize gate positions and wiring routes, but they can take one of three actions to reduce chip size: logic minimization, partitioning, and folding. Traditional logic minimization, which attempts to reduce the number of Boolean functions, does not necessarily reduce PLA area, because the size of the PLA chip depends on the number of inputs, outputs, and minterms.

Instead, since the inputs and outputs are fixed by the chip function, PLA minimization algorithms attempt to reduce the number of minterms. Current software eliminates redundant, unused, and constant minterms and recognizes simple recoding possibilities when a minterm can be eliminated by OR-ing together existing minterms. No efficient minimization algorithms have been developed, so the goal is to find PLAs that are small enough, but do not entail excessive computer time.

Often a space or performance advantage can be gained by separating disjoint or nearly disjoints sets of equations—those with few or no terms in common—into distinct PLAs. Partitioning algorithms collect outputs that require the same minterms and compare the total area of the smaller PLAs thus generated with the area of one large PLA.

Folding algorithms split rows or columns in the PLA so they can be used by more than one signal. A common form of folding allows inputs and outputs at both the top and the bottom of the PLA, with two signals sharing the same AND-plane and OR-plane column. Some layout systems insert logic in front of the inputs to give precalculated terms to the PLA, thereby eliminating some PLA inputs and minterms. An important advantage of PLAs is that their straightforward point-to-point wiring avoids excessively long connections.

PLAs can be made as electrically programmable units in which unwanted connections are burned out by the user after the chip has been fabricated, just as in electrically programmable ROMs. Electrically programmable PLAs can be mass-produced like ROMs and later customized, thereby gaining even greater replication advantages than gate arrays. They are offered by such manufacturers as Monolithic Memories Inc., Sunnyvale, Calif.; Texas Instruments Inc., Dallas; and Signetics Corp., Sunnyvale.

More commonly, though, PLAs are included as parts of larger designs, because of their rectangular shapes. PLAs are very good for functions with inputs numbering in the dozens that require complex operations on those inputs. PLAs are inefficient at handling very simple functions. The function $C = A$ or B, for example, would take two input columns, two minterm rows, and one output column. Also, PLAs implement functions of many inputs and outputs poorly; chip size becomes excessive, and speed is adversely affected by long wiring paths.

Knowing what to look for

Designers who have experience with several processors develop a generalized mental model of a processor. When called on to implement a processor, they use the mental model as the initial structure. Given the logic elements in the processor data path, the fields of the microcode that controls those elements, and the processor word size, the generation of the implementation is routine.

A program that embodies this knowledge is called a standard floor-plan system. Floor plan refers to the overall structure, which is predefined with open slots like empty rooms to be filled as the user sees fit. A standard floor-plan system may produce a layout as efficient as a human's from a very abstract description of a chip. For this reason, such systems are sometimes called silicon compilers.

Researchers at the California Institute of Technology, Pasadena, have developed a standard floor-plan program they call Bristle Blocks, after a toy consisting of modules that can be joined in complicated shapes by meshing bristles. Bristle Blocks is a system for constructing a microprocessor data path—the calculation portion of a processor, which includes registers, arithmetic-and-logic units (ALUs), and shifters on a shared bus. The control portion of the microprocessor, including the microcode and the microsequencer, must be laid out by other means (a PLA, for example).

The Bristle Blocks data path consists of data-processing elements strung along two buses. A processing element is defined as a 1-bit slice of a particular processor function—for example, a shifter, register, I/O port, or an ALU. The elements are stacked vertically to form the complete functional unit. Functional units are strung horizontally along the data buses [Fig. 5]. Externally supplied microcode enters at the bottom, and the fields required by each data-path element are decoded from the microcode and supplied to the core from the bottom.

The elements are parameterized cells; they are defined by a program capable of laying out elements having the required parameters, rather than by the geometrical representation of a functional element, as in standard cell systems. Because the core elements are parameterized, they can be used in various situations for diverse chips.

A designer gives Bristle Blocks the width of a data path in bits, the kinds of elements in the core, the microcode, and the microcode conditions for each element that cause that element

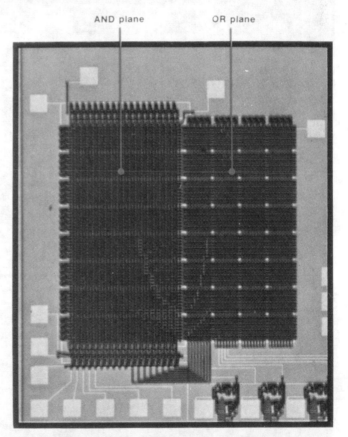

AND plane   OR plane

[4] Programmed logic array built at Caltech contains AND plane and OR plane. Both planes are made up of NOR gates, so the input port inverts inputs and the output port inverts outputs, resulting in AND-OR combinations of signals. Feedback loops linking the output and input ports of the chip appear at bottom.

Trimberger—Automating chip layout

to perform its operation. With that information, the Bristle Blocks system calls each of the core element programs, supplying the parameters to generate the chip layout.

Construction of a data path proceeds in two phases: query and layout. A query algorithm asks the core elements selected by the user for their minimum height and power consumption. The algorithm sums the power consumption to determine the width of power and ground lines. Then it positions the core elements horizontally in the order specified in the input. The algorithm gives core element programs instructions on where to connect the power, ground, and buses. These conventions are determined from the largest minimum size constraint acquired in the query phase. Small elements stretch to make the connection, lengthening vertical wires inside the element as necessary to reach the horizontal bus [Fig. 6].

Because of stretching, there is no need for routing between core elements, and so multiwire routing channels are unnecessary. The speed of the resulting data path is improved, since the connections between elements are shorter. Since each element is given its bit position in the data path, carry chain buffering or carry look-ahead for the ALU can readily be inserted by the program at the appropriate bit positions.

The control signals for each core element are supplied by the externally generated microcode. The proper microcode signal for each input is routed from the pads along the bottom of the chip through a simple decoding array that resembles one plane of the PLA. The decoder is made as large as necessary to generate the signals required by the core elements. The decoded signals are routed to the data-path elements by means of a "river route," a simple route between parallel sets of connectors. The system places the decoder section below the routing area, places bonding pads on the chip, and connects them to the microcode decoder and to the data-path I/O ports.

Bristle Blocks implements processor data paths using no more than 10-percent more area—and sometimes even less area—than hand layout. The routing area, including routing to pads and the area lost by stretching core elements, is typically around 20 percent of the total area. Routing is virtually nonexistent inside the core.

No commercial data-path chips have been produced yet, but Bristle Blocks has been used to lay out 8-, 12-, and 16-bit data paths. Also, IBM and Bell laboratories are developing data-path systems of their own.

Most optimizations of standard floor-plan systems can be implemented as improvements in the core elements, rather than changes in the floor plan itself. With one floor-plan–optimization algorithm, data-path elements can trade off width for length, and the algorithm can find the combination of width and

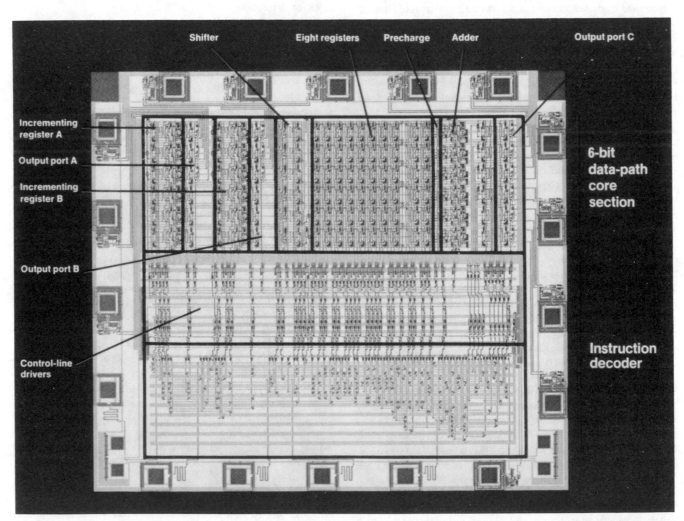

[5] Diagram of a Bristle Blocks layout designed at Caltech shows NMOS mask layers. The chip, which is being built as a demonstration, is called a stock market predictor. It analyzes a stream of 6-bit input numbers to detect trends in real time.

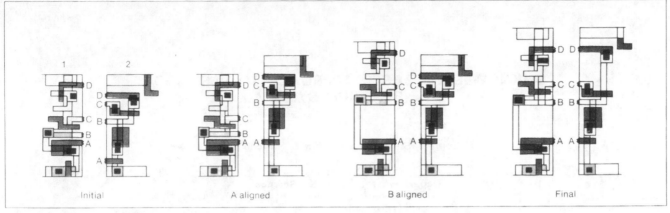

Initial   A aligned   B aligned   Final

[6] Cell stretching is used when two cells in a data path whose connectors are not at the same vertical positions must abut. To avoid generating wires for the interconnections, the Bristle Blocks software converts such cells into cells that can abut directly. For example, to align the A connectors in cells 1 and 2 at left, the system must increase the distance between the bottom of cell 2 and connector A. Accordingly, the layout system stretches cell 2 as shown in the center left diagram. It then stretches cell 1 as shown at center right to align the B connectors. The stretching process continues until all corresponding connectors have the same vertical position, at which stage the connections are made by merely abutting the cells.

length that leads to the smallest chip.

Standard floor-plan layout makes a powerful ally of PLA layout; one produces processors, the other controllers. Together they produce an automatically laid out microcomputer, complete except for memory, either on multiple chips or, recently, on the same chip.

Because some of the data-path elements stretch when they are assembled, there is no good control of an element's electrical properties. Some designers consider this cause for alarm, but stretching has no more serious effect than the routing it replaces. Besides, the electrical properties of a cell can be estimated as a function of the stretching.

A more serious drawback of standard floor-plan systems is their limited applicability. Bristle Blocks can implement only processor data paths and not general logic functions. Many logic operations can be forced into the data-path form, but the conversion can be difficult. However, limitations in applicability can be cured by development of additional special floor plans.

The goals of research in automated layout are to produce tools that can implement any function very quickly, to implement the function as well as a human can, and to get to market as quickly as possible. Toward these ends, much work is going into better general placement and routing techniques, especially for gate arrays, which have the advantage of low fabrication cost. Better medium-scale macros are being developed for standard cell systems, and parameterized cells such as those used in Bristle Blocks, may make for more generally useful standard cells. New floor plan definitions will allow standard floor-plan systems to implement more chips efficiently.

Composite layout systems will also help by allowing designers to implement each function on a chip with the layout tool that makes the best design. Hybrid layout systems eventually may generate paths with a standard floor-plan system, state machines with PLAs, and general logic with a place-and-route system. The latter would allow the system to implement general functions; inefficiencies would be limited to a small part of the chip.

To probe further

Many engineering schools offer courses on LSI and VLSI circuit design, but they usually touch only lightly on automated design tools. A better source of up-to-date, detailed information is the yearly Design Automation Conference, sponsored by the IEEE and next scheduled for June 14–16 in Las Vegas, Nev. Contact Harry Hayman, P.O. Box 639, Silver Spring, Md. 20901. Phone 301-589-3386 for further information.

A new publication, the *IEEE Transactions on Computer-Aided Design*, features "original contributions and tutorial expositions relating to the design, development, integration, and application of computer aids for the design of integrated circuits and systems," according to the editors. Vol. CAD-1, no. 1, was published last January.

For a tutorial article on methods of integrated-circuit design, including automated and hand layout techniques, see J. Tobias' "LSI/VLSI building blocks" in *Computer*, August 1981.

Several proceedings of the Design Automation Conference merit special attention. In those for the 13th conference (1976), G. Persky, O.N. Deutsch, and D.G. Schweiker describe the basic Bell Laboratories standard cell system in "LTX—a system for the directed automatic design of LSI circuits." Other papers by the same authors cover placement and routing of standard cell circuits. In the proceedings for the 14th conference (1977), K.A. Chen and others give an introduction to wiring in gate arrays and describe optimizations in "The chip layout problem: an automatic wiring procedure." The original paper on the Bristle Blocks system, "Bristle blocks: a silicon compiler," by D. Johannsen, is in the proceedings for the 16th conference (1979).

The March 1975 edition of the *IBM Journal of Research and Development* contains three papers on programmed logic arrays, including an introduction to the concept and a description of PLAs for a small system.

A KNOWLEDGE BASED APPROACH TO VLSI CAD
THE REDESIGN SYSTEM

Louis I. Steinberg and Tom M. Mitchell

AI/VLSI Project
Department of Computer Science
Rutgers University
New Brunswick, NJ 08903

Abstract

Artificial Intelligence (AI) techniques offer one possible avenue toward new CAD tools to handle the complexities of VLSI. This paper summarizes the experience of the Rutgers AI/VLSI group in exploring applications of AI to VLSI design over the past three years. In particular, it summarizes our experience in developing REDESIGN, a knowledge-based system for providing interactive aid in the functional redesign of digital circuits. Given a desired change to the function of a circuit, REDESIGN combines rule-based knowledge of design tactics with its ability to analyze signal propagation through circuits, in order to (1) help the user focus on an appropriate portion of the circuit to redesign, (2) suggest local redesign alternatives, and (3) determine side effects of possible redesigns. We also summarize our more recent research toward constructing a knowledge-based system for VLSI design and a system for chip debugging, both based on extending the techniques developed for the REDESIGN system.

I Introduction

Artificial Intelligence (AI) techniques offer one possible avenue toward new CAD tools to handle the complexities of VLSI. This paper summarizes the experience of the Rutgers AI/VLSI group in exploring applications of AI to VLSI design over the past three years. In particular, it summarizes our experience in developing REDESIGN, a knowledge-based system for providing interactive aid in the functional redesign of digital circuits. We also summarize our more recent research toward constructing a knowledge-based system for VLSI design and a system for chip debugging, both based on extending the techniques used by the REDESIGN system.

A. A Knowledge Based Approach to Software Organization

One of the techniques which has arisen from research in AI is the *knowledge based* approach to designing a system which is to achieve some task. The essence of this approach is to ask what knowledge (i.e. what facts and reasoning abilities) is used by a human expert in solving this task, and to develop data-structures and code which

represent this knowledge explicitly, rather than to have the knowledge present in the system only implicitly.

Often the facts which are most useful are a collection of rules of thumb, derived from the expert's experience, which can be represented in a natural way as IF-THEN rules and which can be used in a fairly simple reasoning process. As will be discussed below, we have found it useful to represent design tactics as IF-THEN rules, but to represent other facts about circuits in other ways.

Researchers using the Knowledge Based approach in a number of areas have found it to have several interrelated advantages over more traditional techniques for organizing software:

- *It is easier to make incremental improvements.* Since the knowledge is represented explicitly it is easier to add additional pieces of knowledge and thereby make incremental improvements in the system's capability.

- *It is easier for the system to explain what it is doing and why.* Since the facts and reasoning processes used parallel those used by a human expert, it is often feasible for a knowledge based system to automatically generate an explanation of *how* it reached its conclusions which is understandable by a human domain expert who is not a computer scientist. For example, the program may indicate the chain of IF-THEN rules which was used to make a particular decision about the design of some circuit submodule.

- *It is easier for a human expert to determine what is incorrect or incomplete about the system's knowledge, and explain how to fix it.* Since the system can indicate what knowledge was used and how it was used, it is easier for an expert to determine what is wrong. Since the system is already structured around the kinds of knowledge the expert uses, it is easier to translate the expert's description of how to fix things into an actual change to the program.

- *It is easier to interactively use a human expert's abilities.* Long before a system is capable of handling a task completely automatically, it may be possible to construct a useful interactive system, which aids the user as much as it can given its limited knowledge, and in which the user can take over and do things when he is dissatisfied with the system's recommendations. Since the system's

---

*This material is based on work supported by the Defense Advanced Research Projects Agency under Research Contract N00014-81-K-0394. The views and conclusions contained in this document are those of the authors and should not be interpreted as necessarily representing the official policies, either expressed or implied, of the Defense Advanced Research Projects Agency or the U.S. Government.

Reprinted from *ACM/IEEE 21st Design Automation Conf. Proc.*, 1984, pp. 412-418.

way of doing things is analogous to that of the user, it can be easier to coordinate the system and the user. This ability to interleave user input with the system's processing represents a major difference between knowledge-based approaches and other kinds of approaches, for instance, to silicon compilers.

B. Applying the Knowledge Based Approach to VLSI CAD

A number of research groups are currently exploring knowledge based approaches to various aspects of VLSI CAD (e.g.,[1, 2, 3, 4]). Here we describe a knowledge based system called REDESIGN which addresses the following problem of functional *redesign*: Given an existing circuit, its functional specifications, and a desired change to these specifications, help the user to determine a change in the circuit that will allow it to meet the altered functional specifications without introducing undesirable side-effects.** During this work, it became apparent that providing intelligent assistance for redesign depends on two quite different types of reasoning about the given circuit.

The first essential type of reasoning about circuits is *causal reasoning* about the interrelations among signals within the circuit. This is a generalization of the notions of circuit simulation and symbolic simulation. It involves tasks such as, given a description of the streams of data being input to a component, deriving a description of the output data streams. Or, given a specification of the required characteristics of the outputs of a component, determine what characteristics must be satisfied by the inputs to the component. The subsystem of REDESIGN which has been developed to solve these kinds of problems is called CRITTER.[5]

A second type of reasoning essential to redesign involves reasoning about purposes of components. For example, given a circuit, and its specifications, explain the role played by a particular component in implementing the overall specifications. Or, determine the range of components that can be substituted for this component without violating the overall specifications.

The next section describes our work on the redesign task, and the use of causal reasoning and reasoning about purpose in REDESIGN. Subsequent sections summarize our more recent attempts to extend these ideas and our current research on developing knowledge based systems for VLSI design and chip debugging.

II The Redesign Task

In the functional redesign problem the system is given the schematic of a working digital circuit (e.g., the display controller for a computer terminal), and its functional specifications (e.g., the fact that it displays 80 characters per line, 25 lines per screen, displays the cursor at a programmable address, etc.). The system is also given a data structure called a *design plan*, which relates the circuit schematic to its specifications. Given a desired change to the functional specifications (e.g., require that the terminal display 72 characters per line), the task is then to redesign the circuit so that it will meet these altered specifications***

The formulation of the redesign problem presented here is very similar to planning problems in the AI literature, and the issues addressed in this work are related to those addressed by others working in the areas of planning and design, such as[6, 7, 8, 9, 10, 11, 12] Our work is also related to that of[13] which deals with recognizing circuits rather than designing them, and which addresses the relations among circuit function, structure, and purpose.

The next subsection discusses the representation of circuits, and the notions of circuit behavior and specifications. The following subsection describes the two modes of reasoning about circuits employed by REDESIGN: causal reasoning and reasoning about purpose. We then illustrate the use of these modes of reasoning by REDESIGN, by tracing its use for a specific redesign problem.

A. Representing Circuits, Behaviors and Specifications

The structure of a circuit is represented by a network of *modules* and *data-paths*. A module represents either a single component or a cluster of components being viewed as a single functional block. Similarly, a data-path represents either a wire or a group of wires. The data flowing on a data-path is represented by a *data-stream*, and the operation performed by a module is represented by a *module function*. These representations are described in[14, 5] One aspect of this circuit representation that has been important in REDESIGN is that data-streams represent the entire time history of data values on a data-path, rather than a single value at a single time, as in many circuit simulators. This has proven to allow considerable flexibility in reasoning about circuit behavior over time.

In reasoning about redesign, REDESIGN must distinguish between what happens to be true of the circuit (we refer to this as the circuit *behavior*), and what must be true for that circuit to work correctly (we refer to this as the circuit *specifications*). Therefore, for each module function and data-stream, both behavior and specifications are recorded. For example, the behavior of a particular module may state that its output will be the sum of its inputs, delayed by 100 nanoseconds, while the specifications for that module may simply require that the output be delayed by less that 500 nanoseconds.

B. Two Modes of Reasoning about Circuits

A variety of types of questions arise when redesigning a circuit. REDESIGN uses two separate modes of reasoning to answer these questions -- one to analyze circuit operation based on a causal model of the circuit, and one to reason about the purposes of circuit submodules (i.e. their roles in implementing the global circuit specifications). These two modes of reasoning are combined to provide assistance at various stages of the redesign process.

1. Causal Reasoning

Causal reasoning answers questions such as "If input X is supplied to the circuit module, what will the output be?"

**We chose the redesign problem over the design problem as our first task primarily because it raised a number of important issues about representing and reasoning about circuits in a more tractable context than design from scratch.

***It should be noted that the example circuits used in the work on REDESIGN were not actually VLSI. Rather they were board-level circuits built from standard TTL MSI parts. However, we believe that the same techniques apply directly to VLSI circuits designed with the standard cell approach.

and "If output Y is desired, what must be provided as inputs to the module?" X and Y here may be either complete descriptions or partial descriptions giving, e.g., just the start time or just the value; of course if the question gives only a partial description the answer may also be a partial description.

The question where a completely described input is given is the type of question answered by standard circuit simulators. However, redesigning a circuit requires answering the other kinds of questions as well. For example, if the circuit specifications call for the circuit output to have a certain duration, it is important to be able to determine which properties of the upstream signals will assure this duration. The CRITTER system answers these kinds of questions by a process of propagating full and partial descriptions of data-streams through the circuit, and can test whether a given data-stream's specifications are satisfied by its behavior. CRITTER also maintains a *Dependency Network* that records, for each specification, both its source and the path in the circuit through which it was propagated.

2. Reasoning about Purpose

A second kind of reasoning important in redesign concerns the roles, or purposes, of various circuit modules in implementing the overall circuit specifications. Questions of this sort that arise during redesign include "What is the purpose of circuit module M? and "How are the circuit specifications decomposed into subspecifications to be implemented by separate sections of the hardware?" Questions of this sort can be answered by REDESIGN, by examining the *Design Plan* of the circuit.

The Design Plan is a data structure that shows how circuit specifications are decomposed and implemented in the circuit, as well as the conflicts and subgoals that arise during design. It contains enough information to allow "replaying" the original design, and is characterized in terms of a set of *implementation rules* that embody in executable form general knowledge about circuit design tactics. This Design Plan must be provided as input to REDESIGN, as part of the characterization of the circuit which is to be redesigned.

In order to illustrate the form of the Design Plan, consider the simple Character Generator Module (CGM) circuit shown in figure II-1. This circuit is similar to a standard circuit used in most video computer terminals -- it is the part of the terminal that translates the ASCII character codes into the corresponding dot matrix to be displayed on the screen. This circuit accepts as input (1) a stream of ASCII encoded *Characters*, (2) a stream of binary encoded integers, called *Slice-Indices* that specify which horizontal slice of the character dot matrix is to be displayed, and (3) several clock signals used for synchronization. The circuit must produce a stream of *Character-Slices*, each of which is a bit string corresponding to the dots to be displayed on the terminal screen for the selected horizontal slice of the input Character.

The heart of the CGM design is a read-only memory, the ROM6574. This ROM6574 stores the definition of the character font (the dot matrix to be displayed for each character), one Character-Slice per byte of memory. To retrieve the Character-Slice corresponding to a given Character and Slice-Index, the ASCII code for the character is concatenated with the binary representation of the Slice-Index, and used to address the ROM6574. The other components in this circuit are used to interface the ROM6574 to the desired input and output formats. For example, the CGM specifications require serial output while ROMs produce parallel output. Therefore, a shift register

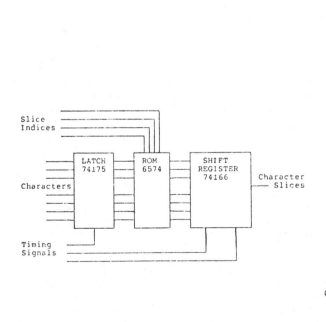

Figure II-1: The Character Generator Module

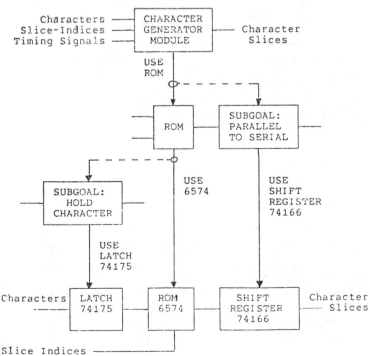

Figure II-2: Design Plan for the CGM

86

(SHIFT-REGISTER-74166) is used to convert the output data to serial. Also, because the address inputs to the ROM6475 must be stable for at least 500 nsec. while the input Characters are stable for only 300 nsec., a latch (LATCH74175) is used to capture the input Characters, and hold these data values stable for an acceptable duration.

The above paragraph summarizes the purpose of each circuit component and the conflicts and subgoals that appear during design. *This is precisely the kind of summary that must be captured in the Design Plan, in order to allow the REDESIGN program to reason effectively about the design and about the purposes of individual circuit components.*

Figure II-2 illustrates the Design Plan used to describe the CGM circuit to REDESIGN. Each node in the Design Plan corresponds to some abstracted circuit module whose implementation is described by the hierarchy below it. The topmost node in this Design Plan represents the entire CGM, and its functional specifications. The bottom most nodes in the Design Plan represent individual components in the circuit. Each solid vertical link between modules in the Design Plan corresponds to some implementation choice in the design, and is associated with some general implementation rule which, when executed, could recreate this implementation step. For example, the vertical link leading down from the topmost module in the figure represents the decision to use a Read-Only Memory (ROM) to implement the CGM. This implementation choice is associated with the implementation rule which states "IF the goal is to implement some finite mapping between input and output data values, then use a ROM whose contents store the desired mapping" (note this leaves open the choice of the exact type of ROM.)

Each dashed link in the Design Plan represents a conflict arising from some implementation choice or choices, and leads to a design subgoal, represented by a new circuit module with appropriate specifications. For example, a conflict follows from the implementation choice to use a ROM, and leads to the subgoal module labelled "Parallel-to-Serial-Subgoal". The conflict in this case is the discrepancy between the known output signal format of ROMs (i.e., parallel) and the required output signal format of the CGM (i.e., serial). The specifications of the new subgoal module are therefore to convert the parallel signal to serial. In a similar fashion, the implementation choice to use the specific ROM6574 leads to another conflict, and to the resulting subgoal to extend the duration of the input data elements.

Not shown are the links between the Design Plan and the Dependency Network, giving the specifications for the various data-streams.

By examining the Design Plan of a circuit, REDESIGN is able to reason about purposes of various circuit modules, and about the way in which the circuit specifications are implemented. The general implementation rules used to summarize the design choices can be used to "replay" the Design Plan for the similar circuit specifications, and thus allow for a straightforward kind of design by analogy.

C. Redesigning a Circuit

This section illustrates the use of both causal reasoning and reasoning about purpose in redesigning a circuit. It traces the actions of the REDESIGN program as it took part in a particular redesign of the Video Output Circuit (VOC)

of a computer terminal. The Video Output Circuit (which contains the Character Generator Module discussed earlier) is shown in figure II-3. It is the part of the computer terminal that produces the composite video information to be displayed on the terminal screen. It produces this output from its combined inputs, which include the characters to be displayed, the cursor position, synchronization information for blanking the perimeter of the terminal screen, and special display commands (e.g., tc blink a particular character).

In this example, we consider redesigning the VOC to display characters in an italics font rather than its current font. Given a redesign problem, REDESIGN guides the user through the following sequence of five subtasks: (1) focus on an appropriate portion of the circuit, (2) generate redesign options to the level of proposed specifications for individual modules, (3) rank the generated options, (4) implement the selected redesign option, and (5) detect and repair side effects resulting from the redesign.

Focus attention on appropriate section(s) of the circuit. In many cases, the most difficult step in functional redesign is determining which portions of the circuit should be ignored. Focusing on relevant details in one locality of the circuit while ignoring irrelevant details in other localities can greatly simplify the complexity of redesign. In order to determine an appropriate focus, REDESIGN "replays" the Design Plan by reinvoking the recorded implementation rules with the changed circuit specifications. During this replay process, whenever an abstract circuit module is produced by some implementation step, its purpose is compared with the purpose of the corresponding module in the original Design Plan. If the purpose is unchanged, then the original implementation of this module will be reused without change in the new design****. If the new module has a different purpose than the corresponding module in the old Design Plan, (e.g., the new CGM must implement a different character font), an attempt is still made to apply the same implementation rule as in the original design (e.g., still try to use a ROM). If this implementation rule is not useful in the new design (as with the rule that suggests using the specific ROM6574), then REDESIGN stops expanding this portion of the Design Plan, and marks the corresponding portion of the circuit as a portion to be focused on for further redesign. The use of the Design Plan as sketched above leads in the current example to a focus on redesigning the abstract ROM module within the CGM within the VOC circuit. This abstract ROM module is implemented in the current circuit by two components as shown in figure II-2 (the ROM6574 and LATCH74175). A second method of focusing is possible, by using the Dependency Network produced by CRITTER. This method involves isolating those points in the circuit that possess specifications derived from the changed specification on the output data-stream. The resulting focus is generally broader than that determined from the Design Plan, because out of the many places in the circuit that can impact any given output specification, only a small proportion of these involve circuitry whose main purpose is to implement that specification.

****One must still make certain that changes elsewhere in the design do not interact dangerously with the implementation of this module. In REDESIGN, this is accomplished without having to directly examining the implementation of the module. Instead, design changes elsewhere in the circuit are checked for consistency with the constraints recorded in the Dependency Network produced by CRITTER.

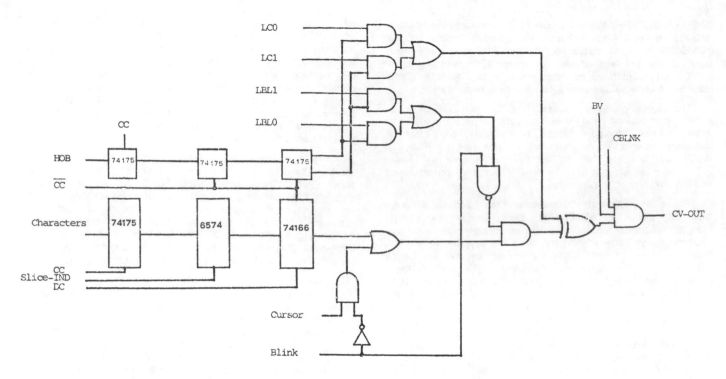

Figure II-3: Video Output Circuit

Generate redesign options to the level of proposed specifications for individual circuit modules. Once an initial focus for the redesign has been determined, redesign options are generated which recommend either altering the specifications of individual modules, or adding new modules with stated specifications. In both cases, only the new functional specifications are determined at this point -- the circuitry to implement these specifications is determined later. The constraint propagation capabilities of CRITTER provide the basis for generating these redesign options. In the current example, once REDESIGN has focused on the section of the VOC including the ROM6574 and LATCH74175, it considers the new output specification for this circuit segment, and propagates it back through this segment. Before each propagation step, REDESIGN considers the option of breaking the wire at that point and inserting a module to transform the values on that wire to values satisfying the required specification. In addition, it considers the option of altering the module immediately upstream, so that it will provide the required signal at that point. For each of the generated options, the new functional specifications are defined in terms of (1) the new specification to be achieved, and (2) a list of unchanged specifications found in the original Dependency Network, which are to be maintained. In the current example, the option generation process produces a list of five candidate redesign options. This list includes redesign options such as "replace the ROM6574 by a module which stores the new character font", and "introduce a new module at the output of the ROM6574, which will transform the output values into the desired font" (these options are described by the program in a formal notation, and the above are only English summaries).

Rank the generated redesign options. Heuristics for ranking redesign options can be based on a variety of concerns: (1) the estimated difficulty of implementing the redesign option (e.g., components with zero delay cannot be built), (2) the likely impact of the implemented redesign on global criteria such as power consumption and layout area, and (3) the likelihood and severity of side effects that might be associated with the redesign[*****]. In the current example, the heuristic that selects the appropriate redesign option suggests "Favor those redesign options that replace existing modules whose purpose has changed." In this case, since the purpose of the ROM6574 has changed, the option of replacing this component is recommended. The recorded Dependency Network and Design Plan also provide very useful information for estimating the relative severity of various changes to the circuit. Because the Design Plan shows the dependencies among implementation decisions (e.g., the purpose for the LATCH74175 is derived from the decision to use the specific ROM74175) it provides a basis for ordering the importance of components and associated constraints in the overall design (e.g., if the ROM6574 is removed, the LATCH75174 may no longer have a purpose for existing). This ordering of circuit modules, and of the data-stream constraints that they impose, provides an important basis for estimating the relative extent of side effects associated with their change.

Implement the selected redesign option. The above steps translate the original redesign request into some set of more local (and hopefully simpler) specification changes. While the implementation rules that REDESIGN possesses might be used for design[******], the REDESIGN system does not make use of this potential. Thus, the user is left to implement the redesign option.

[*****]The current REDESIGN system has only a primitive set of heuristics for ranking redesign options.

[******]In fact they are used this way in the design consultant. See below.

Detect and Repair Side Effects Arising from the Redesign Once the redesigned circuit is produced, REDESIGN checks the new circuit segment to try to determine (a) that it does achieve the desired new purpose, and (b) that it does not lead to undesirable side effects. Undesirable side effects are detected as violations of the Dependency Network specifications at the inputs and outputs of the altered circuit segment. If a specification is violated, the new circuitry might be redesigned, or the specification might itself be modified or removed by redesigning a different portion of the circuit. The Dependency Network can be examined to determine the source of the violated specification, and to determine the locus of circuit points at which the specification could be altered.

D. Conclusions from the Work on REDESIGN

REDESIGN is a research prototype system that demonstrates the feasibility of providing intelligent aids for redesign and design of digital circuits. While the current REDESIGN system has many limitations (e.g., in the size of circuits it can handle, its inability to help with certain classes of redesigns, shortcomings of its causal reasoning methods, incompleteness of its knowledge base of implementation rules, etc.), it demonstrates clearly the importance of reasoning about causality and purpose in circuits when attempting to automate various subtasks involved in redesign and design.

Several features of REDESIGN have been important to its success. The most apparent of these are the means of combining reasoning about *causality* in the circuit, and reasoning about the *purposes* of parts of the circuit to assist in various subtasks of redesign. There are also some important aspects to *how* REDESIGN reasons about causality and purpose. In reasoning about causality, REDESIGN describes both the behavior and the specifications for a data stream, in a way that allows it to describe entire histories, rather than data stream values at single time instants. CRITTER can propagate these descriptions through the circuit, to build a Dependency Network showing how the specifications for each data stream are derived from the behaviors of the modules and the specifications for the circuit as a whole. In reasoning about purposes, we have viewed the original design process essentially as a planning problem, with subgoals derived both from the decomposition of parent goals and from conflicts between other subgoals. The Design Plan provides REDESIGN with an explicit summary of this planning process, with detail enough to replay the process, and to examine the particular relationships among design goals and subgoals.

III An Intelligent Aid for VLSI Design

To follow up on the work on REDESIGN, we are presently developing an interactive intelligent consultant, called VEXED,[15, 5, 16] to aid in designing cells and arrays of cells for VLSI circuits. VEXED begins with the functional specifications of the cells, and constraints on the placement of their interconnections, and is intended to produce a design at the sticks or perhaps layout levels. As an intelligent aid, VEXED is designed to offer advice about alternative methods of decomposing and implementing the desired function, about how to choose among such alternatives, and about detecting and handling interactions and conflicts among implementation choices. By running in background mode inside a graphics-oriented circuit editor, VEXED is intended to provide much the same kind of aid

as that provided by a human expert watching over the shoulder of a designer during an editing session. The user has the ability to focus on a particular portion of the design, and to edit it as he pleases. However, the program may offer advice as it follows the tasks pursued by the user, provided its knowledge base contains expertise appropriate to the task at hand. In such cases, the user may elect to follow the consultant's advice, or to ignore it and implement the portion of the circuit as he wishes.

The design of the VEXED system builds upon our past experience with REDESIGN in several respects. Its design expertise is represented using the same type of If-Then rules used to characterize design steps in REDESIGN, and the two main modes of reasoning about circuits used by REDESIGN are also to be employed by VEXED. However, there are many new issues that must be addressed by VEXED, due to its focus on design rather than redesign, and due to our desire to develop it to the point of a practical tool for VLSI design. One of the major issues lies in building up and managing the knowledge base of design expertise. We expect that, as with many recent expert systems, in order to achieve high levels of performance VEXED may required several thousand If-Then rules. One interesting direction that we intend to pursue is to have VEXED acquire its own rules by observing the user's design steps, much as an apprentice assistant would learn from experience. In particular, in those instances in which the user disregards the advice of VEXED, the system should note the design step that the user takes, and attempt to form a general rule to characterize this step. For example, suppose that the current task is to implement a module that converts parallel to serial signals, and that based on its rule set VEXED suggests using a shift register from its component library. If the user ignores this advice, and instead uses the editor to construct his own circuit, then the system should note the circuit, verify that it accomplishes the desired function, and formulate a new rule that summarizes this new design tactic. Of course the task of formulating new rules in this way can be quite difficult, because such rules should be formulated with an appropriate degree of generality. We plan to base the method for generalizing rules on our previous work on learning heuristics and generalizing from examples[17], and believe that such a capability for acquiring knowledge from interactive problem solving is a crucial direction for research on knowledge based systems during the 1980s.

IV An Intelligent Aid for Chip Debugging

A second current thrust of the AI/VLSI group involves the development of an intelligent aid to assist in debugging VLSI circuits. In particular, we are concerned with the situation faced when the first samples of a newly designed circuit are tested. In the event that the circuit does not perform correctly, the task is to determine whether the failure is due to a design or manufacturing error, and to attempt to localize the cause of the failure. Our goal in this case is to provide an intelligent assistant that is able to generate and rank hypotheses regarding possible sources of the circuit failure, reasoning back from output failure symptoms to plausible internal faults. We find that the kinds of reasoning about the circuit that are essential for providing this kind of assistance in debugging overlap a great deal with the kinds of reasoning essential to design. In particular, the CRITTER system provides one mechanism for tracing output failure symptoms back through the circuit to generate candidate failure hypotheses, and the hierarchical description of the circuit provided by the design plan is essential to controlling the combinatorics of the debugging process (i.e., the circuit is viewed hierarchically, so that the bug is first localized in terms of a small number of possible circuit modules, whose details are then examined in order to further localize the failure within the suspected module).

One thesis of this research is that debugging is best approached by considering design and debugging as interrelated problems. Not only is information from the design plan useful for constraining the debugging process, but the way in which the design is accomplished influences the difficulty of subsequent debugging. One straightforward example of this is the importance of designing VLSI circuits to allow internal signals to be observable at the output pads of the circuit. Furthermore, the result of the debugging process should certainly influence the redesign of the circuit. As our research on design and debugging progresses, we hope to develop ways of assuring closer coupling between these two processes.

References

[1]   Kowalski, T.J. and Thomas, D.E. "The VLSI Design Automation Assistant: Prototype System." In *20th Design Automation Conference*. IEEE, August, 1983, 479–483.

[2]   Stefik, Mark and Conway, Lynn "Towards the Principled Engineering of Knowledge." *The AI Magazine*. 3:3 (1982) 4–16.

[3]   Zipple, R., "An Expert System for VLSI Design", MIT VLSI Memo 83–134

[4]   Kim, J. and J. McDermott "TALIB: An IC Layout Design Assistant." In *Proceedings of the 1983 National Conference on Artificial Intelligence*. AAAI, 1983, 197–201.

[5]   Kelly, V. "The CRITTER System: Automated Critiquing of Digital Hardware Designs", Technical report WP–13, Rutgers AI/VLSI Project, November 1983, to appear in Design Automation Conference, 1984.

[6]   Green, C., et al. "Research on Knowledge–Based Programming and Algorithm Design", Research Report KES.U.81.2, Kestrel Institute, September 1982.

[7]   J. McDermott "Domain Knowledge and the Design Process." In *Proceedings of the 18th Design Automation Conference*. IEEE, Nashville, 1981.

[8]   Mostow, D.J., and Lam, M. "Transformational VLSI Design: A Progress Report", Technical report, USC–ISI, November 1982.

[9]   Rich, Charles; Shrobe, Howard E.; Waters, Richard C. "Computer Aided Evolutionary Design For Software Engineering", AI Memo 506, Massachusetts Institute Technology, January 1979.

[10]   Stefik, Mark Jeffrey, *Planning With Constraints*, PhD dissertation, Stanford University, January 1980.

[11]   Sussman, Gerald Jay; Holloway, Jack; Knight, Jr., Thomas F. "Computer Aided Evolutionary Design For Digital Integrated Systems", AI Memo 526, Massachusetts Institute Technology, May 1979.

[12]   Wile, David S. "Program Developments as Formal Objects", Technical report, Information Sciences Institute, July 1981.

[13]   de Kleer, Johan, *Casual And Teleological Reasoning In Circuit Recognition*, PhD dissertation, Massachusetts Institute Technology, January 1979.

[14]   Kelly, V., Steinberg, L. "The CRITTER System: Analyzing Digital Circuits by Propagating Behaviors and Specifications." In *Proceedings of the National Conference on Artificial Intelligence*. August, 1982, 284–289, Also Rutgers Computer Science Department Technical Report LCSR–TR–30, and Re-Design Project Working Paper #6

[15]   Roach, J. "The Generalization of Symbolic Layout", Technical report WP–12, Rutgers AI/VLSI Project, November 1983, to appear in Design Automation Conference, 1984.

[16]   Steinberg, L. and Mitchell, T. "Artificial Intelligence Aids for VLSI", Technical report WP–9, Rutgers AI/VLSI Project, June 1983.

[17]   Mitchell, T. M., Utgoff, P. E. and Banerji, R. B., "Learning by Experimentation: Acquiring and Refining Problem–Solving Heuristics," in *Machine Learning*, Michalski, R. S., Carbonell, J. G. and Mitchell, T. M., eds., Tioga, 1983.

Section I-C
Design Automation and Simulation

Computer-Aided Design of VLSI Circuits

ARTHUR RICHARD NEWTON, MEMBER, IEEE

Invited Paper

Abstract—With the rapid evolution of integrated circuit (IC) technology to larger and more complex circuits, new approaches are needed for the design and verification of these very-large-scale integrated (VLSI) circuits. A large number of design methods are currently in use. However, the evolution of these computer aids has occurred in an ad hoc manner. In most cases, computer programs have been written to solve specific problems as they have arisen and no truly integrated computer-aided design (CAD) systems exist for the design of IC's. A structured approach both to circuit design and to circuit verification, as well as the development of integrated design systems, is necessary to produce cost-effective error-free VLSI circuits. This paper presents a review of the CAD techniques which have been used in the design of IC's, as well as a number of design methods to which the application of computer aids has proven most successful. The successful application of design-aids to VLSI circuits requires an evolution from these techniques and design methods.

I. INTRODUCTION

THE NUMBER of components which can be implemented on a single-chip integrated circuit (IC) has increased rapidly in recent years [1]. As a result, new approaches to design and verification are necessary for the effective use of very large scale integrated (VLSI) circuits. The evolution of computer aids for IC design has occurred in an ad hoc manner. In most cases, computer programs have written to solve specific problems as they have arisen and very few truly integrated computer-aided design (CAD) systems exist for the design of IC's. Most CAD systems currently in use consist of a loose collection of programs, requiring a large collection of data formats and often requiring manual intervention to move from one program to another.

The use of a particular class of circuit structures is referred to as a *design method*, or design style, and while the development of new algorithms and techniques for CAD continues, the most significant contribution to the design of VLSI circuits will come from the development of new circuit design methods. However, while the implementation of a design method does not *require* the use of computer aids, the most successful design methods will be those designed to take maximum advantage of the computer in both the circuit design and verification phases. The design method must provide the *structure* necessary to use both human and computer resources effectively. For VLSI, this structure also provides the reduction in design complexity necessary to reduce design time and to ensure that the circuit function can be verified and the resulting circuit can be tested. In describing the variety of computer-aids used for IC design, a distinction is made between those techniques used for *design*, or synthesis, of the IC and those techniques used for its *verification*. In both of these categories, a further distinction is made between techniques relating to the *physical*, or topological, aspects of the design process, such as mask layout or the placement of components in a circuit, and *functional* considerations, such as logic description, synthesis, simulation, and test-pattern generation.

Computer aids for design, or synthesis, at both the functional and physical levels, are primarily concerned with the use of *optimization* to improve performance and cost. These design tasks may be formulated as combinatorial optimization problems for operations such as cell placement, routing, logic minimization, and logic state assignment, or as parametric optimization problems for operations such as design at the electrical level. These optimization problems are often too complex to solve directly. Therefore, *partitioning* is often used to reduce the problem to a set of simpler subproblems. The solutions of these subproblems are later combined in a separate step. Both the partitioning task and the solution of each subproblem generally involves the use of heuristics to reduce the complexity further.

This paper presents a review of the CAD techniques which have been used in the design of IC's, as well as a number of design methods to which the application of computer aids has proven most successful. The successful application of design-aids to VLSI requires an evolution from these techniques and design methods. The organization of this paper follows the historical evolution of CAD. Following a brief historical review in Section II, Section III describes the variety of data representations used in the design of VLSI circuits. Sections IV–VII review CAD techniques in use and under development for the design and verification of basic circuits, or cells, circuit building blocks which may consist of many cells, and integrated systems, which may consist of a number of building blocks. After a summary of the major ideas described in this paper, the need for a unified approach to IC design and verification is described in Section VIII.

II. HISTORICAL REMARKS

The first digital IC's were available commercially in the early 1960's. However, it was a number of years before computer-aids were applied to the design and verification of these circuits. In retrospect, it is surprising how little the computer has been used in the design of IC's. Early circuits were sufficiently small that mask patterns could be drawn by hand on rubylith, and then photographically reduced to generate the IC masks directly. However, for the verification of the *function* of the circuit, simulators proved quite useful. Hence initial work in the mid-1960's focused on the development of device analysis [2], [3] and circuit analysis [4]–[9] techniques. These circuit simulators were originally developed for the analysis of nonlinear and

Manuscript received April 21, 1981; revised May 27, 1981.
The author is with the Department of Electrical Engineering and Computer Sciences, University of California, Berkeley, CA 94720.

Reprinted from *Proc. IEEE*, vol. 69, pp. 1189–1199, Oct. 1981.

radiation effects in discrete circuits and it was not until the early 1970's that circuit simulators suitable for IC analysis became generally available [10]–[17].

As the complexity of the circuits increased, industry turned to the computer to store IC layout data and to produce the masks required for manufacture. Systems for layout digitization and interactive correction found extensive use by the early 1970's. However, it was not until the mid-1970's that programs for the physical layout rule checking (LRC) of the circuit began to find widespread use [18]–[20].

By 1975 it had become clear that computer-aids were a necessity in the design of complex IC's, both for physical and for functional design and verification. Until then, the layout of an IC and its transistor-level schematic diagram had been quite separate. In the late 1970's, computer programs became available for such tasks as connectivity verification [21], extraction of transistor-level schematics from IC artwork data [22]–[25] and even extraction of gate-level schematics from the transistor list [26]. These programs are loosely coupled in general and are often incompatible with one another. In fact, the *only* integrated CAD systems that exist today for the design of complex IC's are those for some highly specialized design approaches, such as standard cell and gate arrays.

In parallel with the development of computer-aids for IC design, a great deal of work has been done to aid the digital system designer, particularly as applied to printed circuit (PC) board design using standard components. In particular, algorithms and programs for the optimal placement and routing of cells [72], [73], logic simulation techniques [27]–[29], and test grading [30], [31] have resulted in sophisticated design packages.

III. REPRESENTATIONS OF THE DESIGN

Throughout the IC design process, a variety of different representations or *views* of the design are used. These representations may reflect a particular level of abstraction, such as the functional specification of the circuit or its mask layout, or they may reflect the view required for a certain application, such as the information required for simulation. The choice of appropriate representations for each level of the design process is a key factor in determining the effectiveness of computer aids since it is via these representations that both the structure of the design as well as specific information relating to a particular design level are expressed. The design process then involves transformations between these representations, both for design and verification. In this section, a classification of representations is presented. This classification is used in the following sections to relate different design aids. One particular representation which promises to provide a significant improvement in the design process over the next few years is symbolic layout. For this reason, a number of symbolic layout representations are described in more detail at the end of the section.

While the particular set of representations used in a design depends on the particular design approach being used, the major categories may be defined as shown in Fig. 1. These representations fall into three major categories: behavioral, schematic, and physical. At the behavioral or algorithmic level, functional intent of the design is described independent of a particular implementation. In some cases, programming languages such as concurrent Pascal [99] have been used to represent the design at this level, as well as providing a simulation

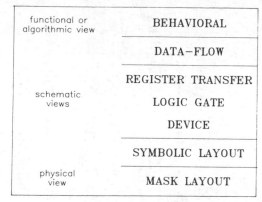

Fig. 1. Representations of the IC design process and their classification.

capability. Languages specifically designed for this task have also been developed [114], [116], [117].

Once a functional implementation strategy has been determined, a schematic view may be generated. At its most abstract level, this schematic view consists of a *chip plan*, illustrating the loose physical placement of the major components and busses, or a register transfer level (RTL) description, defining the functional relationships between the major components of the design. As the implementation is refined further, logic gate level and finally transistor level schematics may be generated. While the nature of the information contained at each level is different, each more detailed view may be considered a different level of "zooming in" on the implementation. With each new level of refinement more information concerned with the detailed physical and functional implementation of the circuit is included in the description. The final transformation consists of the generation of detailed, mask-level geometries from a device-level schematic view.

The transition between functional schematic descriptions and between schematic and mask layout may involve the use of additional views. At the higher level, a data-flow description of the circuit function may serve this purpose, as shown in Fig. 1. At the behavioral level, this description may be viewed as the parse tree generated by a compiler operating on the algorithmic description of the intended function. At the RTL level nodes in the data-flow graph represent an initial configuration of circuit building blocks used to implement the function and branches indicate data paths to- and from- these functions.

At each level, these descriptions must express the structure of the design in such a manner that it can be exploited by the design aids. In particular, *regularity* and *hierarchy* must be exploited. For example, regularity in the form of one- to two-dimensional arrays of similar, e.g., RAM, or iterated, e.g., ROM, components can reduce the design time since only a small number of basic circuit types need be designed by hand. The verification time is reduced also since only one example of each possible spatial combination of this small number of cells need be verified to certify the entire array. Hierarchy can aid the verification process in a similar manner. The components of a circuit block, such as the logic gates used to implement an arithmetic-logic unit, need only be checked in detail once. When the composition of these cells is checked, only the relationships *between* the cells need be verified. A detailed check of the internals of the cells is not necessary. If these cells are used a number of times, this process can provide substantial

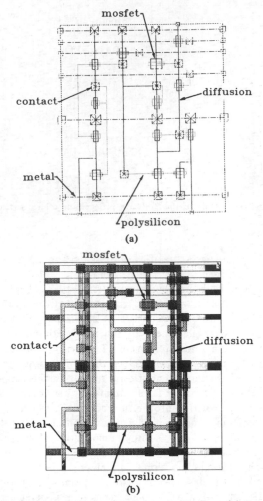

(a)

(b)

Fig. 2. (a) Loose symbolic layout of a simple flip-flop. (b) Corresponding uncompacted mask-level layout.

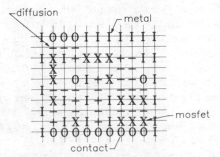

Fig. 3. Simple-grid symbolic layout of an IC cell.

mined by a set of dominant layout rules in such a way that it is impossible for user input to violate spacing rules. A conversion program may then be used to replace each symbol in the user input with its equivalent mask layout description. The compactness of the resulting layout depends on the quality of the user input, since spaces in the layout left by the user are not removed, and the coarseness of the grid. A simple ASCII format can be used to represent fixed-grid symbolic layout, as shown in Fig. 3 [34].

The second type of symbolic layout representation is called *relative grid* since the loose layout plan indicates only the relative placement and interconnection of element symbols [36]–[39]. Fig. 2(a) is an example of a graphical view of such a relative-grid approach.

Once an appropriate set of representations for a particular design method has been determined, it is essential that a self-consistent approach be used to maintain the data for each design.

IV. DEVICE LEVEL

As IC technology scales down to ever-smaller feature sizes, new processing techniques and the circuit devices constructed in the new processes must be characterized accurately for circuit design. If a process has been developed, it may be characterized by extensive measurement. However, if a process is to be designed for specific device characteristics, computer aids can play a major role. For a clear understanding of the process, not only is accurate modeling and analysis of subsilicon surface effects essential, but the characteristics of the lithographic process itself becomes critical as the fundamental limits of mask making, exposure, and pattern-etching technologies are approached. In this section, a number of techniques used for the analysis and design of MOS processes and devices are described. Fig. 4 lists the type of programs used most often for these tasks.

Process simulators, such as the SUPREM program [40], provide one-dimensional impurity profiles and oxide thicknesses for any sequence of process steps, such as oxidation, ion implantation, diffusion, and epitaxial growth. These programs are finding widespread use while work continues to develop more accurate models for each step in the IC manufacturing process.

With this decrease in device size there is also an increasing need to simulate two-dimensional processes, such as line-edge profiles resulting from the various lithography, growth, and etching processes used during IC manufacturing. Programs such as SAMPLE [41], [42] can be used to simulate and model profiles for lithography and etching. SAMPLE allows the user to mimic a large variety of lithographic and etching processes,

savings in computer time. Circuit structure can also be exploited in other areas, such as simulation, circuit synthesis, and testing, as described in the remaining sections of the paper.

Symbolic layout forms a bridge between a schematic view of the circuit and its mask-level layout. A symbolic layout contains explicit connectivity information as well as the relative placement of circuit components, such as transistors, to form a basic circuit cell, cells to form a building-block, and building-blocks to describe the entire circuit. At the transistor level, the symbolic layout is often called a *stick diagram* since interconnections are represented by their center lines and hence resemble sticks. Fig. 2(a) shows the uncompacted, symbolic layout of a cell and Fig. 2(b) shows its corresponding mask level representation [32]. One of the key advantages of a symbolic layout is its ability to maintain explicit electrical connectivity information through to the mask level descriptions. Not only can symbolic layout be used to aid the verification of the circuit, but by separating layout-sensitive cells and interconnections, computer programs can be used to optimize the area utilization of the circuit by modifying only the noncritical interconnections. This is described in Section IV.

There are two common approaches to the use of symbolic layout. In the *fixed grid* or coarse grid method [33]–[35], the layout is drawn on spaced grids. The grid spacing is deter-

	FUNCTIONAL	PHYSICAL
DESIGN	circuit simulation schematic capture	layout digitization symbolic layout compaction
VERIFICATION	schematic extraction electrical-rule check circuit simulation schematic comparison	layout—rule check connectivity verification

Fig. 4. Computer aids used in the design and verification of IC devices and technologies.

	FUNCTIONAL	PHYSICAL
DESIGN	device analysis parasitic analysis interconnect analysis	process simulation lithography simulation process optimization
VERIFICATION	device analysis parasitic analysis interconnect analysis	process simulation lithography simulation

Fig. 5. Computer aids used in the design and verification of basic circuit cells.

including oxide etching, sputter etching, evaporation, and epitaxial growth, and provides cross sections of the line-edge profiles at various stages in the processing.

Once a set of physical processing steps has been established for the design of a circuit device, such as an MOS transistor, the electrical characteristics such as transconductance, breakdown voltage, and parasitic capacitance values must be determined. One, two, or three-dimensional semiconductor device analysis programs can be used here. Most of these programs use finite-element or finite-difference techniques to solve for the static [43] or dynamic [44] characteristics of the device. It is clear that there is a strong interaction between the processing steps and the device characteristics. Recently, work has been carried out on the coupling of process and device analysis to allow the process to be optimized for required device characteristics [45].

A further result of the reduction in size of IC geometries is that parasitic and distributed effects become increasingly significant. Numerical analysis techniques such as finite-difference and finite-element methods may also be used here to determine the electrical characteristics of complex geometrical patterns [46]. These analyses can then be used to produce equivalent electrical models for this geometry [47]. It may soon be necessary to consider detailed coupled-line effects and microstrip propagation in the design of these circuits.

V. BASIC CELL

Once the IC process has been characterized, either by computer analysis or by measurement, basic circuit cells can be designed and verified. These cells may be simple logic gates, such as NAND, NOR, or flip-flops. Alternatively, they may be cells used in regular arrays, such as the one-transistor cells used in a programmable logic array (PLA), which do not perform a complete logic function alone. Computer aids for basic cell design are listed in Fig. 5 and these programs are described in the following.

A. Layout Entry

The most commonly used aid for the physical design of a cell is mask-level digitization and interactive correction program. The user draws the mask layout of each mask layer and enters it into the computer via a digitization process. Once the layout has been entered, it may be plotted and then edited on an interactive graphics display. In some cases the user may enter the initial mask information via the graphics editor directly.

The output from such programs is a file suitable for driving a computer-controlled mask pattern generator. Some of these programs provide rudimentary error detection on input, such as checking for undersized geometry on a particular mask layer or checking the local physical layout rules are not violated as each new piece of geometry is entered [48].

Some programs allow direct symbolic layout entry, using either fixed-grid [33]–[35] or relative-grid [36]–[39] schemes. With the fixed-grid symbolic approach, the grid is designed to ensure all basic layout rules are satisfied upon data entry. For relative-grid schemes, it is necessary to modify the layout such that all layout rules are satisfied. Programs which carry out this operation are often referred to as *compaction* programs since they also attempt to reduce the area occupied by the circuit.

B. Layout Compaction

Once a symbolic layout has been entered into the computer, it may be compacted by adjusting the size of noncritical components, such as interconnections, under the constraints imposed by the physical and electrical layout rules of a given technology. The FLOSS program [36], developed at RCA for the compaction of circuit cells, paved the way for the development of transistor-level compaction programs for IC's. The algorithms used for compaction generally perform x and y axes iterative compaction steps until all layout rules are satisfied and no further area reduction can be achieved. Local modifications to the layout can then be performed to allow further compaction. These modifications generally consist of distortions to interconnect, such as the introduction of "jogs" or "doglegs," or the rotation of transistors and cells [32], [37]. Critical path algorithms and force-directed heuristics are used to determine the best location for the introduction of these layout modifications. To ensure the layout is least sensitive to processing tolerances, noncritical components must then be placed midway between constraints to maximize yield. At each compaction step, a layout-rule analysis must be performed to determine the values of geometrical constraints on the layout. This process is relatively expensive and determines the time complexity of the compaction process. The compaction time for very large layouts can become prohibitive.

The use of an hierarchical description of the circuit can be exploited to reduce the analysis time by compacting the cells independently. The resulting compacted cells may then be combined and compacted to form the circuit. While this may not result in an optimal area utilization, the primary objective

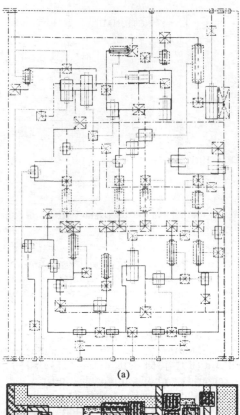

(a)

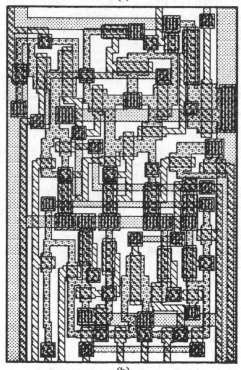

(b)

Fig. 6. (a) Loose symbolic layout of a latch-buffer cell in NMOS depletion load technology. (b) Mask layout of the cell after compaction.

of error-free layout is achieved. Fig. 6(a) shows the loose, symbolic layout of an IC building-block and Fig. 6(b) shows its
physical layout after compaction with the CABBAGE compaction program [32].

C. Circuit Simulation and Modeling

When accurate circuit models are available, circuit simulators
provide precise electrical information, such as frequency response, time-domain waveforms, and sensitivity information,
about the circuit under analysis. The majority of circuit simulators currently in use contain models for a wide range of active
devices and hence are largely independent of technology. For
this reason, these programs must employ general algorithms
for the solution of the set of coupled, nonlinear, ordinary differential equations which describe the integrated circuit and
hence cannot exploit the special characteristics of a particular
technology.

Without models whose accuracy is well matched to the
expected accuracy of a simulation, the results of the simulation
may not reflect the performance of the circuit under analysis.
Recent work on modeling for MOS circuit simulation [49]–
[52] has focussed on the development of empirical models for
MOS transistors which predict the characteristics of the devices
accurately without requiring large amounts of computer time.
With small geometries, signal delays and signal degradation
caused by interconnect can dominate circuit operation. For
this reason, explicit models for interconnect are necessary for
accurate simulation. The parameters of such models may be
provided by the designer interactively or by design programs
directly.

Most circuit simulators are batch-oriented programs and the
input to the program consists of a textual description of the
transistors and their interconnections. In some cases, an interactive graphics editor is used to capture the schematic diagram
and provide simulator input. This description can also be used
for comparison with a schematic diagram extracted directly
from the layout, as described later.

D. Layout-Rule Check

If hand-generated mask-level layout is used for cell design,
compaction programs cannot be used and LRC programs are
necessary to verify that all of the physical layout rules, such
as minimum spacing, minimum size, and minimum enclosure
constraints, are satisfied. Since the introduction of such programs in the mid-1970's [20], considerable improvements have
been achieved in time and memory requirement complexity
[53], [54]. However, if unconstrained polygonal shapes are
permitted in the layout, the computer run times of LRC programs on circuits containing over 10 000 transistors become
prohibitive. LRC programs are under development which
exploit the structure of a circuit, in particular the circuit hierarchy, to reduce the cost of LRC for complex circuits [42].

E. Layout Analysis

If a transistor-level schematic description of a circuit is available, either from simulator input data or a graphical entry system, a comparison between the circuit elements and their
connections in the schematic and the corresponding geometry
in the layout can be performed. A layout extraction program
is required to extract a schematic diagram from the circuit layout and a connectivity verification, or electrical rules check,
program can be used compare it with the intended schematic
captured earlier [21]–[24]. The extraction program can also
be used to determine parameter values for simulation, such as
device and interconnect capacitance, the effective width and
length of MOSFET channels, and the area of source and drain
diffusions. The extracted schematic can also be checked for
simple electrical rule violations, such as a short circuit between
power supply and ground or an isolated transistor, without the
need for a full schematic diagram entered separately. A large

number of errors can be detected by connectivity verification and extraction before any simulations are performed.

VI. BUILDING BLOCKS

Computer aids can be used to combine the cells described earlier and form more complex building blocks for VLSI. While many design methods are used for the implementation of these building blocks, the methods can be classified in four categories: programmable arrays, standard-cell, macrocell, and procedural design.

A VLSI circuit may consist of one large building block or it may consist of a number of building blocks combined either manually, using the techniques described in the previous section, or by a computer program. After a brief description of each of these design approaches, a number of computer aids used for the design and verification of IC building blocks are described in this Section. These programs are listed in Fig. 7.

A. Design Methods

A programmable array is a one- or two-dimensional array of repeated cells which can be customized by adding or deleting geometry from specific mask layers. Since a number of processing steps are completed prior to customization, the locations of components on those layers are independent of a particular circuit implementation. Examples of programmable arrays include the gate array, storage-logic array, PLA, and ROM.

The *gate array* (also referred to as master-slice, or uncommitted logic array) is by far the most common programmable array designed by computer. In this approach, a two-dimensional array of replicated transistors is fabricated to a point just prior to the interconnection levels. A particular circuit function is then implemented by customizing the connections within each local group of transistors, to define its characteristics as a basic cell, and by customizing the interconnections between cells in the array to define the overall circuit. Generally a two-level interconnection scheme is used for signals and, in some approaches, a third, more coarsely defined layer of interconnections is provided for power and ground connections. The interconnections are implemented on a rectilinear grid in the *channels* between the cells. In many cases, channels are also provided which run over the cells themselves and in some arrays, wider channels are provided in the center of the array to alleviate the congestion often found in that area.

Gate arrays are used in many technologies, in particular bipolar and CMOS, and arrays containing many thousands of gates have been used [56], [57]. In the storage logic array (SLA) approach [58], each "gate" consists of a storage element (flip-flop) and a small, uncommitted PLA. This design method has considerable potential for VLSI but effective design-aids for the synthesis of logic functions in SLA form are not yet available.

PLA's may also be used to implement building blocks directly, with storage elements in the feedback path to implement sequential logic in the classical Moore or Mealy style [59]. The PLA consists of a number of transistor arrays which implement logic AND and OR operations. In MOS technology, NOR arrays are used [60]. A conventional PLA consists of two arrays of cells: an input, or lookup, plane followed by an output plane. A *folded* PLA may use an additional plane, since rows and/or columns in the structure may be shared by more than one circuit variable, as described later.

	FUNCTIONAL	PHYSICAL
DESIGN	circuit simulation timing simulation logic simulation mixed—mode simulation logic synthesis testability analysis	cell placement routing direct layout synthesis
VERIFICATION	circuit simulation timing simulation logic simulation mixed—mode simulation gate extraction schematic comparison critical path analysis fault coverage analysis test generation	layout—rule check connectivity verification

Fig. 7. Computer-aids used in the design and verification of circuit building blocks.

The *standard cell* (or polycell) approach refers to a design method where a library of custom-designed cells is used to implement a logic function. These cells are generally of the complexity of simple logic gates or flip-flops and may be restricted to constant height and/or width to aid packing and ease of power distribution. Unlike the programmable array approach, standard cell layout involves the customization of all mask layers. This additional freedom permits variable width channels to be used. While most standard cell systems only permit intercell wiring in the channels between rows of cells or through cells via predetermined "feedthrough" cells, some systems permit over-cell routing. Standard cell systems are also used extensively in a variety of technologies including bipolar and CMOS [61], [62].

It is often relatively inefficient to implement all classes of logic functions in a single design approach. For example, a standard cell approach is inefficient for memory circuits such as RAM and stack. In the *macrocell* method, large circuit blocks, customized to a certain type of logic function, are available in a circuit library. These blocks are of irregular size and shape and may allow functional customization via interconnect, such as a PLA or ROM macro [63], or they can be parameterized with respect to topology as well [64]–[66]. With the parameterized cell, the number of inputs and outputs may be parameters of the cell. In some systems macrocells may also be embedded in gate array or standard-cell designs.

All of the design methods described above may be classified as *data driven*. That is, a description of the required logic function, in the form of equations or an interconnection list, is used as input to a software system which interprets the data and generates the final design. Techniques have been developed recently which can be classified as *procedure*, or program, driven [64]–[66]. In this case, each design is described by a set of procedures and the execution of those procedures is required to generate the design. While this technique provides a flexible and powerful design paradigm, sophisticated verification and debugging techniques must be developed to support it. The use of conditional circuit assembly and loop constructs within the design specification make this verification difficult at other than the final geometrical level.

B. Placement and Routing

All of these techniques involve the choice of an optimal cell *placement* and the interconnection, or *routing*, of the cells to form the final circuit.

The placement problem involves the assignment of specific

locations to building blocks of the layout. This includes the assignment of logic gates within a gate array [67], the placement of cells in a standard cell layout [68], [69] or the placement of macrocells in a nonuniform building-block approach [71], [72]. While a considerable amount of theoretical work has been developed in this area [72], [73], the most successful approaches involve the use of simple, force-directed heuristics and pairwise interchange. Partitioning may be used here to improve the final result by first clustering cells, placing the clusters, and then assigning the cell placements within clusters. In these approaches, either total interconnect length or estimated or exact routing requirements are used to drive the placement algorithms.

Once a cell placement has been determined, either by hand or by computer, an optimal channel routing must be determined. A channel router solves the problem of routing a specified list of nets, or multiport interconnections, between rows and/or columns of terminals across a multilayer channel. In most cases, two layers are available for signal interconnections. Nets are generally routed with horizontal segments on one layer and vertical segments on the other. One objective of channel routing is the minimization of the number of tracks needed to route the nets. Unfortunately, solving the resulting optimization problem exactly, perhaps with a branch-and-bound method [74], may require enormous computing resources for practical problems of even moderate size. This problem can be overcome using heuristic algorithms which generate optimum or near-optimum solutions in a reasonable amount of computer time. As an example, [74] describes new algorithms which allow fast routing for single-channel problems with and without track-sharing by multiple nets.

It is clear that in a methodology which permits only preallocated, fixed-width channels, such as the gate array approach, it is advantageous to determine *a priori* whether a circuit function can be implemented on a given master template and if so what circuit density can be achieved. While techniques are not available to solve this problem exactly, without actually routing the circuit, *estimation* techniques can be used to predict the feasibility of the placement and routing tasks. Estimation algorithms are based on the assumption of equal sized blocks and uniform channel widths [75], or more recent work has extended these techniques to unequal size logic blocks and channels [76].

C. Logic Synthesis

While the above placement and routing techniques involve physical optimization at the topological level, it is also possible to perform optimization at the functional level. Logic synthesis techniques, which carry out this optimization, may be applied to both combinational and sequential circuits.

The generation of a minimal, multiple-output combinational logic implementation of a set of Boolean equations requires the use of heurisitc techniques. For two-level logic, a number of successful programs exist and have been applied extensively to the design of such structures as the PLA [77], [78]. For multilevel logic minimization, a number of heuristic algorithms have been developed for, and applied to, specific classes of problems, such as the implementation of multilevel multiplexor-based logic [79].

The problem of sequential logic design can be divided into the tasks of determining an optimal partitioning at the state-machine level, state assignment, and optimal combinational

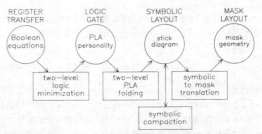

Fig. 8. Data representations (circles) and design aids (boxes) used in the translation of a set of Boolean logic equations to an optimized two-level folded PLA implementation.

logic design techniques. While considerable research has been done in terms of general algorithms for use in this area, these techniques have not yet been successfully applied to the design of integrated digital systems.

Clearly there is often a strong interaction between the topological design of a circuit and its functional counterpart. For example, it may be better to leave additional product terms in a PLA if the resulting increase in sparsity can be used more effectively to increase the overall density of the PLA implementation. A folded PLA is a PLA in which a number of circuit variables share the same row or column in the PLA [80]. In a two-level folding scheme, this is achieved by "splitting" the row or column electrode [81]. This is illustrated in Fig. 9(d) for column folding where the circuit variables A and B share the same input column. Fig. 8 indicates the steps involved in translating from Boolean equations to the IC implementation of a two-level folded PLA and Figs. 9(a)–(d) show the representations of a simple PLA at each stage of the synthesis process. If the objective function of the logic minimizer is simply to provide a minimal set of product terms, the resulting PLA may not be well suited to the folding process. However, if the logic minimization heuristics attempt to increase the sparsity of the PLA with folding in mind, as well as reduce the number of the product terms, the result after folding may be more area-efficient.

D. Simulation

Circuit simulation techniques can provide accurate waveform analysis for circuits of building-block complexity. However, as circuit size increases the time and memory requirements of a circuit simulation become prohibitive. On an IBM 370/168 computer, the average cost of a SPICE [16] analysis is 6 ms/device/clock/timepoint. For a 10 000 device circuit, with 3 clocks and for an analysis of 10 μs at 1 ns steps, the computation time would be in excess of 20 computer days! Nevertheless, the success of circuit simulation in design evaluation has been such that designers wish to continue to simulate large circuits at the level of accuracy provided by this type of program.

By applying node tearing techniques [82], [83] to the interface between cells in the circuit, inactive cells can be bypassed during the equation solution phase. However, it is anticipated that these techniques alone will provide less than an order of magnitude speed improvement. This is not sufficient improvement in performance to permit cost effective device-level analysis of VLSI circuits.

If simulation algorithms are tailored to specific technologies or applications substantial speed improvements can be achieved. Many components of digital MOS or I^2L circuits can be con-

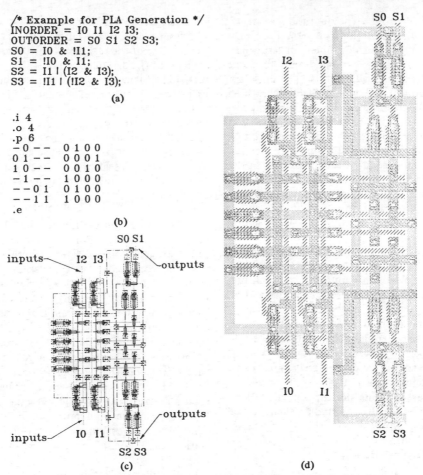

```
/* Example for PLA Generation */
INORDER = I0 I1 I2 I3;
OUTORDER = S0 S1 S2 S3;
S0 = I0 & !I1;
S1 = !I0 & I1;
S2 = I1 | (I2 & I3);
S3 = !I1 | (!I2 & I3);
```

(a)

```
.i 4
.o 4
.p 6
-0--    0100
01--    0001
10--    0010
-1--    1000
--01    0100
--11    1000
.e
```

(b)

(c)

(d)

Fig. 9. (a) Boolean equations for simple combinational logic function. (b) PLA personality after logic minimization. (c) Symbolic layout after topological minimization (folding). (d) Final layout of an NMOS implementation of the PLA.

sidered unilateral in nature. This characteristic, as well as the facts that these families are saturating and hence accumulated voltage errors are lost at the extremes of signal swing, and that large digital circuits are relatively inactive at the gate level, are exploited in *timing simulation*. Timing simulators [84]-[86] can improve simulation speed by up to two orders of magnitude while maintaining acceptable waveform accuracy. These savings are achieved by using node decoupling techniques in conjunction with simplified lookup table models for nonlinear devices.

Where a library of cells is used during the design, or when a group of transistors is used to implement a common function, such as a cell or logic gate, it is often possible to exploit the known structure of the circuit and use a simplified representation which maintains the essential characteristics of the cell at reduced computational expense. Such a reduced representation is called a *macromodel* [87]-[89] and macromodels are used in both circuit and timing level analysis.

The speed of timing simulation is insufficient for the analysis of large circuit blocks. Gate-level logic simulators have been used for over twenty years in digital system design [90] however it was not until the late 1970's that logic simulators applied to IC simulation were capable of dealing with many of the problems associated with MOS logic gates [91]. Logic simulation is considerably faster than circuit simulation since the evaluation of a new logic state involves only Boolean operations, the fastest operations available on most computers, and the logic simulator deals with logic gates directly rather than the

details of their transistor-level implementation. In a logic simulator, the signals propagated between gates are no longer voltages or currents but rather they are discrete logic states [92]-[94]. The most successful MOS logic simulators use as many as nine static logic states to describe the terminal characteristics of a gate or bus [52], [91].

In some simulations it may be necessary to obtain the waveform accuracy of circuit simulation for parts of the circuit while in other parts of the circuit, a more efficient gate-level logic analysis is sufficient. In this situation, *mixed-mode* analysis can be used [95]-[98]. In a mixed-mode simulator, different levels of abstraction are combined in the single analysis. Circuit, timing, and logic analyses [95]-[98] or even behavioral and RTL analyses [99] can be combined in a single program.

E. Testing

With the increasing complexity of IC's, techniques for the cost-effective evaluation of whether a circuit has been manufactured correctly becomes critical to the success or failure of a design [100]. One approach to this problem is to determine an adequate set of tests, after the circuit has been designed, which can qualify it as a functioning part. Even for LSI circuits, the ability to generate test patterns automatically and perform fault simulation is almost impossible unless the testability issue has been considered during the design phase. While a number of ad hoc techniques have been used to ensure cost-effective testability, two very successful and somewhat similar structured approaches have evolved. They are the *level sensitive scan design* (LSSD) [101] and the *scan path* [102] approaches. These techniques reduce the problem of test generation to one of generating tests for combinational logic, for which powerful algorithms are available [105]. Both approaches involve the addition of circuitry to the design, which may increase manufacturing cost and can reduce performance, but for a major reduction in testing cost. A recent extension of these approaches involves the use of *built-in logic block observation* (BILBO) [103]. The BILBO approach uses the separation of combinational and sequential functions, as in the LSSD and scan path approaches, combined with on-chip linear feedback registers to generate input patterns for a signature analysis [104] form of circuit evaluation.

If an ad hoc approach is used, then programs are available which provide a *testability measure* [106, 107] of the design. These programs are quite fast and provide the designer with a heuristic measure of how "difficult" it may be to observe or control nodes in the circuit. While this approach is useful, further research is required to calibrate these techniques and to extend the heuristics.

Once the circuit has been designed, test patterns must be generated which provide an adequate cover of all possible faults. If a structured design approach was used, or if only combinational logic is involved, this task is relatively inexpensive and the required test sequences are generally quite small. If testability was not considered during design, a great deal of fault simulation [108] will be required to find an adequate set of test patterns, if they can be found at all.

VII. System Design

Very few design aids are available to assist the IC designer at the system level. Design at this level involves the translation of a required behavioral-level specification into a register-transfer-level implementation. Once the functional partitioning of the design is completed, estimates of the layout size, power-supply requirements, and speed of the high-level circuit blocks used to implement the various sub-functions are required. A chip-plan must also be constructed to determine the relative placement of these building blocks. This chip plan is then further refined as the design proceeds. These tasks are often performed manually, perhaps with the computer used to perform bookkeeping tasks such as the storage of interconnection data. Research is in progress to develop aids for the designer in this area.

The Carnegie-Mellon University's Design Automation Project [109] aims at the evaluation of a set of high-level tradeoffs involving such parameters as speed, cost, and power supply requirements as a function of *design style*. For example, increasing the priority for speed of the circuit would result in a new design, probably using more parallelism. Once a satisfactory design style has been selected [110], the system will continue through an *allocation* [111] and *module binding* phase [112], where library components are allocated to implement the building blocks of the design, and finally to produce mask layout. Initial results for a restricted class of implementations look promising.

Recent work in the use of constraint-based design [113] has also shown some significant results. In this approach, the relationships between the components of a circuit are defined by a set of *incremental constraints*. These constraints may be physical, such as the required minimum spacing between mask geometries, as well as electrical, such as the size of a bus as a function of its electrical context. A computer program can then act as the *designer's assistant* and remember the reasons for design choices as well as perform simple deductions with them. One of the tasks controlled by the assistant is the maintenance of the design constraints.

Digital system-level simulation can be performed with the use of register transfer level simulators [114], [115]. These simulators generally operate at the synchronous machine level and hence they cannot be used to detect assynchronous timing problems. As mentioned in Section II, behavioral-level simulation can also be performed using programming languages, such as concurrent Pascal [99], as well as special-purpose languages and associated simulators [116], [117].

VIII. Concluding Remarks

In the design and verification of VLSI circuits extensive use of computer aids is required. The effective use of these computer programs necessitates structured design approaches so that the complexity of the design and verification tasks is reduced to a manageable level. Nevertheless, large amounts of data and a variety of design representations will be used for each circuit. For this reason, it is important that an integrated set of computer aids, coupled with a unified approach to data management, be provided to the IC designer [118].

With the decreasing cost of hardware, it also seems apparent that a dedicated special-purpose *design station* can be provided for VLSI designers on a per-designer basis. It is anticipated that such a design station would have the computation power of a high-performance minicomputer of today, coupled to a high-resolution color graphics display, and interfaced to other design stations and common resources via a high-speed computer network.

In the past, design software has often been tailored to a specific design style. To employ a different approach to design would then require a complete rewrite of the software in the design system. As is clear from the previous sections of this paper, an optimal design methodology for VLSI is not known at this time. In fact, it is clear that different design approaches are necessary for different applications. A design style suitable for low-volume custom circuits would not suit the high-volume general-purpose component designer. Hence it is necessary that the underlying operating system software on such a design station be configured to support a variety of different design methods without the need to rewrite common components used in a number of differing approaches. This software framework would involve the use of flexible data management aids which understand such structural metaphors as *hierarchy* and *regularity*. Other key requirements of such a system are the ability to apply integrity constraints to the various data representations, concurrency control to allow multiple access to the data, and incremental update so that the system can be "backed up" to an earlier version of the design. Early experiments with the use of general-purpose data-base systems for VLSI have met with limited success [119, 120] and further research is required in this area.

VLSI concerns the implementation of systems in silicon and the skills required in the design of a single chip span the entire range, from technology to system architecture. The most successful solutions to the VLSI design problem will come from a synergism of skills in all areas.

Acknowledgment

The author wishes to thank D. O. Pederson and A. L. Sangiovanni-Vincentelli for their continuous encouragement and criticism during the development of this paper. He also gratefully acknowledges the assistance of R. S. Tucker for his careful review of the manuscript.

REFERENCES

[1] W. Lattin, "VLSI design methodology: the problem of the 80's for microprocessor design," in *Proc. 16th Design Automation Conf.* (San Diego, CA) pp. 548–549, June 1979.

[2] H. K. Gummel, "A self-consistent iterative scheme for one-dimensional steady-state transistor calculations," *IEEE Trans. Electron Devices*, vol. ED-11, pp. 455–465, Oct. 1964.

[3] C. N. Gwyn, D. L. Scharfetter, and J. L. Wirth, "The analysis of radiation effects in semiconductor junction devices," *IEEE Trans. Nuclear Sci.*, vol. NS-14, pp. 153–169, Dec. 1967.

[4] R. W. Jensen and M. D. Lieberman, *The IBM Electronic Circuit Analysis Program*. Englewood Cliffs, NJ: Prentice-Hall, 1968.

[5] A. F. Malmberg, F. L. Cornwell, and F. N. Hofer, "NET-1 network analysis program 7090/94 Version," Los Alamos Scientific Lab., Los Alamos, NM, Rep. LA-3119, 1964.

[6] L. D. Milliman, W. A. Massera, and R. H. Dickhaut, "CIRCUS-A digital computer program for transient analysis of electronic circuits," Harry Diamond Lab., Washington, DC, Rep. AD-346-1, 1967.

[7] J. C. Bowers and S. R. Sedore, *SCEPTRE: A Computer Program for Circuit and System Analysis*. Englewood Cliffs, NJ: Prentice-Hall, 1971.

[8] E. D. Johnson *et al.*, "Transient radiation analysis by computer analysis (TRAC)," Harry Diamond Lab., Washington, DC, Contract DAA 639-68-C-0041, 1968.

[9] F. F. Kuo, "Network analysis by digital computer," *Proc. IEEE*, vol. 54, pp. 821–835, June 1966.

[10] W. J. McCalla and W. G. Howard, "BIAS-3—A program for the nonlinear dc analysis of bipolar transistor circuits," in *Dig. Tech. Papers, IEEE Int. Solid State Circuits Conf.* (Philadelphia, PA), pp. 82–83, Feb. 1970.

[11] L. Nagel and R. A. Rohrer, "Computer analysis of nonlinear circuits, excluding radiation (CANCER)," *IEEE J. Solid State Circuits*, vol. SC-6, pp. 166–182, Aug. 1971.

[12] T. E. Idleman, F. S. Jenkins, W. J. McCalla, and D. O. Pederson, "SLIC—A simulator for linear integrated circuits," *IEEE J. Solid-State Circuits*, vol. SC-6, pp. 188–204, Aug. 1971.

[13] F. S. Jenkins and S. P. Fan, "TIME—A nonlinear dc and time-domain circuit simulation program," *IEEE J. Solid State Circuits*, vol. SC-6, pp. 188–192, Aug. 1971.

[14] L. W. Nagel and D. O. Pederson, "Simulation program with integrated circuit emphasis," in *Proc. 16th Midwest Symp. Circuit Theory* (Waterloo, Ont. Canada), Apr. 1973.

[15] W. T. Weeks *et al.*, "Algorithms for ASTAP—A network analysis program," *IEEE Trans. Circuit Theory*, vol. CT-20, pp. 628–634, Nov. 1973.

[16] L. W. Nagel, "SPICE2: A computer program to simulate semiconductor circuits," Univ. California, Berkeley, ERL Memo ERL-M520, May 1975.

[17] E. Cohen, "Program reference for SPICE2," Electronics Res. Lab., Univ. California, Berkeley, ERL Memo ERL-M592, June 1976.

[18] H. S. Baird, "A survey of computer aids for IC mask artwork verification," in *Proc. IEEE Int. Symp. on Circuit and Systems* (Phoenix, AZ), Apr. 1977.

[19] L. Rosenberg and C. Benbassat, "Critic: An integrated circuit design rule checking program," in *Proc. 11th Design Automation Workshop* (Denver, CO), pp. 14–18, June 1974.

[20] Design Rule Checking System, NCA Corporation, Sunnyvale, CA.

[21] R. M. Allgair and D. S. Evans, "A comprehensive approach to a connectivity audit, or a fruitful comparison of apples and oranges," in *Proc. 14th Design Automation Conf.* (New Orleans, LA), pp. 312–321, June 1977.

[22] I. Dobes and R. Byrd, "The automatic recognition of silicon gate transistor geometries: An LSI design aid program," in *Proc. 13th Design Automation Conf.* (San Francisco, CA), pp. 414–420, June 1976.

[23] J. Le Charpentier, "Computer aided synthesis of an IC electrical diagram from mask data," in *Dig. Tech. Papers, IEEE Int. Solid State Circuits Conf.* (Philadelphia, PA), pp. 84–85, Feb. 1975.

[24] B. T. Preas, B. W. Lindsay, and C. W. Gwyn, "Automatic circuit analysis based on mask information," in *Proc. 13th Design Automation Conf.* (San Francisco, CA), pp. 309–317, June 1976.

[25] L. Szanto, "Network recognition of an MOS integrated circuit from the topography of its masks," *Computer-Aided Design*, vol. 10, no. 2, pp. 135–140, Mar. 1978.

[26] L. Scheffer and R. Apti, "LSI design verification using topology extraction," in *Proc. 12th Asilomar Conf. Circuits, Systems, and Computers* (Asilomar, CA), pp. 149–153, Nov. 1978.

[27] E. B. Eichelberger, "Hazard detection in combinational and sequential switching circuits," *IBM J. Res. Develop.*, Mar. 1965.

[28] L. Bening, "Developments in computer simulation of gate level physical logic," in *Proc. 16th Design Automation Conf.* (San Diego, CA), pp. 561–567, June 1979.

[29] S. A. Szygenda and E. W. Thompson, "Modeling and digital simulation for design verification and diagnosis," *IEEE Trans. Computers*, vol. C-25, pp. 1242–1253, Dec. 1976.

[30] T. W. Williams, "Design for testability," presented at NATO Advanced Study Institute on Computer Design Aids for VLSI Circuits, Sogesta-Urbino, Italy, July 1980.

[31] P. S. Bottorff, "Computer aids to testing—An overview," presented at NATO Advanced Study Institute on Computer Design Aids for VLSI Circuits, Sogesta-Urbino, Italy, July 1980.

[32] M. Y. Hsueh, "Symbolic layout and compaction of integrated circuits," Univ. California, Berkeley, ERL Memo UCB/ERL M79/80, Dec. 1979.

[33] R. P. Larson, "Versatile mask generation technique for custom microelectronic devices," in *Proc. 15th Design Automation Conf.*, pp. 193–198, June 1978.

[34] D. Gibson and S. Nance, "SLIC—Symbolic layout of integrated circuits," in *Proc. 13th Design Automation Conf.*, pp. 434–440, June 1976.

[35] K. Hardage, "ASAP: Advanced symbolic artwork preparation," *LAMBDA*, vol. 1, no. 3, pp. 32–39, Fall 1980.

[36] Y. E. Cho, A. J. Korenjak, and D. E. Stockton, "FLOSS: An approach to automated layout for high-volume designs," in *Proc. 14th Design Automation Conf.*, pp. 138–141, June 1977.

[37] A. E. Dunlop, "SLIP: Symbolic layout of integrated circuits with compaction," *Computer-Aided Design*, vol. 10, no. 6, pp. 387–391, Nov. 1978.

[38] J. D. Williams, "STICKS—A graphical compiler for high-level LSI design," in *Proc. AFIPS Conf.*, vol. 47, pp. 289–295, June 1978.

[39] A. D. Ivannikov and P. P. Sipchuk, "Computer-aided design of MOS integrated circuit layout," in *Proc. IEE-CADMECCS* (Brighton, England), pp. 47–51, July 1979.

[40] D. Antoniadis, S. Hansen, and R. Dutton, "Suprem II—A program for IC process modeling and simulation," Stanford Electron. Lab., Stanford, CA, Tech. Rep. 5019-2, 1978.

[41] W. G. Oldham, S. N. Nandgaodnkar, A. R. Neureuther, and M. O'Toole, "A general simulator for VLSI lithography and etching processes: Part I—Application to projection lithography," *IEEE Trans. Electron Devices*, vol. ED-26, pp. 717–723, Apr. 1979.

[42] W. G. Oldham, A. R. Neureuther, C. Sung, J. L. Reynolds, and S. N. Nandgaonkar, "A general simulator for VLSI lithography and etching processes: Part II—Application to deposition and etching," *IEEE Trans. Electron Devices*, vol. ED-27, no. 8, pp. 1455–1459, Aug. 1980.

[43] S. Liu, B. Hoefflinger, and D. O. Pederson, "Interactive two-dimensional design of barrier-controlled MOS transistors," *IEEE Trans. Electron Devices*, vol. ED-27, pp. 1550–1558, Aug. 1980.

[44] S. Selberherr, A. Schultz, and H. W. Potzl, "MINIMOS—A two-dimensional MOS transistor analyzer," *IEEE J. Solid State Circuits*, vol. SC-15, pp. 605–615, Aug. 1980.

[45] A. Doganis and R. W. Dutton, "Optimization of IC processes using SUPREM," Stanford Electron. Lab., Stanford, CA, Tech. Rep., 1981.

[46] A. E. Rhueli, "Inductance calculations in a complex integrated circuit environment," *IBM J. Res. Develop.*, vol. 16, no. 5, pp. 470–481, Sept. 1972.

[47] A. E. Rhuehli, P. A. Brennan, "A micromodel for the inductance of perpendicular crossing lines," in *Proc. IEEE Int. Symp. Circuits, and Systems* (New York, NY), pp. 781–783, May 1978.

[48] W. J. McCalla, and D. Hoffman, "Symbolic representation and and incremental DRC for interactive layout," in *Proc. IEEE Int. Symp. on Circuits, and Systems* (Chicago, IL), p. 710–715, Apr. 1981.

[49] "MOS models for micro-computer CAD," MOSAID, Inc.

[50] L. M. Dang, "A simple current model for short-channel IGFET and its application to circuit simulation," *IEEE J. Solid State Circuits*, vol. SC-14, pp. 358–367, Apr. 1979.

[51] S. Liu, "A unified CAD model for MOSFETS," Ph.D. dissertation, Univ. California, Berkeley, Dec. 1980.

[52] A. R. Newton, "Timing, logic, and mixed mode simulation for large MOS integrated circuits," presented at NATO Advanced Study Institute on Computer Design Aids for VLSI Circuits, Sogesta-Urbino, Italy, July 1980.

[53] H. S. Baird, "Fast algorithms for LSI artwork analysis," in *Proc. 14th Design Automation Conf.* (New Orleans, LA), pp. 303–311, June 1977.

[54] P. Wilcox, H. Rombeek, and D. M. Caughey, "Design rule verification based on one-dimensional scans," in *Proc. 15th Design Automation Conf.* (Las Vagas, NV), pp. 285–289, June 1978.

[55] T. Whitney, "Hierarchical design-rule checking," M. S. rep., Dep. Computer Science, California Institute of Technology, Pasadina, June 1981.

[56] Y. Horiba *et al.*, "A bipolar 2500-gate subnanosecond masterslice

LSI," in *Dig. Tech. Papers, IEEE Int. Solid State Circuits Conf.* (New York, NY), pp. 228–229, Feb. 1981.

[57] C. M. Davis *et al.*, "IBM System 370 bipolar gate-array microprocessor chip," in *Proc. IEEE Int. Conf. Circuits and Computers*, pp. 669–673, Oct. 1980.

[58] S. H. Patil and T. A. Welch, "A programmable logic approach for VLSI," *IEEE Trans. Computers*, vol. C-28, pp. 594–601, Sept. 1979.

[59] C. H. Roth, *Fundamentals of Logic Design.* St. Paul, MN: West Publ. Co., 1979.

[60] C. Mead and L. Conway, *Introduction to VLSI Systems.* Reading, MA: Addison-Wesley, 1979.

[61] B. T. Murphy *et al.*, "A CMOS 32b single-chip microprocessor," in *Dig. Tech. Papers, IEEE Int. Solid State Circuits Conf.* (New York, NY), pp. 230–231, Feb. 1981.

[62] M. Watanabe, "CAD tools for designing VLSI in Japan," in *Dig. Tech. Papers, IEEE Int. Solid State Circuits Conf.* (Philadelphia, PA), pp. 242–243, Feb. 1979.

[63] J. W. Jones, "Array logic macros," *IBM J. Res. Develop.*, pp. 98–109, March 1975.

[64] D. Johansen, "Bristle blocks: A silicon compiler," in *Proc. 16th Design Automation Conf.*, pp. 310–313, June 1979.

[65] J. B. Brinton, "CHAS seeks title to global CAD system," *Electronics*, pp. 100–102, Feb. 10, 1981.

[66] G. B. Goates, "ABLE: A LISP-based layout modeling language with user-definable procedural models for storage logic array design," M.S. Rep., Univ. Utah, Salt Lake City, Dec. 1980.

[67] M. Hanan, P. K. Wolff, and B. J. Agule, "Some experimental results on placement techniques," in *Proc. 13th Design Automation Conf.* (San Francisco, CA), pp. 214–220, June 1976.

[68] R. L. Mattison, "Design automation of MOS artwork," *Computer*, vol. 7, pp. 21–28, Jan 1974.

[69] G. Persky, D. N. Deutsch, and D. G. Schweikert, "LTX—A system for the directed automation design of LSI circuits," in *Proc. 13th Design Automation Conf.*, 1976.

[70] U. Lauther, "A min-cut placement algorithm for general cell assemblies based on a graph representation," in *Proc. 16th Design Automation Conf.* (San Diego, CA), pp. 1–10, June 1979.

[71] B. T. Preas and C. W. Gwyn, "General hierarchical automatic layout of custom VLSI circuit masks," *J. Des. Automat. Fault-Tolerant Comput.*, vol. 3, no. 1, pp. 41–58, Jan. 1979.

[72] M. Hanan and J. Kurtzberg, "Placement techniques," in *Design Automation of Digital Systems: Vol. 1, Theory and Techniques*, M. A. Breuer, Ed. Englewood, Cliffs, NJ: Prentice-Hall, 1972, Ch. 5.

[73] M. Hanan, P. Wolff, and B. Agule, "A study of placement techniques," *J. Des. Automat. Fault-Tolerant Comput.*, vol. 1, no. 1, pp. 28–61, Oct. 1976.

[74] T. Yoshimura and E. S. Kuh, "Efficient algorithms for channel routing," to be published in *IEEE Trans. Circuits and Syst.*

[75] W. R. Heller, W. F. Mikhail, and W. E. Donath, "Prediction of wiring space requirements for LSI," *J. Des. Autom. Fault Tolerant Comput.*, vol. 2, pp. 117–144, 1978.

[76] A. A. El Gamal, "Two-dimensional estimation model for interconnections in master-slice integrated circuits," *IEEE Trans. Circuits, and Syst.*, vol. CAS-28, pp. 127–137, Feb. 1981.

[77] S. J. Hong, R. G. Cain, and D. L. Ostapko, "MINI: A heuristic approach for logic minimization," *IBM J. Res. Develop.*, vol. 18, no. 5, pp. 443–458, Sept. 1974.

[78] The PRESTO program was developed by A. Svoboda and D. Brown, Tektronix, Inc.

[79] E. Porter, STC Microtechnology Corp., Private Commun.

[80] R. A. Wood, "A high density programmable logic array chip," *IEEE Trans. Comput.*, vol. C-28, pp. 602–608, Sept. 1979.

[81] G. D. Hachtel, A. L. Sangiovanni-Vincentelli, and A. R. Newton, "Some results in optimal PLA folding," in *Proc. IEEE Int. Conf. Circuits, and Computers*, pp. 1023–1028, Oct. 1980.

[82] A. L. Sangiovanni-Vincentelli, L-K. Chen, and L. O. Chua, "A new tearing approach—Node-tearing nodal analysis," in *Proc. IEEE Int. Symp. Circuits and Systems*, pp. 143–147, Apr. 1977.

[83] P. Yang, "An investigation of ordering, tearing, and latency algorithms for the time-domain simulation of large circuits," *Tech. Rep.* R-891 Univ. Illinois, Urbana, Aug. 1980.

[84] B. R. Chawla, H. K. Gummel, and P. Kozak, "MOTIS—An MOS timing simulator," *IEEE Trans. Circuits and Syst.*, vol. CAS-22, pp. 901–909, Dec. 1975.

[85] S. P. Fan, M. Y. Hsueh, A. R. Newton, and D. O. Pederson, "MOTIS-C: A new circuit simulator for MOS LSI circuits," in *Proc. IEEE Int. Symp. Circuits Syst.*, pp. 700–703, Apr. 1977.

[86] G. R. Boyle, "Simulation of integrated injection logic," Univ. California, Berkeley, ERL Memo ERL-M78/13, Mar. 1978.

[87] E. B. Kosemchak, "Computer aided analysis of digital integrated circuits by macromodeling," Ph.D. dissertation, Columbia Univ., New York, NY, 1971.

[88] N. B. Rabbat, "Macromodeling and transient simulation of large integrated digital systems," Ph.D. dissertation, The Queen's Univ., Belfast, U.K., 1971.

[89] M. Y. Hsueh, A. R. Newton and D. O. Pederson, "The develop-

[90] ment of macromodels for MOS timing simulators," in *Proc. IEEE Symp on Circuits Systems* (New York) pp. 345–349, May 1978.

[90] R. C. Baldwin, "An approach to the simulation of computer logic," 1959 AIEE Conf. paper.

[91] The LOGIS logic simulator, Information Systems Design, Inc., Santa Clara, CA.

[92] S. A. Szygenda and E. W. Thompson, "Digital logic simulation in a time-based, table-driven environment, Part 1. Design verification," *IEEE Comput.*, pp. 24–36, Mar. 1975.

[93] M. A. Breuer, "General survey of design automation of digital computers," *Proc. IEEE*, vol. 54, no. 12, Dec. 1966.

[94] P. Wilcox and A. Rombeck, "F/LOGIC—An interactive fault and logic simulation for digital circuits," in *Proc. 13th Design Automation Conf.*, pp. 68–73, 1976.

[95] A. R. Newton, "Techniques for the simulation of large-scale integrated circuits," *IEEE Trans. Circuits Syst.*, vol. CAS-26, pp. 741–749, Sept. 1979.

[96] G. Arnout and H. De Man, "The use of threshold functions and Boolean-controlled network elements for macromodeling of LSI Circuits," *IEEE J. Solid-State Circuits*, vol. SC-13, pp. 326–332, June 1978.

[97] V. D. Agrawal *et al.*, "The mixed mode simulator," in *Proc. 17th Design Automation Conf.*, pp. 618–625, June 1980.

[98] K. Sakallah and S. W. Director, "An activity-directed circuit simulation algorithm," in *Proc. IEEE Int. Conf. on Circuits and Computers*, pp. 1032–1035, Oct. 1980.

[99] D. D. Hill and W. M. Van Cleemput, "SABLE: Multi-level simulation for hierarchical design," in *Proc. IEEE Int. Symp. Circuits and Systems*, pp. 431–434, (Houston, TX), Apr. 1980.

[100] T. W. Williams and K. P. Parker, "Testing logic networks and design for testability," *Computer*, pp. 9–21, Oct. 1979.

[101] E. B. Eichelberger and T. W. Williams, "A logic design structure for LSI testability," *J. Des. Automat. Fault Tolerant Comput.*, vol. 2, no. 2, pp. 165–178, May 1978.

[102] S. Funatsu, N. Wakatsuki, and T. Arima, "Test generation systems in Japan," in *Proc. 12th Design Automation Conf.*, pp. 114–122, June 1975.

[103] B. Koenemann, J. Mucha, and G. Zwiehoff, "Built-in logic block observation techniques," in *Dig. Papers, 1979 IEEE Test Conf.*, pp. 37–41, Oct. 1979.

[104] H. J. Nadig, "Signature analysis: Concepts, examples, and guidelines," *Hewlett-Packard J.*, pp. 15–21, May 1977.

[105] J. P. Roth, W. G. Bouricius, and P. R. Schneider, "Programmed algorithms to compute tests to detect and distinguish between failures in logic circuits," *IEEE Trans. Electron. Comput.*, vol. EC-16, pp. 567–580, Oct. 1967.

[106] J. Grason, "TMEAS, a testability measurement program," in *Proc. 16th Des. Automat. Conf.* (San Diego, CA), pp. 156–161, June 1979.

[107] L. H. Goldstein and E. L. Thigpen, "SCOAP: Sandia controllability/observability analysis program," in *Proc. 17th Design Automation Conf.* (Minneapolis, MN), pp. 190–196, June 1980.

[108] E. G. Ulrich, T. Baker, and L. R. Williams, "Fault test analysis techniques based on simulation," in *Proc. 9th Design Automation Workshop*, pp. 111–115, June 1972.

[109] A. Parker *et al.*, "The CMU design automation system: an example of automated data path design," in *Proc. 16th Design Automation Conf.* (San Diego, CA), pp. 73–80, June 1979.

[110] D. E. Thomas, "The design and analysis of an automated design style selector," Ph.D. dissertation, Dep. Electrical Engineering, Carnegie-Mellon Univ., Pittsburgh, PA, 1977.

[111] L. Hafer, "Data-memory allocation in the distributed logic design style," M.S. Rep., Dep. Electrical Engineering, Carnegie-Mellon Univ., Pittsburgh, PA, 1977.

[112] G. W. Leive, "The binding of modules to abstract digital hardware descriptions," Ph.D. dissertation, Dep. Electrical Engineering, Carnegie-Mellon Univ., Pittsburgh, PA, May 1980.

[113] G. J. Sussman, J. Holloway, and T. F. Knight, Jr., "Design aids for digital integrated systems, an artificial intelligence approach," in *Proc. IEEE Int. Symp. on Circuits and Computers*, pp. 612–615, Oct. 1980.

[114] M. Barbacci, "The ISPL language," Dep. of Computer Science, Carnegie-Mellon Univ., Pittsburgh, PA, 1977.

[115] A number of references may be found in *IEEE Computer*, vol. 7, no. 12, Dec. 1974.

[116] C. Y. Chu, "An ALGOL-like computer design language," *Commun. ACM*, vol. 8, no. 10, pp. 607–615, Oct. 1965.

[117] *CASSANDRE and LASCAR systems—User's manual.* Grenoble, France: ENSIMAG.

[118] P. M. Carmody *et al.*, "An interactive graphics system for custom physical design," in *Dig. Tech. Papers, IEEE Int. Solid State Circuits Conf.* (Philadelphia, PA), pp. 246–247, Feb. 1979.

[119] L. Rosenberg, "The evolution of design automation to meet the challenges of VLSI," in *Proc. 17th Design Automation Conf.* (Minneapolis, MN), pp. 3–11, June, 1980.

[120] P. B. Weil and L. P. McNamee, "Report of data base workshop," presented at IEEE Computer Science Design Automation Technical Committee, Santa Barbara, CA, Feb. 1980.

Circuit Analysis, Logic Simulation, and Design Verification for VLSI

ALBERT E. RUEHLI, SENIOR MEMBER, IEEE, AND GARY S. DITLOW, MEMBER, IEEE

Invited Paper

Abstract—In this paper, we consider computer-aided design techniques for VLSI. Specifically, the areas of circuit analysis, logic simulation and design verification are discussed with an emphasis on time domain techniques. Recently, researchers have concentrated on two general problem areas. One important problem discussed is the efficient, exact-time analysis of large-scale circuits. The other area is the unification of these techniques with logic simulation and design verification technique in so called multimode or multilevel systems.

I. INTRODUCTION

THE FACT that this special issue is dedicated to VLSI design attests to the importance of the subject area. Many talented professionals are presently striving to find solutions to very challenging VLSI problems. The increase in complexity is the one fundamental issue which confronts both chip and Computer-Aided Design (CAD) tools designers [1]. The techniques and tools and algorithms which were successfully applied to LSI are inadequate for VLSI. Mainly, algorithms which are of $O(N^3)$ or worse in time for practical computations may not be extendable to VLSI with a larger number of subcircuits N.

Fundamental design methodologies like structured design [1], hierarchical, and "divide and conquer" strategies [2] are proposed by many authors. The interaction between the design methodology, the designer, and the CAD techniques is of key importance to the ease with which the VLSI chip design is achieved. This paper is mainly concerned with CAD techniques for circuit design and timing-oriented problems. We assess the status of three important areas, circuit analysis, logic simulation, and design verification. Here, we de-emphasize the historical aspects since review papers and texts are available which include extensive references on design tools and techniques pertinent for the electrical chip and package design [1]–[14]. Also, simulation for testing is beyond the scope of this paper.

We will first define the terminology used before giving an introduction to the three main topics—circuit analysis, logic simulation, and verification or checking.

A. Terminology

We start this paper with the definition of a few frequently used terms to be able to give a clear account of the subject. One source of ambiguity is the fusion of the languages of circuit theorists and computer scientists. The subject of this paper is an illustration of a classical "EECS" subject.

Here, *circuit* is used in two ways. It may be used to describe a collection of gates with possible interconnection models. Unfortunately, this ambiguity is unavoidable since *circuit* can also mean a fraction of a gate or a single gate. We use the term *subcircuit* for a smaller portion of a circuit, especially when we talk about partitioning concepts. Another ambiguity exists in the usage of the term *gate*. We will use gate to mean a logical gate or subcircuit. If we refer to the gate electrode of an MOS transistor, we call it an MOS transistor gate.

Analysis rather than simulation is used to describe techniques which are based on mathematics as opposed to heuristics. The reasons become clear if we consider a dictionary definition for *analysis*: 1) to separate into parts of basic principles so as to determine the nature of the whole; 2) examine methodically; 3) to make mathematical analysis; 4) opposite to synthesize.

The mathematical or physics (basic principles) foundation of analysis is clear from these definitions.

We contrast this definition to the meaning of *simulation* in dictionary terms: 1) to have or take the appearance; 2) assume or imitate a particular appearance or form; 3) counterfeit; 4) to reproduce the conditions of a situation.

The empirical nature of a *simulation model* is apparent from these definitions. Analysis techniques are a subset of simulation approaches since by some lucky insight we may intuitively conceive a model or technique which is mathematically accurate. Simulation techniques span an extremely wide range even if we restrict ourselves to subjects of a technical nature. Here we are interested in topics like *logic* or *circuit simulation* rather than, for example, airplane flight simulation. We use *circuit simulation* to imply that we use approximate, heuristically based models. *Logic simulation* is the conventional process where we replace electrical signals by two or more values, e.g., 0 and 1.

Another specification of a model pertains to its "granularity." Three terms are useful, *micromodel*, *model*, and *macromodel*. A circuit *model* describes the physical situation at an intermediate level. If we desire to analyze at a far more detailed level as, for example, at the level of the transistor physics, we construct very detailed *micromodels* [5], [15], [16]. On the other hand, the need to simulate large VLSI systems has generated much interest in *macromodels* which are designed to trade off internal complexity for speed, e.g., [3], [17], [18]. A macromodel may not necessarily be less accurate than a model, but it may have fewer parameters or exclude unimportant dependencies.

Recently, the usage for the terms *mixed and multilevel (mode) simulation and analysis* techniques has become clear [19], [20]. A *mixed-mode* technique includes two techniques

Manuscript received May 13, 1982; revised October 15, 1982.
The authors are with the IBM Thomas J. Watson Research Center, Yorktown Heights, NY 10598.

Reprinted from *Proc. IEEE*, vol. 71, pp. 34–48, Jan. 1983.

103

in the same system at the same level [19] while a *mixed-level* technique includes different hierarchical levels which interact with each other. Details are discussed in Section IV.

Another term which we use frequently is *verification* which is the following according to a dictionary: 1) evidence that establishes or confirms the accuracy or truth of; 2) process of research, examination required to establish authenticity.

Verification enters in many aspects of the design processes as will be discussed in Section V. In fact, the major purpose of the techniques in Section II is *timing verification*.

B. Circuit Analysis and Simulation

From the sixties until recently, the emphasis in computer-aided circuit analysis was on general-purpose programs. ECAP [21], which was one of the earliest widely available programs, was replaced by many newer ones. Today, the most used programs are SPICE [22]–[24], ASTAP, [25] and ADVICE [26]. These programs employ techniques which were innovated in the last decade such as 1) sparse tableau analysis method (STA) [27]; 2) modified nodal analysis method (MNA) [28]; 3) implicit integration methods; and 4) sparse matrix techniques.

In the early seventies a few researchers devised techniques suitable for larger circuits [29]–[31]. For a while the field of large-scale circuit simulation and analysis (LSSA) was called macromodeling [31], [17], [32]. Today, the term *macro-modeling* is more correctly applied to the *modeling* aspect of large-scale circuits. The field of LSSA has expanded in the last few years due to its importance for *timing verifications* for VLSI circuits.

A multitude of new ideas have been conceived in LSSA in the last few years. MOTIS [33] is one of the first programs where simplified analysis techniques are employed for MOSFET circuits. Other programs include MACRO [34], DIANA [17], and MOTIS-C [35]. The specific techniques employed will be discussed in Section II-A.

We differentiate between *incremental* time techniques which have their origin in circuit analysis and *waveform* techniques which include logic simulation with delay times. In the *incremental* approach the analysis proceeds globally in time steps which are usually smaller than the signal rise times. All the above programs are based on *incremental* time updating techniques. The local circuit waveforms are computed for a sizable time segment in a *waveform* approach [36]–[38]. Two major sources of time saving result from the sparsities present in the space–time relationship of the circuits and from employing *decomposition techniques*. The waveform technique, and *latency*, e.g., [34], [40] are approaches which take advantage of time domain or *temporal* sparsity. Section II-B details these techniques. The other area of *decomposition* pertains to the structure of the logic circuits. The average fan-out for a logic gate is between two and three. Thus the connectivity of most circuits is extremely sparse leading to *structural sparsity*. Usually, for an exact representation of the gates the number of internal nodes of the circuits is much larger than the external nodes. This leads to a *natural* decomposition of the circuits in terms of the gates. Numerous papers referenced give techniques for the exploitation of this structure, e.g., [3], [90], [92]. For MOS devices the MOS transistor gate represents an almost unidirectional node with the exception of the gate-to-drain and -source capacitances. The approach given in [37] exploits this, and the so-called one-way (1-way) scheduling technique is based on this. Here, we discuss mostly

graph-based algorithms, while a recent paper [4] concentrates on equivalent sparse matrix techniques. Some of these techniques are detailed in Section II-B.

A source of inaccuracy may be the approximate inclusion of feedback in the solution. Feedback may be *local* or *global*. Further, the strength of the feedback may be *weak* or *strong*. For example, a flip-flop is locally coupled with strong feedback. The Miller feedback in an MOS circuit is local and weak. Further, the feedback extending over several subcircuits in sequential logic is global and strong. Techniques will be discussed below for the inclusion of feedback. For example, the basic waveform 1-way modeling technique [37] works for the flip-flop case. Two recent general methods are the waveform relaxation (WR) technique for a *waveform* analysis [36] and the symmetric displacement [41] for an *incremental* system.

Logic simulation is performed at the functional gate and recently at the transistor level. In this paper we concentrate on simulation for design verification rather than testing. Gate level simulation is an area which evolved over the last two decades. Even the early work was concerned with efficiency improvement techniques, e.g., [42], [43] since the logic circuits of interest were already large and computers were less powerful. Since then a continuous progress has been made towards the simulation of very large circuits, e.g., [7]–[10], [44]–[48]. Both temporal and structural sparsity is exploited in these techniques. Mainly, logic simulation is an area where the need for large-capacity CAD programs is very evident. Many of these simulators include crude timing information.

Transistor level simulation [49], [52] has recently emerged as an important problem due to the emergence of new complex MOS transistor designs which cannot easily be cast into standard logic gates, e.g., [53]. This new field is presently receiving attention from many researchers.

Another area of importance is mixed-mode and multilevel systems. In a mixed system two techniques are used like circuit analysis and simplified macromodels [54] or logic simulation with an electrical WR technique [19]. Many multimode and/or multilevel systems include several simulation/analysis levels combining many of the above-mentioned techniques. [19], [20], [54]–[64]. Detailed aspects on multilevel systems are given in Section IV.

C. Verification

There are two fundamental processes in VLSI-CAD. *Design* is the process of construction while *verification* ensures the correctness of the design [65]–[78]. The following are general examples of *verification* steps: timing; layout; mask; electrical connectivity; functional correctness; and design rule.

Some of the verification steps involve mask and layout steps, e.g., [65]–[70] for which we refer the reader to another paper in this issue [70] since we want to concentrate on timing aspects in this paper.

Verification of the functional correctness of the logic design is another area which has intrigued computer scientists for more than two decades [71]–[75], [8], [14], [44]–[48]. This topic is obviously closely related to timing verification since functional correctness is a prerequisite and therefore a subset of timing verification, which is discussed in Section V. Simulation for design verification [48], [75] was successfully applied to truly large-scale problems with more than 0.5×10^6 gates more than a decade ago. However, the chips themselves were LSI rather than VLSI.

Timing verification is perhaps the most evasive aspect of verification, e.g., [76], [78]. Specifically, the physical design may be perfect, but still the electrical performance of the design may be incorrect or insufficient. This is why we concentrate on timing design and timing verification in this paper. The spectrum spans from exact circuit analysis to extremely efficient algorithms designed for timing simulation and verification. A discussion of verification techniques is given in Section V.

II. CIRCUIT ANALYSIS AND SIMULATION

A. General-Purpose Circuit Analysis

The most widely used tools for circuit analysis are the *general-purpose* or *standard* circuit analysis programs like SPICE [22]–[24] or ASTAP [25], based on the modified nodal analysis (MNA) approach and the sparse tableau analysis (STA), respectively. They are designed for the detailed analysis of a large variety of circuits and are characterized by having a multitude of features. For VLSI, these tools have a place in the detailed design of special circuits. Specifically, the following applications are a natural for them.

They can serve as an interface to the usually elaborate models used for semiconductor device analysis, e.g., [15], [16]. In this capacity they are used to verify the validity of the semiconductor circuit models from the detailed usually two-dimensional micromodels.

The individual subcircuits or gates are designed and optimized using these programs. The compute time for this type of work is moderate since usually only a few devices are involved.

The correctness of newly developed large-scale programs is verified using these well-proven exact circuit analysis programs.

Thus even for VLSI, they play an important part in the early design phases where the semiconductor devices and the menu of subcircuits are established. Importantly, the process of circuit analysis is exact within the accuracy of the models used and the mathematical numerical techniques employed.

The circuit equation in both the MNA and the STA approaches lead to a set of coupled equations of the form

$$f_1(x, \dot{x}, y, t) = 0 \qquad (1a)$$

$$f_2(x, y, t) = 0 \qquad (1b)$$

where x are the differentiated variables and y the nondifferentiated variables. The formulation of the equations in terms of f_1 and f_2 results naturally from both formulations although the STA formulation is a larger, more sparse matrix while the MNA matrix is smaller and more dense.

We are interested in the solution for the analysis time which usually is $t \in [0, T\mathrm{max}]$ where $T\mathrm{max}$ is the final time. Two main techniques are used today to solve (1). The time derivatives are in the most simple case replaced by the backward Euler (BE) formula where

$$\dot{x}_n \cong \frac{x_n - x_{n-1}}{h_n} \qquad (2)$$

where n represents the "now" time, $n - 1$ the time step before, and $h_n = t_n - t_{n-1}$ is the present time step. Equation (2) is the most simple A-stable formula. However, other integration methods like the second-order backward-differentiation formula [79], [80] and the ACA methods [81] are more desirable for general purpose programs. Numerical damping [82] is severe for the BE formula (2), and misleading results may be

obtained especially for oscillatory solutions which are present in high-performance logic. In this discussion, we will proceed using the BE formula for time integration for clarity. The result of applying (2) to the system (1) is

$$\tilde{f}_1(x_n, x_n - x_{n-1}, y_n) = 0 \qquad (3a)$$

$$f_2(x_n, y_n, t_n) = 0 \qquad (3b)$$

which is a system of nonlinear algebraic equations.

To transform the nonlinear equations into a system of linear equations, we usually apply Newton's method to the system (3) which is written as $g(z) = 0$ for

$$\frac{\partial g}{\partial z} z^i = -g(z)^{i-1} + \frac{\partial g}{\partial z} z^{i-1}. \qquad (4)$$

Here $z = \begin{bmatrix} x_n \\ y_n \end{bmatrix}$ and $g = \begin{bmatrix} \tilde{f}_1 \\ f_2 \end{bmatrix}$ and i is the index of the iteration. Usually, (4) is written in the form of a linear system of equations

$$Az = b \qquad (5)$$

where A is the Jacobian

$$A = \frac{\partial g}{\partial z} \qquad (6)$$

and b is evident from (4) and (5).

The above mathematical procedure for incremental circuit analysis is summarized in ALG. 1. Again, *incremental* refers to the small increments with which time proceeds for all circuits.

ALGORITHM 1: Incremental Circuit Analysis

Inputs:	Circuit	C
	Input Waveforms	IW
	Time Step	h
	Stop Time	$T\mathrm{max}$
Results:	Waveforms of node voltages $v(t)$	
	Waveforms of branch currents	
Note:	$z(n, i)$ Vector of voltages and currents at time n and Newton iteration i	

```
Procedure        CircuitAnalysis (C, I, h, Tmax)
BEGIN
                 Fill in Jacobian Matrix A and
                 right-hand side b with MNA stamps for
                 time invariant circuit elements
                 dc solution at time = 0
                 time = h
                 n = 1
TIMEloop:        for time <= Tmax do
                 BEGIN i = 1
NEWTONloop:          for ABS(z(n, i) - z(n, i-1))
                         < EPSILON do
                     BEGIN Fill in A and b
                         with MNA stamps for time
                         variant and nonlinear
                         circuit elements
                         Solve A z(n, i) = b
                         i = i + 1
                     END
                     time = time + h
                     n = n + 1
                 END
END
```

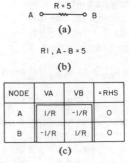

Fig. 1. Processing of resistor R. (a) Graphical representation. (b) Input language statement. (c) Entries into the A-matrix.

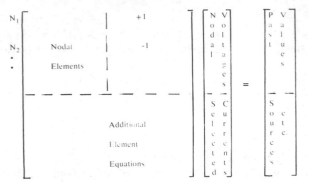

Fig. 2. Circuit analysis matrix $Az = b$.

The relevant aspects of the techniques outlined in ALG. 1 are detailed next. The input data may be supplied to the program in one of three forms. Fig. 1(a) shows a circuit element which may be entered with a graphics system. Fig. 1(b) shows the equivalent language statement where A and B are the connecting nodes. In the third way, the circuit element is extracted from the layout with the assistance of a preprocessing program. As shown in Fig. 1(c) the circuit element data are "stamped" directly into the A matrix in the locations which are identified by the connecting nodes. The linear resistor used in this example is a particularly simple case for the way A is assembled inside the NEWTONloop in ALG 1.

Other circuit elements like transistors, capacitors, inductors, current and voltage sources are stamped into the A matrix in a similar way. A schematic representation of A for the MNA approach is shown in Fig. 2.

Many other algorithms are implemented in an efficient program. In the following, a few are enumerated which are not evident from the discussion above: conversion of node-names into program internal numbers; efficient numerical integration techniques for inductances and capacitances; sparse matrix techniques for storing A and solving the system $Az = b$; efficient assignment of variable time steps h; and convergence improvements for the Newton loop (NEWTONloop) which solves the nonlinear equations.

General-purpose analysis programs of the SPICE and ASTAP type may contain many features which add to their flexibility such as the following: statistical analysis; ac sinusoidal steady-state analysis; built-in source time waveforms; changeable parameters and restart facilities; design centering for design improvement, e.g., [83]; and transmission line analysis.

The growth of the solution time for sophisticated circuit analysis programs is $O(N^m)$ where $1.2 \leqslant m \leqslant 1.5$ and N is the number of nodes. These programs are mostly limited to the applications listed in Section I-B since they are exhausted for most practical cases for 50–100 gates. Further, improvements of perhaps factor 3 can be envisioned in the future from the application of techniques like node tearing [84], and programming improvements.

Another effort to improve the speed of circuit analysis programs is based on *vector processing* [85], [86]. Basically, repetitive subcircuits can be analyzed in parallel, which results in time savings. To evaluate the potential gain with this approach, a comparison has been made between SPICE2 and CLASSIE, a vector processing circuit program. For a sufficiently large and regular circuit like an adder, the speed up is about an order of magnitude compared to SPICE2.

B. Large-Scale Circuit Simulation and Analysis

Clearly, substantial improvements must be made for large circuits in the *simulation/analysis* (S/A) techniques above the analysis approaches presented in Section II-A for VLSI. A new technique must lead to at least an order of magnitude improvement in the number of circuits to be analyzed to offer a distinct advantage above the general-purpose circuit analysis programs. Thus we expect to analyze at least 500 subcircuits in several minutes of compute time on a high-performance machine. In the recent past, the following areas of potential gain have been identified: special-purpose programs; repetitive subcircuit structure (modularity); structural sparsity, approximation and simplification techniques; and time domain or temporal sparsity (latency, etc.).

Special-purpose programs can lead to substantial gains in performance. As an example, the analysis of a digital filter subcircuit is efficiently performed by special techniques, e.g., [87] while a general-purpose program lead to an inefficient, time-consuming solution. Another special-purpose program may be involved in the solution of the linear package or interconnection equations [5], [88], [89]. Efficient techniques may involve the symbolic solution [88] of subcircuits like the interconnections and the exploitation of the special sparsity structure [89]. Specifically, if we employ the general solution in ALG. 1 for interconnection circuits, we can completely eliminate the NEWTONloop since f_1 and f_2 in (1) are linear for this case.

Repetitive substructures or modularity lead to gains in several areas of a S/A system. First, we notice that gains can be made in the DATA specifications of ALG. 1. A large amount of repetitive data is present in the input obtained from a VLSI chip. The essential information is usually a very small repertoire of fundamentally different devices and subcircuits. This is in contrast to the general-purpose circuit analysis program where the inherent data structure must accommodate large amounts of data of a different nature. Internally, each element is individually processed. As an example, in a coarse gate oriented MOS circuit simulator, we may specify a few different gates, the interconnections, and perhaps a single capacitance value per interconnection. In this case, we can severely limit the data handling and tailor the algorithms. So-called *modular* methods are based on the regularity of the subcircuits from an analysis point of view [34], [40], [90]–[92]. The major gain in employing modularity is a reduction in storage requirements and convenience in the implementation of other saving concepts rather than time. One of the modular methods will be discussed below in more detail.

Structural sparsity must be exploited beyond the conven-

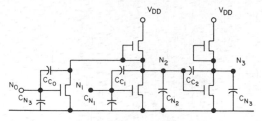

Fig. 3. Example of an MOS circuit.

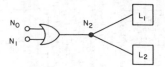

Fig. 4. Illustration of loading.

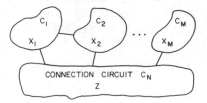

Fig. 5. Circuit with modular subcircuits.

tional sparse matrix techniques for linear systems in circuit analysis programs [93], [94]. Most digital circuits exhibit even more structural sparseness at the global level than analog circuits. The typical fan-out of a logic gate is between 1 and 3 while an analog circuit has about 3 connections per node with other dependent coupled sources. Following this concept, we clearly want to avoid stamping each subcircuit into a single large A matrix as it is done in the general-purpose circuit analysis ALG. 1. Two basic formulations based on the subcircuit structure are widely used. For low- and medium-performance MOS generally connected transistor circuits, a specialized set of nodal equations can be written in the form

$$I(v_N, V_{DD}) + C \frac{dv_N}{dt} = 0 \qquad (7)$$

where $I(v_N, V_{DD})$ corresponds to the MOS transistors which are connected between the nodes as exemplified in Fig. 3. C is the matrix of node capacitances, which are of two types, the node-ground C_N and the node–node capacitances C_C. The node capacitance C_N usually lumps several device and wire capacitances into a single value, and the coupling capacitances C_C is C_{gd} the gate-to-drain capacitance in simple cases. This type of model is used in MOTIS [33], MOTIS-C [35], SPLICE [55], MACRO [34], and MEDUSA [92], and other programs with variations. Usually, the voltage sources are treated as having zero internal resistance in this scheme. The time discretization employed is similar to (2) in Section II-A, and the overall time solution approach is *incremental* like ALG. 1.

In the other approach where more elements are present in a subcircuit model, the number of external connections is usually small or sparse compared to the internal connections. For this case, the modular approach is employed at the subcircuit level. Examples of modular techniques are [40], [61], [83], and [90]. Each of the subsystem of equations for the subcircuit is of the form (1) or (7). If we take the discrete time form of the nonlinear equation (3) for the first subcircuit in Fig. 3, it can be written in the form

$$g(v_{n_0}, v_{n_1}, v_{n_2}, v_{n_3}) = 0 \qquad (8)$$

where n is the "now" index of time. This model is ideally suited to show how the various approximations employed in the different programs operate.

First, we will exemplify the Gauss–Jacobi and Gauss–Seidel iterative solution methods for the simple example of a system of linear equations $Ax = b$. We decompose the matrix A into $A = L + D + U$ with L being strictly lower triangular, D diagonal, and U strictly upper triangular. Then the system of equations can be written as

$$Lx + Dx + Ux = b. \qquad (9)$$

In the Gauss–Jacobi method, we evaluate Dx^k at the present iteration k while Lx^{k-1} and Dx^{k-1} are evaluated at the pre-

vious iteration $k-1$. This leads to a very simple solution. In the Gauss–Seidel method, only Ux^{k-1} is evaluated at iteration $k-1$. Thus the Gauss–Seidel case corresponds to a lower triangular system of equations

$$(L + D)x^k = b - Ux^{k-1} \qquad (10)$$

which is related to levelizing the circuit graph using 1-way models [37] and can easily be solved by back substitution.

Returning to (8) for the linear case, L corresponds to elements in the inputs or the previously iterated variables v_{n_0} and v_{n_1}, v_{n_2} corresponds to the diagonal or variable to be updated while v_{n_3} corresponds to the not-yet-evaluated voltage, or the Ux contribution. Taking the analogous nonlinear case [95], Gauss–Jordan corresponds to the iterative scheme

$$g(v_{n_0}^{k-1}, v_{n_1}^{k-1}, v_{n_2}^k, v_{n_3}^{k-1}) = 0 \qquad (11)$$

while the Gauss–Seidel results in

$$g(v_{n_0}^k, v_{n_1}^k, v_{n_2}^k, v_{n_3}^{k-1}) = 0. \qquad (12)$$

The scheme in (12) needs updated input variables before it can update node 2. A source of confusion exists since the Gauss–Seidel and Gauss–Jordan schemes can be applied to the solution of the nonlinear equation as exemplified here, or to the linear equations, (9) and (10). Both approaches are practiced, and the notation employed is usually hard to understand.

Circuit theorists have a similar way of reducing forward coupling among the subcircuits. A circuit motivated approach to reduce the coupling variables in (8) is *loading*. Fig. 4 gives a loading circuit for the analysis of the two input gates in Fig. 3 with the output node $N2$. $L1$ is an approximate circuit for the coupling capacitance C_{c_2}, which is discussed below, while $L2$ approximates the transistor input. Loading models are employed in many programs, e.g., MACRO [34] and RELAX [38].

Many of these approximation techniques apply to both the incremental and waveform approaches. Clearly, the intent is to save computations. In MOTIS [33], transistor characteristics are approximated by secants which results in a linear local equation which does not require multiple Newton iterations. In MOTIS-C [35], table look-up procedures are used for the nonlinear MOS device characteristics to save function evaluations.

MEDUSA [92] is used as an example of a modulator treatment of the circuit matrix A. The circuit is partitioned into "natural" subcircuits with least connections, as shown in Fig. 5.

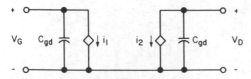

Fig. 6. Equivalent circuit for C_{gd}.

The modular subcircuits C_m with internal variables x_m are embedded in a general interconnection circuit C_N with variables z. The A-matrix (5) can be represented in terms of the modularity as (13)

$$
\begin{bmatrix}
\dfrac{\partial f_1}{\partial x_1} & & & \dfrac{\partial f_1}{\partial z} \\[2mm]
& \dfrac{\partial f_2}{\partial x_2} & & \dfrac{\partial f_2}{\partial z} \\[2mm]
& & \dfrac{\partial f_m}{\partial x_m} & \dfrac{\partial f_m}{\partial z} \\[2mm]
\dfrac{\partial f_n}{\partial x_1} & \dfrac{\partial f_n}{\partial x_2} & \dfrac{\partial f_n}{\partial x_m} & \dfrac{\partial f_n}{\partial z}
\end{bmatrix}
\begin{bmatrix}
\Delta x_1 \\[2mm] \Delta x_2 \\[2mm] \Delta x_m \\[2mm] \Delta z
\end{bmatrix}
=
\begin{bmatrix}
-f_1 \\[2mm] -f_2 \\[2mm] -f_m \\[2mm] -f_n
\end{bmatrix}
\qquad (13)
$$

where the circuit variables are again combinations of voltages and currents like in ALG. 1.

With $\partial f_m / \partial x_m = A_m$, each row of the matrix is easily solved for Δx_m, or

$$
\Delta x_m = -A_m^{-1} f_m - A_m^{-1} \frac{\partial f_m}{\partial z} \Delta z \qquad (14)
$$

for the mth subcircuit. Δz is found by inserting (14) into the last row of (13). Thus the solution of the structurally bordered block diagonal matrix is formally simple. However, to gain decoupling, the second term on the right-hand side in (14) is ignored in [92]. The omission of these terms results in a lower block triangular system which is used in a Gauss–Seidel iteration scheme.

Approximations for the feedback coupling capacitance C_{c_1} in Fig. 3 are discussed next. It is natural for incremental techniques to attempt to achieve decoupling by replacing unknown values at the "now" time $x(t_n)$ by known values $x(t_{n-1})$ at the previous time steps [105]. This is equivalent to having 1-way properties for t_n.

Explicit forward Euler schemes were used by some authors to eliminate the coupling due to the "floating" capacitances by

$$
i_{c,n} = C_{gd} \frac{v_{n-1} - v_{n-2}}{h}. \qquad (15)
$$

Unfortunately, this scheme severely limits the time step h for the response to be stable. Fig. 6 shows an improved model for a floating capacitor where

$$
i_1 = -C_{gd} \frac{dv_2}{dt} \cong -C_{gd} \frac{v_{2_{n-1}} - v_{2_{n-2}}}{h} \qquad (16)
$$

$$
i_2 = -C_{gd} \frac{dv_1}{dt} \cong -C_{gd} \frac{v_{1_n} - v_{1_{n-1}}}{h}. \qquad (17)
$$

Equations (16) and (17) together with Fig. 6 represent an equivalent circuit interpretation of the IIE method proposed in [96]. Note that this model maintains the 1-way coupling property by restricting the forward Euler integration to the

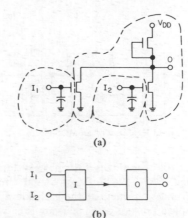

(a)

(b)

Fig. 7. (a) Two-input gate with strongly connected components. (b) One-way model for gate.

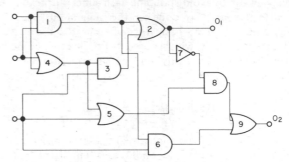

Fig. 8. Example of an adder circuit.

current which represents feedback. A further refinement of this approach is the symmetric displacement method in [41].

In a recent paper a method has been proposed of the form

$$
g(v_{n_0}^k, v_{n_1}^k, v_{n_2}^k, v_{n_3}^*) = 0 \qquad (18)
$$

corresponding to (18) and Fig. 3. Here $v_{n_3}^*$ is computed from

$$
v_{n_3}^* = v_{n-1_3} + h v_{n-1_3} \qquad (19)
$$

which is again a prediction of the feedback variables in the time coordinate.

The "now" time step h_n is more or less synchronized among all subcircuits in the incremental approaches. An attempt has been made at decoupling the time steps by using a master clock with subcircuit running at their own h_n [98]. In a waveform approach, complete decoupling among the subcircuit is achieved and the subcircuits are allowed to run at their own maximum time step [36], [37]. This may result in considerable time saving due to the reduction in the number of computations. This approach is illustrated on the example circuit of Fig. 8, where the model of Fig. 7 is used to represent the 1-way circuits. The interaction graph G of Fig. 9 results if the individual gates or subcircuits are replaced by 1-way models. The key insights gained from this graph are the portions which should be analyzed simultaneously. For example, the subcircuits inside the dashed line B may be strongly connected. Note that the interconnections are conveniently accommodated in this approach. We form a new graph from Fig. 9 which has the strongly connected subcircuits as its nodes. Fig. 10 shows G'' where additionally the nodes have been assigned to levels to further simplify the processing. The levelizing ensures that we analyze the circuit such that all the input information is available for each subcircuit at the processing time. Interpolation must be used in this approach to obtain the input

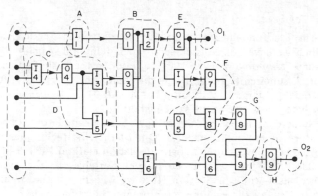

Fig. 9. One-way interaction graph G for Fig. 8.

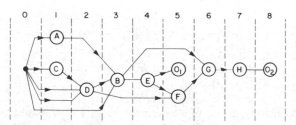

Fig. 10. Levelized graph G'' for circuit.

waveforms at the appropriate time points due to the unequal time steps in the subcircuits [37]. The processing of the subcircuits and the subsequent discarding of this portion of a circuit, is a desirable approach to the time analysis for large-scale circuits. In this process, the waveforms of interest are stored on disk and thus the amount of data present in active store is limited. The accuracy of the solution techniques in the straightforward 1-way technique depends on the size of the gate-to-drain capacitance. Accuracy is always a key issue in the S/A techniques discussed since inaccuracies may lead to faulty chip designs [99]. The WR technique leads to an approach for which the compute time can be reduced at the expense of accuracy and vice versa.

The basic WR approach is easily illustrated for C_{gd} with the assistance of the equivalent circuit Fig. 6. Again, full-time waveforms are involved and we choose

$$i_1(t) = -C_{gd} \frac{dv_2(t)}{dt} \qquad (20)$$

$$i_2(t) = -C_{gd} \frac{dv_1(t)}{dt}. \qquad (21)$$

The left circuit portion is evaluated first for $t \in [0, T]$ where T is the final time. We iterate between the left and right circuits until convergence, with $i_1(t)$ and $i_2(t)$ being the waveforms at the previous iteration. This process is generalized in ALG. 2.

ALGORITHM 2: Waveform Relaxation for Multiple Circuits

Inputs: Circuit C
Start time Tmin
Stop time Tmax
Input waveforms IW
Result: Waveforms for each node
Note: $S(i, j)$ = subcircuit j for iteration i
NS = number of subcircuits
$X(i, j, k, t)$ = voltage of node k in subcircuit j
iteration i at time t

Procedure WaveformRelaxation (Tmin, Tmax, C, I)
BEGIN
        Partition C into subcircuits S
        Schedule the subcircuits
        $X(0, *, *, t) = IW$
        $i = 0$
        Repeat
          For $j = 1$ to NS do
          BEGIN
            Solve each subcircuit $S(i, j)$ for voltages
            $X(i, j, k, t)$ at iteration i using the results
            of iteration $i - 1$
            Note: use the Incremental Circuit Analysis
                 Algorithm to solve each $S(i, j)$
        END
        until max norm $(x(i, j, k, t) - x(i - 1, j, k, t))$
        $<$ ERROR for all j, k, t
END

Different techniques are employed for choosing the subcircuits in ALG. 2. In [100], [19] the decomposition is by *strongly connected component* while in [36], [38] the decomposition is *by circuit* given by the topology of the logical gate.

In the beginning of this section, the desirable goal of at least an order of magnitude speed enhancement over a general-purpose circuit analysis program was stated and a multitude of methods for improving the solution speed are given. A comparison among the methods may be somewhat misleading due to the different implementation, computers, and accuracy demanded in the solution. An indication of the state of the art may be a factor 50 in speed improvement for RELAX [38] as compared to SPICE for a medium-size circuit.

C. Time Waveform Representation

A key aspect of a circuit S/A program and also a logic simulator is the representation of the time waveforms. This is true for incremental as well as waveform techniques. Some of the factors to take into account are the number of circuit, application, intended accuracy, available computer, and storage space, as well as the compute time to be consumed by a typical run. We easily can think of these considerations as well as the waveform representation as more fundamental than the S/A techniques which lead to the waveforms. Fig. 11 illustrates different time waveform representations. In circuit analysis which has the potential of retaining a better than 1-percent accuracy, all of the details of the waveforms are represented as shown in Fig. 11(a). The high accuracy is obtained at the cost of both high storage and time requirements. For example, at the circuit analysis level, 10^3 time points for a single waveform is not unusually high. If we analyze a problem with $N = 10^4$ nodes we may want to retain 10 percent of the node results for future observation of the resultant waveforms. The storage for these waveforms (time, voltage pairs) will require 2×10^6 words of storage. Usually, this will be a secondary store database. Clearly a form of *data compression* is desirable for realistic large-scale systems. In general, point of the amount of information H in a signal with k levels occurring with probability P_i is

$$H = -\sum_{i=1}^{k} P_i \ln P_i. \qquad (22)$$

From this point of view it is clear that the binary simulator (0, 1) employs the minimum waveform representation which

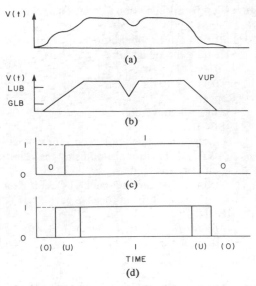

Fig. 11. Different waveform representations.

contains any information at all. However, adding a few additional intermediate levels between the "scaled" levels 0 and 1 is not the most practiced approach to increasing the amount of useful information. Fig. 11 illustrates a few other examples. In (b) a linear rise and fall time has been used as well as a half-signal spike representation. The pure binary signal is given in (c), while in (d), the unknown U- or X-state is added. Both (c) and (d) may ignore the spike shown.

Simulators with 7 or more states are in use, e.g., [102] today attempting to maintain sufficient accuracy while keeping compute time and storage to a minimum. In terms of (22), adding an X-state may increase the useful information content far more than adding an intermediate signal level. It is obvious that the concepts earlier presented unify the techniques employed in analysis and simulation programs. However, new challenges are introduced when different level representations are used simultaneously in a multilevel system. Two fundamentally different types of transitions are necessary: high-to-low information; and low-to-high information.

The first transition is obviously accomplished by discarding information, although challenges remain like the matching of many different waveform representation like the ones with X or high impedance state, etc. The transition from logic variables, Fig. 11(d), to circuit voltages is interesting since information needs to be added. Assume that we want to obtain the constant slope representation of Fig. 11(b). First, the amplitude is scaled by associating a logical 1 with a typical up-level voltage VUP = 4.5 V. Further, the slope of the rise time is

$$SL = \frac{V_{LUB} - V_{GLB}}{t_{LUB} - t_{GLB}}. \tag{23}$$

Thus the constant slope in Fig. 11(b) can be constructed. This obviously only works for X-states corresponding to transitions. Most likely, this waveform will be the input to the circuit analysis portion in a multilevel system obtained from the simulation portion.

The concepts given in this section show the similarities between the fundamental aims of the simulation and analysis techniques which is the computation of waveforms. However, it should be clear at this point that the means of obtaining them may be quite different.

D. Temporal Sparsity

The last fundamental technique mentioned in Section II-B which we have not considered so far is temporal sparsity. We choose *temporal sparsity* [90] to describe the basic fact that not all subcircuits are active at the same time. Typically, for sufficiently large circuits only between 0.01 to 10 percent of the gates are active at the same instant.

Numerous clever techniques have been devised to take advantage of temporal sparsity. One of the first techniques reported is the event-driven logic simulation method [43], [44], [103]. In this section we restrict ourselves to the event mechanism while in Section III-B we will discuss the simulation. Usually the events are represented in a time event queue for the actions to be taken. In the scheme in ALG. 3, the actions taken at the present or "now" time t_n are illustrated.

ALGORITHM 3: Time Event Scheduling

Inputs: Circuit C
            Subcircuit Delays $ScD(i)$
            Event Queue EQ
            Initial events IE
Result: Updating of event queue for circuit scheduling

Note: tn is present time

Procedure EventScheduling(C,ScD,time)
BEGIN
  $EQ = IE$
  While EQ is not empty do
    BEGIN
    tn is smallest new next event time on EQ
      P = Pop from EQ all events at tn
      For all $P(j)$ do
      BEGIN
        S = Successors($P(j)$)
        For all $S(j)$ whose states have changed do
        SelectiveTracing:
        Push onto EQ the event $[S(i),tn+ScD(i)]$
      END
    END
END

The kernel of ALG. 3 is the *selective tracing* which finds all new events which are caused by the present events at t_n. Then the event queue EQ is updated by these events. A further refinement may be if they actually switch and cause new events before we execute the successors. This scheme leads to an excellent exploitation of temporal sparsity since actions are taken only if they lead to activities, or equivalently computations must be performed. It is noted that the number of computations is directly related to how busy the event queue is. Selective tracing usually assumes the logic circuits models to be simple having 1-way properties. It should be noted that special algorithms must be employed for a circuit with pass-transistors which are 2-way [55].

Temporal sparsity was not taken into account in circuit analysis and large-scale circuit simulation until recently when the latency concepts were introduced, e.g., [34], [84], [90]. Before, all subcircuits were analyzed with the same time-step h, whether the subcircuits were latent or not. In a modular or subcircuit oriented program, the inactive subcircuits can be singled out and no computations are done inside the subcircuit. Then, in an incremental analysis, the internal variables are simply [40]

$$z_n = z_{n-1} \qquad (24)$$

which is a zeroth-order integration method with a local truncation error in the variable k is

$$E_{0_k} = \dot{z}_n(t_{n-1})h. \qquad (25)$$

Thus a latency detector keeps track of the derivative of all variables in a subcircuit.

The computations for the subcircuit are resumed if changes in one of the subcircuit variables occur which exceed an error criterion. Thus many of the new incremental programs are taking latency into account, e.g., [35], [61], [84], [90]. The waveform programs have independent time steps in the subcircuits. Therefore, if a subcircuit is inactive, its variable time step algorithm will choose a very large time step which is equivalent to the latency concept [36], [37], [19].

Thus it can be concluded that temporal sparsity can be exploited for both general-purpose circuit analysis, as well as for large-scale programs. In logic simulators, the computations are reduced using event scheduling illustrated in ALG. 3.

III. Logic Simulation

By *logic simulation* we mean the simulation of two $(0, 1)$ or more states like H, X, etc. For many years, logic simulation was performed at the *gate type level* only. However, with the advent of custom VLSI pass-transistor designs, logic simulation at the *transistor level* was devised, e.g., [49]. Simply, these designs cannot efficiently be represented by gates. A logical gate is a higher level representation since usually several transistors are mapped into a single gate. Obviously, for a large-scale circuit this reduction in complexity is important if it can be accomplished for at least a portion of the logic at hand.

A. Gate Level Simulation

Today, the majority of simulators are of the gate or subcircuit level type. Logic design is most easily performed in terms of gates which perform logic functions like AND, OR, NAND, etc. This implies that groups of transistors are easily identifiable as one of a few standard gates like an AND or a NOR subcircuit.

Gate level simulation of high-performance machines is perhaps the only discipline which has so far been executed at a large-scale level. In these machines the number of gates per integrated circuit is small by VLSI standards due to the power dissipation required to obtain the high-performance operation. However, simulation runs involving 0.5×10^6 gates have been made already several years ago, e.g., [48]. The simulation methodology does not strongly depend on whether the physical location of the gates is on VLSI chips or on separate chips at least at the gate level for high-performance logic.

The following factors contribute to the convenience and efficiency by which gate-level simulation can be executed.

The gates almost directly correspond to the logical operators OR, AND, NAND, etc., and these operations can efficiently be executed in most computers.

Delay equations for timing verification may yield accurate answers for bipolar transistors. An accuracy of 10 percent is achievable in some cases especially if the interconnections add a large portion to the delay.

The structure of the circuits is extremely sparse since the number of connections per subcircuits is limited for high performance and since the overall size of the circuit is very large.

Temporal sparsity is large since the percentage of active gates usually decreases with increasing circuit size.

Some of the concepts implemented in gate level simulators which were developed in the last two decades are outlined next. Mainly, the design and verification of large, high-performance machines with up to $\sim 10^6$ gates depend heavily on this simulation capability. While simulation is used for both verification and testing, we consider verification only to limit the scope of this paper. Some of the techniques which make the simulation of large-scale circuits possible are these.

Parallel simulation is a technique where for example each bit in a word is used to represent an input variable, e.g., [42]. Thus this process allows for the simulation of 32 input patterns to a circuit in parallel for a 32-bit word machine. Thus this method has the potential of speeding up the simulation time of almost a factor of 32.

Compiled-code simulation is a technique which takes advantage of the high speed of execution of compiled code. The circuit description is programmed in a suitable language like Pascal or Assembler. Either logical macros or the entire circuit is compiled and then run simultaneously. Clearly, scheduling is required which specifies the sequence in which the logical functions are processed similar to G'' in Fig. 10. Hard problems in compiled-code simulators are the methodology for taking advantage of the temporal sparsity and also to efficiently change the logic configuration represented by the compiled code.

Table lookup is a frequently used technique where the truth table and other important information are stored in a table for the logic gates, e.g., [104]. These models are accessed using the input values to determine the resultant output. This approach is quite flexible and is widely used.

As mentioned above, gate level simulation is widely applied to large-scale, high-performance machines where standard gates are employed. The techniques in this discipline are well established and are sophisticated. Timing delay approximations are relatively accurate for high-performance circuits since the delay is mainly in the interconnections [6]. Thus far more accuracy is obtained by this type of timing simulation for high-performance machines than for MOS-VLSI circuits even if standard logical gates are employed.

B. Transistor Level Simulation

Transistor level logic simulators are a relatively new addition to the CAD tools used especially in the design of MOS transistor circuits. Mainly, new circuit designs have been invented which cannot efficiently be treated by a gate simulator. More function per transistor can be obtained by direct transistor designs rather than gate designs [53].

A good example of an MOS logic simulator is MOSSIM [49], [50], which takes the actual transistor circuit into account. MOSSIM is a three-logic level $(0, 1, X)$ simulator. At each *state* of the circuit the steady-state levels (SSL) are established. Note that the transients from one state to the next state are *not* calculated in contrast to the techniques discussed in the last section. However, the SSL $(0, 1, X)$ must be distinguished from the "*dc*" levels since the SSL may in some case be a dynamic level held by a ratio of capacitances. Dynamic elements like capacitances and inductances are ignored in these simulators unless they are involved in determining the SSL. The fundamental transistor level simulator requires a circuit diagram description where we need to specify the type of each transistor.

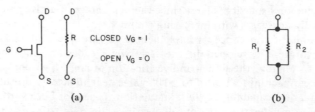

Fig. 12. (a) Simplified transistor model. (b) Two resistances in parallel.

The conventional way to solve for the SSL is to apply ALG. 1 with the time-step loop fixed where the time step h is chosen sufficiently large. This large h assures that the companion resistances corresponding to the capacitances are sufficiently large [93], [94]. This approach is both accurate and time consuming. Approximate solutions are of interest since the approximate knowledge of the logic levels suffices in many cases and since we are interested in simulating very large transistor circuits.

The basis of the technique in MOSSIM is what we can call *approximate circuit theory*. The interpretation of the transistor model is best done in terms of resistances of the device. In Fig. 12(a), we show the model for an MOS transistor where R depends on the type of device. For example, a load device may have $R_1 = 10$, while the active devices may have $R_2 = 1$. The switch for an enhancement load is always closed while the active device in an inverter is controlled by the input. Thus the logic output levels v at the inverter are a logical 1 for a 0 input and for a 1 input

$$v = \frac{R_2}{R_1 + R_2} = 0.0909 \qquad (26)$$

which is equivalent to a logical zero, since we divide the logical range for v

$$v = 0, \quad \text{for} \quad v < v_L$$
$$v = x, \quad \text{for} \quad v_L \leqslant v \leqslant v_H$$
$$v = 1, \quad \text{for} \quad v_H < v. \qquad (27)$$

Computations are saved if we use approximate computations. For example, in Fig. 12(b) the parallel resistance is given by

$$R = \min (R_1, R_2) \qquad (28)$$

which saves a multiplication, an addition, and a division compared to the exact parallel resistor formula. A similar simplification can be worked out to avoid the computation in (26) for the voltage divider, where one of the conditions is

$$v = 0, \quad \text{if} \quad R_1 > \frac{1 - v_L}{v_L} R_2. \qquad (29)$$

A complete set of equations can be worked out for an approximate circuit theory.

A MOSSIM simulator algorithm is outlined below. Here, the nodes are classified as *internal* and *external*. *Clock nodes* are casually called inputs NE which have a state $E(NE)$. The internal nodes NI have a state $I(NI)$ in ALG. 4.

ALGORITHM 4: Transistor Level Logic Simulator

Inputs: Circuit C
    State of each node $E(NE), I(NI)$
    State of each transistor $T(NE, NI)$
Result: Updated States (SSL) $I(NI), T(NE, NI)$

Procedure TransistorSimulation (C, S, T, E)
    For all MOS-transistor gates
    connected to NE's update the state
    $T(NE) = E(NE)$
BEGIN
    For all internal nodes NI do until no changes
    occur
    BEGIN
    Internal NI connected transistors are
    held fixed in state $T(NI)$
        Compute new nodal states $I(NI)$
        using approximate circuit theory
        Update transistor $T(NI)$ states from
        new nodal states $I(NI)$
    END
END

All nodal states are fully determined for each input set or time phase. Since the MOS transistor gates form ideal 1-way connections at dc, they control the MOS transistor source and drain nodes. Groups of MOS transistors are formed by the interconnections [50] between the source and drain nodes. These groups suggest another level of hierarchy which is left out of ALG. 4 for simplicity. Specifically, strong components or groups can be formed which are isolated by FET gates while the interconnections join the members of a group. It is interesting to note that for FET circuits, these groups closely resemble the 1-way components in Section II-B.

The relaxation procedure used in MOSSIM given in ALG. 4 starts from one steady-state level defined by all inputs and clock signals $E(NE)$ at the input or external nodes NE. Then a new input excitation $E(NE)$ is applied and the new internal states $I(NI)$ are computed where $E \in [0, 1]$ and $I \in [0, X, 1]$. This basic transistor level MOS simulation procedure is an iterative procedure to avoid the solution of large, sparse matrices in the form of (4). The transistor states $T(NI)$ connected at internal nodes NI are updated simultaneously only once per iteration while the internal computations are performed with fixed transistor states. The nodal states for internal nodes are computed from the simplified computations of the voltage dividers.

MOSSIM [49] has been applied to circuits with 10^4 transistor on a DEC-20 using about 10 s of compute time per input clock cycle. This suggests that the methodology is applicable to about 10^5 transistors corresponding to approximately 3×10^4 logical gates. Presently, many researchers are active in this general area of MOS transistor level simulation.

The question of approximate time-domain simulation at a far coarser level than the techniques reported in Section II-B is of general interest. In [60], [106] an approximate delay simulation is attempted for MOS circuits. The delays are computed using a table lookup and delay equation-type procedure. In [109] the MOSFET problem is formulated in terms of a switching theory in contrast to the resistive model approach discussed here.

IV. Mixed- and Multilevel Simulation/Analysis Systems

A VLSI circuit may be represented at several levels of complexity depending on the level of the model which may be architectural; functional; gate; device, transistor, circuit; and detailed device macromodel.

In fact, a complete VLSI design of a new technology usually

spans all of these levels. Thus it is very desirable for a design system to encompass as many of these levels as possible.

In the literature a distinction is made between a *multilevel* and a *multimode* system. The multilevel system is based on a hierarchical structure where a common database is employed. Each lower level adds more detail and accuracy to the simulation results like the above list. *Mixed* systems involve two levels and *multi* systems may involve more than two levels.

The first mixed and multilevel systems were different in scope. For example, at the high end logic level simulation was combined with the functional level, e.g., [62]. In the circuit domain, the first macromodels for digital logic were incorporated in a circuit analysis program leading to mixed circuit analysis/simulation systems [32], [55].

The technology of multilevel S/A systems is presently evolving. An early example of such a system is DIANA [87]. Here, the techniques span the digital as well as the analog circuit domain which include analog to digital (A/D), D/A converters as well as filters. With the advent of digital filters, this capability is becoming increasingly important. The internal structure of DIANA is such that even the circuit-analysis-oriented portion can be event driven.

In *mixed-mode* and *multimode* systems, the different programs operate at the same level but the function is different. For example in [19], a logic simulator is employed in the mask design task of the system while the WR time S/A program is used for the timing verification of design. An interesting combined mixed-level and mixed-mode piecewise linear modeling technique is given in [20].

Two recent papers report on CAD systems at Sandia [107] and Hughes [108]. They have a multilevel S/A "engine" as a central part of the system. This approach has many advantages for large-scale VLSI circuits. First, the problems of data conversion from one program to another is avoided in a well-designed system. This is only one reason why the design time is reduced. Other, more important time and storage savings result from the "magnifying glass" technique where, for example, the overall simulation proceeds at a high level while only a few circuits are analyzed accurately at a more detailed level. The potential increase in the number of subcircuits which can be analyzed is substantial. Clearly, a multilevel system provides the environment for these tradeoffs. Further, the danger of losing relevant information by using a too-simple implementation always exists. In a multilevel system the results can always be checked by a lower level technique which ultimately may be the exact circuit analysis level.

V. Verification

As mentioned above, *design* and *verification* are the two key purposes of a CAD system. The former is the "creation" process while the latter insures the "correctness" of the design. Verification is of concern at all levels in the design. Examples of verification steps are the following: functional design from architectural design; correctness of gate level implementation; electrical connectivity; design rules violations; and timing verification.

Verification is not only of importance in the initial design, but also in the engineering design changes which follow each design phase. The verification process must establish the correspondence among all the models used for the representation of the design [76]. The design and verification processes span the entire spectrum from the system architecture to the implementation of the shapes on silicon wafers. Some of the

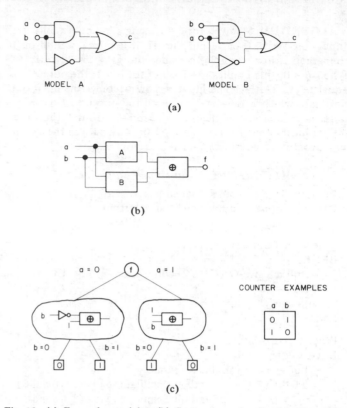

Fig. 13. (a) Example models. (b) Comparison circuit. (c) Expansion procedure.

design procedures at the shapes level involve automatic translation programs [64]–[65], [72] which guarantee correctness. This shifts the burden of verification to the CAD programs. This is a desirable trend since the designs obtained by such a CAD system are automatically correct. Some of the design steps are more amenable to automation, e.g., [71] while others like timing verification are much more difficult to quantize. Here, we give more details on two verification levels, the functional equivalence and timing verification.

A. Functional Equivalence of Logic Circuits

One form of design verification is to determine whether two implementations of the same specification are equivalent. In the IBM 3081 machine, the Static Analysis program [74] was used to find input conditions which would cause two implementations to differ. The two implementations were at the gate level and register transfer level (RTL). Usually the RTL level design is considered correct since a large amount of simulation is possible at this level. However, the 3081 experience demonstrated that errors were found at each level.

The fundamental idea behind this form of design verification is the efficient solution of a set of Boolean equations. The RTL level design is translated to Boolean equations and then each output is exclusive-ored with the corresponding output in the gate level description. All the outputs are ored together to form a single output function f. Solving $f = 1$ is equivalent to finding input patterns which cause the gate level and RTL level designs to differ. If no solutions are found, the two models are declared equivalent. If solutions are found, they represent counter examples which cause at least one of the outputs to be different. Fig. 13 demonstrates the design verification algorithm for comparing two models A and B for equivalence.

ALGORITHM 5 solves for $f = 1$ using symbolic simulation. Inputs are set to 0 or 1, and the effect is propagated through the graph. For example, consider the gate $C = \text{AND}(A, B)$. When $B = 0$, this implies $C = 0$. But for $B = 1$, the output C is equal to A. $C = A$. The input A now becomes the input of a new gate which is one level closer to the output. This procedure is recursively applied by expanding around the other inputs until either $f = 1$ or until $f = 0$. Any path in the expansion process which causes $f = 1$ is a counter example.

ALGORITHM 5: Functional Equivalence

Inputs: Two combinational circuits $C1, C2$
Result: Input patterns which cause output
         values of $C1$ and $C2$ to be different

Procedure BooleanCompare$(C1, C2)$
BEGIN
    $C = $ Exclusive-Or$(C1, C2)$ connections as Fig. 13
    Stack = empty
    Expand(C)
END

Procedure Expand(C)
BEGIN
    Choose input variable X from C
    $CX0 = C(X = 0)$ Symbolic simulate (expand around 0)
    $CX1 = C(X = 1)$ Symbolic simulate (expand around 1)
    Traverse$(X, 0, CX0)$
    Traverse$(X, 1, CX1)$
END

Procedure Traverse$(X, \text{Direction}, C)$
BEGIN
    Push onto Stack $(X, \text{Direction})$
    If C not $= 0$ or 1 then   BEGIN Expand(C)
                                           RETURN
                                    END
    If $C = 1$ then Print Stack
    Pop Stack$(X, \text{Direction})$
END

The main advantage of this form of verification is that no simulation is required. The algorithm formally proves the equivalence of two designs. The computational complexity of this algorithm is known to be NP-complete. However, for most practical problems the algorithm exhibits polynomial time behavior. One reason for this is that during the expansion process, two subgraphs may be identical to one another. In this situation, recursively solving for counter examples in two subgraphs reduces to solving for counter examples in one. This verification step clearly precedes the timing verification step given in the next section.

B. Timing Verification

As discussed above, timing verification is an important aspect of the process unless the circuits are designed for a very low performance technology. Here we again will distinguish between high-performance *bipolar* circuits which are packaged on multichip modules and MOSFET circuits where most of the communication among the circuits is on chip.

Again, the *bipolar* transistor delay may be dominated by the interconnection delay [6] and the subcircuits are usually gates which can be described by delay equations. Thus rather accurate delay equation macromodels can be obtained which can

be added to obtain path delays. Algorithms for timing verification based on this delay model are applicable for large-scale systems.

Timing verification algorithms using path delays are discussed in [13]. They have the capability of selectively disabling unused paths. This is useful in detecting long functional paths which are hidden by unused paths with worse timing problems. Their implementation propagates a null signal through unused paths which no longer participate in the delay calculation. To approximate statistical delay, they use a min. and a max. range for the rising and falling delays. Furthermore, logic gates fall into one of three categories—positive unate(and, or), negative unate(inverter), and nonunate(xor). A rising delay through a negative unate gate becomes a falling signal. Their implementation also includes the analysis of clock skew, pulsewidth, and both long and short paths. Since the algoritm is block oriented rather than path oriented, the performance is linear with the number of gates in the circuit. The implementation is sufficiently general for large problems of 100 000 logic gates and consumes only 2 ms/gate.

The SCALD timing verifier [76] uses timing assertions based on a calculus of 7 values to verify the design. Here a signal can take the value 0, 1, stable, changing, rising, falling, or unknown. During a block cycle, SCALD determines when a signal is changing and when it is stable. By setting inputs to 0 or 1, certain paths are eliminated from the analysis since changing signals no longer propagate. This is necessary to achieve the proper timing for variable length cycles. To verify the correct functioning of memory chips, a setup and hold check is performed. This guarantees that the data are stable before the clock arrives and that it is held long enough before the clock falls. Finally, a minimum pulsewidth is checked for because of the large variations in circuit rising and falling delays. The algorithm is event driven and is computationally efficient on problems as large as 100 000 gates.

The Timing Analysis (TA) technique [78] is the only algorithm which creates slack diagnostics. The slack for a logic gate is defined as the difference between the expected arrival and the actual arrival time. Slack is a measure of how bad a long or short path is. Another feature of the algorithm is that path lengths are statistical rather than worse case. This allows for process variation rather than worse case. A three sigma design allows for process variation which naturally occurs in the manufacturing cycle. The analysis of paths is block oriented rather than by path enumeration so that an entire IBM 3081 processor of 700 000 gates can be handled efficiently. For problems of this magnitude, a software paging scheme is necessary. Another way of avoiding the data explosion problem for 700 000 gates is to use TA to solve small partitions of the machine and then combine the solutions together. Delay modifiers are also a feature of the TA technique. This allows the user to cut paths that are never active or to adjust delays for multicycle paths. All gates are considered as either inverting or noninverting. The rising and falling delays for each gate use this information to accurately predict delay values as is carefully illustrated in Fig. 14 and (30). The meaning of the labels used is as follows:

$a_0, b_0 = $ rising arrival times
$a_1, b_1 = $ falling arrival times
$\quad d_0 = $ delay if output is rising
$\quad d_1 = $ delay if output is falling
$\quad c_0 = $ rising output arrival
$\quad c_1 = $ falling output arrival.

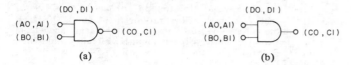

Fig. 14. (a) TA model for an inverting gate. (b) Model for a noninverting gate.

For the late mode the following equations hold for Fig. 14(a)

$$c_0 = \text{MAX}(a_1, b_1) + d_0$$
$$c_1 = \text{MAX}(a_0, b_0) + d_1 \tag{30a}$$

while for the early mode

$$c_0 = \text{MIN}(a_1, b_1) + d_0$$
$$c_1 = \text{MIN}(a_0, b_0) + d_1. \tag{30b}$$

Similarly, for a noninverting circuit in Fig. 14(b) the late mode equations are

$$c_0 = \text{MAX}(a_0, b_0) + d_0$$
$$c_1 = \text{MAX}(a_1, b_1) + d_1 \tag{30c}$$

while for the early mode

$$c_0 = \text{MIN}(a_0, b_0) + d_0$$
$$c_1 = \text{MIN}(a_1, b_1) + d_1. \tag{30d}$$

Using this gate or block model, ALG. 6 gives the timing verification for a large-scale clocked circuit.

ALGORITHM 6: Timing Analysis (TA)

Inputs: Combinational Circuit C
        Primary input arrival times Tin
        Expected Output times $Tout$
        Circuit delays D
Results: Worst case arrival times $tARR$
        Slack times for each circuit $tSLACK$

Procedure TimingVerification($C, Tin, Tout, D$)
BEGIN
    Levelize C by a topological sort
    Schedule gates in levelized order
    For each gate i do
        $tARR(i) = D(i) + \text{MAX}(tARR(\text{predecessors of } i))$
    For all outputs j do
        $tSLACK(j) = Tout(j) - tARR(j)$

    Schedule gates in reverse levelized order k
    For each connection Edge(m,k) do
    BEGIN
        Edge $tSLACK(m,k) = tARR(k) - D(k)$
        $+ tSLACK(k) - tARR(m)$
    END
END

Worst case arrival times are computed in linear time by making only one sweep through the circuit from primary inputs to primary outputs. Slacks are useful as diagnostics for timing verification. They represent how late (or early) a signal is with respect to the expected arrival time at the primary outputs. The computation of slacks is done in linear time by making one sweep through the circuit in the backward direction from primary outputs to primary inputs. However, it needs the arrival times computed during the forward propagation phase.

The critical path in the circuit is the one with the smallest slack. There are many ways to fix these timing problems. One way is to change low-power subcircuits to high power on nets which drive a high capacitance load. For chip-to-chip timing problems, designers may change to IO assignment or driver and receiver types to prevent electrical reflections. This is another example of how innovative algorithms extend the capabilities of the design procedure to VLSI dimensions.

The area of verification tools is presently evolving [62]–[78] as is illustrated by the above examples. For timing verification at the transistor level, the techniques given in Section II are used. Again, the procedures at the transistor level are more complicated than at the gate level. Mainly MOS transistor responses are highly dependent on the interconnections and the capacitances and thus the delay and rise times include much more variability than the corresponding bipolar transistor gates. This forces the verification to a much more refined level especially for passtransistor circuits.

VI. Conclusions

At present, the areas reported on in this paper are receiving considerable attention from industry as well as universities since they are of key importance for the design of realistic VLSI circuits. As is evident from the recent references in this paper, much progress has been made in the last few years across the entire spectrum of VLSI problems. The main task that needs to be performed is to find algorithms and implementations which yield acceptable running times at the VLSI level. Programs are in use today in some areas like logic simulation and timing verification which accommodate circuits at the VLSI level without excessive running time and storage requirements.

Acknowledgment

The authors would like to thank G. Almasi for the valuable suggestions and careful reading of the manuscript. They would also like to acknowledge the contributions made by the discussions with other researchers active in the field. Within the space given here they can only mention a few names: C. Carlin, W. Donath, G. Hachtel, I. Hajj, E. Lelarasmee, F. Odeh, G. Rabbat, A. Sangiovanni-Vincentelli, and V. Visvanathan.

References

[1] C. H. Sequin, "Managing VLSI complexity: An outlook," this issue, pp. 149–166.
[2] C. Niessen, "Hierarchical design methodologies and tools for VLSI chips," this issue, pp. 66–75.
[3] A. E. Ruehli, N. Rabbat, and H. Y. Hsieh, "Macromodeling—an approach for analyzing large-scale circuits," *Comput. Aided Design*, vol. 10, pp. 121–130, Mar. 1978.
[4] G. D. Hachtel and A. L. Sangiovanni-Vincentelli, "A survey of third generation simulation techniques," *Proc. IEEE*, vol. 69, pp. 1264–1280, Oct. 1981.
[5] A. E. Ruehli, "Survey of computer-aided electrical analysis of integrated circuit interconnections," *IBM J. Res. Develop.*, vol. 23, pp. 627–639, Nov. 1979.
[6] E. E. Davidson, "Electrical design of high speed computer package," *IBM J. Res. Develop.*, vol. 26, pp. 349–361, May 1982.
[7] M. A. Beuer, Ed., *Digital System Design Automation: Languages, Simulation and Data Base*. Woodland Hills, CA: Computer, Sc. Press, 1975.
[8] M. A. Breuer and A. D. Friedman, *Diagnosis & Reliable Design of Digital Systems*. Potomac, MD: Comp. Sc. Press, 1976.
[9] M. A. Breuer, A. D. Friedman, and A. Iosupovicz, "A survey of the state of the art of design automation," *Computer* (IEEE), pp. 58–75, Oct. 1981.
[10] L. Bening, "Developments in computer simulation of gate level physical logic," in *Proc. 16th Design Automation Conf.*, San Diego, CA, June 1979, pp. 561–567.
[11] E. Ulrich and D. Herbert, "Speed and accuracy in digital network simulation based in structural modeling," in *Proc. 19th*

Design Automation Conf., Las Vegas, NV, June 1982, pp. 587–593.

[12] R. B. Hitchcock, Sr., "Timing verification and the timing analysis program," in *Proc. 19th Design Automation Conf.*, Las Vegas, NV, June 1982, pp. 594–604.

[13] L. C. Bening, T. A. Lane, and C. R. Alexander, "Developments in logic network path delay analysis," in *Proc. 19th Design Automation Conf.*, Las Vegas, NV, June 1982, pp. 605–613.

[14] J. P. Roth, *Computer Logic, Testing, and Verification.* Potomac, MD: Comp. Sc. Press, 1980.

[15] D. A. Antoniadis and R. W. Dutton, "Models for computer simulation of complex IC fabrication process," *IEEE J. Solid-State Circuits*, vol. SC-14, pp. 412–422, Apr. 1979.

[16] W. L. Engl, H. K. Dirks, and B. Meinerzhagen, "Device modeling," this issue, pp. 10–33.

[17] G. Arnout and H. De Man, "The use of threshold function and boolean-controlled network elements for macromodelling of LSI circuits," *IEEE J. Solid-State Circuits*, vol. SC-13, pp. 326–332, June 1978.

[18] H. De Man, "Computer aided design for integrated circuits: Trying to bridge the gap," *IEEE J. Solid-State Circuits*, vol. SC-14, pp. 613–621, June 1979.

[19] H. De Man, "Mixed mode simulation for MOS VLSI: Why, Where and How?" in *Proc. IEEE Int. Symp. Circuits System*, Rome, Italy, May 1982, pp. 699–701.

[20] W.M.G. Van Bokhoven, "Mixed-level and mixed-mode simulation by a piecewise-linear approach," in *Proc. IEEE Int. Symp. Circuits System*, Rome, Italy, May 1982, pp. 1256–1258.

[21] R. W. Jensen and M. D. Lieberman, *IBM Electronic Circuit Analysis Program.* Englewood Cliffs, N.J.: Prentice-Hall, 1968.

[22] L. W. Nagel, "SPICE2: a computer program to simulate semiconductor circuits," Univ. of California, Berkeley, ERL Memo ERL-M520, May 1975.

[23] E. Cohen, "Program reference manual for SPICE2," Univ. of California, Berkeley, ERL Memo ERL-M592, June 1976.

[24] A. Vladimirescu, K. Zhang, A. R. Newton, D. O. Pederson, and A. Sangiovanni-Vincentelli, "SPICE Version 2G User's guide," University of California, Berkeley, Tech. Memo., Aug. 10, 1981.

[25] "Advanced statistical analysis program (ASTAP)," Program reference manual, Pub. No. SH20-1118-0, IBM Corp. Data Proc. Div., White Plains, NY 10604.
W. T. Weeks, A. J. Jimenez, G. W. Mahoney, D. Mehta, H. Quassemzadeh, and T. R. Scott, "Algorithms for ASTAP—a network analysis program," *IEEE Trans. Circuit Theory*, vol. CT-20, pp. 628–634, Nov. 1973.

[26] L. W. Nagel, "ADVICE for circuit simulation," in *Proc. IEEE Int. Symp. Circuits and Systems*, Houston, TX, Apr. 1980.

[27] G. D. Hachtel, R. K. Brayton, and F. Gustavson, "The sparse tableau approach to network analysis and design," *IEEE Trans. Circuit Theory*, vol. CT-18, pp. 101–113, Jan. 1971.

[28] C. Ho, A. E. Ruehli, and P. A. Brennan, "The modified nodal approach to network analysis," *IEEE Trans. Circuits Syst.*, vol. CAS-22, pp. 504–509, June 1975.

[29] N. B. Rabbat, W. D. Ryan, and S. Q. Hossain, "Computer modeling of bipolar logic gates," *Electron. Lett.*, vol. 7, pp. 8–10, Jan. 1971.

[30] O. Wing and E. B. Kozemchak, "Computer analysis of digital integrated circuits," in *NERM Conf. REC.*, Boston, MA, Nov. 1971, IEEE Cat. No. 71C51, pp. 189–191.

[31] S. C. Bass and S. C. Peak, "Terminal models of digital gates allowing waveform simulation," in *Proc. IEEE Int. Symp. Circuit Theory*, Apr. 1973, pp. 287–289.

[32] N. Rabbat, A. E. Ruehli, G. W. Mahoney, and J. J. Coleman," A survey of macromodeling," in *Proc. IEEE Int. Symp. Circuits Systems*, Apr. 1975, pp. 139–143.

[33] B. Chawla, H. K. Gummel, and P. Kozah, "MOTIS—a MOS timing simulator," *IEEE Trans. Circuit Syst.*, vol. CAS-22, pp. 301–310, Dec. 1975.

[34] N. Rabbat and H. Y. Hsieh, "A latent macromodular approach to large-scale sparse networks," *IEEE Trans. Circuit Syst.*, vol. CAS-22, pp. 745–752, Dec. 1976.

[35] S. P. Fan, M. Y. Hsueh, A. R. Newton, and D. O. Pederson, "MOTIS-C: a new circuit simulator for MOS LSI circuits," in *Proc. IEEE Int. Symp. Circuits Systems*, 1977, pp. 700–703.

[36] E. Lelarasmee, A. E. Ruehli, and A. L. Sangiovanni-Vincentelli, "The waveform relaxation method for time-domain analysis of large-scale integrated circuits," *IEEE Trans. CAD Integ. Circ. Syst.*, vol. CAD-1, pp. 131–145, Jul. 1982.

[37] A. E. Ruehli, A. L. Sangiovanni-Vincentelli, and N.B.G. Rabbat, "Time analysis of large scale circuits containing one-way macromodels," *IEEE Trans. Circuits Syst.*, vol. CAS-29, pp. 185–189, Mar. 1982.

[38] E. Lelarasmee and A. Sangiovanni-Vincentelli, "RELAX: A new circuit simulator for large scale MOS integrated circuits," Electronic Research Laboratory, Univ. of California, Berkeley, Memo UCB/ERL M82/6, Feb. 1982.

[39] M. Tanabe, H. Nakamura, and K. Kawakita, "MOSTAP: An MOS circuit simulator for LSI circuits," in *Proc. IEEE Intl.*

[40] *Symp. Circuits Systems*, Houston, pp. 1035–1038, Apr. 1980.

[40] N. G. Rabbat, A. L. Sangiovanni-Vincentelli, and H. Y. Hsieh, "A multilevel Newton algorithm with macromodeling and latency for the analysis of large-scale nonlinear circuits in the time domain," *IEEE Trans. Circuits Syst.*, vol. CAS-26, pp. 733–741, Sept. 1979.

[41] G. DeMicheli and A. L. Sangiovanni-Vincentelli, "Numerical properties of algorithms for the timing analysis of MOS VLSI circuits," in *Proc. 1981 Europ. Conf. on Circuit Theory and Design*, Aug. 1981, pp. 387–392.

[42] M. A. Breuer, "Techniques for the simulation of computer logic," *Commun. Ass. Comput. Mach.*, pp. 443–446, Jul. 1964.

[43] E. G. Ulrich, "Time sequenced logical simulation based on circuit delay and selective tracing of active network path," in *Proc. ACM Nat. Conf.*, 1965, pp. 437–448.

[44] E. Ulrich, "Exclusive simulation of activity in digital networks," *Commun. Ass. Comput. Mach.*, vol. 12, no. 2, pp. 102–110, Feb. 1969.

[45] S. A. Szygenda, "TEGAS-Anatomy of a general purpose test generation and simulation at the gate and functional level," in *Proc. 9th Design Automation Conf.*, June 1972, pp. 116–127.

[46] E. G. Ulrich and T. Baker, "The concurrent simulation of nearly identical digital networks," in *Proc. 10th Design Automation Conf.*, June 1973, pp. 145–150.

[47] S. A. Szygenda and E. W. Thompson, "Digital logic simulation in a time-based table-driven environment: part 1, design verification," (IEEE) *Computer*, pp. 24–36, Mar. 1975.

[48] H. E. Krohn, "Design verification of large scientific computers," in *Proc. 14th Design Automation Conf.*, June 1977, pp. 354–361.

[49] R. E. Bryant, "MOSSIM: A switch-level simulator for MOS LSI," in *Proc. 18th Design Automation Conf.*, Jul. 1981, pp. 786–790.

[50] R. E. Bryant, "An algorithm for MOS Logic simulation," *Lamda Mag.*, Fourth Quarter, pp. 46–53, 1980.

[51] J. Watanabe, J. Miura, T. Kurachi, and I. Suetsugu, "Seven value logic simulation for MOS LSI circuits," presented at the IEEE Intl. Conf. Circuits and Computers, Port Chester, NY, Oct. 1980, pp. 941–944.

[52] W. Sherwood, "An MOS modeling technique for 4-state true-value hierarchical logic simulation," in *Proc. 18th Design Automation Conf.*, Nashville, TN, Jul. 1981, pp. 775–785.

[53] C. Mead and L. Conway, *Introduction to VLSI Systems.* Reading, MA: Addison-Wesley, 1980.

[54] N. Rabat, A. Y. Hsieh, and A. E. Ruehli, "Macromodeling for the analysis of large-scale networks," in ELECTRO-76 Professional Program 21, May 1976, pp. 1–8.

[55] A. R. Newton, "Techniques for the simulation of large-scale integrated circuits," *IEEE Trans. Circuits Syst.*, vol. CAS-26, pp. 741–749, Sept. 1979.

[56] D. Hill and W. van Cleemput, "SABLE: a tool for generating structural, multi-level simulation," in *Proc. 16th Design Automation Conf.*, San Diego, CA, June 1979, pp. 403–405.

[57] V. D. Agrawal, A. K. Bose, P. Kozak, H. N. Nham, and E. Pacas-Skewes, "A mixed-mode simulator," in *Proc. 17th Design Automation Conf.*, Minneapolis, MN, June 1980, pp. 1–8.

[58] T. Sasaki, A. Yamada, S. Kato, T. Nakazawa, K. Tomita, and N. Nomizu, "MIXS: a mixed level simulator for large digital system logic verification," in *Proc. 17th Design Automation Conf.*, Minneapolis, MN, June 1980, pp. 626–633.

[59] V. D. Agrawal, A. K. Bose, P. Kozak, H. N. Nham, and E. Pascal-Skewes, "A mixed model simulator," in *Proc. 17th Design Automation Conf.*, Minneapolis, MN, June 1980, pp. 618–625.

[60] H. H. Nham and A. K. Bose, "A multiple delay simulator for MOS LSI circuits," in *Proc. 17th Design Automation Conf.*, Minneapolis, MN, June 1980, pp. 610–611.

[61] P. H. Reynaert, H. De Man, G. Arnout, and J. Cornelissen, "DIANA: a mixed-mode simulator with a hardware description language for hierarchical design of VLSI," in *Proc. IEEE Intl. Conf. Circuits and Computers*, Port Chester, NY, Oct. 1980, pp. 356–360.

[62] D. D. Hill and W. M. Van Cleemput, "SABLE: Multilevel simulation for hierarchical design," in *Proc. IEEE Int. Symp. Circuits and Systems*, Houston, TX, Apr. 1980, pp. 431–434.

[63] W.M.G. van Bokhoven, "Macromodeling and simulation of mixed analog-digital networks by piecewise-linear system approach," in *Proc. IEEE Intl. Conf. Circuits and Computers*, Port Chester, NY, Oct. 1980, pp. 361–365.

[64] M. E. Daniel and C. W. Gwyn, "Hierarchical VLSI circuit design," in *Proc. IEEE Intl. Conf. Circuits and Computers*, Port Chester, NY, Oct. 1980, pp. 92–97.

[65] L. Scheffer and R. Apte, "LSI design verification using topological extraction," in *Proc. 12th Asilomar Conf. Circuits and Systems, and Computers*, Nov. 1978, pp. 149–153.

[66] C. R. McCaw, "Unified shapes checker—A checking tool for LSI," in *Proc. 16th Design Automation Conf.*, June 1979, pp. 81–87.

[67] R. Auerbach, "FLOSS: Macrocell Compaction system," presented at the 1979 IEEE Design Automation Workshop, East

Lansing, MI, 1979.

[68] C. S. Chang, "LSI layout checking using bipolar device recognition technique," in *Proc. 16th Design Automation Conf.*, June 1979, pp. 95–101.

[69] T. Mitsuhashi, T. Chiba, M. Takashima, and K. Yoshoda, "An integrated mask artwork analysis system," in *Proc. 17th Design Automation Conf.*, June 1980, pp. 277–284.

[70] J. P. Avenier, "Digitizing, layout, rule-checking—The everyday tasks of chip designers," this issue, pp. 49–56.

[71] H. C. Godoy, G. B. Franklin, and P. S. Bottorff, "Automatic Checking of Logic design structures for compliance with testability ground rules," in *Proc. 14th Design Automation Conf.*, June 1977, pp. 469–472.

[72] C. M. Baker and C. Terman, "Tools for verifying integrated circuit designs," *Lambda Mag.*, Fourth Quarter, pp. 22–30, 1980.

[73] R. N. Gustafson and F. J. Sparacio, "IBM 3081 processor unit: Design consideration and design process," *IBM J. Res. Develop.*, vol. 26, pp. 12–21, Jan. 1982.

[74] G. L. Smith, R. J. Bahnsen, and H. Halliwell, "Boolean comparison of hardware and flow charts," *IBM J. Res. Develop.*, vol. 26, pp. 106–116, Jan. 1982.

[75] M. Monachino, "Design verification system for large-scale LSE designs," *IBM J. Res. Develop.*, vol. 26, pp. 89–99, Jan. 1982.

[76] T. M. McWilliams, "Verification of timing constraints of large digital systems," in *Proc. 17th Design Automation Conf.*, Minneapolis, MN, June 1980, pp. 139–147.

[77] S. Newberry and P. J. Russell, "A programmable checking tool for LSI," in *Proc. IEE Europ. Conf. Electrical Design Automation*, Brighton, U.K., Sept. 1981, pub. no. 200, pp. 183–187.

[78] R. B. Hitchcock, Sr., G. L. Smith, and D. D. Chang, "Timing analysis of computer hardware," *IBM J. Res. Develop.*, vol. 26, pp. 100–105, Jan. 1982.

[79] C. W. Gear, "The automatic integration of ordinary differential equations," in *Proc. Information Processing 68*, A.F.H. Morrel, Ed. Amsterdam, The Netherlands: North-Holland, 1968, pp. 187–193.

[80] L. O. Chua and P. M. Lin, "Computer-aided analysis of electronic circuits: algorithms and computational techniques." Englewood Cliffs, NJ: Prentice Hall, 1975, chap. 10, pp. 410–431.

[81] F. Odeh and W. Liniger, "On A-Stability of second-order two step methods for uniform and variable steps," in *Proc. IEEE Intl. Conf. Circuits and Computers*, Port Chester, NY, 1980, pp. 123–126.

[82] A. E. Ruehli, P. A. Brennan and W. Liniger, "Control of numerical stability and damping in oscillatory differential equations," in *Proc. IEEE Intl. Conf. Circuits and Computers*, Port Chester, NY, Oct. 1980, pp. 111–114.

[83] V. M. Vidigal and S. W. Director, "A design centering algorithm for non convex regions of acceptability," *IEEE Trans. CAD Integ. Circ. Syst.*, vol. CAD-1, pp. 13–24, Jan. 1982.

[84] P. Yang, I. N. Hajj, and T. N. Trick, "Slate: A circuit simulation program with latency exploritation and node tearing," in *Proc. IEEE Intl. Conf. Circuits and Computers*, Port Chester, NY, Oct. 1980, pp. 353–355.

[85] D. A. Calahan, "Multilevel vectorized sparse solution of LSI circuits," in *Proc. IEEE Intl. Conf. Circuits and Computers*, Port Chester, NY, Oct. 1980, pp. 976–979.

[86] A. Vladimirescu and D. O. Pederson, "Performance limits of the CLASSIE circuit simulation program," in *Proc. Int. Symp. on Circuits Systems* (Rome, Italy, May 1982), pp. 1229–1232.

[87] H. De Man, J. Rabaey, G. Arnout, and J. Vandervalle: "DIANA as a mixed-mode simulator for MOS LSI Sampled-data circuits," in *Proc. IEEE Intl. Symp. on Circuits and Systems,* Houston, TX, Apr. 1980, pp. 435–438.

[88] P. Penfield, Jr. and J. Rubinstein, "Signal delay in RC tree networks," in *Proc. 18th Design Automation Conf.*, June 1982, pp. 613–617.

[89] A. E. Ruehli, N. B. Rabbat, and H. Y. Hsieh, "Macromodular latent solution of digital networks including interconnections," *Proc. IEEE Int. Symp. Circuits and Systems*, New York, NY, Apr. 1978, pp. 515–521.

[90] K. A. Sakallah and S. W. Director, "An activity-directed circuit simulation algorithm," in *Proc. IEEE Intl. Conf. Circuits and Computers*, Port Chester, NY, 1980, pp. 1032–1035.

[91] ——, "An event driven approach for mixed gate and circuit level simulation," in *Proc. IEEE Int. Symp. Circuits and Systems*, Rome, Italy, May 1982, pp. 1194–1197.

[92] W. L. Engl, R. Laur, and H. Dirks, "MEDUSA—A simulator for modular circuits," *IEEE Trans. CAD Integ. Circ. Syst.*, vol. CAD-1, pp. 85–93, Apr. 1982.

[93] L. O. Chua and P-M Lin, *Computer-Aided Analysis of Electronic Circuits.* Englewood Cliffs, NJ: Prentice-Hall, 1975.

[94] J. Vlach and K. Singhal, *Computer Aided Circuit Analysis.* New York: Van Nostrand, 1983.

[95] J. M. Ortega and W. Rheinboldt, *Iterative Solution of Nonlinear Equations in Several Variables.* New York: Academic Press, 1970.

[96] A. R. Newton, "The analysis of floating capacitors for timing simulation," in *Proc. 13th Asilomar Conf. on Circuits Systems and Computers*, Pacific Grove, CA, Nov. 1979.

[97] Y. P. Wei, I. N. Hajj, and T. N. Trick, "A prediction—relaxation based simulator for MOS circuits," in *IEEE Intl. Conf. Circuits and Computers* (New York, NY, Sept. 1982), pp. 353–355.

[98] A. L. Sangiovanni-Vincentelli and N. G. Rabbat, "Techniques for the time domain analysis of LSI circuits," *IEE Proc.*, vol. 127, part G, pp. 292–301, Dec. 1980.

[99] R. Bernhard, "Technology '82/82 solid state VLSI/LSI components," *IEEE Spectrum*, pp. 49–63, Jan. 1982.

[100] E. Lelarasmee, A. E. Ruehli and A. S. Sangiovanni-Vincentelli, "Waveform relaxation decoupling (WRD) method," *IBM Techn. Discl. Bulletin*, vol. 24, no. 7B, pp. 3720–3721, Dec. 1981.

[101] P. Goel, H. Lichaa, T. E. Rosser, T. J. Stroh, and E. E. Eichelberger, "LSSD fault simulation using conjunctive combinational and sequential methods," in *Proc. IEEE 1980 Test Conf.*, Nov. 1980, pp. 371–376.

[102] E. M. DaCosta and K. G. Nichols, "MASCOT," *IEE Proc.*, vol. 127, part G, no. 6, pp. 302–307, Dec. 1980.

[103] P. W. Case, H. H. Graff, L. E. Griffith, A. R. LeClercq, W. B. Murley, and T. M. Spence, "Solid logic design automation," *IBM J. Res. Develop.*, vol. 8, pp. 127–140, 1964.

[104] E. Ulrich, "Table lookup techniques for fast and flexible digital logic simulation," in *Proc. 17th Design Automation Conf.*, Minneapolis, MN, June 1980, pp. 560–563.

[105] R. A. Rohrer and H. Nosrati, "Passivity Considerations instability studies of numerical integration algorithms," *IEEE Trans. Circuits Syst.*, vol. CAS-28, pp. 857–866, Sept. 1981.

[106] V. B. Rao, T. Trick, and M. Lightner, "Hazards in a multiple delay logic simulation," *Proc. IEEE Int. Symp. on Circuits and Systems*, Rome, Italy, May 1982, pp. 72–75.

[107] M. E. Daniel and C. W. Gwyn, "CAD system for IC design," *IEEE Trans. CAD Integ. Circ. Syst.*, vol. CAD-1, pp. 2–12, Jan. 1982.

[108] H. W. Daseking, R. I. Gardner and P. B. Weil, "VISIA: A VLSI CAD system," *IEEE Trans. CAD Integ. Circ. Syst.*, vol. CAD-1, pp. 36–51, Jan. 1982.

[109] J. P. Hayes, "A unified switching theory with applications to VLSI design," *Proc. IEEE*, vol. 70, no. 10, pp. 1140–1151, Oct. 1982.

Part II
Design Automation

THIS part consists of eight sections dealing with important aspects of design automation. High-level languages are dealt with in Section II-A. Only a representative sample of the research work in this area is included here. The area of high-level languages for VLSI design will witness substantial changes in the years to come and the reader should follow these changes in current conference proceedings and journals. Section II-B deals with one of the most, if not the most, important areas in design automation: silicon compilation. The use of silicon compilers in the 1990's will be a must, as the use of silicon simulators is a must today. The area will no doubt mature over the next five years and only useful and efficient compilers will survive, impacting the design of digital VLSI systems significantly. The papers presented in this section should be viewed as: "introduction to silicon compilation, more to come." Section II-C deals with circuit (mask) layout, and Section II-D deals with symbolic layout. It is the latter which will make in-roads into future layout design automation systems. The areas of layout verification and layout languages, placement, and routing are the subjects of Sections II-E and F.

Having good design automation tools is necessary, but not sufficient, for successful VLSI designs. What is needed is to integrate these DA tools, along with simulators, into one design environment using one CAD system. Thus, data structures and CAD systems for VLSI design become very important and these are covered in part in Sections II-G and H. Section II-H should be regarded as only representative of the state-of-the-art of this dynamic area.

Section II-A
High-Level Languages

HARDWARE DESCRIPTION SYSTEM DEVELOPMENT

BY DAVE ACKLEY, JOHN CARNEGIE, EDWIN B. HASSLER, JR.
DESIGN AUTOMATION DEPARTMENT
TEXAS INSTRUMENTS, INC.
P.O. BOX 226015, MS 3602
DALLAS, TEXAS 75266

I. ABSTRACT

Hardware Description Languages provide the basis for complete system design and specification. The Texas Instrument's Hardware Description Language provides not only the language framework for hierarchical specification, but a consistent basis for design verification, checkout, and communication. Current hardware design complexities require accurate representation and simulation. Hence, simulation is a required adjunct to a hardware description language. Together the hardware description language and simulator provide the user with a definitive model of the hardware. This paper describes the details of the hardware description language and associated simulator currently in use at Texas Instruments. Finally, a perspective is given on the impact the tool has on complex VLSI design programs.

II. DESIGN PROCESS

Figure 1, Design Process, shows one representation of the steps involved in a system design. The design process starts with a specification for the system. In addition to specifying the inputs and outputs to the system the specification would also define the desired outputs in terms of the inputs and time. This system is then verified or analysed to determine that the system will satisfy the desire intent. Clearly during this verification and analysis process the specification may need to be corrected or revised. Once a satisfactory system specification has been arrived at it is time to decompose this system into a group of interconnected subsystems each with its own specification. Next, the specifications of the subsystems plus their interconnections must be verified against the system specification. During this verification process, corrections and revisions may need to be made to the system decomposition, the subsystem specifications, or even the system level specification. This decomposition and next lower level specification process plus verification process is repeated until a level is reached which is directly implementable with known elements. These known elements may be devices, intergrated circuit chips, programmable components or previously designed modules or subsystems. The key aspect to this design process is the repeated decomposition and lower level specification followed by verification and analysis. This iterative top-down approach allows the designer to concentrate on the function (behavior) to be performed before addressing the implementation (structure plus lower level behavior). Thus, as this iterative process continues down through the lower levels of the design, the designer can focus in with increasing detail over a narrowing portion of the design. Therefore, the amount of information associated with each design block can be held to a managable level. This reduction of breadth with increasing of detail as you go down through the design hierarchy makes it possible to handle large complex designs in parts, each of which is of managable size. These formal design tools and methodology are relatively new and will be compared with historical design methodology in the next section.

III. HISTORICAL DESIGN METHODOLOGY

For discussion purposes I will break the system design process into four levels (system, subsystem, module and device). In actual practice each of these levels may be further subdivided for complex designs.

Figure 2 shows the historical methods used for specifying, simulating, and verifying these different design levels. The system level specifications are typically written in English. Various simulation programs such as the General Purpose Simulation System (GPSS), the Continuous System Modeling Program (CSMP), or SIMULA have been used for the simulation and verification at the system level. At the subsystem level a more formal specification process has been used. Examples of subsystem specification languages are various Register Transfer Languages (RTLs) and the Instruction Set Processor (ISP) language. Both RTL and ISP types of formal specification allow functional simulation for evaluating and verification of subsystem specifications. These subsystems were then broken down into modules. These modules were described in terms of the interconnection of gates or possibly other lower level modules and simulation was performed at the logic level. At the lowest level, these gates were broken down into the interconnection of individual discrete devices. These interconnected devices would then be simulated using an electrical simulator such as SPICE. The primary disadvantage of this previous approach was that there were multiple languages and simulators for the specification and evaluation of the different design levels. This made the designer's task very difficult. He had to be an expert in many different languages and simulators. The designer was thus forced to do many manual translations of his design from one level to another, plus transform the input data and compare the output data in different simulator formats. Clearly what was needed was a unified approach to the specification, simulation, and verification of the different levels of the system design.

IV. HARDWARE DESCRIPTION LANGUAGE (HDL)

If we go back and examine the design process, we see that there are two different types of information that must be dealt with. One type is the description of the functional behavior of a block (system, subsystem, or module) and the other type is the implementation of a block in terms of the interconnections (structure) of lower level blocks. The behavior is described in terms of the inputs and outputs plus the relationship between the inputs and the outputs as a function of time. Figure 3, shows the hierarchical design of a system in a Hardware Description Language (HDL). At the topmost level, the system is specified in terms of its behavior. Then the system is decomposed into the structured interconnection of subsystem blocks. Then the specifications for these subsystems is written in terms of their individual behaviors. If this design step has been done correctly, then for the same set of inputs, the interconnection speci-

Reprinted from *IEEE Int. Conf. Circuits and Computers Proc.*, 1982, pp. 608–611.

fied at the system level combined with the behavior of the subsystems will produce the same outputs as the system behavior. Next, we decompose each subsystem into a set interconnected modules. We then repeat the specification part of the process by specifying the behavior of each of the modules that are interconnected to form the subsystems. As shown in Figure 1, we continue this iterative decomposition and specifications until we reach a level which is a directly implementable with devices, IC chips, programmable components, or some previously designedmodule or subsystem. Note, that during the design process, the behavior of a block serves as the specification for that block. When the design has been completed, the behaviors at each level will have been revised to reflect the actual behavior of the block and thus serves as a data sheet for the block. By separating the behavior of a block from the decomposition (structure) of a block, we divide the HDL language requirements into two parts, behavior and structure. The behavior is a formal specification or model for the functional behavior of a block. In the case of the Texas Instrument's Hardware Description Language, a Pascal like programming language was chosen to describe these functional behaviors. This allows the user to write a program which describes what the outputs are as a function of the inputs and time. The structure of a block is described by listing the lower level block names, their generic block type, and the names of the interconnecting nodes of the structure. By describing the structure in this way we avoid imposing limitations on the behavior of the boxes that are being described. that is, the structure section can be used to describe the interconnection of electrical devices, interconnecting cables in a computer system, or even pipes that carry fluid in a building. It is important to note that for each block the behavior and the structure plus lower level behavior are redundant input/output definitions. If these were not redundant input/output definitions, then the blocks implementation (structure plus lower level behavior) does not match its specification and the design is incorrect.

V. MULTILEVEL SIMULATOR

The key design tool that must be available for use with an HDL is simulator. This simulator must be able to handle any and all levels of the HDL design hierarchary. Such a multilevel simulator would be used to exercise the system level behavior for verification and analysis purposes. This would allow us to verify the systems specification by applying a suitable set of inputs and examining the resulting outputs. We would then use the simulator to verify the system decomposition plus subsystem behaviors. We would apply the same set of inputs that we had used to verify the system level and compare the results from the system level behavior simulation. If the outputs match, then we have, to the limit of our test patterns, verified that our decomposition of the system and specifications of the subsystems are correct. These different levels of simulation are depicted in Figure 4. Thus, we see that the simulator needs to be capable of simulating both the system level behavior and the system level structure plus subsystem behavior. Similarly, when we decompose the subsystems into their structure plus module behaviors, we would use the multilevel simulator to verify the correctness of our decomposition. There are several ways that we can do this. One way is to have recorded the inputs and outputs for each subsystem during the system, subsystem level of verification. These recorded inputs would be used to stimulate the individual subsystems. If the decomposition is correct, these inputs should produce the same outputs. An alternate approach is to do a multilevel simulation to generate and interpret the internal inputs and outputs. Looking at Figure 4, we see that we could use the system decomposition (structure) plus the other subsystem behaviors to provide the environment for testing a subsystem structure and module level behaviors.

Figure 4 shows how we can go down through the hierarchy using the behaviors at one level or the structures at that level and the behaviors from some lower level. Clearly there are economic reasons why it would not be desirable to simulate a system or subsystem in terms of the lowest level devices and higher level structures. The amount of information that would have to be processed would cause the simulation to run very slowly and probably would not give any more information than a more straightforward test of a small portion of the system. The multilevel simulator that we are currently using at Texas Instruments not only allows the mixing of several levels, but also allows functional (ones and zeros) type simulation plus miltistate logic simulation. Multivalue logic simulation allows a signal to be represented as being in one of a number of states other than just one or zero. These additional states include high impedence, rising, falling, ambiguous, uninitialized, etc. Thus we can run portions of the simulation at a functional behavior level for efficiences, while other portions are run at a more detailed logic level. In principal we could run functional, logical, and electrical simulation all as one integrated multilevel simulator. The simulation overhead and the difficulty of correctly converting from logic levels to electrical levels and back make the value of such a simulator questionable. In practice we have found what is needed is a single unified language for describing the different design of the levels is the key feature and that having a separate simulator for the electrical level versus the functional/logic level simulation has proved fully adaquate.

VI. OTHER DESIGN AIDS

While the multilevel simulator driven from an HDL description of the system is the key design aid, other design aids can also be of significant value during the design process. Figure 5 shows some of these and is divided into two categories. One is a set of general capabilities which can be applied to system, subsystem, module, and the IC design levels. The other category of design aids are those that are specific to the IC design. In terms of general capabilities one of the problems with a simulation only methodology, is that it is not practical to apply all of the possible patterns to test for worst case timing paths (or race conditions). Thus, it is appropriate to have a separate timing analyzer which will go through the HDL data base and determine which are the worst case timing paths. Such an analyzer would be able to examine all paths, whereas the simulator would only have shown those paths that had been stimulated by the particular test patterns used. Another design correctness analyzer is a drive/load analyzer. Such an analyzer would compare the drive capability of each source against the loading of the inputs to make sure that the drive capability of individual driven devices has not been exceeded. Next, even if the design is correct, is it testable? Thus there are two additional design aids that are desirable to support testing. One is testability analyzer which will indicate the relative difficulty of stimulating internal nodes and the relative difficulty of observing these internal nodes. This information can be used during the design process to make sure that the system is reasonably testable. The other two test design aids are a test pattern generator and a test pattern grader. For IC design specific capabilities, TI's current system supports a logic array routability analyzer, an automated logic array router, an interactive chip layout system, and a layout to HDL data base verification subsystem. The key to being able to support this board spectrum of design aids is the unified Hardware Description Language data base.

VII. CONCLUSIONS

In this paper we have described the system design process and the role played by a Hardware Description Language, its associated simulator, and other design aids. It is clear that a unified Hardware Description Language coupled to an appropriate design methodology are key to the design of complex systems, subsystems, and VLSI components. The use of Hardware Description Lanaguages and its interfacing to various design aids will continue to evolve and improve as we strive to develop more sophisticated and complex systems.

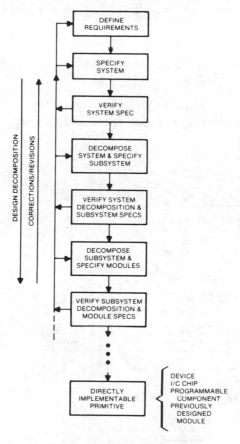

FIGURE 1. DESIGN PROCESS

DESIGN LEVEL	SPECIFICATION	SIMULATION
SYSTEM	ENGLISH	GPSS, CSMP, SIMULA
SUBSYSTEM	RTL, ISP	FUNCTIONAL
MODULE	GATE INTERCONNECTION	LOGICAL
GATE	DEVICE INTERCONNECTION	SPICE

FIGURE 2. PREVIOUS APPROACHES TO DIGITAL DESIGN

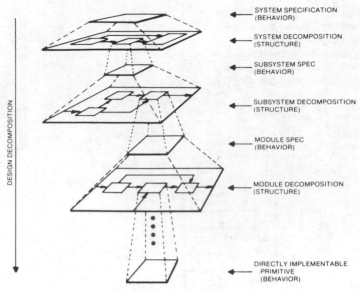

FIGURE 3. HARDWARE DESCRIPTION LANGUAGE (HDL)

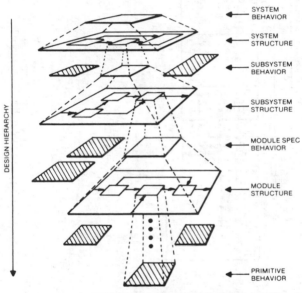

FIGURE 4. MULTI-LEVEL SIMULATION

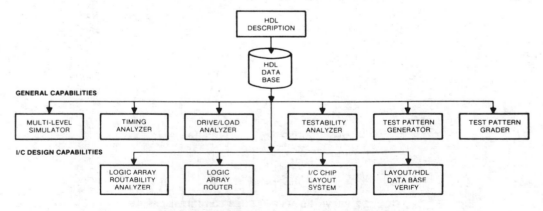

FIGURE 5. INTERFACE OF HDL TO OTHER DESIGN AIDS

LOGAL+ - A HARDWARE DESCRIPTION LANGUAGE
FOR HIERARCHICAL DESIGN AND MULTILEVEL SIMULATION

by Ghulam M. Nurie

Sperry Univac
P.O. Box 43942
St. Paul, Minnesota 55164

ABSTRACT

Structured and hierarchical design methodologies presently being used in the design of VLSI circuits depend heavily on a suitable design language. LOGAL+ is a hardware design language that facilitates a multilevel description of digital systems, to be used in a top-down hierarchical design approach, incorporating multilevel simulation. The behavior and structure descriptions are maintained in separate sections with common terminal information. This facilitates automated consistency checks for different levels of descriptions. In this paper, these concepts and some key features of LOGAL+ are described.

INTRODUCTION

The advent of VLSI has brought with it a major change in the philosophy of designing digital systems. It is almost impossible to design a VLSI circuit without adequate design tools. Various design techniques have been proposed to solve this problem[1,2,3]. Most of these involve a structured design methodology[4], which emphasizes a hierarchical top-down approach to the design. The design starts from the highest level, which is a conceptual level. At this level, the design consists merely of the behavioral or functional specification. This design then progresses down one level of refinement at a time, until the lowest or the final implementation level is reached.

LOGAL+ is a major extension of LOGAL2 (Logic Algorithm Language), which has been in use at Sperry Univac for a number of years[5]. LOGAL+ supports top-down hierarchical design methodology by allowing a multilevel description capability.

TOP-DOWN DESIGN WITH LOGAL+

In recent years many efforts have been made to reduce the complexity of large scale development projects. A number of these efforts are based on the notion of modularization; that is, partitioning the solutions to large complex problems into smaller, more understandable and hence more manageable components. LOGAL+ promotes this modularization concept by allowing a design to be specified in terms of a number of different modules known as "models" in the LOGAL+ environment. The word "MODEL" was chosen instead of "MODULE" because the description of the design is really a model of the actual module, and a number of such models can exist for the same hardware module. A LOGAL+ model consists of the behavior and/or structure description, together with timing and terminal information, about a piece of computer hardware. No restriction is placed on the size of the model. One model can represent the total design, as might be the case at the highest level, or a model can represent a portion of a design.

Figure 1 shows the general format for a LOGAL+ model. Every model has a heading MODEL, followed by the name of the model, which will be the name by which the circuit is recognized in the system. This is followed by a brief description of the type of circuit (e.g., microprocessor, memory, adder, etc.). This information can be used to search the library for models with certain characteristics. Then the terminal information is specified. A model communicates with the external world through its input and output terminals. No global signals are allowed because there is no such thing as a global signal in the computer hardware.

Following the input and output lists is the local declaration section, where the internal states and other temporary variables can be declared. LOGAL+ is a strong type-checking language, and requires all the variables and their types (INTEGER, REAL, REGISTER, IMEDREG, NET) to be explicitly declared. Following this, the model can have up to four classes of data. These are the BEHAVIOR, STRUCTURE, TIMING and PHYSICAL classes. Each of these four sections is optional, and the presence of all of them is not necessary for the compilation of the model.

```
*MODEL;
 name;module type;
  INPUT (a, b, ... c);              @INPUT LIST
  OUTPUT (x, y, ... z);            @OUTPUT LIST
  local declarations
*PHYSICAL;
*TIMING;
 propagation delays and timing constraints
*BEHAVIOR;
 This section describes the behavior of the module. The
 behavior need not have any kind of relationship with
 the structure description other than the input/output
 relationship.
*STRUCTURE;
 This section has the structure description of the module.
 It describes the interconnection of lower level models.
*ENDMODEL;
```

Figure 1. Model Format

BEHAVIOR

The behavior of computer hardware can be described in a procedural manner or in a non-procedural manner. A procedural description usually consists of a sequence of steps, and the order of the statements is important. A non-procedural description is usually not order dependent. In general it has been found that the control section of the computer hardware is best described in a procedural manner, while the data section is best described in a non-procedural manner. Therefore, it was decided to support both forms of description in LOGAL+.

Reprinted from *IEEE Int. Conf. Circuits and Computers Proc.*, 1982, pp. 600-603.

A procedural description is facilitated with common programming constructs such as IF-THEN-ELSE, WHILE-DO, REPEAT-UNTIL, FOR-LOOP, and CASE STATEMENT[6]. These constructs facilitate structured descriptions, which are highly recommended. Alternatively, a design can also be expressed in a non-procedural manner[7].

LOGAL+ has three basic classes of variables: REAL, INTEGER, and LOGIC. Integer and real variables, in addition to being used as indices and loop counters, are very useful in high level behavior models. For instance, the address line on a 256-word RAM can be modeled as an integer with a range between 0 and 255.

Variables belonging to the logic class can have four states: logic-0, logic-1, unknown, and high-impedance. The logic class is a generic class consisting of the types REGISTER, IMEDREG and NET.

A REGISTER variable is analogous to a computer register and has a fixed number of bits associated with it. A REGISTER variable can be declared with a clock constraint. A REGISTER declared with such a constraint can receive new data only under the condition described in the constraint. As an example, consider the following three register declaration statements:

    REGISTER(R1 $<$ 1:4 $>$,R2 $<$ 0:3 $>$,R3);
    REGISTER(/CLK1.EDGE1/SHF_REG $<$ 0:3 $>$);
    REGISTER(/CLK1.LEVEL1/DLATCH_A);

The first statement declares two 4-bit registers, R1 and R2, and one 1-bit register, R3. There are no constraints on these registers; therefore they can receive new data any time.

The second statement declares a 4-bit register called SHF_REG that is clocked on the positive edge of a signal called CLK1. Thus SHF_REG can receive new data only when the signal CLK1 goes through a positive transition. From this it is obvious that SHF_REG is a positive edge triggered register.

The third statement delcares a 1-bit register called DLATCH_A that can receive new data only when CLK1 is high.

IMEDREG (IMEDiate REGister) is a data abstract, similar to a register in concept. However, it does not have any delay associated with it. Its main use is in high level behavior models for register abstraction. It is very useful in describing sequential operation, where the results from one computation are to be used in the next computation.

A variable of type NET is an abstract for combinational elements. Such a variable has zero delay. In a model, all transfer statements whose destinations are NET type variables are iteratively executed until their values stabilize. With this iterative feature, the user does not have to levelize the network description.

LOGAL+ also supports a finite state machine description[8] of a hardware module. This is facilitated by allowing labels and GO TO statements. The labels represent a particular state of the machine. Control flows from one labeled section into another unless altered by a GO TO or RETURN statement.

Two types of labels are allowed. One is where the label is just an identifier for the state. Explicit GO TO statements are required to direct the flow of control. This is similar to the use of labels in common programming languages.

The other type of label is where the state variable itself, together with the state encoding, is used as a label. The state label takes on a Boolean characteristic in the sense that it is either true or false. Therefore, explicit GO TO statements are not required to direct the flow of control.

LOGAL+ also accepts a truth table for the behavior description of a circuit. A truth table description is very efficient for simulation and is highly recommended for simple combinational circuits.

With these different ways available to describe behavior in LOGAL+, the user has much power and freedom with which to describe the design. The behavior description in LOGAL+ can be a very high algorithmic level description, a register transfer level description, or a gate level description. Furthermore, it is also possible, through the macro facility provided, to describe a design in terms of an already existing model.

Multiple Views

The multiple view of a variable or a label is a very powerful construct that allows an object to be looked at from a different view or under some constraint[9]. For example, at one point a designer may want to treat a given variable as an integer, and at some other point he may want to treat the variable as a bit string of logical values. Though special operators are allowed in LOGAL+ to accomplish this simple translation, the view construct gives much more freedom. This multiple view is obtained by using the view declaration:

    VIEW view-suffix .AS. type-definition or
        constraint .FROM. view-source;

The view-suffix is an identifier that is used as a suffix with the originally declared variable or view-source to obtain its multiple views defined by the type definition or constraint.

As an example, suppose A is declared as a 16-bit REGISTER. Then the declaration:

    VIEW I .AS. INTEGER $<$ 0,255 $>$ .FROM. A $<$ 0:7 $>$;

declares I as a suffix to view the lower eight bits of register A as an integer whose range is between 0 and 255. This alternative view is obtained by referencing A.I. Similarly the declaration:

    VIEW P2 .AS. EDGE1.CLK1 .FROM. A;

puts a constraint on A, when used with the suffix P2, of being clocked on the positive edge of a signal called CLK1. Then, if the following two statements exist in the model:

    X .TO. A;
    Y .TO. A.P2;

the first is an asynchronous transfer of X to A, whereas the second statement describes a synchronous transfer, and Y will be transferred to A only if a positive edge of CLK1 occurred.

The view-suffix can also be used with statement labels. This is very useful in finite state machine descriptions, when describing synchronous entry into a new state.

STRUCTURE

In LOGAL+, the structure of a design is kept separate from the behavior. This separation is almost essential for an efficient multilevel simulation. Since a particular behavior can be realized with many different structures, the separation of the two in the LOGAL+ model ensures that any changes in the structure description are not automatically reflected in the behavior part. This, in turn, ensures that any modification made to the structure will not accidentally alter the desired behavior, because any functional discrepancy that exists between the behavior and the structure descriptions will be detected during simulation.

In the structure section of a LOGAL+ model, a structure that realizes the function described in the behavior section can be described. In a multilevel description, the structure is described as the interconnection of lower level models. However, the structure description language is not yet developed. Until the structure description language is developed, the Nodal Design Language (NDL)[10] will be used to describe the structure. NDL is presently used as the Structure Description Language in the CAD data base.

TIMING

LOGAL+ allows for timing to be specified in the timing section. Propagation delays, as well as timing constraints, are allowed. Separate rise and fall times can be specified from any input to any output. There is also a provision to handle state dependent delays. For complex sequential circuits, the input to output delays vary, depending upon the state of the machine. These delays can be modeled within the behavior.

Inward timing constraints such as set up and hold time, minimum pulse width, etc., can also be specified. During simulation of the model these constraints are checked and, if violated, warning messages are generated.

PHYSICAL

This section deals with the information on the I/O pins. The I/O pins are categorized into data, control, clock, address, groups, etc., and their voltage and current requirements are given. The information in this section is non-executable and does not affect simulation. It is used to verify the structural interconnect of the models, and in generation of test lists which contain I/O pin information for the tester.

A DESIGN EXAMPLE

Figure 2 illustrates a hierarchical block diagram for the design of a 4-bit, 1's complement integer adder. In a top-down design methodology, a high-level behavior is initially written for the outermost block. This block has two integer inputs, A and B, and an integer output called SUM, and an OVERFLOW flag. A LOGAL+ model for this block is shown in Figure 3. The behavior is described in an algorithmic manner. No implementation is implied in the behavior section.

After this behavior is simulated and satisfactory results are obtained, the structure section is added. The structure section shows the interconnect of the full adder and the overflow blocks. A language similar to NDL is used to describe the structure. It should be noted that a level transformation of the signals is required to map the integer into logical signals. In order to simulate the system at this lower level, behavior descriptions of the full adder and the overflow blocks are required. Figure 4 shows a LOGAL+ model for the full adder circuit. The behavior is described in a Boolean equation form. A structure description is also provided that uses half adders to design the full adder. Figure 5 shows a LOGAL+ model for the overflow block. Here the behavior is

described in an algorithmic manner. A structure description is also provided which gives a gate-level description for the overflow circuit. The half adders can be decomposed down to the gate level, as done in the model shown in Figure 6.

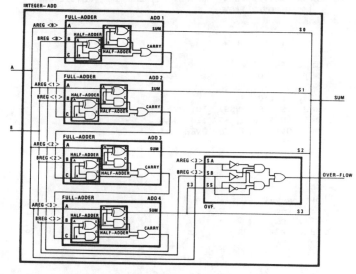

Figure 2. 4-Bit 1's Complement ADDER

```
*MODEL;
 INTEGER_ADD; 4-BIT ADDER
 INPUT (A, B);                    @TWO INPUT TERMINALS
 OUTPUT (SUM, OVER_FLOW);         @TWO OUTPUT TERMINALS
 INTEGER (A <-7,7>, B <-7,7>, SUM <-7,7>);
*BEHAVIOR
 A + B .TO. SUM;
 IF(ABS(SUM) .GT. 7) THEN
   1 .TO. OVER_FLOW
 ELSE
   0 .TO. OVER_FLOW
 ENDIF
*STRUCTURE
 A .INTLOG. AREG <0:3>;           @Transformation of
                                  @integer to logic
 B .INTLOG. BREG <0:3>;           @signals
 S3 & S2 & S1 & S0 .LOGINT. SUM;  @Transformation of
                                  @logical outputs
                                  @ into integer
 .FULL_ADDER. [ADD1] (AREG <0>, BREG <0>, C3)
   .TO. (S0,CO);
 .FULL_ADDER [ADD2] (AREG <1>,  BREG <1>, CO)
   .TO. (S1,C1);
 .FULL_ADDER. [ADD3] (AREG <2>, BREG <2>, C1)
   .TO. (S2,C2);
 .FULL_ADDER. [ADD4] (AREG <3>, BREG <3>, C2)
   .TO. (S3,C3);
 .OVF. (AREG <3>, BREG <3>, S3) .TO. OVER_FLOW
*ENDMODEL;
```

Figure 3. LOGAL+ Model for a Four-Bit Adder

```
*MODEL;
 FULL_ADDER; ADDER;
 INPUT (A, B, C);
 OUTPUT (SUM, CARRY);
*BEHAVIOR
 A .XOR. B .XOR. C .TO. SUM;
 B .AND. C .OR. A .AND. (B .XOR. C) .TO. CARRY;
*STRUCTURE
 .HALF_ADDER. (A,S1) .TO. (SUM,CO);
 .HALF_ADDER. (B,C) .TO. (S1,C1);
 .OR. (CO,C1) .TO. (CARRY);
*ENDMODEL;
```

Figure 4. LOGAL+ Model for a Full Adder

127

```
*MODEL;
 OVF; OVER_FLOW
 INPUT (SA, SB, SS);
 OUTPUT (OVER_FLOW);
*BEHAVIOR
 IF(SA .EQ. SB .AND. SA .NE. SS) THEN       @Result has
    1 .TO. OVER_FLOW;                        @opposite sign
 ELSE;                                       @When two
                                             @operands have
    0 .TO. OVER_FLOW;                        @the same sign
 ENDIF;
*STRUCTURE
 .NOT. SA .TO. NSA;
 .NOT. SB .TO. NSB;
 .NOT. SS .TO. NSS;
 .AND. (SA, SB, NSS) .TO. OV1;
 .AND. (NSA, NSB, SS) .TO. OV2;
 .OR. (OV1, OV2) .TO. OVER_FLOW;
*END_MODEL
```

Figure 5. LOGAL+ Model for an Overflow Circuit

```
*MODEL;
 HALF_ADDER; ADDER;
 INPUT (A, B);
 OUTPUT (S, C);
*BEHAVIOR
 .TRUTH_TABLE.
   (A, B : S, C);
   (0, 0 : 0, 0);
   (0, 1 : 1, 0);
   (1, 0 : 1, 0);
   (1, 1 : 0, 1);
*STRUCTURE
   .AND. (A, B) .TO. C;
   .XOR. (A, B) .TO. S;
*ENDMODEL;
```

Figure 6. LOGAL+ Model for Half Adder

At this point, the entire system in Figure 2 can be simulated at the gate level, even though a gate level structure description for the whole system was not explicitly described. However, the gate level interconnect for the entire system is implicit in the hierarchy, and can be easily obtained and put into a data base. Furthermore, only one model for each type of block is required. If the same type of block is used in more than one place, a new instantiation of the model is automatic. In the example, only one model is needed for each of the full adder, half adder and overflow blocks. The user does not have to worry about multiple uses in the structure description. In the structure description the users can assign a unique name to each block, as done for the full adders in the model of Figure 3 (ADD1, ADD2, ADD3, ADD4). However, the names can be left out and the system will assign unique names.

MULTILEVEL SIMULATION

After the gate level or the implementation level of the design is achieved, it must be verified to ensure that the original functional objective is still intact. One of the ways to do this is to simulate the entire design at the gate level and compare the result against those results obtained at the high level. For the previous example, this is not a problem because the network is small, and the gate level description of the entire system can easily fit in the host computer memory. But in reality, designs are very large, and it is very inefficient, if not impossible, to represent the entire design at the gate level. One solution to the problem is to represent part of the design at a higher level and part of it at a lower level and verify it piece by piece. For instance, in the previous example of the 4-bit adder, the network can be simulated with one full adder, for example ADD1, expanded to gate level and maintaining the other full adders at the high level. In another instance, the overflow module can be verified by expanding it to its gate level and leaving the four full adders at the high level.

MULTILEVEL CONSISTENCY

Though hardware design languages that support hierarchical designs exist[11],[12], none of them solves the problem inherent in the hierarchical approach, that is, the issue of consistency of different levels of design. It is not always easy to verify that a high level behavior description corresponds to the lower level structure description. However, this problem is partially solved by LOGAL+. Due to the fact that the primary input/output structure in a LOGAL+ description is maintained at all levels of the hierarchy (Figure 2), the same input stimulus can be applied at all levels, and the same output response should be observed. This consistency check can be completely automated.

CONCLUSION

LOGAL+ is a hardware design language that facilitates hierarchical design of digital systems. The top-down design concept allows early introduction of high level, relatively abstract behavioral models. The models can be used to evaluate the design early in the design cycle. The decomposition of high level models into successively lower level models, with each lower level model providing additional detail, follows a classical synthesis-analysis loop where each newly created model is analyzed for compliance with the design goals.

However, the use of LOGAL+ is not limited to a top-down design process. It can be used in bottom-up or inside-out designs. The user is not restricted to follow any particular design methodology. But the hierarchical design approach is highly recommended, as it eases the design process and lends itself to multilevel simulation, which in turn eases the verification process.

ACKNOWLEDGEMENT

The author would like to acknowledge the contributions of Charles Loegering and Dr. Bulent Dervisoglu, who provided valuable comments and suggestions in the development of LOGAL+.

REFERENCES

[1] R. Rice, VLSI - The Coming Revolution in Applications and Design, IEEE, New York, 1980.

[2] S. Trimberger, J.A. Rowson, C.R. Lang and J.P. Gray, "A Structured Design Methodology and Associated Software Tools," IEEE Trans. Circuits and Systems, July 1981.

[3] S.W. Director, A.C. Parker, D.P. Siewiorek and D.E. Thomas, "A Design Methodology and Computer Aids for Digital VLSI Systems," IEEE Trans. Circuits and Systems, July 1981.

[4] C.A. Mead and L.A. Conway, Introduction to VLSI Systems, Addison Wesley, 1980.

[5] J.H. Stewart, "LOGAL, A CHDL for Logic Design and Synthesis of Computers," Computer, June 1977.

[6] J. Welsh and J. Elder, Introduction to Pascal, Prentice-Hall Int'l Inc., London, 1979.

[7] D.L. Dietmeyer and J.R. Duley, "Register Transfer Languages and their Translation," in Digital System Design Automation (ed. M. Breuer), Computer Science Press, Woodland Hills, CA, 1975.

[8] D.L. Dietmeyer, Logic Design of Digital Systems, Allyn and Bacon, Inc., Boston, 1971, pp. 453-466.

[9] S.J. Piatz, "VHSIC Transportability for CAD Data," Document no. PX-13557, UNIVAC Technical Report, October 1981.

[10] J.J. Deck and J.B. Dietel, "UCADS Nodal Design Language User Guide," Document no. 4168201, UNIVAC Technical Report, 1980.

[11] W.A. Johnson, J.J. Crowley and J.D. Ray, "Mixed Level Simulation from a Hierarchical CHDL," Proc. of 4th Int'l Symposium on Computer Hardware Description Languages, 1979.

[12] N. Kawato, T. Saito, F. Maruyama and T. Uehara, "Design and Verification of Large Scale Computers by Using DDL," Proc. of 16th Design Automation Conference, June 1979.

TOWARDS A STANDARD HARDWARE DESCRIPTION LANGUAGE

Karl J. Lieberherr

GTE LABORATORIES INC.
40 Sylvan Road, Waltham, Massachusetts 02254

Abstract: General requirements for hardware description languages are defined. The hardware description language Zeus is claimed to satisfy most of these requirements. Zeus is summarized and briefly compared to other hardware description languages. The expressiveness of Zeus is demonstrated on designs for music generation, comparison and unary-to-binary conversion.

1. Introduction

Hardware description languages (HDLs) must assist the system designer/computer scientist in designing, documenting, and validating entire hardware/software systems. A large number of hardware description languages currently exists (see [Nash (1984)]). Unfortunately, none of these languages are widely used. This wide diversity and lack of standardization is a big hindrance to the use, development and the dissemination of CAD tools for hardware design. A versatile and widely used language would considerably increase the efficiency of CAD tool user and developer.

Requirements for hardware description languages are similar to the requirements for programming languages and are not easy to quantify. In general, though, a useful hardware description language should satisfy the following requirements: simplicity, expressiveness, orthogonality, readability, security, extendability, and availability of efficient and user friendly tools (e.g. simulators, silicon compilers) which produce quality results (e.g simulation output, chips). A consequence of these general requirements is that a HDL should have functional, structural, and layout semantics.

The HDL Zeus is an attempt to provide a language which satisfies most of these requirements. Unfortunately there are at this time no finished CAD tools available which use Zeus as their input language. This is a serious limitation because a hardware description language is only as good as the tools which use it as input language. There are, however, several CAD tools, using Zeus as input language, under development and Zeus itself has been benchmarked on more than 40 designs. They range from adders, matrix multipliers, sorters to microprocessors. Three of the designs are shown in this paper.

Zeus is essentially a structural language although it has functional and layout semantics. Zeus uses ALGOL's and Pascal's ideas for specifying structure. Three type concepts (COMPONENT, ARRAY and basic types) and the possibility to instantiate these types are used for describing complex hardware. Types can be generic and recursive. Interconnections are specified in an algorithmic manner with the *assignment*, *for* and *when* statement. The type equivalence problem for Zeus types is undecidable [German (1984a)]. The type equivalence problem asks whether any two parameterized types $t1(n)$ and $t2(n)$ describe functionally identical circuits for all n.

In the following we report on the history and current status of Zeus. Zeus was developed at ETH Zurich, Princeton, MIT and GTE Laboratories, where it is currently being used in various projects. Zeus is a general purpose hardware description language which is designed to serve as a uniform human interface for a variety of design tools. Zeus provides the basic hardware description facilities required by most tools and it has been made extendable so that it can be tailored to meet the specific needs of particular tools.

Work on Zeus was initiated in the summer of 1982 by a paper on the hardware description language Hades by Niklaus Wirth at ETH Zurich [Wirth (1982b)]. In the following we highlight some of the improvements which have been made to Zeus since its first publication in [Lieberherr (1983a,b)].

Zeus was interfaced with the programming language Modula-2 [Wirth (1982a)] without modifying Modula-2. During the hardware description process there are many situations where the need for the facilities of a general programming language like Modula-2 arises. For example: the specification of complicated interconnection patterns, the computation of test vectors, the specification of behavior in procedural form or the description of system software which is to be executed on the described hardware. We decided to use Modula-2 as an associated programming language of Zeus because of its elegance and its efficient implementation. Modula-2 has been used, for example, to program the complete single-user operating system Medos-2 [Knudsen (1983)].

Reprinted from *ACM/IEEE 21st Design Automation Conf. Proc.*, 1984, pp. 265-272.

The specification of complicated interconnection patterns may be accomplished by allowing a call to a Modula-2 defined function procedure in the test of a conditional interconnection statement. Test vectors may be computed in a Modula-2 module and exported to a Zeus component. The behavior of a Zeus component may be specified by a Modula-2 procedure.

Zeus has a separate compilation mechanism similar to the one in Modula-2. A compilation unit is either an IMPLEMENTATION HARDWARE_MODULE or a DEFINITION HARDWARE_MODULE. A HARDWARE_MODULE is a collection of declarations. Import and export lists are used in the same way as in Modula-2. The separate compilation mechanism is useful for describing big hardware systems, including libraries, in a modular fashion.

The parameters which can be given to a Zeus type have been extended. For component types we distinguish between control and signal parameters. The control parameters are used for passing numerical information and for passing types as parameters. The signal parameters describe the wires which exit and enter the component. The type passing mechanism is similar to the procedure passing mechanism in regular programming languages.

Recursive types are now specified more concisely. Type virtual has been eliminated and now constant, type or hardware declarations are allowed in *when* statements. The scope of these declarations is restricted to their respective when statements.

Multiple clocks are now available in Zeus. A component type *C* can have one clock input. All the registers local to *C* are controlled by the clock input. Asynchronous systems can either be expressed as combinational circuits or as synchronous systems with a fast clock.

The Zeus syntax has been modified at several places to incorporate the LR(1) property. A recursive descent parser has been implemented. The modified syntax is given in the appendix of this paper.

Space limitations here prohibit a detailed comparison with other languages, e.g. ADLIB/SABLE/SDL, AHPL, CASL, CSP, HADES, HISDL, KARL, MODEL, OCCAM, VHDL etc. (See [Nash(1984)] for an almost complete list of references to published hardware description languages.) It suffices to mention that Zeus has a combination of features which gives it a versatility rarely found in other hardware description languages.

Zeus has a disadvantage relative to CSP or OCCAM: certain Zeus programs tend to be an order of magnitude longer than the corresponding CSP or OCCAM descriptions. One reason for this blow-up is that in Zeus all sequencing has to be done explicitly by describing a finite state machine. We plan to remedy this situation by permit-ting the definition of a Zeus component in a CSP-like language [German (1984b)]. This will give Zeus the power of CSP while retaining the powerful structuring facilities of Zeus.

In the following sections we demonstrate some of the features of Zeus on several design examples. Here we combine the explanation of the language features with the explanation of the designs.

2. Electronic organs

In this example we apply Zeus to describing a slightly modified version of the MM5891 top octave frequency generator. First we give a generalized form of the frequency generator. Then we specialize the generalized form to describe a circuit that provides 12 semitone outputs like the MM5891.

This example shows Zeus' facilities for design decomposition (components and modules), finite state machines, and parts of the typing mechanism. Our description follows Fig. electronic_organs.

To cope with complexity, large designs must be decomposed into partitions which can be considered one at a time without too much regard for the other partitions. At the lowest level of the decomposition structure are statements, at the next level components, and at the highest level hardware modules.

A basic decomposition rule is that connections or interfaces between partitions are simple and "thin". The thickness of an interface between two parts is the number of items that take part in it. The interface between two hardware modules is sketched in terms of the module's import lists. Hence the goal of modularization is to make the import lists reasonably short. The distinctive property of the hardware module is its ability to hide details and thereby establish a new level of abstraction.

Hardware modules at the outermost level are divided into definition and implementation parts. The definition part contains all the information that a user of the hardware module is supposed to know. The details of the definition part's operation - its realization - are contained in the corresponding implementation hardware module. The definition part contains the list of exported objects and their declarations. Only constants and types can be exported.

The hardware module *electronic_organ* is subdivided into these definition and implementation hardware modules. From the definition hardware module we export two component types. We call the parameterized frequency generator *electronic_organ_type*. It has two control parameters, called *nr_of_tones* and *control*. *nr_of_tones* specifies how many frequencies the chip generates. The array control passes *nr_of_tones* integers which define the frequencies to be generated.

```
DEFINITION HARDWARE_MODULE
    electronic_organs;
  EXPORT electronic_organ_type,
      organ_MM5891_type_modified;
  CONST
    freq= [451,426,402,379,358,338,
         319,301,284,268,253,239];

  TYPE
    electronic_organ_type(nr_of_tones: INTEGER;
     control:
        ARRAY[1..nr_of_tones] OF INTEGER)=
     COMPONENT(CLOCK clock_input: logical;
      OUT tones:
          ARRAY[1..nr_of_tones] OF logical)
      <* frequency generator *> ;
    organ_MM5891 type modified=
      electronic_organ_type (12,freq)
END electronic_organs.

IMPLEMENTATION HARDWARE_MODULE
  electronic_organs;
  FROM functions IMPORT log2, zero_seq;
  FROM adders IMPORT increment;
  TYPE
    period_counter_type(n:INTEGER)=
    COMPONENT(CLOCK clock_input: logical;
      OUT tone: logical) IS
    CONST <* log2(n) is the floor of
          log of n to base 2 *>
    logn = log2(n) + 1;
    critical = binary(n,logn);
    zeros = zero_seq(.logn.);
    HARDWARE
      state: ARRAY[1..logn] OF REG; switch: REG;
    CONNECT
      IF reset THEN
        state.i:=zeros; switch.i:=zero;
      ELSIF EQUAL(state.o, critical) THEN
        switch.i := NOTg switch.o;
        state.i:=zeros;
      ELSE switch.i:=switch.o;
        state.i:=increment(.logn.)(state.o);
      END;
      tone:= switch.o;
    END <* period_counter_type *>;

    electronic_organ_type(nr_of_tones: INTEGER;
     control:
        ARRAY[1..nr_of_tones] OF INTEGER)=
    COMPONENT(CLOCK clock_input: logical;
      OUT tones:
        ARRAY[1..nr_of_tones] OF logical) IS
    HARDWARE
      period_counter: ARRAY[1..nr_of_tones:i] OF
        period_counter_type(control[i])
    CONNECT
      FOR i:=1 TO nr_of_tones DO
        period_counter[i](clock_input,tones[i])
      END
    END <* electronic_organ_type *>;
  END electronic_organs.
```

Fig. electronic_organs

The component type *electronic_organ_type* has two signal parameters called *clock* and *tones*. The signal *clock* represents a 2.00024 MHz clock. The array *tones* describes the *nr_of_tones* wires which carry the generated frequencies. The output *tones*[*i*] represents the frequency of the clock divided by *control*[*i*].

The type *organ_MM5891_type_modified* is a specialization of type *electronic_organ_type*. As actual control parameters we substitute 12 (for the number of tones) and the constant array *freq* (for specifying the output frequencies).

The implementation hardware module contains the details of the implementation of type *electronic_organ_type*. We first import several objects from the modules *functions* and *adders*. Then we define type *period_counter_type* as a component type which has one control parameter, called *n*. The signal parameters are the clock which is called *clock_input* and the signal *tone*. Type *period_counter_type* divides the input frequency of the clock by *n*. Signal *tone* carries this divided frequency.

This frequency division is done by a simple state machine which has *log*(*n*) one-bit registers, called *state*. These registers are declared as an array of element type *REG*. *REG* is a predefined component type which introduces the concept of time into Zeus descriptions. An instance of type *REG* has an input pin called *i* and an output pin called *o*. Whatever bit value is stored at the *i* pin of the register at time *t* is available at the *o* pin at time *t+1*. All the registers local to a component are controlled by the clock of the component.

The register *switch* is used for storing whether the output *tone* is *zero* or *one*. Any instance of type *period_counter_type* has a "reset button" for initialization. If this button is pressed, then *state* and *switch* are initialized to *zero*. The initialization of *state* is accomplished by calling a function component which returns *zeros*.

During normal operation the register *switch* keeps its value and *one* is added to the number stored in the register array *state*. This addition is done by a function component increment which takes as input an *m*-bit number *c* and returns as output the *m*-bit number *c+1*. If state contains the number *n* (in binary) then the bit stored in register *switch* is negated and the number stored in *state* is set to 0. The output *tone* is set to the value of the output port of the register called *switch*.

The type *electronic_organ_type* instantiates *nr_of_tones* copies of type *period_counter_type*. This is done by declaring hardware called *period_counter* as an array of *nr_of_tones* elements. Element *i* is declared to be of type *period_counter_type*(*control*[*i*]).

The first parameter of *period_counter*[*i*] is connected to the clock input and the second parameter to *tones*[*i*]. This completes the description of hardware module *electronic_organs*.

3. Comparator

Here we apply Zeus to describing a combinational circuit for comparing numbers. This example shows Zeus' facilities for defining recursive types. A recursive type describes hardware which has a recursive structure. Recursive hardware structures emerge naturally when problems are solved with the divide and conquer approach. Our description follows Fig. *recursive_comparator*. The floor plan is shown in Fig. *floor_plan_recursive_comparator*.

```
IMPLEMENTATION HARDWARE_MODULE
  recursive_comparator;

TYPE
logic(n:INTEGER)=ARRAY[1..n] OF logical;
comparator(n:INTEGER) =
<* n a power of two *>
COMPONENT (IN a,b:logic(n);
  OUT agreaterequalb,agreaterb:logical)IS
  <* compares the n-bit binary numbers a and b;
    the OUT signals return the outcome of
    the comparison; the hardware is
    specified recursively following
    a divide-and-conquer approach *>
  WHEN n=1 THEN
    CONNECT
      agreaterequalb:=ORg(a[1],NOTg b[1]);
      agreaterb:=ANDg(a[1],NOTg b[1]);
  OTHERWISE
    HARDWARE
      left_c, right_c: comparator(n DIV 2);
    CONNECT
      FOR i:=1 TO n DIV 2 DO
        left_c.a[i]:=a[i];
          right_c.a[i]:=a[i+n DIV 2];

        left_c.b[i]:=b[i];
          right_c.b[i]:=b[i+n DIV 2]
      END;
      agreaterequalb:=
        ORg(left_c.agreaterb,
          ANDg(left_c.agreaterequalb,
            right_c.agreaterequalb));
      agreaterb:=
        ORg(left_c.agreaterb,
            ANDg(left_c.agreaterequalb,
              right_c.agreaterb));
  END;
END <* comparator *>;

END recursive_comparator.
```

Fig. recursive_comparator

The comparator which we are going to describe has as input two numbers, called *a* and *b*, each *n* bits wide. The output of the circuit consists of two signals which tell whether *a* is greater or equal to *b* or whether *a* is strictly greater than *b*.

The body of the component *comparator* consists of a *when* statement. When the control parameter *n* is 1 then the problem has a trivial solution. Oth-

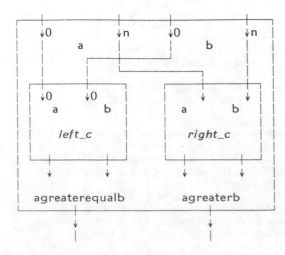

Fig. floor_plan_recursive_comparator

erwise we instantiate type *comparator* twice recursively with half the parameter size. The first instance, called *left_c*, compares the two left halves of *a* and *b*. Similarly, the second instance, called *right_c*, compares the two right halves of *a* and *b*. The splitting of *a* and *b* into two halves is expressed by the *for* statement. Given the comparison results for the two halves, it is a simple matter of a few logical gates to compute the two results of the entire circuit.

4. Unary-to-binary Conversion

Here we describe a combinational circuit for unary-to-binary conversion which has been presented in [Cappello/ Steiglitz (1983)]. This example shows Zeus' facilities for design decomposition and recursion on a more complicated example. Our description follows Fig. UBC. The corresponding floor-plan for n=7 is shown in Fig. floor_plan_UBC.

The unary-to-binary converter has a control parameter *n* which specifies the number of incoming bits. *n* is assumed to be a number of the form $2**k-1$, $n \geq 3$. The output indicates the number of input wires which carry a *one*. The output has to be therefore $log(n+1)$ bits wide since it counts, in binary, the number of input wires which carry a *one*. The addition is performed by a network of fulladders.

When *n=3* then one fulladder performs the conversion. Otherwise we instantiate the type *UnaryToBinaryConverter* twice recursively with half the parameter size. The first instance is called *left_converter* the second *right_converter*. We also declare an array of $log(n+1)-1$ fulladders which is used to add the results of *left_converter* and *right_converter*. In the connection part the details of the interconnection are specified. A special case is shown in Fig. *floor_plan_UBC* for n=7.

```
IMPLEMENTATION HARDWARE_MODULE
  unary_to_binary_conversion;
FROM functions IMPORT log2;
TYPE
  logic(n:INTEGER)= ARRAY[1..n] OF logical;
  UnaryToBinaryConverter(n:INTEGER)=
  COMPONENT(IN input:logic(n);
    OUT output: logic(log2(n+1))) IS
<* this is the unary-to-binary converter
   described in [Cappello/Steiglitz (1983)];
   n is assumed to be of the
   form 2**k-1, n≥3 *>
  CONST
    n1=log2(n+1)-1; n2=n DIV 2;
    n12= (n+1) DIV 2;
  TYPE
    FullAdderType=
    COMPONENT(IN a,b,carry_in: logical;
      OUT carryout,sum: logical) IS
    CONNECT
      <* ... *>
    END;
  WHEN n=3 THEN
    HARDWARE fulladderleaf: FullAdderType;
    CONNECT fulladderleaf(input[1],input[2],
          input[3],output[1] <*carry*>,
          output[2] <*sum*>);
  OTHERWISE
    HARDWARE
    left_converter, right_converter:

      UnaryToBinaryConverter(n2);
    fulladder: ARRAY[1..n1] OF FullAdderType;
  ORDER lefttoright
    left_converter;
    ORDER toptobottom
      FOR i:=n1 TO 1 BY -1 DO fulladder[i]
    END;
    [flip0] right_converter;
  END;
  CONNECT
    <* connect fulladders to
       left and right part *>
    FOR i:=2 TO n1 DO
      fulladder[i](left_converter.output[i],
        right_converter.output[i],
        fulladder[i-1], carryout, output[i])
    END;
    <* first fulladder needs
       special treatment *>
    fulladder[1](left_converter.output[1],
      right_converter.output[1],
      input[n12], output[1]);
    <* last output *>
    output[log2(n+1) ]:= fulladder[n1].carryout;
    <* connect inputs for left and right part *>
    left_converter.input:=input[1..n12-1];
    right_converter.input:=input[n12+1..n]
  END
END <* UnaryToBinaryConverter *>;

END unary_to_binary_conversion.
```

Fig. UBC

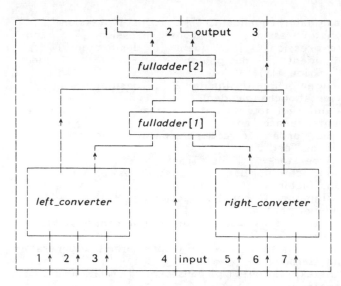

Fig. floor_plan_UBC

Zeus has a facility for specifying relative place-ment with the *order* statement. Each instantiated component type is thought to be contained in a bounding rectangle. The *order* statement puts constraints on the relative placement of the rec-tangles. These constraints are intended as hints to a silicon compiler to improve the quality of the layout.

In this example we might want to specify that *left_converter* is right of *right_converter* and that the fulladders are in between. Furthermore, the fulladders should be on top of each other. These constraints are expressed concisely by the follow-ing order statement

```
ORDER lefttoright
   left_converter;
   ORDER toptobottom
      FOR i:=n1 TO 1 BY -1 DO fulladder[i];
   END;
   [flip0] right_converter
END
```

The orientation change *flip0* flips the right con-verter along a vertical axis.

5. Conclusions

Zeus has been benchmarked on over 40 design ex-amples and has been found to satisfy most of the requirements for HDLs mentioned in section 1. We therefore believe that the language mechanisms contained in Zeus should be included in any hard-ware description language which is proposed as a standard.

Appendix: Zeus syntax

```
directionofseparation=ident.
qualident=ident { "." ident }.
```

133

designator=(qualident {"[" (constexpression [".."
constexpression] {"," constexpression [".." con-
stexpression] } | "*" ident "(" designator ")" {","
"*" ident "(" designator ")" }) "]"| "." ident
[".." ident]})|"*".
designatorlist= designator { "," designator } .
orientationchange=ident.
boundary=("top"|"right"|"bottom"|"left") lbstate-
mentsequence "end".
order="order" directionofseparation lostatementse-
quence "end".
muloperator= "*"|"div"|"mod"|"and".
constfactor=number|"(" constexpression ")"| "not"
constfactor|
designator["(" constexpressionlist ")"].
addoperator="+"|"-"|"or".
constterm=constfactor {muloperator constfactor}.
relation= "="| "<>"| "<"| "<="| ">"| ">=".
simpleconstexpr= ["+"|"-"] constterm { addoperator
constterm }.
constexpression=simpleconstexpr [relation simple-
constexpr].
constexpressionlist= constexpression {"," constexpression }.
types=(arraydeclaration|componentdeclaration| qual-
ident ["(" constexpressionlist ")"| "is" dstate-
mentsequence "end"]).
assignment= designator (":=" | "==") expression.
basic= ["[" orientationchange "]"] designatorlist.
expression=designator ["(." constexpressionlist
".)"] [expression]| "(" expression {"," expres-
sion} ")" ["." designator].
withs= "with" designator "do" statementsequence
"end".
result= "result" expression.
whenstatement= "when" constexpression "then"
statementsequence {"otherwise_when" constexpres-

sion "then" statementsequence } ["otherwise"
statementsequence] "end".
forstatement="for" ident ":=" constexpression "to"
constexpression ["by" constexpression] "do" statementsequence "end".
conditional= "if" expression "then" cstatementsequence { "elsif" expression "then" cstatementsequence } ["else" cstatementsequence] "end".
connection= designator [expression].
connectionspecification= "connect" cstatementsequence .
statement= [connectionspecification| assign-
ment|basic| connection| conditional| forstatement|
whenstatement| result| withs| order| boundary|
declaration].
statementsequence= statement {";" statement }.
lostatement=[basic| forstatement| whenstatement|
withs| order].
lostatementsequence= lostatement {";" lostatement}.
lbstatement=[basic|forstatement|whenstatement|withs].
lbstatementsequence= lbstatement {";" lbstatement
}.
cstatement=[assignment| conditional| connection|
forstatement| whenstatement| result| withs].
cstatementsequence= cstatement { ";" cstatement}.
dstatement= [declaration| connectionspecification|
whenstatement| order| boundary].
dstatementsequence= dstatement {";" dstatement}.
idlist= ident {"," ident }.
parameterlist= idlist ":" types .
fparams= ["in"| "out"| "clock"| "type"] parameter-
list.

subrangetype= "[" constexpression ".." constexpression [":" ident] {"," constexpression ".."
constexpression [":" ident] } "]".
componentheading= "component" ("(" fparams {";"
fparams }] ")" [":" types] | "predefined").
structconstexpression= constexpression| "[" struct-
constexpression {"," structconstexpression}"]".
constantcollection= "(." idlist ".)".
constant= constexpression| structconstexpression|
constantcollection.
hd= idlist ":" types.
hardwaredeclaration= "hardware" hd {";" hd}.
td= ident [("(" fparams {";" fparams} ")")| "pre-
defined"] "=" types.
typedeclaration= "type" td {";" td}.
cd= ident "=" constant.
constdeclaration= "const" cd {";" cd}.
declaration= [constdeclaration| typedeclaration|
hardwaredeclaration| moduledeclaration| hardware-
moduledeclaration].
componentdeclaration= componentheading ["is"
dstatementsequence "end"].
arraydeclaration= "array" [subrangetype {"," su-
brangetype}] "of" types.
block= declaration {";" declaration} "end".
definition= [constdeclaration| typedeclaration|
hardwaredeclaration].
export= "export" ["qualified"] idlist ";".
import= ["from" ident] "import" idlist ";".
hardwaremoduledeclaration= "hardware_module"
ident ";" { import } [export] block ident ["."].
hardwaredefinitionmodule= "definition" "hard-
ware_module" ident ";" { import } [export] definition {";" definition} "end" ident "." .
hardwarecompilationunit= hardwaredefinitionmodule|
["implementation"] hardwaremoduledeclaration.

Appendix: Definition of Zeus

The following is an excerpt from: "Report on the
Hardware Description Language Zeus". The full
version of the defining report on Zeus is being
disseminated through the MIT VLSI Memo series of
the MIT Microsystems program office.

6. TYPE DECLARATIONS

Zeus maintains two classes of types: signal types
and control types. The signal types define a type
for signal constants and hardware instances and
the control types define a type for control
constants, including control parameters. A type
determines a set of values which instances of that
type may assume, and it associates an identifier
with the type. In the case of structured types, it
also defines the structure of instances of this
type.

For signal types, Zeus provides COMPONENT,
ARRAY and basic types which can be arbitrarily
nested.

6.1 BASIC TYPES

6.1.1 SIGNAL TYPES

The following basic types are predeclared and denoted by standard identifiers:

- An instance of type *tri_state* assumes as values the constants *zero, one, undef*, and *high_imp*.

- An instance of type *logical* assumes as values the constants *zero, one*, and *undef*.

Other basic signal types can be defined as enumeration types or subrange types in a special Modula-2 module called *basic_types*.

6.2.1 SIGNAL TYPES

An array is a hardware structure consisting of a fixed number (which might be a function of a control parameter) of elements. These elements are usually all of the same type, in which case the array serves as a hardware replication facility. However, the elements might be of different types.

Example:

    TYPE d=ARRAY[1..n:i,1..n:j] OF t(i,j)

Type *d* represents a chessboard-like processor configuration if $t(i,j)$ is a "white" processor for $i+j$ even and a "black" processor otherwise.

The elements of the array are designated by indices, with values belonging to the index type. The index type is a control type and can be either BOOLEAN, a Modula-2 enumeration type, or a subrange type. A subrange type is defined as a subrange of a basic control type by specifying the least and the highest value in the subrange.

6.1.2 CONTROL TYPES

The following basic types are predeclared and denoted by standard identifiers.

- A constant of type INTEGER assumes as values the integers between MaxInt and MinInt.

- A variable of type BOOLEAN assumes the truth values TRUE or FALSE.

Other basic control types can be defined as enumeration types or subrange types in a Modula-2 module. The exported type identifiers are used in control parameter specifications in Zeus.

6.2 ARRAY TYPES

The array type declaration specifies the index type as well as the element type.

A declaration of the form

    ARRAY T1, T2, ..., Tn OF T

with *n* index types *T1, ..., Tn* must be understood as an abbreviation for the declaration

    ARRAY T1 OF
    ARRAY T2 OF
    ...
    ARRAY Tn OF T

Examples of signal array types:

    ARRAY [0..n-1] OF logical;
    ARRAY [-5..100] OF tri_state;

Signal array types can be parameterized in the same way as component types (see 6.3.2).

A subrange type specification might contain a colon, followed by an identifier. This identifier is a fresh identifier valid within the array type only. The fresh identifier can be used in place of a constant expression. The scope of the identifier starts after its definition. Therefore it is possible to have "shaped" arrays.

Example:

    TYPE d= ARRAY[1..n:i, 1..i:j] OF t(i, j)

represents a triangular array.

6.2.2 CONTROL TYPES

Array control types must be (nested) arrays of the basic control types INTEGER or BOOLEAN.

6.3 COMPONENT TYPES

A component type defines a circuit. (There are no control component types.) The wires which are attached to the circuit are described by formal signal parameters. A signal parameter can be of any signal type. The names of these signal parameters are called parameter identifiers. They are used for connecting the component to its environment.

We distinguish two kinds of signal parameters: *ports* and *fields*. A signal parameter, and in general a type, is called a port if it has any of the following types:

a) A basic signal type.

b) An array type of a) b) c).

c) A linear component type of a) b) c). A linear component type is a component type which does not have a statement part.

All other signal parameters are called fields. A component type is a structure having a fixed number of ports or fields of possibly different types. The component type declaration specifies for each signal parameter its type and an identifier which denotes the parameter. The scope of these parameter identifiers is the component definition itself, and they are also accessible within parameter designators (see 8.) referring to parameters of instantiated component types, and within *with* statements.

A component type declaration consists of a component heading, specifying the ports and fields, and a statement sequence. A component type may be specified by a qualified identifier *x* followed by IS and a statement sequence. The qualified identifier *x* must denote a component type without a body.

There are two kinds of component types, namely proper component types and function component types. Function component types are activated by a designator (see 8.) as a constituent of an expression, and yield a result that is an operand in the expression. Function component types can only have ports as formal parameters. Proper component types have to be instantiated first in a hardware declaration before they can be used (e.g. in an assignment or a connection statement).

The function component type is distinguished in the declaration by an indication of the type of its result following the signal parameter list. Its body must contain a result statement which defines the result of the function component.

All constants, types, hardware objects, and modules declared within a component are local to that component type. The scope of constant, type and hardware declarations within a *when* statement is limited to the *when* statement. Since component types may be declared as local objects too, component types may be nested.

In addition to the parameters and local objects of a component type *C*, the constants and types declared in the declaration environment of *C* are known and accessible in *C* (with the exception of those objects that have the same name as objects declared locally).

The use of the component type identifier T in a type or hardware declaration within the declaration of T implies a recursive definition of component type T.

6.3.1 FORMAL SIGNAL PARAMETERS

Formal signal parameters are identifiers that denote actual parameters specified in a connection statement (or in a function designator in an expression).

The correspondence between formal and actual parameters is established in the connection statement (or, in the case of a function, when the function is called). There are four kinds of parameters, namely IN, OUT, neutral and clock parameters. The kind is indicated in the formal parameter list. An IN parameter is used to transmit a value to a component, and an OUT parameter is used to transmit a value from a component. A neutral parameter is specified by not specifying the IN or OUT reserved words.

Acknowledgements

I would like to thank Steve German, Terry Glagowski and Martin Resnick for their feedback on Zeus and Paul Lindemann for improving the contents and style of this paper.

Parts of this research were done while the author was visiting at MIT and was partially supported by a grant from the Swiss National Science foundation.

References

[Cappello1983a.] P.R. Cappello and K. Steiglitz, "A VLSI layout for a pipelined Dadda multiplier," *ACM Transactions on Computer Systems* Vol. 1(2) pp. 157-174, May 1983.

[German1984b.] S. German and K.J. Lieberherr, "Zeus-CSP," *In preparation*, 1984.

[German1984a.] S. German "Private Communication," *January*, 1984.

[Knudsen1983a.] S.E. Knudsen, "Medos-2: A Modula-2 Oriented Operating System for the Personal Computer Lilith," *Dissertation ETH Zurich*, 7346, 1983.

[Lieberherr1983a.] Karl J. Lieberherr and Svend E. Knudsen, "Zeus: A hardware description language for VSLI," *Tech. Report 51, Institut fuer Informatik, ETH Zentrum, CH -8092 Zurich, Switzerland*, February 1983.

[Lieberherr1983b.] Karl J. Lieberherr and Svend E. Knudsen, "Zeus: A hardware description language for VSLI," *Proceedings of the ACM/IEEE 20th Design Automation Conference*, June 27-29, 1983.

[Lieberherr1983c.] K.J. Lieberherr, "Multi-Level Simulation for VLSI" *Proc. IEEE International Conference on Computer Design: VLSI in Computers, Port Chester, New York*, pp. 441-444, 1983.

[Nash1984a.] J.D. Nash, "Bibliography of hardware description languages," *SIGDA Newsletter*, Vol. 14(1), pp. 18-37, February 1984.

[Wirth1982a.] N. Wirth, "Programming in Modula-2," Springer, 1982.

[Wirth1982b.] N. Wirth, "Hades: A notation for the description of hardware," *ETH Zurich*, August 1982.

Block Description Language (BDL): A Structural Description Language

Eric Slutz, Glen Okita, Jeanne Wiseman

Hewlett-Packard, 1501 Page Mill Rd, Palo Alto California, 94304

Abstract

The Block Description Language (BDL), a language for capturing the structure of an electronic system, is described. The structure of a system may be specified hierarchically in this language. Additional information may be associated with the structural description with general properties. A facility for declaring and using arrays and records of the structure elements is provided.

Keywords: hardware description language, logic design

Introduction

BDL is used to transfer information between design systems, design databases, and application programs. To maximize the portability of this design information, the design language is ordinary text.

BDL captures hierarchical connectivity for a design. Many design systems and application programs take advantage of hierarchical descriptions of designs. Hierarchy makes checking for design errors easier and reduces the volume of data needed to represent a design.

Other information about a design may be captured by using properties in BDL. Properties are a way of attaching additional information to a connectivity description.

Background

In the past two decades, many languages have been proposed for describing electronic hardware [1,2,4,5]. Most of these languages attempt to model a hardware system functionally; they are often called register transfer languages.

Given a functional description, the system being designed can be simulated. Logic syntheses from the high level description is sometimes possible. The primary goal of these languages is to capture design intent at a very high level. The high level register transfer languages have proven useful in the design of complex digital systems. However, most of these languages are of limited value in capturing purely structural representations of a design.

As discussed in the existing literature [3,6], a structure description language allows the separation of structural and functional descriptions. This separation of descriptions can lead to a flexible, more modular, system. For example, one can compare the simulation of a register transfer description with the simulation of explicitly connected components in a structure description.

Structure descriptions of electronic designs are useful for more than design verification. For example, the implementation of a system specified with a register transfer language is usually described with structure because of limitations in logic synthesis. Also, automated physical implementation, such as for PC boards, gate-arrays, or standard cells, often requires a structural design representation. A structural description language also overcomes the inherent limitations of register transfer languages, such as the ability to describe an analog circuit. With the advent of low-cost workstations, structural descriptions are an easy way to pass design data on to mainframe simulators, gate-array vendors, or traditional large CAD installations.

The languages most similar to BDL in philosphy are SDL and HISDL. BDL differs from these languages by providing structural types for connections, similar to the data structures provided in modern high-level languages. The explicit declaration of types preserves information about the connectivity elements which can be useful for analysis tools and physical design systems. For example, a router may choose to route sub-elements of a 32-bit bus together, or type declarations in BDL can be checked for consistency with declarations in a register transfer language. BDL also provides a more general properties mechanism than SDL or HISDL. (Note: it appears that HISDL has no properties at all.)

BDL lacks the iterative and recursive features of HISDL because its primary purpose is to capture concrete structures which do not require further interpretation. The macro expansion facilities in SDL are also absent as a result of this philosophy. The advantage of requiring concrete structure is that BDL can be used directly. Thus, post-processed results, such as SDL "object", need not exist for applications using BDL.

Reprinted from *ACM/IEEE 21st Design Automation Conf. Proc.*, 1984, pp. 81–85.

Hierarchical BDL

BDL describes basic electronic entities called "blocks." A block is a set of interconnected sub-blocks. The connections are described in terms of "signals." A given block can be instantiated any number of times in other blocks. Each instantiation is like a copy of its defining block. A block is composed of instances of other blocks interconnected by signals.

The description is hierarchical because a block is described in terms of other blocks. If a block 'reg16' were instantiated in a block 'alu' then block 'reg16' would be at a lower hierarchical level than block 'alu'. The lowest level blocks may be transistors, gates, or TTL parts.

Block RSFlipFlop example

An example block, 'RSFlipFlop', is shown with its corresponding BDL description. It introduces important connectivity concepts described in following paragraphs.

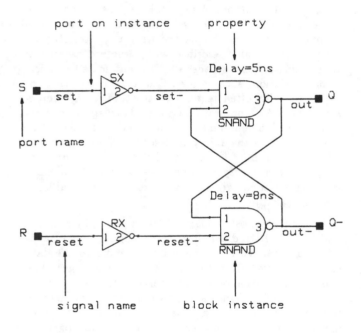

```
BLOCK Inverter;
  PORTS
    NET : 1, 2;
  ENDPORTS
ENDBLOCK
BLOCK Nand;
  PORTS
    NET : 1, 2, 3;
  ENDPORTS
ENDBLOCK
```

```
BLOCK RSFlipFlop;
  PORTS
    NET : R, S, Q, Q-;
  ENDPORTS
  SIGNALS
    NET : reset, reset-, set, set-, out, out-;
  ENDSIGNALS
  INSTANCES
    Inverter : RX, SX;
    Nand     : RNAND(Delay=''8NS''),
               SNAND(Delay=''5NS'');
  ENDINSTANCES
  SIGNALLIST
    reset    : THISBLOCK*R, RX*1;
    reset-   : RX*2, RNAND*2;
    set      : THISBLOCK*S, SX*1;
    set-     : SX*2, SNAND*1;
    out      : THISBLOCK*Q, SNAND*3, RNAND*1;
    out-     : THISBLOCK*Q-, RNAND*3, SNAND*2;
  ENDSIGNALLIST
  INSTANCELIST
    RX        : 1=reset, 2=reset-;
    SX        : 1=set, 2=set-;
    RNAND     : 1=out, 2=reset-, 3=out-;
    SNAND     : 1=set-, 2=out-, 3=out;
    THISBLOCK : R=reset, S=set, Q=out, Q-=out-;
  ENDINSTANCELIST
ENDBLOCK
```

Block Concepts

Ports, signals, and block instances compose a block. The example is used to explain these concepts.

Ports

Ports allow logical connections to blocks and block instances. When a block is defined, instances of other blocks are created and signals are created which connect to the ports of the instances. In the example, signal 'set-' connects to port '1' of instance 'SNAND'.

Within the definition of a block, each port is usually connected to one internal signal. In the example, 'R' is a port with the signal 'reset' connected to it.

Signals

A signal is a "wire" or collection of "wires" carrying electrical information. It connects instances together in a block definition. In the example above, the signal 'out' is connected to port '3' of instance 'SNAND', port '1' of instance 'RNAND', and port 'Q' of the block 'RSFlipFlop' itself.

Block Instances

A block instance is a use of another block. The instantiation of a block is a reference to the definition of the block being used. Each port instance on a block instance may have signals connected to it. In the example, 'RX' is an instance of an inverter block. The signal 'reset' is connected to its input port, 'reset-' is connected to its output port.

Properties

Properties are name-value pairs associated with elements such as signals, ports, block instances, and block definitions. The interpretation of properties is left to the user. It is not part of the BDL definition. In the example, block instance 'SNAND' has a property named 'Delay' with the value of '5NS'. If the block RSFlipFlop were simulated, then this property could be interpreted as a propagation delay of 5 nanoseconds for this instance. Instance 'RNAND' also has the 'Delay' property, but with a value of 8 nanoseconds.

Properties are name-value pairs, where the "name" is an identifier which names the property, and "value" is a text string. They provide a general mechanism to capture aspects of a design which often require special constructs in other languages. For example, ports on a block may have many different characteristics associated with them, such as direction (input, output, bidirectional), technology (CMOS, TTL, ECL), assertion level (active-low, active-high), inversion (inverting, non-inverting), capacitive loading, or physical location.

Properties provide a way to associate information with the basic connectivity description so that all information contained in design databases can be represented. For example, the value of a property may be graphics for a schematic, graphics for a layout, or a functional description for a simulator.

Making the value of a property the name of a file containing the appropriate information is one way to associate sophisticated non-connectivity data with a BDL description. These files could be binary files or database files.

Ports and Signals

One of the features of BDL is that it treats signals and ports as different entities. A signal and a port connected together may have the same or different names.

Consider the signal 'out' which connects to port 'Q' of the block itself. When this block, RSFlipFlop, is used as an instance, the signal connected to that instance need only know about port 'Q', not the signal 'out'. If block RSFlipFlop were edited and a buffer were inserted, the name of the signal connected to port 'Q' could be changed to 'buf_out' without affecting the connectivity description at higher hierarchical levels.

This distinction between signals and ports is also useful for PC boards, where ports are named like edge connectors, e.g. 'J2_12', while signals attached to the connector are named mnemonically. Signals may also connect several ports of a block definition together, making them equivalent.

Connectivity Descriptions

The connectivity for a block is usually described in either a "signal oriented" form or an "instance oriented" form. A signal oriented description lists all signals, and for each signal there is a list of block instances to which they connect. An instance oriented description lists all block instances, and for each instance there is a list of signals to which they are connected. Either description is sufficient to fully capture connectivity information. A description of connectivity is often called a "structural", or "topological" description.

Structure Concepts

Groups of abstract signals, ports, and instances are considered as structures. Given that a signal, port, or instance may have structure, there must be ways to specify the structure and select one or more components of a structure.

Block BigSys example

This more sophisticated example shows block BigSys as a schematic diagram with its corresponding BDL description. It introduces some of the structuring capabilities of BDL.

```
TYPES
   widebus  = BUS[0:31] : NET:
   smallbus = BUS[0:7]  : NET;
   sysbus   = BUNDLE
     AD  : widebus;
     CTL : smallbus;
     ENDBUNDLE;
ENDTYPES
BLOCK CPU;
   PORTS
     sysbus : io;
   ENDPORTS
ENDBLOCK
```

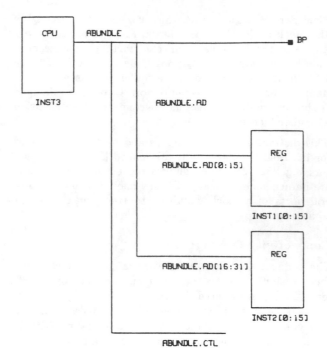

```
BLOCK REG;
  PORTS
    NET : input, output;
  ENDPORTS
ENDBLOCK

BLOCK BigSys;
  PORTS
    sysbus : BP;
  ENDPORTS
  SIGNALS
    sysbus : ABUNDLE;
  ENDSIGNALS
  INSTANCES
    REG : INST1 [0:15],
      INST2 [0:15];
    CPU : INST3;
  ENDINSTANCES
  SIGNALLIST
    ABUNDLE          : THISBLOCK*BP, INST3*io;
    ABUNDLE.AD[0:15]  : INST1*input;
    ABUNDLE.AD[16:31] : INST2*input;
  ENDSIGNALLIST
```

```
  INSTANCELIST
    THISBLOCK : BP=ABUNDLE;
    INST1     : input=ABUNDLE.AD[0:15];
    INST2     : input=ABUNDLE.AD[16:31];
    INST3     : io=ABUNDLE;
  ENDINSTANCELIST
ENDBLOCK
```

Port and Signal Structure

Ports and signals have "net", "bus", or "bundle" structure types. A net type represents a single "wire." A bus type represents an array of types, which are selected by a non-negative integer. The type which makes up the bus is the "base type." A bundle type is a record structure of types. Sub-types in a bundle are selected by a "field" identifier.

All port and signal types are named and declared. Properties may be attached to types and to fields of types.

The example includes a bundle named 'ABUNDLE' which has a field named 'AD' and a field named 'CTL'. The field 'AD' is a 32-bit bus. The first 16 bits of field 'AD' are connected to instance 'INST1', while the second 16 bits are connected to instance 'INST2'.

Block Instance Structure

Block instances may be structured into arrays. Multi-dimensional arrays are permitted.

The example shows two one-dimensional instance arrays, 'INST1' and 'INST2'. Both instance arrays are 16 bits wide.

Type Compatibility

BDL only allows connections between compatible types. Type compatibility is determined by rules similar to most structured programming languages. In the example above, the signal selection 'ABUNDLE.AD' has the type defined by field 'AD' in bundle 'ABUNDLE'. This signal has been further selected to 'ABUNDLE.AD[0:15]', which is a 16 element subset of field 'AD'. The instance array connected to this selected signal, INST1[0:15], is compatible if and only if the type of the port on block 'REG' is compatible with the type of a single element in field 'AD'.

Application of BDL

BDL serves as a general way to communicate connectivity information. Typical uses are to load a database with a connectivity description, or to output the contents of a database. The "network description language" translators mentioned for HISDL to NDL conversion are provided by translation programs running directly off the database.

Similarly, communication to automatic physical design programs is done by loading the connectivity into the database used by the auto route and place programs directly.

Summary

To summarize, the main features of BDL are:

BDL is hierarchical.

BDL is textual.

BDL allows either instance or signal oriented connectivity descriptions.

BDL supports properties associated with elements.

BDL supports structured signals, ports, and instances.

Conclusion

The features of a language for describing the hierarchical structure of electronic circuits has been presented. The language allows for the declaration of busses, bundles, and arrays of instances. A general capability to attach additional information in the form of properties exists. The BDL language is used as an intermediate form to drive application programs, or entire design systems.

References

[1] Chu, Y., "An ALGOL-like computer design language," *Comm. ACM,* Vol. 8, October 1965, pp. 607-615.

[2] Duley, J. R., Dietmeyer, D. L., "A digital system design language (DDL)," *IEEE Trans. Computers,* Vol. C-18, September 1968, pp. 850-861.

[3] Lim, Willie Y.P., "HISDL, a Structure Description Language," *Communications of the ACM,* November 1982, Vol. 25, Num. 11, pp. 823-830.

[4] Piloty, R., Barbacci, M. Borrione, D., Dietmeyer, D., Hill, F., Skelly, P., "CONLAN – A formal construction method for hardware description languages: basic principles," *AFIPS Conf. Proc.,* Vol 49, 1980, pp. 209-217.

[5] Shiva, S. G., "Computer hardware description languages – A tutorial," *Proc. IEEE,* Vol. 67 Num. 12, December 1979, pp. 1605-1615.

[6] vanCleemput, W. M., "A hierarchical language for the structural description of digital systems," *Proc. 14th Design Automation Conf.,* New Orleans, Louisiana, June 1977, pp. 377-385.

A High Level Synthesis Tool for MOS Chip Design

Jean Dussault, Chi-Chang Liaw, and Michael M. Tong

AT&T Bell Laboratories, Murray Hill, NJ 07974

Abstract

This paper describes a design tool called Functional Design System (FDS) that supports high level MOS LSI design. Designers can build circuits at the register transfer level by using a set of high level FDS primitives. FDS then automatically produces in seconds an accurate and efficient polycell implementation for these primitives. Therefore, the design cycle time can be reduced significantly. FDS is an integral part of a larger CAD system [1] which supports other aspects of the design cycle, namely, graphical design capture, simulation, test generation, and layout. The system has proved to be highly successful in helping designers to develop extremely reliable chips in a short time frame.

1. Introduction

In the conventional polycell design approach, designers interconnect library defined function blocks together to form circuits. Each function block, which is called a polycell, performs an SSI function like a logic gate, a transmission gate, or a flip flop. Several versions of a polycell which meet different power and speed requirements are normally available in the library for designers to choose from. The connectivity between polycells is described manually in a circuit description language call LSL [2]. This process is often very tedious, time consuming, and error prone as detailed design information is required during early stages of the design. In addition, circuit changes which require adding, deleting, and rearranging polycells are often cumbersome and vulnerable to mistakes.

Figure 1 shows a new polycell design approach using the Functional Design System (FDS). Designers build their circuits using polycells and FDS primitives. These primitives can model high level function blocks similar to those MSI catalog parts being used in the circuit pack design environment.

Each primitive provides a rich set of variations selectable by a user to meet the exact specifications he wants for a circuit. A user can create an FDS primitive dynamically by describing its functional behavior through a graphical design capture system called SCHEMA [3]. The behavior of primitives is captured in a menu driven interactive process. SCHEMA then produces a graphical symbol for each primitive and its behavior is automatically compiled into a concise textual representation in the Functional Primitive Description Language (FPDL). Figure 2 shows the symbol of a 4-bit shift/load FDS register and its corresponding FPDL description. All FDS primitives thus created are generic functions and their symbols can be interconnected together to form a circuit schematic. SCHEMA then automatically produces the LSL description for the design.

Next, the design is verified by simulating it at the register transfer level [4]. Each FDS primitive in the circuit is treated as a single simulation element. The functional behavior of each primitive is extracted automatically from its FPDL description by the simulator. Most design changes done at this stage can be easily accomplished by changing the behavior of the primitives. When the circuit is verified at the functional level, the FPDL description of each primitive is fed into FDS to produce a polycell implementation. In addition to the LSL description, FDS provides the schematic placement information so that SCHEMA can display the polycell structure in a schematic. The expanded circuit is then verified in greater details by running logic and timing simulation at the polycell level [5,6]. Finally, automatic layout tools [7] are used to layout the chip.

By default, FDS generates area efficient designs. There are options available in these primitives which produce designs that are optimized in terms of operating speed and/or power consumption. Since the response of FDS is very fast, a designer can, in the same interactive environment, try different design strategies on a circuit, compare them, and then choose

Reprinted from *ACM/IEEE 21st Design Automation Conf. Proc.*, 1984, pp. 308-314.

the optimum solution. FDS greatly enhances the efficiency of the design effort significantly reducing the design cycle time.

This paper will describe the different types of FDS primitives and their features. The implementation of their synthesis routines will be discussed. Finally, an example that illustrates the power and the convenient use of FDS will be given.

2. FDS Primitives

An FDS primitive is a high level function block for building circuits. Its logic behavior and its data path size are user definable. Different primitives can be interconnected together to realize a larger and more complex function. A top down design of a circuit using FDS can be done by partitioning its logic into several subfunctions and then mapping each subfunction to an FDS primitive.

A user can get access to two categories of FDS primitives. The first category consists of system defined primitives each of which is dedicated to perform a specific logic function. For example, there is a primitive for adders, and there is a primitive for counters, etc. The second category contains the Random Logic Combinational primitive. The logic function of this primitive is user defined through boolean equations or through a truth table. Each category of primitives will be described in more details in the following sections.

3. Functional Primitive Description Language

The functional behavior of an FDS primitive is defined using FPDL. This language serves as a common medium between the logic simulator, which verifies the design, and the FDS which implements the polycell structure. FPDL can be either written manually, or generated automatically by SCHEMA. Its syntax consists of a set of system defined keywords. There are five statements in the language set:

```
CKTNAME:   circuit name;
TYPE:      FDS primitive type (size parameter);
INPUTS:    input list;
OUTPUTS:   output list;
OPTIONS:   design options;
```

The CKTNAME statement is used to name a primitive. The generic function thus created will be referenced by this name. The primitive type is specified in the TYPE statement. Each type has a keyword, like "ADDER", and "COUNTER", etc. The numeric parameter following the type gives the size of the primitive. For example, ADDER(4) means a 4-bit adder. The INPUTS and OUTPUTS statements are used to name the primitive input and output terminals, respectively. Each primitive type has its own set of keywords for these two statements to define its functional behavior. For example, by choosing adder keywords "CI" and "CO" in the INPUTS and OUTPUTS statement, defines that the adder has a carry input and a carry output. Finally, the OPTIONS statement is used to define the implicit functions of a primitive. For example, the keyword "CARRY_LOOK_AHEAD" means a carry-look-ahead adder is selected over a ripple-carry adder. Each primitive has a convenient set of default values for its behavioral definition like the polarity of its data and control signals, etc.

4. System Defined Primitives

This category of primitives evolves from the MSI catalog parts that are commonly used in the circuit pack design environment. Currently FDS supports seven system defined primitives. They are the Register, Counter, Adder, Decoder, Multiplexer, Parity Generator, and Comparator. New primitives can be easily added onto the list.

Each primitive provides a rich set of features for a designer to choose from. For example, the register primitive has the following features:

- n bit configuration
- left/right/bidirectional shift
- parallel load
- preset/preclear controls
- clock enable control
- single clock/two clock scheme
- static/dynamic implementation
- high/low active controls
- synchronous/asynchronous controls

A user can create registers, each with a different configuration, by choosing different feature combinations. Listing 1 gives the FPDL description of an 8-bit left shift register with parallel load. The register has a falling edge triggered clock, and its synchronous preclear control is low active. "SHLD" is a keyword for the shift/load control. By default, the register will do a shift(load) operation when SHLD is high(low). "RS" is the signal whose values latches into the most significant bit of the register at each shift operation. "PI[7:0]" is a vector notation for the parallel load data lines. PI7 is the most significant bit and PI0 is the least significant bit. The outputs

"Q[7:0]" and "QN[7:0]" are complementary of each other.

```
CKTNAME:    BOB;
TYPE:       UNREG(8);
INPUTS:     CK, RS, PI[7:0], SHLD, PC;
OUTPUTS:    Q[7:0], QN[7:0];
OPTIONS:    FALLING_EDGE_TRIGGER,
            SYNC, LOW_ACTIVE(PC);
```

Listing 1.   FPDL Description of an 8-bit Register.

4.1 Implementation

The polycell structures of these primitives are very regular in nature. Primitives like registers, counters, and ripple carry adders have linear structures, and primitives like carry-look-ahead adders, multiplexers, and comparators have tree-like structures. In addition to a polycell library, FDS maintains an internal library of intermediate building blocks to aid the synthesis effort. Each of these blocks is hand designed using standard polycells. The block size ranges from three to ten polycells. The front end of these primitives is a compiler that compiles the FPDL description. Based on the primitive type, FDS determines the design framework to be either a linear structure or a tree-like structure. Next, intermediate building blocks as well as polycells are picked from the library to fit this structure and they are connected together. Final touches are then applied to adjust the design to meet the exact functional specifications. For example, one technique is to add inverters or to interchange signal connections to adjust the polarity of data and control signals. The output processor of FDS then produces the circuit description in LSL and also provides the schematic placement information of the circuit to SCHEMA for graphic display.

5. Random Logic Combinational Primitive

This primitive is designed specifically for modeling random combinational circuits which do not have a regular structure like an adder or a parity generator which are best handled by system defined primitives. FDS accesses a system of routines to manipulate and transform multiple-output combinational switching functions into a polycell LSL circuit description. The FDS serves as a front end to this system and is used to capture the circuit description using the FPDL. The FDS in turn, produces an intermediate file which is passed to the logic synthesis programs. As far as the user is concerned, this feature looks like any other FDS primitive except that, instead of specifying parameters and switches, one has to specify the function to be realized.

The first step in the logic synthesis is to extract the common subfunctions from the set of output functions so that they can be processed separately. Each subfunction (main and extracted) is then minimized. Using a factoring procedure, each subfunction is implemented using a minimal set of gates obtained from the specified polycell library. The subfunctions are then connected together appropriately and each is logically optimized by migrating inverters and performing symbolic replacements. Finally, the LSL output is produced.

5.1 Logic Function Description

At present, the function description is entered using the input language of DCBE. (Desk Calculator for Boolean Expressions; developed at AT&T Bell Laboratories in North Andover) [8]. DCBE is a general tool to manipulate boolean functions and its equation style is used to specify the logic function of combinational blocks. A function is defined by assigning ($=$) a valid expression to a variable. It is possible to specify whether that expression defines the ones of the output function, its zeros or its don't care cover by suffixing its name with .on, .of or .dc, respectively. A valid expression is any well-formed expression containing input variables, output variables, subfunctions and the operators * (and), + (or), ! (not), % (xor), # (sharp: relative complement), (,) (parentheses).

There is a syntax which may be used to enter expressions in cubic notation [9]. For example, given three inputs x1,x2 and x3, the cube "1x0 represents x1* !x3. A '1' indicates that a variable is part of the product, a '0' that its complement is part of the product and 'x' indicates that the variable does not appear in the product. Listing 2 gives a sample function description file for a small combinational circuit.

5.2 Extraction of Design Parameters

Design parameters for the synthesis of circuits are obtained from the particular polycell library used. Combinational gates like AOI (And-Or-Invert), OAI (Or-And-Invert), ND (Nand), NR (Nor), XOR (Exclusive-Or), XNOR (Exclusive-Nor), and INR (Inverter) are sorted by size into tables, one for each type. Fan-in limitations can be specified and are achieved by populating the tables only with cells satisfying the fan-in constraints.

```
CKTNAME: CCV;
TYPE: COMBINATIONAL;
INPUTS: B4N,B3,(B2,B2N);
OUTPUTS: CCMNR9;
OPTIONS: ROOT_NAME (GATE),MAX_FANIN (3);
<DCBE>
A = !(!B2*B3*B4N)
B = !(B2*B3*B4N)
F = !(E*A)
CCMNR9.ON = "0XX111X + "1100XX1 + F
<END>
```

Listing 2. Sample Function Description File.

These parameters determine the largest size a two-level structure (A(OI), O(AI), and the equivalent structures constructed out of ND, NR and INR) can have. Synthesis uses these two-level structures as templates and pick the "best" implementation.

During synthesis and optimization, AOIs and OAIs will be transformed into one another. In the same routine that extract the design parameters, the fixed costs of such transformations are computed.

5.3 Approach to Multiple-Output Synthesis

The philosophy here is to deal with single output problems as much as possible. Indeed, processes like minimization and synthesis are essentially made to handle single input functions. This constrains the complexity of the problems and increases the speed of the solution. However it is recognized that completely ignoring the possible sharing among multiple output functions is not realistic. For that reason the truth table is pre-processed before any manipulation such as minimization is attempted. The scope of the pre-processing is to extract the common subfunctions from the output array, to augment the array with the extracted subfunctions, and replace the occurrences of those subfunctions in the main functions with don't care conditions.

Two heuristics are used to find the "best" common subfunctions. The first one attempts to find the widest vector of ones which is contained in the largest number of implicants. The second heuristic is somewhat more exhaustive and is better at finding larger subfunctions but which may be shared among fewer outputs.

5.4 Minimization and Implementation Procedures

After the expanded truth table is created, each subfunction (or its complement) is minimized

separately. For the purpose of synthesis we have found it sufficient and efficient to find irredundant covers rather than minimal covers as could be obtained from the extraction algorithm. In the case where more optimal minimization is required, the intermediate file format described above is fully interfaced with RDC [8,9], POP [10] and a version of MINI [11].

After each function is minimized, it is implemented using a factoring based procedure. Both the true function and its complement are examined for implementation. A quick algorithm is used to find the best factors of the function [9]. In this case, a good factor is a wide product term contained in a large number of implicants (represented as cubes). The function is implemented by recursive decomposition and the number of factors at each level is bounded by the limit on fan-in.

5.5 Reconnection and Logic Optimization

Next, each function and its associated subfunctions are reconnected together using the equivalent of an OR gate. When the subfunctions are brought together they are optimized. Synthesis produces redundant inverters and AOI and OAI gates which are not necessarily well formed. A procedure has been devised to eliminate the inverters by removing consecutive ones or by pushing them towards the inputs. A simple cost function drives the process. Other redundant gates are also removed and the degenerate ones replaced by an appropriate equivalent.

5.6 Design Verification

In addition to producing LSL, the synthesis program produces files which can be used to check the implementation using logic simulation. The input cubes used in the synthesis process are expanded and used as inputs to a true value simulation of the synthesized circuit. The outputs produced by simulation are then compared with the output cubes and the discrepancies are reported. The entire verification process is automated by a single command.

6. Example

Since its implementation, FDS has been used to design several production ICs. For the purpose of illustration, a scaled down design is used to show the power of the system. Figure 3 is a schematic of an 8-bit Booth Multiplier. The design is captured at the register transfer level by using SCHEMA. The whole design took seven FDS primitives which include five registers, one multiplexer, and one adder. SCHEMA produces the FPDL of each primitive as well as their interconnections in LSL. FDS expanded the circuit into a polycell design and it is verified using a logic simulator [5]. Figure 5 shows a portion of the output timing diagram. Except for the graphical design capture and functional test generation, the whole logic design process is virtually automatic.

7. Conclusions

A high level MOS design aid has been described. The capabilities of the FDS primitives and their implementations have been discussed. The tool is being used to design several custom MOS production chips that use different MOS technologies. Feedback from users shows that FDS integrates very well into their design process. The primitives are easy to use and the response of the system is very fast. FDS helps remove the design tedium and avoids designer's burnout.

Acknowledgements

We would like to thank A. K. Bose for his original idea on FDS, and his technical guidance throughout the project. J. W. Eddy and the SCHEMA group provided fantastic graphical features for FDS. We are grateful to P. Agrawal and G. Mandeville for their valuable comments to our paper. Finally, special thanks to P. Gagnon for providing the Booth Multiplier example.

References

[1] A. K. Bose, 'A System of Computer Aids for LSI Design', The 3rd International Conference on Semi-Custom ICs, London, England, Nov 1-3, 1983.

[2] H. Y. Chang, G. W. Smith, and R. B. Walford, 'LAMP: System Description', Bell System Technical Journal, Vol. 53, N0.8, October 1974, pp. 1431-1449.

[3] L. A. Fajardo, C-C. Liaw, and M. Tong, 'A System for High Level Design Capture and Synthesis', AT&T Bell Laboratories Technical Journal, to appear.

[4] V. D. Agrawal, A. K. Bose, P. Kozak, H. N. Nham, and E. Pacas-Skewes, 'A Mixed-Mode Simulator', Proc. of 17th Design Automation Conference, Minneapolis, Minn., June 23-25, 1980, pp. 616-625.

[5] H. N. Nham, and A. K. Bose, 'A Multiple Delay Simulator for MOS LSI Circuits', Proc. 17th Design Automation Conf., Minn., June 23-25, 1980, pp. 610-617.

[6] B. R. Chawla, H. K. Gummel, and P. Kozak, 'MOTIS - An MOS Timing Simulator', IEEE Trans. Circuits and Systems, Vol. CAS-22, No. 12, December 1975, pp. 901-910. Also, P. Kozak, H. K. Gummel, and B. R. Chawla, 'Operational Features of an MOS Timing Simulator', Proc. 12th Design Automation Conference, Boston, Mass., June 23-25, 1975, pp. 95-101.

[7] G. Persky, D. N. Deutsch, and D. G. Schweikert, 'LTX - A System for the Directed Automatic Design of LSI Circuits', Proc. 13th Design Automation Conf., San Francisco, June 28- 30, 1976, pp. 399-407. Also G. Persky, D. N. Deutsch, and D. G. Schweikert, 'LTX - A Minicomputer Based System for Automated LSI Layout', J. Des. Aut. & Fault-Tol. Comp., Vol. 1, No. 3, May 1977, pp. 217-255.

[8] P. D. Halkiotis, T. B. Merrick and E. D. Walsh, AT&T Bell Laboratories, North Andover, MA., 1979.

[9] D. L. Dietmeyer, 'Logic Design of Digital Systems', 2nd ed., Chapters 9-11, Allyn and Bacon, 1978.

[10] D. W. Brown, 'A State Machine Synthesizer - SMS,' 18th Design Automation Conference, pp. 301-305. Developed by A. Svoboda, reported by Brown and implemented by P. Simanyi at UC Berkeley.

[11] S. J. Hong, R. G. Cain, and D. L. Ostapko, 'MINI: A Heuristic Approach for Logic Minimization,' IBM Journal of Research and Development, vol. 18, no. 5, Sept. 1974.

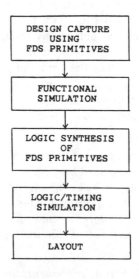

Figure 1. The Design Process Using FDS

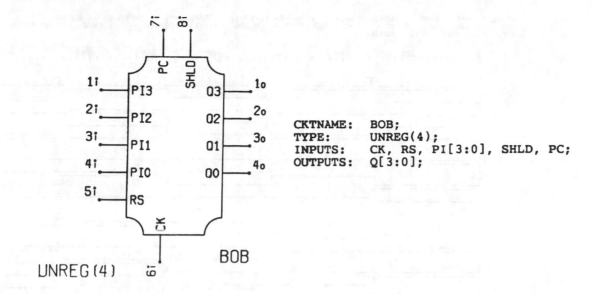

CKTNAME: BOB;
TYPE: UNREG(4);
INPUTS: CK, RS, PI[3:0], SHLD, PC;
OUTPUTS: Q[3:0];

Figure 2. SCHEMA Symbol and FPDL Description
of a 4-bit FDS Register

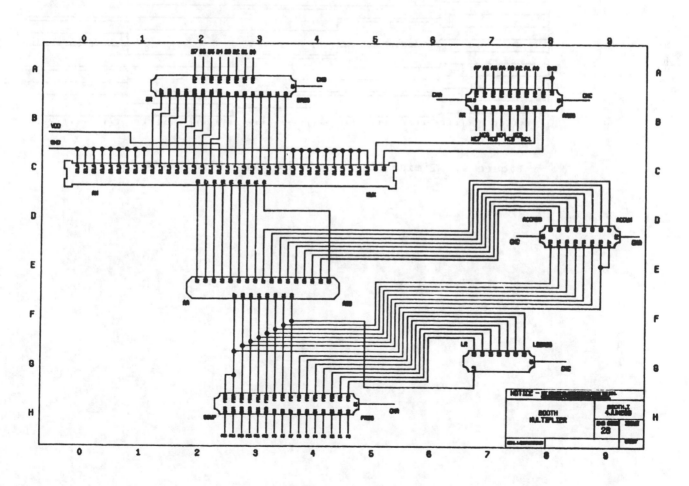

Figure 3. An 8-bit Booth Multiplier

147

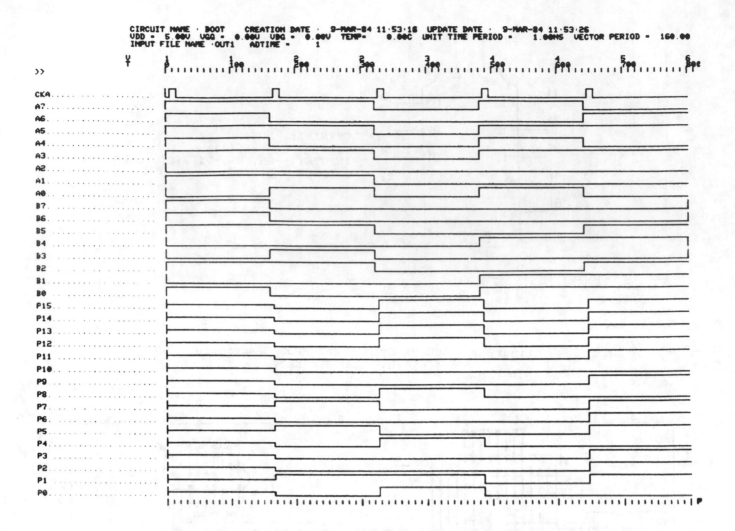

CIRCUIT NAME : BOOT   CREATION DATE : 9-MAR-84 11:53:16 UPDATE DATE : 9-MAR-84 11:53:26
VDD = 5.00U VGG = 0.00U VBG = 0.00V TEMP=   0.00C UNIT TIME PERIOD =   1.00MS VECTOR PERIOD = 160.00
INPUT FILE NAME :OUT1   ADTIME =   1

Figure 4. Timing Diagram of the Booth Multiplier

148

EDIF: A MECHANISM FOR THE EXCHANGE OF DESIGN INFORMATION

John D. Crawford
Tektronix, Inc.
P.O. Box 500
Beaverton, OR 97077

Abstract

The large variety of CAD/CAE tools/systems and foundry services now available to the electronic engineer have brought to a head the long-standing need for a standard format for the exchange of design information. Existing formats have not served this need for a variety of reasons. This paper describes an Electronic Design Interchange Format (EDIF) that has arisen from collaboration of the principals involved in four preceding efforts.

Introduction

Over the past few years, there has been a rapid increase in the number of silicon foundries and CAD/CAE system and workstation companies. As a result, and due to the increasing problem of moving data among in-house design systems, a standard interchange format for electronic design data has become essential.

While many interchange formats and hardware description languages have been developed over the past decade, each has suffered from one or more of the following drawbacks:

- *Narrow Focus:* For example, many formats have been developed to address only a specific aspect of the design data, such as only mask artwork data, or only netlist information, but not both.

- *Proprietary:* Many formats are proprietary to specific semiconductor or CAD companies.[1,2] This severely limits their effective use.

- *Difficult to Implement:* Often, as a format evolves, it becomes increasingly difficult to develop parser/generator software for dealing with it. This is particularly true if upward compatibility is an issue.

- *Difficult to Extend:* As IC design methods rapidly evolve, new facets are added to the data that must be transferred. Many languages or formats cannot grow to meet these new needs.

There are a large number of formats in use, many of which have been quite successful.[1-6] Unfortunately, each one suffers from one or more of the drawbacks listed above.

Benefits of Standardization

The benefits of the general adoption of a standard electronic design interchange format are important and far reaching. The customer receives a much wider variety of feasible choices, and free competition in both CAE/CAD system/tool businesses and foundry businesses is enhanced. Electronic circuit designers and customers of CAE/CAD system and foundry service can expect compatibility among these products. As a result, they can choose equipment and services best suited for a particular task with minimum overhead. Foundries can work with customers regardless of what CAE/CAD equipment the customers are using. The CAD/CAE tool or system vendor can maintain compatibility with a variety of foundries and other systems with a minimum investment in software development for that purpose. Currently, a wide variety of formats must be supported.

Background

In the last one to two years, several formats have been developed to address one or more of the issues previously mentioned. EDIF was started by the principals involved in four of these efforts in order to combine the best features of each and coalesce them into a single, more powerful format. This cooperative approach has been backed by organizations originally responsible for the development of each of the four formats. The organizations involved in the current effort and committed to its success are Daisy Systems Corp.; Mentor Graphics Corp.; Motorola, Inc.; National Semiconductor Corp.; Tektronix, Inc.; and Texas Instruments, Inc. The formats from which EDIF is derived are described briefly below.

CIDF (Common Interchange Description Format) was the result of two years of semi-regular meetings among interested people from several organizations. The effort evolved from UC Berkeley research in this area[7-9] and produced a format for the the transmission of schematic diagrams, symbolic layout, mask-level layout, and connectivity information. The syntax, based on the LISP programming language, is simple and extensible.[10]

GAIL (Gate Array Interface Language) was developed as a proprietary format to specifically address the transfer of information related to gate arrays at certain points in the design cycle.[11]

Reprinted from *IEEE Proc. Custom Integrated Circuits Conf.*, 1984, pp. 446–449

The development of TDF, also targeted at gate arrays, was driven by the need to transfer graphical information on gate array macro libraries from semi-conductor vendor to CAE/CAD system vendor and end customer.[12]

TIDAL is the most comprehensive of the formats, and addresses almost all identifiable design data from behavioral descriptions down through mask geometry. TIDAL addresses areas that none of the other three formats address, such as test data, design rule and technology data, and behavioral description data. TIDAL provided the largest base from which the development of EDIF proceeded.[13]

In late 1983, a committee consisting of the principal developers of the above formats was formed to co-ordinate the development of a single, public domain, interchange format.

Design Data Requirements

Requirements were generated to identify the classes of design data the format was to handle.

- The initial version handles the data required for a designer using a CAD/CAE system and a foundry supplying gate array or semi-custom design services to exchange all needed information.

- All information that a foundry typically sends to customers to enable them to do these types of designs must be expressible in EDIF Version 1.0. Macro libraries, technology information, and substrate information can be sent.

- All information that a designer sends to a foundry at various stages of the design process must be supported. This includes netlists, mask-level geometry, and test data.

- Later versions of the language must, in addition, support ECB information, system-level design information, and procedural layout.

Format Design Requirements

The design data requirements listed above and the need for greater capability over time drove the specification of several format design requirements.

- The format needed to be designed to minimize implementation difficulty of translators into and out of it.

- Since the format will evolve through several versions both to add capability and to improve existing capabilities, extensibility must be designed in from the beginning.

- The initial version of the format must permit the transfer of a useful subset of the design information and must also provide a framework for continuing work.

Scope and Structure of EDIF

EDIF version 1.0 is targeted at the exchange of design information for gate-array and semi-custom IC designs in a multi-vendor environment. This is regarded as a two-way process.

Foundry to Designer

Base arrays, technology information, and macro libraries must be sent to customers. EDIF provides mechanisms for expressing all of this information. To provide a feel for the kinds of information that can be exchanged using EDIF, some the information normally required to describe a cell in a macro library is detailed. A simple example is shown in figure 1.

- *Logic Model:* A connect list of logic primitives with various attributes provides the basis for a logic model. If this model has been generated using a schematic editor, the graphical information may also be useful. The set of logic primitives used in a logic model is often tied to a particular logic simulator. EDIF provides a set of typical primitives (basic gates, etc). However, a richer set may be used and is expressible in the format.

- *Circuit Model:* A circuit model consists of a connect list based on circuit-level primitives such as transistors, resistors, and capacitors. As with the logic model, schematic graphics information may also be included.

- *Macrocell Layout:* A set of graphics primitives is provided to construct the layout view or detailed layout of the cell. These primitives may be easily mapped into popular existing public (such as CIF) or proprietary (such as Calma STREAM) formats.

- *Layout Abstraction:* EDIF is specifically designed to support abstractions. In particular, the information normally needed to design with or produce a layout using a cell is easily differentiated from the detailed information about that cell. A layout abstraction containing routing blockages, port and terminal locations, names, permutability information, etc., is easily constructed.

- *Schematic Symbol:* A schematic symbol ready for use with a schematic entry system is easily expressible. The symbol may have electrical ports with names and attributes. The set of graphics primitives in EDIF is designed to handle both schematic and layout graphics.

- *Base Array:* The base array, including plug and socket information, can be expressed.

The EDIF logic model, schematic symbol, detailed layout, and layout abstraction for placement and routing for a very simple cell is outlined in figure 1.

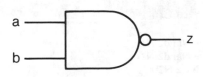

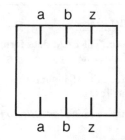

```
(Cell NAND2
  (View Topological Logic
    (Interface
      (Declare (INPUT PORT a b) (OUTPUT PORT z))
      (Body (Layer SchematicSymbol
        (shape (20 0) (arc (40 0) (60 20) (40 40)) (20 40))
        (circle (60 2]) (70 20))
          )
      )
      (PortImplementation a (Layer SchematicPort (Path (0 30) (20 30))))
      (PortImplementation b (Layer SchematicPort (Path (0 10) (20 10))))
      (PortImplementation z (Layer SchematicPort (Path (70 20) (90 20))))
      (Permutable a b)
    )

    (Contents
      (Instance Nand2Primitive)
    )
  )
  (View Physical
    (Interface
      (Declare (INPUT PORT a b) (OUTPUT PORT z))
      (Body (Layer MetalBlockage (Rectangle ....)))
      (PortImplementation b (Layer Poly . . . . .))
            .
            .
            .
      Permutable a b)
    )
    (Contents
      (Layer Metal (Path ...) (Polygon ...) ...)
      (Layer Poly  (Path ...) (Path ...) ...)
            .
            .
            .
    )
  )
)
```

Figure 1

Information from Designer to Foundry

On receipt of the macro library, users read it into their local system. A design may be shipped back in many forms.

- *Netlist:* The design parts-list/net-list information is expressible in EDIF. If the design was entered using a schematic entry system, the schematic graphics may also be expressed. A very simple netlist example is shown in figure 2. Figure 3 illustrates the addition of placement information to the simple netlist.

- *Layout:* A complete mask-level layout may be translated into EDIF either from an internal format or from any of the commonly used layout formats. Additionally, a simplified symbolic layout typical of the output of a gate-array placement and routing package may be translated into EDIF.

Figure 2 shows a latch constructed from two instances of the NAND2 cell described above. The instances and interconnections are expressible in EDIF in a simple parts-list/net-list format as shown. Some of the expressive power of EDIF is shown in figure 3 by adding a little more information to the same description. In this case, placement locations are provided for the instances and the **MustJoin** construct is used (rather than the **Joined** construct) to indicate that the connections must be implemented by place and route software.

As illustrated in the above examples, EDIF supports multiple views of a cell. Full instance hierarchy is also supported. The syntax is based on the LISP programming language for ease of implementation and extensibility.

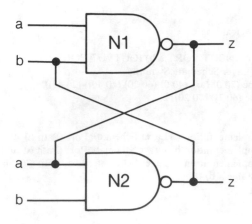

```
(Cell SimpleLatch
  (View Topological Logic
    (Interface
      .
      .
      .
    )
    (Contents
      (Instance Nand2 N1)
      (Instance Nand2 N2)
      (Joined N1:z N2:a)
      (Joined N2:z N1:b)
    )
  )
)
```

Figure 2

```
(Contents
  (Instance Nand2 N1 (0 1000))
  (Instance Nand2 N2 (0 0))
  (MustJoin N1:z N2:a)
  (MustJoin N2:z N1:b)
)
```

Figure 3

Future Directions

While the format currently addresses a large variety of data, it also provides a framework for future work. The area of behavioral descriptions requires a great deal of additional work and is one of the more difficult areas to standardize. Information specific to ECB design will be included in a future version. While the current implementation handles most of the information for full custom IC design, that topic will also be addressed in greater detail. Procedural layout descriptions are growing in popularity. Future versions of the format will address this area using constructs for expression of typical program flow information, e.g., while , do , etc. Several of these constructs are already supported in EDIF.

Acknowledgments

This paper reports the work of the EDIF technical committee, which, at the time of this writing, is composed of:

John Crawford	Tektronix, Inc.
John Eurich	Daisy Systems Corp.
Mark Faust	Tektronix, Inc.
Vandana Gupta	National Semiconductor Corp.
Carl Hage	Daisy Systems Corp.
Penny Herscher	Daisy Systems Corp.
Prof. Richard Newton	U.C. at Berkeley
Mike Ransom	National Semiconductor Corp.
Paul Stanford	Texas Instrument, Inc.
Mike Waters	Motorola, Inc.

These individuals have contributed a great deal of time and effort in the specification and implementation of EDIF. The author also acknowledges the support of the institutions listed above and the additional support and contributions of Jeff Lavell of Motorola Inc., Robert Rozeboom of Texas Instruments Inc., Dick Smith of National Semiconductor Corporation, and Steve Swerling of Mentor Graphics Corporation.

References

1. Proprietary Specification for STREAM data format, Calma Corporation, Santa Clara, California.

2. Proprietary Specification for TEGAS input format, COMSAT General Corporation, Austin, Texas.

3. A number of references may be found in *IEEE Comput.,* Vol. 7, No. 12, December 1974.

4. C.A. Mead and L.A. Conway, *Introduction to VLSI Systems,* Addison-Wesley, 1980.

5. W.A. Johnson, J.J. Crowley and J.D. Ray, "Mixed-Level Simulation from a Hierarchical CHDL," Proceedings of the 1979 International Symposium on Computer Hardware Description Languages and their Applications.

6. W.M. vanCleemput, "A Structural Design Language for Computer Aided Design of Digital Systems," Technical Report No. 136, Stanford Electronics Laboratories, Stanford University, April, 1977.

7. J.D. Crawford, A.R. Newton, D.O. Pederson, and G.R. Boyle, "A Unified Hardware Description Language for CAD Programs," Proceedings of the 1979 International Symposium on Computer Hardware Description Languages and their Applications.

8. S.A. Ellis, "A Symbolic Layout Language and a Database for an Integrated VLSI Design System," MS Report, Dept. of EECS, University of California at Berkeley, December 1981.

9. C.H. Sequin, A.R. Newton, "Standard Interchange Formats for Integrated Circuit Design," NATO Advanced Study Institute on Design Methodologies for Very Large Scale Integrated Circuits, Louvain-la-Neuve, Belgium, July 8, 1980.

10. J. Solomon, J.D. Crawford, M.G. Faust, J. Gordon, W.J. McCalla, A.R. Newton, R. Smith, et al, "CIDF—Common Interchange Description Format Version 1.0," unpublished preliminary specification, October, 1983.

11. "Gate Array Interface Language User's Manual," Daisy Systems, September, 1983.

12. "TDF—Technology Definition File Interface Specification," Motorola Inc., August, 1983.

13. "Provisional Tidal Specification," Report No. 025,002, Texas Instruments, June, 1983.

VLSI circuit design reaches the level of architectural description

by Stephen C. Johnson,
Silicon Compilers Inc., Los Gatos, Calif.

Silicon compiler lets systems engineers design chips quickly without performance loss or inefficient use of chip area

Silicon compilation, the process of automatically synthesizing complex designs and their analysis and verification models from high-level architectural descriptions of systems, is now capable of producing a wide variety of very large-scale integrated circuits that are of merchant-market quality. Virtually any systems designer can participate immediately in the design process. Furthermore, short design turnaround times and design simplicity do not occur at the expense of added silicon area.

Silicon compilation was first described in 1979 by David L. Johannsen (see "What is silicon compilation?" p. 122). Under the guidance of his thesis adviser, Carver Mead, at the California Institute of Technology at Pasadena, Calif., Johannsen demonstrated the practicality of silicon compilation for implementing certain types of VLSI architectures. Johannsen and Mead, along with Edmund K. Cheng, formerly of Intel Corp., Santa Clara, Calif., subsequently founded Silicon Compilers Inc. to

further such research and explore the practical applications. One eminently practical application is the recently introduced MicroVAX from Digital Equipment Corp., Maynard, Mass., which was built on two boards totaling 158 square inches. This space requirement can be compared with the VAX 11/750, which consists of five boards totaling 800 sq in. and has just slightly better performance. A substantial part of the 5 : 1 improvement in functional density came from the MicroVAX's data-path chip, which carries the entire 32-bit processor, except its microcode store and control-store sequencer. The design and implementation of this 37,000-transistor IC took the three-member team and the proprietary silicon compiler only seven months. The n-channel MOS chip, which measures 98,000 sq mils, carries a 32-bit arithmetic and logic unit, a 64-bit barrel shifter, a control decoder, flag logic, a dual-port block of random-access memory housing 47 registers, a timer, a seven-level operand-restore stack, and a 32-word block of read-only memory (Fig. 1).

Until recently, silicon compilation continued to be viewed by many as simply an interesting research topic, with little promise of immediate practical value. The merchant-market IC industry, in particular, has shown little interest in new design methodologies, especially those that seemed to sacrifice silicon area or performance. This point of view is understandable in an industry that competes primarily on cost and performance. As a result, there has been little change in merchant-market IC design methodology in the last 15 years. Even computer-aided design, without which VLSI could not have come into existence, did not change the underlying design

Reprinted with permission from *Electronics*, vol. 57, pp. 121–128, May 3, 1984.

What is silicon compilation?

Silicon compilation is a design-automation methodology that moves the process of realizing a merchant-market-quality very large-scale integrated circuit toward the architectural level and away from classical, microlevel considerations, so that systems engineers can effectively design chips. Silicon compilation encapsulates the knowledge of expert IC designers in the areas of layout, timing analysis, and functional simulation.

Just as software compilation synthesizes complex assembly-language programs from programs written in high-level languages, modern silicon compilation synthesizes critical views of an IC, such as layout (topology), functional description (simulation model), and performance (timing model) from high-level architectural descriptions. These descriptions are in terms of structures "known" to the compiler, such as programmable logic arrays, read-only and random-access memories, arithmetic and logic units, registers, or complete data paths. Structures can be described functionally (shall the ALU multiply?), in size (how wide is the data path?), or in terms of mathematical, logical, or state equations.

The structures themselves are actually synthesized from stored rules that are built into the compiler. The rules represent an attempt to embed the best of the IC designer's art and science, and when invoked by the compiler, yield a layout, a functional simulation model, a timing model for the structure, and ultimately, a timing model for the chip. Silicon compilation typically uses several thousand rules that will yield from 200 to 500 of the most useful high-level VLSI

structures from about 40 different classes. Designs can be provided for different VLSI processes, and a specific compiler usually has a wide range of similar processes for which it can produce circuits with no user input other than process selection.

Architectures may be specified hierarchically, and the layout may be synthesized either hierarchically (wiring blocks into modules and modules into chips) or with the flat approach (working at the block level, wiring directly from one block to another). Interconnection routing is accomplished automatically by the compiler. Silicon compilation also uses an integrated hierarchical data base and provides complete functional simulation at any level of the design hierarchy automatically. It also produces complete performance reports and critical path analysis automatically from the completed layout.

During the design process, which is incremental by nature, silicon compilation provides estimates of a wide variety of circuit characteristics before they are fully determined. Finally, it puts out masking tapes and all supporting plots and documentation, thereby eliminating from the design process such steps as logic design, logic simulation, circuit design, circuit simulation, layout, postlayout parameter extraction and resimulation, electrical rule checks, and topological design-rule checks. Thus, all of the arcane aspects of VLSI design disappear, permitting a competent systems designer (in addition to a chip designer), to utilize silicon compilation to create merchant-market-quality custom or standard integrated systems.

---

methodology. CAD simply speeded up, made more accurate, or changed the form of certain steps of a well-worn design path.

This lack of interest in silicon compilation by the IC industry is compensated for, however, by its potential for use by manufacturers of electronic end products—the systems houses (see "The application-specific IC market starts to grow," p. 124). These manufacturers look forward to the advent of a technology that provides rapid, accurate development by systems engineers of VLSI systems with no sacrifice of silicon or performance.

The MicroVAX project and two others—the development of a single-chip Ethernet data-link controller (EDLC), which is in commercial production by Seeq Technologies Inc., San Jose, Calif., and the RasterOp Controller used in the Sun II work-station products—have proven that such a technology is at hand.

Integrated systems technology

Silicon compilation sprang from efforts to develop an integrated systems technology that not only encapsulated the expertise and methodologies of the best designers in the IC industry, but provided higher levels of design leverage by means of radical new methodologies. The resulting technology bears little resemblance to layout-simplification methodologies now in vogue, such as standard cells or gate arrays. Instead, the technology addresses the entire VLSI design process simultaneously, and

implements an approach practiced only in the more mature arts and sciences: design by successive refinement, or incremental design.

The term design leverage refers to processes or methodologies whereby small inputs of time or effort by a designer result in large amounts of high-quality, useful design output. For example, if a methodology will permit three designers to produce a working 37,000-transistor chip in six months, when classical methods require a combination of 20 engineers and layout draftsmen working for two years, that methodology has leverage.

Current IC design technology, including standard-cell and gate-array practices, has too little design leverage to be able to support the market for user-designed chips. Furthermore, what leverage exists is not the most important type. For example, a gate-array methodology that permits a designer to go from a detailed logic diagram to a metalized 2,000-gate complementary-MOS gate array in two weeks does have some layout leverage, but at the expense of silicon efficiency and versatility (many functions cannot be implemented on a gate array). More important, however, is that this methodology has had absolutely no effect on the many other time-consuming and error-prone tasks in the total design process, such as logic design, modeling, simulation, or architectural exploration.

With respect to IC design technology as it exists, the position of integrated systems technology and silicon compilation can be described by analogy: if producing an integrated system is akin to arranging transportation between two points, then present design technology is a garage full of automobile parts, some power tools (the equivalent of CAD tools), and an assembly manual. Integrated systems technology is the assembled automobile: the user need merely get in and drive.

Designers needing a VLSI implementation of a system can, with great effort, create a collage of systems from all the component parts in the current IC-design-technology market, and get a fair measure of expertise and methodology. But the vital leverage is missing.

Tower of Babel

The development of integrated systems technology involved creating several models of the classical design process. The most important is the "view" model, which holds that during the design process, the designer creates and maintains several views of his circuit, which include a functional view, a functional simulation view, a logical view, a logic-simulation view, and circuit and circuit-simulation views. It includes also a topological view, and performance, power, and testing views.

Each of the foregoing design views historically has been documented in different formats and has had its own CAD support tools. During the course of the design process, the designer is forced to generate views separately and to translate repeatedly from one view to another. Frequently, such translations are generated manually, with great effort. The major differences between these views, and the lack of standardization of the CAD and computer-aided-engineering tools supporting the views (that is, a nearly total lack of common interfaces), create a Tower of Babel syndrome. The lack of a common base for communications means confusion and fragmentation of purpose. It is also one of the reasons that VLSI design is so difficult, and is practiced by so few persons.

The "complexity" model of VLSI design is depicted in Fig. 2. In the classical, top-down design approach, a 20,000-gate circuit may be composed of 50,000 transistors implemented topologically with over 1 million geometries. Perhaps 2,000 cells (of the complexities associated with standard-cell design) would have been required to implement the 100 or so major blocks from the block diagram. Finally, these 100 blocks were functionally part of about 20 architectural-level modules. The classical designer must manage all of these complexities, at each level (or view) of the design.

Virtually all designers today approach VLSI design by such a top-down, hierarchical decomposition. Unfortunately, a VLSI designer, once having specified a behavioral description of an integrated system and having designed a high-level architectural description (at the block level, or above), must then wallow in logic, circuit, and geometric design, often for years on one chip. Design at higher levels of abstraction implies that the design process may stop at perhaps the architectural (or module) level.

Design leverage

Silicon compilation creates design leverage by keeping the user from having to do any design activity below the leverage line denoted in Fig. 2. That is, all details of VLSI design and implementation below the block level are hidden from, and transparent to, the user by the technology. Design leverage is attained by holding design activity to high levels of abstraction. This management of complexity, in turn, is obtained through design synthesis, the process of automatically creating from the architectural-level circuit description all required views of the circuit and all abstractions below the leverage line.

There are several other important points in the VLSI

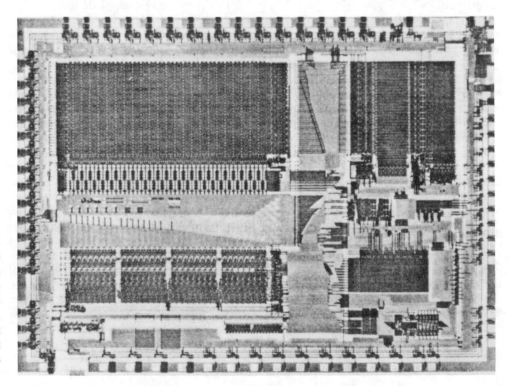

1. 64-bit data path. The main engine of the MicroVAX, less its microcode store and control-store sequencer, is a new architecture developed using a silicon compiler. Containing the control decoder, 32-bit arithmetic and logic unit, a 64-bit barrel shifter, 47 dual-port registers, and more, it measures 98,000 square mils.

The application-specific IC market starts to grow

In recent months, industry watchers such as Dataquest Inc. have begun to predict that a revolution is at hand, fueled by moves made by electronics manufacturers to integrate entire systems or major subsystems on single chips. The San Jose, Calif., market-research firm believes that in 1990, application-specific integrated circuits will account for $20.6 billion of a total IC market of $43 billion, up from a 1980 application-specific market of about $2 billion. More significantly, it predicts that nearly 40% of these will be designed by their end users.

The tremendous payoff of system-level integration has already forced scores of systems manufacturers to learn the difficult and arcane arts of IC design and manufacture. Even then, their expertise is typically concentrated in small groups of IC experts, whom the systems designers treat almost as chip vendors. But it is estimated that only about 2,000 people are employed as IC designers, as opposed to approximately 400,000 systems engineers, worldwide. Therefore, most observers believe that this dramatic growth—especially in user-designed chips—will not occur unless there is either a significant change in VLSI design methodology, or introduction of powerful new design-auto-

mation tools, or both, so as to enable systems engineers to design cost-effective VLSI. Cost-effective implies design-turnaround time measured in months, effective use of the silicon real estate, and design at only the architectural level.

The market for application-specific ICs is causing two submarkets to appear—the IC-design-technology market and the silicon-foundry market. Suppliers in the design technology market range from small houses that hand-design chips for third parties; vendors of computer-aided-design tools, who supply equipment for various parts of the design process; suppliers of work stations for computer-aided engineering, who initially concentrated on design-documentation entry but now seem inexorably drawn into the CAD arena; and several leading-edge firms concentrating on design automation. Organizations that provide standard-cell libraries, gate-array libraries, and either CAD or design-automation tools for using them are also players.

The silicon-foundry market today is principally served by several merchant-market IC makers who offer standard-cell or gate-array methodologies to customers, and hence are participants in both the IC-design and silicon-foundry markets.

---

design process where tremendous leverage can be applied to meet the goal of an integrated systems technology. One of the most obvious and least exploited points in integrated systems design, which can effect a large improvement in the function, performance, quality, or silicon efficiency of a VLSI circuit, is that of the exploration or investigation by the designer of a variety of alternative architectures for a given circuit. Although such exploration is practiced to a small degree today, the process is tedious and inaccurate. Much information that would influence the selection from competing architectures is unavailable until logic design, circuit design, and layout are nearly complete. At this point, the issue of changing architectures is moot. But architectural exploration, when done correctly, can result in final VLSI designs that are significantly more efficient—and hence more cost-effective—than those attainable under the limited exploration typical of classical VLSI design.

Such exploration would be limited without an accurate means of estimating the characteristics of the finished circuit—such features as die size, power dissipation, or critical-path timing. Today, most important features, especially those affecting performance or cost, cannot be determined accurately until layout is essentially complete. The ability to estimate these and other features during architectural exploration means that the trade-offs between competing architectures become clear.

Other design leverage is also manifest. Since IC designers spend a significant amount of time tediously translating each view of the design to the next, such as from the architectural view to the logic view or to the logic-simulation view, a technology that can synthesize all views from one common abstraction will not only reduce time, but will reduce the attendant errors and confusion.

The decoupling of a design from a manufacturing pro-

cess solves one of the most pervasive problems for users of custom or application-specific VLSI chips—that of ensuring an uninterrupted source of ICs by enabling several fabrication lines or manufacturers to produce the same part. Today, however, the common reality is that for a given chip to run in a manufacturing area (other than the one for which it was designed), often it must be redesigned.

Human creativity

Another vital and completely overlooked area where the design process can be significantly enhanced is that of the design flow itself. The classical design process, even when approached with a standard-cell or gate-array methodology, proceeds monotonically from state to state, from view to view.

Silicon compilation, by contrast, addresses all of the circuit-design views listed above at once. It integrates all ancillary support functions such as simulation and timing verification. But architectural design is left to the human; the designer's creativity has not been replaced.

The table lists the aspects of IC design technology that are accommodated by silicon compilation. Those aspects served by CAD or CAE are also shown. Integrated systems technology, in the true spirit of design automation, automates and integrates virtually every aspect of the design process possible, except those tasks requiring human creativity and judgement.

The compilation philosophy, which provides for a compiled layout from an architectural description, has been generalized to include the compilation of all other information germane to the design process, enabling the designer to guarantee that his design can function as desired, within required timing and power constraints.

There is an analogy with software: not only does a

software compiler produce assembly code, but it also concurrently produces cross-reference listings, ancillary debugging modules, and, with modern systems, source-code control aids, data-flow diagrams, program hierarchy diagrams, and so forth. All of these items, produced concurrently by the expanded compilation system, are vital to the program-development process. The fact that they are produced concurrently and automatically, rather than asynchronously and manually, gives the programmer vital leverage to optimize his creative work. In addition, clerical work and the Tower of Babel syndrome are eliminated.

In the same manner, silicon compilation provides designers with integrated design, verification, and documentation. Formatting and translation problems between the various tools are eliminated because the compiler is the only tool required. It concurrently synthesizes, from the same architectural description, functional models at all levels in the design hierarchy, functional simulations, timing models, timing simulation, power estimates, and, of course, layouts. Thus nearly all of the important views of a circuit are produced automatically as part of the compilation process. Two additional views, the logic-simulation and circuit-simulation models, are not required with silicon compilation. (The last view, the high-level architectural description, is always left up to the designer's creativity).

New design process

Because of this concurrent nature of the development of each required circuit view, an incremental or successive-refinement approach to design becomes far easier. Incremental design is quite different from the classical sequential VLSI design process, but it is very similar in nature to other mature, creative pursuits of man. For example, an artist, when approaching a new painting, starts with a sketch and slowly refines the entire canvas, adding more color and more detail, until the work is complete. He designs by successive refinement. Integrat-

ed systems architecture is also an art, and systems designers are artists as well as engineers.

The compilation process is such that at any point, most of the ultimate characteristics of the finished chip can be estimated from whatever data is available. In practice, this means that as soon as an architectural description of a circuit is provided (consisting of major elements such as ALUs, register files, and programmable logic arrays), the compiler will, upon request, estimate die size and power consumption. The architectural description need not be finalized for this to occur. As additional details are filled in (such as whether the ALU has result registers, and if so, how many), the estimates become more accurate. The compiler will inform the designer when sufficient information has been accumulated so that the next major step may be attempted.

With silicon compilation, the designer can lay out the floor plan long before the total details are known about each of the architectural building blocks. However, when the last detail is actually provided, the user can begin functional simulation (the model having been automatically constructed by the compiler), or begin wiring. Similarly, timing estimates may be obtained before wiring is begun.

As the design is refined, the estimates are improved and updated. The designer provides additional refinement in an incremental fashion, as he determines from the estimate whether he wishes to continue to pursue his current approach. At any time, he may change it. But since the net cumulative investment in design time at any point is so small (hours or days), he is encouraged to continue to experiment until the best architecture emerges and all the architectural refinements are specified. The ultimate refinement is, of course, the finished circuit. The data that is available at this point includes the final layout in a form suitable for pattern generation.

Architectural exploration, which is one of the greatest leverage points in the design cycle, occurs spontaneously. Regardless of the innate ability of any silicon compiler to

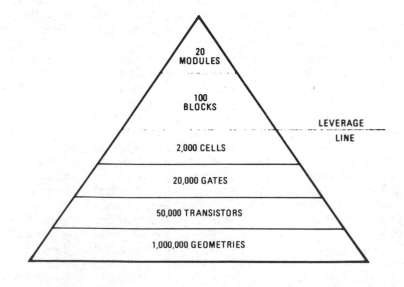

2. Top-level leverage. Putting an automated tool to work at a high level in this design-complexity model gives the systems designer with an architectural viewpoint the ability to work with ideas at that level; yet he or she will still have detailed knowledge of the effects of these ideas in terms of the cost and performance of the silicon-based end product.

IC-design steps	Computer-aided engineering	Computer-aided design	Silicon compilation
Expertise components:			
Intuition			
Creativity			
Experience			X
Discipline			X
Methodology steps:			
Implementation strategies			
Problem breakdown			X
Manpower allocation			X
Toolset selection			X
Analysis limits defined			X
Designer-imposed constraints			X
Design capture	X	X	X
Design style strategies			
Clocking approach			X
Low-level architecture			
High-level architecture			X
Verification strategies			
Circuit modelling			n/r
Logic modelling			X
Stimulation modeling			
Functional verification (simulation)		X	X
Topology verification (design-rule checking)		X	n/r
Modeling for performance verification			X
Performance verification (timing)		X	X
Manufacturing strategies			
Processes			X
Manufacturers			X
Manufacturing preparation:			
Verification/qualification			
Functionality verification (testing)		X	X
Process checking			
Circuit characterization			
Manufacturability qualification			
Manufacturing setup			
Foundry-specific layout additions			X
Test-tape preparation			
Bonding-diagram preparation			X
Layout digitizing		X	n/r
Pattern-generation tape preparation		X	X
Mask preparation			
Manufacturing:			
Fabrication			
Assembly			
Testing			
Documentation	X	X	X
(n/r means this component not required)			

Rather, the designer ends up spending more time on other pursuits, such as architectural exploration, than he did in the classical process.

Silicon Compilers undertook several custom chip-design projects designed to wring out the silicon compilation process. The Ethernet data-link controller project, begun shortly after the firm's founding in 1981, took a total of eight months, with the first three consumed by architectural exploration. This phase culminated in a chip specification, which was then implemented using a prototype silicon compiler in the remaining five months. A significant portion of this time was consumed by the development of the compiler itself. Two engineers were required for the chip architecture and implementation, while the development of the compiler took the remainder of the company's resources.

The resultant chip, which is being manufactured under license from Silicon Compilers, was implemented in a 3-micrometer n-MOS process (Fig. 4).

Input to the compiler was an eight-page high-level description of the circuit written in a proprietary architectural-description language. Among the blocks synthesized by the compiler are two programmable logic arrays, a first-in, first-out buffer, three data-path blocks (registers, comparators, counters, and zero-detection logic), a serial data path, clock generators, a linear-feedback shift register, bonding pads, and assorted random logic. The floor plan was laid out interactively, which the company prefers for architectures containing less than 100 blocks. Interconnection was accomplished with the compiler's intrinsic router.

Rapid turnaround

The second project was a proprietary circuit for bit-mapped-graphics operations. Called the RasterOp Controller (Fig. 5), it was designed for Sun Microsystems of Santa Clara, Calif., for use in their line of work-station products and has been well received by Sun, according to Andrew Bechtolsheim, who is vice president of technol-

produce circuits of a certain density, architectural exploration virtually guarantees that the final circuit will be efficient by comparison with a classically designed circuit. Here, the difference is in the final architecture. In the classical design process, architectural exploration is minimal, because it is so difficult. With silicon compilation as it is today, the mechanics take only a few hours. Virtually all of the designer's time can be spent in the architectural-design phase.

A matter of time

Figure 3 shows the practical results, both the time and the manpower differences for chips at the 40,000-transistor level. The times from the classical design process do not scale to the silicon-compilation example.

Electronics/May 3, 1984

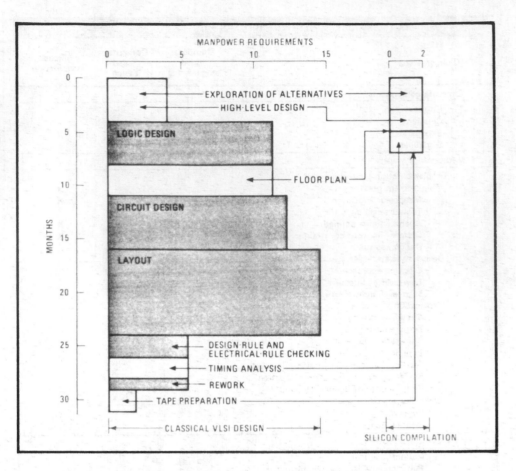

3. Comparing outlays. A comparison of the time and effort put into various phases of IC design as it is generally practiced versus current work levels for Silicon Compilers Inc. at the 40,000-device level includes the use of computer-aided-design and -engineering tools, which have reduced the size of some of these blocks a certain percentage. Silicon compilation leaves the designers in the architectural phase even longer, but then truncates the process, making logic and circuit design, layout, and electrical-rule and design-rule checking unnecessary.

ogy. The entire design required only one engineer working five months.

The rapid turnaround of the MicroVAX I data-path chip was a key element in DEC's strategy to complete the design and implementation of the MicroVAX within one year. Silicon Compilers worked in concert with a VAX design team from Digital's Belleview, Wash., facility (DEC West).

The partitioning of the design into the ROM microcode store, control-store sequencer, and the data-path chip was considered optimum, since it kept the latter chip at a reasonable size and took advantage of the proven, high-density ROM technology. From a project-management standpoint, it permitted DEC West to concentrate on the microcode development for the VAX instruction set, while the compiler was applied to the data-path chip itself. Finally, it offered a simple way of implementing last-minute changes in the microcode.

The internal architecture of the data-path chip was totally new. Although systems manufacturers often attempt to reduce costs by reimplementing old architecture in a new technology, as is often done today with gate arrays, such an approach is usually not optimal, since many of the original trade-offs do not hold for the new technology. Instead, as was the case with the MicroVAX chip, a totally new architecture is more appropriate.

The silicon-compilation technology has already been expanded to provide automatic synthesis of new blocks required for the chip. The IC carries RAM, ROM, a timer, a counter, random logic, programmed logic arrays (inhibit and ALU-decoding logic), and a multiplexer for oper-

and logic, in addition to the 32-bit ALU and 64-bit barrel shifter mentioned earlier.

Since silicon compilation synthesizes large blocks automatically, one of the major challenges in its development is to guarantee that every single possible combination of user-specified options for a given block type generates electrically and topologically correct layouts. To this end, traditional CAD and CAE tools, used to verify the construction and operation of the silicon compiler, and including gate-level simulation, circuit verification, design-rule checking, network analysis, and engineering-change control, have been considerably enhanced.

Functional correctness of the data-path chip was guaranteed by an extensive simulation effort, with both firms involved. Simulators built by each team were run with identical sets of test data until both the simulators and the design converged. Silicon Compilers concentrated on simulation at the functional level by generating models that predicted the performance of each of the blocks. Simulation models are now automatically synthesized at the same time as the block layouts.

Slow to catch on

The performance of the MicroVAX chip was guaranteed by static-timing analysis. Timing models are now also automatically generated for each block, and for an entire chip, or any partitioning. An integral static-timing verifier then yields timing reports indicating critical paths, maximum clock frequency, setup and hold times, and other such performance-related data.

If silicon compilation were simply to become a tool

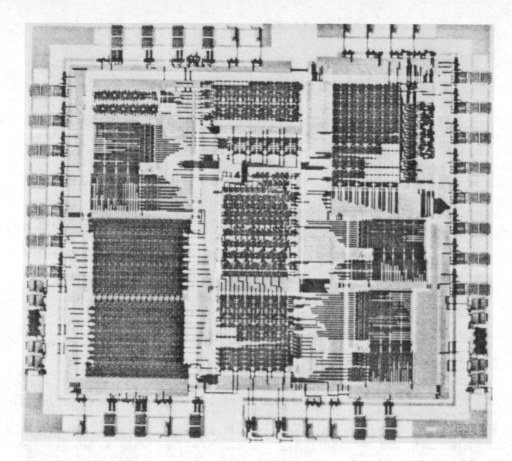

4. First project. The Ethernet data-link controller, built in a 3-micrometer n-channel MOS process, was designed with a prototype compiler during the design phase of the current silicon compiler. The floor plan was laid out interactively, which is preferred for parts with less than 100 blocks.

used by today's VLSI designers, it would fall far short of its ultimate goal, which is to make it possible for the hundreds of thousands of competent systems designers in the electronics industry today to begin designing integrated systems at the VLSI level. The benefits of integration are so great that many electronics manufacturers are beginning to view silicon compilation as the most important single breakthrough in IC design technology in this decade.

Until now, however, silicon compilation has failed to capture the imagination of much of the IC design community. Many pundits have claimed that silicon compilation's ultimate manifestation includes automatic architecture—that is, silicon output given no more than a behavioral description. As a result, many IC designers who already know better end up dismissing silicon compilation out of hand. Others fear it the way an assembly-line worker fears the coming of robots. But in reality, the production silicon compiler such as has now been developed greatly simplifies the job and would captivate the most conservative IC designer today. □

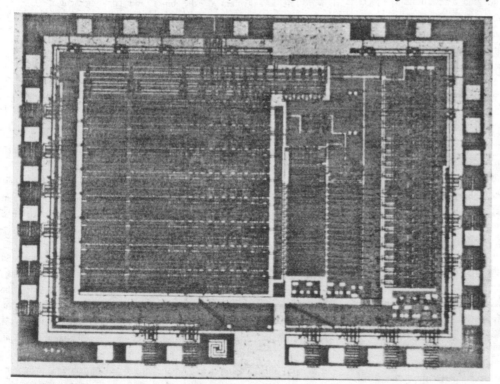

5. Graphics chip. The RasterOp controller for high-performance bit-mapped graphics was designed for Sun Microsystems Inc.'s work stations by a single engineer working only five months, using silicon compilation. The MOS part measures 182 by 151 mils in a 3-micrometer high-performance process.

Bristle Blocks: A Silicon Compiler

Dave Johannsen
California Institute of Technology

Standard LSI Design Automation systems are database management systems that aid the circuit designer by organizing the collection of submodules that comprise a chip. This type of file system usually does not aid in the actual computation of silicon layout, and can hinder a designer with program constraints that have little or nothing to do with silicon constraints. The Bristle Block system is an attempt to create a silicon compiler that will perform the majority of the implementation computation while placing a minimum set of constraints on the designer. The goal of the Bristle Block system is to produce an entire LSI mask set from a single page, high level description of the integrated circuit.

Introduction

After designing a few reasonable sized LSI integrated circuits, it becomes painfully obvious that the current design tools aid little in the every increasing task of automating chip design. Most systems are little more than fancy filing cabinets that help organize the data associated with various elements composing a chip. Even with these advanced systems, 90% of the design is completed in 10% of the time, while the remaining 10% of the design requires the remaining 90% of the time. Too many design decisions are "frozen in concrete" early in the game, before the effects of those decisions become apparent.

What if a person were able to sit down and design a complete chip in a single afternoon? The user would grab a few building blocks, snap them together, and experiment with many radically differing designs before deciding upon the actual implementation. What if that person were given complete mask layouts and simulations for each of his or her experimental configurations with almost no effort? This is the environment that the Bristle Blocks system attempts to provide.

Design Goals

The goal of the Bristle Block system is to allow the user to design LSI integrated circuits with as little concern for the mechanics involved as possible. The system must be capable of designing fairly complicated chips, and the resulting designs should have a high degree of optimization in the layout. The results should be competitive with "hand layout" chips.

Structured Design. The Bristle Block system is intimately dependant on the structured design methodology, as presented in Mead and Conway [1]. This system encourages the design of regular computing structures. These regular computing structures tend not to limit the power or usefulness of the resulting machines, but rather to enhance the performance and generality of the hardware.

Hierarchical Design. The chips designed by Bristle Blocks are hierarchical in nature. This hierarchy imposes a locality on the various sections of the chip that can be exploited when performing design rule verifications and electrical simulations of the chip.

Various Representations. The Bristle Block system provides various representations for the integrated circuit. The representations span the entire range from the physical to the conceptual aspects of the chip. Bristle Blocks is designed to handle the following seven representations:

LAYOUT. The "lowest" level of representation is the Layout level, which produces the actual chip masks.

STICKS. The Sticks level produces a stick diagram of the chip, which has the same topology as the layout, but with all of the features reduced to single-width lines. The resulting diagram is much easier to comprehend than the full layout diagram.

TRANSISTORS. The Transistor level produces a transistor diagram for the chip or subsection of the chip.

LOGIC. The Logic level of the chip can produce a logic diagram of the chip in the TTL style.

TEXT. The Text level prints a hierarchical description of the chip that can be used as a "user's manual" for the completed chip.

SIMULATION. The Simulation level can be used to logically simulate the chip, so that software can be written for the chip to explore the feasability of the design.

BLOCK. The Block level draws a block diagram of the chip, showing the arrangement of the buses and core elements.

Every fundamental element in the Bristle Block system has the capability of containing each of these seven representations for itself. When the system is physically connecting the elements, the logical and textual links are also generated. Other representations may be added to the system at any time.

Bristles and Blocks

The fundamental unit in the Bristle Block system is the cell, which may contain geometrical primitives and references to other cells. These cells to the LSI designer can be equated to the programmer's subroutines: each contain a few primative operations and references to lower levels in the hierarchy. The geometric primitives are instances of lines, boxes, and polygons, each with an associated mask layer.

Bristle Blocks also has a structure for specifying the location and flavor of connection points between cells. These connection points are like bristles along the edges of the cells, and it is upon these bristles that the Bristle Block system builds most of the computable structures. Connection points help keep local data local and global data global, while delaying the binding of many design constraints. For instance, a cell that requires an input from a pad would contain a connection point stating where in the cell the pad should connect and what type of pad is needed. When the chip is compiled, the appropriate pad is automatically placed on the chip and a wire is routed between the pad and the cell. Thus, the local data, which declare where in the cell the pad connects and the type of pad, is kept local to the cell, while the global data, which specify where the pad is

Reprinted with permission from *Proc. 16th Annu. Design Automation Conf. Under Joint Sponsorship ACM/IEEE*, 1979, pp. 310-313.

located and how the wire is routed, is kept global to the cell. Also, none of the data need be supplied by the user at compile time.

The data necessary to specify the various representations for the cells and connection points may be stored in disk files and read in as needed, to allow for the use of common cell libraries and sharing of data. Any low level cell that the user may need for his chip must be entered into either the system or a library before the chip can be compiled. Associated with each cell is the information necessary to extract the cell's definition from a file, if such a file exists.

The low level cells in a library are defined by entering the actual layout of each cell representation in a standard cell design language. The low level cell design task was not given to the compiler, but left to the user, for three reasons:

1) The design of low level cells does not take much time. Each cell usually has a small area, few transistors, and is reasonably well defined.

2) Very few mistakes are made in the design of low level cells. Experience has shown that virtually none of the fatal chip errors occur in the low level cells, but that chips fail because of faulty "glue": the high level cells holding the chip together.

3) Human ingenuity pays off well in the low level cell design. There has yet to be written a computer program that approaches human ingenuity in cell design, which is most noticable in the low level cells.

A distinction between Bristle Block cells and layout cells used in current design practices should be noted: Bristle Blocks uses procedural cells while standard practice makes use of database cells. Database cells are merely static "pictures" of circuits that are described either as rectangles on pieces of mylar or as coordinate pairs in some data file. Each copy of the cell will be identical to every other copy of that same cell. Procedural cells are little programs that can do several things, one of which is to draw itself. These cells may also stretch themselves, compute their power requirements, and, in future versions of Bristle Blocks, simulate themselves.

Chip Format

As a starting point, one style of chip design was selected to explore the possibilities of a Bristle Block system. Although it was initially felt that this style would be rigid and special purpose, the generality of the style became apparent as the potentials of Bristle Blocks were discovered. The style of chip design is stated in the description of three formats of the chips: the physical, logical, and temporal formats.

Physical Format

The physical format of chips built by the current Bristle Block system is shown in figure 1. A chip consists of a central core which is controlled by an instruction decoder, both of which are surrounded by pads. The core is composed of a series of data processing elements, such as memories, shifters, and arithmetic-logic units. Based upon the requirements of the core, the instruction decoder and pads are automatically generated and placed in the final chip.

Logical Format

The logical format of the chips is shown in figure 2. Each of the core elements can communicate with either of two buses that run through the elements. These buses may run the length of the chip, or they may stop anywhere along the chip with new buses servicing the remainder of the chip. The order of placement of elements along the core is irrelevant to the system, with the exception that at most two buses may run through any element. The microcode words which control the operation of the chip enter the decoder twice during each clock cycle. The appropriate control functions are derived from these microcode words and latched by control signal buffers. These buffers then drive the control lines of each core element.

Temporal Format

The chip is driven by a two phase, non-overlapping clock. One phase, refered to as $\varphi 1$, controls the transfer of data between elements via the buses. The alternate clock phase, refered to as $\varphi 2$, controls the operation of the data processing elements. During $\varphi 2$, the buses are

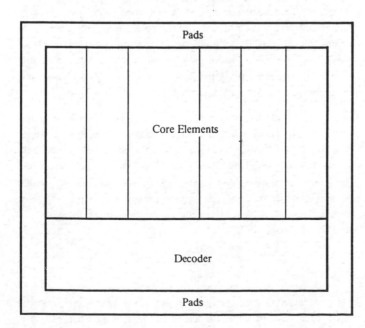

Figure 1. Physical Chip Format

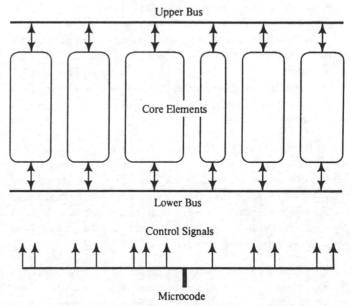

Figure 2. Logical Chip Format

precharged to a high state. Any processing element that is activated during φ1 may pull the bus lines low. Any element may read the data from the bus by the end of φ1. While the buses are transfering data, the data processing elements may be precharged. An example of a precharging pocessing element is an adder, where the carry chain is precharged to a high state. During the following φ2, the various points along the carry chain may be pulled to a low state. Instructions enter the control buffers through the decoder logic on the clock phase preceeding the phase when the instruction is to be executed.

Operation

Bristle Blocks is a three pass compiler, consisting of a core pass, a control pass, and a pad pass. The core pass takes both the user's input and low level cell definitions to construct the core of the machine. The control pass adds the instruction decoder to the core by generating a decoder which fills the constraints posed by control connection points in the core. The pad pass adds the pads to the perimeter of the chip and routes wires between the pads and the corresponding connection points in the core and decoder.

User Input

The input to the compiler consists of three sections. The first section states the microcode instruction width and describes the decomposition of the microcode word into various fields, such as the "Register Select Field" or the "ALU Operation Field." The second section states the data word width for the chip and lists the buses that run through the core of the chip. The final section lists the elements of the chip's core, and provides any parameter values that the elements require.

Pass 1:   Core Layout

The core pass is driven by the user's input. The input contains the list of elements and associated parameters. After all of the elements vote on the values of global parameters, each element is executed in turn, resulting in a heirarchy of cells which implement the core of the chip.

To allow any two elements to plug together, cells must interface either by maintaining a common pitch (width) or by wire routing. To save the space and costly routing needed if cell widths vary, a design constraint states that all cells must be of equal width. To have a uniform width for all cells, every cell must be designed as wide as the widest cell. The width of the widest cell is not known until after all of the cells are designed! And as future cells are designed, they must either be forced to have the same width as current cells, or else all of the cells must be redesigned to accomidate the wider cells.

By introducing stretchable cells, this problem can be avoided. Each of the cells are designed with places to stretch. As the core is being scanned, each element reports the width of its cells, so that when the end of the core list is reached, the widest cell width is known. As the elements produce their cells, each cell is stretched (a painless operation) to fit all other cells. The cells can also be stretched to allow the power lines to expand as power demands increase.

Another restriction arising from the arbitrary element ordering is that each of the cells must meet certain interface requirements. Proper interface standards eliminate intercell problems such as shorting out a neighbor's transitor or power supply line. By agreeing on a standard interface to begin with, any cell can be guaranteed to mesh properly with adjacent cells before the neighboring cells are specified. Boundry conditions like these allow design rule checking to be performed on individual cells as the cells are designed, rather than on fully instantiated artwork.

Along with meshing of cells, the busing requirements between cells must also be met. Buses may need to break or stop, and bus precharge circuits must be added for each bus. Details like these need not be specified by the user, but are added by the compiler.

Pass 2:   Control Design

Given the results of the core pass, the control design and layout proceeds. First, control buffers to drive the control lines are inserted along the edge of the core. The timing is also added to the control signals by the buffers. Then, an text array is constructed which specifies the decode functions needed for each buffer. A two-tape Turing machine operates on one "tape", which contains the text array, and writes the second "tape", producing compiled silicon code. When it has finished operating on the array, the Turing machine will have generated and optimized the instruction decoder, and created pad connections for the inputs to the decoder.

Pass 3:   Pad Layout

The pad layout pass of the silicon compiler begins by collecting all of the connection points which need to be connected to pads. These connection points are sorting in clockwise order, and pads are allocated in the same order. The pads and connection points are examined by a Roto-Router, which rotates the pads around the perimeter of the chip in an attempt to minimize the length of wire between pads and connection points. The Roto-Router spaces the pads evenly around the chip to avoid generating pad layouts that would be difficult to bond. The third pass concludes by adding wires between the pads and the connection points.

Conditional Assembly

Bristle Blocks can allow for conditional assembly of silicon. For example, when designing prototype chips, the internal state of a state machine may need to be routed to pads, but when production chips are produced, the area of the pad and wires may need to be reclaimed. The user may declare a global boolean variable PROTOTYPE, which, if TRUE, will add the connection points for the pads, but if FALSE will not. At any time prior to actually compiling the chip, the user may decide whether this is a prototype chip or not, and properly modify the layout.

Since Bristle Blocks provides a high level programming language to the user, cells can be smart, and perform calculations as they are being added to the core. As an example, there may be several possible cell layouts which implement the desired function. After the cell width has been determined, the possible layouts which fit within the specified width can be judged to find the cell with minimum resulting area. The minimum area cell is then used in the actual chip, which optimizes the length of the chip.

Implementation

Bristle Blocks is implemented in ICL, a high level graphics, integrated circuit design, and general purpose programming language created by Ron Ayers [2]. ICL allows the user to generate new data structures at any time. Functions and coercions can also be added to manipulate the data structures. The coercions are invoked automatically by ICL when the data it receives are not of the type it expects. Also, ICL provides for polymorphic functions, where the operation performed by the function is dependant upon the data supplied to the function. ICL is an extensible language, so that any user defined functions, datatypes, and coercions become a part of the current version of ICL. This also means that the Bristle Block system provides the user with all of the basic ICL operations, and that Bristle Blocks is an extensible system.

The only disadvantage with using ICL is that the current version is an implementation for a PDP10/PDP20, which has a small (18 bit) address space.

Conclusion

In the 2½ man-months that have gone into developing the first Bristle Blocks system, approximately 80% of the system has been implemented. Given a high level description of the chip and definitions for core elements, the system produces a complete layout, sticks diagram, transistor diagram, logic diagram, and block diagram. There are a few special cases in the circuit topology that have still to be considered, but lack of time is the only reason they have not been implemented. Hooks for the circuit simulator and text producing code have been included in the system, but with the constraint that all code must remain in core, these features must wait.

The chips produced by the system are fairly well optimized, having ±10% of the area of a chip produced by hand using the structured design methodology. The compiler takes approximately 4 minutes to generate a small chip, in all five of the current representations. The time needed to generate a fairly large chip should be in the neighborhood of 10-15 minutes.

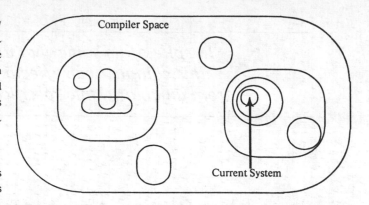

Figure 3. Hierarch of Systems

At present, the Bristle Blocks system is tailored to produce chips of a particular architecture. As the domain of silicon compilers is examined through the use of the current Bristle Blocks, a more general chip architecture will be found that encompasses the current chip architecture. A more generalized Bristle Blocks will be developed to compile the generalized chip architecture. The generalization of the current Bristle Block system may continue through a number of generations. At some point, however, a totally new Bristle Blocks will be generated to handle a completely different style of chip. Thus, there will emerge several classes of Bristle Block systems, each compiling a separate class of chip architectures, rather than one Super Bristle Block system which attempts to produce all chips.

References

[1] Mead, Carver A., and Conway, Lynn   An Introduction to VLSI Systems, limited printing, 1978

[2] Ayres, Ronald, ICL Reference Manual, SSP File #1364, Caltech, 1978

In spite of its somewhat unlikely sounding name,
this complier can effectively design custom ICs
from an algorithmic specification of circuit behavior.

MacPitts: An Approach to Silicon Compilation

Jay R. Southard, MIT Lincoln Laboratory

Silicon compilation is a new technique for integrated circuit design that has recently undergone intense development, scrutiny, and criticism. At the moment, there is even disagreement over the meaning of the term "silicon compilation"[1] itself. In this article, "silicon compilation" will refer to the automatic synthesis of an integrated circuit layout from a description of its behavior. Traditional integrated circuit design techniques, on the other hand, are dependent on the *structure,* rather than the *behavior* of the integrated circuit.

Of course, design is not the only step in the production of useful ICs. Fabrication and testing are also important. In fact, advances in fabrication capability have actually outstripped advances in design, which is why design is a topic of such interest at present.[2] Some testing issues— design for testability and automatic generation of test patterns, for example—are potentially part of the design process. Current research is beginning to show progress in these areas, and there is reason to believe that silicon compiler design methodology is more amenable to test automation than more handcrafted methods.

The disparity between fabrication and design is such that while it is economic to fabricate a moderately complex circuit—one that is to be replicated in production about 1000 times—it is not economic to design such a circuit unless it will be replicated at least 10,000 times.* Thus, most IC vendors will not accept commissions for designs without a commitment to a 10,000-unit production volume. Reducing the design cost of an IC by a factor of two would probably reduce the minimum economic production volume to about 1000 units, and a good silicon compiler can probably reduce design cost by much more than that—probably by a factor of three to five (see Figure 1).

*This is a rough calculation based on quotes from silicon foundries for prototype and low-volume production runs of custom integrated circuits compared to similar runs of printed circuit boards.

The design of integrated circuits

There are at least two separable aspects to IC design. One is the rigorous specification of the integrated circuit's function. Another is the layout of the active devices and interconnections required to instantiate the circuit. This layout description assumes some specific or generic IC fabrication technology. Because going directly from behavior to layout is usually too complicated to accomplish in one step, designers customarily construct a circuit description that lies somewhere between the behavioral and layout descriptions of a specific circuit. The amount of engineering time spent in each of these three design areas is shown in Figure 1.

During the past 15-20 years, great strides have been made in the understanding and utilization of program like behavior specifications. It was quickly discovered that the majority of the functions of even the most complex digital hardware system could be described by a modest program. Thus, the main thrust of research activity must be to improve layout and circuit design productivity.

Below, we will connect these factors to the evolution of the MacPitts[4] silicon compiler. For those desiring a still more comprehensive survey of integrated circuit design methods, one can be found in the literature.[5]

Layout

In simple designs, the layout is often obvious to a good designer. In more complex designs, however, there are a number of reasonable layouts to consider. Additionally, the functions of medium-scale integrated circuits can often be circuit designed in several intuitively reasonable ways. In order to improve productivity, two basic approaches to IC layout design have traditionally been used: computer-aided drafting and layout synthesis from circuit design.

Reprinted from *Computer*, vol. 16, pp. 74–82, Dec. 1983.

Computer-aided drafting. This layout method recognizes the "fine structure" of the layout design task. In order to achieve efficiency vis-a-vis a specific technology, the real design process must be iterative, as shown in Figure 2, rather than follow the straight flow of Figure 1. A computer-aided drafting tool combined with hierarchical design style can thus be valuable for easing the task of reworking a layout. In addition, the layout specification in the computer can be utilized to provide automatic layout and electrical checks and device simulations.

Because the drafting system and the photomask fabrication process deal with the same objects, the designer has, potentially, complete freedom to utilize all the capabilities of any imaginable process. In practice, however, two productivity implications must be taken into account when laying out a circuit. One factor is implied by a hierarchical design style. If some cells already exist cells that can be incorporated into a new design—this can result in greatly reduced design time and greatly increased confidence that those cells will work. The other productivity factor is a consequence of the fact that the time required to design a large-scale IC often spans one or two modifications of the target IC fabrication process. Successful integrated circuit production will span many such changes. A highly optimized layout for one process will be suboptimal for its successor. It is valuable to be able to automatically modify otherwise satisfactory masks so that they match an improved fabrication process. Although the modified layouts are generally still suboptimal, automatic modification is usually an acceptable engineering compromise.

Layout synthesis. New layout tools and techniques generally emphasize the above-mentioned practices, even at the cost of some further reduction in the absolute efficiency of the available fabrication technology.

The standard cell layout method institutionalizes the hierarchical borrowing practice. This method provides a cell library composed of logic gates and complex cells. It is common practice to include automatic place-and-route routines in the cell library, meaning that the designer only needs to specify the connectivity between cells, not lay out

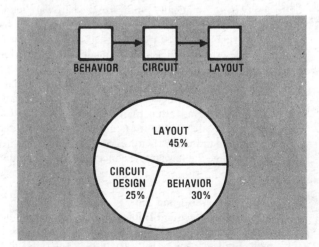

Figure 1. Three levels of IC specification and the time spent in each. The time chart is based on actual measurements.[3]

![Figure 2 diagram: BEHAVIOR, CIRCUIT, LAYOUT]

Figure 2. The iterative design process.

the interconnection.[6] Again, this is a trade-off of efficiency for productivity.

Another approach to layout synthesis is sometimes called "symbolic layout." Here, the layout is not specified down to its absolute geometrical location. Rather, relationships are sketched out from which the layout can be automatically generated.[7] The sketchier the relationship specification, the harder the automatic program must work, and the less efficient the resulting layout will be. However, the relationship sketch is a general specification that need not be altered for reasonable changes in fabrication capabilities. Only the layout program will be modified, and that will be used to resynthesize many designs. Although advantageous, these symbolic layout systems still require considerable skill and effort for them to be used properly.

Circuit design: silicon compilation precursors

The difficulty with the layout methods described above is that none of them provide an underlying systemization for transforming behavior into the required structural, circuit terms. In other words, they do not address the problems of circuit design. In this context, two *subsystem* generation techniques deserve special mention. One involves the use of PLAs (programmable logic arrays) in implementing finite-state machines.[8] The other is an offshoot of the standard cell methodology—the "data-path generation" technique pioneered by Johannsen[9] and also recently used by Shrobe.[10]

Subsystem generators must be used in conjunction with other subsystem generators to create a complete integrated circuit. This means that several decoupled descriptions of the same system must be created—one for each subsystem generator. Also, the generated layouts must be combined, either manually or by some "chip assembler" techniques that are not yet generally available.

Finite-state machines and PLAs. Finite-state machines, or FSMs, are a powerful conceptual technique for specifying some forms of control. PLAs are an efficient and easily laid out implementation of logic. A PLA implements a universal form of logic; all logic functions not requiring storage can be implemented in a large enough PLA. Additionally, efficient algorithms for transforming any logic expression into PLA forms are well-known. Generating a PLA layout from the PLA forms is also a well-understood process, though new wrinkles are always being introduced. A simple addition to a PLA generator will create an FSM implementation.

Finite-state machines are not enough for complete systems, especially when computation must take place. In such cases, a computation subsystem can often be controlled by an FSM. One choice for the computation subsystem is a "data path."

Data-path generation. A data-path generator is a specialized standard cell system that systematizes and takes advantage of typical constructions. In a data-path generator, bit-wise subunits are combined to create functional units, and multiple units are combined to create a data path. A data path is a sequence of operators—addition units, multipliers, comparators, etc.—connected to perform a computation. Thus, a systematic transformation is possible from the behavior/computation specification of an IC directly to its structural, standard cell specification. The connection of operators is done by automatic routing of interunit signals.

The systems that use the generated data paths, however, typically require conditional data flows, looping, and other control constructs. The data paths, therefore, must be capable of being switched during operation by signals generated outside the data path. To do this, the data-path generators leave hooks for control signals. A different tool, and, more importantly, an unrelated description, must be used to create these control signals.*

First

Another precursor to general silicon compilation is First.[12] First (fast implementation of real-time signal transforms) was designed for digital signal processing (DSP) applications. First and MacPitts differ in nearly every way—from philosophy of behavioral specification to circuit implementation. Both are alike, however, in that they produce complete chip layouts, including I/O pads and routing.

There are several interesting facets to First, but the one most important is that many DSP applications can be described strictly in terms of data-flow graphs—for ex-

*Recently, Agre[11] reported on work in which a unified high-level description—one similar to MacPitts—was used to generate the descriptions for both an existing data-path generator and a control signal synthesizer.

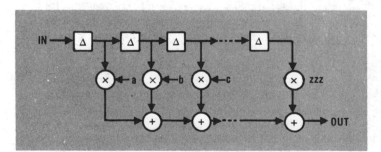

Figure 3. Data-flow graph of a nonrecursive digital filter. Note that there are no options in the flow of data; there is no control flow.

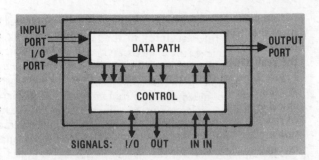

Figure 4. The MacPitts target architecture.

ample, the nonrecursive filter of Figure 3. Note that there are no conditionals of any kind because *there is no need for them.* This is in contrast to the data-path generators typically used for algorithms and systems that do require conditionals, time-varying interconnections, and so on. First is also superior to most data-path generators in that it can interconnect any network of its functional units. First uses bit-serial operations to implement its functions, and this often yields a superior fabrication cost/function ratio. However, it also results in greater design complexity because of the necessity of taking the bit delay timings into account when connecting functions.

The necessity of considering bit delay timings led the First team into a powerful, formal method of bit-level design. Also, the very restrictive operator combination paradigm allows the First designer to guarantee speed performance. In turn, this guarantee has led to methods for time multiplexing units to match the actual performance with specified performance criteria.

MacPitts

MacPitts was designed to be a synthesizable, algorithm description language. By synthesizable, we mean that the language was designed in conjunction with IC synthesis routines. Therefore, during the evolution of MacPitts, proposed language features and concepts were considered both on the basis of possible application and on our ability to automatically synthesize the hardware required for its implementation. We have called this set of hardware structures the "target architecture" (see Figure 4).

By describing MacPitts as an algorithm description language, we mean that it has essential programlike features, notably flow of control. We also mean that the designer has the ability to describe the algorithm that the IC is to perform, rather than one that describes the integrated circuit's structure (standard cell). In order to understand the salient features of this architecture, let us first examine the few simple concepts on which MacPitts is based.

Data structures and storage. MacPitts has only two data types: integer and Boolean. All integer-type objects are of the same length and are implemented in the data path. Boolean-type objects are implemented in the control and flags sections. Our terminology is based on the following:

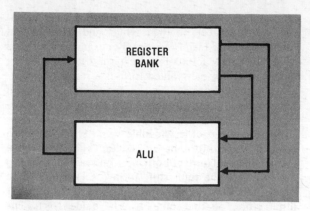

Figure 5. Conventional data-path organization.

Integer	Boolean	
register	flag	stored
port	signal	not stored

Storage is kept in master-slave flip-flops. These flip-flops are called registers and physically exist in the data path if they store integer-type data; they are called flags if they store Boolean-type data. All storage is modified on a single, synchronous "state transition." During a state (between two successive state transitions), the storage devices present a constant output. At the same time, their inputs are allowed to settle. At the state transition, the storage input values are propagated to their outputs. The flip-flops are designed so that the new outputs cannot modify the inputs during the state transition. This is precisely the effect that a programmer expects from the Lisp instruction

(setq a (+ a 1))

or in Fortran

A = A + 1

On the other hand, nonstorage items are designed so that they are modified asynchronously during the state. Thus, the MacPitts instruction

(setq a (+ a 1))

while syntactically valid whether *a* is a register or a port, will only be useful if it is a register.

Data-path operations and transfers. So far, we have only presented the framework in which computation can be accomplished. Because many of the target applications require high throughput and can utilize parallelism, we wanted a design language and a target architecture that would support such parallelism. Our target architecture achieves this parallelism by implementing a high degree of concurrency in data operations and transfers within the data path. For comparison, this concurrency can be considered alongside "standard" and "horizontally microprogrammable" computer architectures.

The conventional computer data path is typically partitioned into a register/memory array and an ALU, as shown in Figure 5. Only one operation can proceed each clock cycle. Efficient algorithm specification in the form

of a program, however, is simplified because of this limitation; the ALU merely needs to be kept busy for full efficiency.

A horizontally microprogrammed machine typically has several buses connecting a small number of functional units that can operate in parallel. The microprogram specifies the connection of units to buses on an instruction-by-instruction basis. Although some parallelism is available, there is usually not enough to execute all the parallelism implicit in the algorithm. Both the number of functional units and their allowed interconnection are usually limited. Such restrictions create both programming complications and execution bottlenecks.

For example, in one typical, high-performance microprogrammed array processor, it is not possible to directly sum a series of values from the main data memory because both the main data memory and the adder output are limited to entering the adder at one and the same input port. Efficient programming demands that many of the units (and hence transfer buses) be kept busy simultaneously. Fitting an algorithm into the Procrustean bed of any specific microprogrammable machine is typically an error-prone and difficult programming chore. This chore is complicated by the usual situation of obtaining partial results in a unit that cannot be directly connected to the unit required for the remaining computation.

MacPitts, however, allows a designer to specify an algorithm as though completely general and sufficient parallelism existed in some general-purpose machine. In other words, any control/data-flow graph can be directly specified in MacPitts. Then the MacPitts compiler "extracts" the minimum-hardware microprogrammed machine which executes that parallel algorithm, with all the bus and unit merging and sharing that that implies. The resulting structure is topologically similar to any microprogrammable machine's architecture but is organized as shown in Figure 6. The data-steering control functions are provided in the control section, which is generated from the same algorithm specification by the extraction process. Thus, MacPitts combines ease of programming with the efficiency and parallelism of the horizontally microprogrammed architecture.

From our knowledge of the data-path generators previously discussed, we were aware of the usefulness of constructing the data path out of bit-wise units "glued" together in one direction to form word-length units, which are then bused together in the other direction. A more detailed picture of the layout scheme for the data path is given in Figure 7. A bit-wise unit is called an *organelle*. A standard library of the usual functions—adder, subtractor, shifters, comparators, etc.—is provided. More sophisticated users can design their own organelles. However, as units become more special-purpose, the MacPitts program becomes more a *structural* than a *behavioral* specification.

MacPitts specification language—forms. There are two fundamental concepts to the MacPitts language. The first of these is "state transition," which we have already discussed, and the second is that of a "form." Form includes the syntax and semantics of logic, arithmetic, and control expressions. Forms are composed of an operator

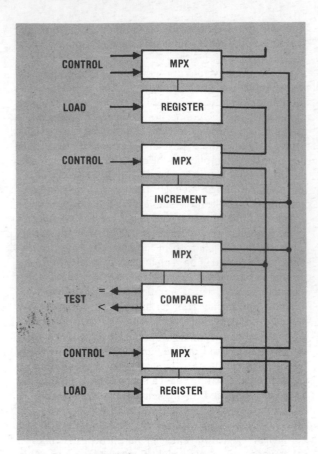

Figure 6. Organization of the MacPitts-generated data path. Note the lines going to and from the control section (not shown). The control section is generated from the same program-like specification.

and argument values, and they produce a value and, possibly, a side effect. In their syntax and intent, forms usually resemble Lisp functions. For example, (+ b 1) is a legal Lisp or MacPitts form if *b* is integer valued. The quintessential integer-valued form is a register or port. However, the above code is also an integer-valued form because it results in an integer value. Therefore, the statement (+ (+ b 1) 1) is also legal.

There are two other possible value types—Boolean and void. A void form has no resultant value. Any form can contain a mixture of both input and resultant values. For example, the right shift form (>>) can use one Boolean and one integer input and produce an integer result. Note, too, that some forms have side effects. The most obvious examples of such forms are assignment statements and control flow branches. These forms accommodate the storage and state transition concepts previously presented. Thus (setq a (+ a 1)) has the desired effect of incrementing *a* after the next state transition.

A compile-time side effect not usually occurring in a standard computer language is that some forms cause other forms embedded in their scope—that is, the forms that make up the arguments of the outer form —to be executed in parallel instead of sequentially. Forms that cannot be executed in parallel can share physical units in the data path. Those that cannot syntactically (at compile time) be determined as mutually exclusive, must be assumed to execute in parallel and thus cannot share physical units. *Parallelizing forms increase the computational throughput of a design at the expense of silicon area.* A later example will demonstrate the use of parallelizing forms in examining the trade-off between throughput and silicon area.

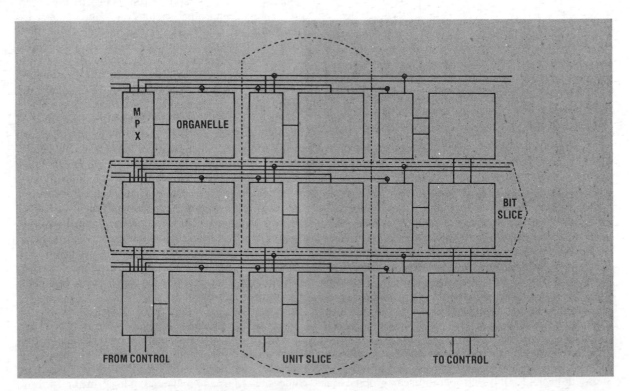

Figure 7. Detailed organization of MacPitts data paths.

The *cond* operation. In MacPitts, the *cond* operation is the method of condition testing for execution and control flow. This operation is probably the most important MacPitts form; it is certainly the most complicated. It is *cond* that distinguishes MacPitts from a data-path generator and makes it an algorithm specification language rather than a data-path specification language. The *cond* operation is also a parallelizing form.

There are several ways of thinking about the *cond* statement. First, it is syntactically identical to the Lisp *cond* statement and may easily be viewed as a program-like "case" statement. This paradigm, however, ignores not only the implicit parallelism of the form, but also a possibly preferable paradigm based on finite state machines.

Consider the following code fragment (assuming that *a* and *b* are registers and *exch* and *incr* are Boolean values):

```
execute
(cond   (exch  (setq a b) (setq b a) (go fetch))
        (incr  (setq a (+ a 1))
               (setq b (+ b 1))
               (go fetch))
        (t     (setq a (− a 1))
               (setq b (− b 1))
               (go fetch)))
```

MacPitts interprets this specification of the state labeled *execute* as follows: If *exch* is true, exchange the contents of registers *a* and *b* and go to the state labeled *fetch*. (Note that *cond* has parallelized the two *setq*s, otherwise this code would just have set *a* to *b*.) If *exch* is not true, but *incr* is, then increment both *a* and *b* simultaneously and go to *fetch*. Finally, if neither conditional is true, *a* and *b* will both be decremented (simultaneously), and, again, *fetch* will be the next state. Notice that MacPitts has assumed that the programmer wanted all the conditions to be mutually exclusive. This is consistent with the Lisp interpretation of the conditions (predicates), but not the consequent actions. The flowchart equivalent of this code is displayed in Figure 8.

In the *execute* sample, the predicates are simple Boolean values. These would be distributed via the control section, commanding the organelle multiplexers to connect the operators and registers to the correct buses. Instead of Boolean values, any form with a Boolean result could have been used as a predicate:

```
(cond ((= a b) (setq a 0))
      (t       (setq a (+ a 1))))
```

In this case, a comparator in the data-path would compare *a* and *b*. The comparison would result in a Boolean signal that would be distributed via the control section to command other multiplexers in the data path.

The crux of the difference between Lisp and MacPitts interpretations is that Lisp's consequent actions are sequential but MacPitts' are parallel. The entire MacPitts *cond* statement can be, and is, compiled into silicon capable of executing it in a single state cycle. The MacPitts code, and its timing, make eminently good sense in the context of a Mealy-type FSM.* First, let us make the distinction between data storage and storage of FSM

state. The *cond* form can then be interpreted as the way MacPitts specifies the next-state and output mappings for the current FSM state. In fact, we can adopt the convention that each FSM state is represented by a set of < condition / actions / transition > triples. Note that, unlike the usual FSM conventions, "actions" may affect data storage as well as specify output. The sample code is thus the equivalent of the triples

$$exch / a \leftarrow b, b \leftarrow a / fetch$$
$$\overline{exch} \cap incr / a \leftarrow a+1, b \leftarrow b+1 / fetch$$
$$\overline{exch} \cup incr / a \leftarrow a-1, b \leftarrow b-1 / fetch$$

for the state *execute*. This is presented graphically in Figure 9.

*The Mealy-type FSM output functions depend on both the current state and the input values.

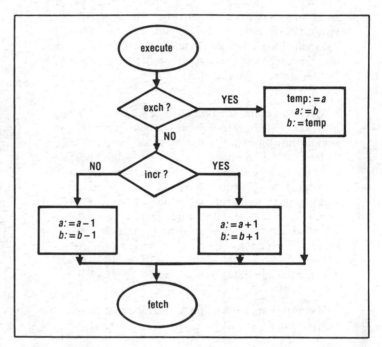

Figure 8. Flowchart of the exchange, increment, or decrement code. Note the awkwardness in describing the exchange due to the implications of the sequence of execution.

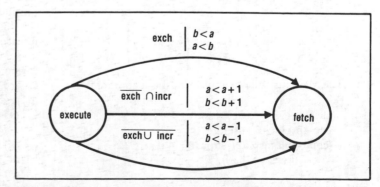

Figure 9. FSM implementation for exchange, increment, or decrement code. Note that FSM concept implies that the actions performed on the selected state transfer arm are simultaneous.

By inspecting the FSM representation, it is clear that actions on different "arms" of the graph can never be executed simultaneously. Therefore, the physical units executing the operations on any arm can be shared with those used for other arms. The only requirement is that the appropriate input signals be switched (the task of the multiplexers and their control) into the proper units for the selected arm. The MacPitts compiler keeps track of what units are available and generates the multiplexing and control required for this "merging."

The interpreter. A MacPitts program is not only an IC specification, it is also an algorithm specification. The MacPitts compiler generates IC layout; however, there is also an interpreter that executes the specification program on a general-purpose computer. This interpreter is invaluable because, unlike software or even wire-wrap hardware boards, it is very difficult to interactively probe and modify the internals of an integrated circuit. By using the same language to drive both the interpreter and integrated circuit compiler, human error is necessarily reduced. The algorithmic MacPitts specification is also more suitable than a structural specification for interactive function design and check.

MacPitts examples. Because of the power of MacPitts, it is very difficult to produce simple examples that demonstrate all of the features of this compiler. Many desired algorithms can be described using only one state! For example, the digital filter presented in the section on First can, in MacPitts, be described in only one state that loops back to itself forever. In fact, all such filterlike systems can be so described, and MacPitts has an *always* construct semantically equivalent to such an FSM, or *process*, as we usually call it. Another powerful feature not often used in simple examples is the ability to specify multiple parallel FSMs or *process*es.

Magnitude approximation.* The following algorithm approximates $\sqrt{a^2 + b^2}$:

$$\text{let}: g = \max \left| \begin{matrix} a \\ b \end{matrix} \right|$$

$$l = \min \left| \begin{matrix} a \\ b \end{matrix} \right|$$

$$\sqrt{a^2 + b^2} \approx \max \left\{ \begin{matrix} g \\ \frac{7}{8}g + \frac{1}{2}l \end{matrix} \right\}$$

The MacPitts code to implement this algorithm is

```
(always
; set a-absolute b-absolute
  (cond ((bit msb a) (setq aab (− 0 a)))
        (t (setq aab a)))
```

*I am indebted to Peter Denyer for recently suggesting this algorithm, since it points up the power of the *cond* construction. The algorithm itself is much older and has a long history at MIT's Lincoln Laboratory.

```
  (cond ((bit msb b) (setq bab (− 0 b)))
        (t (setq bab b)))
; set g and l
  (cond ((unsigned −> aab bab)
         (setq g aab)
         (setq l bab))
        (t
         (setq l aab)
         (setq g bab)))
; sqs: = 7/8 g + 1/2 l
  (setq sqs ( + (−g (3 >> f g))
               (>> l)))
; max of sqs and g
  (cond ((unsigned −>g sqs) (setq res g))
        (t (setq res sqs)))))
```

The first code section computes the absolute values of *a* and *b*. The second uses these to compute *g* and *l*. Next, $7/8\,g + 1/2\,l$ is calculated and compared to *g*. Both *a* and *b* must be specified as input ports and *res* as the output port.

Note, too, that the word size must be defined and the pinout specified. These can be easily changed to accommodate modified requirements. The definitions are located in an initial MacPitts definitions section, shown below for a four-bit word size:

```
(program genmag 4
  (def msb constant 3)
  (def 1 ground)     ; pin number one is ground
  (def a port input (2 3 4 5))  ; pins 2-5 are reserved
                                ; for input a
  (def b port input (6 7 8 9))
  (def res port output (10 11 12 13))
  (def 14 power)
  (def 15 phia)      ; other than for pin definition of
                                these pins,
  (def 16 phib)      ; clocking is implicit in the state
                                concept,
  (def 17 phic)      ; and can otherwise be ignored.
  (always
;     set a-absolute b-absolute
  . . . . . . .
```

The above code does not fully specify the design; the degree of parallelism and pipelining must still be decided. It is a simple matter to specify and modify these characteristics, and we will now investigate some of the possibilities.

If the integers *aab, bab, g, l,* and *sqs* are declared as internal *ports*, then the entire computation will proceed asynchronously and combinationally. We make the declaration in the *def* section of the MacPitts code:

```
(program magasync 4
  (def msb constant 3)
  (def aab port internal)
  (def bab port internal)
  (def g port internal)
  (def l port internal)
  (def sqs port internal)
  (def 1 ground)
  . . . . . .
```

There is really no "state" or cycle here. Rather, the program accepts inputs and produces the outputs. The combinational delay, however, is fairly large, as the inputs must propagate through several cascaded functional operator units.

For slightly greater area, pipelining can be accommodated. For example, *aab, bab, g,* and *l* can be declared as registers instead of ports. If so, the start of the program now looks like the following:

```
(program magpipe 4
    (def msb constant 3)
    (def aab register)
    (def bab register)
    (def g register)
    (def l register)
    (def sqs port internal)
    (def 1 ground)
      . . . . . .
```

With no other change in the code, this becomes a fully pipelined system. The worst-case combinational delay per cycle is shortened. It takes three cycles to obtain a result (latency from inputs to output). However, new inputs can be accommodated each cycle, thus increasing overall throughput.

Finally, by removing the parallelizing *always,* a true multistate sequential loop can be easily formed. Though slower, this design is able to share (time multiplex) some of the actual physical operation units—the number of subtraction units can be reduced from 3 to 1, for example. The determination of which physical operation units can be multiplexed is done automatically, as is the actual multiplexer specification and layout. However, register sharing is not done automatically. In this case, the *aab* and *g* values can share one register, while *bab, l,* and *sqs* share another. An extra input signal, *reset,* must be provided. The MacPitts compiler automatically generates hardware to set the process state to the beginning of the loop if the *reset* signal is high. This ensures that the resulting FSM can be started in a known state. Putting together the definition section and the sequential loop, the code to do all of this now looks like this:

```
(program magseq4
    (def msb constant 3)
    (def aab-g register)
    (def bab-l-sqs register)
    (def 1 ground)
    (def a port input (2 3 4 5))
    (def b port input (6 7 8 9))
    (def res port output (10 11 12 13))
    (def reset signal input 14)
    (def 18 power)
    (def 15 phia)
    (def 16 phib)
    (def 17 phic)
    (process compmag 0
; set a-absolute b-absolute
loop
    (cond ((bit msb a) (setq aab-g (− 0 a)))
          (t (setq aab-g a)))
```

```
    (cond ((bit msb b) (setq bab-l-sqs (− 0 b)))
          (t (setq bab-l-sqs b)))
; set g and l
    (cond ((not ( unsigned −>aab-g bab-l-sqs))
          (setq aab-g bab-l-sqs)
          (setq bab-l-sqs aab-g)))
; sqs: = 7/8 g + 1/2 l
    (setq bab-l-sqs (+ (− aab-g (3 >> f aab-g))
              (>> bab-l-sqs)))
; max of sqs and g
    (cond ((unsigned −> aab-g bab-l-sqs)
          (setq res aab-g))
          (t (setq res bab-l-sqs)))
    (go loop)))
```

These modifications are simple ways of achieving performance goals either for area or speed. The area requirements and number of transistors for each of the three designs (four-micron minimum feature size) is shown in Table 1. Several recently proposed silicon compilers, as well as Agre's work, attempt to automate this process by using resource and timing constraints to guide an extraction of parallelism from a single, fundamentally sequential program representation.

It is also possible to construct custom designed units. For example, a max-min organelle could be created to generate *g* and *l*. This was the solution forced on First. With *cond,* we can compose functions from the primitive library set in minutes (paying area and speed penalties, of course). Without such a construct, however, circuit design and layout must be performed for many new applications.

MacPitts: past, present, and future

Several MacPitts chips have been fabricated and tested. The latest and largest of these is an automatic gain control chip, shown in Figure 10, that is actually in use in a digital vocoder system built at MIT's Lincoln Laboratory.[13] This chip was designed in a few weeks by an engineer with no previous experience in IC design. The large areas of unused space and other layout inefficiencies are continually being reduced by compiler improvements. Note that *the chip does not need to be redesigned to take advantage of any such improvements.*

A more ambitious project, a test controller, is currently being fabricated. Since the original design, which used standard functional units, was initially too large to be fabricated, some improvements to the basic organelles' layouts had to be implemented. Doing this did not require a new MacPitts specification for the test controller, and other designs could benefit from these improvements as well. However, these improvements were still not sufficient. At this point, the designer was confronted

**Table 1.
Area requirements and number of transistors
for each of three designs.**

DESIGN	SIZE (mm×mm)	NO. OF TRANSISTORS
asynchronous	5.1×2.1	867
pipelined	5.4×2.1	1088
sequential	3.4×2.4	806

with three courses of action. First, a speed/area trade-off could be investigated, but this approach was deemed unsuitable for a test controller. A second method was to partition the system into several integrated circuits. MacPitts does not have any facilities for aiding the partitioning of a too-large system into a chip set. By way of comparison, First can easily partition systems because of its bit-serial approach; splitting a word-parallel design usually generates overly large pin counts. The third approach, the one finally used, was to create some special-purpose organelles.

These designs, and others, have pointed out several weaknesses of MacPitts. Performance prediction and automatic trade-off analysis is not a strong point; MacPitts lacks system partitioning facilities and only generates NMOS designs. Finally, the test generation mechanisms of MacPitts still require human interaction. We have no shortage of ideas, but a great deal of work will be necessary to overcome these problems. Nevertheless, MacPitts has demonstrated its fundamental goal: Custom integrated circuits can be effectively designed from an algorithmic specification of the circuits' behavior. ∎

Figure 10. Automatic gain control chip.

Acknowledgments

I thank my colleagues Jeffrey M. Siskind and Kenneth W. Crouch, who, together with the author, invented, developed, and implemented MacPitts. I would also like to thank Allan Anderson and Peter Blankenship for their help in editing this paper.

This work was sponsored by the Defense Advanced Research Projects Agency. The United States government assumes no responsibility for the information presented.

References

1. J. Werner, "The Silicon Compiler: Panacea, Wishful Thinking, or Old Hat?," *VLSI Design*, Vol. 3, No. 5, Sept./Oct. 1982, pp. 46-52.

2. Gordon Moore, "VLSI: Some Fundamental Challenges," *IEEE Spectrum*, Vol. 16, No. 4, Apr. 1979, pp. 30-37.

3. Rex Rice, *VLSI: The Coming Revolution in Applications and Design,* IEEE CS Press, Los Alamitos, Calif., 1980.

4. Jeffrey Siskind, Jay Southard, and Kenneth Crouch, "Generating Custom High-Performance VLSI Designs from Succinct Algorithmic Descriptions," *Proc. Conf. Advanced Research in VLSI,* P. Penfield, ed., MIT, Cambridge, Mass., Jan. 1982, pp. 28-40.

5. *Proc. IEEE* (special issue on VLSI design), Vol. 71, No. 1, Jan. 1983.

6. Borgini et al., *Automated Design Procedures for VLSI,* final report DELET-TR-78-2960-F, Advanced Technical Laboratory, RCA Government Systems Division, 1981.

7. J. D. Williams, "Sticks—A Graphical Compiler for High-Level LSI Design, *AFIPS Conf. Proc.,* Vol. 47, 1978 NCC, pp. 289-295.

8. C. Mead and L. Conway, *Introduction to VLSI Systems,* Addison-Wesley, Reading, Mass., 1980.

9. D. Johannsen, "Bristle Blocks: A Silicon Compiler," *Proc. 16th Design Automation Conf.,* June 1979, pp. 310-313.

10. H. E. Shrobe, "The Data Path Generator," *Proc. Conf. Advanced Research in VLSI,* P. Penfield, ed., MIT, Cambridge, Mass., Jan. 1982, pp. 175-181

11. P. E. Agre, "Designing a High-Level Silicon Compiler," *VLSI Research Review,* MIT, Cambridge, Mass., Dec. 1982.

12. Neil Bergman, " A Case Study of the F.I.R.S.T. Silicon Compiler," *Proc. Third CalTech Conf. VLSI,* R. Bryant, ed., Pasadena, Calif., Mar. 1983, pp. 413-430.

13. J. A. Feldman and E. J. Beauchemin, " A Custom IC for Automatic Gain Control in LPC Vocoders," *Proc. ICASSP 83*, Boston, Mass., Apr. 1983.

PERFORMANCE PREDICTION WITH THE MACPITTS SILICON COMPILER

Jeffrey R. Fox
GTE Laboratories Incorporated
40 Sylvan Road
Waltham, Massachusetts 02254

ABSTRACT

The MacPitts Silicon Compiler is a VLSI circuit synthesis program that accepts a high-level algorithmic description of a VLSI circuit in a language that syntactically resembles LISP. In evaluating the suitability of MacPitts for designing telecommunications circuitry, this author disclosed the need for a system that can make predictions of circuit performance. A circuit performance prediction system has since been added to the MacPitts Silicon Compiler. The performance package operates interpretively and provides the user with information on the location of the critical timing paths within a circuit. This paper describes the MacPitts performance prediction package.

INTRODUCTION

Advances in integrated circuit technology are rapidly increasing the density and complexity of circuits that can be fabricated on a single silicon substrate. One micron feature size is becoming practical, allowing the integration of close to one million devices on a single chip. Complexity of this degree makes it possible to provide the function of a large system on a few devices. Using conventional computer assisted techniques with a designer productivity of one to ten devices per person-hour, designing circuits of this magnitude will require 50 person-years to complete. This may not be a problem for microprocessors that will be used in large quantities, or for memories where individual elements are repeated many times. It is a problem, though, for systems used in quantities of hundreds or thousands, or for parts where time to market is crucial.

The silicon compiler is a means for raising the level of abstraction at which a designer may think, allowing the designer to specify an integrated circuit by describing its function in an algorithmic, rather than a logical description. This technique reduces the amount of detail that the designer must consider; the silicon compiler does the actual logic design, circuit design, and layout from the high-level algorithmic description. This results in an order of magnitude savings in design time, and increased design longevity, since the circuit description is written in a high-level language whose meaning can transcend designer turnover and technology updates. A detailed review of silicon compiler technology can be found in Refs. 1 and 2.

The MacPitts silicon compiler is a VLSI circuit synthesis program that accepts a high-level algorithmic description of a VLSI circuit in a language that syntactically resembles LISP.[3,4] In evaluating the suitability of MacPitts for designing telecommunications circuitry, this author noted the need for predicting the performance of the circuit designed by the silicon compiler and for identifying to the user the source statements that resulted in the synthesis of a critical path.[5] MacPitts, as obtained from MIT Lincoln Laboratories, had no features related to performance prediction. This paper describes an interpretive performance prediction system that has been added to the MacPitts Silicon Compiler at GTE Laboratories.

THE MACPITTS SILICON COMPILER

Figure 1 shows a section of the MacPitts source code extracted from the six-page description of a Dual Tone Multi-Frequency Detector chip (DTMF3), used in a telephone exchange to decode pushbutton telephone tones. A MacPitts circuit description comprises a set of *def* declaration statements that define the chip's datapath width, global storage registers, input and output pins, and one-bit storage units called flags; and one or more *processes* that specify independent finite state machines that may all function in parallel.

The specification of the chip's clock-period as 500 ns in the declaration section of Figure 1 is a capability provided by the performance prediction enhancement to MacPitts made by GTE Laboratories.

Many of the functions available to the user of MacPitts are implemented in the program as macros. All Boolean operations, such as AND, OR, XOR, cond (conditional execution) statements, and certain integer functions (e.g., <0, >0), are implemented as macros. The use of these macros is transparent to the user, but may have an impact on performance.

The MacPitts synthesis process begins by parsing the user's circuit description and processing all the declaration type *def* statements. Next, all macros are identified and expanded. Each *process* specified by the user is then examined to determine the number of state bits required to implement the control-flow sequencer for its corresponding finite state machine. This is a simple process since each top-level MacPitts form is executed on precisely one state. Depending on the chip's control flow, there are a few varieties of state counters that are implemented by the compiler. What happens next is called "logic extraction" and is the heart of the MacPitts logic synthesis algorithm.

During the logic extraction phase, the MacPitts compiler examines each source statement of the chip description to determine what hardware will be necessary to implement the desired function. Word-wide (integer) operations are implemented by functional cells called "organelles" within the chip's datapath. Boolean operations and control logic are implemented with an interconnection of NOR gates within the chip's control section. The layout of the control section is implemented as a Weinberger array. For a given source statement, the compiler chooses the datapath and control array resources needed to execute the current statement and then synthesizes the logic

Reprinted from *IEEE Proc. Custom Integrated Circuits Conf.*, 1984, pp. 351–355.

necessary to transfer control to the next state that the chip will execute. The compiler then proceeds to the next source statement. Once again the compiler will instantiate the required hardware for this state, but this time it will perform certain optimizations based upon the hardware that it has already chosen for previous states. Hardware resources will be shared through the use of multiplexers.

PERFORMANCE PREDICTION

The GTE Laboratories MacPitts circuit performance analyzer operates interpretively and gives the user an estimate of the delay through the hardware generated by the compiler. As each source statement is processed by the MacPitts compiler, the appropriate hardware is instantiated. After MacPitts generates the hardware for each statement, the performance analyzer prints the source statement and the process that it is in, and examines the hardware that was just generated. Each organelle has associated with it a LISP function that describes the propagation delay for that element. The delay function takes into account the circuit design of the organelle, including transistor sizes, capacitance, and number of logic levels. Additionally, the delay function takes into account the environment of the organelle, such as the width of the datapath, and whether the architecture calls for a ripple-carry, wired-or bus, or other configuration-dependent delay.

The performance analyzer identifies the longest delay path for a given set of operations, and adds to it the delay due to the corresponding control circuitry. Consider the following statement:

(setq reg–a (+ (+ reg–b reg–c) (+ reg–d reg–e)))

In this statement register "reg–a" is loaded with the sum of the contents of "reg–b," "reg–c," "reg–d" and "reg–e." Since "+" accepts only two arguments, it is necessary to cascade operators. In this statement,the longest delay path is two levels of "+" deep. The "+" operator for reg–b and reg–c and the "+" operator for reg–d and reg–e are at the same level and can be performed in parallel. If the delay calculated by the performance analyzer for this statement exceeds the user specified clock-width, a warning message is displayed on the user's terminal and the compilation pauses for user acknowledgment. The performance prediction package continues to analyze all of the source statements in this manner, pausing wherever a critical path is identified.

The circuit analyzed in this manner does not take into account sharing of resources that may affect the delay of previously analyzed statements. Organelle sharing, which is designed to save hardware, may result in increased delay in all statements using a shared resource, due to the complexity of the multiplexing hardware. In addition, control-flow branches due to conditional "go" statements may increase the complexity and delay through control logic that has already been extracted from a previously analyzed statement. After each source statement has been analyzed, the performance prediction package makes a second pass over all the hardware generated by the compiler, looking for sharing of resources that has increased the delay of potentially critical paths. This sharing of hardware resources never increases the delay through datapath organelles, but it may increase the delay through the control circuitry that Mac-Pitts generate to operate the multiplexers.

The philosophy of the performance prediction package is to provide the user with a rough approximation of chip performance in an interactive, interpretive manner. It is meant to identify source statements that may cause the compiler to generate critical delay paths. With this information, the designer can make changes to the algorithmic description to eliminate these critical paths. Analysis of the six-page source for the DTMF3 circuit required about ten elapsed minutes on a moderately loaded VAX 11/780 running EUNICE™*/VMS.

The approach taken in this performance prediction program is simpler than that needed for timing verifiers such as TV[7] or Crystal,[8] because it analyzes only logic delays and is used as an aid at the chip algorithmic specification level, rather than after the chip has already been completed. A silicon compiler can use simplified timing verifiers even at the layout level since the intended current flows and interconnect are synthesized and are therefore known. The performance analysis package does not take into account wire lengths that are a function of the physical layout of the chip. Layout synthesis by the MacPitts silicon compiler is a time-consuming procedure. The layout synthesis for the DTMF3 circuit took over six hours to complete. It is clear that it would not be desirable for the designer to wait this long to identify critical paths and get a rough performance estimate. It is important, however, to insure that the chip will operate at the desired clock rate after fabrication. After the designer has shortened the critical paths and synthesized the chip's layout, it would be desirable to extract the circuit — including wire lengths — to verify proper operation with a timing simulator such as RSIM.[6]

EXAMPLE OF MACPITTS PERFORMANCE PREDICTION

Figure 2 shows a fragment of the output of the performance analyzer after compiling the DTMF3 circuit whose source code is partially shown in Figure 1.

As shown in Figure 2, all of the MacPitts executable user source code (after macro expansion) is echoed back during the performance analysis. The first line in Figure 2 is the first executable statement shown in Figure 1 (at state label tp1). Only one critical path is found — the statement marked with the warning message. After each source statement is listed and analyzed, a list of all the organelles that have been instantiated in the chip is displayed. Following the organelles is a list of all of the resources that have been shared, with a warning message given if the delay is greater than the clock period.

In the DTMF3 example in Figures 1 and 2, the only critical path was found in process "tone-proc" state "tp4." The delay time of 570 ns is due to the cascading of three integer operations — < > (not equal), increment (1 +), and decrement (1 –) — all of which have ripple carries. The post-process analysis of organelle sharing reveals that the delay for this statement is actually worse — 590 ns due to sharing of the register named "pc." A solution to this problem is to spread the integer operations over two states. There are at least two ways to do this. One would be to store the incremented value of ltch1 in another register on a prior state, and eliminate the 1 + operation in state tp4. Another would be to set a flag with the result of the conditional test, and execute the 1 – operation on a subsequent state. Without the performance evaluator, the designer would not have known to make these changes until later.

---

*The Wollongong Group.

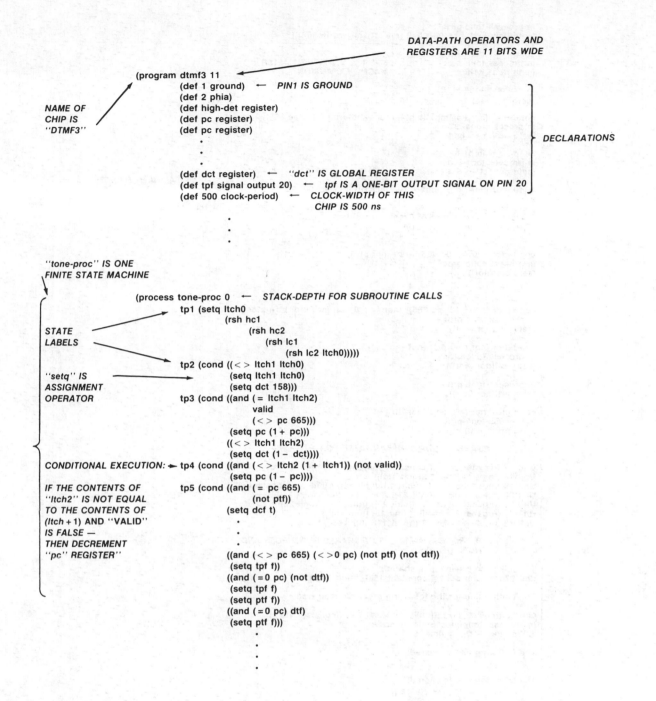

DATA-PATH OPERATORS AND
REGISTERS ARE 11 BITS WIDE

```
                          (program dtmf3 11
                              (def 1 ground)   ←   PIN1 IS GROUND
                              (def 2 phia)
                              (def high-det register)
                              (def pc register)
                              (def pc register)
                                    .
                                    .
                                    .
                              (def dct register)   ←   "dct" IS GLOBAL REGISTER
                              (def tpf signal output 20)   ←   tpf IS A ONE-BIT OUTPUT SIGNAL ON PIN 20
                              (def 500 clock-period)   ←   CLOCK-WIDTH OF THIS
                                                               CHIP IS 500 ns
                                    .
                                    .
                                    .
```

NAME OF
CHIP IS
"DTMF3"

DECLARATIONS

"tone-proc" IS ONE
FINITE STATE MACHINE

```
                          (process tone-proc 0   ←   STACK-DEPTH FOR SUBROUTINE CALLS
                              tp1 (setq ltch0
                                      (rsh hc1
                                           (rsh hc2
                                                (rsh lc1
                                                     (rsh lc2 ltch0)))))
                              tp2 (cond ((< > ltch1 ltch0)
                                          (setq ltch1 ltch0)
                                          (setq dct 158)))
                              tp3 (cond ((and (= ltch1 ltch2)
                                              valid
                                              (< > pc 665)))
                                          (setq pc (1 + pc)))
                                         ((< > ltch1 ltch2)
                                          (setq dct (1 - dct))))
                              tp4 (cond ((and (< > ltch2 (1 + ltch1)) (not valid))
                                          (setq pc (1 - pc))))
                              tp5 (cond ((and (= pc 665)
                                              (not ptf))
                                          (setq dcf t)
                                    .
                                    .
                                    .
                                         ((and (< > pc 665) (< >0 pc) (not ptf) (not dtf))
                                          (setq tpf f))
                                         ((and (=0 pc) (not dtf))
                                          (setq tpf f)
                                          (setq ptf f))
                                         ((and (=0 pc) dtf)
                                          (setq ptf f)))
                                    .
                                    .
                                    .
                                    .
                                    .
```

STATE
LABELS

"setq" IS
ASSIGNMENT
OPERATOR

CONDITIONAL EXECUTION:

IF THE CONTENTS OF
"ltch2" IS NOT EQUAL
TO THE CONTENTS OF
(ltch + 1) AND "VALID"
IS FALSE —
THEN DECREMENT
"pc" REGISTER"

Figure 1. A Portion of the MacPitts Source Code for the DTMF3 Tone Decoder IC

```
in process tone-proc
(setq ltch0 (rsh hc1 (rsh hc2 (rsh lc1 (rsh lc2 ltch0)))))
```

cond-(<> ltch1 ltch0)
in process tone-proc
(setq ltch1 ltch0)

SOURCE STATEMENT AFTER
MACRO EXPANSION

in process tone-proc
(setq dct 158)

cond-(nor (nor (= ltch1 ltch2)) (nor valid) (nor (<> pc 665)))
in process tone-proc
(setq pc (|1 +| pc))

cond-(<> ltch1 ltch2)
in process tone-proc
(setq dct (|1 −| dct))

CRITICAL PATH FOUND

cond-(nor (nor (<> ltch2 (|1 +| ltch1))) (nor valid)))
in process tone-proc
(setq pc (|1 −| pc))
**Warning delay time 570 Greater than clock period 500
Type c to continue, s to stop

cond-(nor (nor (= pc 665)) (nor (nor ptf)))
in process tone-proc
(setq dcf (nor))
        .
        .
        .
cond-(nor (nor (<> pc 665)) (nor (<>0 pc)) (nor (nor ptf)) (nor (nor dtf)))
in process tone-proc
(setq tpf (nor (nor)))

cond-(nor (nor (= 0 pc)) (nor (nor dtf)))
in process tone-proc
(setq tpf (nor (nor)))

in process tone-proc
(setq ptf (nor (nor)))

cond-(nor (nor (= 0 pc)) (nor dtf))
in process tone-proc
(setq ptf (nor (nor)))

LIST OF ORGANELLE UNITS INSTANTIATED FOR THIS CHIP

Unit 17 (register dct t (((constant 158)) ((unit 23))))
Unit 18 (register ltch2 t (((unit 15))))
Unit 19 (organelle = nil (((unit 15) (unit 18)) ((unit 20) (constant 665))))
Unit 20 (register pc t (((unit 21)) ((unit 23)) ((constant 158))))
Unit 21 (organelle |1 +| t (((unit 20)) ((unit 15)) ((unit 24))))
Unit 22 (organelle <> nil (((unit 15) (unit 18))))
Unit 23 (organelle |1 −| t (((unit 17)) ((unit 20))))

POST-PROCESS ANALYSIS OF SHARED RESOURCES AND
CRITICAL DELAYS

The following organelle is shared:
unit 17 (register dct t (((constant 158)) ((unit 23))))

The control line for the following organelle is shared:
Control-line 1
unit 20 (register pc t (((unit 21)) ((unit 23)) ((constant 158))))
In process: tone-proc State number: 3
**Warning — Delay time: 590

STATE "tp4"

The following flag is shared:
dtf

The following flag is shared:
dcf

"load" SIGNAL FOR REGISTER "dt"

The control line for the following organelle is shared:
Control-line 1
unit 24 (register dt t (((constant 0)) ((unit 21))))

The following organelle is shared:
unit 24 (register dt t (((constant 0)) ((uniit 21))))

The following organelle is shared:
unit 19 (organelle = nil (((unit 15) (unit 18)) ((unit 20) (constant 665))))

The following flag is shared:
ptf

Figure 2. Performace Analysis of DTMF3 Circuit

178

CONCLUSION

The technique of silicon compilation is in its infancy. Early silicon compilers had restricted architectures, poor optimization algorithms, and no performance prediction capability. We believe that refinement of the silicon compilation process will be an evolutionary, rather than a revolutionary process. The performance prediction package presented in this paper demonstrates progress towards a commercially viable silicon compiler.

ACKNOWLEDGMENTS

The author acknowledges the cooperation of Jeffrey Siskind and Jay Southard, of MetaLogic Inc., the authors of MacPitts. Paul Lindemann, of GTE Laboratories assisted in the preparation of the manuscript. Prof. Chris Terman of MIT made many helpful suggestions.

REFERENCES

1. J. Werner, "Progress Toward the Ideal Silicon Compiler," *VLSI Design* (September 1983, October 1983).
2. D. Gajski and R. Kuhn, "New VLSI Tools," *Computer Magazine*, pp. 11 – 14 (December 1983).
3. J. Siskind, J.R. Southard and K. Crouch, "Generating Custom High-Performance VLSI Designs from Succinct Algorithmic Description," Proceedings of the Conference on Advanced Research in VLSI, MIT (January 1982).
4. J.R. Southard, "MacPitts: An Approach to Silicon Compilation," *Computer Magazine*, pp. 74 – 82 (December 1983).
5. J.R. Fox, "The MacPitts Silicon Compiler: A View From the Telecommunications Industry," *VLSI Design* (May/June 1983).
6. C. Terman, "RSIM — A Logic-Level Timing Simulator," Proceedings of ICCD '83 (October 1983).
7. N. Jouppi, "TV: An nMOS Timing Analyzer," Proceedings of the Third Caltech Conference of Very Large Scale Integration (1983).
8. J. Ousterhout, "Crystal: A Timing Analyzer for nMOS VLSI Circuits," Proceedings of the Third Caltech Conference of Very Large Scale Integration (1983).

Hardware Compilation from an RTL to a Storage Logic Array Target

FREDRICK J. HILL, MEMBER, IEEE, ZAINALABEDIN NAVABI, MEMBER, IEEE, CHEN H. CHIANG, MEMBER, IEEE, DUAN-PING CHEN, AND MANZER MASUD

Abstract—This paper treats the automatic translation of register transfer level (RTL) descriptions of digital systems to VLSI realization. The target technology is the storage logic array or SLA. The approach is aimed at applications where the emphasis is on reducing engineering effort and design turnaround time rather than maximizing chip area utilization. The paper develops a mapping between the register transfer language, AHPL, and the SLA. It is shown that each primitive explicitly appearing in an AHPL description can be mapped into an area of real estate in an SLA realization. A detailed development of some of the algorithms is presented. The entire process has been successfully implemented and applied to a set of examples. This is accomplished by developing a final stage for an already existing three-stage multi-application compiler for AHPL. Layout and routing are shown to be a single optimization process if the hardware target is an SLA.

I. INTRODUCTION

AUTOMATIC translation of register transfer level (RTL) descriptions of digital systems to VLSI realization has been a topic of discussion and research for several years. The target realizations of RTL translation processes have included gate arrays [7], [16], standard cells [3], [8], or unconstrained custom integrated circuits. These approaches have met varying degrees of success. It is not anticipated that in the near future any automatic design process will generate results competitive in the use of chip area with the largest and more densely packed handcrafted designs. The engineering time required to accomplish these handcrafted designs is prohibitive except for parts

Manuscript received March 22, 1983; revised January 20, 1984. The research for this paper was supported in part by General Instrument Corporation and the National Science Foundation.

F. J. Hill, Z. Navabi, and D-P. Chen are with the Department of Electrical and Computer Engineering, University of Arizona, Tucson, AZ 85721.

C. H. Chiang is at Rockwell International, Newport Beach, CA 92663.

M. Masud is with the Department of Computer Science and Engineering, University of Petroleum & Minerals, Dhahran, Saudi Arabia.

which anticipate a volume market. If the advantages of integrated circuit realization are to be extended to lower volume parts, the emphasis will be on reducing engineering effort and design turnaround time rather than maximizing chip area utilization. This can be accomplished if the design medium is an RTL with an integrated circuit target format arranged to map directly onto the RTL. The early sections of this paper will develop a mapping between the register transfer language, AHPL, and one such target, the storage logic array or SLA [10], [13], [15]. The later sections will detail the algorithms required in the AHPL to SLA translation process which has been sucessfully implemented.

Whether or not the source description is an RTL, the hardware generation process for gate arrays, standard cells, and custom realization all consist of two separate steps. The first step is the layout of a floor plan constraining the position of both logic gates and memory elements. This is followed by a separate step which automatically routes the interconnecting channels. The compactness of the final routing and the size of the resulting chip are, of course, heavily dependent on the initial layout. For gate arrays layout is often manual followed by automatic routing. A second layout is sometimes used, if the first pass at routing yields a grossly inefficient design; but the layout and routing are not jointly optimized. In the standard cell approach an automatic iterative layout and routing process is usually employed. For more complex standard cells a very difficult system partitioning problem must be accomplished by the hardware compiler. Sections of the described system must be mapped onto the available standard cells.

As we shall demonstrate in this paper, layout and routing is a single process, if the hardware target is an SLA. Thus a single optimization process can search for the overall best layout and routing. We shall also show that each primitive

Reprinted from *IEEE Trans. Computer-Aided Des. of ICAS*, vol. CAD-3, pp. 208–217, July 1984.

explicitly appearing in an AHPL description can be mapped into an area of real estate in an SLA realization.

The existence of a hardware compiler capable of translating AHPL to wire list form has been reported previously [4]-[6]. An enhanced AHPL compiler, written in Fortran, is now available in which the output stage can be adapted to produce output format consistent with any particular implementation technology. A final stage of this compiler has been added to complete the design process for the SLA. This paper will introduce the SLA, show the relation between its structural features and AHPL primitives, and make a case for the assertion that the SLA is competitive with other approaches as a target for automatic hardware generation.

II. Symbolizing the SLA Target

A symbolic diagram for an SLA appears very much like that of the more familiar programmed logic array (PLA) with rows of memory elements distributed uniformly between the rows and columns. In working with PLA's of fixed size, part of the mask design is independent of the actual logic functions to be realized. If the PLA's are stand-alone parts, this independence extends to part of the fabrication process. In the case of SLA's no stage of the design or fabrication process can be completed in the absence of a complete specification of the system under design. The primary reason for this is that the number of logic rows between any two rows of memory elements and the number of logic columns between any two columns of memory elements will vary from circuit to circuit. The memory elements are not efficiently configured from groups of logic cells, and the system clock lines must be routed vertically through the columns of memory elements.

To organize the design process the area of the SLA is assumed to be partitioned into a grid of squares, called cells [13]. A cell may be considered to be a unit of occupied or unoccupied real estate. The number of cells required to realize a clocked memory element will vary with the technology in which the SLA is to be realized. In this document we shall assume that a flip-flop will occupy four cells, as shown in Fig. 1(a). The Q and $\bar{Q}$ output lines extend vertically from both top and bottom of the right most half of the device area. The D input line for the flip-flop extends from both top and bottom at the left. Lines connected to the write enable extend from both sides of the device. With spacing as required by a specific design, memory elements are arranged in the array in carefully aligned rows. The row alignment is necessary so that a single horizontal control line can enable a write operation on a vector of memory elements as depicted in Fig. 1(b).

Components other than memory elements will be regarded as implementable in a single cell of SLA area. A column to row logic element (less pull-up), a row to column element, a pull-up resistor, and an inverter are shown in Fig. 2. Each column consists of two wires representing the true and complemented values of the column variable. The column wires may extend from both the top and bottom of a logic element cell, as shown, or from only the top or only the bottom. Thus columns may be shared by more than one variable as in folded PLA's. Similarly a row variable may extend through a cell or

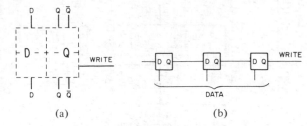

Fig. 1. Clocked memory element.

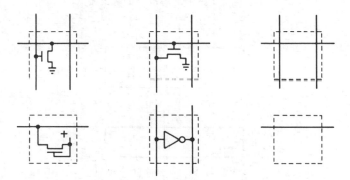

Fig. 2. SLA logic and circuit elements.

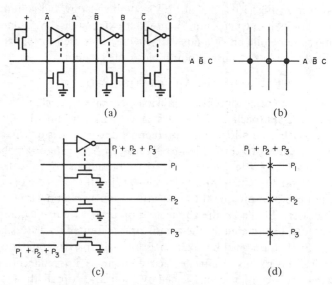

Fig. 3. Symbolizing the OR and AND operations.

be terminated on either the left or the right. A column of adjacent and connected cells implementing a logic variable must somewhere include a column pull-up and an inverter to generate the complement of the variable. A row of adjacent and connected cells must include a row pull-up resistor. Row and/or column wires may extend through a cell with no logic element or pull-up in the cell. Two examples are illustrated in Fig. 2.

In discussing specific SLA realizations, a symbolic representation is clearly preferable to a cell by cell map of the circuit elements in the realization. Fig. 3(a) depicts NMOS implementation of a logical product on an SLA. This realization is symbolized in Fig. 3(b) where a solid dot indicates connection to the true value of a column variable and a white circle indicates connection to the complemented value of the variable. Fig. 3(c) illustrates the effective ORing of three row

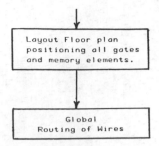

Fig. 4. Usual VLSI design process.

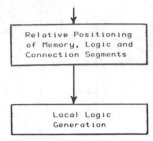

Fig. 5. SLA design process.

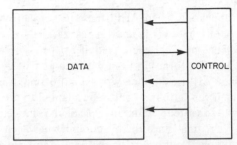

Fig. 6. Physical partition of control and data units.

products. The symbol to be used for the column OR operation is given in Fig. 3(d). As in PLA's combined row and column operations implement sum of product expressions. A column may also be one bit of a bus with row sections used to connect data to the bus at various points along the column.

III. THE SLA GENERATION PROCESS

From the discussion in the previous section we conclude that the SLA approach to VLSI design deliberately blurs the distinction between logic and interconnecting channels or wires. A continuous vertical column of cells can be a gate or a connecting wire. The same is true for a continuous row of cells. The transistor which can implement one input of an AND operation can also provide the connection from a column wire to a row wire. Similarly, the connection of a row wire to a column wire can be effected by the transistor represented by a cross in Fig. 3(d). Care must be taken to assure that an even number of logical inversions occur in each contiguous interconnecting wire. This can be accomplished by choosing properly the true or complemented value in each column to row connection. To limit the scope of the paper we shall not introduce notation for direct ohmic contacts between rows and columns which are usually available as well.

Since there is no physical distinction between logic and wires, no distinction between them is made in the automatic SLA generation process. This feature permits a significant departure from the flow of events in all other integrated circuit design processes including custom, standard cells and gate arrays. The two essential steps of the latter process are depicted in Fig. 4. The primary steps of the contrasting SLA process are shown in Fig. 5. Given a poor layout in Fig. 4, an efficient wiring pattern is not possible. Very often in, for example, the standard cell approach sufficient chip area is set aside for wiring so that the wire routing program will succeed regardless of the layout.

The entire process of Fig. 5 has been automated with only an AHPL description required as input. As will be demonstrated in succeeding sections, the question of efficiency of

chip area utilization is essentially settled by the first of the two steps. Because the SLA makes no distinction between wires and logic, it is quite natural for this first step to jointly optimize layout and interrow connections. Once this is accomplished local logic networks are generated independently with minor impact on chip size.

IV. MAPPING AHPL OPERATORS AND OPERANDS INTO SLA SEGMENTS

Input to an integrated circuit design automation system frequently takes the form of a wire list. The demonstrated capability to compile descriptions written in AHPL to wire list form is not alone sufficient to guarantee that the design process of Fig. 5 will function as claimed. Extraction of information to guide a layout from a wire list is extremely difficult. As we shall show, this necessary insight is much more apparent in a source AHPL description. The task is to establish a framework through which this insight can be carried through the compilation process to be used in step 1 of Fig. 5.

The compiler for AHPL presumes a partitioning of a description into a control and a data portion. The control unit is specified by the branch statements in the description. A typical step containing a branch statement is illustrated below

$$10 \quad A \leftarrow X;$$
$$\rightarrow (a, \bar{a} \wedge b, \bar{a} \wedge \bar{b})/(11, 20, 30).$$

In the process of compiling the AHPL description including this step, control memory elements will be assigned to each of steps 10, 11, 20, and 30; and a branching network will be provided to route the control level to the proper memory element following the execution of step 10.

The same partitioning of data and control extends to the SLA realization. As depicted in Fig. 6, the control unit is located at the right of the SLA data unit. The approach to control unit implementation will be tailored so that its vertical dimension will match that of the data unit. If necessary more than one column of control flip-flops will be used.

The term segment will be used to refer to an element in the domain of the mapping of an AHPL description to areas of real estate within an SLA chip. All chip area not included in a segment will be referred to as unused. One measure of success of any integrated circuit design is the ratio of the unused area to the total chip area. The segment areas are not necessarily physically disjoint. For example, a bus is typically implemented as a vector of parallel columns which route information between row vectors of memory elements. A bus segment might map into some physical area overlapping a combinational

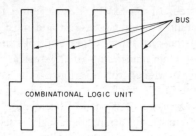

Fig. 7. Overlapping bus and combinational logic unit segments.

```
MODULE: SEQSEL
INPUTS: X[4]; Y[2]; a; b.
MEMORY: A[4]; B[4]; C[4]; F[4].
BUSES: BUS1[4].
OUTPUTS: Z[4].
CLUNITS: INC[4](BUS1); DCD[4](Y).

1  →  (a, b, ā ∧ b̄)/(2, 4, 1).
2  →  (a)/(2).
3  A ← X; B ← A; C ← B;
   →  (1).
4  F ← INC(BUS1);
   →  (1).
ENDSEQUENCE
Z = F; BUS1 = (A!B!C!(1,1,1,1)) * DCD(Y).
END.
```

Fig. 8. Example AHPL description.

logic unit, another type of segment. This situation is illustrated in Fig. 7.

Explicit data segments are the operators, operands, and declared targets of AHPL falling into types, as listed below:

1) register target of a transfer,
2) bus target of a connection,
3) input or output vector,
4) declared memory file,
5) single operator,
6) combinational logic unit,
7) ROM,
8) vector of binary constants,
9) derived segment (combination of declared memory elements).

The declared file will consist of a few rows of memory elements organized as a RAM and treated as a unit by the SLA compiler. Files and ROM's will not be included in any of the examples in this paper to limit the scope of the discussion. Sometimes expressions on the right-hand side of a transfer or connection consist of a single AND or OR operation. Sometimes the unique treatment of these as type 5 segments will result in a more efficient SLA. All other logic expressions on the right hand side of transfers or connections will be treated as combinational logic unit segments.

Not every segment mapped into an area in an SLA appears explicitly in the AHPL description. It has always been necessary for the hardware compiler to generate at the input of registers buses which were not declared in the AHPL description. In the context of the SLA these will be referred to as fan-in-bus (FIB) segments.

The implicit SLA data segments fall into the following three types:

10) FIB,
11) FOV (fanout vertical),
12) connecting row section.

Instances of data segments of types 10, 11, and 12 are generated as needed during the SLA compiling process. Like a bus an FOV is a vector of columns on the SLA. The FOV is typically used when it is necessary to fan-out information from one row of memory elements on the SLA to several other rows. All vertical routing of information between rows is accomplished by I/O lines, declared buses, FIB's, or FOV's. From time to time it is necessary for the hardware compiler to instantiate type-12 row sections to route information from bus to bus or to provide proper column alignment of operand bits. The ultimate use of each cell of chip area is established

by the mapping of explicit and implicit segments from the AHPL description into the SLA. As an illustration of a particular mapping consider the AHPL description of Fig. 8, which accepts a stream of four bit vectors on lines X and delivers any one of the most recent three vectors at its output. The two-bit input vector Y specifies which of the three data vectors or the constant vector $(1,1,1,1)$ is desired at the output. Acceptance of an input vector requires a positive pulse on control input a, and a logical 1 on control line b allows the output to change values.

An inspection of the module SEQSEL reveals the following explicit segments classified into types.

Type 1: A; B; C; F
Type 2: $BUS1$
Type 3: X; Y; Z
Type 6: DCD; INC
Type 8: $(1,1,1,1)$

The hardware compiler is able to obtain this same tabulation of explicit segments as part of the process of parsing the syntax of the AHPL description. Only one implicit segment will be required in the realization of this deliberately uncomplicated system. This will be an FOV which routes the output of the decoder to the rows where it is used to control connections to $BUS1$.

In Fig. 9 we see an SLA realization of the module SEQSEL. Although symbolic notation is used, the layout is constructed to scale. If it were overlaid with an SLA area grid, one would see the symbolic horizontal and vertical lines running through the centers of the grid squares. The leftmost column of memory elements labeled 1 through 4 is the implementation of the control unit which will be discussed in the next section. Segments of types 1, 6, 8, and 12 are always classified as horizontal segments in that each one occupies some number of consecutive rows of the SLA. Each horizontal segment of SEQSEL is listed at the left of the rows occupied by that segment in Fig. 9. The top row of memory elements is the register segment F followed by four rows of combinational logic implementing the segment INC. The three rows of logic below the next register C is input logic to this register and output logic connecting this register to $BUS1$. Since these rows must be adjacent to the memory elements in C they are not named as separate segments but are considered part of segment C. The same can be said for the three rows below B. Below register segment A are four rows of logic implementing the

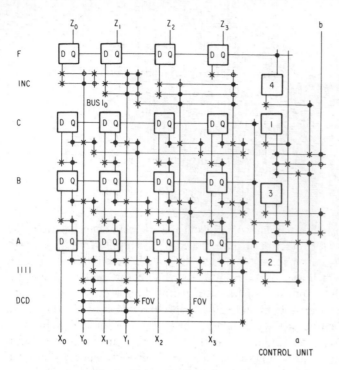

Fig. 9. SLA realization of SEQSEL.

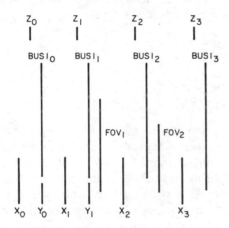

Fig. 10. Vertical segments of SEQSEL.

decoder horizontal segment DCD. The constant segment *BUS1* does not require a separate row but is merely a connection to BUS1 of one row of the decoder.

The segments not discussed above are vertical segments. For clarity the horizontal segments have been removed from Fig. 9 to form Fig. 10. Each of the remaining vertical segments is carefully labelled in this figure. It will be noted that unlike horizontal segments, the columns comprising a vertical segment are not adjacent. Each column of a vertical segment is adjacent to a column of memory elements. Again Fig. 10 is to scale. The column lines shown would run through the centers of squares of an SLA area map.

V. SLA REALIZATION OF THE CONTROL UNIT

The control unit of an SLA is a mapping of horizontal and vertical segments just as is the data unit. None of these control unit segments are explicit in the AHPL description. As always the hardware compiler must assign memory element

states to the steps of the AHPL description. The two state assignments of most practical interest and most easily implemented are 1) the one flip-flop per control state assignment and 2) a direct binary encoding of 2^n steps as the combinations of values of n control flip-flops. The current implementation of the SLA compiler provides for only option 1).

Once a state assignment is made, the memory elements of the control unit must be arranged in register segments. The approach to the state assignment is chosen and the definitions of register segments are made with the goal of matching the vertical dimension of the control unit with that of the data unit. The one flip-flop per control state assignment was used in SEQSEL, and each control flip-flop was defined as a horizontal segment. We see from Fig. 9 that this approach happened to lead to a very efficient realization for this circuit. There is, of course, no guarantee that the length of the vertical dimensions of the control and data unit will be so near the same. If twice as many control states were included with a data unit similar to that of Fig. 9, either two control flip-flops per register segment or an option-2 state assignment would be the choice.

Also defined as a horizontal segment in the control unit is each of the logic functions specifying a transition from a step to any next step. Defined as implicit vertical segments are FIB's which may serve to collect signals from prior control states or FOV's which route activity from present control states to potentially active next states. The process for defining FOV's and FIB's in the control unit is similar to that for the data unit. Once explicit segments are defined the algorithms for generating the control section of the SLA are the same as those for the data section with one constraint. The lines connecting the control unit and data unit are defined as row sections in the control unit. These row sections must be located on the same physical rows in the control SLA as in the data SLA which is generated first.

VI. THE AHPL TO SLA COMPILER

A function level simulator for AHPL together with a hardware compiler capable of translating AHPL descriptions to an abstract wire list have been available for some time. More recently, as the focus of effort became VLSI realization of AHPL designs, it became apparent that more than one form of output would be required from the AHPL compiler. Outputs from the compiler now drive a gate level simulator and an automatic test sequence generation program and will be used for more than one approach to integrated circuit realization. In order to provide for so many separate output formats with a unified set of software, the system of Fig. 11 was conceived and the first two stages implemented.

The first two stages of Fig. 11 are common to every application of the AHPL compiler. The first stage scans and parses the AHPL description and converts it to a set of executable tables. These tables will be used to drive subsequent versions of the function level simulator. Stage 2 is the actual hardware compiler, translating the executable tables into a linked list representation of the circuit wire list. For the benefit of the SLA compiler and other hardware generation approaches which must produce a layout, stage 2 provides a second data

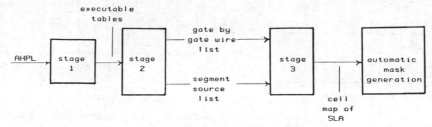

Fig. 11. The three-stage AHPL compiler.

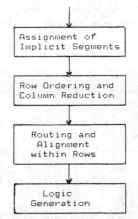

Fig. 12. Major steps in SLA realization.

```
MODULE: STACKBUFFER
  INPUTS: push; pop; complete; look.
  MEMORY: R1[8]; R2[8]; STATUS[2]; STKPNT[8].
  OUTPUTS: ADDRESS[8]; write; read; STATOUT[2].
  COMBUS: OBUS[8]; RAMBUS[8].

  1   STKPNT ← 0,0,0,0,0,0,0,0; STATUS ← 1,0.
  2   STATUS[0]*(push∨ pop) ← 0;
      → (push∨ pop, push, pop)/(2,3,5).
  3   R1 ← OBUS; R2 ← R1; STKPNT ← INC(STKPNT).
  4   RAMBUS = R2; write = 1; STATUS[1]*(∧/STKPNT) ← 1;
      STATUS[0]*complete ← 1;
      → (complete,complete)/(4,2).
  5   STKPNT ← DEC(STKPNT); STATUS[1] ← 0; R1 ← R2.
  6   read = 1; → (complete)/(6).
  7   R2 ← RAMBUS; STATUS[0] ← 1;
      → (2).
ENDSEQUENCE
  OBUS = R1 * look; ADDRESS = STKPNT;
  STATOUT = STATUS.
END.
```

Fig. 13. AHPL description of STACKBUFFER.

base listing all explicit segments in the AHPL description together with their respective sources.

Definition 1: Any segment which serves as an input to explicit segment S is called a *source* of S.

Only one stage 3 is shown in Fig. 11, but the single stage 2 has been designed to drive third stages for a broad spectrum of applications including interconnection of MSI parts as well as integrated circuit realization. The stage 3 for the SLA compiler implements a series of algorithms which translate the segment list and wire list to a cell by cell specification of the SLA. Fig. 12 represents the sequence of major steps included in stage 3 as employed for realization of the data section. The process has been slightly modified for generation of the control section by allowing for position constraints on row sections.

Each of the four blocks in Fig. 12 is composed of a number of procedures. The entire process for both the control and data unit has been implemented. The first two blocks are of most interest and largely determine the efficiency of the final realization. These processes will be described in the next two sections. Section IX will summarize the results obtained from running the program on five circuits of varying complexity.

VII. GENERATION OF IMPLICIT VERTICALS

As pointed out in Section IV, not all connections between row segments of an SLA are accomplished by declared or explicit vertical segments. Before the process of ordering row segments and merging verticals into a minimum number of columns, all vertical segments must be identified. For this purpose row segment data, as supplied by stage 2 of the hardware compiler is assembled into a row-segment connection matrix.

Definition 2: A *row-segment connection matrix* for a given

AHPL description is a matrix whose columns consist of all targets of direct connections between horizontal segments (Types 1, 5, 6, 9), and whose rows consist of all source segments of such direct connections. A mark is placed in element (i, j) if the segment corresponding to column j is a target of the segment listed for row i.

A source may be connected to a target through an FOV, an FIB, or a direct connection, if the two segments are adjacent in the final layout.

Vertical Declaration Algorithm: While there exists a row or column in the row segment connection matrix with more than one entry, select the row or column with the most entries; assign the row as an FOV or column as an FIB; and delete the row or column from the matrix. *Join* selected pairs of row segments to satisfy single remaining connection entries and delete these rows. Assign FOV's to the remaining rows.

If two horizontal segments are joined, they will be treated as a single segment throughout the rest of the process and will be adjacent rows in the final layout. The criterion for joining segments has been left deliberately imprecise. As yet no single criterion has been found which will lead to optimal results in all cases. All implemented versions will join rows which are both targets and sources of each other.

The AHPL description given in Fig. 13 provides a simple illustration of the assignment of implicit verticals. All of the scalar inputs and outputs, indicated in lower case in Fig. 13 are control variables and will be connected directly to the control unit which will not be included in this example. Likewise $\wedge/STKPNT$ must be logically combined with other signals in the control unit. This function is generated in the data unit, but the resulting line is routed horizontally in the same row to the control unit. Therefore, $\wedge/STKPNT$ is not a source in the row-segment connection matrix. Since the outputs of the

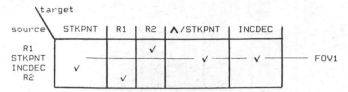

Fig. 14. Row-segment connection matrix.

	FOV1	OBUS	RAMBUS	STATOUT	ADDRESS
R1/R2		1	1		
INCDEC/STKPNT	1				1
∧/STKPNT	1				
STATUS				1	

Fig. 15. Span matrix for STACKBUFFER.

register, *STATUS*, are connected only to a declared output and this register has no sources in the matrix, it is not included. Also not included in the list of sources are constant vectors (e.g., 00000000) which are most easily generated at the row where they are used. The resulting row-segment connection matrix, as assembled from data provided by stage 2 of the compiler, is given in Fig. 14. Since the increment and decrement logic units have the same target and the same source, they have been combined together as a single unit. This can be accomplished by a slight modification of the AHPL description prior to application of the hardware compiler.

Only one row (and no columns) in Fig. 14 includes more than one connection. Thus the fanout vector FOV1 is created to connect *STKPNT* to targets *INCDEC* and ∧/*STKPNT*. Since *R*1 and *R*2 are mutual targets and sources, these rows are joined. Similarly, *STKPNT* and *INCDEC* are joined. Since all rows are thus eliminated from the table, no additional FOV's are required.

VIII. ORDERING OF ROW SEGMENTS

In most SLA implementations adding a column will have greater impact on chip size than adding a row. It is, therefore, important to use as few columns as possible to route information between the various register row sections and combinational logic units. The number of necessary columns vary significantly with the ordering of row segments in the layout. This section will describe a process for ordering rows to permit maximum sharing of columns by the vertical segments. The process will begin with a span matrix whose rows are all individual row-segments or joined row-segments and whose columns are all verticals both explicit and implicit. An element of the span matrix will be 1, if the corresponding row and column are to be connected. The span matrix for STACKBUFFER is given in Fig. 15.

The object of the row ordering process is to merge the verticals of the span matrix into as few shared columns as possible. From the span matrix the connection set of each row and the span set of each vertical may be determined. These terms are defined as follows.

Definition 3: The *connection set of a row segment* is the set of all verticals to which that row must be connected.

Definition 4: The *span set of a vertical* is the set of all row segments which must be connected to that vertical. We say that a vertical must span the rows in its span set.

The final layout of the rows and verticals will be described by a layout matrix. Each input or output vertical must pass through either the top or the bottom of the layout to provide for external connections. For convenience we refer to the top and bottom of the layout as extremes and the set of all inputs passing through an extreme as its extreme set. The two extreme sets are disjoint, and in the absence of external constraints the I/O segments are assigned to the extreme sets by the following simple procedure.

Procedure 1 for Generation of the two Extreme Sets: Let *A* be the set of I/O vertical segments. While $A = 0$, select the row segment whose connection set contains the most members of *A*; enter these I/O vertical segments in the extreme set with the fewest entries; and delete them from *A*.

Definition 5: A *partial layout matrix* is a matrix whose rows are rows from the span matrix. Entries are blank or are identifiers of verticals. A particular vertical may be entered in only one column. If a vertical is entered in a given column, the identifier of the vertical must be placed in each row in the span set of that vertical. All columns are valid layout columns as given by definition 6.

Definition 6: A *valid column* of a partial layout matrix must satisfy the following two conditions.

1) If a vertical entry in the column is a vertical in the upper/lower extreme set, only entries of the same vertical or blank entries may occur above/below that entry in the column.

2) All entries for the same vertical are adjacent or separated by blank entries.

Definition 7: A *layout matrix* is a partial layout matrix which includes every row of the span matrix and includes entries for every vertical in the span matrix.

Procedure 2: A vertical may be added to partial layout matrix to form a new partial layout matrix by entering its identifier in those rows of an existing column which are included in the span set of that vertical, provided a valid layout column will result. If no column will remain valid with the addition of that vertical, the vertical must be entered as a new column. If a row in the span set of the vertical is not already present in the partial layout matrix, the position of the row is defined by the inclusion of the vertical.

Procedure 3: A row segment may be added to a partial layout matrix to form a new partial layout matrix by placing it between any two existing rows or at the top or bottom of the matrix, so that the maximum number of columns remain valid. Any resulting invalid column is replaced by two valid columns.

In Fig. 16 we see a partial example in which row 6 with connection set (F, D) can only be included by adding a column.

A layout matrix may be obtained by beginning with a partial layout matrix consisting of a single row, including a column entry for each vertical in its connection set, or with a single valid column, including an entry for each row in its span set. The partial layout matrix could be extended by adding rows and verticals in any order. Clearly the order in which rows and verticals are added will significantly influence the number of columns in the final layout matrix. In the remainder of this section we shall develop a procedure for establishing an order in which rows and verticals are added to the partial layout matrix. In general the approach will be to identify rows and

Fig. 16. Example of row addition to a partial layout matrix.

columns in the span matrix which, if added to the partial layout matrix at a particular stage in the process, will not increase the number of columns. These rows or verticals are then placed on a stack (and deleted from the span matrix) so that entries are popped from the stack and added to the partial layout matrix in reverse order of their placement on the stack. Usually the rows and verticals with most connections are stacked last and entered first on a relatively sparse partial layout matrix. As rows and columns are deleted from the span matrix, the result will be referred to as a reduced span matrix.

Definition 8: A row *i* of a reduced span matrix is said to *dominate* row *j* of the reduced span matrix if and only if row *i* includes a 1 entry in every column in which row *j* includes a 1 entry.

Theorem 1: If a partial layout matrix is formed from the rows and verticals remaining in a reduced span matrix, a dominated row may be entered after the other rows and verticals without forcing the addition of a column in the partial layout matrix.

Proof: If row *i* dominates row *j*, row *j* may be placed adjacent to row *i* in the partial layout matrix so that each vertical in the connection set of row *j* will be entered immediately below the entry of the same vertical in row *i*.

Theorem 2: If a row of a reduced span matrix includes only a single 1 entry, a partial layout matrix may be formed and the respective row entered last without the addition of a column in the partial layout matrix.

Proof: If the single vertical in the connection set of row *i* is included in another row *j* in the partial layout matrix, row *i* is dominated. If the vertical is in no other row, it is possible to enter the vertical in any column at a point not affecting the validity of the column.

Theorem 3: If a column of a reduced span matrix includes only a single 1 entry, a partial layout matrix may be formed and the respective vertical entered last without the unnecessary addition of a column in the partial layout matrix. (The proof is similar to that of Theorem 2.)

On the basis of the above three theorems we are able to construct the procedure given in Fig. 17 for reducing the span matrix and stacking rows and verticals. Once all rows and verticals are stacked, they are popped one by one to form successive partial layout matrices employing the procedures given above for adding rows and verticals. No proof is offered regarding the effectiveness of the last row deletion operation. It merely allows the process to continue. Empirical evidence gathered thus far and discussed in the next section has been encouraging.

The STACKBUFFER whose span matrix is given in Fig. 15 provides a very simple example of the column merging process. First procedure 1 places *OBUS* and *RAMBUS* in the same extreme set leaving *STATOUT* and *ADDRESS* in the other. The

```
rcnt = number of rows in span matrix
ccnt = number of columns in span matrix
while ( ( rcnt ) 1 ) OR ( ccnt ) 1 ) )
  [ progress = 0
    while (single entry row exixts)
      [ delete row from reduced span matrix;
        stack row and remainder of connection set;
        decrement rcnt;
        progress = 1
      ]
    while (dominated row exists)
      [ delete dominated row form reduced span matrix;
        stack row and remainder of connection set;
        decrement rcnt;
        progress = 1
      ]
    while (single entry column exists)
      [ delete column from reduced span matrix;
        stack column and remainder of span set;
        decrement ccnt;
        progress = 1
      ]
    if (progress = 0)
      [ delete the row which will leave the most single
        entry columns from the reduced span matrix;
        decrement rcnt;
        progress = 1
      ]
  ]
```

Fig. 17. Procedure for reducing the span matrix.

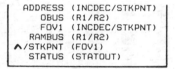

Fig. 18. Stacked rows and verticals.

Fig. 19. Evolution of layout matrix.

procedure of Fig. 17 first stacks the row-segments $\wedge$/*STKPNT* and *STATUS* and deletes them from the span matrix. The four verticals FOV1, *OBUS*, *RAMBUS*, and *ADDRESS* are each now represented by single entries in the reduced span matrix and are thus placed on the stack. The final stack of rows and verticals with their respective connection sets and span sets from the reduced span matrices are given in Fig. 18.

Fig. 19 depicts the evolution of the layout matrix as the segments are successively popped from the stack. For convenience the verticals are abbreviated with their first letters *F*, *O*, *R*, *S*, and *A*. For reference the two extreme sets which are checked at each step to insure column validity, are shown at the top and bottom of the figure. For example, $R1/R2$ is placed above *INCDEC*/*STKPNT* at the second step so that the OBUS is not blocked from the upper extreme set.

IX. APPLICATIONS

So far printouts similar to Fig. 9 have been obtained as a result of running the VAX-11/780 version of the SLA compiler program on five circuits. These include SEQSEL intro-

circuit	# of AHPL statements	# of explicit segments	number of data flip-flops	CPU time	chip area utilization
SEQSEL	10	10	16	4.05s	29 %
HWTEST	10	5	4	3.82s	24 %
SERCOM	12	5	28	7.17s	20 %
STACKMEM	33	17	32	9.38s	14 %
PROCESSOR	112	17	39	148s	12 %

Fig. 20. Summary of applications of SLA compiler.

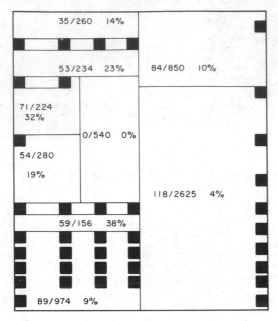

Fig. 21. Map of SLA layout for STACKMEM.

duced in Section IV, two other small circuits, HWTEST and SERCOM, for checking results, and two more substantial examples. One of these was STACKMEM, a stack controller together with a memory module realized as a single SLA. The most challenging description was that of a 4-bit microprocessor, designed specifically as a test for the SLA compiler. The respective printouts were analyzed, and the results summarized as given in Fig. 20.

For the purpose of this analysis the chip area utilization was defined as the number of cells occupied by transistors or memory elements divided by the total number of cells in the chip area. The number of AHPL statements includes both branch and transfer statements, several of which may appear in a single step. This may be the best overall measure of circuit complexity. Each of the examples was analyzed in detail, reasons were ascribed to various instances of wasted chip area. These observations have led to a series of proposed refinements in the stage-3 algorithms.

A map of the layout of the SLA realization of STACKMEM is given in Fig. 21. Only the memory elements are shown individually. The devices used to implement AND and OR operations between row and column wires are represented by numerical summary in each of nine disjoint areas making up the SLA. Left of the slash is the number of cells within the area containing transistors. On the right of the slash is the total number of cells within the area. The cells occupied by memory elements are not counted in either number. The

absence of data in the smaller areas between memory elements may be assumed to indicate that no transistors are found there.

The three deficiencies most commonly responsible for wasted chip area are all evident in the map of Fig. 21. The most significant factor was poor resolution of the tradeoffs between data section logic and control section logic. A bad choice of location of certain combinational logic which could be placed in either the control or data section will result in extra horizontal wires between the control and data unit and extra vertical lines within the control unit. The large area in the lower right corner of Fig. 21 consists primarily of vertical wires with only 4 percent of the cell occupied by transistors. A second factor contributing to waste is lack of uniform register size. Notice that the second row of memory elements in the data section is only a 2-bit register, and the third row consists of only a single memory element. At the right of these registers is a 540 cell area containing no memory elements or logic devices. Currently the program treats the area between data flip-flops as unuseable.

An improved version of the compiler is under development. This program will be sensitive to the location of combinational logic implementing branch functions to minimize the number of lines passing between the control and data units and, thus reduce the number of verticals within the control unit. It will also implement criteria for merging shorter than standard-length registers. Merging must take place early in the process to permit the generation of connecting row sections to carry the memory element outputs horizontally as necessary. Elimination of the wasted area between memory elements will be best approached by broadening our view to include circuit design considerations. A design goal for the circuit set to be developed will be short wide flip-flop areas with vertical wires allowed to pass through.

The summary in Fig. 20 suggests that the overall effectiveness of the program as well as the required CPU time increases dramatically for more complex circuits such as the four-bit processor. While the chip area utilizations for these more complex circuits can be expected to improve significantly as the algorithms are refined, the opposite will probably be the case for CPU time. When compared to alternative approaches, however, this use of CPU time is not a drawback.

X. CONCLUSIONS

The project described by this paper offers evidence that the overall concept of a silicon compiler with an RTL description as its source is valid. In particular it justifies the place of the SLA as a target of such a compiler. The emphasis has been on setting forth the principal algorithms and establishing a terminology for design automation based on aligned rows and columns of memory elements.

The scope of a comparison of results achieved by the AHPL to SLA compiler with other approaches to digital design automation must include circuit considerations. Research is continuing, and the data necessary to make precise quantitative comparisons will be forthcoming. For now it should be recognized that much of the chip area of competitive approaches such as gate arrays and standard cells would not be considered occupied according to our definition of chip area utilization. A utilization factor of 30 percent would probably be competitive and satisfactory for small production quantity applications where minimum turnaround time and engineering cost are the principal criteria. Improvement in chip utilization over the figures given in the previous section may be expected from successive versions of the compiler.

REFERENCES

[1] S. P. Reiss, and J. E. Savage, "SLAP-A methodology for silicon layout," in *Proc. IEEE Int. Conf. on Circuits and Computers* (New York), p. 281, Sept. 1982.

[2] A. A. Szepieniec, "SAGA: An experimental silicon assembler," in *Proc. 19th Design Automation Conf.* (Las Vegas, NV), p. 365, June 1982.

[3] T. S. Hedges *et al.*, "The siclops silicon compiler," in *Proc. IEEE Int. Conf. on Circuits and Computers* (New York), p. 277, Sept. 1982.

[4] F. J. Hill and G. R. Peterson, *Digital Systems: Hardware Organization and Design.* 2nd ed. New York: Wiley, 1978.

[5] F. J. Hill, "Introducing AHPL," *Computer*, p. 28, Dec. 1974.

[6] R. E. Swanson, Z. Navabi, and F. J. Hill, "An AHPL compiler/ simulator system," in *Proc. 6th Texas Conf. on Computing Systems*, Nov. 1977.

[7] L. F. Todd *et al.*, "A multitechnology gate array layout system," in *Proc. 19th Design Automation Conf.* (Las Vegas, NV), p. 792, June 1982.

[8] D. G. Gajski, "The structure of a silicon compiler," in *Proc. IEEE Int. Conf. on Circuits and Computers* (New York), p. 272, Sept. 1982.

[9] Z. Navabi *et al.*, "VLSI design automation using a hardware description language," in *Proc. 1982 Phoenix Conf. on Computers and Communications*, p. 54.

[10] S. S. Patil and T. A. Welch, "A programmable logic approach to VLSI," *IEEE Trans. Comput.*, vol. C-28, pp. 594–601, Sept. 1979.

[11] D. Johansen, "Bristle blocks: A silicon compiler," in *Proc. 16th Design Automation Conf.*, pp. 310–313, 1979.

[12] C. Tanaka *et al.*, "An integrated computer aided design system for gate array masterslices," in *Proc. 18th Design Automation Conf.*, pp. 812–819, June 1981.

[13] K. F. Smith, T. Carter, and C. Hunt, "Structured logic design of integrated circuits using the storage/logic array (SLA)," *IEEE Trans. Electron Devices*, vol. ED-29, pp. 765–776, Apr. 1982.

[14] J. E. Hassett, "Automated layout in ASHLAR: An approach to the problems of general cell layout for VLSI," in *Proc. 19th Design Automation Conf.* (Las Vegas, NV), p. 777, June 1982.

[15] K. F. Smith, "Implementation of SLA's in NMOS technology," in *Proc. VLSI 81 Int. Conf.* (Edinburgh, U.K.), pp. 247–256, Aug. 1981.

[16] J. P. Gray, I. Buchanan, and P. S. Robertson, "Designing gate arrays using a silicon compiler," in *Proc. 19th Design on Automation Conf.* (Las Vegas, NV), p. 377, June 1982.

AN INTERACTIVE PLA GENERATOR AS AN ARCHETYPE FOR A NEW VLSI DESIGN METHODOLOGY

Lance A. Glasser and Paul Penfield, Jr.
Department of Electrical Engineering and Computer Science, and
Research Laboratory of Electronics
Massachusetts Institute of Technology
Cambridge, Massachusetts

ABSTRACT

A group of APL functions for creating programmable logical arrays with NMOS lambda design rules is described. These are used as a vehicle with which to study structured VLSI system design.

INTRODUCTION

Programmable logic arrays (PLA's) have emerged as one of the fundamental building blocks of VLSI systems, and as such are an excellent vehicle for the study of structured integrated circuit design. In this paper we describe a group of APL functions designed to generate artwork and perform other support functions for PLA's. The final artwork specifications, in CIF [1], conform to the Mead and Conway [2] lambda design rules for NMOS.

Structured design generally involves the use of hierarchy and modularity. In structured integrated circuit design we find these concepts useful in at least three domains. Because we are working in a programming environment, a clean programming style which embodies these concepts is clearly useful. In addition, the use of modularity, hierarchy, and a constant attention to interface specification is desirable in the geometrical and electrical domains.

THE SOFTWARE

The APL group consists of three top level functions. The first is used to create the PLA artwork specification. There are five inputs to this function:

- A character array containing the sum of products specification. Also included in the array is the specification of which inputs and outputs are clocked and a designation of feedback terms. In the future, the selection between various input and output driver options will also be included in this array. A specification of a simple PLA is shown below and the corresponding schematic is shown in Fig. 1.

```
IIOOO
NCCCN
OX101
11011
10100
```

- The value of lambda in centimicrons. This is needed so that the program can automatically size the fourteen different power buses in the PLA.

- Whether folding is desired. Folding is a way of changing the aspect ratio (as well as the speed and power) of PLA's which contain a sufficient degree of code repetition in the AND plane. One may think of a folded PLA as containing two OR planes from which one is selected via a 2:1 multiplexer controlled by the high order input bit. We also allow a specification as to whether or not this selector input bit is clocked.

- The output scaling factor. Basically this parameter forces the user to explicitly state whether or not the design is being done in units of centimicrons or lambda.

- Mode selection. We provide several modes to the PLA generator of which only one specifies a request to create the PLA artwork. Other modes include requesting the size of the bounding box, the location of the inputs, the recommended point to connect VDD, and so forth.

During the design of the floor plan for an LSI or VLSI chip we do not necessarily want to fully instantiate the PLA's. What is needed is quick answers to questions such as, "How big would this PLA be if we decided to use it?" Maybe after getting the answer we will decide to try a folded PLA instead. The interactive nature of the APL environment is well suited to building programs which allow this sort of interaction. We have modes of the PLA generator which, without going through the computation involved in actually building the CIF for the PLA, can quickly return some of the PLA's important parameters. The PLA generator function has nine modes. They are to return:

- the artwork specification for the PLA

- the bounding box

- a graphical abbreviation of the PLA showing the important connection points. This is for use in generating quick check plots.

- the locations of the inputs

- the locations of the outputs

- the location of the input clock line

Reprinted from *Proc. IEEE Int. Conf. Circuits and Computers*, 1980, vol. 2, pp. 608–611.

- the location of the output clock line

- the recommended point to connect VDD

- the recommended point to connect GROUND

All of these outputs can, of course, be used as arguments of other APL functions more prominent in the design hierarchy. An example might be a router which needs to know the locations of the PLA outputs.

The second top level function in the group is called PLAFACTS. This function returns a summary of the important geometric, electrical, and logical facts about the PLA, including estimates of the speed, average power dissipation, and peak current requirement. Once again, the ability to quickly get these facts without instantiating the PLA is a great aid to the designer.

The last top level PLA function is a logic simulator designed to take the same input format as the PLA generator. Because it works directly from a high level specification it is computationally efficient.

Both the PLA generator and the PLAFACTS functions call a function which performs all of the global calculations of size, speed, current requirements, etc. Among the duties of this function are an analysis of the required topology (the present version of the PLA group only considers the two cases of folded or nonfolded); two calculations to optimize the speed performance of the PLA (see the section on electrical considerations); and the calculations of the fourteen different sized power buses based on the three criteria of current load, local design rules (metal must be at least three lambda wide), and global design rules (such as how a certain bus must stretch to keep different sections of the PLA from colliding when they grow to accommodate the current budget).

After all of the calculations global to the PLA have been performed, the various modules which make up the parts of the PLA are invoked. To each geometrical object in the PLA (such as the input buffers) there corresponds an APL function. For instance, the PLA generator invokes the function which builds the AND plane, which in turn invokes the functions which build the input drivers and the pullup array. Each of these functions mirrors the organization of the PLA generator. That is, each goes off and does some calculations global to its limited view of the world, and then branches on its input mode to the appropriate section of code. This uniformity makes it easy to connect modules together. When we connect the central VDD bus in the input driver array to the VDD bus in the output driver array we simply run a wire from the point which is returned by calling the input driver function (in the appropriate mode) to the point returned by calling the output driver function (in its corresponding mode).

THE FLOOR PLAN

Figure 2 shows the floor plan of the PLA. A checkplot, together with the source sum of products table, is shown in Figure 3. In this example the folding option was invoked. Notice how the pullups, which are on an 8 lambda pitch, fan out from the 7 lambda pitch of the AND and OR planes into the extra space provided by the ground refresh buses.

ELECTRICAL CONSIDERATIONS

An electrical model for the PLA is shown in Figure 4. Based on this model the software performs two speed optimizations. The first involves sizing the drive transistors T1-T4 in the input superbuffer array. The optimum aspect ratio of these transistors is proportional to the square root of the total load capacitance in the AND plane, which includes the distributed capacitance of the long polysilicon runs and the gate capacitance of the coding straps.

The second optimization is obtained by choosing the optimum width of the pulldown (and hence the pullup) transistors, T5, in the OR plane. Either 2, 3, or 4 lambda wide transistors can fit within the available 7 lambda pitch. The program analyzes both the high and low going transients and picks the pulldown size which minimizes the worst case delay.

Extensions to this PLA could include the placement of drivers between the AND and OR planes, and repeaters to break up the quadradic delay incurred in the polysilicon distributed RC lines. It would also make sense to optimize the PLA speed within a user specified power budget.

To achieve modularity in the electrical domain we allow inputs which may come through pass transistors and require output drivers of sufficient size to drive a reasonable load. (We are presently examining how one might define "a reasonable load." One possible criterion for the output drivers is that attaching a 3 lambda wide metal line to the output that travels halfway around the perimeter of the PLA should add a delay of not more than 15% when compared to the delay through the PLA.)

While we are willing to tolerate the relatively small area penalty incurred by placing inverting superbuffers on the PLA outputs, we did not feel we could accept the large static power dissipation that went with them. We circumvented this limitation by redesigning the superbuffer to have lower steady state power dissipation by including a fifth enhancement mode transistor and increasing the impedance of the depletion mode drive transistor as shown in Figure 4. A SPICE simulation comparing the traditional superbuffer with the low power version is shown in Figure 5. The user will only realize the full speed advantage when pass transistors are not placed in series with the output. (Note that a traditional superbuffer is shown in the Figure 3 checkplot.)

REFERENCES

[1] R. Hon and C. Sequin, A GUIDE TO LSI
    IMPLEMENTATION, Xerox, Palo Alto, CA, 1980.

[2] C. Mead and L. Conway, INTRODUCTION TO VLSI
    SYSTEMS, Addison-Wesley, Reading, MA, 1980.

This work was supported by United States Air
Force Contract AFOSR F49620-80-C-0073 and by the
Department of Electrical Engineering and Computer
Science, MIT.

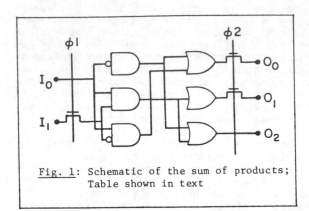

Fig. 1: Schematic of the sum of products;
Table shown in text

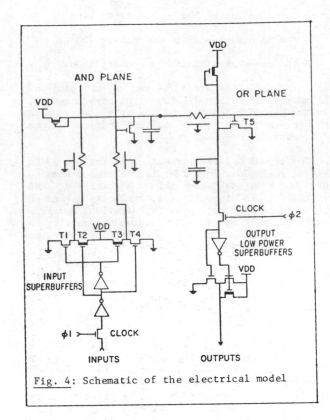

Fig. 4: Schematic of the electrical model

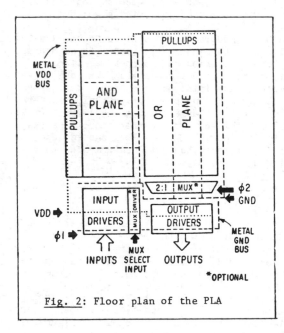

Fig. 2: Floor plan of the PLA

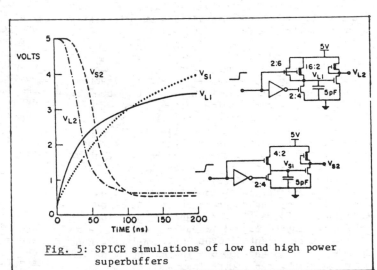

Fig. 5: SPICE simulations of low and high power
        superbuffers

192

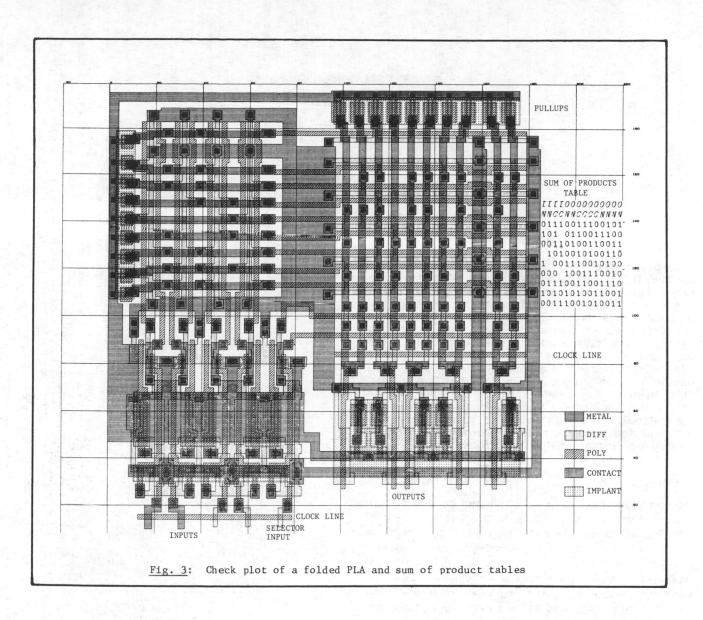

Fig. 3: Check plot of a folded PLA and sum of product tables

193

A VLSI FSM Design System

M. J. Meyer

AT&T Bell Laboratories
Holmdel, NJ 07733

P. Agrawal

R. G. Pfister

AT&T Bell Laboratories
Murray Hill, NJ 07974

ABSTRACT

This paper describes a fully automated finite-state machine (FSM) synthesis system. The FSM is realized as a PLA. This synthesizer accepts a high-level description of the FSM and generates a mask level layout. Several simulation models are produced at different levels of abstraction; these models can be integrated with other modules on the chip to aid in the debugging of the overall VLSI chip design. Valuable information on speed, area, and testability of the PLA can be obtained through a collection of audit programs. This system has been used to design complex controllers for many VLSI chips at AT&T Bell Laboratories. Although a PLA implementation is assumed, the system can be extended to synthesize a random logic implementation of the FSM.

1. Introduction

Modern VLSI chips, such as 32-bit microprocessors [1], require complex control units to process a variety of information. Often, the performance and cost of these chips are dominated by the complexity of the control circuitry. For this reason, the methodology used for designing a VLSI chip's control circuitry needs to be carefully thought out. The important requirement on such a methodology is to define a set of procedures that allow the design of area efficient control units that operate at the required speed. The methodology should be flexible enough to respond quickly to changes in the design and should include effective means for evaluating the area and speed. This paper will describe such a methodology developed at AT&T Bell Laboratories. The procedure is fully automated and permits the designer to describe a FSM in a high level language and results in a mask level layout description. The language permits the complete description of the FSM, including its inputs, outputs, state variables, and transfer functions. Presently the use of the methodology results in a PLA implementation for the FSM. Current work is focussed towards extending this methodology to include microprogrammed as well as random logic implementations.

Figure 1 shows the VLSI design methodology at AT&T Bell Laboratories. The chip is partitioned into control, I/O, and data path. The data path contains regular structures: ALUs, register arrays, etc. The control is

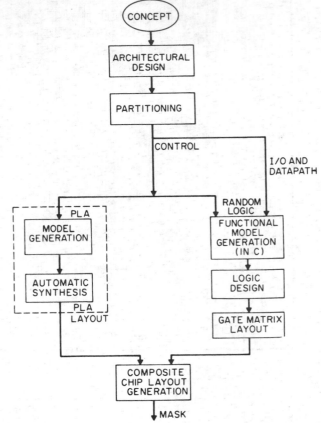

Figure 1: A VLSI Design Methodology

implemented as a collection of FSMs. The FSMs are implemented as clocked PLAs, except in those few cases where speed requirements are better met by a random logic implementation. In cases where a two level logic implementation of a PLA requires more area than a random logic implementation, the latter is used. Random logic is also used for clock generation and some intermodule interfacing.

A functional description of the random logic and the data path is written in C [2]. After the functionality of the random logic has been verified a logic model is manually synthesized. The layout for the random logic is then done in gate matrix [3]. A special language is used to describe the FSMs at a high level. A PLA generation system (shown within dotted lines in Figure 1) converts this FSM description to a PLA implementation, and

Reprinted from *ACM/IEEE 21st Design Automation Conf. Proc.*, 1984, pp. 434–440.

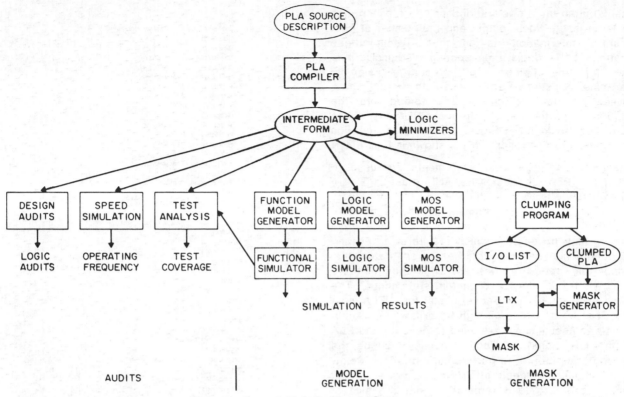

Figure 2:   A PLA Synthesis System

produces models and layout that can be merged with the description of the rest of the chip. This paper will describe the tools used to convert the FSM description to a PLA, its associated models and layout.

The PLA synthesis system (see Figure 2) consists of a collection of programs. A compiler translates the PLA description language, which is C-like in syntax, to an intermediate form, which is used by the other PLA programs. Minimization programs operate on the intermediate form to generate PLAs that are logically and topologically minimized. A set of audit programs give valuable feedback to the designer about the PLA speed, area, testability, and other connectivity information. Several simulation models can be generated to drive simulators that accept different levels of abstraction for the design. An automatic layout program [4] converts the intermediate file to the layout information, which can be combined with similar information for the remaining parts of the chip to produce a composite layout. The following sections describe the methodology and the individual tools that make up the synthesis system.

2. Implementation of the FSM as a PLA

Once the timing relationships of inputs and and outputs and the functionality of a FSM are established, a decision has to be made regarding its implementation. The two choices available for realizing a FSM are, designing it as a PLA or as random logic. If the restrictions of a PLA implementation are not a problem, which are described later, the advantages and

disadvantages of both implementations are examined to decide what approach is best. The PLA implementation has the advantage of providing more flexibility to the designers as the logic changes can be made easily and quickly. Also, due to the increased automation provided by a PLA design system, a PLA is easier to layout and debug and is less error prone compared to the random logic implementation.

A PLA implementation imposes certain restrictions on the FSM. The PLAs in AT&T Bell Laboratories are designed using dynamic CMOS technology and assume the availability of a four phase clock in the logic design. All the inputs to a PLA are latched and are assumed to be valid in phase one. All outputs are assumed to be used only after two additional phases. Thus the outputs are latched in phase three. This allows two phases for computation within the PLA. The fourth phase is allocated for any interfacing random logic external to the PLA. The FSMs that are designed using PLAs are one cycle machines, i.e. only one computation can be made per cycle. The FSM is realized in random logic whenever the needed speed of computation is faster than a cycle.

3. The PLA Description Language

The functionality of a PLA is expressed by the designer in a language that is similar to the *C* programming language. Briefly, this description consists of PLA input/output declarations, which describe the input/output latches, and a statement section, which describes the personality of the wordlines in the PLA.

The input/output latch declarations include a description of the latch input, clock, output and the number of bits. The state information is included within these declarations. The declaration section also includes a facility for describing the precharge signals, programming functions and constants. The state assignment is done externally and folded into the description. A partial description of the PLA description language can be found in Figure 3. A PLA description for the traffic light controller in [5] is shown in Figure 4.

If statements and *PLA case* statements are used to describe the PLA decoder (the AND plane). The *if statement* usually is used to decode one or more inputs signals. The *PLA case* statement usually is used to decode the present state of the FSM. *If* and *case* statements may be nested, which results in an ANDing of the condition expression for the nested statements. The *input* and *output* latches are D latches and are declared with the *input* and *output* statements. The assertion level for the logic level signal for a PLA input or output can be specified as high (h) or low (ℓ). Output functions and assignments are used to describe the PLA ROM (the OR plane). Output functions describe the outputs that must be set to perform an operation. Below is an output function declaration that generates output signal necessary to load register c with the sum of registers a and b. When this function is involved, the outputs RA_BUS1, RB_BUS2, ALUFUNC, and RC_LATCH will be set to latch the result of the addition in c.

$$funct \ c < -a + b \ \{RA\_BUS1 = 0; \ RB\_BUS2 = 0; \ ALUFUNC = 0x4;$$
$$RC\_LATCH = 0;\}$$

The use of the output functions is optional, an assignment can be made directly to an output. Such use is preferred to assignment statements, since output functions offer a more functional description of the result of the outputs.

If a value is not specified for an output, then that output will assume a default value specified in the output declaration. This simplifies a description since the output only has to be specified if it differs from the default. There are cases where it is easier to implement a function's inverse, rather than its true form. The description of the output also describes whether the output should be connected to the inverted or non-inverted latch output, as indicated by the $-B$ or $-Q$, respectively. This can be used with the default value for an output to describe either the output or its complement (which can be inverted by the latch). Figure 4 contains several latch descriptions.

The functional description of the FSM is translated into an intermediate form by the compiler for use by the rest of the programs in the design system. This intermediate form includes a description of all latches and the truth table description of the combinatorial logic.

4. Minimization

After the source has been compiled, the PLA can be

```
PLA:
    planame ( ) { declarations statements }
declarations:
    declaration
    declaration declarations
declaration:
    int      constant-description ;
    input    latch-description ;
    output   latch-description ;
    rompre   signal ;
    decpre   signal ;
    funct    function-description ;
    version = string ;
latch-description:
    latch
    latch , latch-description
latch:
    ( level ) name suffix_{opt} ext-name_{opt} : width # clock default_{opt} inv-flag
level:
    l
    h
name:
    signal
suffix:
    ( constant )
ext-name:
    signal
width:
    constant
clock:
    signal
default:
    constant
function-description:
    function
    function , function-description
function:
    template { assignments }
assignments:
    assignment
    assignment assignments
assignment:
    signal = value ;
value:
    constant
    name
    $unsigned-integer
statements:
    statement
    statement statements
statement:
    signal == constant: statement
    if ( expression ) statement
    { statements }
    output-function-invocation ;
    INVALID ;
expression:
    expression & expression
    ( expression )
    signal == constant
```

Figure 3:   PLA Language Description

minimized using one or both of the available logic minimization programs. These logic minimization programs are based on algorithms found in [6] and [7]. The minimization programs create a new intermediate file containing the minimized PLA. This new intermediate file is in the same format as the unminimized intermediate file, so either file can be used by other parts of the system. The first minimizer [6] is a heuristic minimizer that reduces the number of product terms, this decreases the area required for the PLA. The second minimizer [7], is then used to reduce the PLA density by removing redundant transistors. The second minimizer improves the PLA's speed and fault coverage.

5. Design Audits and Analysis

The PLA design system provides a set of audit programs. These programs estimate the PLA's area and speed for a given technology, and makes suggestions that can be considered to improve either of these.

```
#define red      b'11'
#define yellow   b'10'
#define green    b'01'

#define HiGreen      b'10'
#define HiYellow     b'00'
#define FmGreen      b'11'
#define FmYellow     b'01'

traffic()
{
    input
        (h)SSTATE   [MSTATE]:   2      #clock1 -Q,
        (h)SFCAR    [CAR]:      1      #clock1 -Q,
        (h)STIME(4) [SEC]:      2      #clock1 -Q;
    output
        (h)MSTATE:   2      #clock33      -Q,
        (h)MCLRTIM:  1      #clock31      -Q,
        (h)MFARM:    2      #clock33      -Q,
        (h)MHIWAY:   2      #clock33      -Q;
    decpre    S1PLA;
    rompre    T2PLA;
    funct
        next=_v     { MSTATE = $1;    },
        init_time   { MCLRTIM = 0;    },
        hiway=_v    { MHIWAY = $1;    },
        farm=_v     { MFARM = $1;     };

        SSTATE == HiGreen: {
            hiway = green;
            farm = red;
            if (SFCAR == 0)
                next = HiGreen;
            if (STIME == t'0x')
                next = HiGreen;
            if ((SFCAR == 1) & (STIME == t'1x')) {
                init_time;
                next = HiYellow;
            }
        }
        SSTATE == HiYellow: {
            hiway = yellow;
            farm = red;
            if (STIME == t'x0')
                next = HiYellow;
            if (STIME == t'x1')
                next = FmGreen;
        }
        SSTATE == FmGreen: {
            hiway = red;
            farm = green;
            if ((SFCAR == 1) & STIME == t'0x')
                next = FmGreen;
            if (SFCAR == 0) {
                init_time;
                next = FmYellow;
            }
            if (STIME == t'1x') {
                init_time;
                next = FmYellow;
            }
        }
        SSTATE == FmYellow: {
            hiway = red;
            farm = yellow;
            if (STIME == t'x0')
                next = FmYellow;
            if (STIME == t'x1') {
                init_time;
                next = HiGreen;
            }
        }
}
```

Figure 4:   PLA Description of Traffic Light Controller

There is a program that allows the designer to print both the truth table and the connection array for the compiled PLA. A connection array differs from a truth table in that more than one entry in the connection array may be true at a time. In the case where two lines in the connection array indicate that a result is determined by logical OR'ing the indicated outputs. The connection array reflects the PLA implementation with each line corresponding to a wordline. In the truth table, on the other hand only one line will be true for a particular input. The truth table and connection array are identical for a PLA where one and only one wordline is selected at a time. The truth table is useful in determining the effects of selecting more than one wordline at a time. In the example below, on an input of '111', to produce an output of '00', the truth table indicates the output would be '00', while the two entries in the connection array indicate the output would be '01' or '10' (ANDing these results in '00' as in the truth table). The outputs of multiply selected wordlines are ANDed as the PLA is implemented as a NOR-NOR array. The output of the audit program for the traffic light controller example is shown in Figure 5.

Inputs	Outputs		Inputs	Outputs
10x	01		1xx	01
1x0	01			
111	00		111	10

Truth Table	Connection Array

The designer can also get suggestions for possible ways to reduce the number of inputs or outputs from the PLA. Other audits check for possible design errors, for example, cases where more than one wordline can be selected.

The speed of a PLA is determined by generating a model for the worst case paths in the PLA and doing a circuit simulation [8] to determine its operating speed. The worst case input line, wordline and output line are identified by finding the lines that have the largest capacitive load. The critical paths corresponding to these cases are simulated to see that the wordlines and output lines properly precharge and discharge at a given clock rate. The simulation is done with process files that correspond to best, medium and worst case processing characteristics. The simulation gives the maximum clock rate for the PLA assuming no clock skew. Clock skew is then added to the operating period to give the chip operating frequency. Input set-up requirements and output delays are substituted in the model of the chip, and a critical path analysis is performed [9].

Another program examines the results of a series of tests run on the chip containing the PLA and provides fault and functionality coverage estimates along with recommendations for additional PLA input sequences that would improve these metrics. The functionality coverage is estimated by determining the percentage of valid combinations of outputs that are generated by the set of tests. These estimates, derived from the functional simulation, are available early in the design cycle so that the need for additional tests can be determined before extensive logic or MOS level design verification is possible.

6. Simulation Model Generation

After compilation, and optionally, minimization, PLA model generators are used to build several levels of simulation models automatically. A C language model is

```
Tue Mar 27 14:50:12 1984 TRAFFIC
Source file: houxk!/u1/mjm/pla/examples/traffic.s
Mon Sep 13 13:28:53 1983 Created by @(#)pla-gen 1.4
@(#)pla-aud 1.2
Decoder precharge is S1PLA.
ROM precharge is T2PLA.
10 word lines, 5 input lines, 7 output lines
1 invalid input conditions
15.5078 square mils (assuming all 5-3 clumps and 1.75um technology)


Tue Mar 27 14:50:12 1984 TRAFFIC Loading Analysis
Input line loading analysis

LSSTATE1 contains 6 transistors
HSSTATE0 contains 5 transistors
LSSTATE0 contains 5 transistors
HSSTATE1 contains 4 transistors
HSFCAR contains 2 transistors
LSFCAR contains 2 transistors
HSTIME4 contains 2 transistors
LSTIME4 contains 2 transistors
HSTIME5 contains 2 transistors
LSTIME5 contains 2 transistors

Output line loading analysis

MSTATE0 contains 5 transistors
MSTATE1 contains 5 transistors
MCLRTIM contains 4 transistors
MFARM1 contains 3 transistors
MHIWAY1 contains 3 transistors
MFARM0 contains 2 transistors
MHIWAY0 contains 2 transistors

no dangling inputs
no dangling outputs
no dangling word lines
Tue Mar 27 14:50:12 1984 TRAFFIC Decoder

     S S
     S F T
     T C I
     A A M
     T R E
     E

 0  10 0 xx
 1  10 x 0x
 2  10 1 1x
 3  00 x x0
 4  00 x x1
 5  11 1 0x
 6  11 0 xx
 7  11 x 1x
 8  01 x x0
 9  01 x x1
Tue Mar 27 14:50:12 1984 TRAFFIC ROM

     M M M M
     S C F H
     T L A I
     A R R W
     T T M A
     E I   Y
         M

 0  10 1 11 01
 1  10 1 11 01
 2  00 0 11 01
 3  00 1 11 01
 4  11 1 11 10
 5  11 1 01 11
 6  01 0 01 11
 7  01 0 01 11
 8  01 1 10 11
 9  10 0 11 11
Tue Mar 27 14:50:12 1984 TRAFFIC Multiple Selection Analysis

wordlines        selection input

{0, 1}           10 0 0x
{6, 7}           11 0 1x
```

Figure 5:   Audit of Traffic Controller Example

available for simulation in a functional environment. In the functional simulation model of the chip, each module (PLA, ALU, register array, etc.) is a C language function. A scheduling function then calls each module to simulate the operation of the chip for each phase. To assist in the evaluation and verification of the control design, several monitoring facilities are available that determine which wordlines are exercised by a test.

At the logic level a LAMP [10] compatible model is developed for gate level simulation. At the timing simulation level, a MOTIS [11] compatible model is produced. The AND and OR arrays are modeled as functional ROMs to reduce the simulation complexity while still maintaining timing accuracy. The AND array model's inputs are the PLA inputs and the outputs are the wordlines. The OR array model's inputs are the wordlines and the outputs are the PLA outputs. This is shown in Figure 6. This organization allows a different delay to be associated with each input bit, wordline, and output bit of the PLA. The delays and capacitances associated with the PLA are calculated, for MOTIS simulation, based on technology parameters contained in an auxiliary file, and the loading and length of each line in the PLA.

7. Layout

A collection of programs converts the intermediate file, which describes the PLA's personality, to mask information, and integrates it with the rest of the chip. The first step in this process is to clump the PLA [3]. A clump is a group of wordlines and their associated decoder ground line. A 3-2 clump groups two wordlines with one ground line (Figure 7a). The 5-3 clump groups four wordlines with one ground line (Figure 7b). This reduces the area of the PLA compared to using 3-2 clumps, which need two ground lines for the same four wordlines. The 5-3 clumps place restrictions on the outputs of the wordlines; in Figure 5b, wordlines (1) and (2) must have identical outputs, while the ouputs of wordline (4) must be a subset of the outputs of wordline (3). The clumping program uses 5-3 clumps to reduce the area of the PLA, if that is desirable. The clumping program can also increase the height of the PLA to the allocated size; this is useful since it gives room to make future PLA code changes without changing the size of the PLA.

After the PLA is clumped, the size and I/O of the PLA are used for chip assembly [12]. The inputs and outputs for the clumped PLA are re-ordered to match I/O specified in the chip assembly and the mask information for the resulting chip is generated by another program [3].

8. Results

The FSM synthesis system described in this paper has been successfully used in the development of many VLSI chips [13],[14]. Table 1 contains a list of some of the PLAs that have been designed with the synthesis system. The PLAs range from mainly combinatorial PLAs to those with 66 states, with up to 209 product terms after minimization. The CPU time is the time required to compile the description, minimizing using the algorithm in [7], and generate the models. The fully automated system is a true silicon compiler as it produces a layout from a high level description of the controller. Figure 8 shows the layout plot for a PLA with 49 product terms, 17 inputs, 23 outputs and 8 states.

198

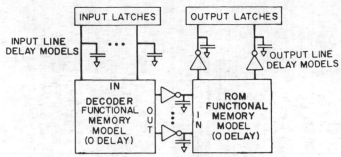

Figure 6:   PLA Timing Model

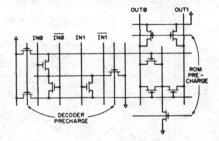

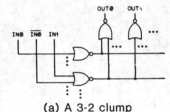

(a) A 3-2 clump

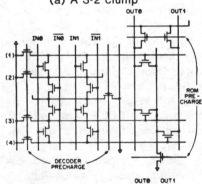

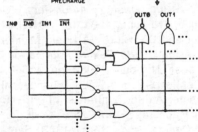

(b) A 5-3 clump

Figure 7:   Clumping of wordlines

9. Conclusions

Several other synthesis systems [15] [16] [17] have been described elsewhere. The system described in this paper offers several important advantages. First, the layout, circuit model, logic model, and functional model are all generated from the same high level description. This results in fewer errors, since the high level, functional description is more understandable than

Table 1

Some Finite-State Machines Designed by the System

Product Terms	Inputs	Outputs	States	CPU time (Min:Sec) (VAX 11/780)
209	30	55	66	1:25
180	57	81	32	5:20
152	30	37	59	1:02
147	32	91	64	1:12
140	55	63	36	7:23
108	31	28	6	2:01
88	45	25	17	0:21*
49	17	23	8	0:13
47	23	16	10	2:13
42	25	12	3	0:11

*   This PLA was not minimized.

a logic or circuit description. Also, since there are fewer steps in the process of generating the layout, the designer is less likely to introduce an error after the functional design is complete. Second, the system generates predictable results when changes are made to a design. This is important since it allows a layout work for the rest of the chip to proceed, while the functionality of the PLA evolves. Third, the system is fast. For example, a designer can compile, minimize, and generate all models for a reasonable size PLA, (around 200 product terms) in less than 10 minutes on a VAX-11/780.

The FSM design system described in this paper is still evolving. There is an effort underway to improve and generalize the description language. Planned enhancements to the FSM design system will include support for automatic minimal state assignment [18]. PLA partitioning [19] and PLA folding [20] are being investigated as an additional technique for producing more area efficient PLAs.

Acknowledgements
The PLA tools are a collection of programs that have been written by many authors. T-W. J. Chou, A. J. Greenberger, H. Schichman, H-F. S. Law, N. R. Miller, and B. M. Swinyer have all written or made substantial improvements in one or more of the PLA tools. The PLA tools would not be the success they are today without the comments and suggested improvements that all the users of the PLA tools have provided.

References
1.   A. Gupta and H. D. Tong, "An Architectural Comparison of 32-bit Microprocessors", *IEEE Micro*, Feb., 1983, pp. 9-22.
2.   B. Kernighan and D. Ritchie, *The C Programming Language*, Prentice-Hall, 1978.
3.   A. D. Lopez and H-F.S. Law, "A Dense Gate Matrix Layout Method for MOS VLSI", *IEEE Trans. on Electr. Devices*, Vol. ED-27, Aug., 1980, pp. 1671-1675.

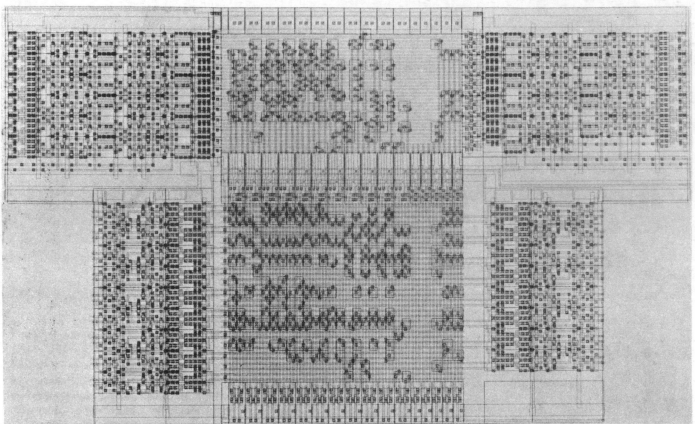

Figure 8:   Layout of a finite state machine with a 49- product term PLA

4. H-F. S. Law and M. Shoji, "PLA Design for the BELLMAC-32A Microprocessor", *IEEE Int. Conf. on Circuits and Computers Proc.*, New York, NY, Sept. 1982, pp. 161-164.

5. C. Mead and L. Conway, *Introduction to VLSI Systems*, Addison-Wesley, 1980, pp. 85-88.

6. S. J. Hong, et al, "MINI: A Heuristic Approach for Logic Minimization", *IBM J. of Res. and Dev.*, Vol. 18, September, 1974, pp. 443-458.

7. D. L. Dietmeyer, *Logic Design of Digital Systems*, 2nd edition, Allyn and Bacon, 1978, pp. 646-655.

8. L. W. Nagel, "ADVICE of Circuit Simulation", *IEEE Symp. on Computers and Systems*, 1980, Houston, TX.

9. V. D. Agrawal, "Synchronous Path Analysis in MOS Circuit Simulator", *19th Design Automation Proc.*, Las Vegas, NV, June 1982, pp. 629-635.

10. H. Y. Chang, et al, "LAMP: System Description", *Bell System Technical Journal*, Vol. 53, Oct., 1974, pp. 1431-1449.

11. H. N. Nham and A. K. Bose, "A Multiple Delay Simulator for MOS LSI Circuits", *17th Design Automation Conference Proc.*, Minneapolis, MN, June, 1980, pp. 610-617.

12. B. W. Colbry and J. Soukup, "Layout Aspects of the VLSI Microprocessor Design", *Proc. of the IEEE Symp. on Circuit and Systems*, Rome, Italy, May 1982, pp. 1214-1228.

13. D. E. Blahut, et al, "Hierarchical Design Methodology for a Single Chip 32-bit Microprocessor", *IEEE Int. Conf. on Circuits and Computers Proc.*, New York, NY, Sept., 1982, pp. 16-20.

14. P. W. Diodato, et al, "CAD Construction of a VLSI Memory Management Unit", *IEEE Int. Conf. on Computer-Aided Design Proc.*, Sept., 1983, Santa Clara, CA, pp. 226-227.

15. S. Kang and W. M. vanCleemput, "Automatic PLA Synthesis from a DDL-P Description", *18th Design Automation Conf. Proc.*, Nashville, TN, June, 1981, pp. 391-396.

16. D. W. Brown, "A State-Machine Synthesizer — SMS", *18th Design Automation Conf. Proc.*, Nashville, TN, June, 1981, pp. 301-305.

17. B. Teel and D. Wilde, "A Logic Minimizer for VLSI PLA Design", *19th Design Automation Conf. Proc.*, Las Vegas, NV, June, 1982, pp. 156-162.

18. G. DeMicheli and A. Sangiovanni-Vincentelli, "Computer-Aided Synthesis of PLA Based Finite State Machines", *IEEE Int. Conf. on Computer-Aided Design Proc.*, Sept., 1983, Santa Clara, CA, pp. 154-156.

19. G. DeMicheli and M. Santomauro, "Topological Partioning of Programmable Logic Arrays", *IEEE Int. Conf. on Computer-Aided Design Proc.*, Sept., 1983, Santa Clara, CA, pp. 182-183.

20. G. DeMicheli and A. Sangiovanni-Vincentelli, "PLEASURE: A Computer Program for Simple/Multiple Constrained/Unconstrained Folding of Programmable Logic Arrays", *20th Design Automation Conf. Proc.*, Miami Beach, FL, June 1983, pp. 530-537.

LAYOUT COMPILER FOR VARIABLE ARRAY-MULTIPLIERS

N.F. Benschop

Philips Research, Eindhoven, The Netherlands

Abstract

A software layout generator for parameterized array-multipliers will be described, of which a large part is technology independent. An example for static NMOS array-multipliers will be elaborated.

Introduction

There is a pressing need for LSI design-time reduction, ever since technology allows integration at the digital system level. The pace of development in for instance digital signal processing (DSP) dictates a corresponding level of design automation, to exploit the possibilities of digitalisation in telecommunication, audio and video, on the professional as well as the consumer market. This holds not only for general purpose 'random' control logic at the gate level, where for instance symbolic layout started, but also for larger functional building blocks in high performance custom design DSP chips.

Dense and regular arrays for logic, arithmetic and memory are well suited to be generated automatically. However, loss of area and performance cannot be tolerated in high throughput DSP chips. The method to be described obtains this by combining optimized cell design with software generation of parameterized arrays. An arithmetic array for fast multiplication will be used as example.

A critical function in custom LSI for real time DSP, such as digital filtering, is fast multiplication. Parallel array multipliers of various precisions and speeds are required, depending on the application. Other variables are the amount of pipelining (latch layers) in the array, single- or double precision output, an extra adder or accumulator, rouding or truncation, and such geometrical features as desired horizontal or vertical pitch of columns and rows for a best fit with the rest of the chip.

The following sections will explain first the general approach taken to merge most easily with existing CAD, and so encourage IC designers to generalize their expertise and gain experience with parameterized arrays. Then follows a description of two's complement array-multipliers with inverting full adders, the required cells, and their NMOS implementation. And finally a simple software implementation (fortran program) of the corresponding layout compiler is outlined.

General Approach

The aim is an efficient method of algorithmic layout generation, in the form of parameterized arrays, with a minimum of changes in existing layout-CAD and a maximum cooperation from IC designers. Only one feature was missing in the in--house layout language: variable coordinates. This was easily solved by a pre-processing subroutine for conversion of parameterized cells, to be described later, so that no change in existing CAD was necessary at all. Just using a normal text editor is not sufficient since arithmetic with coordinates is required.

A layout consists of a hierarchy, or rather a tree, of patterns with the whole chip or array--design as root (top level) and the basic cells as leaves (bottom level), which are defined in terms of rectangles in the various masks. Any intermediate level may be called a pattern or network of patterns. In a parameterized or "variable" layout, some of the coordinates in cells or pattern placings in an array, are parameters or simple arithmetic expressions in the parameters. A parameter may also be a boolean variable, as a selection mechanism for the presence or absence of certain cells or patterns in the layout; or more generally: an integer to select from a number of alternatives.

Coordinate parameters and selection parameters may concern either cells or the array structure, or both. For instance, a cell-pitch parameter influences interconnect lines in the cell, and cell placings in the array. The length of a depletion load transistor in a cell is a local parameter, while the number of operand bits determining the array size is clearly a global array parameter.

For a compact array of good performance the cells are hand layed-out, including interconnect for a close fit. The array generating software can be simple if the number of different cells and cell-variants (along the array edges) is small. The general set-up is now as follows:
1. The array is logically decomposed into as few as possible different types of slices, and these slices again into a minimum number of different subpatterns or cells, with a view on iterative structure. A slice can be a row or column or a pair of rows or columns, for instance.
2. A minimal array containing all necessary types of cells is simulated, to verify proper logic behaviour.
3. The cells are layed out, including interconnec-

Reprinted from *IEEE Proc. Custom Integrated Circuits Conf.*, 1983, pp. 336–339.

tions for supply, clocks and data, to fit with their neighbours in the array. Here the minimal complete array helps to keep an overview and check all necessary connections in the layout: a cell may be used in different contexts, and must fit all of them.

4. Electrical behaviour of cells and critical paths, in a slice or the whole array, is simulated and verified for worst case conditions.

5. Now the desired parameters are built into both parts of the layout description: local parameters into the cells, and global parameters into the (array-)network description where cells and slices are placed.
The network part of the layout is software generated, and is technology independent; it reflects directly the logic design. Once the required cell types are known, they can be redesigned to another technology, without influencing the network generating software.

6. A simple routine, provided with parameter values, converts the parameterized cells with variable coordinates into fixed cells, on a line by line basis, as shown later. These cells are called by the array layout generating program, which uses the desired global parameter values for computing cell- and slice placement coordinates, in order to generate the fixed array layout file:

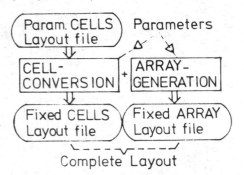

It depends on the available layout primitives whether to have a parameterized array (network-) file and convert it, like the cells, or to generate the required layout text directly in the program, with format statements. The latter is done in the present example, because the desired structural variability could not be provided by the layout language. This will be the case in most non-trivial arrays. Any higher level programming language will suffice for this purpose; in our case Fortran is used.

The sketched method will now be illustrated for a class of two's complement array multipliers.

Linear Array Multipliers

The product of two binary coded n-bit numbers $X = \sum x_i 2^i$ and $Y = \sum y_j 2^j$ is $XY = \sum \sum x_i y_j 2^{i+j}$ with $i,j = 0,..,n-1$.
The bitwise products $x_i y_j$ are the boolean AND of

x_i and y_j. In a parallel multiplier all n^2 product bits are produced simultaneously by as many AND-gates at the crossings of two orthogonal sets of bitlines that distribute the two input operands. Reduction of the n^2 bits to the 2n-bit product is obtained by repeated additions with many full adder cells in the plane. There are parallel array multipliers of various types, depending on the required speed: tree structure (Wallace[1]), array with bit-pair precoding of one input operand (Booth[2]), and linear carry-save array with adder/subtractor cells (Pezaris[3]) or with adders only (Benschop[4]). The latter type is shown in figure 1.

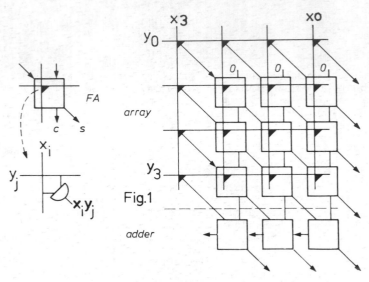

Fig.1

These types are mentioned in the order of increasing regularity and logic depth (decreasing speed). To double the speed of the most regular last type, the full adder cells are made inverting, by deleting the output inverters following the normally inverting sum- and carry- function gates. The consequence is a slight complication in the input AND matrix, where now alternating rows of AND and NAND gates must be used. Alternate rows process then normal and inverted signals respectively. Also, at the array output edges the polarity must be put right in every other bit-position. Without going into much detail here (see reference 4) a two's complement array multiplier is obtained by inverting both sign bits, change gates in the leftmost column into alternating NOR/OR gates, and the bottom row into NOR's, except the left/bottom corner which must be an AND gate. The whole array is very regular and compact, and has only 3 dissipating loads per cell (product-gate, sum, carry). The non-inverting AND and OR are 'source-follower' gates, with zero-threshold driver transistors pulling up, and a depletion load pulling down to ground. For the array-structure, cell logic and transistor circuit see figures 2, 3 and 4.

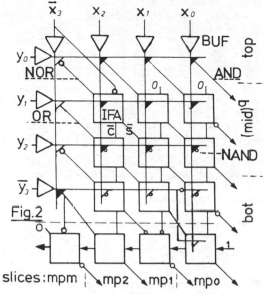

Fig.2

slices: mpm   mp2   mp1   mpo

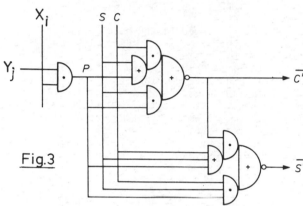

Fig.3

$$\overline{C'} = \overline{(s+p)c + sp}$$
$$\overline{S'} = \overline{(p+s+c)\,\overline{c'} + psc}$$

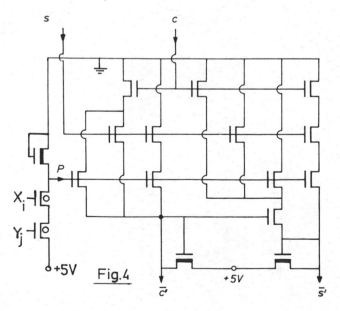

Fig.4

Array Decomposition

The array is partitioned columnwise into four different types of slices: a least significant (rightmost) column slice MP0, a most significant (leftmost) column slice MPM, and intermediate odd- and even type column slices MP1 and MP2 which differ only in their bottom output cell, due to the inverting full adders. Also the alternating rows are taken pairwise. Consequently, the operands must have an even number of bits, although this is not an essential restriction.

Similarly, each slice is decomposed into a TOP, MIDdle and BOTtom pattern . TOP contains two input buffers for x_i and y_i and one AND gate. The input operands are assumed to come from above, but other input connections, for instance as in the figure, can equally well be chosen. BOT contains a bottom row NOR+IFA cell and a carry-ripple adder cell under the array. MID contains cells from two adjacent even and odd rows (AND+IFA and NAND+IFA), so that iteration suffices to generate the middle part of a column slice.

Schematically the array structure can then be described as a slice sequence MPM(MP2,MP1)^{p}MP0 and each slice as a pattern sequence TOP$_i$ (MID$_i$)^{q}BOT$_i$ with i$\in$ {M,2,1,0} and p,q $\geqslant$ 0. This results in a 2(p+1) x 2(q+1) bit array multiplier, with an output precision of 2(p+q+2) bits.

Performance

The inverting full adder cell, implemented in 2 μ static NMOS, measures 60 x 45 μ^2 and has two loads that draw 60 μA when active. So the average dissipation with half the loads active is 0.3 mW at 5V supply voltage. Including the product gate, the basic array cell measures 90 x 45 μ^2 with about 0.45 mW dissipation. The gate delay with a fanout of 4 is typically 2.5 ns, so the carry delay is 2.5 ns and the sum delay is 5 ns. The longest path through an m x n array of m columns and n rows is n sum delays (along a diagonal) and m carry delays along the bottom adder. The total delay is then (m + 2n) x 2.5 nsec. For instance a 12 x 12 bit array multiplier has 90 ns delay, and the dissipation of about 11 x 12 cells, which is some 60 mW, excluding I/O buffers. For non square arrays, the largest precision should determine the number of columns (carry delays of one gate each), and the smallest precision the number of rows (sum delays of two gates each) to obtain the fastest multiplier.

Array multipliers of this type are used in the DFT chip described by v. Meerbergen[5], which contains an 8x8 and a 12 x 12 bit array, and in a digital audio filter chip[6] with a 12 x 16 array. One or two rows of latches have been used for pipelining to increase the speed. The present layout compiler has one pipeline option as a parameter: a row of latches between the array and the bottom adder, decreasing the clock period to the maximum of m and 2n gate-delays.

Software

The layout compiler, apart from the parame-

terized cell library, consists of three main parts (each comprising about one page of Fortran code):
. User interface: parameter input, and array characterization output.
. Cell conversion: replaces variable coordinates by fixed coordinates.
. Array generation: computes cell placements and generates layout text.

— The user interface asks the designer for the required parameters, mentions and tests for their type and range, and generates appropriate error messages if necessary. Default parameters for an 4 x 4 array multiplier can be used to generate a layout plot or color graphics display to familiarize the designer with the array's structure. A compact listing of area, speed and dissipation is produced, with their dependence on the various parameters.

— Cell conversion: variable coordinates in a cell layout description are identified by a special character, for instance "#", and have the following format: $\pm$#i $\pm$ d, where "i" is an integer indicating the i-th parameter, and "d" is a displacement (real) in microns. This format appeared to cover the required variability. For instance, the interconnections between cells, for supply and data, can be stretched in x and/or y-direction according to the parameters XPITCH and YPITCH. This obviates the need for extra "filler" patterns. The routine reads a line, scans for #, and replaces each variable coordinate by the value of the parameter expression, taking into account the resolution.

— Array generation: various parameters can be built in, according to the expected applications. The present layout compiler has 9 parameters (4 integers and 5 booleans). They are:
  2 operand precisions XBITS, YBITS (even, from 2 to 24)
  2 pitches XPITCH, YPITCH (for streching > 90 μ resp. 45 μ)
y/n adder (X.Y + Z)
y/n accumulator (sum Xi.Yi)
y/n full precision output (including least signif. YBITS positions)
y/n pipeline latch layer (just above bottom adders)
y/n rounding (vs. truncation to most signif. XBITS positions)

The described array decomposition into column slices, and the slices into TOP/MID/BOT patterns, makes generation of the required layout text quite simple. Some intermediate variables are computed from the given parameters, for instance the number of copies of the middle patterns and slices, and their placings.

The layout compiler should also generate transistor-circuit and logic network descriptions for circuit resp. logic simulation as part of a larger design. Such an extension, though not trivial, could well benefit from the array decomposition on which the layout compiler is based.

References

1. C.S. Wallace: "A Suggestion for Parallel Multipliers", IEEE Tr. El. Computers, EC-13, p14-17 (Feb 1964).
2. A.D. Booth: "A Signed Binary Multiplication Algorithm", Quart. J. Mech. Appl. Math, V4, Pt2, p236-240 (1951).
3. L.R. Rabiner, B. Gold: "Theory and Application of Digital Signal Processing", p 523, Prentice-Hall, N.J. (1975).
4. N.F. Benschop, L.C. Pfennings: "Compact NMOS Array Multipliers with Inverting Full Adders", Philips J. Res. V36, p 173-194 (1981).
5. J. van Meerbergen, F. van Wijk: "A 2 μ NMOS 256-point Discrete Fourier Transform Processor", ISSCC 1983, Digest p 124-125.
6. D. Goedhart, R. v.d. Plassche, E.F. Stikvoort: "Compact Disc - de omzetting van digitaal in analoog bij het afspelen" (Dutch), Philips Techn. Tijdschrift, V40, p 290 (1982).

A Technology Independent MOS Multiplier Generator

Kung-chao Chu
Ramautar Sharma

Bell Laboratories
Murray Hill, New Jersey 07974

ABSTRACT

A layout generator for technology independent implementation of the MOS multiplier is described. The modified Booth's algorithm with a structured floor plan has been used. The layout has been optimized and described as a program in a high level layout language. The fabrication process related information is maintained in a separate technology database that is coupled with the layout program at the time of execution to generate the mask data. The user can choose from a variety of architectures for speed, area, and power trade-off's. The user can also specify geometric and electrical constraints tailored to his system specification.

1. INTRODUCTION

VLSI technology has provided a great impetus for the design of single chip digital information processors. These processors can be either general purpose or customized, but they generally tend to have modules of memory, arithmetic and logic unit, PLAs, multipliers, and some random logic to glue these modules together. To find an optimal speed and area performance, various architectures involving different modules have to be tried out. Since it is time consuming to provide every possible layout of each module, one approach to provide fast and efficient design is to use parameterized modules. These parameterized modules are created by software packages called generators or silicon compilers.

In this paper, a layout generator for MOS multipliers is described. Benschop also described a generator for NMOS multipliers using parallel linear carry-save array of adders [1]. The basic idea of using dense and regular arrays is good, but his multiplier is slow because of the large logic depth in carry-save adder chain. However, a careful design of data path and floor plan can produce fast multipliers and it is incorporated in our generator.

The circuit design and layout depend heavily on the processing technology, therefore currently available generators also depend on it. In the present generator, technology independence is achieved by combining the process definition database with a layout language. The multiplier architecture is designed to offer many options like sign-magnitude or two's complement, pipelining, different data precision, and data flow. Another novel feature of the generator is the choice of three circuit styles for different area, speed, and power requirements. This is achieved by doing circuit design and layout in a higher level of abstraction.

The multiplier architecture and the corresponding basic cells are described in the Section 2. The details of the layout language and generator architecture are given in the Section 3. Finally, we describe an example and the generator specifications in the Section 4.

2. MULTIPLIER ARCHITECTURE AND BASIC CELLS

The parallel multiplication is done by generating the partial products and adding them together. Thus the speed of

Reprinted from *ACM/IEEE 21st Design Automation Conf. Proc.*, 1984, pp. 90–97.

multiplication depends on the number of additions, and the generation and propagation delays of the sum and carry signals. Although, a generator using conventional carry-save adder scheme is easy to implement because it consists of identical cells, the number of additions (and hence the adders) needed to complete the multiplication is high. To speed up the operation, several techniques such as Wallace's tree structure, and Booth's encoding scheme have been proposed [2]. These schemes require complicated interconnections that introduce more signal delay, and involve more routing for the generator.

The modified Booth's algorithm has been used in the generator because it requires only half the number of adders needed in the conventional carry-save adder array design. The basic cell has been designed in such a way that many variations of interconnections and data flow can be realized without changing the basic architecture and the floor plan of the multiplier in Figure 1. The multiplicand x and the multiplier y are first latched into input registers; the y input register then feeds an encoding circuit to generate control signals. There are $(n + 1)/2$ encoders for n-bit y input. For m-bit x input, the core of the multiplier consists of $(m + 1) \times (n + 1)/2$ identical cells except those at the boundary. A typical cell is shown in Figure 2(a). The encoder generates the control signals one[i], two[i], and sign[i] according to the following equations.

$$one[i] = y[i] \oplus y[i-1] \tag{1}$$

$$two[i] = \overline{y[i+1]} \cdot y[i] \cdot y[i-1] + y[i+1] \cdot \overline{y[i]} \cdot \overline{y[i-1]} \tag{2}$$

$$sign[i] = y[i+1] \tag{3}$$

The operation of the basic cell at ith row and jth column can be described by the Eqns.(4) through (6).

$$dummy[i,j] = sign[i] \oplus (one[i] \cdot x[j] + two[i] \cdot x[j-1]) \tag{4}$$

$$sum[i,j] = sum[i-1,j+2] \oplus dummy[i,j] \oplus carry[i-1,j+1] \tag{5}$$

$$carry[i,j] = (sum[i-1,j+2] + dummy[i,j]) \cdot carry[i-1,j+1] + sum[i-1,j+2] \cdot dummy[i,j] \tag{6}$$

For a 16x16 multiplier 136 such cells form the main body of the multiplier. The carry signal flows through 8 adder stages from top to bottom, except the last carry look ahead circuit. The cell can be modified to provide the multiplication without the carry look ahead circuit, by replacing the term carry[i-1,j+1] with carry[i,j-1] on the right hand sides of Eqns.(5) and (6). The resulting multiplier will be smaller but slower. The new carry signal flows from right to left through each ripple carry adder stage, the flow of sum signal remains unchanged as shown in Figure 2(b). With the ripple carry scheme, two bits of the product come out of each stage. The remaining higher order bits come out of the last stage.

To increase the throughput, pipelining can be used at each stage or at some other convenient places determined by the inherent delay of the stages. For instance, the portions enclosed by dotted lines in Figure 1 show the additional delay elements needed for the synchronization of the data flow.

3. LAYOUT GENERATORS

Recently, generators for custom NMOS microcomputer have been discussed [3]. The multiplier generator extended the methodology to the MOS technology independent domain.

The multiplier generator is a C program that calls other programs hierarchically, and outputs circuit layout descriptions according to user's specification. There are two kinds of information that can be fed into the generator. First, the user can control the architecture and logic design through generator specification files. These files also allow some geometric and electrical constraint specifications. Second, the technology information is described in a separate set of files called the technology database tailored to the details of the processing technology. Currently, there are four different NMOS and CMOS technology databases.

When the generator is executed, it reads both the technology database and the user specification files. The specifications are checked against the technology for violations and the user is advised accordingly. The layout design rules are embedded in the technology database and can be changed without modifying the generator program. This feature makes the generator technology updatable.

The outputs of the generator are in the format of Si language [4] that is a technology independent version of hierarchical circuit layout description language i [5]. Si is integrated in a VLSI design database system called Vdd [4], and shares the same technology database with the generators. The basic primitives in Si include transistors, wires, contacts, cells, and terminals. The wires can be made of compound mask levels. Because Si can specify transistors and compound wires directly, it greatly simplifies the coding of the generators. The Si parser is capable of resolving geometric constraints, and minimizes the need of the absolute coordinates in a Si description. A looping mechanism is also provided through the FOR statements in Si. Table 1 shows an Si language description of a static CMOS inverter. To provide numerical computation and conditional testing, the Si circuit description is embedded in C program to form a language called ic [3]. The ic preprocessor not only recognizes Si text from C program, it also allows variables in C to be inserted in Si description.

To reduce the area and increase the speed, the cell layout is hand packed. The layout of the cells can either be described directly in Si language or first manipulated with the color graphic layout editor vv of Vdd system and translated into Si description at the end of editing. In either case, the Si texts are edited to minimize the coordinate variables and constraints. The remaining variables are then replaced by C variables when ic texts are created.

The C variables are derived from the user specifications and the technology dependent design rules to satisfy the geometric constraints. To satisfy the electrical constraints, the lower level cells and critical paths are simulated with circuit simulator to find empirical formula for relations between user specification and parameters like sizes of transistors, and widths of power lines.

The floor plan of the multiplier is such that the global layout is controlled by architecture instead of circuit style. There are three circuit styles to choose from. For low power consumption, static logic can be used. If the application needs fast multiplication, dynamic circuit with precharged pull up device is the choice. A third variation adds a ground switch to the pull down device in the dynamic circuit to provide high speed and cut down the power consumption.

The coding has been simplified by observing that the design of the circuit can be broken down into two parts, viz., load and logic devices. For NMOS circuit, the load device is either a depletion mode transistor or a clocked enhancement mode transistor or a combination of both. For dynamic CMOS circuit, the load device is a clocked p-channel transistor. In both NMOS and dynamic CMOS circuits, the logic is performed by n-channel devices. Therefore, for NMOS and dynamic CMOS circuit the idea of technology independent can be implemented readily because the circuits are similar and the extra space needed in CMOS for the isolation of p_channel device from n_channel device is used by routing in NMOS layout. For static CMOS circuit, the load is the complement of the logic devices, and this part of the layout is coded separately.

Figure 3 shows the organization of the MOS multiplier generator. It is a hierarchical system of generators, and Unix shell scripts are used to communicate between them. Each generator takes two input files and writes to two output files. One of the input file is the technology file and contains the information about design rules also. Right now, the technology dependent circuit design information is embedded in the generators instead of the technology file. The second input file contains derived user specifications propagated through the generator hierarchies. The generator can derive the same information from the original user specifications, but that would make the generator dependent on the multiplier architecture. By having a separate specification, the same generator can be used for other high level generators. For instance, a single generator was used for the two inverter generators in Figure 3. The third input file, which is also one of the output files from lower level generators, provides feedback from lower level layout. The other output file contains layout descriptions. The mask generator takes layout descriptions and technology database to produce mask data.

Figure 4 illustrates how the geometric and electrical constraints can be embedded in the generators. The two modified Booth's encoders shown have been generated by a Booth's encoder generator. Figure 4(a) is for a 8x8 static CMOS multiplier, and Figure 4(b) for a 16x16 static CMOS multiplier. The driving capability of the control signals are doubled in Figure 4(b). The metal power lines for the drivers are also doubled in width to avoid voltage drop and metal migration problems. The layout of the driver is such that the increase in driving capability will not stretch the vertical dimension of the layout. It is this kind careful layout consideration that enables the generator to satisfy both the geometric and electrical constraints. Another important issue is the substrate contacts for the CMOS technology. Different starting materials will require different types of tubs (or wells) for the transistors. The generator always keep the substrate contacts close to the transistors such that only small islands of tubs are needed.

4. GENERATOR SPECIFICATIONS

The generator specification file has the following parameters:

- technology : process type and design rules.

- power : allowed power dissipation.

- circuit type : static or dynamic with/without ground switch.

- numerical system : 2's complement or sign magnitude.

- inputs : number of bits of the multiplicand and multiplier.

- outputs : truncated or rounded, with/without accumulation.

- pipelining : number and locations.

- mode of operation : ripple or carry look ahead adder.

- geometric constraints : inputs and outputs positions.

- logic simulation options : with/without logic simulation file.

The logic simulation of the multiplier is done by a switch level simulator [6]. Since the switch level simulation is independent of the details of the layout, we are able to test the correctness of the logic design efficiently.

The generator can also be used in an interactive mode where the program helps the user in selecting a set of consistent specifications. The generator checks the parameters like the number of bits in the input and the output, stage delays, and the number and the location of the pipeline, allowed power dissipation and the circuit design style, etc. It is up to the user to choose a consistent set of specifications. The multiplier can be used as a component on a silicon chip, or the frame generator in Vdd can be used to design a single chip multiplier.

A typical layout for the multiplier as produced by the generator with 3 micron normalized CMOS design rules is shown in Figure 5. It is a 8x8 multiplier using ripple adder and static circuit style; the product is neither truncated nor rounded and there is no pipelining. There is also no accumulation of the product, and the inputs and outputs are distributed evenly around its four sides.

5. DISCUSSION

The timing simulations [7] on the circuit and device levels have only been done for the primitive cells and critical paths. Empirical equations are used to estimate the electrical and geometrical constraints. It would be a great improvement if capacitance extraction and circuit simulation could be done more precisely and efficiently inside the generators. It is hard to predict how far technology independent generators can be pushed. On the surface, technology differences only propagate to the circuit design level. In actuality, many architecture decisions have to be made according to what the technology can do. It would be very hard for a silicon compiler to make decisions if the components were not well characterized.

6. CONCLUSION

A generator for automatic layout of multiplier modules has been developed. The multiplier uses modified Booth's algorithm for area and speed efficiency. The generator is technology independent in MOS domain, and user definable through the generator specification files. Beside the usual parameters required for a multiplier, different circuit styles and data flow are provided.

REFERENCES

[1] N. F. Benschop, "Layout Compiler for Variable Array-Multipliers," Proc. Custom Integrated Circuits Conf., May 1983, pp.336-339

[2] K. Hwang, Computer Arithmetics: Principles, Architecture and Design, Wiley, New York, 1979.

[3] T. G. Matheson, M. R. Buric, C. Christensen, "Embedding Electrical and Geometric Constraints in Hierarchical Circuit-Layout generators," Digest of Technical Papers, ICCAD 1983, pp.3-5.

[4] K. C. Chu, J. P. Fishburn, P. Honeyman, Y. E. Lien, "Vdd - A VLSI Design Database System," Proc. ACM/IEEE Database Week, San Jose, CA. May 1983, pp.25-37.

[5] S. C. Johnson, "Hierarchical Design Validation Based on Rectangles," Proc. Conf. on Advanced Research in VLSI, MIT, Jan. 1982, pp.97-100.

[6] T. G. Szymanski, Private Communication.

[7] A. K. Bose, B. R. Chawla, H. K. Gummel, "A VLSI Design system," IEEE 1983 International Symposium on Circuits and Systems Proceedings, pp. 734-739, 1983.

Table 1.

Si Description of a Static CMOS Inverter

```
SYM inv
    {
    TP (4,2) tp;
    TN (2,2) tn;
    tn.gw LEFT 4 TO tp.ge;
    INPUT POLY in tp.gw;
    OUTPUT METAL out;
    tn.dn UP 1 RIGHT 1 MDN LEFT 6 TO
    out LEFT 6 MDP DOWN TO tp.dn;
    MDNS %90 gndx;
    MDPS %-90 vddx;
    XDEF METAL gnd0 gndx.ms S=gnd;
    XDEF METAL vdd0 vddx.ms S=vdd;
    tn.ds DOWN 1 RIGHT 1 TO gndx.d;
    tp.ds DOWN 1 TO vddx.d;
    }
```

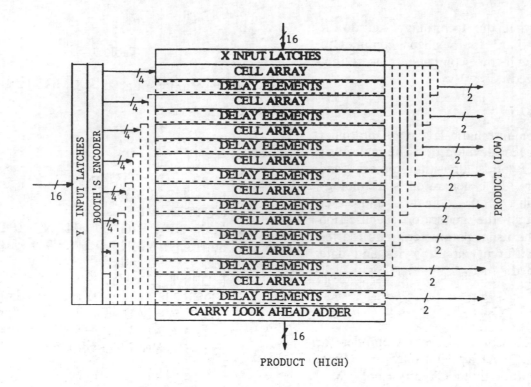

Figure 1. Fully Pipelined m x 16 Multiplier

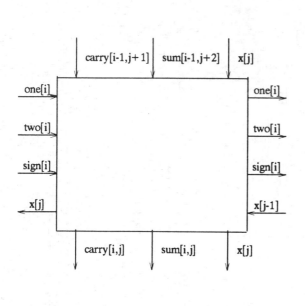

Figure 2(a). Typical Cell for Multiplier

with Carry Look Ahead Adder

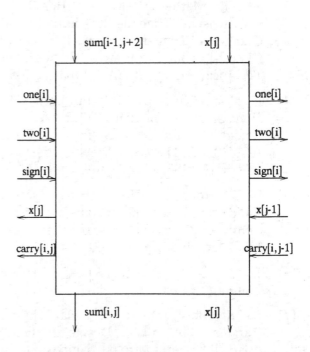

Figure 2(b). Typical Cell for Multiplier

with Ripple Adder

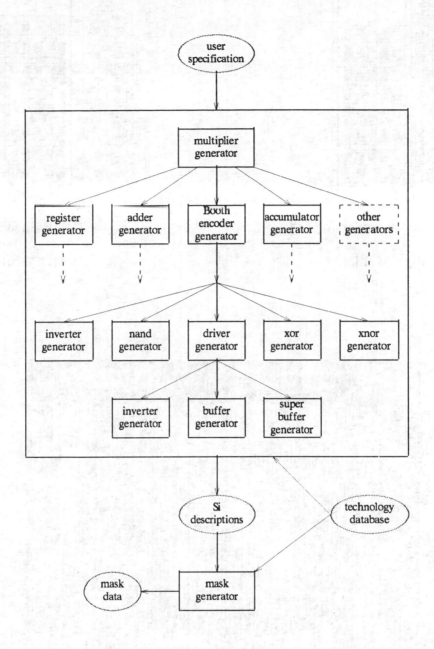

Figure 3. Organization of MOS Multiplier Generator

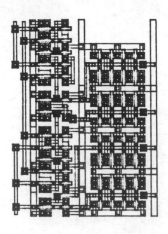

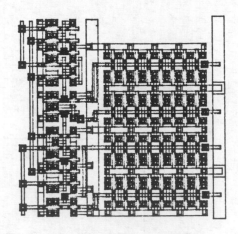

Figure 4(a). Booth Encoder for 8x8

Static CMOS Multiplier

Figure 4(b). Booth Encoder for 16x16

Static CMOS Multiplier

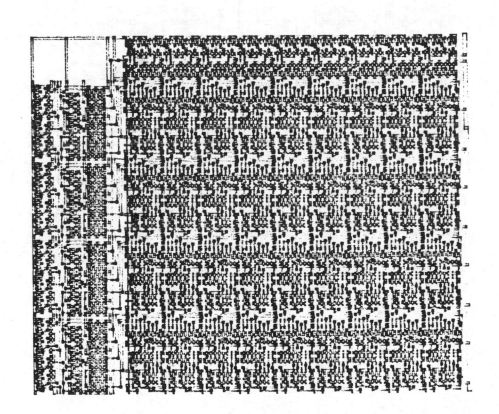

Figure 5. A 8x8 Static CMOS Multiplier

CMOS CELL-LAYOUT COMPILERS FOR CUSTOM IC DESIGN

Duane Edgington, Bill Walker, Scott Nance, Chris Starr,
Suresh Dholakia and Mike Kliment

VLSI Technology, Inc. (VTI)
1101 McKay Drive, San Jose, CA   95131

Abstract

This paper describes the design methodology and software used to implement a family of CMOS cell compilers: computer programs that generate layout of user-configurable logic functions, which are used to build custom integrated circuits.

Introduction

Techniques for standard cell-based design have been in existence for many years, and have begun to gain acceptance in the semiconductor industry. While the technologies used have changed radically over time, the basic concept and structure of standard cell libraries has remained virtually unchanged. In a standard cell-based design, all the physical layout of function blocks comes from previously defined data bases for those blocks (thus, "standard cells"); the custom aspect of the design is in the interconnect between the standard cells. With this approach, a system designer can generate a custom IC quickly with a high probability of first silicon success. Scarce custom engineering and designer talents can be focused on development of the standard cell libraries, which are used in many designs.

A drawback of standard cell libraries is that a large number of cells must be available to do even a limited range of circuit design applications. Since each standard cell has fixed geometry and performance, typically hundreds of cells must be available in a library to handle the range of function, load and speed encountered in an IC. This volume of cells places considerable burden on the engineering staff responsible for building the library; even if advanced CAD tools are used, each standard cell must be independently designed, drawn, characterized and documented [1]. If the data base for each cell is fixed, changes to incorporate technology improvements can be very time-consuming.

As an alternative to standard cells, we at VTI have developed cell compilers. Cell compilers are computer programs written in a high level procedural language [2]. Each program accepts parameters from the user that specify the desired configuration of a particular cell. The program then constructs and outputs the appropriate physical geometry for that cell, in CIF [3] format. The advantages of cell compilers over standard cells are (1) a much smaller number of cells is needed to address a wide range of function, speed and load requirements, and (2) fewer cells means less engineering time is needed to build a useful library, or to incorporate technology upgrades. The only disadvantage of cell compilers is that they are each computer programs, often of considerable sophistication.

This paper describes a new method of implementing "cell-layout compilers" [2] that allow users to generate a large number of customized cell configurations from a small number of cell compilers. We summarize design methods and software development tools that we used to build the VTI CMOS cell library.

Cell Compiler Software Development Tools

Cell compilers in the VTI CMOS cell library were designed by developing a set of fixed-geometry subcells, using a graphics-system layout editor. Cells are formed by placing subcells side-by-side, and by overlaying subcells.

Assembly of subcells is controlled by the cell compiler programs. To simplify coding and testing the programs, we developed a high-level cell-assembly language. The language gathers the user's parameters; instructions to the language control which subcells are drawn, where they are drawn, and when they are drawn, based upon the user's functional and behavioral parameters to the cell.

The cell-assembly language is an extension of VIP [2,5], a procedural language developed at VTI. In VIP, goemetries are described in text; the size and location of geometries can be controlled by parameters passed to VIP at run-time. In a VIP program, one typically encodes small portions of a cell as subcells, defined within a procedure or macro. The cell is constructed by calling these lower-level procedures, and then wiring between

Reprinted from *IEEE Proc. Custom Integrated Circuits Conf.*, 1984, pp. 512–517.

the subcells. Our extension to VIP consists of a set of high-level procedures, optimized for placement of subcells. High level statements control generation of rows or arrays of subcells, and support extensive error checking and reporting.

Features were built into the high level cell-assembly language to allow computation of buffer sizing from user's specification of load and speed requirements, which optimize the area used to achieve the required performance. Each cell has associated with it a performance table, that specifies a set of parasitic capacitances and drive capability for its internal subcells. Standard internal buffer subcells have their own performance tables. A standard program within the library reads these performance tables; any cell compiler can call this program with the user's load and speed requirements - the program returns a parameter that tells the calling cell which standard buffer cell to draw.

Circuit Design Requirements and Methodology

General Design Considerations

Designing any cell based system, whether standard cells or cell compilers, require setting up goals and priorities. These serve as a foundation to decide design trade-offs that occur when conflicts arise. Careful consideration must be given when picking design goals; for example, picking too high a clock rate could cause the cells to be obsessively large. Factors that should be considered include what size die these cells are going to be used in and what performance is necessary.

The following is a list of design requirements and priorities for VTI's CMOS cell compilers:

(1) Development cycle for cells should be short.
(2) Technology will be two layer metal, 3 micron CMOS.
(3) The cell compilers will be integrated with VTI's design tools.
(4) The cells should be simple enough to be used by a novice circuit designer (someone with no previous IC design experience). The methodology should incorporate SCR protection to avoid latch-up.
(5) The cell compilers should allow the more sophisticated designer to make the most efficient use of silicon area and tailor the layout to their needs.
(6) The cell compilers should allow 20 MHz clock operation with a single clock input.
(7) The behavior of the cell compilers should emulate, when feasible, the functions of TTL standard ICs, allowing users to

quickly convert their designs.
(8) The cell compilers must be compatible with a fully automated auto-route system.
(9) Minimize work needed to modify cells for a change in CMOS process.

Design Philosophy

The design methodology we developed was driven by the use of CMOS technology. In our previous HMOS cell compiler family [2] the size of most transistors were parameters to each cell. While parameterization of transistors sizes did not save size in many of the cells, it did allow optimization of speed and power consumption. For the CMOS library we found that the difference in area was small, whether we designed to achieve 20 MHz performance or whether the same logic function was implemented using minimum size transistors. Since there is negligible power savings gained from parameterizing the transistors in CMOS we decided to parameterize transistor sizes only in special cases.

The design methodology we chose was to implement a set of building blocks that met the 20 MHz performance goal, and then use our procedural design language to select and assemble the correct building blocks based upon the user's inputs.

Standard Function Blocks

After the ground rules are defined, cells and functions are implemented. A typical design cycle for a cell could be two to three weeks, with two circuit design engineers and one layout engineer. To meet our goals of short design cycle and building a large library we could not build cells in the classical way. The method we chose was to break up the large functions like counter, serial/parallel shift registers, and so forth, into their basic subfunctions. By making these subfunctions general enough to be used in several cells, they could be designed and drawn only once.

The range of complexity of these Standard Function Blocks varies within the library. Their simplest form is as interface cells, well caps for SCR protection, or inverters; more sophisticated subcells include complete NAND gates and clock buffers that automatically scale their speed and drive performance.

Standard Function Blocks are the lowest-level logic functions that can be isolated. These Standard Function Blocks can be designed and characterized once, then used to build several different complex functions. Similar functions are grouped into family types; all members of a family are designed to interface together. They use the same busing scheme, cellheight and input and output configurations. They are usually directly interchangeable with any other cell that provides the same function but may vary internally depending upon the implementation. An illustration of a Standard Function Block and family group are the shift register bits shown in Figure 1. The Shift Register Bit family consists of four members: static, static with reset and clear, dynamic, and dynamic with reset and clear. They were all designed to meet the same performance goals and they are all interchangeable. Thus, if they were used in a counter, four entirely different versions of the counter could be created by simply substituting the different register bits. Currently roughly 60 standard function blocks are used in 53 cell compilers that expand into over a 1000 unique combinations of physical layout. Some of the types of families of Standard Function Blocks that currently are used include NAND, NOR, latches, registers, internal buffers, external I/O buffers (CMOS and TTL) and pad cells.

Special Function Blocks Expanded into Intelligent Modules

One of the powerful ramifications of using a procedural language to implement our cell compilers is the ability to process equations and direct the layout based upon results from these calculations. Parameters used in the equations may be internal or external to the program, exist in technology files, or may be passed directly from the user to the cell when the cell is invoked.

An example of this feature is the clock buffer Standard Function Blocks. In the VTI CMOS cell compiler library we decided to use a single clock scheme in which one clock is bused throughout the chip; each cell compiler locally generates a buffered clock signal (and clock bar if necessary) for internal logic. This puts strict demands on the clock buffer's performance, to guarantee that all clock buffers generate consistent and compatible timing so no data corruption occurs. Normally the burden of auditing the clock delays for different clocks would be on the user, but the use of equations and parameters that are called when a cell compiler is invoked automatically maintains the desired performance set by the user.

Performance Parameterization by Standard Function Blocks

The performance of the output from any cell compiler can be customized for load and speed requirements. This performance scaling is accomplished by the cell calling a central buffer-sizing program. The equations within this program predict the performance for various loads and select the correct set of Standard Function Blocks from the buffer family of subcells to meet the performance requirements.

Also, the output can be inverted to provide the signal and its compliment. The outputs of all cells are designed to interface to the buffer family of subcells. The performance degrades minimally with the addition of more drive capability. The user can also

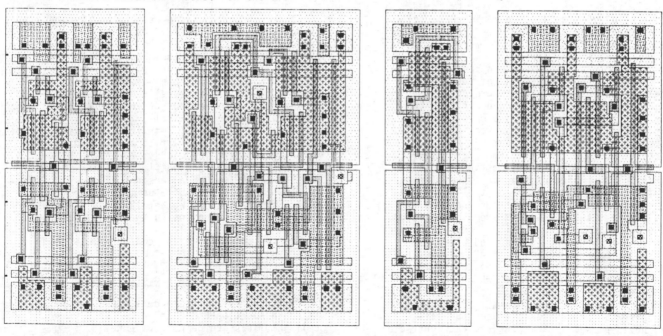

Static   Static with set and clear.   Dynamic   Dynamic with set and clear.

Figure 1. Versions of shift register bit Standard Function Blocks.

manually override the auto sizing and place a specific buffer if desired, or no buffer at all, if the base Standard Function Block cell drive is adequate. The more sophisticated user can use the manual feature to optimize area and speed, relying on performance curves in the data sheet of each cell, or relying on circuit simulation.

Creating Complex Functions from Standard Function Blocks

More complex functions like Counters and serial/parallel shift registers were designed by using as many of the Standard Function Blocks as possible; these Standard Function Blocks were combined with "glue" cells that contain the special logic required to generate the desired function. Signals that are used for control are routed in wiring channels above and below the cell's power buses.

An example of the generation of complex functions is the cell compiler CCTR00, an up counter. The counter uses the static shift register bit which was shown previously. Glue cells consisted of two versions of carry logic, toggle control, and parallel load interface. The carry cells are used to enable the counter and to propagate carry information to the next bit. The toggle control takes the current bit count, and carry information, and decides if the bit will toggle on the next clock, thus incrementing the count value. The parallel load feature allows external data to be loaded directly into the register bit, bypassing the toggle cell. The glue cell also brings connection points into the routing channels above and below the cell to be used as reference points when interconnect wiring is done.

Parameterization is added to the cell by creating different versions of the glue cells: for example, three versions of parallel load subcells were designed. One version brings data into the counter from an external input preset source, the second version ties the preset source internally to ground, and the last ties the preset source to VDD. The counter utilizes these three versions to implement four cell compiler options. The options are controlled by parameters passed to the cell when the compiler is invoked; two parameters are used to control the four options for parallel load. They are CTR00 PARALLEL LOAD FEATURE and CTR00 PARALLEL LOAD SOURCE. The first parameter determines if the option is to be placed, the second parameter determines which version of the Standard Function Block cell for the parallel load is to be drawn. Since the counter can be N-bits long, the parameter value is passed as a string and each character in the string determines which parallel load Standard Function Block is placed. The parameter string "10D", for example, would place the load source to VDD, VSS, and External, respectively, from the least to the most

significant bit.

Other parameterized features can be created in a similiar fashion. Some features that exist in the library are active high or low signal levels, outputs on rising or falling edge of the clock, number of bits, number of inputs and inverting or non-inverting outputs.

Construction of Cell Compilers

Once (1) all the parameters have been decided, (2) logic functions defined, and (3) Standard Function Block cells and glue cells drawn, the cell compiler can be written. As an example, the counter CCTR00 is shown below, with default parameters for a two bit counter (Figure 2). First, the cell compiler draws the front end cells that will provide the clock and control signals. These Standard Function Block cells are selected automatically to meet the performance necessary to drive the specified number of bits. Once the clock buffers and signal buffers are drawn the cell compiler generates the first unique carry buffer. Next the first bit is assembled. The first bit is then repeated N times, with the appropriate parallel load subcell substituted for each bit depending upon the value passed for CCTR00 PARALLEL LOAD SOURCE. Once all the base cells are placed, the interconnect is completed by point-to-point wire statements, and all outputs are labeled to reflect their relative bit placement (eg. "Q1").

The real power of the methodology becomes evident when the counter is expanded to include new functions like down count, and up/down count, because these features can be added by creating new glue cells to perform these functions. The names of the new Standard Function Block cells can be substituted for the previous defined up counter Standard Function Block cells and with little other change a new cell compiler is created. This avoids having to rewrite complex software for similiar functions. Further, by substituting different family types for the register bits four more versions of each counter can be created. Thus from three basic counter designs we get 12 useful counter types (Figure 3) and the software was generated basically only once. This allowed the 12 counter to be designed and coded in less than ten days.

Simulation and Documentation

Performance goals are designed into the Standard Function Block and glue cells when cell compilers are being configured. Using the completed cell compilers, test cases are generated and netlists are extracted using VTIextract; the resulting netlists, which include precise transistor size and stray capacitance information, are converted into

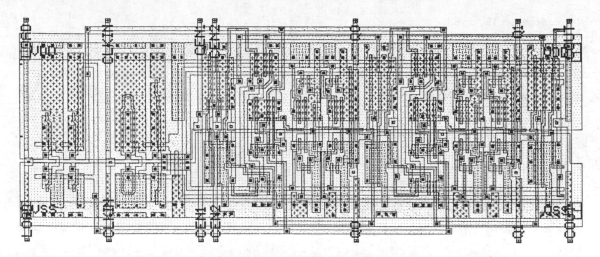

Figure 2. Two bit static up counter.

CCTR00	N bit static up counter
CCTR01	N bit static up counter with set and clear
CCTR10	N bit dynamic up counter
CCTR11	N bit dynamic up counter with set and clear
CCTR20	N bit static down counter
CCTR21	N bit static down counter with set and clear
CCTR30	N bit dynamic down counter
CCTR31	N bit dynamic down counter with set and clear
CCTR40	N bit static up/down counter
CCTR41	N bit static up/down counter with set and clear
CCTR50	N bit dynamic up/down counter
CCTR51	N bit dynamic up/down counter with set and clear

Figure 3. Counters in the CMOS cell compiler library.

ASPEC circuit simulation files using VTIsimasp. VTIextract and VTIsimasp are programs in the VTI design system. Performance is characterized and information is derived to calibrate the equations used by each cell compiler. Data sheets are then generated in a semi-automatic fashion by using software tools to extract parameter lists from technology files; descriptions, performance data and schematic information are added. Due to the consistancy and uniform methodology used to design the cell compilers, cells of the same family type can utilize very similar data sheets. Thus, the documentation effort is greatly reduced.

Ease of Use

Depending on the cell compiler, there could be several parameters that would have to be specified when a compiler is invoked. It would be a burden on the user if they had to specify the value of each parameter every time they used a cell. To make the cells easier to use, the library is integrated into a window inter-face in the VTI design tools [4].

The cell library window is menu driven and allows the user to view and modify parameter before a cell is compiled. The parameters have default values that should meet most needs; the legal range of parameter values is also displayed, as illustrated in Figure 4.

Conclusions

This design methodology enabled us to develop an extensive library of CMOS cells relatively quickly; by focusing effort on developing high level general software and design tools, we were able to minimize the development effort required for any particular cell.

References

1. N. Tosuntikool and C. L. Saxe, "Automated Design of Standard Cells", IEEE Custom Integrated Circuits Conference, Rochester, NY, May 1983.

2. S. Nance, C. Starr, B. Duyn and M. Kliment, "Cell-layout Compilers Simplify Custom-IC Design", EDN,

September 15, 1983.

3. C. Mead and L. Conway, Intro-
    duction to VLSI Systems,
    Addison-Wesley, 1980.

4. B. Duyn and S. Trimberger,
    "Structured-design system takes
    over the complexities in VLSI
    circuits", Electronics, August 25,
    1983.

5. F. Riherd, A. Haines and D.
    Fairbairn, "Mead-Conway design-
    ing: No IC experience needed",
    Electronic Design, March 4, 1982.

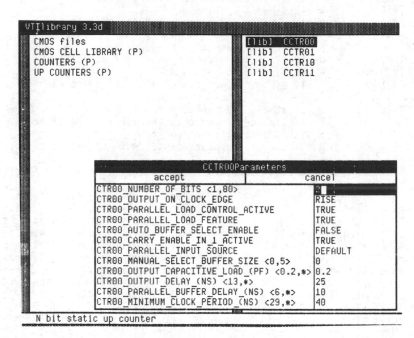

Figure 4. Screen image of a VTI library window showing the parameter menu for CCTR00.

Cell Compilation with Constraints†

Chidchanok Lursinsap and Daniel Gajski

Department of Computer Science
University of Illinois at Urbana-Champaign
Urbana, Illinois 61801

ABSTRACT

This paper describes a cell compiler that translates cell descriptions given in form of Boolean equations including pass transistors into layout descriptions in Caltech Intermediate Form (CIF). The translation process is constrained by given height and width of the cell and the position of each I/O signal on the boundary of the cell. Furthermore, the size of each transistor as well as power consumption can be arbitrarily chosen. This cell compiler allows routing through the cell in any direction. The cell architecture is based on PLA structures.

1. Introduction

One method in solving the VLSI-layout problem is a library of standard cells which are used as standard building blocks. Therefore, the design problem is partitioned into two subproblems: (a) decomposition of design into standard cells and (b) placement and routing of standard cells used in the design. The one disadvantage of standard-cell methodology is that library has to be maintained and updated. Secondly, each cell is usually needed in many different versions, all of them having the same functionality but slightly different electrical characteristics such as time delay, power consumption and geometrical characteristics such as height, width ratio, and position of input/output signal lines. The solution to this problem is a library of parametrized cells.

The second problem with standard-cell approach is the extensive area of the chip used by routing. First, we layout very dense cells and then lose all that gain by extensive routing between the cells. Through-the-cell routing provides a solution.

In this paper, we describe an automatic cell-synthesis program for generation of cells with different electrical parameters and with the custom through-the-cell routing. In section 2, we define our methodology and how the cell compiler fits into it. Section 3 describes the features of our cell-compiler and compares our work with previous work in the field. Section 4 summarizes the features of the layout architecture. Section 5 gives a short

†This research was supported in part by the National Science Foundation under grant no. NSF MCS 83 00981.

description of the compiler structure and algorithms used in translating of cell descriptions into layout. Section 6 describes our environment, while section 7 contains one example. Conclusion and possible extensions of our work are given in section 8.

2. Compilation Methodology

The problem of designing digital systems is a translation process from algorithms or behavioral description into silicon. This process is complex and accomplished in only several steps. Each step refines the design by adding more detail. Any digital system can be represented as a forest of hierarchies, each hierarchy corresponding to one representation such as logic, circuit and layout. In each hierarchy we have basic elements or elementary components which we use to define higher level blocks (macros) which are then used as building blocks on the next level in the hierarchy. Each representation has unique properties which influence the rules of design in each hierarchy. This influence decreases and sometimes disappears on the very abstract levels of the hierarchy. The translation process is an ordered sequence of steps through one or more hierarchies. In our model of translation, we have three steps which involve four levels of design.

(1) Algorithmic description
(2) Register-transfer description
(3) Abstract-cell description
(4) Layout description

Algorithmic description is the instruction set description in a high level language such as ISPS [Barb75]. The register-transfer-level description defines a structure in which basic components are registers, functional units (ALU, shifters), ROMs and RAMs. The instruction set is converted into register-transfer structure by a processor compiler [RaPG84]. Each register-transfer component is compiled into a set of abstract cells by a module compiler. Abstract cells are described by Boolean equations which have been extended to include pass transistor and dynamic-register specification. In this paper, we describe compilation between the third and fourth levels. This compiler is a part of ARSENIC system [Gajs82].

Reprinted from *ACM/IEEE 21st Design Automation Conf. Proc.*, 1984, pp. 103–108.

ARSENIC methodology is characterized by the top-down, hierarchical decomposition from algorithmic description to layout description. On each step of translation, the designer interacts by specifying constraints such as time delay, number of components allowed or the position of input/output signals [Boze84]. This interactive compilation shrinks the design space and improves convergence toward a high-quality design. In this paper, we deal with translation between the last two levels.

3. System Features

Each module compiler breaks up a module into cells and determines the cell functionality, estimates its size, and assigns I/O signals to points on the cell boundary for each cell in the module. The cell synthesis that follows is executed in two steps: symbolic phase, in which a cell is decomposed into a regular array of subcells each represented by a symbol, and geometric phase, in which each symbol is replaced by its parametrized layout macro. Cell synthesis is characterized by the following features derived from the principles of ARSENIC.

1. 4-directional I/O. Any input I/O signal may be assigned by a module compiler or the designer to an arbitrary position on any side of the cell. The cell synthesizer will assign each I/O signal to a row or a column on a virtual grid inside the cell. Two layers of metal are allowed in the cell.

2. Through-the-cell Routing. All routing in our methodology is done through the cells. Routing information is considered to be part of the functional description. The connection between two ports assigned to different sides of a cell is an identity function. This way, routing and function layout in the entire chip are done concurrently. This is quite opposite to the prevailing approach in which function layout inside the cell is done first and routing between cells is added later. The separation of function synthesis and routing makes checking for errors and frequent changes easier and less costly to implement. In our system both of these tasks are automatic. So there is no need for design rule checks and changes are not time consuming, which removes the need for separation of these two tasks. Our cells are larger in size than standard cells, but we expect the entire design to be smaller since there is no routing outside of any cell.

3. Resizable devices. The entire cell is decomposed symbolically into subcells called parametrized layout macros such as contact, transistor, connection, etc. This first step is technology-independent. Layout macros are technology-dependent. However, each macro is parametrized which allows one to resize devices, and the connection (power supply) after cell synthesis is completed. Under time and area constraints, other systems ([SMWP83], [WiSh83], [ChSe83], [Chuq83]) do not offer this feature.

3.1. Cell synthesizer

We expect several different cell synthesizers to be available in ARSENIC system for each different module type such as register file, ALU, ROM. This paper describes a universal cell synthesizer which is based on the well-known PLA structure. However, it differs from other PLA generators in several ways.

First, we have defined an assignment algorithm that assigns each variable to a virtual row inside the cell. The algorithm clusters the variables that are used in the same function. This clustering allows higher packing density of terms during column folding.

Secondly, AND and OR planes are not physically separated but interleaved. This approach can achieve the high utilization of area comparing to the other models of separated AND and OR planes.

Thirdly, we fold pull-up transistors by moving them into the unused holes inside of the cell (a feature not offered by other systems like Berkeley's PLEASURE [MiSV83]). This pull-up folding will produce more compact cells and improve the regularity of power and ground lines in some cases.

Fourthly, we have included a pass transistor in our cell description so that dynamic registers and other structures using pass transistors can be handled by our cell synthesizer. The pass transistor is treated as a product term and placed concurrently with other terms.

Fifthly, the functional description of the cell includes any number of levels of logic. We do not restrict the cell description to standard sum-of-product form. Furthermore, registers and pass transistors can be included in the description.

3.2. Comparison with previous work

We compare our cell synthesizer, which is a part of ARSENIC [Boze84], [Gajs82], with some existing systems. PAOLA [ChSe83], [Chuq83] allows routing inside the PLA. However, routing is performed after PLA folding, not concurrently. PAOLA stretches the PLA structure and inserts extra rows and/or columns whenever a route must be created. PAOLA allows only 2-directional I/O. HOPLA [WiSh83] and APSS [SMWP83] are very similar. HOPLA was a high-level hardware definition language while APSS accepts Boolean equations as an input. They do not allow routing inside the cell, but they do allow folding. Both of them allow only 2-directional I/O. The presently available set of the layout and routing programs at Berkeley has not been integrated into a system. There are many programs on folding [MiSV83], [HNSV82], [MiSa83], and some on layout [MaOu83], [EKNP82]. Routing inside a PLA is not allowed and only 2-directional I/O is allowed. The following table summarizes this comparison.

	PAOLA	HOPLA	APSS	Berkeley	ARSENIC
4-directional I/O	No	No	No	No	Yes
Through-the-cell Routing	Yes	No	No	No	Yes
Folding	Yes	Yes	Yes	Yes	Yes
Pull-up Transistor Folding	No	No	No	No	Yes
Resizable Macros	No	No	No	Yes	Yes
Pass Transistor	No	No	No	No	Yes
Fully Automatic	Yes	Yes	Yes	No	Yes

4. Layout Architecture

The geometrical layout is NMOS-technology dependent and is based on the design rules given in

[MeCo80]. Presently, we have restricted all nets to 2-terminal nets. There two layout models used, V and H models. In V model, all terms are compacted vertically, while in H model, all terms are compacted horizontally. The following are the features of our layout architecture :

1. Two layers of metal, with first layer is for power line, ground line and I/O signals running horizontally and the second layer for routing and connections between the first layer of metal.

2. One layer of polysilicon and diffusion

3. Vertically laid product terms in V model and horizontally laid in H model.

5. Compiler Structure

The input to the compiler is a simple cell-description language which specifies the size (height and width) of the cell, the position of each input or output variable on the boundary of the cell, the routing through the cell and the function of the cell in terms of Boolean expressions. For example, the description of a cell containing one bit register with two OR inputs and a connection through the cell is given in figure 1a while the circuit diagram is given in figure 1b. The output of the compiler is a layout description in CIF. The compiler also produces two different symbolic layouts, routing and term layouts. The following symbols are used in the symbolic layout.

X	for AND operation when a variable is false
x	for AND operation when a variable is true
O	for OR operation
T	for pull-up transistor of rule 1 (section 5.6)
t	for pull-up transistor of rule 2 (section 5.6)
-	for 1st metal layer
|	for 2nd metal layer
o	for connection between 2 layers of metal for routing
#	for connection between 2 layers of metal for inverter
c	for contact cut acting as one end of the pass-transistor term
p	for pass transistor
H	for the combination of pull-up transistor and inverter
I	for inverter

The compiler consists of seven software modules executed in the following order:

1. Translator
2. Net Classifier
3. Row Assignment
4. Column Assignment
5. Term Assignment
6. Pull-up Transistor Assignment
7. Layout Generator

5.1. Translator

This module translates the cell-description language into PLA virtual grids. An example of cell description language is shown in figure 1 (detailed semantics can be found in [Boze83]). Symbol 't' indicates the type of functional description. In our case, the function is described by Boolean equation in standard (not necessarily minimal) sum-of-product form. A pass transistor is considered to be a special term. Symbol 'o' and 's' indicate the origin and size of the cell, respectively. Symbol 'f' indicates all the terms inside the cell. There are four types of terms: product (*), pass transistor (|), inverter (-) and connection (=). In our example, the meaning of * x1 a b ; term is 'x1' equals 'a AND b'. Symbol 'p' indicates the position of an I/O variable on the boundary of the cell.

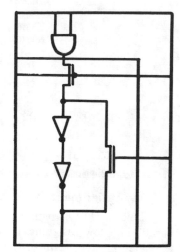

```
t bool
o 0 0
s 87 180
f = f f1 ;
f = c c1 ;
f * x1 a b ;
f | x2 x1 c ;
f | e x2 d ;
f - x3 x2 ;
f - e x3 ;
p f 65 180
p f1 0 20
p a 12 0
p b 35 0
p c 0 30
p c1 87 30
p d 87 55
p e 33 180
```

Figure 1a. Cell description.   Figure 1b. Circuit diagram.

The input to the translator is generated by a higher-level module compiler which partitions register-transfer-level components such as register file, ALUs, shifters, ROMs, and RAMs into cells of different types. The cell description can be modified by the user before the cell compilation is attempted. The design starts with oversized cells, which are more and more constrained on each iteration until a high-density layout is achieved. The output of the translator is unseen by the user. The translator will assign virtual row or column to each I/O signal. In addition, it will determine the side on which the signal enters the cell.

5.2. Net Classifier

This module classifies nets into eight different classes so that conflict can be avoided during the assignment of nets to virtual grid. Classes are processed in ascending order. In this paper, we consider only 2-terminal nets. Each terminal can be on any side of the cell. If T, L, B, R stand for top, left, bottom, and right sides of the cell, then 2-terminal net can be characterized by pair (x, y) where x and y are the sides of the first and second terminals. If one of the terminals is internal to the cell then a net is characterized by (x) where x is a terminal on the boundary of the cell. The following are net classes:

Class 1:   (T,L), (T,R), (B,L) and (B,R) nets
Class 2:   (L,L) and (R,R)
Class 3:   (L,R) with L = R, (L) and (R) nets
Class 4:   (L,R) nets with $L \neq R$
Class 5:   (T,B) nets with T = B
Class 6:   (T,B) nets with $T \neq B$
Class 7:   (T), (B),(T,T) and (B,B) nets
Class 8:   nets with both internal terminals.

Figures 2 and 4a show a set of nets, their classes after the classification process and the corresponding routing inside the cell, respectively.

Nets	Direction	Class
f	T5,L0	1
c	L0,R0	3
x1	-	8
a	T2	7
b	T3	7
x2	-	8
e	B3	7
d	R2	3
x3	-	8

Figure 2. Nets, directions and classes of the cell.

5.3. Row Assignment Heuristic

This module assigns virtual rows to all variables in a cell. The row assignment clusters the variables used in the same product term so that efficient column folding can be achieved. There are four main steps in the row assignment. First, we create a weight table with n rows and n columns, where n is the number of variables. A weight in a row i and a column j as follows:

$$W_{ij} = \begin{cases} number\ of\ terms\ having\ variables\ i\ and\ j & \text{if } i \neq j \\ 0 & \text{if } i = j \end{cases}$$

This table will be used in Algorithm 1 to select and assign virtual rows to all variables in the functional description.

Algorithm 1

1. Mark all rows having two I/O ports, and add those variables to the set of unassigned variables.

2. For each row with only one I/O port, assign the corresponding variable to that row.

3. For all variable v_i already assigned, compute $W_i = \sum_j W_{ij}$.

4. For each assigned variable v_i,
   -find every unassigned variable v_j such that $W_{ij} = \max_k W_{ik}; 1 \leq k \leq n$,
   -randomly select first v_j having max W_j, and
   -set $W(v_i) = W_i + W_{ij}$.

5. Randomly select v_i having max $W(v_i)$.

6. Assign v_j to the first empty row next to the row assigned to v_i.

7. Update the list of all assigned variables.

8. Go to 2 if all variables are not assigned to rows.

9. Specify a jog between the marked and assigned rows of the same variable (to be implemented in column assignment part).

Finally, the row assignment procedure will assign an extra row to one of the variables forming a loop to avoid conflict. The results of row assignment for the description in figure 1a. are illustrated in figures 3a. and 3b., respectively. Figure 3a. gives the weight table, while figure 3b. gives the rows assigned to all the variables.

```
0 0 0 0 0 0 0 0 0 0
0 0 1 0 0 1 0 0 0 2
0 1 0 1 1 1 0 0 0 4
0 0 1 0 1 0 0 0 0 2
0 0 1 1 0 0 0 0 0 2
0 1 1 0 0 0 1 1 1 5
0 0 0 0 0 1 0 1 1 3
0 0 0 0 0 1 1 0 0 2
0 0 0 0 0 1 1 0 0 2
```

Variable	row assigned
f	0
c	1
x1	4
a	5
b	6
x2	3
e	8
d	2
x3	7

Figure 3a. Weight table.   Figure 3b. Rows assigned to the cell.

5.4. Column Assignment

The purpose of column assignment is to complete the routing inside a cell. This module assigns connections between I/O ports on top or bottom sides of the cell and virtual rows assigned to variables in classes 1, 4, 5, 6 or 7. A connection symbol is assigned to the point on the virtual grid where the column assigned to an I/O port meets a row assigned to a variable. In addition, the module assigns a connection between two virtual rows assigned to all variables in classes 1 or 6 and any two variables forming a loop. Figure 4a. illustrates the routing of the cell after column assignment.

5.5. Term Assignment

This module assigns a part of a virtual column to each term. This assignment problem can be solved by using the channel assignment algorithm [HaSt71] which gives an optimal solution.

The module finds the top and bottom rows for each term first, and then it applies the channel assignment algorithm to assign terms to column. The first and bottom rows of each term correspond to the upper and lower bounds of each wire segment interval in the channel, respectively. The terms and columns of the cell described in figure 1a. after the term assignment is shown in figure 4b.

5.6. Pull-up Transistor Assignment

We try to utilize the silicon area as much as possible and improve the regularity of the power and ground lines by filling the unused holes of the layout with the pull-up transistors. There are two rules used to assign the pull-up transistors to the unused holes

Rule 1 assigns the pull-up transistor to the first empty hole in each column.

Rule 2 places the pull-up transistor aside if there are no empty holes in that column.

```
++--o              --T---
++--+              p-i---
++--+              +pi---
+#--+              cci--I
+++--+             c+O--H
o++--+             -+X--I
-+o--+             -+X--I
#o---+             -+---I
o-o--+             -c----
```

Figure 4a. Routing.   Figure 4b. Term layout.

5.7. Layout Generator

This module translates symbolic layout into geometric layout macros. The layout generator replaces each symbol in the symbolic layout by set of rectangles by calling corresponding layout-macro routine. There are 23 different types of layout macros, including AND, OR, VDD and GND layout macros. Each layout macro consists of a set of parametrized rectangles. The width and height of each macro are calculated from the given timing and power consumed by the macro. The height of each row and the width of each column are equal to the height and width of the largest macro. Other macros are stretched to conform to the column width or row height.

The layout generator will add boundary macros including VDD and GND macros to the layout. The size of VDD and GND macros are estimated by the power consumed by the entire cell. Figure 5 gives all macros used the cell described in figure 1a.. The complete layout is given in figure 6. The output of this part is in CIF and in the form which is internally used for plotting program.

```
VVVVVV
--T---
p-i---
+pi---
cci--I
c+O--H
-+X--I
-+X--I
-+---I
-c----
GGGGGG
```

Figure 5. Symbolic layout with VDD and GND macros

6. Environment

The cell compiler is written in C and run on VAX 11/780 under the UNIX† operating system. This program is used by MUSIC [Boze84] for generation and editing of VLSI designs. The output of the program can be displayed on the RAMTEK graphical device. The hard copy of the layout can also be obtained on the HP7585 plotter.

7. Examples

The example in figure 6 illustrates how product terms and pass-transistor terms can be combined in one cell and laid out by a PLA-based structure. The position of each I/O signal as well as the size of the cell can be changed by the designer to achieve better density or better abutment. Figure 7 shows the same cell with different size and some I/O positions changed. Figure 8 shows the layout generated by ARSENIC of the ALU mentioned in [AbGa81].

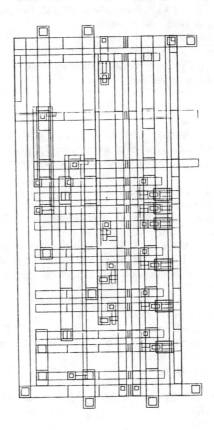

Figure 6. Layout.

8. Conclusion

This paper describes an automatic layout compiler based on PLA structure. Some new features have been implemented, for example 4-directional I/O, through-the-cell routing, resizable devices and row assignment to achieve multiple-column folding while the circuit is laid out. Our further research will include row folding and extension of our techniques to cells with multiterminal nets and partially defined I/O.

9. References

[AbGa81] J.A. Abraham and D. Gajski, "Design of Testable Structures Defined by Simple Loops," IEEE Transaction on Computers, Vol C-30, No. 11, November 1981, pp.875-884.
[Barb75] M.R. Barbacci, "A Comparison of Register Transfer Languages for Describing Computers and Digital Systems," IEEE Transaction on Computers, Vol C-24 No. 2, February 1975, pp. 137-149.

[Boze84] J. Bozek, "MUSIC : Management and Editing of Functional Design," Project ARSENIC Technical Report, Department of Computer Science, University of Illinois at Urbana-Champaign, 1984.

[ChSe82] S. Chuquillanqui and T.P. Segovia, " PAOLA: A Tool for Topological Optimization of Large PLAs," Proceedings of 19th Design Automation Conference, 1982, pp. 300-306.

[Chuq83] S. Chuquillanqui, "Internal Connection Problem in Large Optimized PLAs," Proceedings of 20th Design Automation Conference, 1983, pp. 795-802.

[EKNP82] S.A. Ellis, K.H. Keller, A.R. Newton, D.O. Peterson, A.L. Sangiovanni-Vincentelli and C.H. Sequin, "A Symbolic Layout Design System," IEEE Symposium on Circuits and Systems, 1982, pp. 670-676.

[Gajs82] D.D. Gajski, "The Structure of a Silicon Compiler," Proceeding of the International Conference on Circuits and Computers, October 1982, pp. 272-276.

[HaSt71] A. Hashimoto and J. Stevens, "Wire Routing by Optimizing Channel Assignment within Large Apertures," Proceedings of 8th Design Automation Conference, 1971, pp. 155-169.

[HNSV82] G.D. Hachtel, A.R. Newton and A.L. Sangiovanni-Vincentelli, " Techniques for Programmable Logic Array Folding," Proceedings of 19th Design Automation Conference, 1982, pp. 147-155.

[MaOu83] R.N. Mayo and J.K. Ousterhout, "Pictures with Parenthesis: Combining Graphics and Procedures in a VLSI Layout Tool," Proceedings of 20th Design Automation Conference, 1983, pp. 270-276.

[MeCo80] C. Mead, L. Conway, "Introduction to VLSI Systems," Addison Wesley, 1980.

[MiSa83] G. Micheli and M. Santomauro, " SMILE: A Computer Program for Partitioning of Programmed Logic Arrays," CAD, vol 15, No. 2, March 1983, pp. 89-97.

[MiSV83] G. Micheli and A. Sangiovanni-Vincentelli, "PLEASURE: A Computer Program for Single/Multiple Constrained/Unconstrained Folding of Programmable Logic Array," Proceedings of 20th Design Automation Conference, 1983, pp. 530-537.

[RaPG84] V. Raj, B. Pangrle and D. Gajski, "Microprocessor Synthesis," Proceedings of 21th Design Automation Conference, 1984, This proceeding.

[SMWP83] M.W. Stebnisky, M.J. McGinnis, J.C. Werbickas, R.N. Putatunda and A. Feller, "A Fully Automatic, Technology-independent PLA Macrocell Generator," Proceedings of 20th Design Automation Conference, 1983, pp. 430-435.

[WiSh83] S. Wimer and N. Sharfman, "HOPLA - PLA Optimization and Synthesis," Proceedings of 20th Design Automation Conference, 1983, pp. 790-794.

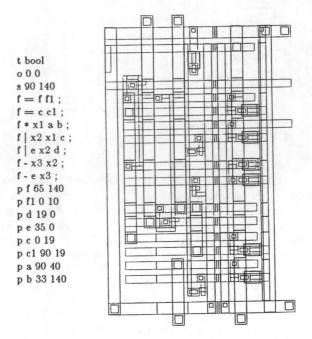

```
t bool
o 0 0
s 90 140
f = f f1 ;
f = c c1 ;
f * x1 a b ;
f | x2 x1 c ;
f | e x2 d ;
f - x3 x2 ;
f - e x3 ;
p f 65 140
p f1 0 10
p d 19 0
p e 35 0
p c 0 19
p c1 90 19
p a 90 40
p b 33 140
```

Figure 7. New description and layout.

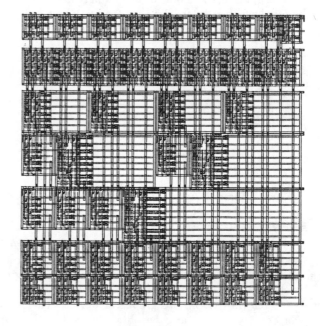

Figure 8. ALU.

† UNIX is a trademark of Bell Laboratories

AUTOMATED IMPLEMENTATION OF SWITCHING FUNCTIONS
AS DYNAMIC CMOS CIRCUITS

Robert K. BRAYTON, Chih-Liang CHEN, Curtis T. McMULLEN, Ralph H.J.M. OTTEN, Yiannis J. YAMOUR

Thomas J. Watson Research Center
IBM corporation
P.O.Box 218
Yorktown Heights, NY 10598
United States of America

Abstract: This paper describes an automated design system for realizing switching functions as dynamic CMOS circuits, suitable for incorporation into a silicon compiler. The system can effectively use several objectives that relate to the speed, area and power consumption of these circuits. Design turn-around times can be extremely short. Though the mask definition for these circuits can be derived completely automatically from the specification of the function, the system allows for human interaction in various stages. The features are illustrated with an automatically designed control logic macro of a 32-bit microprocessor with a complexity equivalent to a 500 4-way-NOR-gates network. An APL implementation of the system can generate the masks for that circuit from an algebraic or tabular specification in less than 10 minutes IBM-3081 CPU-time. The result compares favorably with other realizations of the same switching function.

1. INTRODUCTION

Large systems usually are decomposed into smaller modules which may be decomposed themselves, thus forming a hierarchy. A *hierarchy* is by definition either a *compound* (i.e. a set of hierarchies), or a *cell* (i.e modules that, for the purpose of the actual design procedure, are considered not to be decomposed into modules). Whereas most of the decisions concerning compounds are to a high degree technology independent, cell design is dominated by possibilities and limitations of the target technology and the style imposed by the chosen layout design algorithms. Cells can be classified from several view points. For example in the layout stage we distinguish *inset cells* (cells of which the layout is fixed and stored in a master or user library, and to which a layout design program only can assign a location and an orientation), *stretch cells* (with some shape flexibility) and *algorithmic cells*. The layout of an algorithmic cell is generated by a special procedure from a functional specification. For the purpose of that procedure the cell can have a decomposition of its own in subcells or circuits. In that case we speak of a *macro*.

Since the topic of this paper is designing a particular kind of macro, section 2 will be devoted to the selected logic family and its technology. The consequences of this choice for the logic decomposition part of the system are discussed in section 3 using a small example. The layout style chosen for this type of macro, the *pluricell* style, is introduced in section 4.

A top-down design system is essentially a system that derives properties of the environment of a module before designing that module. It is therefore desirable that certain aspects of a preliminary environment can be taken into account by such a design system. Examples of such aspects are the capacitance of global nets, the shape of the region that the environment leaves for realizing that module, and the location of nets to which the module has to be connected. Ignoring the environment, typical for mainly bottom-up approaches, might lead to unacceptable penalties in speed, size and power consumption of the whole design. For example, implementing switching functions in so-called programmable logic arrays optimized with respect to speed and area by sophisticated programs using row and column folding, very often inflates the space used by the wiring connecting that PLA with the other modules, because of its severe constraints on the sequence in which nets enter the module region. In this paper we describe another way of implementing switching functions in a rectangular region with the same degree of automation as is possible for PLA implementation. Yet it effectively uses estimated data about the environment, such as signal delays, aspect ratio and net sequences.

In the last section we give as an example the result generated by an implementation of the system. This result will be compared with results of other approaches for realizing switching functions.

2. SINGLE-ENDED CASCODE VOLTAGE SWITCH LOGIC

The system uses a very large logic family to implement switching functions. The idea behind the operational principle of member of this family goes back at least to 1976 [1]. It can be easily explained with figure 1[a]. **N** represents a network of which the branches are switches. The state of a switch in the network depends on the value of one of the inputs of **N**. The network has two terminal nodes, s and b in the figure. The double switch **PC** is periodically moved up and down. When it is in the position as depicted in the figure, the capacitor C_s is *precharged*. Just before changing this position the node s will therefore be *high* and, consequently, the output u of the inverter will be *low*. After

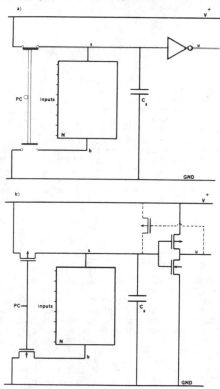

Figure 1. The principle and a CMOS realization of domino circuits.

Reprinted from *IEEE Proc. Custom Integrated Circuits Conf.*, 1984, pp. 346–350.

changing the position of **PC** the capacitor C_s will discharge through the network if there is a path of conducting switches from s to b, and this, of course, depends on the values of the inputs. The switching function realized by such a circuit is a single-output function equivalent to the or-sum of all product terms obtainable by and-ing the input variables associated with the switches on a simple path from s to b in **N**.

In CMOS technology the double switch **PC** can be easily realized by a p-channel and an n-channel device with the same net, the *precharge* or *clock* line, connected to their gates. This avoids dead time necessary for reliable operation under multi-phase clocking. Partitioning the logic of the system in two parts might even enable the designer to hide the precharge phase of one part under the logic evaluation phase of the other part. Thus, extremely fast logic circuitry can be obtained this way. The switches in **N** can be realized by n-channel devices with the corresponding input signals on their gates. A conventional CMOS inverter completes this so-called *single-ended cascode voltage switch circuit* (figure 1[b]). The switches in such a circuit are predominantly n-channel devices, which are more area efficient than p-channel devices under high performance requirements. Power consumption is low as usual with CMOS circuits, and this type of logic can be combined with static and other dynamic CMOS logic. For fairly large instances of **N** some additional devices are necessary for an acceptable noise margin. A simple solution is a feedback device between the sense node s and the voltage supply line as indicated by the dashed lines in figure 1[b] [2]. The size of this device, if added, should be based on a trade-off between noise insensitivity and performance. If the desired noise margin cannot be reached by this modification one may have to resort to precharging all internal nodes of **N**.

An important limitation, inherent to implementing switching functions with circuits as in figure 1, is that the output signal u can only change from *low* to *high* after precharge. Such circuits have been called domino circuits because of this drawback [3]. One way out of this problem is to combine single-ended cascode voltage switch (SCVS) circuits with circuits not limited in this way, such as differential cascode voltage switch (DCVS) circuits [2]. In practice a more attractive solution is often feasible. This case will be discussed in the next section, together with the logic synthesis part of the system making use of those possibilities.

3. INVERT FREE LOGIC SYNTHESIS

The system contains a number of algorithms for decomposing a switching function into switching functions that can be realized as circuits of the chosen logic family [4]. The decomposition can be executed automatically. An experienced user, however, has several means of controlling the synthesis process. He may call the algorithms individually in a sequence of his choice. He can give other values to the parameters controlling the result with respect to the number of elementary circuits, the number of transistors, delay, and other figures of merit. It is also possible to use preliminary data concerning placement and wiring, and results from network simulations to obtain more accurate data for signal delays and critical path identification.

If the family described in section 2 is chosen, the decomposition has to cope with the problem that intermediates are generated in only one form. In this paper, it is assumed that input signals to the macro are available in both the true value as well as the complement value. This enables us to realize any polarity of any intermediate signal by a chain of SCVS circuits only. Moreover, if the circuits into which the output signals of the macro are to be fed, can be properly set by either the true or the complement signal, only one arbitrary polarity of each of these output signals must be generated by the network of SCVS circuits. This leads to the problem of assigning the output polarity in such a way as to realize all intermediate signals with a minimal number of SCVS circuits. This problem is solved using a branch and bound method starting with a heuristically generated feasible bound. This problem is similar to the *optimal output phase assignment* problem for PLA's [5].

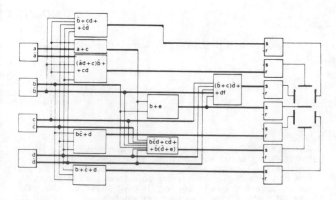

Figure 2. An example of an SCVS implementation of a seven-segment-decoder. The network realizing the associated switching function is bounded by input and output latches. Not only does this make both phases of the input signals available, but also the number of inverted intermediate signals can be reduced by selectively building either the true or the complement phase of the outputs of that network. The figure only shows an example of such a network. It is not optimal in any obvious sense!

4. THE PLURICELL LAYOUT STYLE

Physiognomically pluricells seem to be polycell or standard cell layouts as is immediate when comparing figure 6 with polycell results [6]. Both have their circuits placed in columns, either with two sides accessible for signal nets, or one side with the other abutting cells in a neighboring column. The latter is chosen in the system we are describing. Supply and clock lines are realized by abutting subsequent cells in each column. The signal nets are realized mainly in channels between the (double) columns. Two wiring layers, polysilicon and first-level metal, are used to realize the nets in the channels.

One important difference is in the way of obtaining the individual cell layouts. Whereas polycell layouts find their cells in a master library that is built and maintained independently from the individual applications, pluricell cells are constructed ad hoc. In case of an SCVS circuit the switching network is realized as a one-dimensional transistor array along the edge of the channel. Such an array is formed by crossing a strip of n-diffusion with polysilicon tracks carrying the associated logic signal. This is not always possible without modifying the network. Here, modifying means duplicating transistors, introducing transistors with grounded gates, changing sequences in series connections, etc. As long as the logic function and the electric properties of the result are still acceptable these modifications may be used for making the diffusion strip short and the number of tracks to realize the network topology small. The rest of the circuit of figure 1[b] is realized between the power supply and ground lines that run parallel to the diffusion strip on the side opposite from the channel. The size of those devices can be adapted on the basis of critical path information generated by the algorithms in the logic synthesis part, and in the future by simulation procedures in the system. Between those two parts runs the clock line to which the **PC** gates are connected. Any number of additional first-level metal tracks can be allocated between that line and the ground line of the inverter. One, however, is always there without increasing the circuit area. This one is mostly used for sense node s if an arbitrary position between the the supply lines is allowed for the inverter. Another track that comes for free, because of the spacing rules, is available between the clock line and the transistor array. That one is used, if possible, for making the output u available via first-level metal. If this cannot be used the output will enter the channel via polysilicon. Along the channel side of the array runs a ground line for connecting the part of the **PC** switch between the node b and ground.

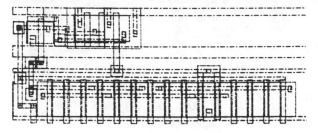

Figure 3. The layout of the circuit of figure 1[b].

Of course other types of circuits can be incorporated in pluricell layouts as long as their periphery is compatible, i.e. pitches of clock and supply lines are matched, and signal pins are on the correct side. Carrying this to the extreme with completely specified circuits (inset cells!) leads to polycell layouts. Pluricell can therefore be seen as a generalization of polycell styles.

To connect parts of nets in different channels, polycell layouts use so-called feed-through cells, in addition to the possibility of leaving the channel at one of its ends and making the required connection outside the macro. In pluricell layouts these connections are made by second-level metal tracks. This also allows signals to enter the macro from all four sides. Each signal requiring a second-level metal connection gets exactly one track assigned to it. If this track is in the interval determined by the possible first-level metal output line in an SCVS circuit an entry on polysilicon in the channel is superfluous.

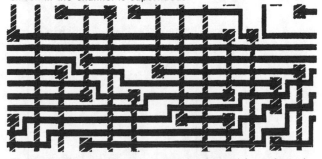

Figure 4. The polysilicon and first-level metal in a channel. Contacts are displayed smaller to show connection paths.

Channels in a pluricell layout are different from channels resulting from a classical channel router, because signals can be *dropped into* the channel by contacts between first-level and second-level metal. Thus, there are three ways of bringing signals into the channel, whereas the classical formulation has only two: the sides and the ends of the channel. The channels created in the design system differ in one more aspect from classical channels. Design rule sets have developed towards smaller and smaller widths for metal and polysilicon tracks. However, the area taken by contacts between polysilicon and first-level metal did not decrease that fast. By now the stage is reached that, with minimum pitch wiring under the presence of these contacts, there still is enough space for a first-level metal track in both directions. The router implemented for the system makes use of this fact by allowing twice as many tracks in the channel as would be allowed by a classical channel router. However, contacts in neighboring tracks should not interfere with each other. Moreover, in lattitudinal direction the router uses the extra space for adjusting the position of a net by first-level metal jogs (figure 4). The wiring space is therefore reduced with a factor of about two.

5. THE MACRO ASSEMBLER

Figure 5 shows the sequence in which procedures are called. Input for the logic decomposition block is a specification of the switching function to be realized. The form can be a number of single output switching functions or a PLA or a set of linked PLA's. The components of the switching function are reduced to their minimal disjunctive forms and then decomposed by finding common subexpressions. The results are tested on their realizability as an SCVS circuit, and further decomposed if necessary. The invert free logic synthesis procedure creates a separate function for each complement signal required in the result of the logic decomposition block. At this point some of these functions can be eliminated, while others may not pass the test for realizability as a single SCVS circuit. Again local decomposition is applied to break up those functions, possibly followed by merging some functions into others. The output of the invert free logic synthesis block is a set of single output switching functions each of which can be implemented as an SCVS circuit. Typically, many different networks (**N**) can be used to implement such a function.

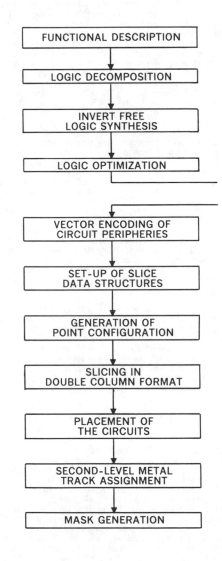

Figure 5. The sequence of routines in the macro assembler

The logic optimization block tries to find a factored and-or switching function of which the corresponding network is expected to be optimal in some sense, for example the least number of transistors or delay [4]. The determination of the delay of a circuit is based on the wiring capacitance in a quick trial placement, or on a worst-case net geometry. Some interaction by an experienced user at this point can often improve the overall design considerably.

The results of the logical synthesis part have to be made available to the layout program. Since information on delay and critical paths is available, also indications about desirable device sizes can be handed to the layout part. For example, functions in a critical path may be marked so that the associated SCVS circuit gets a more powerful inverter (figure 1[6]). All this information is to be filed in a specific format, readable for the layout procedures that have to generate the mask level specification of the macro. After the design of the whole chip has entered the layout stage, this file will be read at some point after the floorplanning and the global net assignment of the higher level modules has been executed to obtain some information about the environment of the macro.

The construction of a circuit in a pluricell layout is executed by first performing a topological analysis of the switching or memory function, and encoding the result in a compact vector from which all periphery data of the cell, such as dimensions and pin positions, can be derived. SCVS circuits are specified as factored and-or switching functions. These have an obvious translation into a two-terminal-series-parallel network of switches (N in figure 1). Consequently, the analysis heavily uses the freedom in the sequences of a series connection. This analysis is followed by a placement procedure consistent with the structure rules of the SLICE system [7]. First a configuration of points, each representing an SCVS circuit, in the plane is created on the basis of proximity data such as net length and delay. Slicing will preserve the relative position within each slice. The determination of the final slicing structure is based on aiming at approximately equal length double columns. Once the distribution of the circuits over the columns is known their position in the column will be fixed. This enables the program to determine the intervals over which the nets have to be present between each pair of double columns in order to connect the corresponding pins along the sides of the channel. If the same net occurs in several channels or enters the macro from one of the directions perpendicular to the columns, a second-level metal track will be reserved for that net. The position of these second-level metal tracks is optimized with respect to net resistance and net size as far as permitted under possible sequence and location constraints in the environment of the macro. After finishing the second-level metal track assignment, the data for the channel routing problems is complete.

Pluricell floorplans are, as a slicing structure, always represented by a five-level slicing tree. The leaves of that tree represent the circuits and the channels of course. By generating the mask data of these leaves in the same sequence as they are visited by a depth first search over that tree the coordinates of the elementary mask elements can be immediately assigned. Consequently, circuit geometry and wire specification are interleaved, and in one pass the mask data is complete.

6. CONCLUDING REMARKS

The system has been implemented in APL on an IBM 3081. Figure 6 shows a control logic macro of a CPU, which is designed automatically by the implemented system. This control logic macro has a total of 81 SCVS circuits. and has been estimated to be equivalent to 500 four-way NOR logic gates, plus inverters. Comparisons with other realizations of the same switching function are summarized in table 1. The macro generated by the system described is not only superior in area and speed to the results of the other approaches, but also is able to use environmental data such as signal delays, aspect ratio and net sequence information to create a macro which is more compatible with global requirements. This makes the macro assembler a useful part of a silicon compiler system.

References

[1]   J.A. Luisi, *High-speed, low-cost, clock-controlled CMOS logic implementation*, Sept 21 1976, US patent 3982138

[2]   L.G. Heller, W.R. Griffin, J.W. Davis, N.G. Thoma, *Cascode voltage switch logic -A differential logic family*, 31st International Solid State Circuits Conference, Digest of Technical Papers, 1984, pp. 16-17

[3]   R.H. Krambeck, C.M. Lee, H.-F.S. Law, *High-speed compact cicrcuits with CMOS*, IEEE Journal of Solid-state Circuits, Volume SC-17, Number 3, June 1982, pp. 614-619

[4]   R.K. Brayton, C.T. McMullen, *The decomposition and factorization of boolean expressions*, Proceedings of the International Symposium on Circuits and Systems, Rome, 1982, pp. 49-54

[5]   T. Sasao, *Input variable assignment and output phase optimization of PLA's*, IBM Research Report RC10003, June 1983

[6]   G. Persky, D.N. Deutsch, D.G Schweikert, *LTX: A minicomputer-based system for automated lsi layout*, Journal of Design Automation & Fault-Tolerant Computing, Volume 1, 1977, pp. 217-255.

[7]   R.H.J.M.Otten, *Layout structures*, Proceedings of the Large Scale Systems Symposium, Virginia Beach, October 1982, pp. 349-353

	pluricell (scvs CMOS)	gate array (CMOS)	PLA (static NMOS)	PLA (dynamic CMOS)
minimum design rule μm)	1.25	1.00	1.25	1.25
area (mm²)	.61	1.67	.82	.82
critical path delay (3V supply) (ns)	13.8 + pcd		30.0	19.0+ pcd
critical path delay (5V supply) (ns)	6.5+ pcd		20.8	13.0+ pcd

Delay time definition

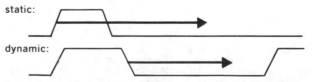

static:

dynamic:

Table 1. Comparison with gate array and PLA results. Delay times were compared in this way because of the possibility of hiding the precharge time. However, not using this possibility and assuming worst-case precharge time (pcd) still leads to a much faster macro than the static NMOS PLA realization.

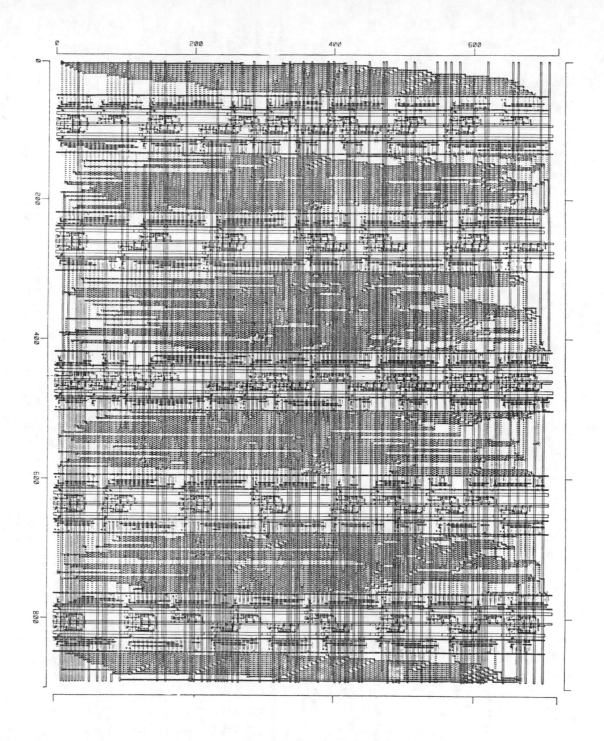

Figure 6. The result of the macro assembler for the control macro of 32-bit microprocessor.

AN AUTOMATIC ASSEMBLY TOOL FOR VIRTUAL GRID SYMBOLIC LAYOUT

Bryan Ackland
Neil Weste

Bell Laboratories
Holmdel, New Jersey 07733

1. INTRODUCTION

The object of any custom integrated circuit design system should be to reduce, in a disciplined manner, the tedious complexity of the design process. Symbolic layout systems accomplish this by removing unnecessary physical geometric constraints (*e.g.* design rules, mask layers) while capturing, more explicitly, the structural intent of the design. Compared to manual (geometric) layout, symbolic techniques provide faster turnaround with less possibility for error. In addition, they generate layout descriptions that are process design rule independent.

To date, most symbolic systems have concentrated on the symbolic representation of leaf cells and their conversion to correct mask equivalents. Few [1-3] have tackled the problems of composition and assembly once these symbolic leaf cells have been designed. Structured IC designs frequently use a composition style based on leaf cell abutment with implicit communication flowing directly across cell boundaries. To haphazardly assemble cells in this fashion without some structured composition discipline is to throw away much of the design integrity which is captured in good leaf cell design. The potential for error during this phase of the design process is particularly acute in symbolic systems where there is at least one level of abstraction between the symbolic cell description and the corresponding mask cell.

MULGA [4-6] is a proven symbolic design system based on the free form placement of symbolic circuit components on a virtual (topological) grid. In its original implementation, the authors concentrated on the capture and verification of good leaf cell designs and their correct transformation to geometric data. Over a period of time, a formal strategy for the assembly of symbolic leaf cells into larger modules has evolved. In this paper we describe our overall design philosophy and its current implementation. In so doing, we advance the concept that a detailed floor plan is a valid structural description of a chip. A set of structural (floor plan) design rules, similar to those used in hierarchical geometric systems, are provided. The paper also describes a proposed assembly tool which directs these compaction, pitch matching, circuit interpretation and mask data generation operations according to the structural floor plan. This tool is currently under development.

2. DESIGN PHILOSOPHY

There are at least two fundamentally different approaches to custom VLSI design. The first is characterized by the automatic generation of mask data from a user supplied functional or structural specification. Design proceeds in a strictly top-down manner converting function into structure and thence into layout with little or no human intervention [7-9].

The second approach seeks to capture, either procedurally or interactively, the creativity of a human designer and then verify the result. It is based on a mixture of top-down specification and bottom-up implementation. *MULGA* is an interactive system based on the top-down/bottom-up approach. The design process begins with a hierarchical floor plan. The floor plan is required to conform to a *restricted hierarchy*, similar in principle to the *separated hierarchy* of Rowson [10]. This restricted hierarchy classifies cells as modules, composition cells and leaf cells as shown in Figure 1. Leaf cells contain symbolic layout primitives - *i.e.* they define the physical layout of the circuit. They are symbolically connected (by abutment) through a hierarchy of composition cells. Composition cells are cells which contain only instances of other cells (leaf and composition). Different modules contain cells which are

structurally unrelated. For example, a PLA controller and a data path would normally be designed as two separate modules. These modules are then connected to each other and to the I/O modules (pads) using conventional place and route techniques.

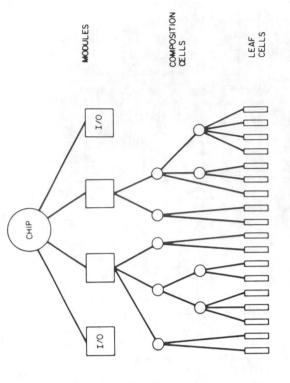

Figure 1 Restricted Hierarchy

It should be noted that within modules, cells are always connected across abutting boundaries. Modules thus form the highly structured, pitch matched portions of the layout. This does not mean that cells within a module cannot be routed together. Intra-module routing is accomplished using symbolic routing cells that also pitch match to their leaf cell neighbors.

Through the use of this restricted hierarchy and some composition design rules that will be described later, the floor plan can be used as a complete structural specification for the chip. This is similar to the technique used by Mudge [11] in which the floor plan is used to subdivide the logic, circuit and layout representations into a set of consistent hierarchies. Because composition cells define only cell placement and connectivity, they are representation independent and can be used to generate other (*e.g.* functional) descriptions of the chip. Composition cells are, therefore, not only process but also (potentially) technology independent.

3. IMPLEMENTATION STRATEGY

The symbolic design techniques used in the *MULGA* system have previously been described in detail [4,6,12]. The implementation of a floor plan begins with symbolic leaf cell design as shown in Figure 2. Leaf cells are generated either interactively (using a graphics editor) or procedurally in terms of a symbolic Intermediate Circuit Description Language (*ICDL*). This hierarchical language defines a cell as a distribution of electrical circuit elements over a virtual grid as shown in the example of Figure 3. Primitive elements of the language (*devices, wires, contacts*) have functional and structural as well as physical properties. *Pins* are named interconnection points that identify particular nodes by name. The only other primitive, *cell instance*, provides for hierarchical cell definition.

ICDL elements are linked both electrically and topologically through the virtual grid. This grid defines a relative layout topology without specifying the actual physical distance between elements. It is the virtual grid, rather than the *ICDL* elements themselves, that is manipulated by subsequent processing in order to arrive at geometrically correct mask description.

Reprinted with permission from *Proc. Int. Conf. Very Large Scale Integration*, F. Anceau and E. J. Aas, Eds., Elsevier Science Publishers B. V., North-Holland, Amsterdam, 1983, pp. 457–466. Copyright © 1983, IFIP.

Once a leaf cell has been symbolically designed, an audit program checks for circuit inconsistencies (e.g. implicit transistors, floating nodes, inappropriate contacts) and assigns node numbers to the elements in the cell. A compaction program uses this node information along with the *ICDL* description to determine minimum physical spacings in the cell. It does this by examining each symbolic grid line in the layout to determine how far it must be spaced from its neighbors in order to satisfy all process design rules for all elements along its length. The compactor stores its numerical results in a *design grid* file.

Because the compactor operates on a per-grid basis, it becomes inefficient (in terms of area) to compact an entire module as one cell. It is also inefficient (in terms of time) to consider the compaction of large regular structures such as memory arrays. We therefore compact leaf cells individually and subsequently pitch match neighboring cells at critical connection points. This involves a restretching of the compacted cells. Because the *ICDL* elements are still "attached" to their virtual grid lines, this stretching is relatively straightforward. A pitch match program simply modifies the entries in the *design grid* file to ensure the correct geometric connection of specified interconnection points.

Virtual grid lines that are not associated with interconnect can be locally adjusted to optimize leaf cell layout. A stretch weight is assigned by the compactor to each pair of virtual grid lines according to the elements present. The pitch match program will then stretch grids having a lower weight (e.g. those connected only by metal) in preference to those of a higher weight (those connected by diffusion).

Once pitch matching is complete, a mask conversion program uses the modified *design grid* spacings to expand the symbolic *ICDL* description into a full geometric mask layout. Leaf cells are then placed in the mask layout according to the original floor plan.

Functional verification of symbolic leaf cells begins with the audit program. A circuit interpretation program then converts the *ICDL* file into a circuit description suitable for simulation on either *SPICE* or *EMU* - a locally resident timing simulator. Exact parasitics are calculated using the spacing information held in the design grid file.

4. MODULE ASSEMBLY

Module assembly is the process of combining leaf cells into a verifiably correct set of module descriptions. This is done by instantiating the composition cell hierarchy that was specified in the floor plan. In the *MULGA* system, this process involves compaction, pitch matching and mask generation. In its original implementation, module assembly was performed manually - *i.e.* the designer manually invoked the compaction, pitch match and mask generation programs according to his target floor plan. Needless to say, this process was time consuming and error prone. As a result, a simple composition language was developed. This language, which allowed the designer to specify relative cell placement and interconnect, was used as input to a chip assembler. An example is shown in Figure 4.

```
PLACE cell 1 A
EAST A cell 2 B
EAST B cell 3 C
CONNECT A.vss.E B.vss.W
CONNECT A.vdd.E B.vdd.W
CONNECT A.do B.di
CONNECT B.vss.E C.vss.W
CONNECT B.vdd.E C.vdd.W
CONNECT B.do C.di
LABEL A.di DIN
LABEL C.do DOUT
```

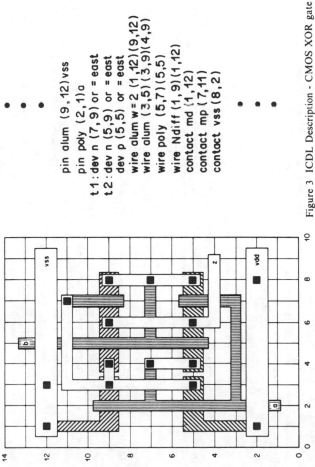

Figure 4 Simple Composition Language

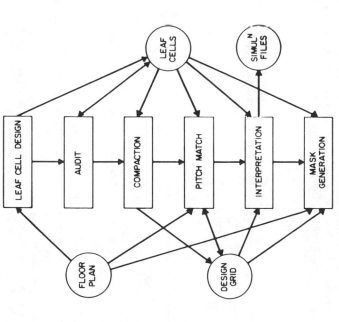

Figure 2 MULGA Implementation Steps

```
pin alum (9,12) vss
pin poly (2,1) a
t 1 : dev n (7,9) or = east
t 2 : dev n (5,9) or = east
     dev p (5,5) or = east
wire alum w= 2 (1,12)(9,12)
wire alum (3,5)(3,9)(4,9)
wire poly (5,7)(5,5)
wire Ndiff (1,9)(1,12)
contact md (1,12)
contact mp (7,11)
contact vss (8,2)
```

Figure 3 ICDL Description - CMOS XOR gate

Whilst this assembly language specification was a considerable improvement over manual composition, there were a number of problems associated with its use. Firstly, being a fairly simple language, there were no constructs to allow hierarchical composition. Secondly, it was found that language entry was a tedious (and hence error prone) technique for capturing the pictorial content of a floor plan. Thirdly, the program did not deal with potential design rule violations at cell boundaries. Finally, there were many special cases where extra manual intervention was required. For example, if one symbolic leaf cell was to be used in two separate portions of the layout, then it was up to the designer to ensure that the derived mask description was compatible with both environments.

At the same time, we noticed that designers invariably used the hierarchical capabilities of *ICDL* to generate graphical symbolic floor plans of their modules. These floor plans were never actually used in the generation of the module. Rather, they served as a pictorial aid to the designer through which he could plan modifications to leaf cells and to the language based assembly specification. It occurred to us that most of the information needed to assemble a module was already contained in this easily generated symbolic floor plan. Leaf cell placement was explicitly defined and interconnect could be derived by examining pin overlap at cell boundaries. We therefore decided to build a module assembly tool which would automatically and correctly compose modules using a symbolic floor plan of leaf and composition cells.

Graphical floor planning has been used by Trimberger and Rowson in their *RIOT* assembly tool [1]. They point out the danger of using implicit connection through placement *viz.* that after an initial floor plan is laid out, connections may be inadvertently made or destroyed through subsequent graphical editing. Another criticism that has been levelled at this graphical approach is that it does not have the versatility of programming language based systems.

Accordingly, we have made two modifications to the *MULGA* system. The first is a set of structural design rules. The second is an enhancement to the instancing capabilities of *ICDL* in order to simplify the composition phase of symbolic layout. These language modifications are based on the *ABCD* language proposed and implemented by Rosenberg [13]. Basically, they allow cells to be placed relative to one another so that the floor plan is not destroyed when leaf cells change size and/or shape. In the example shown in Figure 5, cell B has been bound so that its lower left corner is coincident with the lower right corner of cell A. Note that these cells may, in fact, be empty. This means that the graphical editor can be correctly used as a floor planning tool prior to any leaf cell design.

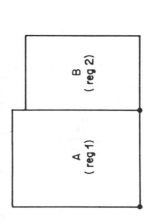

A: instance reg 1   (0, 0)
B: instance reg 2   ll = A(lr)

Figure 5 Relative Cell Placement

Alternatively, a compiler could be used to generate the symbolic floor plan from a higher level composition language. This would allow the designer to explicitly specify the desired connectivity. No matter which approach is taken, the symbolic floor plan becomes a stable human readable specification of module composition.

5. ASSEMBLY RULES

A number of authors of structured geometric systems [14-16] have proposed structural design rules to simplify incremental checking of large designs. Checking, in this context, means process design rules and circuit connectivity. In the *MULGA* system, design rule checks are eliminated through the use of symbolic compaction. A set of structural design rules is still needed, however, to efficiently compact, pitch match and then interpret - especially if this task is to be reliably automated.

5.1 Rectangular Cells

All cells (both leaf and composition) must be rectangular in shape. This simplifies the interpretation of communication across boundaries. Cells are not required, however, to tile orthogonally. Boundaries may be staggered as shown in Figure 6.

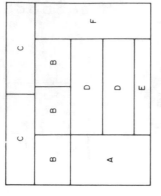

Figure 6 Staggered Tiling

5.2 Leaf Cell Naming

As mentioned previously, all cells within a module are classified as either leaf or composition. Composition cells contain no primitive symbolic elements - only cell instances. A leaf cell, on the other hand, may contain cell instances. This allows the designer to use the hierarchical capabilities of *ICDL* to efficiently construct leaf cells. Note, however, that leaf cells that do contain instances are always fully instantiated prior to compaction and pitch matching.

A cell instance can also be classified as a leaf cell by assigning it a leaf cell name. This name is then used to identify the geometric mask file which is ultimately derived from this instance. A designer may indicate that two instances of the same symbolic cell are to be treated as equivalent by assigning them the same leaf cell name. Sometimes, however, a designer may wish to use one symbolic cell in two places but have each instance locally optimized (in terms of compaction and pitch matching) to its environment. This is achieved by assigning different leaf cell names to the instances.

5.3 Cell Overlap

A generally accepted rule of structured layout is that cells should not be allowed to overlap. This permits the interior of a cell to be checked independently of other cells in the design. There still remains, however, the problem of dealing with components which lie on the boundaries of adjoining cells. In *MULGA*, cells are joined by placing them in such a way that they share a common virtual grid line. Because *ICDL* elements always lie symmetrically over grid lines, skeletal connectivity [14] is achieved. In this scheme, the only elements shared across cell boundaries are those elements which lie directly on the virtual grid bounding box.

There is, however, a class of circuits in which it becomes very inefficient to insist upon non-overlapping cells. These circuits are characterized by large numbers of nearly identical cells. Examples include PLA's, memory and other regular structures such as address decoders. As a result, we have introduced a special class of *ICDL* elements known as *compaction* elements. These elements are treated as normal by the compactor in determining intra-cell spacings. They are, however, ignored by the mask generation program. The correct mask components are generated using a special overlay cell which contains only the required (functional) *compaction* elements. This cell is simply pitch matched to those cells it overlays.

5.4 Boundary Connections

Because boundary components are shared by neighboring cells, they represent potential physical and electrical connection points. We therefore require that all boundary elements be electrically connected, within the cell, to a named *pin*. This serves two functions. Firstly, it ensures that cells are pitch matched at all boundary locations occupied by a physical element. Secondly, it enables all inter-cell connections (intentional or otherwise) to be recorded in terms of named electrical *pins*. These *pin* names are used to generate a net list description of the module floor plan.

In addition to specifying a *pin's* name, the designer may also specify its function type. This is taken from a scheme proposed by Corbin [16]. Allowable *pin* types are *clock, output, input, ioput, generic, vdd* (power) and *vss* (ground). Typed *pins* are readily checked for consistent connection across cell boundaries.

6. AUTOMATIC ASSEMBLY

The module assembler is a program which takes, as input, a symbolic module floor plan and produces a complete hierarchical mask description according to target process design rules. It also produces a hierarchical circuit description suitable for module simulation. The module assembler is responsible for ensuring that all the steps outlined in Figure 2 are performed correctly and in the right order. When changes are made, the assembler is used to incrementally update the design in a disciplined manner.

6.1 File System

The module assembler uses the *UNIX** file system to manage the various representations of a design. Researchers at Berkeley [17] have exploited the hierarchical features of the *UNIX* file system to represent the hierarchical structure of their designs. This mapping is not well suited to the *MULGA* system since there is not always a good one-to-one correspondence between symbolic cell descriptions and mask leaf cells. Accordingly, we manage our design data using a number of functionally distinct subdirectories as shown in Figure 7. All files in a given directory are similar representations of different cells.

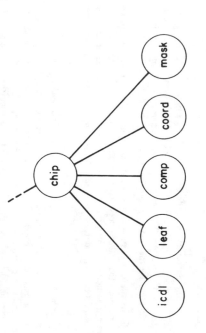

Figure 7 Chip Directory Structure

6.2 Mask Generation

The input to the module assembler is a symbolic floor plan consisting of composition and leaf cells that are stored in the *icdl* directory. Referring to Figure 8, the assembler first examines the structure to determine those cell instances that will become leaf cells in the physical hierarchy. At the same time, cells are checked for correct (non-overlapping) abutment. Cells that have not been previously audited are updated with correct node data. Cells are then examined for the existence of un-named nodes on their boundaries.

_____
* *UNIX* is a trademark of Bell Laboratories

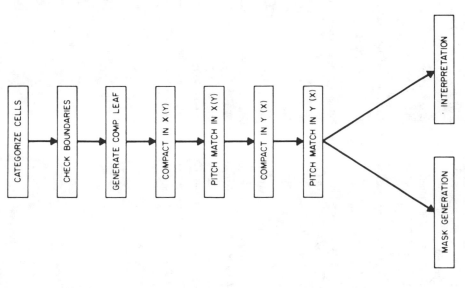

Figure 8 Tasks Performed by Module Assembler

Once these checks have been performed, composite leaf cells are generated and stored in the *leaf* directory. These composite leaf cells are the union of the original symbolic elements describing that cell and all the boundary elements contributed by all neighboring cells. These new elements are marked as *boundary* elements. The composite leaf cell is then audited to assign correct node numbers to these boundary components.

The composite leaf cells are compacted in one direction according to process design rules. This initial direction is specified by the designer (suppose it is the X direction). The results of this X compaction (*i.e.* the physical grid spacings) are stored in a file in the *comp* directory. In compacting composite leaf cells, boundary elements are correctly spaced from normal elements. Boundary elements are not, however, spaced from each other. This is to avoid the situation where boundary components from different boundary cells cause a leaf cell to expand unnecessarily to accommodate an element interference which never physically occurs. Elements which do not lie on the symbolic boundary of the cell are spaced at least half a design rule tolerance from the boundary.

Once all leaf cells have been compacted in the X direction, they can be pitch matched in this direction. Pitch matching is performed by aligning all vertically adjacent cells at those grid points where physical boundary elements have been placed. The module assembler treats this as a graph problem in which the grids to be pitch matched are nodes and the constraints stored in the *comp* directory are weighted directed arcs between these nodes. Stretch weights are used to determine the optimum placement of non-matched grids. The results of this process are a set of modified grid coordinates. These are stored in the *coord* directory.

The assembler then proceeds to compact and pitch match leaf cells in the Y direction. Diagonal tolerances are now taken into account based on X spacings already determined. Overlay cells are then pitch matched directly onto their corresponding leaf cells.

The final step is mask conversion. Leaf cells are generated according to the spacings defined in the *coord* directory. These leaf cells are then composed in the mask language according to the original floor plan. The end result is a hierarchical module mask description ready to be inserted in the overall chip layout.

6.3 Circuit Interpretation

In addition to producing a correct mask description, the assembler also generates a hierarchical circuit description of the module. Leaf cells are analyzed and parasitics assigned according to grid spacings stored in the *coord* directory. The resulting leaf cell circuit description is in the form of a subroutine whose parameters are the signals passed by the boundary pins.

Composition cells translate into subroutines which call other subroutines. Parameters are linked according to the connections that exist across cell boundaries. Pins named in the root composition cell are used to generate a structural summary of the module which lists the name and location of all external electrical connections. This data is subsequently used to place and route the module into the overall chip layout.

6.4 Modifications and Updates

It should be obvious by now that a fair amount of processing is required to convert a symbolic module floor plan into a mask description. In a situation where the designer makes a small change to his circuit, it is important to be able to incrementally update the stored information so as to avoid unnecessarily long processing delays. Like the *UNIX MAKE* program, the module assembler examines file creation dates to determine whether a particular operation needs to be performed. Thus the assembler, and not the designer, takes responsibility for determining which portions of the design need to be modified to reflect the update.

7. CONCLUSIONS

In a symbolic layout system, a structured composition methodology is just as important as the symbolic leaf cell design tools. A convenient way to specify composition is through the use of a hierarchical symbolic floor plan. The interpretation of this floor plan, however, and the subsequent conversion of symbolic leaf cells into correct mask equivalents is too complex a task to be managed manually by the designer. The module assembly program that we have described in this paper performs this transformation in a reliable, repeatable fashion. Design updates generate incremental modifications to the database. More importantly, it is the use of this assembler which enables the symbolic floor plan to be interpreted as a concise, unambiguous structural specification.

REFERENCES

[1] **Trimberger, S. and Rowson, J.** RIOT - A Simple Graphical Chip Assembly Tool, *Proc. 19th. Design Automation Conf.*, Las Vegas NV, June 1982, pp. 371-376.

[2] **Mosteller, R.** An Experimental Composition Tool, *Proc. of Microelectronics '82*, Adelaide Aust., May 1982, pp. 130-135.

[3] **Keller, K., Newton, A.R. and Ellis, S.** A Symbolic Design System for Integrated Circuits, *Proc. of 19th. Design Automation Conf.*, Las Vegas NV, June 1982, pp. 460-466.

[4] **Weste, N.** MULGA - An Interactive Symbolic Layout System for the Design of Integrated Circuits, *Bell System Tech. Journal*, Vol. 60. No. 6. July-Aug. 1981, pp. 823-858.

[5] **Weste, N.** Virtual Grid Symbolic Layout, *Proc. of 18th. Design Automation Conf.*, Nashville TN, June 1981. pp. 225-233.

[6] **Weste, N. and Ackland, B.** A Pragmatic Approach to Topological Symbolic IC Design, *Proc. of 1st International Conf. on VLSI.* Edinburgh UK. Aug. 1981. pp. 117-129.

[7] **Denyer, P., Renshaw, D. and Bergmann, N.** A Silicon Compiler for VLSI Signal Processors, *Proc. of 1982 European Solid State Circuits Conf.*

[8] **Johannsen, D.** Bristle Blocks - A Silicon Compiler. *Proc. of 16th. Design Automation Conf.*, San Diego CA. June 1979. pp. 310-313.

[9] **Siskind, J., Southard, J. and Crouch, K.** Generating Custom High Performance VLSI Designs from Succinct Algorithmic Descriptions, *Proc. of Conf. on Advanced Research in VLSI.* Jan. 1982. pp. 28-39.

[10] **Rowson, J.** Understanding Hierarchical Design, *Ph.D Thesis*, Calif. Inst. of Tech.. 1980.

[11] **Mudge, J.C.** VLSI Chip Design at the Crossroads, *Proc. of 1st. International Conf. on VLSI.* Edinburgh UK. Aug. 1981. pp. 205-215.

[12] **Ackland, B. and Weste, N.** Functional Verification in an Interactive Symbolic IC Design Environment. *Proc. of 2nd. Caltech Conf. on VLSI.* Jan. 1981. pp. 285-298.

[13] **Rosenberg, J.** Vertically Integrated VLSI Circuit Design, *Ph.D. Thesis*, Duke University, 1983.

[14] **McGrath, E. and Whitney, T.** Design Integrity and Immunity Checking, *Proc. of 17th. Design Automation Conf.* Minneapolis MN, June 1980. pp. 263-268.

[15] **Scheffer, L.** A Methodology for Improved Verification of VLSI Designs Without Loss of Area, *Proc. of 2nd. Caltech Conf. on VLSI.* Jan. 1981. pp. 299-309.

[16] **Corbin, L.** Custom VLSI Electrical Rule Checking in an Intelligent Terminal. *Proc. of 18th. Design Automation Conf.*, Nashville TN. June 1981, pp. 696-701.

[17] **Newton, A.R., Pederson, D., Sangiovanni-Vincentelli, A. and Sequin, C.** Design Aids for VLSI: The Berkeley Perspective, *IEEE Trans. on Circuits and Systems* Vol. 28. No. 7. July 1981, pp. 666-680.

THE ICEWATER LANGUAGE AND INTERPRETER

Patrick A. D. Powell and *Mohamed I. Elmasry*

VLSI Research Group, Institute for Computer Research
University of Waterloo, Waterloo, Ontario, Canada N2L 3G1

ABSTRACT

A symbolic circuit design language for describing the topology and topography of a VLSI design in a simple and hierarchical manner is described. The language was intended to provide a simple manner of structuring a VLSI design, based on the Mead and Conway design methodology. Cells may be constructed from other cells and technology specific devices. Terminals for interconnecting cells are explicitly named, and may be accessed in a symbolic fashion from the language. The restriction of methods of interconnecting cells to simple abutment and specific wiring provides a simple and clear method of maintaining the connectivity of the design. The methodology proposed obviates the need for overly complex geometrical design rules. Other tools will provide design compaction, mask generation, and circuit extraction using a technology specific database and the design description.

1. Introduction

The design process can be partitioned into the problems of *definition*, *construction*, and *verifcation* of a solution to a problem. Design automation is aimed at performing simple or well structured procedures of design by computer. The problem of design definition is one of the more intractible parts of the design process, and is being addressed by researchers in the field of artificial intelligence and problem solving systems.

In the VLSI design area, one of the most pressing problems is the construction and verification parts of the design process. The construction of a workable design is pursued by using a design methodology, which enables the designer to structure his solution so that it is *implementable* and *verifiable*. This is a similar problem faced by software engineers, when attempting to construct large programs or systems of interacting programs. The field of software engineering has identified characteristics of software design that enables designers to effectivly build and automatically verify the correctness of the design. As noted by Buchanan, [1] these characteristics can be used to structure VLSI designs. By partitioning a design into interacting modules, and describing the interface and interaction between the modules, the designer keeps the number of elements of a design to a managable level. By implementing a module as a set of interacting smaller modules, the designer can again limit the number of different elements. In addition, in many designs it turns out that the same basic modules are used in many parts of the design, but are arranged or connected in different manners. Many designs are composed of highly regular structures, such as a RAM memory cell, or a PLA. In many processor designs, there is a single bit slice, which is replicated until the neccessary word size is obtained. As pointed out by Sietz [2] a VLSI design can be partition into equipotential regions, where the signal delay between circuit elements is within a minimal amount. By using circuit structures which are small and close, the locality property can be used to ensure correct functionality of operation.

The purpose of the ICEWATER language and methodology is to bring the power of the tools that have been developed for software generation to bear on the problems of VLSI generation. These tools have been characterized by their partitioning of the problem of software generation in distinctly different phases, and by providing managment tools for each phase as well as for translating between phases. For example, a common model of VLSI design partitions the design process into *functional*, *logic block*, *circuit*, *layout*, and *mask* levels (Figure 1.1). One goal of silicon compilers is to

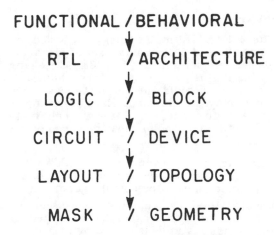

Figure 1.1 Design Representation Hierarchy

The work described in this paper is part of the Systems On Silicon Project at the University of Waterloo, and has been partially supported by a N.S.E.R.C. Strategic Grant.

Reprinted from *ACM/IEEE 21st Design Automation Conf. Proc.*, 1984, pp. 98–102.

provide automatic translation of a functional description into a mask description, which when fabricated will yield a correctly functioning circuit. This may be accomplished by restricting the class of designs to be implemented to a narrow set implementable by a predesigned set of mask level building blocks. The synthesis operation then consists of placement and abutment of the building blocks. This is a powerful synthesis mechanism, as a large class of digital designs fall into the implementable functions. Unfortunately, the design procedure may not always yield an optimal implementation for a particular problem. In the perspective of software engineering, this problem is similar to translation of a high level language to a set of *macros*, which are in turn expanded blindly into machine instructions. This approach usually does not yield high quality code, as very little *optimization* is performed. More importantly, in a VLSI design there may be alternate structures possible for a particular requirement. If the VLSI design system does not provide the facilities for construction or use of these facilities in a design, it may not be possible to produce optimal designs. [3]

The ICEWATER methodology provides a way of introducing layers of representation into the design process that make it possible to generate *optimal* designs by exploiting the modular, hierarchical, and regular features of VLSI designs. In addition, it allows the user to develop tools which can exploit these features, and easily integrate them into a design system.

Finally, the ICEWATER methodology makes easy *design retargetting* possible. If a techonolgy dependent value used in the construction of a design changes, then only the design constructors effected by the value need to be changed. The design can then be easily regenerated, allowing a series of different mask sets for the same design to be generated.

2. ICEWATER

The ICEWATER methodology is based on the representation of a design as a set of hierarchically structured cell instances. The ICEWATER language is used to define cell constructors which are procedures used to build the cell instances. The constructors may be parameterized, allowing a cell constructor to generate different cell instances whose function or topology may be based on parameter values (Figure 2.1).

The designs produced by ICEWATER are hierarchically structured sets of cell instances. A cell instance has topological, geometrical, and electrical attributes which are specified or constructed by the cell constructor. The geometrical model of a cell instance is a rectangular region containing the mask elements of a cell instance (bounding box). The instance has named terminal points which are the connection points for the cell. Terminals are contained in the bounding box, and are adjacent to one of the 4 borders (north, south, east or west), or are interior terminals. A terminal has a set of materials that are compatible with it. By using these properties, it is possible to perform checks to ensure that a cell terminal is connected correctly.

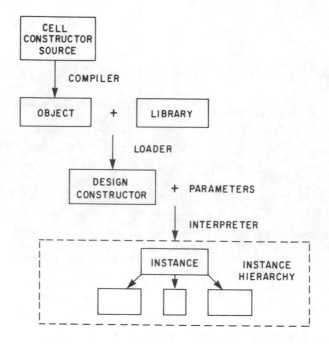

Figure 2.1 ICEWATER Hierarchy

Cell constructors are written in ICEWATER, a simple, block structured language. ICEWATER provides integer, real, and coordinate data types for local and global variables, as well as the ability to create cell instances. A cell instance can be regarded as a particular value of a record (Pascal) or a structure (C programming language). It has two parts: a form part which is initialized by the parameterized cell constructor, and a location part which is assigned by the invoking cell constructor using the place statement.

An example of a cell constructor is given in Figure 2.2. The first part of the cell constructor defines the parameter names, types, and default values. The terminal and attribute statements define the names of terminals and attributes which are part of the created cell instance. The CHAIN cell constructor creates an array of instances which are initialized with the value returned by the N_device cell constructor. Cell constructor parameters are passed by keyword, rather than position as is commonly done in most high level programming languages. One reason for this is to allow functions with relatively large numbers of parameters to be used in a relatively safe manner. In addition, it has been observed that most cells are constructed with a standard or default set of parameters, and only infrequently do the unusual values of parameters get specified. The use of keywords allows type checking to be done, as well as specification of a minimal number of values for instance creation.

When a cell constructor is to be invoked, the cell constructor parameter list is built by overriding default parameter values with specified values. If the cell constructor has already been invoked with the same parameter values, a copy of the previously returned value is used. This limits the execution of

```
define CHAIN( N=1, L=2 ) begin
    int t;
    attribute middle = 0%0;
    terminal s, d, g[N];
    N_device( l=L ) : dev[N];

    if( N < 1 || N > MAX_IN_LINE ) then
        error( NAND: N (%d) out of range, N );
    endif;
    place dev[1] at 0%0;
    name: dev[1].src as s; name: dev[1].gate as g[1];
    for t = 2 to N do
        place N[i].src at N[i-1].dr + 0%L;
        connect: N[i-1].dr, N[i].src;
        name(border=IE|IW): dev[i].gate as g[i];
    endfor;
    name: dev[N].dr as d;
    middle - 0 % (( %< dev[N].dr )/2 )
end
```

Figure 2.2 CHAIN Cell Constructor

cell constructors to those needed to create unique cell instances.

The other statements of the language are used for instance placement, connection, terminal definition, and attributes. Instances have a reference point or local origin, which must be placed at an absolute coordinate or relative to some other object. All coordinates of an instance are taken relative to this reference point. Instances cannot overlap unless they are placed using the abut statement.

Connections between cell instances can be made only at terminals. The terminals of a cell instance have a location, compatibility, and connection direction attribute. Compatibility is used to determine if a connection can be made to a terminal by a wire of a particular material, while the connection direction is used to determine if the direction of connection is legal. This is similar in high level languages to type and bound checking of variables.

The name statement specifies the location and name of a terminal. The locations of terminals are restricted to terminals of cell instances, contacts, or points which are on the path of a wire.

Wires are used to interconnect the parts of the cell instance. A wire has a type which is dependent on the particular technology of the design. Two wire types are compatible if they can cross with no violation of technology design rules. A contact is a technology dependent device which has no active properties, but allows the connection of different wire types. Wires can be parameterized, allowing width, extension, and type to be specified. When a wire is created, it is maintained as an entry in the wire table, an array of instance-like objects with attributes. The wire array is accessed by using the reserved name wire. The wire_number variable contains the index number of the last entry into the wire table.

Example:
```
connect(width=10,ext=HALF):
    N[i-1].dr, N[i].src;
name wire[wire_number].path[1] as g[N];
```

The path of a wire is a list of coordinates and terminals. In addition, the wire_last variable contains the index number of the last entry on the path. The extend statement is similar to the wire statement, but will cause a selected wire to be extended by the path.

As indicated in the above examples, simple arithmetic (vector addition and subtraction) can be performed on coordinates, and the x and y components can be extacted using the %< and %> operators respectively.

Cell instances and wires have associated values which can be accessed by using the attribute mechanism. There are two classes of attributes: user defined and computed. The user defined attributes are declared as part of the instance value returned by the cell constructor. During instance construction, attribute variables are treated as local variables. After the instance has been constructed, the values of the attribute variables are available for use by other cell constructors. Computed attributes are built-in functions of the ICEWATER system. They may be used to provide information about the cell instance. For example, the size attribute returns a coordinate whose value is the size of the bounding box of the instance.

In addition to attributes, ICEWATER supports user defined functions. The user must declare the function type and parameter list, in a similar manner to a cell constructor.

Example:
```
int Larger( coord, coord );
```

In the example above, the Larger function returns an integer value when passed pair of coordinates. Function parameters are passed positionally. The compiler will perform compile time type checking for parameter types. Functions are implemented as part of the ICEWATER interpreter, which is described later.

3. Compilation and Interpretation

The ICEWATER language is supported by a compiler and interpreter. The compiler translates a source file containing cell constructors into intermediate code, which is put into a *library* or an *object* file. Part of the object file is a set of definitions for the objects defined in the object file which the compiler can use in separate compilation.

One of the goals of ICEWATER was to support separate compilation, as well as attempting to provide consistency and type checking similar to high level software languages. To support separate compilation, it is neccessary to know the type and calling sequence of functions or variables used from other separately compiled objects. The *import* and *export* statements are provided to allow the compiler to do compile time checking. The *export* keyword can be attached to a global variable or cell constructor definition, to indicate to the compiler that it is available for use by other constructors or functions. When a variable or cell constructor is used in another module, the compiler must be informed of the *object module* containing the compiled constructors. For example, if the CHAIN constructor was defined in file *CHAIN.ice*, then the object file *CHAIN.obj* would be produced by the ICE-WATER compiler. If the CHAIN constructor was used by another constructor in a different source file, the line

import "CHAIN.obj";

in another source file would cause the compiler to extract the necessary information from the *CHAIN.obj* file.[1]

A *constructor library* may be generated by the compiler. All constructors in the library are assumed to reference a common set of constructors, and may reference other constructors in the library. This is useful to shorten the time needed by the compiler to extract information and object modules.

The ICEWATER interpreter is used to construct the cell instances. When invoked, the interpreter is provided with a list of object files or libraries, and a *root cell constructor*. The interpreter constructs a *design constructor file* containing all of the neccessary cell constructors.

Next, the intepreter executes the cell constructors, starting at the root cell. The interpreter constructs a data structure representing the design, which can then be stored on a design data structure file.

User defined functions are supported in a similar manner to the LISP dynamic loading capability. The user specifies an object library of functions which are to be used by the interpreter. The interpreter will dynamically load these functions, and when encountering a function call in the cell constructor, will call the function in the user library.

4. Design Representation

The output of the ICEWATER interpreter is a data structure representing a hierarchically structured set of cell instances. The format of the data structure patterned on the concept of the LISP *s-expression*. The basic elements of the data structure are *atoms* and lists. Atoms can be integer or real *numbers*, *strings*, *names* (which are short strings), and *pointers* to lists. A list is a sequence of atoms or lists. There is a special *null* list, which is a list with no entries.

The notation adopted for describing the data structure is cribbed from standard LISP notation, with the following simple extensions. A list is represented by a parenthesis, followed by the elements of the list which can be separated by commas for readability, and terminated by a closing parenthesis. A list can be *named*, and a pointer to a named list can be formed by prefacing the name with the ^ (caret or "hat") sign. The word *nil* may be used to represent the empty or null list.

Examples of lists:
```
( )
Def = ( ( name NAND),
          (parms ( (L int 0), (M int 0) ) ))
( ^Def )
```

In the above example, there are three lists, the null list, the list tagged with *Def*, and a list which consists of a pointer to the *Def* list.

A cell instance is constructed by forming a list containing information about the cell instance. The list consists of a sequence of entries of *key* and values. For example, the *Def* list in the above example could be the first couple of entries in a cell instance. The instance name, the parameters and their values, are recorded in the list structure, along with other information.

The data structure is implemented in two forms: an internal or machine level form, and an external or human readable form. There are simple utilities available to translate between the two forms, as well as routines available to perform a large number of LISP-like list manipulations.

5. Rationale

Many of the design and implementation decisions of the ICEWATER system and methodology were based on careful consideration of a desirable environment for an integrated set of VLSI design tools. First, a basic assumption was that the tools would operate in an environment similar to the UNIX Operating System, or a LISP based environment, which would be characterized by a hierarchically structured file system, and powerful task and job control facilities. For example, the UNIX pipe facility, where simple file IO redirection facilities combined with concurrent process execution provide a convenient method of pipelining data between simple software tools.

Another assumption is that the problem of *management* of a design will soon (if not already) become as important as the actual design itself. To this end, tools for managing designs will need to be incorporated. By providing rudimentary facilities for separate compilation, source managment and other *dependency management* tools can be easily incorporated.

Finally, the problems of debugging and extension of the language were considered. The ICEWATER language could have been embedded in a high level language similar to the "i" language [4] or to the ALI language. [5] These languages allow the full power of their *host* language to be applied to a problem. However, the process of program development now encounters an added layer of complexity, in ensuring that the design language can be easily mapped into the host language. Another problem is in the *design turnaround* time. If a modification is made in an ALI program, it is necessary to recompile the ALI program into a Pascal program, then compile the Pascal program, and finally run the compiled Pascal program. If the generated Pascal program is of any size, or there is no separate compilation facility available, this can be a painfully slow process. The implementation of ICEWATER by a simple translation followed by interpretation should be a much faster method. In addition, it is possible to construct *interactive editors* for ICEWATER, in a similar fashion that the *BASIC* programming language supports.

The manner of incorporating user defined functions was carefully considered. First, it was observed that no matter how carefully a language was designed, it would be sure to have a class of algorithms that would not be easily implemented.

Secondly, ICEWATER was not intended to be a general purpose language. It is aimed at the problem of creating hierarchical structures representing a VLSI design. Thirdly, it was observed that the Franz Lisp language [6] provided the capability of interfacing to routines written in other languages. Having already borrowed the data structuring concepts of LISP, it was a simple step to deciding to add user defined functions by a similar mechanism. This facility allows users to implement functions in an appropriate high level programming language.

With the close relationship of the data structures to LISP, as well as inclusion of some advanced LISP facilities, it is surprising that the language of implementation was *not* a dialect of LISP, but the C Programming Language. The reasons for this choice was based on evaluation of the tools available for implementation and the *portability*, *understandability*, and *complexity* of Lisp. Firstly, there were no *compiler writing tools* available for developing LISP based compilers for LR(1) based grammars. This meant that the ICEWATER compiler would have to be written in one of the available languages such as C, Fortran, Pascal, Mainsail or ADA. Secondly, there was a large body of experience in the developement and implementation of interpreters. By using the a block structured language, it would be possible to transfer this experience in a simple manner. Lastly, there was the problem of documenting and understanding a complex LISP program. *Raw* lisp code is very hard to read or understand. For this reason, users extend LISP and use the macro facilities of LISP to provide simple tools. Unfortunately, even with these extensions LISP can be very difficult for the *novice* user to understand. If the ICEWATER system is to be developed and used, it will have to be examined by novices, and then modified and extended. After careful consideration of the software developement environment at the University of Waterloo, a LISP implementation was rejected.

6. Integration With Other CAD Tools

The ICEWATER system provides a data structuring system that allows hierarchical specification of a VLSI design. There are semantics for specifying the connectivity and placement of the cell instances. This may be extended to allow a full mask level specification. For example, the *rectangle* and *polygon* statements could be added to the language, allowing specification of mask level information.

Another approach is to use ICEWATER to specify a design using a *virtual grid* system as in the Mulga [7] system. In this system, there are a basic set of technology dependent devices, with a standard fixed *representation*. The designer specifies his design in terms of these devices and a virtual grid. The design is then *expanded* by substitution of real device sizes, and then *compacted* to occupy the minimum area.

Another possibility would be to extend the language to allow *relative* positioning of devices. Not surprisingly, the *place* and *connect* statements can be extended to support this feature. It is then possible to add an additional stage of design *postprocessing* to place components in optimal locations.

As mentioned in the introduction, *construction* of a design is only part of the designers task. He must be able to verify that the design performs correctly. Currently, there are two major areas of verification: geometrical or mask level and functional or electrical. Geometrical verification (or *design rule checking*) is used to ensure that the mask layout does not violate any design rules of a particular process. If a design is structured hierarchically, and a design methodology is used which composes a design into disjoint modules, it is possible to use a *hierarchical DRC* which can be run quite effectively on even large designs. The interface between a hierarchical DRC and the ICEWATER data structure should be fairly easy to implement.

Another area of verification is the electrical or functional simulation of a design. One class of simulators requires a design to be specified as a *net* of nodes and connections; they make only rudimentary use of the hierarchical nature of the design. However, a new class of simulators is being developed which take advantage of the structure, and allow *hierarchical* or *mixed-mode* simulation. For example, if there is a *nand-gate* in the design, it is possible to specify a *nand-gate* model. When an instance of a nand gate is created, then a model can be created for the gate, which consists of the general nand-gate model, with suitable parameters for the particular instance. This can be extended upwards in the design hierarchy. Similarly, if there is no model for a particular structure, using the connectivity and placement information in the ICEWATER data structures, it may be feasible to automatically *synthesis* a model.

7. Summary

The ICEWATER Methodology provides a tool for the hierarchical description of a VLSI design. Implementation of the compiler and interpreter is under way, and a preliminary version should be available in July 1984.

8. References

[1] Irene Buchanan, Modelling and Verification in Structured Integrated Circuit Design, Ph.D. Thesis dissertation, Silicon Structures Project Report 3642, California Institute of Technology, Pasadena, Calif., 1980.

[2] Carver Mead and Lynne Conway, *Introduction to VLSI Systems*, Addison-Wesley, 1980.

[3] Jeffrey R. Fox, The MacPitts Silicon Compiler: A View from the Telecommunications Industry, *VLSI Design* Vol. 4(No. 3), May-June, 1983.

[4] Steven C. Johnson and Sally A. Browning, The LSI Design Language i, Bell Labs Technical Memorandum 1980-1273-10, Bell Labs, November, 1980.

[5] Richard J. Lipton, Stephen C. North, Robert Sedgewick, Jacobo Valdes, and Gopalkrishnan Vijayan, ALI: a procedural language to describe VLSI layouts, *Proceedings of the 19th Design Automation Conference*, 1982.

[6] John K. Foderaro, The FRANZ LISP Manual, Computer Systems Research Group, University of California at Berkeley, 1980 [Franz Lisp Reference Manual, Liszt Lisp Compiler, and Joseph Lister Trace Package].

[7] Neil H. Weste, Mulga - an interactive symbolic layout system for the design of integrated circuits, *Bell Sys. Tech. J.* Vol. 60(No. 6), July-August, 1981.

A TARGET LANGUAGE FOR SILICON COMPILERS

Robert Mathews, John Newkirk, and Peter Eichenberger

Stanford University

Abstract

A general-purpose layout description language can serve both for direct use and as a target for higher-level "silicon compilers". We discuss our experiences with LAVA, a sticks-based layout language that guarantees design-rule-correct layouts for general designs. We conclude that such a constraint-based system provides the designer considerable leverage on the problems of designing and laying out integrated circuits, and in fact can form a viable "code generator" for a silicon compiler.

Introduction

Integrated circuit descriptions may be broadly classified along two dimensions: level of abstraction, and generality of application.

Greater abstraction permits the designer to describe a given chip more concisely without the possibility for lower-level errors. However, no one knows how to compile an abstract description into a practical layout without severely restricting the class of chips that the designer may specify. Currently, the toolmaker must decide how to trade off abstraction and generality to achieve acceptably efficient compilation.

We decided to investigate what gains we could make in abstraction while retaining full generality. Thus, we are in effect exploring a code generator for silicon compiling, in the guise of a language practical for direct use. In contrast, the archetypical silicon compiler, Bristle Blocks[1], accepts a moderately abstract description of a datapath, but is restricted to producing layouts of a fixed floorplan. Such a compiler would be much easier to construct if targeted into the chip "assembly language" we shall discuss presently.

LAVA

LAVA is a general-purpose layout language for nMOS intended to increase the designer's productivity. Using LAVA, a general design may be created more quickly and modified more easily than with the current generation of geometry-based tools.

Language Features

LAVA descriptions are not geometric, but rather, electrical and topological. A circuit consists of gates or stick diagrams with explicit interconnections; a layout consists of cells with pins around their boundaries, their placement described in relative terms ("to the left of", "below"), and their positions controlled by constraints. Geometric details are hidden from the designer, precluding design-rule violations, inadvertent transistors, and similar geometric mistakes. A design is described as a nesting of cells. The base-level, or leaf, cells contain all components. Leaf cells are composed and interconnected in the higher-level, non-leaf cells. This separation, together with inviolate cell boundaries, enforces a design style that permits the LAVA compiler to perform fast circuit extraction.

LAVA is descriptive, not algorithmic: the order of statements within a cell body is immaterial. Neither assignment nor recursion is permitted, so general computations cannot be performed either (this restriction has not caused any problems). However, LAVA does provide passed parameters and simple control structures, and they play the same role here as in conventional programming languages: to permit a single, generic description to take the place of many fixed, specific descriptions. LAVA control structures are the if-then-else conditional and the foreach and iterate "looping" constructs.

Reprinted from *24th IEEE Computer Soc. Int. Conf. Dig. Papers*, Spring 1982, pp. 349-353.

LAVA also includes electrical checking based on typing mechanisms. V_{dd} and ground signals are declared and checked, but more importantly, clocking may be declared and checked as well. Thus, the compiler can enforce the validity of a logic signal with respect to a clock. The typing system is only weakly developed in the current compiler, however.

```
/*
 * Test example for circuit extraction
 *
 */
cell toplevel()
 {
      c();
 }

cell c(l1, l2, l3, b1, b2, t1, t2, t11)
 l1:left[1];
 l2:left[2];
 l3:left[3];
 b1:bottom[1];
 b2:bottom[2];
 t1[3]:top[1];
 t2[2]:top[2];
 t11[3]:top[3];
 {
      top_group = b (tl1, tbarray, tr1, tr2; 0);
      b (bl1, bbarray, br1, br2; 1)
              flip ud below top_group;
      tl1 # bl1;
      l1 # bl1;
      b1 # tr1 # br1 # t1[1];
      b2 # tr2 # br2 # t2[0];
 }

cell b(l1, barray, r1, r2; flag)
 l1:left[1];
 barray[3]:bottom[1];
 r1:right[1];
 r2:right[2];
 {
      Iterate (3) right a(cl1, cb1, cr1, ct1; 0)
         {    cr1 # cl1;      };
      l1 # cl1[1];
      if(flag == 0)
              r1 # r2;
      else
              barray # cb1;
 }

cell a(l1, b1, r1, t1; flag) leaf
 l1:left[1];
 b1:bottom[1];
 r1:right[1];
 t1:top[1];
 {
      tran1 = tran h [10,10] (1:1);
      l1 # tran1.chan2;
      r1 # tran1.chan1;
      b1 # tran1.gate;
 }
```

Figure 1. A simple LAVA example.

A simple logical description appears in Figure 1. The top level is effectively the composition cell "c", which consists of two instances of cell "b" interconnected and connected to c's pins. (Connections are declared by the "#" operator.) Cell b in turn consists of an array ("iterate (3)") of "a" leaf cells, containing in this case a single, minimum-sized, enhancement MOSFET connected to a's pins.

The Language Processor

The LAVA language processor performs 5 major functions:

(1) circuit extraction from logic diagrams, stick diagrams, and connectivity information;
(2) signal type checking;
(3) sticks compaction, i.e., the conversion of stick diagrams to layouts;
(4) automatic constraint generation and solution to stretch and abut cells; and
(5) routing to complete the interconnections.

Functions (1) and (2) permit LAVA to function as a hardware description language (HDL) for verifying a logic design. (3) - (5) are necessary for reducing the description to a layout. We shall explain further by describing the design flow for an actual chip.

A Serial Memory Chip

The example we shall discuss is a serial memory subsystem for signal-processing applications. The core of the design is an array of 3-transistor memory cells arranged in 50 columns, each column containing 6 9-bit words (2700 bits total). The array is addressed serially by a shift-register-based selector, yielding a serial memory, but one with comparable area and much lower power consumption than a simple inverter-based shift register. Additionally, one bit of each word is used to provide 1-bit defect tolerance per word[2].

The floorplan for the memory chip is shown in Figure 2. There are 7 generic cells, excluding pads and wiring, and a grand total of

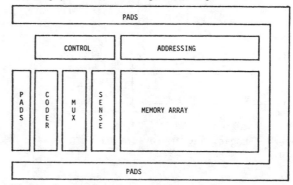

Figure 2. Floorplan of serial memory.

241

nearly 10,000 transistors. Simple control is included; the routing required is simple as well. However, since each of the generic cells is used in many different contexts, the chip has proved a good test case for LAVA.

Logical Design

First we verified the logical design of the chip using the hardware description facilities of LAVA, as illustrated in Figure 3. Leaf cells were specified in mixed notation; connectivity, in text (Figures 4 and 5). This description was compiled to extract the circuit, which we then checked for proper ratio and control of pass transistors[3].

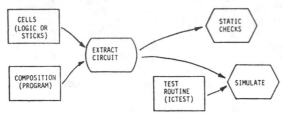

Figure 3. Logical design flow.

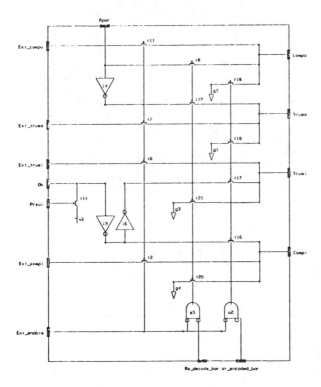

Figure 4. Logic diagram for serial memory control.

To check the functional correctness of the design, we wrote a suite of test programs in the testing language ICTEST[4] and ran them against a switch simulation of the chip[3]. This step was important not only for verifying the design and assessing its testability, but also for developing the test procedure. Hereafter, these tests served to check the subsequent refinements of the design, and because the same ICTEST programs can drive our testers, also served ultimately to test fabricated chips.

```
cell core (Data, E, Sel, Sense, Drive, Prebit,
          Wr_encoded_bar, Re_decode_bar,
          Enable, Preok, A, Phi1, Phi2, PWR,GND)
    Data[TOTALWORD_SIZE]: blue: width 3.0: left;
    Sel[WORDS_COLUMN]: blue: width 3.0: up;
    Sense: blue: width 3.0: up;
    Drive: blue: width 3.0: up;

    • • •
        .
        .
    {
        mem (bio, Sense, Drive, Prebit, A,
             Vddl1, Vddl2, Vddl3, Gndl1, Gndl2, Gndl3);
        io (Data, E, Preok, Wr_encoded_bar,
            Re_decode_bar, Enable, bior, Sel);
        bior # bio;
        PWR # Vddl1 # Vddl3;
        GND # Gndl1 # Gndl2 # Gndl3;
        PWR # VddE1;
        GND # GndE1 # GndE2;
        PWR # VddA1 # VddA2 # VddA3 # VddA4;
        GND # GndA1 # GndA2 # GndA3
           # GndA4 # GndA5 # GndA6;
        control_sim(Phi2, Phi1);
    }
}
```

Figure 5. Fragment of LAVA code.

Layout

The verified design was then converted to a layout as illustrated in Figure 6. We described the leaf cells as sticks (Figure 7) and compacted them (Figure 8). Using the existing connectivity description, these were assembled by abutting and stretching, and together with some wiring cells formed the core of the chip. Because automatic routing was not available, we were forced to finish the layout using a simple, geometry-based language[5]. The layout was circuit extracted and verified using the test battery developed previously.

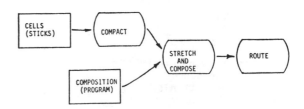

Figure 6. Layout flow.

Result

The finished layout is shown in Figure 9. Using a 5μ process and λ design rules[6], the die is essentially 5 mm x 4 mm. Of 4 dice available to us, one is completely functional, two have defects in the memory plane corrected by the defect tolerance circuitry, and one has a total failure in the precharging circuitry for the memory bit lines. The design is functionally correct, and is now undergoing further parametric and timing characterization.

Conclusions

The LAVA compiler is in an early, experimental stage. Leaf cell handling, including sticks compaction, is completely functional; however, the abut/stretch and HDL portions are just barely working, and routing remains an unsolved problem. Nevertheless, we have been able to draw several major conclusions:

(1) Even the partial system that now exists is a very effective design tool. It affords the designer a means to verify his design and tests early, and the compact/abut/stretch system eliminates concerns about design-rule violations and minimizes the trauma of adjusting pitches of cells. (In the serial memory, the only design-rule violations were—of course!— in the manual routing.)

(2) Despite previous problems with sticks compactors, LAVA demonstrates that sticks compaction works, provided it is properly integrated into a larger composition tool. A standalone sticks compactor is less useful, especially if it attempts to introduce jogs and component reorganizations automatically - the resulting cells tend to be small, but not what the designer had in mind. On the other hand, a sticks system that treats whole chips as one large sticks problem will be hopelessly inefficient.

(3) The critical question for constraint-based tools is whether the constraint system can be generated and solved efficiently. Automatic generation of reasonably well-behaved systems of constraints seems practical; results from LAVA indicate that graph-based constraint solving, exploiting the sparse problem structure, can also be done efficiently. Compiling VLSI designs (100,000+ devices) is still likely to be time consuming, however, on a par with recompiling an operating system and its utilities. Separate compilation of pieces of a design will be essential.

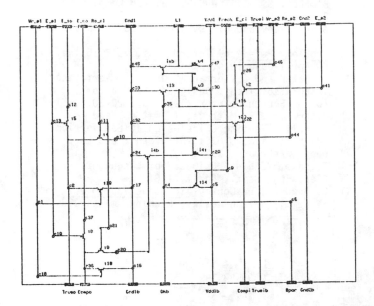

Figure 7. Stick diagram for serial memory control.

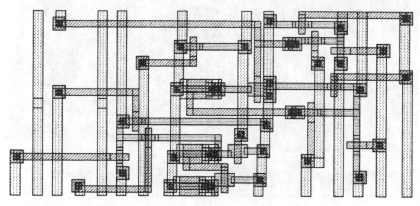

Figure 8. Compacted and stretched layout corresponding to sticks shown in Figure 7.

243

(4) The language is unnecessarily clumsy, primarily owing to bulkiness due to enumerating connections and pins. These problems are syntactic, not fundamental.

(5) A simple graphics editor is invaluable for describing non-parametric leaf cells, primarily because LAVA uses a pseudo-coordinate system to represent logic and stick diagrams. Since non-parametric leaf cells are common, their graphical description saves time. Despite (1), however, composition cells are best described as text so as to capture easily their parametric and conditional aspects.

Summary

We have developed and performed some initial experiments with LAVA, a general-purpose nMOS layout language with a non-geometric basis. Sticks compaction and constraint solution are a promising foundation for higher-level silicon compilers, making a language like LAVA good as a target for them. In the interim, LAVA is an effective direct-use language.

Acknowledgements

We gratefully acknowledge the crucial roles played by Tim Saxe, who wrote the LAVA graphics editor and carried through much of the design and most all of the testing of the serial memory, and Dan Perkins, who (re)designed, implemented, and contributed ideas toward the LAVA compaction, abutting, and stretching subsystem. This research was sponsored by DoD ARPA under contract MDA 903-79-C-0680.

References

(1) D. Johannsen, "Building a Chip with Bristle Blocks", SSP File #3519, CalTech Comp. Sci. Dept., February 1980.

(2) A Kushetsov and B. Tsybakov, "Coding in a Memory with Defective Cells", translated from Problemy Peredachi Informatsii, vol. 10, no. 2, April-June 1974, pp. 62-65.

(3) C. Baker and C. Terman, "Tools for Verifying Integrated Circuit Designs", Lambda vol I, no. 3, Fourth quarter, 1980, pp. 22-30.

(4) J. Newkirk, R. Mathews, I. Watson, and W. Wolf, "Testing Chips Using ICTEST version 1", VLSI File #020281, Stanford Elec. Eng. Dept., February, 1981.

(5) T. Saxe, "CLL - A Chip Layout Language", Elec. Eng. Dept., April, 1981.

(6) C. Mead and L. Conway, Introduction to VLSI Systems, Addison-Wesley, Reading, Mass. 1980.

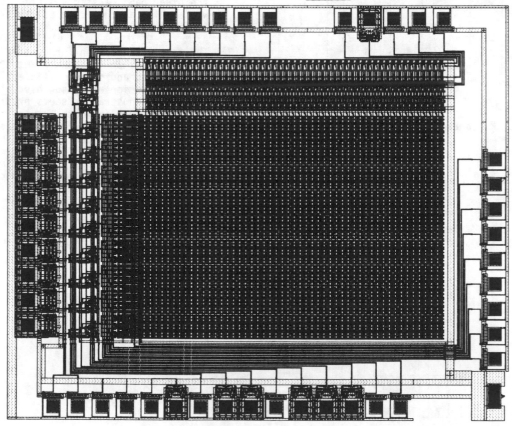

Figure 9. Complete layout of the serial memory.

Digitizing, Layout, Rule Checking—The Everyday Tasks of Chip Designers

JEAN PIERRE AVENIER

Invited Paper

Abstract—The layout phase is most critical in the design of integrated circuits (IC's) because of the cost of the phase itself, since it involves expensive tools and a large amount of human intervention, and also because of the consequences for production cost. Several approaches are used that need more or less computer and/or man time.

The compromise is difficult because of the number of parameters to be taken into account. This paper presents the methods most commonly used with their advantages and disadvantages.

I. INTRODUCTION

THE DESIGNER who has the task of drawing the layout of an integrated circuit (IC) is confronted with a large set of methods. The main compromise is related to the ratio between design cost and production cost. An IC with a high-density layout will be long and expensive to draw, but the small area will induce a higher production yield and a higher number of dies per wafer as well. So the main criterion to be taken into account for the choice of a layout methodology is the production volume. The design duration is next. Today, the size and complexity of circuits impose hierarchy and symbolic approaches. These techniques advocated by MEAD and CONWAY are not yet fully established in the industrial world, but they prefigure the state of the art. It is important, also, for high-cost, complex circuits to have the opportunity of following the technology improvements. The symbolic methods generally allow such transformation, but other tools offer the ability to shrink the layout with some intelligence, sizing the patterns in accordance with their electrical functions.

As the cost and duration are increasing rapidly, it is of prime necessity to run a maximum of verification programs on the layout before the mask generation. These checks refer to the design rules, the electrical rules, and the consistency with a reference wiring diagram.

II. HAND-DRAFTED METHOD

A. Hand-Drafted Layout

1) Presentation: Although the size of VLSI circuits has made the use of manual drawing for a complete layout almost impossible, these methods are widely used and their limits have been pushed away by the new graphic systems available on the market. The growing number of systems installed in various semiconductor companies is proof of a large success. But the kind of work is changing at the same time. They are used as the front end of more complex CAD systems—for the

Manuscript received June 22, 1982; revised October 24, 1982.
The author is with Thomson-EFCJS, F-38019 Grenoble Cedex, France.

definition of the basic cell library, for example. However, in spite of the increasing part of regular structures (ROM, RAM, PLA) in the designs, a great deal of the layout is still done manually. Why? The main reason is the good density of the hand-drafted layout and, above all, its aptitude to fill the remaining areas left by the regular blocks. There is no doubt, again, that for VLSI circuits the shift to more sophisticated methods (that we shall examine later) is compulsory.

B. Methodology

1) Digitizing: It is the oldest approach. The work is split into two phases. The first is drawing of the several layers of the artwork on mylar at a large scale (1000, 2000 for up-to-date technologies). This requires a very good knowledge of the design rules. Generally, the designer himself has to do the major part of the job. A second phase of coding on a large-scale (more than 1 m × 1 m) and accurate (0.1-mm) digitizer inputs the layout into the computer. Associated is a screen that allows a feedback control of the operation.

This process is obviously very cheap in computer resources. The coding phase represents only 5 percent of the total time. Human interaction with the computer is very poor. The drawing of the artwork is long and extremely tedious, and the digitizing itself requires sustained care.

2) Interactive Input on a Graphic Terminal: The availability on the market of high-quality color displays for a relatively low price and the wide spread of computer use have incited people to go to more interactive methods. The layout is done directly on the screen, often starting from a manual sketch [4].

Some functions (design rules checking (DRC), electrical values extraction) can be used for a quick feedback. The computer system investment is naturally far more important than for the previous method, but the realization times are shorter and the job is more pleasant.

3) Description Language: There is another way to define the geometrical patterns; it is to use a description language instead of a graphic input. This method, seldom developed, has the great advantage of offering all the wealth of a procedural programming language, that is, parameters, conditional and inconditional branches, labels, do-loops computed variables, and so on. This allows the description of parametrizable cells that can be rather independent of the technology.

The drawback of this method is its batch mode process. It is, however, very interesting for the description of the basic elements of a technology (transistors with their W/L ratio as a parameter, for example). It is the essential accessory of a more sophisticated approach for the definition of bristle blocks or

Reprinted from *Proc. IEEE*, vol. 71, pp. 49–56, Jan. 1983.

stretchable cells used in interactive methods. Another advantage is the trace that is automatically kept, which limits the effect of any computer failure.

III. Symbolic Approach

A. Introduction

Symbolic layout methodologies are a means of abstracting the detailed and often laborious task of mask design of IC's. They offer the advantages of hand-packed mask design with regard to density of layout, while also having advantages over manual layout for design duration and correctness. In essence, the use of symbology reduces the complexity of the IC design process, and, in addition to the advantages mentioned, allows experienced designers to undertake more complicated circuits than would otherwise be possible, and, more importantly, allows novice designers to complete designs with a high degree of confidence. This last point is regarded as especially important as system designers move to using silicon as an implementation medium, rather than more conventional techniques.

B. Symbolic Layout Methodologies

Symbolic layout methods attempt to abstract the detailed task of designing IC masks to clarify this operation. Normally, this is achieved by simplifying the design rules for a given process. These design rules include the minimum spacings and widths of the mask layers used in the technology. They also include electrical rules for interconnecting layers and the formation of active devices. These simplified rules ideally result in a quicker turnaround of designs and a reduction in errors compared to manual layout.

1) Fixed-Grid Layout: Fixed-grid layout systems divide the chip surface into a uniformly spaced grid in both the x and y directions. The grid size represents the minimum feature or placement tolerance that is desired in a given process. For each combination of mask layers that exists at a grid location, a symbol is defined. Given a particular design system, these symbols are then placed on the grid to construct the desired circuit much in the same way one would tile a floor. Symbol sets may be defined as characters, or perhaps graphical symbols, if a graphics display is used for design.

American Microsystems International (AMI) and Rockwell International have made use of character-based symbolic layout for some time.

The Symbolic Interactive Design System (SIDS) uses a color character terminal as a design station which provides a high degree of user feedback. In addition to these character-based systems, Hewlett–Packard has developed an interactive graphics system (IGS), which is capable of accepting symbolic input on a fixed grid. The IGS also uses symbolic representations to reduce the time to display hand-designed layouts.

The design process in these systems consists in laying symbols down on the coarse grid. The use of fixed-size symbols simplifies geometric design rules but does not totally alleviate them. SIDS, therefore, provides on-line DRC for geometric design rules violations, and a "trace" facility to trace circuit nets to visually check for electrical connectivity. Similarly, the IGS provides "bumpers" which surround symbols to aid designers in placing them.

2) Sticks Layout: The term "sticks" [1], [3] is a generic term given to symbolic design systems that do not necessarily

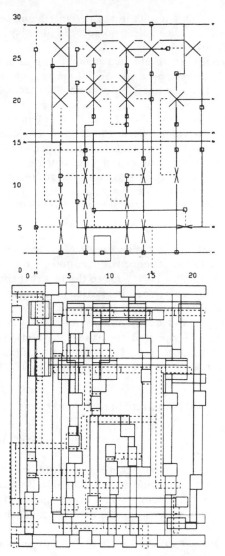

Fig. 1. Input and output of the symbolic program with compaction "TRICKY" (CMOS technology).

constrain the designer to a grid and generally require the designer to enter a free-form topological description of a layout via an IGS (digitizer, tablet, color work station, etc.).

Graphical symbols are placed relative to each other rather than in an absolute manner. Following the definition phase, the symbolic descriptions are converted into valid mask descriptions using a variety of compaction strategies designed to space symbols in accordance with the process design rules.

The TRICKY system [2], developed by Thomson-EFCIS, Grenoble France, uses a grid-based placement scheme. Grid-based placement allows grid entry of geometric topology by "snapping" elements to the grid (and also aids the capture of the circuit details and the subsequent processing needed prior to preparing a valid set of masks). TRICKY accepts free composed and arranged input sketches and results in good layout density and reliability, using a spacer for this purpose; a one-direction algorithm is used successively in both directions, starting in a user-selected one, and can be repeated in alternate directions either a given number of times or until no more space is saved (Fig. 1).

Moreover, STICKS from CALMA removes the grid as a design

consideration. This results in the designer being totally freed from geometric design rules and accessing the layout through a symbolic editor.

Unfortunately, there are two major disadvantages in these methods. The first one is the lack of diagonals because of the complexity of the compaction process. The second one is the fact that the spacer is a batch processed program.

3) Benefits of Symbolic Layout: The simplification of geometric design rules relieves the designer of details that can cloud more global and important issues, such as achieving the correct circuit or communication requirements.

Transparent design rules also make designs relatively process independent. If a process design rule changes, the mask descriptions for a circuit may be regenerated with a minimum effort.

The more recent sticks systems have, in addition to the perceived benefits of a design rule free environment, the basis for capturing circuit connectivity, although few have treated this benefit in detail. This is due to the fact that specific problems have been addressed, in particular compaction, rather than the complete design cycle and the relevant tools required. The time saving and quick turnaround allow a better layout optimization. As opposed to manual methods, DRC is not necessary for this approach. Finally, a loss of 10 to 20 percent of density has to be compared to a reduction of 50 to 75 percent of layout time.

IV. BUILDING-BLOCK APPROACH

This design method is very fast and secure and especially well suited for logic circuits of medium size; density and performances are not critical.

For each technology and range of applications, the designer builds his circuit with standard cells stored in libraries.

The building-block approach consists of a set of blocks arranged in regular bands separated by channels dedicated to interconnections.

Thus the regular layout is able to support automatic placement and routing algorithms.

An important investment of this method is associated with the definition of the standard cells library, a necessary condition for an efficient method.

A. Definition of the Library

For a given technology, a set of standard cells is selected: a two-input NOR gate, a three-input NOR gate, elementary latches and flip-flops with and without asynchronous inputs, input–output buffers, and so on.

Then the layout shape of each cell is chosen; it is generally rectangular with fixed height and variable width.

The global density of the circuit depends strongly on the position of the input–output pads of each cell as follows:

a) I/O pad on each side: important loss of density inside the cell (Fig. 2);

b) no duplication of I/O pad: high density inside the cell but necessity of foreseeing cross cells for interchannel connections (Fig. 3);

c) lateral I/O pad for regular cell structure (register–counter) self-connection (D–Q) just by cell juxtaposition: important gain of density in interconnection area, but necessity to design three types of cells: first, current, and last one (Fig. 4).

Once these basic choices are made, each cell is laid out and

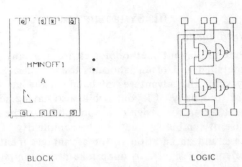

NOR LATCH

Cell Name : HMNOFF1   Iss:A

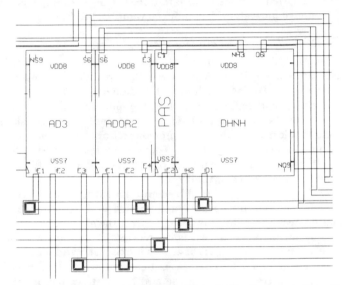

BLOCK   LOGIC

Fig. 2. Example of case a) in Section IV-A (extracted from *COMIC User's Manual*, Hughes Me. Ltd.).

Fig. 3. Example of case b) in Section IV-A. PAS cells are pass-through.

the library contains: logical features—function or logical diagram; electrical features—propagation delay, maximum fan-out; and graphical features—outline rectangle, I/O pads level and coordinates, geometrical details of masks.

B. Layout

The starting point is a logic diagram containing only standard cells. The layout consists in the placement and interconnection of these cells. This task may be performed by means of a graphical station (display and tablet); in this case, the area dedicated to interconnection may be used by the designer to insert some parts of random logic (for instance, transmission gate logic). More frequently, this task is performed by a CAD program that, from a wiring list of cells and from some constraints (critical path and proximity), determines the cells' placement and the coordinates of each interconnection wire. According to the performances of the algorithms used, this process can be either a fully automatic approach, or a semi-automatic tool with manual modifications of automatic placement and completion by hand of unsuccessful wirings.

A main advantage of this method is that the same circuit

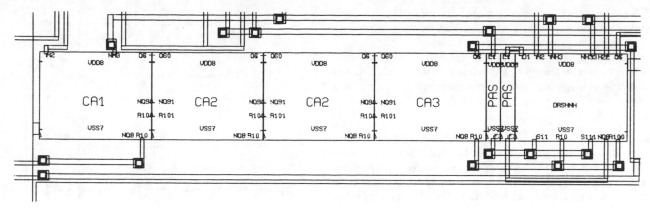

Fig. 4. Example of case c) in Section IV-A, *CA* 1—first cell; *CA* 2—current cell; *CA* 3—last cell.

description (wiring list) is used for layout and also for logic simulation and test pattern generation.

C. Enhancement of this Approach

Many building-block CAD systems offer the opportunity to take into account nonstandard cells, such as blocks of any rectangular shape designed by other methods (PLA's ROM's, RAM's) [5], [6]. Thereby, the placement and routing algorithm had to be modified to support these non-regular structures.

V. GATE ARRAY APPROACH

In the custom design area, it is of prime importance to get very short design delays. So the enhancement of the quality of process has updated gate array approach. The basic concept is to stock nearly fully processed wafers, waiting for the last mask layers, the ones which correspond to the interconnections.
The following different kinds of arrays are possible.

1) *Transistors arrays:* Basic elements (transistors) are first interconnected to build elementary gates or blocks (these interconnections can often be stored in library), then the gates themselves are interconnected to realize the whole circuit.

2) *Gate arrays (ULA):* Basic elements are already logic elements (elementary gate, D flip-flop, etc.) and design work consists only in blocks wiring.

Once the logic description has been simulated to obtain the layout of the circuit, it is possible to use either an interactive display station or automatic tools such as partitioning, placement, and routing.

The loss of silicon area due to the gate array approach in comparison with the hand-drafted method has been reduced by the development of optimization tools. A great amount of work and money has been invested in this area by large companies (IBM, Motorola, Fujitsu, AMI, Signetics) and CAD software companies such as Silvar Lisco, V-R, Information Systems Inc., Scientific Calculations Inc., Compeda, etc.

It is possible to find on the market a wide range of tools starting with fully hand-drafted layout up to fully automated process with 100-percent placement and routing performed by CAD. This type of "always successful" algorithm requires special gate arrays arrangement with sufficient size of routing channels. But the absence of manual operation and graphical associated tools may justify such an approach. Generally, a semiautomatic technique is used, with only about 5 to 10 percent of unsuccessful wiring completed by the user on a graphical display.

Main Remarks: The CAD systems associated with the gate array method, according to the type of concerned circuits, can be very highly integrated (schematic entry, simulation, test generation, automatic layout).

VI. AUTOMATED PLACEMENT AND ROUTING

A. The Problem

The goal is to place the elements (transistors, gates, functions) and to route the wires corresponding to the equipotentials, taking into account the design rules and using a minimal chip area. The task is done automatically, using algorithms. When problems arise, human intervention tries to solve them. The man–machine, manual–automatic interaction is the key to good tools able to work on various technologies.

B. The Placement

This problem is mathematically hard to define. Several measures are taken for optimization. The initial goal may be total length of the wires, number of contacts, or channel density. Moreover, the placement is simplified by the suppression of trees in the equipotentials (only pairs are treated).
There are generally two steps, as follows.

1) Automatic Elaboration of an Initial Placement: Various methods are used.

Static baycenter: Starting from the position of some elements, forces between modules (generated by the connections) are computed and a balanced position is searched.

Branch and bound: Given an evaluation function, the tree of possible solutions is tracked.

Linear order: In the case of linear placements, the problem is to find an optimal order.

Heuristic and constructive methods: The modules are placed one by one, by maximal conjunction or minimal disjunction with others already placed. The associated measures vary. They are the most used.

Min-cut algorithms: The set of modules is cut in two parts to satisfy a law (minimum number of wires between the two parts, equal size of the subsets. . .). These methods are best suited for modules of various size.

2) Automatic Improvement of the Placement: Local modifications are tried. All the measures described before may be improved. However, starting from the initial placement, the routing is simulated to introduce real constraints in the algorithms.

The static baycenter is computed module by module to allocate new positions.

Pair exchange is the most used method, with various measures according to the exchange test. Some difficulties arise when the exchanged modules are of different sizes.

3) Interactive Placement: The designer may preplace some modules or choose between algorithms. He can also, during the improvement phase, modify the placement of a module, specify its form, and change the choice for an I/O allocation.

C. The Routing

The equipotentials are drawn between the modules. Some choices such as the following must be made as a preliminary in accordance with the density or the technology: ON or OFF grid routing (OFF grid is mandatory for high density); EQUI-POTENTIAL TREE cutting; routing order (depending on the length or the strategic position of the wires); and number of interconnection levels.

As for the placement, the routing is now done in two phases.

1) Global Router: This phase results in the allocation of the connections to the various areas of the chip.

Channels definition: The area available between the modules is divided into channels. The wiring is done afterwards channel by channel. The various methods cut the area in rectangles defined by the block boundaries. Association between rectangles depends on the local router.

Connections allocation for each channel: The methods vary with the router. A Lee-type algorithm may be used. A graph of contiguous channels is used to find the wires. First of all, the wires are allocated to some channels. Then the wires of the most chocked channels are redrawn.

Channel wiring order: This point is fundamental because the preliminary allocation of the wires to the channels does not solve the communication problem between channels. The graph of contiguous channels can be used.

2) Local Router: Three approaches exist.

Expansion algorithms on a grid: A wave is propagated from the starting to the ending point of a connection.

Aim algorithm: Two orthogonal lines are drawn from a point until they reach an obstacle. An escape point is then chosen on the line. The process is restarted from this new point. The same process is run from each end of the wire. When the lines cut, the routing is done. This method is much faster than the previous one but the solution is not always found.

Channel algorithms: These methods are widely used. Given a number of connections, several tracks defined in the channel, and I/O positions on each side of it; the routing is done inside the channel. Some connections may be impossible because of the positions on the I/O on the side of the channel.

The order of the routing must be defined previously for the two first algorithms and is given by the third one.

3) Interactive Router: Automatic routers often follow a manual process, and this eases this understanding of the task by the designer. In an interactive system, the user can intervene by commands before or during the routing step for several decisions: allocation of I/O pads; allocation of wires to the channel; modification of the already-drawn connections to relieve some channels; and introduction of underpasses or modification of mask layer for a wire.

VII. SILICON COMPILATION

A. Regular Parametrizable Blocks

If one observes present VLSI circuits, a large amount of silicon area is dedicated to blocks with a repetitive structure, such as PLA's, ROM's, RAM's, Registers stacks, and ALU.

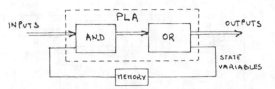

Fig. 5. Example of a synchronous automation using PLA.

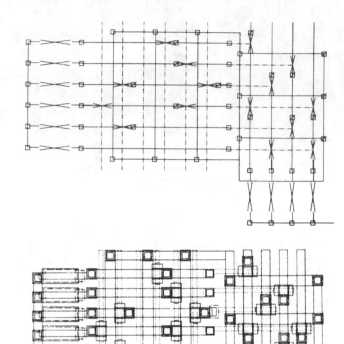

Fig. 6. Example of PLA automatic generation.

Thus in order to save time and to increase reliability, it is very interesting to automatize the design of such elements, that will be called macrofunctions.

To speak about this concept, we take the example of a PLA because it can be taken into account from functional level to layout level.

PLA's are very often used to implement synchronous automation: combinational two-levels logic (AND–OR) and feedback-loop state variables register (see Fig. 5).

The functional description can be described by a language or a flowchart. The logic part consists in deriving a logic equation from flip-flop type and state assignment. This set of equations may be minimized. The final layout is obtained by fixing the sizes of transistors and by specifying the order and position of inputs/outputs (Fig. 6).

B. Silicon Compilers

The trend is to continue the automation process towards the definition of the macrofunctions and the final layout of the whole circuit. The design time can become very short with high reliability. Unfortunately, critical circuits for area, timing, or consumption reasons are not subject to such methods. Silicon compilation starts from the functional description of the circuit. Automatic synthesis and automatic definition of the macrofunctions give the description of the major blocks.

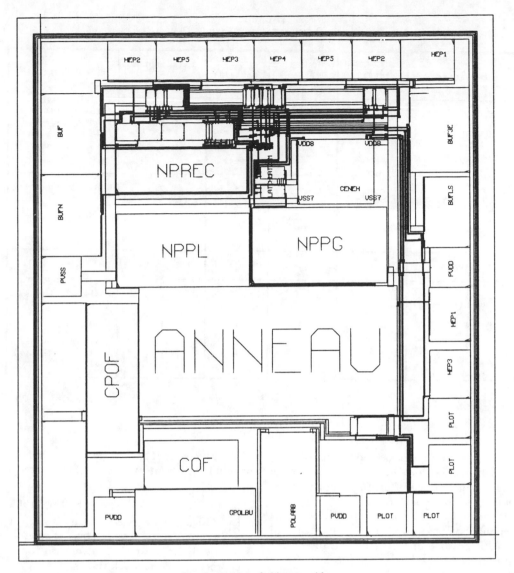

Fig. 7. Output of chip assembler.

A chip assembler (automated placement and routing) draws the final layout. The loss of silicon area is generally high [8].

C. Advantages

These automatic techniques of implementation for complex and programmable blocks allow one to speed up the design and to increase the reliability due to minimal human intervention.

The increase of abstraction level allows one to push the limits in design area (analog to software engineering).

VIII. Hierarchical Approach

Although several methods of structured design have been known for a long time (simultaneously with structured programming), they were far slower to reach an industrial use. The optimization constraints and the need of a high performance level were a powerful brake to this evolution that will be mandatory for VLSI.

A. Introduction

The design methodology must be roughly top-down with stepwise refinement; the layout by itself, on the contrary, has to be bottom-up to reach a good result.

The first step is a general floor plan at a jumbo cell level with

their global connections (buses). The number of elements manipulated must remain under thirty. The shape, the area, the I/O positions are already known for several blocks such as ROM, RAM, PLA, ALU. This first approach will allow the definition of remaining unknown parameters using the general floor plan. The allocation of the PLA lines to variables is made, for example. The remaining area is allocated to the logic not yet implemented. Reshaping of some block of main importance may be decided to fit with its proximity. For example, an $8 \times 2n$ ROM will be better than a $16 \times 1n$. Starting from this point, the work can be divided among several teams for which a more accurate layout process is defined. The same approach may also be used at a lower level.

B. Tools

1) Database: All this process must be supported by only one database for the whole circuit to allow consistency checking between the several parallel tasks. To improve the design security, the layout has to be associated with the previous design steps (functional and logic synthesis) that must be themselves structured. It is interesting to speak about a database that accepts all the data of the various design levels: functional, register-transfer, logic, electric, and layout. This single

TABLE I
ADVANTAGES AND DISADVANTAGES OF VARIOUS DESIGN STYLES

Evaluation / Implantation methods	DENSITY	SPEED	SECURITY	DATA PROCES. COSTS implantation	COMPANIES
DISPLAY	+ + +	- - -	- - -	- - -	CALMA.APPLICON
DIGITIZER	+ + +	- - -	- - -	- - -	CALMA.APPLICON
LANGUAGE	+ + +	- - -	- - - DRC obligatory	(with DRC) + + + +	
MANUAL	- - -	+	-	0	COMIC (Hughes)
AUTO ROUTING	- - -	+ +	+ +	+ +	PHILIPS
FULL AUTOMATIC	- - -	+ + + 200 blocks a week	+ + +	+ + +	MPS2D (RCA) CALMOS (LISCO)
GATE ARRAYS	- -	+ +	+ +		
GRID Max.	-	+	+	0	MASKS (ROCKWELL)
GRID Min.	0	+	-	0	MASKS (ROCKWELL)
Multi step	0	+	0	0	SLIC ou SIDS (AMD)
DEFORMATION	0	+	+	+	ABRAITIS
COMPACTION 1 DIR	+	0	+ + +	+ +	STICKS (CALMA)
2 DIR	+ +	0	+ + +	+ + +	BERKELEY (TRICKY) SLIP (BELL)

+ Good
0 Medium
- Poor

description of the circuit under a blocks + connections form, whatever the level of abstraction, and the memorization of the correspondence between these levels offers the opportunity of an accurate control on the project evolution.

2) Evaluators: In order to avoid a number of loops in the definition of the circuit layout, it is of prime importance to have an idea as soon as possible of the shape and the size of the blocks that will be drawn.

It is the goal of the evaluators, which, given a technology (that is design rules and circuitry) and a specific function to design, will compute some characteristics of the block.

3) Chip Planning:

Chip planning and chip assembler: In connection with the database, two programs are useful for the layout: the chip planning program that helps in the establishment of a general floor plan during the top-down phase and a chip assembler for the accurate definition of the block positions.

The first one must be highly interactive to allow a large number of tries of the placement of the cells. An automatic placement program may be used to propose some solutions. At the current state of the art, only rectangular blocks may be handled by these programs. The connections must be taken into account for the placement because of the large impact they have on the total area of the chip. (Half of the circuit is devoted to connections.)

The chip assembler [7], on the contrary, is involved in the bottom-up phase to get the layout. Each time a block is placed, all the connections are made with respect to the I/O positions and the design rules. Automatic routers can be used to ease the work (to run buses, for example). The lack of diagonals in such routers unfortunately greatly reduces their utility (Fig. 7).

Main Remarks: This approach requires a parallel between the cutting up at the various levels.

Advantages: The process structuration allows the design of very large circuits, preserving realistic security and delays.

IX. ARTWORK ANALYSIS

The design of integrated circuits is not fully automated but contains more or less manual steps. Therefore, error-free design is practically impossible. If an error is found after the fabrication process, it costs much time and money because new masks must be prepared again. Thus complete checkout for design errors becomes a necessary step of the design. Several types of errors may appear in a mask, such as violations of geometrical rules, connectivity errors, or topological errors. So artwork analysis computer programs must be able to perform a wide range of functions. These functions include verifying design rules, discovering electrically conducting paths, and recognizing devices. Several approaches are possible to realize these functions: a "figure-based" method, an "edge-based method," or a "scanning" method. Hierarchical design of the artwork allows checks on basic cells and avoids checks on multiple repetitions of a cell. Using special representation of the border region of the cells, a check can then be performed on all the artwork.

A. "Figure-Based" Approach

This is the most commonly used method. The smallest entity is a closed figure, and various operations are performed on these figures to check for design violations. These operations are Boolean combinations, contraction or expansion, calculation of area and perimeter, and labeling of figures (partitioning the artwork to avoid useless comparisons) or bit map procedures. With this method, one rule can be checked at a time, since the Boolean operations (and expansion or contraction) can be performed on a single layer or between two layers. In order to increase the efficiency of the process, "layer splitting" can be used; this means the elements of one layer are split into several sublayers, each with logically identical components. The bit map procedure presents the difficulty of dealing with the nonorthogonal edges of figures. This approach is quite general and is being used in many programs, with little differences about the partitioning of the artwork. It tends to be expensive.

B. "Edge-Based" Approach

Another checking method is based on "line segments." This approach is more economical since less data must be processed (vertical edges are not part of the data). Each segment is oriented, and the procedure is based on comparisons between segments (check for intersection, test if a point is within a figure, etc.). This method is similar to the "figure-based" one and seems to have the same efficiency.

C. "Scanning" Method

This third method assumes that the artwork is in the form of segments (or points) sorted in lexicographic order. All mask levels are merged into one file before the sort, so all layers will be checked in parallel. Then the artwork can be scanned from left to right by a virtual vertical line drawn across the mask, searching the data to locate all segments which cross this line. The position of the scan line is incremented from the left side of the mask to the right side. It can be stepped by a user-defined increment or by variable increment auto-scan based on "when the data changes" (each point in the data represents a data change). The design rules to be checked or the devices to be recognized are described in technology-dependent subroutines or are compiled separately before the scan begins. This method has two main advantages. All mask levels can be checked in one pass. Besides, interlevel checks can be performed simultaneously.

X. Conclusion

Until today the design methods have evolved rather slowly, the layout remaining the larger part of the work. For VLSI circuits, the trends are evolving rapidly because of the growing need for security and control. The compromise between the various parameters may be done using Fig. 8 and Table I.

More academic approaches are used for the synthesis at the functional and logical levels. Naturally, the layout follows the same evolution. Hierarchic methods are mandatory for large circuits for two main reasons. First, the need of reliability offers the possibility to split the work into several subtasks.

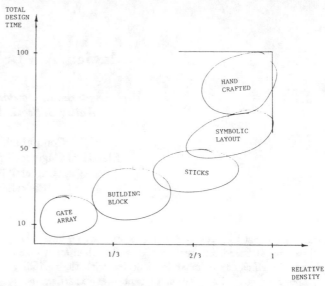

Fig. 8. A diagram of the density versus design time for the various design styles.

The first point refers to the delays. Second, the need of security in the conduct of a process is more and more complex due to the large number of parameters to optimize at the same time. But old methods are still in use and will remain in use for a while each time a difficult problem arises. The evolution towards silicon compilation will cut the design time and the design cost, allowing a large number of small companies to design VLSI circuits mainly for small production volume.

References

[1] N. Weste, "MULGA: An interactive symbolic layout system for the design of integrated circuit," *Bell Syst. Tech. J.*, vol. 60, no. 6.
[2] A. Hanczakowski, "TRICKY symbolic layout system for integrated circuits," in *Proc. Spring Compcon 81* (San Francisco), pp. 374–376.
[3] J. Williams, "STICKS: A new approach to LSI design," thesis, Massachusetts Institute of Technology, Cambridge, MA, June 1977.
[4] J. Ousterhout, "CEASAR: An interactive editor for VLSI layouts," *VLSI DESIGN*, Fourth Quarter, 1981.
[5] A. Feller and R. Noto, "A speed oriented, Fully R automatic layout program for random logic VLSI devices," presented at the Nat. Computer Conf., 1978.
[6] H. Beke, W. Sansen, and R. Van Overstraeten, "CALMOS: A computer aided layout program for MOS/LSI," *IEEE J. Solid-State Circuits*, vol. SC-12, pp. 281–282, June 1977.
[7] S. Trimberger, "RIOT: A simple graphical chip assembly tool," presented at the 19th Design Automation Conf.
[8] A. Szepieniec, "SAGA: An experimental silicon assembler," presented at the 19th Design Automation Conf.

Magic: A VLSI Layout System

John K. Ousterhout, Gordon T. Hamachi, Robert N. Mayo,
Walter S. Scott, and George S. Taylor

Computer Science Division
Electrical Engineering and Computer Sciences
University of California
Berkeley, CA 94720

Abstract

Magic is a "smart" layout system for integrated circuits. The user interface is based on a new design style called *logs*, which combines the efficiency of mask-level design with the flexibility of symbolic design. The system incorporates expertise about design rules and connectivity directly into the layout system in order to implement powerful new operations, including: a continuous design-rule checker that operates in background to maintain an up-to-date picture of violations; an operation called *plowing* that permits interactive stretching and compaction; and routing tools that can work under and around existing connections in the channels. Magic uses a new data structure called *corner stitching* to achieve an efficient implementation of these operations.

1. Introduction

Magic is a new interactive layout editing system for large-scale MOS custom integrated circuits. The system contains knowledge about geometrical design rules, transistors, connectivity, and routing. Magic uses its knowledge to provide powerful interactive operations that simplify the task of creating layouts. Moreover, once a layout has been entered, Magic makes it easy to modify it; this permits designers to fix bugs easily, experiment with alternative designs, and make performance enhancements.

Magic provides several new operations for its users. Design rules are checked continuously and incrementally during editing sessions to keep up-to-date information about violations. When the layout is finished, so is the design-rule check. A new operation called *plowing* allows layouts to be compacted and stretched while observing all the design rules and maintaining circuit structure. Routing tools are provided that can work under and around existing wires in the channels (such as power and ground routing) while still providing the traditional efficiency of a channel router.

Two aspects of Magic's implementation make the new operations possible. First, the system is based on a data structure called *corner stitching* which is both simple and efficient for a variety of geometrical operations [6]. Without corner stitching, most of Magic's new operations would be too slow for interactive use. Second, designs in Magic are specified using abstracted layers we call *logs*, rather than actual mask layers. The logs design style represents circuit structures such as contacts and transistors in a form that appears somewhat like symbolic layout [10,14,15] except that objects are seen in their actual sizes and positions. The logs design style incurs no density penalty over mask-level layout, but it simplifies the designer's view of the system and provides more explicit information about the circuit structure.

This paper gives an overview of the Magic system. Section 2 describes the specific problems Magic attempts to solve, and the overall approach of the system. Sections 3 and 4 describe Magic's internal data structures and the logs design style. Sections 5-11 discuss Magic's new operations, and Section 12 presents the implementation status of the system. Three companion papers discuss design-rule checking, plowing, and routing in detail [2,11,12].

2. Background and Goals

Our previous layout editing systems, Caesar [5,7] and KIC2 [3], have been used since 1980 for a variety of large and small designs in several MOS technologies. They are similar to systems currently in use in industry. Although our systems have proven quite useful, there are a few areas where they (and most other existing layout systems) are inadequate. The most severe inadequacy is in the area of routing, where most systems provide little support. We estimate that between 25% and 50% of all layout time for our circuits is used for hand-routing the global interconnections. The task of routing is tedious and error-prone.

A more general problem is one of flexibility. Once a design has been entered into the layout system, it is hard to change. This makes it difficult to fix bugs found late in the layout process, and almost impossible to experiment with alternative designs. If designers cannot experiment with and evaluate alternatives, it is hard for them to develop intuitions about what is good and bad. Routing is the most extreme example of the flexibility problem. It takes so long to route a circuit that it is out of the question to re-route a chip to try a new floorplan. Even small cells are difficult to change: modest changes to the topology of a cell often require the entire cell to be re-entered. In many industrial settings, layouts are so difficult to enter and modify that designs are completely frozen before layout begins.

Our overall goal for Magic is to increase the power and flexibility of the layout editor so that designs can be entered quickly and modified easily. When the system is complete, we hope it will provide order-of-magnitude speedups for three different parts of the design process:

1) Once a large circuit has been routed, it should be possible

Reprinted from *ACM/IEEE 21st Design Automation Conf. Proc.*, 1984, pp. 152–159.

to remove the routing and re-route in a few hours. Even the initial routing should not require more than a few days for a large custom circuit. With our current systems, routing requires a few weeks to a few months.

2) The turnaround time for small bug fixes should be less than 15 minutes. For example, if a bug is found while simulating the circuit extracted from a layout, it should be possible to fix the layout, verify that the new layout meets the design rules, and re-extract the circuit, all in 15 minutes. This process currently requires several hours of CPU time and at least a half-day of elapsed time.

3) It should not take more than 30 seconds to 1 minute to re-arrange a cell to try out a different topology. With our current systems this requires anywhere from tens of minutes to several hours.

Magic meets these goals by combining circuit expertise with an interactive editor. It understands layout rules; it knows what transistors and contacts are (and that they must be treated differently than wires); and it knows how to route wires efficiently. Magic uses the circuit knowledge to provide interactive operations that re-arrange a circuit as a circuit rather than as a collection of geometrical objects. It also performs analysis operations, like design-rule checking, *incrementally*, as the circuit is created and modified. This means that only a small amount of work must be done each time the circuit is modified.

3. Corner Stitching

In Magic, as in most other layout editors, a layout consists of cells. Each cell contains two sorts of things: geometrical shapes and subcells. Magic represents the contents of cells using a technique called *corner stitching*. Corner stitching is a geometrical data structure for representing Manhattan shapes. It provides the underlying mechanisms that make possible most of Magic's new features. Corner stitching is simple, provides a variety of efficient searching operations, and allows the database to be modified quickly. What follows is a brief introduction to corner stitching. See [6] for a more complete description.

The basic elements in corner stitching are *planes* and *tiles*. Each cell contains a number of corner-stitched planes to represent the cell's geometries and subcells; each plane consists of a number of rectangular tiles of different types. There are three important properties of a corner-stitched plane, illustrated in Figures 1, 2, and 3:

Coverage: Each point in the x-y plane is contained in exactly one tile (Figure 1). Empty space is represented as well as the area covered with material.

Strips: Material of the same type is represented with horizontal strips (Figure 2). The strip structure provides a canonical form for the database and prevents it from fracturing into a large number of small tiles.

Stitches: The records describing the tile structure are linked together in the database using four links per tile, called *stitches*. The links point to neighboring tiles at two of the tile's four corners (Figure 3).

The stitches permit a variety of search operations to be performed efficiently, including: finding the tile containing a

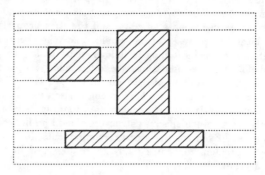

Figure 1. Every point in a corner-stitched plane is contained in exactly one tile. In this case there are three solid tiles, and the rest of the plane is covered by space tiles (dotted lines). The space tiles on the sides extend to infinity. In general, a plane may contain many different types of tiles.

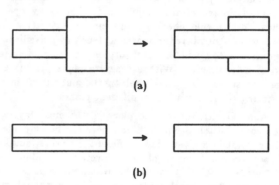

Figure 2. Areas of the same type of material are represented with horizontal strips that are as wide as possible, then as tall as possible. In each of the figures the tile structure on the left is illegal and is converted into the tile structure on the right. In (a) it is illegal for two tiles of the same type to share a vertical edge. In (b) the two tiles must be merged together since they have exactly the same horizontal span.

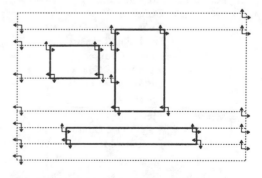

Figure 3. The record describing each tile contains four pointers to other tile records. The pointers are called *corner stitches*, since they point to neighboring tiles at the lower-left and upper-right corners. The corner stitches provide a form of two-dimensional sorting. They permit a variety of geometrical operations to be performed efficiently, such as searching an area or finding all the neighboring tiles on one side of a given tile.

given point; finding all the tiles in an area; finding all the tiles that are neighbors of a given tile; and traversing a connected region of tiles. The coverage property makes it easy to update the database in response to edits, and the strip property keeps the database representation small. To the best of our knowledge, corner stitching is unique in its ability to provide these efficient two-dimensional searches and yet permit fast updates of the kind needed in an interactive tool. The only disadvantage of corner stitching in comparison to less powerful data structures is that it requires more storage space (about three times as much space as structures based on linked lists of rectangles). Even so, the storage requirements do not appear to be a problem for chips likely to be designed in the next several years.

4. Logs

There are several ways in which corner-stitched planes might be used to represent the mask geometries in a cell. One alternative is to use a separate plane for each mask layer; each plane contains space tiles and tiles of one particular mask type. The disadvantage of this approach is that many operations, such as design-rule checking and circuit extraction, require information about layer interactions (such as polysilicon crossing diffusion to form a transistor, or implants changing the type of a transistor). With a separate plane per mask layer, these operations will spend a substantial amount of time cross-registering the information on different planes.

Another alternative is to place all the mask layers into a single corner-stitched plane. Since there can be only one tile at a given point in a given plane, different tile types must be used for each possible overlap of mask layers. This eliminates the registration problem, but results in a large number of small tiles where several mask layers overlap. Even though many of the layer overlaps are not significant (such as metal and implant), separate tile types have to be used to represent them. As a result, the database becomes fragmented into a large number of tiles, and the overheads for all operations increase.

Plane	Tile Types
Poly-Diff	Polysilicon
	Diffusion
	Enhancement Transistor
	Depletion Transistor
	Buried Contact
	Poly-Metal Contact
	Diffusion-Metal Contact
	Space
Metal	Metal
	Poly-Metal Contact
	Diffusion-Metal Contact
	Overglass Via to Metal
	Space

Table I. The corner-stitched planes and tile types used to represent the mask information for an nMOS process with buried contacts and single-level metal. Since polysilicon and diffusion have design-rule interactions, they are placed in the same plane. Metal interacts with polysilicon and diffusion only at contacts, so it is placed in a separate plane. Contacts between metal and diffusion or polysilicon are duplicated in both planes.

The solution we chose for Magic lies between these two extremes. We decided to use a small number of planes, where each plane contains a set of layers that have design-rule interactions. If layers do not have direct design-rule interactions (such as poly and metal), they may be placed in different planes. Some layers, such as contacts, may appear in two or more planes. In our single-metal nMOS process there are two planes: one for polysilicon, diffusion, transistors, and buried contacts; and one for metal (see Table I).

We also decided not to represent every mask layer explicitly. Instead of dealing with actual mask layers, Magic is based around abstracted layers. We call the resulting design style *logs*; it is similar to symbolic layout except that circuit elements are seen in their actual sizes and locations. Designers do not draw implants, wells, buried contact windows, or contact vias. They draw only the primary interconnect layers (polysilicon, diffusion, metal). Certain overlaps between the primary layers are drawn in special layers that represent transistors and contacts of different types. Magic generates the implant, well, and via mask layers automatically when it creates CIF files for fabrication. Table I gives the planes and abstract layers used in Magic, and Figure 4 illustrates how they are used in a sample cell. The logs design style changes the way a circuit looks on the screen, but it doesn't incur any density penalty.

The Magic design style is similar to symbolic systems such as Mulga [14] and VIVID [10], except that the geometries are fully fleshed. As in symbolic design, there are simple operations for stretching and compacting cells. The advantage of the logs approach is that designers can see the exact size and shape of a cell while it is being edited, and they only work with a single representation of the circuit. In symbolic design, designers usually go back and forth between the symbolic and mask representations; the final size of the cell may be hard to determine until it has been compacted and fleshed out. The following sections will show how the abstract layers simplify design-rule checking, plowing, and circuit extraction.

In addition to the planes used to hold mask geometry, each cell contains another plane to hold information about its subcells. Subcells are allowed to overlap in Magic; each distinct subcell area or overlap between subcells is represented with a different tile in the subcell plane. Each tile contains pointers to all of the subcells that cover the tile's area. By using corner-stitching in this way, it is easy to find subcell interactions and to determine which (if any) subcells cover a particular area.

5. Basic Commands

The basic set of commands in Magic is similar to the commands in Caesar [5,7]. Mask geometry is edited in a style like painting: a rectangle is placed over an area of the layout, and mask layers may be painted or erased over the area of the rectangle. Additional operations are provided to make a copy of all the "paint" in a rectangular area and copy it back at a different place in the layout. The corner-stitched representation is invisible to users.

Magic also provides commands for manipulating subcells. Subcells may be placed in a parent, moved, mirrored in x or y, rotated (by multiples of 90 degrees only), arrayed, and deleted. Subcells are handled by reference, not by copying: if a subcell is modified, the modifications will be reflected everywhere that the subcell is used.

6. Incremental Design-Rule Checking

Design-rule checking is an integral part of the Magic system. Our main goal was to make the checker very fast, particularly for small changes: the cost of reverifying a layout should be proportional to the amount of the layout that has been changed, not to the total size of the layout. To achieve this, Magic's design-rule checker runs continuously in the background during editing sessions. When the layout is changed, Magic records the areas that must be reverified. The design-rule checker then rechecks these areas during the time when the user is thinking. For small changes, error information appears on the screen instantly (and also disappears instantly when the problem has been fixed). For large changes (such as moving one large subcell on top of another), it may take seconds or minutes for the design-rule checker to complete its job. In the meantime, the designer can continue editing. If reverification hasn't been completed when an editing session ends, the areas still to be reverified are stored with the cell so that reverification can be completed the next time the cell is edited. Error information is also stored with cells until the errors are fixed. With this mechanism, there is never a need to check a layout "from scratch."

Magic's basic rule-checker works from the edges in a design. Based on the type of material on either side of an edge, it verifies constraints that require certain layers to be present or absent in areas around the edge. There are several reasons why corner stitching and the logs design style allow edge rules to be checked quickly. Each corner-stitched plane can be checked independently. All the "interesting" edges are already present in the tile structure, so there is no need to register different mask layers. Logs make it unnecessary to check formation rules associated with implants and vias. Lastly, corner stitching provides efficient algorithms for locating all the edges in an area and for searching the constraint areas.

In addition to a fast basic checker, the incremental rule checker contains algorithms for handling hierarchy. When a cell in the middle of a hierarchical layout is changed, Magic checks interactions between this cell and its subcells, and also interactions between this cell and other cells in its parents and grandparents. More details on the basic DRC mechanism and on Magic's hierarchical approach can be found in [12].

7. Plowing

Plowing is a simple operation that is used to rearrange a layout without changing the electrical circuit that it represents. To invoke the plow operation, the user specifies a vertical or horizontal line segment (the *plow*) and a distance perpendicular to it (the plow distance). See Figure 5. Magic sweeps the plow for the specified distance, and moves all material out of the area swept out by the plow. The edges of this material are likewise treated as plows, pushing other material in front of them. Mask geometry in front of the plow is compacted as it is moved. Jogs are inserted at the ends of the plow. The plow operation maintains design rules and connectivity so that it doesn't change the electrical structure of the circuit. Most material, such as polysilicon, diffusion, and

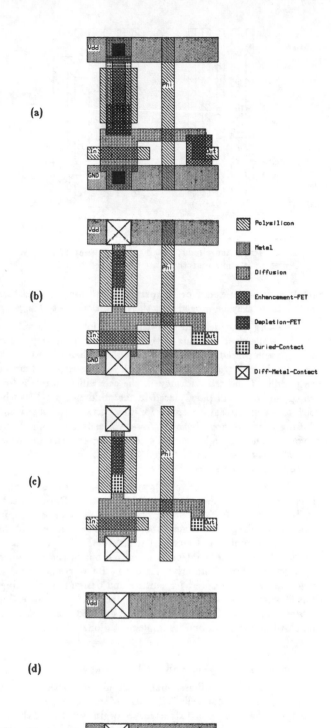

Polysilicon

Metal

Diffusion

Enhancement-FET

Depletion-FET

Buried-Contact

Diff-Metal-Contact

Figure 4. In Magic, transistors and contacts are drawn in an abstract form called logs: (a) a three-transistor shift-register cell, showing actual mask layers; (b) the same cell as it is seen in Magic; (c) the information in Magic's poly-diff plane; (d) the information in Magic's metal plane. Contacts are duplicated in each plane.

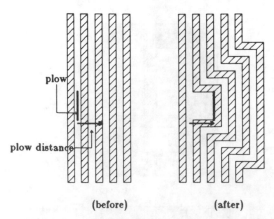

Figure 5. In plowing, a horizontal or vertical line is moved across the circuit, pushing material out of its way. Design rules and connectivity are maintained.

metal, may be stretched or compacted by plowing; transistors and contacts may be moved, but their shape will not change.

Plowing provides most of the operations of a symbolic system, while still working with fully-fleshed geometry. If a large plow is placed to one side of a cell and then moved across the cell, the cell will be compacted. If a plow is placed across the middle of the cell and moved, the cell will be stretched at that point. A small plow placed in the middle of a cell can be used to open up empty space for new transistors or wiring. Plowing may be used both on low-level cells containing only geometry, and on high-level cells containing subcells and routing. Plowing moves each subcell as a unit, without affecting the contents of the subcell.

The implementation of plowing is dependent on corner stitching, logs, and the edge-based design rules. Corner stitching provides the fast geometric operations used to search out plow areas. Logs tell Magic about materials that cannot be stretched or compacted (such as transistors). The edge-based design rules indicate what must be moved out of the way when a particular edge of material is moved. By working from the same data structure used for editing and design-rule checking, the plowing operation avoids the overhead of converting between representations. See [11] for a detailed presentation of the plowing operation and its implementation.

8. Circuit Extraction and Cell Overlaps

The Magic database makes circuit extraction almost trivial for individual cells. Because of the logs and corner stitching, the circuit is extracted to begin with. All that is needed is to traverse the tile structure and record information about what connects to what. There is no need to register layers or infer the structure and type of transistors and contacts: all this information is represented explicitly.

For hierarchical designs, the situation becomes complicated when cells overlap. Each cell uses a separate set of corner-stitched planes, so information from the separate planes must be combined in order to find out what connects to what. If arbitrary overlaps are allowed, then transistors may be split between cells, or may be formed or broken by cell overlaps. In this case, circuits cannot be extracted hierarchically, since the structure of a cell may be changed by the way it is used in its parents.

One approach to the overlap problem is to prohibit cell overlaps. This has two drawbacks, however. First, it makes for clumsy designs, since overlap areas must be placed in separate cells. This makes it harder to understand designs and harder to re-use cells. Second, it doesn't eliminate the problems in circuit extraction, since information will still have to be registered along the boundaries of abutting cells. For example, a cell abutment can cause two separate transistors to join together.

Instead of prohibiting overlaps, we decided to restrict them. In Magic, cells may abut or overlap as long as this only connects portions of the cells without changing their transistor structure. Overlaps and abutments may not change the type or number of transistors from what it would be without the overlap (e.g. polysilicon from one cell may not overlap diffusion from another cell, since this would create a new transistor). These restrictions will be verified by using a special set of design rules in the part of the design-rule checker that deals with cell overlaps.

The Magic approach still requires information to be registered between subcells, but it allows the extracted circuit to be represented (and extracted) hierarchically. The extracted circuit for any cell consists of the circuits of its subcells, plus the circuit of the cell itself, plus a few connections between the subcells.

9. Routing

Routing is the single most important area where we hope Magic will speed up the design process. Most of the Magic routing effort has been spent in two areas: a) creating a channel router that can work around obstacles in the channel (such as critical signals or power and ground routing); and b) modifying the routing tools so that they can work for custom designs that are not laid out on a fixed routing grid.

Magic uses the standard three-phase approach to routing. In the first phase, called *channel definition*, the empty space of the layout is divided up into rectangular channels. In the second phase, called *global routing*, nets are processed sequentially to decide which channels will be crossed by each. In the third phase, called *channel routing*, each channel is considered separately and wires are placed to achieve the necessary connections within the channel. Channel definition is not yet implemented in Magic. At present the system just uses the space tiles in the subcell plane as routing channels, which results in unsatisfactory long skinny channels. Eventually, the channel definer will re-arrange the space tile structure on the subcell plane into larger, more natural channels, perhaps using the bottleneck approach of the BBL system [1], or the approach of the PI system [8]. Global routing uses a wavefront approach [4], slightly modified so that it searches the most direct route to the destination first. Since corner-stitching is used to keep track of the channel space, there is no need to build a special graph for global routing, and the global router runs very fast. The channel router is an extended version of Rivest's greedy router [9]. Magic does not provide placement tools.

In order to make the routing tools more useable in a custom design environment, we have developed a channel router that can work around obstacles in the channels. It is important for designers to be able to wire critical nets by hand, and to have the automatic routing tools route the less critical nets without affecting the hand-routed ones. It is also convenient to run power and ground routing tools as a separate step before signal routing, and have the signal router work around the power and ground wires. Where there are obstacles in the channels, Magic will route under short obstacles if possible; it will route around those that are block both routing layers or are large. Where there is insufficient space to route around single-layer obstacles, Magic can make interconnections under them using river-routing. See [2] for details on how Rivest's greedy router has been extended to handle obstacles. Figure 6 shows an example of results produced by the Magic router.

The router, combined with plowing and the other editing features, provides designers with considerable flexibility. Critical signals and power and ground can be routed by hand. Then the router can be invoked to complete the rest of the interconnections. If the router is unable to make all connections, the final ones can be placed by hand. Or, the designer can provide hints to the router by placing a few key connections by hand and letting the router fill in the rest. Yet another alternative is to use plowing to re-arrange the placement, and then re-run the router. The plowing operation will maintain the existing connections.

We have also extended the standard routing approach to handle designs that are not based on a uniform routing grid. Most channel routers assume a uniform grid based on the minimum contact-contact spacing: channel dimensions must be an integral number of grid units, and all wires must enter and leave channels on grid points. Unfortunately, custom cells are not usually designed with the router's grid in mind, so the cell boundaries and terminals do not line up on a master grid. We are experimenting with two approaches to this problem, called *sidewalks* and *flexible grid*.

The sidewalk approach is illustrated in Figure 7, and involves a pre-routing step where all cells are expanded so that their dimensions are integral grid units. This additional cell area is called its *sidewalk*. In addition, wires are added to connect the terminals of the cell to grid points on the outer edge of the sidewalk. After the sidewalk generation stage, everything is grid aligned so standard routing tools can be used. Magic currently implements the sidewalk approach. Sidewalks are inefficient because the sidewalk areas cannot be used for channel routing, even though they usually contain little material. Sidewalks typically cause the channels to be reduced in size by 2-3 tracks and 2-3 columns.

The flexible grid approach distributes the sidewalks among the channels by jogging the track and column structure at the ends to match up with connection points that don't fall on grid lines. This is illustrated in Figure 8. In the flexible grid approach, wasted space occurs within the channel because some columns and tracks cannot extend all the way across the channel. There appears to be less wasted space in this approach than in the sidewalk approach. In the worst case, the wasted space is equivalent to two tracks and two columns per channel. If connection points are sparse, however (and this appears to occur frequently), the flexible grid approach has almost zero wasted space. We are still in the early stages of exploring this alternative.

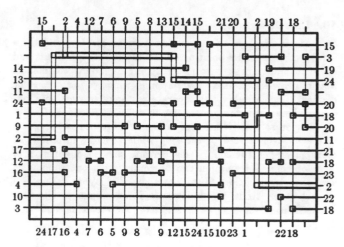

Figure 6. An example of routing with obstacles in the channel. The Magic router was initially unable to route this switchbox. However, after careful hand-placement of one net, shown with double lines, the router was able to complete the remaining connections while working around the hand-placed wiring.

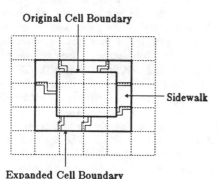

Figure 7. In the sidewalk approach, each cell is enlarged so that its boundary is grid-aligned. Then connections on the edge of the original cell are routed to grid points on the outside of the sidewalk.

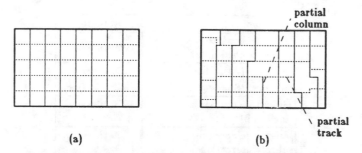

Figure 8. Rather than expand cells to grid points as in the sidewalk approach, the flexible grid approach modifies the track and column structure of the channel. The channel is grid-based in the center, but the grid lines jog at the edges to meet up with non-gridded connections. (a) shows the standard orthogonal channel structure, and (b) shows a channel whose grid structure has been flexed. The flexible grid approach can result in tracks or columns that don't extend all the way across the channel.

10. User Interface

Magic displays the layout on a color display, and users invoke commands by pointing on the display with a mouse and then pushing mouse buttons or typing keyboard commands. Magic provides multiple overlapping windows on the color display. Each window is a separate rectangular view on a layout. Different windows may refer to different portions of a single cell, or to totally different cells. Windows allow designers to see an overall view of the chip while zooming in on one or more pieces of the chip; this permits precise alignments of large objects. Information can be copied from one window to another. In addition to windows on the layout, there are several other kinds of windows used for menus and system maintenance functions such as color map editing and icon design.

11. Technology Independence

Although Magic contains a considerable amount of knowledge about integrated circuits, the information is not embedded directly in code. All the circuit information is contained in a textual technology file. This file defines the abstract layers for a particular technology, the corner-stitched planes used to represent them, and the assignment of abstract layers to planes. It tells how to display the various layers and defines the semantics of the paint and erase operations from Section 5 (for example "if poly-metal-contact is painted over diffusion, erase the diffusion and place poly-metal-contact tiles on both the poly-diff and metal planes"). The technology file contains the design rules used in design-rule checking and in plowing, and indicates which tile-types will be electrically connected when they are adjacent. Lastly, it tells how to fill in the structural details of transistors and contacts when generating CIF for circuit fabrication. The technology file format is general enough to handle a variety of nMOS and CMOS processes. Our technology file for an nMOS process with buried contacts and single-level metal contains about 460 lines, of which about half are comments. Our technology file for a P-well CMOS process contains 750 lines.

Operation	Speed
Painting tiles into corner-stitched database	1000 tiles/sec.
Redisplay (AED display)	100 tiles/sec.
Redisplay (Sun workstation)	400 tiles/sec.
Design-rule checking (single-cell)	1900 tiles/sec.
Design-rule checking (subcell interactions)	600 tiles/sec.
Plowing	600 tiles/sec.
Channel routing ("Deutsch's difficult example," 60 nets)	3 sec.

Table II. Some sample measurements of the speed of the Magic system. All measurements except the Sun redisplay time were made on a VAX-11/780. For most operations, the limiting factor is the redisplay time.

12. Implementation

The implementation of Magic was begun in February of 1983. By early April 1983, a primitive version of the system was operational. Although the first system was based on corner stitching and logs, it provided user features only equivalent to Caesar. During the summer of 1983 implementation was begun on the subsystems for routing, multiple windows, plowing, and design-rule checking. As of this writing, most of the advanced features are either operational or expected to be operational in the near future. The design-rule checker is complete; plowing runs in only a single direction; the routing tools are complete except for channel definition; and circuit extraction has not been started. The system has been in use since April 1983 by the designers of a 32-bit microprocessor [13], and was used during the Fall of 1983 by several dozen students in an introductory VLSI design class.

Magic is written in C under the Berkeley 4.2 Unix operating system for VAX processors and Sun workstations. The current VAX implementation works only with AED color displays with special Berkeley microcode extensions. Altogether, Magic contains approximately 64000 lines of code. Table II gives a few sample performance measurements of pieces of the system, and Table III gives a breakdown of code size.

13. Conclusions

We have not yet had enough designer experience with Magic to evaluate the system thoroughly, but the initial response has been favorable. The greatest problem encountered so far concerns the logs design style: if designers are accustomed to working with actual mask layers, then the abstract layers in Magic are confusing and irritating at first. This problem was exacerbated in the early versions of the system because the design-rule checker wasn't implemented. With continuous feedback from the checker, we hope that it will be much easier for designers to learn the abstract layers. We expect that logs will be easier for designers to work with than the actual mask layers, since they hide irrelevant details.

The pieces of the Magic system work well together. Corner stitching appears to be a complete success: it provides all the operations needed to implement Magic's advanced features, and results in simple and fast algorithms. The design-rule checker's edge-based rule set meshes well with the corner-stitched data, and is used also for plowing. The logs

Component	Size (lines)
Database (corner stitching and hierarchy)	12500
Graphics, windows, redisplay	13700
Command interpreter	6700
Design-rule checker	2900
Plowing	7200
Routing	12200
Miscellaneous	9000
TOTAL	64200

Table III. Code sizes of various elements of the Magic system. The table reports total lines, of which about half are comments or blank lines.

design style simplifies the design rules, provides information needed for plowing and circuit extraction, and simplifies the designer's view of the layout.

We hope that Magic's flexibility will change the VLSI layout process in two ways. First, we hope that it will enable designers to experiment much more than previously. At the cell level, they can use plowing to rearrange cells quickly and easily. Cells can be designed loosely, then compacted. At the chip level, plowing and the routing tools can be used together to re-arrange the floorplan, route the connections, compact or stretch, and try again. The ability to experiment means that students will be able to develop better intuitions about how to design chips; it also means that designers will be able to fix bugs and enhance performance more easily.

Second, we hope that Magic will make it easier to reuse pieces of designs. To design a new chip, a designer will select cells from a large library, use plowing and painting to make slight modifications in their shape or function to suit the new application, and perhaps design a few new cells. Then the routing tools will be used to interconnect the cells. We hope that this approach will result in a substantial reduction in design time for large circuits.

14. Acknowledgements

As tool builders, we depend on the Berkeley design community to try out our new programs, tell us what's wrong with them, and be patient while we fix the problems. Without their suggestions, it would be impossible to develop useful programs. The SOAR design team, and Joan Pendleton and Shing Kong in particular, have been invaluable in helping us to tune Magic; they were willing to use the system even in the dark early days when Magic was a worse tool than its predecessor. Randy Katz was courageous enough to use Magic in his VLSI design class. Randy Katz, David Patterson, and Carlo Séquin all provided helpful comments on this paper. The Magic work was supported in part by the Defense Advanced Research Projects Agency (DoD) under contract N00034-K-0251, and in part by the Semiconductor Research Cooperative under grant number SRC-82-11-008.

15. References

[1] Chen, N.P., Hsu, C.P., and Kuh, E.S. "The Berkeley Building-Block Layout System for VLSI Design." Memorandum No. UCB/ERL M83/10, Electronics Research Laboratory, University of California, Berkeley, February, 1983.

[2] Hamachi, G.T. and Ousterhout, J.K. "A Switchbox Router with Obstacle Avoidance." *Proc. 21st Design Automation Conference*, 1984.

[3] Keller, K.H. and Newton, A.R. "KIC2: A Low-Cost, Interactive Editor for Integrated Circuit Design." *Proc. Spring COMPCON*, 1982, pp. 305-306.

[4] Lee, C. Y. "An Algorithm for Path Connections and Its Applications." *IRE Transactions on Electronic Computers*, September 1961, pp. 346-365.

[5] Ousterhout, J.K. "Caesar: An Interactive Editor for VLSI Layouts." *VLSI Design*, Vol. II, No. 4, Fourth Quarter 1981, pp. 34-38.

[6] Ousterhout, J.K. "Corner Stitching: A Data Structuring Technique for VLSI Layout Tools." *IEEE Transactions on CAD/ICAS*, Vol. CAD-3, No. 1, January, 1984, pp. 87-99.

[7] Ousterhout, J.K. "The User Interface and Implementation of Caesar." *IEEE Transactions on CAD/ICAS*, Vol. CAD-3, No. 3, July, 1984, to appear. Also appears as Technical Report UCB/CSD 83/131, Computer Science Division, University of California, Berkeley, August 1983.

[8] Rivest, R.L. "The 'PI' (Placement and Interconnect) System." *Proc. 19th Design Automation Conference*," 1982.

[9] Rivest, R.L. and Fiduccia, C.M. "A Greedy Channel Router." *Proc. 19th Design Automation Conference*," 1982, pp. 418-424.

[10] Rosenberg, J. et al. "A Vertically Integrated VLSI Design Environment." *Proc. 20th Design Automation Conference*, 1983, pp. 31-36.

[11] Scott, W.S. and Ousterhout, J.K. "Plowing: Interactive Stretching and Compaction in Magic." *Proc. 21st Design Automation Conference*, 1984.

[12] Taylor, G.S. and Ousterhout, J.K. "Magic's Incremental Design Rule Checker." *Proc. 21st Design Automation Conference*, 1984.

[13] Ungar, D., et al. "Architecture of SOAR: Smalltalk on a RISC." *Proc. 11th Symposium on Computer Architecture*, 1984, to appear.

[14] Weste, N. "Virtual Grid Symbolic Layout." *Proc. 18th Design Automation Conference*, 1981, pp. 225-233.

[15] Williams, J. "STICKS - A Graphical Compiler for High Level LSI Design." *Proc. 1978 National Computer Conference*, pp. 289-295.

Virtual Grid Symbolic Layout

Neil Weste

Bell Laboratories
Holmdel, New Jersey 07733

ABSTRACT

Free form or "stick" type symbolic layout provides a means of simplifying the design of IC subcircuits. To successfully utilize this style of layout, a complete design approach and the necessary tools to support this methodology are required. In particular, one of the requirements of such a design method is the ability to "compact" the loosely specified topology to create a set of valid mask data. This paper presents a new compaction strategy which uses the concept of a *virtual grid*. The compaction algorithm using the virtual grid is both simple and fast, an attribute which allows the designer to conveniently interact with the algorithm to optimize a layout. In addition to the compaction algorithm, methods used to create large building blocks will be described. The work described here is part of a complete symbolic layout system called MULGA which is written in the C programming language and resides on the UNIX operating system.

1. INTRODUCTION

A proven method of designing integrated circuits is by reducing functions to manageable subcircuits, implementing these subcircuits, verifying their performance and then using automatic or structured techniques to build larger subcircuits until the complete chip is specified. Based on this design philosophy, there is a need for systems which allow the design of cells in an error free manner and the subsequent documentation of their performance. One method for reducing the complexity of the physical design of cells is to use symbolic layout techniques. These methods reduce or completely eliminate the many geometric design rules generally needed to specify an IC mask set, and provide additional benefits at the *capture* stage of a design.

_____
* UNIX is a Bell System Operating System

This paper describes the compaction strategies that are used in MULGA [1], a UNIX-based interactive symbolic layout system. The components used in this task consist of a compaction program and various methods of converting the symbolic description into a valid mask set. All software is written in the C programming language and runs under the UNIX operating system on a microcomputer (DEC PDP 11/23) based design station.

2. MULGA

Symbolic layout methodologies use symbolism to reduce the complexity of custom mask design. Apart from offering densities comparable to manually designed mask layouts, symbolic layout methods provide potentially shorter design times, a reduced liklihood of errors and a technology tracking library. MULGA is a UNIX-based interactive symbolic design system consisting of a suite of programs residing on a high performance color display station. The MULGA system has been designed to provide the smooth man-machine dialogue necessary to support the tasks required during design capture and performance verification of IC subcircuits. Figure 1 summarises the software modules that constitute MULGA and the paths along which information is passed to complete a design.

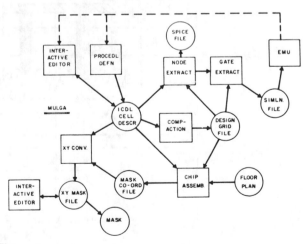

Figure 1.

Software Modules in the MULGA Design System

Reprinted from *ACM/IEEE 18th Design Automation Conf. Proc.*, 1981, pp. 225–233.

The system uses as a central data base, a symbolic Intermediate Circuit Description Language (ICDL), which uses a derivation of the *coordinode* notation used by Buchanan [2]. The language combines the physical and structural aspects of a design by using data elements which have predefined geometric properties in conjunction with designated connection points which are used to establish electrical connectivity. In this way the language aids the capture of designer intent with respect to circuit topology and geometric placement.

The basic structure in ICDL is a *cell* which consists of a variety of elements placed in the X-Y plane as shown in Figure 2. These elements may be **devices** (transistors), **wires, contacts, pins** or other cell **instances.** Devices are the active elements in a process. Wires, which may exist on any interconnection layer, serve to interconnect devices in conjunction with contacts. Pins are used to name internal nodes in a circuit or to specify connection points on the boundary of a cell. Although they have no physical significance in the final mask representation of the circuit, they are an important attribute of the language being used in cell placement and circuit verification. Cell instances allow cells to be defined hierarchically and in addition allow arrays to be easily specified. Figure 2 depicts a CMOS 2-input nand gate in a graphical representation, with the

ICDL textual description for that cell. It is evident from this diagram that each line of text represents a circuit element rather than a geometrical shape as in conventional physical mask description languages. Each component is snapped to a grid as shown. Note, however, that this grid does not initially have any physical mask related spacing. Subsequent processing of the ICDL description yields the spacing between grid points. The attribute of having a topology driven spacing for the grid is thought of as a *virtual grid* which is the name given to this type of layout. This paper will deal with the methods of processing the ICDL to yield a valid mask set.

Before dealing with the compaction methods, the overall design methodology used in MULGA will be reviewed. ICDL cell descriptions may be generated interactively via a color graphics/text editor or alternatively, procedurally defined using the C programming language. Once physically designed, the cell may be checked structurally using a circuit extractor and a gate extractor [3], which produce text descriptions suitable for a circuit analysis program or timing simulator, respectively. The timing simulator EMU [3] runs on the design terminal and provides initial estimates of timing performance. When the cell has been verified behaviourally, the final mask description is generated using a *chip-assembly* program, which uses as guidance, a chip floor-plan.

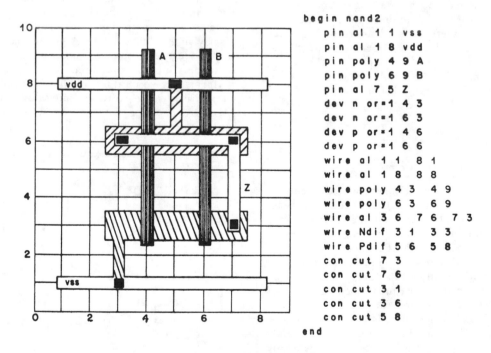

Figure 2. A 2-input Nand gate
- graphical representation
- ICDL description

3. COMPACTION

3.1 Introduction

Once a topology has been generated, it is necessary to *compact* the design to produce a mask description. One of the first examples of a compaction algorithm was by Akers et.al.[4]. This scheme was based on a fixed-grid system where the layout was represented as entries in a rectangular matrix. Each entry in the matrix represented a fixed sized mask element. The matrix thus constructed was searched to determine a path of blank cells across the matrix. Paths were allowed to be straight lines as shown in Figure 3a or could follow what were termed *shear lines* as shown in Figure 3b. Rules applied to paths to protect the topology when space was extracted from the matrix. Compaction proceeded by repeatedly scanning in the X then Y directions until no paths could be found or some other terminating condition was satisfied.

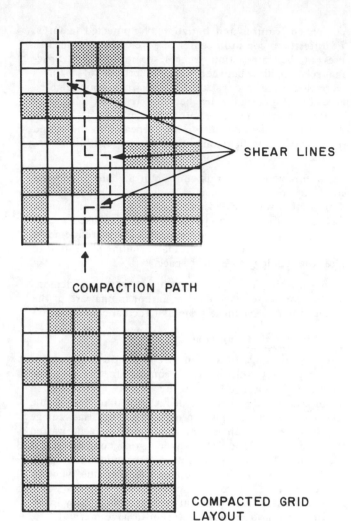

SHEAR LINES

COMPACTION PATH

COMPACTED GRID LAYOUT

Figure 3b.

Compaction on a Coarse Grid - Use of Shear Lines

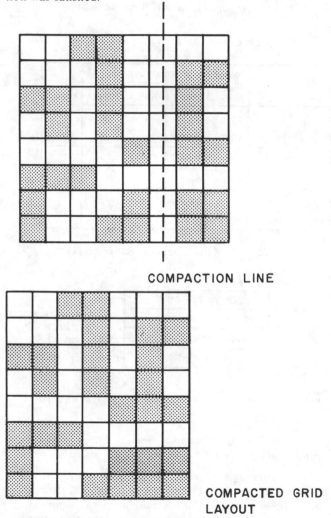

COMPACTION LINE

COMPACTED GRID LAYOUT

Figure 3a.

Compaction on a Coarse Grid - Simple Example

The general idea of compacting a "sticks" type of symbolic layout was first described in [5]. Subsequent to this, Dunlop [6] proposed a detailed scheme where the fixed grid of a coarse grid compactor is traded for a "node and line" symbolic representation. In this scheme, the symbols existed as objects which were not fixed to any grid. These objects were manipulated in much the same way as design rule checking programs work (i.e. by comparing element to element spacings using lists of the elements) to arrive at a set of allowable geometric spacings between objects. The same general search techniques as used in Akers' algorithm were used. The advantage of using the "node and line" representation was potentially improved packing density at the expense of execution time.

A compaction algorithm which used a graph theory approach was proposed by Hseuh et.al. [7]. In this algorithm the symbolic layout was partitioned into vertically connected features and horizontally connected features. Features on a common center line were then grouped to prevent the compaction process separating them. Spacing relationships between such collections of features were then used to construct a graph which was weighted according to design rules between features. The most costly path was then determined and used to derive final mask coordinates. Again this was repeated in the X and Y directions until some termination condition was satisfied.

A combination of this compactor and the "node and line" compactor was used in a system called SLIM[8]. Compaction was divided into "local" and "global" phases. The minimum cost graph compactor was used locally and the "node and line " globally. Other refinements such as jog insertion and automatic routing were used to alter the topology to extract space.

While all the above approaches have all demonstrated compacted layouts, the author is unaware of the application of any these systems to actual chip designs.

3.2 Virtual Grid Compaction

The advantage of a fixed or coarse grid compactor is that adjacency information required for design rule checks is locally available in the matrix. This results in rapid design rule checking, a fact noted in design rule checkers which use this approach [9]. The disadvantage of such an approach is that mask space may be wasted by the fixed grid. Non-fixed grid compactors as described above achieve better compaction but spend most of their time in conventional design rule type calculations (i.e. list comparisons). Thus a system which does not use a fixed grid but does treat adjacency information on a grid basis would seem to have merit.

The ICDL data base consists of elements or "objects" such as transistors and wires that exist on a conceptual grid. For instance, in Figure 4a a transistor is shown (in symbolic graphic format) connected to four wires using three contacts. Because the data structure is based on a grid, the data structure shown in Figure 4b may be constructed, which indicates the elements located at each grid point in the symbolic layout. For instance at point [2,3], there exists the drain or source of a device, a contact and a wire designated *wire1*. To the right of this point is the gate of the same device and wire *wire2*. Below the point [2,3], *wire1* exists. Thus by using a suitably constructed data representation at each grid point one may examine the adjacent neighbourhood of a grid point to determine local constraints. The data structure used in the current system is shown in Figure 4c. At each grid location a pointer may point to the ICDL data structures for a device, a contact and two wires. In practice, to date, this has been sufficient for the fabrication processes used, but may of course be modified to suit different technologies (i.e. two-level metal). Considering compaction in the X direction firstly then, the matrix shown in Figure 4b is examined column-wise. Thus, when travelling up column 2 and looking to the

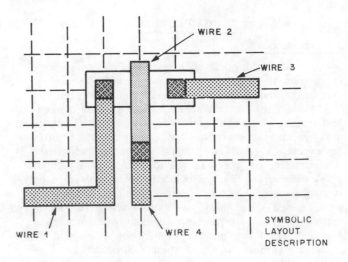

Figure 4a. Symbolic Graphical Representation

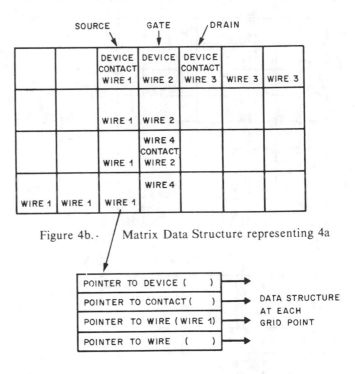

Figure 4b. Matrix Data Structure representing 4a

Figure 4c. The data structure at each matrix element

right, *wire4, wire2,* a contact and a device would be found adjacent to *wire1*.

Thus, the first step in compaction is to plot the ICDL description into the array using the data structure shown in Figure 4c. On small machines only two columns or rows need be plotted at once, thus conserving memory space. Wires are regarded as stretchable in the direction of compaction. To achieve this, the breakpoints (or bends) in a wire are plotted when compacting in both directions, but only interior elements on wires perpendicular to the direction of compaction are plotted. For instance, Figure 5a shows the actual data structure used for the X compaction of Figure 4a. Note that grid point [1,0] is not plotted as part of *wire1*. Similarly point [5,3] of *wire3* is not plotted. The data structure for Y compaction is shown in Figure 5b. Here grid points in the vertical sections of *wire1* and *wire4* have been omitted. This procedure ensures that wires collapse to a minimum length if topology permits. Apart from wires, the only other element treated specially is the device (or transistor). A device is plotted as three points, representing the drain, gate and source. Contacts are plotted as single entries on the grid.

		DEVICE CONTACT WIRE 1	DEVICE WIRE 2	DEVICE CONTACT WIRE 3		WIRE 3
		WIRE 1	WIRE 2			
		WIRE 1	WIRE 4 CONTACT WIRE 2			
WIRE 1		WIRE 1	WIRE 4			

Figure 5a. Actual matrix during X compaction

		DEVICE CONTACT WIRE 1	DEVICE WIRE 2	DEVICE CONTACT WIRE 3	WIRE 3	WIRE 3
				WIRE 4 CONTACT WIRE 2		
WIRE 1	WIRE 1	WIRE 1	WIRE 4			

Figure 5b. Matrix during Y compaction

When the circuit has been plotted into the array, adjacent columns are examined pairwise from left to right and from bottom to top for non-zero entries. Upon finding two adjacent points which reference non-zero elements, the relative mask dimension of each point is then calculated. This is achieved by examining the elements and applying a procedure to each element to find its contribution to the dimension on each available mask level. For instance Figure 6a shows in symbolic graphic representation a vertical polysilicon wire

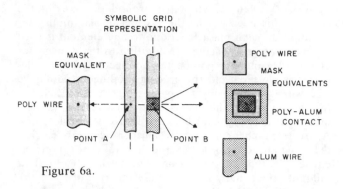

Figure 6a.

Symbolic Graphic example for spacing calculation

"A" WIDTHS	DEVICE	CONTACT	WIRE
POLY	0	0	4
ALUM	0	0	0
DIFF	0	0	0

"B" WIDTHS	DEVICE	CONTACT	WIRE
POLY	0	5	4
ALUM	0	7	4
DIFF	0	0	0

Figure 6b. Width contributions of each element

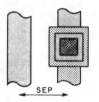

SPACING DESIGN RULES APPLICABLE : —

$$\text{POLY} - \text{POLY} \quad 4$$
$$\text{ALUM} - \text{POLY} \quad 0$$
(WINDOW RULES ACCOUNTED FOR BY CONTACT CONSTRUCTION)

$$\text{SEP} = \text{WORST CASE} \; ((\text{WIDTH}_A + \text{WIDTH}_B)/2 + \text{SPACING})$$
$$= ((4+5)/2+4)$$
$$= 8 \cdot 5 \; \text{UNITS}$$

Figure 6c. Spacing calculation

adjacent to a contact joining a vertical polysilicon wire to an aluminum wire. The points of interest, A and B, have been designated. Considering compaction in the X direction, the mask contribution at A would be the width of the polysilicon wire. At point B, there are three contributions, due to the contact, aluminum wire and polysilicon wire. A table may be constructed for the contributions to the "width" of each point from each coincident element. Figure 6b summarises these for both points. At point A there is a single polysilicon contribution from the wire, while at B there are aluminum and polysilicon contributions from the contact and the two wires. To calculate the separation required between these two points, the allowed interlayer spacings for the process are combined with the worst case widths on each layer to arrive at a design rule correct value. Figure 6c summarises this calculation, indicating for the widths and spacings shown, these two grid points could be at minimum, 8.5 units apart. Normally, mask contributions are viewed as symmetric around the axes of the grid but this need not be the case. The maximum contribution on each mask level due to any element is used to ensure correct spacing. In addition to the dimensions of the adjacent grid points, the connectivity of the elements is checked and the spacing modified accordingly. The worst case spacing for comparing the two columns for the height of the cell is then used as the notional minimum mask spacing between these two grid values. Apart from checking strictly adjacent column entries, the program backtracks to the left to check previous entries. In this manner design rules which exert their influence over many grid units are incorporated. Figure 7 shows this effect. By obeying a strict adjacency scheme the critical spacings between column m-1,m and m,m+1 would appear to be spacings A and B due to aluminum spacing. In fact the correct spacing is the P-device to N-device spacing shown at C. As columns are scanned from left to right (in the case of X compaction) the row position of the worst case spacing elements is stored for future reference.

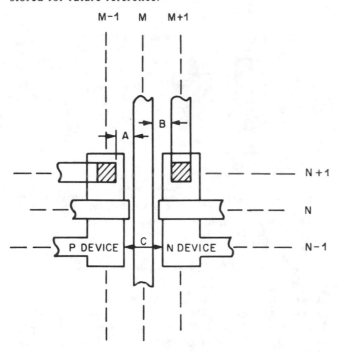

STRICT ADJACENCY WOULD INDICATE SPACINGS A AND B ARE DOMINANT

HOWEVER SPACING C — N DIFFUSION TO P DIFFUSION IS DOMINANT

Figure 7.

Backtracking to allow for non-adjacent critical points

The termination of this process completes what is termed the X compaction. At this stage each column has a mask dimension associated with it in addition to the design rule tolerance to the adjacent column to the right. The same procedure is then repeated for the rows

that comprise the cell. In addition to the backtracking step described above, an arc is swept out by the grid point in question to account for oblique design rule spacings. This is shown graphically in Figure 8. For instance during the X compaction, *wire1* may be moved arbitrarily close to *wire2*. The oblique checking causes *wire2* to be vertically spaced from *wire1* during the Y compaction to obey design rules. When the Y compaction is completed the spacings of rows and columns are stored in the *design grid* file shown in Figure 1.

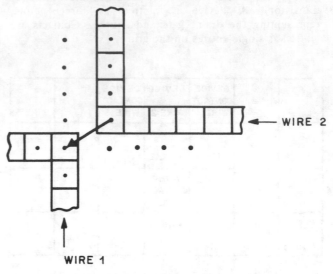

Figure 8. Oblique tolerance checking during Y compaction

It was mentioned that the critical compaction points were stored during compaction. These may now be used as an overlay on the original ICDL graphical representation to allow the designer to interactively change the topology. An important philosophical point emerges here. Previous compaction algorithms have included "jog" insertion as a method of changing the topology to improve compaction. This is not done in this system for a number of reasons-

— It increases the complexity of the compactor

— It introduces unforseeable topology changes

— Jogs form a minor role in space saving topology moves

The first point is minor, but of some import on a small machine. The second point is rather more serious in situations where cells have been designed to abut to complete a larger building block. In these instances, connection points which have been assigned along boundaries may be disrupted by jog insertion. Finally, perhaps the most important reason comes from the extensive experience gained using this system. It has been found that when a particular cell does not meet some predefined mask dimension, rather drastic topology alterations are usually necessary. This ranges from re-routing a wire, to interchanging directions of the aluminum and polysilicon layers. Thus, with this in mind, it is a better strategy to allow the designer to interactively make *all* topology changes and then run a compactor, the output of which is predictable. This allows the designer to interact more smoothly with the algorithm. The MULGA system supports this style of design by providing compaction feedback to the designer and an efficient man-machine dialogue supported by the color graphics editor.

3.3 Performance

Figure 9 shows the relation between compaction time and cell size in grid units. The algorithm has essentially a linear execution time with respect to the area (in grid units) of a cell. Thus if the density of cells is comparable, this constitutes a linear dependence with respect to the number of elements. Approaches in [7] and [8] indicate complexity of N(1.5). Again it must be stressed that this performance greatly enhances the interactive design style.

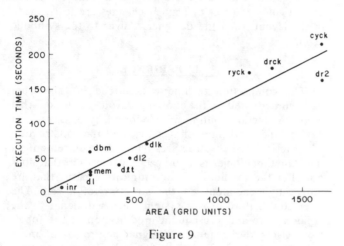

Figure 9

Virtual grid compaction algorithm performance

4. PHYSICAL MASK GENERATION

The MULGA system and the compaction strategy described may be used to design cells for various styles of chip design. For instance in a standard cell system, the advantage of using a symbolic approach is that physical design-rule independent cells may be constructed. This is especially advantageous as rapid technology advances force changes in processes and hence design rules. Another style of design that is supported is what amounts to custom design. However in contrast to other approaches, the use of symbology reduces the complexity of cell design to the point where a large number of performance verified cells may be constructed by an individual designer. To date, this is the style of design for which the system has been used, and this section illustrates how the construction of large cells is supported by the compaction scheme.

4.1 Structured Cell Composition

A designer starts the design of a chip or section thereof by first generating a *floor plan*. This takes into account the data and control flow of the network to be designed, and in addition partitions the circuit into manageable subcircuits. This may be repeated recursively until the designer decides the complexity of a sub-circuit is appropriate for design. Taking into account the global communication dictated by the floor-plan and the circuit itself, the cell is then designed. This is repeated for all cells at the bottom of the hierarchy. Cells are then combined by either abutment, repetition or simple routing. Once the top of the hierarchy is reached, the designer may determine the size of the layout and the critical portions of the design with respect to final mask size. Such critical areas may receive additional attention to improve packing density. Once this optimisation step has been completed a mask set may be generated.

The principle of defining performance verified cells and then combining them through the techniques described above has worked extremely well in practice. Once large functional blocks have been designed, conventional automatic techniques may be used to interconnect them[10].

4.2 Macro-Cell Construction

As a first step in constructing larger cells, which are pitch-matched to each other a simple language defines the relative placement of the cells. The following example is illustrative,

    PLACE linty 0 0 a
    NORTH a sl2y b
    EAST a lintg c
    NORTH c gl2 d

Translated, this means,

— Place cell *linty* at coordinate 0,0 and call this instance *a*

— North of *a* put cell *sl2y* and call it *b*

— East of *a* put cell *lintg* and call it *c*

— North of *c* put cell *gl2* and call it *d*

With the component cells shown in Figure 11a this leads to the placement shown in Figure 11b.

In addition one may then specify the nets that are to be connected (or pitch-matched) which are illustrated by the following examples,

    a.vss.N b.vss.S
    a.yi.N b.si.S
    a.cp1.W c.cp1.E
    c.ld.N d.ld.S
    b.a0.W d.a0.E

For example, the first command means,

Find all the North facing *vss* pins on *a* and match them to the South facing *vss* pins of *b*

269

A program checks for the correct matching of these specified connection points reporting any errors. Once this process has been completed the virtual grid points at which adjacent cells have to match in the physical domain have been defined.

Correct matching of virtual grid coordinates in the mask domain may be achieved in two ways. Firstly, the composite cell shown in Figure 10. may be split into "pseudo-cells" as shown in Figure 11. These have been designated *ab, ac, cd* and *bd*. Each of these cells is compacted. The mask file for cell *a* is then generated by using the X design grid file of *ab* and the Y design grid file of *ac*. This method is similarly applied to *b* (X- *ab* Y- *bd*), *c* (X- *cd* Y- *ac*) and *d* (X- *cd* Y- *bd*). By using this strategy all grid points on a common boundary match in the mask domain.

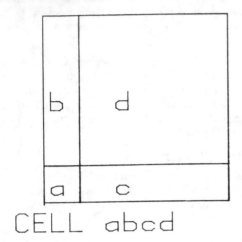

Figure 10. Composite cell abcd

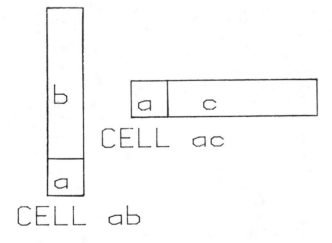

Figure 11   Pseudo cells used for pitch-matching

A refinement of this technique uses the connectivity specified previously. Here each cell to be considered is individually compacted. A program then examines the design grid files of each compacted cell at the boundaries and the grid points specified by the connection rules, and constructs a new file called a *mask coordinate* file which in effect combines the design grid files of the cells concerned according to the connection constraints.

Cells may be defined at the mask level only. These cells have fixed design grid files to which stretchable cells may be pitch matched. If this is impossible, either the variable cell is redesigned or a routing channel is employed. An example of such a cell would be the connection of a symbolically defined cell to an I/O pad.

Once the design grid and mask coordinate files have been adjusted to reflect to communication between adjacent cells, a program uses the original ICDL description and these files to generate XYMASK, the Bell System mask description language. Generation of other languages such as CIF is also possible. When mask conversion is complete the mask output may be viewed on the color graphics display. If changes are to be made in the layout, then the designer returns to the symbolic level.

5. EXPERIENCE

The compaction technique described in this paper has worked well with evolving layout methods that favor completing routing first followed by placement of devices and local interconnect[11], and the construction of generic bit-slice structures [12]. A requirement in these methodologies is a compaction algorithm that does not alter the topology of communication paths that are inherently dependent on the structure of a cell. The interactive nature of this system leads a designer to the rapid capture and subsequent modification of the topology. Using the system, a designer aquires a far more global view of his circuit at a far earlier stage than with more conventional methods. This allows for the efficient planning of effort on areas of the design which will yield the biggest payoff. In the experience gained so far with the system, one or two iterations seem to yeild an effective layout. The design rule free environment tends to encourage experimentation with the layout of complete structures prior to a decision being made about a final embodiment. For example, when constructing a bit-slice, a decision may be made to determine whether each cell should be long and thin or short and fat. In this case two layouts would be created and compared in area and aspect ratio to determine the most suitable for inclusion in the design.

It is important to note that although the system works well with more structured design styles, it is equally at ease dealing with less regular styles. The tendency in these situations is that the system guides a designer towards a more structured layout than might have been initially entered.

A chip photograph of the data path of a special purpose 16 bit CMOS processor is shown in Figure 12. This chip was designed on the MULGA system in 2 man-months. It contains 5000 transistors contained in approximately 30 cells. In addition to the CMOS chip two smaller NMOS chips have been designed and successfully fabricated. Currently, a 11000 transistor 16-bit CMOS special purpose processor is being fabricated.

6. SUMMARY

The concept of a virtual grid has been introduced as a means of describing and manipulating symbolic "stick" type layouts. A compaction algorithm that is a component of the interactive symbolic layout system that supports virtual grid layout has been described. It combines fast execution with predictability. The compaction method has been found to be very efficient in evolving structured layout styles. The interactive nature of the complete system compliments this efficiency. Additionally, inherent methods of pitch matching cells to build macro-cells have been described. The efficacy of this system has been demonstrated during the design of chips in both CMOS and NMOS technologies.

7. ACKNOWLEDGEMENTS

Bryan Ackland provided many fruitful discussions and ideas in addition to valuable feedback as the first "user" of this system.

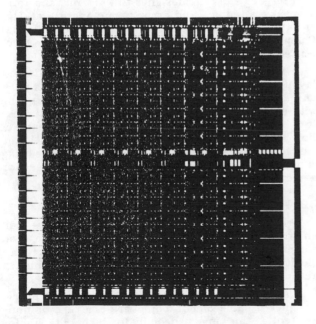

Figure 12. Chip photograph of 5000 transistor
CMOS circuit designed with MULGA

REFERENCES

[1] Weste, N.,"MULGA- An Interactive Symbolic Layout System for the Design of Integrated Circuits," accepted for publication in the Bell System Technical Journal.

[2] Buchanan, I.,"Modelling and Verification in Structured Integrated Circuit Design," PhD Thesis,University of Edinburgh,Scotland, 1980.

[3] Ackland, B. and Weste, N.,"Functional Verification in an Interactive Symbolic IC Design Environment," Proceedings of the 2nd. Caltech Conference on VLSI, Pasadena, CA, 1981.

[4] Akers, S.B., Geyer, J.M. and Roberts, D.L.,"IC Mask Layout with a Single Conductor Layer," Proceedings 7th. Design Automation Workshop, San Francisco, 1970, pp. 7-16.

[5] Williams, J., "STICKS- A Graphical Compiler for High Level LSI Design," Proceedings of the 1978 NCC, May 1978, pp. 289-295.

[6] Dunlop, A., "SLIP: Symbolic Layout of Integrated Circuits with Compaction," Computer Aided Design, Vol. 10, No. 6, November 1978, pp. 387-391.

[7] Dunlop, A., "SLIM- The Translation of Symbolic Layouts into Mask Data," Proceedings of the 17th. Design Automation Conference, June 1980, pp. 595-602.

[8] Hsueh, M. Y. and Pederson, D. O., "Computer-Aided Layout of LSI Circuit Building Blocks," Proceedings of the 1979 International Symposium on Circuits and Systems, July 1979, pp. 474-477. (also University of California, Berkeley Memo No. UCB/ERL M79/80, 10 Dec, 1979 "Symbolic Layout and Compaction of Integrated Circuits)

[9] Losleben, P. and Thompson, K., "Topological Analysis for VLSI Circuits," Proceedings of the 16th Design Automation Conference, San Diego, 1979, pp. 461-473.

[10] Persky, G., Deutsch, D.N. and Schweikert, D.G., "LTX - A Minicomputer-Based System for Automated LSI Layout," Journal of Design Automation and Fault-Tolerant Computing, Vol. 1, No. 3 (May 1977), pp. 217-255.

[11] Lopez, A.D. and Law, H.F.,"A Dense Gate Matrix Layout Style for MOS LSI," IEEE Journal of Solid State Circuits, Vol. SC-15, No. 4, Aug. 1980, pp. 736-740.

[12] Johannsen, D., "Bristle- Blocks", Proceedings of the 16th Design Automation Conference, San Diego, 1979, pp. 310-313.

SYMPLE: A PROCESS INDEPENDENT SYMBOLIC LAYOUT TOOL FOR BIPOLAR VLSI

Kevin S. B. Szabo
M. I. Elmasry

VLSI Group, Institute For Computer Research
University Of Waterloo,
Waterloo, Ontario
Canada N2L-3G1

Abstract

Bipolar VLSI has the potential to be the dominant process for high performance VLSI submicron devices. Bipolar processes and mask layout tend to be much more complicated than their MOS counterparts and hence are prime candidates for computer aided layout.

The automated design approach described here relies on symbolic layout techniques which hide most of the process complexities from the designer. Mechanisms for compaction, mask generation and circuit extraction are included. The symbolic approach has a number of interesting features; it contains schematic, electrical and topological information, and the designer has instant feedback regarding the final layout of the design.

A video taped demonstration of the prototype system will be shown.

1. Introduction

Bipolar VLSI technology, while not today as popular as MOS processes, continues to be very useful for high speed high performance circuits. The technology also performs well in mixed analog/digital and low voltage circuitry. As the advance in submicron device processing continues we may see bipolar become a competitive general purpose technology[1,2].

For a given system complexity a bipolar layout is more complicated than a MOS one. This is due to the greater number of basic devices, as well as their layout geometries and number of masks required. The NMOS technology has two basic devices, enhancement and depletion transistors; CMOS has both enhancement and depletion NMOS and PMOS, but usually only enhancement devices are used. The bipolar technology has resistors, junction and Schottky diodes, npn, pnp, and upside-down mode transistors. *Nesting* of multiple mask layers is necessary for all but the simplest of bipolar circuit structures while devices in MOS processes are generated by simple *overlapping* of two layers (diffusion and polysilicon). The much higher information content results in a higher cognitive load on the designer. This leads us to believe that automating bipolar layout will result is a higher net gain than the automating of MOS has, and that bipolar technology will be more acceptable to the VLSI community in this form.

2. The Question of Automation

2.1. Low Level Automation: The Mask Editor

Today, many interactive mask level editors are available. Caesar is one such editor that is becoming well known for its succinct and powerful command interface[3]. Caesar is process independent and can easily be used to generate bipolar devices. It unfortunately restricts the designer to work with the actual masks, a tedious and error prone operation. This situation is worse in bipolar VLSI considering that Complexity and number of mask levels is high in bipolar technologies.

Many tricks must be used to obtain an efficient layout. Compaction and design rule checking is difficult since the number of masks that are interacting with each other is non trivial. Coupling with simulators is definitely not straight forward; circuit extraction is a very expensive process, as it is with the simpler MOS technologies. Hence, mask level editing is not practical for bipolar VLSI and a need arises for a more powerful CAD layout tool.

2.2. Full Automation: The Silicon Compiler

The advantages of silicon compilers are well known. Applying the structuring found in high level computer languages affords good control over complexity, and many software management techniques automatically become available.

Although frequently mentioned in the literature, designer-useable Silicon compilers are rare or nonexistent at this time, especially for bipolar technology. Compilers that are available (or in the process of being constructed) tend to use one logic circuit type, eg. static or dynamic in MOS implementations. In bipolar technology there are many logic families, eg. IIL, ECL, EFL and STTL. These families are fundamentally different in both their physical layout and their electrical parameters. Hence, most compilers would have to be split into separate modules to generate

Reprinted from *Proc. IEEE Int. Conf. Computer Design: VLSI in Computers*, 1984, pp. 474–479.

the different circuits. Compilers that can intelligently switch between circuit types for design synthesis have yet to be demonstrated.

Structural Compilers which operate on a "mask" level are more difficult to use since they require a time consuming "write procedure; compile; view the output" loop[4]. This would be further aggravated by the many possible orientations of the circuits for the different logic types. Admittedly this occurs only with the low level cells where fine placement of components is absolutely necessary, but construction of the lower level cells occurs quite frequently. This kind of iteration is more appropriate on an interactive graphical editor, not in the 'batch' environment of an editor + compiler.

Restricted system architecture exhibited by current compilers is also detrimental, especially when one considers the mixed logic family and hybrid analog/digital circuits available with the bipolar technologies. Thus at this time we consider silicon compilers too immature for bipolar VLSI.

2.3. Symbolic Layout

Symbolic layout seems to offer ideal features as required by the bipolar system designer. High level circuit/layout representations convey the most significant information about device and interconnection, yet they are easily converted to the intricate and convoluted geometries frequently found in bipolar layouts. Elmasry[5, 6] has recently proposed a bipolar notation which defines a small set of symbols which can describe most useful constructs in bipolar technology. The symbolic diagram is part layout, and part schematic. It does not differentiate between different modes of bipolar circuit operation. The symbology hides the complexity of the actual layout, and stores the context of the circuit in the data. Immediate conversion to masks would throw the context information away, requiring the expensive pattern analysis techniques of present day design rule checkers. Coupling with simulators is straightforward and compaction calculations are also simplified.

3. The Bipolar Stick Notation

The notation is based on the MOS stick notation publicized by Mead and Conway[7], which was originally suggested by J. Williams[8]. The symbols fall into two categories, wires and points. Wires are used to interconnect points, thus denoting the connectivity of the circuit. One special wire type is the N-plus buried layer (See table 1). It interconnects point symbols, but these connections create devices. Thus N-plus is not used to route signals, it is used to provide an area for devices. The symbol layout is quite similar to the device placement on the mask, without the fine mask detail. There are no constructs which denote nesting of the symbols. The fact that the base region actually extends under the emitter region is understood from the context of the symbols. Due to the number of different ways of actually making a transistor we would not want our symbols to contain this information.

The other two symbols of the wire type are first and second metal. They provide all the actual signal carrying capabilities. A complete treatment of the symbols, and their conversion to two common process types, is discussed by Elmasry[6]. An example of a stick layout is shown in Figure 1.

		B/W	COLOR
1	n⁺ UNDERLAYER	▭	ORANGE
2	n⁺ COLLECTOR AND COLLECTOR CONTACT OF npn (OR BASE AND BASE CONTACT OF pnp)	▽	▽ GREEN
3	n⁺ EMITTER AND EMITTER CONTACT OF npn	○	○ GREEN
4	p BASE AND BASE CONTACT	✕	✕ RED
5	SCHOTTKY CONTACT (+VE TERMINAL)	✳	✳ BROWN
6	VIA	◨	▢ BLACK
7	FIRST METAL	—	— BLUE
8	SECOND METAL	▬	▬ PURPLE
9	I²L INJECTOR AND INJECTOR CONTACT OR EMITTER/COLLECTOR OF pnp AND CONTACT	⊠	⊠ RED

Table 1. Bipolar Stick Notation, Elmasry[6].

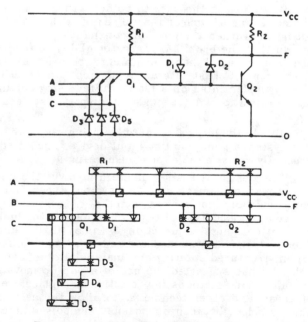

Figure 1. A simple STTL gate described using symbolic notation, Elmasry[6].

4. Goals Of Symple

We believe that a useful design tool for bipolar VLSI should incorporate a number of contemporary ideas in man-machine communication. Foley and Wallace[9] compiled a paper which addresses a number of issues in interaction. A prominent requirement is that of consistency and continuity in the interface. The system must not do unexpected things for certain input, there must be a way to undo an operation (to reduce the fear of making a mistake). There must be a clearly defined home state which is returned to after the requested action has been performed. The techniques they describe are sadly lacking in many interfaces used in the computer aided environment of today. In our work we have tried to reverse this state of affairs by using the techniques described below.

4.1. User Interface

Symple allows the user to naturally create and modify symbolic diagrams. Since Symple has very little knowledge of what the designer really wants to do, it presents a powerful editor which allows the designer to input symbols in the order most comfortable to the designer (i.e. unstructured). If we wish to perform design capture early in the evolution of the circuit we should allow the designer to enter the circuits by him/herself. In this case we do not have an experienced 'operator' in charge of the system, hence we must allow the casual user to feel comfortable with this system.

The system is menu driven, with the menu presented as light buttons on a high resolution display. This makes the system essentially prompt for input by allowing the user to point to the command choice. Giving a command menu requires the user to recognize rather than recall commands. It is well known that the brain has a greater ability to recognize something it has encountered before, rather than recall it. Fast interactive menus can be efficiently implemented on a frame buffer using colour table animation and segmented display list techniques.

Due to the large amount of 'pointing' required in this system a good pointing input device is required. Presently we use a tablet or mouse for input.

The design of a stick diagram involves the unstructured placement of many symbol types, both point symbol and various lengths of wires. This 'chose and place' type of task will dominate the bulk of the designers interaction with SYMPLE. Ron Baecker[10] has introduced a number of techniques for unstructured entering and manipulation of symbols. He has suggested the use of a pop-up moving menu, various methods for selection and placement of areas, and other techniques to allow the user to quickly select between all possible symbols without the gross arm movements required to return to a main menu area. The integration of these and other techniques into an interactive CAD environment will be the subject of another paper.

4.2. Functionality

Symple allows rectangular cells to be created and used in a hierarchical fashion. Currently cells must be convex and non-overlapping. Of course one may make odd shaped cells by interconnecting a number of rectangular cells.

Initially we believed that the compact/space operation would be used in a global sense, i.e. a design would be generated, compacted as a whole, then modified in an attempt to optimize the layout. This is actually incorrect since it is usually only necessary to compact the circuit of present interest. Thus the cycle is usually restricted to compaction of a single cell. Normally the lower level cells have already been spaced, but it is not an error if they have not been. Unspaced cells are guaranteed to not violate design rules, they are assumed to be spaced for the worst case design rules.

We can consider Symple to be simply a powerful editor with a "secondary" output form (the mask data). This distinction solves a number of logistical problems, such as distinguishing between the compact and uncompacted cells. Since compaction is an editing function there no longer needs to be this distinction. We shall just apply the normal text editor practice of allowing the designer to save/abort the current session. This approach is experimental and we are studying subjects to determine whether it is a natural and efficient method of symbolic VLSI layout and design.

The compactor is capable of compacting a whole circuit at a time, this is necessary for conversion of basic layouts between differing processes. Inefficient layouts may result due to design rule differences, and these will have to be fixed using the editor. This would be the case in any other symbolic system, and in a number of silicon compilers as well.

Mask generation is a straightforward operation since the symbols map cleanly into mask features. Circuit extraction is far simplified on the symbolic level since there is no pattern recognition required, and the context of the layout is always available.

It should never be necessary to manually tailor the mask output file from Symple. In order to provide a 'correct by construction' environment we have provided a rich enough symbol set that all necessary structures can be created. This does not take into account the analog/digital case where one wishes to mate these two families. Symple does not allow the fine control over device structures required for low noise, high sensitivity circuitry. The designer has the option of including a raw mask file created by one of the many available mask level editors. Hopefully this exercise will be rarely required since it requires that the mask design be checked for design rule violations and some of the advantages of using SYMPLE are lost. However, the designer gains the ability to have very fine control over geometries of space and performance critical cells, or to include output from other geometry generating tools (such as a PLA generator).

5. Data Structures

Choice of data structures for a data intensive program such as Symple is of very high importance. The structure needs to support rapid insertion/delete and it should be trivial to find one's nearest neighbour. The former requirement allows the structure to support the editor, the latter allows the compaction process to work efficiently. Alternatively there could be two structures that are easily transformable, but the normal use of a tool of this kind results in much redundant conversion between the two representations.

The structure we are using is based on neighbour pointers. A simple symbolic layout is shown in Figure 2a. The schematic represents a multi-emitter transistor with a base pull up resistor. The stick symbolic layout is shown immediately below. The data structure generated for this layout is in Figure 2b. Each node in the structure has the type definition of Table 2. Pointers are bidirectionally linked (shown singly linked for clarity) allowing quick insertion and removal of nodes, plus bidirectional searching. This structure is similar to that used by Neil Weste in his Mulga symbolic layout system[11].

With this structure it is simple to find one's immediate neighbour, and thus determine the minimum spacing to that neighbour. This structure also allows one to quickly determine connectivity between elements. Neighbour pointers determine what nodes are topologically residing beside a node, whereas wire information (nplus, first and second metals) describes the connectivity between nodes. The wire

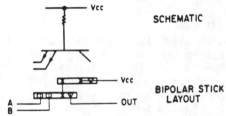

SCHEMATIC

BIPOLAR STICK LAYOUT

Figure 2a: A Simple Bipolar Circuit

information is actually a compact structure (an array of four bytes per wire type, eg 32 bits) and thus only a small overhead will be incurred if a designer defines a new wire type. This new wire type would be an additional interconnect level.

The actual wire size information is held as separate structures. Each wire type (nplus, first and second metal) can have one of 1 ... max_wire_size wire sizes. We have chosen 255 different wire size for this implementation, thus allowing each wire type to fit into a 32 bit word. The wire size indexes into a structure which stores the actual wire width (for metal) or resistance/transistor-size (for N-plus). This indirection not only results in compact storage, but simplifies alterations to the design. One may attach names to the different 'sizes' eg. Pullup-Tran, Load-Res, Base-Res1, Base-Res2 and design with these names. If a change to all load resistors is required one simply changes the resistor value that corresponds to 'Load-Res'.

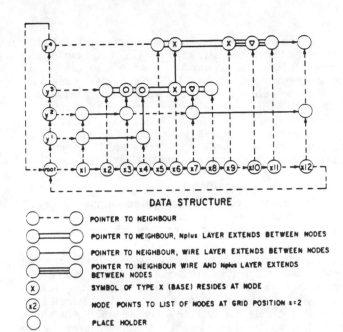

DATA STRUCTURE

- ○----○ POINTER TO NEIGHBOUR
- ○——○ POINTER TO NEIGHBOUR, Nplus LAYER EXTENDS BETWEEN NODES
- ○——○ POINTER TO NEIGHBOUR, WIRE LAYER EXTENDS BETWEEN NODES
- ○——○ POINTER TO NEIGHBOUR WIRE AND Nplus LAYER EXTENDS BETWEEN NODES
- (X) SYMBOL OF TYPE X (BASE) RESIDES AT NODE
- (x2) NODE POINTS TO LIST OF NODES AT GRID POSITION x=2
- ○ PLACE HOLDER

Figure 2b: The Data Generated For The Circuit

Cells are represented simply as a node in the data which points to the root of the 'called' cell. Wires cannot extend into the cell, they must terminate on 'pins'. The cell will fix the spacing of these pins and thus we can perform pitch matching between cells. This restriction may be lifted at a later date if it is warranted, at this time we find this implementation

TYPE

```
direction      = ( up, down, left, right );

wire_info      = record
                 size          : integer;
                 aspect_ratio  : real;
                 resistance    : integer;
                 end;

wire_index     = ( 0 .. max_wire_sizes );

wire_info      = array [ wire_index ] of wire_info;

wire           = array [ direction ] of wire_index;

ndtype         = ( root, yaxis, xaxis, data );

symboltype     = ( emitter, base, collector, schottky,
                 injector, via, label, pin, cell );

nodeptr        = ^ node;

node           = record
                 nodetype      : ndtype;
                 symbols       : set of symboltype
                 label         : ^ char;
                 nplus         : wire;
                 first_metal   : wire;
                 second_metal  : wire;
                 up_ptr        : nodeptr;
                 down_ptr      : nodeptr;
                 left_ptr      : nodeptr;
                 right_ptr     : nodeptr;
                 x_position    : integer;
                 y_position    : integer;
                 end;
```

Table 2. Data Structure For A Node

adequate.

Overhead in pointer space is significant, but not excessive. This data arrangement is much easier to extract information from than bins or a simple list of nodes (equivalent to a netlist). Also the data does not have to be reformatted into a structure suitable for compaction for each invocation of the spacer. Graphical operations such as deleting wires and moving areas are also easily implemented since it is straightforward to isolate the extent of the object in question, and perform an operation on it.

6. Scaling and Compaction

The names scaling, spacing and compaction are used interchangeably in this paper, they refer to conversion of topological (relative placement) data to a correctly spaced mask layout.

The symbolic layout has no absolute coordinates held within its data, these must be generated by the scaling process. The ability to properly scale the layout is the most rewarding aspect of Symple. If we were not able to generate a layout which efficiently utilized the real estate that it consumed our tool would be quickly discarded.

The compaction process can be arbitrarily complex. As implemented at the time of writing the process is based on the scheme used in Mulga, where symbols are never moved from grid points. The inter-grid spacing is non uniform however, and is calculated by examining neighbouring symbols on each grid line.

This scheme was straightforward to implement, and is quite reliable. The inability of the compactor to move structures off a grid line is a shortcoming and we expect the next versions of the compactor to overcome this. Initially we do not want to include jog insertion, since this is a design decision and the designer should have full control over it. Jog insertion increases the complexity of the compactor's code greatly since it is now attempting to synthesize a layout, as opposed to merely compacting it. We believe that given a good graphical editor the designer can place the jogs he/she requires much more effectively than the machine can. The design of more intelligent compactors will be the basis of further research.

The compactor is table driven. This table is derived from a process file at execution time. Symple allows the designer to specify what process a specific cell should use, thus allowing the user to mix and match design rules according to circuit function. This is specifically useful when designing pad drivers and memory cells. At the present time the creation of a process/design-rule file is not straightforward, however we are designing a rule compiler which will convert a list of design rules directly into the appropriate table to check a circuit. This should demystify the task of setting up SYMPLE for new processes.

7. Mask Generation, Circuit Extraction and Simulation

Once the data has been correctly spaced the generation of masks is straight forward. Taking into account the process that is being used, we expand each symbol into a series of rectangles on different mask levels. For most symbols the process ends here, but the program has to do additional work for the 'base' symbol. A base must extend underneath all neighbouring emitters that are part of the same device. This is quickly determined by inspecting the nearest neighbours that are on the same Nplus wire. If they hold emitters we extend the base diffusion underneath them.

Circuit extraction is of somewhat similar a process, except instead of generating masks we determine the mask size (hence parasitics) and the connectivity. It is not a difficult matter to generate a net list where each node is a point on the netlist. Another pass over the data can be used to remove redundant nodes.

8. Environment And Portability

The program Symple works in conjunction with the UNIX† operating system. The code is portable; it presently resides on a VAX11/780 (running Berkeley 4.2bsd) and an Orcatech 3000 (running System III). Graphical input/output is handled by a device independent graphics package developed at the University of Waterloo by the Computer Graphics Lab and the VLSI Group. The program is written in C[12] and is approximately 10,000 lines long.

Acknowledgments

We would like to thank the members of the Computer Graphics Lab for their help in implementing Symple. Specifically John Beatty and Baldev Singh had many good comments and insight from a human factors point of view, as well as some useful software which they donated to the project. Jim Leask has contributed much to SYMPLE, he is currently implementing a MOS version of the system.

References

1. J. S. T. Huang, "Bipolar Technology Potential For VLSI," *VLSI Design*, pp. 64-67 (1983).

2. B. Stehlin, "CMOS and Bipolar Technologies Are More Parteners Than Rivals," *Electronics*, pp. 95-98 (1984).

---

†UNIX is a Trademark of Bell Laboratories.

3. J. Ousterhout, "Caesar: An Interactive Editor For VLSI Layouts.," *VLSI Design*, pp. 34-38 (1981).

4. S. C. North, "Molding Clay: A Manual for the Clay Layout Language," VLSI Memo #3, Princeton University, Princeton, New Jersey (1983).

5. M.I. Elmasry, "Stick-Layout Notation For Bipolar VLSI," *VLSI Design*, pp. 65-69 (1983).

6. M.I. Elmasry, *Digital Bipolar Integrated Circuits*, John Wiley and Sons, New York, USA (1983).

7. C.A. Mead and L.A. Conway, *Introduction to VLSI Systems*, Addison-Wesley, Reading, Mass. (1980).

8. J. D. Williams, *Sticks - A New Approach To LSI Design.*, M.I.T. (1977).

9. J. D. Foley and V. L. Wallace, "The Art of Natural Graphic Man-Machine Conversation," *Proceedings of I.E.E.E.*, pp. 462-470 (1974).

10. R. Baecker, "Towards An Effective Characterization Of Graphical Interaction," *Seillac II Workshop On Man-Machine Interaction*, North-Holland Publishing Co. IFIP, (1980).

11. N. Weste, "Virtual Grid Symbolic Layout," *Proceedings 18th Design Automation Conference*, pp. 225-233 (1981).

12. B. W. Kernighan and D. M. Ritchie, *The C Programming Language*, Prentice-Hall, Englewood Cliffs, New Jersey (1978).

An Algorithm to Compact a VLSI Symbolic Layout with Mixed Constraints

YUH-ZEN LIAO, MEMBER, IEEE, AND C. K. WONG, SENIOR MEMBER, IEEE

Abstract—A popular algorithm to compact VLSI symbolic layout is to use a graph algorithm similar to finding the "longest path" in a network. The algorithm assumes that the spacing constraints on the mask elements are of the lower bound type. However, to enable the user to have close control over the compaction result, a desired symbolic layout system should allow the user to add either the equality or the upper bound constraints on selected pairs of mask elements as well. This paper proposes an algorithm which uses a graph-theoretic approach to solve efficiently the compaction problem with mixed constraints.

I. Introduction

THIS PAPER describes an algorithm to compact the symbolic layout of VLSI circuits. The algorithm accepts the symbolic layout of an IC design and a set of user-specified constraints on the spacing of the mask elements as input. Using a graph-theoretic approach, it then outputs a physical mask layout with the minimum possible chip area and satisfying both the user-defined and the design-rule constraints. Because the algorithm allows the user to define the spacing constraints, it greatly facilitates the designer to closely control over the layout progress in an interactive design environment.

A symbolic layout system is an automated design tool for laying out the masks of a VLSI circuit. In this system, circuit elements like transistors, contacts, capacitors, etc., which are composed of overlapped mask levels, are represented by their respective graphic symbols. Interconnecting mask polygons used as wires are also symbolically represented by their center lines. When a designer puts the symbol of a circuit element at a place on the layout plane, it signifies that all the mask levels composing the element exist there. As an example, Fig. 1(a) shows the schematic circuit diagram of a carry-chain circuit for the arithmetic logic unit [1]. Fig. 1(b) is its symbolic layout. Fig. 1(c) is the geometric mask layout represented by Fig. 1(b). A symbolic layout of this type is also called a "stick diagram" as it contains mostly simple line drawings. We will use the terms "stick diagram" and "symbolic layout" interchangeably. A stick diagram conveys the information of a circuit's topology, as the symbolically represented circuit elements are fully routed but loosely placed.

After a stick diagram is drawn through a computer-aided graphics system, a system called *compactor* automatically spaces the circuit elements and interconnections to obtain a mask physical layout. The compactor considers the connect-ing wires as stretchable and packs the circuit elements as tightly as possible without violating the spacing constraints defined by the user and the design rules. With the aid of an efficient compactor, the user can concentrate on the circuit design at the topological level rather than at the geometrical level. As indicated in [1], alternative topologies often lead to very different chip areas after compaction. This suggests that the designer can iteratively modify the circuit's topology until an effective geometrical layout is generated by the compactor. Various papers [2]-[7] report that the stick or symbolic-layout programs achieve faster turnaround time and slightly inferior area-saving performance, as compared to the totally hand-layout approach. In addition, one of the symbolic layout tools in our facility has computed the layout of a 34-bit ALU. Description of its layout and simulation results is given in [11].

To ease the effort of developing a compaction algorithm, the stick system requires that only orthogonal structures be allowed. This means that the edges of the geometrical shape of every mask element are either in the horizontal (X) or in the vertical (Y) direction. With such a restriction, each of the spacing constraints is also either in the horizontal or vertical direction. Ideally, an algorithm which can simultaneously compact the layout in the two orthogonal directions is most desirable, as the tradeoff of moving an element in either direction can be determined at one time. However, such an algorithm with reasonable speed has not been found yet. The general approach first compacts the layout optimally in one direction and then compacts the result in other direction in a similar way. Thus the layout compaction becomes two one-dimensional problems rather than a two-dimensional X-Y coupled problem. The result is in general not a global minimum in chip area. However, such a simplification can rapidly generate the compacted result, enabling the designer to repeatedly modify the stick diagram until a satisfactory result is obtained. Therefore, the approach is generally regarded as acceptable by the users. This approach is also taken in the present paper.

The constraints imposed by the design rules are of the "lower bound" type, in the sense that the spacing between two elements must be greater than or equal to certain prescribed values. If the mask elements are only subject to this type of constraints, the compaction operation becomes quite simple. However, a useful stick system should allow the user to add mixed constraints. This means that for a selected pair of elements (not necessarily adjacent) the designer can define an upper bound, or lower bound, or fixed constraint between them. One reason to include the mixed constraints is to make the compaction

Manuscript received August 26, 1982; revised December 15, 1982.
The authors are with IBM Thomas J. Watson Research Center, Yorktown Heights, NY 10598.

Reprinted from *IEEE Trans. Computer-Aided Des. of ICAS*, vol. CAD-2, pp. 62–69, Apr. 1983.

278

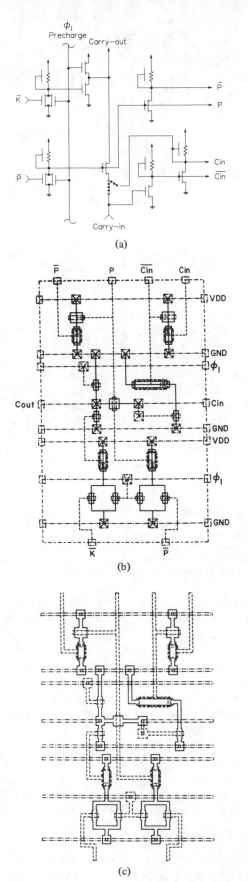

Fig. 1. Carry-chain circuit for the arithmetic circuit. Control signals K and P are for "carry-kill" and "carry-propagate", respectively. (a) The circuit schematic diagram. (b) The stick diagram. (c) The uncompacted geometrical layout represented by (b).

result predictable, so the designer can interact smoothly with the compactor. There are more practical reasons for doing this, and we cite several examples in the following.

In a hierarchical layout approach, it is often desirable to abut two cells directly without generating wires for inter-cell routing. For such cases, the designer can set fixed constraints on the relative distances of the ports so as to align the ports to be connected together.

When the drawing of a stick diagram is not completed yet, the designer may want to view the possible compacted layout. Before invoking the compactor, the designer can block out the chip areas reserved for the unplotted circuits and keep the blocks' external interconnections stuck to the blocks, by using the fixed constraints.

Signals take longer to travel across long wires, so a placement of the circuit elements with shorter interconnections may run faster. However, a placement which results in minimum chip area does not guarantee minimum wire lengths. When the length of a wire connecting two circuit elements is critical to speed, the designer can impose an "upper bound" constraint on the two elements to limit the length of the wire.

The idea of using symbolic system to layout VLSI circuits has appeared for a few years. But the compaction algorithm in most of the related papers can correctly handle the design rule and lower bound constraints only. The CABBAGE system [2] allows the user to define the fixed constraints, but it simply replaces each fixed constraint with a lower bound constraint during the calculation. Consequently, in many cases it returns an "over-constrained" error-message, through there exists a solution satisfying all the constraints. Thus the aim of this paper is to find a solution which satisfies not only the design rule constraints but also the user-defined mixed constraints.

In the proposed algorithm, the first stage in compacting the layout along one axis is to map a stick diagram to a directed graph. Mask elements are converted to the nodes and spacing constraints in the direction of compaction are converted to the weighted edges. If all the constraints solely come from the design rules, i.e., all constraints are of lower bound type, the graph obtained is an acyclic graph. Then with a method similar to finding the longest path in a network, the positions of the mask elements can be derived. In fact, this is the method adopted by the CABBAGE system [2]. When the user-defined mixed constraints are also converted to the edges in the graph, the resultant graph is not necessarily acyclic, and simply applying the above method cannot solve the problem. Instead, the proposed algorithm takes an iterative approach. Each iteration is a variant of the longest path search algorithm. When the iterative terminates, it returns the desired optimal solution satisfying all the spacing constraints.

Detailed description of the algorithm is in Section II. Proof of the correctness of the algorithm and timing analysis is in Section III. Section IV is a discussion of further work.

II. THE ALGORITHM

This section describes the solution process of the layout compaction algorithm. The layout is first compacted along either the vertical or the horizontal direction, and then the other direction. In the sequel, we only discuss the compaction in the horizontal direction. The vertical compaction is identical algorithmically, and its discussion is omitted.

Conceptually, to obtain a layout with minimum overall width, the circuit elements and the vertical interconnections are moved horizontally to the leftmost possible positions, while the attached horizontal interconnection are stretched accordingly. To implement this concept, we translate the stick diagram and the spacing constraint set to a graph, then apply a suitable graph algorithm to yield the coordinates of the mask elements in the compacted layout. This graph is called the *constraint graph* and denoted by G.

We use the same approach as the CABBAGE system to define the nodes in G and translate the spacing design rules to the edges in G. Following is a brief summary of the method. The readers can find the details in [2].

The nodes of G are obtained by a grouping operation. All the circuit elements and vertical interconnections, which are geometrically connected on a common center-line in the stick diagram, are held together as a group. Assume that there are $n + 1$ groups, labeled $V_0, V_1, \cdots, V_n$, in the stick diagram. In particular, the left and right boundaries of the stick diagram are represented by V_0 and V_n, respectively. The nodes in G are defined by mapping each V_i, $i = 0, \cdots, n$, to a node. As the mapping is one-to-one, we will use the same notation V_i, $i = 0, \cdots, n$, to designate the corresponding nodes in G. Fig. 2(a) shows the vertical groups of the stick diagram in Fig. 1.

Next, the relevant spacing requirements among the groups are converted to the edges in G. Assume that the X-axis is in the horizontal direction and $V_0(X), V_1(X), \cdots, V_n(X)$, represent the X-coordinates of the center lines of the groups. After the groups are sorted in the order of their center lines' X-coordinates in the stick diagram, a routine called *design-rule analyzer* examines each group from left to right to find the minimum spacing requirements between this group and its neighboring groups on the right. Specifically, assume the group V_a is being accessed and V_b is a group lying to the right of V_a. If the design rules require a minimum spacing of value p between the center lines of V_a and V_b, then $V_a(X)$ and $V_b(X)$ will be subject to the inequality constraint of the following form of

$$V_a(X) + p \leqslant V_b(X).$$

In such a case, a directed edge from V_a to V_b with edge weight p is inserted in G.

It should be noted that there is always a maximum value of spacing design rules in a given technology. In general, such a maximum value is close to the minimum linewidth in that technology. Consequently, checking the spacing requirement between two groups may become unnecessary when the intervening groups keep the two groups in question far enough apart. For a typical layout, except for a few groups which contain long wires, most groups have relevant spacing requirements with only a limited number of neighboring groups on the right. Thus the number of edges in G is almost linearly proportional to the number of nodes in G, if a carefully devised design-rule analyzer is used.

After the design-rule analysis is finished, user-supplied con-

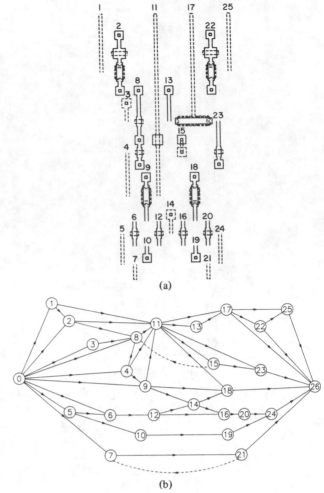

(a)

(b)

Fig. 2. Mapping vertical groups to a constraint graph. (a) The vertical groups of the stick diagram in Fig. 1. The labeling numbers designate the groups. (b) The constraint graph for the groups in (a). The two edges connecting V_7 and V_{21} represent a fixed constraint, for cell abutting. The left-directed edge (V_{15}, V_8) represents an upper bound constraint for limiting the data path crucial to timing.

straints are then translated to edges of G. Again, assume V_a and V_b are two groups and V_b is either on the right or the same horizontal location of V_a. If the user requests that V_b be at least p units to the right of V_a, then the type of inequality constraint is the same as that implied by the design rules. So, a directed edge from V_a to V_b with weight p is added to G. If V_b is required to be no more than q units to the right of V_a, then the inequality constraint on $V_a(X)$ and $V_b(X)$ is

$$V_b(X) - q \leqslant V_a(X).$$

For such a case, a directed edge from V_b to V_a with edge weight $-q$ is added to G. Now, if the user requires that V_b be exactly r units to the right of V_a, then $V_a(X)$ and $V_b(X)$ are subject to the equality (or fixed) constraint:

$$V_a(X) + r = V_b(X).$$

This equality is equivalent to two joint inequalities $V_a(X) + r \leqslant V_b(X)$ and $V_b(X) - r \leqslant V_a(X)$. Thus we add two edges to G. One is from V_a to V_b with weight r, the other is from V_b to V_a with weight $-r$.

Thus in general an edge (V_i, V_j) of weight l_{ij} in G, $i, j = 0, \cdots, n$, represents an inequality constraint in the form of

$V_i(X) + l_{ij} \leqslant V_j(X)$. If in the input stick diagram, the group corresponding to V_j is on the right of V_i, we tag the edge (V_i, V_j) as a right-directed edge. Conversely, if V_j is on the left, then (V_i, V_j) is tagged as a left-directed edge. When V_i and V_j have the same horizontal location, the edge is tagged as a right-directed edge, if V_i is above V_j; else, it is a left-directed edge. Note that all the edges derived from the design-rule analysis are right-directed edges and all the left-directed edges are due to the user-defined constraints. Thus if only design-rule constraints are present, the resulting graph is acyclic. Fig. 2(b) shows the constraint graph constructed from Fig. 2(a).

At this stage, the resulting graph G may have multiple edges from one node to another. For a given pair of nodes V_i and V_j, if there are more than one edge from V_i to V_j, we delete all the edges except the one which has the largest weight, as the constraint implied by this edge overrides other constraints. For example, the effect of the two constraints $V_i(X) + a \leqslant V_j(X)$ and $V_i(X) + b \leqslant V_j(X)$ are equivalent to the constraint $V_i(X) + \max(a, b) \leqslant V_j(X)$. Thus the final constraint graph has at most one directed edge from one node to another. Self-loop on a node will not appear by the construction of the graph.

We define a solution set of G as the set $\{V_i(X), 0 \leqslant i \leqslant n\}$ such that it satisfies all the inequality constraints implied by the edges of G. Since V_0 represents the left boundary of the stick diagram, we normalize $V_0(X)$ in all the solution set to 0. A solution set $\{V_i(X)\}$ of G is defined as the minimum solution set, if, for any solution set $\{V_i'(X)\}$, we have $V_i(X) \leqslant V_i'(X)$. In particular, the value of $V_n(X)$ in the minimum solution set is the minimum possible width of the compacted layout. Note that the constraints may have conflicts and the solution set may not exist. For example, $V_1(X) + 3 \leqslant V_2(X)$ and $V_2(X) + 4 \leqslant V_1(X)$ are two conflicting constraints. When this occurs, we say that the constraints are inconsistent.

Let V and E represent the sets of nodes and edges of G, respectively. Then $G = (V, E)$. Also, let E_f and E_b designate the set of right-directed and left-directed edges of G, respectively. Denote the subgraph (V, E_f) of G as G_f. One property of G_f is that G_f is a single-source single-sink acyclic graph. It is single-source and single-sink because all the stick elements are in-between the left and right boundaries V_0 and V_n of the stick diagram, and cannot cross over the boundaries. It is acyclic because an edge (V_a, V_b) of E_f implies that V_b is below or on the right of V_a in the input stick diagram. Thus there is a topological order for the nodes in the graph G_f. We will exploit this very useful property to design an efficient compaction algorithm.

Based on the graphs G and G_f, the compaction algorithm is performed by iteratively increasing X-coordinates of the nodes in V until all the constraints are satisfied. Each iteration consists of a call to Procedure LongestPath followed by adjusting the positions of the head nodes of the left-directed edges in G.

Procedure LongestPath takes the graph G_f and the initial values of $V_i(X)$, $i = 0, \cdots, n$, as input, and calculates each $V_i(X)$ as

$$V_i(X) = \max_{0 \leqslant j \leqslant n} \{\text{initial value of } V_j(X) + \text{length of the}$$

$$\text{longest path of } G_f \text{ from } V_j \text{ to } V_i\}$$

Here the length of a path is the sum of the edge weights on this path. This procedure is similar to solving the critical path problem in a single-source acyclic network [8], except the initial position of a node may be larger than the length of the longest path from the source V_0 to that node. A formal description of Procedure LongestPath is listed in the following:

Procedure LongestPath(G_f)
    create a stack;
    push V_0 on stack;
    for $i := 1$ to n do
        In_Degree $(V_i) :=$ the in-degree of V_i in G_f;
    while stack not empty do
    begin
        pop the stack's top node V_t from the stack;
        for each edge (V_t, V_j) in G_f do
            begin
                $V_j(X) = \max(V_j(X), V_t(X) + l_{tj})$; $\{l_{tj}$ is the edge weight$\}$
                In_Degree $(V_j) :=$ In_Degree $(V_j) - 1$;
                if In_Degree $(V_j) = 0$ then
                push V_j on stack;
        end;
    end;
  End(LongestPath)

After each call to Procedure LongestPath, the resultant $V_i(X)$, $i = 0, \cdots, n$, satisfies all the constraints implied by the edges in G_f. Next, the algorithm successively accesses each left-directed edge in G to test if the constraint implied by this edge is also satisfied. If it is not true, then the head node of the edge will be moved to the right by a least amount so that the constraint is satisfied. Specifically, assume (V_i, V_j) is a left-directed edge and $V_i(X)$, $V_j(X)$ are the current values when this edge is being tested. Now, if $V_i(X) + l_{ij} > V_j(X)$, $V_j(X)$ will be set to $V_i(X) + l_{ij}$. (Recall that l_{ij} must be negative since (V_i, V_j) is a left-directed edge.)

After all the negative edges in G are accessed, the new setup of the node positions may still not satisfy all the constraints. So another pass of Procedure LongestPath and reallocation of the left-directed edge heads will be executed. This iteration is repeated until all the constraints are satisfied. Assume there are b left-directed edges in G. We will prove in the next section that the algorithm will yield a minimum solution set and terminate by executing at most $b + 1$ iterations, if the constraints are consistent.

The compaction algorithm appears as Procedure Compaction (G), and is listed in the following:

Procedure Compaction (G)
1   Count := 0;
2    $V_0(X) := 0$;
3   for $i := 1$ to n do
4    $V_i(X) = -\infty$;
5   Repeat
6   Flag := ture;
7   call Procedure LongestPath (G_f);
8   For each left-directed edge (V_i, V_j) in G do
9   if $V_j(X) < V_i(X) + l_{ij}$ then
10   begin

```
11              V_j(X) := V_i(X) + l_ij;
12              Flag := false;
13            end;
14         Count := Count + 1;
15         if Count = b + 1 and Flag = false then
16            begin
17              print (inconsistent constraints);
18              Return;
19            end;
20      until Flag := true;
21  End (Compaction).
```

As an example, Fig. 3 illustrates the action of Procedure Compaction on a constraint graph G. The graph and its edge weights are shown in Fig. 3(a). In G, the edges $(V_1, V_2), (V_3, V_4)$, and (V_6, V_1) are left-directed edges; other edges are right-directed edges. As the action begins, the first call to Procedure LongestPath returns the X-coordinates of the nodes, as shown in Fig. 3(b). At this time, the coordinates of each node is equal to the length of the longest path from V_0 to that node, going through the right-directed edges in G only. Further, all the inequality constraints implied by the right-directed edges are satisfied. The subsequent testing of the inequalities represented by the left-directed edges finds that the inequalities of the edges (V_1, V_2) and (V_3, V_4) are not satisfied. Consequently, $V_2(X)$ and $V_4(X)$ are increased to meet the inequalities, as shown in Fig. 3(c). Nevertheless, this operation makes the spacing requirement indicated by the right-directed edges (V_2, V_3) and (V_4, V_6) unsatisfied. In an attempt to correct this, the nodes' X-coordinates shown in Fig. 3(c) are fed to the second run of Procedure LongestPath. Fig. 3(d) shows the result. Now, only the inequality implied by the left-directed edge (V_3, V_4) is unsatisfied. So the subsequent left-directed edge examination increases the value of $V_4(X)$ again, and the result is shown in Fig. 2(e). With the current nodes' coordinates as the initial values, the third call to Procedure Longest-Path returns a setup of the nodes' coordinates which meets all the inequality constraints implied by the left-directed edges. Consequently, Procedure Compaction terminates. The final result is shown in Fig. 2(f). Note that every call to Procedure LongestPath always causes the nodes' coordinates to satisfy all the inequality constraints of the right-directed edges.

III. ANALYSIS

In this section, we shall analyze Procedure Compaction which attempts to obtain a minimum solution set of the constraint graph G. To begin with, Theorem 1 gives a graph-theoretic interpretation of the minimum solution set. That is, in the minimum solution set of G, a node's X-coordinate is the length of the longest path from V_0 to that node. In Procedure Compaction, lines 7 through 13 are one pass of the iteration which involves a call to Procedure LongestPath followed by adjusting the X-coordinates of the head nodes of the left-directed edges in G. Theorem 2 shows that such an iteration moves each node in G to reach the end of a longer path from V_0 to that node. (Given two nodes V_a and V_b, we say that V_b reaches the end of a path from V_a to V_b, if $V_b(X)$ is equal to $V_a(X)$ plus the length of that path.) Applying

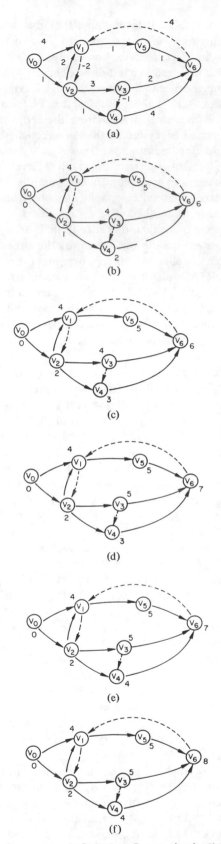

Fig. 3. Iteration progress in Procedure Compaction leading to the minimum solution set of a constraint graph. (a) A constraint graph. $(V_1, V_2), (V_3, V_4)$ and (V_1, V_6) are left directed edges. (b) Results of the first call to Procedure LongestPath. Values associated with the nodes' X-coordinates. (c) $V_2(X)$ and $V_3(X)$ are increased after testing the left-directed edges. (d) Results of the second call to Procedure LongestPath. (e) Results of testing the left-directed edges. (f) Results of the third call to Procedure LongestPath.

Theorems 1 and 2, Theorem 3 shows that Procedure Compaction will terminate and yield a minimum solution set, if there indeed exists a solution satisfying all the inequalities represented by the edges of G. Corollary of Theorem 3 shows that the procedure can correctly detect the inconsistence of the inequalities. We also estimate the time and space requirement for the procedure at the end of this section.

Define a *positive cycle* of G as a cycle in G such that the sum of the edge weights on that cycle is positive. As V_0 is a source of G, there is at least a path from V_0 to any given node in G. Evidently, if G has no positive cycles, the longest path from V_0 to any given node exists and has finite length.

Theorem 1: Assume the constraint graph G has no positive cycle. For $i = 0, 1, \cdots, n$, let $\tilde{V}_i(X)$ be the length of the longest V_0-to-V_i path in G. Then $\{\tilde{V}_i(X), 0 \leqslant i \leqslant n\}$ is a minimum solution set of G.

Proof: We first show by contradiction that $\{\tilde{V}_i(X)\}$ is a solution set. That is, for any edge (V_i, V_j) of edge weight l_{ij}, we have $\tilde{V}_i(X) + l_{ij} \leqslant \tilde{V}_j(X)$.

Assume there is an edge (V_i, V_j) in G such that $\tilde{V}_i(X)$ and $\tilde{V}_j(X)$ do not satisfy the constraint, that is

$$\tilde{V}_i(X) + l_{ij} > \tilde{V}_j(X).$$

The above inequality implies that the longest path from V_0 to V_j does not pass through the edge (V_i, V_j), otherwise $V_i(X) + l_{ij}$ would be equal to $V_j(X)$. Then the V_0-to-V_j path consisting of the longest path from V_0 to V_i and the edge (V_i, V_j) is longer than the longest path from V_0 to V_j. This contradicts the definition of $\tilde{V}_j(X)$. So the above inequality is not true. As $\{\tilde{V}_j(X)\}$ satisfies all inequalities, it is a solution set.

It remains to show that $\{\tilde{V}_i(X)\}$ is minimum. Let $\{V_i'(X), i = 0, 1, \cdots, n\}$ be any solution set of G. For a given node V_i, assume the edges in the longest path from V_0 to V_i are $(V_0, V_{i_1}), (V_{i_1}, V_{i_2}), \cdots, (V_{i_{s-1}}, V_{i_s})$ where $V_{i_s} = V_i$. As the solution set $\{V_i'(X)\}$ satisfies the inequalities represented by these edges, we have

$$V_0'(X) + l_{0,i_1} \leqslant V_{i_1}'(X)$$
$$V_{i_1}'(X) + l_{i_1,i_2} \leqslant V_{i_2}'(X)$$
$$\vdots$$
$$V_{i_{s-1}}'(X) + l_{i_{s-1},i_s} \leqslant V_{i_s}'(X).$$

Adding the above inequalities, we get

$$V_0'(X) + \sum_{k=1}^{s} l_{i_{k-1},i_k} \leqslant V_i'(X).$$

By the definition of

$$\tilde{V}_i(X), \sum_{k=1}^{s} l_{i_{k-1},i_k} = \tilde{V}_i(X).$$

As mentioned in Section II, we assume that the X-coordinate of V_0 is normalized to 0 in any solution set. Thus we have $\tilde{V}_i(X) \leqslant V_i'(X)$. □

Corollary: A constraint graph G has a solution set if and only if there is no positive cycle in G.

Proof: The sufficient condition is proved in Theorem 1. We now prove the necessary condition.

Without loss of generality, assume the edges (V_1, V_2), $(V_2, V_3), \cdots, (V_{s-1}, V_s)$, (V_s, V_1) form a positive cycle. The inequalities represented by these edges are

$$V_1(X) + l_{1,2} \leqslant V_2(X)$$
$$V_2(X) + l_{2,3} \leqslant V_3(X)$$
$$\vdots$$
$$V_s(X) + l_{s,1} \leqslant V_1(X).$$

Adding the above inequalities, we have

$$V_1(X) + \text{(sum of the weights of the edges in the cycle)}$$
$$\leqslant V_1(X).$$

Since the sum of the edge weights is positive, $V_1(X)$ cannot satisfy the above inequality. Thus if a solution set of G exists, G has no positive cycles. □

We now consider the effect of an iterative application of Procedure Compaction. For any integer r, $r \geqslant 0$, let $V_i^r(X)$, $1 \leqslant i \leqslant n$, denote the position of node V_i after the rth call to Procedure LongestPath. In particular, for $r = 0$, the value of $V_i^0(X)$ is $-\infty$ for $i > 0$, and $V_0^0(X) = 0$, as defined in lines 2 through 4 in Procedure Compaction.

Theorem 2: For any r, $r \geqslant 1$, and i, $1 \leqslant i \leqslant n$, $V_i^r(X)$ is equal to the length of a path from V_0 to V_i in G, and $V_i^r(X) \geqslant V_i^{r-1}(X)$.

Proof: We prove the theorem by induction. After the first call to Procedure LongestPath, any node V_i, $0 \leqslant i \leqslant n$, in G reaches the end of the longest V_0-to-V_i path in G_f, and $V_i^1(X) \geqslant 0$. Recall that G_f is obtained by deleting all the left-directed edges in G. Since G_f is a subgraph of G and $V_i^1(X) \geqslant V_i^0(X)$, the assertion is true for $r = 1$.

Assume the assertion is true for $r = k$. So each node is at the end of a path from V_0 after the kth call to Procedure LongestPath is executed. Before the next call, lines 10 through 15 in Procedure Compaction successively examines each left-directed edge to see if the inequality constraint represented by that edge is satisfied. Assume (V_i, V_j) is a left-directed edge of weight l_{ij}. If $V_i^k(X) + l_{ij} > V_j^k(X)$, then $V_j^k(X)$ will be set to $V_i^k(X) + l_{ij}$. This means that V_j is moved right to reach the end of a path from V_0 to V_j via a V_0-to-V_i path of length $V_i^k(X)$ and then via the edge (V_i, V_j). Thus the increased $V_j^k(X)$ becomes the length of the new path to V_j. Note that V_j may move right more than once, if there are more than one left-directed edge leading to V_j. However, after each move, V_j is still at the end of a path from V_0. By the same reasoning, after all the left-directed edges are examined, every left-directed edge's head node either remains unchanged or jumps right to the end of a longer path from V_0. The adjusted $V_i^k(X)$, $0 \leqslant i \leqslant n$, are then input to Procedure LongestPath by the $(k + 1)$th call. As mentioned in Section II, the $(k + 1)$th call to Procedure LongestPath sets the position of node V_i, $0 \leqslant i \leqslant n$, to $V_i^{k+1}(X)$, such that

$$V_i^{k+1}(X) = \max_{0 \leqslant j \leqslant n} \{V_j^k(X)$$

$$+ \text{length of the longest } V_j\text{-to-}V_i \text{ path in } G_f\}.$$

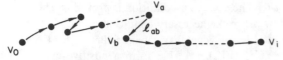

Fig. 4. The longest path t connecting V_0 and V_i.

As a $V_j^k(X)$ in the above expression can be $V_i^k(X)$, we have $V_i^{k+1}(X) \geqslant V_i^k(X)$. Furthermore, $V_i^{k+1}(X)$ is still equal to the length of a path from V_0 to V_i, since every $V_j^k(X)$, is equal to the length of a path from V_0 to V_j.

This completes the proof of Theorem 2. □

By Theorem 2, a node in a constraint graph without positive cycle will not be moved by further iterations, once the node reaches the end of the longest path from V_0 to itself.

For a constraint graph G, $G = (V, E)$, without positive cycles, we define S_k, $k \geqslant 0$, to be the subset of V such that a node is in S_k if, among all the longest paths from V_0 to that node, the one with the smallest number of left-directed edges has exactly k left-directed edges. By definition, S_k can be an empty set and two sets S_i and S_j are disjoint if $i \neq j$. Assume there are b left-directed edges in G. Then for any $k > b$, S_k is an empty set. Furthermore, $\{S_0, S_1, \cdots, S_b\}$ is a partition of V, and $V = \cup_{i=0}^{b} S_i$. Define a number L as follows:

$$L = \min \{u: S_i \text{ is an empty set if } i > u\}.$$

This means that L is the least number such that, for any node in G, one of the longest paths from V_0 to this node should have no more than L left-directed edges. Evidently, $L \leqslant b$. The following theorem proves the correctness of Procedure Compaction when the constraints are consistent.

Theorem 3: Given a constraint graph G without positive cycle, Procedure Compaction yields the minimum solution set of G and terminates, after at most $L + 1$ iterations.

Proof: We will prove by induction that after the rth call to Procedure LongestPath, $r \geqslant 1$, each node in the union $\cup_{i=0}^{r-1} S_i$ reaches the end of its longest path from V_0 in G. If this assertion is true, then Theorem 1 implies the Procedure will terminate and return a minimum solution set, by taking at most $L + 1$ iterations.

For those nodes in S_0, their longest paths from V_0 is in G_f. As the first call to Procedure LongestPath sets them at the end of their longest paths from V_0, the above assertion is true for $r = 1$.

Now assume the assertion is true for $r = k$. By Theorem 2, after the kth call to Procedure LongestPath. The positions of all the nodes in $\cup_{i=0}^{k-1} S_i$ remain unchanged when the iteration operation proceeds further. As shown in Fig. 4, let V_i denote a node in S_k and Γ the longest V_0-to-V_i path with k left-directed edges. Also, let (V_a, V_b) denote the last, i.e., the kth, left-directed edge on Γ with edge weight l_{ab}. The V_b-to-V_i part of Γ does not have any left-directed edge and is a path in G_f. Among all the longest paths from V_0 to a given node on Γ, the one which goes along Γ has the least left-directed edges. Thus $V_a \in S_{k-1}$. By induction hypothesis, after the kth call to Procedure LongestPath. $V_a^k(X)$ becomes the length of the longest V_0-to-V_a path. When the $(k+1)$th call begins, $V_b^k(X)$ has already been set to $V_a^k(X) + l_{ab}$. This value is also the

length of the longest path from V_0 to V_b. After the $(k+1)$th call to Procedure LongestPath, we have

$$V_i^{k+1}(X) \geqslant V_b^{k+1}(X)$$

+ length of the longest V_b-to-V_i path in G_f.

The value of the right-hand expression above is the length of the longest V_0-to-V_i path in G. By Theorem 2, $V_i^{k+1}(X)$ equals the length of the longest path from V_0 to V_i.

The proof is complete. □

Corollary: If the inequality constraints represented by the constraint graph G are inconsistent, Procedure Compaction will terminate and return the inconsistence message.

Proof: Assume that the constraints are inconsistent.

As Procedure LongestPath always forces the X-coordinates of the nodes in G to satisfy the inequalities represented by the right-directed edges, at least one of the inequalities of the left-directed edges will not be met by the output of Procedure LongestPath. Thus the iteration operation will repeat until the end of the $(b+1)$th iteration, where Procedure Compaction is ready to terminate and conclude that the constraints are inconsistent. By Theorem 3, the conclusion is correct.

The proof is complete. □

We now turn to the analysis of the time and space requirement for Procedure Compaction. Let f and T denote the number of right-directed edges and the total number of edges in G, respectively. So, $T = b + f$. Since the operations involved are mainly the traverse of a graph, the natural data structure representing the constraint graph G is the adjacency lists [9]. The space required for this structure is $O(T + n)$. In the practical case encountered in symbolic layouts, T is almost always linearly proportional to n.

Procedure LongestPath traverses each edge in G_f once and then returns the output; so the time it takes is $O(f)$. Each time Procedure LongestPath is executed, every left-directed edge is tested once; this takes time $O(b)$. So each iteration in Procedure Compaction requires $O(b + f) = O(T)$ time. By Theorem 3 and its Corollary, the total time required is $O((L + 1) T)$ if the solution exists; otherwise, it is $O((b + 1) T)$. In the worst case, L is equal to b. This happens when the longest path from V_0 to a node has to pass through all the left-directed edges. As mentioned in Section II, left-directed edges result from user-specified constraints, and right-directed edges are mainly from the design rules. Thus we can regard the speed of the compaction algorithm as linear in T, and the value of the proportional constant is under the control of the user. Practically, in a hierarchical layout, T is rarely more than 10^4. But even when this large value occurs, the user can still specify a sufficient number of constraints and then obtain a desired layout within a few minutes, assuming that the algorithm is running in a typical minicomputer with 1 MIPS speed or higher. We regard the algorithm's speed efficiency as acceptable.

IV. CONCLUSION

A symbolic-layout compaction algorithm which can solve mixed types of constraints has been presented in this paper. The algorithm takes full advantage of the properties inherent in the constraint graph of a symbolic layout for VLSI circuits so as to significantly reduce the computational effort. It is conceptually very simple and is ready to be implemented as part of a layout system.

One problem that needs further study is how to detect and correct the inconsistent constraints which result from the user-specified constraints. Here, we give a brief description of the current strategy. When the program concludes that the constraints are inconsistent and terminates, there are left-directed edges in the constraint graph whose corresponding inequalities are unsatisfied. For each of these left-directed edges, we apply an exhaustive search to find all the positive cycles [10], if any, passing through it. Then all the user-specified constraints on each positive cycle as well as the length of the cycle are listed, and the user can decide which constraints should be relaxed. We are currently looking for a more efficient way to solve this problem.

ACKNOWLEDGMENT

The authors are grateful to the referees whose comments and suggestions help make the presentation of the paper much clearer.

REFERENCES

[1] C. Mead and L. Conway, *Introduction to VLSI Systems.* Reading, MA: Addison-Wesley, 1980.

[2] M. Y. Hsueh, "Symbolic layout and compaction of integrated circuits," ERL Memo. UCB/ERL M79/80, Univ. of California, Berkeley, Dec. 1979.

[3] M. W. Bales, "Layout rule spacing of symbolic integrated circuit artwork," Master thesis, Univ. of California, Berkeley, May 1982.

[4] A. E. Dunlop, "SLIM—The translation of symbolic layout into mask data," in *Proc. 17th Design Automation Conf.*, pp. 595–602, June 1980.

[5] R. Auerbach, B. Lin, and E. Elsayed, "Layout aid for the design of VLSI circuits," *Computer Aided Design*, vol. 13, no. 5, pp. 271–276, Sept. 1981.

[6] N. Weste, "Virtual grid symbolic layout," in *Proc. 18th Design Automation Conf.*, pp. 225–233, June 1981.

[7] K. H. Keller, A. R. Newton, and S. Ellis, "A symbolic design system for integrated circuits," in *Proc. 19th Design Automation Conf.*, pp. 460–466, June 1982.

[8] R. Lipton, S. North, R. Sedgewick, J. Valdes, and G. Vijayan, "ALI: A procedural language to describe VLSI layouts," in *Proc. 19th Design Automation Conf.*, pp. 467–475, June 1982.

[9] E. Horowitz and S. Sahni, *Fundamentals of Data Structures*, Computer Science Press, Inc., 1976.

[10] D. B. Johnson, "Finding all the elementary circuits of a directed graph," *SIAM J. Comput.*, vol. 4, no. 1, pp. 77–84, Mar. 1975.

[11] R. K. Montoye and P. W. Cook, "Automatically generated area, power, and delay optimized ALUs," to appear in Dig. Tech. Papers, *IEEE Int. Solid-State Circuits Conf.*, Feb. 1983.

Section II-E
Layout Verification

SYMBOLIC REPRESENTATION AND INCREMENTAL DRC FOR INTERACTIVE LAYOUT

W. J. McCalla and David Hoffman

Hewlett-Packard Company

HP Design Aids

Cupertino, California

ABSTRACT

User-definable multi-level symbolic representation and incremental design rule checking have been integrated into IGS, an HP-developed, in-house Interactive Graphics System for IC layout. The extensions beyond conventional graphics systems required to support symbolic layout and the DRC algorithms are described.

INTRODUCTION

While capable of being used in a conventional fashion similar to many commercially available systems, IGS, an Interactive Graphics System for IC layout previously described in (1), supports an integrated set of features allowing design and layout in an abstract or symbolic format. In this mode, detailed mask geometries are created concurrently. In addition, a capability for incrementally checking and verifying design rule correctness of each component as it is placed has been included in IGS. The above features and their supporting algorithms are described below.

PROCESS DESCRIPTION

In order to provide cohesion for the material presented in this paper, a five mask NMOS process (Diffusion, Implant, Polysilicon, Contact and Metal) as described in Mead and Conway's Introduction To VLSI Systems (2) is used as the basis for examples throughout. An IGS Process File created via Process Subsystem commands is shown in Fig. 1. The included information consists of mask layer, type, name, color and design rule declarations.

Several considerations influenced the decisions as to both the number and types of masks to be declared. Different and unique sets of mask layers are required for the detailed definition of a device and for its symbolic definition. Further still, a subset of the detailed mask set consisting of those masks used for interconnecting devices or their symbols is distinguished for reasons which will become more apparent later.

Specifically, of the five masks in the process at hand, two, Implant and Contact, are used only in the detailed definition of devices and are declared to be of type 'Detail' by default. Three masks (Diffusion, Polysilicon, and Metal) are used both in the detailed definition of devices as well as for the interconnection of devices or their symbols and therefore were declared to be of type 'Interconnect'. The key concept here is that even when device symbols are displayed it is both desirable and necessary that they be interconnected via 'Detail' mask layers. Thus Interconnect masks are characterized by the property that they are displayable in both a detail display mode or a symbolic display mode.

The second set of masks, layers 11, 12, 13 and 15 are 'Symbolic' masks corresponding to 'Detail' masks 1, 2, 3 and 5, respectively. These Symbolic mask layers are used for building symbols and for indicating locations and types of wiring tabs around or accross a device symbol. Later examples illustrate the use of these masks more fully. The final mask layer, 20, also 'Symbolic' is used to outline and label functional block symbols where the external boundary and functional name are important.

The indicated choice of colors was made in conformance with (2) while mask names were arbitrarily assigned. Explicit bumpers, to be described later, are allowed for masks 1 through 5 while minimum required widths, areas, intra-mask and inter-mask spacings are also defined.

MACRO-COMMANDS AND LOGICAL NESTING

IGS provides a flexible user-defined macro-command capability which can be used to greatly simplify the implementation of a symbolic design methodology. Before describing the interaction between macro-commands, primitive IGS commands and the Process File, it is necessary to describe, as illustrated in Fig. 2, the concept of logical cell nesting as opposed to physical cell nesting. Here REGCELL is assumed to be the device currently being edited so that physically all display references are down from REGCELL. The device REGCELL is a shift register cell including instances of PULLUP, NENH, DIFFCON and BUTTCON. The device PULLUP includes instances of NDEP and, again, BUTTCON. As a consequence, when viewed from REGCELL, one instance of BUTTCON is physically nested at the second level down and another is physically nested at the third level down. The problem created by this situation in a symbolic design environment is that if symbols are to retain a consistent level of representaion, it is desirable to see all occurrences of a given device or no occurrences. For this example, if the display is

Reprinted from *IEEE Int. Symp. Circuits and Systems Proc.*, 1981, vol. 3, pp. 710–715.

controlled only on the basis of physical nesting, then BUTTCON will be seen once in outline form in REGCELL if only the first level is displayed, once in outline and once in detail if the second level down is also displayed, and finally, twice in detail if all three levels are requested.

The basic idea behind logical nesting is to have the designer define a logical hierarchy of devices and device instances which overlays the physical hierarchy. For the example of Fig. 2, the devices BUTTCON, DIFFCON, NENH and NDEP are all low level devices and hence can arbitrarily be assigned a logical function level of 1, the lowest possible, and also default, level. The device REGCELL, which provides a true logical function, is therefore assigned a higher functional level of 2. What about PULLUP? The device PULLUP, consisting of a butting contact BUTTCON which shorts the gate and source of a depletion mode transistor NDEP, typically serves as a two terminal, active load for inverters, etc. and therefore is also a primitive device. It is thus assigned a logical function level of 1, the same as its constituent parts.

To a degree, level control is maintained and enforced by IGS. Level 2 devices cannot be added to level 1 devices but can be added to other level 2 or higher level devices. Thus, there is no conflict in assigning PULLUP the same logical function level as its constituent parts. For the example of Fig. 2, in a logical nesting sense, display is from REGCELL down to logic level 2 devices (i.e. REGCELL itself) or down to logic level 1 devices, the primitive transistors and contacts. As a result, the desired effect of having no instances of BUTTCON or all instances displayed is achieved. As a further point, when the symbols for level 1 devices are displayed, the Interconnect masks for all level 2 and above devices are automatically displayed in order to indicate how the symbols are interconnected. This aspect illustrates the true significance of Interconnect masks and the reason for distinguishing them as a subset of the detail masks.

To summarize, the advantage of working with logical nesting levels is the preservation of a consistent display and multiple symbolic representaions. The multiple representations derive from the fact that each time the logical function level is increased, a new symbolic representation may be defined.

With the concept of logical nesting now in mind, the use of the IGS macro-command facility to provide the basic mechanism for moving back and forth between various display representations can now be described. Two macro commands, 'DETAIL' and 'SYMBOL' can be defined as follows:

DEFINE DETAIL <LV '1'> 'WINDOW :D10,<LV> ';
DEFINE SYMBOL <LV '1'> 'WINDOW :S<LV>,<LV> ';

These macro commands provide the basic mechanism for moving back and forth between various display representations and presume that the editing session has already been toggled into logical display mode. The 'DETAIL' macro uses a WINDOW command to display detail from an arbitrary logical function level of 10 down through a logical function level set by the parameter 'LV' which has a default value of 1.

The 'SYMBOL' macro command is similar to the 'DETAIL' macro. The WINDOW command establishes the logical function level for which symbols are to be displayed. In particular, in a symbolic display mode, it is presumed that new symbols are defined at each new logical function level starting from stick-like symbols for level 1 devices and proceeding to block-like symbols for level 2 devices and above. Most of the time, it is desirable to view only a single level of symbols as well as all interconnections between those symbols. Thus, for the 'SYMBOL' macro, the parameter 'LV' is used for both arguments of :S in the WINDOW command. The interconnections between symbols are presumed to be on masks of type 'Interconnect' and are automatically displayed for all logical device levels from that of the device currently being edited down through the level above that whose symbols are being displayed.

PRIMITIVE DEVICES AND SYMBOLS

The following paragraphs describe the creation, both detail and symbols, of six level 1 primitive devices as illustrated in Fig. 3. The first of these devices, DIFFCON, a diffusion-to-metal contact, consists of a 4 lambda diffusion rectangle on Mask 1, a 2 lambda contact rectangle on Mask 4 and a 4 lambda metal rectangle on Mask 5 as seen at the top of Fig. 3(a). The symbol defined for this is a 2 lambda rectangle on Mask 11, the symbolic diffusion mask, and a 1 lambda rectangle on Mask 15, the symbolic metal mask. This choice of symbol was based on the following observations: In symbolic display mode, lines of any arbitrary width are displayed as centerlines. The minimum width for a diffusion line is 2 lambda. Thus in symbolic mode, any diffusion line which touches the symbolic metal rectangle is also a valid, design-rule-correct connection.

Not previously mentioned is the fact that IGS provides a facility for the definition of stretchable or parameterized cells. For a cell such as DIFFCON, two reference crosses arbitrarily labeled 'LL' for lower-left and 'UR' for upper-right can be declared and positioned at the lower-left and upper-right vertices, respectively, of the symbolic diffusion rectangle. The x- and/or y-coordinates of any vertex of any rectangle, line, polygon or circle on any mask can arbitrarily be assigned to any declared reference cross. By default, all vertices of all primitives are initially referenced or assigned to the device origin. In the case of DIFFCON, the x-coordinates of the lower-left vertices of each rectangle on each mask are assigned to the x-coordinate of the reference cross 'LL'. The remaining lower-left and upper-right vertex coordinates are assigned in similar fasion to 'LL' and 'UR' as appropriate. The result is that whenever an instance of DIFFCON is used in a higher level device definition, its size, both detail and symbol, may be parametrically stretched as needed by appropriately locating the reference crosses 'LL' and 'UR'. In general, up to 5 reference crosses may be declared in any given device or cell definition.

The polysilicon-to-metal contact, POLYCON, is constructed similarly to DIFFCON with the exception that Mask 1, the diffusion mask, and its symbolic counterpart, Mask 11, are replaced by the polysilicon masks, Mask 3 and Mask 13, respectively, as shown in Fig. 3(b).

The butting contact of Fig. 3(c) differs from the previously described contacts in two respects. First, it has a specific orientation associated with it and second, it allows connections between diffusion, polysilicon, and metal concurrently. It consists of a 4 lambda by 3 lambda rectangle on the polysilicon mask which overlaps a 4 lambda by 4 lambda rectangle on the diffusion mask to form a 4 lambda by 6 lambda structure. Centered in the middle is a 2 lambda by 4 lambda contact cut. A 4 lambda by 6 lambda metal rectangle covers the entire structure. The default orientation is with the polysilicon up and diffusion down. The symbol is composed of 3 rectangles on symbolic masks corresponding to diffusion, metal, and polysilicon from bottom to top, respectively. The same considerations were used in defining the size of the rectangles in the symbol as for DIFFCON and POLYCON.

In practice, an enhancement-mode pull-down transistor can be formed simply by crossing polysilicon over diffusion. However, in order to be sure that necessary minimum widths and extensions are maintained and that all design rules described in (2) can be checked, it is useful to define an explicit, stretchable cell, NENH. As shown in Fig. 3(d), it consists of a 2 lambda by 6 lambda horizontal polysilicon rectangle which serves as the gate contact and a similar vertical rectangle on the diffusion mask. Both rectangles are stretchable with their lower left and upper right vertices assigned to two reference crosses 'LL' and 'UR' located at the corners of the active region (i.e. the region of intersection of the two rectangles). The symbolic representation is simply a symbolic zero-width line on Mask 13 horizontally crossing a vertical symbolic zero-width line on Mask 11 and a 2 lambda by 2 lambda rectangle also on Mask 13. The end points of the lines as well as the vertices of the rectangle are appropriately assigned to the reference crosses.

A depletion mode NMOS transistor NDEP is constructed similarly to the enhancement mode transistor with the exception that a 6 lamda by 6 lambda rectangle on Mask 2, the implant mask, is included and appropriately assigned to reference crosses. Also, a 2 lambda by 2 lambda rectangle on Mask 12, the symbolic implant mask, replaces the rectangle on Mask 13. Note that the size of the symbolic rectangles accurately reflects the size of the active gate area of the transistors.

The final level 1 primitive device is a depletion mode NMOS transistor combined with a butting contact for use as a pullup or active load in gates and inverters. An actual default-sized instance of the butting contact device BUTTCON is used together with primitive rectangles and lines as in the device NDEP in order to provide the direct assignment of vertices to achieve stretchability. An additional zero-width line is used in the symbolic view to reflect the shorting of gate to source as shown in Fig. 3(f).

BUILDING A SHIFT-REGISTER CELL AND ITS SYMBOL

A simple, single-bit shift register can be constructed from a chain of inverters gated by pass transistors driven by two alternate, non-overlapping clock signals. Fig. 4(a) shows a symbolic stick representation and Fig. 4(b) a detailed representation for a basic shift register cell, REGCELL, assumed to be a level 2 device. The representations are the level 1 displays as created via the previously described macros 'SYMBOL' and 'DETAIL' respectively. The output of an inverter, to be described more fully below, is taken from the butting contact in diffusion and, after being gated by a polysilicon clock line, is returned back to polysilicon with a second butting contact. The power, ground and output signal lines of one cell align with the power, ground and input signal lines of the next. Thus the cells can simply be concatenated for as many bits as desired with two cells required for each bit.

Fig. 4(c) shows a new block symbolic view of the same shift register cell as created for this new, level 2 device. It consists of a boundary rectangle on Mask 20 together with various labels for signal names including power, ground, clock, input and output. Two zero-width lines on Mask 15, the symbolic metal mask, and a single zero-width line on Mask 13, the symbolic polysilicon mask, are extended completely through the cell boundary in order to indicate the direct, continuous path of these signals through the actual device. Short stubs, also on Mask 15, are used to indicate the exact location of the actual input and output signals. This symbolic block view is displayed whenever level 2 cells are requested via the 'SYMBOL' macro. It contains the essential information required to use this device to create still higher level functional blocks such as a complete shift register array. It conveys more clearly the functional organization of that higher level cell. Finally, it contains less graphic detail thereby resulting in a faster display.

Previously, the description of the inverter within the basic shift register device was ignored. That inverter is now considered. It consists of a depletion mode MOS transistor with its gate and source shorted via a butting contact, the PULLUP device described previously, and an enhancement mode MOS transistor, the NENH device also described previously, used as a pulldown. Diffusion-to-metal contacts, DIFFCON, are used to connect the structure to power and ground lines. Within IGS it is quite possible to simply describe an inverter as another level 2 (or level 1) device and use an instance of this device within the shift register cell. In practice, however, it has been found to be more convenient to write a parameterized macro command which builds an inverter directly out of the primitive level 1 devices of Fig. 3. The reason is that NMOS technology represents a ratio type of logic. Zero and one voltage levels are determined by the relative aspect ratios of the pullup and pulldown transistors with overall aspect ratios of

4:1 to 8:1 being typical. Consequently, a family of inverter cells or a single parameterized cell is required with the latter probably being preferable. However, IGS restricts a reference cross associated with an instance of an included device from itself being x- or y-assigned to another reference cross. Thus, the reference crosses 'LL' and 'UR' in PULLUP and NENH cannot be assigned to reference crosses in such an inverter cell. Hence, the use of parameterized macro commands to build an inverter of proper aspect ratio directly in REGCELL.

INCREMENTAL DESIGN RULE CHECKING

Several types of design rules may be specified and checked for in IGS. Design rules are either entered in the IGS Process subsystem (Implicit Bumpers) or directly into device definitions (Explicit Bumpers). Both types, described more fully below, are applied in the Edit subsystem incrementally as the user modifies the database.

Design rule checks in IGS fall into two basic types: design rules which impose constraints on component geometry, and those which impose constraints on component topology (i.e. placement of components). Constraints of the former type include specification of minimum area and width rules for mask layers. Algorithms for checking geometric components against these constraints are well understood and will not be described here.

Topological design constraints are specified in IGS through the use of 'Bumpers'. Bumpers are pseudo layers which specify regions which may not be overlapped by actual layers in the chip. Each actual layer 'm' has a corresponding bumper layer '-m'. When a component, a metal line on mask 5 for example, is placed in a cell, it may not overlap components on its bumper layer, mask -5. In IGS, bumpers are primarily used to enforce minimum spacing rules. Bumpers are not intended to take the place of complete off-line design rule checking procedures. Rather, they are intended to shorten the design cycle on a large class of errors during the interactive phases of the design.

Bumpers are created by the designer in two ways. First, they may be explicitly created by adding geometric components on bumper layers directly. In this case, standard IGS graphic editing commands (e.g. add, delete, move, stretch, copy, etc.) are employed to manipulate these components. Components created in this fashion are termed 'explicit' bumpers. Second, bumpers may be indirectly created as a result of the mask layer minimum spacing rules specified in the IGS Process subsystem as indicated in Fig. 1. Conceptually, implicit bumper components are automatically created on negative reference mask layers whenever components are added or modified on relative mask layers. The bumper components are automatically oversized by the minimum distances to effectively implement the spacing rules. In IGS, bumpers created as a result of spacing rules are called 'implicit' bumpers. Spacing rules and any resulting implicit bumper violations are ignored whenever components touch or overlap so that components may be electrically connected.

Both explicit and implicit bumpers usually will coexist in a design. However, if an explicit bumper is defined in a cell, then IGS ignores all corresponding implicit bumpers with respect to that cell and any included cells. This characteristic allows the designer to override the globally defined spacing rules with custom rules when necessary.

The decision to define topological constraints in terms of geometric components has the advantage that bumpers can be treated like any other geometric component in IGS. Not only does this allow manipulation with the standard set of commands, but it also allows parameterization as well. Stretchable cells, therefore, can have stretchable bumpers. In addition, the ability of the user to explicitly create bumpers or have them created based on a set of global rules widens their range of potential application. Implicit bumpers are normally defined with a conservative set of design rules. Typically, they find their greatest application when routing interconnects between cells and in intracell designs where cost considerations do not justify optimal designs. Explicit bumpers, on the other hand, may be employed to tailor design rules to cells where more aggressive design approaches have been taken (for reasons of speed, area, etc.).

REFERENCES

1. B. Infante, D. Bracken, W. McCalla, S. Yamakoshi, and E. Cohen "An Interactive Graphics System For Design of Integrated Circuits", 15TH DESIGN AUTOMATION CONFERENCE PROCEEDINGS, pp. 182-187:June 1978.

2. C. Mead and L. Conway, INTRODUCTION TO VLSI SYSTEMS, Addison-Wesley, 1980.

3. I. E. Sutherland and G. W. Hodgman, "Reentrant Polygon Clipping", CACM, 17(1):32, January 1974.

APPENDIX

The implementation of bumper design rule checking in IGS relies heavily on device boundary rectangle information and is able to detect overlap without comparing component data directly. The polygon overlap algorithm described below, which compares disjoint sets of polygons for overlap, forms the core of the bumper checking system.

Strictly speaking tests b and c are not necessary but are included for reasons of effeciency. Other tests can be incorporated here also, but were left out since they were either too complex or not sufficiently general. The primary information used by this algorithm (i.e. boundary rectangles, corner occupancy data, and a simple polygon indicator) is either extracted directly from the IGS database or is obtained as a side effect of the clipping process. A modified reentrant polygon clipper (3) is used to clip polygons to rectangular regions. The clipper is quite efficient and is well suited to this application.

OVERLAP(A,B). Where A and B are disjoint sets of polygons with boundary rectangles [A] and [B], respectively. This algorithm will determine if at least one polygon of A overlaps at least one polygon of B.

Let [AB] be the intersection of [A] and [B].

Sets A and B do not overlap if either:
  a) A or B is null.
  b) Region [AB] is null.

Sets A and B overlap if either:
  c) [A] and [B] share a common corner physically occupied by at least one polygon in both A and B as in Fig. 5(a).
  d) Both A and B are composed of a single simple (non-intersecting) polygon and [A] encloses [B] in the horizontal (vertical) direction and is enclosed by [B] in the vertical (horizontal) direction as in Fig. 5(b).

Otherwise, if [A]<>[B] then A and B are clipped to [AB], giving A' and B', respectively. The algorithm is now recursively applied to A' and B'.

Otherwise, if [A]=[B] then either A or B contains more than one simple polygon. For each simple polygon S in set A (B), the algorithm is recursively applied to S and B (S and A). This continues until an overlap is found or until all simple polygons in the set have been tested.

IGS uses the OVERLAP algorithm to detect violations between actual components and both explicit and implicit bumpers. When testing for explicit bumper violations, OVERLAP can be directly applied to the database.

Detecting implicit bumper violations is more complex. First, implicit bumpers are not actually saved in the database. They must be dynamically generated when required. Second, touches and overlaps of actual components are permitted even if their implicit bumpers overlap. Implicit tests are basically performed by testing a polygon on the reference mask layer for overlaps with oversized polygons generated from the relative mask layers. If no violations are found, then the component on the reference mask is at least the minimum spacing apart from components on all relevant layers (i.e. no violation exists.) However, if an overlap exists, IGS must determine whether actual components touch or overlap before a violation can be reported. This is performed by 1) oversizing the component on the reference mask by one unit and 2) performing the OVERLAP test with this component and each actual component (on the relative masks) whose generated implicit bumper originally caused a violation. If the OVERLAP test succeeds, actual components overlap or touch, so the implicit bumper violation is ignored. Otherwise, the violation is reported to the user.

Since the OVERLAP algorithm depends only on boundary rectangle and corner occupancy information, there is no need to scan component data unless it can possibly affect these two parameters. Both the explicit and implicit bumper violation procedures use this fact to significantly reduce processing time and temporary storage requirements for large tests. When collecting the polygon lists for OVERLAP, the database is accessed only for those components whose boundary rectangles share common corners with or are partially intersecting region [AB] (i.e. the boundary rectangle of polygon list intersection). Components totally within [AB] appear in polygon lists as unprocessed components — to be scanned at some future time only if a new, smaller [AB] should dictate it. Components totally outside [AB], of course, are ignored. By implication, generation of implicit bumpers (and the resulting oversizing operation) is limited to those components requiring a database access. This 'shortcut' effectively eliminates many of the I/O and CPU intensive operations that would result from full application of the OVERLAP algorithm.

EXPLICIT-TEST(deviceA,deviceB)

Let [AB] be the boundary rectangle of intersection between deviceA and deviceB.

FOR each mask m:

    Search for explicit bumpers on deviceA.
    Let A' be the list of polygons intersecting [AB], in deviceA on mask m.
    Let B' be the list of polygons intersecting [AB], in deviceB on mask -m.

    If OVERLAP(A',B') Then an explicit bumper violation exists.

    Otherwise, search for explicit bumpers on deviceB.
    Let A' be the list of polygons intersecting [AB], in deviceA on mask -m.
    Let B' be the list of polygons intersecting [AB], in deviceB on mask m.

    If OVERLAP(A',B') Then an explicit bumper violation exists.

    Else no explicit bumper violation exists between deviceA and deviceB on layers m and -m.

291

```
PROCESS FILE:    PROVLSI.VLSIDEMO.CAD
CREATION DATE:   JANUARY 15, 1981
LAST MODIFIED:   JANUARY 31, 1981  1:56 PM

RESOLUTION:      100 UNITS PER UM

MASK INFORMATION
```

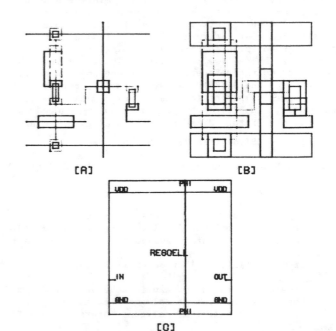

Fig. 2   (a) Physical Nesting as viewed from REGCELL and
         (b) Logical Nesting as viewed from REGCELL.

MSK	TYPE	MNEMONIC	COLOR	EXPL. BUMP.	MIN. AREA	MIN. WIDTH	COMM MASK
1	INT	DIFUSION	GREEN	YES	4	2	4
2	DET	IMPLANT	YELLOW DOT	YES	36	6	
3	INT	POLYSIL	RED	YES	4	2	4
4	DET	CONTACT	WHITE	YES	4	2	1,3,5
5	INT	METAL	BLUE	YES	9	3	4
11	SYM	DIFFCON	GREEN	NO			
12	SYM	IMPLSYM	YELLOW	NO			
13	SYM	POLYCON	RED	NO			
15	SYM	METALCON	BLUE	NO			
20	SYM	OUTLINE	WHITE	NO			

IMPLICIT BUMPER INFORMATION

REFERENCE MASK	DISTANCE	RELATIVE MASK	
1 DIFF	3	1	DIFF
	1	3	POLY
3 POLY	1	1	DIFF
	2	3	POLY
4 CONTACT	2	4	CONTACT
5 METAL	3	5	METAL

Fig. 1 NMOS Process File for Symbolic Design

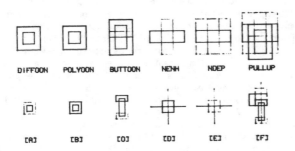

Fig. 3 Primitive Devices with Accompanying Symbols

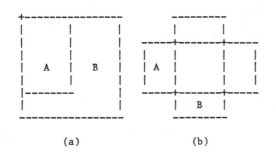

Fig. 5 (a) Common corner test and
        (b) Overlapping boundary rectangle test.

Fig. 4   Stick, Detailed and Block Symbolic
         Views of Shift Register Cell.

Lyra: A New Approach to Geometric Layout Rule Checking

Michael H. Arnold
John K. Ousterhout

Computer Science Division
Electrical Engineering and Computer Sciences
University of California
Berkeley, CA 94720

Abstract

Lyra is a layout rule checking program for Manhattan VLSI circuits. In Lyra, rules are specified in terms of constraints that must hold at certain corners in the design. The corner-based mechanism permits a wide variety of rules to be specified easily, including rules involving asymmetric constructs such as transistor overhangs. Lyra's mechanism also has locality, which can be exploited to construct incremental and/or hierarchical checkers. A rule compiler translates symbolic rules into efficient code for checking those rules, and permits the system to be retargeted for different processes.

1. Introduction

Each process used to fabricate integrated circuits is characterized by a set of layout rules that must be observed when using the process. These rules specify minimum widths and spacings for mask features and also specify the form of transistors, contacts, and other circuit constructs. Lyra is a program that examines VLSI circuit designs to verify that they satisfy a given set of layout rules. Layout rule checking in Lyra is *corner-based*: each rule specifies a class of corners where the rule applies, and a set of constraints that must hold at all corners of that type. Lyra's mechanism has three attractive properties:

Expressive Power. It is possible to specify a wide variety of layout rules in a simple fashion. In particular, rules governing the form of asymmetric constructs, such as transistor gate overhangs, are readily expressed.

Adaptability. Rules are specified in a symbolic language, from which the code for doing the checks is generated by a rule compiler. Lyra can be targeted to a new process by writing and compiling the appropriate rule specifications.

Locality. The corner-based mechanism of Lyra can be used to check small areas of a design independently. Locality has two advantages. First, it permits incremental checkers to verify modified designs by rechecking only those portions that have changed. Second, locality permits checkers to exploit repetitive and hierarchical designs by checking only key pieces; from these key pieces the correctness of the whole can be inferred, based on the repetitive structure [7].

2. Existing Systems

Two major methods for performing geometric rule checks are currently in use. One is raster scan based [2], and the other is organized around operations on polygons [1].

2.1. Raster Scan

In the raster method, the design is cast onto a fine square grid whose spacing is determined by the minimum feature size of the design. Then a small window is systematically stepped over the grid. At each window position, the pattern of mask layers at the grid points under the window is checked for validity using a combination of stored patterns and process-dependent code.

One of the strong points of this method is the inherent simplicity of the algorithm. Another advantage is the locality of its processing: only the information under the window is used to decide whether a layout rule violation has occurred. The raster method appears to be amenable to implementation in special-purpose hardware.

However, the embedding of layout rules in both templates and code means that the checking program is tied closely to a single process. Furthermore, the cost of examining windows grows rapidly as the window size gets larger. This combinatorial explosion limits the raster method to small window sizes (4 grid points × 4 grid points in the Baker implementation). The small window size in turn restricts the power of the mechanism, since only a limited amount of context can be used to decide whether the window area is in error.

The work described here was supported in part by the Defense Advanced Research Projects Agency (DOD), ARPA Order No. 3803, monitored by the Naval Electronic System Command under Contract No. N00039-78-G-0013-0004.

Reprinted from *ACM/IEEE 19th Design Automation Conf. Proc.*, 1982, pp. 530–536.

2.2. Polygon Algebra

This method treats each mask layer as a collection of polygons, and defines an algebra for manipulating and combining the layers. The algebra includes such operations as growing and shrinking the polygons on a single layer, and boolean operations (AND, OR, NOT, etc.) that combine layers. Using the polygon algebra, new "derived" layers are created from the original mask layers by sequences of operations. For example, a derived nMOS layer "transistor" can be created using the operation:

transistor = polysilicon AND diffusion.

Layout rule errors are found by performing built-in operations such as width, spacing, and enclosure checks on the original and derived layers.

The strength of this method is its algebraic rule language. The system can be readily adapted to new or modified rule sets by coding the appropriate sequence of polygon operations and checks. Complex rules can be captured accurately, although they may require long sequences of operations.

The algebraic method of rule checking has several drawbacks. It is inherently global in nature: operations apply to whole mask layers at a time and cannot easily be confined to small areas of the design. Without some method for local operations, it will be difficult to use this method in incremental or hierarchical systems. Also, the algebra is *isotropic* in nature: operations like "grow" apply uniformly to all parts of a design. This leads to complex and clumsy specifications of rules for asymmetric constructs. A third problem with this method is that the creation of intermediate layers requires large amounts of secondary storage. Processing is slowed by the frequent need to access the large data base maintained on disk. This problem becomes acute when dealing with large designs, or when using complex layout rule sets involving many derived layers.

3. Specifying Layout Rules in Lyra

Lyra's processing is based on corners in the input design. By corners we mean not only the corners appearing on individual mask layers, but also corners of regions defined by any combination of mask layers. In an nMOS process for instance, the vertices of a gate region qualify as corners although they do not correspond to corners on any single mask layer. Figure 1 shows the corner points of a sample design.

Each rule that Lyra checks is specified in two parts. The first part of a rule is a *context pattern* indicating where the rule applies. Context patterns are defined in terms of the layers present or absent in each of the four quadrants about a corner. The second part of a rule contains one or more *constraints*. Each constraint specifies a rectangular area, adjacent to the corner, in which certain layer combinations are required or prohibited. Layout rule checking is accomplished in Lyra by comparing context patterns against the actual mask layers present at each corner, applying constraints where context patterns are matched, and outputting error information for each constraint that is violated.

We illustrate how this works with a spacing check. To verify that all features on a mask layer M are separated by some minimum distance, it is sufficient to check that there is space around the corners of features on that layer (see Figure 2). These constraints can be specified by the rule shown graphically in Figure 3 below. The left hand side of the rule indicates that it

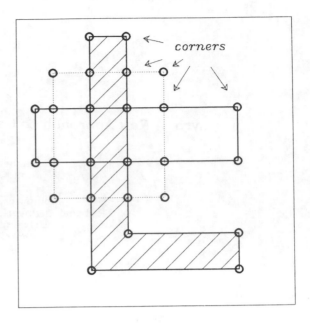

Figure 1. Corners in a design.

applies at all corners where layer M is present in one quadrant (the top right) but absent in the adjacent two quadrants, i.e., the rule applies to convex corners on layer M. The right hand side gives the three constraints to be tested at corners where the rule applies. Each constraint specifies an area where there must be no M. Although the rule explicitly specifies only one corner orientation, it is automatically applied to all possible orientations by Lyra. Similar rules can be used to check spacing on regions consisting of combinations of mask layers, and to do width and enclosure checks. Taken together, such spacing, width, and enclosure conditions constitute the great majority of geometric layout rules.

In addition, asymmetric checks on the form of circuit constructs can also be specified. For example the rules shown in Figure 4 specify constraints on the form of transistor overhangs. These two rules will detect errors of the form shown in Figure 5. In the top rule, the two quadrants on the right indicate a diffusion wire coming out of the top of a transistor region. The condition on the bottom left quadrant indicates that there is no corresponding polysilicon overhang (the situation we wish to detect). The condition on the top left quadrant insures that this rule is not erroneously applied to bends in transistors. If the pattern is matched, we know a polysilicon overhang is missing. Since there is no need to check any constraints, the right hand side of the rule specifies a constraint that is unsatisfied by definition. The second rule checks for missing diffusion overhangs, in a similar way. Separate width checks are required to make sure the overhangs are large enough. The overhang rules may seem somewhat complicated, but they correctly handle missing overhangs, missing chunks of overhangs in large transistors, and bends in transistors.

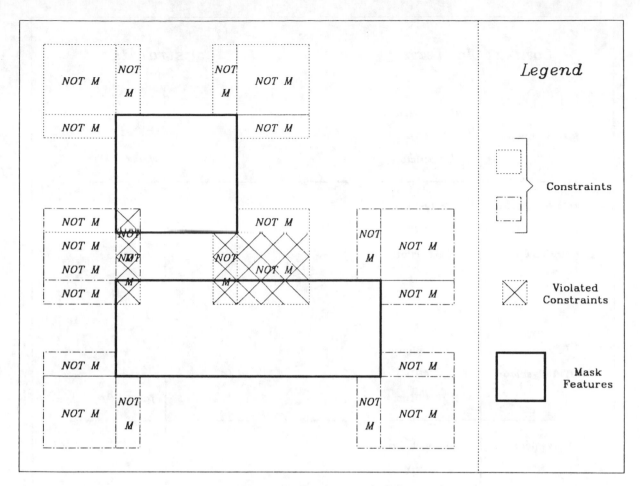

Figure 2. Spacing constraints.

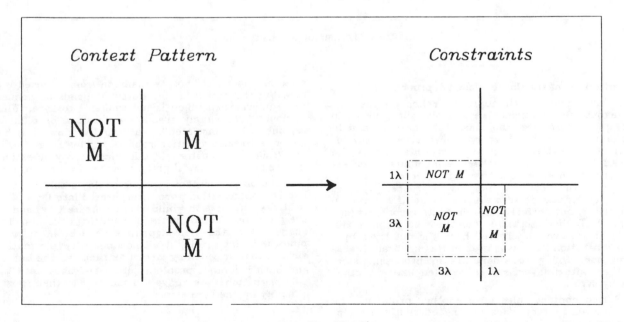

Figure 3. Spacing rule.

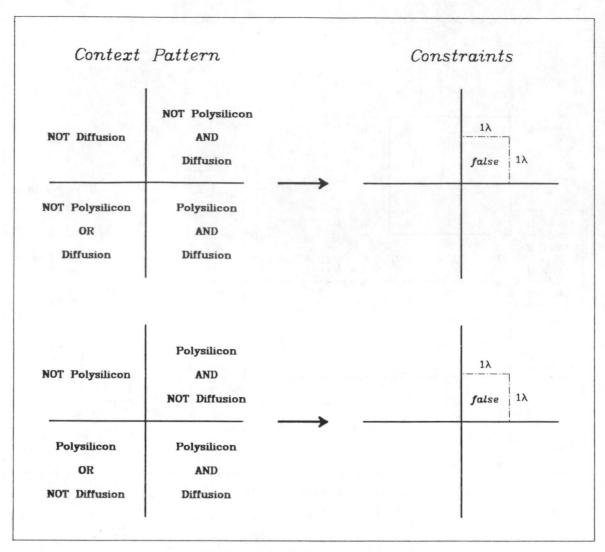

Figure 4. Transistor overhang rules.

4. Limitations of the Corner-Based Approach

Lyra's corner-based approach relies on the fact that layout rule violations always involve some corner in a design. In practice, the mask information immediately adjacent to the corner almost always provides sufficient context to generate constraints. However a few situations arise where there is insufficient context available at corners to process them correctly. For instance, in the Mead & Conway nMOS process [5] there is a corner context pattern which is illegal if it occurs in a transistor, but perfectly legal in butting contacts (see Figure 6). To distinguish between these cases, an auxiliary "context" layer is created from the cut layer by a "grow" operation such as is used in the algebraic checking method. This grow is done in a preprocessing step, and the context layer is then handled like any other layer by Lyra.

Buried contact rules pose another problem. The nMOS buried contact rules we are currently using at Berkeley require that unrelated polysilicon or diffusion must be spaced at least two lambda from a buried window. But the notion of "unrelated" depends on connectivity information which is not available to Lyra. When in doubt Lyra simply assumes polysilicon or diffusion regions to be "unrelated" to a buried window. This can give rise to false violation reports as illustrated in Figure 7. Designers can avoid this idiosyncrasy by extending the buried window slightly, as shown in the figure.

The buried contact problem could be solved if connectivity information were incorporated into Lyra. Connectivity information would also eliminate the need for the grow operation to distinguish butting contacts. This suggests that the Lyra algorithm could be made more powerful by integrating it with connectivity information, such as that provided by circuit extractors. The butting and buried contact problems described above are the only situations we have encountered that posed difficulty for the Lyra algorithm.

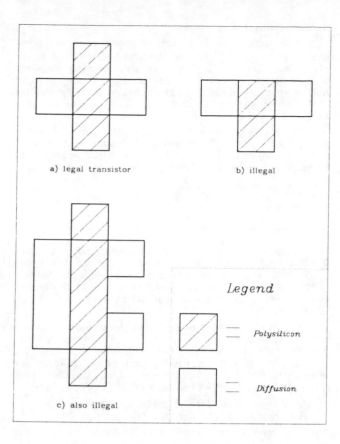

a) legal transistor   b) illegal

Legend

Polysilicon

Diffusion

c) also illegal

Figure 5. Transistor overhang errors.

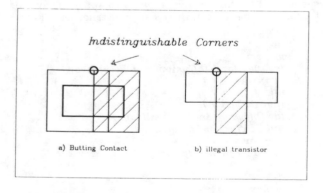

Indistinguishable Corners

a) Butting Contact   b) illegal transistor

Figure 6. An ambiguity due to insufficient context.

The use of corner-based rules in Lyra has additional implications. Corner based checking can only work for designs with corners. Other methods must be used to check designs where curved boundaries are allowed. Furthermore, there is no obvious way to generalize the system to handle arbitrary angles effectively. The current implementation of Lyra, and the rules described above, are *Manhattan*: they assume that all edges are either horizontal or vertical.

Finally, the rectangular nature of Lyra's constraint regions results in slightly over-conservative layout rules, as illustrated in Figure 8. In the figure, the two rectangles are actually more than 3λ apart, but Lyra considers them to be only 2.5λ apart. In practice, slightly conservative diagonal distances are not a problem in a Manhattan Mead & Conway environment such as the one at Berkeley. In some industrial environments, it might be desirable to use a filter which rechecks violated constraints more carefully to obtain the true tolerances.

5. Implementation

Lyra's processing divides into the following principal steps:

(1) Corner detection: all corners in the design are identified by comparing rectangles in the data base to find all feature intersections.

(2) Rule application: the context portions of rules are compared against the corners and, where they match, constraints are generated.

(3) Constraint checking: the constraints are checked against the mask information, and diagnostic output is generated for each violated constraint.

The rule application process is of special interest. Since a typical rule set contains dozens of rules, a linear search through the entire set for each corner in the design is unacceptably slow. Instead, each rule is associated with a corner on some combination of layers. Using this rough indexing scheme, rule application for a corner proceeds in three stages:

(1) Do a 2^n way branch (where n is the number of mask layers in the design) based on the layers present at the corner. Most corners have only one or two layers present, so this branch eliminates most of the mask layers, and hence most of the rules, from further consideration.

(2) For each layer combination that may still be relevant, do a 16 way branch based on the presence or absence of that layer combination in each of the four quadrants about the corner. This determines whether or not a corner of the given layer is actually present. If a corner is present, the branch also determines the orientation of the corner and whether it is concave or convex.

(3) Apply rules specific to the corners found (i.e., those which are indexed under the given layer combination and are of appropriate convexity and orientation).

This strategy works well. The processing time required for rule application is about equal to the time required by each of the other major processing steps (i.e., corner detection and constraint checking).

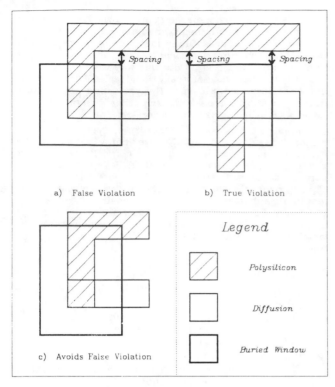

a) False Violation   b) True Violation

Legend

Polysilicon

Diffusion

Buried Window

c) Avoids False Violation

Figure 7. A situation requiring connectivity knowledge.

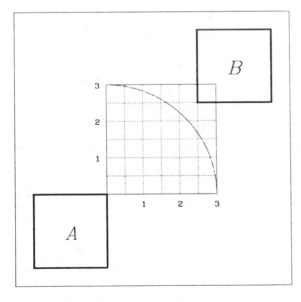

Figure 8. Manhattan distance metric.

6. Rule Compiler

Rule application as described above involves a great deal of code (the 2^n arms of the first branch, and multiple 16 way branches). Furthermore, the code depends heavily on the rules in use. Fortunately, this code need not be specified directly. Lyra contains a rule compiler that classifies the rules and generates the rule application code from a symbolic description of the rule set.

Rules are specified textually to the compiler using a Lisp-like syntax. For example, the spacing check for the metal layer in the Mead & Conway nMOS process can be specified using the following rule:

```
(rule
  (corner: (convex Metal))
  (constraints:
    (extended-outside 3 (not Metal) "Metal spacing")))
```

This is equivalent to the rule shown graphically in Figure 2. The "extended-outside" construct generates three constraints that do not permit metal around the outside of the corner for a radius of 3λ. The string at the end of the constraint gives the diagnostic message with which errors are flagged.

Spacing and width checks like the one above occur quite frequently in rule sets. To simplify the specification of rules, the Lyra rule compiler provides several higher-level macros for such cases. Macros are automatically expanded to equivalent simple rules. Examples from the Mead & Conway nMOS process are:

```
(self-separation Metal 3 "Metal spacing")
(width Metal 3 "Metal width")
```

The *self-separation* macro expands to two rules: the spacing rule outlined above, and a similar rule applying to concave corners which checks for too small holes in mask features. Similarly, the *width* macro expands to a pair of rules: one for convex corners, and one involving concave corners.

As a further example, consider the transistor overhang rules shown in Figure 4. These rules are too specialized for higher-level macros to apply. They must be coded using the primitive rule construct as follows:

```
(rule
  (corner:
    (convex (and Diffusion (not Polysilicon))))
  (also:
    nil
    (not Diffusion)
    (or Diffusion (not Polysilicon))
    (and Diffusion Polysilicon))
  (constraints:
    (inside 1 false "Malformed transistor")))

(rule
  (corner:
    (convex (and Polysilicon (not Diffusion))))
  (also:
    nil
    (not Polysilicon)
    (or Polysilicon (not Diffusion))
    (and Diffusion Polysilicon))
  (constraints:
    (inside 1 false "Malformed transistor")))
```

The *corner* clauses indicate what sort of corners this rule applies to. The four arguments of the *also* clause define additional conditions for each of the four quadrants; these conditions must be satisfied for the constraints to be checked. The word *false* is used in the *constraint* clauses to indicate unsatisfiable constraints (i.e., errors are always reported).

To generate a layout rule checker for a new rule set, the Lyra rule compiler reads in the textual rule specifications, classifies the rules, generates the rule application code, and combines the code with standard corner generation and constraint checking code that is used for all rule sets.

7. Status

Lyra has been implemented in Lisp and runs under the Berkeley version of VAX Unix. The system has been used to check the entire RISC chip (a 44,000 transistor microprocessor developed at Berkeley) and numerous smaller projects. Several rule sets are in use, including:

(1) The original Mead & Conway nMOS rules.

(2) A Berkeley variation on the Mead & Conway nMOS rules with buried contacts and the Lyon implant rules [4].

(3) Berkeley CMOS rules.

Lyra works in conjunction with both the Caesar and KIC layout editing systems [5,3]. Designers can invoke Lyra interactively from inside the editors, specifying rectangular regions of their designs to be checked. Violations are displayed graphically as small rectangular regions with associated text. Lyra can also be used non-interactively, to run larger checks in batch mode.

Lyra, including the rule compiler, but not the code it generates, contains about 3000 lines of code. The specification of the Mead & Conway nMOS rules consists of approximately 130 lines of text (half of these are comments) and expands to about 40 rules. The Berkeley and CMOS rule sets are somewhat shorter and simpler, since they need not deal with butting contacts. The current Lisp version of Lyra processes about 50 corners/CPU-second on a VAX-11/780; at this rate it takes about two CPU-seconds to check a simple shift register cell.

A number of improvements are planned for Lyra. The locality property of the algorithm is being exploited to develop hierarchical and incremental versions of Lyra. A more efficient C-coded implementation of the algorithm is also planned. An efficient incremental hierarchical implementation of the algorithm should be able to provide instantaneous feedback to designers as they enter and modify their designs.

8. Comparison with Existing Systems

As discussed above, existing rule checking programs are generally of two types, being based either on polygon operations or on a raster scheme. Viewed from the standpoint of data representations, Lyra's corner-based processing takes a middle position between these two methods; it doesn't deal with mask features as indivisible entities like the polygon method does, nor does it obliterate them entirely and manipulate bit patterns, as the raster method does. This compromise retains the useful locality property lost by the polygon method while avoiding the nasty combinatorics that plague the raster method.

Lyra and the algebraic programs share the common property of being rule based. Both are readily adapted to new processes. Lyra's rule language, augmented with the "grow" operation, seems to have expressive power equivalent to the algebraic rule languages. However there is an important difference between Lyra's rules and the rules in algebraic systems. The rules of algebraic systems are procedural, indicating a sequence of layer creating operations and checks. Lyra's rules, by contrast, are declarative, simply stating conditions which must be met by an input design. There is no need to create a large database of additional layers on secondary storage, and there is no notion of sequencing through the rules.

9. Summary

Lyra's corner-based context sensitive processing represents a new approach to geometric layout rule checking. The system has a number of strong points. The use of a rule compiler to translate symbolic rule descriptions into the appropriate code makes the system easily adaptable to new rule sets while avoiding the overhead of re-interpreting rules dynamically. Lyra's rules provide enough expressive power to capture a variety of layout rules, including rules for asymmetric constructs. Since the algorithm is corner-based it is intrinsically local, making it suitable for hierarchical and incremental implementations.

10. References

[1] Baird, Henry. "Design of a Family of Algorithms for Large Scale Integrated Circuit Mask Artwork Analysis." Master's Thesis, Princeton University, 1976.

[2] Baker, Clark and Terman, Chris. "Tools for Verifying Integrated Circuit Designs." *VLSI Design*, Vol. 1, No. 3, Third Quarter 1980.

[3] Keller, Kenneth and Newton, Richard. "KIC2: A Low Cost, Interactive Editor for Integrated Circuit Design." Digest of Papers for COMPCON Spring 82, IEEE Catalog No. 82 CH1739-2, 1982.

[4] Lyon, Richard. "Simplified Design Rules for VLSI Layouts." *VLSI Design*, Vol. 2, No. 1, First Quarter 1981, pp. 54-59.

[5] Mead, Carver and Conway, Lynn. *Introduction to VLSI Systems*. Addison-Wesley, 1980.

[6] Ousterhout, John. "Caesar: An Interactive Editor for VLSI Layouts." *VLSI Design*, Vol. 2, No. 4, Fourth Quarter 1981.

[7] Whitney, Telle. "A Hierarchical Design-Rule Checking Algorithm." *VLSI Design*, Vol. 2, No. 1, First Quarter 1981, pp. 40-43.

A New Automatic Logic Interconnection Verification System for VLSI Design

TAKASHI WATANABE, MAKOTO ENDO, AND NORIO MIYAHARA

Abstract—A new VLSI checking program, LIVES (Logic Interconnection VErification System), has been developed. LIVES verifies the geometrical layout data to determine whether or not it correctly reflects the original logic level description. An excellent LIVES feature, which no other programs have possessed yet, is that it can perform the check even when there are no identification marks on each logic gate in layout level description data. This feature is realized by employing new algorithms based on graph theory. In the verification procedure, LIVES adopts graph isomorphism, which uses the new partitioning methods for vertices, especially tailored for the logic diagram. In the error detection procedure, a graph is divided into subgraphs, based on a new concept, "route-subgraph," and tests for graph isomorphism between subgraphs are performed. These algorithms enable precise and rapid search for any error points in the VLSI design.

These new algorithms have been implemented and examined for practical feasibility. Experimental results have ensured that this program can detect errors in the VLSI layout patterns very quickly and precisely. LIVES will be a highly efficient tool in VLSI logic design by using it systematically combined with other CAD programs.

I. INTRODUCTION

VLSI technology has made great advances during the last ten years. This progress makes it more and more difficult to design a VLSI within a reasonably short turn around time. Design automation systems will be indispensable in order to obtain an error free VLSI design within a practical time. However, a general system will not be realized in the near future. Hierarchical and systematic use of well-designed CAD programs currently realizes the high density and high performance VLSI design. This kind of design method includes a man-made interactive phase at various design stages. As a consequence, errors cannot help being included throughout the design process. Therefore, highly efficient checking tools are used to detect and remove errors.

Checking for geometrical layout data is one of the most fundamental procedures. Geometrical layout data have to be checked from three standpoints: their consistency with layout design rules, with circuit level descriptions and with original logic level descriptions. A number of checking programs have been developed in this field [1]-[7]. There are two conventional basic checking methods for comparing geometrical layout data with logic design: one which adopts the logic simulation technology, as used at the logic design stage [7], and one which takes advantage of identification marks attached in advance to the layout description data [4]. The disadvantage

Manuscript received September 16, 1981; revised December 15, 1982.
The authors are with Musashino Electrical Communication Laboratory, Nippon Telegraph and Telephone Public Corporation, Musashino-shi, Tokyo, 180 Japan.

of the first method is that it cannot guarantee complete error detection. The second method puts the burden on the layout designer of including a number of marks in the layout description data. There is the possibility of giving marks incorrectly as well.

LIVES (Logic Interconnection VErification System) is a new automatic checking program, which eliminates the disadvantages of the existing programs. LIVES verifies the consistency between the geometrical layout data and the original logic level descriptions without any marks which identify each individual logic gate. LIVES checks for the existence of errors and then, if errors are detected, goes on to investigate detailed information about errors (their locations, categories and so on). These new LIVES capabilities are realized through the introduction of new algorithms, based on graph isomorphism.

II. LIVES OVERVIEW

LIVES is composed of four routines: data input, verification, error-point detection, and logic diagram output routines. A functional diagram is shown in Fig. 1.

In the input routine, LIVES receives two kinds of data. One kind of data is the original logic level description, specified at the logic design stage, and is used for reference. The other kind of data is the geometrical layout description to be checked. The latter category is changed into a circuit level description by the same procedure as that used in Pattern Analysis System (PAS) [5]. From the circuit level description, LIVES synthesizes logic level description, which is composed of primitive logic gates.

In the verification routine, those two bits of data are compared. Comparison procedure is performed by employing a graph isomorphism algorithm based on partitioning methods. This procedure indicates whether or not the layout description agrees with the logic design. If no errors are found, LIVES reports the correctness of the data. If any errors are found, an error point detection routine is initiated.

In the error-point detection routine, detailed information about errors (category, location, number and so on) is investigated. This investigation is performed by a new algorithm composed of two steps. The first step is the division of graphs into subgraphs, the second is the examination for isomorphism between subgraphs.

In the output routine, LIVES draws a logic diagram with detected error points on it and reports detailed information about errors.

Detailed algorithms about the logic level description synthe-

Reprinted from *IEEE Trans. Computer-Aided Des. of ICAS*, vol. CAD-2, pp. 70–81, Apr. 1983.

300

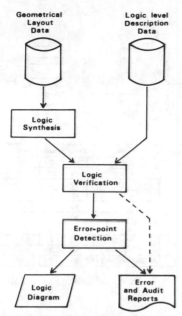

Fig. 1. LIVES functional diagram.

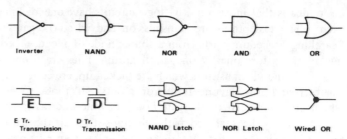

Fig. 2. Primitive logic gate category.

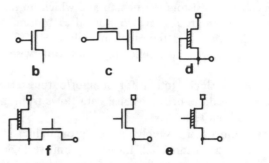

Fig. 3. Input/output terminal.

sis, the verification and the error-point detection procedures are described in Sections III, IV, and V.

III. LOGIC LEVEL DESCRIPTION SYNTHESIS ALGORITHM

In advance of verification, a geometrical layout description to be checked is changed into a logic level description. This procedure consists of two steps, a conversion from a geometrical layout description to a circuit level description and a conversion from a circuit level description to a logic level description. The first step is executed in PAS [5]. A detailed PAS explanation is given in the reference. The PAS input data was restricted to an E/E-MOS layout description at its first version. It can accept E/D-MOS and C-MOS layout descriptions at present. Data which are needed for checking, such as terminal names and node names, are placed in the original layout description. The second step depends on the device technology, because the correspondence between a logic gate category and its circuit expression differs between differing device technologies. In order to accept layout description data for several kinds of device technologies, analysis routines for as many device technologies as possible must be prepared. The routine for E/D MOS circuits is currently in use. A new routine for C-MOS circuits is going to be developed. The output data for the second step is formated according to the logic simulator syntax, with which the logic level description to be referenced is written. The categories for primitive logic gates to be synthesized are also restricted, in order to be fit for this logic simulation format. These categories are shown in Fig. 2. In the following, the detailed logic synthesis algorithm for E/D-MOS is described.

A. Transistor Classification

In this step, according to their roles in circuits, transistors are classified into three categories: normal load transistors (depletion loads), push-pull load transistors, and others.

Normal load: A depletion transistor, whose drain node is connected to a power supply line, and whose source node is connected directly to its own gate node.

Push-pull load: A transistor whose drain node is connected to a power supply line.

All other transistors, which do not satisfy the above conditions, are not considered to be load transistors.

B. Input/Output Terminal Recognition

Nodes which satisfy the following conditions are recognized as input or output terminals. The results are of use in the logic diagram drawing procedure in order to obtain easy-to-see drawings.

(a) A node specified to be an input/output terminal by designers.

(b) A node connected only to driver or transmission gates. (Transistors other than load transistors are thought to be driver transistors (drivers) or transmission transistors (transmissions).)

(c) A source or a drain node of a transistor in the condition that another side node of that transistor is connected only to driver or transmission-gate nodes.

(d) A node connected to load source nodes, but not connected to driver or transmission gate nodes.

(e) A source node of a push–pull load.

(f) A source or a drain node of a one enhancement transistor, under the condition that another side node of that transistor is a (d)-type node.

In the logic diagram drawing procedure, the number of gates between an input and an output terminal are counted for every signal path. The longest path among them is selected for a drawing guide line. Input/output terminal examples are shown in Fig. 3.

C. Transistor Merging into Groups

In this step, transitor groups which correspond to primitive logic circuits, such as NOR and NAND, are constructed. The basic idea is that most primitive logic circuits have one load in each. The exceptions are transmission and wired-OR gates. Searching procedure starts from a source node of a load, picking up transistors one by one, until ground lines are found. The load, and all transistors which are picked up, are collected together and form a one transistor group. If those nodes which are connected to the gate node for a non-normal load are found on the searched path, all transistors on the path from the source to that gate are recognized as transmissions. This procedure is executed repeatedly for all loads. After this procedure, transistors left behind and which do not belong to any transistor groups, are considered as transmissions. The signal direction in a transmission is decided if it is evidently clear according to the following rules, otherwise, it may be bidirectional.

(a) A rule similar to that for an input/output terminal recognition procedure, because signal data flows from input terminals to output terminals.

(b) From a power supply line to a ground line.

Circuits that cannot fall into the primitive logic categories are considered as the collection of transmissions. This convention is applied to the input data for logic simulator, as well. Generally speaking, a circuit block, which is defined explicitly as a block, should be considered as one element. This problem is discussed in Section VI.

D. Gate Category Classification

Transistor groups constructed in the previous step are classified into the primitive logic gates categories. At first, the number of nonload transitors (m) in each group is counted. A classification is executed as follows:

$m = 1$: Invertor
$m > 1$, and all m transistors are placed in parallel between a source node for a load and a ground line: m-input NOR gate
$m > 1$, and all m transistors are placed in series between a source node for a load and a ground line: m-input NAND gate.

All other groups that do not match the above conditions may be AND-OR-INVERTOR gate. For this category, its precise logic function expression is to be decided. Decision procedure is continued in the following way.

Step 1: Making a transistor table. —List transistor numbers, drain node numbers and source node numbers for all transistors in a table, as shown in Fig. 4.

Step 2: Parallel path search. —Find a couple of transistors between which both source node numbers and drain node numbers are the same. Then, those lines which correspond to both transistors are merged into one line. At that time, instead of a transistor number, a logic function is indicated. This operation is repeated until no such transistor pairs are left.

Step 3: Series path search. —Find a couple of transistors, one of whose source node number is the same as the drain node number of the other. Those two lines are merged into one line, then the intermediate node for two transistors is deleted from a list and a logic function is indicated instead of a transistor number. This operation is repeated until no such transistor pairs are left.

Steps 2 and 3 are iterated in turn until only one line is obtained. This procedure is illustrated in Fig. 4.

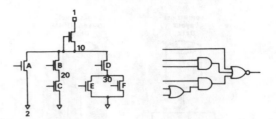

Step	Tr No	Drain	Source	Function
0	A	10	2	——
	B	10	20	——
	C	20	2	——
	D	10	30	——
	E	30	2	——
	F	30	2	——
1	E + F	30	2	OR
2	B ⋈ C	10	2	AND
	D ⋈ (E + F)	10	2	AND
3	A + (B⋈C) + (D⋈(E+F))	10	2	OR

Fig. 4. Gate classification procedure.

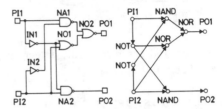

Fig. 5. Relation between logic diagram and graph.

E. Latch and Wired-OR Gate Generation

The NAND/NOR latch gate is composed of a couple of NAND/NOR gates. A Wired-OR gate is generated based on the logic gate interconnection network.

1) NAND/NOR latch gate generation—A couple of 2-input NAND/NOR gates are combined and merged into a NAND/NOR latch gate, when an output node for a NAND/NOR gate is connected to an input node of the other NAND/NOR gate. (Fig. 2.)

2) Wired-OR gate generation—The necessity for a Wired-OR gate generation arises from the convention for a logic simulator. Output nodes, which have the same node number, cannot be connected directly to input nodes of other gates. If there are many such output nodes, they are merged into one node through the Wired-OR gate. Then, this Wired-OR gate is connected to input nodes of other gates. In this step, output nodes that have the same node number are searched for and are connected to a newly generated Wired-OR gate. (Fig. 2.)

IV. LOGIC VERIFICATION ALGORITHM

The algorithm developed for logic verification only makes use of the topological information represented by logic diagrams. This makes it possible for a verification procedure to accept a geometrical layout description which has no identification marks on logic gates.

In order to make it easy to be dealt with, both bits of data, one to be checked and one to be referenced, are expressed as graph representations. A simple example of the relationship between a logic diagram and its graph representation is presented in Fig. 5. Logic gates and input/output terminals are expressed as vertices in the graph. Interconnecting wires are

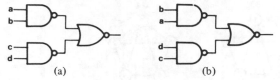

(a)　　　　　　　　　(b)

Fig. 6. Circuit equivalence (example 1).

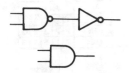

Fig. 7. Circuit equivalence (example 2).

expressed as directed edges, denoting the signal data flow. For bidirectional signal paths, two edges, which have opposite directions from each other, are placed. A bus line will become a vertex connected to many bidirectional edges. Logic gate categories are expressed as labels attached to the vertices. Input/output terminal names are attached as labels to the correspondent vertices. The verification procedure is now reduced to the problem concerning whether or not two graphs have the same topological structures. This problem is known as a graph isomorphism problem in graph theory [8]-[10].

At this point, the information about the geometrical placement order for gates, pins, and terminals is lost. For an example shown in Fig. 6, the placement order for terminals (a) and (b) differs between circuits A and B. However, from a topological point of view, they are equivalent. So are terminals (c) and (d). LIVES treats such gates, pins, and terminals as being equivalent logically, but cannot know about the geometrical placement order. At present, LIVES does not take advantage of primitive gate names or node names, except for input/output names, even when they have been defined at the layout design stage. They are ignored at the conversion procedure described above. As far as the verification efficiency is concerned, it has been reported that the partial correspondence of vertices between two bits of data is of little use [11]. User defined names become very important, when hierarchical structure data are handled. This problem is discussed later in the block handling section.

On the other hand, an equivalence check in logic between different graph representations cannot be performed. For an example shown in Fig. 7, two circuits are judged different, though they are actually equivalent in logic. In order to avoid this kind of misjudgement, two bits of input data for LIVES should have the same expressions in logic.

A. Graph Isomorphism

At this time, no graph isomorphism algorithms have been found which are efficient for all graphs. Furthermore, it is not known if the problem is *NP*-complete (12).

Proposed heuristic algorithms [8], [10] are composed of two steps: a partitioning step and a mapping step. Assume that there are two graphs, G_1 and G_2. At the partitioning step, the vertices in G_1 and G_2 are characterized by their topological features. Then, the vertices are partitioned into groups, each of which contains vertices having the same topological features. These groups are called a "class." At the mapping step, one-to-one mapping of vertices is attemped between graph G_1 and graph G_2. Classes are utilized as a "mark" in order to recognize the topological features each vertex has. A vertex V_a in G_1 is mapped to a vertex V_b in G_2, only if V_a and V_b are marked similarly. For each mapping instance, consistency checking is performed. If current mapping preserves adjacency and nonadjacency for the vertices, the isomorphism of two graphs is verified at that time. If not, backtracking occurs and the next mapping is tried. When all mappings fail, the two graphs prove not to be isomorphic. Fig. 8 shows an example of the above procedure. In this case, vertices are characterized by their edge density, that is, input degrees and output degrees. Vertices are partitioned into three classes for both G_1 and G_2. There are six possible mappings in all. They form a searching tree for consistency check, as shown in Fig. 8(c). In this figure, the mapping results are shown in an adjacency matrix style. If searching activities are executed in order from left to right in this tree, after four backtrackings have occurred, the two graphs prove to be isomorphic and the correct correspondence of vertices between two graphs is obtained. In the case of G_1 and G_3, the partitioning result looks similar to the above example. However, mappings fail for all possible combinations, so that the two graphs are not isomorphic. In the case of G_1 and G_4, it is clear, just from the partitioning result, that the two graphs are not isomorphic.

Assume that L_i is the number of vertices contained in class i. The sum of L_i, $\Sigma_i L_i$, is the total number of vertices (N) in G_1 or G_2. The number of possible mappings (M) amounts to $\Pi_i (L_i!)$. Note that this number strictly depends on the result of the partitioning. In the worst case, only one class is generated and M becomes $N!$. This means that the isomorphism problem is far from solved in practice. However, if all the vertices are characterized uniquely, M will be "one" and computing time is small. Therefore, in the partitioning method, the ability to distinguish among others is a practical necessity.

B. New Partitioning Methods

Apart from general discussions, the isomorphism problem will be investigated from a practical point of view. The algorithm will focus on the limited case to be expected, rather than on the worst case in general. The graph generated from the logic diagram reflects the signal flow in the logic diagram. It has several source and sink vertices, each of which corresponds to input and output terminals. The logic function is realized through the combinations of individual signal flows, and each logic gate may be characterized by its location on each signal flow. Therefore, it seems a very reasonable way to build partitioning methods, which rely on the distances between vertices [10], and on the paths from source vertices to sink vertices. Through these considerations, three partitioning methods are adopted: (1) Partitioning based on the distance distribution. (2) Partitioning based on the distance from terminals. (3) Partitioning based on the label. A detailed description is as follows.

(1) Partitioning based on distance distribution:

In this partitioning method, a vertex is characterized by a vector $(N_1, N_2, \ldots, N_m)$, where N_i is the number of vertices at distances $1, 2, \ldots, m$ from this vertex [10]. The distance between two vertices is defined as the number of edges on the

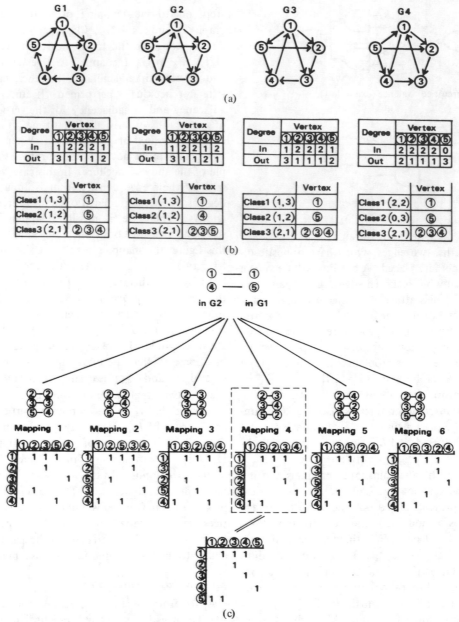

Fig. 8. Isomorphism procedure.

shortest path between those vertices. By convention, the distances from any vertices to a current vertex are negative. The distances from a current vertex to any other vertices are positive. The collection of the number of vertices stated above forms a matrix called "characteristic matrix" [10].

An example of this matrix generated from the graph in Fig. 5 is shown in Fig. 9(a). Vertices are partitioned into five classes. Each class has its characteristic vector: class 1—(0, 1, 2, 2, 1,), class 2—(0, 1, 1, 1, 1,), class 3—(1, 2, 1, 1, 0), class 4—(1, 2, 1, 0, 0), and class 5—(4, 2, 1, 0, 0). Class 3 has 2 vertices, (NA1 and NO1), while others have only one vertex.

(2) Partitioning based on the distance from terminals:

In this partitioning method, a vertex is characterized by the distance from each terminal. This method is introduced because of the poor capability inherent in the first method to distinguish a specific vertex from among others in parallel processing circuits. In such circuits, almost the same circuit unit

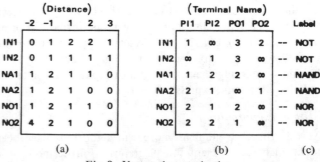

Fig. 9. Vertex characterization.

is used repeatedly along the signal flows. Those parallel circuit units can be distinguished from each other by terminal identification. In order to adopt this method, terminal names have to be given at the geometrical layout design stage. However, since the number of terminals is much smaller than the num-

TABLE I
PARTIONING RESULTS

Circuit	Number of Gates	Partitioning Method							
		(1)		(2)		(1)+(2)		(1)+(2)+(3)	
A	116	116	1	78	2.4×10^3	116	1	116	1
B	123	112	2.8×10^4	123	1	123	1	123	1
C	315	310	32	310	32	310	32	315	1
D	577	559	5.9×10^5	567	1024	567	1024	577	1
E	1360	1268	1.1×10^{28}	1275	1.9×10^{25}	1275	1.9×10^{25}	1360	1

In each Partioning Method column below:
Left-hand side: Number of generated columns
Right-hand side: Possible number of backtrackings

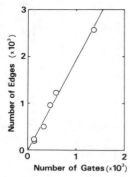

Fig. 10. Edge density for a graph which represents the logic diagram.

ber of logic gates, this penalty is not contrary to the initial LIVES purpose.

A matrix example for the same graph as that used in method (1) is shown in Fig. 9(b). There are 6 classes in all. Each vertex is characterized uniquely.

(3) Partitioning based on the label:

The previous partitioning methods take into account only topological information. In this partitioning method, a vertex is characterized by its label. It is useful to partition the vertices in more detail by their labels. A check on the logic gate category is performed by this method. An example is shown in Fig. 9(c).

Three partitioning methods have been tested by employing them on actual logic circuits. The results are shown in Table I. Circuit *A* is an example of a large gate/terminal ratio circuit. Circuits *B* and *D* are examples of a parallel processing circuit. For each circuit, the more the number of classes increases, the smaller the number of possible backtrackings becomes. At their best, the former is equal to the number of vertices and the latter becomes only "one." Each partitioning method has its favorite circuits. Method (1) is effective for circuit *A*, but not for circuit *B*, and method (2) is effective for circuit *B*, but not for circuit *A*. Complete partition of vertices for all circuits has been achieved, only when three methods are employed together. As described before, the high efficiency of this algorithm strictly depends on the logic diagram interconnection property. The results shown in Table I verify that these three partitioning methods match the property very well.

C. Time Complexity

The verification procedure time complexity depends upon both partitioning and mapping steps. As mentioned before in Section IV, the mapping step has $O(N!)$ time complexity, in general. The partitioning step has several factors. Calculation of distances between the two vertices requires $O(N^3)$ time [13], [14]. The composition of characteristic matrices requires $O(N^2)$ time and other operations be realized in $O(N)$ time.

LIVES partitioning methods are very suitable for logic networks and the mapping step is expected to consume very little time in practice. The distance calculation's practical time complexity depends on graph edge density. A graph which represents the logic diagram does not have as many edges, compared to its vertices, as a complete graph has. As shown in

Fig. 10, the number of edges has a tendency to be in proportion to the number of vertices. This fact reduces the time complexity to $O(N^2)$, in practice. As a whole, the verification procedure is processed in $O(N^2)$ time or less.

V. ERROR–POINT DETECTION ALGORITHM

Error-point detection procedure investigates detailed information about errors. This information is indispensable for the designer in order to correct errors as quickly as possible. However, to obtain this information, one-to-one mapping of vertices between graphs is needed. From this point of view, the isomorphism algorithm has a contradiction, because the one-to-one mapping is accomplished only when no errors exist. One example, in which two graphs are not isomorphic, is the relation between G_1 and G_2 in Fig. 8. The result of a partitioning shows that class 1 and class 2 are not the same. Thus it is obvious that two graphs are not isomorphic. However, there is no information about the cause for this mismatching. In this example, class 1 and class 2 differ between the two graphs. However, this fact does not mean that vertices contained in class 1 and class 2 are incorrect nor that the rest of the vertices are correct. The concept "class" is not suitable for finding error parts.

Attention should be paid to the fact that a graph contains error-free parts as its subgraphs. If error-free subgraphs can be discovered successfully, one-to-one mappings of vertices will be obtained for those subgraphs. Moreover, the error point will be detected by extracting those error-free subgraphs from a total graph.

A. Graph Isomorphism Between Route Subgraphs

The problem of finding a specified subgraph out of a total graph is being studied as a subgraph isomorphism problem [9], [15]. Subgraph isomorphism is discussed on the premise that there are given specified subgraphs which are correct. However, this is not the case here. The correctness of the subgraph is not certain before the procedure, but is to be judged afterwards. Before finding error-free subgraphs, some kind of graph division should be performed. At this point, as there is not any knowledge available related to errors, this division process has nothing to do with what or how actual errors are. In order to guarantee a correct and precise error detection, the following two conditions are to be taken into consideration: (1) The collection of all subgraphs covers an entire original

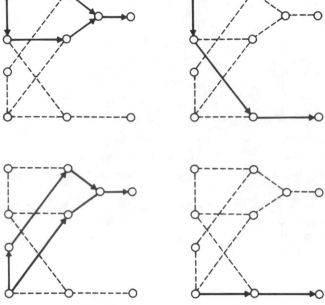

Fig. 11. Route-subgraph example.

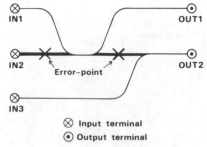

Fig. 12. Error-detection diagram.

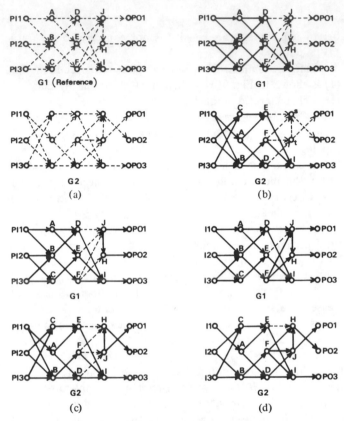

Fig. 13. Detailed error-detection procedure.

graph. (2) Each subgraph has an identification mark to distinguish itself among others.

A new subgraph concept, called "route subgraph," is defined here, which satisfies the above conditions. A route subgraph consists of the vertices and edges which belong to all routes from one input terminal to an output terminal. Input and output terminal names can be used as identification marks for subgraphs. Fig. 11 shows an example of route subgraphs for the graph in Fig. 5. The method to find this route subgraph is very simple. Those vertices which lie on a route subgraph have finite distances from an input terminal and to an output terminal concerned. In the example shown in Fig. 9(b), vertices: (IN1, NA1, NA2, NO1, NO2) have finite distances from terminal PI1, and vertices: (IN1, IN2, NA1, NO1, NO2) have finite distances to terminal PO1. As a result, vertices which belong to a route subgraph from PI1 to PO2 are IN1, NA1, NO1 and NO2. In general, ($p \times q$) route subgraphs are generated at most from one graph, where p and q are the number of input and output terminals, respectively. One vertex may be assumed to belong to several route subgraphs at the same time. There are two subgraph groups: one is generated from the graph to be checked, one from the reference graph. As do the vertices between the two graphs, subgraphs of a group have their partners (subgraphs) in another group. This one-to-one correspondence for subgraphs is recognized by their associated input and output terminal names. The tests for isomorphism are executed between those pairs of subgraphs, according to the same methods described in Section IV. As a result of procedure, subgraphs for which isomorphism have not been established are detected, and are called "error subgraphs." For vertices which belong to the isomorphic subgraphs, one-to-one mapping is established, that is, the vertices are identified.

Fig. 12 is a diagram used to explain the error region limitation procedure. The error points should exist in the error subgraphs, for example, subgraph G(IN2–OUT2). By eliminating isomorphic subgraphs G(IN1–OUT1) and G(IN3–OUT2) from the total graph, the error region appears. This error region contains the error points. However, not all the vertices and edges in this region are errors. The error regions are surrounded by the identified vertices. Therefore, error points are searched for in the error region, by examining correctness of vertices adjacent to the identified vertices. Details of this procedure are as follows. Fig. 13 shows an example. For simplicity, logic categories are omitted from the explanation. G_1 is a reference graph and G_2 is a checked graph. At the initial condition, correspondence between input and output terminals is known (Fig. 13(a)). Through the test for route subgraph isomorphism, it becomes known that subgraphs G(PI1–PO3) and G(PI2–PO3) do not contain errors. Also, vertices which belong to these subgraphs are identified (Fig. 13(b)).

The following procedure is divided into several steps, which find error points as precisely as possible. In these steps, only those vertices in G_2 which are not identified yet are dealt with.

Step 1: The purpose of this step is to distinguish those vertices from others, all of whose fan-in side vertices are identified already. Fan-in side vertices of vertex "*s*" in the example are vertices "*E*," "*F*" and "*t*." So, vertex "*s*" does not meet the above condition. If a vertex on this condition has a unique combination of fan-in side vertices and there is only one such vertex in G_1, which has the same combination of fan-in side vertices, correspondence between them is established and this vertex is considered to be identified.

This step is repeated as long as new identified vertices have been discovered in the current repetition cycle.

Step 2: In this step, fan-out side vertices are checked. Except for this point, the procedure is exactly the same as that for Step 1. The fan-out side vertex for "*s*" is PO2. Then, "*s*" corresponds to "*H*" in G_1. This result leads to another correspondence between "*t*" and "*J*" in the second repetition cycle (Fig. 13(c)).

Step 3: In this step, both fan-in side and fan-out side vertices are checked at the same time. Except that point, the procedure is the same as the previous one.

Step 4: The following two steps utilize the information about vertices within feedback loops. So, this step checks the distance from oneselves to oneselves for all vertices and finds vertices on feedback loops.

Step 5: This step is the same as Step 1, except that fan-in side vertices, which lie on feedback loops, are excluded from checking.

Step 6: This step is the same as Step 2, except that fan-out side vertices, which lie on feedback loops, are excluded from checking.

The checking cycle, composed of the above six steps, becomes the outer repetition loop and whole cycle is repeated as long as newly identified vertices have been discovered.

Step 7: According to the result of above steps, edges are classified into three groups: correct, incorrect, and uncertain vertex groups. When vertices of both ends are identified and there is an edge in G_1 which connects between the same vertices, it is a correct edge. If there is no such an edge in G_1, which connects between the same vertices, it is an incorrect edge. Otherwise, it is treated as an uncertain edge. The last two kinds of vertices are regarded as errors. In G_2 of the example, there is no uncertain edge. It is easily recognized that an edge between vertex "*E*" and "*H*" is extra and that an edge which has to connect vertex "*E*" to "*J*" is missing (Fig. 13(d)).

This algorithm detects error points, except when plural error points exist in clusters or when error regions contain feedback loops within themselves. Even when precise error points are not detected, an error-contained region is easily pointed out as a small limited region.

B. Time Complexity

The error-detection procedure time complexity is under consideration. The subgraph generation step is simply performed by tracking paths from input or from output terminals successively. Then, it requires $O(E)$ time, where E is the number of edges involved. E is smaller than N^2, in general. Test for subgraph isomorphism has the same time complexity as that for the verification procedure. It requires $O(N^2)$ time. Therefore, this procedure is considered to require $O(N^2)$ time, as a whole.

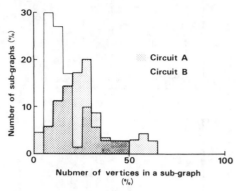

Fig. 14. Route-subgraph distribution according to the contained vertices number.

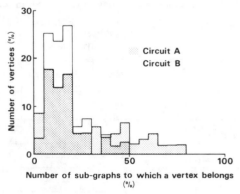

Fig. 15. Vertex distribution according to the included subgraph number.

C. Algorithm Efficiency

Two questions rise concerned with the algorithm efficiency. The first one is whether or not the isomorphism test for a complete graph is needed in advance as a verification procedure. Provided that vertices do not belong to a number of subgraphs at the same time, and, as a consequence, that the total number of vertices of all subgraphs is equal to that of the undivided original graphs, tests for isomorphism of the entire graph would only be redundant and can be omitted. In that case, the verification procedure can be contained in the error detection procedure. However, this is not true in the case of the route subgraphs. As shown in Fig. 14, each vertex belongs to several route subgraphs at the same time. As a result, it is shown in Fig. 15 that subgraphs contain considerably more vertices than the original graph. The total number of vertices for subgraphs is from about 10 to 10^2 times larger than the number of vertices in the original undivided graphs. Because of $O(N^2)$ expected time complexity in the isomorphism algorithm, rough computing time estimation can be accomplished with the expression, $\sum_i (N_i^2)$, where N_i is the number of vertices which are contained in each subgraph. A summary of this calculation is shown in Table II. The test for graph isomorphism in route subgraphs takes longer computing time than the test of an entire graph does. Its factor falls in the range of one to several tens. Thus it is worth processing the verification procedure independently.

The second question is whether or not the expected error-region size is small enough to isolate faults. A larger error region is considered to have a higher possibility of containing

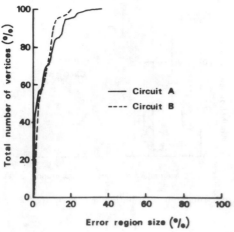

Fig. 16. Error region size when one vertex error is introduced.

TABLE II
COMPUTING TIME ESTIMATION

		Circuit A	Circuit B	Circuit C
A :	Number of Vertices (N)	123	315	1360
B :	Number of Subgraphs (P)	70	138	453
C :	$\sum^{P} \left(\begin{array}{c}\text{Vertices for}\\ \text{a Subgraph}\end{array}\right)$	1450	16565	158695
	$\dfrac{C}{A}$	11.79	52.59	116.69
D :	$\sum^{P} \left(\begin{array}{c}\text{Vertices for}\\ \text{a Subgraph}\end{array}\right)^2$	4.08×10^4	2.69×10^6	7.25×10^7
	$\dfrac{D}{N^2}$	2.70	27.15	39.22

clustered errors or feedback loops. Therefore, the larger the error region size becomes, the more often the precise error-point search fails. Considering the route-subgraph definition, the independent route length is the most essential factor which decides error region sizes. The existence of a long independent route means that a long signal path exists on which no interaction occurs between various kinds of signals. In general, the independent route length is expected to be short because of the logic diagram property.

In order to reveal this situation more clearly, one experiment has been performed. A single error, introduced intentionally, must result in the appearance of an error region. This is repeated for all vertices in the graph. The sizes of the error region for each execution have been summed up and are shown in Fig. 16. In the example circuits, most errors generated smaller error regions than 10 percent of an entire graph. These sizes are reasonably small for the procedure efficiency.

VI. MACRO HANDLING

In the discussion so far, only a primitive gate handling is described. "Cell" or "Block" is a widely used concept involved in constructing LSI data in order to obtain compact and hierarchical data. Depending on the device technology and design methodology, the depth of hierarchy and the number of elements in each level or in each block varies widely. However, in any case, the way to take advantage of this hierarchical structure is one of the best ways to reduce the size of the data to be dealt with. In order to handle this kind of data in LIVES,

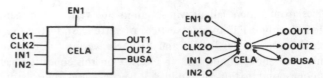

Fig. 17. MACRO graph representation.

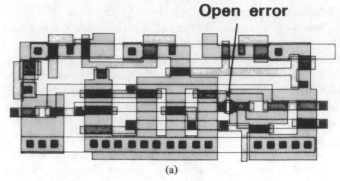

(a)

Fig. 18. Experimental result 1. (a) Input data at layout level description.

MACRO handling capability has been introduced. In the first place, blocks which are placed in one level higher block are considered as a single entity, which has input/output terminals but does not have its detailed content. In the second place, its block name and its terminal names defined by designers remain as node labels in its graph representation in order to distinguish each block and each terminal from among others. The graph representation of a block is shown in Fig. 17. In a hierarchical design method, the number of elements in each hierarchical level ususally amounts to several thousand or less. Therefore, this block handling capability is expected to reduce the practical checking time required for VLSI design.

In order to use this block handling capability, two bits of input data should have the same hierarchical structures. From this point of view, there are two problems. The first problem is the circuit synthesis program capability. Unfortunately, at present, PAS cannot offer hierarchical circuit descriptions, even when the original geometrical descriptions are hierarchical. A new version of PAS, which will be able to process and output hierarchical structure data, is now under development. The second problem is the mismatching of the hierarchical structure between a logic description and a geometrical description. This may often occur when a compact chip area is of the first importance. In these cases, the structure of a logic description may be changed according to a geometrical description modification. For this purpose, a general LSI database manipulator [16] is used, which has a capability to modify the hierarchical structure of a logic level description into any other structures desired.

VII. IMPLEMENTATION AND EXPERIMENTAL RESULTS

LIVES is written in Fortran. Its size is 7K steps in input and output routines, which include logic description synthesis procedure, 1K steps in verification routine and 1K steps in error-detection routine. The LIVES feasibility and practical use were confirmed by applying it to various kinds of actual layout patterns. An experimental result for an ADDER circuit layout pattern is shown in Fig. 18, in which 21 transistors (11

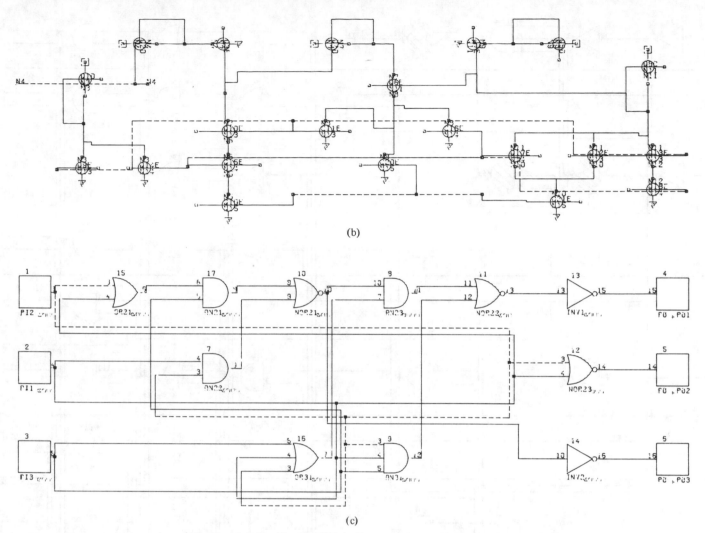

(b)

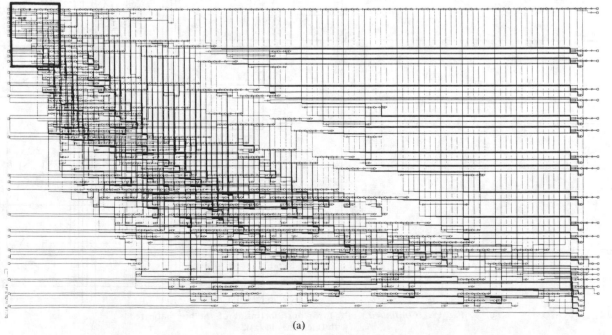

(c)

Fig. 18. (*Continued.*) (b) Circuit diagram output. (c) Logic diagram output.

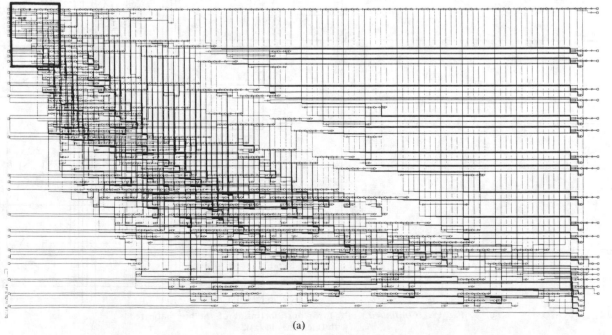

(a)

Fig. 19. Experimental result 2. (a) Whole logic diagram output.

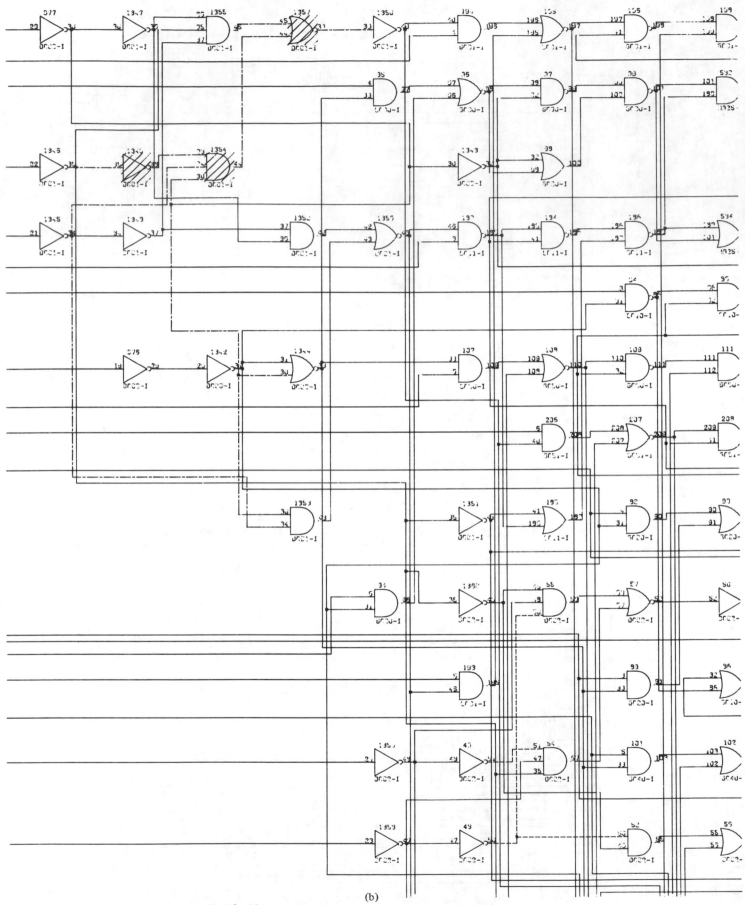

(b)

Fig. 19. (*Continued.*) (b) Error portion (inside of the box indicated with thick lines in (a)).

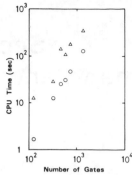

Fig. 20. LIVES computing time.   ○: Error-free case.   △: Error-contained case.

gates) are involved and a connecting wire open-error is intentionally induced. Error-point detection result is indicated with broken lines in Fig. 18(c). A distinctive color is also used to make an error portion easily identified. The precise expression for this error should be broken lines only between PI2 terminal and the nearest gate. Currently, some other lines connected to error wirings are liable to become broken lines because of the drawing algorithm. A designer can backtrack from a logic diagram to a layout description with the help of a circuit diagram shown is Fig. 18(b) and the correspondent audit report. Various size circuits have been examined, as well. Fig. 19 is the result of a rather complicated circuit composed of 1.4K gates. Broken lines are used for incorrect edges and dot-dash lines for uncertain edges. The area shaded with slanted lines indicate that those gates are not identified.

The relation between the computing time and the number of logic gates is shown in Fig. 20. It is found that the computing time is in proportion to about the square of the number of logic gates. The difference between error-free case and error-contained case suggests that the error-detection procedure takes a longer time than the verification procedure by a factor of several times to ten times. The entire computing time for verification is only a few minutes for an erroneous 1K gate circuit on 1 MIPS machine and will be 1 h for a 10K gate circuit. After implementing the macro handling capability into PAS, the computing time will be expected to become short enough for efficient whole VLSI level data processing.

VIII. CONCLUSION

A new VLSI checking program LIVES has been developed. LIVES can check the consistency between the geometrical layout data and the original logic level descriptions without any marks which identify each individual logic gate. For error detection, a graph isomorphism algorithm has been employed, which is composed of three partitioning methods: partitioning based on the distance distribution, partitioning based on the distance from terminals, and partitioning based on the label. These partitioning methods take such good advantage of the logic diagram property that the computation can be realized within a practical order time. The investigation for the detailed information about errors is accomplished by introducing a new concept, "isomorphism between route subgraphs." This new concept allows error-point detection to be accomplished successfully.

LIVES is implemented by using these algorithms. Experimental results show that logic verification and error-point detection can be performed within a practical computing time, even for several kilo-gate layout patterns. This fact has ensured the LIVES capability and practical feasibility.

In order to improve the LIVES performance, several problems have to be solved. The first one is hierarchical structure handling in PAS. Without this improvement in PAS, a general advantage of a macro usage in VLSI design is of no use for LIVES. In the second place, the number of primitive logic gate categories is not large enough to express complicated and ever-progressing VLSI circuits. The requirement for logic simulator input data is one problem and the capability of logic synthesis procedure is another problem. After the implementation of those capabilities, LIVES will become a more efficient tool in logic design, using it in systematic combination with other CAD programs.

ACKNOWLEDGMENT

The authors wish to thank H. Mukai, M. Kondo, and T. Asaoka for their support in this work. They are also grateful to S. Yamada and T. Yano for useful discussions and comments.

REFERENCES

[1] D. Alexander, "Technology independent design rule checker," in *3rd USA–Japan Computer Conf.*, pp. 412–416, Oct. 1978.
[2] C. R. McCaw, "Unified shapes checker–A checking tool for LSI," in *16th Design Automation Conf.*, pp. 81–87, June 1979.
[3] P. Losleben and K. Thompson, "Topological analysis for VLSI circuits," in *16th Design Automation Conf.*, pp. 461–473, June 1979.
[4] K. Aritoyo, H. Kawanishi, H. Yoshizawa, H. Ohno, Y. Fujinami, and K. Kani, "An interconnection check algorithm for mask pattern," in *Int. Symp. Circuits and Systems*, pp. 669–672, July 1979.
[5] S. Yamada and T. Watanabe, "A mask pattern analysis system for LSI (PAS-1)," in *Int. Symp. Circuits and Systems*, pp. 858–861, July 1979.
[6] T. Ozaki, J. Yoshida, and M. Kosaka, "PANAMAP-1: A mask pattern analysis program for IC/LSI-Centerline extraction and resistance calculation-," in *Int. Symp. Circuits and Systems*, Apr. 1980.
[7] T. Mitsuhashi, T. Chiba, M. Takashima, and K. Yoshida, "An integrated mask artwork analysis system," in *17th Design Automation Conf.*, pp. 277–284, June 1980.
[8] D. G. Corneil and C. C. Gotlieb, "An algorithm for graph isomorphism," *J. Ass. Comput. Mach.*, vol. 17, pp. 51–64, Jan. 1970.
[9] A. T. Berztiss, "A backtrack procedure for isomorphism of directed graphs," *J. Ass. Comput. Mach.*, vol. 20, pp. 365–377, July 1973.
[10] D. C. Schmidt and L. E. Druffel, "A fast back-tracking algorithm to test directed graphs for isomorphism using distance matrices," *J. Ass. Comput. Mach.*, vol. 23, pp. 433–445, July 1976.
[11] N. Kubo, I. Shirakawa, and H. Ozaki, "A fast algorithm for testing graph isomorphism," in *Int. Symp. Circuits and Systems*, pp. 641–644, June 1979.
[12] R. M. Karp, "On the computational complexity of combinatorial problems," *Networks*, vol. 5, pp. 45–68, Jan. 1975.
[13] R. W. Floyd, "Algorithm 97, shortest path," *Commun. Ass. Comput. Mach.*, vol. 5, p. 345, June 1962.
[14] S. E. Dreyfus, "An appraisal of some shortest distance algorithms," *Operations Res.*, vol. 17, pp. 395–412, 1969.
[15] J. R. Ullmann, "An algorithm for subgraph isomorphism," *J. Ass. Comput. Mach.*, vol. 23, pp. 31–42, Jan. 1976.
[16] Y. Sugiyama, M. Suzuki, Y. Kobayashi, and T. Sudo, "A subnanosecond 12K gate bipolar 32-bit VLSI processor," in *Custom Integrated Circuits Conf.*, pp. 73–77, May 1981.

An Integrated Aid for Top-Down Electrical Design

Steven M. Rubin

Fairchild Laboratory for Artificial Intelligence Research
4001 Miranda Avenue
Palo Alto, California
U.S.A.

An electrical design aid is described which integrates any number of analysis tools, multiple and mixed technologies, a flexible user interface, and the capability for top-down circuit layout. The system, called Electric, currently has design-rule checking and switch-level simulation, and it handles nMOS, CMOS, Bipolar, and printed circuit board design. Circuits are represented as hierarchical networks of electrically connected and geometrically described components. The wires that connect components have attributes that determine what will happen when the components change. This information is used to ensure that the circuit remains properly connected throughout the hierarchy even when low-level components are modified. Thus, top-down design techniques are encouraged by the hierarchically consistent nature of the database.

INTRODUCTION

Computer-aided electrical design has existed for years and many different systems have been built. Nevertheless, as circuits become more and more complex they swamp the capabilities of less powerful design aids. Frequently changing technologies are creating a need for design aids that are not bound to rigid environments. The growing number of analysis aids must be accommodated in the design environment. Also, the user interface is rarely flexible enough to suit designers. For true flexibility of design, top-down circuit specification must be available so that large pieces of a design can be planned before their contents are specified. Then, as lower-level objects change, the effects are consistently propagated from the bottom-up. The system described here handles multiple technologies, arbitrary analysis aids, has a powerful user interface, and supports top-down design.

One of the problems with today's design aids is their inability to accommodate the ever-increasing number of circuit analysis tools that are in use. Typical design systems consist of many components. After layout, the design is post-processed by a design rule checker, a node extractor, a simulator, a static analyzer, a compacter, a test vector generator, fabrication preparation, and even experimental analysis tools such as circuit and timing verification. In addition to these post-processing steps, there are many pre-processing steps that occur before and during layout. Examples are programmable logic array (PLA) and gate array generators, floor planning systems, cell library searching, wire routing, and many more. To design effectively, the designer needs to understand a complicated set of programs. Each program typically has a different user interface, making the design process difficult to learn. In addition, each program is often un-related to other parts of the system, making user feedback difficult and cryptic.

Another aspect of design systems that is a problem for designers is the user interface and the manner of interaction that it provides. Besides basic human engineering issues, electrical design aids can be categorized along two orthogonal dividing lines that break the systems into four categories. These dividing lines are the text vs. graphics distinction and the connectivity vs. geometry distinction.

Textual design languages are typically used in batch environments where a textually specified circuit is compiled into a layout and plotted for the user's verification. Graphic design languages are more interactive and link the user with a graphics display for immediate visual feedback of the design as it is created. Textual design languages have the advantage that they are a superset of a known programming language [5, 26]. Textual languages can be easier to document, parameterize, and transport. In addition, textual design is cheaper because the need for graphic output is reduced and the plotting devices can be shared. Graphic design languages, however, are typically easier to learn and faster to use in producing a circuit. This is because of the immediate feedback provided to the user and because graphics is a closer representation of the designed circuit than is text. Some design systems provide both text and graphics [5, 8, 9, 17, 26].

The other dividing line in electrical design aids is whether the user describes the circuit in a "sticks" form of connectivity or a geometric form of placement. Sticks systems require that the circuit be described as a network of dimensionless components connected to each other with wires. The actual size of the components and wires is considered later when the design aid produces a layout. Geometric systems, on the other hand, treat the entire circuit as explicit areas of silicon, metal, etc. There are no components or wires, just the geometry that describes both. Sticks systems have the advantage that electrical connectivity information is available to the design aid for analysis of the circuit. The disadvantage is that the user must rely on the design aid to properly generate an efficient layout of the circuit (or must spend much additional effort describing the exact size and placement of each component). Geometric systems require less intelligence because they are simply managing collections of polygons that describe the circuit without understanding their use.

EXISTING SYSTEMS

Current design systems can be viewed in light of the above classifications. Early design aids were textual and geometry based (for example PAL [2], ICLIC [3], and CIF [12]). Many subsequent systems remained geometry based but presented graphics interfaces. Examples are ICARUS [19], Caesar [22], Chipmunk [23], and AGS [1]. Geometry based systems also exist that allow both text and graphics input (Daedalus/DPL [5], SILT [8], and CADDS II [9]).

Printed circuit design systems are typically graphics and connectivity based (for example SUDS [21] and Scald [18]). More recently, the connectivity approach has been used as a basis for textual VLSI layout languages (EARL [15], ALI [16], Sticks & Stones [7], and I [13]). There are even connectivity based systems that allow text and graphics (LAVA/Sedit [7]).

A rare system combines text and graphics design with both connectivity and geometry. Mulga [26] acts like a sticks system but displays and manipulates the fully instantiated geometry. It has a graphics editor and can be used textually with a superset of the C programming language [14].

Todays electrical design aids cover a wide range of interaction styles and design functions. There is little consistency due to the vast complexity of the field: constantly changing technologies and ever-new approaches to design. An electrical design aid is needed that can grow along all of these lines and provide a uniform design environment.

ELECTRIC

A new design aid named Electric is described here. It can incorporate all of the analysis aids described above in a uniform manner. Some of these have already been integrated into Electric and others are available simply by programming the correct interface format. To date, there is a design rule checker and two simulators. The user interface, which graphically manipulates the circuit, is treated as an analysis aid. In addition, there is an input/output system that handles textually described circuits and it, too, acts as an analysis aid.

Electric is like Mulga in its use of both connectivity and geometry in the description of a circuit. All electrical components are treated as nodes in a network and they are connected with wires that are the arcs of the network. In addition, the nodes and arcs of this circuit network have geometric information so that proper graphic display is always available during design. Collections of nodes and arcs can be treated as single complex nodes in higher levels of the design hierarchy. A collection of hierarchically connected cells describes a circuit and is contained in a library. Electric provides top-down design by performing bottom-up propagation of changes so that the entire library always remains properly connected. Multiple libraries can be handled for the purpose of updating old designs and collecting cells of a given class.

At the bottom of a circuit hierarchy there are primitive nodes and arcs which are described by the technologies. Technologies, therefore, are the building blocks from which designs are made. They are collections of primitive components with information about how to use them in design. The components vary greatly with the technology: in printed circuit board design the components are logic elements; in MOS the components are single transistors; and in Bipolar the components are the parts of transistors bases, emitters, and collectors.

Electric allows a hierarchy of cells to be mixed in technology. For example, the VLSI designer can layout a chip in nMOS and then treat an instance of that chip as a node in a printed circuit board design.

Reprinted with permission from *Proc. Int. Conf. Very Large Scale Integration*, F. Anceau and E. J. Aas, Eds., Elsevier Science Publishers B. V., North-Holland, Amsterdam, 1983, pp. 63–72. Copyright © 1983, IFIP.

Simulation and other appropriate analysis aids will cross this technology boundary to provide proper analysis from the computer system level down to the chip transistor level. This is especially useful to VLSI designers who need to consider the pin requirements and proper configuration of their chips.

The remainder of this paper will discuss the three important aspects of Electric: the technologies, the circuit database that provides for top-down design, and the integrated analysis aids. Electric is written in the C programming language and runs on DEC VAX computers. It is expected that the system can be ported to smaller computers for use in a design workstation.

TECHNOLOGIES

Technologies provide the building blocks of electrical design. In addition to the primitive nodes and arcs, they describe all of the attributes of these components for the analysis aids. Examples of information provided for the nodes and arcs are their functional descriptions, their default sizes, and their connectivity. Frequently the nodes and arcs are described in terms of their common components: geometric layers. Examples of the information provided about layers are design rules and graphical attributes.

Electric currently supports the nMOS, Bulk CMOS, I²L Bipolar, and printed circuit board technologies. To show how technology flexibility provides a useful design environment, the salient aspects of each of these technologies will be discussed.

nMOS

N-channel Metal Oxide Semiconductor (nMOS) is the technology that has been used most for Electric's applications. This technology has wires that can run in metal, polysilicon, or diffusion. Both enhancement and depletion transistors exist as primitive nodes in the nMOS technology. It is not proper to create a transistor by crossing polysilicon and diffusion wires: a transistor node must be used. This allows Electric to know the functionality of the circuit more precisely. If polysilicon and diffusion wires are crossed, the design rule checker will issue an error. However, since the design rule subsystem is integrally connected to the design process, it can be programmed to insert a transistor node in the proper place.

In addition to the transistors, the nMOS technology provides nodes for connecting layers. There are nodes for connecting metal to diffusion, metal to polysilicon, diffusion to polysilicon (buried contacts) and all three (butting contacts). The three layers on which arcs can run (metal, polysilicon and diffusion) have dummy nodes, called pins, for making arc corners. All layers have pure nodes associated with them so that arbitrary structures can be built. Figure 1 shows the primitive nodes in the nMOS technology. These have been shown to be sufficient in the design of moderately large (over 10,000 transistor) chips.

CMOS

Complementary Metal Oxide Semiconductor (CMOS) is much like nMOS except that instead of enhancement and depletion mode transistors, there are N-channel and P-channel transistors. These transistors are located in P-well and P⁺ implant regions on the chip. Metal and polysilicon can run in and out of these areas, but the diffusion layer is sensitive to implant presence. In order to relieve the designer of correct implant placement, this technology provides two types of diffusion wires: N⁺ diffusion in P-well and P⁺ diffusion in N substrate. Also provided are two types of metal-to-diffusion contacts and two types of diffusion pins for these cases.

The existence of nodes and arcs that include their surrounding implant means that these implants never have to be explicitly described to the design aid. In connecting a N-channel transistor (in P-well) to a metal-diffusion contact in P-well, the connecting arc and both nodes will include the proper amount of P-well surround. Figure 2 shows the primitive nodes for the CMOS technology, including metal-diffusion split contacts that cross implant boundaries to connect with the substrate and P-well.

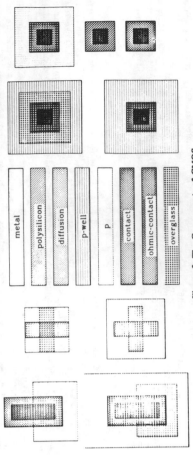

The different layers of CMOS are shown down the center here. To the left are the two transistor types and the split contacts for substrate connection. On the right are the layer contacts: four metal-to-diffusion contacts for the two implant combinations and a metal-to-polysilicon contact in the center right.

Figure 2: The Components of CMOS

Bipolar

Integrated Injecter Logic (I²L) is one of the bipolar technologies in use today. Unlike the MOS technologies, bipolar transistors may have multiple collectors, gates, and emitters of varying shape and may share these components among multiple transistors. Thus, it is not feasible to provide a limited number of transistors as primitive nodes in the technology. Rather, the transistor must be reduced to its components to provide flexible design. The primitive transistor node is therefore an empty region of buried implant. Attached to that can be collector, base, and emitter nodes of arbitrary shape and size.

This particular technology provides two layers of metal and a "via" node for connecting them. In addition to carte-blanch transistors, there are resistors of a similar nature. Figure 3 summarizes the components of this technology.

Printed Circuit

The printed circuit technology can do logic design or actual printed circuit layout. For logic design, the technology provides logic components (NAND, NOR, NOT, etc.) and wires as primitive components. For printed circuit layout, a library of over 700 chips is available which describes many of the most common chips in use today. Information on the multiple layers of a printed circuit board can also be specified. Cells from other technologies can be incorporated as packages and used here. Because the basic unit of layout is measured in true distance (centimicrons), technology mixing gives proper size information. So, when a VLSI circuit is placed inside a TTL package, the true size of that circuit relative to the package bounds is displayed. Very little use has been made of this technology because there are no available wire list generators, placement and routing aids, or other support tools.

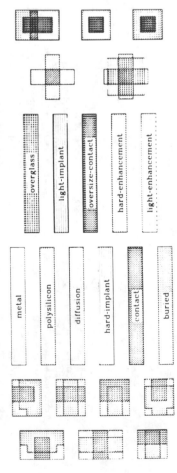

The different layers of nMOS are shown in the center here. To the right are the two transistor types and three contacts between layers. On the left are the seven buried contact configurations.

Figure 1: The Components of nMOS

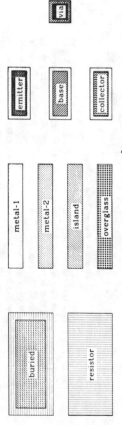

Figure 3: The Components of I²L

The different layers of I²L are shown on the left here. On the right are the three components of transistors: emitters, bases, and collectors; and a via for connecting the different metal layers. Transistors are constructed by placing the components on the buried layer.

THE CIRCUIT DATABASE

At the heart of Electric is a model of circuit representation and modification. This model treats a circuit as a hierarchical network of electrical components, connected with wires. A set of operations can be performed on the components resulting in electrically consistent changes in the network. The centralization of circuit management simplifies the addition of new technologies and analysis aids.

Database Structure

Each technology provides the database with a set of primitive circuit components, called nodes. The primitive nodes have ports for arc connections in fixed locations, relative to the size of the node. For example, the MOS transistor node has four ports at the source, drain, and two gates. If the transistor is scaled to have a larger channel length, the source and drain ports move farther apart and the gate ports elongate to allow connection anywhere along the side. Also provided by the technology are a set of wire types, called arcs. Each arc connects exactly two different ports and each port is typed to indicate the class of arcs to which it may connect. To bend an arc or run it between more than two points, multiple arcs and intermediate pin nodes must be used. An arc cannot exist without making a connection. Therefore if a node is removed from the database, all arcs connected to it are also removed.

Hierarchy of circuit design is achieved by packaging a group of nodes and arcs into a cell which acts as a complex node. Like primitive nodes, cells have ports for arc connections. These are simply exported ports from nodes inside the cell. Ports on complex nodes retain all of the characteristics of the ports on the primitive nodes from which they came, even when the ports are further exported up the hierarchy. As the internal nodes change and their exported ports move, the cell, its ports, and the arcs connected to it change. If an internal port is "unported", then all arcs connected to instances of the cell at the exported port are deleted.

Cell Change

Modification of the circuit network is restricted to transformation of nodes. Nodes can be moved, rotated, mirrored, or scaled. When the ports of a node move due to a transformation, the arcs attached at those ports are also modified. Depending on the attributes of the arcs, the nodes at the other end may also be transformed. The circuit database manager converts every node transformation into a series of database changes that consistently updates the circuit.

There are two arc attributes that affect circuit modification: orthogonality and rigidity. An orthogonal arc (sometimes called a Manhattan arc) is one that must remain horizontal or vertical through all changes. If two nodes are connected with an orthogonal arc that is vertical and the bottom node moves to the left, then the top node (and the connecting arc) will also move left. If the arc were non-orthogonal, then the top node would remain where it is and the arc would rotate to make the connection (see Figure 4).

The other arc attribute that affects circuit modification is rigidity. A rigid arc cannot stretch and connects its two nodes in a fixed configuration. If two nodes are connected with a rigid arc and one moves to the lower-right, then the other node and the arc will move the same amount. If one of the nodes rotates, the other node will spin about the center of the modified node, causing this other node to rotate and translate (see Figure 5).

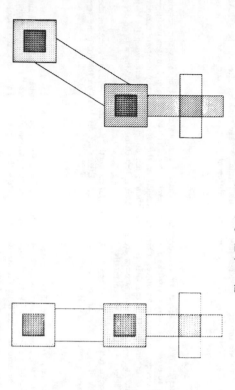

Figure 4: The Orthogonality Attribute of Arcs

On the left, three nMOS nodes are connected with two arcs. The transistor at the bottom is connected with a polysilicon arc to a metal-polysilicon contact which is then connected with a metal arc to a metal-diffusion contact. The polysilicon arc is orthogonal but the metal arc is not. The right side shows the result of translating the transistor node to the left. The same effect could be achieved by translating the metal-polysilicon contact.

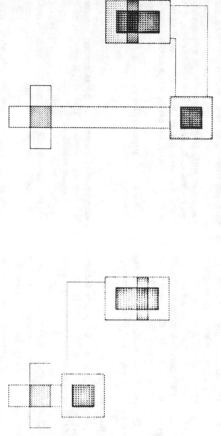

Figure 5: The Rigidity Attribute of Arcs

On the left, four nMOS nodes are connected with three arcs. The butting contact at the bottom is connected with a metal arc to a metal pin which is connected with another metal arc to a metal-polysilicon contact. The contact is then connected with a polysilicon arc to a transistor. Both metal arcs are rigid, but the polysilicon arc is not. The right side shows the result of mirroring the butting contact about the horizontal axis. Both the metal pin and the metal-polysilicon contact also get mirrored, but they are symmetric and therefore show no effects.

315

ANALYSIS AIDS

Electric provides a useful environment for writing analysis aids. Since both connectivity and geometric information are available from the database, all commonly used forms of circuit description are present. Node extraction and rasterization pre-processors are unnecessary as are input and output facilities. The analysis aid writer need only program the "inner loop" which is a fraction of the amount of code typically needed.

Although not currently implemented, background processing is an important consideration for analysis aids. As the number of these systems increases, their activity will overload Electric, causing slow response. An analysis aid should be able to run asynchronously with the design activity. In a completely distributed system, each analysis aid could have its own processor. In single-processor applications, it is sufficient to have only two partitions: the foreground design activity and, in the background, all of the long-running analysis aids. Regardless of the implementation, distributed database updates becomes a non-trivial problem beyond the scope of this paper.

The rest of this section outlines the existing analysis aids and their capabilities. At the end is a list of possible analysis aids that do not presently exist in Electric.

The User Interface

The user interface is a highly flexible and powerful control mechanism. It can support a color display with a pointing device (tablet, mouse, etc.) and a command terminal. However, it does not need all of these devices to function and can tolerate the absence of any component.

Commands to the user interface can be issued via tablet buttons, menu entries, single keyboard letters, or full keyboard commands. All commands are really full keyboard commands, but can be dynamically bound to buttons, menu entries, or keystrokes. Thus, the interface can be tailored to operate in any manner and can mimic other design aids.

The command terminal supports many useful features. To help with command input, Electric requires only as many characters as are unique to a keyword and fills in the rest. The system will also provide a list of keywords that match what has been typed. An on-line help facility explains all commands individually and by category. Parameterized macros and command files are available.

The graphics display also has much of interest. It draws on a scalable lambda grid [19] and permits multiple windows onto the design. The menu entries can be located anywhere on the display and can be scaled to fit any number of commands. Arbitrary symbols can be placed in the entries to help identify their associated commands.

Perhaps the most interesting aspect of the user interface is its portability. The color display and tablet control is done through a device-independent subroutine package modeled after the SIGGRAPH Core System [24]. This package currently supports ten different graphics devices and is adaptable to a wide variety of displays.

Design Rule Checker

The design rule checker is an incremental analysis aid that watches all changes and responds immediately to errors. It currently works at only one level of hierarchy at a time and thus is useful only for leaf-cell design. Whenever an object is created or modified, the layers that comprise it are compared with layers of other objects in the vicinity. Rasterization is not done: instead of examining the object point-by-point, bounding box comparisons are made. This is less efficient when checking a large design, but more efficient when done incrementally.

The design rule checker also keeps track of electrical connectivity so that it will not complain about the proximity of objects that are really allowed to overlap. Errors are displayed on the command terminal, but currently no correction is done.

Simulation

Simulation is an example of an analysis tool that does not have to be written. Instead, two existing simulators, MARS, a hierarchical switch-level simulator [25] and ESIM, an event-based switch-level simulator [4] have been interfaced. When simulation is desired, the analysis aid writes the circuit description suitable for the simulator and helps the user to communicate with the program. Thus, an analysis tool can be integrated without major re-writing.

There are other arc attributes that are useful in design. The notion of temporary rigidity is available for overriding specified constraints in order to effect once-only change. For example, when compacting (or spreading apart) a circuit, the arcs that cross the line of compaction are made temporarily non-rigid and all other arcs are made temporarily rigid. Then, moving a single node will compact properly and leave the arc attributes as they were.

Although not implemented now, it would be useful to have minimum and maximum length attributes that allow arcs to be non-rigid within these limits. Also eight-way orthogonality, in which 45° arcs were allowed, would be useful.

Most arcs begin as non-rigid and orthogonal so that Manhattan designs can easily be made. However this and other default attributes are set by the technology and changeable by the user.

Hierarchical Change

Because of the hierarchy, a single database modification can manifest itself as a large number of changes. If a node transforms, the cell in which it resides may change size which will cause all instantiations of this cell to scale. If the moved node has an exported port (or is connected to a node with an exported port in such a way that the port moves) then the ports on instances of the cell will also change. If, for example, instances of the changed cell are connected to each other with rigid arcs, then their positions will change because the rigid arcs hold them a constant distance apart. This translation will cause the cell with the instances to scale, and the entire network will recursively alter. Figure 6 shows an example of such a hierarchical change. In overconstrained cases, an arc may have to be jogged to retain electrical connectivity. The database will insert new pin nodes and arcs to make this jog.

Every change to the database comes from one of the analysis aids. The change, and all associated side-effects, are packaged into a change-batch that is preserved by the database. When the analysis aid finishes issuing changes, the collected modifications are broadcast to all of the analysis aids. It is possible to request that the database retract a batch of changes. A variable sized list of batches is retained and can be used to undo any modifications. Because of this facility, users are less timid about trying circuit modifications when they know that whatever is done can be undone.

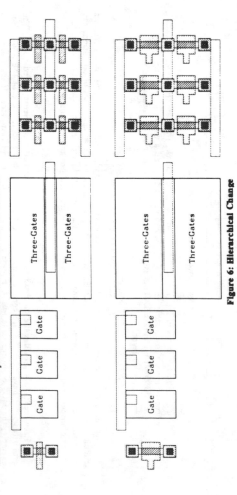

Figure 6: Hierarchical Change

The top half of this figure shows three levels of hierarchical design. At the left is a "gate" cell; next to that is a cell with three instances of the gate tied together; and to the right of that is two instances of the previous cell tied together (one is mirrored horizontally). The top right shows all three levels of the hierarchy fully instantiated. The bottom half shows the same views of these hierarchical levels after a single change has been made: the transistor in the gate cell has been scaled from 2×2 to 2×6 in size. Because of the attributes of the connecting arcs, the entire hierarchy expands to accommodate this change.

Input/Output

The input/output system is viewed as an analysis aid because of the wide variety of description formats that it can potentially handle. Textual languages can be more powerful than graphical ones, so a separate analysis aid is needed to handle them. The current system can read and write binary forms of the circuit for maximum efficiency. It can also read and write textual forms of the circuit that are more portable.

The input/output system can currently write CIF files [12] but cannot read them. CIF input requires node extraction which makes it much more difficult than CIF output.

Unimplemented Analysis Aids

In addition to filling in the missing pieces of the existing analysis aids, it would be useful to have the following facilities:

- A wire router for connecting busses and making optimal use of the channel space between cells.
- A compacter for reducing cell size.
- A test vector generator for exercising all paths through the circuit.
- PLA and gate array layout for regular circuit design.
- Circuit simulation such as SPICE [20].
- Fabrication facilities such as layer compensation, peripheral design (scribe lanes, alignment targets, etc.) and mask making conversion.
- Static analysis for power estimation, transistor ratio checking, and short circuit detection.
- Static verification for comparing high-level intended functionality with actual circuit layout [10].
- Timing verification for static delay analysis [11].
- Natural language interface for cell library search to find existing cells that meet design requirements [6].
- A "novice user" aid to teach those unfamiliar with circuit design, the Electric system, or both.
- Floor plan optimization systems.

The list goes on.

CONCLUSION

Electric is a flexible design aid that can accommodate a diverse range of needs. In its current state, it has enough functionality for the design of VLSI chips and has been so used. Electric also provides a workbench for testing new circuit analysis algorithms. Most important is its model of circuit representation and change that allows top-down design and ease of circuit modification. Combined with a powerful user interface and a variety of technologies, Electric is a valuable system for circuit design.

REFERENCES

[1] "AGS/860 V3.0 User's Guide", Applicon, Inc., Burlington, Mass. (September, 1981).

[2] "Precision Artwork Language (PAL)", Automation Technology Inc. (1971).

[3] Ayres, R.F., "IC Design Under ICL, Version I", Caltech SSP Report #1366 (revised #4031), California Institute of Technology (1978).

[4] Baker, C.M. and Terman, C., "Tools for Verifying Integrated Circuit Designs", Lambda Magazine 1, 3 (fourth quarter 1980), 22-30.

[5] Batali, J. and Hartheimer, A., "The Design Procedure Language Manual", AI Memo 598, Massachusetts Institute of Technology (1980).

[6] Brachman, R.J. and Levesque, H.J., "Research in Knowledge Representation in Support of Consultation About Complex Artifacts", internal memo, Fairchild (June 1982).

[7] Cardelli, L., "The Sticks & Stones Painter's Manual", technical report, University of Edinburgh (1980).

[8] Clark, J. and Davis, T., "SILT, A VLSI Design Language", technical report, Stanford University (1982).

[9] "CADDS 2/VLSI Product Specifications", Computervision Corporation, Bedford, Mass. (February 1982).

[10] Gordon, M., "A Very Simple Model of Sequential Behavior of nMOS", in: Gray, J. (ed.), VLSI 81, (Academic Press, New York, 1981), 85-94.

[11] Hitchcock, R.B., "Timing Verification and the Timing Analysis Program", proceedings, ACM IEEE 19th Design Automation Conference (June 1982), 594-604.

[12] Hon, R.W. and Sequin, C.H., "A Guide to LSI Implementation", technical report SSL-79-7, Xerox Palo Alto Research Center (January 1980).

[13] Johnson, S.C., "Hierarchical Design Validation Based on Rectangles", Conference on Advanced Research in VLSI, MIT (January 1982) 97-100.

[14] Kernighan, B.W. and Ritchie, D.M., "The C Programming Language", (Prentice-Hall, Englewood Cliffs, NJ, 1978).

[15] Kingsley, C. and Williams, G., "Earl - interpreted circuit design system", technical report, California Institute of Technology (1981).

[16] Lipton, R.J., North, S.C., Sedgewick, R., Valdes, J., and Vijayan, G., "ALI: a Procedural Language to Describe VLSI Layouts", proceedings, ACM IEEE 19th Design Automation Conference (June 1982), 467-474.

[17] Matthews, R., Newkirk, J., and Eichenberger, P., "A Target Language for Silicon Compilers", Twenty-Fourth IEEE Computer Society International Conference (Compcon) (February 1982), 349-353.

[18] McWilliams, T.M. and Widdoes, L.C. Jr., "SCALD: Structured Computer-Aided Logic Design", Proceedings of the 15th Annual Design Automation Conference (June 1978).

[19] Mead, C. and Conway, L., "Introduction to VLSI Systems", (Addison-Wesley, Reading, Mass., 1980).

[20] Nagel, L.W. and Pederson, D.O., "Simulation Program with Integrated Circuit Emphasis", Proc. 16th Midwest Symp. Circ. Theory, Waterloo, Canada (April 1973).

[21] Newcomer, J.M. "SUDS Users' Manual", technical report, Carnegie-Mellon University Computer Science Department (1979 and 1980).

[22] Ousterhout, J., "Caesar: An Interactive Editor for VLSI Layouts", VLSI Design, II-4 (1981), 34-38.

[23] Petit, Phil, private communications.

[24] "Status Report of the Graphic Standards Planning Committee", Computer Graphics 13-3 (August 1979).

[25] Singh, N., "MARS: A Multiple Abstraction Rule-Based Simulator", technical report 17, Fairchild Laboratory for Artificial Intelligence Research (1983).

[26] Weste, N. and Ackland, B., "A Pragmatic Approach to Topological Symbolic IC Design", in: Gray, J (ed.), VLSI 81, (Academic Press, New York, 1981), 117-129.

COMBINING GRAPHICS AND A LAYOUT LANGUAGE IN A SINGLE INTERACTIVE SYSTEM

Stephen Trimberger

California Institute of Technology, Computer Science Department, Silicon Structures Project
Pasadena, California

XEROX Palo Alto Research Center
Palo Alto, California

ABSTRACT

Layout languages provide users with the capability to algorithmically define cells. But the specification language is so non-intuitive that it is impossible to debug a design in that language, one must plot it. Interactive graphics systems, on the other hand, allow the user to debug in the form in which he sees the design, but severely restrict the language he may use to express the graphics. For example, he cannot express loops or conditionals. What is really needed is a single interactive system that combines layout language and graphic modifications to the data. This paper describes just such a system.

INTRODUCTION

Two primary methods for generating integrated circuit mask layout data are *interactive graphics* and *layout languages*. Each has tasks which it does well and those which it does not. The result is that users of both kinds of systems are dissatisfied.

When dealing with graphic data, such as integrated circuits, it is necessary to view the data graphically. Often the limiting factor in the speed of design is the time it takes to plot the data. Interactive graphics systems provide "instant plotting", enabling the designer to iterate extremely quickly on the design.

Interactive graphics systems also provide a powerful "language" for handling the data. For example, the user may point to the object of his attention or to a desired position, rather than search for certain numbers in a program printout or type numbers in a program oriented system. But interactive graphics systems do not allow graphic objects to be positioned with respect to other objects, except occasionally, in a most rudimentary adjacency manner. Positions are given in some absolute coordinate space and are independent of one another. Many systems give the ability to replicate a piece of geometry. This is a looping construct, but it is severely limited by the capabilities of the graphics system. Graphics systems do not allow the expression of conditional geometry or relative positioning. Much more powerful language-style operations are needed.

Layout languages attempt to resolve these problems. Layout languages usually fall into the "plotter driver" category. Features are described by a sequence of commands to draw geometry at absolute coordinates. More advanced languages, usually embedded in an existing programming language, have all the powerful control structures that such languages provide, such as loops and conditionals. The power gained by the addition of true programming language facilities to the layout language provides the designer with the ability to *algorithmically define* a circuit or a piece thereof.

Algorithmic definition is the specification of a piece of a layout with an algorithm. The algorithm that generates the layout can be parameterized, giving the ability to define, for example, an *n*-input NAND gate, a line driver with exactly the required power, a Programmed Logic Array (PLA), or even an *n*-bit processor. Such cells are much more versatile than typical "hard" standard cells. This algorithmic design is not possible with current interactive graphic design aids.

Unfortunately, languages specify graphic positions in an awkward fashion, by numbers. A user of a layout language system has a separation between the graphics specification and the graphics viewing. Current languages force the user to go through a tedious and time consuming edit-compile-plot cycle. Interpreted languages get rid of the explicit compilation, but have a corresponding lengthy program execution and evaluation cycle, which achieves the same effect, that of slowing down the design cycle. Interactive techniques have attempted to get rid of this lengthy cycle, but have been usually aimed only at the graphic form and not at the language form.

The major disadvantages of each kind of system correspond to the strong points of the other. Graphics systems are easy to use, but severely limited in their expressability, language systems are versatile but tedious to use. Therefore an attractive idea is to combine both representations in one system which allows modification of the integrated circuit data in both forms. This is called parameterized graphics by graphics system users and instant plotting by language system users.

This paper deals with the design, implementation and evaluation of the ideas for combining graphical and textual data representations.

OVERVIEW OF SAM

Sam is the name of a system which combines the two data representations. The work on Sam was done at the Xerox Palo Alto Research Center. Sam was written on the *Alto*, a personal minicomputer, the key features of which are a high-resolution black and white video monitor used for both graphics and text output, and a "mouse" pointing device for graphic input, as well as a keyboard and facilities for printing and file storage. Sam runs in the *Smalltalk* environment, an object-oriented system with very powerful programming and debugging aids [Ingalls 1977]. Smalltalk is a virtual memory system with its own memory

Reprinted from *ACM/IEEE 18th Design Automation Conf. Proc.*, 1981, pp. 234–239.

```
Def SRcell | GNDy | VDDy | INPy | LEFTx |
RIGHTx |
    |Note: Default Note.
    |Box. Layer: 5. ll: ¯6+LEFTx,12+VDDy
ur: 13+RIGHTx,16+VDDy.
    |Box. Layer: 2. ll: ¯3,12+VDDy ur:
1,16+VDDy.
    |Box. Layer: 4. ll: ¯2,13+VDDy ur:
0,15+VDDy.
    |Box. Layer: 3. ll: ¯4,5 ur: 2,11.
    |Box. Layer: 4. ll: ¯2,3 ur: 0,7.
    |Box. Layer: 2. ll: ¯3,2 ur: 1,5.
    |Box. Layer: 5. ll: ¯3,2 ur: 1,8.
    |Box. Layer: 1. ll: ¯4,3 ur: 2,13.
    |Box. Layer: 3. ll: ¯6+LEFTx,¯1+INPy ur:
3,1+INPy.
    |Box. Layer: 3. ll: 5,¯6+GNDy ur:
7,16+VDDy.
    |Box. Layer: 3. ll: 5 + 4 + 2,¯1+INPy ur:
7+RIGHTx + 4 + 2,1+INPy.
    |Box. Layer: 5. ll: ¯6+LEFTx,¯6+GNDy +
2 + ¯2 ur: 13+RIGHTx,¯2+GNDy.
    |Box. Layer: 4. ll: ¯2,¯5+GNDy ur:
0,¯3+GNDy.
    |Box. Layer: 2. ll: 0,3 ur: 11,5.
    |Box. Layer: 3. ll: 9,¯1+INPy ur: 13,2.
    |Box. Layer: 2. ll: 9,1 ur: 13,5.
    |Box. Layer: 4. ll: 10,0 ur: 12,4.
    |Box. Layer: 5. ll: 9,¯1 ur: 13,5.
    |Box. Layer: 3. ll: ¯3,¯6+GNDy ur: 1,3.
    |Box. Layer: 2. ll: ¯2,¯6 ur: 0,16.
```

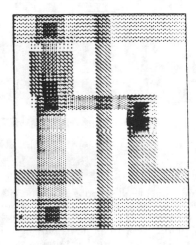

Figure 1. Snapshot of the Sam Display.

manager and garbage collector.

Sam was meant to be an experiment, a quick implementation to test the ideas for combining graphic and language systems. For this reason, many decisions were made to facilitate the implementation at the cost of execution speed and fullness of the user interface. Smalltalk was chosen as the implementation language because of its virtual memory, automatic garbage collection, excellent debugging facilities and powerful language constructs which facilitated code sharing. These features were necessary in order to finish the system in the three months allotted for the project. The price of these features was the inability to perform low-level hacking to improve performance. Since no such hacking was to be done, there was essentially no cost for the powerful language environment.

Sam provides the user with a two-part viewing window on the display as seen in figure 1. The left side shows the program view of the design under edit, the right side shows the graphics view. The user may move the viewing location in either window and may make edits to the data in either window. When the design is changed in either window, the change is reflected immediately in both windows.

The data displayed in the windows are *pictures* of the data structure. The data structure is the base form, the program view and the graphic view are merely different ways of looking at the base form of the data. When either the graphic bitmap form or the program character-string form is needed for display, it is generated from the data

structure. When the user makes what appears to be a modification of the data in either window, the commands are translated into calls on procedures in the data structure to carry out the action. The data structure makes the modification and causes both displays to be updated. The two views are kept consistent because they are both refreshed from the same data in memory.

Internally, Sam consists of four major pieces (see figure 2). The first piece is the data structure, which is more than a conventional design automation database, consisting as it does of objects which have both data and code attributes. Two more major pieces are the Graphic Editor and the Program Editor, which display data and convert inputs to commands to the data structure. The fourth piece is a small coordination piece, which holds together the two editors.

DATA STRUCTURE

The heart of the Sam system is the data structure. The data structure is modelled after the *parse tree* of a simple programming language as seen in figure 3. Each node in the data structure corresponds to one statement in the simple programming language. For this reason, I use the phrase *data structure language* when referring to the operation of the data structure. The parse tree form facilitates the viewing and editing operations. It is more convenient and faster to keep the data in this form than it is to re-construct it from a character-string or token-string base language form when it is needed for graphic operations.

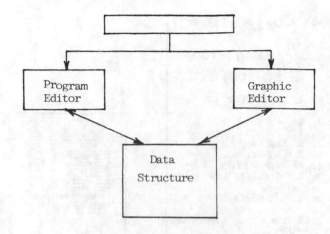

Figure 2. Sam Block Structure.

The data structure language includes loops, conditionals, and variables. Procedure definition in the language provides the cell definition facility for the integrated circuits. Thus, cells defined in Sam can have parameters passed to them, just like procedures in programming languages.

The data structure contains eight kinds of entries: *Box*, *Instance*, *Cell Definition*, *Loop*, *If*, *Assignment*, *Block* and *Note*. Besides the major data structure entries, there are entries for *name*, *expression*, and *comment text*. Each kind of entry is defined as a Smalltalk *Class*, which is a construct consisting of some data and some procedures for manipulating that data. Each statement in the data structure is one *instance* of a Class, which has its own data fields, but shares the procedure code with all other objects of its class.

The statements below are the text representations of the data structure entries. These correspond to the text view of the design. The underlined portions of the statements correspond to the data fields of each class of objects. The data fields are the portions of each object which may be manipulated by the editors. Commands from the editors to modify the design are translated into commands to change one or more of the underlined fields in a data structure entry.

The *Box* entry is the graphic primitive, and is described by a layer, and expressions for the lower left and upper right x-y positions of the corners of the rectangle.

Box. Layer: Polysilicon. ll: 8,1 ur: 10,10.

FOR i := 1 to 7 DO

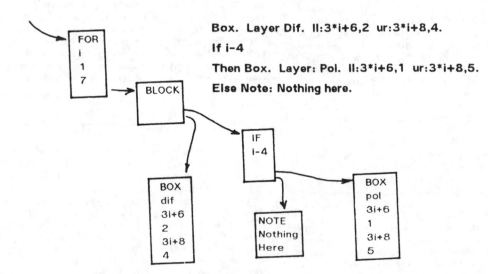

Box. Layer Dif. ll:3*i+6,2 ur:3*i+8,4.

If i-4

Then Box. Layer: Pol. ll:3*i+6,1 ur:3*i+8,5.

Else Note: Nothing here.

Figure 3. The Sam Data Structure

320

The data structure contains entries for programming language constructs. The first is an assignment entry with a destination variable name and source expression. There is a conditional entry with three fields: the conditional expression and two pointers to other Sam data structure entries, one for the *THEN*-branch and one for the *ELSE*-branch. Sam has an entry for iteration, which consists of loop variable name, starting and ending loop value expressions, and the entry for the body of a loop. The *Note* entry is a comment, and is used for annotation.

PLAsize = PLAdrivSize + (minterms * PLAandSize).

If firston
   Then: Box. Layer: Polysilicon. ll: 8,1 ur: 10,10.
   Else: Note: Don't connect the switch

For buscount = 1 to bussize do:
   Note: Connect the busses
   Box. Layer: Metal. ll: LeftSide.bottom + 10*buscount
   ur: RightSide.bottom + 3 + 10*buscount.

Note: Tricky stuff: Be sure DEI and CFS are never both high..

There are entries for building the cell instantiation hierarchy. The cell definition entry has a cell name, a list of parameter names and an entry for the body of the cell. The cell instance entry has the name of the cell to be instantiated, the transformation matrix entries to specify the position and orientation of the instance, and a list of parameters.

Def andPlane (inputs, minterms, code)
   Note: The stuff for the andplane goes here.

Inst PLAcellpair t11:1, t12:0, t21:0, t22:1,
   tx:(1.7*incount), ty:-4+(14*mincount) |
   Params: (code(mincount*2-1, incount)),
   (code(mincount*2, incount)).

The *Block* entry allows many statements to be grouped into one for inclusion in a loop, for example. *Blocks* show indented.

   Note: There is nothing in this loop.
   Note: Except these comments.

The procedures recognized by the data objects define the interface to the data structure. In particular, each class has procedures to update each of its data fields shown underlined above. In addition, each class has procedures to show graphically and print textually in the respective windows. These two procedures provide the pictures of the data structure that the user sees when he manipulates the data.

Commands from the two editors, one textual, one graphical, to alter the data, are translated into calls to entries in the data structure to change a certain field, giving a common interface for both representations. The calls may be passed down the tree if necessary. An operation on an *If* statement may be one on the statement itself, in the case of modifications to the conditional expression, or may be passed down the *THEN*-branch or *ELSE*-branch, in the case of a textual *select* operation.

EDITING THE DESIGN

Sam provides a *syntax-directed editor* for the program view. This is similar in philosophy to interactive graphics editors, since the user may not alter arbitrary pieces of the picture of the data, be it individual bits in the raster of the graphics or, in this case, individual characters in the text. Instead, the user may only manipulate *complete syntactic pieces* of the data, such as whole *Boxes* or complete expressions. Complete syntactic objects are whole statements, expressions, and names. These are, by definition, the data structure entries, shown underlined in the list of data structure entries, above.

Therefore, the syntax of the program view need never be checked. It is always correct because it is impossible to make it incorrect. The editing features do not allow the "o" to be deleted from the *For* keyword, for example. Only meaningful pieces of the data can be changed. When editing an expression, a variable name, or the comment text in a *Note* statement, the user modifies the actual text, which is re-compiled when an attempt is made to terminate the edit. This gives full generality and ease of expression when editing at the lowest level.

The program editor allows the user to *select* a statement or subfield of a statement by pointing to it. When this is done, the selected entry shows video inverted in the program window, and outlined in the graphic window. Selected items may be deleted or modified by commands to either editor. The program editor has commands to create any statement, delete the selected statement, move and copy textually, and edit expressions, names, and comment text.

The graphic editor commands are very similar to those available in commercial systems. In the graphic editor, the user may select a box by pointing to it. The selection works exactly the same in the graphic editor as it does in the program editor. The user may manipulate boxes with commands to create, destroy, stretch, and move and copy graphically. The graphic modifications are interpreted as changes in the expressions that make up the position of the *Box*, for modifications of existing objects, and as changes in the *Block* that contains the *Box* statements, for creating and destroying elements.

The changes from both editors to entries in the data structure are translated into calls on the procedures of the data objects to effect the change. When a data entry is changed, both pictures of the data are immediately updated to reflect the new data structure.

UPDATING PROBLEMS

There are problems that arise in a system of this sort where changes can be made in two different forms which must remain consistent. There are two problems of particular importance because of their frequency: expression update and iteration update.

Expressions: Suppose the x-position of a *Box* is given by the equation "3*w+4" and suppose further that the *Box* was moved graphically. How should the x-position be represented now?

Let us make this an example. Assume w=2. "3*w+4" is 10. In the graphics window, the user sees the x-position as 10 and moves it to 13 (see figure 4). The resulting expression could be any of the following expressions which evaluate to 13:

13	destroy the parameterization
3*w+7	add a constant (translate)
(13/10)*(3*w+4)	multiply by a constant (scale)
3*w+4 {w=3}	change the value of the identifier

The first choice, the most simple, destroys the parameterization. The parameterization may still be relevant and, in any case, is useful to the user in understanding the design, so this may not be very wise. The second and third choices preserve the parameterization, but there is no assurance that this is what the user wanted, either. The last solution is fairly tricky. Since w could itself be defined as an expression, we are faced with this same problem again when updating w. The result is a constraint satisfaction problem. Small changes in the design could have far-reaching and non-obvious effects on the circuit.

None of the solutions can give the correct result every time. The program cannot know the mind of the user. One option is to give the user several different graphic editing modes, one for each of the choices above. This leads to a cluttered user interface, increasing the chance for subtle errors if the user accidentally modifies something with the wrong mode. Another solution could be used where expressions of the form "aX+b" are translated, because the expression already has a translation; expressions of the form "aX" are scaled because the variable is already scaled and expressions of the form "X" modify the variable. Or the system could translate all positions and scale all dimensions. But these guesses could still be wrong, and the user would have to remember all the special conditions. In general, a blatantly naive, but consistent system is better than a clever, but inconsistent one.

Sam translates all changes. This keeps the effects of modifications localized and preserves the parameterization. In use, this was found to be the proper choice in every case. It seems to be a reasonable solution, preserving parameterization in a simple, straightforward manner.

Iteration: When one graphically edits the graphics corresponding to one iteration of a loop, should all the iterations be changed, or just the one?

Typically, language systems modify all iterations, changing the object of a step and repeat, while graphic systems either do the same or disallow the operation. Sam modifies all iterations of the loop. This seems to work well, but there are clear cases when the other choice is preferred. This may be a situation in which two different editing modes would work. The iteration problem has not yet been fully investigated.

GENERAL EVALUATION OF SAM

The individual editors used in Sam were made intentionally weak in order to simplify the programming task so that the project could be completed quickly. These weaknesses were easy to identify and ignore when evaluating the new ideas in Sam and they will not be discussed here. Instead, this section covers problems arising from combining these two data representations.

The evaluation of Sam consisted of the design of cells of varying complexity: an inverter, a totally graphical task; a simple, parameterized *stretchable* shift register cell, which could change its pitch depending on input parameters; and a PLA, a predominantly algorithmic task. The interaction of the two data representations indicated that even the simple Sam system was unusually powerful. The details of the evaluation will not be discussed, but the results are reported here.

Box. Layer: Pol. ll:3*w+4,6 ur: 3*w+12,9.

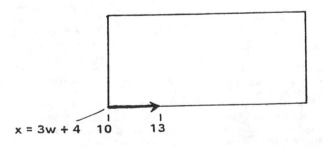

x = 3w + 4   10   13

Figure 4. Expression Update.

With the larger designs, Sam swamped the minicomputer on which it ran. The problem was due mostly to the limited memory space. Since Sam was meant to be a quick and dirty experiment, no coding tricks were used to speed the execution. Sam ran acceptably fast on a more powerful processor with a larger address space.

The language model used for Sam's data structure was inadequate. There were two major problems with the language. First, the Sam data structure language was modelled after a very simple Algol-like language without data scoping or type checking. This made the interpreter simple and obviated the need for error messages. However, structured data types such as points, rectangles and arrays were needed. This need was anticipated and the problems were bypassed in the evaluation, but a real system would have to address these data structure issues.

The Sam data structure language could not properly handle incremental data updates because the language model did not provide a facility for expressing *dependence* of statements. A *Box* statement with a variable in one of its expressions *depends* on the assignment statement that sets the value of that variable. In programming languages, independence is expressed by concurrency, since two pieces of program can run concurrently if they are independent. A piece of the design must be refreshed only when it is dependent on something which has changed. Therefore, a proper language model for the Sam data structure would have to be able to express concurrency. This concurrency would never be seen by the user, since is only used by the system internally.

One deficiency of the Sam data language was not a problem. Since loop termination could not be affected inside the loop, the language is equivalent to a finite state machine, and less powerful than a Turing machine. During the evaluation of Sam, this limitation was not a problem. This would seem to imply that a finite language is sufficient to describe a finite object, such as an integrated circuit chip. This question is still unresolved, however. Some evidence exists that data-dependent recursion or iteration is necessary to produce some designs.

The power of parameterization in Sam's cells was very good. When placing an instance of a cell, one could supply parameters to alter the internal structure of the cell as desired. This is the same as passing parameters to a procedure in a programming language, and is done for the same reason: it allows the procedure/cell to be used in many more situations.

A cell should have connection points on it, which could be used as variables in expressions. These could be used in the program view to connect wires. Attributes of *Boxes* should be accessible also, for the same reasons. This implies that *Instances* and *Boxes* should exhibit attributes in the program view, like SIMULA class instances [Birtwistle 1973].

One graphic editing feature that would have simplified many operations is one which would position new features relative to a point. Then, all items relative to that point could be moved just by moving the point. This would allow huge pieces of the design to be moved quickly and easily by parameters in the program.

Perhaps the most powerful single feature of Sam is the selection operation. The selection shows the relationship between program statements and graphic elements, enabling the designer to move quickly and easily between representations. Not only does Sam give an instant plot, but the plot is an active entity, telling the user where each graphic feature comes from in the program.

CONCLUSIONS

Sam is fundamentally different than any commercially available design system. The Sam system gives the user the ability to pass quickly between the graphics and language worlds. The mating of graphics and language in this fashion provides the user with a vast increase in expressive power without losing the fast iteration of graphic systems. The Sam system was developed in a deceptively short amount of time. A full, usable Sam system requires a good graphics system and a good language system as well as facilities relating the two.

Complex designs can be created algorithmically without giving up the rapid feedback of graphics systems. Such a system can be used to generate a parameterized cell library, which could be used to design a large number of chips. The cells in the Sam library could be parameterized in such a fashion that calls to them would provide a behavioral design language. Thus, large designs could be created from a behavioral description, and the layout could still be optimized graphically. Work is progressing along both these paths.

REFERENCES

[Birtwistle 1973] G. Birtwistle, O.J. Dahl, et. al., *Simula Begin*, Petrocelli/Charter, 1973

[Ingalls 1977] D. Ingalls, "The Smalltalk-76 Programming System Design and Implementation", Proceedings of the Fifth Annual Symposium on Principles of Programming Languages, January, 1977

Automatic Placement and Routing Program for Logic VLSI Design based on Hierarchical Layout Method

Hidekazu Terai*, Michiyoshi Hayase*, Tatsuki Ishii*, Chihei Miura*,
Tokinori Kozawa*, Kuniaki Kishida**, Yohsuke Nagao***

* Central Research Laboratory, Hitachi Ltd.   (Kokubunji,Tokyo,Japan)
** Device Development Center, Hitachi Ltd.   (Kodaira,Tokyo,Japan)
*** Kanagawa Works, Hitachi Ltd.   (Hadano,Kanagawa,Japan)

ABSTRACT

A hierarchical lauout program being applicable to both custom (building blocks) and master-slice LSIs is described. A new objective function is presented for improving iterative placement which is suitable for wiring patterns that are made by channel router. By using this function, the automatic placement and routing program reduced the block area by 5% on an average compared with the manual layout. A custom VLSI including 74K transistors was designed through this program. This chip was fabricated with no error at the first trial. Several kinds of 6K gate master-slice LSIs whose gate utility is more than 90% have also been designed by using this layout program, and full automatic wiring (no un-connected pin-to-pins) was performed in every case.

1. Introduction

We have developed an automatic chip design system being applicable to both custom (building blocks) and master-slice design method.[1]

This system can support a hierarchical design approach, in which the lowest level is the functional cell, such as a single or complex gate, and macro logic unit.[2]

The specific features of our system include:

(1) that the object of design can be partitioned into an arbitary size;

(2) that each partitioned unit on the same level can be concurrently designed;

(3) that functional assignment of automatic and manual design is optimized; and

(4) that various verification programs, such as logical/physical/delay check, are prepared around the design data base.

In this paper, we describe automatic placement and routing programs which are a main part of this system. Application results with a custom VLSI and master-slice LSIs are also presented.

2. Hierarchical chip structure

The hierarchical structure and the layout model for a VLSI chip is shown in Figure 1. The cell is a minimum logical unit consisting of 1 to about 50 gates of as large as SSI (Small Scale Integration). The block is a basic unit of logic

design, and has about 500 to 1000 gates. The chip consists of blocks (inner logic parts) and I/O buffer array (peripheral logic parts).

The whole layout for a VLSI whose number of gates exceeds 10K to 100K gates causes an extreme increase in the drawing size of the XY plotter and memory capacity of the CPU.

In our system, when manipulating the design data base, it is possible to partition the chip into certain subsets assembled of blocks, in order to keep the design object within a reasonable limit.

We name such a subset "a block assembly" (or "assembly" in brief). By creating virtual edge pins on each boundary of the block assembly when partitioning a chip into assemblies, the routing can be done concurrently in each block assembly. By joining the assemblies together like a jigsaw puzzle, the whole layout for the chip is completed.

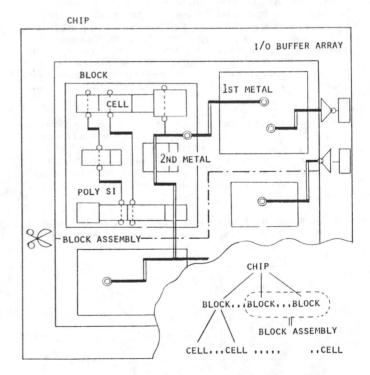

FIG.1 HIERARCHICAL STRUCTURE OF VLSI CHIP

Reprinted from *IEEE Int. Conf. Circuits Computers Proc.*, 1982, pp. 415-418.

3. Automatic placement and routing

Placement and routing process for a LSI layout are as follows.

3.1 Application to custom VLSI

3.1.1 Block design

--Placement Process--

Step 1: Clustering.
Cells are grouped according to the degree of each logical connection to make clusters. We name such clusters "macro cells". The number of macro cells, K , is related as

$$K=M*C .$$

Here, M is the number of cell rows composing the block, and C is a control parameter.

Step 2: Macro cell placement.
Macro cells are placed with a M x C matrix. This procedure can be summarized as two processes, that is, making a binary tree which represents the degree of connection between macro cells, and repeating the decomposing of the tree and positioning of each node (macro cell) on a plane accompanied by mini-cut inter change, from the top to the bottom of the tree.

Step 3: Cell placement.
The order of cells in a macro cell is decided. Cells are arranged one-dimensionally in the macro through a linear clustering method.[3]

Step 4: Iterative Improvement.
Through Step 3, a two dimensional cell placement is generated in accordance with the layout model shown in Figure 1. In this step, the location of each cell is improved iterativelly in a cell row. We have introduced a new objective function for the iterative improvement process. This function makes it possible to realize an optimal location for a specified cell. This function can be used to effectively estimate wiring patterns produced in the routing process and is more effective in reducing wiring area than are existing functions. Details are presented in Section 4.

--Routing Process--

Step 1: Pin assignment.
As shown in Figure 1, a cell has dual equi-potential terminals on its upper and lower sides. Therefore, it is necessary to decide on the terminal to be wired. Between them, that terminal is selected which narrows the wiring space (the number of the first metal channels between cell rows).

Step 2: Layer assignment.
Two kinds of layers (Poly-Si and 2nd metal) are available for y-direction wiring. As is well-known, the resistance value for Poly-Si is 1,000 times that of metal. Therefore, with respect to a specified net, such as a timing signal, the layer to be used is assigned according to the net length (y direction) at this step.

Step 3: Feed-through assignment.
Feed-through points in a cell row are decided for connecting cells located far away over more than two rows.

Step 4: Channel assignment.
In this step, each terminal within a wiring space between cell rows is routed. The algorithm is a channel assignment method employing trunk-division.[4] We have enhanced this method so as to apply it to the wiring for the three layers.

3.1.2 Chip design

Step 1: Relative positioning of blocks.
In accordance with the logical structure of the chip, such as the data/control flow, the relative location between blocks is decided by a designer.

Step 2: Determination of absolute coordinates.
Strict coordinates for blocks on the chip are calculated through a channel capacity estimation program, under constraints of a given chip size.

Step 3: Block assembly partitioning.
A chip is partitioned into block assemblies if concurrent design is necessary, or CPU resources are limited. In this case, the coordinates for points shared jointly by neighbouring assemblies are decided by an edge generation program.

Step 4: Automatic routing.
Each block/assembly edge pin is wired automatically. The algorithm is a maze router.[5] Our maze router can wire not only the block/assembly edge pins, but also the equi-potential paths existing in the previously routed blocks. Such a capability is effective in increasing wireability.

Step 5: Completion of wiring.
Manual routing is performed, using a digitizer, if some portions can not be wired automatically.

3.2 Application to master-slice LSI

For master-slice LSIs, the block and cell model shown in Figure 1 is to be modified partially. The main portions of the modification are as follows.
(1) Two metals (1st and 2nd) are used for connecting cell terminals in a block and Poly-Si is only used for intra-cell wiring.
(2) The number of 1st metal channels between cell rows is fixed before the layout design.
When the scale of the layout is not so large as 10K gates and the computer resources are sufficient for designing such a object, it may be considered that a chip is composed of only one block. In this case the chip design is equivalent to the block design. By only excepting Step 2 of routing process described in 3.1.1, the placement and routing procedure for the block design can be employed for the master-slice LSI. Moreover, if unconnected pin-to-pins are left as the result of the channel routing, the remainders are wired by using the maze router. This combination of two kinds of routing algorithm is effective to perform a high speed and high wireability processing.

4. Algorithm for iterative improvement

The iterative improvement procedure for cell rows is summarized in the following procedures.

"Repeat Steps 1 to 3 until all cells in the block remain at their own location forever."

Step 1: Select a cell row.

Step 2: Compute the optimal location for a specified cell.

Step 3: Make the cell move to its new location only if this will cause a decrease in the sum of the estimated wire length.

The objective function which was employed in order to compute the optimal location has ever been formalized in the following manner.[6]

<Problem 1>

Let { a_i |i=1,N } be a given set whose elements consist of x number of coordinate values of the pins of all nets connected to cell "X" and let $g(x) = \sum |a_i - x|$ be a distance function when cell "X" is located at point x. Then, derive solution x0, which will make g(x) minimum.

Solution x0 is:

$$x0 = a_{\frac{N+1}{2}} \text{ (N is odd) or } x0 = a_{\frac{N}{2}} \text{ (N is even),}$$

where { a_i } is sorted in ascending order.

It is evident that x0 provides an optimal location for cell "X" if the routing path is radial, as shown in Figure 2(A). However, the "channel router" makes a "Steiner tree" as shown in Figure 2(B). Therefore, it was concluded that g(x) is not always a suitable function for the channel router. Thus, we introduce a new function.

<Problem 2>

Let $a_1, a_2, \ldots a_N : a_i > 0$ be N points (=pins) on the x-coordinate, and let these points be partitioned into M groups, $S_1, S_2, \ldots S_M$ (=nets). Also, let $l(S) = \max(S) - \min(S)$ be a mapping function for set S, where max(S) and min(S) are maximum and minimum values for the elements of set S. Then, arrive at a solution, x0, which makes $f(x) = \sum l(S_j \cup \{x\})$ minimum, when cell "X" is located at point x.

The solution is:

"x0 is a median point of sequence $p_1, p_2, \ldots p_{2M}$, which is sorted in ascending order, where p is an element of set $P = \{\max(S_j), \min(S_j) | j=1,M\}$". (*1)

This solution represents the fact that the location is optimal where the number of "right edge nets" equals the number of "left edge nets" with respect to cell "X". Here, right/left edge nets are such that cell "X" occupies the right/left-most location in them, as is shown in Figure 3. We have named this method "net balance placement".

(*1) The description of the proof is omitted due to the limit of the paper volume.

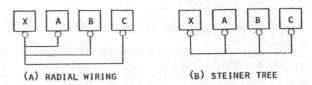

FIG. 2 WIRING PATTERN TO OBJECTIVE FUNCTION

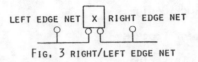

FIG. 3 RIGHT/LEFT EDGE NET

5. Results of layout design

5.1 Evaluation of block design

Table 1 compares block size between manual and automatic layout. Column (A) is the value using the function f(x) --Net Balance Placement--, and column (B) is that for g(x) as described in Section 4.

Table 1 Comparison of block size

	Manual	Automatic	
		(A)	(B)
	mm²	mm²	mm²
BLK1	2.02	1.86	1.87
BLK2	1.29	1.13	1.16
BLK3	1.09	1.19	1.22
BLK4	1.05	1.08	1.09
BLK5	1.72	1.56	1.56
Total Ratio	100%	95%	96%

As shown in Table 1, the Net-balance placement described above results in being more effective for reduction of block size.

Figure 4 shows the computing time necessary for block design. In our program, placement time (Tp) and routing time (Tr) have the following relation to the approximate number of gates.

$$Tp \sim G^{1.4} \qquad Tr \sim G^{1.2}$$

In the case of 5K gate block, Tp=102 sec and Tr=150 sec, on a 3 MIPS computer.

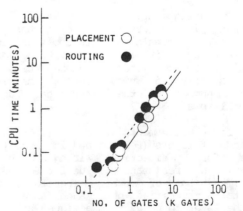

FIG. 4 PROCESSING TIME FOR AUTOMATIC LAYOUT

5.2 Example of chip design

Figure 5 shows an example of a VLSI including 74K transistors designed by using this automatic layout system. This VLSI is a part of the CPU of small/medium scale computers, and is mainly composed of 32 bit ALU, registers and some control units.

Design features of the chip are:

Chip size	12mm x 12mm
No. of transistors	74,000
No. of random gates	17,000
No. of cells	5,700
No. of blocks	40
RAM	2k bits
No. of package pins	200
Technology	2micron CMOS

The total CPU time for routing blocks on this chip (partitioned into 4 block assemblies) was 256 minutes, and the number of un-connected pin-to-pins was 9, where the sum of pin-to-pins to be connected is 2726. The channel utility was 49.1% at 1st metal and 47.0% at 2nd metal, respectively. The total length of wiring in the chip was 10,892mm. The average wiring length was 0.46mm for the intra-block net, and 7.6mm for the inter-block net. This chip was fabricated with no error on the first trial.

Figure 6 is another example of automatic design for 6K gate master-slice LSI.[7] The computer processing time for the whole layout design of this chip was 28 minutes. The average net length is 1.4mm and more than 90% of the nets are wired within a half length of one side of the chip. For the present, this layout system has been applied to several kinds of master-slice LSIs of which gate utility is more than 90%, and full automatic wiring (no un-connected pin-to-pins) was performed in every case.

6. Conclusion

A hierarchical method for designing the layout of a VLSI chip, and automatic placement and routing programs have been described. In this system, the chip can be partitioned into an arbitrary number of block assemblies for concurrent layout design. This makes it possible to reduce design time. Also, a new objective function has been presented for improving iterative placement, which is suitable for wiring patterns made by channel router. By applying these programs to both custom VLSI (74K transistors) and master-slice LSI (6K gates) design, we have confirmed their effectiveness.

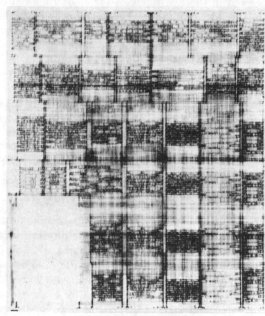

FIG. 5 EXAMPLE OF LAYOUT DESIGN
(74K TRS. CUSTOM VLSI)

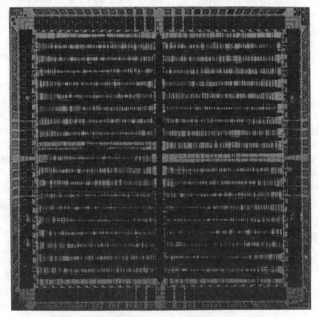

FIG. 6 EXAMPLE OF LAYOUT DESIGN
(6K GATE MASTER SLICE)

7. References

1) Y. Ohno, et. al. ; Integrated Design Automation System for Custom and Gate Array VLSI Design, Proc. of Conf. ICCC ('82)

2) T. Adachi, et. al. ;Hierarchical Top-down Layout Method for VLSI Chip, Proc. of 19th DA Conf. ('82)

3) D. M. Schuler, et. al. ; Clustering and Linear Placement, Proc. of 9th DA Workshop ('72)

4) T. Kozawa, et. al. ; Advanced LILAC-an Automated Layout Generation System for MOS/LSIs, Proc. of 11th DA Workshop ('74)

5) R. Kamikawai, et. al. ; Placement and Routing Program for Master slice LSIs, Proc. of 13th DA Conf. ('76)

6) T. Kozawa, et. al. ;Block and Track Method for Automated Layout Generation of MOS/LSI Arrays, Proc. of ISSCC ('72)

7) T. Itoh, et. al. ;A 6000 Gate CMOS Gate Array, Proc. of ISSCC ('82)

Wire-Routing Machines—New Tools for VLSI Physical Design

SE JUNE HONG, FELLOW, IEEE, AND RAVI NAIR, MEMBER, IEEE

Invited Paper

Abstract—Interconnection of components in a VLSI chip is becoming an increasingly complex problem. In this paper we examine the complexity of the wire routing process and discuss several new approaches to solving the problem using a parallel system architecture. The machines discussed range from compact systems for highly specialized applications to more general designs suited for broader applications. The process speedup due to parallelism and the cost advantage due to the use of large numbers of identical VLSI parts make these new machines practical today.

I. INTRODUCTION

THE PROCESS of designing a chip becomes more challenging as the number of components that it can accommodate increases. Many design automation problems, e.g., logic synthesis, testing, partitioning, placement, and wire routing, are known to be NP-complete or worse in complexity. The optimal solution to these problems requires computation times which could grow exponentially with the number of components. Practical algorithms, therefore, use heuristic techniques with polynomial complexity which lead to near-optimal solutions. Unfortunately, in most aspects of VLSI design, the heuristics for even these suboptimal solutions involve polynomials with degrees higher than 1 that make them prohibitively more costly to apply as the number of components increases.

Physical design of a chip refers to the process of placing the component elements on the chip and determining wire routes to interconnect the terminals of the components. A component may occasionally be a large functional macro (ALU, PLA, ROM, RAM, register, etc.) with terminals on the periphery, a logic gate/circuit (NAND, NOR, driver, flip-flop, etc.), or even an isolated transistor or resistor. Integrated circuits today have components ranging up to a few hundred thousand transistor equivalents in custom designed chips or up to ten thousand logic gates in *gate array* or *master-slice* chips.

The interconnections of terminals are made through one or more wiring planes where wire tracks generally run in either horizontal or vertical direction on alternating layers. A physical wire changes direction usually by means of a *via* connecting the two wiring planes at the intersection of a horizontal wire track with a vertical one. The density of wires today approaches several hundred tracks per millimeter, implying thousands of wire tracks on a chip of size 5 mm × 5 mm. The number of grid points (via sites) defined by such dense wire tracks on a chip can be well over a million.

Manuscript received May 10, 1982; revised September 28, 1982.
The authors are with IBM Thomas J. Watson Research Center, Yorktown Heights, NY 10598.

The active devices and wires compete for space on the chip surface, although multilayer wiring may alleviate this problem. It has been observed that wiring can occupy more than half of the chip area. As the device size becomes smaller, the wiring space would dominate the space needed for the devices by far. A good placement and wiring technique that minimizes the number of wire tracks used is essential for the VLSI chip density, except for highly repetitive structures with regular interconnections such as PLA chips, memory chips, or arrays of identical processing elements. Placement of components and wiring of interconnections between them are conventionally carried out in successive phases for computational convenience. In the "custom" approach which occupies one end of the VLSI design spectrum, the process of physical design also determines the final dimensions of the chip. It is desirable in this case to closely couple the process of positioning the components on the chip with the routes of the interconnecting wires. Often, in order to reduce the number of parameters involved in the optimization process, the positions of the active area of devices and service terminals are fixed before the routing is done. On the other end of the spectrum is the "master-slice" approach where the size of the chip is fixed, components occupy prespecified positions, and wires are limited to predefined tracks. In such cases, not much is sacrificed by separating the processes of placement and routing, provided appropriate wireability measures are considered during the placement. In discussing the wire routing process in this paper we will assume that the precise physical locations of the points to be interconnected are known.

Consider a master-slice chip as shown in Fig. 1. The number of horizontal and vertical tracks associated with a gate/circuit cell is fixed by the chip designer guided by some theoretical analysis of wire demands [1]–[3]. For a given size of chip area, the more the components, the less the available area for wiring. It has been observed that in designing these master-slice chips the channel demands are generally underestimated, at least in certain areas of the chip. This puts a strain on the physical design automation programs especially when the logic designer expects to use most of the active devices on the chip. It is not unusual then to find systems which specify that only a certain fraction (between 50 and 90 percent) of the active devices on the chip should be used by the logic designer. It is also not unusual that a small fraction of connections remain unconnected, as *overflow wires*, by the automatic wiring programs for LSI chips. The overflow connections are then manually routed by rerouting some of the existing wires if necessary, which may take days, weeks, and sometimes months.

Reprinted from *Proc. IEEE*, vol. 71, pp. 57–65, Jan. 1983.

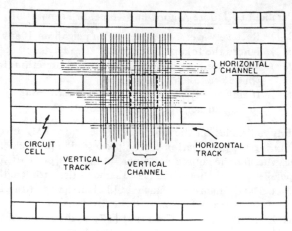

Fig. 1. A typical master-slice chip.

```
                5
            5   4   5
        5   4   3   4   5
    5   4   3   2   3   4   5
5   4   3   2   1   2   3   4   5
    5   4   3   2   3   4   5
        5   4   3   4   5
            5   4   5
                5
```

Fig. 2. The waves of forward propagation.

If the current wire routing programs were employed for complex VLSI chips, the hours of computation that it would require would be intolerable. Furthermore, the percentage of overflows is not likely to decrease. Even 1 percent of interconnections for a ten thousand gate chip amounts to hundreds of wires. Manual embedding of even a single wire segment becomes progressively more difficult as the chip size gets denser and larger. Therefore, what is required for VLSI chip wiring is a new technique that is both fast and reduces, or preferably eliminates, overflows.

There are two approaches to meet this challenge. One is to develop better serial algorithms to run on bigger and faster conventional mainframe computers. The other is to construct a parallel processing machine with effective parallel algorithms. The latter seems more promising on two accounts. 1) The complexity of serial computation is too high even for the fast new mainframes. 2) The economy of VLSI chips, especially that of microprocessors, makes it cost effective to build a large parallel machine.

Before we discuss these special-purpose machines, let us examine some of the characteristics of wire routing problems in more detail in the next section. We refer to Soukup [4] for a comprehensive overview of current practices in VLSI physical design in general.

II. THE WIRE ROUTING PROBLEM

The signal lines in the schematic diagram of a logic network are now to be embedded as connected wires along the given wire tracks on the chip. (Many of the concepts to be discussed apply not only to chips but also to higher levels of packaging like cards and boards.) The grid of wire tracks in a chip contains numerous blockages due to internal wiring of the components (component personalization). Wires of different nets obviously cannot occupy the same wiring space. Further, technology rules often limit the position of vias adjacent to one another or even the routing of wires on adjacent tracks in certain directions. These restrictions transform the wire routing problem to an allocation problem with limited track resources and complex constraints. The quality of routing is measured primarily by the overflow count. Secondary considerations include the total wire length, the number of vias used, and the maximum length of a net.

An important and often employed simplification is to find a shortest wiring path on the rectilinear grid between two given points. The length of a shortest path may be and often is longer than the rectilinear distance between the two points because of blockages. The basic technique by Moore [5] and Lee [6], commonly referred to as the Lee–Moore (LM) algorithm can be informally stated as follows. For simplicity of argument we take a single wiring plane. The technique can be adapted easily to situations where multiple planes are involved. Let one of the points be called the source and the other the sink. We make use of two lists of nodes called OLD and NEW and two status markers per node for recording whether the node has been visited, and if so from which direction.

Lee–Moore Algorithm

LM1) Initialization:
   Mark source node as visited
   OLD ← source node.

LM2) Propagation: Starting with empty NEW list
   For each node in the OLD:
      new ← the neighbor nodes of the current node that are not visited and not blocked.
      Mark new nodes as visited and the direction visited from.
      If sink ∈ NEW, go to LM3, else append new to NEW
   Let OLD ← NEW (if NEW is empty, the path does not exist, i.e., an overflow)
   GO to LM2).

LM3) Backtrace: Starting from the sink node, follow the directions noted on the nodes to the source node.

This description of the shortest path finding algorithm is not unique and many variations of the above have been implemented. The important aspects are as follows: 1) the propagation step, LM2, is iterated p times where p is the shortest path length, and 2) the time taken for the backtrace step, LM3, is linear in the length p, but processing at each step during the backtrace is much simpler than that during the propagation.

During the propagation iteration, nodes are processed in a wave-like fashion where each NEW list represents a fresh wavefront. When there are no blockages, each node within the diamond shaped area with rectilinear radius p must be processed once for propagation. Fig. 2 shows the nodes labeled by the numbered iteration of the propagation step, LM2, from a given source, assuming no blockages. The wire length p is easily identified as the total number of times LM2 is invoked.

The number of nodes within the diamond area of radius p is asymptotically $2p^2$ and hence a serial computer takes computation time that grows as the square of p. Of course, if the chip is known to have no blockages, a greedy progression from source to sink would take just $p = q$ steps where q denotes the rectilinear distance between the two points. A routing tech-

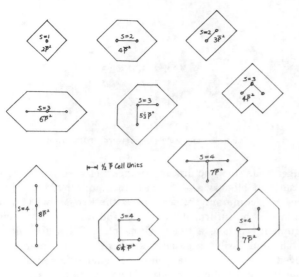

Fig. 3. Examples of sources that are wire segments and their forward propagation areas.

nique based on this greedy approach, called the line probe technique, performs well when wire routes are not constrained by existing blockages or by wires already routed on the chip.

Many techniques [7]-[11] have been developed to reduce the number of nodes processed in finding a path, of which the line probe technique is just one example. Reduction is often accomplished by expanding only within the minimum rectangle area[1] or by target directed depth first expansion. While these techniques tend to reduce the average processing time, the worst case computational complexity remains $O(p^2)$.

The basic Lee–Moore algorithm can be extended to find a shortest path between one piece of wire to another simply by marking initially all nodes representing one wire as sources and all nodes of the other as sinks. The propagation waves in this case emanate from the entire length of the source wire. Suppose a wire source contains s of the t terminals in a net already connected. Assume, for simplicity, that all connections are of length $p = \bar{p}$. The total area within distance $\bar{p}$ of such a wire of length $(s - 1)\bar{p}$ ranges between $(s/2)\bar{p}^2$ to $2s\bar{p}^2$ depending on the wire configuration. Fig. 3 shows some of the configurations and the corresponding areas. It can be seen that if one uses the method of wiring a t terminal net by wiring to one additional terminal at a time from a partially connected wire [7], the total serial time complexity becomes $O(t\bar{p})^2$. The same can be accomplished in $O(t\bar{p})$ time using a parallel machine.

Consider a mesh connected complex of simple processing elements, one per track intersection of the chip being wired. Suppose each processing element can perform the following basic operations (among some other equally simple operations) in parallel:

[1] Minimum rectangle refers to the smallest rectangle enclosing the given nodes. Given two points distance q apart, it can be enclosed in minimum rectangle of size $a \times b$ where $q = a + b$. If propagation is restricted to this area, the average number of nodes processed would be asymptotically

$$\frac{1}{q}\sum_{a=0}^{q} a(q - a) = (q^2 - 1)/6 \approx \frac{1}{6}q^2$$

in contrast to $2q^2$ of the full diamond area. However, in the presence of blockages, a path may not be found inside the area or the path found may not be of the shortest length.

Parallel Lee–Moore Algorithm (Kernel of Propagation)

PLM1) Receive wavefront tokens from neighbors, if any.

PLM2) Ignore the token if blocked or already visited.

PLM3) Mark itself visited and mark the direction from where the token was received.

PLM4) If sink node, signal halt, else send wavefront tokens to all four neighbors simultaneously.

Clearly, all nodes on the wavefront (the entire LM2 invocation) during the propagation can be processed in parallel. This is true whether the source is a single point or a piece of wire. The parallel propagation time is the same as the path length p. The Breuer and Shamsa's L-machine [12] designed at the University of Southern California and the SAM machine proposed by Blank, Stefik, and van Cleemput [13] at the Stanford University both implement the parallel Lee–Moore algorithm in efficient hardware. These machines will be discussed in more detail later on.

Let us now return to the wire routing problem with all the constraints mentioned before. It cannot be solved by a series of path finding steps alone. One approach is to judiciously employ a fast path finding machine as a subroutine in conjunction with a more sophisticated routing technique using a conventional computer as the main processor. Another approach is to route each wire more carefully with due considerations from the beginning to the wire congestions on the chip. That is, when a path is to be found, the goal is not an arbitrary shortest path, but the path with the best cost among all acceptable paths. Here the cost may include an estimate of the congestion, number of available wire tracks in the local area, the length, etc. Other heuristic measures that help steer the path to the "right" one in the overall wire allocation task can also be utilized. Because the cost measures are of heuristic nature, it may be necessary to have rip-and-rerouting feature in any VLSI automatic routing schemes to get rid of overflows. The Lee–Moore algorithm addresses a special case of minimum cost path finding: namely, uniform cost per unit length over the whole area. The algorithm employs a breadth-first search technique which is efficient for uniform costs. (The labels of nodes in Fig. 2 also represent the partial path cost from the source.) Akers [7] presented a modified algorithm that finds the minimum cost path in rectilinear geometry with arbitrary nonnegative cost distribution. His algorithm is a mixture of breadth-first and depth-first search where the propagation initially spreads out like the Lee–Moore algorithm until the sink is reached. The propagation then continues only if some partial path cost is less than the current best cost seen at the sink node. The serial complexity of this scheme again remains the square of the path length. The path length for this case is usually longer than the shortest path length p which, in turn, is greater than or equal to the distance between the two points.

The best cost path algorithm requires much more complex processing per node. Besides, it is extremely difficult to compute a reasonable cost function for track-to-track fine resolution. An effective strategy is to attack the problem in two phases: divide the chip into an array of cells, where each cell contains a reasonable number of horizontal and vertical wiring tracks. During the first phase, routes are determined in cell resolution, this operation is called *global wiring* [14], [15], or *loose wiring* [16]. The number of wiring tracks at the boundary between two adjacent cells is called the channel capacity. Global wiring ensures that all wires are routed without violating the channel capacities of the cells. The second phase deter-

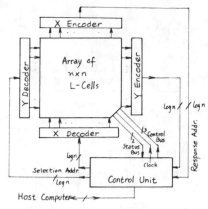

Fig. 4. An overview of the L-machine.

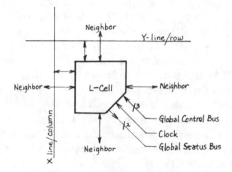

Fig. 5. I/O signals of an L-Cell.

mines the fine track routes within the global route. This is called *exact embedding*. This division into cells comes naturally in the case of master-slice chips. Similar to the machines alluded to earlier, a mesh connected structure of more general-purpose processing elements (microprocessors), one per cell of the chip being wired, can be employed to carry out these two phases efficiently. The Wire Routing Machine (WRM) [17] and associated wiring algorithms being developed at IBM Thomas J. Watson Research Center is aimed at this and other broader applications.

III. THE L-MACHINE

Breuer and Shamsa's L-machine [12] is the first published design of this nature. The L-machine consists of a control unit that communicates with a host computer and sequences the operations of an array of simple processing elements, called L-Cells (see Figs. 4 and 5). This is expressly designed to implement the parallel Lee–Moore (PLM) algorithm. The L-Cells are simple (about 75 gates) and hence many of them can be laid out on a VLSI chip to form a subarray. The machine is capable of performing the following tasks:

1) Initialization: This involves the loading of source, sink, and blockages information into the L-Cells.

2) Parallel Propagation: This essentially implements the PLM described earlier. In addition a BUSY-status signal is raised by those L-Cells which are active during a step of processing the wavefront. The controller receives the wired-OR of BUSY signals from all the L-Cells. Thus the BUSY signal is high as long as some cell is active. If the controller sees the BUSY signal go down before the sink is reached, an overflow is indicated.

3) Backtrace: This process determines the wire path by following the stored direction flags from an activated sink back to the source. The X, Y coordinates of each node on the path of the wire are output by the machine.

4) Clear: Cells along the backtraced wire path are marked as blockages for subsequent wire routing. The internal status of all other L-Cells is cleared to an idle state.

Each L-Cell communicates with its four neighbors through bidirectional lines, one per neighbor. These lines are used during the propagation and the backtrace. In addition there are 7 more signal I/O's and a clock input line per cell:

a) Global Broadcast Bus (3 bits of control signals) from the control unit to all cells in the array.

b) Global Status Bus (2 bits) from all cells to the control unit in wired-OR lines.

c) X and Y (1 bit each) Select/Response wired-OR bidirectional lines along each column and each row of L-Cells.

The control signal lines are used to sequence the loading of blockage, source and sink status (in conjunction with X, Y selection), the forward propagation, the backtrace, and clearing operations. The two status bits inform the control unit of the busy status of the wavefront, whether the sink is reached during the propagation phase, and whether the source is reached during the backtrace phase. The X and Y select/response lines are used for communicating the cell addresses between the control unit and the array. The control unit selects a specific L-Cell, through the X and Y addresses of the cell. These addresses are decoded by X and Y decoders, the outputs of which activate the appropriate column and row signal lines. When an L-Cell raises its X and Y lines the column and row encoders translate these signals to X and Y addresses for the control unit or for other I/O lines of the system.

The machine is capable of processing an entire wavefront of propagation in one clock cycle. Contrast this with tens of instructions necessary to process just one node of the wavefront for the Lee–Moore algorithm in a conventional serial computer. For two wiring planes, the number of L-Cells double and each L-Cell has five neighbor connection lines. To accommodate the two wiring planes in three dimensions, the authors propose to interweave rows of L-Cells of top and bottom layers in a plane. While the arrangement is perfectly capable of finding a shortest path, the current design can neither accommodate preferred wire directions in the wiring planes nor treat vias as anything other than one unit of length, should such technology constraints exist.

For an $n \times n$ track wiring surface, the wiring array contains n^2 L-Cells each of which contains about 75 gates. The total number of pins of the array including two power pins and the row and column decoder and encoder is $4 \log_2 n + 8$ (this can be reduced to $2 \log_2 n + 8$ if the row and column select/response address signals share one bidirectional bus). Breuer and Shamsa envision using one VLSI chip for the entire array of size ranging from 64×64 (300K gates 32 I/O pins) to 256×256 (5M gates and 40 pins) depending on the available VLSI chip technology. In general, when the problem size is in the range of a VLSI chip or of a large board (say 1024×1024), a single chip array is not realizable. The array must then be partitioned into subarrays, each of which is a VLSI chip. One subarray of size $m \times m$ would have $75m^2$ gates for the L-Cells. Due to their desire to incorporate the decoders and encoders, the authors resort to the following three chip types; a corner chip, a side chip, and a mid-array chip. The corner and side chips realize portions of encoder and decoder structure besides the subarray. The mid-array chip now needs $4m$ neighbor lines, $4m$ X, Y lines plus six global wires.

To reduce the pin counts, Iosupovicz [18] suggests a scheme involving serial transfer of information between subarray boundaries. That is, all neighbor communication between a subarray boundary are serially transmitted by encoding the active cell positions, and then passed on to the next chip which, in turn, serially decodes the correct neighbor cell position on the boundary. A substantial reduction in the chip pin count, as well as a complete modularity of subarray chips is accomplished at the expense of lengthening the operation of the machine. The processing time is still more or less proportional to the length of the path.

The array size of the L-machine must be as large as the wiring surface (twice as large for two-layer wiring). If a large enough machine is built, perhaps all practical problems can be handled. But no problem larger than the physical size of the machine can be processed. The L-Cell processor is compact and very fast. However, it is inflexible and limited to a very narrow field of applications; namely, finding a shortest path between two points.

IV. THE SAM-MACHINE

The machine proposed by Blank, Stefik, and van Cleemput [13] is also aimed at a compact design suitable for subarray packing in a VLSI chip. One main difference between this machine and the L-machine is that this was designed with a somewhat broader range of applications in mind. Examples include most of the bit map oriented spatial processing algorithms such as the Lee–Moore algorithm, image processing algorithm, and many elements of design rules checking. This emphasis is expressed in the name given to the machine as Synchronous Active Memory (SAM) machine. The node processor is called a SAM-Cell. It supports 20 assembly level instructions operating mainly on the data width of 1 bit. Similar bit-oriented parallel computing structures had been proposed earlier by Unger [19] and more recently by Reeves [20] and Duff [21].

A SAM-Cell consists of a local control unit, a 2-bit Boolean logic unit, a multiplexor feeding a 1-bit accumulator, a neighbor masking unit, and 16 1-bit registers, as shown in Fig. 6. A cell is enabled by X and Y selection lines from the array control logic. A global data line, five register address lines, and five control lines also enter the chip from the control logic. The cell communicates to each of its four neighbors through four separate bidirectional lines which are not explicitly shown in the figure. The global data output line connects to a wired-OR bus running through all the SAM-Cells. There is a total of 18 signal I/O lines per cell. The first prototype design used approximately 350 transistors in $900 \times 900\ \mu m^2$ in NMOS technology.

The SAM-machine operates by the controller broadcasting each instruction which every cell obeys.[2] The program control and storage can be provided either by the host computer or by the SAM system depending on implementation. The SAM array control logic works as an interface between the host and the array.

One of the most powerful SAM instructions is called NEIB, which masks the 5-bit data (4 neighbor input and self data in the accumulator) by the 5-bit address and ORs the result into the accumulator in one clock cycle. This instruction enables a fast implementation of node processing for the Lee–Moore propagation. The authors programmed a variant of propagation labeling technique known as the 1–1–2–2 scheme (see

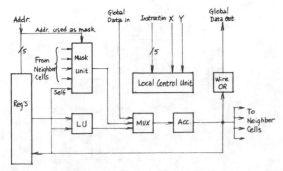

Fig. 6. A SAM-Cell overview.

[7]), where successive wavefronts are labeled 1 twice and 2 twice alternately.

They envision packaging a 16×16 subarray of SAM-Cells on a VLSI chip. To process a 1000×1000 grid wiring, it would require about 4000 such chips in the SAM system. A more cost-effective way of realizing such a large machine is to use a folding technique as proposed by the SAM and as practised with the WRM to be discussed later. In folding, the chip area is divided into the machine size frames and the machine serially processes each frame. In such a scheme, the run time and the memory size of node processors will increase, but the fixed size machine can act as a virtual machine of much larger size, limited only by the node memory capacity. Some of the overhead for this flexibility are

1) optional but worthwhile addition of end-around neighbor connections between the opposing edges of the array;
2) software overhead of managing frame by frame computation;
3) much of the communications may require tags to indicate the frame identification, especially the array boundary cells communicating across the frame boundary to their (folded) neighbors.

The SAM proposal chose to fold actually like an accordion without the end-around neighbor connections. As a consequence, the cells at the boundary of the array have themselves as neighbors across the frame, and the neighborhood directions change to their mirror images between adjacent frames. For the SAM virtual machine design, the SAM-Cells are modified by replacing the accumulator and the registers with 1K and 16K RAM, respectively. This amount of memory would allow the machine to process problems 1000 times larger in size.

The speed advantage of the SAM-machine is illustrated by the following run-time estimate. Unlike the L-machine, it now takes many clock cycles to process one wavefront (hard-wired L-Cell versus assembly level instructions in SAM-Cell). Assuming 300-ns instruction cycle time (100 ns for virtual machine), an estimate is made for routing 1000 two-point nets on a 512×512 grid wiring surface, with average connection length of $\overline{p} = 200$ units. (M refers to the time for initial loading of the problem estimated between 1 and 5 s.)

Machine	Run Time
1K × 1K full size SAM	0.4 s + M
32 × 32 virtual SAM	15 s + M
Conventional computer	5 h

Although it was not specified in [13] which conventional computer was used for comparison, the contrast is clear. The

[2] This mode of parallel processing is called the Single Instruction Multiple Data (SIMD). On the other hand, if each processor node follows its own instruction stream, it is called the Multiple Instruction Multiple Data (MIMD) [32].

same example on a more specialized L-machine with the same clock cycle would take only 0.12 s + M.

Conceptually, the SAM machine can be used for general-purpose applications. However, the limited instruction set, small data width (1 bit), and SIMD operation together restrict the effective range of other applications to bit-map problems arising in certain image processing, bit-vector operations, and simple design rules checking.

V. THE WIRE ROUTING MACHINE (WRM)

The machine described by Hong, Nair, and Shapiro [17] was originally conceived for two-phase wire routing strategy (global wiring and exact embedding). It is also a mesh-connected multiprocessor complex, as the two machines just described. The major difference in the design philosophy between this and the others lies in the power and scope of the node processing element. It is clear that even for the path finding problem alone, the compact L-Cells and SAM-Cells are not flexible enough if there are complex technology constraints on the wire routes. How general and powerful should the node processor be may be a difficult question to answer. Our feeling is that the processing elements must support almost the full range of a general-purpose instruction set and perhaps even some additional special hardware enhancements. The total speed gain of the machine comes not only from the parallelism but also from hardwired instructions for often used kernel operations (for example, NEIB instruction of the SAM-Cell, or even the whole L-Cell mechanism). Each node of WRM requires one or more VLSI chips or even a circuit board for implementation, trading versatility for compactness.

The WRM system consists of three major parts: the *processing array*, the *control processor*, and the *array control unit* that interfaces the array and the control processor. The control processor connects to the host computer, disk units, console, printer, and the array. In the experimental machine at IBM Thomas J. Watson Research Center, the processing elements are implemented using commercial microprocessors. The machine was first constructed in 1979 and has been upgraded in memory and the disk capacity during the past two years.

All communication within the system occur by a memory map mechanism invoked by the microprocessors. All I/O operations and some control operations are performed by accessing special addresses in the memory. The node processors in the array operate in a slave mode to the controller. The controller can select any subset of rows and columns (X, Y selection lines) of node processors and communicate with them. Besides the global I/O ports, the controller has Direct Memory Access (DMA) capability to all node memories, for fast bulk loading, unloading, and broadcasting.

The node processor has a *local neighbor bus* to communicate with neighbors. It has individual *strobe* lines to each neighbor, which latch the local neighbor bus contents to the appropriate neighbor's *receiver latch*. The action of strobing a neighbor also sets his *neighbor flag*. The neighbor's action to read the latch resets the flag. This mail box flag protocol provides for an asynchronous communication. A similar protocol arrangement is made for global port communication. It is envisioned that a more ambitious design would have individual neighbor output ports as well as banks of neighbor input buffers. Fig. 7 shows the I/O ports of the node processor. Global communication is enabled by the X, Y selection mechanism.

An overview of the node processor is shown in the Fig. 8. The local control logic intercepts all memory mapped special

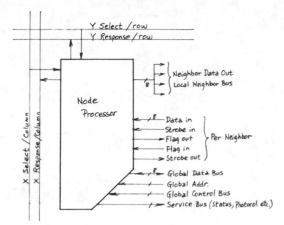

Fig. 7. I/O signals of WRM node processor.

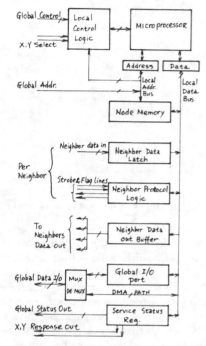

Fig. 8. An overview of WRM node processor.

addresses and controls the local buses, I/O buffers, and the service status register which includes the X, Y reponse bits. The node memory capacity in the current machine, capable of folding operation, is 15K bytes. The whole node processor, except the local memory, can be realized in one VLSI special processor chip.

The machine operates mainly in MIMD fashion, although identical programs are loaded into all the node processors. The node processor programs are sufficiently complex and general so that depending upon the data, the node status, and the node position, each node often processes different program sequences. An SIMD machine would lose the advantage of parallelism in such applications.

During the global wiring process the nets are wired one at a time. For each net, one of the terminals are designated as the initial source, and all the rest of the terminals are designated as sink cells. The WRM finds the optimum cost route from the source to a sink cell within a *deferential detour limit*, that is, the allowed detour length from the shortest possible connection. The connected wire now becomes the source, and the

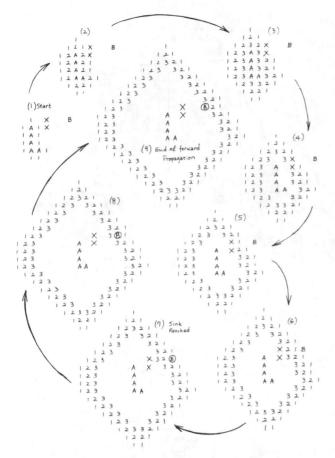

Fig. 9. An example of forward propagation in WRM.

process repeats. Each time, a connection is made to the lowest score sink cell that is within the detour limit D from the nearest one. Similar to the Lee–Moore process, a connection is established by forward propagation and backtrace. The propagation of partial path costs emanate out from the source in a wave fashion. This forward propagation differs from that of simple Lee–Moore in that the active nodes are not just the thin wavefronts but a band of cells of width D. The computation that is required for each node in the active propagation zone is quite a bit more complex also (see [22] for details).

The advantage of parallel processing in the WRM can be illustrated by the following small example. Fig. 9 shows the forward propagation at each time step of parallel operation. The numbers in the cells denote the number of times the cell was processed for the detour limit of 2. In the example, a source cell is denoted by A a sink cell by B, and a cell having no wiring tracks at the boundary by X. The shortest connection length between A and B is 7 cell units. For the given detour limit of 2, the parallel forward propagation takes nine steps as shown in the figure. However, the total number of nodes processed during these nine steps sums up to 602, which will be the number of steps required for a serial processor. The speed advantage is even higher for a large detour limit or for a larger body of source wire segment.

Let us now consider the computational complexity of the global process. For simplicity, assume that all connections made are of the same length $\overline{p}$ and that all nets have the same number of terminals $\overline{t}$. Furthermore, assume that the detour limit is given as 0, that is, the forward propagation proceeds as in the Lee–Moore case except with more involved node processing.

For one net, the complexity is of $O(\overline{tp})^2$ as discussed earlier. The average number $\overline{t}$ of terminals per net is in the range of 2 to 4. The average connection length $\overline{p}$ is a slow growing function of N, the number of circuits, the number of nets, or the size of the chip, all of which are proportional quantities. Feuer derives [23] that $\overline{p}$ is proportional to $N^{1/6}$. Therefore, the serial complexity for N nets would be $O(N^{4/3})$.[3] On the other hand, the parallel processing complexity is $O(N^{7/6})$. The difference here seems small. However, if the detour limit is about the same as $\overline{p}$, the serial complexity increases to $O(N^{3/2})$ while the parallel complexity remains the same. We shall discuss later how the parallel complexity can be reduced even further. Additionally, the simplifying assumption of identical connection length $\overline{p}$ causes an underestimation of the serial complexity, which is not the case for the parallel complexity because each connection takes linear time in p, the actual connection length.

Often the quality of wiring can be improved by iterating on the global wiring already obtained. In some dense examples iteration is a powerful measure to remove overflows. The number of iterations can be a constant so as to not change the complexity of the computation. The need for a high-speed wiring process is more acutely felt when iterative procedures are used.

The exact embedding algorithm currently implemented on the WRM has two major components, *cell boundary assignment* and *cell embedding*. The nets globally routed through the cells are first assigned to exact tracks at all cell boundaries considering the terminal locations and eventual destinations. This process takes $O(N^{1/2})$ time on the WRM, mostly involving sweeping of various information in two orthogonal directions. The remaining task is for each cell to honor the assigned boundary and the internal terminals of the net. This represents a small area (at most about 16 tracks in each direction) wiring problem with small number of internal nets (typically less than 10). During the cell embedding process each node processor wires its own cell in track-to-track resolution considering all the technology restrictions. It is during this cell embedding that the WRM achieves its maximum parallelism, i.e., all node processors are busy.

The degree of restrictions imposed by the technology varies greatly from chip to chip. It is not uncommon to find a master-slice chip that necessitates some representation of restriction status for every track intersection and every track interval. Hence, minimally, the entire length of an embedded wire segment must be examined at every track interval. For an average net of length $(\overline{t}-1)\overline{p}$ cell units, the length is multiplied by the number of tracks per cell in one direction (assuming square cells). The total serial complexity for the exact embedding is then bounded from below by $O(N^{7/6})$. The WRM complexity for the same depends on the serial time taken by a node processor to do the cell embedding. The number of nets that has to be internally wired is of the order $\overline{p}$, estimated roughly by dividing the total net length in cell units, $N(\overline{t}-1)\overline{p}$, by the total number cells, N. We conclude that the parallel com-

[3] $O(N^{4/3})$ is higher than some empirical complexities obtained from experiences on wiring programs run on conventional machines. For instance, Feuer *et al.* [24] reports $O(N^{1.1})$ for combined global and exact embedding. The algorithm they use does not seek the best cost route during global phase as does the algorithm used in the WRM. The serial algorithm uses many techniques to reduce the average computation and the result of wiring may not necessarily be inferior to the exhaustive method used in the parallel WRM algorithm. The $O(N^{4/3})$ would result if the serial computer is to do exactly all the computations done by the parallel algorithm.

plexity is bounded from below by $O(N^{1/6})$. This ignores the boundary assignment complexity of $O(N^{1/2})$, as the multiplicative constant is negligible compared to that of cell embedding even for the numbers (N) involved in VLSI range. Similarly, $O(N)$ process of problem loading is ignored. We also ignored the fact that the number of tracks per cell is a function of chip size for the simplicity of arguments (see [1]–[3]).

The experimental WRM has an array size of 8×8. The memory size of 15K bytes per node is adequate to process wire routing for a chip of 24×24 cells, making use of the folding technique. The array has end-around neighbor connections in both directions for a convenient program implementation of the folding. (Recall the accordion folding of SAM virtual machine without the end-around connections.) The speed penalty due to folding will not be discussed in much detail here. Briefly, for any foreseeable VLSI chip, if the size of the array, $m \times m$, is about 32×32, the time penalty during global wiring is about a factor of 2 over a full size machine. For the exact embedding process, full parallelism penalty of $(n/m)^2$ would incur to process for a chip size $n \times n$ cells. In general, the machine size can be smaller with increased folding, but the processing time would increase, while the total memory capacity of the system would stay constant, that is, the amount of memory per node would have to increase. We feel that the most cost-effective system would have $16 \leqslant m \leqslant 32$.

The main purpose of the experimental machine was to develop efficient parallel wiring algorithms. Performance was not the design objective. The microprocessors used in the machine have 8-bit operands and 250-ns cycle time, the fastest instructions taking 1 μs. Today's technology provides much more powerful microprocessors with 16-bit operands and faster cycle times. Additional hardware features to implement special instructions (such as 4-port compare or NEIB-like operations), and to facilitate neighbor communications, would also contribute to the speedup. It is expected that a production machine with a reasonable array size would improve the speed relative to our experimental system by a factor of 50 or more.

The global wiring experiments performed on 19×23 cell chip examples ($n \approx 400$, folded nine times on the experimental 8×8 machine) took about 1 min of CPU time. An almost identical algorithm was programmed in PL/1 and run on IBM 3033. The run time was about 45 s. Considering the order of computation time involved, the projected speed advantage at the VLSI range ($N \approx 10\,000$) is significant, though not so dramatic as simple Lee–Moore processors for path finding applications.

The global wiring algorithm of WRM uses the parallel speed advantage to compute a more elaborate cost function for the routes as mentioned earlier (see [22] for details). The WRM computation for each node during the forward propagation is more involved than that performed by existing global wiring programs for serial machines. The objective here is to reduce the overflows. Experimental results are encouraging in that more uniform wiring track usage have been observed and fewer overflows have resulted.

Let us return to the complexity of parallel global wiring. Wiring one net at a time, the computational complexity is of $O(N^{7/6})$. It is indeed possible to wire more than one net at a time provided the propagation areas of the nets do not overlap. Suppose all the terminals of a net are enclosed within a minimum rectangle which, for the sake of simplicity, is assumed to be of size $l \times l$. The propagation activity for the entire net would be confined to within $(D + l) \times (D + l)$ area where D is

the detour limit. The side l is proportional to the average connection length $\bar{p}$ which, in turn, is proportional to $N^{1/6}$. Therefore, the propagation activity of the net requires an area proportional to $N^{1/3}$. If the machine size is $m \times m$ and $m \gg \bar{p}$, we may assure that some number of nets, proportional to $m^2/N^{1/3}$ can be simultaneously wired on the machine with only a small probability of overlap in their propagation areas. The controller can schedule groups of nets in such a fashion using a heuristic packing algorithm. If the machine size matches the chip size, $n \times n = N$ and $n \gg \bar{p}$, some number of nets proportional to $N^{2/3}$ can be wired simultaneously. Of course, there is a programming overhead as well as a performance overhead in implementing such a scheme. It is also true that packing of the nets into such groups may not be efficient. However, one can argue that the complexity of global wiring in this manner has to be at most linear in N for the range of VLSI. Taken together with exact embedding complexity, we see that the total WRM routing with simultaneous global wiring would take time at most linear to the size of the chip N. Similar simultaneous path finding tasks can be implemented on a full size L-machine or a SAM-machine.

VI. Summary and Discussions

We have examined the complexity of various wire routing stages. It is shown that special-purpose machines with a rectangular array of mesh connected processors can significantly reduce the wire routing time. These machines range from the highly specialized L-machine, and slightly more general SAM-machine to the WRM designed with fully general microprocessors in the array. These machines can be constructed using today's VLSI parts at a fraction of cost that a mainframe computer would require. These machines gain speed over a conventional machine by special-purpose hardware and by use of parallel computation. High-speed wire routing in VLSI would allow fast feedback to the designers and even enable the designer to interactively improve the design through a series of applications.

The WRM has been discussed in greater detail than the others because it is more general and it addresses the total wire routing problem. Although, master-slice chip wiring is obviously a natural application for WRM, the machine could also route wires on any carrier that has regular wiring tracks and placed net terminals.

Parallel processing machines are organized for many different levels of generality. On the one end, there are general-purpose machines with general-purpose interconnection schemes. On the other end, there are machines with highly specialized simple processing elements that are packed in one (or more) VLSI chip, such as the L-machine or the systolic processors [25]. The WRM and the SAM-machine lie somewhere in between these two ends of the spectrum. So are many two-dimensional array machines designed originally for other applications such as BAP [20], CLIP4 [21], ILLIAC IV [26], DAP [27], MPP [28], FEM [29], CHiP [30], and WISPAC [31]. Most of these machines could be programmed for wire routing and other physical design algorithm, with varying degrees of effectiveness. Problems that require two-dimensional processing, in general, benefit from the obvious match with machine structure. Image processing, matrix operations, and structural analysis of surfaces are examples of such problems. So are specific physical design automation problems in routing, placement, shape generation and checking, timing analysis, chip simulation, rules checking, etc.

Analyses and experiments to date point to practical usefulness of wire routing machines. Whether there should be many machines each of which specializes in some aspect of VLSI physical design automation, or a "consensus" machine that can be used for most of the critical computational needs is an emerging question. For the complexities involved in VLSI physical design, it is clear that some form of parallel processing tool will come to be used. Much of the issues concerning data width, instruction capability, neighbor communication mechanism, MIMD/SIMD, and local memory organization need further research and development. There is also the exciting area of parallel algorithm development for problems solved by such machines.

Acknowledgment

The authors gratefully acknowledge the leadership and encouragement given them by E. Shapiro during the course of WRM development. They also thank H. M. Brauer, M. Denneau, S. Liles, and R. Villani for the invaluable contributions to the design and development of the machine and programs. Finally, they wish to thank M. Dietrich for the typing of this manuscript.

References

[1] W. R. Heller, W. F. Mikhail, and W. E. Donath, "Prediction of wiring space requirements for LSI," *J. Design Automation and Fault Tolerant Computing*, pp. 117–144, 1978.

[2] W. E. Donath and W. F. Mikhail, "Wiring space estimation for rectangular gate array," in *VLSI 81 Conf. Proc.*, J. P. Gray, Ed. New York: Academic Press, 1981.

[3] A. ElGamal, "Two-dimensional stochastic model for interconnections in master-slice integrated circuits," *IEEE Trans. Circuits Syst.*, vol. CAS-28, pp. 127–138, Feb. 1981.

[4] J. Soukup, "Circuit layout," *Proc. IEEE*, vol. 69, no. 10, pp. 1281–1304, Oct. 1981.

[5] E. F. Moore, "Shortest path through a maze," in *Annals of Computation Laboratory*, vol. 30. Cambridge, MA: Harvard Univ. Press, 1959, pp. 285–292.

[6] C. Y. Lee, "An algorithm for path connections and its applications," *IRE Trans. Elec. Comput.*, vol. EC-10, pp. 346–365, Sept. 1961.

[7] S. Akers, "Routing," in *Design Automation of Digital Systems: Theory and Techniques*, vol. 1, M. A. Breuer, Ed. Englewood Cliffs, NJ: Prentice Hall, ch. 6, pp. 283–333, 1972.

[8] F. Rubin, "The Lee path connection algorithm," *IEEE Trans. Comput.*, vol. C-23, no. 9, pp. 907–914, Sept. 1974.

[9] S. Akers, "A modification of Lee's path connection algorithm," *IEEE Trans. Elec. Comput.*, vol. EC-16, pp. 97–98, Sept. 1967.

[10] J. Soukup, "Fast maze router," in *Proc. 15th Design Automation Conf.* (Las Vegas, NV, 1978), pp. 100–101.

[11] J. Soukup and U. W. Stockburger, "Routing in theory and practice," in *Proc. 1st Annual Conf. on Computer Graphics in CAD/CAM Systems* (MIT, Apr. 1979), pp. 126–146.

[12] M. A. Breuer and K. Shamsa, "A hardware router," *J. Digital Syst.*, vol. 4, no. 4, Computer Sci. Press, 1980, pp. 393–408.

[13] T. Blank, M. Stefik, and W. van Cleemput, "A parallel bit map processor architecture for DA algorithms," in *Proc. 18th Design Automation Conf.* (Nashville, TN, 1981), pp. 837–845.

[14] K. A. Chen, M. Feuer, K. H. Khokhani, N. Nan, and S. Schmidt, "The chip layout problem: An automatic wiring procedure," in *Proc. 14th Design Automation Conf.* (New Orleans, LA, 1977), pp. 298–302.

[15] H. Shiraishi, and F. Hirose, "Efficient placement and routing technique for master-slice LSI," in *Proc. 17th Design Automation Conf.* (Minneapolis, MN, 1980), pp. 458–464.

[16] J. Soukup and J. C. Royle, "On hierarchical routing," *J. Digital Syst.*, vol. V, no. 3, Computer Sci. Press, 1981, pp. 265–289.

[17] S. J. Hong, R. Nair, and E. Shapiro, "A physical design machine," in *VLSI 81 Conf. Proc.*, J. P. Gray, Ed. New York: Academic Press, 1981.

[18] A. Iosupovicz, "Design of an iterative array maze router," in *Proc. IEEE Int. Conf. on Circuits and Computers*, pp. 908–911, Oct. 1980.

[19] S. H. Unger, "A computer oriented toward spatial problems," *Proc. IRE*, pp. 1744–1750, Oct. 1958.

[20] A. P. Reeves, "A systematically designed binary array processor," *IEEE Trans. Comput.*, vol. C-29, pp. 278–287, Apr. 1980.

[21] M.J.B. Duff, "CLIP 4: A large scale integrated circuit array parallel processor," in *Proc. Pattern Recognition and Image Processing Conf.*, pp. 728–733, Nov. 1978.

[22] R. Nair, S. J. Hong, S. Liles, and R. Villani, "Global wiring on a wire routing machine," presented at the 19th Design Automation Conf., Las Vegas, NV, 1982.

[23] M. Feuer, "Connectivity of random logic," *IEEE Trans. Comput.*, vol. C-31, no. 1, Jan. 1982.

[24] M. Feuer, K. H. Khokhani, and D. Mehta, "Computer-aided design wires 5000-circuit chip," *Electronics*, pp. 144–145, Oct. 9, 1980.

[25] H. T. Kung, "The systolic (VLSI) system: A powerful computing engine," presented at the ICASE Workshop Array Architecture for Computing in the 80's and 90's, Hampton, VA, 1980.

[26] G. H. Barnes *et al.*, "The Illiac IV computer," *IEEE Trans. Comput.*, vol. C-17, pp. 746–757, Aug. 1968.

[27] S. F. Reddaway, "DAP architecture and algorithms," presented at the ICASE Workshop Array Architectures for Computing in the 80's and 90's, Hampton, VA, 1980.

[28] C. Michelson, "MPP architecture and system software," presented at the ICASE Workshop Array Architectures for Computing in the 80's and 90's, Hampton, VA, 1980.

[29] D. Loendorf, "The finite element machine: an array of asynchronous microprocessors," presented at the ICASE Workshop Array Architectures for Computing in the 80's and 90's, Hampton, VA, 1980.

[30] L. Snyder, "Overview of the CHiP Computer," in *Proc. VLSI 81 Conference*, J. P. Gray, Ed. New York: Academic Press, 1981.

[31] M. J. Redmond and S. D. Smith, "Permutation function simulation on the Wisconsin parallel array computer (WISPAC)," presented at the ICASE Workshop Array Architectures for Computing in the 80's and 90's, Hampton, VA, 1980.

[32] M. J. Flynn, "Some computer organizations and their effectiveness," *IEEE Trans. Comput.*, vol. C-21, no. 9, pp. 948–960, Sept. 1972.

A LOW COST, TRANSPORTABLE, DATA MANAGEMENT SYSTEM

FOR LSI/VLSI DESIGN

by David C. Smith and Barry S. Wagner

Advanced Technology Laboratories
RCA Corporation
Camden, New Jersey

ABSTRACT

The philosophy, structure, and design of a data-management system that is applicable to a wide range of organizational environments - from production groups to research and development groups - are introduced and discussed. The system is in operation in two locations and has been used to design numerous LSI/VLSI chips. This low cost system is transportable, extendable, and suitable for use by designers having a diversity of CAD-related skill levels. User reaction to the system and future plans are discussed.

INTRODUCTION

The availability of a comprehensive set of design aids, including a centralized data management system, is absolutely essential for the efficient design of very large scale integrated (VLSI) circuits. RCA's Advanced Technology Laboratories (ATL) has developed a set of integrated design aids called the Computer Aided Design and Design Automation System (CADDAS).[1] At the heart of this system is a data-management system known as DAMACS, which stands for Data Management and Control System. A block diagram of the integrated CADDAS-DAMACS system is shown in Figure 1. The DAMACS system is discussed in this paper.

The DAMACS system is applicable to a wide range of design environments. In addition to being applicable to a production environment, the system is also oriented towards use in research and development and contains features that aid the R&D process. The system presents a "user friendly" environment to the designer who may be unsophisticated in the use of computers.

Some other important features of this system are its high degree of transportability and expandability. This system was developed in a short period of time and at a low cost.

---

*This system and major programs within it received major support from the U.S. Army Electronics Research and Development Command, Adelphi, Maryland, under various contracts. The work on this system was performed under the technical direction of Mr. Randy Reitmeyer, at the Electronics Technology and Devices Laboratory, Ft. Monmouth, New Jersey.

The DAMACS system has been in use since the middle of 1980, and since that time eleven chips have been designed and/or fabricated with this system. A representative chip from this group is shown in Figure 2.

This paper provides a short summary of CAD development in our organization, briefly describing previous versions of the current system. Following this, we describe the philosophy behind the current system. Next, the underlying structure of the system is described, along with some of the significant design features. The paper concludes with a discussion of the system's advantages and a summary of the system's success to date.

BACKGROUND

The Advanced Technology Laboratories (ATL) of RCA Government Systems Division (GSD) provides technical guidance and recommendations to the business units of GSD. Within this charter, ATL performs on government technical contracts and applies IR&D funds to establish new technical areas for future growth. One of these technical areas has been the development of the standard cell concept, including both the cell families and the required software used for the design of LSI and VLSI arrays.

The literature contains many papers on data-management systems developed for management of data used in the design and manufacture of complex systems containing many circuit boards. While ATL saw a need for some kind of data-management system, the implementation or acquisition of a sophisticated system could not be justified due to our mode of operation. Our hardware products are not major systems but are VLSI circuits and arrays whose purpose is to demonstrate the applicability of new technologies to the requirements of future major government systems. Additionally, the components of our array designs, the standard cells themselves, are frequently being added to our cell families or are being updated, as the technology advances or as requirements change. Because of this, the existing systems did not fit well into our mode of operation, and would have little chance of being accepted.

Reprinted from *ACM/IEEE 19th Design Automation Conf. Proc.*, 1982, pp. 283-290.

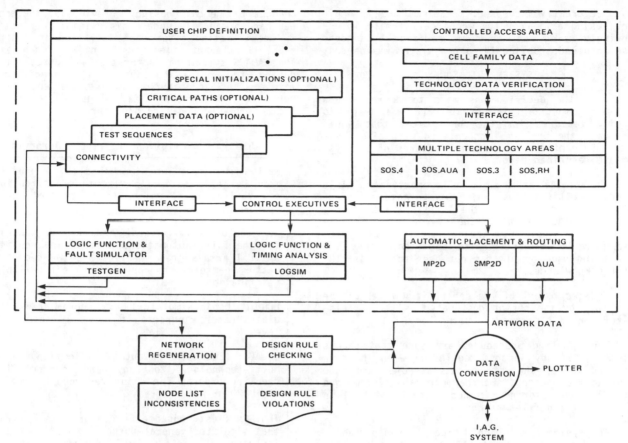

Fig. 1. Integrated CADDAS-DAMACS system.
(Present version of DAMACS lies within dashed lines)

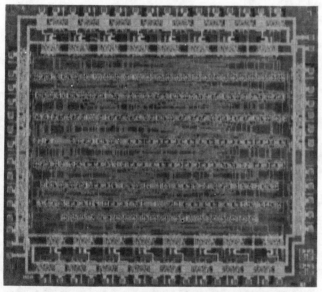

Fig. 2. Representative chip designed and fabricated with the DAMACS system. Chip size: 225 x 204 mils; transistors: 2900.

Computer programs for the simulation, placement and routing, and test sequence generation of chips have been used in ATL since 1966. However, it was not until 1975 that the first attempt at a data-management system was made. This attempt was precipitated by user frustration and incorrectly designed chips caused by differences between the input data to our computer programs.

Because of a scarcity of disk space on a batch-oriented computer system, the resulting data-management system was a set of punched-card-oriented data-translation programs. However, since this early version of the system did eliminate the need to code the same connectivity information in three different formats, it was used on a limited basis.

In 1978, we acquired a system having a large amount of disk space that also supported both batch processing and time-sharing. The translation programs were installed on this system. For the next two years, ATL needs were satisfactorily met by the system.

However, certain events occurred which revealed weaknesses in the system. Because of increased work load, a large number of engineers, not skilled or particularly interested in the use of computers, started to use the system. Consequently, they did not acquire sufficient skill to efficiently use programs which are not the most "user friendly". Additionally, both the number of different cell families being used and the number of chips being designed greatly increased. Problems such as engineers using data from the wrong cell family or unknowingly using data that was modified by someone else occurred. Obviously, a new system was needed.

SYSTEM PHILOSOPHY

When the decision was made to design a new data-management system, great care was taken to specify features for the new system that would eliminate the problems encountered with older systems. Some of the significant features of our system are described below.

The system is capable of supporting multiple standard cell families, which may be mature, undergoing enhancement, or in the initial stages of development. Additionally, the system is easily expandable to handle new technologies.[2] The system also provides the user with confidence that the data being used by him has not been modified without his knowledge.

Updates to any of the centralized files can be made only by the person designated as the administrator for that particular technology.

The updating procedures are such that checks are made for any inconsistencies between the centralized files, and such inconsistencies are made known to the updater. Furthermore, the system does not allow the storing of inconsistent data within the centralized files of the system.

The system has a high degree of transportability. Under past contracts ATL has developed software which has eventually been distributed within and outside RCA. Due to tradition and the expectation that this system will be distributed outside ATL, transportability was emphasized and all software was written in standard FORTRAN IV.

The data-management system and the programs tied into it are "user friendly".[3] A system or program can be made "user friendly" by reducing the data manipulation required of the designer. This reduction was specified in order to improve the productivity of those designers who lack skill in using computer programs. Typical of the programs tied into this system is LOGSIM* which requires input data to be in fixed format. While desiring to make LOGSIM more "user friendly", it was decided not to make coding changes to it. The resultant system and all of its associated programs are now so "user friendly" that a designer needs to know only the very basics of the computer's text editor in order to use the entire system.

The system contains certain closed-loop feedback capabilities that facilitate the use of certain applications programs such as MP2D[4,5] in an iterative manner. Specifically, the use of certain options within the MP2D program causes additional cells to be added to the chip design. With the use of MP2D in an iterative manner, each additional execution of the MP2D program would cause additional

---

*A brief description of acronyms is in Table 1.

changes to the chip design. Previously, the engineer had to manually update his original connectivity specification after each iteration in order to keep all relevant files consistent with each other. The DAMACS system has automatic procedures for updating the original connectivity specification file. These procedures relieve the engineer of performing this manual task and guarantee the correctness of the updating procedure.

Table 1. Description of Program Acronyms

> MP2D (Multi-Port Two Dimensional) - A program for the fully automatic placement and routing of VLSI arrays using the standard cell approach.
>
> AUA (Automated Universal Array) - A program for the fully automatic placement and routing of VLSI arrays that utilizes a fixed geometry master pattern and one layer of variable metalization.
>
> LOGSIM (Logic Simulator) - A program that simulates general purpose logic at a logic gate level.
>
> ENLAVE (Enhanced Layout Verification) - A multi-technology topology and connectivity checking program for chips designed by either the MP2D or AUA program.
>
> TESTGEN (Test Sequence Generator) - A program that simulates general purpose logic under the influence of various fault conditions at the logic gate level.
>
> SOS.4, SOS.AUA, SOS.3, SOS.RH are four different standard cell families which are based on the silicon-on-sapphire technology. These families differ from each other in cell height and performance.

For its operation, the data-management system does not require the use of any special hardware, such as graphic or smart terminals. This feature stems from both the desire to keep the system as portable as possible and the desire to keep the capital investment as low as practical.

SYSTEM DESIGN

To implement these features, we designed a system which executes application software through the use of control executives. These executives automatically gather the appropriate data, perform any necessary manipulations, and cause the execution of the application program. The system supports chip design goals through the use of integrated fully automatic layout, logic simulation, and fault simulation programs.

The system maintains three types of data: the user's chip definition, which includes all of the user's chip-related specifications; the standard-cell technology information; and the various system created data such as the chip artwork and simulation results. To minimize discrepancies and inconsistencies, each type of data is maintained in its own location within the DAMACS structure.

An engineer with the task of designing a chip simply supplies the DAMACS system with the specifications for the chip by creating a user chip definition in the easy-to-learn-and-use Common Data Base Language (CDB). This is a free formatted language for specifying the chip connectivity and logic function relationships. It also contains user specifications as described later in this paper. The design software references this single set of specifications throughout all phases of the design.

The standard-cell technology data required for an application is stored in the central database. This data is maintained in a secure location and is not supplied by the user; it is specified by the administrator, who maintains complete control over the standard-cell family. The administrator supplies this data when the cell family is generated for use and updates it only as the need arises.

Chip designs often have special requirements. "Non-standard" cells may be designed and incorporated into the standard-cell approach to meet these special requirements. This requires the system to support both standard and "non-standard" cells. The system is capable of maintaining both a standard cell and numerous "non-standard" cells, all of the same type and name. The proper version of the cell is supplied by the DAMACS system each time this cell name is requested.

This task is accomplished by allowing the user to supply "non-standard" cell data which will supersede the standard cell data stored in the database only for his design. Multiple users can create their own "non-standard" cells, each applicable only to his design. This procedure has worked well in environments where 80-100 percent of the cell types used on a chip design are standard and 0-20 percent are specially designed. Of course, any "non-standard" cells, which are found useful on multiple chip designs, and which have been properly verified for functionality and reliability, can be placed into the standard-cell database for general use.

In any design system, the designer is the most important component. It is his expertise that the system must be capable of incorporating into the design. It is his time and the amount of data manipulation he must perform which must be minimized.

This is done in the DAMACS system by numerous optional specifications which the user can supply to the application software through the chip definition. Special placement for cells can be designated, critical nodes on a chip can be specified, special routing characteristics identified, special simulation values given, in addition to other options. All of this is accomplished using the Common Data Base Language.

The system, through its control executives, manages all of the data, requiring less of the engineer's time, allowing him to concentrate on more

formidable tasks and to design a better chip in less time. By minimizing his contact with the data, costly human errors are minimized, thereby reducing costs and redesign time.

Figure 3 shows how the designer's interaction with the system during the layout cycle is minimized and structured. He places all of his specifications into his chip definition, in the CDB language, and executes the layout cycle, employing either MP2D, SMP2D, or AUA to perform the layout. After inspection of the layout, the designer either accepts the results and continues his design cycle, or modifies his chip definition and reexecutes. Similar cycles exist for logic and fault simulation operations. This method leaves the engineer in control, while minimizing the amount of time he spends manipulating the data and executing the problem.

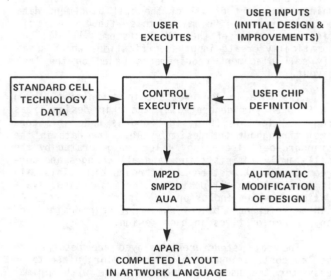

Fig. 3. DAMACS user cycle for fully Automatic Placement and Routing (APAR) using MP2D/ SMP2D/AUA.

DAMACS STRUCTURE

The engineer using the DAMACS system employs two types of input data: the database standard-cell technology data; and the user specified chip definition including chip connectivity, logic function, and all optional specifications such as cell placement data and critical path information. The DAMACS system provides both the structure and the processes for gathering this data into a format processable by the various application programs, for executing the desired program, and for managing the resulting output data.

Each of these types of input data and the output data are maintained in their own distinct area minimizing user difficulties and organizing his design procedure. As the data is required, the DAMACS control system automatically references it from the correct location.

As shown in Figure 1, the information required by the DAMACS control executive is automatically gathered from the various input files. Both user-specified data and standard-cell data are accessed, including both required and optional files. A software interface between the DAMACS control executive and each input data type resolves all file-access protocols, manages the files, and provides the data to the control executive in a structured manner. The result of the control executive is then presented directly to the application program for processing. The output of the application program is then sent to the appropriate location in the DAMACS structure for user inspection and evaluation.

If the operation is a layout using MP2D, SMP2D, or AUA where the cell placement was either totally or partially done by the layout program, the user may wish to cycle through the layout procedure again with modified input specifications but maintaining part or all of the cell placement data he just created. He may do this with the respecification portion of the DAMACS system. It will automatically create input specifications which specify cell placement requirements based on the last layout.

The data in the user's chip definition, shown in Figure 1, is generated by the chip designer, as required for his IC design, and is under his control throughout the design cycle. The data in the standard-cell technology files is generated by the cell-family administrator, and all changes and corrections to it must be made through him. This data is available to all designers for inspection, evaluation, and/or inclusion in his chip design. It can be referenced by any number of chip designs and any number of times in each design.

The data structures employed are designed to be low cost, transportable, and are organized to be easy for the users to learn and use with minimal effort. These structures are designed to provide data security against human errors and unknowledgeable users, and to be flexible, maintainable, and expandable. The structures support all of our CADDA software - layout, simulation, and testing - while providing easy access methods for data retrieval.

The data structure takes advantage of standard multiple-user interactive operating system features and minicomputer nested directories. Since personnel must learn these features to function with the computer, user learning time and software development costs are minimized. To keep the system relatively transportable, during the development phase caution was exercised to use features generally available on standard machines.

Each IC design is assigned a separate Login directory, which is used for maintaining all user data related to that chip. This directory is maintained on the system during the design cycle for the chip and, upon completion, is archived on magnetic tape for any future reference. Advantages of each chip design having its own directory are data and design integrity and cost accounting. The computer system's accounting software apportions design costs to the proper account for each chip.

The cell-technology data is structured to take advantage of the computer's directory-subdirectory feature. The Login directory DATABASE contains all of the technology data. A separate subdirectory is maintained for the technology data of each cell family. At present, four such technology subdirectories have been incorporated into the DAMACS system. Figure 4 shows a two-technology example. Each of these subdirectories contains all of the data related to that particular cell family. It is stored in files, each specifying a different type of data about the cell family.

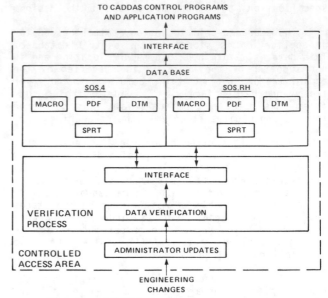

Fig. 4. DAMACS-database structure.

At present, this includes the Pin Data File (PDF), which is a file for specifying the cell and cell pin geometries, pin reassignment characteristics, and transistor count. The MACRO file contains the logical function of the cell in terms of logic gates, and includes cell input/output pins, pin capacitances, and timing information for each gate or combination of gates in terms of a reference to a rise/fall delay information file. The DTM file contains the rise/fall delay timing information which is specified as rise/fall delay slopes for each delay type as functions of capacitance. Specific rise/fall delay times are calculated based on the actual capacitive loading of the gate's output pin.

In addition, the administrator's update verification control procedures are stored here for his convenience, as shown in Figure 4.

Each technology subdirectory is protected against undesired access by passwords on the directory and subdirectory and read-write locks on each file. In this way, all alterations and updates to the technology data can be made only by the authorized administrator. The update procedure requires that the administrator make the changes to the data and that the new data be checked for correctness and consistency. If these checks are passed and the administrator is satisfied with the results, the updates are automatically placed into the master version of the technology data. During the update the old data is available to any user operating the system.

This update procedure verifies that all technology data is free from syntax errors and unreasonable values and is consistent with all of the other data pertaining to that cell and that technology. Only correct and consistent data is released by the verification software for inclusion in the master technology data, and only the data in the master is available for chip design.

Figure 4 also shows the update process with a two-technology database. The expansion to additional technologies is trivial. All work is done in a controlled access area where only the administrator has privileges, and where the data verification software must be passed before inclusion of data in the master. The software interface for accessing the database from either the user's or administrator's side is also shown.

The update system is designed so that, if the computer crashes at any time during the process, data is never lost. At all times, there is backup data, even during the actual replacement of the data in the master.

The transportable nature of the system is partially derived from the files being stored in standard ASCII form with standard 80-character records in sequential access format. The files are completely compatible with standard editors, and can be transferred to other systems with no alterations to the files.

The structure is easily expandable to new technologies, a very important feature in an environment where new technologies are continually being developed and used. The only task to be performed is for the new cell family's administrator to place the data in a new subdirectory created for the family, and have the data pass all data verification checks. The new technology is then ready for use. The data files have already grown from the two technologies which were initially designed into the system to four technologies which are presently available.

Extendability is also an important feature of this system. New types of information can be added into the system. Currently, the cell artwork, both detailed and outlines, as well as cell circuit simulation characteristics, are being added to the data stored by the system. This includes the storage of the data in appropriate files, the additional cross-checking of these files with the existing data, and the automatic accessing of this data for software requiring it.

The execution speed of the system is good. Representative execution times may be seen in Table 2. The directory-subdirectory file structure gives the user and the system software a specific path to follow to locate data types, only requiring searching techniques to locate the desired cell. Searching is minimized because the system automatically stores all data of the same data type by sorted cell names.

Table 2. Representative execution times of DAMACS system - Common Data Base Language to MP2D format

Chip	Clock time (min)	CPU time (min:sec)	I/O time (min:sec)	No. of standard cells
MAU	30	14:32	0:28	1352
IPU	19	17:50	0:31	2636
RALU	20	18:42	0:40	2691

The system is designed so that the unsophisticated user may better utilize valuable application programs with minimal effort, while the sophisticated user may modify the system to meet his unique requirements.[3] The user is the person in control and he can tailor the system's procedures to fit his needs.

PERFORMANCE

To date eleven chips have been designed and/or fabricated using the DAMACS data-management system as an integrated part of the CADDAS system.

These chips, all of which are random logic, are shown in Table 3. They include four different technologies, which differ from each other in performance and cell size. Transistor counts range up to 30,000 transistors.

Table 3. Random logic ICs designed using DAMACS

IC NAME	Technology	No. of Transistors	Size (mils)
Memory Controller	SOS.RH	3,676	261x225
Status Bus	SOS.RH	2,647	226x219
I/O Buffer	SOS.RH	4,491	278x266
Combus	SOS.4	2,900	225x204
Combus	SOS.AUA	2,900	208x208
FIFO	SOS.4	9,500	270x270
FIFO Control	SOS.AUA	1,280	208x208
Program Memory Controller	SOS.4	3,399	253x258
RALU	SOS.3	30,000	-
Memory Address Unit (MAU)	SOS.3	19,000	-
Instruction Prefetch Unit (IPU)	SOS.3	12,877	-

In addition to the DAMACS system being operational on ATL's PRIME 750 computer, it is also operational on a PRIME 400 computer in a second location at RCA. In August 1981 the DAMACS system was provided to the Electronics Technology and Devices Laboratory of the U.S. Army Electronics Research and Development Command, Fort Monmouth, New Jersey, for evaluation. It was installed on their computer, integrated with their CADDAS system, and made operational in less than a day.

The COMBUS chip, shown in Figure 2, designed and fabricated in 4-micron-gate SOS technology, is an example of a chip designed with DAMACS. This random-logic design contains 2900 transistors and measures 225 x 204 mils. Two "non-standard" cells were used; standard cell numbers 8080 and 8090 were modified and "non-standard" cells were created for each. The DAMACS system was used to handle all data, gathering each data type as required, integrating the "non-standard" cells with the required standard cells, providing all data to the application programs, and executing the application programs.

Initially, only the chip connectivity and a preliminary set of test sequences were put into the chip definition. The chip layout was done using DAMACS and MP2D. The first MP2D run had no special requirements in the user chip definition, all cells were standard cells, cell placement was completely free and all routing was unconstrained. The size was 260 x 260 mils.

In further iterations, the "non-standard" 8080 and 8090 cells were created and placed into the chip definition, with "non-standard" pin data and logic descriptions. Also, some constraints were placed on the MP2D program in the areas of cell placement, critical paths and the relative importance of nodes. These constraints were specified by entering appropriate data in free format into various optional files. Upon execution of the control executive, the DAMACS system recognized the invocation of these MP2D options, organized the input data into the fixed format required by MP2D, and executed the MP2D program. These modifications brought the chip size down to the final 225 x 204 mils and fulfilled all layout requirements.

The TESTGEN program was used to verify logic function, analyze race hazards and spikes, and produce fault simulation data. The DAMACS system provided TESTGEN with all required input based upon the specifications in the user chip definition.

The final chip design was fabricated and has been found to be 100 percent functional. It has now been integrated into the U.S. Army's MASS system, an electronic-warfare signal-processing system. It replaces 17 MSI chips.

USER FEEDBACK

Feedback from users of the new system has been very positive.

The system has successfully notified users of numerous mistakes which may have been costly and embarrassing if they were not found prior to chip fabrication. Whether or not all of these mistakes would have been detected by manual means is unknown, but, as chip complexity increases, it is unlikely that manual checking will be considered practical.

Experienced designers have shown a willingness to forsake their past but well-understood methods to learn and adopt this new system, which shows they feel it is a tool worth their investment in learning. Designers began using the system even before it was fully operational in order to take advantage of the features which were available. Since it was completed, every random logic chip designed within ATL has been with the use of the DAMACS system.

ADVANTAGES OF DAMACS

The DAMACS system provides numerous advantages for CADDAS users:

1) DAMACS has brought many isolated application programs for IC design into an integrated design system.
2) The system is flexible enough to operate in a range of environments, from production to R&D.
3) The system is capable of supporting both standard and multiple "non-standard" cell designs regardless of cell type and name.
4) It works from a single input source, simplifying the input data requirements.
5) The DAMACS system requires less user interaction and data manipulation, allowing the user to concentrate on important tasks.
6) The system provides data security. The standard cell technology data is stored in a secure directory on the system, and user-specified chip-design specifications are completely under his control.
7) The integrity of the technology data is improved by having a single administrator who is responsible for a specific technology and through whom all updates and modifications must be channeled.
8) All data to be entered into the master technology data files must pass the data verification software which checks for syntax errors and for consistency with all other data known about that cell and technology.
9) The DAMACS system is easily expandable to include new technologies.

SUMMARY

DAMACS is the Data Management and Control System that is the heart of the RCA-ATL developed Computer Aided Design and Design Automation System. DAMACS is in use and provides the user with a structured approach to the use of the CADDAS system. It is a "user-friendly" system which separates the user from the idiosyncrasies of various application programs and thereby increases his productivity. DAMACS has been designed to function not only in a production-oriented environment, but also in an R&D-oriented environment where numerous modifications to technology oriented data are a normal mode of operation. It is a system which aids the user in his experimental approaches by keeping track of standard and non-standard data for him.

Taking advantage of the pluses and avoiding the minuses of past CAD work at RCA-ATL, DAMACS was designed in less than a single man-year of effort. Yet it is very powerful and has paid back its investment costs many times over. All of this has been done while still being relatively transportable because of the use of standard FORTRAN IV for all software and using generally available multiuser operating system features. The system has received good reception from our user community, and efforts are under way to increase the value of the system by extending it into those areas where DAMACS can provide the greatest benefit.

Future work on DAMACS will be concentrated in two major areas: 1) integration of new and existing programs into DAMACS; and 2) addition of more data within the centralized technology files.

Some of the newer programs which will be integrated into DAMACS are HYPAR[6] (a program for the automatic layout of hybrid microcircuits) and MACPAR (a program for the automatic layout of VLSI devices utilizing random-size rectangular blocks). This represents an expansion of DAMACS into an area of CAD that is not based on the standard-cell technique, which was the original area of use for DAMACS. One of the additional datums presently being added to the centralized technology files is artwork information for the standard cells. This information presently resides on an interactive graphics system. By transferring artwork into DAMACS, the user will be assured of both the integrity and security of his artwork data.

By keeping pace with users' needs, DAMACS will continue to be a system which is eagerly used by our designers.

References

1. R. Noto et al., "Automated Design Procedures for VLSI," Research and Development Technical Report to U.S. Army Electronics Research and Development Command, Fort Monmouth, NJ, March 1981.
2. S. Wong and W. A. Bristol, "A Computer-Aided Design Data Base," Proc. 16th Design Automation Conference, June 1979, pp. 398-402.
3. L. A. O'Neill et al., "Designers Workbench - Efficient and Economical Design Aids," Proc. 16th Design Automation Conference, June 1979, pp. 185-199.
4. R. Reitmeyer Jr.,"CAD For Military Systems, An Essential Link to LSI, VLSI and VHSIC Technology," Proc. 18th Design Automation Conference, June 1981, pp. 3-12.
5. R. Noto et al., "CAD For Hybrid and Monolithic Microcircuits," Research and Development Technical Report to U.S. Army Electronics Research and Development Command, Fort Monmouth, NJ, August 1, 1981.
6. D. C. Smith and R. Noto, "Automatic Layout Program for Hybrid Microcircuits (HYPAR)," Proc. 1981 International Microelectronics Symposium, October 1981, pp. 106-110.

A Database Approach to Communication in VLSI Design

GIO WEDERHOLD, ANNE F. BEETEM, AND GARRETT E. SHORT

Abstract—This paper describes recent and planned work at Stanford University in applying database technology to the problems of VLSI design. In particular, it addresses the issue of communication within a design's different representations and hierarchical levels in a multiple designer environment. We demonstrate the heretofore questioned utility of using commercial database systems, at least while developing a versatile, flexible, and generally efficient model and its associated communication paths. Completed work and results from initial work using DEC DBMS-20 is presented, including macroexpansion within the database, and signaling of changes to higher structural levels. Considerable discussion regarding overall philosophy for continued work is also included.

Keywords and Phrases: database, communication, hierarchical design, DBMS-20, engineering change orders, commercial databases, macroexpansion.

INTRODUCTION

THE DESIGN OF A VLSI device involves the manipulation of a large volume of diverse interrelated data. The data involved may be generated in various ways including recording of interactively made human designer decisions or through computation. The first occurs in graphical entry systems. Computation may be used to generate data for timing and from circuit extraction programs.

In current design methodologies, these design tasks operate on different representational levels of the same device. We may envisage such levels as high level functional specifications, register transfer description, a logical level, circuit level, and detailed layout or artwork. A single task may carry out verification within one level, may generate lower level units by expansion of more abstract definitions, or otherwise necessitate movement between different representational levels.

In addition, if complex designs are to be produced in a tolerable time frame, then several design specialists must be allowed to work concurrently on distinct sections or at distinct levels. All this causes the management of the design data to become increasingly difficult since not only will data requirements overlap, but one design task will often be dependent on the data being manipulated in another area. This data communication effort probably contributes significantly to the increase in effort from 4 to 30 man-years for microprocessor designs, and makes single designer-oriented methodologies infeasible [11]. The exponential growth of design cost while production costs diminish—as stated by Faggin and Moore in [10]—is changing the outlook of the microprocessor industry, pointing to the need for more effective handling of design data.

Database technology has dealt with both the management of large volumes of data and the problems caused by concurrency of data access. Many existing database systems support inter-area communication functions. Instances are found of systems dealing with inventories, where production and sales issues interface, or patient management, where individual care and global health care concerns come together. Databases are used as well to resolve the multidisciplinary design issues in aircraft design [4] and to manage the problems of engineering changes in computing systems[8]. Given this background, we foresee that databases may provide tools to serve the VLSI designer; however, the tool is complex and will not be effective unless well understood and adapted to the demands of this application.

In this paper we discuss our ideas and initial work on tackling the problem of applying database experience to serve in the VLSI design environment. We propose that, despite popular belief, commercial database systems are a reasonable alternative, particularly during the research phase. Finally, we discuss the problem of communication between design levels and a solution using a commercial database system.

REQUIREMENTS FOR A VLSI DATABASE

We have identified several issues that have to be addressed if databases are to become effective tools in VLSI design. (The interested reader may wish to compare these to [7] which has a somewhat different approach.)

1) The database system must support a variety of design methodologies. The support of a hierarchical top-down approach is most easily achieved, but processes at a lower level may also contribute information to higher levels. Also, a designer may need to make changes at any level, which have to be reflected throughout the design levels. An example of the first exception to a pure top-down method is the handling of timing data that can be produced only after a detailed layout is available, while gates have been dimensioned using simple assumptions of intergate transit times. A second case arises when a particular gate of a series is changed to satisfy special interface demands or terminating conditions.

2) Explicit storage for all attributes of all elements at lower levels must be avoided in order to permit effective management of change emanating at a higher level. Excessive replication of lower level elements vitiates the benefits of hierarchical design methods, and creates excessive storage demands. This implies that the database system must be able to fetch actually

Manuscript received February 10, 1981; revised August 6, 1981.

The authors are with the Department of Computer Science, University Stanford, Stanford, CA 94305.

Reprinted from *IEEE Trans. Computer-Aided Des. of ICAS*, vol. CAD-1, pp. 57–63, Apr. 1982.

instantiated data or compute potential, noninstantiated data using stored algorithms.

3) The database must be able to provide a convenient interface to a wide and changing variety of programs. This interface should not change as the database is developed and extended.

4) The performance of the database system has to be such that the degradation of performance relative to an isolated design file operation is proportional to the benefits gained.

COMMERCIAL DATABASE SYSTEMS

The approach we are using in our work includes the use of commercial database systems to assess the general suitability of database approaches to VLSI design data. One objective is to determine whether commercial database systems, used knowledgeably, can perform adequately, or if they do not, where the bottlenecks are. Another is to demonstrate that commercial database systems can satisfactorily model the complex interrelationships and communication paths required.

Use of commercial database systems for VLSI design has promoted some discussion. It has been rumored and is generally believed that a commercial database system would have prohibitively poor performance. This is due to the overhead involved with the generality of the commercial systems. This generality can also cause the modeling means to be nonobvious. The disadvantages of commercial database systems as applied to VLSI design are discussed in [12].

On the other hand, there are many advantages to using a commercial database. The most obvious is avoiding the monumental effort required to custom design a complete, secure, and reliable database system. For these vital issues [18] is a good reference. Commercial systems have built in facilities for maintaining backups and regenerating data. They also incorporate means for enforcing access protection as well as handling the complex storage problems of growth, updates, and locality. A Data Manipulation Langauge (DLM) is provided for high level access. This allows for querying the database without concern for the details of storage. In addition, system support is provided by the vendor, freeing the maintenance staff from this additional difficult and time consuming chore. Perhaps more important is the generality and ease of modification provided. A small change in representation or other aspect of the system need not entail a complete system overhaul.

The current set of experiments uses a network system, DEC's DBMS-20. This is a system based on the published CODASYL database definition [3], which has a strong orientation to well-understood business applications, such as inventory management. The database designer can, through schema specification, select which of the logically appropriate linkages should be implemented. For a thorough discussion of the uses and types of schemas see [18].

An important structure provided by CODASYL databases and which we will be referring to is the *set*. This is represented by notations such as in Fig. 1. Here *Manager* and *Employee* are the record names for collections of data pertaining to the entities, manager and employee. In this example, each has a single field of data called **name**. *Manages* is the name of a set. The notation indicates that for each *Manager* there is an associated

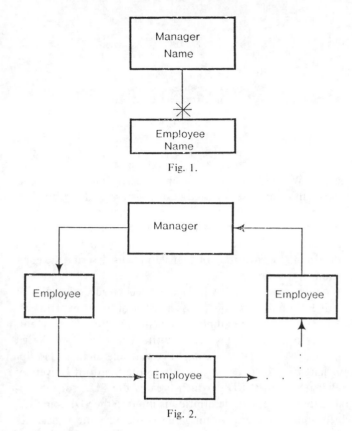

Fig. 1.

Fig. 2.

group of employee records. Each *Manager* record is said to *own* the associated *Employee* records. Sets may have various properties ascribed to them. An excellent discussion is found in [9]. We will also be referring to sets as *rings* or *linksets* as this is how they are implemented and accessed (see Fig. 2).

Relational database systems are also being investigated for their applicability to VLSI design. The use of Ingres is discussed in [5]. Their modeling approach is quite different from ours, being much more tied to the specific representation chosen. We plan to investigate the use of RIMS, a relational system with strong automatic query optimization capabilities [13]. These two experiments should provide an instructive comparison.

COMPARISON OF DBMS-20 AND SPECIALIZED DESIGN FILES

As was stated previously, we wish to either refute or support the notion that the performance and modeling capabilities of a commercial database system cannot compare favorably with a custom specialized design file system. Before we discuss what actual work we have done, it must be noted that due to availability of data, programs, and otherwise existing material at hand, we have been using data from conventional circuit board design to model the VLSI design process. Even so, many of the problems are similar and as such, we believe that the results are validly applicable to the VLSI design process as well. More importantly, by using this data, we were able to make benchmark comparisons with existing specialized design files and programs on the same data. This is essential for measuring the relative performance of commercial database

Level Name	Component Names at Level
cpu	PDP-11
reg	BREG, PROCOUNT, BUSCOUNT, RAM16X16, ALU, STATUS, AMUX
rf1	PRIOARB, TIMER
rf2	TCFF2, RAM16, 40181, 40182
ff	TCFF, LFF16, DFF4, BUSD16, MUX16, RAM
gate	DFF, LFF, XOR, BUSD, MUX
trans	TRANSP, INV, NAND, 3NAND, 4NAND, NOR, 3NOR, 4NOR
bottom	TN, TP, PLA-SMALL, PLA-MED, PLA-LARGE

Fig. 3.

systems. We feel that we have gained insight into the problems involved in a database-oriented design system, and in particular, have demonstrated the viability and efficiency of using commercial database systems at least until the complete model of the design process and its communication requirements is well understood. So, without further ado let us describe the comparison experiment.

A design environment which uses specialized design files of circuit information was used as the control to the experiment. See [16] for a more complete discussion of this environment. The user begins with his design written in Structural Design Language [15] (SDL), a straightforward hierarchical description language, or may be input using the Stanford University Drawing System (SUDS) which creates the SDL for the diagram input. The SDL description is input to the SDL compiler, which then produces a dump file containing the logical description of each component described in the design. This design methodology is strictly hierarchical: the logical description of one component is described in terms of smaller lower level components. For example, gates are described in terms of transistors, flip-flops are described in terms of gates, registers are described in terms of flip-flops, etc. Each description level is given a unique name in the design file. The dump file from the compiler is loaded into the SPRINT database [14], the specialized control subject, via a 'hardwired' schema. Limited Fortran subroutines exist for accessing the data.

Fig. 3 gives a breakdown of the components at the eight different levels of description. Naturally, the PDP-11 itself is the top level component and is described in terms of a 16×16 RAM, an ALU, program counter, status word, etc. Near the bottom, NAND gates are described in terms of transistors, or bottom level components. In this case, the constituents are TN and TP, or N and P type transistors. Fig. 4(a) and (b) show the SDL for NAND, TN, and TP, and the SUDS input for the NAND.

Initially the CODASYL schema was modeled very closely to the SPRINT schema, and due to SPRINT's hierarchical structure, CODASYL network capabilities were not utilized in the first iteration. Later the schema was modified to take advantage of network structures: this effort provided linked access to the library file of components. In addition, we made some changes to more closely model VLSI design. The decomposition method used is excellently described in [17].

Fig. 5 shows a subset of the network diagram of the DBMS-20 schema. For each component described in the database there is a Logical Description record containing its *name*. This field is used as the key to find the record directly via a hashing

```
NAME: NAND;
PURPOSE: ICDESIGN;
LEVEL: TRANS;
EXT::A,B,O,V,GND;
INPUTS: A,B;
OUTPUTS: O;
TYPES: TP,TN;
TN: NA,NB;
TP: PA,PB;
N1=,A,PA.G,NA.G;
N2=PA.D,PB.D,NA.D,,O;
N3=,GND,NB.S;
N4=,V,PA.S,PB.S;
N5=,B,PB.G,NB.G;
N6=NA.S,NB.D;
END;
NAME:TN;
PURPOSE:ICDESIGN;
LEVEL:END;
EXT::G,S,D;
END;
NAME:TP;
PURPOSE:ICDESIGN;
LEVEL:END;
EXT::G,S,D;
END;
```

(a)

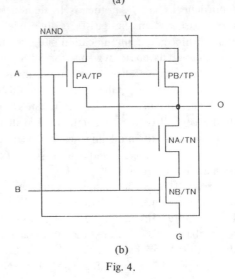

(b)

Fig. 4.

function. Records are connected to related records through three distinct rings of pointers, forming a complex network. One of the rings off of the Logical Description record describes each external pin, with a unique name and its function (i.e., input, output, tristate). The Equivalent-group ring describes the equivalence between pins and sets of pins. The remaining ring off of the Logical Description record supplies the Internal Description of the part. For each Internal Description record, a General Information record is kept describing general characteristics. The attribute fields, *creator,* a *time* stamp, *level*, *purpose*, and *version* number are meant to fully describe the corresponding internal description.

After completing the schema, a loader program was then written to load the same dump file produced by the SDL compiler into the DBMS-20 database. Comparison of loading times showed the SPRINT database to be quicker, as expected, but DBMS times were quite acceptable. It has been rumored that the designer's mode of operation would be drastically affected and that the difference factor would be on the order of a hundred. This was most certainly not the case. The initial loading of the PDP-11 example took SPRINT about one minute and required about four minutes for DBMS-20.

For the actual comparison of the retrieval times, a macroexpander program was used. The macroexpander already op-

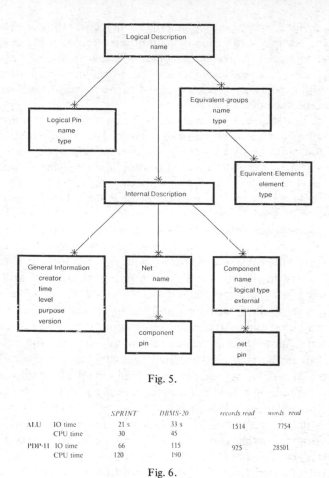

Fig. 5.

		SPRINT	DBMS-20	records read	words read
ALU	IO time	21 s	33 s	1514	7754
	CPU time	30	45		
PDP-11	IO time	66	115	925	28501
	CPU time	120	190		

Fig. 6.

erated on the SPRINT database. We simply substituted the appropriate DBMS-20 DML for the SPRINT Fortran calls. This program reads the logical description of a high level component from the database. The user then specifies the levels to be expanded. Thus if the user specified the program to expand all levels down to the transistor, then the resulting output will be the logical description of the original component described totally in terms of transistors. This form of description may not be very useful to the designer, but could be the input to, for example, a simulator program. Also, in the VLSI environment, this could be the first step in producing the layout diagram.

The macroexpander program needs to do a tremendous amount of random read operations, especially if the component that is being expanded is large and described at a high level. Fig. 6 contains the results of expanding the ALU and the PDP-11. These results show less than a factor of two degradation in performance of the DBMS-20. This is an acceptable tradeoff for the increased flexibility and generality of CODASYL, and disputes the original theories that there would be a dramatic difference in performance, as mentioned above.

We next evaluated DBMS-20 performance in writing back into the database. In order to simulate a real application of VLSI design, we considered how the database would handle instantiations. It is helpful to refer to the schema diagram to see how this is done.

In order to exercise the representation, the macroexpander program was modified so that it could write back into the database. When working in this mode, after the program expands a component, it stores the new representation as another Internal Description along with a General Information record with the appropriate data. In the design atmosphere, suppose one is working on a component at another level or perhaps at the same level. The user makes some changes from the original description and stores his version into the database. This may be done to several or all the components. Now when a higher level component is expanded, the program can selectively choose the appropriate internal description to use. The above scenario was tested on the ALU with following results:

The upper level description of the ALU contains the following pieces and levels:

40181	rf2
40181	rf2
MUX	gate
Xor	gate
INV	trans
NAND	trans
4NAND	trans
NOR	trans
4NOR	trans.

The following is some of the data on expanding the pieces of the ALU:

40181: Total read time = 17.4 s

record name	quantity
net	152
component	127
comp-pin	564
Total read	843 (4432 words)

Total write time = 41.3 s

record name	quantity
net	154
net-pin	898
component	293
comp-pin	898
Total written	2243 (10136 words)

40182: Total read time = 8.78 s

record name	quantity
net	79
component	67
comp-pin	268
Total read	414 (2240 words)

Total write time = 8.82 s

record name	quantity
net	38
net-pin	186
component	59
comp-pin	186
Total written	469 (2128 words).

The MUX and the Xor pieces were also expanded. Then the

ALU was expanded again with the expanded version of the lower level pieces in the database.

Total read time = 48.0 s

record name	quantity	
net	341	
component	441	
comp-pin	1575	
Total read	2357	(12745 words)

Total write time = 203 s

record name	quantity	
net	721	
net-pin	4384	
component	1441	
comp-pin	4384	
Total written	10930	(49364 words).

The current state of the SPRINT design system does not allow writing instantiations back into the database, so no parallel experiments were done, but again the DBMS-20 implementation showed an acceptable performance.

MODELING COMMUNICATION PATHS

After having demonstrated replacement of a custom design by an equivalent database representation, we explored implementation of novel data management facilities not provided by specialized design files. Specifically, as is necessary in order to operate in the mode for the VLSI design environment, the database system has to be augmented with communication paths between levels. These paths may ultimately be oriented in any combination of directions. We are investigating the constituent directions, viz., down hierarchical levels, up to higher level units, and sideways. Current CAD systems generally do not address these communication issues.

In the downward direction, a method must exist to create the effect of lower level instantiation using procedures and higher level descriptions. It is desirable that the query interfaces which access the lower levels do not have to distinguish between actual or computed data elements. Along these same lines, we intend to look at the issues and possibilities of when and how we wish to store redundant or computed data. Hierarchical CAD systems such as SPRINT enforce this sort of communication by providing the names of constituent lower level components at the higher level. This type of communication was used in the macroexpander program.

Communication in the upward direction relates detail to more abstract specifications. The creation or modification of lower level instances has to be bound to the appropriate higher level elements. An initial approach we consider promising and have implemented is signaling such changes to the next levels up in the hierarchy. Multiple structures may become involved because an element may be defined by an association of several higher level entities [17]. A simple example is an element that is defined from the expansion of a functional component and a library description. The signal creates an exception flag at the higher structures. At a later time, when the level to which the signal was directed is accessed by its owner, the system can

provide a warning. An appropriate action could then be taken; for example, verification of continued correctness of the design at that level, or the introduction of a new version of a component, or a new parameterization of the library descriptions. With experience, selected types of changes could trigger automatic updates.

While the need and techniques for up and down passing of information are relatively clear, travel in the orthogonal direction is not nearly so, yet may be essential in the future design system. To explain the need consider the following scenario. Suppose we are designing a microcomputer chip with an ALU composed primarily of registers, dense RAM and ROM, a finite state PLA controller, some random logic, etc. Designing this chip with only one methodology would be a terrific waste of time and energy. Although there are design tools which can fairly efficiently model, for example, random logic, it would be very inefficient to design a regular structure like a PLA using random logic techniques. It would be far more effective to allow each submodule to be designed or simulated in its most appropriate manner when simulating the entire project. Allowing for generalized sideways communication could then greatly increase the overall flexibility and efficiency of a design system. Another extremely important use of sideways communication is in maintaining logical equivalence between representations. This includes transmission of changes made to one representation of a module to its other representations. Current working design systems give no consideration to this sideways communication issue, but the research has begun (see [2], for example). We hope to learn from it and incorporate these new ideas into our database system.

Signaling Changes in the Database

To test upward communication in our DBMS-20 database we devised an experiment to flag the necessary upper level pieces regarding changes made to lower level components. A straightforward approach to this problem would be to have each component keep track of the upper level parts that use that component. To implement this in CODASYL, we would like to have each Logical Description record own other Logical Description records in an *owners* set, as in Fig. 7.

Besides causing cyclic errors, this structure is not allowed in the CODASYL definition. To circumvent this we added another record of the logical description to indirectly find the owners of the module, or similarly the pieces used in the module. This is illustrated in Fig. 8. PointerBox contains a reference to the correct module.

The following example illustrates how upper level pieces are warned of lower level changes. Suppose the multiplexer (MUX) has been changed and we wish to flag all the components that use the multiplexer. In the database there are several parts that use the multiplexer, and for each one of these there is a PointerBox record in the multiplexer's *owners* linkset. Some of these include the ALU, status register, and a 16-input multiplexer (MUX16), see Fig. 4. These same PointerBox records are in the *pieces* linkset of the parts that use the multiplexer. The multiplexer itself is made up of inverters and transistor

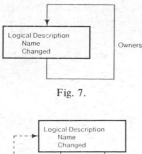

Fig. 7.

Fig. 8.

pairs (TRANSP). Thus there are two PointerBox records in the *pieces* linkset of the multiplexer.

The flagging process begins by marking the first record of the *owners* linkset, in this case say it is the ALU. Here the field *marked* of the PointerBox is incremented. Next, the owner record of the *pieces* linkset of the current PointerBox is found. Each PointerBox record has a pointer directly to the owner record of both the *pieces* linkset and the *owners* linkset. At this time, the Logical Description record of the ALU is located and the field *change* is incremented. Flagging continues from the ALU in a recursive manner. When a Logical Description record with a empty *owners* linkset is found, the upper most level has been reached. The flagging must then move back to the previous level by finding the next PointerBox record of the *owners* linkset.

This method of flagging upper level components is best characterized as "height-first." This method requires more record accesses than a possible level or breath-first flagging, since many of the same records are retrieved many times, but the height-first method allows more information to be stored.

Consider the following example: Suppose the multiplexer and the inverter are changed and their upper level pieces are notified. As shown in Fig. 3, the multiplexer and the inverter are described on different levels of the hierarchy, and a change to the inverter will also affect the multiplexer. If you now query the PDP-11 data record as to whether any changes have been flagged, only the inverter will show up as an altered component. If the upper level pieces of the inverter are unflagged (decrement the *change* and *marked* fields), and the PDP-11 data record is queried again, the multiplexer will be identified as the modified component since correcting for the inverter does not resolve the other multiplexer modification. Now, suppose the multiplexer and the NAND gate, which is described on the same level as the inverter, are changed. A subsequent query from the PDP-11 record will reveal that both the NAND gate and the multiplexer have been changed. The multiplexer is caught in this situation because none of its lower level pieces were changed. Thus this flagging algorithm allows easy detection of all the lowest level independent modifications. Also, every component can be unflagged uniquely by reversing the flagging procedure. The time to carry out this flagging algorithm depends on the level at which the process is started. Here are some sample flagging times:

Piece	Level	Time (seconds)
ALU	reg	0.13
MUX, XOR	gate	0.9
INV	trans	7.1
TN	bottom	21.

CONCLUSION

With the successful implementation of flagging, we have taken time out to write this paper and so we shall summarize our findings and discuss our next moves. Looking back over our work, we believe our initially stated approach to be most promising: the commercial database system allowed us to implement both initial designs and modifications relatively quickly and easily. Coupled with the discovery that performance is not badly affected, this has allowed us to experiment productively. This ability to experiment is important to obtain the knowledge needed for designing useful database systems for VLSI design. As we, and designers themselves, gain more insight into the design process, we hope to develop a database supporting all the types of communication paths for many different kinds of interlocking design tasks. Indeed, a new version of the schema has been written which addresses this problem as well as the requirements for a VLSI design database outlined early in this paper [1]. The final long range goal is, of course, a usable, flexible, expandable, and efficient database-oriented VLSI design system.

REFERENCES

[1] A. Beetem, "A database approach to communication in VLSI design: Part II," internal Rep., Stanford Univ., Computer Science Dep., May 1981.
[2] J. Beetem, "Structured design of electronic systems using isomorphic multiple representations," Ph.D dissertation, Stanford Univ., Elect. Eng. Dep., CA, Dec. 1981.
[3] CODASYL Database Task Group, Apr. 1971, Rep. ACM, New York, 1971.
[4] R. Fulton, "National meeting to review IPAD status and goals," *Astronautics and Aeronautics*, July/Aug. 1980.
[5] A. Guttmann, "Using the Ingres database management system for VLSI design data," internal Rep., Univ. of California, Berkeley, Dep. Elect. Eng. and Computer Sci., Apr. 1981.
[6] D. Hill, ADLIB User's Manual, Stanford CSL Technical Rep. no. 177, CA, Aug. 1979.
[7] L. W. Leyking, "Database Considerations for VLSI," in *Proc. Caltech Conf. on Very Large Scale Integration*, pp. 275-301, Jan. 1979.
[8] F. Mallmann, "The management of engineering changes using the Primus system," *The Seventeenth Design Automation Conf. Proc.*, pp. 348-366, June 1980.
[9] T. W. Olle, *The CODASYL Approach to Database Management*. New York: Wiley, 1978.
[10] A. Robinson, "Giant corporations from tiny chips grow," *Sci.*, vol. 208, no. 4443, pp. 480-484, May 2, 1980.
[11] L. Scheffer, "Database Considerations for VLSI Design," *Design Automation at Stanford*, Ed. W. vanCleemput, Stanford University, CSL (Computer Systems Lab.) internal Rep., 1979.
[12] T. Sidle, "Weaknesses of commercial data base management systems in engineering applications," *The Seventeenth Design Automation Conf. Proc.*, pp. 57-61, June 1980.
[13] J. Simpson et al., *User's Manual for the Rims/FPDB System*, Science Applications Inc., Palo Alto, CA, Mar. 1980.
[14] K. Stevens and W. vanCleemput, Design File Organization in the Sprint System, Stanford CSL Tech. Rep. no. 133, 1979.

[15] W. vanCleemput, "SDL: A structural design language for computer aided design of digital systems." Stanford CSL Tech. Rep. no. 136, 1979.

[16] W. vanCleemput and E. A. Slutz, "Initial considerations for a hierarchical IC design system," Stanford CSL Technical note #132, 1977.

[17] G. Wiederhold, and R. El-Masri, "The structural model for database design," in *The Entity-Relationships Approach to Systems Analysis and Design*, Amsterdam, The Netherlands: North Holland, 1979.

[18] G. Wiederhold, *Database Design*. New York: McGraw-Hill, 1977.

CAD Systems for IC Design

MARVIN E. DANIEL AND CHARLES W. GWYN, SENIOR MEMBER, IEEE

Abstract—As integrated circuit (IC) complexities increase, many existing computer-aided design (CAD) methods must be replaced with an integrated design system to support very large scale integrated (VLSI) circuit and system design. The framework for a hierarchical CAD system is described. The system supports both functional and physical design from initial specification and system synthesis to simulation, mask layout, verification, and documentation. The system is being implemented in phases on a DECSystem 20 computer network and will support evolutionary changes as new technologies are developed and design strategies defined.

INTRODUCTION

COMPUTER-AIDED DESIGN (CAD) of integrated circuits (IC's) has had various interpretations as a function of time and definition source. These interpretations range from use of simple, interactive graphics and digitizing systems to individual programs used for circuit or logic simulation, mask layout, and data manipulation or reformatting. In many instances, the word "aided" is deemphasized to imply nearly automatic design. Within the context used in this paper, CAD refers to a collection of software tools to provide the designer with design assistance during each phase of the design. Although many decisions are made by the software during the design process, important decisions are the designer's responsibility. The computer aids or tools provide the designer with a rapid and orderly method for consolidating and evaluating design ideas and relieve the designer of numerous routine and mechanistic design steps.

Background

A brief review of some of the techniques and terminology that evolved into today's CAD is instructive to viewing the collection of tools available today. CAD had its origins in the mid-to-late 1950's. With the advent of high-speed digital computers. some of the pioneering work of Kron [1], [2] was applied to the simultaneous solution of network equations. This formulation was used, in part, by computer codes such as NET-1 [3] (Network Analysis Program) and ECAP [4] (Electronic Circuit Analysis Program). From a physical point of view, solution of the network problem predicts the behavior of a system in terms of the element characteristics and interconnections. Viewed as a mathematical problem, the properties of a topological structure (linear graph) and a superimposed algebraic structure (interrelations of the nodes, branches, and meshes of the graph) are determined.

Manuscript received April 10, 1981; revised September 8, 1981. This work was supported by the U.S. Department of Energy.

The authors are with Sandia National Laboratories, Integrated Circuit Design, Dept. 2110, Alburquerque, NM 87185.

This early work of solving steady-state and transient network problems was the genesis of circuit simulation and was designated CAD. By the early 1970's, the growing need to simulate large circuits and thus obtain the detailed solution of a large number of partial differential equations forced the development of optimization methods and higher levels of abstraction to represent physical systems. Whereas very detailed models are used to simulate the behavior of individual transistors, more abstract representations are used for timing and logic simulation. A timing simulator uses only current–voltage tables for transistor models, capacitive loading, and circuit connectivity to determine the signal waveforms at each circuit node. Logic or gate-level simulation solves the equivalent Boolean equations with delay elements inserted between gates to account for signal timing.

The use of computer aids in the layout of IC masks essentially proceeded along two approaches: interactive graphic systems and automatic layout based on standard cells. Early interactive graphic systems provided a method for capturing the design by recording coordinate information by manually digitizing and editing data. Methods for superimposing mask layers, scaling, enlarging, contrasting, and reviewing the results were expanded to provide fast, sophisticated drawing commands for constructing, editing, and reproducing complex figures; performing dimensional tolerance checks; selective erase, expand, move, and merge; symbolic input; pattern generation; etc.

Automatic layout methods have classically relied on a library of circuit components or cells in the form of mask geometries defining logic gates. Most software required cells with standard heights and varying width. The layout software places cells in rows, attempts to optimize the cell position in the row, and interconnects the cells in wiring channels between the cell rows. The automatic standard cell layout programs have evolved from simple linear cell placement in a single row, which was subsequently folded to fit a square area [5], to complex placement in two dimensions. The early standard cell layout aids supported the use of single entry (connections on one side of the cell) cells placed in the row in a back-to-back configuration. Power was distributed to the cells through the common connection on the backside of the cell. Newer standard cell layout programs support cells containing terminals on both sides of the cell. Instead of placing cells in a back-to-back configuration, each cell row contains a linear placement of the cells [6], [7] and is separated by wiring channels.

The early use of CAD for physical design verification consisted of performing simple design rule checks for width and spacing violations on mask artwork files. Functional integrity was verified through circuit simulations.

Reprinted from *IEEE Trans. Computer-Aided Des. of ICAS*, vol. CAD-1, pp. 2–12, Jan. 1982.

Initial Sandia Design Aids

After the initial decision to establish a CAD capability for integrated circuits at Sandia, software was obtained from universities and private industry wherever possible to provide a basic capability. For example, the SPICE [8] circuit analysis code was obtained from the University of California at Berkeley, and the PR2D [6] standard cell layout program for metal gate cells was obtained from RCA. Since Sandia had previously developed a simple logic analysis capability, this work was accelerated to provide gate-level simulation. Software was developed to postprocess the mask layout data to pattern generator formats for mask generation and plot file information for the Xynetics flatbed plotter. A software package was purchased from Systems, Science, and Software, Inc., and installed on an existing interactive graphic system. In addition, translator subroutines were written to convert design information data formats to minimize the manual data translation and to allow the design information to be added only once in the system. The above process is oversimplified, since a substantial commitment of staff was required to: (1) understand the acquired software and debug and correct errors, (2) perform modifications to support the software on Control Data computers, (3) identify and develop required translator software, and (4) develop additional design aids. These problems and the subsequent software modifications required a large manpower investment, since most of the software was not documented, had evolved over many years with several authors, used nonstandard Fortran, and was unstructured.

CAD System Requirements

Although the initial design aids provided a valuable capability for simple metal gate CMOS custom IC design, many deficiencies were identified. These deficiencies coupled with the need to support new technologies with shrinking design rules (and thus rapidly increasing circuit complexities), new design styles, and changing design objectives, required a new approach in the development of aids for IC design.

Several general objectives for new aids were identified. The new aids must (1) be user oriented, (2) use modular software, (3) be evolutionary to meet changing design needs, and (4) be integrated into a complete design system rather than exist as an independent collection of disjoint tools.

Computer aids must support both functional and physical design. Functional design aids include synthesis, verification, simulation, and testing at architecture, system, logic, circuit, device, and process levels. Physical design aids support partitioning, layout, and topological analysis at all design levels. Functionality, testability, and physical design must be considered in parallel throughout the design process.

All CAD software must be as technology independent as possible and support various levels of designer sophistication from inexperienced first-time users to state-of-the-art system designers. To meet these goals, the interaction between individual programs, data base, and design engineer must be through a single, consistent interface. This interface should support monitoring the progress of a design, supply options at any stage in the design process, and assist in design documen-

TABLE I

Complexity	Examples
VLSI ($\sim 10^5$ devices)	Microcomputers, cryptographic circuits, ROMs, RAMs
LSI ($\sim 10^4$ devices)	Microprocessors, A/D & D/A converters, ALUs, FFT circuits, ROMs, RAMs
MSI ($\sim 10^3$ devices)	Adders, complex gates, multiplexers, ROMs, RAMs
SSI (~ 100 devices)	AND, OR, NAND, NOR gates, buffers, memory cells
Primitive elements	Transistors, resistors, capacitors

tation and maintenance.

A system design language—or Hardware Description Language (HDL)—must be available for describing all levels of system behavior and structure. Organized in a hierarchical manner, the language should support functional and physical descriptions and the relationships among entities.

Synthesis aids must facilitate the addition of sufficient detail to generate a complete system description. Design verification software monitors internal consistency and completeness of system specifications. Final system specifications should be retained in a dynamic data base providing files in the proper format for input to each of the design aids and for documentation.

A complete CAD system satisfying the above requirements has been designed and implementation is in progress. The basic system description has been divided into three sections. The first section describes design flow and the concept of top-down design with bottom-up implementation. A hierarchical design approach is used with appropriate merging of levels to accomplish the design of VLSI systems. Requirements for specific computer aids are outlined in the second section. The final section describes the implementation philosophy, computer hardware, and present software development status.

HIERARCHICAL DESIGN

The CAD system supports a number of functions at each level in the design hierarchy. Design proceeds in a top-down sequence with bottom-up detailed implementation and addresses both functional and physical problems at each level.

Architecture, the top level of the design hierarchy, contains elements for a broad functional system description, requirements for interfacing units specifying performance and compatibility, and methods for partitioning the system into major functional blocks such as processors, memory, and I/O. The architectural specification can be expanded at the system level to generate a more detailed description consisting of register transfer level subsystem functions.

At the logic level, system functions are defined as interconnections of fundamental gates or modules. Logic modules can be further decomposed into circuit primitives (e.g., transistors, resistors, capacitors). Finally, in order to construct circuit elements and determine their behavior, the physical implementation and process technology may be considered.

To support a variety of technologies in a hierarchical design structure, a data base consisting of a library of elements of varying sophistication must be maintained. A distinct library must exist for each technology used. Typically, the hierarchi-

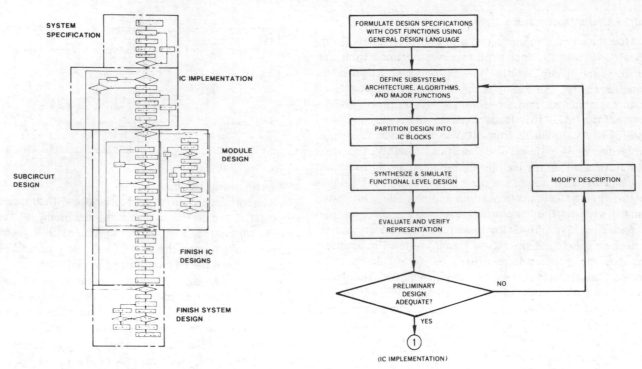

Fig. 1. System design sequence.

Fig. 2. System specification and partitioning.

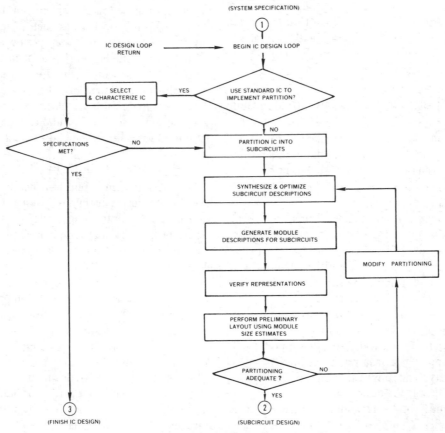

Fig. 3. IC design planning.

cal library will contain circuits and subcircuits at various levels of complexity (VLSI, LSI, MSI, SSI, circuit primitives and/or discrete devices) as defined in Table I.

Electronic system design using computer aids can be divided into six major design sections: (1) system specification and partitioning, (2) IC design planning and initial implementation, (3) subcircuit design, (4) module design, (5) completion of individual IC design, and (6) completion of system design. A design flow diagram summarizing these major functions is shown in Fig. 1.

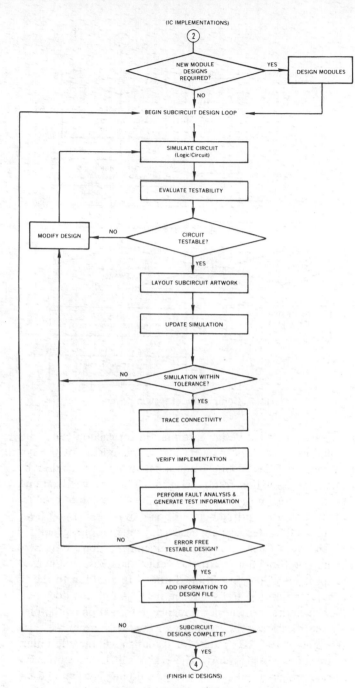

Fig. 4. Subcircuit design.

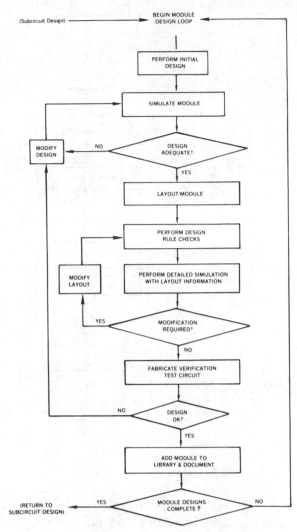

Fig. 5. Module design.

The flow diagram in Fig. 1 is subdivided and enlarged to show the individual design steps in Figs. 2–7. The first major step consists of developing system specifications and partitioning guidelines (Fig. 2). After developing general specifications, subsystems, architecture, algorithms, and major functions are defined. Each subsystem is partitioned into IC blocks which are synthesized and simulated at the system design level. This preliminary design is evaluated and modified if necessary by repeating the above steps.

The next major step in the design is implementation of each IC based on the system specification (Fig. 3). For a general electronic system, the first decision to be made is whether custom or commercial circuits will be used. If a characterized commercial circuit is available in the library, the circuit is se-

lected. If the required commercial circuit is not in the library, the specifications must be obtained and compared with the system specification. If the system specifications are met, the commercial circuit can be used; if not, a custom circuit must be designed.

The first step in designing a custom IC is partitioning the IC into subcircuits. Each subcircuit is synthesized and optimized. Module descriptions for each subcircuit are developed and verified, and a preliminary layout using the estimated module sizes is performed. At this point, a decision can be made concerning the adequacy of the partitioning. If the preliminary layout does not conform to the size restrictions for the circuit or if during the circuit verification, signal-delay paths are longer than desired, the partitioning must be modified and the circuit design repeated.

If partitioning is acceptable, the subcircuit design step can be initiated (Fig. 4). For each subcircuit design, a decision must be made concerning the adequacy of modules in the library for implementing the design; new modules must be designed and chracterized as required (Fig. 5).

During subcircuit implementation, each circuit can be simulated at various individual or combined simulation levels, the testability evaluated, and the artwork generated. After art-

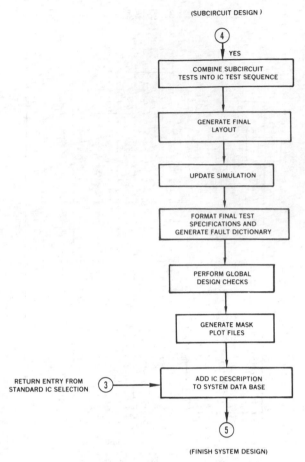

Fig. 6. Completion of IC design.

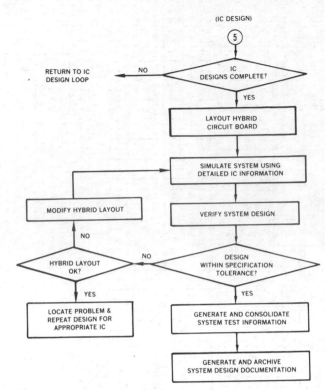

Fig. 7. Completion of system design.

work design, the simulation net list must be updated to include circuit parasitics associated with the specific layout. This simulation also provides information concerning circuit output drive and power required for subcircuit operation. If the simulation is within tolerance, static design verification is initiated; connectivity is traced; and the layout is checked for conformance to design rules. The layout and associated simulation characteristics are compared with the initial circuit specifications. Finally, test information is generated for the subcircuit.

After each subcircuit design has been completed and verified, the circuit design information is added to a design file in the data base. The next major step consists of completing the IC design by combining subcircuit design information (Fig. 6). The IC layout is assembled by placing and interconnecting subcircuit layouts. The simulation is updated to assure correct circuit operation, test specifications are generated, and a fault dictionary is developed for system diagnosis. Physical design verification, including global rule checks and connectivity checks, is performed, and the mask information for each IC is generated. Finally, design information for each IC is added to the system data base.

The last major step in the design sequence consists of completing the system design (Fig. 7). After selecting commercial circuits or designing custom IC's, layouts are performed for the system circuit boards. After the printed- or hybrid-circuit board layouts have been completed, the entire system is simu-

lated, using detailed design information for each circuit. This simulation determines the electrical characteristics of the system and provides a final functional design verification by comparing simulation results with initial design specifications. Test sequences for the individual IC's are combined to produce system tests at the printed-circuit board or hybrid-circuit level. During the final step, system design documentation is generated and archived. Generating system documentation consists of obtaining information from the design data base and consolidating it into the appropriate format. In addition to the IC mask plot files and test-sequence information, the documentation includes system performance information with tolerance specifications for acceptance of the final system.

The complete CAD system for IC design will include a number of computer programs. These aids can be categorized as follows: design specification and partitioning, system and circuit synthesis, system partitioning, simulation at various levels, IC mask layout, design verification, testability evaluation, test sequence generation, data base, and design documentation.

SPECIFIC COMPUTER AIDS REQUIREMENTS

In addition to a number of specific computer programs for supporting a hierarchical design, an extensive data base and support software for monitoring the design flow and communication are required. Requirements for specific programs and interfaces are outlined below.

Design Executive

Currently, most design automation programs exist as independent entities with idiosyncratic user interfaces and incompatible I/O structures. A Design Executive can provide a single consistent interface between a user and the complete set of computer aids available on the CAD system.

The Design Executive supports the complex CAD system structure by mediating interactions between the user and the system, the system and the data base, and the user and the data base. The executive system user interface supports comprehensive "HELP" capabilities, the HDL, and a logically integrated command language. Documentation describing the system and the CAD codes is maintained by the Design Executive in a hierarchical configuration. User access to this documentation structure is facilitated by the intelligent intervention of the Design Executive.

HDL—Design Specification

The structure and behavior of a digital system must be described in various ways at many levels to completely characterize a design. Existing "languages" such as ISP [9], DDL [10], SDL [11] and are usually associated with specific descriptions of architecture, system behavior, system structure, logical structure, circuit structure, logical behavior, or physical structure. These descriptions lack the generality required to support VLSI design, since most are applicable to one level of design and a particular design style and are not integrated into a complete design system.

Ultimately, a comprehensive HDL must be used to describe all levels of system behavior and structure; including normal and faulty electrical operation, logical and functional behavior, and physical structure. Organized in a hierarchical manner, the HDL should allow description of functional and physical entities and relationships among entities.

Synthesis and Functional Verification

Functional synthesis begins at the architecture level with a description of the overall behavior of a large system and interaction with the environment. As synthesis proceeds, more structural detail is added at system, logic, and circuit levels in the form of interconnection structures. Concurrently, behavior specifications are developed at each level of the design hierarchy.

During synthesis, the actual behavior of the system must be forced to match the specifications. To effect the match, models are required for computational algorithms and interpreters, and procedures for mapping one into the other. Models must be expressible in a computer-readable form in order to manage complexity with an interactive computer support system, permit the separation of structure from associated behavior, and support direct fabrication of the synthesized system.

Simulation

Simulation, or dynamic design verification, is the process of calculating the behavior of a system within an environment specified by the designer. The objective of simulation is to verify that the system will perform correctly in an operational environment.

Ideally, simulation should be multilevel; i.e., consider larger circuits at a less detailed level with the capability of simultaneously simulating subcircuits more exactly. This concept complements the hierarchical design implementation. In a hierarchical simulator, the basic elements are more complex than simple modules or gates. Models are formulated for entire IC's and include gate, function, and transistor models. Parts of a system are simulated in great detail while other parts are simulated abstractly. Models are written in high-level languages, with models for MSI or LSI blocks only slightly more complex than the model for a simple NAND gate. Thus a system of many thousands of devices or gates can be reduced to a few hundred hierarchical models, making it feasible to simulate entire VLSI systems.

Testability

Testing is the process of verifying that a system is operating properly. Because of the increased complexity of VLSI circuits, the difficulty of testing is substantially increased. To ensure that a circuit can be tested, design-for-testability techniques must be used throughout the design process.

Approaches to design-for-testability fall into two major categories: testable design styles and testability measure analysis. The use of a testable design style guarantees that generating tests for a circuit will not be impossible. Testability measure analysis computes the difficulty of controlling and observing each internal node from primary input or output pads. This information is used in the design process to locate potential circuit testing problems and provide feedback about the effect of circuit modifications on testability.

In developing a test sequence, knowing that a good test exists is not the same as knowing the test. Deriving a test is the task of the test sequence generation phase. Given a digital circuit and a set of possible faults, a series of input signals (vectors) is generated that will force any faulty circuit to behave differently from the fault-free circuit. The test sequences depend not only upon the intended function of the circuit but also upon the fault set assumed. Test-sequence generation must proceed concurrently with the hierarchical design process.

Physical Design Aids

Physical design aids include tools to partition, place, and interconnect circuit components and verification tools to ensure the synthesis has been performed properly. To provide an optimum system, physical design must be considered in parallel with functional design and testability.

Partitioning—Partitioning programs operate on both functional and physical entities. Partitioning techniques must function in both the initial design and detailed implementation phases. During top-down design, partitioning aids are needed for hierarchical decomposition. During bottom-up implementation, partitions are modified based on lower level implementations.

In addition to optimizing a circuit element assignment based on size and external pin connections, parameters such as circuit speed, power dissipation, and functional groupings must be considered. The partitioning aids must be capable of optimizing any of these quantities as a function of the others.

Symbolic layout—Provides a shorthand method for manually sketching a circuit layout using specified symbols to represent various types of transistors and interconnections. The initial layout can be performed on grided paper or directly on an interactive graphics system. During layout, the designer is not concerned with the geometric design rules. Symbolic layouts are postprocessed using computer aids which expand and relocate all symbols and interconnects based on the geometric design rules for the specified technology to provide mask artwork files.

Physical layout—The physical layout phase of IC design involves positioning and interconnecting electrical components. In hierarchical design, the concept of a component is generalized. Components may range in complexity from primitive transistors and fundamental gates to microprocessors that can be combined to form a microcomputer.

Hierarchical layout is applicable to a wide range of physical design levels. At the highest design level, the shape of LSI components may be adjusted and combined to form a VLSI circuit. At the lower design levels, the same algorithms may be used to combine gates into registers or transistors into gates.

Topological analysis—Because of the high costs and long lead times involved in the fabrication of IC's, it is important to verify the correctness of a circuit design before manufacture. Verification tools must ensure correct physical mask layout, functional operation, and that electrical characteristics are within specified tolerances.

Topological analysis tools can be used to verify correctness of physical layout data by examining the artwork data and the circuit inputs and outputs. Design-rule checking codes perform geometrical, logical, and topological operations on artwork data and compare the results with design rules for the specified technology. Connectivity and electrical parameter data are used to reconstruct a detailed circuit including parasitic electrical components. This reconstructed circuit can be used in performing an accurate functional and timing analysis.

Circuit Design Documentation

After a circuit or system design has been completed, the design can be documented by collecting information from the data base generated during the design process. Design documentation includes information for fabricating the IC's and system as well as information for archiving the completed design to support later modifications. Computer aids should collect manufacturing and archival information and generate appropriate files, specifications, and reports.

Data Base

In general, the distinction between programs and data is that programs are active, data is passive (i.e., programs operate on data). The necessity for supporting the initial design, design changes, and providing software transportability dictates consistency in data handling procedures and mechanisms. Therefore, an efficient data base is vital for coordinating the various functions. As systems become more complex, it is imperative that an overall data base be established and maintained to ensure consistency in design and to eliminate duplication of information.

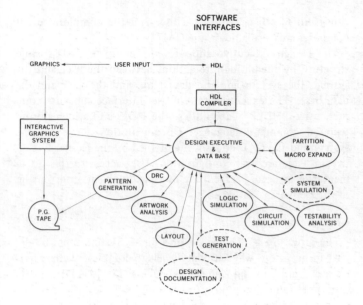

Fig. 8. Sandia hierarchical CAD system.

The data base must be designed in a flexible and modular fashion to provide upward compatibility with computer hardware and the evolving CAD software. Modularity is important where various portions of the data base may reside at more than one computer node. In addition to the actual data used by CAD programs, the data base should contain all necessary information to properly document elements being designed.

SANDIA CAD SYSTEM IMPLEMENTATION

A CAD System meeting the criteria outlined above is being implemented in phases at Sandia National Laboratories to provide design capability—for LSI circuits initially, with evolutionary expansion and enhancement as required to support VLSI designs.

The design procedures consist of: (1) identifying critical aids in the design process based on user needs, (2) acquiring existing software when available and subsequently modifying it to meet system and user needs, (3) developing new aids with appropriate expansion capabilities to support hierarchical design, and (4) integration of all aids into the design system framework. A block diagram outlining the present system implementation is shown in Fig. 8. The solid blocks represent areas where design support is presently available, and the dashed blocks indicate work in progress. Brief descriptions of the hardware and software are given below.

Hardware

The software is supported on a dedicated DECSystem 20 computer system containing 1 million words of main memory, 160 million words of disk files, and two 9-track 800/1600 bit/in, 75 in/s tape drives. Two Applicon 860 interactive graphics systems and a Versatec 36" electrostatic plotter are connected to the DECSystem 20. A VAX 11/780 computer containing 4 megabytes of main memory, 323 megabytes of disk files, and 2 9-track 800/1600 bit/in, 125 in/s tape drives is interfaced in a network configuration to provide additional computing capability. Communication among the computers and the Applicon systems is supported by DECnet. Both dial-up and direct-wired access to the system provide support for alphanumeric and graphics terminals.

Design Software

A Universal HDL (UHDL) is being developed in essentially three stages consisting of: (1) language formation by phrase concatenation, (2) redundancy removal, and (3) consolidation and rewrite of the compound language. During the first step, the SDL [11] language (a PASCAL-like description) is used to describe system structure. Other languages are appended in phrase form to describe functional behavior. After gaining experience with the composite languages and after the deficiencies and required enhancements have been identified, a language for the initial UHDL will be written. Continuous refinement and enhancement of the UHDL will be required to meet design and documentation needs.

Translation from UHDL descriptions (initially SDL) to the required formats for simulation, layout, and verification programs is performed by the Design Executive. The Design Executive also manages the data flow between the Data Base files (outlined below) and each of the design aids.

Design synthesis aids currently available include MINI [12] and DDA [13] (Digital Design Aids). MINI, a Programmable Logic Array (PLA) minimization program uses a branch-and-bound algorithm to produce the optimal two-level realization of a set of Boolean equations. DDA provides synthesis of a sequential-state machine based on flow diagram, state assignment, and flip-flop type.

Logic-level design consists of defining, in complete detail, all components (gates, cells, modules, etc.) and interconnections required to implement a specified function. The SALOGS [14] (SAndia LOGic Simulator) program performs true-value 4-state logic simulation (true, false, undefined, and high impedance) and 8-state timing simulation (four states plus transition to each of the four states) as well as states-applied and gate activity analysis and fault simulation. SALOGS is used for dynamic design verification, test sequence evaluation, and fault coverage calculations. Race and hazard analysis is available and the program is capable of using library or user-defined models.

A typical input/output for SALOGS is shown in Fig. 9 for a small logic circuit. The input consists of a connectivity description of logic gates or cells contained in the model library. The plotted output describes the logic signal changes as a function of time for each of the nodes indicated.

Although approximate timing simulation is accomplished in SALOGS through the use of extra states to represent logic values in transition, more exact timing simulation is possible with MOTIS-C [15] and SIMPIL [16] for MOS and I^2L circuits, respectively. These programs use tables to describe active devices in a circuit instead of models with complex equations. Both programs provide more accurate analysis than logic simulators and run more efficiently than circuit simulation codes such as SPICE.

A more exact circuit analysis is obtained by using the SANCA (SANdia Circuit Analysis) program circuit simulation program based on SPICE. This program simulates MOS and bipolar

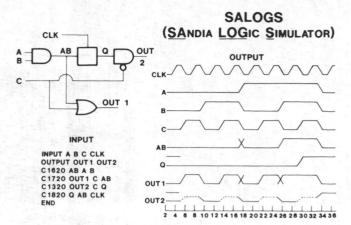

Fig. 9. Typical logic simulation input and output using SALOGS.

circuits, including analog and digital circuits, by numerically solving the network equations to calculate the desired voltages and currents. The cost of increased accuracy (over timing simulation) is a decrease in both execution speed and circuit size that can be analyzed. Model libraries [17] for circuit analysis are maintained to reflect varying levels of accuracy and different modeling philosophies. Circuit models based on device physics, empirical behavior, and hybrid approaches are all used in the circuit simulation environment. SANCA is an enhanced, quasi-interactive, topological analysis program containing model information for specific Sandia technologies and output plot routines.

Physical mask layout is performed using the SICLOPS [18], [19] (Sandia IC Layout Optimization System) and SLOOP [20] (Standard cell LayOut Optimization Program) programs to automatically place and interconnect generalized circuit elements. Two different forms of mask layout are used. A hierarchical mask layout system has been developed and used for specific designs. Hierarchical layout is applicable to a wide range of design levels. At one extreme, LSI components can be combined to form a VLSI circuit. At the other extreme, the same algorithms can be used to combine gates into registers based on a standard cell layout. This hierarchical approach relies on the use of SICLOPS to perform an automatic layout of an integrated circuit using rectangular-shaped blocks. The blocks must be rectangular and can be of arbitrary size and aspect ratio. The program automatically places the blocks and performs the interconnections. Each of the blocks can contain nested subblocks to any depth to provide a hierarchical layout. SICLOPS can also be used to layout a thick film hybrid circuit, as shown in Fig. 10.

The more conventional layout using standard cells placed in rows with interconnection between cell rows in routing channels is supported by SLOOP. In addition to designing modules used by SICLOPS, conventional standard cell IC layouts can be designed. The basic program can be used in an interactive mode to optimize the design, will perform a 100-percent completion of all interconnections by expanding the chip size as required to complete interconnections, contains hierarchical interfaces for use with the SICLOPS code, and supports multiport gates or cells. A typical IC design using SLOOP is shown in Fig. 11.

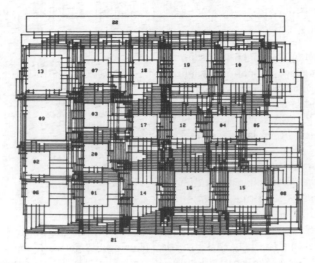

Fig. 10. Thick-film hybrid circuit layout obtained using SICLOPS.

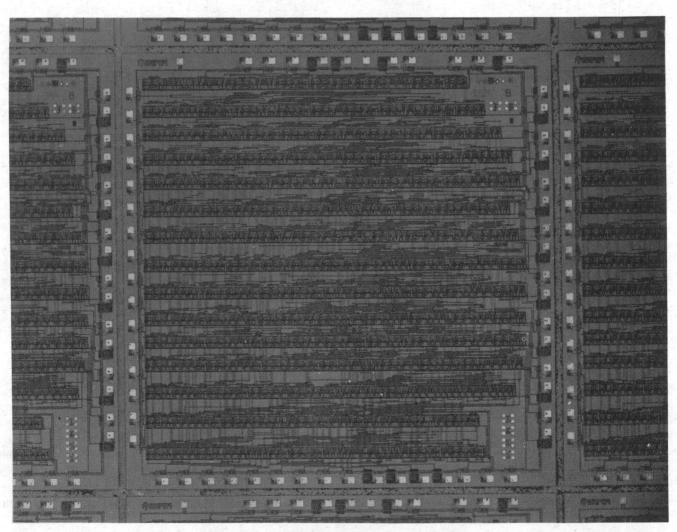

Fig. 11. Typical standard cell IC design using SLOOP.

The LOGMASC (LOGical MASk Checking) [21] program is a tool for design rule checking of IC masks. This program produces Boolean logic combinations of mask levels and is based on pattern recognition techniques. LOGMASC provides essentially all design rule checks which can be performed manually and provides pattern identification and extensive topological and geometrical information concerning the interrelationships of entities in the file. False errors are minimized by selectively applying design rule checks based on the circuit use of the particular patterns begin checked. Many of the algorithms can be used to solve other mask analysis problems in addition to the design rule checks.

During the circuit mask design process, transistor current or voltage gains, resistance, and capacitance values are scaled and converted to area definitions based on the nominal circuit parameter values used for circuit analysis. The actual sizes of the circuit elements and devices are adjusted to fit layout constraints based on minimum spacing, relative element partitioning, and size requirements. The actual circuit defined by the mask layout may be quite different than the original circuit schematic. For example, the interconnection distances between transistors introduce resistances and capacitances into the circuit which were not included in the initial circuit simulation. Additional parasitic transistors and diodes can be unintentionally introduced into the circuit which can cause improper operation. The CMAT [22] (Circuit MAsk Translator) code operates on circuit mask information, using pattern recognition to recognize and transcribe the mask areas into the circuit elements. CMAT employs the basic Boolean operations algorithms contained in LOGMASC and performs the required scaling operations to obtain circuit element values.

SCOAP [23] (Sandia Controllability Observability Analysis Program) is used to evaluate the testability of a circuit after the basic circuit design has been completed. This analysis is performed prior to layout and is based on a simple analysis of the logic interconnectivity. The SCOAP program calculates the ease or difficulty of setting an internal node from a primary input to a zero or one and the difficulty of observing a logic value at any internal node from a primary output. The zero and one values for controllability may be different depending upon the exact circuitry and interconnection of the logic gates. For nodes having very high values for controllability or observability changes are usually required in the logic circuitry to reduce the controllability and observability numbers. This reduction is required, since the numbers are proportional to the testing difficulty. For nodes with high controllability/observability measures, additional input or output test points or signal multiplexing may be required to set logic gates or observe node logic values to achieve circuit testability. At present, a structured design for testability, such as the IBM LSSD [24] approach, is not required. However, since the designer is responsible for designing a testable circuit, techniques equivalent to the LSSD approach may be used in the design based on the testability analysis information.

The testability analysis program is used for both combinational and sequential circuits. Combinational measures basically relate to the number of different line assignments or input assignments which must be made to set an internal node to a

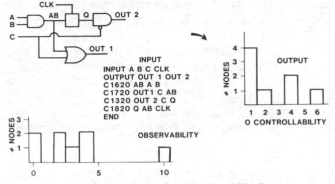

SCOAP
(SANDIA CONTROLLABILITY
OBSERVABILITY ANALYSIS PROGRAM)

Fig. 12. Testability analysis using SCOAP.

given value. For sequential circuits, the measures relate to the number of clock cycles or time frames required to propagate a specified logic value from an input pad to a node or from a specific node to an output terminal.

An example at testability analysis is shown in Fig. 12. All input nodes are directly controllable as indicated for the four nodes with a "zero controllability" value of one. The most difficult node to control is usually an output node; for this example, output 2 has a zero controllability value of six. The observability is shown in the lower left of the figure. Again, the most observable nodes are the output nodes as indicated by the two nodes with an observability value of one. Ordinarily the most difficult nodes to observe are the input nodes and, therefore, the nodes A, B, C, and CLK should be the least observable. In this case, the CLK signal is input to a transmission gate between nodes AB and Q; a change of state at node AB must occur to observe the clock signal and, therefore, the CLK signal has the large value of 10 for observability.

Data Base

Because of the initial need for integrating loosely related CAD software and evolving design system specifications, a data–file base has been developed. The Sandia CAD data structure [25] is in the form of a "deciduous tree"; that is, one that can lose its leaves. This structure is a two-dimensional network with several restrictions. In one dimension the data structure forms a shallow tree one level deep. That is, any number of data types can be associated with a given node, but they may not be broken down further. In the other dimension, the structure is a network of nodes and subnodes to any depth. The leaves on the tree are the equivalent of small sequential files of data which can be retrieved by the user application program. This data file approach forces the designer to conform to a hierarchical structure while maintaining the freedom to store test files, documentation, etc., in the same file.

SUMMARY AND FUTURE PLANS

Although a basic computer-aided LSI design capability has been established and is used by design engineers, the system is continuously evolving as new capabilities and enhancements are added. The modular programs and data structures, as well as the flexibility designed into the overall system framework, tend to minimize the cost for system modification as requirements change. In addition to continuous enhancement of existing programs, new aids for design synthesis, hierarchical simulation, symbolic layout, and test sequence generation will be developed. A continuing emphasis is placed on integrating the CAD software into a complete design system.

REFERENCES

[1] G. Kron, *Tensor Analysis of Network*. New York: Wiley, 1939.
[2] —, "A set of principles to interconnect the solutions of physical systems," *J. Appl. Phys.*, vol. 24, pp. 965–980.
[3] A. F. Malmberg, F. L. Cornwell, and F. N. Hofer, "NET-1 network analysis program," Los Alamos Rep. LA-3119, 1964.
[4] R. W. Jensen and M. D. Lieberman, *The IBM Circuit Analysis Program*. Englewood Cliffs, NJ: Prentice-Hall, 1968.
[5] A. Feller and M. D. Agostino, "Computer-aided mask artwork generation for IC arrays," in *Dig. 1968 IEEE Computer Group Conf.*, pp. 23–26.
[6] A. Feller, "Automatic layout of low-cost quick-turnaround random-logic custom LSI devices," in *Proc. 13th Design Automation Conf.*, pp. 79–85, June 1976.
[7] G. Persky, D. N. Deutsch, and D. G. Schweikert, "LTX–A system for the directed automatic design of LSI circuits," in *Proc. 13th Design Automation Conf.*, pp. 399–407, June 1976.
[8] E. Cohen, Program reference for SPICE 2, Memo ERL-M592, Electronics Research Laboratory, Univ. California at Berkeley, June 1976.
[9] M. R. Barbacci *et al.*, The ISPS computer description language, Tech. Rep Dept. of Computer Science, Carnegie-Mellon Univ., Pittsburgh, PA, Mar. 1978.
[10] W. E. Cory *et al.*, An introduction to the DDL-P language, Tech. Rep. 163, Computer Systems Lab., Stanford Univ., Stanford, CA, Mar. 1979.
[11] W. M. van Cleemput, A structural design language for computer-aided design of digital systems, Tech. Rep. 136, Digital Systems Lab., Stanford Univ., Stanford, CA, Apr. 1977.
[12] J. Abraham, MINI–A logic minimization program, Univ. Illinois, Urbana, IL.
[13] DDA–Digital design aids for state machine synthesis, Tektronix, private communication.
[14] J. M. Acken and J. D. Stauffer, "Logic circuit simulation," *IEEE Circuits Syst. Mag.*, vol 1, no. 2, pp. 3–12, June 1979.
[15] S. P. Fan *et al.*, "MOTIS-C: A new circuit simulator for MOS LSI circuits," in *Proc. IEEE 1977 Int. Symp. on Circuits and Systems*, pp. 700–703, Apr. 1977.
[16] G. R. Boyle, *Simulation of Integrated Injection Logic*, Memo. UCB/ERL M78/13, Electron. Res. Lab., Univ. California at Berkeley, Mar. 1978.
[17] D. R. Alexander, R. J. Antinone, and G. W. Brown, *SPICE 2 Modeling Handbook*, BDM/A-77-071-TR, May 1977.
[18] B. T. Preas and W. M. van Cleemput, "Placement algorithms for arbitrarily shaped blocks," in *Proc. 16th Design Automation Conf.*, June 1979.
[19] B. T. Preas and W. M. vanCleemput, "Routing algorithms for hierarchical IC layout," in *Proc. Int. Symp. Circuits and Systems*, July 1979.
[20] SLOOP (Standard-Cell Layout Optimization Program), Sandia Lab., to be published.
[21] B. W. Lindsay and B. T. Preas, "Design rule checking and analysis of IC mask designs," in *Proc. 13th Design Automation Conf.*, June 1976.
[22] B. T. Preas *et al.*, "Automatic circuit analysis based on mask information," in *Proc. 13th Design Automation Conf.*, June 1976.
[23] L. H. Goldstein, "Controllability observability analysis of digital circuits," *IEEE Trans. Circuits Syst.*, vol. CAS-26, Sept. 1979.
[24] E. B. Eichelberger and T. W. Williams, "A logic design structure for LSI testability," in *Proc. 14th Design Automation Conf.*, pp. 462–468, June 1977.
[25] J. D. Stauffer, "SADIST (The Sandia Data Index Structure)," Sandia National Lab., Albuquerque, NM, SAND80-1999, Nov. 1980.

IBM's Engineering Design System Support for VLSI Design and Verification

Larry N. Dunn, IBM

VLSI has fundamentally changed the relationship of the design automation tool developer and the product designer. IBM's design and verification subsystem responds to that change.

Many authors and design automation practitioners have postulated a design system that can describe a large, complex design at a behavioral level, rapidly verify its functions, and then provide the data required to automatically generate the physical implementation. This article presents IBM's first implementation of such a design and verification, or DAV, system.

The IBM DAV system is the result of prior research, which took place both within and without IBM. Also, user requirements studies, specific research, and advanced development projects were important sources of concepts incorporated into DAV. These ranged from design modeling and RTL support to verification and synthesis.[1-14] DAV concepts previously published were system overview and design entry, verification, and design transformations.[15-19]

VLSI and productivity

VLSI promises lower cost per circuit and better product performance. However, an effective product strategy requires the developer to provide more functions with higher performance on a shorter design schedule, while holding development costs constant. In other words, VLSI requires increased productivity.

DAV objectives. The scope of our development effort was to extend the existing Engineering Design System, or EDS, support from physical design and test generation to engineering design, which absorbs more than 50 percent of designer manpower (Figure 1). The stated goal for DAV is: get the product to the marketplace in 30 percent less time with 40 percent lower design cost and with 98 percent of the errors removed prior to hardware prototyping.

The original, 1978 DAV system objectives defined extensive database support and a tightly coupled system environment. Our objectives were scaled down, in recognition of the problems inherent in implementing a tool set, a system structure, and a database even as we educated future users in the benefits of structured design. Currently, most of our effort is in high-leverage tools loosely coupled to a common model with hooks for future migration to a relational database.

Tool development progressed from Boolean comparison (1978), to functional simulation (1979), to logic transformation (1981), to prototype entry/system functions (1982). The experience of developing this initial set of tools will prove invaluable as we continue toward our original objective.

Philosophy

Where verification emphasizes efficient error removal, DAV emphasizes avoiding injection of errors into the logic design. Figure 2 illustrates IBM's migration from verification to design and verification. It shows we are approaching the

An earlier version of this article appeared in the *ACM/IEEE 20th Design Automation Conference Proceedings*, June 1983.

Reprinted from *IEEE Des. Test Comput.*, vol. 1, pp. 30–40, Feb. 1984.

goal: the ability to bring up hardware that works correctly the first time, every time.

Three principles. Three basic principles guided our design.

Error prevention. The first was that *not* injecting errors into the design is preferable to error detection. Key requirements, then, were to reduce the amount of manual designer input required to define a digital logic product and to ensure error-free input. We also concentrated on new applications that could eliminate tasks, rather than obtaining subsecond response times on current applications.

We needed a way to describe a design at a level higher than the register-transfer level employed by the IBM 3081 project—a way that did not require programming experience on the part of the designer, yet could be "synthesized" to a gate-level network description.

At IBM, it is extremely difficult to define and enforce a set of standard components or macros. Differences in performance, space, and function leave room for cost-justified variations requiring thousands of predefined macros.

This led us to define generic operators. The generic operator resembles a predefined macro backed by a library of behaviors, but there is a difference. The DAV system provides a basic set of operators representing logical functions. The engineer selects parameters that represent a unique usage (a maximum of 10 keystrokes). The system input automatically provides a Polish string representation for simulation, as well as an equivalent logic model for early logical and physical checking and for eventual mapping to a design implementation.

Besides the generic operators, the user can define any other desired function as a functional entity. This entity can then be defined by another logic specification, which details its structure in terms of other functional entities or generic operators. This hierarchy can be carried to any depth (Figure 3), although eight levels is the

practical limit requested by our designers.

Logic specs can contain up to 999 logic sheets, each of which can contain any desired mix of generic operators and functional entities.

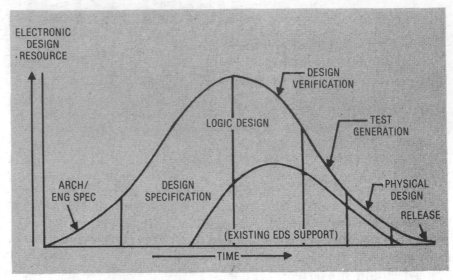

Figure 1. Logic design resource distribution.

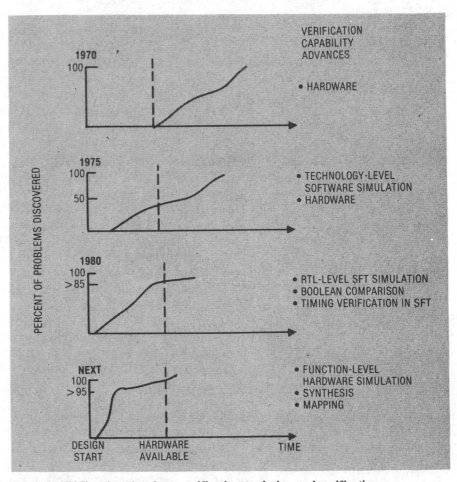

Figure 2. IBM's migration from verification to design and verification.

However, the lowest level in the hierarchy can contain only generic operators. At each level, several different specs can describe a given functional entity; each spec represents an alternative design or implementation.

No one has all the answers to the complete set of design problems at any point in time.

The use of hierarchy reduces the model-build storage requirements. Generic operators significantly reduce the time-consuming task of behavior model definition; direct handling of these operators facilitates designer use of hierarchy during the verification process and relieves the designer of the burden of manually redefining these functions at the primitive logic level.

Analytical proof of correctness. The second principle stated that analytical proof of correctness should be used whenever possible. We decided to continue to try to obtain definitive design verification rather than use heuristic techniques, such as

nonexhaustive simulation, at the detailed logic level.

The IBM 3081 project design verification strategy demonstrated that this concept is practical, if functional and timing verification are separated. The effect of this separation simplifies the verification process by ensuring that a correct functional design is in place before the analytic timing analysis tool is run. The designer can isolate a timing problem, fix it, and use the Boolean analysis tool to verify that the change did not inject a functional error. Note that successful application of Boolean analysis requires use of a structured design methodology that allows for practical comparison of two representations of the design for a given product.

No one has all the answers. The third principle is more an assumption: No one has all the answers to the complete set of design problems at any point in time.

We recognized early that we would not be able to satisfy the wide range of requirements within IBM. Therefore, we provided a modular architecture that allowed addition of unique, project-developed tools and local modification of job setup and display menus. Within the application tools, we established user-program inter-

faces allowing extensions to specific applications. In several cases, these interfaces were developed as command languages, so engineers without programming experience could add specific functions not provided by the system.

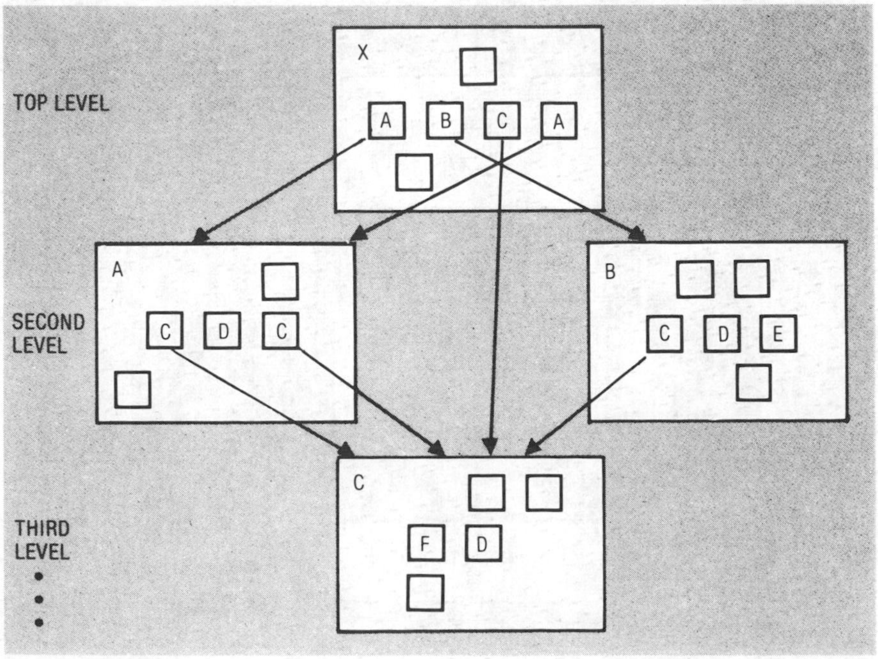

Methodologies

Figure 4 is a simplified view of the design methodology. The emphasis is on structured design, be it top-down, middle-out, or bottom-up. Computer-aided synthesis and design transformation coupled with definitive design verification get preference. Each step in the Figure 4 flowchart has some key characteristics.

Step 1. Specification of the functional-level design:
- graphical entry via dual-screen terminal
- hierarchic structured design
- generic operators
- bundle notation

Step 2. Verification of the functional-level design:
- functional simulation
- timing verification

Step 3. Translation to a specific technology:
- automatically by the logic transformation system
- manually by the designer
- feedback from physical design system

Step 4. Verification of the technology-level design:
- Boolean comparison between the two levels
- timing analysis using technology defaults and constraints

Step 5. Checking for chip testing and physical design:
- physical design rules checks
- level-sensitive scan design rules and compatibility checks

Step 6. Transmitting the design to the physical design system:
- partitioning, placement, wiring, etc.
- test pattern generation
- timing analysis based on physical implementation
- manufacturing process controls

Figure 3. Hierarchy.

367

The first five steps are within the DAV set of programs. Step 6 is performed by other subsystems of the IBM Engineering Design System.[6,20] All proposed tools are now implemented, in production, or in prototype programs.

Functional elements

Step 1. Design specification. The primary entry form in DAV is graphic.[16,17] Entry is via the IBM 3277 GA workstation, which consists of an IBM 3277 alphanumeric terminal, a medium-function graphics terminal, and a joystick, with optional light pen, data tablet, and electrostatic printer. We chose this configuration for its cost, early availability, and ability to handle very detailed images. The dual-screen terminal allows display of the full picture of the logic sheet on the graphics screen while an option menu, tutorial, or alphanumeric edit menu appears on the alphanumeric screen. The 3277 can be used stand-alone in applications that do not require the graphics screen. Graphics operations include, among many others, adding, moving, copying, editing, and deleting blocks; adding, deleting, and modifying nets, using either manual or automatic line routing; adding, deleting, and changing intersheet or external connections.

Bundling. Bundling indicates that one line on the diagram represents many connections in the circuitry. Connections can be numerically related, logically related, or unrelated.

For numerically related data, such as the bits in an integer operand of an ALU, the bundles can be dimensioned. Up to three dimensions are provided for the data dimensions; up to two dimensions are provided for parity.

For example, a bundle could have the dimensions 2 256 32 P 4 16, meaning that it is a bus with $2 \times 256 \times 32$ data bits and 64 parity bits in a 4×16 configuration. For logically related bits, such as the control bits of a register, or for unrelated bits, an alphanumeric strand notation is provided. The bits are given names, numbers, or ranges of numbers. For example, 15, +NC,T0:3 could describe a six-strand bundle whose strands are named 15, +NC, T0, T1, T2, and T3.

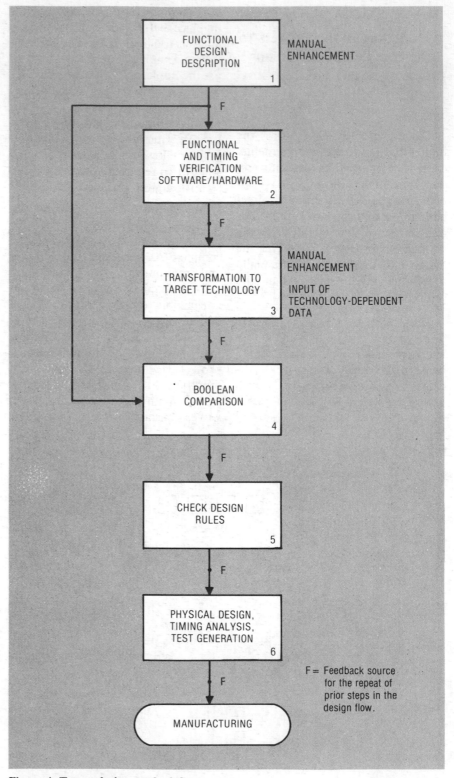

Figure 4. Target design methodology.

368

The bundling notation reduces routing clutter and allows the engineer to enter all desired interconnections very quickly.

Generic operators. DAV supplies generic operators that run the gamut from the most elementary, such as AND and INVERT, to complex, such as registers, multiplexers, decoders, and ALUs. These operators do not comprise a prestored set of definitions, but a set of macros that can be tailored to any need.

The advent of VLSI poses new challenges, in that the verification goal now aims for 100 percent error removal prior to hardware commitment.

For example, if an engineer wants an operator that adds a 12-bit input to a 15-bit input to produce a 17-bit output, the dimensions on the block pins can be edited to these values, and the system will generate the desired behavior or equivalent logic model. Through bundling notation, even the elementary operators become high-level. For example, a NAND block with its inputs and outputs dimensioned 16 represents 16 separate NAND blocks operating in parallel.

The behaviors of the generic operators are either built into the various DAV application programs or generated by a generic operator processor. Generated behaviors can take the form of Polish strings or equivalent logic models. An ELM is a network of unbundled usages of generic operators in which the level of abstraction depends on which primitives the application accepts.

Step 2. Functional simulation. During the past two decades, logic simulation systems have been the workhorses that met the major portion of the logic designer's verification needs. However, the verification process invariably moved to a hardware-debug phase, to remove the re-

maining set of design errors. The advent of VLSI poses new challenges, in that the verification goal now aims for 100 percent error removal prior to hardware commitment.

As with most changes in any way of doing business, there is resistance to this new approach to hardware design. Some of the resistance to top-down, multilevel design comes from those who do not accept the idea that more resources applied early result in significant savings in the field. However, there is growing recognition that multilevel modeling can reduce manpower requirements in the implementation phase and shrink the overall design cycle.

Three simulators. Three key areas in a logic simulation system for VLSI designs are hierarchy, interactivity, and performance.

Hierarchy is clearly required at IBM. We support top-down, middle-out, and bottom-up methodologies that allow the user to simulate at different levels of detail.

Flexible and rapid interaction is especially needed early in the verification cycle, when frequent design changes are being made, but it is also needed later in the cycle, for solving subtle design problems.

Finally, performance becomes more critical as the verification cycle progresses, because larger experiments must run longer before each subsequent error can be found.

We recognized that such diverse goals required three simulators.

The first was the experimental functional simulator, which provided specialized support for synchronous designs involving rapid model-build and simulation of major system components. EFS also provided the environment for experimenting with interactive functions.

The techniques developed from this experimentation led to the interactive functional simulator. IFS addressed the requirements for interactivity for asynchronous designs.

The third simulator, the engineering verification engine, provided total system verification with the performance capability of a hardware en-

gine: EVE is based on the Yorktown Simulation Engine—YSE[11]—which, in turn, is an enhancement of IBM's first logic simulation machine—LSM.[10]

EFS. The experimental functional simulator is a stand-alone tool used to simulate a logic design described in the BDL/CS form of functional design specification. BDL/CS stands for basic design language for cycle simulation. As the name implies, it is a cycle simulator for logic that is clocked on a cycle boundary. EFS was based on the experience gained in using the variable mesh simulator.[13,21]

Basically, cyclic simulation consists of determining the next state of the machine as a function of the machine's present input values and its present state. Determination of the next state is wholly combinatorial; hence, simulation of parallel processing is an essential part of the simulator.

The BDL/CS language allows generation of hardware simulation models that closely match the design algorithms. Programming sequence considerations are transparent to the user. The cycle of the simulation model corresponds directly to the real machine cycle; signals are generated in proper sequence, but no attempt is made to match internal logic delays.

In the early stages of design, most designers are responsible for specific areas or subunits. The designer is most interested in simple test cases that allow basic test patterns to feed the logic; he then observes the actions on a cycle-by-cycle basis. In most cases, this functional testing leads to problem definition and analysis testing. At this design stage, it is most important to pinpoint failures rapidly so that a functioning subunit can be delivered for integration with other subunits for gradual construction of an entire functioning unit.

Subunit-level test cases should, therefore, be designed so as to be easily generated and understood by the designer. In EFS, it is possible to display the outputs of the logic on a cycle-by-cycle basis, so the actions of the logic are visible to the designer. Later on, the results of individual cycles are less interesting than the final result.

As the design proceeds, the subunits are integrated with other subunits to form entire processors or control units, and the need for broader tests becomes apparent. This is when instruction- or command-level tests are needed and the speed and capacity of a hardware engine are required. Although the need for subunit level test cases diminishes when the subunits are integrated into a full unit, they can still be a valuable tool for problem analysis.

Since the simulator executes directly from a hierarchically arranged set of stored tables, modifications to the design can be incremental.

IFS. IFS is a versatile, interactive functional simulator.[18] Input and output are shown as waveform diagrams. The simulator provides a variety of breakpoint and tracing facilities.

One of the simulator's most interesting human-factors features is its use of the logic diagram. During simulation, the user can select any sheet of any logic spec in the hierarchy. Simulation results can be displayed directly on the diagram with the value at each output pin superimposed on the net. Use of the hierarchy and the incremental model-build processor allow quick invocation of the simulator after a logic change is entered. Since the simulator executes directly from a hierarchically arranged set of stored tables, modifications to the design specification can be incorporated incrementally. That is, if only one node in the hierarchy is changed, only that node's table need be replaced in the model. If no design change has been made since the last experiment, the model is simply reloaded.

Interactive stimulus and debugging aids are addressed via a command

and waveform entry function, which operates during user breakpoints. This function presents an interactive display of previously entered test patterns and user command fields. Patterns can be changed by keying over the existing pattern image. Shorthand notation is supported, to ease the process of entering repetitive patterns. The user can insert, change, and delete any pattern during any breakpoint. These pattern sets can be stored, retrieved, and modified to drive simulation experiments.

Line commands can be entered, providing a standard text-editing capability for the patterns. This capability is important, because it gives the user maximum flexibility during an interactive experiment. The user can change patterns during any experiment, without restart or cleanup before resuming simulation. These patterns can be applied to any net in the model. Thus, different levels in a hierarchical design can be simultaneously verified during one experiment.

Simulation efficiency in such an environment depends on the representation of the logic selected. The range of performance is 25,000 to 35,000 (primitive logic) blocks per second to 2000 (eight-bit counter) blocks per second on an IBM 3081.

EVE. Architecturally similar to YSE, the engineering verification engine performs simulations by executing instructions representing gates on a number of processors connected in parallel. Additional hardware units, including array processors for simulating (embedded) arrays, support these logic processors.

The set of hardware units is referred to as the "engine." The basic simulated unit is a four-input, one-output gate. This gate can accept and produce four-valued signals represented by two bits. The values are: zero, one, undefined, and three-state high. Function codes defined by truth tables determine the computation of output values for each gate. Users can modify and extend these truth tables from the standard set pro-

vided. Typical activities supported by EVE are

- high-level simulation, for verification and system-level testing utilizing the same models and test cases as EFS;
- detailed logic design verification at the unit, subsystem, and system level; and
- verification of diagnostic programs by inserting bugs and running the diagnostic programs on the simulated machine.

The primary job of the EVE host software is converting the user's design data to a form the special-purpose hardware can recognize. This conversion occurs in several steps.

The first is to translate the design data into a flat model format, called a nodes file. In this format, logic is described as a network of nodes (gates). Each node is described in terms of its simulation function and interconnecting net names. EVE can generate a nodes file from a technology-dependent or technology-independent design language base.

The compiler program performs the next step, which is actually a double step.

First, each node file record (gate) must be converted to one or more logic processor instructions, which compute an output value based on the gate's assigned function code and the input values. Second, the instructions must be spread evenly across available processors. The output of this step is a load module containing the collection of data to be entered into the various memories of the EVE hardware.

An optional step is to combine load modules. This is helpful when the user anticipates considerable incremental updating. Test cases are then translated to virtual logic and effectively merged with the logic under test.

The EVE system can compute outputs of gates based on function names (AND, OR, etc.) from a set of commonly used functions defined within the system. However, if the designer requires additional function definitions, he can invoke a function compiler program to enter them. To actually run the simulation, it is necessary to invoke the simulation control program, SIM.

SIM is activated through user commands and performs such tasks as: loading the hardware memories from the files created by the software, specifying breakpoints, setting net values, issuing the run command (which starts the simulation processors), and transferring results to host files. After completion of the simulation run, the results can be selectively displayed, filed, or printed by the output program.

Experience. Table 1 summarizes the design differences among the LSM, YSE, and EVE and compares them to the reference software simulator EFS. The simulators are optimized toward specific design points, but the choice of those points is left to the designer. Table 2 illustrates the range of capability. The

table provides a view of one design point—performance—across EFS and EVE simulators. As the data in Table 2 indicates, there is a definite cut-over point between EFS and EVE at about 1500 cycles.

The engineering verification engine is currently undergoing final testing. It will be in production use later this year.

Step 3. Logic transformation. A key component in DAV is the logic transformation subsystem.[19] Using LTS, the designer can automatically synthesize and map a higher-level logic description into a design implementation under cost, performance, and technology constraints. The function of transforming a design described in BDL/CS into a logic network form was adopted from the logic synthesis system.[14]

Two fundamental problems. Transformation aids give the designer the freedom to fully exploit the

benefits of structured design. High-level design specification and functional verification approaches have been successful in the past,[2] but were constrained by two fundamental problems.

Design pressures have tended to push designers prematurely into low-level design.

First, design pressures have tended to push designers into low-level design prematurely. The saying, "There's never enough time to design it right the first time, but always enough time to do it over," seems to have held consistently true. This low-level bias results in the carrying out of gate-level design activities in parallel with functional verification.

As such designs progress, it becomes increasingly difficult to affect functional changes, due to the considerable investment in the design implementation. Moreover, the functional model begins to trail the implementation model, as engineers begin to make quick fixes.

Second, technology constraints (primarily power, area, and timing) can be assessed only after fairly detailed and timely trial implementation. Once an implementation is roughed out, there is little motivation to return to the functional level.

Transformation, with a goal of reducing an implementation's cycle time, can make it more plausible for the designer to perform implementation experiments and still rely on functional-level design as a source specification. The transformation approach also helps justify a design group deferring its commitment to a specific technology. Thus, design and functional verification can be proven in the absence of a technology, and then the cost/performance benefits of the latest technology development can be exploited.

A sample LTS methodology. The sample methodology described below will clarify the LTS system design.

Table 1.
Comparison of simulators.

	EFS (3081K)	LSM	YSE (PROTOTYPE)	EVE
GATE INPUTS	N/A	5	4	4
SIMULATION VALUES	4	3	4	4
HARDWARE INTERACTIVE	N/A	NO	YES	YES
NOMINAL DELAY	NO	YES	NO	NO
CAPACITY (GATES)	800K	63K	64K	2.0M
CAPACITY (MEMORY)	0	512KB	N/A	12.8M
I/O SPEED (BYTES/SEC)	N/A	16.5K	5K	50K
SIMULATION SPEED (370 INSTRUCTIONS/SEC)	1	100	560	450

Table 2.
Elapsed time for selected run duration.

	RUN DURATION (CYCLES)		
SIMULATOR/ELAPSED TIME	1	10^2	10^6
SUBSYSTEM (EFS)	3.4 MIN.	4.5 MIN.	250 HRS.
SYSTEM (EVE)	49 MIN.	49 MIN.	66 MIN.

Environment:
- Elapsed time is total user perceived response time, including all model-build, data transfer, simulation, and presentation of results.
- Host CPU is 3081K.
- Model simulated is 500K gate design.

Starting with an entity described by a BDL/CS or generic operator design specification, the process of SYNTHESIS creates a functionally equivalent logic network. Primitive logic blocks are created to represent control structures, such as an IF-THEN-ELSE sequence (Figure 5). Signals of interest, such as primary inputs, outputs, and memory element outputs, are directly associated with nets in the logic. Blocks of corresponding functions replace arithmetic and logical operations. Memory elements are explicitly defined and are replaced with generic register blocks. The resulting network contains a set of register blocks, along with both simple and complex combinational function blocks.

At this point, local transformations, applied to one block at a time, begin to reduce the number of blocks in the network. Each block is then analyzed with respect to its relationship to the relatively small logic neighborhood surrounding it. Based on this analysis, the subnetwork is altered, according to a particular objective, to a functionally equivalent subnetwork. An example of a transformation applied at this stage traces the block inputs to see whether they are identical to inputs to a block of the same function. If so, the block, now considered redundant, is removed and its output is reconnected as an additional fan-out of the remaining block.

If the entity had originated as a structural specification containing generic operators, a compiler called ELM, for equivalent logic model generator, is used for synthesis. ELM provides an expansion that takes variations in operator usage into consideration. The resulting expansion contains primitive Boolean and register blocks. If a PLA format is the original entity specification, the synthesis operation is the simple creation of an AND (plus required input inverters) for each product term and an ORing-applicable format. Minimization then reduces the number of logic cubes. Factorization transforms the cubes back into a random logic

representation, which is substituted into the design model as an alternate subnetwork.

During the second mapping stage, the closest match to an available technology book replaces each block. Local transformations then modify the network according to design (e.g., LSSD rules), technology (e.g., the use of pull-down terminators), and performance (e.g., timing) objectives. Some transformations attempt multiple strategies and then select the least costly.

For example, in repowering a block output, another block of the same function might be tied in parallel and the fan-outs redistributed among the two functionally equivalent outputs. Alternatively, a block might have an out-of-phase output available, which, with an additional inverter, can provide another source of fan-out. A third alternative might be to select a higher-power book. Each alternative is evaluated in terms of the overall objectives of cell count, area, power, connection count, and delay.

Finally, the network is partitioned into pages to create a set of person-readable logic diagrams. Results are then fed fack into the entry subsystem of DAV as an alternate structural specification. There, it can be subsequently edited or verified.

Another class of transformation. Although the emphasis in LTS is on local transformation, there is another general class available, the global or two-level transformation. Compression takes a user-specified, combinational, random-logic subnetwork and converts it into an equivalent of sum-of-products product terms for each output function.

MAPPING can then be applied to the network in two stages. During the first, the model is further optimized without considering technology, design, or performance constraints. Remaining complex combinational blocks in the model are replaced by an equivalent network of ANDs, ORs, and INVERTs, via macro expansion from a predefined library. Local transformations are again run at this level, performing straightforward Boolean algebra reductions. Transformations eliminate the logical constant blocks through a process of constant propagation. The logic is then converted to a set of NANDs (or NORs, depending on target technology) and register blocks. Local transforms are again applied to this technology-independent NAND net-

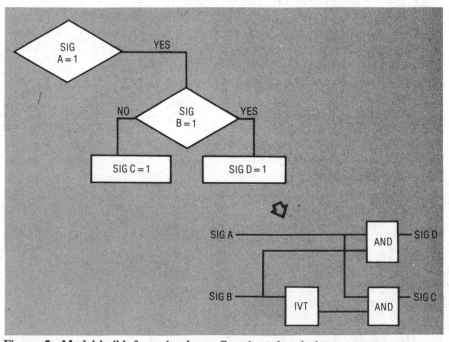

Figure 5. Model-build from hardware flowchart description.

work, to reduce both block and connection counts. During the reduction, adjustments are made for the generic technology fan-in constraints.

Examples of local transformations at this level include common-term elimination (Figure 6) and logical subsumption. To date, approximately 12 different transformations can be applied at this level.

Experience. Initial results of LTS use have been encouraging. In one experiment, six pages of a BDL/CS-like language were processed through LTS in 19 minutes of CPU time on a 3081, resulting in approximately 3000 logic gates. When reviewed by designers, LTS-generated logic produced results that equaled those of an experienced logic designer. Table 3 summarizes the productivity results.

Step 4. Boolean comparison. The methodology produced two logic representations that purportedly describe equivalent sequential machines

of the same state assignments. However, the flexibility of the system allows user-written transforms and user editing of the technology-level

---

The methodology of functional simulation and Boolean comparison has resulted in less than 15 functional design errors in thermal conduction modules on IBM 3081Ds in the field to date.

---

logic. This could inject errors into the design.

The Boolean comparison programs determine whether or not the functions driving each corresponding pair of latches or outputs in two representations are Boolean equivalent. This capability provides a complete verification when the DAV

system is used to map a design between technologies. Thus, one can validate functional equivalence of two design description levels or validate an engineering change.

Although Boolean comparison programs can be found within IBM in unpublished documents from as early as 1965, it was only with the differential Boolean analyzer,[4] that Boolean comparison became practical. Table 4 compares the DBA and the verify algorithm,[9] which utilizes a modified D-algorithm for a range of small test cases.

DBA was enhanced to identify identical logic groups. This function is especially effective when a degree of order is imposed across the design and a high degree of replication exists. The enhanced DBA was named ESP, for equivalent sets of partial differential equations; its effect is shown in Figure 7.

Experience. The Boolean comparison runtime for a 500K gate machine (60K gates at a time) is about 45 CPU hours on a 3033, compared to verification via conventional techniques estimated at thousands of CPU hours over 12 months. Boolean comparison on the design of the IBM 308X family of computers has been outstanding. The methodology of functional simulation (EFS) and Boolean comparison has resulted in less than 15 functional design errors in thermal conduction modules on IBM 3081Ds in the field to date.

Step 5. Check design rules. The DAV methodology includes two types of checks: physical and LSSD checks.

Physical design checks. The designer must perform a number of physical design checks before inputting the design to the physical design system. These checks cover important physical design errors, mainly those that can be corrected only in the logic design. They include checks on the number of input/output pins used in a package, the number of cells of different types, legal prespecified circuit locations, power consumption,

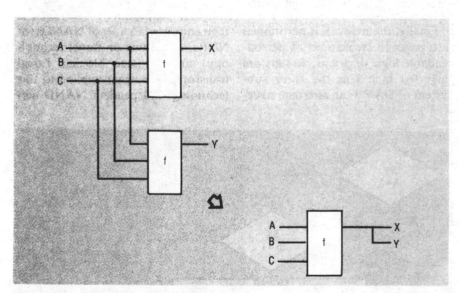

Figure 6. Common-term elimination example.

Table 3.
The LTS experiment.

	LTS	DESIGNER
INITIAL DESIGN		
Functional description	3 weeks	–
Technology-level description	19 min.	6 months
ENGINEERING CHANGE	3 days	3 weeks

the number of tie-down resistors needed, net fan-out and dotting constraints, net loadings, estimated connection count, and various internal names and flags that are checked for both validity and consistency of use.

These early checks reduce the number of iterations between DAV and physical design, since most errors in the logic are eliminated before they get into the physical design.

The physical design checks cover all package types designed at IBM: chips, modules, cards, and boards. Only one set of programs is needed, since most of the checking information is contained in the IBM corporate rules library. This library contains information on the package being designed, such as the number of input/output pins available and the maximum power consumption allowed.

It also contains information on the circuits used in the design, based on the logic block selected by the designer. The logic block identifies the selected technology and the associated rule. The rule contains such information as the number of pins in each circuit, the amount of "real-estate" used by the circuit, the loading of each circuit pin, and the specification that tells which drivers can be dotted.

The checks are performed for the user, in the foreground. Approximately 100 circuits per second are checked, using an IBM Model 3033. The success or failure of the checks is presented to the user, followed by the informational and error messages. These messages are organized by circuit and net and are cross-referenced

for each unique message. They can be browsed in the foreground and, optionally, routed to hard copy for later use. Updates to the logic, to correct

---

Only one set of programs is needed, since most of the checking information is contained in the IBM corporate rules library.

---

any errors, can be made in the same session, and the physical design checks can be repeated to verify the updates.

LSSD rules checks. Design rules checks help produce designs for which efficient test patterns are easily generated. Level-sensitive scan design rules place testability constraints on the design,[22,23] including constraints on the way latches are constructed and clocked. They require separation of clock lines from data lines and separation of the clocks controlling certain registers and latches from those controlling others. Taken together, the rules assure success of

test-pattern generation algorithms, which generate test patterns with a high degree of test coverage. The process is very efficient; the associated cost is 1.5 seconds of CPU (3081) time per 1000 blocks, where each block is assumed to be an AND invert with 2.5 inputs and one output.

In the past, testability problems could not be detected until late in the design process. If these checks are made early in the design process, there is no need to rerun the physical design programs due to lack of conformance to LSSD design rules.

The software capability described here has had restricted distribution and limited use, though several tools within the system have had significant usage, notably EFS and Boolean comparison. The full database support[17] remains to be implemented, along with changes based on user feedback.

One such change is the separation of design description and simulation. The current system tightly couples these capabilities, providing rapid update/verification while maintaining design integrity. However, in response to the variety of user design descrip-

**Table 4.
DBA/verify comparison.**

TEST CASE (#blocks/#PI)	VERIFY (min:sec)	DBA (min:sec)
100/20	1:53	:19
200/20	6:03	:55
200/25	4:14	:23
500/80	1:53	1:38

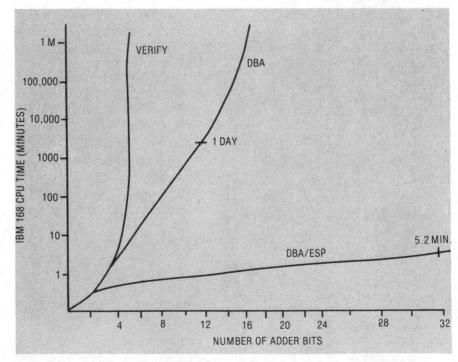

Figure 7. Verify, DBA, and DBA/ESP computation time for Boolean comparison on adders with *N* bits.

tion preferences, we are in the process of separating design capture and simulation.

Additional changes include migration of interactive functions to an intelligent workstation and coupling EVE more directly to the design specification functions.

VLSI has significantly changed the relationship of the DA tool developer and the product designer. No longer is design automation something nice to have or something that merely makes the job easier. Now, DA tools are critical to the success of any VLSI-based product. In fact, the tool developer will significantly influence how VLSI products will be developed, and the product designer will overtly or covertly enforce a methodology far different than that of the past. The system outlined in this article represents a major first step in establishing one such VLSI methodology at IBM. ∎

Acknowledgments

The Design and Verification project is the result of many contributions of my development team, which was headed by Don Cottrell and Hillel Ofek. Innovative ideas, programs, and valuable feedback from both design and research organizations throughout IBM have been most valuable. This article benefitted from the input of many of the developers, including Dick Price, Bruce Raymond, Charlie Rimkus, John Bendas, Frank Rubin, and the late Maureen Kane.

References

1. A. E. Fitch, "Will Your Bridge Stand the Load?" *Proc. 17th Design Automation Conf.*, 1980.

2. M. Monachino, "Design Verification System for Large Scale LSI Systems," *Proc. 19th Design Automation Conf.*, June 1982, pp. 83-90.

3. R. B. Hitchcock, Sr., G. L. Smith, and D. D. Cheng, "Timing Analysis of Computer Hardware," *IBM J. Research and Development*, Vol. 26, No. 1, Jan. 1982.

4. G. L. Smith, R. J. Bahnsen, and H. Halliwell, "Boolean Comparison of Hardware and Flowcharts," *IBM J. Research and Development*, Vol. 26, No. 1, Jan. 1982, pp. 106-116.

5. R. B. Hitchcock, "Timing Verification and the Timing Analysis Program," *Proc. 19th Design Automation Conf.*, June 1982, pp. 594-604.

6. P. W. Case et al., "Design Automation in IBM," *IBM J. Research and Development*, Vol. 25, No. 5, Sept. 1981, pp. 631-646.

7. C. J. Evangelista, G. Goertzel, and H. Ofek, "Designing With LCD—Language for Computer Design," *Proc. 14th Design Automation Conf.*, 1977, pp. 369-376.

8. L. I. Maissel and D. L. Ostapko, "Interactive Design Language: A Unified Approach to Hardware Simulation, Synthesis, and Documentation," *Proc. 19th Design Automation Conf.*, 1982, pp. 193-201.

9. J. D. Roth, "VERIFY: An Algorithm to Verify a Computer Design," IBM Technical Disclosure Bulletin 15, 2646-2648, 1973.

10. J. K. Howard, R. L. Malm, and L. M. Warren, "Introduction to the IBM Los Gatos Logic Simulation Engine," *Proc. IEEE ICCD*, Oct. 1983, pp. 580-583.

11. G. E. Pfister, E. Kronstadt, and M. M. Denneau, "The Yorktown Simulation Engine," *Proc. 19th Design Automation Conf.*, 1982, pp. 51-64.

12. H. Ofek et al., "Structured Design Verification of Sequential Machines," Research Report RC7037, IBM Thomas J. Watson Research Center, Yorktown Heights, N.Y., 1978.

13. G. Parasch and R. Price, "Development and Application of a Designer Oriented Cyclic Simulator," *Proc. 13th Design Automation Conf.*, June 1976, pp. 48-53.

14. J. A. Darringer et al., "Logic Synthesis Through Local Transformations," *IBM J. Research and Development*, Vol. 25, No. 5, July 1981, pp. 272-280.

15. L. N. Dunn, "An Overview of the Design and Verification Subsystem of the Engineering Design System," *Proc. 20th Design Automation Conf.*, June 1983, pp. 237-238.

16. F. Rubin, "A Logic Design Data Entry System," *Proc. Int'l Conf., Circuits and Computers*, June 1980, pp. 107-110.

17. P. Horstmann and F. Rubin, "A Logic Design Front-End for Improved Engineering Productivity," *Proc. 20th Design Automation Conf.*, June 1983, pp. 239-245.

18. D. Cheng et al., "Structured Design Verification: Function and Timing," *Proc. 20th Design Automation Conf.*, June 1983, pp. 246-252.

19. J. B. Bendas, "Design Through Transformation," *Proc. 20th Design Automation Conf.*, June 1983, pp. 253-256.

20. J. L. Sanborn, "Evolution of the EDS Data Base," *Proc. 19th Design Automation Conf.*, June 1982, pp. 214-218.

21. P. N. Agnew and M. Kelly, "The VMS Algorithm," Technical Report TRO1.1338, IBM System Products Division Laboratory, Endicott, N.Y., 1970.

22. E.B. Eichelberger and T.W. Williams, "A Logic Design Structure for LSI Testability," *Proc. 14th Design Automation Conf.*, June 1977, pp. 462-468.

23. H. C. Godoy, G. B. Franklin, and P. S. Bottorff, "Automatic Checking of Logic Design Structures for Compliance with Testability Ground Rules," *Proc. 14th Design Automation Conf.*, June 1977, pp. 469-478.

INTEGRATED DESIGN AUTOMATION SYSTEM
FOR CUSTOM AND GATE ARRAY VLSI DESIGN

Yasuhiro Ohno*, Norio Yamada*, Katsuya Sato*, Yoshinori Sakataya*,
Makoto Endo*, Hisashi Horikoshi**, Yuichi Oka***

  * Kanagawa Works, Hitachi Ltd.   (Hadano,Kanagawa,Japan)
 ** Central Research Laboratory, Hitachi Ltd. (Kokubunji,Tokyo,Japan)
*** Device Development Center, Hitachi Ltd.   (Kodaira,Tokyo,Japan)

Abstract

Integrated design automation system has been
developed which allows rapid and error-free design
of both standard cell approach VLSI's and gate
array VLSI's. Around a central design data base
this system provides integrated DA system which
has logic verification, test pattern generation,
etc. Many standard cell approach VLSI's ranging
from 15K to 20K gates and gate array VLSI's of 6K
gates have been designed with this system. Layout
DA system for standard cell approach VLSI's is
mainly described in this paper including design
philosophy, Floor Planning, wiring method
interconnecting different levels of hierarchy,
various automated functions and layout results.

1. Design philosophy of DA system

The number of new VLSI types to be designed
increases drastically as a wide variety of
electronic equipment has begun to use VLSI's. It
is required that many types of VLSI's of high
density can be designed concurrently and with ease
like conventional printed circit boards. To meet
these design needs, the following design
philosophy was adopted for layout DA system.

(1) The design approach capable of partitioning
the design object into arbitrary size and the
hierarchical design approach are adopted in order
to be able to lay out VLSI's of up to several tens
of thousands of gates.
(2) New Floor Planning method is adopted to
achieve as small chip area as possible satisfying
delay constraints.
(3) Wiring method interconnecting different
levels of hierarchy is employed to achieve high
wiring density.
(4) A wide variety of automated functions in
addition to automatic placement and routing[3] are
provided in order that many types of new VLSI's
can be designed in a short period and with small
layout effort.
(5) Great care is taken in order that both
standard cell approach VLSI's and gate array
VLSI's can be designed with the same interface to
a designer. For example, the same data base, the
same logic input language and the same design
procedure are used for both types of VLSI's, and
also for PCB's and logic equipment.
(6) The layout DA system is embedded in the
integrated DA system [1] shown in Figure 1.

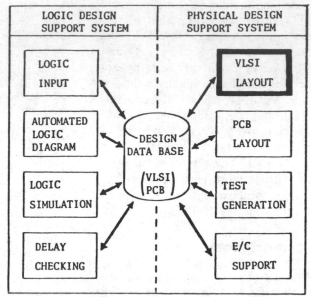

Figure 1. Overview of Integrated DA System

2. Organization of VLSI layout system

VLSI layout system has seven subsystems as shown
in Figure 2.

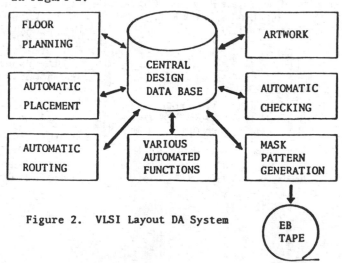

Figure 2. VLSI Layout DA System

3. Chip layout model

Figure 3 illustrates the hierarchical chip layout
model employed. The cell is a minimum logical
unit consisting of 1 to several tens of gates.

Reprinted from *IEEE Int. Conf. Circuits Computers Proc.*, 1982, pp. 512-515.

The block is a basic unit of logic design with several hundreds of cells in it. RAM, ROM, etc. called a macro block are predefined in a library. The chip consists of these blocks and I/O buffer array. Two layer metals and one layer Poly Si are used for intrablock and interblock wiring. The chip can be partitioned into subsets called "a block assembly" in order to keep the design object within a reasonable size.

CHIP

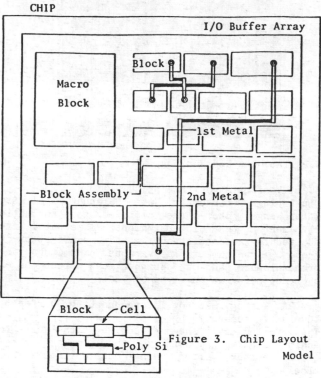

Figure 3. Chip Layout Model

4. Features of VLSI layout system

4.1 Floor Planning

It is necessary to fix the layout image of an entire VLSI chip prior to detailed layout design in order to achieve as small chip area as possible with its shape close to square. The detail of this process called Floor Planning will be described according to the design flow shown in Figure 4.

Step 1. Estimation of shape and size of blocks
Shape and size of random logic blocks can be obtained after placement and routing are performed. For this reason shape and size data for each block are obtained by performing automatic placement and routing programs changing the number of cell-rows in a block. An example is shown in Table 1. In Figure 5 shown is an example of block area vs the number of cell rows corresponding blocks in Table 1. Figure 5 shows the following:
(1) When the number of cell rows is in the range between 2 and 4, the block area is reduced drastically as the number of cell rows becomes larger.
(2) When the number of cell rows is larger than four, the block area fluctuates but remains almost on the same level.

Step 2. Relative positioning of blocks
A designer makes a plan of the relative location between blocks so as to take the following into consideration.

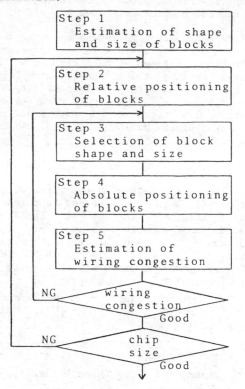

Figure 4. Design Flow of Floor Planning

Table 1. Estimation of Block Shape and Size [grid]

Number of Cell Rows	2		...	7		8		9		...	15	
Size \ Block No.	x	y		x	y	x	y	x	y		x	y
1	998	110		308	266	(270)	(280)	244	326		164	536
2	758	96		236	232	210	252	(194)	(284)		128	456
3	800	112		(248)	(270)	220	318	202	354		140	504
4	884	104		274	252	(242)	(278)	220	300		144	504
5	968	116		(300)	(280)	266	326	238	364		154	532

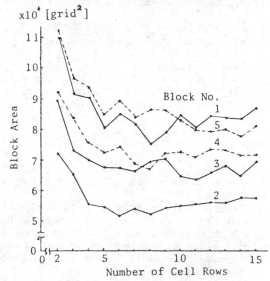

Figure 5. Example of Block Area vs the Number of Cell Rows

(1) Locations of VLSI I/O buffer pins.
(2) Data flow.
(3) Delay constraints(blocks which have tight delay specifications such as critical paths are placed in close proximity to each other).

Step 3. Selection of bolck shape and size
Average area $\overline{Si}$(i means block No.) is obtained for each block from a group of cell rows in which the block area remains almost on the same level in Figure 5. Next the target horizontal length L of a chip is determined. The target vertical length Y of blocks is calculated for each block row of the relative block positioning plan obtained at Step 2.

$$Y = \sum_i \overline{Si} \div L$$

The block size and shape with the vertical length close to the above Y is selected from the block area table obtained at Step 1. In Figure 6 illustrated is an example of a block row obtained in this way. The same block No. means the repetition of the block. Block No's in this example corresponds to those of Table 1 and block sizes circled in Table 1 are adopted.

Block No.1	2	3	3	3	4	5	Block No.5

|←————————————————— L —————————————————→|

Figure 6. Example of Block Shape Estimation

Step 4. Absolute positioning of blocks and Step 5. Estimation of wiring congestion
An absolute positioning plan of blocks is made by determining the location of each block row, obtained at Step 3, on a chip. An estimation program of wiring congestion is performed for this plan and the following two indices are calculated.

(1) Total track demand index
This index indicates macro wiring congestion for the entire wiring space and is defined as follows.

$$\text{Total track demand index} = \frac{\text{Estimated total wire length}}{\text{Total available track length}}$$

(2) Local track demand index
This index is defined as follows when the wiring space is cut by a vertical or horizontal crossing line and indicates local wiring congestion.

$$\text{Local track demand index} = \frac{\text{Estimated number of tracks}}{\text{Available number of tracks}}$$

The chip size and the block positions are adjusted so that these two indices do not exceed 50% and 80% respectively.
Vertical length and horizontal length of a VLSI chip are adjusted by iterating Step 3 through Step 5 or Step 2 through Step 5 changing the shape and size of each block to achieve one of the purposes of Floor Planning, "as small chip area as possible with its shape close to square."
In Figure 7 illustrated is the chip layout result with Floor Planning method.

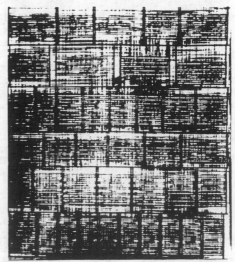

Figure 7. Example of Chip Layout

4.2 Placement and Routing

4.2.1 Various algorithms [3][4]
Clustering, Min-cut, Net Balance, pairwise interchange, etc. are employed for automatic placement. Channel assignment, maze, etc. are adopted for automatic routing. The combination of these algorithms and various parameters are precatalogued and a designer can select one of them depending on the type of VLSI.

4.2.2 Wiring method interconnecting different levels of hierarchy
The followings show effective techniques to achieve high wiring density.
(1) Interblock wiring using unused tracks in a block
When an upper level module is laid out with this system, a lower level module is not treated as a black box and terminal pins to be connected between lower level modules are not placed on boudaries of these modules. Shown in ① and ② of Figure 8 the interblock automatic router is capable of using unused tracks in a block to achieve high wiring density.

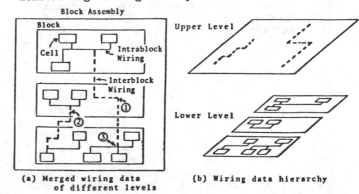

(a) Merged wiring data of different levels   (b) Wiring data hierarchy

Figure 8. Interblock wiring method

(2) Interblock wiring utilizing intrablock wiring patterns (equi-potential wiring method)
As shown in ③ of Figure 8, the automatic router is permitted to wire not only points in a block but also any part of wiring patterns previously

378

connected in a block, which is called equi-potential wiring method. This is effective in increasing wirability. For example, in a VLSI chip whose number of interconnections was 2726, there were 25 overflows when this method was not used but only 9 overflows with this method.

Maze algorithm is also employed for interblock wiring to attain the above (1) and (2).

(3) Multi-level checking
Since manual interventions are permitted to modify wiring patterns that belong to different levels, multi-level checking is required. As illustrated in Figure 8, equi-potential signal nets are merged between upper and lower levels when there are wiring patterns passing through lower level modules. Connectivity checking with logic file and layout rule checking are performed based on this merged layout data.

4.2.3 Gate Array VLSI[2]
Gate array VLSI's as well as standard cell approach VLSI's are supported by the same layout DA system. Shown in Table 2 is the comparison between both types of VLSI's.

Table 2. Comparison between both types of VLSI's

NO	item	standard cell	gate array
1	The number of tracks between cell rows	not fixed	fixed
2	The number of signal layers	3 (2 metals+ 1 Poly Si)	2 (2 metals)

Particular attention is paid to meet the needs which occur when both types of VLSI's are used in the same electronic equipment;
(1) common data base,
(2) the same design procedure and language,
(3) the design can be converted from gate array to standard cell approach or vice versa.

4.3 Various automated functions

This section describes a wide variety of automated functions provided other than automatic placement and routing programs to reduce the layout effort, and to be capable of iterating the design process from Floor Planning to detailed layout in a short turn-around-time to obtain optimal chip size. These automated functions saved 80% from the layout effort required other than Floor Planning, and automatic placement and routing.
(1) Interboundary pins between partitioned areas
The block assembly can be partitioned to facilitate concurrent layout design. Interboundary pins are required on boundaries to interconnect signals between partitioned areas. These interboundary pins are determined automatically.
(2) Power supply and ground nets
(3) Probing pads
Probing pads are special pads which are touched with a small probe and the potential is measured to analyze a VLSI chip for improving yield. One probing pad is generated for each signal net on a chip.

5. Results of layout design

The layout DA system has been applied to the design of many types of standard cell approach and gate array VLSI's, and none of them have had layout errors. As shown in Figure 9, layout effort and layout design time are significantly reduced with this system as compared with the conventional method.
The total design effort from the start of the logic design to release to manufacturing were five to seven man years and one man year, for a standard cell approach VLSI of 20K gates and for a gate array VLSI of 6K gates, respectively.

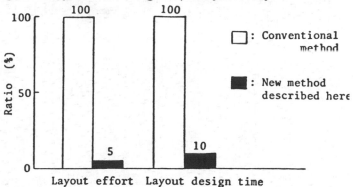

Figure 9. Layout effort and design time

6. Conclusions

The VLSI layout DA system embedded in the integrated DA system has been put to practical use and proved to be effective. More work will be done to enhance the capability of this system.

7. Acknowledgements

The authors would like to thank Kozo Kayashima, Masayoshi Tsutsumi, Yuichiro Oya, Kenji Taniguchi and Kisaburo Nakazawa who gave a chance to develop the system and helpful encouragements. They also wish to thank Mitsugu Edagawa, Yoji Tsuchiya, Toshiyuki Yamagami, Kyoji Chiba, Yasuhiro Ikemoto and Masao Kato for their efforts in working together to develop the system.

8. References

[1] Y. Ohno, et al., "Logic Verification System for Very Large Computers using LSI's," Proc. of 16th DA Conference, pp367-374, June 1979.
[2] T. Itoh, et al., "A 6000 Gate CMOS Gate Array," Digest of Technical Papers of ISSCC 82, pp176-177, February 1982.
[3] H. Terai, et al., "Automatic Placement and Routing Program for Logic VLSI Design based on Hierarchical Layout Method," Proc. of ICCC 82, September 1982.
[4] T. Adachi, H. Kitazawa, M. Nagatani and T. Sudo, "Hierarchical Top-Down Layout Design Method for VLSI Chip," Proc. of 19th DA Conference, pp.785-791, June 1982.

Part III
Simulation

THIS part consists of two sections: one deals with mixed-mode simulators and the other with performance evaluation and optimization. Mixed-mode simulators covering a wide range of simulation levels will become a focus of research attention in the next five years. These levels include processing at the lower end and behavior simulation at the upper end. One important topic which is not covered in this part because of lack of space is modeling for these different levels of simulations. The reader is referred the third section of the bibliography at the end of this book for further reading. The second section of this part deals with simple performance evaluation and optimization techniques which could be used, for example, in silicon compilers to make design decisions. These techniques should lead to fast but accurate evaluation of performance. The topic is relatively new and will be enhanced by researchers in the future as part of design automation.

MIXED-MODE ANALYSIS AND SIMULATION TECHNIQUES FOR TOP-DOWN MOSVLSI DESIGN.

H. De Man

Katholieke Universiteit Leuven
Kardinaal Mercierlaan 94
3030 Heverlee

Abstract

Based on a structured top-down design strategy a number of principles to certify overall behavioral correctness of MOSVLSI circuits are derived. It is shown that Boolean verification can best be performed based on a stick diagram rather than a logic schematic. A mixed register transfer-logic-switch simulator is necessary. Electrical behaviour is checked by mixed-mode circuit-timing-logic simulation of functional blocks after compaction of sticks into layout. The concept of guided simulation involving automatic partitioning in unilateral subnetworks, assignment of abstraction levels, transformation of parametervectors and test generation for simulation of subnetworks, is introduced.

1. THE PROBLEMS

Actual MOSVLSI chips contain over 50.000 transistors. To the user, they behave as complete systems such as microcomputers, digital filters, graphic processors etc. but to the designer they are to be realized as a geometrical mapping of transistor networks into silicon. Therefore the design spectrum ranges from high level system description down to a mask set containing of the order of 10^6 rectangles. Two major design problems are :

a) The conception of a suitable system architecture which, when translated into a network, gives rise to a minimum layout area with maximum ratio of total transistor count to designed transistor count. Recently a lot of attention has been paid to this problem as inspired by the Mead and Conway [1] structured design methodology.

b) The verification of design correctness, which can be considered as two separate problems :

b-1) The correct mapping of high level behaviour into the behaviour of a transistornetwork.

b-2) The correct mapping of the transistornetwork into its geometrical representation.
There is some overlap between b-1) and b-2) as the non-Boolean i.e. electrical problems can only be verified after the layout phase has been completed. In this paper we will be concerned with the problem of verification of behavioral correctness which usually is performed by computer simulation. For IC designers for a long time this has been limited to circuit simulation at the transistor level.
When faced with VLSI two severe problems arise :

1) CAD tools for simulating higher than circuit levels such as system, register and logic level are necessary. Such tools exist for discrete digital systems but have no unified language, compatible data-structures or algorithms. As a result their use is inefficient and error prone. Logic simulators do not cope with MOS transistornetworks which behave more like "relay" networks [24] with capacitive storage nodes than as traditional Boolean operators. Register transfer languages are too much like computerlanguages than representations of actual hardware e.g. by lacking operators representing bilateral data flow as occurs regularly in MOS networks.

2) Simulation is basically an ad hoc technique which does not guarantee design correctness unless stimuli and abstraction level are "well" chosen.

Recent efforts to cope with higher complexity belong to two classes :

a) Extend circuit analysis to much higher complexity either by new algorithms or macromodeling [2] [3] [4] [5] or by exploiting parallelism in advanced computer systems [6] .

b) The use of simultaneous simulation at different abstraction levels by unifying language, data-structures and algorithms. This is called mixed-mode simulation [7] [8] [9] [10] . Although both efforts are of great importance, two major problems remain :
1) Tools under a) can handle very large scale (VLS) problems at a given abstraction level but they also require VLS input data (which stimuli ?) and produce VLS output data (where and what to look for?). They are likely to produce the "Garbage in - Garbage out syndrome", but do not guarantee that critical problems are indeed identified.
2) Mixed mode simulation reduces the complexity problem in principle by assigning an "appropriate" abstraction level to different circuit parts. However the existing programs leave the decision on abstraction level to the intuition of the designer who tends to simulate everything at the lowest (i.e. most inefficient) level.
Nearly all these problems can be traced back to the fact that CAD tools can only be efficiently conceived after the definition of a design methodology. In what follows we will propose such a methodology, mainly oriented towards, but not restricted to n-MOS technology. It will then become clear how the data-structure and algorithms of a mixed mode simulator allow to shift the emphasis away from simulation to analysis. Analysis of the MOS networks will allow to select the optimum abstraction level at a given stage of the design which will insure to a much higher degree that design correctness is preserved during the design cycle with optimum use of computer and designers resources.
The methods and ideas suggested in this paper are be no means exhaustive nor complete. They represent a reflection based on design experience and status of actual simulation tools and are intended to stimulate further discussion and research in this direction.

2. DESIGN METHODOLOGY AND VERIFICATION SOFTWARE.

Similar to structured programming techniques, design of complex digital systems is usually done by a so called "top-down design" and "bottom-up imple-

Reprinted from *Proc. 1981 European Conf. Circuit Theory and Design*, R. Boite and P. Dewilde, Eds., North-Holland, Amsterdam, 1981, pp. 5-10.

mentation" procedure as illustrated in Fig. 1.
A concept is translated from its behavioral or algorithmic description in a high level language such as PASCAL, LISP, ISP etc... into detailed layout by a sequence of <u>hierarchical decompositions</u> into lower forms of abstraction.
Hereby behaviour is decomposed into <u>structure</u> i.e. <u>interconnection of functional blocks</u>, which by preference can be mapped into layout blocks with a structured interconnection pattern (chip plan). Clearly an expansion of information volume (complexity) of ca. 6 orders of magnitude occurs, which can only be managed if design has a <u>repetitive structure in time and space</u>.

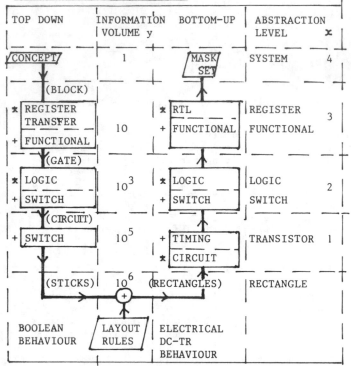

Fig. 1. Design Methodology & simulation.

Time partitioning is achieved by the use of predominantly <u>synchronous</u> systems whereas space partitioning and repetitivity occurs as a natural property of bit-slice data-path structures [11]. In VLSI so called random control logic tends to be regularized by the use of programmable matrixlike <u>transistor-structures</u> which often may not easily be recognized as classical gate networks (e.g. PLA's, carry chains, logic trees etc...).
Within this context Fig. 1. shows how the "classical" simulation tools (∗) as well as some more recent tools (+) are superimposed on such a design methodology. It can be seen that during the decomposition process 5 transformations taken place i.e.

$\underline{1}$ concept → {functional blocks}
$\underline{2}$ block → {gates}
$\underline{3}$ gate   → {circuit}
$\underline{4}$ circuit → {stick (symbolic layout)}
$\underline{5}$ stick → {rectangles} → {masks}

A stick diagram is a symbolic representation of an MOS circuit. Different symbols exist for different transistortypes, contacts and interconnection lines. Interconnections have zero width. A stick diagram contains the same graph as the electric circuit representation but has the topography reflecting the final placement of real components in a layout. Its importance will become clear later on. Fig. 2b) and e) illustrate the concept. In principle each of the transformations introduces risks for errors which

can be avoided or decreased by :

a) the use of synthesis techniques [13] [14]
b) the use of formal correctness proving [15];
c) verification by modeling, computer simulation and analysis ;
d) use of a minimum number of design representations (languages). The ideal would be a single topological description whereby abstraction levels differ only in the parameters associated with the elements of description.

Solutions a) and b) are, for the time being, not able to beat "fuzzy" human intelligence in terms of economical solutions (except for small scale problems) and c) and d) remain necessary and will be discussed below.
Based on the above and the fact that verification software at abstraction level x operating on information volume y (Fig. 1) tends to have a cost of the form:

$$C = 10^{-x}\, y^{\alpha} \quad (\alpha > 1) \tag{1}$$

the following additional principles can be formulated :

e) verification of a given behavioral characteristic should be done at the <u>highest</u> possible <u>abstraction</u> level which still guarantees correctness of that characteristic.
f) verification at lower abstraction levels can only be done on a set of <u>subnetworks</u> with <u>decoupled local behaviour</u> which can be represented at different abstraction levels the parameters of which are mathematically linked. This partitioning should maximally exploit <u>time and space repetitivity</u> resulting from hierarchical decomposition.

The principles d), e) and f) are the historical basis for <u>mixed-mode (MM) simulators</u> such as SPLICE [8] and DIANA [9] which allow for simultaneous simulation and analysis of subnetworks at logic (i.e. gate),timing [3] [16] and circuit simulation level. With respect to Fig. 1. actual MM simulators are all to be located in the implementation branch i.e. they are used to verify electrical DC and transient (timing) behaviour which only makes sense <u>after completion of the layout</u>.
Using them during the top-down phase for <u>logic verification</u> (since no electrical information present) would violate principle e). Furthermore transformations between abstraction levels are left to the designer and thus artificial and errorprone. In what follows we propose, based on existing algorithms and programs, a potential architecture of a behavioural verification system for MOSVLSI which obeys closely principles c) to f) and supports the design methodology of Fig. 1.

3. ARCHITECTURE FOR A GUIDED SIMULATION SYSTEM.

Under <u>guided</u> simulation system we understand a simulation system which, based on a <u>preanalysis</u> phase of a network :
1) partitions the network into appropriate subnetworks according to principles e) and f);
2) transforms automatically one level of abstraction in another wherever possible (principle d);
3) if 2) is not possible the system can interactively generate local "test" patterns to compare behaviour between different levels (or leave it to the designer).
4) for a given verification always the highest abstraction level is automatically selected unless this is overwritten by the designer.

Fig. 1. indicates a clear distinction between the prelayout phase where only Boolean behaviour can be verified, and the post layout phase where electrical behaviour needs to be verified. We will therefore make a clear distinction between both phases.

3.a. The prelayout phase.

During this phase the system concept is translated into the interconnection of functional (synchronuous) blocks. This requires a functional (register transfer) simulator which allows to simulate the system from one clock phase $\emptyset_i$ to the next based on a nested hierarchical description. By preference the simulator should be able to handle tristate bidirectional buses (1,0,X,Z system) and have parametrizable primitives for the most widely used blocks such as RAM, ROM, REG, MPX, DEC, ENC, COUNT etc...

To our knowledge no such systems, integrated with lower level tools to be described below, exist today although many desirable features can be found in such systems as CASSANDRE [17], ADLIB-SABLE [18] or SAMSON [19].

CASSANDRE for example allows for a good nested hierarchical description and in such case each functional block then has to be translated into a logic gate representation which in term is translated into a transistornetwork and/or a stick-diagram reflecting layout topography.

However for MOS logic, in view of the importance of layout structuring and exploiting the voltage controlled switch character of MOS devices, designers tend to think more directly in terms of logic transistorstructures rather than logic gates [21].

Therefore traditional logic simulation is of little value since too many transformations are involved and logic simulators cannot handle the bilateral character of MOS switches.

In view of the principle d) therefore it looks much more natural to accept the stick diagram as the decomposition level of functional blocks for the following reasons :

1. Stick diagrams can algorithmically be verified with respect to circuit schematics.
2. Stick diagrams can, by use of compaction algorithms, interactively be converted into rectangle layout with layout rules as input (e.g. CABBAGE [20]). This eliminates error risks due to design rule violations.
3. Extraction of electrical circuit description from layout generated from stick diagrams is, in view of 1., extremely simple to achieve, which provides an exact link between Boolean and electrical representation.

All this however is only valuable if it is possible to use stick diagrams :
a) for subnetwork partitioning
b) Boolean verification (logic-switch simulation)
c) verification of first order DC logic levels.

Recent work at MIT [21] and actual research by the author indicates that a), b) and c) are indeed possible. Fig. 2. illustrates some of the principles. Fig. 2a. is a simple register-transfer functional block. Fig. 2i. is its logic representation for a logic simulator but contains in this case not more information than Fig. 2i. In contrast Fig. 2b, e are two stick diagrams of typical MOS transistorstructures which contain final topographic as well as circuit information. Figures in brackets are aspect ratios (W/L) of the MOS devices.

Bryant [21] defines a circuit graph of which the vertices are the circuit nodes and the edges sourcedrain branches of transistors named after their gate names (Fig. 2c, f). Nodes are partitioned into forced nodes (f) such as inputs, supply, ground), pull-up nodes (p) and storage nodes (s) (all nodes which can become isolated when incident switches are off). Nodes are ordered according to "logic strength" as f > p > s. Partitioning the circuit graph into its connected subgraphs automatically partitions the network into DC unilateral subnetworks ($\mathcal{N}_u$) with gates as inputs and whereby outputs can be pull-up or storage nodes when driving an input. Notice that the MOS gate is indeed unilateral DC-wise but may have non unilateral AC effects to be discussed later. This partitioning is the same as used in timing simulators based on submatrix scheduling such as MOSTAP [22], MOTIS [5] and DIANA .

This partitioning also allows to apply circuit analysis algorithms exploiting unilaterality [23]. Bryant [21] proposes algorithms to logically simulate such a switch network based on an event-driven data-structure. Simulation proceeds by first detecting connected subgraphs in scheduled $\mathcal{N}_u$'s depending on the state of the inputs. Then, within these $\mathcal{N}_u$'s, node values $\in \{1,0,X\}$ are assigned based on the strength of the connected (or isolated) nodes.

Actual research by the author further shows that an $\mathcal{N}_u$, containing a single pull-up node, can be partitioned into a transmission gate network ($\mathcal{N}_{ut}$) and a pull-down ($\mathcal{N}_{up}$) network.

The transmission gate network ($\mathcal{N}_{ut}$) is formed by all paths from a non-pull up output node ending into the first encountered p or f node.

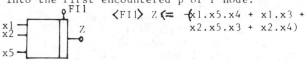

⟨FI1⟩ Z <= (x1.x5.x4 + x1.x3 + x2.x5.x3 + x2.x4)

a) REGISTER-FUNCTIONAL BLOCK

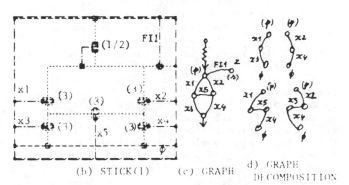

(b) STICK(1)   (c) GRAPH   d) GRAPH DECOMPOSITION

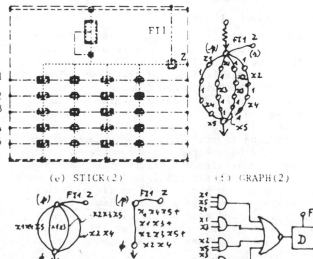

(e) STICK(2)   (f) GRAPH(2)

(g) AND Collapse   (h) OR Collapse   (i) LOGIC DIAGRAM

385

The remaining network is the pull-down network. This is illustrated in Fig. 2c) and f) where FI1 belongs to $\mathcal{N}_{ut}$ and all the other branches to $\mathcal{N}_{up}$. One can show that, for $\mathcal{N}_{up}$, it is always possible to derive a minimal inverted sum of products representation. This is done by a sequence of "AND collapses" and "OR collapses" first. In an "AND" collapse all edges x_i in series are substituted by a single branch with logic value $\prod_{i=1}^{k} x_i$ and an $(W/L)^{-1} = \sum_{i=1}^{k} (W_i/L_i)^{-1}$.

In an "OR" collapse, parallel edges x_i are substituted by a single branch of logic value $\sum_{i=1}^{k} x_i$ and a $W/L = MIN(W_i/L_i)$. If no more collapsing is possible it is possible that a single branch from p to ϕ is found resulting in a sum of products representation in which, to each product, a resulting worst case $(W/L)_j$ ratio is associated. We define the resulting W/L of $\mathcal{N}_{up}$ as the minimum of all $(W/L)_j$ over all product terms.

If however more than one branch remains, then the resulting graph contains floating branches of connectivity in excess of two such as in Fig. 2c). In this case all acyclic paths from p to ϕ have to be found. Each path can then be expanded in a sum of products and again a worst case equivalent W/L can be found. Further literature on MOS switching networks, including synthesis, can be found in [24] [25]. We have now shown that, from a stick diagram :

a) Boolean equations can be reconstructed for all $\mathcal{N}_{up}$ and be compared to intended logic or used in classical logic simulations,
b) Switch simulation can be applied to $\mathcal{N}_{ut}$,
c) Worst case resulting W/L ratios can be found and compared to design standards.

In this way functional blocks in RTL can be translated into circuit schematics or stick diagrams with local check of Boolean behaviour as well as by substituting the RTL block by its expansion in a mixed mode simulator. Also mixed-mode can be used to test the circuit or stick representation with the RTL description one by one using e.g. randomly generated test patterns to both representations and use EXORING or transition count [26] on the outputs.

Using this system, the prelayout phase results in stick diagrams for all functional blocks with :

a) All aspect ratios for DC levels verified;
b) All Boolean logic verified and all Boolean functions for all $\mathcal{N}_{up}$ defined;
c) All logical bus contentions resulting from logical errors are checked.

3.b. The postlayout phase.

When a stick diagram is available it can automatically be transformed into a layout respecting layout rules using compaction algorithms [20]. At this point detailed electrical DC and transient behaviour can be checked since now transistor-characteristics and capacitive loads are known and can easily be derived from the stickdiagram and generated layout.

However according to principle e) only non-Boolean behaviour should be checked and according to f) it can only be done by adequate partitioning in subnetworks each at the highest possible level of abstraction according to the purpose of verification.

In what follows we propose a guided simulation technique using mixed mode at circuit, timing and logic level whereby the unit to be verified is at the functional block level and whereby the goal of verification is :

1. Logic level errors due to dynamic capacitive effects (signal feedthrough, charge redistribution);

2. Timing errors due to gate and interconnection delays;
3. Dynamic errors due to spikes in local asynchronous circuits (e.g. data-latches).

To guide the designer through mixed mode simulation we propose the following six CAD procedures :
INPUT : rectangle layout from stick diagram. DC unilateral subnetworks.

1. Extract : circuit from layout through stick.
2. Partition : into AC unilateral networks $\mathcal{N}_{ua}$.
3. Identify : identical unilateral networks from structural description.
4. Assign : abstraction level based on network topology and time partitioning.
5. Transform : parameter vector from low to high abstraction level.
6. Analyse : dynamic and threshold related level disturbance, charge redistribution spikes; timing errors.

We will now discuss each procedure separately and refer to existing algorithms and programs to show its feasability :

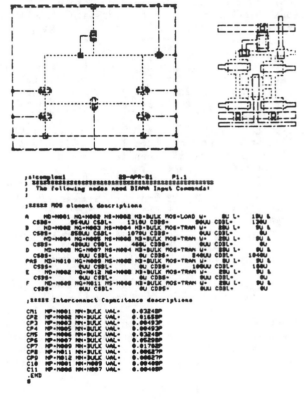

Fig. 3.

1. Extract : Fig. 3. shows a CABBAGE [20] stick, its compacted layout and the input for the circuit-simulation part in DIANA [9] language. Procedure extract produces a full circuit description including all parasitic capacitance which is guaranteed to correspond to the produced layout.

2. Partition : start from the DC unilateral models ($\mathcal{N}_u$) in the prelayout phase and compute the capacitive loads of all nodes. Assign fan-out capacitance C_{FO} to driving network.
Merge $\mathcal{N}_u$'s where floating gate-(drain)-source capacitors violate unilaterality rules.
E.g. in Fig. 4. when Ⓢ is a storage node and $\Sigma C_F > C_L$.

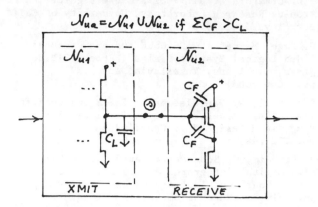

$$N_{u2} = N_{u1} \cup N_{u2} \text{ if } \Sigma C_F > C_L$$

Fig. 4.

This defines a new set of unilateral subnetworks N_n partitioned into (non merged) DC networks (N_{DC}) and (merged) AC networks (N_{AC}).

3. Identify : uses the structural hierarchical description to identify identical copies (e.g. set of identical full adders) in N_n as well as their interconnection net. Therefore the language should contain a provision for parametrizable nested "FOR" loops such as in CASSANDRE [17].

4. Assign : networks of N_n now have to be assign an abstraction level for time domain simulation in a mixed mode simulator. Although research for Gauss-Seidel-Newton timing simulation [8] of floating capacitors in N_{AC} is going on [27], it does not seem very reliable for the time being and we propose to assign a single submatrix to every N_{AC}. These submatrices can then be solved at the circuit level using single- or multilevel [2] Newton-Raphson techniques (the latter only if the networks are large enough). Since now all networks are unilateral, analysis can take place by exploiting space ordering and time scheduling methods such as proposed in [23]. Simplifications are possible by cutting feedback loops and allowing for a minimum resolvable time step [8] delay as in [28]. Otherwise strongly connected components [23] would create too large submatrices (violation of principle f). The remaining N_{DC} are now to be assigned either to logic or timing level.
Timing simulation is event driven simplified circuit simulation either based on single iteration Gauss-Jacobi [3] [9] or Gauss-Seidel-Newton [8] [16] methods either on the full timing network [16] [8] or on macromodels [9] [3] and/or submatrices of macromodels and transmission-gate trees [9] [22]. All these techniques do not allow for floating capacitances except when driven from low impedance (for p) nodes to model e.g. clock-feedthrough in transmissiongates in DIANA [9]. An interesting new perspective for more accurate timing simulation is the use of explicit (or mixed explicit-implicit) integration rules [29] which allow to avoid iteration completely (or partly).
Timing simulation is useful for all N_{DC} containing transmission gate trees, single transmission gates when their source-bulk to drain-bulk capacitance ratio is in excess of 3 [21] (potential charge redistribution or spike problem) or when pull-down networks (N_{up}) contain internal grounded capacitors larger than the pull-up to ground capacitance (internal spike danger). If the latter does not occur pull-down networks can be macro modeled by an "equivalent" driving transistor with worst case (W/L) and driven by Boolean controlled network

variables [4] [9].
All other N_{DC} are static Boolean gates and are assigned to the logic simulation part.
Notice that the timing models are potential candidates for internal spikes and dynamically disturbed logic levels. Based on the output report designers can simulate these simple circuits separately to identify pathological behaviour.

5. Tranform : once the assignment is done, according to principle d), it is necessary that parameters are automatically transformed from low to high abstraction levels. This is especially important when macromodels are used in timing or gate-delays have to be extracted for logic simulation. A particular problem occurs because in MOS logic, the falling edge delay t_{pHL} depends on the driving edge risetime t_{rLH} [30] and the impedance of the selected driver path in the pull-down graph. Therefore, to extract meaningfull t_{pHL}, t_{pLH} for logic models, the following worst case procedure is proposed :
1. Compute all t_{rLH} of logic gates by a single step response circuit simulation.
2. From the sum of products representation derived in the prelayout phase : derive the worst case input test for t_{pLH} of N_{up} as follows :

let $Z := \sum_{j=1}^{\Sigma} P_j$ and $P_j = \prod_{i=1}^{\pi} x_i$ be the logic function and $(W/L)_j$ the resulting $(W/L)_j$ for the selected product j. Let $t_{rj} = \underset{\forall k \bullet P_j}{MAX} (t_{rLH}, k)$.
Then search j such that $t_{rj}/(W/L)_j = \underset{\forall k}{MAX} \{t_{rk}/(W/L)_k\}$.
This identifies the input test pattern for the worst case t_{pHL} as follows : $x_i = 0 \; \forall_i \neq j$ and x_j is assigned the rising input waveform j. One single analysis of N_{up} then generates t_{pHL} for logic simulation.

6. Analyse : it is to be noticed that a lot of information is gathered during procedures 1-5 which allow for a lot of potential troublespot indication, worst-case delay information and charge redistribution effects. One particular problem of interest is the timing verification of static logic subnetworks occuring between two transmission gate registers in a multiphase clocked MOS system. Static logic blocks belong, by definition, to Boolean logic gates with worst case t_{pLH} en t_{pHL} obtained during the transform procedure.
Instead of doing a complete simulation only the networks partitioned between e.g. clocks $\emptyset_i$ and $\emptyset_j$, with T_{ij} as rising edge time difference have to be checked for their longest delay path which must be smaller than T_{ij}. The logic network can be translated into a graph in which vertices are gates and edges are assigned delay values. Using graph or PERT techniques it is possible to identify the longest path in $O((V+E)^2)$ operations. Since logic behaviour has already been checked in the prelayout phase, this analysis on small subcircuit certifies logic timing errors without need for simulation.

4. CONCLUSION

The above are only a few indications how, based on a stick-diagram representation and a hierarchical compaction system, the concept, algorithms and data-structures of a mixed mode simulator can be used as a guided simulation and analysis tool for n-MOS VLSI. Such a system is based on, but also imposes, a structured top-down design methodology.
This paper does not claim that such a system exists yet but is written to stimulate research, developments and ideas in this direction.

[1] C. Mead, L. Conway :"Introduction to VLSI systems", Addison-Wesley, 1980.

[2] N.B. Rabbat et al. :"A Multilevel Newton Algorithm with Macromodeling and Latency for Analysis of Large-Scale Nonlinear Circuits in the Time Domain", IEEE Trans. Circuits & Systems, CAS-26, No. 9, pp. 733-741, Sept. 1979.

[3] B.R. Chawla, H.K. Gummel and P. Kozak :"MOTIS - An MOS timing simulator", IEEE Trans. Circuits Syst., Vol. CAS-23, pp. 301-310, Dec. 1975.

[4] G. Arnout and H.J. De Man :"The use of threshold functions and Boolean-controlled network elements for macromodeling of LSI circuits", IEEE J. Sol. St. Circuits, Vol. SC-13, pp. 326-332, June 78.

[5] P. Yang et al. :"SLATE : A Circuit Simulation program with latency exploitation and node tearing", Proc. IEEE ICCC Conference, Oct. 1980, New York, pp. 353-355.

[6] A. Vladimirescu et al. :"A computer program for the simulation of LSI circuits", Proc. 1981 IEEE ISCAS Conference, April 1981, Chicago.

[7] H. De Man, R. Newton :"Hybrid Simulation", Proc. 1979 IEEE ISCAS Conference, July 1979, pp. 249-259.

[8] A.R. Newton :"Techniques for the Simulation of Large Scale Integrated Circuits", IEEE Trans. Circuits & Systems, CAS-26, No. 9, pp. 741-749, Sept. 1979.

[9] Ph. Reynaert et al. :"DIANA - A mixed-mode simulator with a hardware description language for hierarchical design of VLSI", Proc. 1980 IEEE ICCC Conference, Oct. 1980, New York, pp. 356-360.

[10] N.H. Nham, A.K. Bore :"A multiple delay simulator for MOSLSI circuits" and :
V.D. Agrawal et al. :"A mixed-mode simulator", Proc. 17th Design Automation Conference, June, 1980, pp. 610-625.

[11] D. Johannsen :"Bristle Blocks; A silicon compiler", Proc. 16th Design Automation Conference, pp. 310-313, June 1979.

[12] J.D. Williams :"STICKS - A graphical compiler for High-Level LSI design", AFIPS Conference Proceedings, Vol. 47, pp. 289-295, June 1978.

[13] M.R. Barbacci :"Introduction Set Processor (ISPS): The notation and its application", IEEE Trans. Computers, C-30, no. 1, pp. 24-40, Jan. 1981.

[14] H.M. Lipp :"Current trends in the design of digital circuits", in "Computer-Aided Design of digital electronic circuits and systems", North Holland, Amsterdam, 1979, pp. 91-102.

[15] W.M. Van Cleemput, W.E. Cory :"Developments in Verification of Design Correctness", Proc. 17th Design Aut. Conf., June 1980, Minneapolis, pp. 156-165

[16] J.D. Crawford et al.:"MOTIS-C users guide", Electronics Research Laboratory, University of Calif. Berkeley, June 1978.

[17] G. Bogo, Mermet :"CASSANDRE and the Computer Aided Logical Systems Design", Proc. IFIP Congress, Ljubljana, Yugoslavia, Aug. 1971, pp. 1056-1063.

[18] D.D. Hill, W. Van Cleemput :"SABLE : A tool for generating structured, multi-level simulations", Proc. 16th Design Aut. Conf., June 1979, pp. 272-279.

[19] P. Draheim, B. Schendel :"Verification for complex digital integrated circuits", Proc. 4th Europ. Conf. on Electrotechnics, EUROCOM 80, pp. 151-155, March 1980.

[20] M.Y. Hsueh, D.O. Pederson :"Computer Aided Layout of LSI Circuit Building Blocks", Proc. 1979 Int. Symp. Circuits & Systems, Tokyo, pp. 474-477, July 1979.

[21] R.E. Bryant :"An Algorithm for MOS Logic Simulation", Lambda Magazine, A Fourth Quarter 1980, pp. 46-53.

[22] N. Tanabe et al. :"MOSTAP : An MOS Circuit Simulator for LSI", Proc. IEEE Int. Symp. Circ.& Syst., pp. 1053-1079, Houston, April 1980.

[23] A. Ruehli et al.: "Time Analysis of Large Scale Circuits Containing One-Way Macromodels", Proc. IEEE Int. Symp. Circ.& Syst., pp. 766-770, Houston, April 1980.

[24] L. Weinberg :"Optimal MOS Logical Design Using Minimal Partitioning and Graph Theory", Proc. 1980, Int. Conf. Circuits & Computers, New York, pp. 1036-1039, Oct. 1980.

[25] T.K. Liu :"Synthesis algorithms for 2 level MOS networks", IEEE Trans. Computers, C-24, No. 1, pp. 72-79, Jan. 1975.

[26] J.P. Hayes :"Testing Logic Circuit by Transition Counting", International Symposium Fault Tolerant Computing, pp. 215-222, June 1975.

[27] G. De Micheli et al.:"New Algorithms for timing analysis of Large Circuits", Proc. IEEE Int. Symp. Circ. & Systems, pp. 439-444, Houston, Texas, April 1980.

[28] A.E. Moselhi et al. :"A macormodel and a macrosimulator for GaAs FET Digital IC's", Proc. IEEE Int. Symp. Circuits & Computers, pp. 366-373, New York 1980.

[29] R. Rohrer, H. Nosrati :"A Modified Forward Euler Approach to Stable Step Response Computation", IEEE Trans. Circuits & Systems, CAS-28, no. 3, pp. 180-185, April 1980.

[30] I. Ohkura et al.:"A New Exact Delay Logic Simulation for ED MOS LSI", Proc. 1980, IEEE Int. Conf. Circuits & Computers, pp. 953-956, Oct. 1980.

THE SECOND GENERATION MOTIS MIXED-MODE SIMULATOR

C.F. Chen, C-Y Lo, H.N. Nham, and Prasad Subramaniam

AT&T Bell Laboratories, Murray Hill, New Jersey 07974

ABSTRACT

This paper describes the second generation MOTIS mixed-mode simulator. In particular, it extends the current modeling capabilities to include resistors, floating capacitors, and bidirectional transmission gates. It employs a relaxation algorithm with local time-step control for timing simulation, and a switch level approach for unit delay simulation. It provides logic and timing verification for general MOS circuits in a mixed-mode environment. The new simulator is being used for production chips, and it is more accurate, flexible, and efficient than the existing MOTIS mixed-mode simulator.

1. INTRODUCTION

In VLSI design verification, mixed-mode simulators [1,2,3] have proved to be flexible and cost-effective. In general, the simulation level varies from architectural, register transfer, unit delay, multiple delay, timing, to circuit level. The MOTIS mixed-mode simulator covers unit delay, multiple delay, and timing simulation. The timing simulator [15] produces continuous voltage waveforms within a finite resolution for each signal by integrating the net charging current flowing into the signal capacitance. The multiple delay simulator [16] is considerably faster but produces less accurate timing waveforms. In multiple delay mode, transistors are simulated logically with transitions at the output nodes being scheduled with integer rise and fall delays. These delays are calculated automatically from the I-V curves and the loading capacitances. The unit delay simulator is a special case of the multiple delay simulator, its primary function is to verify the logic function of a circuit.

The present MOTIS mixed-mode simulator can only handle driver-load structures such as NAND, NOR, AND-OR-INVERT, etc., and limited configurations of bidirectional components. For example, a configuration comprising more than two transmission gates in series is not allowed. In addition, polysilicon resistance and bootstrapping cannot be modeled. Furthermore, the accuracy of timing simulator degrades for small gate delays.

In this paper we will discuss the second generation MOTIS mixed-mode simulator, which handles general MOS circuits with resistors and floating capacitors. A circuit is first decomposed into unidirectional blocks by grouping all the related bidirectional components together. Then these blocks are simulated either in unit delay, multiple delay, timing, or mixed-mode. While the unit delay uses switch level simulation techniques [7,8,9] to evaluate transistors, the timing simulator uses relaxation techniques [4,5,6,13] with local time-step control [6] to calculate and propagate signals. The new simulator is now being used for production LSI chips. The models, algorithms, implementation, and performance of the simulator are addressed here.

2. CIRCUIT DESCRIPTION, MODELS, AND EQUATION FORMULATION

The new mixed-mode simulator allows users to describe their circuits in a hierarchical fashion. A circuit is typically described as an interconnection of several subcircuits, which could in turn be described similarly. Each individual subcircuit communicates with another through a set of input, outputs, and ioputs. Ioputs are those nodes that can be driven from both inside as well as outside a subcircuit. The input processors first compile the subcircuits individually and then expand (link) the circuit into a flat structure. Before simulation, this flat circuit structure is regrouped into unidirectional sections called superblocks. In this section, we will discuss the models used in the simulator and the procedure for setting up the superblocks.

Reprinted from *ACM/IEEE 21st Design Automation Conf. Proc.*, 1984, pp. 10–17.

2.1 Circuit Models

The basic primitives in the simulator and their directionalities are shown in Figure 1. They consist of resistors, floating capacitors, transistors, logic gates, and functional elements. Based on the nature of these primitives, we can classify them into the following four different simulation models:

2.1.1 MOS Normal Gates

A MOS normal gate is a device with multiple inputs and one output. It consists of two sets of series-parallel transistors, one of them (the load) connected between VDD and the output, and the other (the driver) connected between VSS and the output. Figure 1(a) is an example of a CMOS two-input NAND gate. Extensive use of normal gates in a circuit reduces its complexity significantly.

2.1.2 Bidirectional Components

To be able to model circuits more accurately, the simulator also accepts resistors for modeling polysilicon resistance, and floating capacitors for modeling bootstrapping effects. In addition, transistors which cannot be clustered into normal gates are modeled as bi-directional transmission gates. Figure 1(b) shows the bidirectional components within the simulator. For efficiency, a pair of p-channel and n-channel transistors in parallel is treated as a single CMOS transmission gate.

2.1.3 Logic Gates

A logic gate can be described by its logic function with associated integer rise/fall delays. Commonly used logic gates such as AND, OR, NAND, NOR, etc. are supported in the simulator. A two-input logic NAND gate is shown in Figure 1(c).

2.1.4 Functional Primitives

Functional primitives are, in general, easier to describe and more economical in terms of storage and simulation time. They help to reduce the circuit complexity. Figure 1(d) shows an example of a memory primitive.

2.2 Superblock and Sparse Matrix Set-ups

In order to decouple a circuit into unidirectional blocks, bidirectional components and their related driving normal gates are grouped together into sections called superblocks. This is an automatic process which is completely transparent to the user. The primary inputs, the input sides of primary ioputs, and the gate terminals of transistors form the borders of a superblock. A superblock can be considered as a multiple input and multiple output device, which is internally bidirectional but unidirectional externally. For efficiency reason in timing simulation, the superblocks are divided into two types. If a superblock contains resistors and/or floating capacitors, it is a type 1 superblock, otherwise, it is a type 0 superblock. Different techniques are used for these two types of superblocks.

Table 1 shows the superblock statistics for some production circuits. In this table, circuits A and B are MOTIS compatible, but circuits C through F can only be analyzed in the new simulator. It is observed that the average number of equations per superblock gets larger when more transmission gates are used in the circuit design. Also, the number of superblocks (nsupb) tends to decrease if more resistors and floating capacitors are included. This is because, as the number of bidirectional components in the circuit increases, there is a higher chance for more nodes to interact with one another, resulting in a few large superblocks. Although the average number of equations per superblock is less than ten, some can be as large as four hundred. Sparse matrix methods are therefore necessary to evaluate the superblocks efficiently.

To facilitate the superblock evaluation in timing mode, a threaded linked-list sparse matrix is used to store and manipulate the nonzero elements. For a type 1 superblock, the Markowitz reordering scheme [17] is also used to minimize the fill-ins during the LU decomposition process. The sparse matrix pointers are updated accordingly; no equation swapping tables are used during the actual simulation.

ckt name	# trs	# r,c	#gates	#normal gates	nsupb	eqn/supb
A	8768	0	3440	2632	769	2
B	4562	0	965	939	26	1
C	3423	328	2719	1004	190	5.2
D	3586	211	3074	1056	488	3.2
E	12246	29	8242	2820	988	3.8
F	28807	43	21406	6190	2633	3.8

Table 1 Superblock statistics

The superblock can either be simulated in timing or in unit delay. At this moment, no multiple delay simulation is allowed for superblocks due to the lack of an accurate delay model. In unit delay simulation, only a subset of superblock data structure is used.

3. SIMULATION ALGORITHMS AND IMPLEMENTATION

In this section the algorithms and implementation of the generalized unit delay and timing simulation will be discussed. The multiple delay simulation technique is similar to the one used in MOTIS [16], and will not be described here.

3.1 Generalized Unit Delay Simulation (Switch Level Simulation)

In unit delay simulation, the MOS normal gates can be evaluated fairly efficiently. This is achieved not only by the use of an efficient transistor structure representation that facilitates early terminations during evaluation, but also by the enhanced latency technique that effectively looks ahead and avoids unnecessary evaluations. It has been shown that a savings in speed by as much as 40% can be obtained [14]. Superblock evaluations are triggered only by external input changes. For internal node changes, the simulator iterates until all nodes have settled to a steady state. Oscillations are detected if the number of iterations exceeds the number of nodes in the superblock. The value of a node is represented by 5 strengths, and 3 states 0, 1, and X. The strengths in decreasing order are: forced, input, pull-up, big, and small, following Bryant's notation[7]. These strengths are used for internal evaluation, and subsequently ignored outside the superblock boundaries. It has been shown previously [14] that the logic evaluation of a MOS normal gate produces a value of 3-strengths, namely, driving, unknown, and high impedance (Z). The interface between normal gate strengths and

superblock strengths will be discussed later. Since the switch level simulation works on the same data structure as in timing simulation, information about transistor and gate characteristics are readily available. Therefore, it is very easy to increase the accuracy of the simulation by handling more signal strengths.

The superblock evaluation consists of 3 phases, namely, node initialization, node interaction, and event generation. A node list is used for the first phase, and a switch list for the second phase. In the node list, each node entry contains a set of unilateral elements whose outputs are tied to the node. Unilateral elements include MOS normal gates, logic gates, primary inputs, and input sides of primary ioputs. The switch list is a list of bilateral elements, and their connected node pairs (e.g. source, and drain).

In the first phase, all nodes in the superblock are initialized. Nodes driven by primary inputs or active primary ioputs assume a forced strength with the input states since primary inputs are treated as independent voltage sources. Nodes driven by normal gates and logic gates have their outputs as the initial values. The rest of nodes are driven by switches and are assigned a high Z strength with previous states. Except for nodes forced by primary inputs, the initial node values are then converted to superblock node strengths according to Table 2.

In the second phase, the interaction between adjacent nodes connected by ON, or X switches is evaluated. As a result, nodes in the superblock are partitioned into mutually exclusive equivalent classes. Nodes are in the same ON (X) equivalent class if they are connected by some ON (ON, and/or X) switches. During evaluation, each node contains the values of its ON, and X classes. Unlike MOSSIM, equivalent classes are created incrementally. Initially, every node is a class by itself. As the conduction

normal gate strength	superblock strength	comments
driving	input	non-depletion load or state = 0
driving	pull-up	depletion load and state = 1
unknown	input	pessimistic
high Z	big, small	depends on node sizes

Table 2 Strength Conversion

states of switches are determined, the equivalent classes of the connected nodes are merged into larger ones. The new class value is obtained by merging the two old class values, the merged value being the stronger of these two. In case the two values are of equal strengths but conflicting states, the resulting state is X. The ON switches are processed before the X switches. Then the values of X classes are compared against those of the ON classes to produce the steady state node values. The conduction state of a switch is determined by evaluation of the switch element. For an NMOS transmission gate, the conduction state is determined by both its transistor type and the voltage at its gate terminal. For a CMOS transmission gate, the conduction state is ON if either transistor is ON, is OFF if both transistors are OFF, and X otherwise. For simplicity, resistors are treated as ON switches, and capacitors as OFF switches. The last phase of evaluation is to generate events if the steady state node states differ from previous states (not strengths).

The complexity of this implementation is linear with the number of transmission gates (T) in the superblock. This is because that for each switch, only connected nodes in the current equivalent classes need to be traversed, where such number has been observed to be independent of T [8] and is typically less than five. In addition, through the extensive uses of normal gates, the number of nodes in a superblock is usually small (less than 10). It should also be noted that by storing the previously evaluated value for each element in a superblock, an element (normal gate or transmission gate) is re-evaluated only when there is a change at its input. Furthermore, the enhanced latency analysis technique is applicable to normal gates inside superblocks. Input changes to these normal gates may be ignored if the newly evaluated values are the same as before. In this case, the evaluation of the entire superblock may be avoided.

The utilization of normal gates also facilitates the race analysis of cross-coupled gates. Before simulation, all cross-coupled normal gate pairs are identified and classified as one of the following three types, namely, NOR, NAND, and GENERAL. A NOR (NAND) type race pair is one where a 1 (0) on one output will force the other to 0 (1). All other types of cross-coupled pairs are categorized as GENERAL. During simulation, race analysis is invoked whenever both outputs of a race pair change at the same time. A race condition is detected if such simultaneous changes cause the circuit to oscillate when other inputs are held constant. For example, in a NOR type race pair, when both outputs change from 0 to 1 at the same time, the outputs will return to 0 and then oscillate between 0 and 1. In this case, a race condition is reported and the transitions will be set to unknown. For a GENERAL race pair, assuming one output changes from a to b and the other changes from c to d simultaneously, the simulator will look ahead and evaluate both gates with new inputs values b and d, and if the new output values change back to a and c again, a race condition is detected. In short, the race analysis not only avoids excessive run time, it can also points out potential design problems, and has been found to be a very useful feature in this simulator.

3.2 Timing Simulation

Timing simulation is a special kind of simulation which uses simplified circuit simulation techniques with less accurate transistor models to evaluate the signals in the circuit, and logic simulation techniques to propagate their values. The simulation output is a continuous voltage with a certain resolution rather than a logic value (0,1,X). The simulation speed is about 100 times faster than conventional circuit simulation as in SPICE2 [12], and is about 100 times slower than logic simulation.

The original timing simulator in MOTIS is designed to handle circuits containing only MOS normal gates with some limited configurations containing transmission gates and busses [15]. As technology heads toward the submicron era, the polysilicon resistance and bootstrapping effect due to floating capacitors become more important. Also, designers favor the use of more transmission gates in their circuits to optimize the silicon area. These constraints become a limiting factor in using MOTIS. In addition, the accuracy of timing simulation deteriorates as the device feature size gets smaller. The situation can only be remedied by a reduced time-step which, unfortunately, dramatically increases the simulation time [6].

In order to remove the topological and modeling constraints, increase the accuracy, and have comparable or better speed performance than MOTIS, a new timing simulator has been developed. The detailed algorithms of this simulator are presented elsewhere [6]. Here we highlight some of the techniques used in the simulator.

3.2.1 Relaxation techniques

Among all unidirectional blocks in a circuit, only MOS normal gates and superblocks can be simulated in timing mode. Since these blocks are fully decoupled and unilateral, nonlinear Gauss-Seidel relaxation can be employed at the network level. For each active block, a mixture of line origin and regular two point secant method [11] is used to linearize the nonlinear transistor curve, and to iterate the solution until it converges. Because a normal gate has a special structure and a single output, the output voltage at the $(k+1)$th iteration at t_{n+1} can be derived explicitly as shown in Eqn.(3-1)[6], where g_d^k is the equivalent conductance of the driver transistors at the kth iteration, and g_l^k is the equivalent conductance of the load transistors. The Backward Euler integration method is used in the simulator, and the equivalent loading conductance due to the ground capacitance C is given by $\frac{C}{h_n}$ where h_n is the time-step.

$$V_{n+1}^{k+1} = \frac{g_l^k\ VDD\ +\ \frac{C}{h_n}\ V_n}{g_d^k\ +\ g_l^k\ +\ \frac{C}{h_n}} \quad (3\text{-}1)$$

For superblock evaluation, the nodal formulation technique with the secant method is first employed to set up the equations for the node voltages. Then to solve these equations, Gauss-Seidel relaxation at the linear level is used for a type 0 superblock, and LU decomposition is used for a type 1 superblock. We have observed that 10% CPU time is saved by adopting this strategy, because the coupling within the type 0 superblock is much weaker than that within the type 1.

The relaxation techniques used here are quite different from that in RELAX [5,13] and SPLICE [4]. RELAX uses waveform relaxation to decouple and iterate the circuit at the differential equation level. SPLICE iterates every node at the nonlinear equation level with each node being evaluated only once for every outer loop iteration. SPLICE does not have an extra level of hierarchy like the superblock. The technique used in SPLICE is similar to the one we use for evaluating a type 0 superblock except that the relaxation level is different. For mixed-mode simulation, we feel that the relaxation techniques used in the new MOTIS timing simulator are much cleaner and easier to interface with the different levels of simulation.

The average number of iterations per superblock per event depends on the time-step, the accuracy desired, and the coupling within the superblock. The larger the time-step, the more the number of iterations needed for convergence. We have observed that for a given accuracy, the time-step in the new simulator can be larger than that of MOTIS because the latter does not iterate.

3.2.2 Local Time-step Control

The objective of local time-step control is to allow every active gate or superblock to step at its own pace. In other words, every active gate or superblock can have a different time-step as opposed to the global fixed time-step in MOTIS. This is more efficient because every active gate or superblock may have a different delay due to difference in loading, and even for the same gate, the rise delay may be different from the fall delay (e.g. NMOS technology).

Since iteration count is a good predictor for the time-step size, it is used to control local time-steps. The problem can be described as follows: Given the upper and lower bounds of time-step (h_{max} and h_{min}), and the upper and lower thresholds of iteration counts (U and D), the next time-step (h_{n+1}) is estimated based on the current iteration count (IC) and the current time-step (h_n). A simple solution of this problem is to double the time-step if IC is less than D, and to halve the time-step if IC is greater than U. Care is taken to ensure that h_{n+1} is bounded between h_{max} and h_{min}. h_{min} in general is equal to UT unless an event collision occurs, where UT is the basic simulation time unit or the so called minimum resolution time(MRT).

In case a gate is scheduled for evaluation again before its pending event gets updated, the pending event is canceled, and the gate is evaluated and updated right away. In this situation, an event collision occurs (like a spike in logic simulation) and h_{min} is equal to the current simulator time minus the time when the canceled pending event was generated. In order to minimize the number of event collisions and synchronize active gates or superblocks, h_{max} should be specified. This is a technology dependent parameter and is related to gate delays.

For a superblock, every member within the block has the same time-step, and the entire superblock is scheduled with a single superblock event whenever it is active. Our experiments show that the local time-step control scheme is about two to three times faster than the global fixed time-step [6].

3.2.3 Table Models

A new table model has also been developed to model the transistor drain current as a function of both the device geometry and the terminal voltages including the back-gate bias effect. Each model uses about 1000 words of memory and is about two times faster but 3 to 5% less accurate than the equivalent analytical model. The detailed derivation and performance of this new model is described elsewhere [10].

	superblock	normal gate	logic gate	functional primitive
unit delay	X	X	X	X
multiple delay	-	X	X	X
timing	X	X	-	-

Table 3 Simulation modes and models

3.3 Mixed-Mode Simulation

The new MOTIS mixed-mode simulator uses an event-directed, table-driven approach. While the scheduling of the timing blocks is a little different due to the local time-step control scheme, the control structure of the simulator and the interface between timing gates and non-timing gates are similar to that of MOTIS. Table 3 shows the possible mode settings among the different levels of models.

At this moment, superblocks cannot be evaluated in multiple delay mode due to the lack of an accurate delay model. They can only be evaluated in timing or unit delay mode (switch level simulation). For a MOS normal gate, the rise and fall delays are computed automatically from the transistor I-V curves and loading capacitances as described in [16]. For logic gate and functional primitives, the rise and fall delays are computed from user specified delay coefficients, information on the gate types, routing capacitances, and the number of fanouts.

The mixed-mode simulation also allows a user to change the mode dynamically during simulation. For example a user can use the timing mode to initialize a circuit, and then change to unit delay mode to verify its logic function. For timing verification, the user can use timing simulation to evaluate all superblocks, and multiple delay simulation to evaluate the rest of the circuit. This is extremely cost-effective.

4. PERFORMANCE

The new MOTIS mixed-mode simulator is being used for production chips which include both MOTIS compatible and incompatible circuits. For a MOTIS compatible circuit, the new simulator has always shown more accurate results than MOTIS. In this section the speed performance of two production circuits (A and C in Table 1) on a VAX 11/780 running under the VMS operating system is discussed.

4.1 Circuit A --- MOTIS compatible circuit

This circuit consists of 8768 transistors, with 769 superblocks. There are no resistors or floating capacitors. The simulation speed is about 16 vectors per hour for the MOTIS timing simulator; 53 vectors per hour for the new timing simulator; 1614 vectors per hour for the new unit delay simulator; and finally 127 vectors per hour for the new mixed-mode simulator. In the mixed-mode simulation run, all the superblocks (1542 gates) were simulated in timing and the rest of the circuit (1898 gates) was simulated in multiple delay. For this example, the new timing simulator is about three times faster than the MOTIS timing simulator and the new mixed-mode simulator is about 2.5 times faster than the new timing simulator.

4.2 Circuit C -- MOTIS incompatible circuit

This circuit consists of 3423 transistors, 328 resistors and floating capacitors, and 190 superblocks. The simulation speed is about 70 vectors per hour for the new MOTIS timing simulator, and 3000 vectors per hour for the new MOTIS unit delay simulator. Portions of this circuit have also been simulated using ADVICE [18] which is an in-house circuit simulator, and the results show that the new timing simulation is only three to five percent less accurate than the circuit simulation.

We can see that MOTIS incompatible circuits are more cpu intensive than MOTIS compatible circuits because they are more tightly coupled. This means that more simulation time can be expected if there are more bidirectional components in the circuit. However, for MOTIS compatible circuits, the new simulator is always faster than MOTIS.

5. CONCLUSION

A generalized MOS mixed-mode simulator has been described. The simulator not only handles general MOS circuits with resistors and floating capacitors which MOTIS cannot accommodate, but it also improves the accuracy and speed performance. It provides circuit designers a flexible and useful means to perform logic and timing verification. The algorithms used in the timing simulator are quite general and can be extended to circuit simulation as well.

ACKNOWLEDGEMENTS

The authors gratefully acknowledge the contributions of F. Fazal, E. Pacas-Skewes, and W.J. Dai for their work on the input and output processors.

REFERENCES

[1] V.D. Agrawal, A.K. Bose, P. Kozak, H.N. Nham, and E. Pacas-Skewes, 'A Mixed-Mode Simulator,' *Proceedings of 17th Design Automation Conference*, Minneapolis, Minn., June 23-25, 1980, pp. 618-625.

[2] A.R. Newton, 'The Simulation of Large Scale Integrated Circuits,' **Memorandum No. UCB/ERL M78/52**, Electronics Research Lab., University of California, Berkeley, CA., July 1978.

[3] H. De Man, L. Darcis, I. Bolsens, P. Reynaert, and D. Dumlugol, 'A Debugging and Guided Simulation System for MOS VLSI Design,' *Digest of Technical Papers, ICCAD - 83*, Santa Clara, CA., Sept. 12-15, pp. 137-138.

[4] R.A. Saleh, J.E. Kleckner, and A.R. Newton, 'Iterated Timing Analysis in SPLICE1,' *Digest of Technical Papers, ICCAD - 83*, Santa Clara, CA., Sept. 12-15, pp. 139-140.

[5] E. Lelarasmee, A.E. Ruehli, and A.L. Sangiovanni-Vincentelli, 'The Waveform Relaxation Method for Time-Domain Analysis of Large Scale Integrated Circuits,' *IEEE Trans. on Computer-Aided Design of Integrated Circuits and* Systems", **Vol. CAD-1, No.3, July 1982**, pp. 131-145.

[6] Chin Fu Chen, and Prasad Subramaniam, 'The Second Generation MOTIS Timing Simulator - An Efficient and Accurate Approach for General MOS Circuits,' *Proceedings of the 1984 International Symposium on Circuits and* Systems", Montreal, Canada, May 7-10, 1984.

[7] R.E. Bryant, 'An Algorithm for MOS Logic Simulation,' *Lambda*, **Vol. 1, No.3, 1980**, pp. 46-53.

[8] R.E. Bryant, 'A Switch-level Model and Simulator for MOS Digital systems,' **Caltech Report 5065:TR:83**, 1983.

[9] D. Dumlugol, H.J. De Man, P. Stevens, and G.G. Schrooten, 'Local Relaxation Algorithms for Event-Driven Simulation of MOS Networks Including Assignable Delay Modeling,' *IEEE Trans. on Computer-Aided Design*, **Vol. CAD-2, No.3, July 1983**, pp. 193-202.

[10] Prasad Subramaniam, 'Table Models for Timing Simulator,' *Proceedings of the 1984 Custom Integrated Circuits Conference*, Rochester, New York, May 21-23 1984.

[11] R. Rohrer, 'Successive Secants in the Solution of Nonlinear Network Equations,' *SIAM-AMS Proceedings*, **Vol. III**, Symposium on Mathematical Aspects of Electrical Analysis, 1971.

[12] L.W. Nagel, 'SPICE2, A Computer Program to Simulate Semiconductor Circuits,' **ERL-M520**, University of California, Berkeley, CA., May 1975.

[13] J. White, and A.L. Sangiovanni-Vincentelli, 'RELAX2: A New Waveform Relaxation Approach for the Analysis of LSI MOS Circuits,' *Proc. 1983 International Symposium on Circuits and Systems*, May, 1983.

[14] C-Y Lo, H.N. Nham, and A.K. Bose, 'A Data Structure for MOS Circuits,' *Proceedings of the 20th Design Automation Conference*, Miami Beach, FL. June 27-29 1983, pp. 619-624.

[15] B.R. Chawla, H.K. Gummel,and P. Kozak, 'MOTIS - An MOS Timing Simulator,' *IEEE Trans. on Circuits and Systems*, **Vol. CAS-22, No.12, Dec. 1975**, pp. 901-910.

[16] H.N. Nham, and A.K. Bose, 'A Multiple Delay Simulator for MOS LSI Circuits,' *Proceedings of 17th Design Automation Conference*, Minneapolis, Minnesota, June 23-25 1980, pp. 610-617.

[17] H.M. Markowitz, 'The Elimination Form of the Inverse and Its Application to Linear Programming,' *Management Science*, **Vol. 3, April 1957**, pp. 255-269.

[18] L.W. Nagel, 'ADVICE for Circuit Simulation,' *Proceedings of the 1980 International Symposium on Circuits and* Systems", Houston, Texas, April 28, 1980.

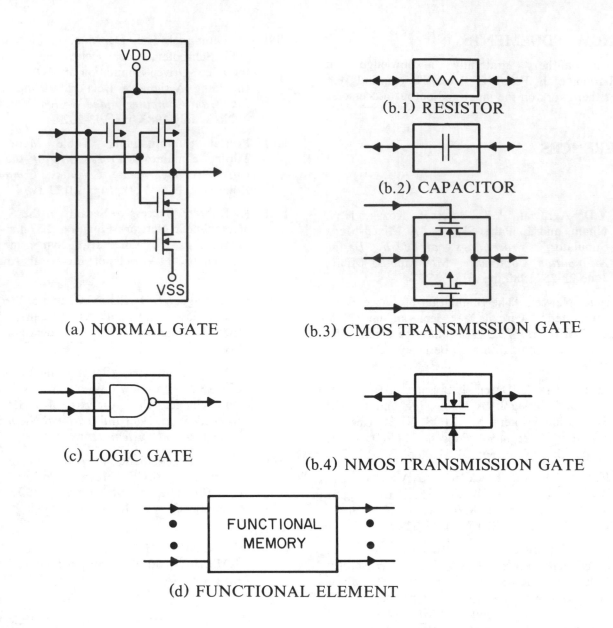

(a) NORMAL GATE

(b.1) RESISTOR

(b.2) CAPACITOR

(b.3) CMOS TRANSMISSION GATE

(c) LOGIC GATE

(b.4) NMOS TRANSMISSION GATE

(d) FUNCTIONAL ELEMENT

Figure 1. Basic primitives used in the second generation MOTIS simulator

THEMIS LOGIC SIMULATOR - A MIX MODE, MULTI-LEVEL, HIERARCHICAL, INTERACTIVE DIGITAL CIRCUIT SIMULATOR

Mahesh H. Doshi, Roderick B. Sullivan and Donald M. Schuler

Prime Computer, Inc.
Framingham, Massachusetts

ABSTRACT

A new logic simulator called THEMIS™ Logic Simulator for the design of LSI, VLSI and PCBs is described. THEMIS supports design verification and test development from initial specification in behavioral and RTL languages to analysis of the final layout at the gate and switch level. To allow the simulation of an entire system or check the correctness of a single circuit, the different modeling techniques can be easily intermixed. THEMIS is a highly interactive simulator that minimizes a hardware engineer's time and effort to debug logic. This paper gives an overview of THEMIS and its use by design engineers.

INTRODUCTION

The expanding use of increasingly complex gate array, semi-custom and full custom LSI/VLSI digital circuits has forced system designers into using new design techniques and new design tools. It is no longer possible to debug a wire-wrap breadboard version of a system in the laboratory when errors may be in the custom designed ICs that comprise the system. For designs that only use standard parts, bread boarding may not be possible if high speed logic is used. In high speed ECL logic, wire delay significantly effects circuit operation and the wire-wrap version is not an accurate model of the printed circuit version. Finally the increase in the gate count of digital systems is making it more difficult to detect and correct errors during hardware prototyping.

The increase in complexity is causing hardware designers to turn to the same structured design and step-wise refinement techniques that have been used by the software designers. Essential to new design methods for complex systems are a new generation of logic simulation and verification programs. Traditionally, the logic simulation tools detect logic and timing errors after a system has been described at the logic level. Designers now require more sophisticated simulation tools that can evaluate a design at earlier stages in the cycle.

The functionality required of these new simulators include:

- The ability to simulate very large circuits. The simulator must handle the entire design of VLSIs, as well as the logic on the PCB which will include the VLSI under design. This means that the simulator should handle tens of thousands of gates in one session.
- The ability to simulate interactively, as opposed to batch simulation. The batch mode simulators generate tremendous amounts of data which is impractical, if not impossible, to digest. The designers want the ability to detect bugs, fix them, and then verify that they work without having to leave the simulation session.
- The speed of simulation is critical. The turnaround time in batch simulators is not acceptable. In an interactive environment, the simulator must respond in a reasonable time.
- A variety of modeling techniques must be included to satisfy the diverse needs of the designers.
- The ability to define the behavior of incomplete designers.
- The ability to handle mixed levels of simulation.
- The ability to provide for various structured design methodologies including top-down, bottom-up and mixed.

A new logic simulator has been developed in an attempt to solve these problems. THEMIS stands for The Hierarchical Multi-level Interactive Simulator. It performs simulation at the behavioral, register transfer, logic and switch levels of description. This paper describes the overall features of the THEMIS simulator and the variety of ways it can be used by design engineers. Finally the benefits derived from actual use at PRIME are discussed.

CREATING SIMULATION MODELS

THEMIS supports a range of modeling techniques at different levels of detail that can be combined to form more complex models. Models are created from:

- PL/1, Fortran, Pascal and C subroutines,
- An RTL language called Themis Architectural Design languages (TAD),
- Boolean equations,
- Builtin primitives, and
- Networks of existing models

Common to all modeling methods is a definition of pins.

Reprinted from *ACM/IEEE 21st Design Automation Conf. Proc.*, 1984, pp. 24–31.

The pins define electrical characteristics of the components, and provide an interface through which models can communicate with other models. This interface allows the simulator to use different modeling techniques for different components in the same simulation run. The pin definition includes pin name, its type and optional data to specify the following:

- Pin number
- Type
- A float value for the pin when left unconnected
- A capacitance value seen by an external output driving the pin
- A signal name used to describe internally connected pins

Figure 1 shows some examples of pin definitions.

```
ADD PINS
   1 = A               INPUT
   2 = B               INPUT        FLOAT 1 CAPACITANCE 3.9
   {3:6},{9:12} = DATA{0:7}  BIDIRECT   FLOAT FF:H
   Q1                  OUTPUT       SIGNAL DUAL__OUT
   Q2                  OUTPUT       SIGNAL DUAL__OUT
```

Figure 1 Pin Definitions

Once the pins of a component are defined, the system accepts a definition of the component behavior, i.e. simulation model, in any one of several available modeling languages. Each model is compiled in an interactive environment. Errors can be immediately corrected without leaving THEMIS. The models are then linked together with the necessary component models from either central libraries or a local library.

Boolean Modeling

Many simulators provide a fixed set of primitives such as AND, OR, NAND and NOR for describing logic. The disadvantages with this approach is that complex primitives must be constructed out of the fixed set of primitives thereby increasing the number of elements to be simulated. In THEMIS, primitives are built from user defined boolean equations. A NAND is simply described as OUT = $\sim$(A & B & C & D ...). Complex devices can be described as well. Efficient table lookup techniques are used for device evaluations. Even the most complex boolean primitive simulates at the same speed as a two input AND gate.

The boolean models can be defined for any element that has up to eight inputs and one output. The output can be used as an internal feedback. The limitation of eight inputs is only parametric and can easily be extended to permit more input pins. The output is either two-state (binary) or tri-state. Figure 2 shows an example of a boolean model for a tri-state AND-OR-INVERT gate. The two equations specified are compiled into a truth table which is used by the boolean evaluator.

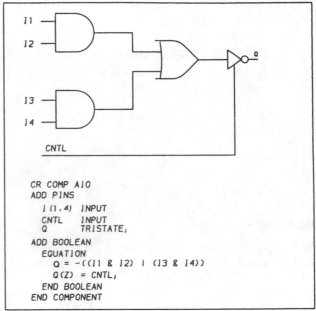

```
CR COMP AIO
ADD PINS
   I (1,4)  INPUT
   CNTL     INPUT
   Q        TRISTATE;
ADD BOOLEAN
   EQUATION
      Q = ~((I1 & I2) | (I3 & I4))
      Q(Z) = CNTL;
   END BOOLEAN
END COMPONENT
```

Figure 2 AND-OR-INVERT modeled as a single boolean element.

The THEMIS Architectural Design Language

To verify the logical correctness of a design at the earlier stages of the design cycle, THEMIS supports register transfer language (TAD) and programming language models. TAD and programming language models allow models to be developed before the details of the implementation are known. Some of the advantages to this approaches are:

(1) High level models can serve as functional documentation,
(2) Reduces the time spent in creating models,
(3) Designers can explore design alternatives prior to committing to a particular approach or strategy, and
(4) Rather than wait for the detailed, component-level description, the simulation can begin immediately.

Using TAD, the designer encodes the behavioral description of any element such as flip flops, ALU, register files, stacks and microprocessors. The component behavior is described using statements that reference and manipulate registers, RAMs, ROMs, PINs, and CLOCKs. The TAD compiler compiles these statements into a Reverse Polish Notation (RPN) code, much like the assembly language for microcomputers. This code is then interpreted when the model is evaluated.

The component's delay may be a function of the states of both inputs and outputs. The TAD operators include logical, arithmetic, relational, and sequencing operators. Table 1 is a listing of TAD operators. To control the execution of the code, TAD supports conditional statements, such as IF-THEN-ELSE and DECODE-CASE. TAD is a non-procedural language. By default all statements not explicitly sequenced by the NEXT operator will execute in parallel. This has proven to be a very natural medium for the expression of hardware algorithms. Internal multiphase clocks, as well as new operators can be specified and defined.

Operator	Symbol	Syntax Typical	Description
Concatenation	‖	A‖B	a set consisting of all elements of A and B
#Complement	~	~A	bit-by-bit complement of A
Negate	-	-A	negate A
Reduction		oA	A1 o A2 o A3...where o is any operator AND, XOR or OR
*Divide	/	A/B	arithmetic division of A and B
Add	+	A + B	arithmetic sum of A and B
Subtract	-	A-B	arithmetic difference of A and B
Less than	⌄	A⌄B	true if A is smaller
Greater than	⌃	B⌃A	number than B
Equal	=	A = B	true if A and B are same number
Less than Equal	⌄=	A⌄=B	true if A is less than
Greater than or equal	⌃=	B⌃=A	or equal to B
Not equal	⌄=	A⌄=B	true is A and B are different numbers
#AND	&	A&B	bit-by-bit logical product
#Exclusive-or	#	A#B	bit-by-bit exclusive-or
#OR	I	AIB	bit-by-bit logical sum

#also available in BOOLEAN EQUATIONS.
*available in future release.

Table 1 TAD operators

The extensive modeling features of TAD allow for a concise representation of the timing and functional behavior of a component. For example, Figure 3 shows a model of an eight-bit arithmetic logic unit (ALU). The instruction bits f2, f1 and f0 are decoded to select one of the eight different functions the alu can perform. In each case, the result is calculated using arithmetic and logical operators. After the calculation, the result is assigned to a nine bit temporary register. The status flag (ZERO), carry out (COUT-) and the output (OUT{7:0}) are then calculated.

TAD supplements, but does not replace, the Boolean models. While the Boolean models simulate faster, TAD models are easier to write and include a larger class of operations.

Behavioral Modeling

Standard programming languages available on PRIME computers, such as FORTRAN, PL/I PASCAL and "C" are used to create behavior level models. Unlike many simulators that interface with these types of language models, THEMIS does not allow user access to the THEMIS internal data structure. All calls to obtain logic values, and/or set logic values are by the PIN, REGISTER or MEMORY name defined by the user. The advantage is that users cannot accidently modify or overwrite other data structure in THEMIS. It also allows special efficient techniques to store and track the contents of these variables.

Language models access the values of REGISTERS, RAMS, ROMS and PINS, using the supplied subroutine set. The REGISTERS, RAMS, ROMS and PINS are declared with the same command as TAD. The variables are manipulated using the standard features of the programming language. Figure 4 shows an example of eight-bit arithmetic logic unit modeled at behavioral level using PL/1 programming language.

```
create component alu
    add pins
        cin-             input
        count-           output
        ain{0:7}         input
        bin{0:7}         input
        out{0:7}         output
        zero             output
        f{0:2}           input
    add function
    registers
        temp[8:01];
    tad
    decode f{2:0}
        case (0) : do    /* a + cin
                         temp = ain {7:0} + ~cin-
                         end
        case (1) : do    /* b + vin
                         temp = bin {7:0} + ~cin-
                         end
        case (2) : do    /* transport a
                         temp = 0:b ‖    ain{7:0}
                         end
        case (3) : do    /* transport b
                         temp = O:b ‖    bin{7:0}
                         end
        case (4) : do    /* a + b + cin
                         temp = ain{7:0} + ~cin-
                         end
        case (5) : do    /* a xor b
                         temp = ain{7:0} # bin{7:0}
                         end
        case (6) : do    /* a and b
                         temp = ain{7:0} & bin{7:0}
                         end
        case (7) : do    /* a or b
                         temp = ain{7:0} bin{7:0}
                         end
    end decode
    zero =  (out0 ‖ out1 ‖ out2 ‖ out3 ‖ out4 ‖ out5 ‖ out6 ‖ out7)
    next do
        cout - = ~ temp [8]
        out{7:0} = temp [7:0]
    end
    exit
    end function
    end component
```

Figure 3 A TAD model for 8-bit ALU

Special Built-in modeling

Built into THEMIS, as standard models, are some components of digital circuits which are difficult to model using any of the modeling techniques described so far. These standard models include:

- UNIDIRECTIONAL MOS transmission gate,
- BIDIRECTIONAL MOS transmission gate,
- PULL-UP and PULL-DOWN resistors,
- DELAY line, and
- TIMING blocks (to detect violations of setup time, hold time race condition, etc)

Transmission gates, pull-up/down resistors and 3 logic state strength provide a switch level simulation capability in THEMIS. Switch level simulation allows acurate modeling of transistor level MOS circuits.

Delay Modeling in all Primitive Models

THEMIS permits timing to be determined by loading, function performed (e.g. different count and clear delays for counters), or by the type of analysis (nominal, minimum, etc.). Furthermore, separate elements can be simulated at different levels. This allows the designer to specify an element that is to have minimum delay, and another to have maximum delay. Each output can have up to six different delay specifications based on the transition it makes (RISE, FALL, Z→0, Z→1, 1→Z and 0→Z).

Network Modeling

To form new models, network models interconnect existing behavioral, functional, boolean, network and builtin primitive models. There is no limit to the nesting of network models. Moreover, any circuit that has been simulated can be referenced as a network model in other circuit models. The delay definition and initial values at lower levels can be overridden in the network models.

```
#define BIT8 0x100

lml()

{

long    cin__,count__,ain,bin,out,zero,f;
long    temp;
long    rddp__b(),rdpin2();
/*Read inputs.*/
ain = rddp__b("AIN{0:7}");
bin = rddp__b("BIN{0:7}");
cin__ = rdpin2("CIN-");
f   = rddp__b("F{0:2}");

/* Perform ALU function.*/
switch(f) {
    case 0:      temp = ain + !cin__;      /*a + cin      */
                 break;
    case 1:      temp = bin + !cin__;      /*b + cin      */
                 break;
    case 2:      temp = ain;               /* transport a  */
                 break;
    case 3:      temp = bin;               /* transport b  */
                 break;
    case 4:      temp = ain + bin + !cin__;  /* a + b + cin  */
                 break;
    case 5:      temp = ain^bin;           /* a xor b     */
                 break;
    case 6:      temp = ain & bin;         /* a and b     */
                 break;
    case 7:      temp = ain | bin;         /* a or b      */
                 break;
}
/* Calculate new outputs.*/
cout__ = !(temp & BIT8);
out  = rddp__b("OUT{0:7}");
zero  = (out = = 0);
out  = temp;

/* Schedule outputs, with default (unit) delay. */
setdp__b("OUT{0:7}", out," ");
setpin__b("ZERO",zero," ");
setpin__b("COUT-",cout__," ");

}
```

Figure 4 Behavioral level model for 8-bit ALU

LINKING

The LINK command is used to link the network, specified at compile time, into a data structure that drives the simulation. The linker resolves component references by searching libraries in a user specified order. The libraries used by THEMIS are the same central libraries used by PRIME's Electronic Design Management System™. As illustrated in Figure 5, the library (LMS) and design management system (DMS) are optional. If the EDMS library is not used, then only component definitions local to THEMIS are searched.

The library is used to store component or marco-cell information that includes physical characteristics of the device (e.g. pin capacitance). A powerful feature of THEMIS allows multiple simulation models for a single component. Initially a component might be modeled in TAD and later a network model may be added. The choice of which model to use for simulation is specified by a parameter at compile time.

Unresolved component references are reported, along with questionable circuit interconnects, such as connected components inputs not driven. A detailed statistical report of system usage, including component counts, number of hierarchy levels, etc. is generated before control is passed to the simulation interface.

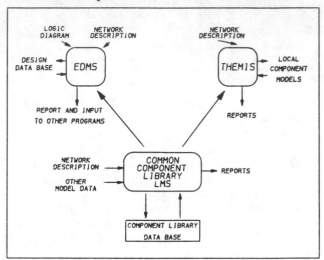

Figure 5 THEMIS and EDMS interface

SAVE and RESTORE

At any point after linking, the entire state of the simulation can be saved for later rollback. Saving the compiled and linked model in a file eliminates the need to recompile for future simulation session. Saving the entire state also enables checkpointing during a simulation session.

For purposes of efficiency, implementation of SAVE and RESTORE relies heavily on PRIME's segmented virtual memory architecture. The implementation of SAVE and RESTORE requires all state information, including static external and dynamic data, to be saved. This is accomplished by confining all external data to a fixed segment, and all dynamically allocated data to a known set of segments. The static segment and as many dynamic segments as necessary are then directly written to disk by a call to a PRIMOS routine. Typically, it takes about a half minute of elapsed time to save or restart a 5000 gate simulation on a moderately loaded Prime 750.

SIMULATION OVERVIEW

THEMIS is a ten state simulator, utilizing driven, resistive, and capacitive strengths for the values 0,1, and X. The high impedance state (Z) is implicitly the weakest state. Z is used for nodes that are driven to any particular value, and have no stored charge. Boolean and functional devices recognize only the four fundamental states 0,1,X, and Z at their inputs, while MOS switches and bussing elements recognize the full set of ten states.

Since interaction was a major design goal of this simulator, it was necessary to allow user access to any circuit object at any time. This required the implementation of a permanently resident symbol table. The symbol table, organized as a tree, reflects the hierarchy of the circuit model. Though a component type may be used many times in a network, its symbol table is represented only once in the hierarchy.

FEATURES OF THE SIMULATION USER INTERFACE

THEMIS supports a high degree of user interaction. Test patterns can be generated, modified, and applied at any time. Any circuit value can be examined and modified, checkpoints can be taken and simulation time rolled forward or backward to the checkpoint. The interactive user interface also provides a full range of simulation control, breakpoint setting and editing commands.

Applying Patterns

There are a number of ways to input patterns to the simulated circuit. A simple language for generating test patterns on signals and datapaths can be used by the designer at any time. Clock signals, periodic waveforms, and a wide variety of test vector types can be quickly generated and applied using this technique. Test patterns can also be stored in a file and read onto circuit inputs manually, or automatically as a result of actions triggered by circuit events, clocks, or circuit stability.

Breakpoints

A breakpoint facility exists which allows simulation to be halted on a number of conditions. The following is a list of possible conditions:

- Signal and datapath changes
- Signal edges
- Datapaths, registers or memory cells changing to a specified value
- Circuit stability
- Event counts
- Memory writes
- User controlled events on "dummy signals"
- Register, memory cell, and datapath increments or decrements

An extension of this facility allows **action lists**, i.e. a series of simulation commands, to be associated with a breakpoint. These commands are executed whenever the event(s) that trigger the breakpoint occur. If the action list is terminated by a CONTINUE command, then simulation does not actually halt. Use of the CONTINUE command allows events to trigger the execution of simulation commands, such as value printing or pattern reading, on an ongoing basis without interrupting the simulation. By allowing action lists to enable and disable other breakpoints, the user can set up breakpoints that are triggered by a **series** of events. For example, a simulation can be stopped when the second operand is being fetched within an ADD instruction.

The same facility allows a designer to trace signal changes during windows of time, such as during the active region of a clock pulse. Action lists have proven to be extremely powerful in debugging logic.

Breakpoints are implemented by wiring "event detectors" into the circuit. These devices, in conjunction with the normal event tracing mechanism of the simulator, signal a breakpoint if the user specified event occurs on its inputs. If the breakpoint has an action list associated with it, then, when event processing for the step is complete, the action list is loaded into an input buffer and control is returned to the command processors. The action list is then interpreted in the same manner as any other command typed in by a user.

Tracing

Closely related to the breakpoint facility is a TRACE command. The TRACE command allows ongoing circuit activity to be sent to an output file, or to be monitored at the users terminal. The TRACE ON WINDOW command formats the traced data and prints it on the upper portion of the screen in fixed locations. This command makes it very easy to follow the activity on circuit objects of interest. An example output of a TRACE ON WINDOW command is shown in Figure 6. The bottom portion of the screen is still available for command entry and scrolling. The state information on the right is updated dynamically. Thus the user can watch state and time values change.

```
                                                          Time [4227]
(1)   A{1:32} + .........................    [CCFD0001]   H
(1)   B{1:32} + .........................    [A0000000]   :H
(2)   MBC.MAR...............................    [000FC36A]   :H
(2)   MBC.DAR...............................    [FFFFFFFF]   :H
(2)   MBC.DIN{1:32}- .......................    [00000000]   :H
(2)   MBC.DOUT{1:32} + ...................    [0CFFFA34]   :H
(2)   SUNIT.STO ..........................    [000CAA10]   :H
(2)   DTAR[0].............................    [170000]   :0
(2)   DTAR[1].............................    [170000]   :0
(3)   DTAR[2].............................    [170227]   :0
(3)   DTAR[3].............................    [170244]   :0
(3)   E1.PC ...............................    [00000000000001111]   :B
(3)   SUNIT.CACHE[125] .................    [00100223C0003FF1]   :H
```

Figure 6 Example of TRACE ON WINDOW output

Logic Analyzer Emulation

One important feature of the user interface is the ability to emulate a logic analyzer - a device familiar to most logic prototype debuggers. THEMIS allows any number of "virtual logic analyzers" to be connected to the circuit. Each logic analyzer supports up to 64 probes. Sampling of the probes is controlled by the previously discussed breakpointing mechanism, and is more versatile than most logic analyzers on the market. Sampling can be done synchronously, asynchronously, or triggered by changes in the sampled signals. Sampled values are stored in a push-down stack of user specified depth which can be examined at any time. The contents of the stack can be displayed in a list format in any of the standard radixes. The stack can also be displayed as waveforms on the screen as shown in Figure 7. The designer can scroll through the waveforms, and zoom in or out.

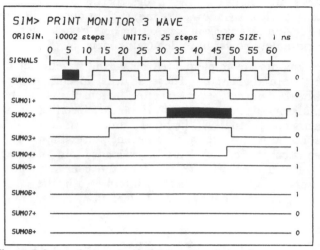

Figure 7 A waveform output of 8 monitored signals

Examining The Results Of Simulation

The values of registers, nets, datapaths, and memories can be examined using the PRINT command. The output of this command can be tailored to meet the needs of the user. This command has a number of options, including spacing, radix, logical inversion, and VERTICAL/HORIZONTAL format selection. The output shown in Figure 8 was produced by triggering a PRINT VERTICAL whenever any of the printed circuit values changed.

STEPSIZE (1 NS)	A + 1:32	B + 1:32	S C U M I N +	C O U T +
PMODE	H	H	B H	B
0	FFFFFFFF	000000000	0 FFFFFFFF	0
2	FFFFFFFF	000000000	1 FFFFFFFF	0
3	FFFFFFFF	000000000	1 FFFFFFFFE	0
5	FFFFFFFF	000000000	1 FFFFFFFEC	0
7	FFFFFFFF	000000000	1 FFFFFFFC8	0
9	FFFFFFFF	000000000	1 FFFFFEC80	0
11	FFFFFFFF	000000000	1 FFFFEC800	0
13	FFFFFFFF	000000000	1 FFFEC8000	0
15	FFFFFFFF	000000000	1 FFEC80000	0
16	FFFFFFFF	000000000	1 FE8000000	0
17	FFFFFFFF	000000000	1 EC8000000	0
19	FFFFFFFF	000000000	1 C80000000	1
21	FFFFFFFF	000000000	1 800000000	1
23	FFFFFFFF	000000000	1 000000000	1

Figure 8 Example of PRINT VERTICAL output

The contents of RAMs and ROMs can be displayed on the screen or, for later reading, dumped into a file. Hazard reporting of output spikes, bussing conflicts, memory addressing errors, and device timing violations can be selectively enabled or disabled at any time. In general, output can be sent to any number of files simultaneously. A dynamic entry to the PRIMOS editor allows examination and modification of these files while report generation is in progress, without leaving THEMIS.

FAULT SIMULATION

The same circuit model used for simulation is also used for fault simulation. In fact, at any time during a simulation session a user can insert faults into the circuit and thereby initiate fault simulation.

Likewise, at any point during fault simulation all faults can be dropped with the DROP command. Fault origins and fault effects are cleared from the circuit and normal, fault-free simulation resumes.

Concurrent fault simulation is used because of its flexibility, accuracy, and efficiency. A wide variety of faults can be simulated with this technique, including the standard stuck-at-1, and stuck-at-0 faults. THEMIS also allows the insertion of stuck-at-Z faults to simulate the effect of a wire with its driver disconnected. Unlike parallel simulation, the concurrent algorithm accurately simulates the rise and fall delays in faulted circuits.

Experience at Prime indicates that concurrent fault simulations runs 10 to 20 times faster than parallel for modest size circuits (500 gates) on a Prime 750 with 2 megabytes of main memory.

GENERAL FEATURES OF THE USER INTERFACE

Because it is impossible to foresee the needs of all users, it is desirable to include features that allow users to customize the interface to meet their own individual needs. THEMIS allows users to abbreviate commands or sequences of commands to almost any short string of characters. This same facility allows commands to be bound to program function keys.

An input facility exists to redirect the input stream from the terminal to a file. This very useful feature enables commonly executed commands to be stored in a file. Also, file nesting, up to some fixed number of levels, is allowed. File input can be temporarily halted by pressing the BREAK key, and resumed at a later date.

THEMIS'S USE IN HIERARCHICAL DESIGN PROCESS AT PRIME

THEMIS has been used at Prime for about 2 years. It has been successfully used in a production capacity to simulate PCBs with TTL parts, ECL gate array chips, and NMOS custom chips. Acceptance of the simulator has been very good and its use within Prime continues to grow.

To give a feel for how THEMIS can be used in a top-down design environment, we discuss its use in the development of several datapath chips implemented in high density ECL gate arrays. The first simulations of these chips

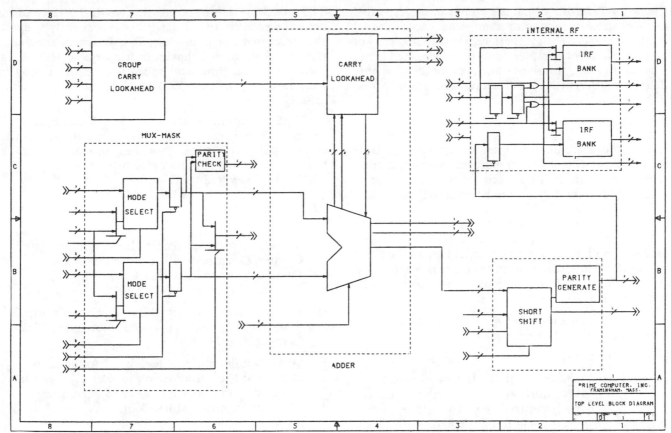

Figure 9 An eight bit slice arithmetic unit (AU)

used TAD models for the functional elements of the datapath (selectors, registers, ALU, register files, etc.). The purpose of this stage of the design verification process was to make sure that the datapath elements played together properly, and that the required functionality was present and implemented correctly. A secondary, and not originally forseen, purpose was to provide detailed documentation of the behavior of each element in the datapath. This proved beneficial for the cases in which gate level implementation was done by different people than had originally designed the datapaths.

The modeling time for each element was low, once the TAD language had been learned and the function of the logic being modelled was well understood. Most models were of the complexity shown in Figure 9, and took a few hours each to design, program, and debug. Most of the problems encountered in modelling were in understanding the TAD documentation, which at that time was rudimentary.

The initial simulations were done interactively, at the chip level, to iron out first order bugs in the models and the interconnections. Later runs, to thoroughly test the logic, were run in a batch mode. Input stimuli and clocks were applied using the waveform language, rather than being read as vectors from a file. This was possible because of the functionally structured nature of the datapath inputs. Many patterns were generated using just a few lines of code.

The results of the functional level chip simulations were saved by triggering output value printing on the leading edge of the system clock, when all results from the previous cycle had stabilized. This approach to saving output values is independent of the delays of the logic, provided that all activity has settled by the time the next clock pulse arrives. Thus, TAD models of the datapath elements could be replaced directly by the gate level implementations, and the results file generated would be the same as that generated by the functional simulation, provided, of course, that the gate level models match the functional models.

The gate level logic was designed along the same boundaries as the TAD models, using library cells that had been modelled as interconnections of boolean gates. In some cases, due to implementation considerations, the TAD models were changed to reflect design simplifications, or alternate approaches. The functional simulation was then rerun, and a new results file generated. Gate level verification was considerably simplified by this approach, making it possible to identify implementation bugs almost immediately.

As a result of in-house usage, several areas were identified for improvement in THEMIS. One of these, which has been subsequently addressed, was the slow execution time of some of the interpreted TAD instruction codes. Another area in which work is progressing is the extension of the fault simulation user interface to simplify the tasks of (a) partitioning faults for large fault simulations and (b) discarding test patterns that don't catch faults.

The areas of THEMIS which are consistently identified by users as greatly enhancing their productivity are (a) the highly interactive user interface, (b) the simplicity of modelling relatively complex functions using TAD, and (c) the ability to simulate mixtures of components modelled at different levels of detail (TAD, gate, MOS, etc.)

SUMMARY

An interactive, hierarchical digital circuit logic and fault simulator was developed to support mixed-mode modeling and simulation at the functional, register transfer, and logic levels from design specification to final fabrication. THEMIS is capable of simulating very large circuits (100,000 plus gates and a million equivalent gates) to support simulation of the entire system for VLSI designs. Its useage in an interactive environment greatly reduces the amount of time taken up in traditional batch mode simulators. It is extremly efficient in simulation, as illustrated in Figure 10.

All benchmarks done on a PRIME 750 with 2 Mbytes of memory, under moderate user loads.

Circuit Descriptions:

1) RALU-G   - 32 Bit register file, arithmetic and logic unit, with accumulator register implemented at the gate level.

2) RALU-T   - RALU-G implemented with TAD models.

3) B-RALU   - A network of RALU bit slices implemented at the gate level.

4) VLSIbnch   - A cruise control circuit used in VLSI magazine benchmark. (VLSI magazine, November 1983)

5) EALU   - Extended Arithmetic Logic Unit implemented at gate level.

CIRCUITS	#ELEMENTS	#PATTERNS	CPU TIME	#EVENTS	#EVENTS/SEC	#ELEMENTS EVALUATED	EVALUATION PER SEC
RALU	4672	8085	1006	1914853	1903	3422348	3402
RALU (TAD)	7	8085	281	70891	252	33981	121
B-RALU	50136	384	604	658351	1090	1275732	2112
EALU (9950)	4735	9486	1996	3839060	1924	10330535	5176
VLSI BENCH	100	715	5	4713	942	13993	2799
EALU (750)	4735	4347	5330	3791600	711	6895212	1294

Figure 10 Some statistics on THEMIS runs

The modeling technique provided in THEMIS gives the desired flexibility to the designers and directly supports the structured design process. The designers are able to model and simulate behavior of incomplete designs, explore new alogrithms and speed up the simulation by concentrating on their portion of design with detailed models.

THEMIS is extremely useful for LSI, VLSI, multi-chip and multi-board systems. It is easy to use and enables accurate timing modeling. Providing interactive simulation, the THEMIS simulator is fully compatible with EDMS, an integrated CAD system.

REFERENCES

1) D. M. Schuler, et al, "A Program for the simulation and Concurrent Fault Simulation of Digital Circuits Described with Gate and Functional Models," **Cherry Hill Test Conf. Proc.**, pp. 203-207, Cherry Hill, NJ (1979).

2) E. Ulrich and T. Baker, "The Concurrent Simulation of Nearly Identical Digital Networks," **IEEE Computer** (1974).

3) D. M. Schuler and R. Cleghorn, "An Efficient Method of Fault Simulation for Digital Circuits Modeled from Boolean Gates and Memories," **14th Design Automation Conf. Proc.**, pp. 230-238 (1977)

4) Melvin A. Breuer and Alice C. Parker, "Digital System Simulation: Current Status and Future Trends," **18th Design Automation Conf. Proc.**, pp. 269-275 (1981)

5) M. Abramovici, M. A. Breuer and K. Kumar, "Concurrent Fault Simulation and Functional Level Modeling," **14th Design Automation Conf. Proc.**, pp. 128-137 (1978)

6) R. Cleghorn, "PRIMEAIDS: An Integrated Electrical Design Environment," **18th Design Automation Conf. Proc.**, pp. 632-638 (1981)

7) R. McCann, "An Electronic Design Management System," **AF-SD/INDUSTRY/NASA Conference and Workshops on Mission Assurance Proc.**, pp. F185-F192 (1983)

MIXS: A MIXED LEVEL SIMULATOR FOR
LARGE DIGITAL SYSTEM LOGIC VERIFICATION

Tohru Sasaki, Akihiko Yamada, Shunichi Kato,
Terufumi Nakazawa, Kyoji Tomita and Nobuyoshi Nomizu
Computer Engineering Division
Nippon Electric Co., Ltd.
Fuchu City, Tokyo, 183 JAPAN

ABSTRACT

A mixed level simulator, MIXS, is a logic verification tool which has multiple simulation capabilities. Main MIXS techniques are time wheel and selective trace algorithm for functional level simulation based on 'node' model concept and the linkage function of functional models, described in different detail, with network information. The mixed level simulation for large digital systems can be achieved very efficiently by using the above techniques.

INTRODUCTION

Due to rapid progress in semiconductor technology, the importance of design phase for large digital circuits and systems has increased dramatically, and so has the need for tools that help designer's efforts.

One useful tool is a logic simulator. Therefore, the important factors in regard to the simulator should be clarified.

1. Real network verification. High level architecture simulation is not able to guarantee logic behavior in a physical digital system. The primary simulator debugging objective is network information verification, which is subsequently supplied to the manufacturing phase. In this respect, the simulation which covers IC level or cell level, which is the lowest verification level, is important.

2. Total digital system simulation. Design period, manpower and machine resources are limited. So, logic verification by simulation should be more efficient than that of manual verification, such as checking logic diagrams and manual simulation. The key point in efficient simulation is to simulate a total digital system, such as a central processing unit. The total digital simulation results in the following merits.

   2.1 For functional verification. Partial logic simulation for a package or a card could detect careless mistakes, such as coding errors or logic diagram drawing errors. However, these mistakes can be detected by manual checking. The bugs detected should include functional errors due to a designer's omissions in preparing specifications. Total digital system simulation results in functional verification.

   2.2 For logic test pattern generating ease. The second merit in total digital system simulation is the decrease in manpower necessary for test pattern preparation. At partial simulation, such as card level, preparing bit patterns for physical card terminals is time consuming. In the total system simulation, software or firmware test programs can be used as simulation test cases. Therefore, designers can concentrate their efforts on simulation results analysis.

3. Top down design support. Some important logic bugs depend on human factors, especially human communication. Early documentation for designer's communication about the logic behavior is important. These documents, which are specified by designers, should be simulated at an early phase in the design. In this respect, functional description language, linkage functions for those modules and swapping design information for different detail levels are important.

To attain the above factor needed for simulation, a simulation reconstruction strategy is needed. First, larger digital system verification in a limited memory size precedes other factors. Therefore, a primitive element technique in conventional gate level simulators is not adequate, and a more abstractly described model simulation technique using high level language should be taken up. Accuracy in timing check, such as hazard and race detection, is not an essential factor. Second, a high level digital system description language should be developed which supports descriptions from chip level to architecture level. Third, verifying network information on the design automation database is important. So, linking models with this network is necessary for simulation. Last, top down design approach is necessary in addition to bottom up verification approach.

Here, it is necessary to review past work in regard to these simulation respects. The concept and its implementation for block behavior description in high level language and linking them with block interconnectivity is seen in papers by Caplener [1] and Weber [2] in 1974. The concept includes mixed level simulator and multilevel simulator for top down design support. However, implementation technique was not disclosed. SABLE [5] succeeds in the concept presented by Caplener and Weber. However, it is a compiler driven simulator, and the language is an expansion of PASCAL program language. Anyway, these simulators objective is not to replace conventional gate level simulator for large digital system with them.

Lamp System [3] and F/LOGIC [4] aim to take the place of gate level simulator. Their other main purposes are fault simulations. Therefore, they do not cover high level design support area. The other approach [6], [7], reported recently can simulate a mixture mode in gate level model and high level language model. However, their technique is one that runs conventional gate level simulator and compiler object in a memory. None of them is flexible enough to execute for any combination model at different detail levels.

This paper describes a mixed level simulator, MIXS, which is efficiently applicable to design verification for large digital systems. This simulator has been developed to satisfy the above mentioned designer needs.

MIXS objectives are:

1. To support a specification language from architecture level function to integrated circuit level function.
2. To verify logic network from board level to chip or cell level on the design automation database.
3. To simulate a logic network at higher speed and with less memory capacity requirement than a conventional simulator in gate level.
4. To simulate a total digital system, with a mixture mode for the lowest level networks and the high level language model, by the firmware or software test patterns.
5. To simulate firmware effectively by architecture level model.
6. To simulate continuously from architecture level to integrated circuit level.

In order to achieve these objectives, several techniques have been developed.

First, a new language concept, suitable for a mixed level simulator, is taken up. The language has "node" concept which enables the simulator to handle different level models and to use selective trace.

Second, the two time wheel and selective trace algorithm are taken up. MIXS proves these techniques are also efficient in a high level language model as well as in conventional gate level simulators.

Last, new linkage functions are adopted. By this technique, models with different detail level can be linked together in multinesting level.

SYSTEM OVERVIEW

MIXS concept

MIXS accepts hardware block functional description and hardware block interconnectivity information. A hardware block can be a functional one or a physical component block. MIXS combines hardware block functional description with hardware block interconnectivity information. Hardware block functional descriptions are written in Functional Description Language (FDL), and stored in the hardware block area on a design automation database. The hardware block can be at any nesting level. Hardware block interconnectivity information is described in Logical Description Language (LDL), and stored into the database.

MIXS loads these two kind of information from

the database to execute simulation. MIXS system image is shown in Fig. 1.

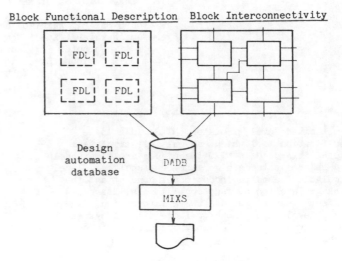

Fig. 1 MIXS SYSTEM CONCEPT

MIXS also can simulate a model in FDL description without block interconnectivity information. This capability is useful for a preliminary model debugging or architecture level model verification.

System configuration

MIXS consists of five parts; a component compiler, a network compiler, a linker, a simulator and a monitor. The system configuration is shown in Fig. 2.

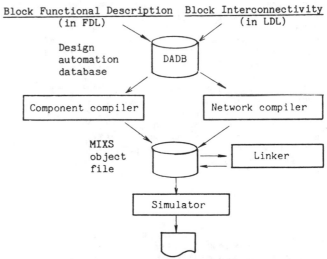

Fig. 2 MIXS SYSTEM CONFIGURATION

To use the system, a user must provide the database with block functional description in FDL and block interconnectivity information. The user first runs the component compiler and the network compiler, and makes up MIXS object file.

The component compiler accepts a block functional description from the database or card image file, generates a model module and stores it into MIXS object file. There are two kinds of modules, one is functional model and the other is component model. Both kinds of modules consist of mainly a node table (NT), functional description table (FDT) and functional signal table. The node table mainly expresses node connections in a component, while the functional description table expresses a node behavior by reverse polish notation. The "node" is a basic element in MIXS simulation.

The network compiler accepts block interconnectivity, generates a network model module, and stores it into the MIXS object file. The network compiler also generates linkage information for the linker.

The linker loads model modules according to their digital system configuration, links them to each other by linkage information, generates a loading module, and stores it into MIXS object file.

The simulator loads a loading module on MIXS object file, simulates them under the control of MIXS Control Language (MCL), and outputs time chart.

The monitor controls all the simulation activities under the conrol of MCL description.

Test Pattern

As test patterns for simulation, users can use software program object, microprogram object, a piece of bit pattens in chips such as RAMs, ROMs and PLAs on database. They are loaded into the memory defined in FDL model by the LOAD command in MCL description. MCL can also specify bit patterns directly.

MIXS Control Language

The monitor reads MIXS Control Language (MCL) description from a terminal, and controls simulation sequence interactively in accordance with the description.

MODELING LANGUAGE

MIXS provides a facility for describing digital systems in two kinds of languages at two levels. One language is FDL (Functional Description Language), for describing a hardware block logical function. The other language is LDL (Logical Description Language), for describing network and physical components interconnectivity.

FDL (Functional Description Language)

Description for hardware block. FDL is capable of describing a hardware block model. A hardware block is an arbitrary circuit assembly constructed with functional circuits, such as data path control circuits, status control circuits, timing signal control circuits and so on. This block can be a block to coincide with logical function or a block of physical components, such as a board, a card, a multi chip package or a chip. This block can be described as a low level block or logic node.

Logic node. A logic node is a basic element for describing a hardware block. The kinds of logic nodes are combinational circuits, register, memories and registers or memories following the combinational circuits.

Logic node examples are shown in Fig. 3.

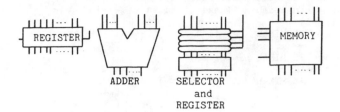

Fig. 3 LOGIC NODE EXAMPLES

A node is expressed in a statement. A statement format example is as follows:

Keyword node name (length) <delay> = expression;

DELAY.   Each node may have a set of delay expression, as shown below:

< minimum delay, nominal delay, maximum delay > or <nominal delay>

One of three kinds of delays is selected at model linking process, with which the node is simulated.

OPERATOR. The following operators may be used in an "expression".

1. LOGICAL operator; NOT OR AND EOR REDUCTION AND/OR GATING COINCIDENCE

2. FUNCTIONAL operator; SHIFT RIGHT/LEFT ROTATE RIGHT/LEFT CONCATENATE ADJUST SIZE COPY

3. ARITHMETIC AND RELATION operator; ADD SUBTRACT MULTIPLY DIVIDE CARRY EQUAL/NOT EQUAL GREATER THAN/EQUAL LESS THAN/EQUAL

4. TIMING operator; GO UP GO DOWN This operation is used to support edge trigger function

EXPRESSION.

1. FUNCTION statement: This statement is used for description with "OPERATOR"s and braces.

2. "IF" statement: This statement is used for description with IF, THEN and ELSE.

3. CASE statement: This statement is used for description with CASE and OF.

Statement

Hardware block model descriptions are composed of five kinds of statements.

1. INPUT/OUTPUT statement: This statement defines interface signals through which this hardware block can be connected to other hardware blocks in the same level or in a higher level.
    FORMAT
        INPUT   a list of input signal names;
        OUTPUT   a list of output signal names;
        INOUT   a list of in/out signal names;
    EXAMPLE
        INPUT   A,B,CREG(0:31);
        OUTPUT   DREG(0:7),E,F;

2. TERMINAL statement: This statement is used to describe combinational circuits.
   FORMAT
        TERMINAL   terminal name (length) <delay>
        = expression;
        TERMINAL can be omitted.
   EXAMPLE
        TERMINAL AOUT (0:4) = IF ACNT.EQ.1
                        THEN LI.ADD.RI
                        ELSE LI.SUB.RI;

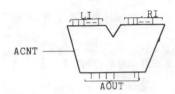

        TERMINAL   SELOUT (0:7)
        = CASE SCNT OF X1, X2, X3, X4;

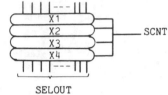

3. REGISTER statement: This statement is used to describe registers and combinational circuits.
   FORMAT
        REGISTER   register name (length) <delay>
        = expression;
   EXAMPLE
        REGISTER ROUT (0:32) = IF RST.EQ.1 THEN 0
                        ELSE IF CL.UP.
                        THEN RIN;

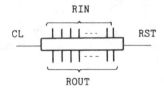

4. MEMORY statement: This statement is used to describe memories and combinational circuits.
   FORMAT
        MEMORY
        Memory name (address width, word width)
            <delay>= expression;
   EXAMPLE
        MEMORY   MDT (0:255, 0:7)
                    = IF WENB.EQ.1 THEN WRITE MIN
                    AT MADR & READ AT MADR;

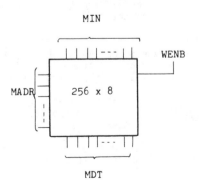

5. MODULE statement: This statement is used to call the model already described in the library, such as an LSI.

   FORMAT
        MODULE
        Module name = library name (LOAD input name
                    SOURCE output name);

   EXAMPLE
        MODULE   CHIP1 = C2901 (LOAD PATHOUT
                    (0:8),A,B,C SOURCE
                    AREG(0:7), DBUS(0:7));

Gate level logic node. For supplementary description to functional description by REGISTER, MEMORY and TERMINAL node description, gate level node description may be used.

   EXAMPLE
        OR   A <17, 19, 21> = B, C;
        AND   B <15, 17, 18> = D, F;
        AND   C <15, 17, 18> = E, G;
        NAND   G <10, 12, 14> = F;

LDL (Logical Description Language)

     There are two kinds of statements. One is for block names and their function names. The other is for pin names and assigned network names for pins.

   FORMAT
   ALOC   block ID, a list of block name and its
               function name.
   CONN   block ID, a list of pin name and
               network name.

   EXAMPLE

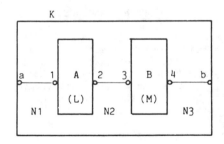

     A,B; block name
     K; block ID
     L, M; block function name
     1,2,3,4,a,b; pin name
     N1,N2,N3; network name

   (LDL DESCRIPTION)

        ALOC   K,A,L,B,M,TER,TER-1;
        CONN   K,A,1,N1,2,N2;
        CONN   K,B,3,N2,4,N3;
        CONN   K,TER,a,N1,b,N3;

A group of external pins is named a block name and a block type function name as TER and TER-1.

LINKER

Database Consideration

A functional description in FDL can communicate with other descriptions only through functional signals in INPUT, OUTPUT, INOUT statement. It is essentially necessary for a mixed level simulator to link a functional model to hardware physical information. For this purpose, functional signal names are defined in pin areas on database. A functional model can communicate with other model through pins and network information assigned to them.

Models for linker

There are three kinds of models for linker on MIXS object file, as shown in Fig. 4.

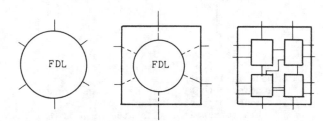

Functional model   Component model   Network model

Fig. 4   THREE MODELS

Functional model is FDL model only. Component model is FDL model, which is linked to block pin information. Network model is pin to pin infomation in a block.

Linkage mode

Both functional model and component model have node table. The linker fundamentally generates one Node Table by using three kinds of models. The linker has three linkage modes, as shown in Fig. 5, Fig. 6 and Fig. 7.

Functional signal linkage uses functional signal name in INPUT, OUTPUT, INOUT statements. Pin linkage links component models through network models by using the pin name in models. Argument linkage is for MODULE statement in FDL model.

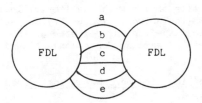

a, b, c, d, e: functional signal name

Fig. 5   FUNCTIONAL SIGNAL LINKAGE

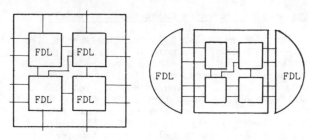

Pin Linkage   Pin Linkage and Functional
                                    signal linkage

Fig. 6   PIN LINKAGE

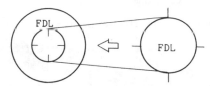

Fig. 7   ARGUMENT LINKAGE

SIMULATION TECHNIQUE

Total storage structure at simulation is shown in Fig. 8. The selective trace algorithm is based on the storage structure.

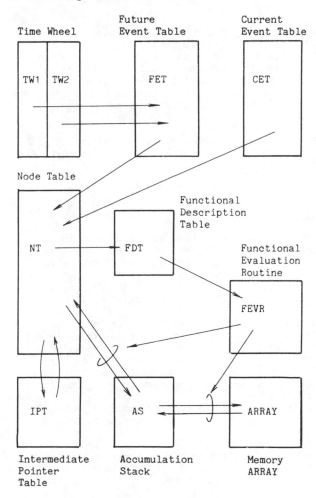

Fig. 8   MIXS STORAGE STRUCTURE

Time mapping mechanism using two time wheels.

The time mechanism uses two time wheels, TW1 and TW2. Each wheel has a pointer. When the TW1 pointer turns around the time wheel once, the TW2 pointer steps forward, much like the minute and hour hands on a clock. A time wheel slot includes a pointer to event linked list in the Future Event Table. When the TW2 pointer steps forward, the event linked list by the new TW2 slot pointer is mapped to new event linked lists, which are pointed by TW1 slot new pointers. There are 512 slot numbers for each time wheel. Multi wheel time mapping technique was proposed by Ulrich (8), for gate level simulator.

Selective trace algorithm

The selective trace algorithm treats nodes like primitive gates in conventional gate level simulator. The algorithm is shown in the following.

```
Begin for TW2 pointer:=1 to 512 do
 Begin for TW1 pointer:=1 to 512 do
  Get an event linked list in Future Event Table
    through TW1 pointer;
  For every event in the event linked list do
   Begin
    Propagate logic value from a node at the event to
      fanout nodes on Node Table;
    Put the fanout nodes to Current event Table as
      events;
   End;
  For every event in Current Event Table do
   Activate evaluation routine;
    If evaluation result is not equal to the value
      previously on the Node Table, put the event
      node to Future Event Table through a slot at
      (current time + a delay time in the node)
      event linked list and change node logical
      value to a new one on Node Table;
  End;
   Move all events from a event linked list at (TW2
     pointer + 1) to event linked lists through TW1
     slot pointers;
 End
```

Functional evaluation

The selective trace algorithm activates the evaluation routine. The evaluation routine evaluates the event node using lists of evaluation codes on Functional Description Table and routines in Functional Evaluation Routine. The evaluation codes could be categorized into functions to:

1. Obtaining logical value; push down the Accumulation Stack and move logical value from Node Table (NTI) to the topmost Accumulation Stack.
2. Calculating; calculate values between that of the topmost area and the next topmost area for Accumulation Stack.
3. Branching; check the topmost Accumulation Stack area and branch to a code according to this result.
4. Array accessing; move logical value from Accumulation Stack to Memory Array or vice versa.
5. Exiting; exit from code list and return to selective trace algorithm.

There are about 60 evaluation codes.

Node Table

A Node Table includes information and areas for network, logical status, an FDT pointer, a delay time, and simulation control flags. A Node Table has the capability to treat multi bit connection and multi bit logical status with maximum 72 bit width.

Node table format is shown in Fig. 9.

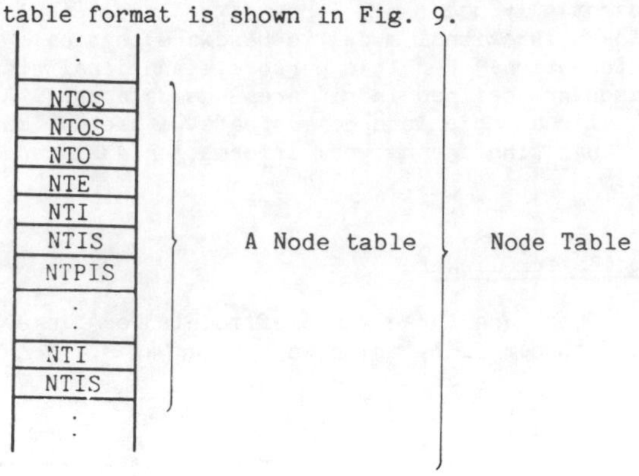

NTOS (Node Table Output Status)
    multiple output logical status area
NTO (Node Table Output)
    a pointer to NTI for one of fanout nodes
    simulation control flag
    1 bit output logical status area if 1 bit
    output
NTE (Node Table Entry)
    a Functional Description Table pointer
    node attribute
    node output bit width
NTI (Node Table Input)
    a pointer to NTI for one of nodes connected
    in series by the same wire. 1 bit input
    logical status
NTIS (Node Table Input Status)
    multiple bit input logical status
NTPIS (Node Table Previous Status) if needed,
    previous input logical value

Fig. 9 NODE TABLE FORMAT

Each information and its format are fixed at each 4 byte.

APPLICATION

MIXS proposes many simulation forms.

Logic network verification at IC level

A common library of integrated circuits (IC) is prepared in FDL forms. Designers can verify networks consisting of ICs with the library logic. The capabilities for network visibility and setting logical value to flip-flop and register at the IC level are almost the same as for a conventional gate level simulator.

Network verification with peripheral logic described in FDL

To simulate logic networks in as much of an actual circumstance manner as possible, MIXS is able to simulate logic with peripheral logic described in FDL.

IC level simulation with peripheral logic is shown in Fig. 10.

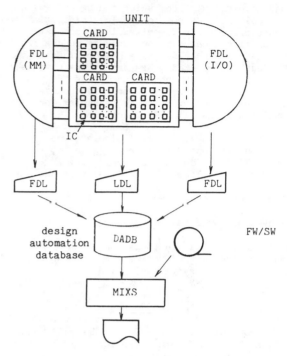

Fig. 10   IC LEVEL WITH PERIPHERAL LOGIC SIMULATION

The mixed level simulation is also applicable to LSIs simulation and digital system simulation including them. LSIs can be described as a single model in FDL. They can also be described as blocks or cells plus their interconnection. In this case, blocks or cells are described in FDL as a library and their interconnectivity information is given in LDL for each LSI circuit.

Microprogram verification

Microprogram simulation is able to participate in the simulation form, as described before. For more efficient simulation, a higher level model can be selected from architecture description of digital systems or the hardware description translated from microprogram specifications.

Top down design support

MIXS is a very useful tool to support top down design. The top down design flow for a large digital system is shown in Fig. 12. MIXS is also able to support LSI chip top down design.

Network verification at mixed level

By preparing the components model, described in FDL, designers can simulate, at mixed levels, a mixture of IC chip level and the functional level shown in Fig. 11.

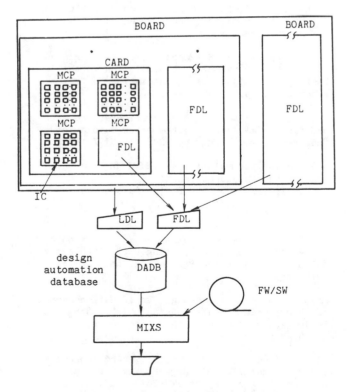

Fig. 11   MIXED LEVEL SIMULATION

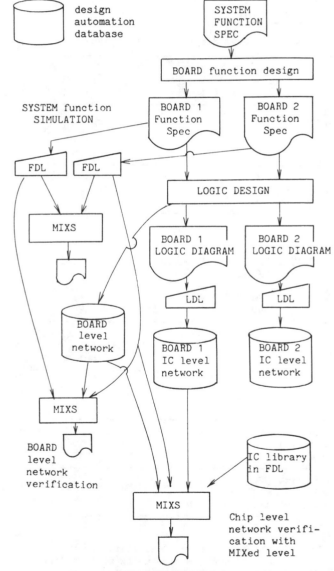

Fig. 12   TOP DOWN DESIGN FLOW FOR A LARGE DIGITAL SYSTEM

RESULT

MIXS has been implemented on NEC ACOS computer using about 37 K steps mainly structured Fortran, including assembler statements. MIXS has been used for large computer development. Performance example on ACOS system 700 (1.1 MIPS machine) is shown in the following.

Functional level result.

Table 1 Functional Level Simulation Performance

Condition	
- Model size (steps)	3,000
(100,000 gates HW model)	
- Simulation time	5,800
(micro program proceeding words)	
Processing time (seconds)	
- Simulator	2,300

IC level result

Table 2 IC Level Simulation Performance

Condition	
- Model size (ICs)	216
- Simulation time (clocks)	1,320
Processing time (seconds)	
- Compiler	84
- Linker	182
- Simulator	627
Memory size (kw)	
- Model size	14
(NT + FDT)	

Comparison IC level result beween MIXS and a conventional simulator

Table 3 Comparison between MIXS and Conventional gate level simulator

	MIXS	Conventional gate level simulator
Condition (using same HW model)		
- Model size (ICs)	57	57
- Simulation time (clocks)	380	380
Processing Time (seconds)		
- Modeling	64	206
- Simulator	49	101
Memory size (KW)		
- Model size	9	14
(Connection Table + (FDT))		

The conventional gate level simulator is implemented on NEC ACOS computer using Fortran like language by primitive gate technique.

CONCLUSION

Mixed level simulation, which can cover a wide range of application forms for logic verification, has been described. In the mixed level simulation method, functional description language models are linked to each other by hardware network information and efficiently simulated. New techniques have been developed, which use the node concept for the functional description language oriented connection table, time wheel selective trace mechanism for functional models and linkage functions.

MIXS, which is a mixed level simulator using the above mentioned techniques, is running on NEC ACOS computers. MIXS can be used to verify logic networks for ICs in less than half the time and with one fifth the memory capacity requirement than conventional gate level simulator without sacrificing visibility and accuracy needed for digital systems logic verification.

The logic verification forms for which MIXS is proposed are also very effective for top down design for VLSI and digital system using VLSI.

ACKNOWLEDGEMENTS

The authors wish to thank Dr. H. Kanai, Mr. A. Kobayashi and Mr. T. Kitamura for their guidance and advice.

REFERENCES

1. H. D. Caplener and J. A. Janku "Top-Down Approach to LSI System Design", Computer design, August, 1974.

2. H. Weber, "A Tool for Computer Design", 11th Design Automation Conference Proceeding, pp.200-208, 1974.

3. S. G. Chappell, P. R. Menon, J. F. Pellegrin and A. M. Schowe, "Functional Simulation in the LAMP system", 13th Design Automation Conference Proceeding, pp. 42-47, 1976.

4. P. Wilcox "Digital logic simulation at the gate and functional level", 16th Design Automation Conference Proceeding, pp. 242-248, 1979.

5. William vanCleemput, et. al., "SABLE: A Tool for generating structured, multi-level simulation", 16th Design Automation Conference Proceeding, pp. 272-279, 1979.

6. Kawato, N. et. al., "Design and Verification of Large-Scale Computers by using DDL", 16th Design Automation Conference Proceeding, pp.360-366, 1979.

7. Miyoshi, M. et. al., " Logic Verification System for Very Large Computers using LSIs", 16th Design Automation Conference Proceeding, pp.367-374, 1979.

8. Ulrich, E. G., "Serial/Parallel Event Scheduling for the Simulation of Large System", ACM National Conference, pp.279-287, 1968.

Switch-Level Delay Models for Digital MOS VLSI

Computer Science Division
Electrical Engineering and Computer Sciences
University of California
Berkeley, CA 94720

Abstract

This paper presents fast, simple, and relatively accurate delay
models for large digital MOS circuits. Delay modeling is organized around chains of switches and nodes called *stages*, instead of logic gates. The use of stages permits both logic gates
and pass transistor arrays to be handled in a uniform fashion.
Three delay models are presented, ranging from an RC model
that typically errs by 25% to a slope-based model whose delay
estimates are typically within 10% of SPICE's estimates. The
slope model is parameterized in terms of the ratio between the
slopes of a stage's input and output waveforms. All the models
have been implemented in the Crystal timing analyzer. They
are evaluated by comparing their delay estimates to SPICE,
using a dozen critical paths from two VLSI designs.

1. Introduction

This paper describes the techniques used to estimate
delays in Crystal [8], a timing analyzer for MOS integrated circuits. Crystal's overall function is to locate the critical paths
that limit the performance of a circuit. It breaks the circuit
up into pieces, estimates delays through the pieces, and uses
the delay estimates to search for a critical path. In order to
process large circuits, both the delay estimation and the
searching must execute quickly. At the same time, the delay
estimates must be accurate; otherwise, the program may overlook the performance-limiting paths.

The traditional approach to delay modeling is the one
used in circuit simulation programs such as SPICE [6]. The
primary concern in those models is *accuracy*: the ability to
simulate exactly the real-world behavior of devices. A circuit
is modeled by a collection of differential equations, and the
equations are solved to predict the circuit's behavior. This
approach produces highly accurate results, and has resulted in
widespread usage of circuit simulators as a design tool.

Unfortunately, the accuracy of the circuit models comes
at a high price in execution time: circuit simulators typically
require several seconds of CPU time per transistor. As a
result, the programs are impractical for today's state-of-the-art
VLSI circuits, which contain tens or hundreds of thousands of
transistors. Although there have been recent improvements in
the speed of circuit simulators [4], they still require too much
time for VLSI circuits.

Crystal uses a different approach to transistor modeling,
where speed is the primary consideration. The models treat
each transistor as a perfect switch in series with a resistor.
Instead of solving differential equations, the switch-level models

use tables to compute the value of the series resistance. The
switch-level approach results in four orders of magnitude
improvement in speed: only a few hundred microseconds of
execution time are needed per transistor. In spite of their simplicity, the switch-level models provide delay estimates for digital MOS circuits that are typically within 10% of what SPICE
would estimate for the same circuits.

The switch-level models achieve their accuracy and speed
by capitalizing on the uniform design style used in large circuits. For example, digital VLSI circuits tend to have only a
few different sizes of transistor and a few pullup-pulldown
ratios, used over and over. Since the pieces of the circuit have
about the same structure, they also have about the same delay
properties. The delay properties of the basic constructs are
measured by running SPICE on small examples and distilling
the results down to a few tables. When analyzing large circuits, delay estimates are computed quickly using the tables. If
there isn't much variation in the structures used in the circuit,
small tables will produce accurate results. The approximate
models tend not to work as well for sensitive analog components or circuits with large variation in design style.

This paper describes and evaluates three switch-level
delay models that have been implemented in Crystal. Crystal's
models contain two features that permit fast and accurate
delay estimates. First, circuits are decomposed into chains of
transistors called *stages*. Each stage is independent for purposes of delay calculation. The stage decomposition permits
Crystal to handle both logic gates and pass transistors in a uniform fashion. Second, each transistor is modeled by an
effective resistance whose value depends on the shape of the
transistor, the waveform driving its gate, and the load being
driven by the transistor. For any given transistor, all of these
factors are combined together into a single ratio that determines the effective resistance. The ratio approach means that
small tables can be used to handle a large variety of actual
situations.

Section 2 describes the stage decomposition and the
advantages it provides over a logic-gate decomposition. Sections 3-5 present the three delay models. The lumped
resistor-capacitor model of Section 3 is the simplest and
fastest, but is only accurate to within about 25%. Section 4
describes the more complex slope model, which is usually accurate to within 10%. Section 5 applies the Penfield-Rubinstein
models for distributed capacitance [10,12] to get still greater
accuracy. Section 6 summarizes our experience in using Crystal for actual chips. Section 7 discusses the limitations of the
models, and Section 8 compares this work to previous work in
the area.

Reprinted from *ACM/IEEE 21st Design Automation Conf. Proc.*, 1984, pp. 542–548.

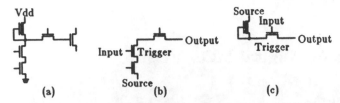

(a) (b) (c)

Figure 1. A stage is a chain of transistors analyzed together for delay calculations. During different phases of analysis, the stages in (b) and (c) might be extracted from the circuit in (a) for delay calculation. The trigger transistor is the last one in the stage to turn on.

2. Stages

At any given time, Crystal's delay modeler considers a collection of transistors called a stage. A stage consists of a chain of nodes and transistor channels forming an electrical path from a strong signal source (such as Vdd, Ground, or a highly capacitive bus) to some other node, called the *output* of the stage. As shown in Figure 1, a single transistor may be associated with different stages during different phases of the analysis. Stages generally correspond to logic gates, except that pass transistors are lumped together with the logic gates that drive them. See [8] for information on how Crystal extracts stages from a circuit and uses the delay information to locate critical paths.

The delay modeler is given the sizes and types of the transistors in the stage, and the parasitic resistances and capacitances of the nodes along the stage. One of the transistors is identified as the *trigger*: it is assumed to be the last transistor to turn on in the stage. Its gate is called the *input* of the stage, since it controls the activation of the stage. The modeler is also given information about the waveform at the input. Its job is to use this information to compute the waveform at the output of the stage, assuming that all the transistors in the stage except the trigger are turned on.

Waveforms are described at different levels of precision in different delay models. In the simplest delay model, each waveform is described by a single value: the time at which its voltage crosses the logic threshold (the input voltage for a standard inverter where the input and output voltages are equal in the dc transfer curve). This is called the *inversion time* for the waveform. In the slope-based models each waveform is also characterized by its *rise-time* (in ns/volt) at the logic threshold voltage.

Crystal's delay modeler only considers information on the direct path between signal source and output. All side transistors connecting to the path are assumed to be turned off: their gate-source capacitance is included in the parasitic capacitance, but information on the far side of side transistors is ignored (see Figure 2). This approach is used in Crystal because the program does not have specific information about whether side transistors are turned on or off; if it automatically included all side capacitance, its delay estimates would be unrealistically high. Busses and pass transistor arrays account for most of the situations with many side paths, and in these cases only a single path is usually active through the structure at once. (If it becomes necessary to include side paths, the results of Section 5 can be extended to handle them). Most stages in VLSI circuits are simple: in the critical paths found by Crystal, 80-90% of all stages contain only a single transistor.

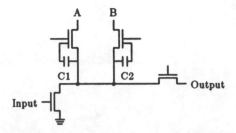

Figure 2. In computing delays, information from side paths is not included. For example, in this figure the capacitance at A and B is not included. However, the gate-source overlap capacitance from the side transistors (C1 and C2) is included in the parasitic capacitance of the stage.

Using stages as the basis for delay computation has worked out well, because both normal gates and pass transistors can be handled in a uniform fashion. Most previous timing analyzers have been organized around logic gates. They have generally had difficulty dealing with pass transistors because the non-linear pass transistor effects cannot be separated cleanly from the rest of the gate. The stage approach handles gates with or without pass transistor structures uniformly as chains of switches. It also accommodates complex gates such as AND-OR-INVERT.

3. The RC Model

The RC model is the simplest and least accurate of Crystal's models, and is a slight generalization of the "tau" models proposed by Mead and Conway [5]. The RC model computes a resistance and capacitance value for each node and transistor along the stage. All of the resistances are summed separately, as are all the capacitances, and the product is used as the delay through the stage:

$$delay = \left[\sum R\right]\left[\sum C\right]$$

If a stage contains several transistors, a separate resistance computation is made for each transistor, and the values for the different transistors are summed, along with the resistances of the interconnect. No information about waveforms is used by the RC model: the delay time for the stage is added to the inversion time at the input to calculate the inversion time at the output.

The effective resistance of each transistor in the stage is computed using a table based on the transistor's type (enhancement, depletion, etc.) and geometry. For each type of transistor there are two effective resistance values, each expressed in ohms per square. The first resistance value is used if the transistor is transmitting a logic one and the second is used if the transistor is transmitting a logic zero (this distinction results in more accurate modeling of pass transistors). The values currently used in Crystal are given in Table I. The effective resistance of a transistor is computed by selecting the appropriate parameter value and multiplying it by the length/width ratio of the device.

The type of a transistor is not determined solely by its physical structure (enhancement, depletion, p-channel, etc.) but also by the way it is used in the circuit. For example, enhancement transistors driven through pass transistors have

Transistor Type	Ohms/square (transmitting 1)	Ohms/square (transmitting 0)
Enhancement	30000	15000
Enhancement driven by pass transistor	---	48000
Depletion Load	22000	---
Super-buffer (depletion, gate 1)	5500	5500
Depletion (gate 0 or unknown)	50000	7000

Table I. The effective resistances used by the RC model for a 4-micron nMOS process. The missing entries are for situations that do not ever occur (for example, depletion loads are used only to transmit 1's).

different characteristics than enhancement transistors driven directly by depletion loads. When reading in circuits, Crystal distinguishes these two uses of enhancement transistors and uses different types for each. Three different uses of depletion devices are also distinguished by Crystal, since they can result in different behavior. Users can add new types of their own and label transistors as being of the new types; this provides a crude facility for dealing with special circuit constructs such as bootstrap drivers.

The table values are generated by running SPICE simulations of simple stages. In the SPICE simulations a step function is used to drive the input, and the effective resistance is computed by dividing the delay time by the load capacitance. This generally results in an underestimation of effective resistance (see below).

Capacitance includes parasitics from the nodes, gate-channel capacitance from transistors in the path, and gate-

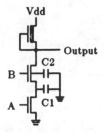

Figure 3. A switch-level approach to delay analysis automatically accounts for different delay characteristics at different inputs of a NAND gate. The capacitance at *C1* and *C2* will be included when calculating delays from A, but not from B.

source capacitance from side transistors attached to the path. When summing the capacitances along the path, only those between the trigger transistor and the output are used. The trigger transistor is the last to turn on, so all the capacitance between it and the signal source is assumed to have discharged. This means that different delays will be computed from each input of a NAND gate (see Figure 3).

Two large circuits, a microprocessor [3] and an instruction cache [9], were used to compare the RC model to SPICE. Out of the 40000-50000 transistors in each chip, Crystal used the RC model to extract 12 critical paths containing a total of 157 stages. SPICE was used to simulate each critical path, and the SPICE delays were compared with Crystal's estimates. Figure 4 makes a stage-by-stage comparison and Figure 5 compares total delays through the critical paths. Although the RC model often errs by a factor of 2 or more on individual stage calculations, the errors of successive stages tend to cancel. On average, it can usually estimate the total delay through a path to within 25% of SPICE (see Table II).

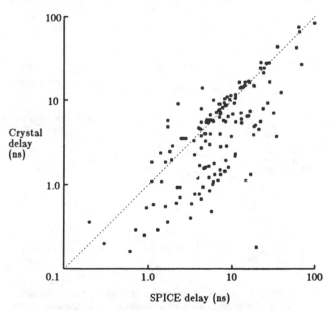

Figure 4. A comparison between the RC model and SPICE, using 12 critical paths (157 stages) extracted by Crystal from two large chips. Each point compares Crystal's delay estimate for a stage with the corresponding SPICE time, measured from a simulation of the critical path. Ideally, all points should fall along the diagonal.

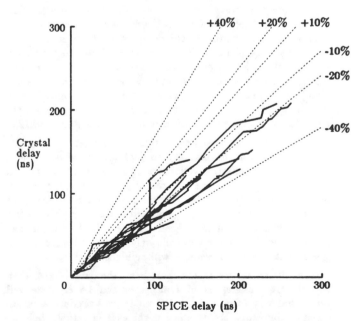

Figure 5. A comparison between the RC model and SPICE, using total delays through critical paths. Each line corresponds to one critical path, and each point in the line corresponds to a stage. The point compares Crystal's and SPICE's estimates for the total delay in the critical path up through that stage.

Model	Overall Error	Av. Stage Error	Av. Path Error	Std. Deviation in Path Error
RC Model	-22%	45%	24%	10%
RC Model with 1.3 Scale Factor	-3%	51%	15%	16%
Slope Model	4%	23%	8%	9%
PR-Slope Model	2%	20%	6%	7%

Table II. A comparison of the delay models. "Overall Error" compares the sum of all the delays computed by Crystal with the sum of all delays computed by SPICE, to point out consistent underestimates or overestimates. "Av. Stage Error" is the average of the absolute value of errors for individual stages (the points in Figures 4 and 7). "Av. Path Error" is the average absolute error in estimating total delays in the critical paths up to each stage (the points in Figures 5 and 8). The rightmost column gives the standard deviation in the errors from the "Av. Path Error" column.

Much of the RC model's error is due to consistent underestimation: the sum of all the RC delays is 20% less than the sum of all SPICE delays. This means that the overall accuracy of the results can be improved by simply scaling each of Crystal's estimates upward. To see what effect a scale factor might have, I tried several different values. The best value for the test data was 1.3: when each of Crystal's estimates was multiplied by 1.3, the the average stage error increased to about 50% but the average path error decreased from 25% to about 15%. The scale factor approach is used in the TV timing verifier [2] and other programs. Unfortunately, it isn't obvious whether or not such a scale factor depends on the circuit constructs or fabrication parameters, nor is it obvious how to characterize such a dependency if one exists. For these reasons, Crystal has not used the scale factor approach.

There are two sources of error in the RC model. One source of error is the lumping of resistances and capacitances. This tends to overestimate the delays since it assumes that all the capacitance must be discharged through all the resistance. Fortunately, most stages contain only a single transistor and small parasitic resistances, so lumping introduces only a small error and is not the major problem with the RC model. Section 5 improves on the lumped model by applying the Penfield-Rubinstein models for distributed capacitance [10,12].

The second and most significant source of error in the RC model comes from its inability to deal with waveform shape. In practice, the effective resistance of a transistor depends on the waveform on its gate. If the trigger transistor turns on instantaneously, then its full driving power is used to drain the output capacitance and the transistor has a relatively low effective resistance. If the trigger turns on slowly, then it may do much or all of its work while only partially turned-on. In this case its effective resistance will be higher.

If all waveforms in a circuit have the same shape, then the effective resistances of transistors can be characterized using that waveform and the RC model will produce accurate results. Unfortunately, this is not the case in actual VLSI circuits. Although almost all waveforms have an exponential shape, they vary by more than three orders of magnitude in their slopes. As a result, the effective resistance of the transistors varies by more than a factor of ten and the RC model produces only a rough estimate for delays.

4. The Slope Model

The slope model incorporates information about waveform shape in order to make more accurate delay estimates. It assumes that all waveforms are exponential in overall shape but vary in their slopes. Each waveform is represented by its inversion time and its rise-time, in ns/volt at the logic threshold. A rise-time of zero corresponds to a step function, and a large rise-time corresponds to a slowly rising or falling signal.

Unfortunately, the effective resistance of a transistor depends not just on the rise-time of its gate, but also on the load being driven by the stage and on the sizes of the transistors in the stage. If a stage is driving a large load, or has very small transistors, then only very slow input rise-times will affect the stage's delay. If a stage is driving a small load or has very large transistors, its delay will be more sensitive to the rise-time of its input. The important issue is whether or not the trigger transistor is fully turned-on when it does most of its work, and this depends on the input rise-time, the output load, and transistor size.

The key to implementing the slope model was the discovery that all of these factors can be combined into a single ratio, which alone determines the transistor's effective resistance. First, the output load and transistor size are combined into a single value called the *intrinsic rise-time* of the stage; this is the rise-time that would occur at the output if the input were driven by a step function. The input rise-time of a stage is then divided by the intrinsic rise-time of the stage's output to produce the *rise-time ratio* for the stage. The rise-time ratio gives an estimate of how fully turned-on the trigger transistor is when it is doing its work. SPICE simulations showed that the rise-time ratio is an accurate predictor of the effective resistance of a transistor, independent of the specific input rise-time, transistor size, or output load. Pilling and Skalnik were apparently the first to suggest the ratio approach [11].

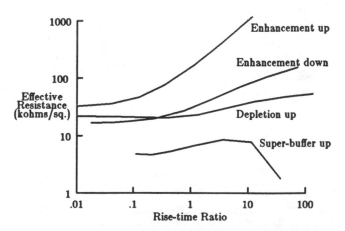

Figure 6. A plot of the tables that characterize the resistance of different types of transistors as a function of the rise-time ratio of the stage. A rise-time ratio of zero means the stage's input rises very quickly in comparison to the output. Different tables are used when transmitting zero (e.g. the "enhancement down" curve) and transmitting one (e.g. the "enhancement up" curve). The curves represent a few of the transistor types for a 4-micron nMOS process.

In the slope model, each transistor type is characterized by two resistance tables, one used when the transistor is transmitting a logic 0 and one used when it is transmitting a logic 1. Each table gives effective resistance values as a function of rise-time ratio. During timing analysis, the effective resistance of the trigger transistor is computed by interpolating in the appropriate table. When a stage contains several transistors, the slope model is applied only to the trigger transistor. The other transistors are assumed to be fully turned-on, so the RC model is used to compute their resistances. Once a resistance has been computed for each transistor, the separate values are summed as before, and multiplied by the total capacitance to compute the delay.

The parameter tables for the slope model were generated in much the same way as for the RC model: SPICE simulations were run on simple stages and parameters were extracted from the output. Each table contains six to ten values. Figure 6 plots the contents of a few of the tables for our 4-micron nMOS process. As with the RC models, non-standard circuit structures can be handled by giving their transistors different types and generating special parameter tables for them.

The ratio approach is important because it allows transistors to be parameterized with one-dimensional tables, instead of three-dimensional tables based on input rise-time, load, and transistor size. The three-dimensional approach would require large amounts of CPU time to generate the tables and would also make the delay analysis slower by requiring three-dimensional interpolation.

The accuracy of the slope model depends on the accuracy with which rise-times can be calculated. It is the responsibility of the delay modeler to estimate the rise-time at the output of each stage; this value is used as the input rise-time of later stages. The current implementation of the slope model approximates actual output rise-times by using intrinsic rise-times everywhere. The intrinsic rise-time is computed under the assumption that the input is a step function and the output has an exponential shape. For any given voltage, the slope of an exponential waveform at that voltage is proportional to the delay time to reach that voltage. This means that the intrinsic rise-time of a stage is proportional to its intrinsic delay, which is the delay computed by the RC model.

Figures 7 and 8 compare the slope model to SPICE with the same data as in Figures 4 and 5. These figures show that the slope model is substantially more accurate than the RC model. The average error for individual stages was reduced from 45% to 23%, and the average overall error over several consecutive stages dropped from 24% to 8%. Only rarely does the estimate for a critical path differ from SPICE by more than 20%. Table II summarizes this data and the corresponding data for the additional enhancement described in Section 5.

5. Complex Stages and the PR-Slope Model

Although most stages in MOS circuits contain only a single transistor, the delay modeler must still be able to make reasonable delay estimates for more complex stages. In the RC and slope models, complex stages are handled by lumping resistances and capacitances. The lumped approach is pessimistic for complex paths with distributed capacitance, since it assumes that all the capacitance must be discharged through all the resistance (see Figure 9 for an example).

In order to handle such situations more accurately, a new model called *PR-slope* was implemented. It is similar to the slope model except that it uses the results of Penfield and Rubinstein [10,12] to avoid lumping the capacitance. [12] provides upper and lower bounds for the delay; Crystal uses the average of the two. Since each stage is a linear chain of transistors with no side trees, the average of the upper and lower bounds is

$$delay = \sum_i R_i C_i$$

R_i is the resistance from the signal source to point i, and C_i is

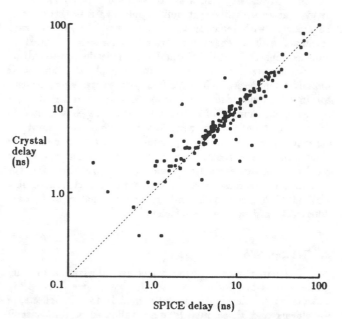

Figure 7. A stage-by-stage comparison between the slope model and SPICE, for the same critical paths as in Figures 4 and 5.

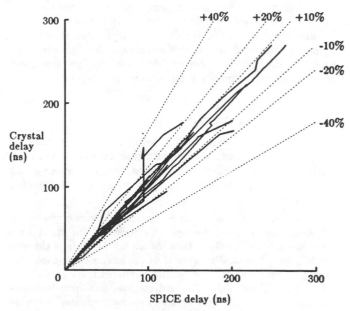

Figure 8. A total-delay comparison between the slope model and SPICE, using the same critical paths as in Figures 4 and 5.

417

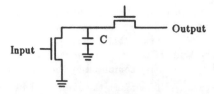

Figure 9. The basic slope model will compute delays under the assumption that the capacitance at C must discharge through both transistors, whereas in fact it only discharges through one.

the capacitance at point i. This means that instead of weighting all the capacitance by all the resistance, each separate capacitor is weighted only by the resistance between it and the signal source. Aside from this difference in weighting, the PR-slope model is identical to the slope model.

The addition of the Penfield-Rubinstein models made a small additional improvement in accuracy, which is summarized in Table II. The average error for a single-stage estimate dropped from 23% to 20%, and the average error over paths containing several stages dropped from 8% to 6%. The small improvement supports the conclusion that complex stages are rare in large circuits. However, the PR-slope model is almost as efficient as the basic slope model, and can avoid gross overestimates that will occasionally happen in the basic slope model (for example, the two largest vertical jumps in Figure 8 were due to lumping errors, and disappeared with the PR-slope models).

6. Experiences with Crystal

The initial implementation of Crystal was developed in the summer and fall of 1982, at about the same time that the two test circuits were being completed. Crystal was used for analysis of both those circuits, although only the RC models were available at the time. See Table III for a summary of the experience with the circuits. In one case (the RISC II cache) Crystal located about a half dozen performance bugs that would have slowed the cycle time by nearly a factor of two. In the other case (the RISC II CPU), Crystal located several performance bugs and one functional error. Both of the chips have been fabricated and tested. Both are fully functional, and both run at approximately the predicted speeds.

7. Limits of the Models

The models presented here are nearly as accurate as SPICE over a wide range of digital circuit constructs and loading characteristics. There appear to be three limitations to the models: rise-time estimation, complex stages, and the number of transistor types.

Crystal estimates rise-times by using intrinsic rise-times, which is only a first-order approximation. If the input to a stage has a large rise-time, then the actual rise-time of the output will be substantially larger than the intrinsic rise-time. On average, Crystal's rise-time estimates differed from SPICE's by 30% over the test data. To find out the maximum improvement one can expect with better rise-time estimates, I ran an experiment on the test data using both Crystal and SPICE. Crystal's PR-slope model was used to compute delays over the 12 critical paths, but SPICE provided all the rise-time esti-

	CPU [3]	Cache [9]
Size (transistors)	41000	46500
Clock phases	4	4
Crystal processing time (CPU minutes, VAX 11/780)	35	15
Stages processed	154000	72000
Estimated cycle time (ns, RC model)	480	450
Actual cycle time (ns)	500	600

Table III. Experiences with Crystal on two nMOS circuits. The processing time includes circuit pre-processing and searching functions as well as delay modeling; delay modeling accounts for only a small fraction of the total time.

mates. There was no significant improvement in Crystal's delay estimation with "perfect" rise-time estimation; in several cases, the delay estimates actually got worse! This suggests that rise-time estimation is not a major source of errors.

Although the PR-slope model makes a first-order attempt to deal with complex stages, there are still several situations that can cause it to produce inaccurate results. One source of error in complex stages is the assumption that only a single transistor is turning on or off at once. If two transistors turn on simultaneously in a NOR gate, the gate's delay will be less than predicted; if two transistors turn on simultaneously (and slowly) in a NAND gate, its delay will be greater than predicted. Another source of error in complex stages is the additive treatment of different transistors and resistors: total resistances and rise-times for stages are computed by summing the contributions of each device along the stage. This is an accurate approximation for resistors, but it is less accurate for non-linear devices such as pass transistors.

The third potential limitation of the models has to do with the number of transistor types. When transistors are used in different ways, such as bootstrap drivers, super buffers, or even gates with different pullup/pulldown ratios, they must be modeled with separate transistor types. In general, each construct with a separate dc transfer characteristic must be modeled separately. SPICE simulations must be run to extract the parameters for each type. In principle this allows an unlimited number of different circuit constructs: every transistor in the circuit could ostensibly be given a different type. However, the number of different transistor types is limited in practice by the human and computer time that must be expended to extract all their parameters. If a large number of transistor types is required to model a circuit accurately, too much parameter extraction time will be required for the approach to be reasonable. Fortunately, our current design style at U.C. Berkeley is uniform enough that a half dozen different transistor types is sufficient.

8. Related Work

The importance of using waveform information in computing delays has been recognized for some time [7,11,13,14]. Tokuda et al. developed analytic models for the effects of waveforms and those models were validated against circuit simulation, but only over a relatively small range of rise-time ratios. As a consequence, they concluded that load transistors were insensitive to waveform. Tamura et al. have also developed an analytical model for waveform effects [11], but their model appears to apply only to simple gates without pass transistors. A recent thesis by Mark Horowitz also explores simplified analytical models for MOS circuits [1]. The ratio approach was suggested in 1972 by Pilling and Skalnik [11], although they also did not deal with pass transistors, and the idea doesn't appear to have been used since then.

9. Conclusions

Crystal demonstrates that simple models can estimate delays in MOS digital circuits to within 10% of SPICE. Other factors, such as variations in processing, are likely to cause errors at least as great as this, so the accuracy of the simple models appears to be acceptable for a wide variety of applications. The models are fast: delays can be estimated for typical stages in less than one millisecond on a VAX 11/780 using the PR-slope model.

Two factors contribute to the success of Crystal's models. The first factor is the switch-level approach based on stages: it enables the system to handle a variety of different circuit constructs in a uniform fashion. The second factor is the use of rise-time ratios, which results in small parameter tables and fast interpolation.

10. Acknowledgements

Mark Horowitz gave me several valuable pieces of technical advice and corrected some misconceptions on my part. Gordon Hamachi, Bob Mayo, Walter Scott, Carlo Séquin, and George Taylor all provided helpful comments on early drafts of this paper. The work was supported in part by the Defense Advanced Research Projects Agency under contract N00034-K-0251.

11. References

[1] Horowitz, M. "Timing Models for MOS Circuits." Technical Report SEL83-3, Stanford University, December 1983.

[2] Jouppi, N.P. "Timing Analysis for nMOS VLSI." *Proc 20th Design Automation Conference*, 1983, pp. 411-418.

[3] Katevenis, M., Sherburne, R. Patterson, D., and Séquin, C.S. "The RISC II Micro-Architecture." *Proc. IFIP TC10/WG10.5, International Conference on VLSI*, North Holland, 1983, pp. 349-359.

[4] Lelarasmee, E. and Sangiovanni-Vincentelli, A. "RELAX: A New Circuit Simulator for Large Scale MOS Integrated Circuits." *Proc. 19th Design Automation Conference*, 1982, pp. 682-691.

[5] Mead, C. and Conway, L. *Introduction to VLSI Systems.* Addison-Wesley, 1980.

[6] Nagel, L.S. "SPICE2: A Computer Program to Simulate Semiconductor Circuits." ERL Memo ERL-M520, University of California, Berkeley, May 1975.

[7] Okazaki, K., Moriya, T. and Yahara, T. "A Multiple Media Delay Simulator for MOS LSI Circuits." *Proc. 20th Design Automation Conference*, 1984. pp. 279-285.

[8] Ousterhout, J.K. "Crystal: A Timing Analyzer for nMOS VLSI Circuits." *Proc. Third Caltech Conference on VLSI*, R. Bryant, ed., Computer Science Press, 1983, pp. 57-70.

[9] Patterson, D.A., et al. "Architecture of a VLSI Instruction Cache for a RISC." *Proc. 10th International Symposium on Computer Architecture*, 1983 (*SIGARCH Newsletter*, Vol. 11, No. 3), pp. 108-116.

[10] Penfield, P. Jr. and Rubinstein, J. "Signal Delay in RC Tree Networks." *Proc. 18th Design Automation Conference*, 1981, pp. 613-617.

[11] Pilling, D.J. and Skalnik, J.G. "A Circuit Model for Predicting Transient Delays in LSI Logic Systems." *Proc. 6th Asilomar Conference on Circuits and Systems*, 1972, pp. 424-428.

[12] Rubinstein, J., Penfield, P. Jr., and Horowitz, M.A. "Signal Delay in RC Tree Networks." *IEEE Transactions on CAD/ICAS*, Vol. CAD-2, No. 3, July 1983, pp. 202-211.

[13] Tamura, E., Ogawa, K. and Nakano, T. "Path Delay Analysis for Hierarchical Building Block Layout System." *Proc. 20th Design Automation Conference*, 1984, pp. 411-418.

[14] Tokuda, T., et al. "Delay-Time Modeling for ED MOS Logic LSI." *IEEE Transactions on CAD/ICAS*, Vol. CAD-2, No. 3, July 1983, pp. 129-134.

Section III-B
Performance Evaluation and Optimization

MICRO-COMPUTER ORIENTED ALGORITHMS FOR DELAY EVALUATION OF MOS GATES

D. ETIEMBLE[*] - V. ADELINE[**] - NGUYEN H. DUYET[**] - J. C. BALLEGEER[**]

[*] UNIVERSITE PIERRE ET MARIE CURIE
4 place Jussieu, 75230 Paris, France
[**] THOMSON-CSF LABORATOIRE CENTRAL DE RECHERCHES
GROUPE INTERACTION HOMME-MACHINE, 91401 Orsay, France

Abstract

We present microcomputer-oriented algorithms to calculate very quickly the switching times and propagation delays of basic CMOS and NMOS gates. The results show less than 10% difference with model 1 SPICE 2-G results.

I. INTRODUCTION

For several years, many authors have proposed analytic expressions for calculation of propagation delays of basic gates for several logic technologies: ECL, ISL and STL, NMOS [1] and CMOS [2].
With VLSI circuits, quick calculation of switching times and propagation delays demand new methods. Standard circuit simulators, as SPICE [3] require great amounts of CPU time and storage to simulate complex circuits. To improve performance, new simulators as MOTIS [4] or RELAX [5] bypass one or more features of standard simulators. For instance, MOTIS uses bachward Euler integrated and relaxation techniques to decompose and solve the system of equations and tables to avoid the use of transistors equations. RELAX applies relaxation techniques directly to the system of nonlinear algebraic differential equations. We present another approach, based on the evaluation of propagation delays and switching times of basic gates with expressions or simple algorithms.
The advantage of using expressions or simple algorithms to calculate the propagation delay through devices and connections versus complete simulation with analog simulators is obvious as these simple expressions or algorithms can be executed efficiently by micro or mini-computers and give the same results as analog simulators (less than 10% difference). With a building block design methodology for VLSI, the designer must specify the timing characteristics of the basic gates and blocks (inverters, NANDs, NORs, transmission gates, super buffers,

etc.) and check the critical delay time paths. We present an algorithm for logic gates to be used with microcomputers (workstation approach). This algorithm uses the Shichman and Hodges model [6]. It is the same as the level 1 model used in SPICE 2-G. Electrical parameters are deduced from layout (capacitive load corresponding to interconnections and gate capacitances). In this paper, we present the algorithms for the simulation of CMOS and NMOS inverters, NOR gates, NAND gates.

II. METHOD

We use the following abbreviations for transistor equations

$$K_i = 1/2 \; Mu \; C_{ox} \; W_i \; / \; L_i \quad \text{with} \; Mu = Mu_n \; \text{(NMOS)}$$
$$Mu = Mu_p \; \text{(PMOS)}$$

V_{tn} = N-transistor threshold voltage
V_{tp} = P-transistor threshold voltage

It is assumed that the input signals can be linearized according to Figure 1. The rise and fall times are proportional to the slope of the linearized part of the rising and falling signals.

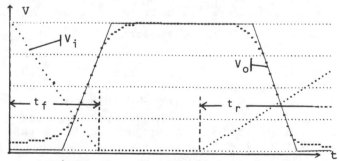

Figure 1 : Signals modelization

The equations employed are derived from the Shichman and Hodges model. This equations may include bulk effect on threshold level and channel length modulation. The input rising signal for NMOS gates is expressed as:

$$V_i - V_{tn} = V_{dd} \; t \; / \; t_r \quad \text{for} \; V_t < V_i < V_{dd} \quad (1)$$

Reprinted from *ACM/IEEE 21st Design Automation Conf. Proc.*, 1984, pp. 358-364.

For CMOS gates, the input falling signal is expressed as:

$$V_i = - V_{dd} t / t_f + V_{dd} - V_{tp} \qquad (2)$$

$$\text{for } 0 < V_i < V_{dd} - V_{tp}$$

The switching characteristics of the MOS inverter correspond to the schematic in Figure 2, where i_1 is the NMOS transistor current and i_2 is the PMOS transistor (CMOS) or depletion mode NMOS transistor (NMOS) current. The corresponding equation is given in (3).

$$C_c \, dV_i / dt = i_1 - i_2 + (C_c + C_L) dV_o / dt \qquad (3)$$

C_c corresponds to the Miller capacitor derived from the NMOS and(or) PMOS transistor capacitors. C_L corresponds to the transistor load capacitances (interconnection and gate capacitors).

The problem is more complicated when considering NMOS NAND gates and CMOS NOR-NAND gates because of the presence of transistors in series. The schematic in Figure 3 corresponds to a two-input NMOS NAND gate and a two-input CMOS NAND gate. The circuit leads to a system of two differential equations. However, a good approximation of the output response can be obtained by the following simplification.

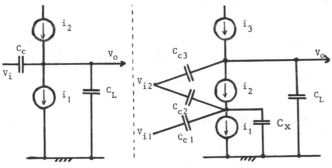

Fig. 2. Inv. schematic   Fig. 3.

When $V_{i1} = V_{dd}$, the series of two transistors is replaced by an equivalent transistor with $L = 2.L_{nand}$ and with equivalent capacitances as shown in Figure 4. When $V_{i2} = V_{dd}$, the equivalent schematic is shown in Figure 5. More generally, the equivalent schematic is that of Figure 4 when $V_{i1} < V_{i2}$ and that of Figure 5 when $V_{i1} > V_{i2}$. These simplifications are based on results of SPICE NAND simulations. An alternate solution takes into account the drain-source voltages of the series transistors in the "on" state without changing transistor sizes. The first is used with the CMOS version, the second with NMOS version.

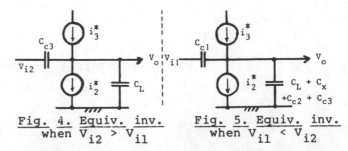

Fig. 4. Equiv. inv. when $V_{i2} > V_{i1}$ Fig. 5. Equiv. inv. when $V_{i1} < V_{i2}$

For NAND gates, SPICE simulations show that the output signal and some of the intermediate points in the transistor chain (according to the used input) cover each other in the linear part of the switching curves.

According to these assumptions, the switching characteristics of the inverter, the NOR gate and the NAND gate correspond to the resolution of the differential equation (3). This equation could be solved by numerical methods (simulator approach). But, expressions or simple algorithms can be deduced from this equation to calculate the propagation delay and the fall time (rise time) in the linearized part of the output signal. This approach has been used in [7] to calculate the propagation delays of NMOS inverters with some drastic assumptions, such as considering the depletion mode current i_2 to be negligible. (This assumption is valid only when capacitive loads are very large as in a PLA).

In this paper, we show how this method can be generalized and applied to most of the basic CMOS and NMOS gates, with more precise equation and(or) algorithms. We first calculate the propagation delay between $t = 0$ (when $V_i = V_t$) and t^* when the output signal reachs some given value in the linear region. With this value t^* in the linear region, the falling time can be computed as $(dV_o / dt)(t^*)$.

We illustrate this method with CMOS and NMOS gates.

III. CMOS GATES

1. INVERTER

The CMOS inverter is shown in Figure 2, with i_1 corresponding to the NMOS transistor, and i_2 corresponding to the PMOS transistor. The transistor parameters are given in Table 1. Figure 6 shows SPICE results for a CMOS inverter driven by a rising input and Figure 7 shows the results corresponding to a falling input. In these cases, the transistor sizes have not been chosen to equalize p and n currents. In some cases, $i_1 < i_2$ (falling input) or $i_2 < i_1$ (rising input).In the worst cases, $i_1 \ll i_2$ or $i_2 \ll i_1$.

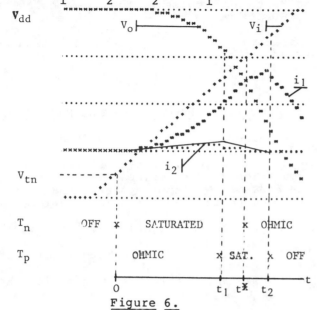

Figure 6.

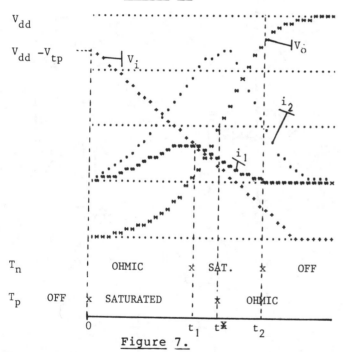

Figure 7.

The equation (3) can be solved with the assumptions $i_2 = 0$ (rising input) and $i_1 = 0$ (falling input). These assumptions are valid when the load capacitance are greater than the device capacitance. This corresponds to the actual operation of the circuits and to the cases for which it is important to check critical time data path. However, it is possible to improve the evaluation precision by considering the effect of the other transistor currents and

second-order effects such as channel length modulation.

1.1 ALGORITHMS FOR RISING INPUT

In this paragraph, we use the following definitions and approximations :

- Let V_o be the output corresponding to the threshold between saturated and ohmic mode for the NMOS transistor. It corresponds to a point where the output signal can be easily linearized

$$V^*_i - V_{tn} = V^*_o = (t^* / t_r) V_{dd}$$

- A_n and A_p corresponds to channel length modulation, in Shichman and Hodge models. They are considered as constant values in differential equation (4) (second order effects)

$$A_n = 1 + Lambda_n . V^*_o \quad ; \quad A_p = 1 + Lamda_p . (V_{dd} - V^*_o)$$

- Figure 6 shows the PMOS current i_2

- t_1 is the time corresponding to the switching between satured and ohmic modes for the PMOS transistor. Many SPICE simulations shows that for t_1
$V_{tp}/2 < V_i(t_1) - V_i(t^*) < 3.V_{tp}/2$.
So, we approximate t_1 as:

$$t_1 = t^* - (t_r / V_{dd}) . V_{tp}$$

- t_2 is the time corresponding to the switching between saturated and off modes for the PMOS transistor

$$t_2 = t_r - (V_{tn} + V_{tp}) \, t_r / V_{dd}$$

- in the ohmic mode, i_2 is expressed as:

$$i_2 = i_2[t_1] . t / t_1$$

with $\quad i_2[t_1] = A_p \, K_p \, (V_{dd} - (V_{tn} + t_1 V_{dd} / t_r) - V_{tp})^2$

- in the saturated mode, i_2 expressed as:

$$i_2 = i_2[t_1] . (t_2 - t) / (t_2 - t_1)$$

The equations (3) becomes :

$$C_c V_{dd} / t_r = A_n K_n (V_i - V_{tn})^2 - i_2 + \quad (4)$$
$$(C_c + C_L) \, dv_o / dt$$

$$C_c (V_{dd}/t_r) \, dt = A_n K_n (V_{dd} / t_r)^2 t^2 dt$$
$$- i_2 \, dt + (C_c + C_L) \, dv_o$$

Integrated equation from $t = 0$ to $t = t_1$

and $t = t_1$ to t^* (V_o from V_{dd} to V_o^*)

$$1/3\ A_n\ K_n\ (V_{dd}/t_r)\ t^{*3} + C_L\ V_{dd}\ t^*/t_r$$
$$-\ i_2[t_1].t_1/2 + 1/2\ i_2[t_1]\ \{\ -(t_2-t_1)$$
$$+\ (t_2-t^*)^2/(t_2-t_1)\} = (C_c + C_L)\ V_{dd}$$

This equation is solved by numerical methods, yielding t^*. We can then find dV/dt as follows :

When $t = t^*$, $dV_o^*/dt = \{\ 1/\ (C_c + C_L)\ \}$

$$\{A_p\ K_p\ (V_{dd}-V_i^*-V_{tp})^2 - A_n\ K_n\ (V_i^*-V_{tn})^2\}$$

When knowing t^* and dV_o^*/dt, $V_o(t)$ and the propagation delay can be calculated.

1.2 ALGORITHM FOR FALLING INPUT

According to CMOS duality, similar equations can be derived for falling input.

1.3 RESULTS

The corresponding programs have been written in FORTRAN 77 on a DEC VAX computer. The corresponding results are shown in Table 1, where they are compared with SPICE 2-G results for the same inputs.

PARAMETERS VALUES USED IN TABLE 1 AND 3

$V_{tn} = 0.75$ V $\qquad\qquad$ $V_{tp} = -0.75$ V

Channel length of transistor : 6 mm
Channel width of transistor : 6 mm
$Mu_n = 600$ cm**2/V.S $\quad$ (NMOS transistor)
$Mu_p = 200$ cm**2/V.S $\quad$ (PMOS transistor)
$C_{ox} = 5.6$ pF/mm**2

Time t^*, for falling inputs are presented on line below for rising inputs

SIM	C_L pF	P_{ri} ns/v	P_{fi} ns/v	t^* ns eva	t^* ns spi	$-P_{fo}$ ns/v eva	$-P_{fo}$ ns/v spi	P_{ro} ns/v eva	P_{ro} ns/v spi
1	0.1	4.0	4.0	8.8	8.6	0.88	1.15	1.3	1.6
				12.8	12.0				
2	0.2	4.0	4.0	10.8	10.0	1.32	1.8	2.0	2.3
				13.7	13.0				
3	0.3	4.0	4.0	11.6	11.0	1.66	1.9	2.6	2.8
				14.4	14.0				
4	0.6	4.0	4.0	13.2	13.5	2.47	2.4	5.3	5.0
				15.8	15.5				
5	0.9	4.0	4.0	14.2	14.5	3.16	3.0	6.82	6.8
				16.6	17.0				

Channel length of transistors : 12 mm (PMOS)
$\qquad\qquad\qquad\qquad\qquad$: 6 mm (NMOS)
Channel width of transistors : 6 mm (PMOS)
$\qquad\qquad\qquad\qquad\qquad$: 6 mm (NMOS)

SIM	C_L pF	P_{ri} ns/v	P_{fi} ns/v	t^* ns eva	t^* ns spi	$-P_{fo}$ ns/v eva	$-P_{fo}$ ns/v spi	P_{ro} ns/v eva	P_{ro} ns/v spi
6	0.1	4.0	4.0	8.95	9.0	0.75	1.3	1.9	2.2
				13.1	13.5				
7	0.5	4.0	4.0	11.9	12.0	1.85	2.7	7.1	8.4
				15.44	16.5				

Table 1

The results are similar to SPICE results, with less than 10% differences for propagation delays. For the slopes, the results are less precise. It should be noticed that the evaluation of output slope is not as precise as the slope calculated by SPICE but that a less precise slope evaluation can be tolerated. A first approximation deduced from equation (4) shows that the delay time depends on $(t_r/V_{dd})^{2/3}$.

2. ALGORITHM FOR CMOS NOR and NAND GATES

As previously explained, the CMOS NOR and NAND gates delay time and slopes can be deduced from the results of the inverter simulation. The method is presented for the case of two input gates, and can be extended to the case of n input gates.

The NAND gate shematic is shown in Figure 8

The NOR gate shematic is shown in Figure 9.

The equivalent inverter shematic is shown in Figure 10.

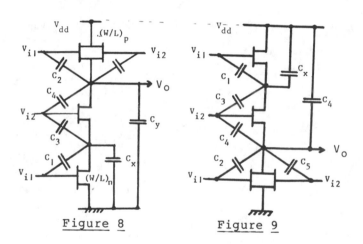

Figure 8 $\qquad\qquad$ Figure 9

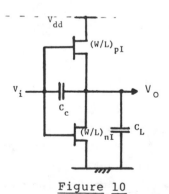

Figure 10

The transformations are given in Table 2, with the notations used in Figures 8, 9,10.

The comparaisons between our evaluation and simulations with SPICE 2-G are given in table 3.

NAND GATE		NOR GATE	
$(W/L)_{pI} = (W/L)_p$		$(W/L)_{pI} = (W/2L)_p$	
$(W/L)_{nI} = (W/2L)_n$		$(W/L)_{nI} = (W/L)_n$	
$v_i = v_{i2}$ $v_{i1} = v_{dd}$		$v_{i1} = 0$	
	$C_c = C_4 + C_5$		
	$C_L = C_y + C_2$		
$v_i = v_{i1}$ $v_{i2} = v_{dd}$		$v_{i2} = 0$	
	$C_c = C_1 + C_2$		
	$C_L = C_y + C_x + C_3 + C_4 + C_5$		

Table 2

SIM	C_L pF	P_{ri} ns/v	P_{fi} ns/v	t^* ns eva	t^* ns spi	$-P_{fo}$ ns/v eva	$-P_{fo}$ ns/v spi	P_{ro} ns/v eva	P_{ro} ns/v spi
1	0.1	4.0	4.0	8.95 13.1	9.0 14.0	0.75	1.03	1.9	2.7
2	0.5	4.0	4.0	11.9 15.44	12.0 16.8	1.85	2.4	7.1	9.1

TABLE 3 : NOR Gate CMOS

IV. NMOS GATE

1.   INVERTER GATE

For NMOS gates, the depletion mode transistor cannot be ignored (except for very large load capacitance [5]). We model the depletion mode transistor current i_2 in order to obtain a simple equation with either an explicit or an implicit solution.

We present the algorithm for the depletion load NMOS inverter. The NMOS NOR gates can be calculated by the same algorithm with just one capacitance modification (C_L is replaced by $C_{cg} + C_L$ where C_{cg} is the Miller capacitance of the non active input). The NMOS NAND gate can be calculated with the inverter algorithms by using transformations similar to the NAND CMOS transformations, and by considering the drain-source voltages of the "on" transistors.

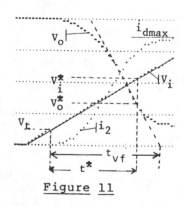

Figure 11   Figure 12

1.1 ALGORITHM FOR RISING INPUT

Figure 11 shows a SPICE 2-G result corresponding to a rising input. The input equation is :

$$v_i - V_t = V_{dd} \cdot t/t_r \text{ for } 0 < t < t_r$$

With various capacitive loads, SPICE simulations have shown that the depletion load current i_d can be linearized as shown in Figure 11.

$$i_d = i_{dmax} \cdot t / t_{vf} \text{ for } 0 < t < t_{vf} \quad (5)$$

$$\text{Let } V^*_o = V^*_i - V_t = V_{dd} \cdot t^* / t_r \quad (6)$$

At time t^*, the enhancement mode transistor is switching from saturated to ohmic mode. From $t = 0$ to $t = t^*$, equation (3) can be written as (7) with solution (8)

$$i_{dmax}/t_{vf} \int_o^{t^*} t.dt - K_e(V_{dd}/t_r)^2 \int_o^{t^*} t^2 dt$$
$$= C \int_{Vdd}^{V^*_o} dV_o \quad (7)$$

$$i_{dmax} \cdot t^{*2}/2t_{vf} - K_e(V_{dd}/t_r) t^{*3}/3 \quad (8)$$
$$= C \cdot V_{dd} \cdot (t^*/t_r - 1)$$

At t_{vf} , $V_o = 0$

Let S_f be the output falling slope.

$$S_f = dV_o/dt$$

$$V_o = S_f(t - t^*) + V^*_o \quad (9)$$

$$t_{vf} - t^* = -V^*_o / S_f \quad (10)$$

S_f is calculated for $V_o < V^*_o$ (in this case, the transistor operates in the ohmic mode and the slope is less sensitive to V_o variations). Let V_{il} be the value of input corresponding to $V_{ol} < V^*_o$.

$$t_1 = (V_{il} - V_t).t_r/V_{dd} = t^* \quad (11)$$
$$+ (V^*_o - V_{ol})/S_f(V_{ol})$$

$$\text{with } S_f(V_{ol}) = dV_o/dt \quad (12)$$
$$= 1/C.\{ i_d(V_{ol}) - i_e(V_{il},V_{ol})\}$$

by using (12) in (6), we obtain a second order linear equation that gives $V_{il}(V_{ol})$. By replacing $V_{il}(V_{ol})$ in (12), We obtain $S_f(V_{ol})$ to be used with equation (10). From this we calculate the value t_{vf} to be used in equation (8). The algorithm can be summarized :

STEP 1 : with $i_d = 0$ (t_{vf} very large),
evaluate an initial value t^*_{max}
and $V^*_o(t^*_{max})$

let $k = 1$ and go to STEP 3

STEP 2 : if $t^*(k) - t^*(k-1) < eps$, then END
STEP 3 : with equations (11) (12) (10),
calculate $t_{vf}(k)$
STEP 4 : with equation (8), calculate $t^*(k)$
STEP 5 : $k = k + 1$ and go to STEP 2

The convergence is fast and gives t^*
(delay time between $V_i = V_t$ and
$V^*_o = V^*_i - V_t$) and the falling slope S_f.

This method can be used with more
complicated transistor equations.

- with linearized bulk effect, V_t is
replaced by $V_t = V_{to}+B_1 V_{sb}$ where $V_{sb}=V_o$
for depletion mode transistors.
- with channel lenth modulation, transistor
currents are multiplied by :

$1 + lambda\ V_o$ (enhancement mode
transistor) or

$1 + lambda\ (V_{dd} - V_o)$ (depletion mode
transistor)

in both cases, $V_o(t) = V^*_o + S_f\ (t-t^*)$.

Table (4) compares the calculated delay
times and slopes with corresponding SPICE
simulations for inverters of varying
geometry, capacitive loads and rise times.

1.2 ALGORITHM FOR FALLING INPUT

Figure 12 shows a SPICE 2-G result
corresponding to a falling input. The
input is expressed as $V_i = V_{dd} - V_{dd}t/t_f$.

We consider time t^* when $V^*_i - V_t = V^*_o$,
when the enhancement mode transistor
switches from ohmic to saturated mode, and
time t_I , when $V_i = V_o$. We may observe
that V_o (t) can be linearized from $t = 0$ to
$t = t^*$ with $dV_o/dt \approx 2.V_t/t_f$. Between
t^* and t_I , the current is considered as
constant :

$$i_d = K_d\ V_{td}^2 \{ 1 + Lambda\ .\ (\ V_{dd} - V_{oI}\ ) \}$$

The algorithm can be summarized the way :

STEP 1 : evaluation of t^*

acording to equation (3) with :

$$i_1 = i_e = (1 + Lambda_e\ .\ V_o)\ K_e \quad (13)$$
$$.\ \{2\ (V_i - V_t)\ V_o - V_o^2\}$$

$$i_2 = i_d = \{(\ 1+Lambda_d\ .\ (V_{dd} - V_o)\} \quad (14)$$
$$.\ K_d\ V_{td}^2$$

$$dV_o/dt \approx 2\ V_{te}/t_f \ ;\ dV_i/dt = -\ V_{dd}/t_f$$

By using $V_i(t^*)$ and $V_o(t^*)$ with equation
(9), (14) and (3), we obtain an equation
whose solution is t^* .

STEP 2 : Between t^* and t_I , the two
transistors operate in the saturated mode.
We evaluate the channel length factor by :

$1 + Lambda\ .\ (\ V_{dd} - V_{te}\ )$ for $t = t^*$.

Equation (3) is used with :

$$i_e = (\ 1+Lambda_e\ .\ V_o)\ K_e\ (V_i - V_{te}) \quad (15)$$
$$i_d = \{\ 1+Lambda_d\ .\ (V_{dd}-V_{te})\}K_d\ V_{td}^2 \quad (16)$$

Solving this equation gives $t_I - t^*$.

PARAMETERS VALUES USED IN TABLE 4 AND 5

V_{te} = 0.6 V V_{td} = -1.4 V
Channel length of transistor : 4 mm
Mu   = 800   (enhancement) and
         780   (depletion)   cm**2/V.S
Lambda = 0.1 ; gamma = 0.32 (bulk effect)
C_{ox} = 5.6 pF / mm**2
t_p are measured at V = 1.25 V

SIM	WD	WE	C	t_r/V_{dd}	t^*	t^*	$1/S_f$	$1/S_f$	t_p	t_p
			pF	ns/V	ns cal	ns spi	ns/V cal	ns/V spi	ns cal	ns spi
1	5	14	0.1	0.2	3.610	0.600	0.10	0.11	0.67	0.68
2	5	14	0.1	2	3.600	3.500	0.44	0.66	2.54	2.50
3	5	14	0.1	6	8.310	8.500	0.96	1.22	4.54	4.70
4	5	14	0.1	12	14.29	14.70	1.68	2.0	6.39	6.50
5	5	14	1.0	0.2	0.87	0.80	0.61	0.55	2.63	2.6
6	5	14	0.5	0.2	0.81	0.80	0.25	0.34	1.38	1.6
7	5	22	0.5	60	61.90	62.0	6.36	8.80	21.5	20.0
8	5	14	1.0	6.0	14.43	13.0	1.95	2.60	12.8	13.0

TABLE 4 : Falling output

SIM	WD	WE	C	t_r/V_{dd}	t^*	t^*	$1/S_f$	$1/S_f$	t_p	t_p
			pF	ns/V	ns cal	ns spi	ns/V cal	ns/V spi	ns cal	ns spi
1	5	14	0.1	2.0	8.3	8.3	1.83	2.14	1.45	1.5
2	2	14	0.1	6.0	23.4	24.5	2.81	2.49	1.33	1.3
3	5	14	0.1	4.0	15.9	15.8	2.40	3.20	1.36	1.2
4	5	14	0.1	10	38.5	38.3	3.36	3.50	1.30	1.0
5	5	14	0.5	1.0	4.6	4.7	7.52	7.80	7.52	8.0
6	5	14	0.5	20	79.3	80.0	11.82	15.20	6.88	7.0
7	5	14	1.0	1.0	4.7	4.5	15.16	15.80	14.75	16.5
8	5	14	1.0	2.0	9.3	9.5	15.04	15.80	15.01	15.5

TABLE 5 : Rising output

STEP 3 : Point I is located below the line
tangent to the rising output. This slope
dV_o/dt when $t = t_I$ is raised to calculate
the slope of the rising output.

$$(dV_o/dt)_{t=t_I} = i_d\ (t_I)-\{(1 + Lambda_e\ V_o(t_I)\}$$
$$.\ K_e\{V_i(t_I)-V_{te})^2\} \quad (17)$$

where

$$V_i(t_I) = V_{dd} - V_{dd}\ t_I\ /\ t_f$$

$$i_d(t_I) = \{ 1 + Lambda_d (V_{dd} - V_i(t_I))\}$$

Table 5 gives the corresponding results.

2. NAND GATE RESULTS

Table 6 and 7 presents comparisons between the results of our programs and SPICE results for four inputs NAND gates.

PARAMETERS VALUES USED IN TABLE 6 AND 7

Channel length of transistor : 6 mm
Others parametres are the same as in Table 4 and 5
For simulation 1, V_{te} = 0.6 V and V_{td} = -1.4 V
For simulation 2, V_{te} = 0.9 V and V_{td} = -2 V

SIM	WD	WE	I	C	t_r/V_{dd}	t^*	t^*	$1/S_f$	$1/S_f$	t_p	t_p
				pF	ns/V	ns cal	ns spi	ns/V cal	ns/V spi	ns cal	ns spi
1	6	60	1	0.1	12.0	14.34	13.0	1.46	2.0	6.46	6.0
1	6	60	2	0.1	12.0	13.71	12.80	1.35	2.4	5.76	5.5
1	6	60	3	0.1	12.0	12.84	12.10	1.31	2.4	4.78	5.0
1	6	60	4	0.1	12.0	11.40	11.20	1.23	1.8	3.23	3.5
2	6	60	1	1.0	12.0	18.88	19.20	2.20	3.0	15.35	17.1
2	6	60	2	1.0	12.0	19.72	21.0	2.37	3.4	16.45	19.0
2	6	60	3	1.0	12.0	20.56	21.50	2.54	3.6	17.54	21.0
2	6	60	4	1.0	12.0	21.41	23.50	2.72	3.8	18.60	22.5

TABLE 6 : 4 input-NAND gate. Falling output

SIM	WD	WE	I	C	t_r/V_{dd}	t^*	t^*	$1/S_f$	$1/S_f$	t_p	t_p
				pF	ns/V	ns cal	ns spi	ns/V cal	ns/V spi	ns cal	ns spi
1	6	60	1	0.1	6.0	27.0	26.5	7.65	9.20	10.18	9.8
1	6	60	2	0.1	6.0	26.3	26.0	5.71	7.80	7.48	7.8
1	6	60	3	0.1	6.0	25.5	25.2	4.38	5.40	5.26	5.6
1	6	60	4	0.1	6.0	24.6	24.5	3.93	4.0	3.42	3.8
2	6	60	1	1.0	6.0	26.0	25.0	10.52	11.0	9.71	7.0
2	6	60	2	1.0	6.0	25.8	24.8	10.11	10.20	8.85	6.8
2	6	60	3	1.0	6.0	25.5	24.8	9.71	11.0	7.82	6.8
2	6	60	4	1.0	6.0	25.1	24.8	9.42	11.20	6.57	6.8

TABLE 7 : 4 input-NAND gate. Rising output

* I = input

V. CONCLUSION

We have presented simple algorithms to calculate propagation delays for CMOS and NMOS inverters, NAND and NOR gates. These algorithms are simple for CMOS gates and more complicated for NMOS gates. In both cases, the corresponding programs give results with less than 10% difference from SPICE results for propagation delays, and 10% difference in most cases for output slopes to be used for calculation of propagation delays of cascaded gates.
Similar algorithms for pass transistors in dynamic logic are under developpment.

The programs have been written in FORTRAN on a VAX 750 computer; they are now being reformulated in the 'C' language, to be used in workstations (68000 + UNIX + C + Ethernet network). These programs will be used:

- for propagation delay calculations to search for critical data paths. They will be used in the EMILIE-2 system, of which different parts have already been presented [8], [9].

- for automatic sizing of MOS gates according to speed, power dissipation, chip area. This approach appears promising. This work will be continued by considering short-channel length transistor equations.

REFERENCES

[1] Ed SEEWANN "Switching Speeds of MOS Inverters" in IEEE J. Solid State Circuits. Vol SC-15 N 2. april 80.

[2] Sung Mo KANG "A Design of CMOS Polycells for LSI Circuits" in IEEE Trans. on Circuits and systems. Vol CAS-28 N 8. August 81

[3] L. W. Nagel, "SPICE2: A computer program to simulate semiconductor circuits", Electronics Res. Lab. Rep. ERL-M520, Univ. California, Berkeley, May 1975.

[4] B. R. Chawla, H. K. Gummel, and P. Kozak, "MOTIS-An MOS timing simulator", IEEE Trans. Circuits Syst., Vol. CAS-22, pp. 901-910, Dec. 1975.

[5] Ekachai Lelarasmee, Albert E. Ruehli, and Alberto L. Sangiovanni-Vincentelli, "The waveform relaxation method for time-domain analysis of large scale integrated circuits", IEEE Trans. on Computer-aided design of integr. circuits and systems, Vol. CAD-1, NO. 3, July 1982.

[6] H. SHICHMAN and D.A. HODGES "Modeling and Simulation of Insulated Gate Field Effect Transistor Switching Circuits" in IEEE J. of Solid State Circuits. Vol SC-3. September 1968.

[7] M. OBREBSKA "Etude comparative de differentes methodes de conception des parties controle des microproesseurs" These de Docteur Ingenieur. Grenoble. June 1982

[8] K. BARABAN, M. ISRAEL and J.L. KORS "An interactive Design Rules Checker" in ICCAD-83

[9] G. NOGUEZ "Interactive Design Checking Based On Behavior Simulation" in ICCAD-83

CHIP LAYOUT OPTIMIZATION USING
CRITICAL PATH WEIGHTING

A. E. Dunlop
V. D. Agrawal
D. N. Deutsch*
M. F. Jukl
P. Kozak
M. Wiesel

AT&T Bell Laboratories
Murray Hill, New Jersey 07974

ABSTRACT

A chip layout procedure for optimizing the performance of critical timing paths in a synchronous digital circuit is presented. The procedure uses the path analysis data produced by a static timing analysis program to generate weights for critical nets on clock and data paths. These weights are then used to bias automatic placement and routing in the layout program. This approach is shown to bring the performance of the chip significantly closer to that of an ideal layout which is assumed to have no delay due to routing between cells.

1. INTRODUCTION

Polycell layout [1] of chips consists of two steps. First the cells are placed and then the routing is completed. A good placement and routing procedure will normally try to minimize the total routing area. However, from the viewpoint of speed performance of the circuit, the critical nets should be compacted, if necessary, at the expense of the non-critical ones which may be allowed to stretch. This suggests the use of critical net data for cell placement and routing.

Several synchronous path delay analysis programs have been reported in the literature [2,3]. In this paper we use the synchronous path delay (SD) analysis [2] of the MOTIS simulator [4]. This analysis determines clock skews and data path delay between clocked elements. The basic inputs are the clock signals, their waveforms and the

desired operating frequency. The analysis is topological. An exhaustive path search is performed without the need for a pre-defined vector set. The gate delays computed from device and routing information give SD the information necessary to provide circuit performance as accurately as the multiple delay simulator [5].

Figure 1 shows how MOTIS-SD and the layout program LTX2 [1], can be used together to obtain high performance circuit layouts automatically. An SD analysis is performed first with an estimated delay based on fanout. Nets that are found to be on failing paths, near-failing paths or having large clock skews are assigned priorities. These are critical nets. The net priority data are then passed to LTX2. The priorities are used to bias the placement and routing procedures so as to improve performance. After layout, actual routing delay data are passed from LTX2 back into MOTIS. A second SD run at this time will indicate the circuit's performance taking layout loading into consideration. This process can be stopped when a layout is found to have no failing paths for the desired frequency.

In Figure 1a, the first MOTIS-SD analysis is done with estimated delays. Alternatively, an LTX2 layout can be generated without any critical net priorities and the corresponding routing delay data could then be used in the first pass of SD (Figure 1b). It is also possible to determine a theoretical upper limit to the operating frequency of the chip layout by applying SD to the circuit using transistor drive and gate loading capacitance but no delay due to interconnections.

* Presently with Bell Communications Research, Murray Hill, N. J. 07974

Reprinted from *ACM/IEEE 21st Design Automation Conf. Proc.*, 1984, pp. 133–136.

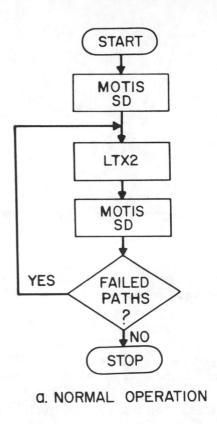

a. NORMAL OPERATION

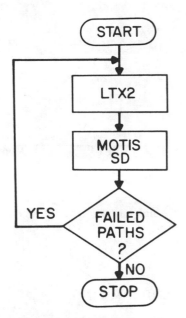

b. ALTERNATE OPERATION

Figure 1.   MOTIS-SD-LTX2 Chip Layout Procedure

2. SYNCHRONOUS PATH DELAY ANALYSIS

Synchronous path delay (SD) analysis in MOTIS [2] consists of two phases. First, the given clock signals are traced from their inputs forward to the clock terminals of flip-flops. The delays on these paths are used to determine the clock signal skews. Second, combinational paths are searched between the flip-flops. Propagation delays for these paths are examined for correct operation with the given clock waveforms and the clock skews. For each combinational path, the analysis determines the maximum frequency of operation. If the skew of a clock path is found to be greater than 10 percent of the clock period corresponding to the given frequency or the maximum frequency of proper operation for a combinational path is found to be less than 1.2 times the given clock frequency, then all the signals on the path are assigned greater than zero priorities. These priorities are used during placement and routing in order to minimize the interconnection delay between components.

3. NET PRIORITY

Normally every noncritical signal net is assigned a zero weight. Higher weights are assigned to the nets on critical paths. MOTIS-SD analysis assigns the weights to clock signal nets as shown in Figure 2. Weights for signals on a clock path are determined from the skew that the delay of that path introduces in the clock signal. If the clock goes through gating logic, then these gates may receive a weight if they induce a large skew. The weight assignment for latch-to-latch combinational data path nets is shown in Figure 3. The maximum frequency of correct operation f_{max} for a path and the clock frequency f_0 determine the weight for all signals on the path. If f_{max} is greater than 1.2 f_0, where f_0 is the desired operating frequency, then the signals are considered noncritical. When $f_{max} = f_0$, a medium weight is assigned. For $f_{max} < f_0$ the weight rapidly increases to a maximum value.

There are several priorities for critical signal nets that are passed from MOTIS-SD to LTX2. For each net these include a placement priority, a loose routing priority, a maximum number of feedthroughs, a fine routing priority, and whether 'doglegs' are permitted.

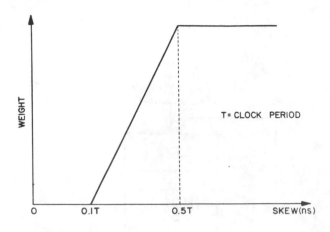

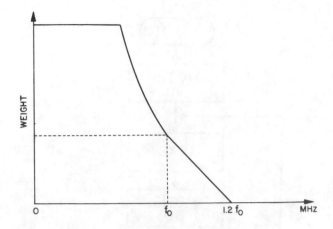

Figure 2.   Priority weight assignment for signal nets on clock paths.

Figure 3.   Priority weight assignment for signal nets on combinational paths.

4. PLACEMENT AND ROUTING

LTX2 uses the critical net data to bias automatic placement and routing procedures [1]. The information passed from MOTIS-SD to LTX2 is a net priority file. The file specifies placement priorities, loose routing priorities and fine routing priorities. Normally the net priority file is read into LTX2 before any geometric layout has been made. The priority data can also be read in at a later time to augment any existing priorities. Subsequent automatic placement and routing will be biased by these priorities.

In general, the placement procedure [6] uses the net priority number to indicate how many duplicates of the original net are to be considered when determining placement. The greater the number of nets that connect two modules together, the closer they will be placed.

The loose router [7] considers the loose routing priority and the maximum number of feedthroughs. Critical nets get routed before less critical nets. The sooner a net is routed the straighter the routing path will be resulting in a smaller propagation delay. MOTIS-SD sets the maximum feedthrough count on all critical clock nets to 0 so no polysilicon feedthroughs will be used. Other critical nets receive a small maximum feedthrough count. The channel router, a new version of [8], uses the fine routing priority. Nets with the highest priority are routed before nets with a lesser priority. The rule is to minimize total capacitance. If a prioritized net has more connections on the top side of a horizontal routing module than the bottom, the horizontal connection is biased toward the top.

Since the placement and routing procedures are being biased by the critical net data, some tradeoffs will be made to favor circuit performance instead of layout area. In general, increasing the number of nets that have priority, and giving individual nets higher priorities tends to increase the layout area.

5. EXPERIMENTAL RESULTS

The critical net data can make a substantial impact on the performance of a chip. An upper limit for the chip performance will be considered to be the maximum working frequency, assuming no delay due to routing (no capacitance and no resistance). This is not a tight upper bound since some routing will be required, but it is the best known. The improvement due to critical net data can be measured by how much closer to the upper limit of chip performance a layout using critical net data comes compared to a layout without critical net data.

For the circuit in Table 1 with 89 polycells, the performance is 15 percent closer to the upper

TABLE 1

Circuit with 89 polycells

Delay Description	Working Frequency
Without Net Priority	4.37 MHz
With Net Priority	4.43 MHz
No Routing Delay	4.77 MHz (upper limit)

TABLE 2

Circuit with 558 polycells

Delay Description	Working Frequency
Without Net Priority	6.75 MHz
With Net Priority	6.99 MHz
No Routing Delay	7.50 MHz (upper limit)

bound. For the circuit in Table 2 with 558 polycells, the performance is 32 percent closer to the upper limit. For all the layouts in the above examples, the placement and routing were done automatically.

6. CONCLUSION

The procedure employing critical net data from MOTIS-SD path analysis in LTX2 automatically improves circuit performance. Using this automatically generated critical net data in an automatic layout program is a new concept. It represents a deviation from the conventional layout approaches in which the chip area is unconditionally minimized. The illustrative examples show that the circuit performance obtained through the use of critical net data is

closer to the theoretical upper limit of the chip operating frequency than would be achieved by automatic layout without the additional net priority data.

REFERENCES

[1] A. E. Dunlop, "Automatic Layout of Gate Arrays", *Proceedings of IEEE International Symposium on Circuits and Systems*, Newport Beach, CA, May 2-4, 1983, pp. 1245-1248.

[2] V. D. Agrawal, "Synchronous Path Delay Analysis in MOS Circuit Simulator", *Proceedings of 19th Design Automation Conference*, Las Vegas, Nevada, June 14-16, 1982, pp. 629-635.

[3] J. K. Ousterhout, "Crystal: A Timing Analyzer for nMOS VLSI Circuits", *Third Caltech Conference on Very Large Scale Integration*, (Rockville, MD: Computer Science Press), 1983, pp. 57-69.

[4] V. D. Agrawal, A. K. Bose, P. Kozak, H. N. Nham, and E. Pacas-Skewes, "Mixed-Mode Simulation in the MOTIS System", *Journal of Digital Systems*, Vol. V, pp. 383-400, Winter 1981.

[5] H. N. Nham and A. K. Bose, "A Multiple Delay Simulator for MOS LSI Circuits", *Proceeding of the 17th Design Automation Conference*, Minneapolis, Minnesota, June 23-25, 1980, pp. 610-617.

[6] A. E. Dunlop and B. W. Kernighan, "A Placement Procedure for Polycell VLSI Circuits", *IEEE International Conference on Computer Aided Design*, Santa Clara, CA, September 13-15, 1983, pp. 51-52.

[7] M. Wiesel, "Automatic Routing of Gate Arrays", *European Conference on Circuit Theory and Design*, Stuttgart, W. Germany, September 5-9, 1983.

[8] D. N. Deutsch, "A 'Dogleg' Channel Router", *Proceedings of the 13th Design Automation Conference*, San Fransico, CA, June 1976, pp. 425-433.

DELAY and POWER OPTIMIZATION in VLSI CIRCUITS

Lance A. Glasser
Electrical Engineering and Computer Science Department
and the
Research Laboratory of Electronics
Massachusetts Institute of Technology
Cambridge, Massachusetts 02139

Lennox P. J. Hoyte
Prime Computer, Inc.
Framingham, Massachusetts 01701

Abstract: The problem of optimally sizing the transistors in a digital MOS VLSI circuit is examined. Macromodels are developed and new theorems on the optimal sizing of the transistors in a critical path are presented. The results of a design automation procedure to perform the optimization is discussed.

1. Statement of the Problem

The circuit optimization techniques and algorithms appropriate for use in VLSI design automation systems must have low computational complexity. In this paper we investigate efficient optimization techniques for expediting the automatic design of high performance digital integrated circuits.

The automatic generation of LSI and VLSI circuits is already mundane. These synthesis systems are typically one of four types: gate array, polycell, macrocell, or silicon compiler systems. Definitions vary in the literature, but roughly speaking a gate array is a fixed matrix of gates or gate parts which is customized to implement a particular logic function by means of a customized metalization layer or layers. In a polycell system fixed cells are chosen from a library and then automatically placed and routed. A macrocell system typically embodies the placement and routing functions of the polycell system, but the cells are not fixed. The exact implementation of the cells is a function of the logic the chip must perform. A software program which implements programmable logic arrays of different sizes depending on the input sum of products specification is an example of a macrocell generator. Silicon compilers transform the specification of a chip from a hardware description language into a layout specification. In these systems cells are generally not fixed. In many silicon compilers the cells can be stretched; they generally have other variable attributes as well. Gate array and polycell systems are not readily modifiable to include automatic circuit optimization in which the cell design is dependent on the function of the LSI chip because the cell design is bound before the functionality of the chip is specified. Thus the design of the cell cannot depend on the source and load impedance

context of the cell in the actual chip implementation (as usual there are exceptions, for instance cases where drivers of different sizes can be used). In both macro-cell and silicon compiler systems the implementation of a circuit, or at least its layout, is dependent on the function that circuit will be performing in the context of the total chip design. Such systems provide, at least conceptually, the hooks necessary to do global circuit optimization. For our purposes it is not necessary to differentiate between macro-cell and silicon compiler systems since, from a circuit optimization standpoint, they are substantially the same.

We will be examining algorithms which run in near linear time. In order to achieve this degree of efficiency we will need to develop macro-models for the digital logic gate which are both simple and accurate.

We limit our discussion to digital MOS integrated circuits. In nMOS integrated circuits there are two typical circuit optimizations we might wish to perform. The first is to optimize a critical path for minimum delay without regard to power dissipation. This may be done for one of two reasons. We may want that path to run at maximum speed, or we may wish to obtain some measure of the difficulty of optimizing that path compared to optimizing some other path. The second, and more common, optimization problem in nMOS is to optimize a path such that it meets a delay specification (say related to the clock period) and dissipates minimum power. Power is a global scarce resource on an nMOS LSI chip. The optimum speed problem is a special case of the second problem. Prior work in this field includes that of [Kang81, Lin75, Moshen79].

2. Circuit Optimization Techniques

Given a critical path, there are six general circuit optimization techniques.

(i) Change the widths and lengths of various transistors in the schematic. This is the most straight-forward technique and has the minimum impact on the layout.

(ii) Insert one or more buffer stages between a high impedance source and a low impedance load. There is an optimum number of buffer stages.

Reprinted from *ACM/IEEE 21st Design Automation Conf. Proc.*, 1984, pp. 529–535.

(iii) Decouple the loads on a node with a large load capacitance and optimize each of the resulting paths separately. By this technique one can, for instance, expend significant power driving only loads in the critical path. This technique is illustrated in Fig. 1.

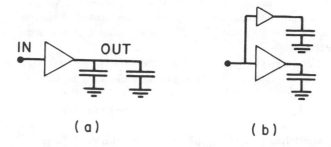

Fig. 1. The driver in (a) is split to allow separate path optimizations as seen in (b).

(iv) Use clocked buffers to take advantage of the low impedance characteristics of a strong system clock. This technique can only be applied in cases when an appropriate clock edge exists. This technique is illustrated in Fig. 2 using nMOS technology. (The usefulness of clocked circuit techniques is one of the motivations for having a multiplicity of clock edges in the architectural specification of a chip.)

(v) Precharge heavily capacitive nodes to achieve both better speed performance and, in nMOS, lower static power dissipation. A precharged circuit is illustrated in Fig. 3. Precharging improves the speed performance of an MOS circuit when the delay is significantly different between a rising and falling output transition. As with technique (iv), precharging requires the existence of appropriate clock edges, though they can sometimes be generated with timing chains if the system architecture allows.

(vi) Use specialized circuits. These circuits are particularly useful for high fan-in (large input capacitance) or fan-out (large output capacitance) situations. Common techniques are differential sensing, cascode connections, and superbuffers.

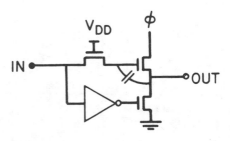

Fig. 2. A clocked bootstrap buffer.

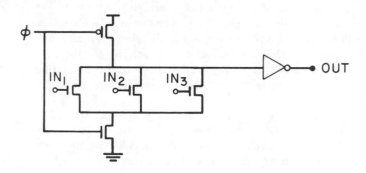

Fig. 3. A clocked precharged cMOS OR gate.

In addition to these circuit domain techniques, there are optimizations which can be performed in other domains of abstraction including the domains of machine organization, logic, and layout. Of the six circuit domain techniques enumerated above we will concentrate on only the first.

3. The Macro-model

The classical model of the delay through a string of logic gates is simply that the delays through the individual logic gates add and that these delays do not depend on the shape of the input waveform. This is the model which we will use in this paper, with a couple of minor exceptions. One exception is that we consider the delay through a logic gate to be a function of the sense of the input transition. That is, the delay through an MOS logic gate depends on whether the output is pulling up or pulling down. If used with care, this simple model can yield reasonably good results [Terman83], in the ±30% range. On the other hand, in order to understand the limitations of this model, its regions of validity, and directions for future research, we must investigate the delay characteristics of MOS logic gates in finer detail.

TTL gates behave much closer to the ideal than do MOS logic gates. The major differences between MOS and TTL logic gates are first that the voltage rise and fall times seen in digital MOS integrated circuits are much slower in relation to the gate propagation delay than in TTL, and second that the maximum differential gain of MOS logic gates is smaller than that of TTL gates. The ramifications of these facts are that one must look more carefully at both the definition of delay and the assumption that gate delay is independent of the input transition time.

Let us first examine the shortcoming of expressing delay as the sum of the individual gate delays. Figure 4 illustrates a typical collection of voltage waveforms representing the delay through a series of N inverters. The delay times $\tau_1...\tau_N$ represent the individual gate delays. We have picked a reference voltage V_{REF}, in this case 1.57V which is the DC voltage at which the voltage input to the gate equals the output, as the reference point at which to perform the delay measurements. Notice that we do not account for the last part of the delay τ_E which represents the tail of the transition of the last gate. While one might decide to use a different measure of gate delay, all commonly used definitions have this problem of not accounting for the edge conditions correctly (this is really the fence post counting problem in disguise). A first order correction to the classical theory can handle this problem in a rather trivial fashion. Our approach is that, if the output is terminated with a conventional latch rather than another gate or a transparent latch, we increase τ_N to include τ_E. This can be easily modeled by multiplying the load capacitance by a constant larger than one.

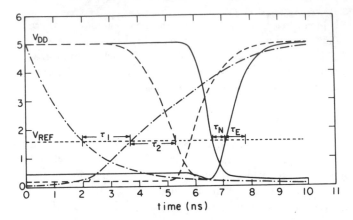

Fig. 4. A collection of voltage waveforms.

The second exception deals with is the dependence of gate delay on the shape of the input waveform. It is instructive to look at two regimes of operation—very fast inputs and very slow inputs. For very fast inputs we examine the gate's step response, which consists of two components. The first component of the response is governed by the capacitive division of the Miller capacitance with all other capacitances on the output node including the load and self capacitance terms. This capacitive feedthrough is negligible in almost all optimized circuits. After the input step is over, the response of the network is determined only by the current sourcing capabilities of the logic gate output and the total effective capacitance of the load. Assuming that the capacitive load on the output can be approximated by a linear effective load capacitance then we can express the stage delay t_D as

$$t_D = C \int \frac{1}{I(V_{OUT})} dV_{OUT}. \qquad (1)$$

We can rewrite (1) as

$$t_D = R(C_{LOAD} + C_{SELF}) \qquad (2)$$

where the effective charging resistance R is defined as

$$R \equiv \int_{V_{START}}^{V_{REF}} \frac{1}{I(V_{OUT})} dV_{OUT} \qquad (3)$$

and where C has been decomposed into load and self capacitance terms. Thus we see that, as expected, the delay increases linearly as the output load grows. This has been corroborated by SPICE [Vladimirescu80] circuit simulations.

For very slow inputs the situation is quite different. In this case the gate dynamics are determined only by the DC transfer characteristics of the logic gate. Thus the output is actually slowed by the slowness of the input transition. The output transition is constrained to be no faster than the input transition multiplying the maximum differential gain of the logic gate.

Figure 5 shows simulation results which illustrate the effective charging resistance as a function of the input transition time. We define transition times as the time it takes the voltage to transition between the 20 and 80% levels. We define the slew rate S as the 80% voltage minus the 20% voltage ($0.6V_{DD}$ where V_{DD} is the supply voltage) divided by the transition time. The output load capacitance is fixed in this experiment. In circuits where the load capacitance on the output (which includes the self capacitance) dominates over the gate output to input capacitance—which is almost always—we can use the ratio of the output transition time to input transition time as a single metric to characterize the effective charging resistance R_i of logic gate i. We can do this because the only time dependent terms in the equations describing the dynamics of the logic gate come from the time dependence of the input and the $C\frac{d}{dt}$ terms. If the dominant capacitance comes from the

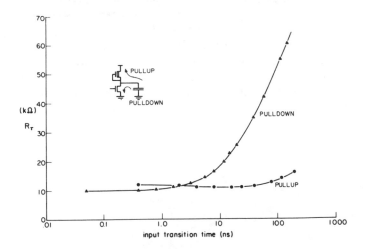

Fig. 5. The effective charging resistance as a function of input transition time.

load as we have suggested, then scaling the capacitive load and the time parameter of the voltage waveforms does not change the equations of the system, implying that the effective charging resistance remains unchanged. This has been verified through SPICE simulations. Figure 6 illustrates the dependence of R_τ on the ratio of the input transition time over the output transition time for an nMOS inverter. We define R_τ as the effective charging resistance of a square pulldown device. In a typical circuit context the rising transient of an nMOS logic gate is 5 to 20 times slower than the falling transition time. We see that for the case of an output pulling high, and this is the critical transition as far as the total delay is concerned, the logic gate typically operates in a region where R_i is independent of the input transition rate. For the pulldown case, which luckily is not as critical, the reverse is true and so the pulldown R_i is sharply dependent on even slight changes in the output/input transition time ratio (or equivalently, the input/output signal slew rate ratio S_{in}/S_{out}). In this case, the accuracy of the model is much poorer. In cMOS the constant R approximation is generally valid for both transitions. In the remainder of the paper we will assume that R_i is independent of the input transition times and use a typical value. This is a good first order approximation, but it will turn out to be the limiting accuracy issue.

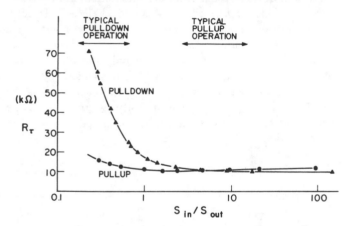

Fig. 6. The effective charging resistance as a function of the slew rate ratio S_{in}/S_{out}.

The final modeling problem is to relate the input capacitance of a logic gate to its drive for different device widths. Neglecting narrow width effects the current sourced by a transistor (and hence R_i^{-1}) is proportional to its effective width. The input capacitance C_i of a transistor at a given voltage is also proportional to the effective width of the device. (We are neglecting the differences in the Miller capacitance due to changing the width of a transistor though we are not neglecting the Miller capacitance totally.) Thus the RC time parameter of a MOS logic gate is constant. We can then define, for a given technology, a time constant τ such that

$$\tau \equiv R_i C_i. \qquad (4)$$

This τ is essentially the same as the one presented by Mead and Conway. [Mead80] It is roughly the time it takes one transistor of width W to discharge a second device of identical dimensions.

4. Theoretical results

As we discussed in the last section, the delay through a cascade of logic gates can be represented as a sum of RC's. We write the total delay T as

$$T = \sum_{i=0}^{N} t_i \qquad (5)$$

where

$$t_i \equiv R_i \left(C_{i+1} + Y_i + \frac{\alpha_i C_i}{n_i} \right) n_i \qquad (6)$$

and where R_i is the equivalent pulldown resistance of the the ith stage, C_{i+1} is the effective input capacitance of the $i+1^{st}$ stage, and Y_i is the total effective interconnect capacitance of the wire connecting the ith and $i+1^{st}$ stages. Y_i also includes the equivalent input capacitances of stages not in the critical path. α_i characterizes the self capacitance loading of a stage. As the width of a transistor increases, its parasitics also increase. To first order, these increases are captured in α. α_i is usually independent of stage number. That is, $\alpha_i = \alpha_0$ for all i. Figure 7 illustrates these various parameters. The parameter n_i is called the fan-out parameter and has a number of distinct interpretations depending on the problem being solved. It can be used for accounting for edge conditions on the delay summation in the first and N^{th} stages; the difference between pullup transistor networks, and fanout. Using (4) we can rewrite (5) as

$$T = \tau \sum_{i=0}^{N} \left(R_i \left(\frac{1}{R_{i+1}} + \frac{Y_i}{\tau} \right) + \frac{\alpha_i}{n_i} \right) \qquad (7)$$

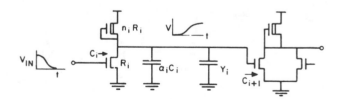

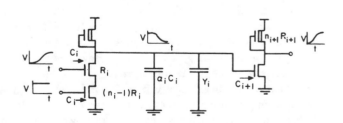

Fig. 7. Interpretation of the circuit parameters.

where $n_0 \equiv 1$. R_0 is the source equivalent resistance and $C_{N+1} = 0$. In (6) we have define $Y_N \equiv 0$ and $R_{N+1} \equiv \tau/C_{LOAD}$ where C_{LOAD} is the output capacitance. This is purely a notational convenience. We define the delay t_i of the ith stage as the difference between the time the input voltage passes V_{REF} and the time the output voltage passes V_{REF}.

Equation (7) is the central equation in this article. We will first solve it in closed form in a number of special cases. The most important special case is that of all the wire capacitances Y_i much much smaller than the respective input capacitances $C_{i+1} = \tau/R_{i+1}$ (with the exception of C_{LOAD}). This case is likely to occur under conditions of minimum delay where the drive transistors are very wide. We find the conditions of optimum speed by taking the gradient of T with respect to R_i and setting it equal to zero. We have

$$ 0 = \nabla_{R_i} T = -\frac{R_{i-1} n_{i-1}}{R_i^2} + \frac{n_i}{R_{i+1}} \quad \text{for} \quad i = 1 \text{ to } N. \quad (8) $$

Because in this case we have

$$ t_i = n_i \tau \left(\frac{R_i}{R_{i+1}} + \frac{\alpha_i}{n_i} \right), \quad (9) $$

we can see that (8) leads to all of the t_i being equal in the special case of $\alpha_i = \alpha_0$. We have

$$ t_i = t_{i+1} \equiv t_0. \quad (10) $$

We define

$$ t_{00} \equiv t_0 - \alpha_0 \tau. \quad (11) $$

Equation (8) can be solved in closed form. The results are

$$ R_i = \tau \frac{n_{i-1}}{t_0} R_{i-1} \quad (12) $$

$$ R_i = R_0 \left(\frac{t_{00}}{\tau} \right)^{-i} \Pi_{j=i}^{i-1} n_j \quad (13) $$

$$ t_{00} = \tau \left(\frac{R_0}{R_{N+1}} \Pi_{j=1}^{N} n_j \right)^{\frac{1}{N+1}} \quad (14) $$

and

$$ T_{MIN} = (N+1) t_0. \quad (15) $$

These formula, which can clearly be computed in linear time, can provide a simple first guess of the optimum device sizes and delay for a string of simple gates. For the simple case of all $n_i = 1$ these results reduce to the familiar geometric device scaling in which each device is a constant times the size of the preceding device.

We have not been able to find closed form solutions in the cases in which wiring capacitance can not be neglected or in the cases when we wish to consider power dissipation or area as an additional metric. Let us write the total power or area in the form

$$ B = \sum_{i=1}^{N} b_i. \quad (16) $$

That is, we associate with each stage a quantity which captures the resources used by that stage. For the case of power dissipation we interpret the b_i as the average power dissipation of each stage. In the case of active area we would interpret b_i as the area of each stage. For both the cases of power dissipation and active area, b_i is of the form

$$ b_i = m_i C_i \quad (17) $$

where m_i is a constant which depends on the topology of the logic gate. For instance, the active area of a cMOS logic gate is proportional to the number of transistors in the logic gate times the area of each transistor. Similarly the power dissipated in an nMOS logic gate depends on the width of the pulldown transistors (which is directly proportional to the capacitance C_i) times the β-ratio times the number of pulldown transistors connected in series. We therefore have

$$ B = \sum_{i=1}^{N} m_i C_i = \tau \sum_{i=1}^{N} \frac{m_i}{R_i}. \quad (18) $$

The new problem which we wish solve is given T, minimize B. That is, given a delay specification, minimize either the area or power dissipation of the total circuit. To solve this problem we invoke the marginal utility argument which states that to minimize B (the global scarce resource) we require that locally each must stage make the very best use of that scarce resource. This is true if the sensitivity of the total delay to the global resources used at each stage is the same for all stages. This marginal utility argument is the equivalent to a Lagrange multiplier technique formulation of the problem. We have

$$ 0 = \nabla_{b_i} B + \lambda \nabla_{b_i} T \quad (19) $$

which reduces to

$$ -\lambda^{-1} = \nabla_{b_i} T. \quad (20) $$

Solving, we obtain

$$ \lambda^{-1} = \frac{R_i}{\tau m_i} \left(\frac{n_i R_i}{R_{i+1}} + \frac{n_i R_i Y_i}{\tau} - \frac{n_{i-1} R_{i-1}}{R_i} \right) \quad (21) $$

for $i = 1$ to N, where λ is the Lagrange multiplier.

In the next section we will discuss our algorithms for numerically solving (21). Before going on to do this there are a number of useful observations which may be made. Equation (21) reduces to (8) in the special case of optimum speed by setting $\lambda^{-1} = 0$. λ^{-1} represents the negative of the derivative of delay with respect to area or power. Optimum solutions will always occur for non-negative λ because a negative λ implies that we are operating in a region where we can simultaneously decrease the area and decrease the delay. We are hardly at an optimum if we are in a region were such happy circumstances exist. We can rewrite (21) in terms of the t_i as

$$t_i = t_{i-1} + \frac{\tau^2 m_i}{\lambda R_i} - n_{i-1} R_{i-1} Y_{i-1} + \tau(\alpha_{i-1} - \alpha_i) . \quad (22)$$

τ, n_i, m_i, R_i, Y_i, and λ are all non-negative. When α is constant this equation teaches us that for optimum speed $(\lambda^{-1} = 0)$ in the presence of interconnect wiring parasitics, the stage delay is a monotonically non-increasing function of position in the chain i.e., $t_i \geq t_{i+1}$. That is, the greatest delay occurs at the beginning. Physically, this occurs because the drive capability one has at the beginning of a chain of gates is limited, but by the end of the chain one has had a chance to telescope that drive to a higher level and hence make the interconnect capacitance more negligible. On the other hand, if there are no interconnect capacitances but power or area is important, then the stage delay is a monotonically non-decreasing function of position i.e., $t_i \leq t_{i+1}$. We can understand this by considering the problem of driving a large load capacitance through many stages. If λ^{-1} were equal to zero then the delay in every stage would be the same. The last stages would clearly draw the greater percentage of the power. Thus if we were willing to have the chip run slower it would be most advantageous to draw the power from the last stages, slowing them down.

5. Experimental results and Conclusions

A procedure for optimally sizing transistors in a single critical path has been written in C. While more sophisticated techniques are possible [Matson82, Vergnieres80], the straightforward relaxation techniques we used resulted in a program which is computationally very fast. In an example contrived to illustrate the speed of the program, a chain of 500 stages (1000 transistors) was sized in 50 seconds of CPU time on a DEC System 20/60. Further experiments showed the program to run in near linear time. The inner loop of the program uses a simple relaxation technique to size the transistors given λ, Y_i, n_i, R_0, C_{LOAD}, and α_i. After an arbitrary initial assignment of the R_i, the program refines the R_i based on (21). The program proceeds down the critical path from R_1 to R_N. Let R_i^k represent the kth iteration on R_i. The relaxation equation we use is

$$R_i^{k+1} = \sqrt{\frac{\tau^2 m_i \lambda^{-1} R_{i+1}^k + n_{i-1} R_{i-1}^{k+1} R_{i+1}^k \tau}{n_i \tau + n_i Y_i R_{i+1}^k}} \quad (23)$$

This procedure typically converges to an acceptable accuracy within two dozen iterations.

Since the relaxation loop takes as input λ, not a delay specification, the correct value of λ must be found. Binary search is used because of its simplicity of implementation. The value of λ, if one exists, needed to produce the specified optimum delay can be found in just a few iterations. If a solution is not possible, the fastest possible circuit is found.

Figure 8 plots optimization time as a function of the number of stages.

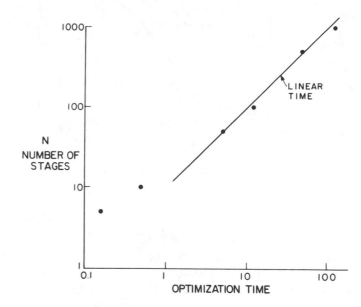

Fig. 8. Computer time required to perform optimization versus number of stages.

The results of circuit simulations to verify our theory must be evaluated in terms of the accuracy of the results compared to circuit simulation and also in terms of the optimality of the result.

The accuracy of our results were observed to lie within 3 to 70% of SPICE circuit simulation. This wide variation was shown to depend directly upon the value of effective charging resistance R. Recall that R_τ was shown to be sharply dependent on S_{in}/S_{out}. The reason we sometimes produce errors much larger than those quoted by other workers using the same model is that, unlike in unoptimized circuits, fall times in, for instance, the odd numbered stages of some optimized circuits can be comparable to rise times in the even numbered stages. Since the fall time accuracy has a tendency to be poor, the total delay accuracy is degraded in this class of circuits. Note however that one must go to some lengths to get such inaccuracies and any reasonable estimate of the circuit's regime of operation will result in reasonable accuracy.

We show the results of optimizing a single 5-stage circuit in Figs. 9 and 10. Figure 9 shows the circuit to be optimized. Despite the circuit's seeming simplicity, this circuit is actually quite difficult because there are no parasitic

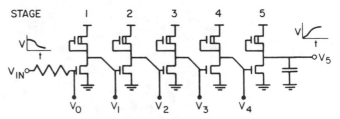

Fig. 9. Example circuit to be optimized.

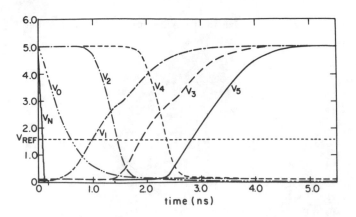

Fig. 10. SPICE waveforms for optimized circuits.

Stage number	0	1	2	3	4	5	Total
t predicted	.442	.442	.442	.442	.442	.442	2.65
t observed	.51	.40	.495	.40	.49	.465	2.76
% error	13.3	−10.5	10.7	−10.5	9.8	4.95	4.0
S_{in}/S_{out} assumed	.4	.3	.4	.3	.4		
S_{in}/S_{out} observed	1.72	.35	3.23	.39	2.3		
% error	77	14	88	23	83		

Table 1. Macromodel accuracy in the example.

capacitances to mask errors. A value of R_τ is chosen based on the curves in Fig. 6, the circuit is optimized, and the optimal circuit is compared to SPICE. We show a plot of the SPICE waveforms in Fig. 10 and compare the predicted stage delays to the actual (SPICE) stage delays in Table 1. Clearly, the error is introduced wherever the observed S_{in}/S_{out} ratio differs from the assumed S_{in}/S_{out} ratio. The pulldown devices are more volatile because of the region in which they tend to operate (see Fig. 6). For the pullup a 77% error in the slew rate ratio produces only a 10% error in delay, while for the pulldown a 14% error in slew rate ratio produces a 10% error. Thus while the key second order effect is signal slope dependencies, reasonable accuracies are attainable in most situations. Indeed, the results are most accurate in cases where wiring parasitics dominate, as they would in most present macrocell or silicon compiler

systems. For these systems we have demonstrated a way of improving the circuit performance in a computationally inexpensive way. (Exploiting the delay inequalities in a more knowledge based approach is an alternative.) We believe that much of this computational efficiency will carry over to implementations exploiting more complex macro-models.

Research is continuing on multipath optimization techniques, the application of more accurate and general macro-models, and the use of more general circuit techniques such as buffer insertion.

6. Acknowledgements

It is our pleasure to acknowledge helpful and inspiring discussions with Paul Penfield, Jr., Dan Dobberpuhl, Mark Matson, Bob Yodolowski, and Isaac Bain.

This work was supported in part by Air Force Contract F49620-84-C-0004 and in part by Defense Advanced Research Projects Agency under contract N00014-80-C-0622.

References

[Vladimirescu81] A. Vladimirescu, K. Zhang, A. R. Newton, D. O. Pederson, and A. Sangiovanni-Vincentelli, "SPICE Version 2G User's Guide," August 10, 1981.

[Vladimirescu80] A. Vladimirescu and S. Liu, "The Simulation of MOS Integrated Circuits using SPICE2," revised October, 1980.

[Mead80] C. A. Mead and L. Conway, *Introduction to VLSI Systems*, Addison-Wesley, 1980.

[Tokuda83] T. Tokuda, K. Okazaki, K. Sakashita, I. Ohkura, and T. Enomoto, "Delay-Time Modeling for ED MOS Logic LSI," *IEEE Trans. Computer-Aided Design*, vol. CAD-2, pp. 129-134, 1983.

[Lin75] H. C. Lin and L. W. Linholm, "An Optimized Output Stage for MOS Integrated Circuits," *IEEE J. Solid-State Circuits*, SC-10, pp. 106-109, 1975.

[Mohsen79] A. M. Mohsen and C. A. Mead, "Delay-Time Optimization for Driving and Sensing of Signals on High-Capacitance Paths of VLSI Systems," *IEEE Trans. Electron Devices*. ED-26, pp. 540-548, 1979.

[Bandler79] J. W. Bandler and M. R. M. Rizk, "Optimization of Electrical Circuits," *Engineering Optimization*, Mathematical Programming Study 11, North-Holland, Amsterdam, 1979.

[Brayton80] R. K. Brayton and R. Spence, *Sensitivity and Optimization*, Computer-Aided Design of Electronic Circuits, vol. 2, Elsevier Scientific Publishing, 1980.

[Terman83] C. J. Terman, *Simulation Tools for Digital LSI Design*, VLSI memo 83-154, MIT, 1983.

[Matson83] M. Matson, "Circuit level optimization of Digital MOS VLSI designs," Ph.D. Thesis Proposal, MIT, June 1982.

[Kang81] S. Kang, "A Design of CMOS polycells for LSI circuits," *IEEE Trans. Circuits and Systems*, vol. CAS-28, 1981.

[Verngnieres80] B. Verngnieres, "Macro Generation algorithms for LSI Custom Chip Design," *IBM J. of Research and Development*, vol. 24, pp. 612-621, 1980.

Part IV
Testability and Fault-Tolerant Design

THIS part consists of two sections: one deals with testability and self testing and the other with fault-tolerant and error detection. Because of space limitations, the two sections can only briefly address these topics, although a full volume could be devoted to the testing of VLSI chips. The reader is referred to the bibliography for further reading.

Section IV-A
Testability and Self Testing

Design for Testability—A Survey

THOMAS W. WILLIAMS, MEMBER, IEEE, AND KENNETH P. PARKER, MEMBER, IEEE

Invited Paper

Abstract—This paper discusses the basics of design for testability. A short review of testing is given along with some reasons why one should test. The different techniques of design for testability are discussed in detail. These include techniques which can be applied to today's technologies and techniques which have been recently introduced and will soon appear in new designs.

I. INTRODUCTION

INTEGRATED Circuit Technology is now moving from Large-Scale Integration (LSI) to Very-Large-Scale Integration (VLSI). This increase in gate count, which now can be as much as factors of three to five times, has also brought a decrease in gate costs, along with improvements in performance. All these attributes of VLSI are welcomed by the industry. However, a problem never adequately solved by LSI is still with us and is getting much worse: the problem of determining, in a cost-effective way, whether a component, module, or board has been manufactured correctly [1]–[3], [52]–[68].

The testing problem has two major facets:

1) test generation [74]–[99]
2) test verification [100]–[114].

Test generation is the process of enumerating stimuli for a circuit which will demonstrate its correct operation. Test verification is the process of proving that a set of tests are effective towards this end. To date, formal proof has been impossible in practice. Fault simulation has been our best alternative, yielding a quantitative measure of test effectiveness. With the vast increase in circuit density, the ability to generate test patterns automatically and conduct fault simulation with these patterns has drastically waned. As a result, some manufacturers are foregoing these more rigorous approaches and are accepting the risks of shipping a defective product. One general approach to addressing this problem is embodied in a collection of techniques known as "Design for Testability" [12]–[35].

Design for Testability initially attracted interest in connection with LSI designs. Today, in the context of VLSI, the phrase is gaining even more currency. The collection of techniques that comprise Design for Testability are, in some cases, general guidelines; in other cases, they are hard and fast design rules. Together, they can be regarded essentially as a menu of techniques, each with its associated cost of implementation and return on investment. The purpose of this paper is to present the basic concepts in testing, beginning with the fault models and carrying through to the different techniques associated with Design for Testability which are known today in

the public sector. The design for testability techniques are divided into two categories [10]. The first category is that of the ad hoc technique for solving the testing problem. These techniques solve a problem for a given design and are not generally applicable to all designs. This is contrasted with the second category of structured approaches. These techniques are generally applicable and usually involve a set of design rules by which designs are implemented. The objective of a structured approach is to reduce the sequential complexity of a network to aid test generation and test verification.

The first ad hoc approach is partitioning [13], [17], [23], [26]. Partitioning is the ability to disconnect one portion of a network from another portion of a network in order to make testing easier. The next approach which is used at the board level is that of adding extra test points [23], [24]. The third ad hoc approach is that of Bus Architecture Systems [12], [27]. This is similar to the partitioning approach and allows one to divide and conquer—that is, to be able to reduce the network to smaller subnetworks which are much more manageable. These subnetworks are not necessarily designed with any design for testability in mind. The forth technique which bridges both the structured approach and the ad hoc approach is that of Signature Analysis [12], [27], [33], [55]. Signature Analysis requires some design rules at the board level, but is not directed at the same objective as the structure approaches are—that is, the ability to observe and control the state variables of a sequential machine.

For structured approaches, there are essentially four categories which will be discussed—the first of which is a multiplexer technique [14], [21], Random Access Scan, that has been recently published and has been used, to some extent, by others before. The next techniques are those of the Level-Sensitive Scan Design (LSSD) [16], [18]–[20], [34], [35] approach and the Scan Path approach which will be discussed in detail. These techniques allow the test generation problem to be completely reduced to one of generating tests for combinational logic. Another approach which will be discussed is that of the Scan/Set Logic [31]. This is similar to the LSSD approach and the Scan Path approach since shift registers are used to load and unload data. However, these shift registers are not part of the system data path and all system latches are not necessarily controllable and observable via the shift register. The fourth approach which will be discussed is that of Built-In Logic Block Observation (BILBO) [25] which has just recently been proposed. This technique has the attributes of both the LSSD network and Scan Path network, the ability to separate the network into combinational and sequential parts, and has the attribute of Signature Analysis—that is, employing linear feedback shift registers.

For each of the techniques described under the structured approach, the constraints, as well as various ways in which

Manuscript received June 14, 1982; revised September 15, 1982.
T. W. Williams is with IBM, General Technology Division, Boulder, CO 80302.
K. P. Parker is with Hewlett-Packard, Loveland Instrument Division, Loveland, CO 80537.

Reprinted from *Proc. IEEE*, vol. 71, pp. 98–112, Jan. 1983.

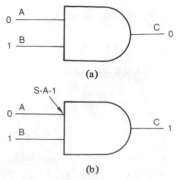

Fig. 1. Test for input stuck at fault. (a) Fault-free AND gate (good machine). (b) Faulty AND gate (faulty machine).

they can be exploited in design, manufacturing, testing, and field servicing will be described. The basic storage devices and the general logic structure resulting from the design constraints will be described in detail. The important question of how much it costs in logic gates and operating speed will be discussed qualitatively. All the structured approaches essentially allow the controllability and observability of the state variables in the sequential machine. In essence, then, test generation and fault simulation can be directed more at a combinational network, rather than at a sequential network.

A. Definitions and Assumptions

A model of faults which is used throughout the industry that does not take into account all possible defects, but is a more global type of model, is the Stuck-At model. The Stuck-At model [1]–[3], [9], [11] assumes that a logic gate input or output is fixed to either a logic 0 or a logic 1. Fig. 1(a) shows an AND gate which is fault-free. Fig. 1(b) shows an AND gate with input "A," Stuck-At-1 (S-A-1).

The faulty AND gate perceives the "A" input as 1, irrespective of the logic value placed on the input. The pattern applied to the fault-free AND gates in Fig. 1 has an output value of 0 since the input is 0 on the "A" input and 1 on the "B" input, and the AND'ing of those two leads to a 0 on the output. The pattern in Fig. 1(b) shows an output of 1, since the "A" input is perceived as a 1 even though a 0 is applied to that input. The 1 on the "B" input is perceived as a 1, and the results are AND'ed together to give a 1 output. Therefore, the pattern shown in Fig. 1(a) and (b) is a test for the "A" input, S-A-1, since there is a difference between the faulty gate (faulty machine) and the good gate (good machine). This pattern 01 on the "A" and "B" inputs, respectively, is considered a test because the good machine responds differently from the faulty machine. If they had the same response then that pattern would not have constituted a test for that fault.

If a network contained N nets, any net may be good, Stuck-At 1 or Stuck-At 0; thus all possible network state combinations would be 3^N. A network with 100 nets, then, would contain 5×10^{47} different combinations of faults. This would be far too many faults to assume. The run time of any program trying to generate tests or fault simulate tests for this kind of design would be impractical.

Therefore, the industry, for many years, has clung to the single Stuck-At fault assumption. That is, a good machine will have no faults. The faulty machines that are assumed will have one, and only one, of the stuck faults. In other words, all faults taken two at a time are not assumed, nor are all faults taken three at a time, etc. History has proven that the single

Stuck-At fault assumption, in prior technologies, has been adequate. However, there could be some problems in LSI—particularly with CMOS using the single Stuck-At fault assumption.

The problem with CMOS is that there are a number of faults which could change a combinational network into a sequential network. Therefore, the combinational patterns are no longer effective in testing the network in all cases. It still remains to be seen whether, in fact, the single Stuck-At fault assumption will survive the CMOS problems.

Also, the single Stuck-At fault assumption does not, in general, cover the bridging faults [43] that may occur. Historically again, bridging faults have been detected by having a high level—that is, in the high 90 percent—single Stuck-At fault coverage, where the single Stuck-At fault coverage is defined to be the number of faults that are tested divided by the number of faults that are assumed.

B. The VLSI Testing Problem

The VLSI testing problem is the sum of a number of problems. All the problems, in the final analysis, relate to the cost of doing business (dealt with in the following section). There are two basic problem areas:

1) test generation
2) test verification via fault simulation.

With respect to test generation, the problem is that as logic networks get larger, the ability to generate tests automatically is becoming more and more difficult.

The second facet of the VLSI testing problem is the difficulty in fault simulating the test patterns. Fault simulation is that process by which the fault coverage is determined for a specific set of input test patterns. In particular, at the conclusion of the fault simulation, every fault that is detected by the given pattern set is listed. For a given logic network with 1000 two-input logic gates, the maximum number of single Stuck-At faults which can be assumed is 6000. Some reduction in the number of single Stuck-At faults can be achieved by fault equivalencing [36], [38], [41], [42], [47]. However, the number of single Stuck-At faults needed to be assumed is about 3000. Fault simulation, then, is the process of applying every given test pattern to a fault-free machine and to each of the 3000 copies of the good machine containing one, and only one, of the single Stuck-At faults. Thus fault simulation, with respect to run time, is similar to doing 3001 good machine simulations.

Techniques are available to reduce the complexity of fault simulation, however, it still is a very time-consuming, and hence, expensive task [96], [104], [105], [107], [110], [112]–[114].

It has been observed that the computer run time to do test [80] generation and fault simulation is approximately proportional to the number of logic gates to the power of 3;[1] hence, small increases in gate count will yield quickly increasing run times. Equation (1)

---

[1] The value of the exponent given here (3) is perhaps pessimistic in some cases. Other analyses have used the value 2 instead. A quick rationale goes as follows: with a linear increase k in circuit size comes an attendant linear increase in the number of failure mechanisms (now yielding k squared increase in work). Also, as circuits become larger, they tend to become more strongly connected such that a given block is effected by more blocks and even itself. This causes more work to be done in a range we feel to be k cubed. This fairly nebulous concept of connectivity seems to be the cause for debate on whether the exponent should be 3 or some other value.

$$T = KN^3 \qquad (1)$$

shows this relationship, where T is computer run time, N is the number of gates, and K is the proportionality constant. The relationship does not take into account the falloff in automatic test generation capability due to sequential complexity of the network. It has been observed that computer run time just for fault simulation is proportional to N^2 without even considering the test generation phase.

When one talks about testing, the topic of functional testing always comes up as a feasible way to test a network. Theoretically, to do a complete functional test ("exhaustive" testing) seems to imply that all entries in a Karnaugh map (or excitation table) must be tested for a 1 or a 0. This means that if a network has N inputs and is purely combinational, then 2^N patterns are required to do a complete functional test. Furthermore, if a network has N inputs with M latches, at a minimum it takes 2^{N+M} patterns to do a complete functional test. Rarely is that minimum ever obtainable; and in fact, the number of tests required to do a complete functional test is very much higher than that. With LSI, this may be a network with $N = 25$ and $M = 50$, or 2^{75} patterns, which is approximately 3.8×10^{22} Assuming one had the patterns and applied them at an application rate of 1 μs per pattern, the test time would be over a billion years (10^9).

C. Cost of Testing

One might ask why so much attention is now being given to the level of testability at chip and board levels. The bottom line is the cost of doing business. A standard among people familiar with the testing process is: If it costs \$0.30 to detect a fault at the chip level, then it would cost \$3 to detect that same fault when it was embedded at the board level; \$30 when it is embedded at the system level; and \$300 when it is embedded at the system level but has to be found in the field. Thus if a fault can be detected at a chip or board level, then significantly larger costs per fault can be avoided at subsequent levels of packaging.

With VLSI and the inadequacy of automatic test generation and fault simulation, there is considerable difficulty in obtaining a level of testability required to achieve acceptable defect levels. If the defect level of boards is too high, the cost of field repairs is also too high. These costs, and in some cases, the inability to obtain a sufficient test, have led to the need to have "Design for Testability."

II. Design for Testability

There are two key concepts in Design for Testability: controllability and observability. Control and observation of a network are central to implementing its test procedure. For example, consider the case of the simple AND block in Fig. 1. In order to be able to test the "A" input Stuck-At 1, it was necessary to control the "A" input to 0 and the "B" input to 1 and be able to observe the "C" output to determine whether a 0 was observed or a 1 was observed. The 0 is the result of the good machine, and the 1 would be the result, if you had a faulty machine. If this AND block is embedded into a much larger sequential network, the requirement of being able to control the "A" and "B" inputs to 0 and 1, respectively, and being able to observe the output "C," be it through some other logic blocks, still remains. Therein lies part of the problem of being able to generate tests for a network.

Because of the need to determine if a network has the attributes of controllability and observability that are desired, a number of programs have been written which essentially give analytic measures of controllability and observability for different nets in a given sequential network [69]–[73].

After observing the results of one of these programs in a given network, the logic designer can then determine whether some of the techniques, which will be described later, can be applied to this network to ease the testing problem. For example, test points may be added at critical points which are not observable or which are not controllable, or some of the techniques of Scan Path or LSSD can be used to initialize certain latches in the machine to avoid the difficulties of controllability associated with sequential machines. The popularity of such tools is continuing to grow, and a number of companies are now embarking upon their own controllability/observability measures.

III. Ad Hoc Design for Testability [10]

Testing has moved from the afterthought position that it used to occupy to part of the design environment in LSI and VLSI. When testing was part of the afterthought, it was a very expensive process. Products were discarded because there was no adequate way to test them in production quantities.

There are two basic approaches which are prevalent today in the industry to help solve the testing problem. The first approach categorized here is Ad Hoc, and the second approach is categorized as a Structured Approach. The Ad Hoc techniques are those techniques which can be applied to a given product, but are not directed at solving the general sequential problem. They usually do offer relief, and their cost is probably lower than the cost of the Structured Approaches. The Structured Approaches, on the other hand, are trying to solve the general problem with a design methodology, such that when the designer has completed his design from one of these particular approaches, the results will be test generation and fault simulation at acceptable costs. Structured Approaches lend themselves more easily to design automation. Again, the main difference between the two approaches is probably the cost of implementation and hence, the return on investment for this extra cost. In the Ad Hoc approaches, the job of doing test generation and fault simulation are usually not as simple or as straightforward as they would be with the Structured Approaches, as we shall see shortly.

A number of techniques have evolved from MSI to LSI and now into VLSI that fall under the category of the ad hoc approaches of "Design for Testability." These techniques are usually solved at the board level and do not necessarily require changes in the logic design in order to accomplish them.

A. Partitioning

Because the task of test pattern generation and fault simulation is proportional to the number of logic gates to the third power, a significant amount of effort has been directed at approaches called "Divide and Conquer."

There are a number of ways in which the partitioning approach to Design for Testability can be implemented. The first is to mechanical partition by dividing a network in half. In essence, this would reduce the test generation and fault simulation tasks by 8 for two boards. Unfortunately, having two boards rather than one board can be a significant cost disadvantage and defeats the purpose of integration.

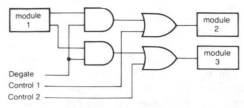

Fig. 2. Use of degating logic for logical partioning.

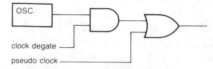

Fig. 3. Degating lines for oscillator.

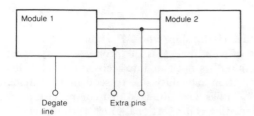

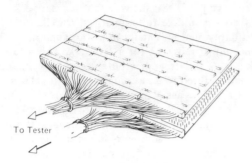

Fig. 4. Test points used as both inputs and outputs.

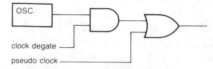

Fig. 5. "Bed of Nails" test.

Another approach that helps the partitioning problem, as well as helping one to "Divide and Conquer" is to use jumper wires. These wires would go off the board and then back on the board, so that the tester and the test generator can control and observe these nets directly. However, this could mean a significant number of I/O contacts at the board level which could also get very costly.

Degating is another technique for separating modules on a board. For example, in Fig. 2, a degating line goes to two AND blocks that are driven from Module 1. The results of those two AND blocks go to two independent OR blocks—one controlled by Control Line 1, the other with Control Line 2. The output of the OR block from Control Line 1 goes into Module 2, and the output of Control Line 2 goes into Module 3. When the degate line is at the 0 value, the two Control Lines, 1 and 2, can be used to drive directly into Modules 2 and 3. Therefore, complete controllability of the inputs to Modules 2 and 3 can be obtained by using these control lines. If those two nets happen to be very difficult nets to control, as pointed out, say, by a testability measure program, then this would be a very cost-effective way of controlling those two nets and hence, being able to derive the tests at a very reasonable cost.

A classical example of degating logic is that associated with an oscillator, as shown in Fig. 3. In general, if an oscillator is free-running on a board, driving logic, it is very difficult, and sometimes impossible, to synchronize the tester with the activity of the logic board. As a result, degating logic can be used here to block the oscillator and have a pseudo-clock line which can be controlled by the tester, so that the dc testing of all the logic on that board can be synchronized. All of these techniques require a number of extra primary inputs and primary outputs and possibly extra modules to perform the degating.

B. Test Points

Another approach to help the controllability and observability of a sequential network is to use test points [23], [24]. If a test point is used as a primary input to the network, then that can function to enhance controllability. If a test point is used as a primary output, then that is used to enhance the observability of a network. In some cases, a single pin can be used as both an input and an output.

For example, in Fig. 4, Module 1 has a degate function, so that the output of those two pins on the module could go to noncontrolling values. Thus the external pins which are dotted into those nets could control those nets and drive Module 2.

On the other hand, if the degate function is at the opposite value, then the output of Module 1 can be observed on these external pins. Thus the enhancement of controllability and observability can be accommodated by adding pins which can act as both inputs and outputs under certain degating conditions.

Another technique which can be used for controllability is to have a pin which, in one mode, implies system operation, and in another mode takes N inputs and gates them to a decoder. The 2^N outputs of the decoder are used to control certain nets to values which otherwise would be difficult to obtain. By so doing, the controllability of the network is enhanced.

As mentioned before, predictability is an issue which is as important as controllability and observability. Again, test points can be used here. For example, a CLEAR or PRESET function for all memory elements can be used. Thus the sequential machine can be put into a known state with very few patterns.

Another technique which falls into the category of test points and is very widely used is that of the "Bed of Nails" [31] tester, Fig. 5. The Bed of Nails tester probes the underside of a board to give a larger number of points for observability and controllability. This is in addition to the normal tester contact to the board under test. The drawback of this technique is that the tester must have enough test points to be able to control and observe each one of these nails on the Bed of Nails tester. Also, there are extra loads which are placed on the nets and this can cause some drive and receive problems. Furthermore, the mechanical fixture which will hold the Bed of Nails has to be constructed, so that the normal forces on the probes are sufficient to guarantee reliable contacts. Another application for the Bed of Nails testing is to do "drive/sense nails" [31] or "in situ" or "in-circuit" testing, which, effectively, is the technique of testing each chip on the board independently of the other chips on the board. For each chip, the appropriate nails and/or primary inputs are driven so as to prevent one chip from being driven by the other chips on the board. Once this state has been established, the isolated chip on the board can now be tested. In this case, the resolution to the failing

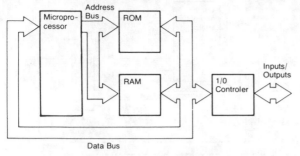

Fig. 6. Bus structured microcomputer.

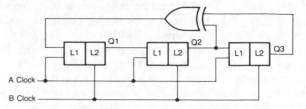

Fig. 7. Counting capabilities of a linear feedback shift register.

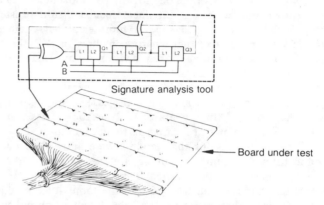

Fig. 8. Use of signature analysis tool.

chip is much better than edge connector tests, however, there is some exposure to incomplete testing of interconnections and care must be taken not to damage the circuit when over-driving it. Design for testability in a Bed of Nails environment must take the issues of contact reliability, multiplicity, and electrical loading into account.

C. Bus Architecture

An approach that has been used very successfully to attack the partitioning problem by the microcomputer designers is to use a bus structured architecture. This architecture allows access to critical buses which go to many different modules on the computer board. For example, in Fig. 6, you can see that the data bus is involved with both the microprocessor module, the ROM module, the RAM module, and the I/O Controller module. If there is external access to the data bus and three of the four modules can be turned off the data bus—that is, their outputs can be put into a high-impedance state (three-state driver)—then the data bus could be used to drive the fourth module, as if it were a primary input (or primary output) to that particular module. Similarly, with the address bus, access again must be controlled externally to the board, and thus the address bus can be very useful to controlling test patterns to the microcomputer board. These buses, in essence, partition the board in a unique way, so that testing of subunits can be accomplished. A drawback of bus-structured designs comes with faults on the bus itself. If a bus wire is stuck, any module or the bus trace itself may be the culprit. Normal testing is done by deducing the location of a fault from voltage information. Isolating a bus failure may require current measurements, which are much more difficult to do.

D. Signature Analysis

This technique for testing, introduced in 1977 [27], [33], [55] is heavily reliant on planning done in the design stage. That is why this technique falls between the Ad Hoc and the Structured Approaches for Design for Testability, since some care must be taken at the board level in order to ensure proper operation of this Signature Analysis of the board [12]. Signature Analysis is well-suited to bus structure architectures, as previously mentioned and in particular, those associated with microcomputers. This will become more apparent shortly.

The integral part of the Signature Analysis approach is that of a linear feedback shift register [8]. Fig. 7 shows an example of a 3-bit linear feedback shift register. This linear feedback shift register is made up of three shift register latches. Each one is represented by a combination of an $L1$ latch and an $L2$ latch. These can be thought of as the master latch being the $L1$ latch and the slave latch being the $L2$ latch. An "A" clock clocks all the $L1$ latches, and a "B" clock clocks all the

$L2$ latches, so that turning the "A" and "B" clocks on and off independently will shift the shift register 1-bit position to the right. Furthermore, this linear shift register has an EXCLUSIVE-OR gate which takes the output, $Q2$, the second bit in the shift register, and EXCLUSIVE-OR's it with the third bit in the shift register, $Q3$. The result of that EXCLUSIVE-OR is the input to the first shift register. A single clock could be used for this shift register, which is generally the case, however, this concept will be used shortly when some of the structured design approaches are discussed which use two nonoverlapping clocks. Fig. 7 shows how this linear feedback shift register will count for different initial values.

For longer shift registers, the maximal length linear feedback configurations can be obtained by consulting tables [8] to determine where to tap off the linear feedback shift register to perform the EXCLUSIVE-OR function. Of course, only EXCLUSIVE-OR blocks can be used, otherwise, the linearity would not be preserved.

The key to Signature Analysis is to design a network which can stimulate itself. A good example of such a network would be microprocessor-based boards, since they can stimulate themselves using the intelligence of the processor driven by the memory on the board.

The Signature Analysis procedure is one which has the shift register in the Signature Analysis tool, which is external to the board and not part of the board in any way, synchronized with the clocking that occurs on the board, see Fig. 8. A probe is used to probe a particular net on the board. The result of that probe is EXCLUSIVE-OR'ed into the linear feedback shift register. Of course, it is important that the linear feedback shift register be initialized to the same starting place every time, and that the clocking sequence be a fixed number, so that the tests can be repeated. The board must also have some initialization, so that its response will be repeated as well.

After a fixed number of clock periods—let's assume 50—a particular value will be stored in $Q1$, $Q2$, and $Q3$. It is not necessarily the value that would have occurred if the linear feedback shift register was just counted 50 times—Modulo 7.

The value will be changed, because the values coming from the board via the probe will not necessarily be a continuous string of 1's; there will be 1's intermixed with 0's.

The place where the shift register stops on the Signature Analysis Tool—that is, the values for $Q1$, $Q2$, and $Q3$ is the Signature for that particular node for the good machine. The question is: If there were errors present at one or more points in the string of 50 observations of that particular net of the board, would the value stored in the shift register for $Q1$, $Q2$, and $Q3$ be different than the one for the good machine? It has been shown that with a 16-bit linear feedback shift register, the probability of detecting one or more errors is extremely high [55]. In essence, the signature, or "residue," is the remainder of the data stream after division by an irreduceable polynomial. There is considerable data compression—that is, after the results of a number of shifting operations, the test data are reduced to 16 bits, or, in the case of Fig. 8, 3 bits. Thus the result of the Signature Analysis tool is basically a Go/No-Go for the output for that particular module.

If the bad output for that module were allowed to cycle around through a number of other modules on the board and then feed back into this particular module, it would not be clear after examining all the nodes in the loop which module was defective—whether it was the module whose output was being observed, or whether it was another module upstream in the path. This gives rise to two requirements for Signature Analysis. First of all, closed-loop paths must be broken at the board level. Second, the best place to start probing with Signature Analysis is with a "kernel" of logic. In other words, on a microprocessor-based board, one would start with the outputs of the microprocessor itself and then build up from that particular point, once it has been determined that the microprocessor is good.

This breaking of closed loops is a tenant of Design for Testability and for Signature Analysis. There is a little overhead for implementing Signature Analysis. Some ROM space would be required (to stimulate the self-test), as well as extra jumpers, in order to break closed loops on the board. Once this is done, however, the test can be obtained for very little cost. The only question that remains is about the quality of the tests—that is, how good are the tests that are being generated, do they cover all the faults, etc.

Unfortunately, the logic models—for example, microprocessors—are not readily available to the board user. Even if a microprocessor logic model were available, they would not be able to do a complete fault simulation of the patterns because it would be too large. Hence, Signature Analysis may be the best that could be done for this particular board with the given inputs which the designer has. Presently, large numbers of users are currently using the Signature Analysis technique to test boards containing LSI and VLSI components.

IV. STRUCTURED DESIGN FOR TESTABILITY

Today, with the utilization of LSI and VLSI technology, it has become apparent that even more care will have to be taken in the design stage in order to ensure testability and produce-ability of digital networks. This has led to rigorous and highly structured design practices. These efforts are being spearheaded not by the makers of LSI/VLSI devices but by electronics firms which possess captive IC facilities and the manufacturers of large main-frame computers.

Most structured design practices [14]-[16], [18]-[21], [25], [31], [32], [34], [35] are built upon the concept that if the

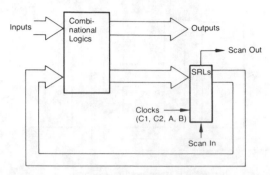

Fig. 9. Classical model of a sequential network utilizing a shift register for storage.

values in all the latches can be controlled to any specific value, and if they can be observed with a very straightforward operation then the test generation, and possibly, the fault task, can be reduced to that of doing test generation and fault simulation for a combinational logic network. A control signal can switch the memory elements from their normal mode of operation to a mode that makes them controllable and observable.

It appears from the literature that several companies, such as IBM, Fujitsu Ltd., Sperry-Univac, and Nippon Electric Co., Ltd. [14]-[16], [18]-[21], [31], [32], [35] have been dedicating formidable amounts of resources toward Structured Design for Testability. One notes simply by scanning the literature on testing, that many of the practical concepts and tools for testing were developed by main-frame manufacturers who do not lack for processor power. It is significant, then, that these companies, with their resources, have recognized that unstructured designs lead to unacceptable testing problems. Presently, IBM has extensively documented its efforts in Structured Design for Testability, and these are reviewed first.

A. Level-Sensitive Scan Design (LSSD)

With the concept that the memory elements in an IC can be threaded together into a shift register, the memory elements values can be both controlled and observed. Fig. 9 shows the familiar generalized sequential circuit model modified to use a shift register. This technique enhances both controllability and observability, allowing us to augment testing by controlling inputs and internal states, and easily examining internal state behavior. An apparent disadvantage is the serialization of the test, potentially costing more time for actually running a test.

LSSD is IBM's discipline for structural design for testability. "Scan" refers to the ability to shift into or out of any state of the network. "Level-sensitive" refers to constraints on circuit excitation, logic depth, and the handling of clocked circuitry. A key element in the design is the "shift register latch" (SRL) such as can be implemented in Fig. 10. Such a circuit is immune to most anomalies in the ac characteristics of the clock, requiring only that it remain high (sample) at least long enough to stabilize the feedback loop, before being returned to the low (hold) state [18], [19]. The lines D and C form the normal mode memory function while lines I, A, B, and $L2$ comprise additional circuitry for the shift register function.

The shift registers are threaded by connecting I to $L2$ and operated by clocking lines A and B in two-phase fashion. Fig. 11 shows four modules threaded for shift register action. Now note in Fig. 11 that each module could be an SRL or, one level up, a board containing threaded IC's, etc. Each level of pack-

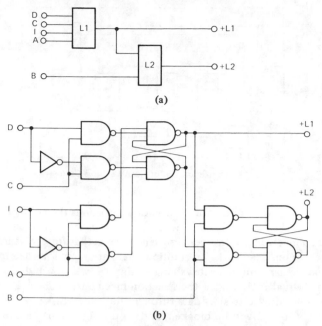

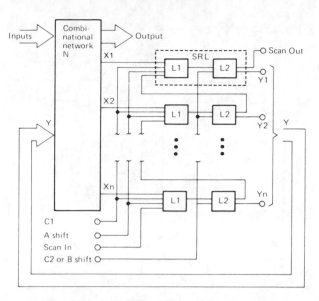

Fig. 12. General structure of an LSSD subsystem with two system clocks.

Fig. 10. Shift register latch (SRL). (a) Symbolic representation. (b) Implementation in AND-INVERT gates.

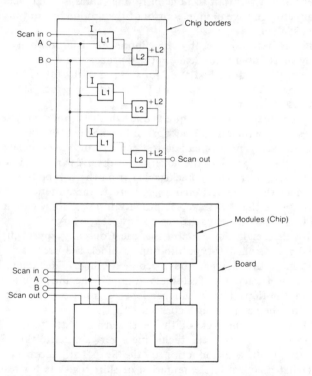

Fig. 11. Interconnection of SRL's on an integrated circuit and board.

aging requires the same four additional lines to implement the shift register scan feature. Fig. 12 depicts a general structure for an LSSD subsystem with a two-phase system clock. Additional rules concerning the gating of clocks, etc., are given by Williams and Eichelberger [18], [19]. Also, it is not practical to implement RAM with SRL memory, so additional procedures are required to handle embedded RAM circuitry [20].

Given that an LSSD structure is achieved, what are the rewards? It turns out that the network can now be thought of as purely combinational, where tests are applied via primary inputs and shift register outputs. The testing of combinational circuits is a well understood and (barely) tractable problem. Now techniques such as the D-Algorithm [93] compiled code Boolean simulation [2], [74], [106], [107], and adaptive random test generation [87], [95], [98] are again viable approaches to the testing problem. Further, as small subsystems are tested, their aggregates into larger systems are also testable by cataloging the position of each testable subsystem in the shift register chain. System tests become (ideally) simple concatenations of subsystem tests. Though ideals are rarely achieved, the potential for solving otherwise hopeless testing problems is very encouraging.

In considering the cost performance impacts, there are a number of negative impacts associated with the LSSD design philosophy. First of all, the shift register latches in the shift register are, logically, two or three times as complex as simple latches. Up to four additional primary inputs/outputs are required at each package level for control of the shift registers. External asynchronous input signals must not change more than once every clock cycle. Finally, all timing within the subsystem is controlled by externally generated clock signals.

In terms of additional complexity of the shift register hold latches, the overhead from experience has been in the range of 4 to 20 percent. The difference is due to the extent to which the system designer made use of the $L2$ latches for system function. It has been reported in the IBM System 38 literature that 85 percent of the $L2$ latches were used for system function. This drastically reduces the overhead associated with this design technique.

With respect to the primary inputs/outputs that are required to operate the shift register, this can be reduced significantly by making functional use of some of the pins. For example, the scan-out pin could be a functional output of an SRL for that particular chip. Also, overall performance of the subsystem may be degraded by the clocking requirement, but the effect should be small.

The LSSD structured design approach for Design for Testability eliminates or alleviates some of the problems in designing, manufacturing and maintaining LSI systems at a reasonable cost.

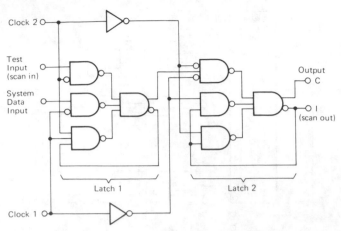

Fig. 13. Raceless D-type flip-flop with Scan Path.

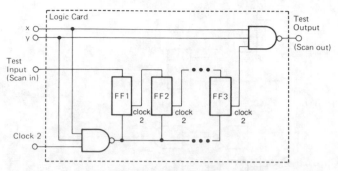

Fig. 14. Configuration of Scan Path on Card.

B. Scan Path

In 1975, a survey paper of test generation systems in Japan was presented by members of Nippon Electric Co., Ltd. [21]. In that survey paper, a technique they described as Scan Path was presented. The Scan Path technique has the same objectives as the LSSD approach which has just been described. The Scan Path technique similarities and differences to the LSSD approach will be presented.

The memory elements that are used in the Scan Path approach are shown in Fig. 13. This memory element is called a raceless D-type flip-flop with Scan Path.

In system operation, Clock 2 is at a logic value of 1 for the entire period. This, in essence, blocks the test or scan input from affecting the values in the first latch. This D-type flip-flop really contains two latches. Also, by having Clock 2 at a logic value of 1, the values in Latch 2 are not disturbed.

Clock 1 is the sole clock in system operation for this D-type flip-flop. When Clock 1 is at a value of 0, the System Data Input can be loaded into Latch 1. As long as Clock 1 is 0 for sufficient time to latch up the data, it can then turn off. As it turns off, it then will make Latch 2 sensitive to the data output of Latch 1. As long as Clock 1 is equal to a 1 so that data can be latched up into Latch 2, reliable operation will occur. This assumes that as long as the output of Latch 2 does not come around and feed the system data input to Latch 1 and change it during the time that the inputs to both Latch 1 and Latch 2 are active. The period of time that this can occur is related to the delay of the inverter block for Clock 1. A similar phenomenon will occur with Clock 2 and its associated inverter block. This race condition is the exposure to the use of only one system clock.

This points out a significant difference between the Scan Path approach and the LSSD approach. One of the basic principles of the LSSD approach is level-sensitive operation—the ability to operate the clocks in such a fashion that no races will exist. In the LSSD approach, a separate clock is required for Latch 1 from the clock that operates Latch 2.

In terms of the scanning function, the D-type flip-flop with Scan Path has its own scan input called test input. This is clocked into the $L1$ latch by Clock 2 when Clock 2 is a 0, and the results of the $L1$ latch are clocked into Latch 2 when Clock 2 is a 1. Again, this applies to master/slave operation of Latch 1 and Latch 2 with its associated race with proper attention to delays this race will not be a problem.

Another feature of the Scan Path approach is the configuration used at the logic card level. Modules on the logic card are all connected up into a serial scan path, such that for each card, there is one scan path. In addition, there are gates for selecting a particular card in a subsystem. In Fig. 14, when X and Y are both equal to 1—that is the selection mechanism—Clock 2 will then be allowed to shift data through the scan path. Any other time, Clock 2 will be blocked, and its output will be blocked. The reason for blocking the output is that a number of card outputs can then be put together; thus the blocking function will put their output to noncontrolling values, so that a particular card can have unique control of the unique test output for that system.

It has been reported by the Nippon Electric Company that they have used the Scan Path approach, plus partitioning which will be described next, for systems with 100 000 blocks or more. This was for the FLT-700 System, which is a large processor system.

The partitioning technique is one which automatically separates the combinational network into smaller subnetworks, so that the test generator can do test generation for the small subnetworks, rather than the larger networks. A partition is automatically generated by backtracing from the D-type flip-flops, through the combinational logic, until it encounters a D-type flip-flop in the backtrace (or primary input). Some care must be taken so that the partitions do not get too large.

To that end, the Nippon Electric Company approach has used a controlled D-type flip-flop to block the backtracing of certain partitions when they become too high. This is another facet of Design for Testability—that is, the introduction of extra flip-flops totally independent of function, in order to control the partitioning algorithm.

Other than the lack of the level sensitive attribute to the Scan Path approach, the technique is very similar to the LSSD approach. The introduction of the Scan Path approach was the first practical implementation of shift registers for testing which was incorporated in a total system.

C. Scan/Set Logic

A technique similar to Scan Path and LSSD, but not exactly the same, is the Scan/Set technique put forth by Sperry-Univac [31]. The basic concept of this technique is to have shift registers, as in Scan Path or in LSSD, but these shift registers are not in the data path. That is, they are not in the system data path; they are independent of all the system latches. Fig. 15 shows an example of the Scan/Set Logic, referred to as bit serial logic.

The basic concept is that the sequential network can be

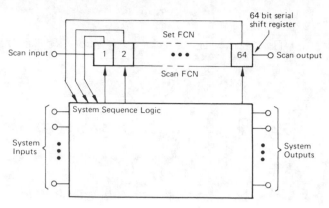

Fig. 15. Scan/Set Logic (bit-serial).

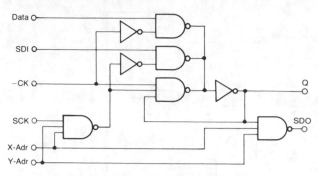

Fig. 16. Polarity-hold-type addressable latch.

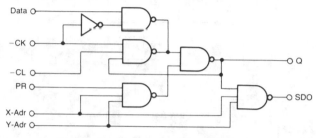

Fig. 17. Set/Reset type addressable latch.

sampled at up to 64 points. These points can be loaded into the 64-bit shift register with a single clock. Once the 64 bits are loaded, a shifting process will occur, and the data will be scanned out through the scan-out pin. In the case of the set function, the 64 bits can be funneled into the system logic, and then the appropriate clocking structure required to load data into the system latches is required in this system logic. Furthermore, the set function could also be used to control different paths to ease the testing function.

In general, this serial Scan/Set Logic would be integrated onto the same chip that contrains sequential system logic. However, some applications have been put forth where the bit serial Scan/Set Logic was off-chip, and the bit-serial Scan/Set Logic only sampled outputs or drove inputs to facilitate in-circuit testing.

Recently, Motorola has come forth with a chip which is T^2L and which has I^2L logic integrated on that same chip. This has the Scan/Set Logic bit serial shift registers built in I^2L. The T^2L portion of the chip is a gate array, and the I^2L is on the chip, whether the customer wants it or not. It is up to the customer to use the bit-serial logic if he chooses.

At this point, it should be explained that if all the latches within the system sequential network are not both scanned and set, then the test generation function is not necessarily reduced to a total combinational test generation function and fault simulation function. However, this technique will greatly reduce the task of test generation and fault simulation.

Again, the Scan/Set technique has the same objectives as Scan Path and LSSD—that is, controllability and observability. However, in terms of its implementation, it is not required that the set function set all system latches, or that the scan function scan all system latches. This design flexibility would have a reflection in the software support required to implement such a technique.

Another advantage of this technique is that the scan function can occur during system operation—that is, the sampling pulse to the 64-bit serial shift register can occur while system clocks are being applied to the system sequential logic, so that a snapshot of the sequential machine can be obtained and off-loaded without any degradation in system performance.

D. Random-Access Scan

Another technique similar to the Scan Path technique and LSSD is the Random-Access Scan technique put forth by Fujitsu [14]. This technique has the same objective as Scan Path and LSSD—that is, to have complete controllability and observability of all internal latches. Thus the test generation func-

tion can be reduced to that of combinational test generation and combinational fault simulation as well.

Random-Access Scan differs from the other two techniques in that shift registers are not employed. What is employed is an addressing scheme which allows each latch to be uniquely selected, so that it can be either controlled or observed. The mechanism for addressing is very similar to that of a Random-Access Memory, and hence, its name.

Figs. 16 and 17 show the two basic latch configurations that are required for the Random-Access Scan approach. Fig. 16 is a single latch which has added to it an extra data port which is a Scan Data In port (SDI). These data are clocked into the latch by the SCK clock. The SCK clock can only affect this latch, if both the X and Y addresses are one. Furthermore, when the X address and Y address are one, then the Scan Data Out (SDO) point can be observed. System data labeled Data in Figs. 16 and 17 are loaded into this latch by the system clock labeled CK.

The set/reset-type addressable latch in Fig. 17 does not have a scan clock to load data into the system latch. This latch is first cleared by the CL line, and the CL line is connected to other latches that are also set/reset-type addressable latches. This, then, places the output value Q to a 0 value. A preset is directed at those latches that are required to be set to a 1 for that particular test. This preset is directed by addressing each one of those latches and applying the preset pulse labeled PR. The output of the latch Q will then go to a 1. The observability mechanism for Scan Data Out is exactly the same as for the latch shown in Fig. 16.

Fig. 18 gives an overall view of the system configuration of the Random-Access Scan approach. Notice that, basically, there is a Y address, an X address, a decoder, the addressable storage elements, which are the memory elements or latches, and the sequential machine, system clocks, and CLEAR function. There is also an SDI which is the input for a given latch, an SDO which is the output data for that given latch, and a scan clock. There is also one logic gate necessary to create the preset function.

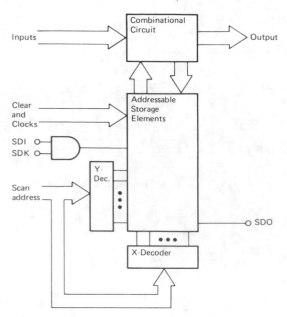

Fig. 18. Random-Access Scan network.

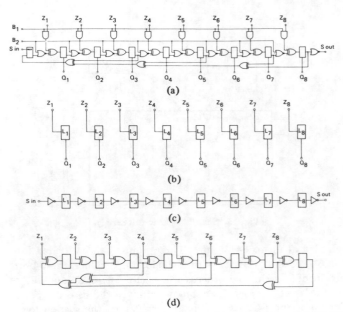

Fig. 19. BILBO and its different modes. (a) General form of BILBO register. (b) $B_1B_2 = 11$ system orientation mode. (c) $B_1B_2 = 00$ linear shift register mode. (d) $B_1B_2 = 10$ signature analysis register with m multiple inputs $(Z_1, Z_2, \cdots, Z_8)$.

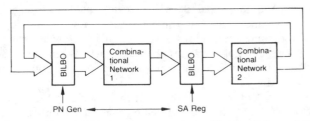

Fig. 20. Use of BILBO registers to test combinational Network 1.

The Random-Access Scan technique allows the observability and controllability of all system latches. In addition, any point in the combinational network can be observed with the addition of one gate per observation point, as well as one address in the address gate, per observation point.

While the Scan Path approach and the LSSD approach require two latches for every point which needs to be observed, the overhead for Random-Access Scan is about three to four gates per storage element. In terms of primary inputs/outputs, the overhead is between 10 and 20. This pin overhead can be diminished by using the serial scan approach for the X and Y address counter, which would lead to 6 primary inputs/outputs.

V. SELF-TESTING AND BUILT-IN TESTS

As a natural outgrowth of the Structured Design approach for "Design for Testability," Self-Tests and Built-In Tests have been getting considerably more attention. Four techniques will be discussed, which fall into this category, BILBO, Syndrome Testing, Testing by Verifying Walsh Testing Coefficients, and Autonomous Testing. Each of these techniques will be described.

A. Built-In Logic Block Observation, BILBO

A technique recently presented takes the Scan Path and LSSD concept and integrates it with the Signature Analysis concept. The end result is a technique for Built-In Logic Block Observation, BILBO [25].

Fig. 19 gives the form of an 8-bit BILBO register. The block labeled L_i ($i = 1, 2, \cdots, 8$) are the system latches. B_1 and B_2 are control values for controlling the different functions that the BILBO register can perform. S_{IN} is the scan-in input to the 8-bit register, and S_{OUT} is the scan-out for the 8-bit register. Q_i ($i = 1, 2, \cdots, 8$) are the output values for the eight system latches. Z_i ($i = 1, 2, \cdots, 8$) are the inputs from the combinational logic. The structure that this network will be embedded into will be discussed shortly.

There are three primary modes of operation for this register, as well as one secondary mode of operation for this register. The first is shown in Fig. 19(b)—that is, with B_1 and B_2 equal to 11. This is a Basic System Operation mode, in which the Z_i values are loaded into the L_i, and the outputs are available on Q_i for system operation. This would be your normal register function.

When B_1B_2 equals 00, the BILBO register takes on the form of a linear shift register, as shown in Fig. 19(c). Scan-in input to the left, through some inverters, and basically lining up the eight registers into a single scan path, until the scan-out is reached. This is similar to Scan Path and LSSD.

The third mode is when B_1B_2 equals 10. In this mode, the BILBO register takes on the attributes of a linear feedback shift register of maximal length with multiple linear inputs. This is very similar to a Signature Analysis register, except that there is more than one input. In this situation, there are eight unique inputs. Thus after a certain number of shift clocks, say, 100, there would be a unique signature left in the BILBO register for the good machine. This good machine signature could be off-loaded from the register by changing from Mode $B_1B_2 = 10$ to Mode $B_1B_2 = 00$, in which case a shift register operation would exist, and the signature then could be observed from the scan-out primary output.

The fourth function that the BILBO register can perform is B_1B_2 equal to 01, which would force a reset on the register. (This is not depicted in Fig. 19.)

The BILBO registers are used in the system operation, as shown in Fig. 20. Basically, a BILBO register with combinational logic and another BILBO register with combinational logic, as well as the output of the second combinational logic network can feed back into the input of the first BILBO regis-

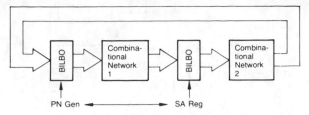

Fig. 21. Use of BILBO registers to test combinational Network 2.

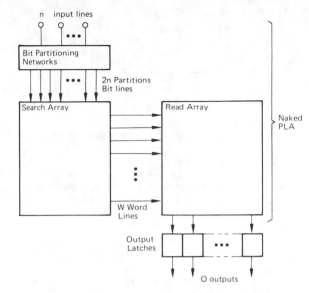

Fig. 22. PLA model.

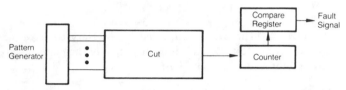

Fig. 23. Syndrome test structure.

ter. The BILBO approach takes one other fact into account, and that is that, in general, combinational logic is highly susceptible to random patterns. Thus if the inputs to the BILBO register, $Z_1, Z_2, \cdots, Z_8$, can be controlled to fixed values, such that the BILBO register is in the maximal length linear feedback shift register mode (Signature Analysis) it will output a sequence of patterns which are very close to random patterns. Thus random patterns can be generated quite readily from this register. These sequences are called Pseudo Random Patterns (PN).

If, in the first operation, this BILBO register on the left in Fig. 20 is used as the PN generator—that is, its data inputs are held to fixed values—then the output of that BILBO register will be random patterns. This will then do a reasonable test, if sufficient numbers of patterns are applied, of the Combinational Logic Network 1. The results of this test can be stored in a Signature Analysis register approach with multiple inputs to the BILBO register on the right. After a fixed number of patterns have been applied, the signature is scanned out of the BILBO register on the right for good machine compliance. If that is successfully completed, then the roles are reversed, and the BILBO register on the right will be used as a PN sequence generator; the BILBO register on the left will then be used as a Signature Analysis register with multiple inputs from Combinational Logic Network 2, see Fig. 21. In this mode, the Combinational Logic Network 2 will have random patterns applied to its inputs and its outputs stored in the BILBO register on the far left. Thus the testing of the combinational logic networks 1 and 2 can be completed at very high speeds by only applying the shift clocks, while the two BILBO registers are in the Signature Analysis mode. At the conclusion of the tests, off-loading of patterns can occur, and determination of good machine operation can be made.

This technique solves the problem of test generation and fault simulation if the combinational networks are susceptible to random patterns. There are some known networks which are not susceptible to random patterns. They are Programmable Logic Arrays (PLA's), see Fig. 22. The reason for this is that the fan-in in PLA's is too large. If an AND gate in the search array had 20 inputs, then each random pattern would have $1/2^{20}$ probability of coming up with the correct input pattern. On the other hand, random combinational logic networks with maximum fan-in of 4 can do quite well with random patterns.

The BILBO technique solves another problem and that is of test data volume. In LSSD, Scan Path, Scan/Set, or Random-Access Scan, a considerable amount of test data volume is involved with the shifting in and out. With BIBLO, if 100 patterns are run between scan-outs, the test data volume may be reduced by a factor of 100. The overhead for this technique is higher than for LSSD since about two EXCLUSIVE-OR's must be used per latch position. Also, there is more delay in the system data path (one or two gate delays). If VLSI has the huge number of logic gates available than this may be a very efficient way to use them.

B. Syndrome Testing

Recently, a technique was shown which could be used to test a network with fairly minor changes to the network. The technique is Syndrome Testing. The technique requires that all 2^n patterns be applied to the input of the network and then the number of 1's on the output be counted [115], [116].

Testing is done by comparing the number of 1's for the good machine to the number of 1's for the faulty machine. If there is a difference, the fault(s) in the faulty machine are detected (or Syndrome testable). To be more formal the Syndrome is:

Definition 1: The *Syndrome S* of a Boolean function is defined as

$$S = \frac{K}{2^n}$$

where K is the number of minterns realized by the function, and n is the number of binary input lines to the Boolean function.

Not all Boolean functions are totally Syndrome testable for all the single stuck-at-faults. Procedures are given in [115] with a minimal or near minimal number of primary inputs to make the networks Syndrome testable. In a number of "real networks" (i.e., SN74181, etc.) the numbers of extra primary inputs needed was at most one ($<$5 percent) and not more than two gates ($<$4 percent) were needed. An extension [116] to this work was published which showed a way of making a network Syndrome testable by adding extra inputs. This resulted in a somewhat longer test sequence. This is accomplished by holding some input constant while applying all 2^k inputs ($k < n$) then holding others constant and applying 2^l input patterns to l inputs. Whether the network is modified or not, the test data volume for a Syndrome testable design is extremely low. The general test setup is shown in Fig. 23.

The structure requires a pattern generator which applies all possible patterns once, a counter to count the 1's, and a com-

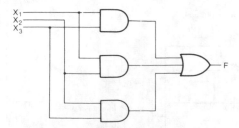

Fig. 24. Function to be tested with Walsh coefficients.

TABLE I
EXAMPLES OF WALSH FUNCTIONS AND WALSH COEFFICIENTS

$X_1 X_2 X_3$	W_2	$W_{1,3}$	F	$W_2 F$	$W_{1,3} F$	W_{ALL}	$W_{ALL} F$
0 0 0	−1	+1	0	+1	−1	+1	+1
0 0 1	−1	−1	0	+1	+1	−1	−1
0 1 0	+1	+1	0	−1	−1	−1	−1
0 1 1	+1	−1	1	+1	−1	+1	−1
1 0 0	−1	−1	0	+1	+1	−1	−1
1 0 1	−1	+1	1	−1	+1	+1	−1
1 1 0	+1	−1	1	+1	−1	+1	−1
1 1 1	+1	+1	1	+1	+1	−1	+1

$$C_{ALL} = 4$$

pare network. The overhead quoted is necessary to make the CUT Syndrome testable and does not include the pattern generator, counter, or compare register.

C. Testing by Verifying Walsh Coefficients

A technique which is similar to Syndrome Testing, in that it requires all possible input patterns be applied to the combinational network, is testing by verifying Walsh coefficients [117]. This technique only checks two of the Walsh coefficients and then makes conclusions about the network with respect to stuck-at-faults.

In order to calculate the Walsh coefficients, the logical value 0 (1) is associated with the arithmetic value −1(+1). There are 2^n Walsh functions. W_0 is defined to be 1, W_i is derived from all possible (arithmetic) products of the subject of independent input variables selected for that Walsh function. Table I shows the Walsh function for W_2, $W_{1,3}$, then $W_2 F$, $W_{1,3} F$, finally W_{all} and $W_{all} F$. These values are calculated for the network in Fig. 24. If the values are summed for $W_{all} F$, the Walsh coefficient C_{all} is calculated. The Walsh coefficient C_0 is just $W_0 F$ summed. This is equivalent to the Syndrome in magnitude times 2^n. If $C_{all} \neq 0$ then all stuck-at-faults on primary inputs will be detected by measuring C_{all}. If the fault is present $C_{all} = 0$. If the network has $C_{all} = 0$ it can be easily modified such that $C_{all} \neq 0$. If the network has reconvergent fan-out then further checks need to be made (the number of inverters in each path has a certain property); see [117]. If these are successful, then by checking C_{all} and C_0, all the single stuck-at-faults can be detected. Some design constraints maybe needed to make sure that the network is testable by measuring C_{all} and C_0. Fig. 25 shows the network needed to determine C_{all} and C_0. The value p is the parity of the driving counter and the response counter is an up/down counter. Note, two passes must be made of the driving counter, one for C_{all} and one for C_0.

D. Autonomous Testing

The fourth technique which will be discussed in the area of self-test/built-in-test is Autonomous Testing [118]. Autonomous Testing like Syndrome Testing and testing Walsh coefficients requires all possible patterns be applied to the network inputs. However, with Autonomous Testing the outputs of

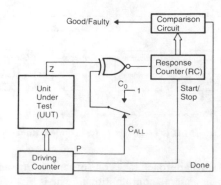

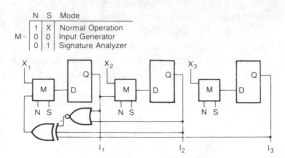

Fig. 25. Tester for veryfying C_0 and C_{all} Walsh coefficients.

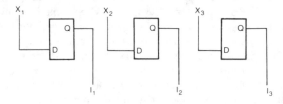

Fig. 26. Reconfigurable 3-bit LFSR module.

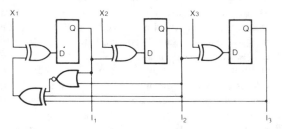

N = 1: Normal Operation

Fig. 27. Reconfigurable 3-bit LFSR module.

N = 0, S = 1: Signature Analyzer

Fig. 28. Reconfigurable 3-bit LFSR module.

the network must be checked for each pattern against the value for the good machine. The results is that irrespective of the fault model Autonomous Testing will detect the faults (assuming the faulty machine does not turn into a sequential machine from a combinational machine). In order to help the network apply its own patterns and accumulate the results of the tests rather than observing every pattern for 2^n input patterns, a structure similar to BILBO register is used. This register has some unique attributes and is shown in Figs. 26–29. If a combinational network has 100 inputs, the network must be modified such that the subnetwork can be verified and, thus, the whole network will be tested.

Two approaches to partitioning are presented in the paper "Design for Autonomous Test" [118]. The first is to use

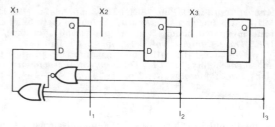

N = 0, S = 0: Input Generator

Fig. 29. Reconfigurable 3-bit LFSR module.

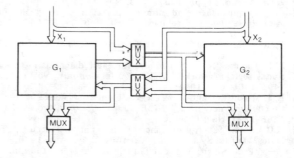

Fig. 30. Autonomous Testing—general network.

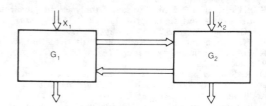

Fig. 31. Autonomous Testing—functional mode.

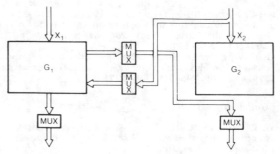

Fig. 32. Autonomous Testing—configuration to test network G_1.

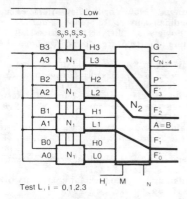

Test L_i, i = 0,1,2,3

Fig. 33. Autonomous Testing with sensitized partitioning.

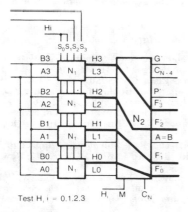

Test H_i, i = 0,1,2,3

Fig. 34. Autonomous Testing with sensitized partitioning.

multiplexers to separate the network and the second is a Sensitized Partitioning to separate the network. Fig. 30 shows the general network with multiplexers, Fig. 31 shows the network in functional mode, and Fig. 32 shows the network in a mode to test subnetwork G_1. This approach could involve a significant gate overhead to implement in some networks. Thus the Sensitized Partitioning approach is put forth. For example, the 74181 ALU/Function Generator is partitioned using the Sensitized Partitioning. By inspecting the network, two types of subnetworks can be partitioned out, four subnetworks N_1, one subnetwork N_2 (Figs. 33 and 34). By further inspection, all the L_i outputs of network N_1 can be tested by holding $S_2 = S_3 = $ low. Further, all the H_i outputs of network N_1 can be tested by holding $S_0 = S_1 = $ high, since sensitized paths exist through the subnetwork N_2. Thus far fewer than 2^n input patterns can be applied to the network to test it.

VI. CONCLUSION

The area of Design for Testability is becoming a popular topic by necessity. Those users of LSI/VLSI which do not have their own captive IC facilities are at the mercy of the vendors for information. And, until the vendor information is drastically changed, the Ad Hoc approaches to design for testability will be the only answer.

In that segment of the industry which can afford to implement the Structured Design for Testability approach, there is considerable hope of getting quality test patterns at a very modest cost. Furthermore, many innovative techniques are appearing in the Structured Approach and probably will continue as we meander through VLSI and into more dense technologies.

There is a new opportunity arriving in the form of gate arrays that allow low volume users access to VLSI technology. If they choose, structured design disciplines can be utilized. Perhaps "Silicon Foundries" of the future will offer a combined package of structured, testable modules and support software to automatically provide the user with finished parts AND tests.

ACKNOWLEDGMENT

The authors wish to thank D. J. Brown for his helpful comments and suggestions. The assistance of Ms. B. Fletcher, Ms. C. Mendoza, Ms. L. Clark, Ms. J. Allen, and J. Smith in preparing this manuscript for publication was invaluable.

REFERENCES

General References and Surveys

[1] M. A. Breuer, Ed., *Diagnosis and Reliable Design of Digital Systems.* Rockville, MD: Computer Science Press, 1976.
[2] H. Y. Chang, E. G. Manning, and G. Metze, *Fault Diagnosis of*

Digital Systems. New York: Wiley-Interscience, 1970.

[3] A. D. Friedman and P. R. Menon, *Fault Detection in Digital Circuits.* Englewood Cliffs, NJ: Prentice-Hall, 1971.

[4] F. C. Hennie, *Finite State Models for Logical Machines.* New York: Wiley, 1968.

[5] P. G. Kovijanic, in "A new look at test generation and verification," in *Proc. 14th Design Automation Conf.*, IEEE Pub. 77CH1216-1C, pp. 58–63, June 1977.

[6] E. I. Muehldorf, "Designing LSI logic for testability," in *Dig. Papers, 1976 Ann. Semiconductor Test Symp.*, IEEE Pub. 76CH1179-1C, pp. 45–49, Oct. 1976.

[7] E. I. Muehldorf and A. D. Savkar, "LSI logic testing—An overview," *IEEE Trans. Comput.*, vol. C-30, no. 1, pp. 1–17, Jan. 1981.

[8] W. W. Peterson and E. J. Weldon, *Error Correcting Codes.* Cambridge, MA: MIT Press, 1972.

[9] A. K. Susskind, "Diagnostics for logic networks," *IEEE Spectrum*, vol. 10, pp. 40–47, Oct. 1973.

[10] T. W. Williams and K. P. Parker, "Testing logic networks and design for testability," *Computer*, pp. 9–21, Oct. 1979.

[11] IEEE, Inc., *IEEE Standard Dictionary of Electrical and Electronics Terms.* New York: Wiley-Interscience, 1972.

Designing for Testability

[12] "A designer's guide to signature analysis," Hewlett-Packard Application Note 222, Hewlett Packard, 5301 Stevens Creek Blvd., Santa Clara, CA 95050.

[13] S. B. Akers, "Partitioning for testability," *J. Des. Automat. Fault-Tolerant Comput.*, vol. 1, no. 2, Feb. 1977.

[14] H. Ando, "Testing VLSI with random access scan," in *Dig. Papers Compcon 80*, IEEE Pub. 80CH1491-OC, pp. 50–52, Feb. 1980.

[15] P. Bottorff and E. I. Muehldorf, "Impact of LSI on complex digital circuit board testing," *Electro 77*, New York, NY, Apr. 1977.

[16] S. DasGupta, E. B. Eichelberger, and T. W. Williams, "LSI chip design for testability," in *Dig. Tech. Papers, 1978 Int. Solid-State Circuits Conf.* (San Francisco, CA, Feb. 1978), pp. 216–217.

[17] "Designing digital circuits for testability," Hewlett-Packard Application Note 210-4, Hewlett Packard, Loveland, CO 80537.

[18] E. B. Eichelberger and T. W. Williams, "A logic design structure for LSI testability," *J. Des. Automat. Fault-Tolerant Comput.*, vol. 2, no. 2, pp. 165–178, May 1978.

[19] —, "A logic design structure for LSI testing," in *Proc. 14th Design Automation Conf.*, IEEE Pub. 77CH1216-1C, pp. 462–468, June 1977.

[20] E. B. Eichelberger, E. J. Muehldorf, R. G. Walter, and T. W. Williams, "A logic design structure for testing internal arrays," in *Proc. 3rd USA-Japan Computer Conf.* (San Francisco, CA, Oct. 1978), pp. 266–272.

[21] S. Funatsu, N. Wakatsuki, and T. Arima, "Test generation systems in Japan," in *Proc. 12th Design Automation Symp.*, pp. 114–122, June 1975.

[22] H. C. Godoy, G. B. Franklin, and P. S. Bottoroff, "Automatic checking of logic design structure for compliance with testability groundrules," in *Proc. 14th Design Automation Conf.*, IEEE Pub. 77CH1216-1C, pp. 469–478, June 1977.

[23] J. P. Hayes, "On modifying logic networks to improve their diagnosability," *IEEE Trans. Comput.*, vol. C-23, pp. 56–62, Jan. 1974.

[24] J. P. Hayes and A. D. Friedman, "Test point placement to simplify fault detection," in *FTC-3, Dig. Papers, 1973 Symp. on Fault-Tolerant Computing*, pp. 73–78, June 1973.

[25] B. Koenemann, J. Mucha, and G. Zwiehoff, "Built-in logic block observation techniques," in *Dig. Papers, 1979 Test Conf.*, IEEE Pub. 79CH1509-9C, pp. 37–41, Oct. 1979.

[26] M. D. Lippman and E. S. Donn, "Design forethought promotes easier testing of microcomputer boards," *Electronics*, pp. 113–119, Jan. 18, 1979.

[27] H. J. Nadig, "Signature analysis-concepts, examples, and guidelines," *Hewlett-Packard J.*, pp. 15–21, May 1977.

[28] M. Neil and R. Goodner, "Designing a serviceman's needs into microprocessor based systems," *Electronics*, pp. 122–128, Mar. 1, 1979.

[29] S. M. Reddy, "Easily testable realization for logic functions," *IEETC Trans. Comput.*, vol. C-21, pp. 1183–1188, Nov. 1972.

[30] K. K. Saliya and S. M. Reddy, "On minimally testable logic networks," *IEEE Trans. Comput.*, vol. C-23, pp. 1204–1207, Nov. 1974.

[31] J. H. Stewart, "Future testing of large LSI circuit cards," in *Dig. Papers 1977 Semiconductor Test Symp.*, IEEE Pub. 77CH1261-7C, pp. 6–17, Oct. 1977.

[32] A. Toth and C. Holt, "Automated data base-driven digital testing," *Computer*, pp. 13–19, Jan. 1974.

[33] E. White, "Signature analysis, enhancing the serviceability of microprocessor-based industrial products," in *Proc. 4th IECI*

Annual Conf., IEEE Pub. 78CH1312-8, pp. 68–76, Mar. 1978.

[34] M.J.Y. Williams and J. B. Angell, "Enhancing testability of large scale integrated circuits via test points and additional logic," *IEEE Trans. Comput.*, vol. C-22, pp. 46–60, Jan. 1973.

[35] T. W. Williams, "Utilization of a structured design for reliability and serviceability," in *Dig., Government Microcircuits Applications Conf.* (Monterey, CA, Nov. 1978), pp. 441–444.

Faults and Fault Modeling

[36] R. Boute and E. J. McCluskey, "Fault equivalence in sequential machines," in *Proc. Symp. on Computers and Automata* (Polytech. Inst. Brooklyn, Apr. 13–15, 1971), pp. 483–507.

[37] R. T. Boute, "Optimal and near-optimal checking experiments for output faults in sequential machines," *IEEE Trans. Comput.*, vol. C-23, no. 11, pp. 1207–1213, Nov. 1974.

[38] —, "Equivalence and dominance relations between output faults in sequential machines," Tech. Rep. 38, SU-SEL-72-052, Stanford Univ., Stanford, CA, Nov. 1972.

[39] F.J.O. Dias, "Fault masking in combinational logic circuits," *IEEE Trans. Comput.*, vol. C-24, pp. 476–482, May 1975.

[40] J. P. Hayes, "A NAND model for fault diagnosis in combinational logic networks," *IEEE Trans. Comput.*, vol. C-20, pp. 1496–1506, Dec. 1971.

[41] E. J. McCluskey and F. W. Clegg, "Fault equivalence in combinational logic networks," *IEEE Trans. Comput.*, vol. C-20, pp. 1286–1293, Nov. 1971.

[42] K.C.Y. Mei, "Fault dominance in combinational circuits," Tech. Note 2, Digital Systems Lab., Stanford Univ., Aug. 1970.

[43] —, "Bridging and stuck-at faults," *IEEE Trans. Comput.*, vol. C-23, no. 7, pp. 720–727, July 1974.

[44] R. C. Ogus, "The probability of a correct output from a combinational circuit," *IEEE Trans. Comput.*, vol. C-24, no. 5, pp. 534–544, May 1975.

[45] K. P. Parker and E. J. McCluskey, "Analysis of logic circuits with faults using input signal probabilities," *IEEE Trans. Comput.*, vol. C-24, no. 5, pp. 573–578, May 1975.

[46] K. K. Saliya and S. M. Reddy, "Fault detecting test sets for Reed-Muller canonic networks," *IEEE Trans. Comput.*, pp. 995–998, Oct. 1975.

[47] D. R. Schertz and G. Metze, "A new representation for faults in combinational digital circuits," *IEEE Trans. Comput.*, vol. C-21, no. 8, pp. 858–866, Aug. 1972.

[48] J. J. Shedletsky and E. J. McCluskey, "The error latency of a fault in a sequential digital circuit," *IEEE Trans. Comput.*, vol. C-25, no. 6, pp. 655–659, June 1976.

[49] —, "The error latency of a fault in a combinational digital circuit," in *FTCS-5, Dig. Papers, 5th Int. Symp. on Fault Tolerant Computing* (Paris, France, June 1975), pp. 210–214.

[50] K. To, "Fault folding for irredundant and redundant combinational circuits," *IEEE Trans. Comput.*, vol. C-22, no. 11, pp. 1008–1015, Nov. 1973.

[51] D. T. Wang, "Properties of faults and criticalities of values under tests for combinational networks," *IEEE Trans. Comput.*, vol. C-24, no. 7, pp. 746–750, July 1975.

Testing and Fault Location

[52] R. P. Batni and C. R. Kime, "A module level testing approach for combinational networks," *IEEE Trans. Comput.*, vol. C-25, no. 6, pp. 594–604, June 1976.

[53] S. Bisset, "Exhaustive testing of microprocessors and related devices: A practical solution," in *Dig. Papers, 1977 Semiconductor Test Symp.*, pp. 38–41, Oct. 1977.

[54] R. J. Czepiel, S. H. Foreman, and R. J. Prilik, "System for logic, parametric and analog testing," in *Dig. Papers, 1976 Semiconductor Test Symp.*, pp. 54–69, Oct. 1976.

[55] R. A. Frohwerk, "Signature analysis: A new digital field service method," *Hewlett-Packard J.*, pp. 2–8, May 1977.

[56] B. A. Grimmer, "Test techniques for circuit boards containing large memories and microprocessors," in *Dig. Papers, 1976 Semiconductor Test Symp.*, pp. 16–21, Oct. 1976.

[57] W. A. Groves, "Rapid digital fault isolation with FASTRACE," *Hewlett-Packard J.*, pp. 8–13, Mar. 1979.

[58] J. P. Hayes, "Rapid count testing for combinational logic circuits," *IEEE Trans. Comput.*, vol. C-25, no. 6, pp. 613–620, June 1976.

[59] —, "Detection of pattern sensitive faults in random access memories," *IEEE Trans. Comput.*, vol. C-24, no. 2, Feb. 1975, pp. 150–160.

[60] —, "Testing logic circuits by transition counting," in *FTC-5, Dig. Papers, 5th Int. Symp. on Fault Tolerant Computing* (Paris, France, June 1975), pp. 215–219.

[61] J. T. Healy, "Economic realities of testing microprocessors," in *Dig. Papers, 1977 Semiconductor Test Symp.*, pp. 47–52, Oct. 1977.

[62] E. C. Lee, "A simple concept in microprocessor testing," in *Dig.*

Papers, 1976 Semiconductor Test Symp., IEEE Pub. 76CH1179-1C, pp. 13–15, Oct. 1976.

[63] J. Losq, "Referenceless random testing," in *FTCS-6, Dig. Papers, 6th Int. Symp. on Fault-Tolerant Computing* (Pittsburgh, PA, June 21–23, 1976), pp. 81–86.

[64] S. Palmquist and D. Chapman, "Expanding the boundaries of LSI testing with an advanced pattern controller," in *Dig. Papers, 1976 Semicondctor Test Symp.*, pp. 70–75, Oct. 1976.

[65] K. P. Parker, "Compact testing: Testing with compressed data," in *FTCS-6, Dig. Papers, 6th Int. Symp. on Fault-Tolerant Computing* (Pittsburgh, PA, June 21–23, 1976).

[66] J. J. Shedletsky, "A rationale for the random testing of combinational digital circuits," in *Dig. Papers, Compcon 75 Fall Meet.* (Washington, DC, Sept. 9–11, 1975), pp. 5–9.

[67] V. P. Strini, "Fault location in a semiconductor random access memory unit," *IEEE Trans. Comput.*, vol. C-27, no. 4, pp. 379–385, Apr. 1978.

[68] C. W. Weller, in "An engineering approach to IC test system maintenance," in *Dig. Papers, 1977 Semiconductor Test Symp.*, pp. 144–145, Oct. 1977.

Testability Measures

[69] W. J. Dejka, "Measure of testability in device and system design," in *Proc. 20th Midwest Symp. Circuits Syst.*, pp. 39–52, Aug. 1977.

[70] L. H. Goldstein, "Controllability/observability analysis of digital circuits," *IEEE Trans. Circuits Syst.*, vol. CAS-26, no. 9, pp. 685–693, Sept. 1979.

[71] W. L. Keiner and R. P. West, "Testability measures," presented at AUTOTESTCON '77, Nov. 1977.

[72] P. G. Kovijanic, "testability analysis," in *Dig. Papers, 1979 Test Conf.*, IEEE Pub. 79CH1509-9C, pp. 310–316, Oct. 1979.

[73] J. E. Stephenson and J. Grason, "A testability measure for register transfer level digital circuits," in *Proc. 6th Fault Tolerant Computing Symp.*, pp. 101–107, June 1976.

Test Generation

[74] V. Agrawal and P. Agrawal, "An automatic test generation system for ILLIAC IV logic boards," *IEEE Trans. Comput.*, vol. C-C-21, no. 9, pp. 1015–1017, Sept. 1972.

[75] D. B. Armstrong, "On finding a nearly minimal set of fault detection tests for combinational logic nets," *IEEE Trans. Electron. Comput.*, vol. EC-15, no. 1, pp. 66–73, Feb. 1966.

[76] R. Betancourt, "Derivation of minimum test sets for unate logical circuits," *IEEE Trans. Comput.*, vol. C-20, no. 11, pp. 1264–1269, Nov. 1973.

[77] D. C. Bossen and S. J. Hong, "Cause and effect analysis for multiple fault detection in combinational networks," *IEEE Trans. Comput.*, vol. C-20, no. 11, pp. 1252–1257, Nov. 1971.

[78] P. S. Bottorff *et al.*, "Test generation for large networks," in *Proc. 14th Design Automation Conf.*, IEEE Pub. 77CH1216-1C, pp. 479–485, June 1977.

[79] R. D. Edlred, "Test routines based on symbolic logic statements," *J. Assoc. Comput. Mach.*, vol. 6, no. 1, pp. 33–36, 1959.

[80] P. Goel, "Test generation costs analysis and projections," presented at the 17th Design Automation Conf., Minneapolis, MN, 1980.

[81] E. P. Hsieh *et al.*, "Delay test generation," in *Proc. 14th Design Automation Conf.*, IEEE Pub. 77CH1216-1C, pp. 486–491, June 1977.

[82] C. T. Ku and G. M. Masson, "The Boolean difference and multiple fault analysis," *IEEE Trans. Comput.*, vol. C-24, no. 7, pp. 691–695, July 1975.

[83] E. I. Muehldorf, "Test pattern generation as a part of the total design process," in *LSI and Boards: Dig. Papers, 1978 Ann. Semiconductor Test Symp.*, pp. 4–7, Oct. 1978.

[84] E. I. Muehldorf and T. W. Williams, "Optimized stuck fault test patterns for PLA macros," in *Dig. Papers, 1977 Semiconductor Test Symp.*, IEEE Pub. 77CH1216-7C, pp. 89–101, Oct. 1977.

[85] M. R. Page, "Generation of diagnostic tests using prime implicants," Coordinated Science Lab. Rep. R-414, University of Illinois, Urbana, May 1969.

[86] S. G. Papaioannou, "Optimal test generation in combinational networks by pseudo Boolean programming," *IEEE Trans. Comput.*, vol. C-26, no. 6, pp. 553–560, June 1977.

[87] K. P. Parker, "Adaptive random test generation," *J. Des. Automat. Fault Tolerant Comput.*, vol. 1, no. 1, pp. 62–83, Oct. 1976.

[88] ——, "Probabilistic test generation," Tech. Note 18, Digital Systems Laboratory, Stanford University, Stanford, CA, Jan. 1973.

[89] J. F. Poage and E. J. McCluskey, "Derivation of optimum tests for sequential machines," in *Proc. 5th Ann. Symp. on Switching Circuit Theory and Logic Design*, pp. 95–110, 1964.

[90] ——, "Derivation of optimum tests to detect faults in combinational circuits," in *Mathematical Theory of Automation*. New York: Polytechnic Press, 1963.

[91] G. R. Putzolu and J. P. Roth, "A heuristic algorithm for testing of asynchronous circuits," *IEEE Trans. Comput.*, vol. C-20, no. 6, pp. 639–647, June 1971.

[92] J. P. Roth, W. G. Bouricius, and P. R. Schneider, "Programmed algorithms to compute tests to detect and distinguish between failures in logic circuits," *IEEE Trans. Electron. Comput.*, vol. EC-16, pp. 567–580, Oct. 1967.

[93] J. P. Roth, "Diagnosis of automata failures: A calculus and a method," *IBM J. Res. Devel.*, no. 10, pp. 278–281, Oct. 1966.

[94] P. R. Schneider, "On the necessity to examine D-chairs in diagnostic test generation—An example," *IBM J. Res. Develop.*, no. 11, p. 114, Nov. 1967.

[95] H. D. Schnurmann, E. Lindbloom, R. G. Carpenter, "The weighted random test pattern generation," *IEEE Trans. Comput.*, vol. C-24, no. 7, pp. 695–700, July 1975.

[96] E. F. Sellers, M. Y. Hsiao, and L. W. Bearnson, "Analyzing errors with the Boolean difference," *IEEE Trans. Comput.*, vol. C-17, no. 7, pp. 676–683, July 1968.

[97] D. T. Wang, "An algorithm for the detection of tests sets for combinational logic networks," *IEEE Trans. Comput.*, vol. C-25, no. 7, pp. 742–746, July 1975.

[98] T. W. Williams and E. E. Eichelberger, "Random patterns within a structured sequential logic design," in *Dig. Papers, 1977 Semiconductor Test Symp.*, IEEE Pub. 77CH1261-7C, pp. 19–27, Oct. 1977.

[99] S. S. Yau and S. C. Yang, "Multiple fault detection for combinational logic circuits," *IEEE Trans. Comput.*, vol. C-24, no. 5, pp. 233–242, May 1975.

Simulation

[100] D. B. Armstrong, "A deductive method for simulating faults in logic circuits," *IEEE Trans. Comput.*, vol. C-22, no. 5, pp. 464–471, May 1972.

[101] M. A. Breuer, "Functional partitioning and simulation of digital circuits," *IEEE Trans. Comput.*, vol. C-19, no. 11, pp. 1038–1046, Nov. 1970.

[102] H. Y. P. Chiang *et al.*, "Comparison of parallel and deductive fault simulation," *IEEE Trans. Comput.*, vol. C-23, no. 11, pp. 1132–1138, Nov. 1974.

[103] E. B. Eichelberger, "Hazard detection in combinational and sequential switching circuits," *IBM J. Res. Devel.*, Mar. 1965.

[104] E. Manning and H. Y. Chang, "Functional technique for efficient digital fault simulation," in *IEEE Int. Conv. Dig.*, p. 194, 1968.

[105] K. P. Parker, "Software simulator speeds digital board test generation," *Hewlett-Packard J.*, pp. 13–19, Mar. 1979.

[106] S. Seshu, "On an improved diagnosis program," *IEEE Trans. Electron. Comput.*, vol. EC-12, no. 1, pp. 76–79, Feb. 1965.

[107] S. Seshu and D. N. Freeman, "The diagnosis of asynchronous sequential switching systems," *IRE Trans, Electron. Compat.*, vol. EC-11, no. 8, pp. 459–465, Aug. 1962.

[108] T. M. Storey and J. W. Barry, "Delay test simulation," in *Proc. 14th Design Automation Conf.*, IEEE Pub. 77CH1216-1C, pp. 491–494, June 1977.

[109] S. A. Szygenda and E. W. Thompson, "Modeling and digital simulation for design verification diagnosis," *IEEE Trans. Comput.*, vol. C-25, no. 12, pp. 1242–1253, Dec. 1976.

[110] S. A. Szygenda, "TEGAS2—Anatomy of a general purpose test generation and simulation system for digital logic," in *Proc. 9th Design Automation Workshop*, pp. 116–127, 1972.

[111] S. A. Szygenda, D. M. Rouse, and E. W. Thompson, "A model for implementation of a universal time delay simulation for large digital networks," in *AFIPS Conf. Proc.*, vol. 36, pp. 207–216, 1970.

[112] E. G. Ulrich and T. Baker, "Concurrent simulation of nearly identical digital networks," *Computer*, vol. 7, no. 4, pp. 39–44, Apr. 1974.

[113] ——, "The concurrent simulation of nearly identical digital networks," in *Proc. 10th Design Automation Workshop*, pp. 145–150, June 1973.

[114] E. G. Ulrich, T. Baker, and L. R. Williams, "Fault test analysis techniques based on simulation," in *Proc. 9th Design Automation Workshop*, pp. 111–115, 1972.

[115] J. Savir, "Syndrome—Testable design of combinational circuits," *IEEE Trans. Comput.*, vol. C-29, pp. 442–451, June 1980 (corrections: Nov. 1980).

[116] ——, "Syndrome—Testing of 'syndrome-untestable' combinational circuits," *IEEE Trans. Comput.*, vol. C-30, pp. 606–608, Aug. 1981.

[117] A. K. Susskind, "Testing by verifying Walsh coefficients," in *Proc. 11th Ann. Symp. on Fault-Tolerant Computing* (Portland, MA), pp. 206–208, June 1981.

[118] E. J. McCluskey and S. Bozorgui-Nesbat, "Design for autonomous test," *IEEE Trans. Comput.*, vol. C-30, pp. 866–875, Nov. 1981.

AUTOMATION IN DESIGN FOR TESTABILITY

V. D. Agrawal
S. K. Jain
D. M. Singer

AT&T Bell Laboratories
Murray Hill, New Jersey 07974

Invited Paper

ABSTRACT — Through a well-defined set of design rules, the scan approach allows complete automation in chip testability. This paper describes a design automation system, TITUS, which automatically checks the circuit for design rules, implements the testability hardware, and generates tests. Special features in this CAD system, provided to support custom designs, include handling of on-chip memory, chip layout techniques to optimize hardware overhead and performance penalties, and test generation in the presence of tristate devices, buses and bidirectional devices.

INTRODUCTION

Design for testability means designing in such a way that testing can be guaranteed. The methods of implementing testability include serial scan, random scan, and built-in test [1]. All of these require extra hardware over that needed to accomplish the given function of the circuit. In our design automation system we have used serial scan because it requires minimum added hardware and allows the use of the conventional test technology.

To best take advantage of automation, the design process is divided in two stages. In the first stage the circuit is designed only to implement the desired function. One requirement imposed upon the designer is a set of design rules. A computer-aided *design audit* tool guides the designer in adhering to the rules. The required circuit functions are verified through simulation. The second stage of the design is completely automatic. In this stage the testability hardware is implemented, design is reverified, and tests are generated.

This paper presents a design automation system - TITUS (Testability Implementation and Test-generation Using Scan). The TITUS system consists of programs for automatic *design audits*, scan implementation, and test vector generation. Special layout features that minimize the hardware overhead in custom polycell designs have also been developed. On-chip memory blocks are specially designed to support scan testability. MOS circuits contain tristate devices, buses, and bidirectional devices with memory states. Since the test generation in TITUS is done at combinational logic level, special logic models without memory have been developed for these devices such that the tests generated for the models remain valid for the MOS devices.

TITUS fulfills the testability needs of an integrated CAD environment that includes an MOS simulator, MOTIS [2], a fault simulator [3], a layout system, LTX2 [4], and a variety of other CAD tools [5].

SCAN DESIGN

The scan concept [6] tries to solve the general problem of test generation for sequential circuits. It takes advantage of the fact that if all the flip-flops (memory elements) can be controlled to any specific value, and if they can be observed with a straight-forward operation, then the test generation task can be reduced to that of test generation for a combinational logic network. Scan design allows the state (value) of all the internal flip-flops to be both completely controllable and observable by connecting the flip-flops into one or more shift registers when the circuit is in a test (scan) mode.

The memory elements are implemented using D-type master-slave flip-flops. Consider the master-slave latch shown in Fig. 1 where master clock (MC) and slave clock (SC) are two non-overlapping clock signals. Two modifications of this flip-flop will be discussed that allow scan design. The first modification as suggested in [6] is shown in Fig. 2. A double throw switch is inserted in the data path and the mode-switch input selects the normal or scan shift mode (Mode switch = 1 for normal mode). The same master and slave clocks are used in both scan and normal modes. Furthermore, the slave clock is generated locally from the master clock requiring only one clock signal to be routed over the whole chip. This is important in chip layouts since routing of several clocks can introduce time skews between the clock signal. Even in a single clock design special care is usually taken in routing of the clock signal. Routing of more than one clock can introduce major layout and timing problems. Hence, from layout point of view, this scheme is advantageous. However, the double throw switch introduces an additional delay of two gates (actually it is only one gate delay in MOS implementation). This gate delay can be kept to a minimum if the switch is designed within the flip-flop in the cell-based design approach. The D-type master-slave flip-flop and its equivalent flip-flop for the cell based approach is shown in Fig. 3. The actual increase in cell area was only 21%, even though increase in transistor count is 38.9%.

There is an alternative modification of the flip-flop to allow for scan design. This, as suggested in [7], is shown in Fig. 4. In the normal operation inputs MC and SC are used as master and slave clocks, respectively, while the input MCP is held low. For the scan operation, the input MCP and SC are used as active master and slave clocks, while the clock line MC is held low to detach the normal data line D. The area overhead due to the added gates is about the same as the previous case. In this scheme, there is a smaller increase in delay in the data path due to increased fanin of gates G1 and G2. However, the slave clock input SC must be a primary input. Hence, in this scheme three clock signals have to be routed over the whole chip. (In the previous scheme only two signals, clock and mode switch, need to be routed.) This will result in higher routing overhead. In addition, routing of three clock signals involves potential for skews and timing problems.

Reprinted from *IEEE Proc. Custom Integrated Circuits Conf.*, 1984, pp. 159–163.

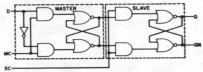

Fig. 1   D type master-slave flip-flop.

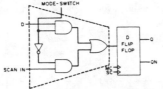

Fig. 2   Scan modifications of D flip-flop.

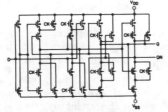

Fig. 3   (a) Single clock D flip-flop polycell (26 transistors).

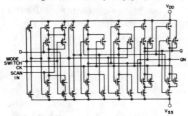

Fig. 3   (b) Scan polycell of the D flip-flop (36 transistors).

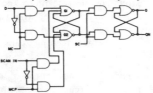

Fig. 4   Scan register flip-flop of reference [7] which requires two clocks, MC and SC, plus a scan clock, MCP.

In TITUS, the first approach to scan modification was chosen. Fig. 5 shows the modification of a synchronous circuit using the flip-flop of Fig. 3(b). A non-scan design can be converted automatically into a scan design by TITUS. A designer using TITUS can complete the design verification without worrying about testability. However, the designer must work with a set of design rules. Fig. 6 shows the flow of scan design implementation and test generation. First, the circuit is checked for adherence to design rules by an audit program. If the circuit, does not have any violation of the design rules, then scan design is automatically incorporated in the circuit. Two options exist for scan implementation. These will be discussed later. An automated scan register audit can be performed at this stage. The last step consists of automatic test generation to cover all stuck type faults in the scan circuit.

DESIGN RULES FOR SCAN CIRCUITS

In this section, a specific set of design rules will be described that will result in a design suitable for scan implementation. The rules are simple to follow and can be checked automatically by a CAD tool. These rules result in a hazard and race-free sequential design and still provide considerable flexibility to the designer.

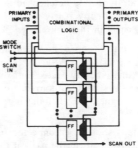

Fig. 5   Application of scan flip-flops. For the state of MODE-SWITCH signal that selects SCANIN through multiplexers, the flip-flops form a shift register. All flip-flops have common clocks (not shown). Scan overhead consists of multiplexers (shaded blocks) and the wiring shown in bold.

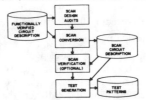

Fig. 6   Automated scan design and test generation.

Rule 1:   All internal memory elements must be implemented in D-type master-slave flip-flops.

Rule 2:   One primary input pin must be allocated for specifying the mode (scan or normal mode). Normal input and output pins can be used as scan-in and scan-out of the shift register. Hence, even for multiple scan chains, only one extra primary input is required.

Rule 3:   It must be possible to identify a set of clock primary inputs from which the clock inputs to flip-flop are controlled. In addition (a) clock inputs should not be controlled through logic that is gated by flip-flop output and/or other primary inputs,
(b) Clock input should trigger the flip-flops on the same edge, and
(c) Clock primary inputs may not feed the data input to flip-flop, either directly or through combinational logic, but may only feed the clock input to the flip-flop.

Rule 4:   The edge triggered flip-flop can have only one asynchronous input (clear/preset). Furthermore, this asynchronous input must be controllable by non-clock primary inputs.

The above design rules can be effectively audited for compliance by a CAD tool One such tool described in [8] checks logic structure for compliance to design rules through logic simulation of the behavioral model consisting of *primitive gates* (AND, OR, NAND, NOR) and a shift register latch. This type of tool could not be used efficiently for the circuit structure described at MOS gate level.

An audit algorithm was developed which does not require a logic simulator. This algorithm identifies the violation of design rules by analyzing the structure of the circuit rather than examining its function. This approach simplifies and speeds up the audit algorithms. A topological analysis of the circuit detects the situations that may violate the scan design rules. The audit algorithm initiates the path tracing from the primary clock inputs or other primary inputs of the circuit. During these path tracings, the logical functions of gates in the path are ignored. Often this results in a conservative check. The audit is implemented as a depth first search algorithm.

Very large logic designs may be checked quickly by this tool. Hence, this provides an effective early warning tool for the logic designer and is used early in the logic design phase.

A sequential circuit designed in accordance with Rules 1 through 4 will be scan testable. After the scan audits, the flip-flops in the circuit must be connected into one or more shift registers. This can be done automatically by TITUS or manually by the designer. The following rules should be followed.

Rule 5:   All flip-flops should be connected in a shift register. There *can* be logical inversions in the path of the scan chain.

Rule 6:   Each shift register should have a primary input and a primary output in the scan mode.

Rule 7:   When the mode specification line is in scan mode then the output of a flip-flop or scan-out primary output should be a function of only the preceding flip-flop output or scan-in primary input of the shift register.

Compliance with Rules 5 through Rules 7 is checked by a scan verification audit. This audit also detects the inversions in the scan chain. The technique used is similar to logic simulation (at MOS gate level) performed locally on the scan portion of the circuit.

AREA AND PERFORMANCE OVERHEAD

The polycell layout style consists of standard cells placed on grids in the rows of the layout (see Fig. 7). The polycells contain simple boolean or memory functions. One dimension (height) of the cells is fixed to allow for an arrangement in rows. The width of the polycells varies. The rows of polycells are separated by routing space (Fig. 7). The routing space consists of routing channels. Routing is greatly facilitated by CAD programs [4]. Routing is mainly done in channels between the adjacent rows of cells.

The implementation of scan design increases the area of the chip in two ways. First, the width of a scan flip-flop is larger than that of an ordinary flip-flop (height of both flip-flops remain the same). The larger flip-flop size is reflected by the increase in width of polycell row (X-direction). Secondly, the scan design requires *at least* two additional routing channels per pair of polycell rows. One of these channels is for the scan data signal and the other channel is for the mode specification line (scan or normal mode). This is shown in Fig. 8. This will increase area in the Y−direction.

The increase in area, due to larger scan flip-flops, is dependent on the fraction of chip area that is occupied by the flip-flops. The total increase in area can be theoretically calculated as follows:

Let

x = Fractional combined width of flip-flop polycells in pre-scan circuit
S_1 = Size of scan flip-flop
S_2 = Size of non-scan flip-flop.

Increase in X direction = $(\frac{S_1-S_2}{S_2}) \cdot x$

y = Fraction of routing area
k = Number of extra channels per polycell row required for scan design
T = Polycell height in terms of number of routing tracks

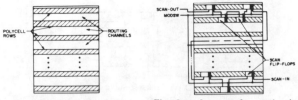

Fig. 7   Polycell design layout.   Fig. 8   Layout of scan circuit.

Increase in Y direction = $\frac{(1-y)k}{T}$

Hence,

Percent overhead = $\left[(1 + \frac{S_1-S_2}{S_2} \cdot x)(1 + \frac{(1-y)k}{T}) - 1 \right] \times 100$

Fig. 9 shows the computed area overhead versus percentage of flip-flops in the circuit. A 50-percent routing area is assumed. It should be noted that area overhead is a function of flip-flop percentage and extra channels required for scan routing.

In TITUS, the scan design can be implemented automatically in one of two different manners. The circuit designs are described hierarchically through a connectivity language [9]. Primitives in this description are the polycells which are pre-laid out blocks of commonly used functions having complexity up to 30 to 40 MOS devices. Flip-flop polycells are automatically replaced by their scan versions. In the first approach, the flip-flops are connected in the shift register such that the hierarchy in the design is preserved. The shift register connection, obtained in this manner, will be similar to that implemented by a designer. With this approach, the impact of scan conversion on a circuit's interconnection description hierarchy is minimal. However, the overhead due to scan wiring cannot be easily estimated. Furthermore the layout program does not differentiate between normal and scan wiring. The normal wiring may be stretched out due to scan wiring. This directly translates into larger routing capacitances for the normal paths.

The second approach takes advantage of the fact that the order of flip-flops in the shift register is not important. In this approach the circuit is first laid out by the layout program [4] without the shift register wiring. The order of the flip-flops in the shift register is then selected to minimize the number of extra routing channels required for scan wiring. The layout program can differentiate between normal and scan wiring at this point. Hence, normal wiring is optimized before the scan wiring is added.

The overhead due to the increased size of scan flip-flops is the same in both approaches. The overhead due to extra routing channels for scan wiring will be different in the two approaches. The second approach gives more optimized scan wiring. A CMOS polycell design was laid out in three different manners. First, the layout of the design *without* scan was made. Then, two more layouts, with the approaches discussed, were made. The actual area overheads were determined and compared with the theoretical values. The results of this experiment are presented in TABLE 1. It can be seen that the flip-flop ordering decided by the layout program is better than the ordering used for the hierarchical design.

The operating frequency for the three layouts was determined by a path analysis technique [10]. As expected, the scan design also increased the delay of normal data paths. Speed of operation, i.e., clock rate is reduced due to this extra

458

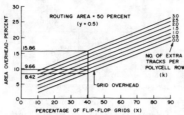

Fig. 9   Estimated overhead of scan testability.

delay. The increased path delay in MOS polycell circuits consists of two components. First, an extra gate delay is caused by the multiplexer on the data input of the flip-flop. Second, the output of a flip-flop is connected to the scan input of the next flip-flop in the scan register chain. This results in extra capacitive loading due to scan wiring at the flip-flop output. The delay overhead due to the multiplexer will be the same in both of the scan implementation approaches. But the delay due to extra capacitances can be minimized if the scan wiring is optimized through proper ordering of flip-flops in the scan register. From the TABLE 1, it can be noted that the scan implementation during layout gives better performance than scan implementation during the logic design phase.

TABLE 1 Overhead of Scan Design
Number of CMOS gates = 2,000
Percentage width of flip-flop polycells = 47.8%
Percentage of routing area = 47.1%

SCAN IMPLEMENTA-TION	PREDICTED OVERHEAD $k = 1.5$	ACTUAL AREA OVERHEAD	NORMALIZED OPERATING FREQUENCY
NONE	0	0	1.0
HIERARCHICAL	14.05%	16.93%	0.87
OPTIMIZED	14.05%	11.9%	0.91

TEST GENERATION

Figure 10 shows the complete test generation system. The circuit is described using a connectivity language [9]. In MOS environment, the circuit is described using MOS gates. Fig. 11 shows a CMOS gate and its equivalent logic gate model. Efficient algorithms [11] are available for test generation (for stuck-at-faults) for combinational networks described with primitive logic gates (AND, OR, NAND, NOR and INVERT). In order to use these algorithms the circuits described with MOS gates should be modeled using only primitive logic gates. Thus, the first step in test generation consists of obtaining the circuit's logic gate model. This is achieved by using a logic gate library for MOS gates. The MOS circuits often contain tristate and bidirectional buffers at input/output, and internal buses using tristate bus drivers. Special logic gate models are substituted for these gates for combinational test generation. These models will be discussed later in this section. The logic model preparation step also prepares a list of stuck-at-faults for test generation. It translates a fault of a CMOS gate to an equivalent fault on the logic gate. Fig. 11 shows an example of such fault translation. The fault "input A stuck-at-1 in gate Q" gets translated to the fault "input A stuck-at-1 in AND gate AB."

The next step in test generation consists of removing flip-flops to obtain the remaining combinational network. The scan flip-flop consists of a multiplexer and a D flip-flop. The

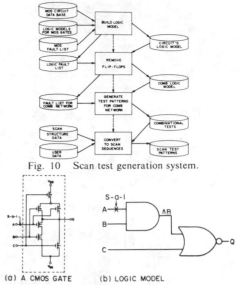

Fig. 10   Scan test generation system.

Fig. 11   Logic model for AND-OR-INVERT polycell.

multiplexer is retained as part of combinational network. Fig. 12 shows the flip-flop removal procedure. The hatched gates are removed and the inputs of the flip-flops become primary outputs and the outputs of the flip-flops become primary inputs of combinational network. This step produces a circuit logic model for combinational test generation. In this manner, any existing combinational test generator can be used for test generation. The faults related to the flip-flops are removed from the fault list for combinational test generation.

An automatic test generator is used for generation of test patterns for stuck-at-faults in the combinational network. The test generator is based on an algorithm which is similar to the D-algorithm [11]. "Easy" faults are first detected by a few random vectors. The logic test generation models for tristate and bidirectional buffers are shown in Fig. 13 and 14. These models will appropriately propagate the faults through the data input A and set the value of control input C. The faults through the control input will not be properly propagated in this model. The faults on the control input often result in sequential behavior at the output and therefore cannot be properly represented by using primitive logic gates. Usually, the control input is a primary input and does not have combinational logic in the path. In the model the input UNKNOWN is tied to a common primary input. This is a specially added primary input signal which remains fixed to *UNKNOWN(X)* value that can not be changed by the test generator.

Figure 15 shows the test generation model for a tristate bus. Here also the faults are propagated properly through data inputs D_1, D_2 and D_3, while fault propagation through control inputs C_1, C_2 and C_3 might not be right.

In general, a test pattern for a fault may contain *don't care* states. For a combinational circuit, two patterns can be merged into one as long as the corresponding bits are identical or align with a don't care. The test patterns obtained are compacted, so as to obtain a smaller set of patterns. The last step consists of translating the test patterns for combinational network into scan tests patterns for the sequential circuit. The combinational network inputs corresponding to flip-flop outputs are converted into scan sequences. This step also includes the automatic generation of test patterns for the shift registers to verify the

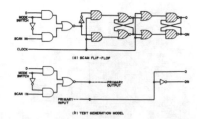

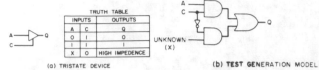

Fig. 12 Test generation model of scan flip-flop.

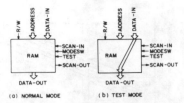

Fig. 16 Memory for scan design.

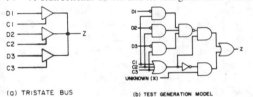

Fig. 13 A tristate device and its test generation model.

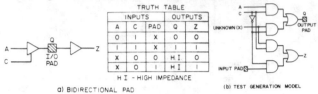

Fig. 14 A bidirectional device and its test generation model.

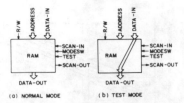

Fig. 15 A tristate bus and its test generation model.

correct shifting operation. The test consists of a shift test in which a 010101 pattern is shifted through every flip-flop of the scan register. Analysis has shown that this test pattern is sufficient to detect faults in the shift path and in the master-slave flip-flops.

SPECIAL FEATURES

There are several special situations in circuit designs which need careful handling. Some chips contain RAM and / or ROM. New techniques were developed for implementing scan design around the memory blocks. Figure 16 shows the block diagram of a RAM. In order to incorporate scan design around the RAM, the data, address and read-write (R/W) inputs need to be observed and data outputs need to be controlled. If all these signals are latched into master-slave flip-flops, then there is no problem in incorporating scan design, as these can be put in a shift register. However, usually the data outputs are not latched into master-slave flip-flops. Fig. 16 shows a modification which makes data-out controllable. In test mode the data inputs are shorted to data outputs. The test mode signal of memory, shown as TEST in Fig. 16, is a separate signal from the scan mode signal of the chip. The memory itself is not tested in this mode and only the logic around the memory is tested. Often a different strategy is adopted for testing RAM/ROM. Designer makes sure that memory inputs are controllable and outputs are observable so that memory tests can be applied directly. This is necessary because memory tests are usually long and their application through a scan register might be inefficient.

A circuit may contain more than one clock. If two flip-flops are driven by different clocks, then they should be put in separate scan chains. The scan implementation program automatically creates separate shift registers for separately clocked flip-flops.

Some circuits contain functional shift registers as part of chip architecture. Scan design take advantage of these preconnected shift registers and adds scan only to the first flip-flop of the shift register. This results in very little overhead.

CONCLUSION

Through the use of the TITUS system, testability implementation and test generation have been made automatic. The designer can utilize more of the design time for circuit optimization and functional verification. A 100-percent fault coverage is guaranteed in almost all cases. This approach produces not only a high quality design but also a high quality product.

REFERENCES

[1] T. W. Williams and K. P. Parker, "Design for testability — A Survey," *IEEE Trans. Comput.* Vol. C-31, pp. 2-15, January 1982.

[2] V. D. Agrawal, A. K. Bose, P. Kozak, N. H. Nham, and E. Pacas-Skewes, "Mixed-mode simulation in the MOTIS system," *J. Digital Systems*, Vol. V. pp. 383-400, Winter 1981.

[3] A. K. Bose, P. Kozak, C-Y. Lo, H. N. Nham, E. Pacas-Skewes, and K. Wu, "A fault simulator for MOS LSI circuits," *Proc. 19th Design Automation Conf.*, Las Vegas, Nevada, June 1982, pp. 400-409.

[4] A. E. Dunlop, "Automatic layout of gate arrays," *Proceedings of 1983 IEEE Symp. on Circuits and Systems*, Newport Beach, CA., May 1983, pp. 1245-1248.

[5] A. K. Bose, B. R. Chawla, and H. K. Gummel, "A VLSI design System," *Proceedings of IEEE 1983 Symp. on Circuits and Systems.* Newport Beach, CA., May 1983, pp. 734-739.

[6] M. J. Y Williams and J. B. Angell, "Enhancing Testability of Large Scale Integrated Circuits via Test Points and Additional Logic," *IEEE Trans. Comput.*, Vol. C-22, pp. 44-60, January 1973.

[7] E. B. Eichelberger and T. W. Williams, "A logic design structure for LSI Testability," *Proc. 14th Design Automation Conf.*, New Orleans, Louisiana, June 1977, pp. 462-468.

[8] H. C. Godoy, C. B. Franklin, and P. Bottorff, "Automatic checking of logic design structures for compliance with testability ground rules," *Proc. 14th Design Automation Conf.*, New Orleans, Louisiana, June 1977, pp. 469-478.

[9] H. Y. Chang, G. W. Smith, and R. B. Walford, "LAMP: System Description," *B. S. T. J.*, Vol. 53, pp. 1431-1449, October 1974.

[10] V. D. Agrawal, "Synchronous path analysis in MOS circuit simulator," *Proc. 19th Design Automation Conf.*, Las Vegas, Nevada, June 1982, pp. 629-635.

[11] J. P. Roth, W. G. Bouricius, and P. R. Schneider, "Programmed algorithms to compute tests to detect and distinguish between failures in logic circuits," *IEEE Trans. Electronic Computers*, Vol. EC-16, pp. 567-580, October 1967.

CHIP PARTITIONING AID: A DESIGN TECHNIQUE FOR PARTITIONABILITY AND TESTABILITY IN VLSI

S. DasGupta
IBM Corporation
P.O. Box 390
Poughkeepsie, NY 12602

M. C. Graf
IBM Corporation
Route 52
Hopewell Junction, NY 12533

R. A. Rasmussen
IBM Corporation
Route 52
Hopewell Junction, NY 12533

R. G. Walther
11400 F.M.R.D. 1325
IBM Corporation
Austin, Texas 78759

T. W. Williams
IBM Corporation
P.O. Box 1900
Boulder, CO 80314

ABSTRACT

This paper presents a structured partitioning technique which can be integrated into the design of a chip. It breaks the pattern of exponential growth in test pattern generation cost as a function of the number of chips in a package. In one of its forms, it also holds the promise of parallel chip testing, as well as migration of chip-level tests for testing at higher package levels.

INTRODUCTION

Level Sensitive Scan Design (LSSD) [1, 2] is one method to solve controllability and observeability problems in sequential networks and hence, ease the problem of test pattern generation. This is achieved by incorporating all memory elements in a sequential network in shift register latches (SRLs) and then connecting all SRLs into one or more shift registers so that the internal state of the network can be controlled or observed at any time through the shift register path. LSSD also permits software-based partitioning techniques [3] to divide a large network into manageable, independent networks, each of which is separately addressed by test pattern generators. This LSSD-based approach to partitioning will be discussed in the next section prior to the main topic of this paper. While this partitioning approach was adequate for large scale integration (LSI), it is inadequate for networks in very large scale integration (VLSI) [4] due to the rapid increase in test generation complexity. Several solutions have been suggested to solve this partitioning problem. Hsu, et al [5], and Tsui [6] have recommended ad hoc techniques for controlling and observing the outputs of chips, and, hence, inputs of other chips fed by the former on a common package. Goel and McMahon [7] have proposed another method where extra circuitry in system latches and multiplexors on chip outputs are required to control and observe chip boundaries.

This paper presents a structured, logical partitioning technique called Chip Partitioning Aid (CPA) that can be designed into a VLSI chip technology. In its simplest form, called Half-CPA (HCPA), it is a structured technique that partitions a network into nearly disjoint, physical segments that are approximately a chip's worth of logic in size. Thus, test generation cost, at higher package levels, drops from the normal exponential cost function to a straight multiple of the number of chips in the package. In the complete version of CPA, called Full-CPA (FCPA), the logic network is partitioned into subnetworks virtually along chip boundaries with built-in latch isolation around chip inputs and outputs (I/O's) that allows the potential reapplication of chip level test data and, more importantly, the potential for simultaneous testing of the internal logic of all chips. This latter version is, of course, the ultimate in the "divide and conquer" approach to test generation. The only addition at each package level is the incremental set of tests for interconnection faults at that package level which can be derived from a considerably simpler model of the network.

Next, we will define the rules associated with the two versions of CPA, and, finally, there will be a discussion on the effect of CPA on sytem design and how it can be mitigated by proper implementation of CPA.

HALF-CPA (HCPA)

Figure 1 shows a conceptual diagram of HCPA. It shows that in this version of CPA, all logic outputs of a chip are buffered by shift register latches (SRLs), called CPA-SRLs here, before being driven off-chip. CPA-SRLs are similar to standard SRLs, an example of which is shown in Figure 2. However, control outputs, such as clocks, are treated differently. They feed off-chip drivers unimpeded as required by system function. However, to ensure that this control function can be tested properly, it is required to also feed a CPA-SRL which is left to the side, out of the system path.

The HCPA structure at the chip boundary described above does not play any significant role in test pattern generation at the chip level. The only difference from a chip without CPA-SRLs is that measurements at chip outputs can be made only after the data has been clocked into the CPA-SRLs. The HCPA structure, however, has a considerble influence in test pattern generation at higher package levels, forcing creation of partitions of approximately a chip's worth of logic.

Reprinted from *ACM/IEEE 21st Design Automation Conf. Proc.*, 1984, pp. 203-208.

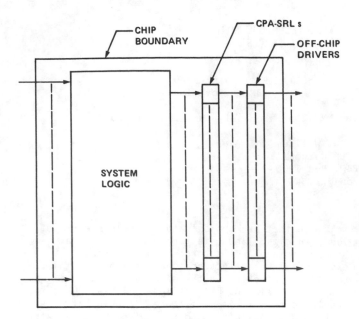

Figure 1. Half-CPA (HCPA) Structure

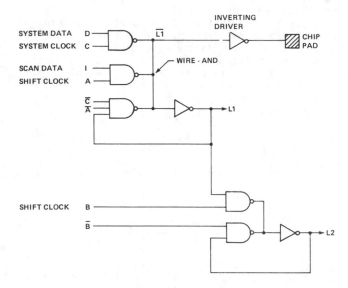

Figure 2. Example of CPA-SRL

An understanding of some partitioning concepts [3] is necessary here to appreciate the effect of HCPA. In an LSSD environment, SRLs, like package outputs, are considered observable nodes since the values in SRLs can be shifted out and observed. Therefore, to divide a network into smaller, independent subnetworks, a back-trace is performed from each observable node, stopping only at package inputs or SRLs, since the latter can be considered as a controllable input. All logic encountered in this back-trace constitutes an independent partition since it contains all the logic that can ever affect this SRL or primary output (PO). An example of this partitioning approach is shown in Figure 3. The unfortunate problem with this approach is that once the design is done, there is no way to bound or change the sizes of these partitions without a redesign; in fact, experience has shown that in many cases a significant segment of the entire network may be accounted for in a single partition.

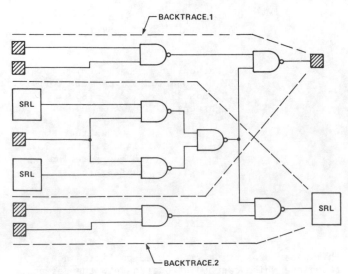

Figure 3. Examples of Partitioning Backtraces

A particularly good example of this is a bus-architected design where backing up from the bus, one can pack up just about the entire network in a single partition. A second problem is partition overlap in which a gate appears in more than one partition back-trace. This gate is considered at least for signal propagation during test generation and fault simulation, thus, effectively increasing the total number of gates that are evaluated by test generation/fault simulation programs.

The HCPA structure of Figure 1 changes the above situation. Figure 4 shows a module with several chips with HCPA structure. Since the CPA-SRLs also satisfy the property that they are controllable/observeable points, they serve both as "start points" and "stop points" of partitioning back-traces. Thus, starting from any HCPA-SRL, a

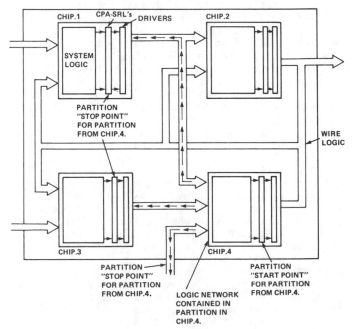

Figure 4. Partitioning with HCPA

back-trace propagates backwards through the logic on that chip and, in the worst case, stops at CPA-SRLs on the outputs of chips feeding the chip from where the back-trace started. Thus, each partition contains a network about the size of a chip's worth of logic, hence, putting an upper bound on the size of the partition. The question now is: how does this concept break the trend of an exponential rise in test generation cost as a function of chip count? To answer that, consider a package of n chips with m circuits on each chip. Assume that without HCPA, the worst case partition is approximately the size of the entire package. Also, assume that test generation cost is proportional to the square of the circuit count. Then, for a package without HCPA, test generation cost for a chip is:

$$T_c = km^2$$

i.e., $\quad m^2 = T_c/k$

Test generation cost at package level, T_p, is given by,

$$T_p = k(nm)^2$$

$$= kn^2m^2$$

or, $\quad T_p = kn^2 T_c/k$

$$= T_c n^2$$

For a fixed chip size, T_c can be considered to be fixed. Hence, T_p varies as the square of n, i.e., the square of the number of chips.

With HCPA, partitions are limited to approximately a chip's worth of logic. Hence,

$$T_p = (km^2)n$$

$$= (kT_c/k)n$$

$$= T_c n$$

Once again, the assumption that T_c is fixed for a fixed chip size makes T_p directly proportional to the number of chips. Thus, HCPA creates, for a given chip size, a linear relationship between the cost of test generation and the number of chips and hence, network size on a high level package.

Once these HCPA partitions are determined, test generation is done as in ordinary LSSD networks [8,9] with package wiring being tested along with on-chip circuitry.

The only exception to what has been said about

partitioning in HCPA relates to control outputs of chips. It is possible to back-trace through multiple chips, starting from a control output, but experience has shown us that these paths, while they may traverse multiple chips, are sparse in logic content. These outputs, therefore, are not expected to have large partitions, even in dense VLSI networks. The CPA-SRLs, that are fed by control outputs and sit on the side, aid in testing the logic since these SRLs act like intermediate observation points for the logic.

FULL-CPA (FCPA)

This is the complete version of CPA and is built on the benefits of HCPA. Unlike in HCPA where only system logic outputs are buffered by CPA-SRLs, in FCPA, both system logic inputs and outputs are buffered by CPA-SRLs, as shown in Figure 5. The only exceptions are control inputs and outputs. In the case of control inputs, they are required to feed CPA-SRLs on the side along with the system logic they are designed for, while control outputs are treated the same way they are treated in HCPA. The FCPA structure has two benefits over HCPA:

1. Though a FCPA chip needs at least two test clocks, they can be shared with all chips at higher package levels.

2. Latches on all system logic inputs/outputs effectively isolate the internal logic of chips allowing all chips with the potential to be tested simultaneously (hence, saving time on the tester) along with the potential to apply tests that were generated for the individual chip.

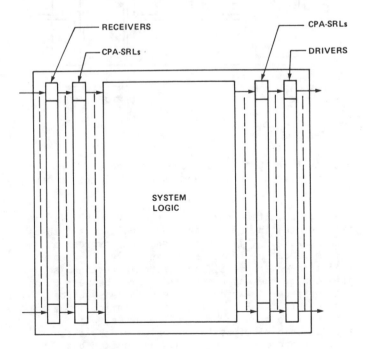

Figure 5. Full-CPA (FCPA) Structure

Test generation in the FCPA environment is now done in two stages:

1. Test generation of the internal logic of chips which is done either at the chip level and migrated up through the packaging levels or are generated again at the package level.

2. Test generation for stuck-at-faults in the package wiring and drivers/receivers on chips for which a simple model is created (see Figure 6) since they are bounded by SRLs. If any logic is performed at this package level, with wire-ORs or wire-ANDs, the number of test required is (r+1) where r is the maximum number of wires that is tied together to perform the largest AND or OR. If no such functions are performed, the package wiring can be tested with two tests.

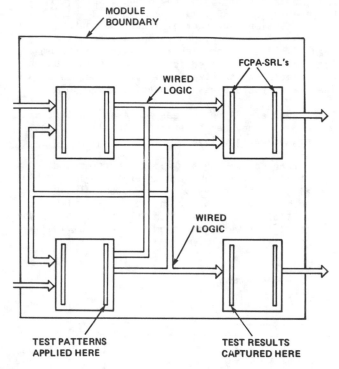

Figure 6. Simplified Model for Wiring Test on FCPA Module

Note that in FCPA, as in HCPA, test generation cost can be shown to be a linear function of the number of chips.

The real advantage offered by FCPA over HCPA is in parallel testing of the internal logic of all chips in a package. While in an idealized environment, all chips can be tested simultaneously, in a more realistic environment, parallel chip testing is affected by the way clocks are shared between chips and the order in which they need to be sequenced during a test. For example, if a particular test for one chip requires a C1-C2 sequence, while another chip requires the opposite order, these patterns cannot be merged into one common pattern for the package, even if everything else in the two patterns match. This limitation, however, is not expected to be a serious problem.

CPA RULES

The rules for HCPA are as follows:

1. Each chip "data output" signal feeding a chip output driver must be fed directly from a single CPA-SRL.

2. Each chip control output (for example, RAM control, shift clocks, tri-state inhibits) must feed a CPA-SRL, as well as it's off-chip driver.

In addition to the above rules, FCPA has the following additional rules:

1. Each chip "data input" must directly feed a CPA-SRL.

2. Each chip control input must feed directly to a CPA-SRL, as well as the system logic that it normally drives.

SYSTEM CONSIDERATIONS

From a system viewpoint, the choice of HCPA and FCPA is dependent upon density, system architecture and performance. At sufficient densities, latches naturally migrate to chip boundaries. In an LSSD environment, these latches would be embodied in SRLs, thus, satisfying the CPA requirements. However, there will be situations where an SRL will be required for CPA only, that is, a test-only SRL with no system application. In this situation, the clock(s) to that SRL will be used to control or observe a chip boundary during test.

When test-only SRLs are required for CPA, several steps can be taken to mitigate the real-estate and delay penalty of CPA-SRLs. In the case of HCPA, the CPA-SRL can be merged with its output driver to minimize both real-estate and delay. Also, the output from the CPA-SRL is taken from its L1 latch to the driver, thus saving the delay of the L2 latch. Figure 7 shows an example of an integrated CPA-SRL and driver where the performance detractor is the loading of the wired-AND function in the CPA-SRL. Estimates have shown that the above techniques can be used to limit real-estate overhead to less than 10% of the chip area and the delay penalty to a fraction of the delay of a logic circuit. And, finally, the test clock that sets data into the L1 latch of the CPA-SRL can be held "on" during system operation so that data can be flushed through it. Note that this test clock would constitute an overhead and at higher level packages, in a worst case situation, each chip might require a separate test clock for race-free testing. However, in a typical multi-chip package, it is possible to have many chips share the same test clock and still have race-free testing.

In the case of FCPA, the above ideas can be applied for the CPA-SRLs on both chip inputs and outputs. Additionally, the latches at the inputs and outputs can be merged into a single SRL with the L2* latch [10] , as shows in Figure 8, so that the L1 latch

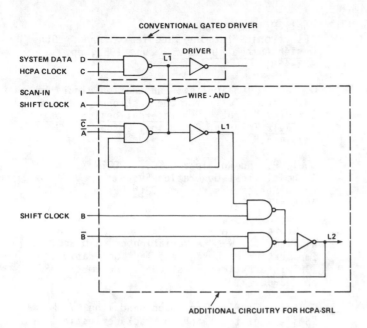

Figure 7. Example of Integrated HCPA-SRL/Driver

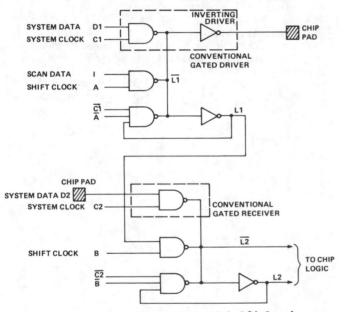

Figure 8. Example of CPA-SRL with L2* Latch Integrated with Receiver and Driver

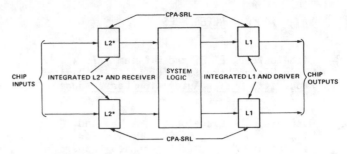

NOTE: SCAN PATHS & CLOCKS NOT SHOWN

Figure 9. FCPA Chip with CPA-SRL Built with L2* Latch Integrated with Driver and Receiver

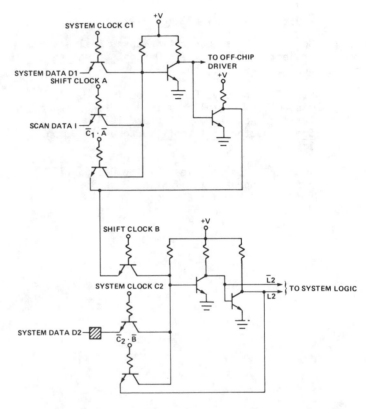

Figure 10. Example of CPA-SRL Implementation for FCPA

could serve as the CPA boundary for an output and the L2* latch could serve as the CPA boundary for an input (see Figure 9). This provides a further reduction in the real-estate overhead for FCPA. The two test clocks that set data into these CPA latches are now part of the CPA overhead. However, at higher package levels, these two clocks can be shared between other chips. Figure 10 shows an example of a CPA-SRL implementation from [11] to show the delay impact on a system data path.

One final note on CPA! Whether the latches at chip boundaries are system usable or not, CPA-SRLs can be used to trap machine states when desired and in the event of an error/fault, can, in most cases, be used to pinpoint the failing chip [12].

CONCLUSIONS

In this paper, we have described a partitioning technique that removes the uncertainty of partitioning sizes, since each partition is forced around chip boundaries and contains approximately a chip's worth of logic. Test generation, at higher package levels, now increases as a linear function of the number of chips and, in one of the versions, allows parallel chip testing which saves time on the tester during manufacturing. We have also defined the design rules and discussed system design aspects of CPA.

REFERENCES

[1] Eichelberger, E. B. and Williams, T. W., "A Logic Design Structure for LSI Testability," Proc. 14th Design Automation Conf., June 1977, pp. 462-468.

[2] DasGupta, S., Eichelberger, E. B. and Williams, T. W., "LSI Chip Design for Testability," Digest of Technical Papers, 1978 International Solid-State Circuits Conference, February 1978, pp.216-217.

[3] Bottorff, P. S., France, R. E., Garges, N. H. and Orosz, E. J., "Test Generation for Large Logic Networks," Proc. 14th Design Automation Conf., June 1977, pp. 479-485.

[4] Goel, P., "Test Generation Costs Analysis and Projection," Proc. 17th Design Automation Conf., June 1980, pp. 77-81.

[5] Hsu, F., Solecky, P., and Zobniw, L., "Selective Controllability: A Proposal for Testing and Diagnosis," Proc. 1978 Semiconductor Test Conf., October 1978, pp. 170-175.

[6] Tsui, F., "In-situ Testability Design (ISTD) - A New Approach for Testing High-Speed LSI/VLSI," Proc. IEEE, Vol. 70. No. 1, January 1982, pp. 59-78.

[7] Goel, P. and McMahon, M. T. "Electronic-Chip-In-Place Test," Proc. 19th Design Automation Conf., June 1982, pp. 482-488.

[8] Goel, P., "An Implicit Enumeration Algorithm to Generate Tests for Combinational Logic Circuits," Proc. 10th International Symposium on Fault Tolerant Computing, October 1980, pp. 145-151.

[9] Goel, P. and Rosales, B. C., "PODEM-X: An Automatic Test Generation System for VLSI Logic Structures," Proc. 18th Design Automation Conf., June 1981, pp. 260-268.

[10] DasGupta, S., Goel, P., Walther, R. G. and Williams, T. W., "A Variation of LSSD and Its Implication on Design and Test Generation," 1982 International Test Conf., November 1982, pp. 63-66.

[11] Culican, E. F., Diepenbrock and Ting, Y. M., "Shift Register Latch for Package Testing in Minimum Area and Power Dissipation," IBM Technical Disclosure Bulletin, Vol. 24, No. 11A, April 82, pp. 5598-5600.

[12] DasGupta, S., Walther, R. G., Williams, T. W. and Eichelberger, E. B., "An Enhancement to LSSD and Some Applications of LSSD in Reliability, Availability and Serviceability," Proc. 11th International Symposium on Fault Tolerant Computing, June 1981, pp.32-34.

AN INTEGRATED DESIGN FOR TESTABILITY AND AUTOMATIC TEST PATTERN GENERATION SYSTEM: AN OVERVIEW

Erwin Trischler *

Siemens Corporate Research and Support, Inc.
Research and Technology Laboratories
Princeton, N.J. 08540

ABSTRACT

A general overview on an Integrated Design for Testability and Automatic Test Pattern Generation System (IDAS) is given. The major components of IDAS include: heuristic controllability/observability (C/O) analysis, prediction of testing costs, tools for evaluation, display and improvement of testability, and C/O guided automatic test pattern generator. The IDAS system includes also the logic and concurrent fault simulator CADAT. A brief description of major components with a scenario how to use IDAS is given. Future research activities are discussed.

1. INTRODUCTION

Because test costs are rising exponentially with growing complexity and density of digital circuits, design for testability (DFT) is receiving considerable attention in the electronics community. Design for testability refers to design methods, procedures, and/or rules which, when applied, result in circuits that are substantially easier, and consequently less expensive to test. A testable design is both highly controllable and highly observable, at least in test mode. It is generally agreed that testability should be considered a design parameter. Over the past few years, interest has focused on the development of computer aided testability analysis programs as an attempt to measure the testability of a circuit, so that, when necessary, improvements may be made as early as possible in the design cycle. Having a measure of testability makes it possible to:

- accept or reject new designs or engineering changes with respect to testability
- choose among alternative designs
- identify problem areas within a design to provide feedback to designers, test engineers, etc.

Computer aided testability analysis is going to be an integral part of future CAD systems, especially with users who are not using structured DFT techniques [14].

In general there are two approaches to DFT:

1) Ad hoc design rules which are largely a matter of common sense and a design engineers' awareness of DFT methodologies, fundamental limitations of test

---

* The author is now with
Siemens AG, Corporate Information Technology
Munich, West Germany

equipment, and test software tools.

2) Structured DFT where a certain structure and other constraints are imposed upon the designer. The classic example is Level Sensitive Scan Design (LSSD) as used at IBM.

In general, DFT requires a certain amount of additional hardware overhead and thus, an increase in circuit size. For a given amount of circuit board or semiconductor chip area and a given number of available external signal connections (pins), testability, circuit functionality and performance may be viewed as arch rivals. Until now, industry has traditionally been unable to justify the additional hardware area (silicon real estate) associated with the current DFT techniques.

Because additional hardware overhead should not exceed practical limits in real-world circuits, a technique called "incomplete scan path" was developed which requires the inclusion of a relatively low percentage of flip-flops into a scan path [15]. The philosophy of incomplete scan is based on the fact that automatic test pattern generation (ATG) is relatively easy and inexpensive, not only for combinational but also for "weak" sequential circuits where the sequential depth does not exceed four. It was shown in [15] that:

(1) The incomplete scan path technique can be used effectively to reduce ATG and fault simulation costs.
(2) The selection procedure based on controllability and observability analysis can be used successfully to determine which flip-flops to include into an incomplete scan path.
(3) Incomplete scan path minimizes total costs - the sum of the additional hardware costs and ATG and simulation costs.
(4) The use of testability analysis in conjunction with incomplete scan path opens a way to achieve testability optimization, i.e. minimization of the additional hardware overhead so as to reduce the overall test cost as much as possible.

Based on these conclusions, the introduction of a customized DFT methodology based on scan structures was proposed. The necessary tools for achieving this and for answering the question "How much DFT is necessary?" were developed and are now included into a system called **Integrated Design For Testability and Automatic Test Generation System** (IDAS) at Siemens' Research and Technology Laboratories in Princeton, NJ. This paper indicates the background of the IDAS and summarizes the status of current work and enhancements that are planned as part of the ongoing research program.

Reprinted from *ACM/IEEE 21st Design Automation Conf. Proc.*, 1984, pp. 209–215.

2. TESTABILITY ANALYSIS

Testability measures can be divided into intrinsic and extrinsic types [1]. Extrinsic measures, such as fault coverage and fault resolution, require consideration of both a circuit and its associated test patterns. Intrinsic or inherent testability is a property of a circuit that excludes input/output test pattern considerations.

Several algorithms for measuring testability have been proposed [2] - [12]. Most of them assume that testability is an inherent property of the circuit based on its topology. A testability analysis program should meet two basic criteria:

1) The computational complexity of the testability measures should be significantly below that of test generation. In general, this is achieved by simplifying the assumptions on which test generation is based.

2) The testability measure should reflect the potential difficulties for a certain type or strategy of test generation (manual, random, automatic). For example, if random test pattern generation is applied, the measures should help to identify difficult-to-test nodes and areas from the viewpoint of random testing; or if an automatic test pattern generator (ATG) is to be applied, the measure should help to identify difficult-to-test areas from this viewpoint. In general, testability values will differ for the same circuit, depending on the test generation strategy or methods applied.

All mentioned testability analysis programs [2] - [10] are basically heuristic approaches which means that the correlation between the testability analysis results and actual test generation efforts may differ in some cases. Some testability measurement algorithms (e.g. [2] - [6], [10], [12]) are linear algorithms, whereas ATG even for combinational circuits is NP-complete [23], i. e., it appears that the computation is, for the worst case, exponential with the size of the circuit. It is obvious that a problem which is, in the worst case, an exponential one cannot be solved using a linear algorithm. Nevertheless, the use of heuristic approaches in ATG can be quite useful; previous [18] - [19] and recent work [19] - [21] in ATG approaches have proven that.

An early attempt to measure testability was undertaken by Rutman [1] who introduced a heuristic cost analysis in order to improve the efficiency of an automatic test pattern generation program for sequential circuits. Rutman's three cost measures for each node N of a circuit are: costs of controlling a node from primary inputs to 0 and 1 and the cost observing it on a primary output. If there is more than one alternative to control or to observe a node, the minimal cost will be calculated and assigned to that node (e.g. to set the output of a two input NAND gate to 1, three alternatives 00, 01, 10 will be evaluated).

Rutman's work was further extended by Breuer [3] who introduced the cost analysis for 'side effects' of fan-out nodes, by Goldstein [4] who introduced the distinction between combinational and sequential controllabilities and observabilities, and recently by Ratiu [12] whose calculation of controllabilities and observabilities is based on primary input variables and not on intermediate variables as in [2] - [4].

The testability of a digital circuit in Goldstein's SCOAP [5] is quantified by performing a topological analysis of a circuit without input patterns being required and by assigning six numbers to each node of the circuit that reflect the degree of difficulty associated with controlling or observing it. Three of the six measures are associated with combinational nodes, while the remaining three are associated with sequential nodes. The meaning of the measures are as follows:

(1) C_{C0} (C_{C1}): Combinational controllability 0 (1) of a node N is related to the minimum number of combinational nodes in the circuit that must be assigned in order to set node N to the 0 (1) state.

(2) C_{S0} (C_{S1}): Sequential controllability 0 (1) of a node N is related to the minimum number of sequential nodes (i.e. flip-flops) in the circuit that over a period of time must be assigned in order to set the node N to the 0 (1) state.

(3) O_C: Combinational observability of a node N is related to the minimum number of node assignments in the circuit required to propagate the state of N to a primary output.

(4) O_S: Sequential observability of a node N is related to the minimum number of sequential nodes that must be set, over a period of time, in order to propagate the state of N to a primary output.

The higher the calculated number, the worse the controllability/observability. The generated measure can range from 0 to infinity, where 0 means direct access and infinity means impossible to access or untestable.

Goldstein's approach [5] was chosen as the basis for this work because its algorithm can be viewed as an emulation of a path sensitization type of ATG. We have further extended this type of heuristic testability analysis.

3. GENERAL OVERVIEW ON IDAS

IDAS was created as an experimental system to provide an interactive logic design and test environment with the purpose of exploring new ways to solve different testing problems. IDAS research and development objectives were to study testability analysis, testability improvement, and ATG approaches and to create an integrated system which would help to study and to solve the following problems in the design and test phase:

- Determining whether the given circuit is testable with respect to the ATG
- Quantification of the testability cost
- Identification of what causes problems
- Suggestions as to how the testability can be improved
- Verification by generating test patterns automatically

Although the system includes logic and concurrent fault simulation, the research was restricted to the following aspects of testing:
- Testability analysis
- Prediction of testing (ATG) cost

- Automatic testability improvement (automation of DFT)
- Controllability/Observability guided ATG

One additional problem in the design process is the evaluation of the large amount of data produced by various analysis tools. An engineering workstation (DAISY Logician) was used to help simplify the evaluation.

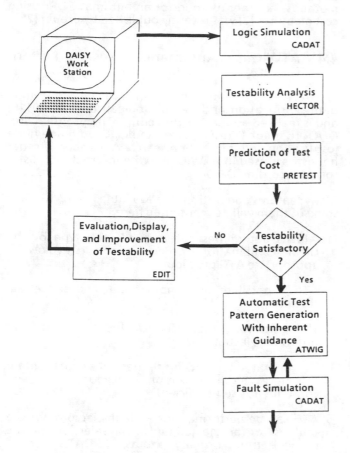

Figure 1. IDAS: An Integrated Design for Testability and Automatic Test Pattern Generation System

IDAS includes:

1. **CADAT**: A logic and concurrent fault simulator [26].

2. **HECTOR**: An heuristic controllability/observability tree search program for testability analysis.

3. **PRETEST**: A program that predicts the total cost of ATG (ATWIG).

4. **EDIT**: A number of programs for Evaluation, Display, and Improvement of Testability that assist the designer through graphical means to isolate testability problems in his circuit and to identify the causes of these problems. One of these programs also automatically selects the flip-flops to be included into an incomplete scan path for a given circuit.

5. **ATWIG**: An Automatic Test Pattern Generator With Inherent Guidance for combinational and sequential digital circuits that is based on the HECTOR testability analysis.

6. Several utility programs (not shown in the picture).

The advantages of this system include:

1. The designer is provided with immediate access to test design tools with minimal human interaction.

2. The designer is provided with an interactive environment to improve testability that is enhanced by graphical feedback from the analysis tools.

4. IDAS MODULES

The IDAS system uses the non-hierarchical CADAT circuit description language as its input. It uses also the same type of primitives as CADAT - simple gates, multiplexers, demultiplexers, adders, latches, flip-flops, etc. All IDAS programs are written in Fortran-77 and C and are running on the VAX11/780 with UNIX operating system.

4.1 HECTOR: Heuristic Controllability and Observability Analysis

The HECTOR program performs a similar analysis as proposed in [2] - [6], [10], [11], [14]. These testability analysis programs assign six numbers to each node of the circuit. Each of these numbers represents the minimal cost to control or to observe a certain node ($C_{C0/1}$, O_C, $C_{S0/1}$, O_S).

HECTOR, however, assigns more information to each node of the circuit. Besides the minimal number (which is identical to the number generated by SCOAP program), HECTOR also assigns six **ordered lists** to each node; each of these lists contains information about the type of node (combinational/sequential), local connectivity, test pattern etc. For example the combinational observability O_c of node N O_c (N) will be represented as follows:

SCOAP: $(O_c (N))_{min}$

HECTOR: $(O_c (N))_{min} \leq (O_c (N))_2 \leq (O_c (N))_3 \leq \ ..... \ \leq (O_c (N))_n$

where n = number of alternative test patterns for observing the node N.

Similar ordered lists exist for other C/O measures. These not only allow the search for a test pattern in ATWIG to be easier and faster, but also automatic DFT to be very efficient. The intention of HECTOR was to use the **same principles** in ATG (ATWIG program) and in automatic improvement of testability (EDIT program).

The complexity of the algorithm is linear in time , i.e. it is O(N), where N = number of nodes in the circuit. The typical run time of 20 faults/sec on a VAX 11/780 could be further accelerated using techniques such as selective trace, etc. A more detailed description of HECTOR is in preparation.

4.2 PRETEST: Prediction of ATG Cost

Based on the fact that the heuristic testability analysis cannot always predict accurately the actual ATG effort or identify difficult-to-test areas, we have chosen to use a sampling technique to estimate the expected ATG costs. For a given number of randomly selected faults the ATWIG program will generate test patterns. Based on the required CPU time for ATG and the number and type of conflicts occurred during the ATG, it is possible to estimate the overall ATG time for the whole circuit. A system for estimation of detectability of faults in logic circuits presented recently in [18] employs similar principles.

4.3 EDIT: Evaluation, Display and Improvement of Testability

EDIT consists of a number of separate programs which can be used independently. Some of the programs are:

TYNODE: This program for the evaluation and display of $C_{C0/1}$, O_C, $C_{S0/1}$, O_S, fault related and mean square testability measures of circuit nodes can be performed either on the VAX or DAISY. The generated controllability/observability (C/O) and testability numbers can be downloaded from the VAX and displayed directly on the circuit schematic on the DAISY screen. This feature eliminates the manual correlation of the testability results and allows the user to identify the difficult-to-test nodes very easily.

TRACE: This program highlights a whole C/O tree (i.e. sensitized path) for a particular fault on the DAISY schematic. This feature is of special importance if the designer wants to know why a node is difficult to test. Using the color terminal, the observability path can be distinguished from controllability paths using different colors.

SEARCH: This program is similar to the TRACE program; it allows the evaluation of the C or O paths using signal names on a non-graphic terminal.

SELECT: This program automatically selects flip-flops for an incomplete scan path. Available options include:

(1) Only flip-flops that are closest to primary outputs of a circuit will be selected for an incomplete scan to be included into a boundary scan.

(2) Flip-flops based on desired sequential depth will be selected for an incomplete scan.

(3) Options 1 and 2 combined.

(4) Flip-flops closest to a given fault or undetected fault list will be selected for an incomplete scan.

Option 4 is of special importance if an existing test program doesn't have satisfactory fault coverage. In this case, an incomplete scan path can be introduced in areas in which the fault coverage is relatively low. This option helps to bridge the gap between intrinsic and extrinsic testability measures.

The EDIT program contains two different procedures for flip-flop selection for an incomplete scan path. The first procedure is indicated above and realized in program SELECT, the second procedure is as proposed in [15].

The methods developed here can be viewed as an **automatic improvement of testability** or **automatic DFT**. The present SELECT version is restricted to automatic selection of flip-flops for an incomplete scan path, but it can be easily extended to other DFT methods. Another method for an automatic DFT using cost measures [2] was recently published by Breuer [25].

4.4 ATWIG: Automatic Test Pattern Generation With Inherent Guidance

The ATWIG program is based on principles of single path sensitization and partial Boolean difference [16] and is realized as a search procedure. The program uses O_C/O_S to guide D-drive search and C_C/C_S to guide node justification and implication search. A number of other heuristics is used in ATWIG; e.g. the internal fault list is sorted according to O_C/O_S.

Many features and options are available in ATWIG. Only some of them will be mentioned here:

(1) ATWIG determines automatically the type of the circuit (combinational or sequential) and invokes the appropriate ATG procedure.

(2) ATG is possible for combinational and sequential circuits.

(3) The search for a test pattern can be performed using different selection criteria

(4) All generated test sets for sequential circuits contain the necessary initialization procedure. They are consequently self-sufficient.

(5) ATG can be performed for all faults, according to a fault list, or for selected faults. The selections can be derived from primary inputs, CADAT's list of undetected faults, etc.

(6) The generated test patterns can be compressed. This saves simulation and fault simulation time.

(7) The generated test patterns are translated into CADAT's Design Simulation Language (DSL) and fault simulated with CADAT.

The experimental results on over 20 combinational and low sequential LSI circuits (300 - 650 gates) have shown that ATWIG can quickly generate tests with a fault coverage of 95% or more (Table 1). If the O_S or C_S of the circuit is, however, very high, i.e. it was designed without DFT, ATWIG cannot always generate tests with reasonable fault coverage (95%) in a reasonable time (> 1000 sec on VAX 11/780). Other heuristics are to be considered for ATG of highly sequential circuits.

No. of investigated circuits	Type of circuits	Indicated C_S/O_S	Fault Coverage %	Runtime VAX 11/780 sec
11	combinat.	0	96 - 100	20 - 40
7	sequential	1 - 7	95 - 100	50 - 120
4	sequential	15 - 30	< 90	> 200

Table 1. ATWIG: Experimental Results

4.5 Utility Programs

Different utility programs are available in IDAS:

- TRADE: A program that allows file and data transfer between the DAISY and the VAX computer with UNIX operating system.

A collection of programs which allow conversion from a given description language into another:

- LASCAD: LASAR to CADAT (non-hierarchical circuit descriptions) conversion programs
- MKDSL: ATWIG test patterns to CADAT (DSL) format
- HILLO: ATWIG test patterns to LASAR (TECO) format
- DAISY graphical representation to CADAT circuit descriptions

5. A SCENARIO FOR THE USE OF IDAS

An engineer designs a circuit using the interactive capabilities at an engineering workstation (e.g. DAISY). Let us assume that a designer has finished the design as shown in Fig. 2 which represents the schematic from DAISY, and intends to perform logic simulation. By typing in the appropriate commands, automatically a CADAT compatible circuit description is extracted from the DAISY Logician data base and is transferred to the VAX computer. Test patterns for logic simulation have to be specified and CADAT can be started. If the results are not satisfactory, the logic schematic on the DAISY can be modified and the whole process repeated. It is assumed that the modification will always be done on the DAISY.

After the logic simulation is performed, testability analysis can be started. The HECTOR program is invoked with the CADAT circuit description as input. After completion of HECTOR, the designer specifies which testability measures are to be evaluated and displayed on the DAISY. The appropriate program on the VAX creates a file of DAISY commands that are forwarded (downloaded) to the DAISY. The circuit schematic is then **annotated with testability numbers** and displayed to the designer. Figure 2 shows such an example where the C_{C0} of each node is back annotated and displayed on the schematic. The information about a particular controllability and/or observability path can be downloaded as well to DAISY and displayed. Figure 3 shows the observability path between node A0 (primary input) and node O2 (primary output). This path is equivalent to the **sensitized path** for input A0 s-a-0/1 faults if it does not contain any conflict. In such a way we can evaluate and verify the assertion of testability analysis. The EDIT program allows the display of a

sequence of paths after each other. Using automatic selection of flip-flops for an incomplete scan path, only recommendations about what could be done will be given; the final modification of the circuit on the DAISY, however, has to be done by the designer.

ATWIG can be invoked from the DAISY console to automatically generate the test patterns for the circuit. Due to identical circuit description for all tools, the simulation program CADAT can be called upon to determine the overall fault coverage; if unsatisfactory, ATWIG can be restarted to generate additional test patterns.

The whole cycle (beginning from extraction of circuit description from DAISY created database, uploading to VAX, running HECTOR, downloading testability numbers and paths, running ATWIG until generated test patterns are converted into CADAT's DSL format) takes less than 5 minutes real time for a circuit as shown on Fig. 2.

6. CONCLUSIONS AND FUTURE WORK

1. The IDAS system is a research tool which allows one to perform:
   - heuristic controllability/observability analysis
   - prediction of ATG cost
   - evaluation, display and automatic improvement of testability
   - controllability/observability guided automatic test pattern generation
   - logic and concurrent fault simulation

Methods developed in IDAS bridge the gap between intrinsic and extrinsic testability measures.

2. Further acceleration is possible in
   - HECTOR using e. g. selective trace technique.
   - ATWIG using multiple path sensitization [22] - [24], other heuristics and refined strategies.

3. The following extensions are planned:
   - HECTOR: The library will be extended towards high level sequential primitives.
   - PRETEST: With the ATG cost and actual hardware cost it is possible to determine 'How much DFT is necessary?'. The future work and extensions will go in this direction.
   - EDIT: The graphic display of EDIT will be extended to provide the highlightening of flip-flops to be included in the incomplete scan path directly on the screen of the engineering work station.

- SELECT: The program will be extended to DFT techniques other than incomplete scan path.

4. Further display of different circuit properties directly on the circuit schematic on the screen of the engineering work station seems to be unreasonably difficult. Based on our present knowledge, none of existing engineering workstations allow easy display of information alphanumeric, highlighted or in color. Future work will require close collaboration with vendors of engineering workstations to make the design and test task easier.

Acknowledgements

The author would like to thank Dr. A. Ashkinazy (now with RCA, Solid State Technology Center, Sommerville,NJ) for his collaboration and many discussions during the development of the first version of HECTOR (called ESCOAP and written in Fortran on DEC/10 computer), and B. Ladendorf for developing the second and third version of HECTOR written in C on VAX 11/780 (UNIX). Thanks also to P. Sochacki, who wrote the extraction program for generation of CADAT and ESCOAP circuit description from DAISY Logician data base.

A number of coop students from Drexel University in Philadelphia,Pa participated in the years 1982-83 on IDAS program development. In particular, the author would like to thank M. Yoo, L. McDowell, V. Kim, and S. Rosen for their excellent collaboration and programming. The contribution of D. Anderson and S. Bahrampour, who have created CADAT and HECTOR library is also appreciated.

Finally, the IDAS system would be impossible without continuous encouragement of K. Anderson and Dr. S. Daniels; their support is greatly appreciated.

References

[1] Keiner, W., et al., "Testability Measures", Proc. 1977 IEEE AUTOTESTCON, pp. 49 - 55

[2] Rutman, R. A., "Fault Detection Test Generation for Sequential Logic by Heuristic Tree Search", IEEE Computer Society Repository, Paper No. R-72-187

[3] Breuer, M. A., "New Concepts in Automated Testing of Digital Circuits", Proc. Symp. on CAD of Digital Electronic Circuits and Systems, 1978, pp. 57 - 80

[4] Goldstein,L. H., "Controllability/Observability Analysis of Digital Circuits", IEEE Trans. on Circuits and Systems, Vol. CAS-26, No. 9, Sept. 1979, pp. 685 - 693

[5] Goldstein, L. H. et al., "SCOAP: Sandia Controllability/Observability Analysis Program", Proc. 17th Design Automation Conference, 1980, pp. 190 - 196

[6] Kovijanic, P. G., "Testability Analysis", Proc. 1979 IEEE Test Conference, pp. 310 - 316

[7] Grason, J., "TMEAS, A Testability Measurement Program", Proc. 16th Design Automation Conference, 1979, pp. 156 - 161

[8] Dussault, J. A., "A Testability Measure", Proc. 1978 Semiconductor Test Conference, pp. 113 - 116

[9] Bennetts, R. G. et al., "CAMELOT: A Computer Aided Measure for Logic Testability", Proc. 1980 IEEE ICCC, pp. 1162 - 1165

[10] Goel, D. K. et al., "An Interactive Testability Analysis Program - ITTAP", Proc. 19th Design Automation Conference, 1982, pp. 581 - 586

[11] Fong, J. Y. O., "A Generalized Testability Analysis Algorithm for Digital Logic Circuits", Proc. 1982 IEEE Int. Symp. on Circuits ans Systems, pp.1160 - 1163

[12] Ratiu, I. M. et al., "VICTOR: A Fast VLSI Testability Analysis Program", Proc. 1982 IEEE Int. Test Conference, pp. 397 - 401

[13] Agrawal, V. D., et al., "Testability Measures - What do they tell us?", Proc 1982 IEEE Int. Test Conference, pp. 391 - 396

[14] Berg, W. C. et al., "COMET: A Testability Analysis and Design Modification Package", Proc. 1982 IEEE Int. Test Conference, pp. 364 - 378

[15] Trischler, E., "Testability Analysis and Incomplete Scan Path", Proc. 1983 IEEE Int. Conf. on CAD, pp. 38 - 39

[16] M. A. Breuer et al., "Diagnosis & Reliable Design of Digital Systems", Computer Science Press Inc., 1976

[17] Kovijanic, P. G., "Single Testability Figure of Merit", Proc. 1981 IEEE Test Conference, pp. 521 - 529

[18] Miyamoto, S. et al., "TRUE: A Fast Detectability Estimation Program", Proc. 1983 IEEE Int. Conf. on CAD, pp. 36 - 37

[19] Hill, F.J. et al., "SCIRTSS: A Search System for Sequential Circuit Test Sequences", IEEE Trans. on Computers, Vol. C-26, No. 5, 1977, pp. 490 - 502

[20] Strebendt, R. E., "Heuristic Enhancement of an Algorithmic Test Generator", Proc. 14th Design Automation Conference, 1977, pp. 84 - 87

[21] Breuer, M. A., "TEST/80 - A Proposal for an Advanced Automatic Test Generation System", Proc. 1979 IEEE AUTOTESTCON, pp. 305 - 312

[22] Benmehrez, C. et al., "The Subscribed D-Algorithm - ATPG with multiple Independent Control Paths", Proc. 1983 IEEE ATPG Workshop, pp. 71 - 80

[23] Benmehrez, C. et al., "Measured Performance of a Programmed Implementation of the Subscribed D-Algorithm", Proc. 20th Design Automation Conference, 1983, pp. 308 - 315

[24] Fujiwara, H. et al., "On Acceleration of Test Generation Algorithms", Proc. of 13th Annual Int. Symp. on Fault Tolerant Computing, 1983, pp. 98 - 105

[25] Breuer, M. A., "The Automatic Design of Testable Circuits", Proc. 1983 IEEE ATPG Workshop, pp. 3 - 6

[26] CADAT Manual, HHB-Softron, Inc., 1983

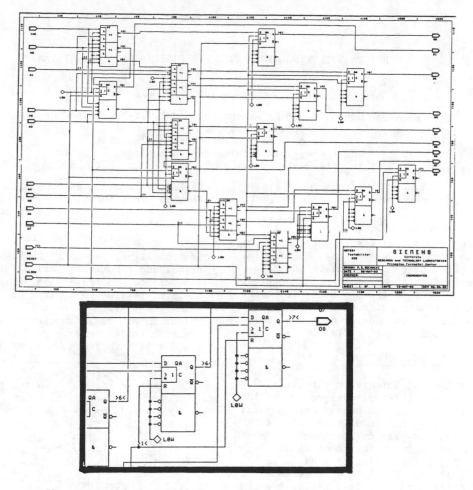

Figure 2. Backannotated circuit schematic with selected
testability numbers

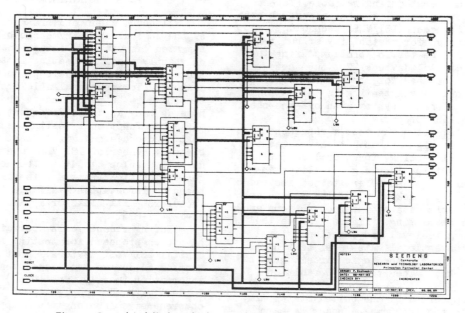

Figure 3. A highlighted observability path displayed on
the DAISY Logician

FUNCTIONAL TESTING TECHNIQUES FOR DIGITAL LSI/VLSI SYSTEMS*

Stephen Y.H. Su and Tonysheng Lin

Research Group on Design Automation and Fault-Tolerant Computing
Department of Computer Science
Thomas J. Watson School of Engineering,
Applied Science and Technology
State University of New York, Binghamton, N.Y. 13901

ABSTRACT

Functional testing is becomilng more important due to the increasing complexity in digital LSI/VLSI devices. Various functional testing approaches have been proposed to meet this urgent need in LSI/VLSI testing. This paper presents the basic ideas behind deterministic functional testing and concisely overviews eight major functional testing techniques. Comparisons among these techniques and suggestions for future development are made to meet the challenges in this fast growing testing field.

1.INTRODUCTION

The rapid growth of integrated circuit (IC) technology has allowed LSI/VLSI devices to provide more powerful functions and to become more popular in commercial, educational, industrial and military applications. These complex devices must be tested for correct functional operation to obtain certain assured reliability before being adopted in an individual application. Due to the increased design complexity and integrated circuit density, testing problems become much more difficult than ever. The high complexity of VLSI makes gate-level testing very difficult and expensive [1]. The need for both manufacturers and users to find reliable, low-cost, comprehensive testing techniques has already become a major problem in modern LSI/VLSI technology. Techniques for design for testability or built-in-self-test consider the testing problem during the design stage of digital devices [2]; but these approaches cannot be used for the off-the-shelf components. Functional-level testing is the natural approach to assure that a system with off-the-shelf components is working properly.

Functional-level testing uses a representation of a digital system higher than gate-level testing. In functional-level testing, functional faults with respect to the specification (e.g., addition instruction in a microprocessor) are tested instead of signal faults (e.g., stuck-at-0) at the inputs and the output of a logic gate or interconnections among gates in gate-level testing. Figure 1 shows the contrast between gate-level and functional-level testing. When conventional gate-level testing is performed, every line in the logic circuit must be individually checked. The stuck-at-0 fault on ℓ_1

* This work is supported by the United States Army Communication Electronics Command under Research Contract No. DAAB 07-82-K-J056.

in Figure 1, is an example of a gate-level fault. Whereas, when functional-level testing is to be performed, only the functions of modules and their interactions are tested. The functional faults of f_1 (e.g., incorrect addition) and f_2 (e.g., incorrect control signal decoding) in Figure 1, are examples of assumed functional faults for data operation and control modules respectively. For example, functional-level testing in testing microprocessors usually means the testing of each instruction in the instruction set based on its specification. Functional-level testing may be performed without knowing the detailed implementation of the system-under-test.

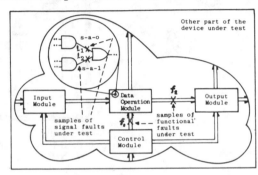

Fig.1 Gate-level testing vs. functional-level testing.

Although manufacturers have detailed knowledge of the logical and parametric behavior of the integrated circuit (IC) chips produced, they usually only apply specific testing instead of comprehensive, thorough testing to their products due to economic reasons. To insure reliable operations, additional testing must be performed by the users. Users may have as many as three different testing environments -- ad hoc functional testing, assembly production test, and occasional field test. Due to the increasing number of LSI/VLSI users, the need is evident to develop "good" (to be defined later) test generation algorithms for testing these devices based on information available to users in manufacturers' data books or application notes [3-5, 11-13, 31-32].

During recent years, certain kinds of techniques have been proposed for testing functional faults in digital LSI/VLSI devices. Most of them have attempted to derive functional tests using the limited information (mostly functional) available to users. To deal with the complexity of VLSI circuits in a comprehensive fashion, these approaches tend to use abstract, formal functional descriptions of VLSI behavior as opposed to traditional gate/circuit-level description [6].

Reprinted from *ACM/IEEE 21st Design Automation Conf. Proc.*, 1984, pp. 517-528.

A complete functional testing effort includes, in general, the following steps:
(1) Describe the digital system under test in terms of a well-defined description medium such as the Register Transfer Language (RTL) or graph description language.
(2) Set up a suitable functional-level fault model (or models).
(3) Develop a test generation algorithm to derive a comprehensive set of functional test patterns (or test procedures).
(4) Validate the effectiveness (e.g., fault coverage and test generation complexity) of the generated functional tests by a fault simulator or an automatic test equipment (ATE).

The purpose of functional testing is to validate correct functional operations of digital systems according to their specifications. Using functional testing techniques, one cannot only reduce the test generation complexity, but also obtain a test set for testing the IC chips with the same functions but different circuit design/implementation (e.g., parallel adder vs. serial adder) which is common among manufacturers today. In reality, manufacturers of LSI/VLSI chips will not release the implementation details of their chips to users lest they be copied by other competitors. Thus the user of LSI/VLSI chips has little other alternative but performing functional testing of these IC chips.

In this paper, we summarize the current state of the art on the development of deterministic functional testing techniques for digital LSI/VLSI devices with special emphasis on the important area of microprocessor testing. Random testing techniques such as that proposed by David, Thevenod-Fosse and Fedi [28] are therefore not included in this paper. The second section briefly discusses the types of digital LSI/VLSI chips from a system organizational point of view and their functional testing approaches. Section three describes functional fault models and fault coverage measures. Section four describes major functional testing techniques described in the current literature and a brief comparison is made among them in section five. The last section includes trends and suggestions for future development in VLSI functional testing.

2. TYPES OF DIGITAL LSI/VLSI DEVICES AND THEIR FUNCTIONAL TESTING

In general, the digital LSI/VLSI devices available to users can be categorized from the system organizational point of view into the following five types:
(1) Microprocessors
(2) Peripheral supporting devices
(3) Special-purpose functional devices
(4) Memory devices
(5) Basic logic devices
Functional-level testing receives different degree of attention among these five types according to their diverse functional complexity. Microprocessors usually have the highest functional complexity among these five types. Peripheral supporting chips include chips which perform interfacing and inter-module communication. They usually have limited specific functions imbedded within and can be acti-

vated under program control by a microprocessor. Special-purpose functional chips, such as devices designed for fast execution of certain functions (e.g., hardware multiplier, fast fourier transform, etc.) or devices designed for enhanced features (e.g., cyclic redundancy check (CRC) chips for increasing system reliability) have medium complexity. The memory chips are used for program or data storage purposes and receive the least emphasis in functional complexity. Basic logic chips include logic circuits such as gate arrays and programmable logic arrays (PLA). They also have much lower functional complexity comparing to microprocessors as depicted in Figure 2.

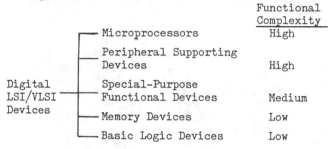

Fig. 2   Digital LSI/VLSI devices and their estimated functional complexity.

Testing of LSI/VLSI devices can be performed using different philosophies by different people based on how much implementation information is available and how much confidence of fault detection is required. Manufacturers who know the implementation details of the devices under test may perform implementation-dependent structural testing to obtain more accurate and more practical fault coverage. Whereas, for users and those chips vendors who do not design the circuitry, only operational/functionl testing strategy can be used in general cases since the structural details are unknown. Both these two approaches can be performed at the functional level [7].

In various kinds of applications, especially those requiring critical reliabiity, users must assure the correct functional operations of purchased LSI/VLSI chips even without knowing their detailed implementation. This usually left them with several uncomfortable alternatives. An often adopted way for testing microprocessors, for example, is to buy a set of test vectors to test the purchased chips. However, most of these test vectors were derived based only on certain sets of faults (e.g., stuck-at faults in input/output pins) considered by vendors and usually do not test other faults (e.g., bridging fault in internal connections) in a comprehensive way. Another possible approach is to run a pseudorandom test sequence of test vectors for simplicity and minimum cost. For example, to test a microprocessor, test engineers may generate random streams of instructions and compare the results with a pretested good module. But, to derive reliable measures of fault coverage using this method is a nondeterministic problem. The most reasonable/conceivable way to test digital LSI/VLSI chips, from chip users' point of view, is to use those techniques attempting to derive conceivable fault coverage functional tests using the limited information available (e.g., those contained in data books or user's manuals). Formal, abstract descriptions of device behaviors should be used to enable the proof of comprehensiveness. To attack the problem of high functional complexity of LSI/VLSI devices (e.g., microprocessors), "Divide and Conquer" techniques (e.g., modular decomposition) are usually used to subdivide the devices into functional modules whose behaviors could be individually verified using appropriate test procedures. For example, testing of a microprocessor can be subdivided into testing data processing and control modules. But two situations may often be observed:

(1) Testing is focused on data processing function The fault types of the control part are usually simplified.

(2) Modules partitioning and tests are developed on an ad hoc basis.

The order of test sequence is of major concern during functional testing. The order is usually derived from the functional description of the device (e.g., observability and controllability of function outputs and inputs) and associated parameters (e.g., type of functional fault under test) which may be structural or behavioral. Two well-known test order mechanisms can be applied: the "start-small" or the "start-big" method [8]. The start-small method tests only a small portion of hardware and then uses the tested part to detect faults in other parts. Each additional test adds a small portion of hardware to the previously tested parts until all are tested. This method requires an ordering among the set of instructions used for testing. The start-big test starts with the verification of the whole system to determine whether any subsystem is faulty and proceeds to sequentially narrow down the region in which fault tends to occur. The start-big method is most useful for reliability failure analysis.

The testing of memory chips only concerns itself with read and write processes of the individual storage element and can usually be done by various memory testing strategies [9,10]. Due to the high density of VLSI memory chips (and hence the extremely tiny distance between cells), data-dependent fault (e.g., pattern sensitivity faults) is the kind of fault receiving more emphasis in practical applications which employed LSI/VLSI memory chips.

3. FUNCTIONAL FAULT MODELS AND FAULT COVERAGE MEASURES

To simplify the evaluation of fault coverage in testing digital systems, a fault model is developed and then the fault coverage is calculated based on this model. In gate-level testing, the popular stuck-type fault model in which any faulty line is modeled as logically stuck-at-0 or stuck-at-1 is used. Whereas in a functional-level fault model, only those faults which can be modeled and/or observed in functional-level are included.

A functional fault model is set up with respect to certain sets of functional faults considered. The more functional faults included in the fault model, the closer the model will approximate the faults in real implementation. Functional faults are actually circuit-level faults manifested at the functional level. Usually any one of several circuit/gate-level faults will cause the occurrence of one functional fault, and hence the number of faults in a functinal fault model will naturally be much

smaller than that of the equivalent circuit/gate-level fault model. Figure 3 shows the relationship between these two levels of fault models. The universal fault, F_u, represents the set of all possible physical faults in circuit/gate-level. After a selective, direct inclusion of faults, a circuit/gate-level fault model can be established. The functional level fault model, on the other hand is developed through selective inclusion of functional level fault types F_i's. The F_i's are in turn originated from circuit/gate-level fault cases by fault equivalence and/or fault dominance [9].

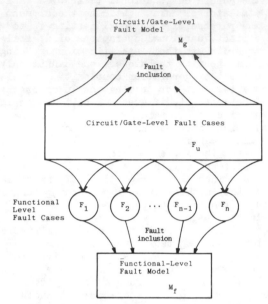

Fig. 3 Close relation between functional-level and circuit-level fault models.

There are usually two types of fault models for testing complex digital devices (e.g., microprocessor). One is called the universal fault model which takes all faults into account. The other is called funcitonal fault model which partitions all possible faults under consideration into several types according to different functions. Typical examples of functional faults are data transfer faults, data storage faults, data manipulation faults, instructon decoding faults, or instruction execution faults. Conceptually, the microprocessor testing approach based on microprogramming concept [11] (to be described in section 4-5) implies the use of universal fault model, whereas the RTL method [12-14] (to be described in section 4-4), for example, adopts the functional fault model.

Fault coverage is calculated based on the fault model used. Fault coverage may be defined based on broad sense or narrow sense. The broad sense of fault coverage is defined as the ratio of faults detected with respect to the set of all existing faults. Whereas, the narrow sense of fault coverage is the ratio of faults detected with respect to the set of all faults considered. The latter is usually used along with a fault model to simplify the measurement of fault coverage. Figure 4 shows the inclusive relation of real world faults and faults detected by testing. Funcitonal fault coverage in this case is expressed as F_d/F_c which is always less than or equal to 1.

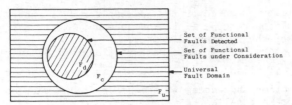

Fig. 4 Fault domain and fault coverage.

Fault coverage is used as a reference to show the effectiveness (i.e., how good) of a test generation algorithm. But fault coverage alone cannot indicate the quality of its corresponding test generation algorithm since the measurement of fault coverage is relative to the fault model used. For simplicity, most fault coverages are measured using simplified simulation which simulates at the gate level all single-occurring faults which are to be detected by test input at a higher level (e.g., RTL level). The more functional faults included in a funcitonal fault model, the more physical faults (e.g., bridging faults, multiple faults and technology dependent faults) will be covered, and hence the closer the fault coverage will reflect the actual faults detected.

The evaluation of fault coverage data is usually done by simulation. Essentially, for a system with n modeled faults, the conventional simulator models the behaviors of n+1 devices: the fault-free device and the n faulty devices. In most cases, only single faults are considered to reduce the complexity. Due to the inherent limitations of simulators (e.g., embedded assumptions on the simulation model, distortion of simulated cases, internal design errors of simulator, etc.), the fault simulator results are often only approximations to those actually observed during actual device testing. This is true even if the simulator models all reasonable types of faults [15]. There is a new statistical fault analysis technique proposed recently by Jain and Agrawal [16]. Based on the experimental results in [16], this kind of fault simulation approach will exhibit the proper fault measuring capability while it may save a lot of time-consuming operations in conventional fault simulation by making use of a promising statistical technique.

4. CURRENT FUNCTIONAL TESTING APPROACHES

In this section, we describe most of the current major approaches and technques appearing in literature for functional testing of digital LSI/VLSI systems. The scope of devices which can be tested by these techniques includes single-chip microprocessors, peripheral supporting chips, special-purpose functional chips and memory chips. The sequence for describing each approach/technique is roughly chosen based on the nature of the described content and does not imply the author's order of preferrance. For clarity, each of them is described within one subsection.

4.1 Abstract Execution Graph Technique

This technique is for the funcitonal testing of microprocessors. Roboch and Saucier [8] proposed that, each instruction of a microprocessor is represented by an "abstract execution graph". Test generation of the microprocessor can be performed based on the set of abstract execution graphs derived from

its instruction set. In such a graph, memory elements, including source and destination registers, are represented as circle-shape type-1 nodes, and the microoperations performed by the instruction are represented as square-shape type-2 nodes. The archs represent data flows. An example of "ADDA n, x" from the MC6800 microprocessor is given in Figure 5. This instruction first computes the effective address by adding n to x, using this effective address for fetching the operand. Then the operand is added to register A and the sum is stored in register A. The abstract execution graph for every instruction in MC6800 can be derived similarly. All instructions in a microprocessor are classified into several groups according to their graph structures. Then exhaustive testing of all nodes within each instruction group is performed. The results obtained from this technique are sets of test procedures.

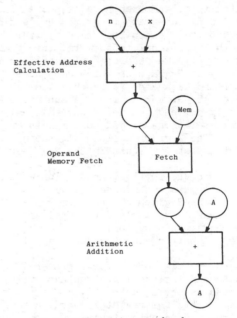

Fig. 5 An illustrative example of "abstract execution graph" of instruction.

The test generation procedure using this technique is given below:
(1) Describe the microprocessor-under-test as a set of abstract execution graphs (AEG) using its functional specification (e.g., instruction set).
(2) Verify that each AEG actually represents the specified function.
(3) Deduce all AEG's into minimum groups, each group contains the AEG's with the same graph structure.
(4) Find an order to test each member of the reduced graphs set using the complexity of each corresponding graph based on either start-small or start-big approach.
(5) Perform the testing of each memory element (type-1 node) and each microoperation (type-2 node) in each AEG-under-test.

Both start-small and start-big strategies (as described in section 2) can be used in generating test procedures. The start-small testing is implemented by testing instructions from class C_i before instructions from class C_{i+1}, where i denotes the order of the class. Each AEG in the reduced graph set obtained in step (3) above represents a class of instructions. There exists an order among these classes of instructions. The order of the instrucion class is derived from the degree of complexity of its corresponding "abstract execution graph". To make the testing complete, two stages of test are proposed: the structure of the abstract execution graph, and the correct function of each node. Memory elements are checked by running appropriate (heuristic) test patterns and every microoperation is checked using every possible operand combination. Essentially stuck-type faults and microoperation-level functional faults are considered.

4.2 System Graph Technique

Wtih a different graph theoretic approach, Thatte and Abraham [17] proposed another method which makes use of register transfer level description of microprocessors. Instructions are classified into three types: transfer (T) type, manipulation (M) type, and branch (B) type. A directed graph called system-graph (S-graph) quite different from the "abstract execution graph" is derived based on the functional, structural information and instruction set. Therefore, a microprocessor is expressed by an S-graph in which each node represents a register and each edge represents the execution of instruction or subinstruction. A subinstruction can be logically defined as one of several register transfer level operations needed to complete an instruction. The S-graph depicts all the derived register transfer paths within the device-under-test.

The procedure for test generation using system graph technique is summarized below:
(1) Derive the S-graph of the microprocessor-under-test using available funcitonal and internal structural information.
(2) Assign label to each node and edge in the S-graph.
(3) Set up the test order for register decoding faults and instruction decoding faults based on results obtained in (2).
(4) Derive tests for register decoding faults and instruction decoding faults.
(5) Derive test sequences to test data transfer and data storage faults among register transfer paths.

An example of an S-graph based on part of a hypothetical microprocessor is shown in Figure 6. The superscript in each edge indicates the subinstruction sequence. I_2^1, for example, represents the first subinstruction of I_2.

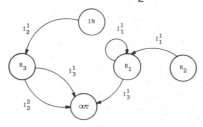

Fig. 6 An illustrative example of S-graph.

I_1: add the contents of R_1 and R_2 and store the results in R_1(M-class); I_1^1 represents both register transfer operations from R_1 to R_1 and from R_2 to R_1.

I_2: jump instruction (B-class); I_2^1 is the first sub-instruction to fetch the address for JUMP instruction from memory; I_2^2 is the second subinstruction to access the referenced memory.

I_3: store R_1 into main memory using implied addressing (T-class); I_3^1 indicaters that data transfer operation and access to the referenced memory both occur in parallel.

R_1, R_2: general-purpose registers

R_3: program counter

IN,OUT: external memory at input/output devices (including memory unit).

The microprocessor functions considered in this method are register decoding, instruction decoding and control, data storage, data transfer, and data manipulation. Individual functional fault model is derived from each of these primary functions. Stuck-type and bridging faults are considered in data storage and data transfer faults. Whereas three types of faults: f_i/ϕ, f_i/f_j, and f_i/f_i+f_j are considered in register decoding and instruction decoding faults. Fault f_i/ϕ means that operation f_i is not executed, f_i/f_j denotes that f_j is executed instead of f_i and f_i/f_i+f_j means that f_j is executed in addition to f_i. In [17], separate fault testing procedures are proposed to test these functional faults (except data manipulation) according to a testing order decided by the label of the associated edges and nodes in that specific function under test. The label of a node or an edge is a number representing the degree of ease of observability from OUT node to that node or edge in S-graph. Under this definition, all register "read" instructions which have smallest label (since they are directly connected to OUT node) are tested first.

The functional faults specified for data storage and transfer functions are stuck-type and bridging faults. A valid test corresponds to the execution of a sequence of instructions which transfer complementary data patterns from the input port to the output port. The faults for register and instruction decoding are failures to address the correct register or to execute the correct instruction. Tests for these faults involve tracing the sequence of instructions from input to output ports. A set of test procedures will be obtained using this technique.

4.3 State Transformation Graph Technique

Lai [18] proposed an ambitious functional testing methodology for general digital systems. A new graph language called state transformation graph (STG) language was designed to describe the digital system to be tested.

The test generation procedure is as follows:
(1) After the graph description of a digital system is prepared, it is inputted to the "functional analyzer" along with a set of graph level fault models chosen by the user. For each graph level fault model and each covered graph primitive (a graph primitive is defined as a basic graphical symbol used to describe the system-under-test at a certain graph level), the functional analyzer generates a parameterized test that detects the fault at that primitive. A parameterized test

consists of a pair of symbolic assertions.
The first assertion states the condition that
must exist before the test cycle. The second
assertion states the expected changes in the
machine state after the test cycle.

(2) The "test case synthesizer" is called to substitute the formal parameters by bit patterns obtained from the test pattern database. The
purpose of the test case synthesizer is to
bring together those functional tests and test
patterns that has been developed through lower-
level, more implementation-oriented fault
models.

(3) Finally, the "test program synthesizer" is
called to map these test vectors into test
program segments. The output of the test
case synthesizer are actual test cases expressed as a pair of assertions which are the
same as the parameterized test they come from
except that the parameters in both assertions
have now been substituted by numbers.

The STG has its root in data flow graphs and
can describe digital systems at all levels ranging
from user-defined primitives of arbitrary complexity
down to logic gates. An STG graph is a directed
graph in which edges represent logical data/control
paths while nodes represent data transformation
operators. Data passes along in the form of
tokens with at most one token per path. A node
can "fire" when it has all the required tokens on
its input paths. The firing of a node results in
the removal of input tokens and the placement of
new tokens on its outputs.

In Figure 7, a rough STG description of the
PDP-8 minicomputer is shown. A detailed STG description of logical AND instruction is shown in
Figure 8. When the logical AND instruction is
executed, the content of accumulator AC is logically
ANDed with the content of the memory location specified by the effective address and the result is
placed in AC. The notation AC-> and M[]-> represent reading from accumulator and memory respectively
whereas the notation AC<- stands for writing into
the accumulator.

The functional analyzer is the heart of this
technique. It first produces parameterized tests
for faults considered by consulting the graph level
fault model chosen to select a set of primitives
and corresponding faults. Then, test generation is
performed for each fault case using techniques similar to those used in gate-level D-algorithm such as
forward drive, backward propagation, and path justification. A complete set of parameterized tests
which detect those faults considered can be obtained
by successively processing each fault case individually. Some heuristic techniques may be used to
recover from failure in the "trial and error" path
traversing. The test program synthesizer produces
test program segments according to the numeric contents of test cases and instructions of the system
under test. The results obtained using this technique is a set of test program segments.

4.4 Register Transfer Language (RTL) Technique

The RTL description of a digital system can be
reasonably derived from the function description
and user-available information (e.g., user manuals
and application notes) of the system. Using RTL,
the behavior of an LSI/VLSI device is comprehensively described, and functional faults derived from

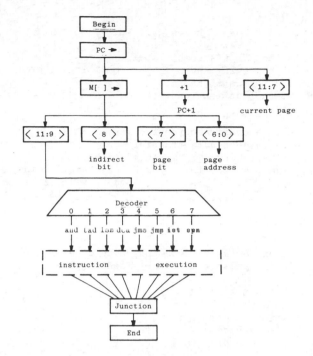

Fig. 7 State Transformation Graph of PDP-8 :
instruction decoding.

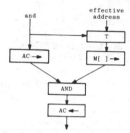

Fig. 8 STG of a PDP-8 instruction : AND.

them is studied. Su and Hsieh [12] proposed two
approaches for functional testing based upon the RTL
description. The first approach constructs a data
graph from the RTL description and uses the existing
algorithm such as the D-algorithm [19] or path sensitizing method to generate the test pattern for functional faults. In the second approach, the symbolic
execution technique [20] is employed and a test
generation algorithm called S-algorithm is developed
[14] to generate test input patterns for detecting
functional faults in the RTL description. Symbolic
execution is like normal program execution except
that symbolic values are used instead of actual variable values during the process of symbolic execution.
The test input pattern of a specific RTL-level fault
is derived by comparing results of symbolic executions for a "good" (fault-free) machine and a "bad"
(fault-injected) machine. First, the test patterns
for data storage and data transfer faults are derived using heuristic test generation procedures.
Then, every remaining functional fault under consideration is enumerated based on the RTL fault model
and is functionally injected into the fault-free RTL
description. The set of functional faults is derived
from a fault analysis for all distinct RTL statements
in the device-under-test. The comprehensive set of
test inputs is obtained after each member in the
functional fault set is processed.

A formal definition of a RTL statement is

defined as [13]:

$$k:(T,C)\ R_d \leftarrow f(R_{S1}, R_{S2}, \ldots, R_{SM}), \rightarrow n$$

where

k is the RTL statement label

T is the timing and C is the condition to execute this statement

R_d is the destination register

R_i is the i-th source register

f is an operation on source registers

$\leftarrow$ represents data transfer

$\rightarrow$ n represents a jump to statement n

For example, the following RTL statement means that when T5=C8=1, the sum of R_1 and R_2 will be stored in R_3, and then the program jumps to RTL statement numbered 12.

$$9:(T5,C8)\ R_3 \leftarrow R_1 + R_2, \rightarrow 12$$

Based on the above notation, nine categories of functional faults can then be identified as follows:

(1) label faults (k/k')
(2) timing faults (T/T')
(3) condition faults (C/C')
(4) register decoding faults (R_i/R_i')
(5) function decoding faults (f/f')
(6) control faults ($\rightarrow n/\rightarrow n'$)
(7) data storage faults ((R_i)/(R_i)')
(8) data transfer faults ($\leftarrow/\leftarrow'$)
(9) data manipulation (function execution) faults ((f)/(f)')

This set of derived functional faults is comprehensive because the internal functional behavior of any digital systsem can be described by a sequence of RTL statements. Functional fault dominance and fault collapse analysis may be applied to shrink the size of the fault set.

It can be justified that the above functional faults are the manifestation of physical faults (e.g., stuck-at faults and bridging faults (i.e., signal lines shorted together), etc.) at circuit level into functional faults at the RTL level. Using the above funcitonal fault model, the RTL technique can be comprehensively developed to consider more practical funcitonal faults. The test generation procedure using symbolic execution involves mainly the following six steps:

(1) Derive or prepare the register transfer description of the digital system under test using the RTL described above.
(2) Derive test generation order among function submodules (e.g., an instruction of a microprocessor represents a function submodule) in whole RTL description.
(3) Using the order obtained in last step for every function submodule, set up its symbolic execution tree for terminated paths.
(4) Perform heuristic test procedures derived from results of (3) for data storage and data transfer faults in current function submodule.
(5) Inject an RTL-level fault which has not been tested along the selected path.
(6) Set up the symbolic execution subtree of the fault-injected machine and choose one terminated path for faulty symbolic results.
(7) Derive a test pattern for current fault by comparing the symbolic results and path constraints of good and bad machines.
In testing complex digital devices (e.g.,

microprocessors), an order of testing based on the partition of function modules (e.g., instructions) is needed to efficiently perform the systematic test generation. To test instruction execution of microprocessors, for example, one may test instructions in the order decided by the number of distinct RTL statements involved in each instruction. This order may be systematically derived from the interconnection graph of the RTL description of microprocessor. Each untested RTL statement in each instruction is tested comprehensively based on RTL fault model. The fault due to the partial execution of instructions in addition to current instruction can be detected by heuristic techniques developed using the RTL fault model. The RTL technique is a promising approach for functional testing of LSI/VLSI digital systems including microprocessors.

4.5 The Microprogram-Oriented Technique

This technique is proposed by Annaratone and Sami [11] mainly for functional testing of microprocessors with microprogrammed control. It functionally describes a microprocessor as a set of microprograms derived from information available to users. The basic information required includes:

(1) The set of internal registers and functional units.
(2) The set of instructions and asynchronous control signals.
(3) The behavior of signals at external pins and internal operations at each clock semicycle which can generally be derived from the timing charts in the user manuals.

Using this technique, testing of a microprocessor is performed by testing each instruction at the microoperation level and each independent asynchronous signals observable at external pins. No specific fault model is assumed. The order of testing instructions is based on the cardinality of each instruction using the "start-small" principle. Instruction cardinality is defined as the number of independent (subsequent) access to registers. The more access to registers a microprogram has, the larger the cardinality its corresponding instruction will be. A complete test for possible microoperation-level faults requires that all instructions with all addressing modes be included in the test sequences.

The steps for test generation using this technique are:

(1) Derive the microprograms description of the microprocessor-under-test using user available information.
(2) Set up a test hierarchy among all microprograms based on the number of access to registers in each microprogram.
(3) Generate test procedures for microoperations and their sequencing within each microprogram.

As an illustrative example, the instruction "NOP" in Z-80 microprocessor will be tested first since the microprogram of this instruction, as shown in Figure 9, only consists of instruction fetching and decoding phase, and hence its cardinality is among the least in the whole instruction set. Then, basic "write" or "read" instructions like "LD A,n" (LD means "Load") will be the next group of instructions to test because such instructions have smaller cardinality among the remaining instructions.

Each microprogram corresponding to an instruction

```
FETCH:  µload AB, PC;  µread
        µload IR, DB;  µload AB, RF;  µrefresh
        µload IR, DB;  µdecode
        µinc  PC
        if  INTREQ then INT else FETCH
```

Fig. 9 An illustrative microprogram of "NOP"
      in Z-80 microprocessor

must be tested. Since the instruction decoder outputs
a starting address of the set of microinstructions
in the microprogrammed store corresponding to the incoming instruction, the length of the starting address is a critical factor because it eventually
determines the efficiency of the final test sequence.
If one can generate all possible combinations of the
starting address pattern, all instruction executions
will be fully exercised. The sharing of microinstructions among different instructions evidently
tends to reduce the number of different starting addresses because several instructions can now be integrated into a microprogram. The result obtained
using this technique is a set of test programs.

4.6 The Extended D-Algorithm Technique

The extended D-algorithm [21,22] technique is
designed for the functional-level testing of general
digital systems. It is a direct application of the
D-algorithm to digital systems described by a high
level hardware description language.

The approach proposed by Levendel and Menon [21]
is to use computer hardware description language
(CHDL) constructs (e.g., if-then-else, case) and
functional operators (e.g., addition) to describe
the behavior of a general logic network. Then
switching algebraic expressions are used to set up
D-cubes for each functional block. To derive tests
for circuits containing CHDL blocks, one has to perform D-propagation from outputs. A CHDL block may
contain different types of functional expressions
(e.g., shift, addition etc.) and cause-effect structures (e.g., if-then-else). The block may contain
memory elements, in such case, a sequence of inputs
to propagate a D or a $\overline{D}$ to some of its outputs would
be required.

The major steps of test generation algorithm
contains the following steps:
(1) Describe the device-under-test as an inter-
    connection of CHDL blocks.
(2) Insert a fault effect to a CHDL construct or a
    CHDL block containing CHDL constructs.
(3) Propagate the fault effect to an observable
    point.
(4) Justify all the decisions made in the above
    two steps by applying sequences at the primary
    inputs of the circuit.
Both switching and non-switching algebraic expressions can be included as part of CHDL constructs.
As an example of the former, we consider a JK flip
flop with output function $Q = J\overline{q} + \overline{K}q$, where q is the
present output and Q is its next output. To derive
the set of D cubes of this flip-flop, we make use of
the simple equations of

$$C^D = (a+b)^D = a^D b^0 + a^0 b^D + a^D b^D$$
$$C^D = (ab)^D = a^D b^1 + a^1 b^D + a^D b^D$$

along with simple expansion technques of Boolean expression and compute:

$$Q^D = (J\overline{q} + \overline{K}q)^D = (J\overline{q})^D(\overline{K}q)^0 + (J\overline{q})^0(\overline{K}q)^D + (J\overline{q})^D(\overline{K}q)^D$$

$$= J^0 K^0 q^D + J^D K^0 q^D + K^D \overline{q}^1 + J^0 K^D \overline{q}^D + J^D K^D q^D + J^1 K^1 q^{\overline{D}} +$$
$$+ J^D K^1 q^1 \overline{D} + J^D q^0 + J^1 K^{\overline{D}} q^{\overline{D}} + J^D K^{\overline{D}} q^{\overline{D}}$$

For the example of non-switching algebraic
expressions, let us consider the "addition" function. Analogous to the regular addition, we may
use SUM (exclusive-OR) and CARRY(AND) table shown in
Figure 10 to derive the set of D-cubes of the "addition of A and B". For example, in Figure 10(a),
we obtain the set of D-cubes of SUM part as $A^0 B^D +$
$A^D B^0 + A^1 B^{\overline{D}} + A^{\overline{D}} B^1$. The x's in the table denote unspecified values. As an example for CHDL control
structures, let us consider the following binary
cause-effect expression:
    If A Then Z ←B Else Z ←C
where A, B and C are switching expressions. This
control statement is equivalent to the following
switching expression:
    Z ← AB + $\overline{A}$C
Then, the D-drive expression can easily be derived as:
$$Z^D = (AB + \overline{A}C)^D$$
$$= (AB)^D (\overline{A}C)^0 + (AB)^0 (\overline{A}C)^D + (AB)^D (\overline{A}C)^D$$
If single D propagation is desired, then we obtain:
$$Z^D = A^1 B^D + A^0 C^D + A^D B^1 C^0 + A^{\overline{D}} B^0 C^1$$

SUM	0	1	D	$\overline{D}$	X
0	0	1	D	$\overline{D}$	X
1	1	0	$\overline{D}$	D	X
D	D	$\overline{D}$	0	1	X
$\overline{D}$	$\overline{D}$	D	1	0	X
X	X	X	X	X	X

(a)

CARRY	0	1	D	$\overline{D}$	X
0	0	0	0	0	0
1	0	1	D	$\overline{D}$	X
D	0	D	D	0	X
$\overline{D}$	0	$\overline{D}$	0	$\overline{D}$	X
X	0	X	X	X	X

(b)

Fig. 10 Table representation of primitive
       function of additon.

The fault model includes conventional data
stuck-at faults and two functional faults -- the
control fault and the general fault. To test the
control fault, each path in the control graph corresponding to the control construct must be tested.
The general faults are defined as faults whose effects are known, but cannot be modeled by stuck
variables or control faults. This kind of fault
can be tested only when their behaviors are expressed explicitly.

In [22], Breuer and Friedman proposed a system
which may generate tests for complex sequential circuits composed of functional primitive logic elements
such as counters or shift registers. Using this concept, a digital circuit may be described as interconnections of basic logic gates, flip flops, and
functional primitive elements such as counters.
Then, a path sensitization type test generation algorithm like D-algorithm can be applied for testing
each test point within the "mixed-component" circuit.
The major problems in this technique are the implication, D-drive, and line-justification of every
functional primitive element in the circuit. To
solve these problems, the concept of mapping the
control input values into an algorithm is utilized.
Primitive elements are modeled by a set of algorithms,
including primitive funcitonal algorithms and unions
of them. A functional solution language (FSL) was
designed, based on the functions of the primitive
functional elements, to specify solutions to the
implication, D-drive, and line-justification problems of each primitive functional element. The
implication problem may be considered as consisting

of forward implication and backward implication. In the former case, the values of control inputs, the present state of current basic funcitonal element and the data inputs are used to determine the next state while the latter is the inverse operation. The D-drive process consists of specifying appropriate values on all inputs in order to "propagate" an error signal (D or $\bar{D}$) on one (or more) of these signals to the next state. The line-justification process is just to keep consistent assignment of values to all inputs with respect to a certain present state.

The test generation procedure is summarized below:
(1) Describe the circuit-under-test as a mixture of basic logic gates and primitive functional elements.
(2) Map the functinal behaviors of each functional element into a set of algorithms.
(3) Enumerate and insert faults (mainly stuck-fault) in the circuit.
(4) Apply a path sensitization algorithm (e.g., D-algorithm) to choose at least one sensitizing path for each fault.
(5) For each funcitonal element along this sensitizing path, derive solution sequences for the implication, D-drive, and line-justification problems using the proposed functional solution language. Tables of control input values and tables of equations representing union and primitive algorithms of each functional primitive element will be employed in implication and line-justification process. D-drive process may produce a single-vector or a sequence of vectors.

This technique is developed in [22] using a shift register and a counter as examples. To illustrate this technique, consider a shift register shown in Figure 11 with the following functional operations: L (left shift), R (right shift), P (parallel load), H (hold; do nothing), and C (clear). Then, using the FSL description, PH^2L represents a sequence of four functinoal operations, namely a parallel load followed by two hold operations followed by a left shift. We can also specify solution for the desired operation encountered in test generation. For example, to justify $Q_1 = 1$ in Figure 11, one possible solution is: first set $A_{n+1} \leftarrow 1$, and shift the register to the right n times. This operation can be expressed using FSL as the following solution sequence

$$\{A_{n+1} \leftarrow 1, R\}H*(RH*)^{n-1}$$

where H* means that we separate each right shift by zero or more hold operations.

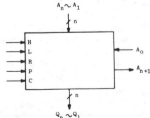

Fig. 11 A parallel load shift register.

Based on the possible input combinations of inherent control lines (e.g., L,R,P,H and C input in Figure 11), a set of potential primitive operations is formed. Then a set of union algorithms is derived from the union of the potential primitive operations. Each of the union algorithms is actually equivalent to a set of constraint equations which can be used to derive input, output and state values during the process of implication. D-drive and line justification of the basic functional element. For example, using the shift register in Figure 11, the union algorithm LPH which means "left shift" or "parallel load" or "halt" is implemented by the following set of equations:

$$Q_i \leftarrow A_i, \text{ if } A_i = q_i = q_{i-1}, \text{ for } 1 < i \le n$$

$$Q_i \leftarrow X, \text{ otherwise, for } 1 < i \le n$$

$$Q_1 \leftarrow A_0, \text{ if } A_0 = A_1 = q_1$$

$$Q_1 \leftarrow X, \text{ otherwise}$$

Where q_i is the value of output signal before operation, Q_i is the value of output signal after operation, A_i is the value appearing at input, and X is don't care.

No matter which approach is used, basically, a fault-model similar to the stuck-type fault will be used. Once the functional D-cubes of each basic functional block is derived, the classical stuck-type testing technique can still be applied without major changes. By building the basic functional blocks from elementary components, modular composition and recursive extension can then be applied during the procedure of test generation.

4.7 Behavioral-Level Techfnique

Behavioral-level description is the abstract description of digital systems using high level hardware description statements like "if-then-else", "do-while", "repeat-until", and "case". Using behavioral-level model, the total device function is described by a relatively abstract behavioral description. Behavioral-level testing is to test digital systems by considering faults which may only occur in behavioral-level model.

Behavioral-level testing is abstract in nature and there are flexible techniques employed. Two approaches are developed in GTE laboratories [23] and Texas Instrument Incorporated [24] respectively. The technique developed at GTE is to devise a behavioral test generation algorithm which is based on functional tables translated from combinations of statements of a high-level hardware description language. Every functional table maintains all information needed to define a functional block in the high-level hardware description. It consists of two parts: one specifies the conditions that determine corresponding actions to be executed and the other describes those actions in the form of algebraic expressions. Using a table-driven funcitonal extraction method, specific, enumerated functional faults can be injected by users. There are two procedures needed in GTE's behavioral test generation algorithm to detect a fault: forward driving and backward tracing. The main steps in forward driving procedure include:
(1) Derive the funciton table, generating the fault list and initializing the functional block.
(2) Identify the location of each fault related table, and then insert composite values that

differentiate a good machine from a bad machine.

(3) Scan the appropriate table at the locations identified in a digital system, testing is then ready to commence. Techniques to monitor Step (2) to select tests are dependent on the type of functional faults (data faults, control faults, operation faults, and user-defined faults are considered here). This step will be performed exhaustively until a test is derived.

The procedure for backward tracing is, for each functional block, to find inputs to justify its outputs.

As an example for this technique, let us consider testing of a simple presettable binary counter. The test generation can be carried out in following five steps:

(1) Describe the functional primitives (e.g., direct clear, count/load control etc.) of this binary counter in a functional table.

(2) Perform initialization that just clears the output.

(3) Exercise the load operation with proper inputs to test data faults at input and output lines.

(4) Apply composite values in a condition expression with the control variables to see any discrepancy at the output to test control faults, (e.g., reaching the count limit and reset to initial count).

(5) Apply a transfer sequence to bring the state that verifies the operation to be performed for testing operation faults, (e.g., increment the counting operation).

The behavioral-level test technique developed in Texas Instrument Incorporated is quite an abstract one. It is intended to be integrated into the total design process at the early stage of the design cycle. Behavioral-level testing then is to insure that the device-under-test is behaviorally exercised. A digital system is considered as composed of a set of behavioral-level circuit models described by a high-level, Algol-like language in the form of a computer program. The language is specifically developed for hardware description and is hierarchical in form. Using this behavioral-level circuit model, circuit arithmetic functions such as addition, subtraction, and logical functions such as AND and OR are represented as program operators. Program variables may represent single or vectored signals or internal state values in the circuit. After the behavior of a digital system is described, techniques to monitor the completeness of the behavioral exercise are used instead of fault modelling. For the simple example of high-level assignment statement: S := A + B, test generation software attempts to behaviorally exercise the variables S, A and B, and the "add" and "assignment" operators.

Various tasks of the behavioral-level automatic test generation software are:

(1) To determine which functional components of the behavioral model are to be exercised to assure adequate test coverage.

(2) To sensitize a path from the primary inputs and existing internal state values through the behavioral description to the chosen component.

(3) To sensitize a path from the site of fault to a primary output.

(4) To resolve value conflicts along the path and to determine the necessary input patterns to cause the predetermined local exercise sequence to be applied to the target fault, and the results to be propagated to a circuit port.

Since failure modes are not explicitly modelled during test development, failed responses are not predicted, and the process yields only an "OK' or "NOT OK" test response.

4.8 Memory Functional Testing Techniques

The functional testing of memories from users' point of view is the application of a sequence of "read" and "write" operations to check the storage funciton of each memory cell. A lot of work has been done in the area of static functional testing of LSI/VLSI semiconductor memories. Interested readers are encouraged to find more detailed discussions and references from [9-10].

5. COMPARISON AMONG DIFFERENT TECHNIQUES

Each of the functional testing techniques described in the last section has its specific feature. Since most of them are still under development, it is not easy to make a decisive judgment among them. However, a brief comparison among them is given in Figure 12 based on the present "appearance" of each technique.

Several observations can be made about the first approach in graph-theoretic technique. First, its fault coverage rests on the exhaustive testing of particular functions and memory elements. It is an ad hoc method and will be specific to the devices-under-test (e.g., MC6800). Moreover, it presumes a black box view of the microprocessor in which each function is independently tested. These factors lead to excessive test length. However, this technique is easily understood.

The S-graph approach is attractive for several reasons: It is based on the minimal available information about the microprocessor (e.g., instruction and register sets). Microprocessor operations are decomposed into a set of functions. For each function, a specific single fault model is defined and a comprehensive test is rigorously derived. Finally, the fault coverage is evaluated through a stuck-type fault simulator. However, the S-graph is derived manually. No automatic generation procedure for S-graph is presented. Furthermore, the derived S-graph should be verified before it is put into use in test generation. No testing procedure for data manipulation fault is presented. The estimated length of test sequence of this method is $O(n_R^3 + n_I^2)$ where n_R and n_I are the number of registers and instructions respectively. To improve the computational efficiency, Abraham and Parker [25] proposed a simplified fault model with revised test sequence length of $O(n_R * n_I)$. One major shortcoming of this approach is that it does not include partial instruction execution and instruction manipulation faults which can be attacked using the RTL technique.

The ideas in state transformation graph technique are explicit and practical. A framework to systematically attack functional testing of general digital systems has been laid down in an ambitious fashion. This technique uses multi-level faults for multi-stage test generation. Modularity and flexibility are taken as basic design considerations.

Item to Be Compared	Abstract Execution Graph	System Graph	State Transformation Graph	Register Transfer Language	μ-program-Oriented	Extended D-algorithm	Behavioral Level	Memory
Major Application Field	μp	μp	GDS	GDS	μP	GDS	GDS	Memories
Description Media	instruction execution graph	register level system graph	state transformation graph	register transfer language	μ-program	HDL	HDL	human language
Functional fault Model Level	μ-instruction	assembly instruction	user-defined multi-level	register transfer instruction	μ-instruction	user-defined functional models	abstract or none	cellular 0/1 fault model
Include Comprehensive Fault Types?	NFI	Yes	NFI	Yes	NFI	Yes (partial)	Yes (partial)	Yes
Algorithmic Test Generation	Yes	Yes	NFI	Yes	No	NFI	NFI	Yes
Algorithm Features	device dependent	high complexity	NFI	high complexity	NFI	NFI	NFI	low complexity
Performed Experiment?	Yes	Yes	Yes	undergoing	NR	NR	NR	Yes
Estimated Fault Coverage	NFI	High	NFI	High	NFI	NFI	NFI	High
Result Generated	TP	TP	TP	TD	TP	TD/TP	TD/TP	TP

Abbreviations

GDS-General Digital Systems;
NFI-Need Further Investigation;
NR-No Report;
TP-Test Program;
TD-Test Data;
HDL-Hardware Description Language;
μp-microprocessor

Fig. 12 Brief comparison among major functional testing techniques.

Users are allowed and are supposed to decide many control factors on the test generation process such as defining fault models of various levels and defining details of graph primitives. To demonstrate the practicality and efficiency of this technique, tests were generated for the DEC PDP-8 minicomputer and compared against test programs supplied by the computer manufacturer using the ISPS fault simulation facility available in Carnegie-Mellon University. The claimed experimental data using this technique shows better result. The present version of this technique deals only with loop-free graphs. Most of difficult test generation issues such as functional level fault modelling or test pattern data base are left for individual user to solve. So there is still a lot of work to be done to make this ambitious technique more complete and more valuable.

The benefits of the RTL technique are generality, clarity, easy understandability, and precise description of the behavior of a given digital system. More research work is still being done for this technique [14]. The algorithm currently under development in RTL technique is comprehensive, and more funcitonal faults than other approaches (e.g., graph approach) are covered in the fault model developed. Like graph approach, the automatic generation of RTL description from the available information of the device-under-test is nontrivial from the viewpoint of practical applications. The way a digital device was actually implemented usually cannot be uniquely determined by user-derived RTL sequence due to lack of detailed product information (e.g., gate-level circuit). In this case, a little redundancy for derived RTL descriptions may be unavoidable. Though RTL technique provides a comprehensive approach to define functional faults, it still needs to be exploited in more detail on the problem of test complexity along with algorithmic test generation using the promising symbolic execution technique [12].

The microprogram-oriented approach is attractive because microinstructions are basic to the functions of the control unit. For control functions, savings in test length are possible for microprogrammed control paths. The fault coverage remains to be argued from the practical point of view. The major shortcomings of the technique are no analysis of the scheme, no algorithmic test procedure and hence hard to automate in practical situations.

The basic foundation to use D-algorithm in functional level has been developed in the extended D-algorithm technique. The derivation of D-cube of basic functional block in high-level hardware description language is explored. The two techniques mentioned here are making use of the existing technique of gate-level D-algorithm. However, no existing technique can easily construct D-cube for each funcitonal block, mainly because even a simple funcitonal block may need several different functional representations. More work is needed to explore and to automate this technique. Also, this technique fails to test the faults performing extra or fewer operations.

Behavioral-level techniques are highly abstract and flexible. There are diverse interpretations of the behavioral-level techniques and little reported work has appeared in literature. The two behavioral-level techniques presented in the last section are proposed and experimented by professionals in industry under the "do and improve" experimenting principle. More study and analysis works are needed to develop this technique.

6. CURRENT TRENDS

Due to the small number of published research papers with solid results in the functional testing of digital LSI/VLSI field and the increasing needs of industry for additional techniques, functional testing is still one of the most active areas in the design automation (DA) development and quality control (QC) process [26]. The techniques presented here are representative selection of recent papers and this paper is not intended to be a complete survey on this subject. In most cases, detailed description of each technique is omitted (partly because

484

most of them are still under development).

After overviewing most of the related research works appearing in published literature we would like to briefly point out in this concluding section, several further development suggestions for functional testing development.

Although various functional testing techniques have been proposed by researchers, most (if not all) of them need further development and/or improvement to meet the following criteria:

(1) From the practical point of view, every functional testing technique should be designed for the ease of automation (i.e., implemented by computer program). This suggests the use of explicit, well-defined hardware description language to describe the complex devices, the use of compact algorithmic expressions to describe the derived test generation procedures, and the consideration of test generation strategies in early system design phase.

(2) Test generation algorithms should be proposed along with detailed analysis of the algorithms themselves.

(3) Complete theoretic analysis of fault coverage and comprehensive time/space complexity of the test generation algorithms should be performed.

(4) Quantitative fault coverage measures by computer experiment/simulation or by hardware tester (automatic test equipment) should be performed to demonstrate the effectiveness of the approach proposed.

(5) The functional fault model should be general enough to meet the future challenges in VLSI technology.

With higher abstraction level than functional level, behavioral-level techniques suggest a feasible way to solve the problem of increasing functional complexity of future digital VLSI systems. Yet, comprehensive behavioral-level fault modelling, fault insertion and fault propagation techniques still need to be developed.

One interesting trend to enable the efficient testing of complex devices is the development of "hybrid" approaches combining functional-level technique and behavioral-level technique. Bellon et al presented in [27] a set of test principles for automatic generation of microprocessor test programs using this idea. To make the functional testing more practical, it is a good idea to consider method employing more than one technique in testing complex systems. The development of the 32-bit microprocessor in Bell Laboratory [29] serves as a good practical example to this approach.

Furthermore, efficient algorithms should be developed for testing digital VLSI (including bit-sliced microprocessors) as well as general digital systems containing mixed logic (e.g., mixture of VLSI, LSI, MSI, and SSI integrated chips). Basic work on the analytical functional test approach for AMD2900 bit-sliced microprocessor has been done [30]. But a general functional test procedure independent of specific products is more desirable. Solid, comprehensive definition of fault coverage in functional level and its practical measures should also be developed to give criteria in judging the quality of the developed algorithms.

REFERENCES

[1] E. Muehidorf & A. Savkar, "LSI logic testing - an overview", IEEE Trans. Computers, Jan. 1981.
[2] T. Williams & K. Parker, "Design for testability - a survey", IEEE Trans. on Computers, Jan. 1982.
[3] S. Su, "Computer-aided design and testing of digital systems and circuits", (invited paper), Proc. Jordan Int'l Elect. & Electronic Eng. Conf., 1983.
[4] D. Siewiorek & R. Swartz, "The theory and practice of reliable system design, Digital Press, 1982.
[5] D. Siewiorek & K. Lai, "Testing of digital systems", Proc. IEEE, Oct. 1981, pp. 1321-1333.
[6] M. Lin & K. Rose, "Applying test theory to VLSI testing", Digest of papers, 1982 Test Conf.
[7] T. Middleton, "Functional test vector generation for digital LSI/VLSI devices", Digest of papers, 1983 Test. Conf., pp. 682-691.
[8] C. Robach & G. Saucier, "Microprocessor functional testing", Digest of papers, 1980 Test Conf.
[9] M. Breuer & A. Friedman, Diagnosis and reliable design of digital systems, Computer Sci. Press 1976.
[10] J. Crafts, "Techniques for memory testing", Computer, October 1979, pp. 23-31.
[11] M. Annarartone & M. Sami, "An approach to functional testing of microprocessors", Proc. 12th Symp. on Fault-Tolerant Computing, June 1982, pp. 158-164.
[12] S. Su & Y. Hsieh, "Testing functional faults in digital systems described by register transfer language", J. of Digital Systems. Also, Digest of papers, 1981 Test Conf., pp. 447-457.
[13] Y. Min & S. Su, "Testing functional faults in VLSI", Proc. of 19th Design Auto. Conf., 1982.
[14] T. Lin, "Functional test generation of digital LSI/VLSI systems using machine symbolic execution technique", Ph.D. thesis proposal, Comp. Sci., SUNY Binghamton, Apr. 1984.
[15] R. Wadsack, "VLSI: how much fault coverage is enough?", Digest of papers, 1981 Test Conf.
[16] K. Jain & V. Agrawal, "STAFAN: An alternative to fault simulation", Proc. of Design Automation Conf., 1984.
[17] S. Thatte & J. Abraham, "Test generation for microprocessors", IEEE Trans. on Computers, June 1980.
[18] K. Lai, "Functional testing of digital systems", Ph.D thesis, Comp. Sci. Dept., Carnegie-Mellon Univ. Dec. 1981.
[19] J. Roth, "Diagnosis of automata failures: a calculus and a method", IBM J. of R&D, July 1966.
[20] J. Oakley, "Symbolic execution of formal machine descriptions", Ph.D. thesis, Comp. Sci. Dept., Carnegie-Mellon Univ., 1979.
[21] Y. Levendel & P. Menon, "Test generation algorithms for computer hardware description language", IEEE Trans. Computers, July 1982, pp. 577-588.
[22] M. Breuer & A. Friedman, "Functional level primitives in test generation", IEEE Trans. Computers, March 1980, pp. 223-235.
[23] K. Son & J. Fong, "Automatic behavioral test generation", Digest of papers, 1982 Test Conf..
[24] W. Johnson, "Behavioral-level test generation", Proc. 16th Design Auto. Conf., 1979, pp. 171-179.
[25] J. Abraham & K. Parker, "Practical microprocessor testing: open and close loop approaches", 1981 Test Conf., pp. 308-311.
[26] M. Breuer, A. Friedman & A. Iosupovicz, "A survey of the state-of-the-art of design automation", Computer, Oct. 1981, pp 58-75.
[27] C. Bellon et.al., "Automatic generation of microprocessor test programs", Proc. 19th Design Automation Conf., 1982, pp. 566-573.
[28] R. David, P. Thevenod-Fosse & X. Fedi, "What about random testing of microprocessors?", Research report, submitted to Computer Review, 1984.
[29] R. Wadsack, "The BELLMAC CPU's: Design verification, debugging, and fault coverage tests", 1982 Bell System Conf. on Electronic testing.
[30] T. Sridhar and J. Hayes, "A functional approach to testing bit-sliced microprocessors", IEEE Trans. Computers, Aug. 1981, pp. 563-571.
[31] K. Saluja, L. Shen & S. Su, "A simplified algorithm for testing microprocessors", 1983 Test Conf., pp. 608-675.
[32] L. Shen and S. Su, "A functional testing method for microprocessors", Proc. 14th International Symp. on Fault-tolerant Computing, June, 1984.

Section IV-B
Fault-Tolerant and Error Detection

Fault-Tolerant Design for VLSI: Effect of Interconnect Requirements on Yield Improvement of VLSI Designs

TÜLIN ERDIM MANGIR, MEMBER, IEEE, AND ALGIRDAS AVIŽIENIS, FELLOW, IEEE

Abstract—In order to take full advantage of VLSI, new design methods are necessary to improve the yield and testability. Designs which incorporate redundancy to improve the yields of high density memory chips are well known. The goal of this paper is to motivate the extension of this technique to other types of VLSI logic circuits. The benefits and the limitations of on-chip modularization and the use of spare elements are presented, and significant yield improvements are shown to be possible.

A mathematical model for yield is developed, and the effects of interconnect densities and logic module complexities in yield improvement are investigated for regularly designed VLSI circuits. The chip contains redundant modules which can be substituted for flawed modules at the probe testing stage. Because of the interconnection complexity involved in the reconfiguration, redundancy is employed only for the logic areas, while their interconnections remain nonredundant. It is shown that yield improvements of two or more over the yields of nonredundant designs can be expected even though redundancy is provided only for the logic areas. It is also observed that the yield improvement saturates above a limited amount (10 percent) of redundancy. This indicates that the chips can be designed around an optimal amount of additional logic on the chip to improve the yields.

Index Terms—Interconnect area estimates, redundancy partitioning, redundancy placement, regular designs, VLSI yield improvement, VLSI fault tolerance.

I. Introduction

PAST improvements in circuit integration have been reductions of defect densities to achieve economically viable yields of perfect devices. This is neither possible nor necessary for VLSI. The proposed structures are so complex ($>10^5$ gates per chip) and the catastrophic defect size is so small (μm), that there is a great likelihood that we will never be able to make them perfect. On the other hand, we can now afford to put redundancy on the chip to make fault-tolerant design viable, by allocating some of the chip area for this purpose. However, a careful study of how and where this redundancy should be placed is necessary to avoid excessive penalty in added area and possibly more complex processing. Currently, there is no analytical model available to optimize the placement of redundancy.

Manuscript received October 8, 1981; revised January 12, 1982. This work was supported in part by the Office of Naval Research under contract N00014-79-C-0866.

T. E. Mangir is with Mangir Associates, Los Angeles, CA 90066.

A. Avižienis is with the Department of Computer Science, University of California, Los Angeles, CA 90024.

The following methods have been suggested to improve the processing yields of VLSI circuits:

1) modifying the design rules, thereby reducing the probability of yield loss due to critical spacings, such as contact to poly layers (defect avoidance);

2) modifying the designs and the design methodologies, so that redundant elements (spares) can be introduced on the chip to compensate for defective areas. Redundancy can be introduced into the design in different forms such as coding or replication.

In previous work redundant elements to improve the processing yield have been proposed and implemented for high density memories [2], [3], [6], [7], [16]. The approach is to substitute spare word or bit lines (address lines) for defective ones during the testing of the chips. The information which identifies good and bad locations can be stored on the chip by one of several of the following alternatives:

1) latches,
2) laser personalization [2], [6], [7], [10],
3) electrically programmable storage elements [8],
4) electrically programmable links [13].

Latches and other electrically alterable memory elements can be reprogrammed in the field, if necessary. Therefore, the redundancy included on the chip for yield improvement can also be used for field maintainance/reliability improvement.

Laser personalization involves blowing the fuses to spare elements [6], disconnecting the contacts to faulty elements [2], or joining interconnect layers together [10]. Thus it is permanent, and once done cannot be changed. In [6] laser pulses are used to burn fuses to isolate the faulty elements from the power supply. This technique is particularly suited to cases where interconnections between modules are done in serial or in pipeline fashion.

The evolution of yield with chip area has been extensively studied in the literature. Various theories have been presented and analytical expressions derived to fit statistical data based on defect density distributions [1], [15], [17], [18]. All work is based on random defect distributions, and the papers differ in their treatment of various defects being distinguishable or indistinguishable from each other. Generally, it is assumed that the yield falls off exponentially with increasing chip area. This occurs when a Poisson distribution of defect densities is assumed. However, it has been observed that in practice yield

Reprinted from *IEEE Trans. Comput.*, vol. C-31, pp. 609–615, July 1982.

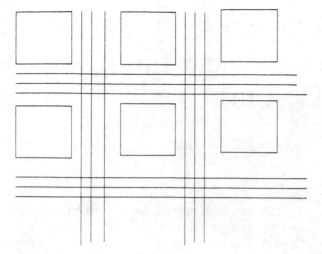

Fig. 1. Two-dimensional array of modules and interconnects.

falls off less sharply [1], but nevertheless significantly with increasing chip area.

The other attempts at yield calculations which take redundancy into account have been reported for memory circuits by Schuster *et al.* [16] and Stapper *et al.* [18]. The model presented in [16] is also based on the exponential dependence of yield on the active chip area. The defects are separated as correctable, uncorrectable, and gross imperfections, and the net yield is calculated as the product of these three independently calculated yields. Stapper *et al.* [18] have described a yield model with redundancy based on the Gamma distribution of defects for RAM chips. Mixed Poisson statistics are used to arrive at a yield expression describing yield of the redundantly designed memory chips. That approach is based on the data obtained from a specific process line for a specific product (i.e., RAM chips), and therefore may not be applicable to other circuit types or process facilities. Also, for our purposes we are interested only in the cumulative effect of the defects in the logic areas, rather than the individual defect types. In this work we have used an average random defect density and Bose–Einstein statistics to derive the yield expressions. This approach is readily extendable to yield calculations with redundancy, and also preserves the generality of the expressions as long as defects per chip can be described with an average defect density. Neither of the references [16] and [18] accounts for the effects of the complexities of areas, connectivities between different areas, or the effect of regularity of interconnections which would affect the processing tolerances, and thus yield [19].

It is important to note that the redundant areas introduced are also as susceptible to defects as other parts of the chip, since the defects are assumed to be randomly distributed. However, the probability of the added (redundant) and the nonredundant areas both being defective at the same time is very small. Also, studies done on high density memories [16], [18] indicate that a large percentage of the failures occurs because of localized defects which affect single bits, rows, or columns of the array, suggesting that this assumption is a valid one.

In Section II a yield model with redundancy is described. In Section III the effects of module complexities and inter-connect densities are studied. Section IV discusses the effect of other factors on yield improvement, and Section V presents a summary of the conclusions of this work.

II. YIELD MODEL WITH MODULE REDUNDANCY

The idealized model of the chip, depicted in Fig. 1, consists of a two dimensional arrangement of N modules, of which $N-R$ are required to be functional, with intermodule connections running vertically and horizontally between the modules. It is assumed that all modules are identical; therefore any faulty module can be replaced by one of the R redundant (spare) modules. Examples of this type of design are: on-chip bit-slice designs, master-slice designs, array processors, and distributed-logic memories. A specific design developed at UCLA that comes very close to the model is the signed-digit arithmetic building block [20].

For the nonredundant case the yield is given by the probability of getting a defect-free chip. What redundancy allows us is to reclaim the chips with nonzero defects which would normally be considered as waste. For example, if we have one extra module on the chip (i.e., $R = 1$), by successfully substituting this extra module for the defective one, we should be able to add the yield of all the chips with a single defect to the yield of defect-free chips to improve the total yield. To describe the expected yield we assume a random defect density which has an average D_0, and statistical partitions to obtain the yield of redundantly designed VLSI chips as follows [12]:

$$Y_R = \sum_{k=0}^{\infty} \frac{(A_R D_0)^k}{(1 + A_R D_0)^{k+1}} (1 - r)^k P_{k,N,R} \qquad (1)$$

where

$$P_{k,N,R} = \begin{cases} C_{N,R} & \text{for } k \leq R, \\ \sum_{j=1}^{R} \left(\frac{j}{N}\right)^{k-j} \prod_{i=0}^{j-1} \left(\frac{N-i}{N}\right) C_{N,R} & \text{for } k > R. \end{cases}$$

In practice, the contribution of terms with $k > R$ to total yield Y_R is quite small; therefore for practical purposes the infinite sum in (1) can be replaced by a finite sum with the upper limit of R.

The bypass coverage $0 < C_{N,R} \leq 1$ is defined as the conditional probability that a module will be bypassed, given that it is faulty. The inclusion of $C_{N,R}$ explicitly allows the consideration of less than perfect bypass mechanisms and tests employed to identify the defective logic modules. Its use here is analogous to the use of the coverage parameter in the modeling of fault-tolerant computer systems [14]. Even though it can be expressed as an explicit function of complexity, interconnection densities, amount of redundancy, and switching mechanisms, this dependency is treated implicitly in this paper.

Equation (1) provides us with the necessary analytic tool to investigate the yield improvement and cost due to the use of redundancy as a function of the parameters of (1). Specific attention will be given to the effect of interconnect requirements and module complexity in the following section. An elementary illustration of the use of (1) is given in Fig. 2, in

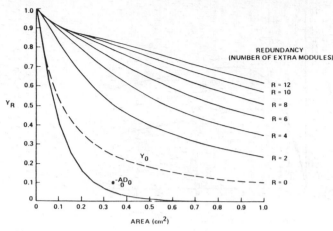

Fig. 2. Yield versus area for bypass coverage $C_{n,R} = 1$ and $D_0 = 9/$ cm^{-2}.

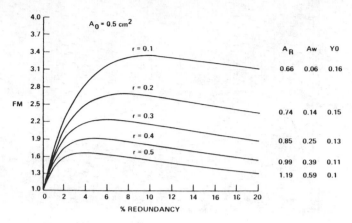

Fig. 3. Figure of merit (FM) versus percent redundancy for varying interconnect area ratios for 10^5 gates/chip $D_o = 9/$cm^{-2}, and active nonredundant chip area $A_o = 0.5$ cm^2. $N = 100$, $n_g = 1000$, $a_g = 500\lambda^2$, $y = 0.5$, $b = 0.5$, $\lambda = 1 \mu$m.

which Y_R is plotted versus the total area A_R, for several values of R and for $D_0 = 9/$cm^2, with the simplifying assumptions of $r = 0$ and $C_{N,R} = 1$. This is an upper bound for improvement, assuming essentially independent modules and perfect bypass coverage. These assumptions will be eliminated in subsequent discussions.

For cost considerations a figure of merit FM is defined as follows. To be able to produce a working chip, on the average $1/Y_0$ nonredundant chips, and $1/Y_R$ chips designed with redundancy, must be produced, respectively. (Y_0 and Y_R are the yields of nonredundant and redundant chips, respectively.) Working chips are defined to be those which perform their intended functions with respect to observable input/output relationships, when accessed at the probe level, even though there may be undetected internal faults.

When the ratio $Y_R/Y_0 > 1$, then redundancy will have increased the yield of VLSI chips. However, redundancy also increases the total area of the chips from A_0 to $A_R = NA_m + A_w$. This assumes that there is no significant increase in the area of pads, pad drivers, etc. The figure of merit (FM) is defined as

$$\text{FM} = (A_0/A_R) \times (Y_R/Y_0)$$

and we consider that a cost advantage is attained by the use of redundancy when FM > 1.

III. EFFECT OF INTERCONNECTS AND MODULE COMPLEXITIES

To incorporate the interconnect areas into the model the following two assumptions are made:

1) the number of defects (faults) in the interconnections is a function of the area of the interconnections,

2) since redundancy is provided only for the logic modules any defect falling into the interconnect area is catastrophic.

The key parameter for the study of the interconnection effect is the ratio r, defined as $r = A_w/A_R$, where A_w is the intermodule interconnect area. The term $(1 - r)^k$ in (1) represents the probability that, given k defects on the chip, A_w remains defect free. In Fig. 3 the FM is plotted versus the percentage of redundancy for a chip of nonredundant active logic

area of 0.5 cm^2, for varying values of r, with a fixed defect density $D_0 = 9/$cm^2. Note that the maximum improvement shifts to a higher percentage of redundancy for lower values of r. This is explained by the interconnect dominance of yield improvement for large ratio of interconnect areas.

In recent papers, Donath [4], [5] and Heller [9] have studied the interconnection problem for LSI, with respect to average interconnection lengths and wireability, with the objective of finding results which reflect the characteristics of the logic design process. The approach used in this paper to estimate the interconnect requirements is based on an extension of the model described in [9]. Briefly, the model considers a doubly infinite array of equal size modules. The number of interconnections originating from each module is taken to be distributed according to a Poisson distribution with parameter y and is assumed to be independent between different modules. Each interconnection length is assumed to be independently chosen according to a geometric distribution, with the mean L'. In interconnecting the digital designs, interconnects in the same net are observed to connect more than two modules. In spite of this correlation it has been observed *a posteriori* that the assumption of independent origination and termination of single two module interconnections does not introduce serious error.

The following relationship has been observed for the number T of terminals (connections) required per module [5]:

$$T = y\, n_g^b$$

where

n_g is the number of blocks (gate equivalents) in the module

y is the average number of interconnects per gate

b is Rent's exponent, $0.5 < b < 1$.

It is observed that $b = 0.5$ when the interconnections are mostly internal to the module. Larger values are observed for modules with high numbers of external interconnections.

Given some regular geometric arrangement of logic areas, one can model the interconnect requirements as a stochastic process. Interconnects will originate and terminate at modules

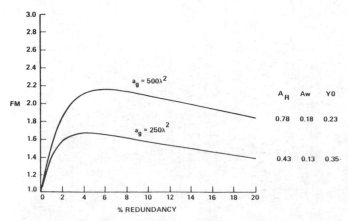

Fig. 4. Effect of size of the blocks on FM for $\lambda = 1$ μm. $n = 10^5$ gates/chip, $D_o = 5$ cm^{-2}. $N = 100$, $n_g = 1000$, $y = 2.5$, $b = 0.75$.

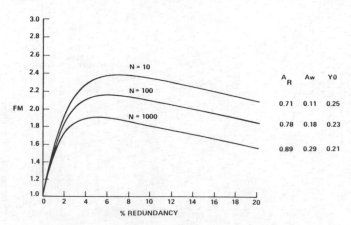

Fig. 5. Effect of number of blocks on FM for $\lambda = 1$ μm. $n = 10^5$ gates/chip, $D_o = 5$ cm^{-2}, $a_g = 500\lambda^2$, $y = 2.5$, $b = 0.75$.

according to some probabilistic models. For a given number of modules, interconnect demand is a function of the average number of interconnections per block. Another factor that affects the demand is the average length of interconnections. Consider the two-dimensional arrangement of modules as in Fig. 1.

The length of interconnects required for N modules, where the average length of interconnections is L' and $(\sqrt{A_m} + wP)$ is the distance between modules, is given by

$$L = N L' y n_g^b (\sqrt{A_m} + wP)$$

where

A_m is the module area
$\quad$ w is the average number of wires in the intermodule interconnect area
$\quad$ P is the distance between centers of the wires, i.e., pitch, which is defined by technology.

For this paper horizontal and vertical pitches are assumed to be the same and $P = 5\lambda$, where λ is the linewidth.

Then assuming that the horizontal and the vertical pitches are the same, the interconnect area between the modules is

$$A_w = L_h L_v - nA_m$$

where L_h and L_v are the total horizontal and vertical interconnect length, respectively. (Here, for simplicity we will assume that the number of interconnect channels between the modules is $\sim\sqrt{N}$. This is a good assumption for $N > 10$.) Then using the equation for A_w and Fig. 1, we get

$$A_w = 2wNP \sqrt{A_m} w 2P + Nw^2P^2. \tag{2}$$

As expected, the interconnect area is a strong function of the average number of wires in the interconnect channel. Depending on the assumptions made on w, the effect on the yield improvement will vary.

A. Effect of the Size of the Blocks

To examine the effect of the size of the individual blocks that make up the modules, we use $a_g = 500\lambda^2$ and $a_g = 250\lambda^2$, where a_g is the individual block area and each module has n_g blocks. The size of the individual blocks is an indication of how

"tight" the design is. Note that this area is assumed to include the wiring area associated with the individual block. From Fig. 4 we make the following observations.

1) For larger block areas, the wiring area and the logic area ratios are smaller than the small size block areas. This results in

$\quad$ a) lower yields because of the increased chip area

$\quad$ b) increased improvement in yield with redundancy, for the larger block area case. We also observe that the peak improvement shifts toward less added redundancy for small block areas.

2) The same relative effect is observed for different linewidths λ. This can be explained, since as block areas are decreased actual critical logic area is reduced, therefore less redundancy is needed. Also, for smaller block areas overall yield becomes more interconnect dominated. In other words, within the assumptions of the model the amount of hardcore is increased.

3) Peak yield improvement is depressed for smaller linewidths. This results from smaller chip area, and therefore higher nonredundant yields.

B. Effect of the Number of Blocks

To observe the effect of the number of blocks within a module we have plotted the yield curves for a fixed total block count n, from $n = 10^4$ to $n = 10^6$ blocks (gate equivalents) per chip. For 10^5 blocks/chip with $y = 2.5$, $b = 0.75$, $a_g = 500\lambda^2$, and $\lambda = lu$, we have

N	n_g	r	Y_0
10	10^4	0.15	0.32
100	10^3	0.22	0.24
1000	10^2	0.32	0.21.

Yield improvement for $D_0 = 5/$cm^2 is shown in Fig. 5. We see that the most significant improvement is for $N = 10$ case, which has the minimum intermodule area and highest nonredundant yield. The maximum improvement can be reached by just adding one spare module. This is encouraging for those cases where a chip is composed of multiple areas of different regularly designed blocks, where design of each area can be maximized for yield with an extra logic module.

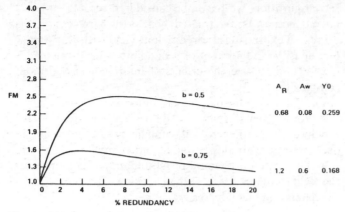

Fig. 6. Effect of complexity of logic modules for Case 1, with $n = 10^5$ gates/chip, $D_o = 5$ cm^{-2}. $\lambda = 1$ μm, varying Rent's exponent b. $N = 100$, $n_g = 1000$, $a_g = 500\lambda^2$, $y = 0.5$.

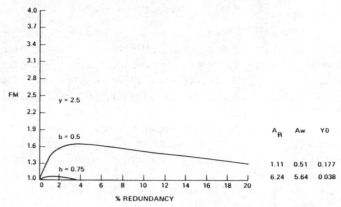

Fig. 8. Effect of module complexity for Case 1, for $y > 1$ with $n = 10^5$ gates/chip, $D_o = 5$ cm^{-2}, $\lambda = 1$ μm. $N = 100$, $n_g = 1000$, $a_g = 500\lambda^2$, $y = 2.5$.

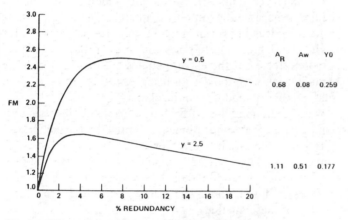

Fig. 7. Effect of complexity of logic modules for Case 1, with $n = 10^5$ gates/chip, $D_o = 5$ cm^{-2}, $\lambda = 1$ μm, varying average connections per block y. $N = 100$, $n_g = 1000$, $a_g = 500\lambda^2$, $b = 0.5$.

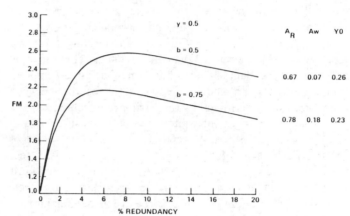

Fig. 9. Effect of complexity of logic modules for Case 2, with $n = 10^5$ gates/chip, $D_o = 5$ m^{-2}, $\lambda = 1$ μm, varying Rent's exponent b. $N = 100$, $n_g = 1000$, $a_g = 500\lambda^2$, $y = 2.5$.

C. Effect of Complexity of the Logic Modules

For our purposes complexity will be defined in two different ways. First, by the block (equivalent gate) count of each module, i.e., number of blocks/module; and second, by studying the external interconnect requirements, i.e., the number of connections required by the module. The first case has been addressed under the section for size of blocks, so in this section only the second case is considered. We examine two cases for the dependence of average number of interconnects in a wiring channel w on the average connection length per module $L'T$.

Case 1: $w \sim O(L'yn_g^b)$ in (2). This results from direct use of the equation for the number of terminals T. For the same block count and y, it is observed that b has a pronounced effect on the overall yield improvement. This can be interpreted as follows: the lower the external information required by the block (i.e., the more it is self contained), the better the improvement in yield. In Fig. 6 it is observed that even though nonredundant yields are close because of the extremely high intermodule wiring required, the maximum improvement in yield is much lower for $b = 0.75$. In fact, FM becomes less than

one for $b > 0.6$ being the crossover exponent for $y > 1$ (Fig. 8). In Fig. 7 FM is plotted for the same case, for varying y. Even though this (Case 1) dependency has been used widely to approximate the wiring area required for LSI chips, we have observed that this dependency overestimates connection requirements, especially for large logic blocks, which tend to be more self contained.

Case 2: $w \sim O(L'yn_g^b)^{1/2}$ in (2). This is a more realistic dependence for actual circuits of LSI and above complexities, where gate-to-pin ratios (or gate to connection ratios) are high. In this case, the dependence of yield improvement on both y and b is observed to be weak. However, for self contained modules a higher yield improvement is possible (Fig. 9).

It is concluded that, for modules with simpler interconnect requirements (Case 2), the number of modules becomes the dominant factor in determining the yield improvement even for large module sizes (Fig. 5). For designs where a high degree of connectivity is required between modules (low gate-to-pin ratios), module complexity becomes the limiting factor for yield improvement using spare modules, and the order of the curves in Fig. 5 is reversed, i.e., the least amount of improvement is possible for $N = 10$.

IV. Other Considerations

1) Effect of Minimum Linewidth λ*:* In general, smaller linewidths mean a smaller overall chip area for the same function capability. In this paper the total number of blocks is used as a measure of function capability. Overall, the yields are observed to be higher for narrower linewidths, as expected for the same defect density. However, from past experience it has been conjectured that the defect density increases inversely proportional to the linewidth, with a ratio $\sim(\lambda_1/\lambda_2)^2$ for scaled technologies, going from a process with λ_1 to a scaled process of λ_2, with $\lambda_1 > \lambda_2$. Also, to avoid high current density effects such as metal migration, the cross sections of the interconnects are not scaled with the process. It is experienced that thick layers of interconnects are prone to higher density of defects. Therefore, to interpret the effect of the linewidth, appropriate curves for defect should be used corresponding to these conditions. Another point is that during the lifetime of a process, depending on the maturity and changes made, both defect density and defect distribution varies. Therefore, actual yield improvement will shift up and down accordingly.

2) Effect of Extent of Failures: In deriving the yield equations we have assumed that the effects of defects in logic areas are independent of each other. Note that this assumption is not required. For a module of n_g gates, the distance between the modules is determined by the wiring requirements of the modules w. For $w = 10$ and a pitch $P = 5\lambda$, this distance is ~ 50 μm which can usually be detected during visual inspection of the chips. Also any defect of this size or larger will also affect the interconnect area. Since we have assumed that any fault in the intermodule interconnection area is catastrophic, defects extending to more than one logic area are also taken into account in the model.

3) Extension to Multiple Regions: On a logic chip which consists of multiple types of logic modules, each area can be designed to improve the yield of the overall design. In this case some areas may utilize the approach addressed here, and others may use error coding. Each type of module would be optimized for yield of that area and the overall chip yield. If we generalize the mathematical approach presented here, then the overall yield with redundancy can be expressed as [12]

$$Y_R = \frac{1}{(1 + A_T D_0)}\left[1 + \sum_{k=1}^{R} (1 - r)^k \sum_{\substack{\text{All possible} \\ \text{combinations of} \\ k \text{ defects in } m \text{ regions}}} \prod_{i=1}^{m} I_i\right]$$

where

A_T is the total chip area

R is the least amount of redundancy

r is the area ratio of common connections between m regions (i.e., hardcore) to the total active chip area

I_i is the yield improvement for each region.

V. Conclusions

In this paper we have presented a model to study the fault tolerance techniques applied in regulary designed VLSI circuits to bypass manufacturing flaws by using on-chip redundant logic modules. Because of the complexity involved in reconfiguration, we have not considered redundancy in the interconnect areas, and treated them as the hard core. We have presented expressions for redundant yield with chip area, defect density, and amount of redundancy as parameters. We have also discussed the effects of interconnect and module complexities.

Our main conclusions are as follows.

• For a given chip size and defect density, yields increase rapidly at first with increased redundancy to a maximum, then decrease relatively slowly due to added area.

• The optimum amount of redundancy which maximizes the yield increases slowly with increasing defect density, while the maximum yield decreases.

• When yields are dominated by interconnect area, yield mprovement is not as large.

• Randomness of the intermodule connection pattern increases the interconnect area, therefore decreasing the yield improvement.

• For large modules with low intermodule interconnect requirements, a smaller number of modules on the chip results in higher yield improvements. To put it another way, for modules with high gate-to-pin (connection) ratio, redundancy is more effective if the extra modules are provided at large module level, with the design being partitioned into a small total number of modules on the chip.

• For logic modules with high intermodule interconnect requirements, yield improvement is limited by the complexity of the module. (Module complexity is defined by the interconnections.) This is to say, if modules have low gate-to-connection ratio, then redundancy must be provided as small modules, with the total design being partitioned into a large number of these modules.

• For this approach to be effective a high bypass coverage (defined as the product of testing and replacement coverages) is necessary. This means that the probability of detecting defective modules (testing coverage), and substituting good ones in their places (replacement coverage) must be high.

In this work we have assumed arbitrary interconnections. If interconnections are restricted to special cases such as nearest neighbor or tree connections between the logic modules, yield improvements will be higher. However, more research is needed in design and reconfiguration techniques inorder to exploit the full potential of this approach for both higher yielding and more reliable VLSI circuits.

List of Symbols

a_g: average area of individual blocks comprising the modules,

A_0: area of chip without redundancy,

A_R: area of the chip with redundancy added,

A_m: area of individual module on the chip $a_g n_g$,

A_w: area of the intermodule interconnections,

b: Rent's exponent,

$C_{N,R}$: bypass coverage,

D_0: average random defect density,

FM: figure of merit = $(Y_R/Y_0) \times (A_0/A_R)$,

I_y: yield improvement Y_R/Y_0,

k: number of defects,

L': average intermodule interconnection length (in units of module to module distance),

L_h: total length of interconnect running in orizontal direction,

L_v: total length of interconnect running in vertical direction,

λ: linewidth, minimum dimension used in a technology,

N: number of modules on the chip,

n: total number of blocks (i.e., gate equivalents) on the chip,

n_g: number of blocks in a module,

P: pitch, separation between the centers of interconnect wires,

$P_{k,N,R}$: probability of distributing k defects into R out of N modules,

r: ratio of intermodule interconnection area to total active chip area with redundancy A_w/A_R,

R: number of redundant modules,

T: average number of connections required per module,

w: average number of interconnect wires in the intermodule interconnect channel,

y: average number of external connections required per block (gate) of module,

Y_0: yield without redundancy (i.e., $R = 0$),

Y_R: total yield with redundancy.

REFERENCES

[1] J. Bernard, "The IC yield problem: A tentative analysis for MOS/SOS circuits," *IEEE Trans. Electron Devices*, vol. ED-25, pp. 939–944, Aug. 1978.

[2] R. P. Cenker *et al.*, "A fault-tolerant 64K dynamic RAM," in *Dig. ISSCC*, vol. 22, Feb. 1979, pp. 150–151.

[3] R. R. DeSimone *et al.*, "FET RAM's," in *Dig. ISSCC*, vol. 22, Feb. 1979, pp. 154–155.

[4] W. E. Donath, "Placement and average interconnection length of computer logic," *IEEE Trans. Circuits Syst.*, vol. CAS-27, pp. 272–277, Apr. 1979.

[5] ——, "Equivalence of memory to random logic," *IBM J. Res. Develop.*, pp. 401–407, Sept. 1974.

[6] B. R. Elmer, "Fault tolerant 92160 bit multiphase CCD memory," in *Dig. ISSCC*, vol. 20, Feb. 1977, pp. 116–117.

[7] B. F. Fitzgerald and E. P. Thoma, "Circuit implementation of fusible redundant addresses on RAMs for productivity enhancement," *IBM J. Res. Develop.*, vol. 24, pp. 291–298, May 1980.

[8] D. Frohman-Bentchkowsky, "A fully decoded electrically programmable MOS-ROM," in *Dig. ISSCC*, vol. 14, Feb. 1971, pp. 80–81.

[9] W. R. Heller, "Prediction of wiring space requirements for LSI," in *Proc. 17th Des. Automat. Conf.*, 1979, pp. 2–42.

[10] L. Kuhn, "Experimental study of laser formed connections for LSI wafer personalization," *IEEE J. Solid-State Circuits*, vol. SC-10, pp. 219–228, Aug. 1975.

[11] T. E. Mangir and A. Avižienis, "Effect of interconnect requirements on VLSI circuit yield improvement by means of redundancy," in *Proc. IEEE COMPCON 1981*, Feb. 1981, pp. 322–326.

[12] T. E. Mangir, "Use of on-chip redundancy for fault-tolerant VLSI design," Ph.D. dissertation, Univ. of California, Los Angeles, 1981; and Dep. Comput. Sci., Univ. of California, Los Angeles, Tech. Rep. CSD 820201, Dec. 1981.

[13] T. Mano *et al.*, "A fault tolerant 256K RAM fabricated with molybdenum-polysilicon technology," *IEEE J. Solid-State Circuits*, vol. SC-15, pp. 865–872, Oct. 1980.

[14] Y. W. Ng and A. Avižienis, "A unified reliability model for fault-tolerant computers," *IEEE Trans. Comput.*, vol. C-29, pp. 1002–1011, Nov. 1980.

[15] J. E. Price, "A new look at yield of integrated circuits," *Proc. IEEE*, pp. 1290–1291, Aug. 1970.

[16] S. E. Schuster, "Multiple word/bit line redundancy for semiconductor memories," *IEEE J. Solid-State Circuits*, vol. SC-13, pp. 689–703, Oct. 1978.

[17] C. H. Stapper, "Yield modelling and process monitoring," *IBM J. Res. Develop.*, vol. 20, pp. 228–234, 1976.

[18] C. H. Stapper *et al.*, "Yield model for productivity optimization of product," *IBM J. Res. Develop.*, vol. 23, pp. 298–409, May 1980.

[19] J. T. Wallmark, "A statistical model for determining the minimum size in integrated circuits," *IEEE Trans. Electron Devices*, vol. ED-26, pp. 135–142, Feb. 1979.

[20] A. Avižienis and C. Tung, "A universal arithmetic building element (ABE) and design methods for arithmetic processors," *IEEE Trans. Comput.*, vol. C-19, pp. 733–745, Aug. 1970.

Concurrent Error Detection and Testing
for Large PLA's

JAVAD KHAKBAZ AND EDWARD J. McCLUSKEY, FELLOW, IEEE

Abstract—A system of checkers is designed for concurrent error detection in large PLA's. This system combines concurrent error detection with off-line functional testing of the PLA by using the same checker hardware for both purposes. The result is a significant saving in hardware cost. For a case example, the total hardware cost is estimated at about 37 percent of the original PLA area. The system is almost totally self-checking and, although the test patterns are not function-independent, their generation algorithm is simple. The total test time for the entire system is within the range of that of some recent PLA design schemes which were specifically aimed at simplifying off-line testing, but which have no provisions for concurrent error detection.

Key Words—programmable logic array, PLA, nonconcurrent PLA, concurrent error detection, testing, two-rail code checker, 1-out-of-n code checker.

I. INTRODUCTION

THE ADVANTAGE of the programmable logic array (PLA) as a flexible and regular structure within a digital system is now well established, [12], [7]. For example, the use of the PLA as the driver of the control section of today's complex microprocessors has become a common practice. Such a PLA might have as many as 40 output lines, 30 input lines, and a few hundred product lines. A PLA of this size constitutes a significant portion of a digital system. This paper addresses the issue of error detection and testing of such large PLA's.

Manuscript received September 30, 1981; revised November 28, 1981. This work was supported in part by the US Army Electronics Research and Development Command under Contract DAAK-20-80-C-0266.

The authors are with the Center for Reliable Computing, Computer Systems Laboratory, Departments of Electrical Engineering and Computer Science, Stanford University, Stanford, CA 94305.

There are two modes of fault detection for any digital system. On one hand, we have off-line testing for either initial acceptance or for periodic testing of the system. In this mode, the normal operation of the system is interrupted, a selected set of inputs is applied, and the corresponding outputs are checked for possible error indication. In this paper we refer to this mode as simply *testing*. The second mode of fault detection, hereafter called *concurrent* or *on-line* error detection or checking, concerns monitoring the system under normal operation (i.e., with normal inputs).

Much work has been done in recent years on the problem of PLA fault detection: [14], [21], [6], [8], [10], [15], [18], [22], and [17]. However, most of this work has been concentrated on the question of testing, and not concurrent checking, of PLA's. For example, Fujiwara [8] suggests extra hardware to augment the PLA so that function-independent test of the PLA would be possible. This augmentation includes a set of *cascaded* XOR gates on the product lines. For a large PLA with a few hundred product lines, the delay corresponding to such an arrangement renders this extra circuitry useless for concurrent checking. In the present work, however, we integrate the two modes of testing so that any additional piece of hardware is used for both of these modes. A price is paid in terms of the generation of test patterns, as the test set for our design will be larger than that of, for example, Hong [10] or Fujiwara [8]. Further, the test set will not be function-independent, although the dependency is simple.

Section II of this paper lays the background, describes the fault model, and provides the definitions and assumptions that will be used in later sections. In Section III, the circuitry for concurrent error detection will be developed. In Section IV, we will show how the PLA can be tested using the detec-

Reprinted from *IEEE Trans. Electron Devices*, vol. ED-29, pp. 756–764, Apr. 1982.

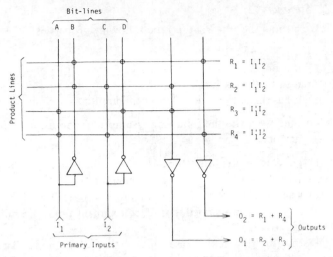

Fig. 1. A NOR–NOR PLA.

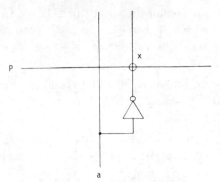

Fig. 2. Portion of the AND plane.

tion circuitry developed in Section III. Also, in Section IV we will address the question of testing the checkers. Finally, in Section V an example will be provided for measuring the cost of hardware incurred by our design. Following this example a comparison is made with three other proposed designs for testing PLA's.

II. THE MODEL, DEFINITIONS, AND ASSUMPTIONS

A. The PLA

We assume the PLA is realized in NOR–NOR logic, as in most MOS implementations, [7], [12]. However, the proposed checking circuitry can be transformed readily to accommodate other implementations. An example of a NOR–NOR PLA is given in Fig. 1. We are interested mainly in large PLA's, as described in the Introduction.

Since this paper is concerned with the structural properties of PLA's, we assume that the primary (external) inputs to the PLA have a separate error-checking mechanism, such as a parity bit, and we exclude the primary PLA inputs from our discussion. If the PLA has n inputs, it need not necessarily see all 2^n possible input vectors during normal, fault-free operation. That is, if we conceive of a truth table corresponding to the PLA, there may be some possible input vectors for which no output is defined in the truth table. We call these the *don't-care* input vectors. If a don't-care input vector is applied to the PLA, the output would be undefined. Any input vector for which a corresponding output vector is defined is called a *normal input*.

Definition: A PLA is said to be *nonconcurrent* iff, under fault-free operation, any normal input vector selects exactly one product term.

We assume that the PLA is nonconcurrent. This property has been shown to be very desirable for testing purposes, [21], [18]. The remaining sections of this paper once again attest to this fact. Furthermore, the nonconcurrency property of the PLA is not necessarily as restrictive as it may appear to be. For example, a PLA has been used for the control section of a special-purpose microprocessor, [5]. This PLA has about 400 product terms, and, of its output lines, eight are used to

specify the state of the PLA and are fed back to the AND plane. This means that the machine has 256 states, or, on the average, it has less than two product terms per state. This implies a great degree of nonconcurrency that naturally exists in the PLA, as the states of the machine are mutually exclusive. It also implies a very easy and cheap search method for any possible concurrencies since concurrencies may occur only between the product lines of the *same* state. Then one can remove these occasional concurrencies by, for example, a method given in [21]. Such a process requires an increase in the size of the PLA, the degree of which depends on the number of concurrencies present.

Next consider a section of the AND plane of the (nonconcurrent) PLA, as shown in Fig. 2. P is a product term and a is an input literal. Then $P = M \cdot a$ where M is the product of some other input literals. Thus the input vector $I = M \cdot a$ is a normal input that selects product line P. We say that the device x in the AND plane (Fig. 2) is *irredundant* if input vector $J = M \cdot a'$ is also a normal input vector. Otherwise, we say that device x is *redundant*. A redundant device may be removed without affecting the function of the PLA or its nonconcurrency property. This is so because, by removing x, product line P will be selected by input M, irrespective of input literal a. But since x is redundant, J is not a normal input and it cannot occur in normal, fault-free operation. So M always occurs with $a = 1$ (never with $a = 0$). Thus device x may be removed without affecting the normal operation of the PLA. However, one may wish to have redundant devices for, say, a fault-tolerant design. Thus, in general, we may have redundant devices in the PLA.

Next, we state the definitions that will be used in the remainder of this paper. Consider a combinational circuit C with input code space S and output code space S' that implements a function Z. Let F be the set of faults that may occur in this system. Let $Z(s, f)$ denote the response of the circuit to input s in S in the presence of fault f in F. Let $Z(s, 0)$ denote the response of the system to input s in S when no fault is present. Then define the following [20]:

Definition: A combinational circuit C is said to be *fault-secure* with respect to input code space S and fault set F iff for all f in F and for all s in S either $Z(s, f) = Z(s, 0)$ or $Z(s, f)$ is not in S'.

Definition: A combinational circuit C is said to be *self-testing* with respect to input code space S and fault set F iff for all f in F, there is an s in S such that $Z(s, f)$ is not in S'.

Definition: A combinational circuit C is said to be *totally self-checking* (TSC) with respect to input code space S and fault set F iff it is both fault-secure and self-testing with respect to S and F.

Definition: A bit-line B_1 of a PLA is said to be *complementary* of another bit-line B_2, iff both B_1 and B_2 correspond to the same input literal.

Note that, under nonfaulty operation of the PLA, any bit line carries the complement logic value of its complementary bit line.

B. The Fault Model

There have been a number of studies conducted that were aimed at modeling physical failures in different technologies by logical faults, [19], [9]. A general conclusion from these works is that the classical stuck-at model is not sufficient for the present MOS technologies, [13]. Furthermore, these works exclusively deal with the so-called solid faults, and, to the best of our knowledge, there has not been any experimental result that has yielded a logic model for the transient or intermittent failures which, by their very nature, are not amenable to off-line testing.

Therefore, in this work, for both solid and nonsolid failures, we use the following logical fault models which have been used widely in recent works, e.g., [14], [21], [6], [8], [10], [17], [22]:

1) stuck-at-0 (sa0) and stuck-at-1 (sa1)
2) short between two adjacent parallel lines
3) extra/missing device at PLA crosspoints.

Further, we assume that a short between two lines always results in ANDing the logic values of the two lines.

In this work we base our design on the single-fault assumption. It has been shown that a set of test patterns that detects all single faults in a PLA also detects most of the multiple faults, [1]. However, for concurrent error detection, where we have no control over the inputs, we need to be more careful about multiple-fault situations. Therefore, our approach will be that of detecting *all single faults* and as many multiple faults as possible. To this end, we will distribute checkers throughout the PLA, as described in Section II; although there will be significant overlaps between the sets of errors detected by these checkers.

For modeling the multiple fault situations, we assume that, at any instance of time, any circuit element type may display at most one failure mode. For example, we assume that we may not have a sa1 on one product line and, at the same time, a short between two other product lines. Or, we may not have a missing device at one crosspoint and, simultaneously, an extra device at a different crosspoint. We refer to this assumption as *the single-mode fault assumption*. This is a reasonable assumption because one expects that the cause of the failures, be it a fabrication problem, extreme environmental condition, or otherwise, may affect similar circuit elements in similar manners.

C. Summary

We assume that the PLA has the following properties:

1) it is large
2) it is nonconcurrent
3) it is NOR–NOR implemented
4) it has single-input decoders at the input.

Further, we make the following assumptions about the faults that may occur:

1) possible faults are stuck-at, extra/missing device, and short between two adjacent parallel lines;
2) a short fault AND's the lines involved;
3) single fault is assumed, but detects as many multiple faults as possible;
4) for multiple-fault case, make the single-mode fault assumption;
5) primary PLA inputs are excluded from this work.

III. CONCURRENT ERROR DETECTION

We will divide the PLA into the AND and OR planes, [12], and will consider them separately for concurrent error detection.

A. The AND Plane

The AND plane consists of

1) the input inverters
2) the bit lines
3) the crosspoints on the bit lines
4) the product lines.

A fault in an input inverter can be modeled as a stuck-at fault at its outputs, which is equivalent to a stuck-at fault on the corresponding bit line. Thus we will ignore the input inverters for fault analysis.

Lemma: For any normal input to the PLA, any number of faults on *one* of the following circuit element groups either cause no error, or desensitize the (only) sensitized product line, or sensitize one or more *extra* product lines. The element groups are

1) the bit lines
2) the crosspoints in the AND plane
3) the product lines.

Proof: We just give the proof for the bit lines. The proofs for the other two cases are similar. There are three possibilities:

1) The faults are all sa1's. An examination of Fig. 1, and the fact that the PLA is NOR–NOR, reveals that these faults can at most desensitize the (only) selected product line.
2) The faults are all sa0's. Again Fig. 1 reveals that these faults do not affect the already-selected product line. They, however, may cause activation of more product lines.
3) The faults are all shorts. If shorted lines carry the same logic value, no change would occur. If two shorted lines carry opposite logic values, they will both become 0, which would put us back in case 2).   Q.E.D.

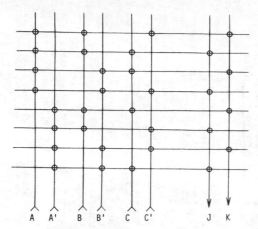

Fig. 3. A TSC two-rail checker.

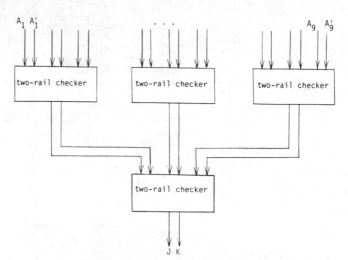

Fig. 4. A two-rail checker tree.

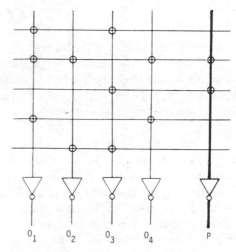

Fig. 5. Even output parity for a nonconcurrent PLA.

The conclusion from this Lemma is that, if there are r product lines, a 1-out-of-r code checker on the product lines detects all single and many multiple faults in the AND plane of the PLA. There are several implementations of a TSC 1-out-of-n checker: [2], [16], [11]. The approach in [2] and [16] is that first the 1-out-of-r code is translated into a k-out-of-$2k$ code, and then a TSC checker for the resulting k-out-of-$2k$ code is designed. In [11], the 1-out-of-r code is first translated into a two-rail code, and then a TSC two-rail checker is used to check the output of this translator. In [11] it is shown that for some values of r, the first method is more efficient, and for the others, the second method results in a better design. Both approaches yield a TSC 1-out-of-r code checker. Here we denote the code translator by L and the checker of the translated code by $C1$. We will choose the design method in accordance with the result of Khakbaz [11].

Another structural regularity of the PLA that can be utilized is that the bit lines form a two-rail code. Thus we can put a two-rail code checker, called $C2$, on the bit lines. The general approach for building a TSC two-rail checker for n pairs (A_i, A_i') is to generate a parity tree with two output lines, [4], [20]. The first output line is the parity of, say, the A_i lines. The other output line is simply the complement of the first. Thus the output of the two-rail checker is a 1-out-of-2 code. A PLA implementation of such a tree is described in [21], and an example of it is shown in Fig. 3 (for three input pairs). The checkers of the type shown in Fig. 3 can be used to make a two-rail checker tree with many input pairs. For example, Fig. 4 shows a TSC two-rail checker with nine input pairs.

B. The OR Plane

The OR plane consists of the following circuit elements:

1) the output lines
2) the crosspoints on the output lines.

As before, we model the failures of the output inverters by stuck-at faults at the corresponding output lines.

Under the single-fault assumption in the OR plane, at most one output line can be altered. Thus for single-fault detection in the OR plane one output parity line suffices. Moreover,

since the PLA is nonconcurrent, the generation of such output parity would be trivial. To generate the even (odd) output parity, we just add one output line and put devices on the crosspoints with those product lines which have odd (even) number of devices on them. An example is shown in Fig. 5. Then a parity tree can concurrently check for any single error on the $m + 1$ output lines.

For better error detection, other encoding schemes may be used for the output lines. For example, we may add one parity line for every three output lines. Any such encoding scheme requires some redundant output lines (denoted by D), and an output code checker $C3$ on the resulting output lines. For example, for the case of a single output parity, D consists of a single output line and $C3$ is a parity tree.

C. Summary

We have suggested the following set of checkers for concurrent (on-line) error checking of the PLA:

1) A TSC 1-out-of-r checker on the product lines. This consists of a code translator L and a checker $C1$ to check the output code of L.

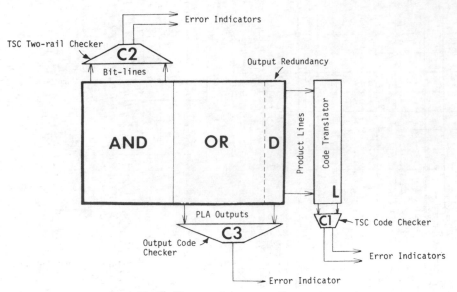

Fig. 6. The complete system diagram.

2) A two-rail code checker $C2$ for the bit lines.

3) An output encoding scheme (e.g., single parity) which requires redundant output lines D and a corresponding output code checker $C3$.

Fig. 6 shows a complete diagram of the system.

IV. TESTING

In this section we will consider the question of test pattern generation for the system of Fig. 6. Each subcircuit will be considered separately. The fault model is described in Section I. We assume that we have full external control over only the primary PLA inputs. This is a realistic limitation since direct external access to points within the PLA or the checkers, or even to the connections between the PLA and the checkers, generally requires extra hardware and/or extra pins on the chip. We will show that even with this limitation, the entire system of Fig. 6 is testable, except the internal parts of $C3$ and the redundant devices in the AND plane of the PLA.

Notation: Let us number the product lines, from top to bottom, 1 through r. Let $I_0(p)$ be the input vector that selects product line p, with all don't-care input literals set to 0. Similarly, let $I_1(p)$ be the input vector that selects product line p, with all don't-care input literals set to 1. Also denote the all-0 input vector by $I0$ and the all-1 input vector by $I1$.

Pin Requirement: We assume that the outputs of $C1$, $C2$, and $C3$ of Fig. 6 are directly connected to external pins. But as mentioned earlier, and as will be discussed in the following, since, in general, the inputs of $C3$ are not controllable from the PLA inputs, a TSC implementation of $C3$ may not be necessary. Thus one output line could suffice for $C3$. Therefore, the system of Fig. 6 requires *5 extra pins*.

A. Testing the AND Plane

A portion of the AND plane is shown in Fig. 7. Table I specifies the input patterns required for testing various kinds of faults in the AND plane. Note that since device x is irredundant, by the argument given earlier, the input vector "M with

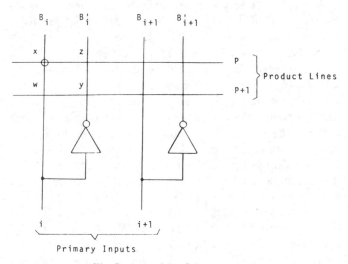

Fig. 7. A portion of the AND plane.

$i = 1$" is also a normal input and thus it selects a product line q. With x missing, this input also selects p, and hence the product lines no longer form a 1-out-of-r code.

B. Testing the OR Plane

A portion of the OR plane is shown in Fig. 8. Table II lists the PLA input patterns that test for single faults in the OR plane.

C. Testing $(L, C1)$

Either of the two implementations of the 1-out-of-r checker $(L, C1)$, one in [2] and the other in [11], is completely tested by applying $I_0(p), p = 1, 2, \cdots, r$; see [11].

D. Testing $C2$

The general two-rail checker tree organization is exemplified in Fig. 4. Any such implementation of $C2$ can be exhaustively tested by 2^t input patterns, where t is the number of input pairs to the largest block of such a tree, [3]. For example, the following is the list of the (PLA) input patterns that are needed

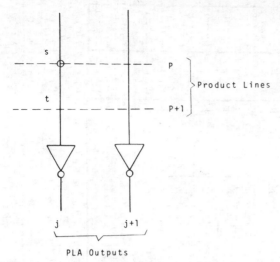

Fig. 8. A portion of the OR plane.

TABLE I
TEST PATTERN FOR THE AND PLANE INPUT

Fault	Test Pattern	Error Indicator
B_i sa0	$I1$	$C2$
B_i sa1	$I0$	$C2$
B_i short to B_i'	any input	$C2$
B_i' short to B_{i+1}	$I0$	$C2$
p sa1	$I_0(p+1)$	$C1$
p sa0	$I_0(p)$	$C1$
p short to p+1	$I_0(p)$	$C1$
extra device at z	$I_0(p)$	$C1$
extra device at w	$I_1(p+1)$ (*)	$C1$
extra device at y	$I_0(p+1)$ (*)	$C1$
missing device at x	M with i=1 (**)	$C1$

(*) Note that input literal i is a don't-care for product line p+1.

(**) Here product line p equals M.i', and device x is irredundant.

TABLE II
INPUT TEST PATTERN FOR THE OR PLANE

Fault	Test Input	Error Indicator
j sa0	$I_0(p)$	$C3$
j sa1	$I_0(p+1)$	$C3$
j short to j+1	$I_0(p)$	$C3$
missing device s	$I_0(p)$	$C3$
extra device at t	$I_0(p+1)$	$C3$

for testing $C2$, for a PLA with nine input lines, and for $C2$ implemented as in Fig. 4.

```
000  000  000
001  011  011
011  001  101
010  010  110
```

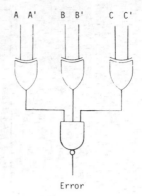

Fig. 9. A simple two-rail checker.

```
101  101  001
100  110  010
110  100  100
111  111  111
```

We denote such set of test patterns by $T(C2)$.

E. Testing C3

Since, in general, we do not have full control over the inputs of $C3$ (outputs of the PLA) from the inputs of the PLA (Fig. 6), it may not be possible to test $C3$ fully. The general approach to this problem is to use a simple implementation of $C3$. For example, if a two-rail output encoding is used for the PLA, we may use the implementation of Fig. 9 rather than that of Figs. 3 or 4. This implementation of a two-rail checker is not totally self-checking. However, since it is simple and it contains less circuitry than the TSC implementations, the possibility of a fault occurring in it is smaller.

F. Summary

In this section we have shown that the system of Fig. 6 can be tested for all single faults, except the missing redundant devices in the AND plane and the internal faults of $C3$. The test patterns are applied to the primary inputs of the PLA and the error indicators are the outputs of $C1$ (two lines), $C2$ (two lines, and $C3$ (one line). Let $I_0(p)$, $I_1(p)$, $I1$, $I0$, and $T(C2)$ be defined as before. Also, let $J(x)$ be the test pattern needed for testing missing device x in the AND plane, as described in Table I. Then Table III summarizes the complete testing scheme for the system of Fig. 6.

V. AN EXAMPLE

Consider a large PLA with 25 inputs, 300 product terms, and 40 output lines. Furthermore, assume that there are 6000 devices in the AND plane. We would like to get an estimate of the cost in area for implementing our design. To this end, we use the standards in [12, p. 103] for area calculations. For simplicity, we separately calculate the cost incurred by circuits $C1$, $C2$, $C3$, L, and D. Then, to estimate the total cost, we will add these numbers together. Thus the details of spacings, interconnections, and the actual layouts are ignored. Further, we assume PLA implementations for $C1$, $C2$, and $C3$. This usually is not the most compact implementation; thus the cost estimates for these circuits are upper limits for these values.

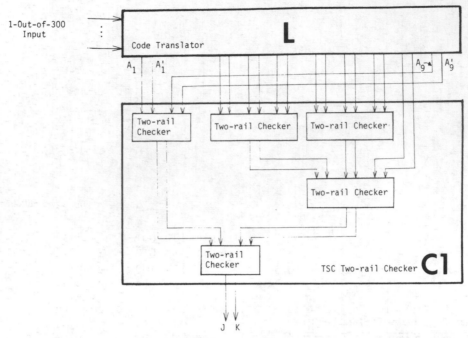

Fig. 10. A TSC 1-out-of-300 code checker.

TABLE III
COMPLETE TEST SET FOR THE SYSTEM OF FIG. 6

Test Input	Error Indicator	Faults Detected	Number of Tests
I0, I1	C2	single faults on bit-lines	2
$I_o(p)$	C1, C3	faults on product lines faults in the OR plane some extra devices in the AND plane, faults in L, faults in C1	r
$I_l(p)$	C1	other extra devices in the AND plane	r
J(x)	C1	missing devices in the AND plane	d
T(C2)	C2	faults in C2	2^t

r : number of product lines;
d : number of devices in the AND plane;
t : number of inputs to the biggest block in the tree implementation of two-rail checker C2.

TABLE IV
AREA COST

Circuit	Area
Original PLA	27,000
L	5,400
C1	264
C2 (Fig. 11)	1,152
D	300
C3 (Fig. 12)	2,740

The unit of area is immaterial, since the purpose here is a relative comparison.

We assume a single output parity for the PLA, i.e., D consists of only one line. Finally, the 1-out-of-r checker $(L, C1)$ for $r = 300$ is implemented using the method of Khakbaz [11]. Such an implementation of $C1$ is shown in Fig. 10. Table IV shows the estimate of the areas of these circuits. From Table IV we conclude that the cost of the excess circuitry is about *37 percent of the original PLA area*.

Next consider the question of error-detection delay for this example. Here we define detection delay as follows:

Definition: The delay between the time the output of the PLA is ready and the time that error indicators can be sampled safely is called *error-detection delay*, or simply, delay.

To calculate the delay of this circuit, we assume that each of the PLA input decoders, the AND plane, and the OR plane has one gate delay. Note that the PLA's used for the two-rail checkers have no input decoders. Also, we assume that each XOR gate has two gate delays. With these assumptions, for the above example, we have

delay of $C1$ (Fig. 10): 6 gate delays:
delay of $C2$ (Fig. 11): 5 gate delays:
delay of $C3$ (Fig. 12): 9 gate delays.

Since the PLA itself has 3 gate delays, for each test pattern applied, the tester must wait at least 12 gate delays before it samples the error indicators. From Table III we conclude that $2^t + 2r + d + 2$ test patterns are required. For our example, this number is 6618. Therefore, the total test time is, at least, 6618 * 12 or about 79 000 gate delays.

It would be interesting to compare the characteristics of our design with those of some other proposed schemes. In particular, the designs of Hong [10], Fujiwara [8], and Yajima [22] have been selected for this purpose, since, in our opinion, they represent the most novel and the state-of-the-art ideas in PLA testing. Table V gives a general comparison of these works, denoted by H, F, and Y, respectively. The present work is denoted by K. In the calculations that resulted in this table, in addition to the assumptions made earlier, we have assumed that one gate delay is required for entering a bit into a shift register.

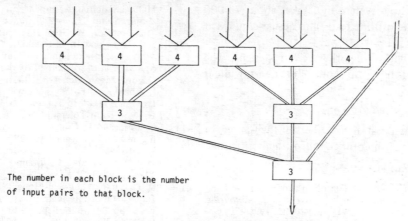

The number in each block is the number
of input pairs to that block.

Fig. 11. A TSC two-rail checker tree with 25 input pairs.

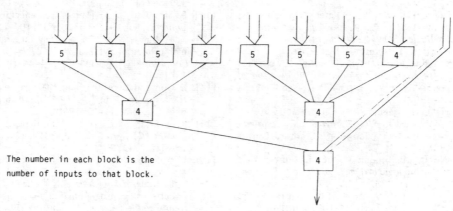

The number in each block is the
number of inputs to that block.

Fig. 12. Parity tree with 41 inputs.

TABLE V
COMPARISON WITH OTHER METHODS

	Concurrent Detection	Function Indep.	No. Test Patterns	Delay per Test Response	No. Extra Pins
K	x		$2r+d+2^t+2$	12	5
H		x	$3n+2r+5$	$7 + 2\lceil \log(r) \rceil$	$7 + \lceil \log(n+1) \rceil$
Y		x	$n+2r+8$	$2m + 7$	3
F		x	$2n+3r$	$2r + 3$	4

TABLE VI
A COMPARISON, NUMERICAL VALUES

	Number of Test Patterns	Total Test Time in Gate Delays	Number of Extra Pins	Area Cost (Percent of PLA)
K	6,618	79,000	5	37
H	680	16,500	12	21
Y	633	56,000	3	27
F	950	573,000	4	21

n : number of PLA inputs;
m : number of PLA outputs;
r : number of PLA product terms;
d : number of devices in the AND plane;
t : max. number of inputs to a block in C2 tree.

To get numerical values, again consider the PLA of the above example. To get an estimate of the cost in area, we assume every block of combinational circuits is PLA implemented. Table VI shows the result.

Note that, of these designs, only this one has combined concurrent error checking with off-line testing. In fact, all of the extra circuitry of Fig. 6 is used for both modes of testing. Thus on the average, over 20-percent savings in area is made over the other systems, as the other designs still need a complete set of circuitry for concurrent error detection. The price for this saving is in terms of the number of test patterns required, which, for our case, is an order of magnitude larger than for

the other three cases. However, the total test time for our method is within the same range as the total test time for the other methods, since the response time to each test pattern is much smaller for our design than for the others.

VI. CONCLUSION

A system of checkers is designed for concurrent error detection in large PLA. This system combines concurrent error detection with off-line testing by using the same circuits for both modes of testing. This results in a significant saving in hardware cost.

An example for our proposed error-detection design is provided. For this example, the cost in area is estimated at 37 percent of the area of the original PLA. Concurrent error indication lags the PLA outputs by 9 gate delays. To test the entire system, we need to apply 6618 test patterns, each requiring 12 gate delays before its corresponding error indicators may be sampled. The whole system thus takes about 79 000 gate delays for a complete off-line test. This design requires at most 5 extra pins.

A comparison is made with three other existing designs. The result is that our system is better in terms of area cost, while it requires a bigger test set.

ACKNOWLEDGMENT

The authors wish to thank all the members of the Center for Reliable Computing for their helpful comments and criticisms.

REFERENCES

[1] V. K. Agarwal, "Multiple fault detection in programmable logic arrays," in *Proc. 9th Annu. Symp. Fault-Tolerant Computing* (FTCS-9), pp. 227–234 (Madison, WI, June 20–22, 1979).

[2] D. A. Anderson and G. Metze, "Design of totally self-checking check circuits for *m*-out-of-*n* codes," *IEEE Trans. Comput.*, vol. C-22, no. 3, pp. 263–269, Mar. 1973.

[3] D. C. Bossen, D. L. Ostapko, and A. M. Patel, "Optimum test patterns for parity networks," in *Proc. American Federation of Information Processing Societies 1970 Fall Joint Computer Conf.*, vol. 37, pp. 63–68 (Houston, TX, Nov. 17–19, 1970).

[4] W. C. Carter and P. R. Schneider, "Design of dynamically checked computer," in *Proc. 4th Congress International Federation of Information Processing Societies*, vol. 2, pp. 878–883 (Edinburgh, Scotland, August 5–10, 1968).

[5] J. H. Clark, "The geometry engine: A VLSI geometry system for graphics," to be published in *SIGGRAPH 82*; also a private conversation, Computer Systems Laboratory, Stanford University, Stanford, CA, 1981.

[6] E. B. Eichelberger and E. Lindbloom, "A heuristic test-pattern generation for programmable logic arrays," *IBM J. Res. Develop.*, vol. 24, no. 1, pp. 15–22, Jan. 1980.

[7] H. Fleisher and L. I. Maissel, "An introduction to array logic," *IBM J. Res. Develop.*, vol. 19, no. 2, pp. 98–109, Mar. 1975.

[8] H. Fujiwara and K. Kinoshita, "A design of programmable logic arrays with universal tests," *IEEE Trans. Comput.*, vol. C-30, no. 11, pp. 823–828, Nov. 1981.

[9] J. Galiay and Y. Crouzet, "Physical vs. logical fault models in MOS LSI circuits, impact on their testability," in *Proc. 9th Annu. Symp. Fault-Tolerant Computing* (FTCS-9), pp. 195–202 (Madison, WI, June 20–22, 1979).

[10] S. J. Hong and D. L. Ostapko, "FITPLA: A programmable logic array for function-independent testing," in *Proc. 10th Annu. Symp. Fault-Tolerant Computing* (FTCS-10), pp. 131–136 (Kyoto, Japan, Oct. 1–3, 1980).

[11] J. Khakbaz, "Totally-self-checking checker for 1-out-of-*n* using two-rail codes," to be published in *IEEE Trans. Comput.* (Special Issue on Fault-Tolerant Computing), vol. C-31, no. 7, July 1982.

[12] C. Mead and L. Conway, *Introduction to VLSI Systems*. Reading, MA: Addison-Wesley, 1980.

[13] V. V. Nickel, "VLSI–The inadequacy of the stuck at fault model," in *Dig. 1980 Test Conf.*, pp. 378–381 (Philadelphia, PA, Nov. 11–13, 1980).

[14] D. L. Ostapko and S. J. Hong, "Fault analysis and test generation for programmable logic arrays," *IEEE Trans. Comput.*, vol. C-28, no. 9, pp. 617–626, Sept. 1979.

[15] D. K. Pradhan and K. Son, "The effects of untestable faults in PLAs and a design for testability," in *Dig. 1980 Test Conf.*, pp. 359–367 (Philadelphia, PA, Nov. 11–13, 1980).

[16] S. M. Reddy, "A note on self-checking checkers," *IEEE Trans. Comput.*, vol. C-23, no. 10, pp. 1100–1102, Oct. 1974.

[17] K. K. Saluja *et al.*, "A multiple fault testable design of programmable logic arrays," in *Proc. 11th Annu. Symp. Fault-Tolerant Computing* (FTCS-11), pp. 44–46 (Portland, ME, June 24–26, 1981).

[18] K. Son and D. K. Pradhan, "Design of programmable logic arrays for testability," in *Dig. 1980 Test Conf.*, pp. 163–166 (Philadelphia, PA, Nov. 11–13, 1980).

[19] R. L. Wadsack, "Technology dependent logic faults," in *Proc. COMPCON Spring 78*, pp. 124–129 (San Francisco, CA, Feb. 28–Mar. 2, 1978).

[20] J. Wakerly, *Error Detecting Codes, Self-Checking Circuits and Applications*. New York, NY: Elsevier–North-Holland, 1978.

[21] S. L. Wang and A. Avizienis, "The design of totally-self-checking circuits using programmable logic arrays," in *Proc. 9th Annu. Symp. Fault-Tolerant Computing* (FTCS-9), pp. 173–180 (Madison, WI, June 20–22, 1979).

[22] S. Yajima and T. Aramaki, "Autonomously testable programmable logic arrays," in *Proc. 11th Annu. Symp. Fault-Tolerant Computing* (FTCS-11), pp. 41–43 (Portland, ME, June 24–26, 1981).

Part V
Design Examples

THIS part deals with digital VLSI design examples in two areas: information and signal processing. It is the latter area which will benefit from high-level DA design tools since we deal with many types of algorithms that call for custom design VLSI chips. Although only the digital aspects of signal processing are dealt with in this section, the digital-to-analog and analog-to-digital conversion, the sampled data and analog circuit and system design, and design automation are all related aspects, and they will witness substantial progress in the years to come.

*Novel VLSI processor architectures, some implemented by
only a few different types of simple cells, are leading
the way towards a new generation of computers.*

VLSI Processor
Architectures

Philip C. Treleaven, University of Newcastle upon Tyne

As an illustration of the rapidly increasing complexity of integrated circuits, Charles Seitz of the California Institute of Technology developed the following analogy:

> In the mid-1960's the complexity of a chip was comparable to the street network of a small town. Most people can navigate such a network by memory without difficulty. Today's microprocessor, using a five-micron technology, is comparable to the entire San Francisco Bay Area. By the time a one-micron technology is solidly in place, designing a chip will be conparable to planning a street network covering all of California and Nevada at urban densities. The ultimate one-quarter micron technology will likely be capable of producing chips with the intricacy of an urban grid covering the entire North American continent.[1]

VLSI microprocessors containing 100,000 transistors, such as the recently announced 32-bit microprocessors from Intel (the iAPX 432) and Bell Laboratories (the Mac-32), are becoming more and more commonplace.[2,3] In fact, the term "VLSI processor architecture" is normally viewed as being synonymous with such designs.[4]

But as Gordon Moore, President of Intel Corporation, said at a relatively recent conference, "Beyond memory, I haven't the slightest idea of how to take advantage of VLSI . . . . How to best make use of the processing technology is what the problem is."[5] The reason for this pessimism is the escalating cost of designing and testing such complex VLSI processors.

However, Mead and Conway's *Introduction to VLSI Systems*,[6] and the companion Multiproject Chip (MPC)[7] courses and silicon foundries[8] they helped to establish, have together stimulated the rapid development of a new VLSI proccessor architecture "culture." A brief survey of these novel VLSI processor architectures will illustrate the exciting work going on in the area, most of which is still at the research stage.

A "good" VLSI architecture in the context of this article should have one or more following properties[9]:

(1) It should be implementable by only a few different types of simple cells.

(2) It should have simple and regular data and control paths so that the cells can be connected by a network with local and regular interconnections. (Long-distance or irregular communication is thus minimized.)

(3) It should use extensive pipelining and multiprocessing. In this way, a large number of cells are active at one time so that the overall computational rate of the simple cells is high.

As we shall see, a whole spectrum of processor architectures is under development based on this philosophy of simplicity and replication.

Special-purpose processors

A special-purpose VLSI processor is usually designed as a peripherial device attached to a conventional host computer. It may be a single chip built from a replication of simple cells, or a system built from identical simple chips, or it may be a chip that uses both of these techniques. Special-purpose processors have a number of advantages[9] that help reduce their cost:

- Only a few different, simple cells need to be designed and tested since most of the cells and chips are copies of a few basic models.
- Regular interconnection implies that the design is modular and extensible, so one can create a large

Reprinted from *Computer*, vol. 15, pp. 33–45, June 1982.

processor by combining the designs of small cells and chips.

- Many identical cells and chips that use pipelining and multiprocessing can meet the performance requirements of a special-purpose processor.

Examples of these special-purpose processor designs can be found in the literature.[5,10-12]

Systolic arrays. H. T. Kung and his co-workers at Carnegie-Mellon University are investigating the relationship of algorithm design to special-purpose chip architecture.[9,13] They state that in designing such chips the most crucial decision is the choice of the underlying algorithm, since it is the suitability of the algorithm that largely determines chip design cost and performance. In other words, they feel the algorithm should receive the largest part of the design effort. They also argue that low-level optimizations at the circuit or layout design levels are probably not worthwhile, as these will lead to only minor improvements in the overall performance and an increase in design time. Borrowing from physiology, Kung's group coined the term "systolic" array for algorithms and architectures with good VLSI properties. Figure 1 shows a number of simple and regular structures for interconnecting processors for VLSI algorithms. Each tends to be good for a particular function—e.g., linear arrays for real-time filters, trees for searching, and shuffle exchanges for Fourier transforms.

A specific VLSI chip[9]—one that performs on-line pattern matching of strings with "wild card" characters—illustrates the design philosophy and methodology of Kung's group. The chip is based on a linear array, the simplest possible geometry, which is good for pipelined operations. It accepts two streams of characters from the host machine and produces a stream of bits, as shown in Figure 2. One of the input streams, the text string, is an endless string of characters over some alphabet. The other, the pattern input stream, contains a fixed-length vector of characters over the alphabet, with X being a "wild card" character. The output is a stream of bits, each of which corresponds to one of the characters in the text string.

The chip is divided into character cells, each of which compares two characters and accumulates a temporary result. The pattern and string follow a preset path of cells from the time they enter the chip until the time they leave it. On each beat, every character moves to a new cell.

It took only about two months to design the pattern-matcher chip and its fabrication was completed in the spring of 1979. Preliminary analysis showed that the chip achieved a data rate of one character every 250 ns, which is higher than the memory bandwidth of most conventional computers. This seems to justify Kung et al.'s claim that careful design of the underlying algorithm leads to high performance.

RSA cipher chip.[14] This chip, designed by Ronald Rivest and his colleagues at the Massachusetts Institute of Technology, implements a public-key encryption algorithm. This operation is computationally demanding, requiring up to several hundred multiplications of several hundred bit numbers. The chip was designed as a general-purpose, big-number processor, using a bit-slice architecture for the ALU. It is a good example of a chip design based on the duplication of simple cells.

Externally, the chip is configured as a memory chip that can be read or written at one of four eight-bit word positions. For instance, one of these positions is the "window" for data I/O, while another is for receiving commands such as "encrypt." To support encryption, the RSA chip has a 512-bit ALU organized in a bit-slice manner. It has eight general-purpose 512-bit registers, as well as up-down shifter logic and a multiplier. Other subsystems include control logic (containing a PLA of 224 72-bit microcode words), a stack/counter array for subroutines and loops, and an array of powerful super-buffers to drive the signals that control the ALU.

Internally, the ALU is only capable of performing the operations $(A*B) \pm C$, shift-left, shift-right, and test least-significant bit. The remaining required operations are implemented by microcode control subroutines. These operations include RSA encryption/decryption (modular exponentiation); generating a large prime

COMMUNICATION GEOMETRY	EXAMPLES
1—DIM LINEAR ARRAYS	MATRIX—VECTOR MULTIPLICATION RECURRENCE EVALUATION
2—DIM SQUARE ARRAYS	DYNAMIC PROGRAMMING PATTERN MATCHING
2—DIM HEXAGONAL ARRAYS	MATRIX MULTIPLICATION GAUSSIAN ELIMINATION
TREES	SEARCHING ALGORITHMS PARALLEL FUNCTION EVALUATION
SHUFFLE-EXCHANGE NETWORKS	FAST FOURIER TRANSFORMS BITONIC SORT

Figure 1. Examples of VLSI algorithms.

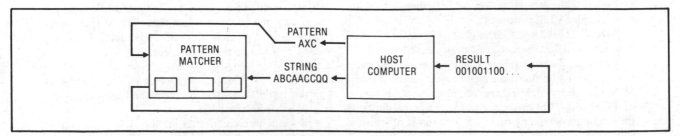

Figure 2. Data to and from the pattern matcher.

number; generating a complete RSA key-set; greatest common divisor; and input or output of a large number through the eight-bit window.

The "floor plan" of the RSA cipher chip is shown in Figure 3. The left side contains a block of 320 slices of the ALU and the upper-right block contains the remaining 192 slices. The central spine carries control signals to the ALU from the superbuffer driver array at its lower right. The microcode PLA occupies the right-center area of the chip, and the remaining logic (stack, pads, etc.) occupies the lower-right portion. In Figure 3, S denotes a stack, X an eight-bit "window," C a small bus-control PLA, and D some debugging logic.

The entire chip contains about 40,000 MOS transistors, has 18 pins, runs at four MHz, and uses a little more than one watt of power. Rivest estimates that the project required about five man-months of effort, with the chip finally being fabricated in the fall of 1979. It was primarily a programming project because he and his colleagues almost exclusively wrote programs in Lisp, which when executed, created the desired Caltech Intermediate Form output file. Altogether, they wrote about 75 pages of Lisp code. The largest pieces specified the final placement and interconnection of the modules (14 pages), the description of the ALU (11 pages), and the microcode assembler and simulator (10 pages). Estimates are that the chip can perform RSA encryption faster than 1200 bps (even faster if shorter keys are used), and the designers predict speeds of 20,000 bps within a few years.

Optical mouse chip. Dick Lyon is investigating new VLSI architecture methodologies for applications such as signal processing[15] and smart digital sensors.[16] A novel example of the latter is the optical mouse chip,[16] designed in the VLSI system design area at Xerox Corporation's Palo Alto Research Center.

The optical mouse (Figure 4a) is a pointing device for controlling the cursor on a personal workstation display, such as the one found on the Xerox Star 8010 information system. The design was motivated by the desire for a highly reliable mouse with no moving parts except button switches, and it was realized through the innovative use of electro-optics, circuit design, geometric combinatorics, and algorithms—all in a single special-purpose sensor chip.

This chip reports the motion of visible spots relative to its coordinate system by combining two techniques. One

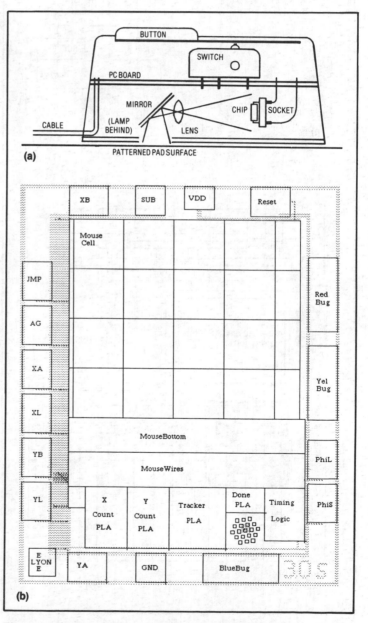

Figure 4. Diagram of optical mouse (a) and chip floor plan (b).

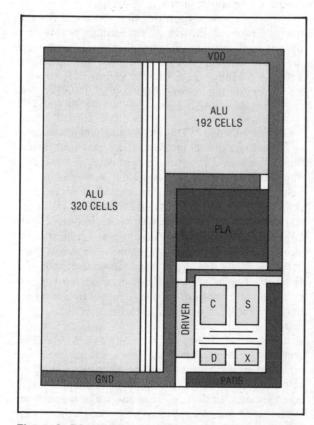

Figure 3. RSA cipher chip floor plan. (This floor plan originally appeared in the fourth quarter 1980 issue of *Lambda* [now *VLSI Design*].)

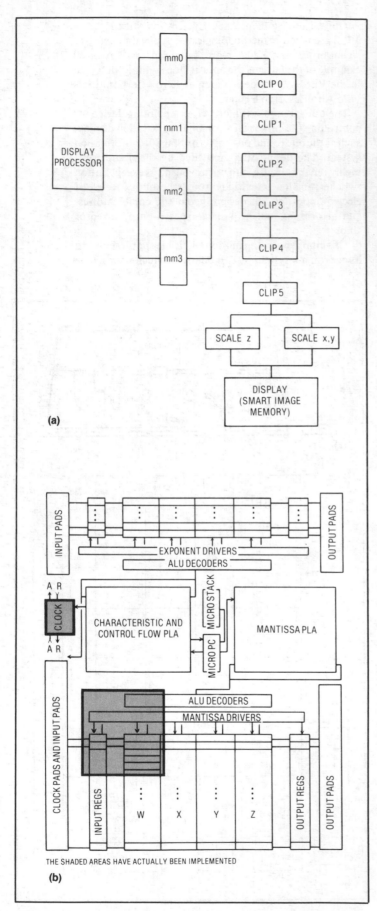

(a)

(b) THE SHADED AREAS HAVE ACTUALLY BEEN IMPLEMENTED

Figure 5. The Geometry Engine: (a) system data flow; (b) chip floor plan.

is a simple, "mostly digital" circuit that produces digital image (bitmap) snapshots of bright features on a dark field using self-timed circuit concepts and mutually inhibiting light sensors. The other technique uses a tracking algorithm with an easy-to-track contrasting pattern, a detector array and inhibition network matched to the pattern, and a digital machine that inputs images of that pattern and tracks relative image motion. (An especially interesting aspect of the design is the integration of sensors, memory, and logic in a single array using standard MOS technology.)

The basis of the sensors is that in NMOS, light striking the circuit side of the chip converts photons to hole-electron pairs; the holes are attracted to negative-biased p-type silicon substrates, while the electrons are attracted to n-type regions. A so-called dynamic node that has been positively charged will detect light by collecting a negative charge (electrons) and "leaking" to a lower voltage. An imager is simply an array of subcircuits, each consisting of a dynamic node, a transistor to reset the node to "high" and isolate it, and an inverter circuit to sense the voltage of the node and communicate it to other circuits.

Figure 4b shows the floor plan of the optical mouse chip. "Mouse cells" represent the four-by-four, two-dimensional sensor array, and "Tracker PLA" tracks spots by comparing images and outputting X and Y movements, as well as outputting "Any-Good" and "Jump" counter control and test signals. The "X counter PLA" and the "Y counter PLA" control up/down counting and transmit "XA XB XL" and "YA YB YL" coordinates to the host computer.

The layout style used in this first version of the chip treats a sensor cell with its logic and memory as a low-level cell and constructs the array by selectively programming the cells in different positions. The resulting chip is about 3.5 mm × 5.4 mm in a typical NMOS process (Lambda = 2.5 microns, or five-micron lines). A second version of the chip has also been designed to improve light sensitivity.[16]

Geometry Engine.[17] This vector function unit, designed by James Clark of Stanford University, performs three of the common geometric functions of computer graphics: transformation, clipping, and scaling. A single-chip version is used in 12 slightly different configurations to accomplish 4 × 4 matrix multiplications; line, character, and polygon clipping; and scaling of the clipped results to display device coordinates. This unit is an excellent example of a special-purpose device built from identical chips.

When configured as a four-component unit, the Geometry Engine allows simple operations on floating-point numbers. Each of its four identical function units has an eight-bit characteristic and (currently) a 20-bit mantissa. It operates with a simple structure of five elements: an ALU, three registers, and a stack. This basic unit can perform parallel additions, subtractions, and other similar two-variable operations on either the characteristic or the mantissa. Since one register can shift down and one can shift up, it can also multiply and divide at the rate of one step per microcycle. The 12-chip system consists of 1344 copies of a single bit-slice layout composed of the five elements. Four pins on the chip are wired to indicate to the microcode which of the 12 functions to carry out, according to the chip's position in the subsystem organization.

Figure 5a shows the geometry subsystem described above. The terms "mm," "clip," and "scale" denote matrix multiply, clipping, and scaling. Each of the four "mm" blocks, for instance, is a separate Geometry Engine chip that does a four-component vector dot product. Although each multiplication is done at the rate of one partial product per microcycle, the matrix multiplier has 16 of these products simultaneously active. Thus, the total transformation time, which is the bandwidth limiting system operation, is about 12 microseconds.

Figure 5b is a plan view of the Geometry Engine. The shaded parts have been implemented as two separate MPC projects. (The very small rectangles are copies of the principal bit-slice.) These projects are five-bit versions of the main function unit and the "pipelined" clock. Almost all the Geometry Engine's complexity is in the microcode that drives it. This microcode represents the logic equations for its finite-state machine, which will be implemented in a PLA. Writing this microcode and making minor additions to the principal bit-slice to accommodate it took up approximately 50 percent of the total design time. Estimations are that the Geometry Engine is capable of performing about four million floating-point operations per second; 48 identical units, four per chip, will each do a floating-point operation in about 12 microseconds.[17] Also note that the Geometry Engine is designed to work in conjunction with an image memory processor whose design is described by Clark and Hannah.[18]

Simple microprocessors

Traditionally, the trend in designing microprocessors, and even mainframe computers, has been towards the use of increasingly complex instruction sets and associated architectures. However, the judicious choice of a simple set of instructions and a corresponding simple machine organization can achieve such a high instruction rate that the overall processing power can exceed that of processors implementing more complex instructions.[19] An additional advantage is that the microprocessor has a short design time. (Patterson and Ditzel[20] give the arguments for such an approach to computer design, with a rebuttal by the designers of the Digital Equipment Corporation VAX computers.) In the following, we examine two contrasting designs for "simple microprocessors."

RISC I.[19,21] The reduced instruction set computer is being developed by Dave Patterson and his co-workers at the University of California, Berkeley. This type of microprocessor combines a small set of often-used instructions, with an architecture tailored to the efficient execution of these instructions. In addition, a single-chip implementation of a simpler machine makes more effective use of the limited resources of present-day VLSI chips—such as the number of transistors, area, and power consumption. The RISC I project has also shown that simplicity of the instruction set leads to a small control section and shorter machine cycle, as well as to a reduction in design time.

RISC I is a register-oriented, 32-bit microprocessor that has 31 operation codes, uses 32-bit addresses, and supports eight-, 16-, and 32-bit data. It has 138 general-purpose 32-bit registers and executes most instructions in a single cycle. The LOAD and STORE instructions—the only operations that access memory—violate this single cycle constraint; they add an index register and the immediate offset during the first cycle, performing the memory access during the next cycle to allow enough time for main-memory access.

Figure 6 shows the machine organization of RISC I. The machine naturally subdivides into the following function blocks: the register-file, the ALU, the shifter, a set of program counter (PC) registers, the data I/O latches, the program status word (PSW) register, and the control section (which contains the instruction register, instruction decoder, and internal clock circuits). In addition, the register file needs at least two independent buses because two operands are required simultaneously. In Figure 6, buses A and B are read-only and bus C is write-only.

An unusual feature of RISC I is the so-called overlapped window registers, a fast and simple procedure calling mechanism using the general-purpose registers. A procedure has access to 32 registers: global (registers 0-9

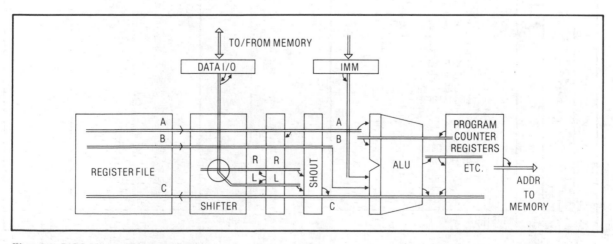

Figure 6. RISC I data path organization.

are shared by every procedure); low (registers 10-15 contain result parameters); local (registers 16-25 are used for local working); and high (registers 26-31 are input parameters). Each time a procedure is called, new registers are allocated in which the low registers of the calling procedure overlap the high registers of the called procedure.

Two different NMOS versions of RISC I are currently being pursued.[21] The "gold" version, whose data path is shown in Figure 6, has actually been fabricated. But the "blue" version is still under development. It's organization is similar to gold's but a more sophisticated timing scheme shortens the machine cycle and reduces chip area. Details of the gold design[21] include a chip size of 406 × 350 mm, 44K devices, a design time of 15 man-months, and a layout time of 12 man-months. Power consumption for the chip is estimated at between 1.2 and 1.9 watts. One very notable impact of the reduced instruction set approach is that the chip area dedicated to control dropped from the typical 50 percent in commercial microprocessors to only 6 percent in RISC I.

Scheme-79. This single-chip microprocessor[22,23] developed by Gerry Sussman and his colleagues at MIT directly interprets a dialect of the Lisp language. As noted by the designers,[23] Lisp is a natural choice among high-level languages for implementation in hardware. It is simple but powerful and, as in traditional machine languages, represents programs as data uniformly. All compound data in the system are built from list nodes, consisting of a CAR pointer and a CDR pointer. Each pointer is 32-bits, comprising a 24-bit data field, a seven-bit type field, and a one-bit field used by the storage allocator.

The Scheme-79 chip, whose architecture is shown in Figure 7, implements a standard von Neumann architecture in which a processor is attached to a memory system. The processor is divided into two parts: the data paths and the controller. The data paths are a set of special-purpose registers, with built-in operators interconnected with a single 32-bit bus. The controller is a finite-state machine that sequences through microcode, implementing both the interpreter for the Lisp subset and the garbage collector that supports an automatic storage allocation system. At each step it performs an operation on some of the registers (for example, transferring the address of an allocated cell in NEWCELL into the STACK register) and selects a next state based on both its current state and the conditions developed within the data paths.

Ten registers in the chip have specialized characteristics. On each cycle, these registers can be controlled so that one of them is gated onto the bus and selected fields of the bus are gated to another register. (The bus is extended off the chip through a set of pads.)

The finite-state controller for the Scheme-79 chip is a synchronous system composed of a state register and a large piece of combinational logic—the control map. From the current state stored in the state register, the control map develops control signals for the register array

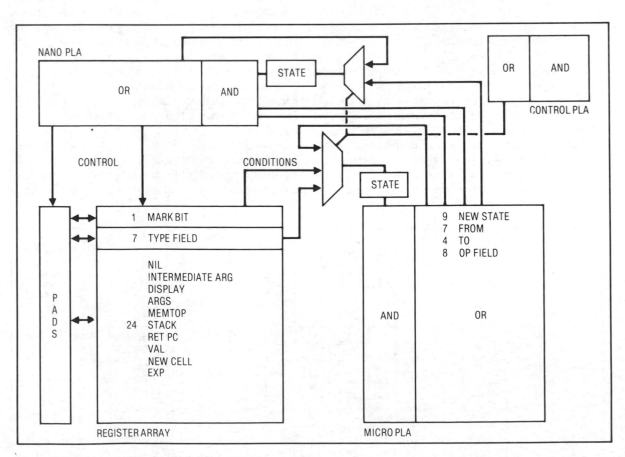

Figure 7. The major blocks of the Scheme-79 chip.

and pads, the new state, and controls for selecting the sources for the next sequential state. The Scheme-79 chip interfaces to the external world through a 32-bit bidirectional data bus that specifies addresses, reads and writes (heap) memory, references I/O devices, reads interrupt vectors, and accesses the internal microcode state during debugging.

To estimate the Lisp chip's performance, Sussman's group computed the values of Fibonacci numbers. For example, they calculated (fib 20) = 6765 with two different memory loadings, a clock period of 1595 nanoseconds (not the chip's top speed), and a memory of 32K-Lisp cells. If the memory was substantially empty (when garbage collection is most efficient) the Scheme-79 chip took about one minute to execute the program. With memory half-full of live structure (a typical load for a Lisp system), the Scheme-79 chip took about three minutes. (A MacLisp interpreter running on a DEC KA10 took about 3.6 minutes for the same calculation.)

The Scheme-79 chip, using a process with a minimum line width of five microns (Lambda = 2.5 microns), was 5926 microns wide and 7548 microns long, a total area of 44.73 mm^2. The entire project,[22] including prototype tool building and chip synthesis, was completed in five weeks and is frequently cited to justify the power of the Mead and Conway VLSI design philosophy. Areas for future work include improving the electrical characteristics of the Scheme-79 chip and implementing a multiprocessor system composed of Lisp chips.

Tree machines

As Seitz[24] has observed, the existence of communication problems makes scaling large single processors to submicron dimensions self-defeating. Communication assumes a progressively more dominant and limiting role in VLSI as chip area, signal energy, and propagation time become increasingly expensive. As these costs rise, im-proving performance with an ensemble of concurrently operating small processors becomes more attractive than using a larger single processor. The principle of locality has a number of effects on VLSI. First, it encourages the repeated use of identical computing elements, each with capabilities for processing, communication, and memory. Second, it becomes desirable for a group of elements to be functionally equivalent to a single element (to overcome problems of increasing miniaturization). Third, it motivates the use of concurrency to counteract the simplicity of the individual computing elements. And lastly, it encourages designers to utilize locality of reference to reduce system-wide communications.

A tree machine architecture combines the above VLSI properties with a general-purpose computing environment. A tree machine is a collection of simple computing elements connected as a binary tree. There is no global communication, only communication between a parent and its child in the tree, and between the root of the tree and the external world. This architecture gives rise to integrated circuits with regular interconnections, local communication, and many repetitions of a single cell. These integrated circuits, in turn, can be assembled into regular patterns at the printed-circuit board and backplane levels to construct machines with thousands of processors. Examples of such tree machines are being investigated at Berkeley,[25,26] Caltech,[24,27] and Carnegie-Mellon.[28]

The X-tree. This University of California, Berkeley, project[25,26] is developing a tree machine built of modular components known as X-nodes. As shown in Figure 8, this is a binary tree enhanced with additional links to form a half- or full-ring tree. These links further shorten the average path length by distributing message traffic more evenly throughout the tree, and they provide the potential for fault-tolerant communciation if a few nodes or links are removed. Notice in Figure 8 that the children of node n have node addresses $2n$ and $2n + 1$, respectively.

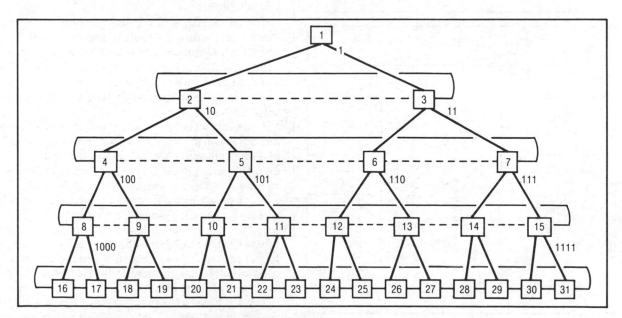

Figure 8. The X-tree—a binary tree with full-ring connections.

All communication throughout the tree is in the form of messages. For effective routing, the complete address is divided into a node address and a second part identifying a particular memory location, if that node has any memory belonging to the global address space. (In the X-tree, secondary memory as well as input/output is restricted to the frontier (i.e., the leaves) of the tree.)

Message routing is based on the binary address. As seen in Figure 8, the root node is assigned address 1; the address of a left child is formed by appending a 0 to the root node address and a right child's address is formed by appending a 1 to it. To address one node from another, the message has to move up the tree to their common ancestor—that is, to the node where the address matches all leading bits in the target address. From there the message moves down to the destination.

On first glance, each X-node (see Figure 9) is simply a computing element that communicates with four or five

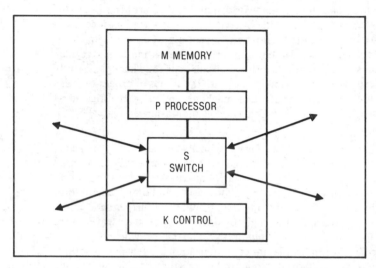

Figure 9. The X-node—a computing element shown in PMS notation.

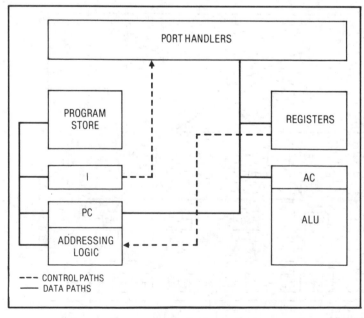

--- CONTROL PATHS
— DATA PATHS

Figure 10. Block diagram of the Caltech processor.

neighbors. However, normal microprocessor input/output techniques are inadequate because of bandwidth requirements. Computation must occur in parallel with communication. Thus, each X-node contains a self-controlled switching network with its own I/O buffers and controllers. The heart of this switching network is a time-multiplexed bus. Since this bus is completely contained within one chip, its parasitic capacity is low. The resulting bandwidth for a given amount of drive power is thus an order of magnitude higher than that running through the package pins associated with the input/output ports. Each of these ports consists of a set of input and output buffers and the necessary finite-state machines to control them.

By the mid 1980's, it appears likely that approximately 64K bytes will be able to be implemented on the same chip with the X-node processor if dynamic RAM or charge-coupled devices are used. Prototype TTL circuitry for the communication system, consisting of I/O ports and a routing controller, has already been implemented. The RISC I processor is definitely a key stage in the development of a complete X-node.

Caltech tree machine. The California Institute of Technolgy tree machine is based on Sally Browning's doctoral dissertation,[27] and the description of the machine presented here is largely taken from this source.

The Caltech tree machine is programmed in tree machine programming language (TMPL), a high-level language resembling Hoare's communicating sequential processes notation.[29] Computing elements of the machine have four main parts: a program store, a bank of registers for storing data, an ALU, and some communications handlers. The control and data paths run between these components, aided by three special-purpose registers. The I register holds an instruction, the PC register is the program counter, and the AC is an accumulator providing the source and destination for the ALU. Figure 10 is a block diagram of one of these computing elements.

The ALU, AC, and registers comprise a functional block that performs the usual arithmetic and logical operations. Communications handlers are related components, and there is one of them for each of the ports. They interface with the outside world, handle message traffic, load the program store, and pass code to their subordinates.

A possible layout of a binary tree of computing elements is shown in Figure 11. There, each element consists of a communication (C), a processor (P), and a memory (M). This arrangement is repeated until an entire silicon chip is covered by the computing element hierarchy.

As discussed in Mead and Conway,[6] the longer the wires in a computing element, the larger the drivers they need. Giving the highest-level computing element large drivers (to drive off-chip without suffering a severe performance penalty) can extend such a machine to many individual chips and yet maintain the full speed of the individual processors within it.

The first Caltech tree machine was designed during the summer of 1980 and has one computing element per chip. As finer design geometries become available, several pro-

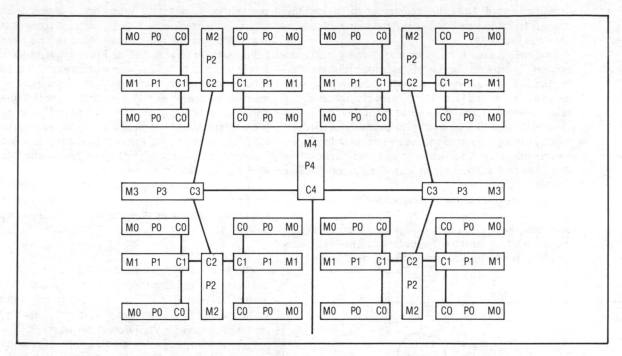

Figure 11. Layout for a binary tree of computing elements.

cessors and their memories will be placed on a single chip. Caltech's goal was to build and test a tree machine by the fall of 1981 and have a machine of at least 1023 processors working in 1982.[24]

Non-von Neumann computers

The final area of VLSI processor architecture covered in this article also concerns the design of decentralized computers built from identical computing elements. Some of these architectures are even tree machines. However, what distinguishes the two areas is the design approach. In the tree machines discussed previously, the design centered on the hardware—i.e., the configuration of computing elements to exploit increasing levels of integration and also to handle problems such as communication. In non-von Neumann computers, though, the design centers on the software. In other words, the design concentrates on the way the computing elements are programmed to exploit parallelism.

The VLSI architectures discussed below are not based on traditional, sequential control-flow program organization, but on alternative naturally parallel organizations such as reduction[30] or data flow.[31-33] All three are specifically designed to exploit VLSI. These program organizations are distinguished by the form of instructions, by the way instructions manipulate their arguments, and by the patterns of control. At present, over 30 non-von Neumann computers are being developed worldwide, and a number of machines are currently operational.[34]

The cellular reduction machine. This machine[30,35] is being developed by Gyula Mago and his colleagues at the University of North Carolina, Chapel Hill. It uses the class of FP functional programming languages designed by Backus[36]; its program organization is string reduction; and its machine organization is a binary tree structure.

Figure 12 illustrates the string reduction form of program organization for a statement $a = (b + 1)*(b - c)$. In this example, each instruction is a "name:(expression)" pair, where the expression is either constant (4) or defines an operator and its input operands. Notice that an instruction has no explicit output operands, but always returns the result to the invoking instructions. Thus, in reduction, instruction execution is often viewed as "demand driven." An instruction is executed when the result it generates is required as input by the invoking instruction. In string reduction, each instruction invoking another will take a copy of the instruction and store the copy in place of the operand address. These copies can then be manipulated independently and in parallel.

A program for the cellular reduction machine is a linear string of symbols mapped onto a vector of memory cells in the computer, one symbol per cell, possibly with empty

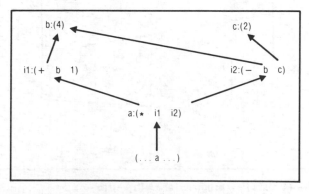

Figure 12. Reduction program for a = (b + 1)*(b − c).

513

cells interspersed. The cellular computer's organization is a binary tree structure with two different kinds of cells, as shown in Figure 13.

Leaf cells (called L cells) serve as memory units, and nonleaf (T) cells are capable of dual processing and communication. An expression is mapped onto the tree, storing each symbol in an L cell. A subtree of symbols (i.e., a subexpression) is linked by some dedicated T cells, as shown in Figure 13. To simplify the operation of the computer, an integer is stored with every symbol indicating its own nesting level. A particular set of L and T cells will be dedicated to a subtree for at least the duration of one machine cycle.

When the expression to be executed has been partitioned into a collection of cells, itself a cellular computer, microprograms handle the interaction of these cells in the reduction of an innermost application. Microprograms normally reside outside the network of cells and are brought in on demand. When one is demanded, it is placed in registers in the L cells. Each cell receives only a fraction of the microprogram needed to make its contribution to the total reduction. For example, if one of the L cells wants to broadcast information to all other L cells involved in reducing a subexpression, it executes a SEND microinstruction,[35] explicitly identifying the information item. As a result, this information is passed to the root of the subexpression and broadcast to all appropriate L cells. The operation of the cells in the network is coordinated, not by a central clock, but by endowing each cell with a finite-state control and letting the state changes sweep up and down the tree. This allows global synchronization, even though the individual cells work asynchronously and only communicate with their immediate neighbors.

The Irvine data flow machine. The Irvine data flow project was originally conducted by Arvind and Gostelow at the University of California, Irvine. It now continues under the direction of Arvind at MIT and primarily concerns the design of a VLSI data flow computer.[31]

Figure 14 illustrates the data flow form of program organization for the statement $a = (b + 1)*(b - c)$. In this example, each instruction consists of an operator, two inputs that are either constants or "unknown" operands defined by empty bracket symbols, and an address such as i3/1 defining a consumer instruction and operand position for the result. In data flow programs, copies of a result are logically stored by the producer instruction directly into each consumer instruction. These so-called data tokens also provide a control signal to the consumer instruction, which is executed when it has received all of its input operands. Data flow program execution is *data driven* and as such is highly parallel.

The Irvine data flow machine, like other data flow computers, is based on a packet communication machine organization.[33] For a parallel computer, packet communication is a very simple strategy for allocating "packets of work" to resources. Each packet to be processed, such as an executable instruction, is placed with similar packets in a "pool of work." When a resource becomes idle, it takes a packet from its input pool, processes the packet, and places a modified packet in an output pool.

Figure 15 illustrates a computing element of the proposed Irvine data flow machine. The machine consists of N computing elements and an $N \times N$ packet communication network. Each computing element is essentially a complete computer with an instruction set. These computing elements have up to 16K words each of program storage and data structure storage, and also contain certain "special" elements, which are divided into sections. These specialized elements include an input section, which accepts input from other computing elements; a waiting-matching section, which forms data tokens into sets for a consumer instruction; an instruction fetch section, which fetches executable instructions from the local program memory; a service section containing a floating-point ALU such as an Intel 8087 that executes instructions; and an output section, which routes data tokens containing results to the destination computing element.

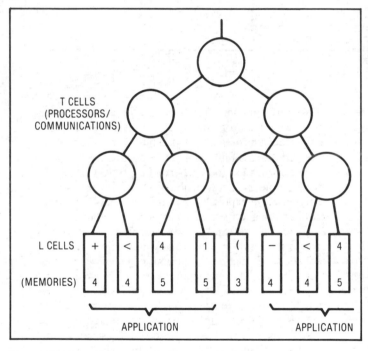

Figure 13. The cellular tree machine.

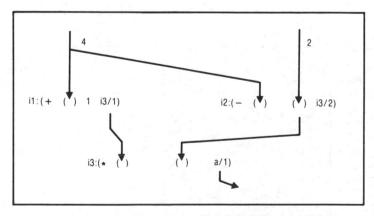

Figure 14. Data flow program for a = (b + 1)*(b − c).

Arvind is currently designing a data flow machine with 64 computing elements at MIT, and he hopes to complete the design for MOS fabrication by the end of 1982.

The recursive machine project.[37] This project, conducted by Wayne Wilner at Xerox Corporation's Palo Alto Research Center in California, builds on earlier work by Bob Barton and Wilner at Burroughs Corporation. This is one of the most sophisticated non-von Neumann computers currently under investigation. The organization of both its program and machine are based on the principle of recursion.

Information is represented in terms of fields, which are recursively defined to be either bracketed strings of characters or bracketed strings of fields

$$field ::= \text{``('' char 1 \quad char 2 \quad ... \quad char } n \text{ '')''}$$
$$field ::= \text{``('' field 1 \quad field 2 \quad ... \quad field } n \text{ '')''}$$

where the character alphabet is disjoint from the bracket alphabet. For example, the number "six" can be represented as (6), a 2×2 matrix as (((a11)(a12))((a21)(a22))), and an empty stack as ( ). Likewise, a machine instruction is recursively defined to be either a string of characters or an n-tuple of machine instructions. The statement $a = (b+1)*(b-c)$ could be represented as $((b + 1)(*)(b - c) ...)$.

Program execution is based on an "actor," or message passing model, such as logically underlies the Xerox PARC Smalltalk[38] programming language. In this model, execution of the instruction $(4 + 1)$ is viewed as the message " + 1" being sent to the integer value 4, which executes and is replaced by the result 5. Messages can be either data driven or demand driven and are sent from one field to another. The addresses used to route messages are the logical addresses of the fields, not their physical designations.

A recursive machine (RM) can be either a single element or a system of recursive machines, as shown in Figure 16.

Several RM elements can be configured into a network by joining their point-to-point connections serially, by busing their hierarchical connections, and by using another recursive machine element to couple the buses to the next higher level of storage and communication. The resulting configuration has exactly the same outward appearance as a single RM element.

A recursive machine element is a small general-purpose computer, forming the bottom level of the storage, control, and communication hierarchy. It consists of a microprogrammable processor with various functional units, or FUs, and writable control store. It has variable-length registers, which are accessed through field-

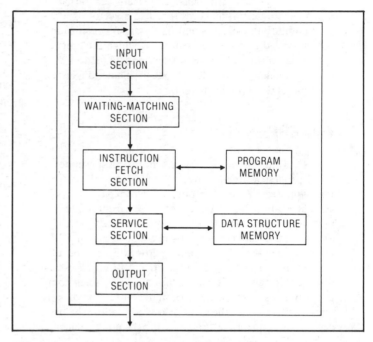

Figure 15. An Irvine data flow computing element.

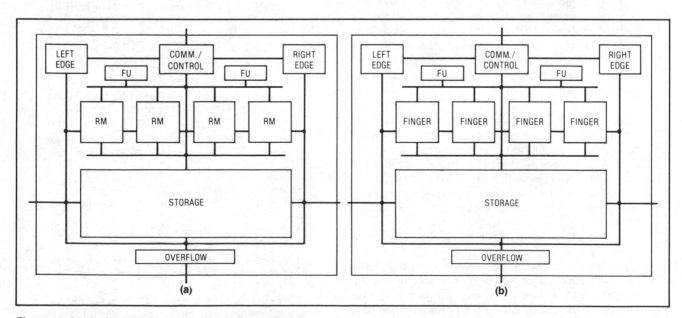

Figure 16. Recursive machine system (a) and element (b).

oriented machines called "fingers." It has input and output ports through which it transmits messages on behalf of fields located within its storage. And it also has point-to-point connections to neighboring elements through which fields migrate when algorithms create more data than can fit inside a single element. Lastly, each recursive machine has two "boundary registers" to hold the logical addresses of the characters at the left and right edges of the RM. These locate the approximate position of an addressed field.

In closing, Treleaven and Hopkins[39] are currently investigating another recursive machine that is based on the work of Wilner[37] and Barton.[40] There are some distinct differences between Wilner's recursive, or actor, machine and Treleaven and Hopkins'. For example, the latter machine's organization is restricted to a vector of identical computing elements and its program organization uses so-called recursive control flow, which attempts to synthesize control flow, data flow, and reduction.[34] A simple, single-chip building block implementing these concepts is presently being designed.

Future directions

The veritable explosion of current research into novel VLSI processor architectures[5,10-12,41] is radically affecting computers. Various groups are interested in incorporating special-purpose VLSI processors into computer systems to increase performance and functionality. Others are more interested in identifying the next era of VLSI building blocks[42,43] that will follow the current LSI microprocessor. To use all of the VLSI processors discussed in this article as building blocks in larger computer systems, they must conform to some common system architecture. If they conform, it will allow the processors to cooperate at both the program and machine organization levels.

Thus, there is a belief among some computer architects that the next generation of computers will be based on a non-von Neumann architecture capable of exploiting both general-purpose and special-purpose VLSI processors.[44] Perhaps the best discussion of the attributes of these so-called fifth generation computers is in a document produced by the Japan Information Processing Development Center.[45] It states that

(1) There will be considerable diversity of functions, types, and levels in fifth generation computers. This will be seen in everything from machines for very high-speed processing, to processors with specialized functions and applications, and personal and built-in computers.

(2) There will be a decline in the former trend toward very general-purpose orientation and, instead, we will see an increase in specialization.

(3) The systems will be based on non-von Neumann architecture.

(4) The importance of individual new microarchitectures will increase; there will be considerable use of composite systems formed by combining a number of processors and hardware, software, and firmware elements; and the importance of system architecture will grow.

Just as Mead and Conway's VLSI design philosophy[6] provides a whole new way to "do electronics," fifth generation architectures are likely to provide a whole new way to "build computing systems." Although we are starting to understand the strengths and weaknesses of various decentralized computer architectures,[34] it is not yet clear just which sets of concepts will contribute to fifth generation computers. It seems likely that they will be significantly influenced by both tree and recursive machines, which are essentially equivalent. Lastly, however, the programming of these decentralized architectures remains a largely unsolved problem. ∎

References

1. "About the Cover," *Lambda,* 1st Qtr. 1980.

2. R. Bernhart, "More Hardware Means Less Software," *IEEE Spectrum,* Vol. 18, No. 12, Dec. 1981, pp. 30-37.

3. R. C. Johnson, "32-bit Microprocessors Inherit Mainframe Features," *Electronics,* Vol. 54, No. 4, Feb. 1981, pp. 138-141.

4. *Very Large Scale Integration (VLSI) Fundamentals and Applications,* D. F. Barbe, ed., Springer-Verlag, New York, 1980.

5. *Proc. Caltech Conf. VLSI,* C. Seitz, ed., Jan. 1979.

6. C. A. Mead and L. Conway, *Introduction to VLSI Systems,* Addison-Wesley, Reading, Mass., 1980.

7. L. Conway, et al., "MPC79: A Large-Scale Demonstration of a New Way to Create Systems in Silicon," *Lambda,* 2nd Qtr. 1980, pp. 10-19.

8. W. D. Jansen and D. G. Fairbairn, "The Silicon Foundry: Concepts and Reality," *Lambda,* 1st Qtr 1981, pp. 16-26.

9. M.J. Foster and H. T. Kung, "The Design of Special-Purpose VLSI Chips," *Computer,* Vol. 13, No. 1, Jan. 1980, pp. 26-40.

10. *Proc. VLSI 81,* J. P. Gray, ed., Academic Press, New York, 1981.

11. *Proc. CMU Conf. VLSI Systems and Computations,* H. T. Kung, R. F. Sproull, and G. Steele, eds., Springer-Verlag, New York, Oct. 1981.

12. *Proc. Caltech Conf. VLSI,* C. Seitz, ed., Jan. 1981.

13. H. T. Kung, "Let's Design Algorithms for VLSI Systems," *Proc. Caltech Conf. VLSI,* Jan. 1979.

14. R. L. Rivest, "A Description of a Single-Chip Implementation of the RSA Cipher," *Lambda,* 4th Qtr. 1980, pp. 14-18.

15. R. F. Lyon, "A Bit-Serial VLSI Architecture Methodology for Signal Processing, *Proc. VLSI 81,* Aug. 1981, pp. 131-140.

16. R. F. Lyon, "The Optical Mouse, and an Architecture Methdology for Smart Digital Sensors," Technical Report VLSI-81-1, Xerox Corporation Palo Alto Research Center, Aug. 1981.

17. J. H. Clark, "A VLSI Geometry Processor for Graphics," *Computer,* Vol. 13, No. 7, July 1980, pp. 59-68.

18. J. H. Clark and M. R. Hannah, "Distributed Processing in a High-Performance Smart Image Memory," *Lambda,* 4th Qtr. 1980, pp. 40-45.

19. D. A. Patterson and C. H. Sequin, "RISC I: A Reduced Instruction Set VLSI Computer," *Proc. Eighth Int'l Symp. Computer Architecture,* May 1981, pp. 443-457.

20. D. A. Patterson and D. R. Ditzel, "The Case for the Reduced Instruction Set Computer," *Computer Architecture News*, Vol. 8, No. 6, Oct. 1980, pp. 25-32.

21. D. T. Fitzpatrick et al., "A RISCy Approach to VLSI," *VLSI Design,* 4th Qtr. 1981, pp. 14-20.

22. J. Holloway et al., "The Scheme-79 Chip," MIT AI Memo 559, Cambridge, Mass., Jan. 1980.

23. G. J. Sussman et al., "Scheme-79—Lisp on a Chip," *Computer,* Vol. 14, No. 7, July 1981, pp. 10-21.

24. S. A. Browning and C. L. Seitz, "Communication in a Tree Machine," *Proc. Caltech Conf. VLSI,* Jan. 1981.

25. A. M. Despain and D. A. Patterson, "X-TREE: A Tree Structured Multiprocessor Computer Architecture," *Proc. Fifth Int'l Symp. Computer Architecture,* Apr. 1978, pp. 144-151.

26. D. A. Patterson et al., "Design Considerations for the VLSI Processor of X-TREE," *Proc. Sixth Int'l Symp. Computer Architecture,* Apr. 1979, pp. 90-100.

27. S. A. Browning, "The Tree Machine: A Highly Concurrent Computing Environment," PhD dissertation, Department of Computer Science, California Institute of Technology, 1980.

28. J. L. Bentley and H. T. Kung, "Two Papers on a Tree Structured Parallel Computer," Technical Report CMU-CS-79-142, Department of Computer Science, Carnegie-Mellon University, Aug. 1979.

29. C. A. R. Hoare, "Communicating Sequential Processes," *Comm. ACM,* Vol. 21, No. 8, Aug. 1978, pp. 666-677.

30. G. A. Mago, "A Cellular Computer Architecture for Functional Programming," *Proc. Compcon Spring 80,* pp. 179-185.

31. Arvind et al., "A Processing Element for a Large Multiple Processor Dataflow Machine," *Proc. Int'l Conf. Circuits and Computers,* Oct. 1980.

32. A. L. Davis, "A Data-Driven Machine Architecture Suitable for VLSI Implementation," *Proc. Caltech Conf. VLSI,* Jan. 1979, pp. 479-494.

33. J. B. Dennis, "Data Flow Supercomputers," *Computer,* Vol. 13, No. 11, Nov. 1980, pp. 48-56.

34. P. C. Treleaven et al., "Data Driven and Demand Driven Computer Architecture," *ACM Computing Surveys,* Vol. 14, No. 1, Mar. 1982.

35. G. A. Mago, "A Network of Microprocessors to Execute Reduction Languages," *Int'l J. Computer and Information Sciences,* Vol. 8, No. 5, pp. 349-385, and Vol. 8, No. 6, pp. 435-471, 1980.

36. J. Backus, "Can Programming be Liberated from the von Neumann Style? A Functional Style and its Algebra of Programs," *Comm. ACM,* Vol. 21, No. 8, Aug. 1978, pp. 613-641.

37. W. Wilner, "Recursive Machines," internal report, Xerox Corporation Palo Alto Research Center, 1980.

38. "The Smalltalk-80 Programming Language," *Byte,* Vol. 6, No. 8, Aug. 1981.

39. P. C. Treleaven and R. P. Hopkins, "A Recursive Computer Architecture for VLSI," *Proc. Ninth Int'l Symp. Computer Architecture,* Apr. 1982.

40. B. Barton, "On Modular Machines," *Proc. Arhus Workshop Software Eng.,* May 1978.

41. P. C. Treleaven, "VLSI: Machine Architecture and Very High Level Languages," *Proc. Joint UK SRC/University of Newcastle upon Tyne Workshop,* Technical Report 156, Computing Lab., University of Newcastle upon Tyne, summarized also in *Sigarch,* Vol. 8, No. 7, Dec. 1980.

42. D. A. Patterson and C. H. Sequin, "Design Considerations for Single-Chip Computers of the Future," *IEEE Trans. Computers,* Vol. C-29, No. 2, Feb. 1980, pp. 108-116.

43. J. R. Tobias, "LSI/VLSI Building Blocks," *Computer,* Vol. 14, No. 8, Aug. 1981, pp. 83-101.

44. *Proc. Int'l Conf. Fifth Generation Computer Systems,* Japan Information Processing Center, Oct. 1981.

45. "Interim Report on the Study and Research on Fifth-Generation Computers," Japan Information Processing Development Center, 1980.

16b CPU Design by a Hierarchical Polycell Approach

T.Tokuda, I.Ohkura, H.Shimoyama, Y.Kazuma, T.Enomoto

Mitsubishi Electric Corporation
Itami, hyogo, 664 Japan

I. Introduction

Custom VLSI chips are available by gate array or standard cell approaches in a short turn-around time. In some fields of LSI application, these approaches are not sufficient in the point of operation speed, power dissipation and packing density. A hierarchical standard cell approach seems to be effective for the increase of LSI performances.

An integrated CAD system which permits the hierarchical LSI design was developed. The system was architected by polycell-based layout program(1) and the precise timing verification(2), to optimize the physical design. Utilizing the design system, high-speed LSIs for a 16b CPU were developed for small business computers.

This paper describes the hierarchical design technique used for the LSI chips by the CAD system with N-ED/CMOS standard cells.

II. CAD System

The CAD system used in the LSI design has been presented in a companion paper by K.Sato(3).
Its salient features are:

Hierarchical LSI Design: As the layout results (block size, peripheral pin positions and net lengths) of layout blocks constructed of standard cells are automatically stored into the data-base, the hierarchical chip design is easily executed in short turn-around time. Also, logic description, logic simulation and layout verifications are carried out hierarchically.

Layout Procedure: In the polycell-based layout program, functional blocks such as ROMs, PLAs and RAMs are handled as layout elements the same as standard cells, and can be automatically laid out concurrently.

Interactive Design Facilities: In order to improve the layout design, modification of the layout results can be carried out interactively using the GDS(Graphic Design System). Updated design modifications are stored automatically into the data-base, after connectivity and spacing checks.

Precise Timing Verification: Timing of LSI circuits can be checked in a short time and on a large scale by precise delay-mode logic simulation.

III. Standard Cell Design

Most of the standard cells are constructed of N-ED (N-channel Enhancement-Depletion) gates, and internal drivers or I/O buffers are designed by CMOS gates, taking advantage of both gates' features (high-packing density of N-ED gates and low power of CMOS gates). N-well CMOS with double level metallization technology is used in the fabrication.

N-ED Gate Cells have three power levels (low, standard and high-power). They can be selected to optimize the delay times of the gates and the power dissipation of the chip. Besides the multi-input NOR, NAND and functional gates (AND-NOR, AND-OR-NAND, etc.), transmission gates are used in the standard cell such as latch and multi-input selector, to decrease the number of LSI elements.

CMOS Driver Cells are prepared for an inverter, a 2NOR, a 2NAND and a tri-state driver. To enable the exact analysis of the timing relationship in the circuits containing both N-ED and CMOS gates, the logical threshold voltage (V_T) of the gates is designed to be same. The DC characteristics of the N-ED and CMOS inverter are shown in Fig.1 as a function of R_I, which is the saturation current ratio of load and switching transistor. This figure shows that the CMOS inverter is suitable for the driver gates because of approximately double capacity drivability compared with N-ED's when the output voltage is rising.

After pattern layout, the art-work data of the standard cells are stored in the data-base of the GDS, while the cell size, peripheral pin position and logic data are stored in the data-base of the host computer. These standard cells are validated completely before LSI design.

IV. Precise Delay-Mode Logic Simulation

Gate Delay Expression

The delay time t_{PLH} and t_{PHL} are defined by the intervals between the two points at which the input and output voltages become equal to the V_T. An N-ED inverter chain was evaluated concerning the delay properties. C_n and C_{n-1} are the load capacitances normalized by the gate size of the "n"th and "n-1"th inverter respectively. Analytical results by circuit simulation with varied values of C_{n-1}, are shown in Fig.2. The delay times of the CMOS inverter was also evaluated replacing the "n"th inverter by the CMOS gate. These are shown by the triangles in Fig.2. Since t_{PLH} is only slightly dependent on C_{n-1} in both inverters, it is not necessary to take C_{n-2} into account in calculating the delay time (2). Then, the delay time of "n"th inverter is shown as a simple function of C_n and C_{n-1}, as follows.

$$t_{PLH}=C_n(K_{0r} + K_{1r}(C_{n-1}/C_n)^{d_r}) \qquad (1.a)$$

$$t_{PHL}=C_n(K_{0f} + K_{1f}(C_{n-1}/C_n)^{d_f}) \qquad (1.b)$$

Parameters K_{0r}, K_{1r}, K_{0f}, K_{1f}, d_r and d_f are empirical constant obtained from the circuit

Reprinted from *IEEE Int. Conf. Circuits and Computers Proc.*, 1982, pp. 102-105.

simulation results. The delay times calculated by equation (1) are also shown in Fig.2.

In this design method, CMOS and N-ED inverter which have the same gate size and same loading condition, have almost the same t_{PHL}, while the t_{PLH} of the CMOS inverter is about 0.7 times smaller than that of the N-ED inverter. Therefore, when the "n-1"th inverter is CMOS, the input waveform of the "n"th inverter is improved. This effect was taken into account by adjusting the normalized capacitances C_{n-1}.

Application to the CAD System

The delay time expression was applied to the CAD system. Multi-input gates have various delay times depending on the input node that is logically active, since each input node has a different capacitance. To simulate this, the input side delay model (2) could be easily applied to the conventional event driven logic simulator without any modification of the logic simulator itself.

The data-base contains the data of gate size, stray capacitances of the input/output node and coefficients. The interface program extracts these data and calculates the delay times, and assigns the delay elements to each input pin of each gate as well as expanding the hierarchy of the network data for the logic simulator.

Besides the precise delay-mode, average delay-mode in which the delay value is implemented in a logic element itself was prepared considering the case that the element counts dealt with the simulator greatly increased in number.

Comparison with Circuit Simulation

The logic simulation of this mode was compared to a circuit simulation in a simple circuit constructed of a D flip-flop and inverters(Fig.3).

The simulation results by this mode are in good agreement with the results by the circuit simulation, as shown in Fig.4.

V. LSI Design for 16b CPU

LSI chips for 16b CPU were designed by the CAD system. The CPU was constructed of a data path chip, controller chip, memory management chip, etc. This section describes the hierarchical and structural methods used in the LSI design of the data path chip and also shows the evaluation results of the chip.

Logic Description

After architectural and logic design, the LSI was designed with the standard cells in the library.

In this process, usage of multi-input N-ED gate cells and multi-input selectors constructed of transmission gates was advantageous in reducing the transistor numbers. The avarage fan-in was increased to 2.7 in the chip and total gate count (regarded the three input N-ED NOR gate as one gate) was decreased to about 5 K.

The hierarchical description and the registration to the data-base were carried out in "logic block" units which correspond to a functional module. The logic blocks are mainly classified into three types according to the gate counts in the LSI. Table 1 shows the scale and number of blocks. The description was checked by format

check programs in each block level, and verified by logic simulation which was carried out in 37 blocks, mainly above the medium logic block.

The total block count amounted to 104, including the chip level, and the logic hierarchy had 10 levels. In spite of large amount of data, data-base registration and logic verification were completed in short period (one month with three people) on account of the hierarchical procedures.

Layout Design

The LSI chip was partitioned into four "layout blocks", considering the manageable amount of data for the layout program or the GDS. In this system, a logic block is easily designated to a layout block by a simple code (3) of the description. The layout hierarchy of the chip had 6 levels, as shown in Fig.5.

As the block(A) which contained a 16bit Register-ALU had high logic regularity, a structural layout design was applied utilizing the regularity as well as a RAM. The 4bit-sliced ALU block could be realized in such regular interconnected structures as the control lines ran in a horizontal straight line from across each block to the next and the data lines ran vertically, as shown in Fig.6. The block was laid out in this way with the automatic layout assisted by mannual improvement using the GDS. After removing the design errors pointed out by the connectivity and spacing verifier, the new layout data were automatically restored into the data-base. A 16bit ALU block was hierarchically constructed from the four blocks plus additional gates. Packing density of the block was improved to 552 transistors/mm^2, and the net lengths along the critical paths from the register output to the register input through the ALU were shortened by this method.

After completing the design of four layout blocks, the chip was automatically laid out with the blocks and I/O buffers. The optimization of the inter-block routing was executed by modifying the peripheral pin positions of the blocks interactively on the GDS.

The layout design of the LSI was accomplished in a short period (2 months with three people) due to the flexible design system.

Timing Verification

Before the layout design, the timing relationship of the logic description was verified by the average delay-mode logic simulation based on predicted wiring length. The power level selection of the standard cells was optimized by the simulation. After layout, precise delay-mode logic simulations were executed at the four layout blocks to verify the precise timing, including analysis of circuit race and hazard. Then, the timing was also verified in the chip level by the average delay-mode. After fabrication of the chip, the simulated delay times in some paths in the LSI show a good agreement with the measured values, as shown in Fig.7.

Fabrication Results of the Chip

The LSI chip, fabricated using the 3 μm-rule silicon gate N-well CMOS technology with double level metallization, contains approximately 20K transistors (including 1K RAM and I/O buffers) in 8.78x9.30 mm^2 die area. The photomicrograph and the characteristics of the LSI are shown in Fig.8 and Table 2. The logic design using multi-input N-ED gates reduced the total transistor number, while the total power dissipation was decreased by using CMOS circuits for the I/O buffer and the internal driver. High speed operation (machine cycle = 250 ns) and high density of 271 transistors/mm^2 were obtained in the LSI resulting from the device technology and flexible design system.

VI. Conclusions

High-performance LSIs for 16b CPU were designed using the integrated CAD system. The design system permits hierarchical logic/layout design and precise timing verification, to meet the requirements of VLSI design. The standard cell library was constructed of N-ED and CMOS gates to utilize the advantages of both gates.

In designing the data path chip, the 16bit Register-ALU block was hierarchically laid out with a RAM and dense 4bit-sliced ALU blocks. These blocks were structurally designed utilizing the regular structure of the ALU.

The LSI chip, fabricated using the 3 μm-rule N-well CMOS technology with double level metallization, contains about 20K transistors/mm^2 in 8.78x9.30 mm^2 die area. High speed operation (machine cycle = 250 ns) and high density of 271 transistors/mm^2 were obtained in low power consumption(1.2 W) owing to the use of N-ED/CMOS devices and the CAD system.

Acknowledgment

The authors wish to thank Drs.H.Oka, Y.Gamoh, T.Yahara, S.murai and S.Takeuchi for their encouragement and especially H.Kimura, J.Hinata, Y.Nishiwaki, S.Asai, Dr.O.Tomisawa, K.Sakashita, J.Korematsu and T.Arakawa, for their contributions.

References

(1) K.Sato et al., "MILD - A Cell-Based Layout System for MOS-LSI", Proceedings of 18th Design Automation Conference, p.828-836, July 1981
(2) I.Ohkura et al.,"A New Exact Delay Logic Simulation for ED MOS LSI", Proceedings of 1980 ICCC, p.953-956, Oct. 1980
(3) K.Sato et al., "An Integrated Custom VLSI Design System", Preceedings of 1982 ICCC

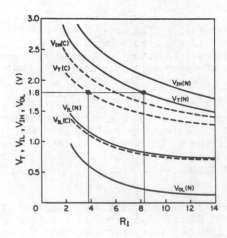

Fig.1 DC characteristics of N-ED and CMOS inverter

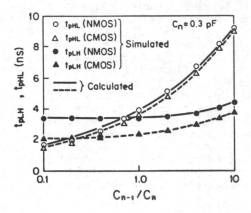

Fig.2 Delay times of N-ED and CMOS inverter

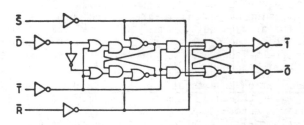

Fig.3 Master-Slave D flip-flop with inverters

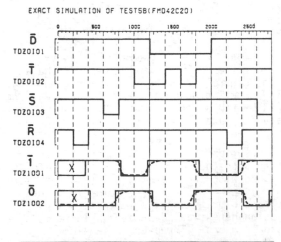

	$\bar{T} \to \bar{1}$		$\bar{S} \to \bar{1}$	$\bar{R} \to \bar{1}$
	t_{PLH}	t_{PHL}	t_{PHL}	t_{PLH}
Precise Delay-Mode Logic Simulator	17.2	23.8	22.4	15.5 (ns)
Circuit Simulator	14.0	20.5	22.5	17.5 (ns)

Fig.4 Comparison of precise delay-mode logic simulation and circuit simulation

Table 1 Scale and number of logic blocks

Block Type	Block Size (Gate Count)	Block Count
Large Block	500 ~ 5000	8
Medium Block	50 ~ 500	23
Small Block	< 50	72

Table 2 LSI characteristics

Function	16 bit data path
Machine Cycle	250 (ms) (typ.)
Active Power	1.2 (W)
Supply Voltage	5 . (V)
Package	124 pin plug — in type
Total Transistor Count	20K (including RAM and I/O)
	Random Logic ····· 11391
	RAM (1Kbit)····7843
	I/O ··· 624
Chip Size	8.78 x 9.30 (mm²)
Packing Density	271 (Transistor/mm²)
Cell Area Ratio	15.9 (%)

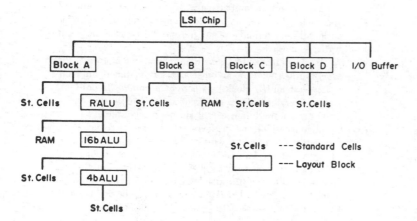

St. Cells --- Standard Cells
[ ] --- Layout Block

Fig.5 Layout block hierarchy of the LSI

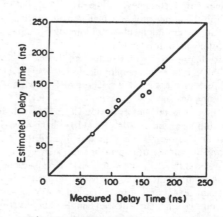

Fig.7 Delay times in the LSI

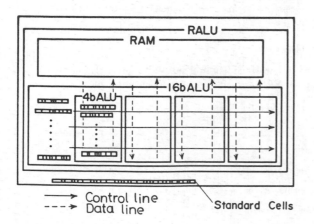

→ Control line
--→ Data line
Standard Cells

Fig.6 Block(A) layout structure

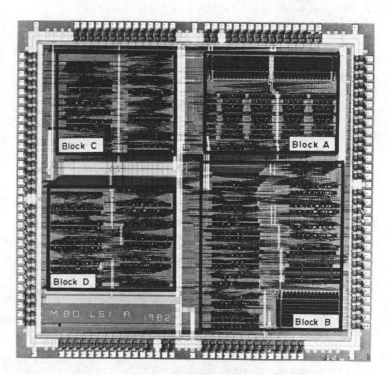

Fig.8 LSI photomicrograph

An Automatically Designed 32b CMOS VLSI Processor

Shoji Horiguchi, Hiroshi Yoshimura, Ryota Kassai and Tsuneta Sudo

NTT Musashino Electrical Communication Laboratory

Tokyo, Japan

AN AUTOMATICALLY designed 32b CMOS VLSI processor chip with mainframe computer architecture, fabricated using 2μm rule silicon gate technology, will be described. The design was centered on the application of a high density fabrication technology, and the development of random logic VLSIs with a quick turn-around-time through CAD.

The density feature was achieved through the use of 2μm-rule silicon gate CMOS fabrication employing double metal layers; Table 1. Full complementary CMOS circuits have been adopted to simplify timing design.

By using the CMOS technology, a 32b CPU chip, including 17K gate random logic and a 2304b RAM have been developed. The CPU chip architecture, Figure 1, was dedicated to a mainframe computer for information processing. Parity generator/checkers, doubled 32b ALUs and scan IN/OUT registers were also included on the chip. This was found to enhance the chip testability, and enable execution of the advanced system diagonostics by the operating system. The CPU can handle data of various types, from one to eight bytes, providing a logical address space of 2^{40} bytes and a physical space of 2^{24} bytes.

The hierarchical and structural design automation system, which processes Hierarchical Specification Language (HSL)[1] as a common design language, is used throughout the VLSI chip design. Figure 2 illustrates the automated design flow. The HSL describes not only all of the structural design data, but also the information needed for the database management. Design methodology is based on a hierarchical polycell approach. A cell pattern example is shown in Figure 3. Several pairs of N and P channel transistors are connected with common gate polysilicon. For basic P and N channel transistors, except for I/O drivers and receivers, the channel widths of 20μm are employed throughout the chip.

Functional descriptions and physical patterns of logic cells are perfectly pre-defined and stored in the data base. The logic hierarchy has 3 levels, namely, the cell, *block* and chip. Each *block* includes 100 to 300 cells; i.e., 300 to 1000 gates. The conventional layout design method that strictly holds the original hierarchical logic design down to pattern layout, usually results in a considerable increase in the inter-block wiring areas. To minimize the areas, we have used a super-block concept, in which the original hierarchical logic design is rearranged for layout. Construction of a super-block, which consists of a *block* array, involved: (*1*)—Arrangement of each horizontal *block* to provide an identical number of cell rows, (*2*)—interchanging the cells within the *block* as well as neighboring *blocks* to obtain the

optimum cell placement, (*3*)—clearing of initial *block* wirings, and redefining of the global super-block wiring.

The super-block method could improve the packing density by as much as 35% in comparison with the conventional inter-block approach. Moreover, a reduction in manpower could also be achieved, since terminal positionings for all *blocks* can be automatically defined.

As shown in Figure 4, one RAM block, one *block* and three super-blocks constitute an inner part, automatically interconnected with an I/O block by a peripheral router. In our CAD system, every complete polysilicon wiring length from one output to another input can be automatically limited to a certain specified length during the layout stage, so as to reduce the CR time constant caused by wiring lengths. In this case, the polysilicon wiring was limited to 140μm in length or about $1k\Omega$. Logic path delay and timing skews, which depend on the chip layout, were verified through logic path sensitizing, and delay evaluating routines. The layout design from the defined logic files to the final mask pattern was accomplished with a two man-months effort.

Figure 5 shows a photomicrograph of the 32b CPU chip. The total transistor count, network count and average network length were 78k, 17125 and 0.59mm, respectively. The resulting average propagation delay was 1.6ns/gate; fanin = 1.7, fanout = 2.5. The chip measured 12×12mm^2, and was housed in a 208-pin plug-in package. Design features are listed in Table 2.

The CAD approach has resulted in a shorter design turn around time with high packing density (542 transistors/mm^2), even for a random logic VLSI. Two more chips for the information processing system have been designed by the same methodology discussed.

Acknowledgments

The author would like to thank M. Watanabe, T. Suzuki and I. Toda of NTT for their direction and encouragement.

[1] Sugiyama, Y., Suzuki, M., Kobayashi, Y., and Sudo, T., "A Subnanosecond 12k Gate Bipolar 32-bit VLSI Processor", *Proceedings of Custom Integrated Circuits Conference*, p. 73-77; 1981.

Substrate Doping	lightly doped n
Poly Doping	n
Channel Length	2µm (effective length=1.4µm)
Gate Oxide	350Å
Threshold Voltage n	0.7V
p	-0.8V
Field Threshold n	>15V
p	<-15V
Wires	double Aluminum layers plus single polysilicon

TABLE 1—Main features of fabrication technology.

Reprinted from *IEEE Int. Solid-State Circuits Conf. Dig. Tech. Papers*, 1982, pp. 54, 55, and 291.

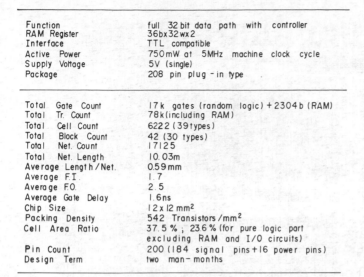

Function	full 32 bit data path with controller
RAM Register	36b x 32w x 2
Interface	TTL compatible
Active Power	750 mW at 5MHz machine clock cycle
Supply Voltage	5V (single)
Package	208 pin plug-in type

Total Gate Count	17k gates (random logic) + 2304b (RAM)
Total Tr. Count	78k(including RAM)
Total Cell Count	6222 (39 types)
Total Block Count	42 (30 types)
Total Net. Count	17125
Total Net. Length	10.03m
Average Length/Net.	0.59mm
Average F.I.	1.7
Average F.O.	2.5
Average Gate Delay	1.6ns
Chip Size	12 x 12 mm²
Packing Density	542 Transistors/mm²
Cell Area Ratio	37.5 % ; 23.6 % (for pure logic part excluding RAM and I/O circuits)
Pin Count	200 (184 signal pins + 16 power pins)
Design Term	two man-months

TABLE 2—Design features for CPU chip.

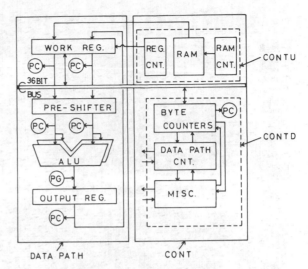

FIGURE 1—CPU chip block diagram.

[Right]

FIGURE 3—Cell pattern example; data latch register with reset input. Either vertical or horizontal scales are normalized by a corresponding wiring pitch.

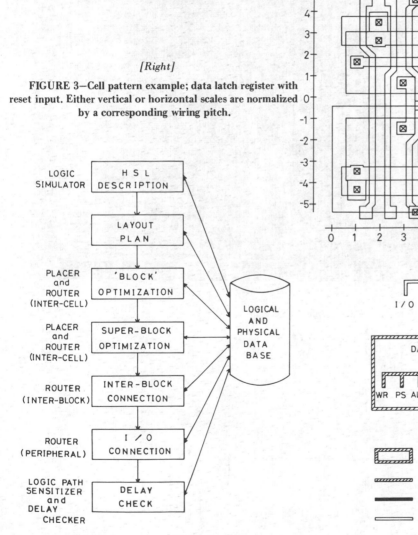

FIGURE 2—Automated design flow.

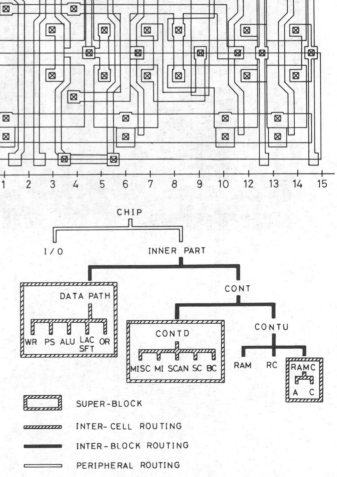

FIGURE 4—Hierarchy used in the layout design.

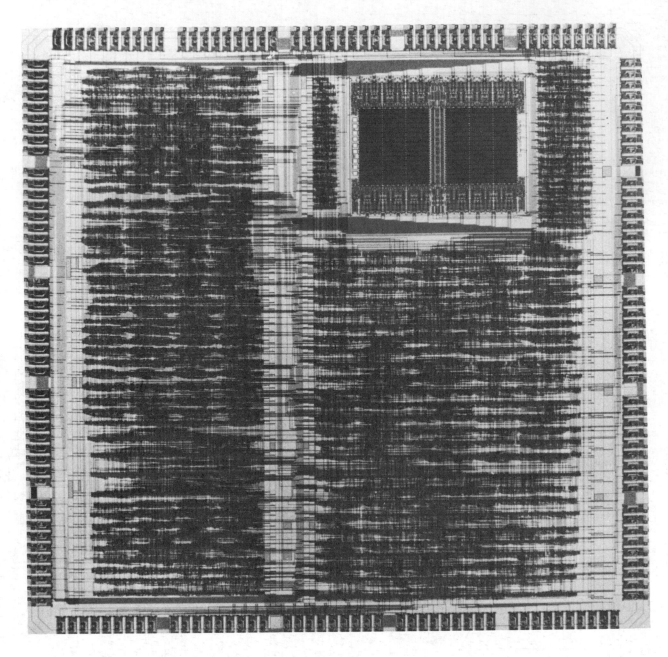

FIGURE 5—Photomicrograph of CPU chip.

RIC, Texas Instruments' restructurable integrated circuit, could be a major step in VLSI technology. Its flexibility promises custom-design performance at off-the-shelf cost.

A Restructurable Integrated Circuit for Implementing Programmable Digital Systems

Rob Budzinski, John Linn, and Satish Thatte, Texas Instruments

Development of a highly programmable, restructurable VLSI IC is an extremely important step toward effecting maximum impact of VLSI. Not only do programmability and flexibility provide new creative opportunities for the system designer, they help overcome two major obstacles to pervasive use of VLSI: design cost and design cycle time.

Flexibility, reliability, and cost

Typically, a state-of-the-art LSI custom design consisting of 5K gates and 5K bits of read-only memory costs approximately $500 thousand and takes about 18 months to design and lay out. This comes to about $100 per gate in the design. If design costs fall an order of magnitude to $10 per gate over the next five years, a typical state-of-the-art VLSI custom system consisting of 50K gates and 50K bits of read-only memory will still cost $500 thousand to design. A very flexible VLSI chip that can be restructured to provide a wide range of capabilities and state-of-the-art performance presents a viable alternative to custom designs in low-volume applications.

Reliability, testing, and maintenance considerations are extremely important in complex VLSI systems. Because poor reliability is reflected in the cost of service calls and returned products, it can, over the life of the system, incur costs greater than those of initial manufacturing. Traditionally, reliability has improved with each increase in the level of integration. However, reliability also benefits from accumulated learning, as shown in Figure 1. A generic programmable chip, in this case the Texas Instruments TMS-1000 microcomputer, increases in reliability as more are manufactured. Although specific programmations of the chip might not be made in significant volume, each programmation benefits from improvements in the reliability of the generic device.

Custom designs, on the other hand, even those using identical processes, do not benefit significantly from the

high-volume reliability learning that applies to other chips. A restructurable VLSI circuit will provide a high degree of reliability learning, even if the volume of most programmation types is small.

Overview of the RIC

The RIC is a semicustom IC that serves much the same purpose as gate arrays and masterslices. The gate-array approach allows a logic diagram to be translated into silicon through software. The software creates an interconnect pattern among the gates so that the logic diagram is implemented in silicon. The gate array approach is very flexible, but provides no special structure for implement-

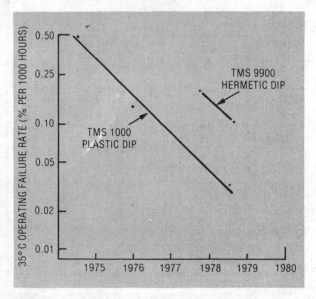

Figure 1. Microprocessor failure rate history.

Reprinted from *Computer*, vol. 15, pp. 43–54, Mar. 1982.

ing programmable digital systems. The RIC approach uses the large number of gates on a VLSI IC to build a chip that is highly flexible in implementing programmable digital systems. The RIC approach differs from gate arrays in that it commits the vast majority of its silicon to a specific design. RIC's flexibility is achieved through the design of a programmable mechanism for controlling the hardware resources on the chip. A block diagram of the restructurable IC is shown in Figure 2.

The RIC is a multimicrocomputer that contains four 16-bit processors called microprogrammable slices, or MPSs. The MPS resources can be controlled at two basic levels. The first level of control is in the coordination of MPSs. The four MPSs can be dynamically configured at runtime into any combination of three fundamental structures. One is the *lockstep,* in which two or more MPSs are structured to form a wider-word computer. This structure is formed by directing the same microinstruction stream to all of the MPSs in the lockstep and structuring the arithmetic status, carry chain, and shift/rotate linkage to configure the MPSs into a wider-word computer. In the second fundamental structure, MPSs are *independent;* each has its own microinstruction stream. A set of array processors can be structured into the independent configuration by directing the same microinstruction stream to the MPSs, without coordination of arithmetic signals between MPSs. In the third fundamental structure, MPSs are *pipelined.* Each forms a stage in the pipeline, and the microinstruction streams are different for each stage. An internal data bus within the RIC provides for simultaneous sending and receiving of data between adjacent stages (MPSs) in the pipeline.

The other basic level for controlling MPSs lies in language interpretation. The language interpretation structure is, in turn, programmable at two levels. The first is a user-definable microcode and/or assembly code. The other is a PLA that interprets instructions through finite-state machines. The microcode and/or assembly code are defined through the contents of the PLA. Microcode can be contained in the on-chip ROM or RAM. If microcode is contained in the RAM, MPSs become user-microprogrammable.

The RIC also provides for coordination of MPSs contained across multiple RICs. (The concept of external coordination of microprocessors has been developed elsewhere.[1,2]) The coordination of MPSs on multiple chips is done by means of status ports. A status port contains four signals for ALU status, plus a carry chain, a shift/rotate linkage, and a synchronization signal. The external status port interconnection to internal MPS structures is programmable. The RIC status ports provide for implementation of computers with word widths greater than 64 bits. The status ports also provide for flexibility in implementing pipelines with multiple RICs. The external interface of the RIC provides the capability for coordinating independent processors implemented on multiple RICs.

The RIC external interface is composed of a data port section, a status port section, and an interrupt port section. The two data ports are 16 bits wide. The 16 lines are bidirectional and carry addresses and data. Each data port has a general-purpose arbitration mechanism. There are three arbitration modes: round robin, master-slave, and a general arbitration method implemented by external hardware. These methods allow great flexibility in the communication among RICs, memories, and I/O devices. The interrupt port on the RIC uses the same arbitration methods as the data port. Thus, the interrupt port has great flexibility in the topology of the interrupt network. The information protocol is very flexible and is user-definable. The interrupt port can be used for a range of applications, from conventional receive-only vectored interrupts up to a user-defined, interrupt-driven interchip communication.

The internal memory system of the RIC supports the internal structures of MPSs. There are four memory modules, which can be accessed in parallel when each MPS accesses its own memory module. Also, the internal RAM bus allows sharing of memory modules among the MPSs and can be structured to support pipelined MPSs. In the pipeline structure, each MPS can send and receive data simultaneously. The internal memory system has a capability for memory mapping. Memory mapping allows the internal RAM to serve as a cache or a member of a virtual memory hierarchy.

Single-chip configurations

The MPSs of a single RIC can be configured into several modes: independent processors, lockstepped processors, and pipelined processors. The configuration is determined by directing the microinstruction stream. In the independent and pipelined configurations, each MPS receives its own microinstruction stream. In the lockstepped configuration, all MPSs in the lockstep receive the same instruction stream.

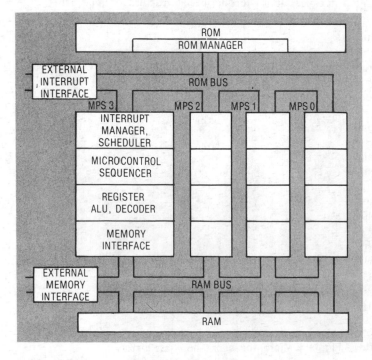

Figure 2. Restructurable IC block diagram.

Independent mode. In the independent mode, there are up to four independent instruction streams on a RIC chip. They operate on four different data streams. The data streams can be completely independent, or they can communicate with one another by sharing memory (on-chip or off-chip RAM). Figure 3 illustrates this configuration.

Lockstep mode. The internal lockstep mode (Figure 4) uses a single microinstruction stream to control multiple MPSs. This mode of operation allows design of a wide-word machine—as wide as 64 bits, when all MPSs have the same instruction stream. Operation in this mode is similar to that of today's bit-sliced microcomputers.[3,4]

Pipeline mode. The pipeline mode uses multiple instruction streams and multiple data streams. Each MPS implements one stage of the pipeline. The data normally flows unidirectionally between neighboring MPSs. In Figure 5, for example, MPS 3 can be programmed to prefetch the machine instructions from the off-chip main memory, MPS 2 decodes the machine instructions, MPS1 is programmed to perform address computation and fetch operands from the main memory, and MPS 0 is programmed to do computation specified by machine instructions.

Combinations. In addition, various combinations of these configurations are possible within a single RIC chip. For example, MPSs 0 and 1 can form one internal lockstep, and MPSs 2 and 3 can form another. The lockstep of MPSs 2 and 3 can emulate the CPU of a 32-bit machine, while the lockstep of MPSs 0 and 1 can be programmed as a graphics processor. Thus, the system can be made suitable for a high-bandwidth graphics applications.

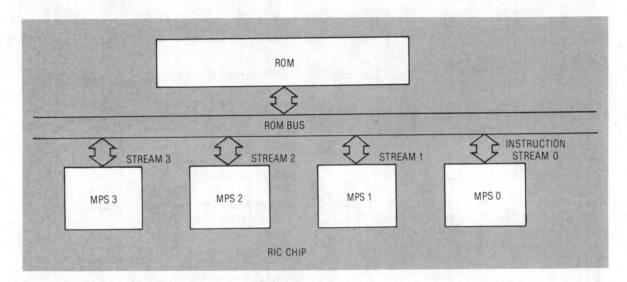

Figure 3. RIC in independent mode.

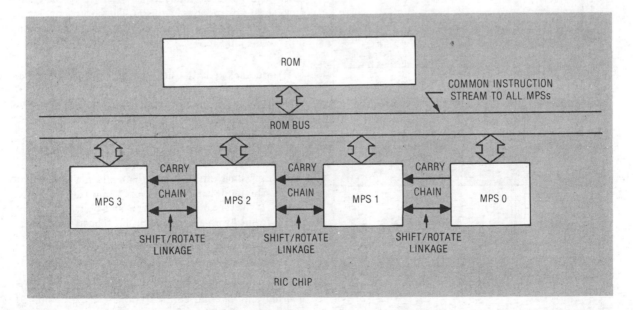

Figure 4. RIC in internal lockstep mode.

March 1982

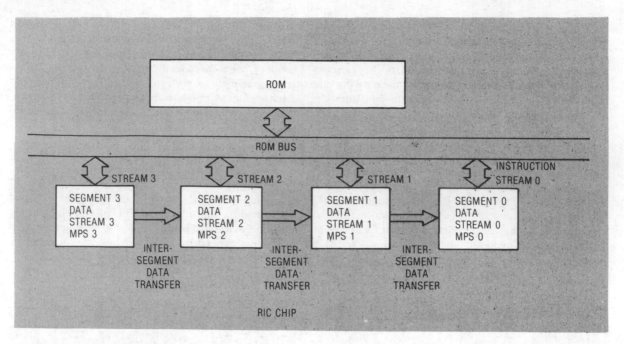

Figure 5. RIC in pipelined mode.

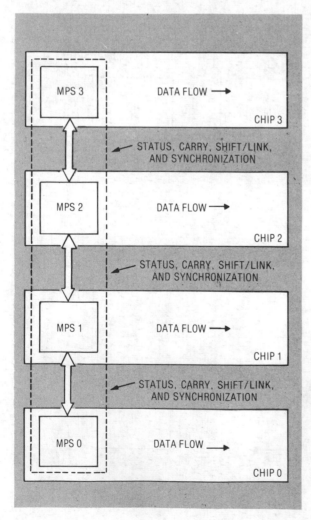

Figure 6. External lockstep.

Multichip structures

Multichip configurations are used to achieve functionality and performance beyond that possible in a single-chip configuration. For example, a multichip structure might employ one or more RICs as the CPU of a machine, another as an I/O processor, and another as a floating-point processor.

External lockstep. An external lockstep connects two or more MPSs, each on a different RIC, to form a lockstep. An external lockstep is illustrated in Figure 6. On each cycle, each MPS in the lockstep executes the same microinstruction, but operates on its own data stream. Since each MPS resides on a different chip, each must fetch its own microinstructions. The status connection synchronizes externally lockstepped MPSs such that they execute instructions in unison. The status connection also contains ALU result status, a carry linkage, and a shift/rotate linkage.

Hybrid lockstep. In the hybrid lockstep structure, MPSs are lockstepped within one RIC, as well as externally to MPSs on one or more other RICs. An example hybrid lockstep is shown in Figure 7.

Pipelines. Two basic types of pipelines can be made with multichip structures. The *internal lockstep pipeline* forms each stage of the pipe with an internal lockstep of MPSs. This mode can be used for pipeline widths of up to 64 bits. If the pipe must be wider than 64 bits, each stage of it is formed with a hybrid lockstep. The *external lockstep pipeline* forms each of its stages with an external lockstep of MPSs.

Combinations. The RIC is designed so that the various internal and external configurations can be combined among various RICs.

Microprogrammable slice design

The MPS is the processing element of the RIC. Each MPS contains six major blocks:
- the data path (computation hardware),
- the PLA for interpreting instructions for controlling the data path,
- the ROM address sequencer,
- the interrupt manager,
- the scheduler, and
- the programmable interconnect.

The data path in an MPS is 16 bits wide. It contains a dual port register file of twenty-four 16-bit-wide registers. The data path contains a high-performance ALU. Two registers in the register file can be simultaneously accessed from two 16-bit-wide buses. The data path also has a hardware unit to shift, extract, and rotate data; it is called the SERU. In addition to its application to the usual shift and rotate operations in the data path, the SERU can be used to extract the fields of a machine instruction as it is emulated and then pass the extracted fields as parameters to the PLA generating control signals for the data path. The data path contains two flag registers which are directly readable and writable through the PLA. The flag registers are mainly used for storing status values or constants generated by the PLA. This PLA also generates signals to coordinate the ROM sequencer. The ROM sequencer is used to generate addresses for the ROM and provides for loop control, subroutine calls, branches, and repetition of the execution of an instruction. The hardware in the MPS is similar in function to that contained in processors implemented with bit-sliced ICs. The MPS differs from conventional bit-sliced processors in that it uses a regular structure, a PLA, to control and coordinate the hardware.

Four techniques can make the MPS restructurable to an architecture:
- ROM programming,
- PLA code stored in each MPS,
- interrupts at the microcode level, and
- programmable interconnect.

The PLA within each MPS is programmed to translate (interpret) the information within an instruction to the control signals that accomplish the operation implied by the instruction. Microcode and low-complexity machine code (assembly code) instructions are interpreted directly through the PLA. More complex machine code instructions are interpreted in terms of microinstructions or microroutines. The instructions that are interpreted by an MPS can be contained either in on-chip ROM, on-chip RAM, or in an external RAM. This feature allows user microprograms to be contained in either ROM or RAM. The ROM can contain system microprograms for a variety of tasks:
- controlling programs for interrupts,
- managing internal memory,
- self-testing,
- initiating internal MPS structures, and
- initiating external RIC structures.

The ROM can also contain microprograms for interpreting machine languages. A RIC can interpret multiple microcode languages and/or multiple machine languages. A PLA within an MPS is programmed to interpret a particular language. Depending upon the languages involved, this PLA can be programmed to interpret more than one language. Since there are four MPSs on a RIC, at least four different languages could be interpreted. We expect that RICs will interpret existing languages (emulation) as well as interpret languages created for particular applications. For example, a language could be created for instruction prefetch, instruction decode, address calculation, self-testing, memory management, or any other computation task.

The information in the ROM is accessible to all MPSs through a shared ROM bus. Instead of a separate ROM for each MPS, we use a centralized ROM because it allows easier code sharing and flexibility in the amount of code that can be dedicated to an MPS. The bus is arbitrated according to a round-robin scheduling discipline. When an MPS issues a ROM access, the ROM manager buffers the tag(s) associated with the sending MPS. Using the tag(s), the ROM manager also routes the microinstructions to the appropriate MPS(s). Multiple tags are sent by an MPS when it is in an internal lockstep mode, and the same microinstruction is routed by the ROM manager to all MPSs involved in the lockstep. The ROM bus is also used to send interrupts.

Interrupts. There are two basic categories of interrupts: *Internal* interrupts travel between MPSs within a single RIC; *external* interrupts travel between an MPS on one RIC to one or more MPSs on one or more other RICs. Each MPS has an *interrupt manager,* which sends and receives both types. The interrupt manager gains control of the ROM bus in order to send an interrupt. It sends the following information:
- identification of the interrupt source MPS,
- the destination MPS(s),
- the priority of the interrupt, and
- runtime information.

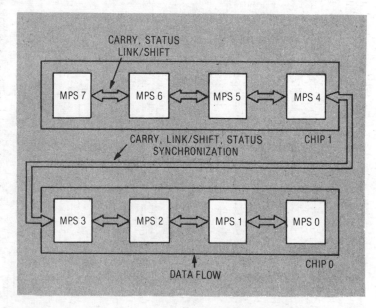

Figure 7. Hybrid lockstep forming a 128-bit lockstep.

An internal interrupt is sent to multiple MPSs to initiate either a lockstep process or a pipeline process. The receivers of an interrupt respond as to whether or not it is to be acted upon immediately. An external interrupt is sent to the external interrupt manager. The external interrupt manager uses the priority of the interrupt to access a message block, which is sent to external RICs and/or other interruptable devices.

A *process* within the RIC is initiated by an interrupt. A process is defined as an *instruction stream,* which is composed of microinstructions, macroinstructions, or a combination of the two. Each process is associated with a priority. Within a RIC, there are 256 priority levels. The priority of a process and the priority of the interrupt that initiates this process have the same value. When an MPS receives an interrupt, its interrupt manager compares the received interrupt priority with that of the currently executing process. If the interrupt manager determines that the interrupt priority exceeds the current process priority, it signals that this interrupt process will cause a context switch. Otherwise, the interrupt manager indicates that the interrupt process is of lower priority. This type of feedback from the interrupt receiver to the sender is needed in the case of multiple receivers. If only a subset of the receivers can perform a context switch, MPSs would be needlessly idle while waiting for other MPSs to finish their higher-priority processes. If, on the other hand, only a subset of the multiple receivers of an interrupt can per-

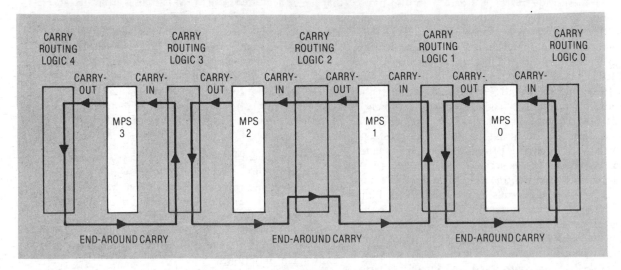

Figure 8. Carry chain configuration when MPS 1 and MPS 2 are working in an internal lockstep and MPS 0 and MPS 3 are working as independent processors.

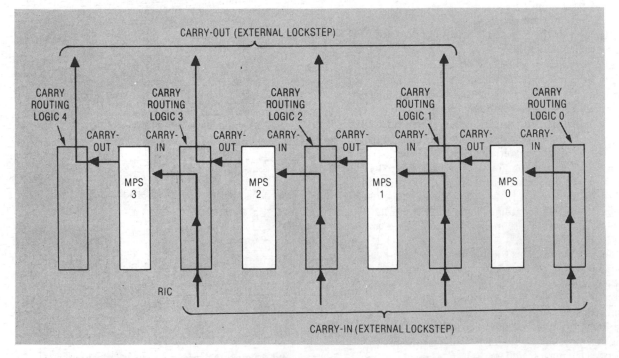

Figure 9. Carry chain configuration when all four MPSs are working as independent processors.

form a context switch, the interrupt is withdrawn and sent again later. This scheme prevents deadlock and unnecessary idling of resources.

If an interrupt is sent to a single receiver and the interrupt is of lower priority than the current process, it is buffered by the receiver MPS's scheduler. The scheduler buffers interrupts by priority in a 256-bit shift register. When a process is active, the scheduler scans the shift register to find the process with the next highest priority. When the current process is finished or timed out, the scheduler uses the priority of the next-highest priority process to access a table that contains a pointer to the process's context.

Programmable interconnect. Programmable interconnect is used for routing the carry chain, the shift/rotate linkage, and the ALU result status flags. The routing of these signals depends upon the single or multichip structure being used. In the following, we discuss only the programmable carry chain.

The carry chain is unidirectional, flowing from MPS 0 to MPS 3 and then looping back to MPS 0. The carry routing logic is also responsible for handling carry-in and carry-out signals in the external and hybrid lockstep modes. We expect the routing logic to be implemented with pass transistors and set up in a particular mode at the beginning of a structure by appropriate signals from the PLA. It should remain set up in this way until the next restructuring.

For example, in the independent mode where all four MPSs are working as four independent processors, the routing logic isolates the carry chain into four independent segments. In the internal lockstep mode, the routing logic establishes a separate carry chain for each lockstep on the chip. Figure 8 shows the carry chain when MPSs 1 and 2 are working in a lockstep and MPSs 0 and 3 are working as independent processors (one's complement arithmetic demands the end-around carry). Figure 9 shows the carry chain when all four MPSs are involved in four different external locksteps. Programmation of the shift/rotate linkage and the ALU result status signals are similar. The ALU status signals differ slightly, in that these signals from lockstepped MPSs are individually connected to a bus via a wired-AND configuration.

Internal RAM

The internal RAM of the RIC is organized as four independent, byte-addressable memory modules. An NMOS RIC with a minimum geometry feature of one micron (lambda = 0.5 micron) could contain about 16K to 32K bytes of dynamic RAM. This RAM would occupy one-third to one-half of the chip area. The internal RAM subsystem of the RIC includes four independent memory modules and a data bus interconnecting the RAM to the four MPSs. The data bus is designed to support the three basic internal MPS structures (independent, lockstep and pipeline). The memory subsystem contains a memory mapper, which automatically directs memory accesses to internal locations when the data is resident internally and to external locations when it is not. The memory subsystem is illustrated in Figure 10.

The data bus and memory modules. The data bus supports four concurrent accesses to memory, provided

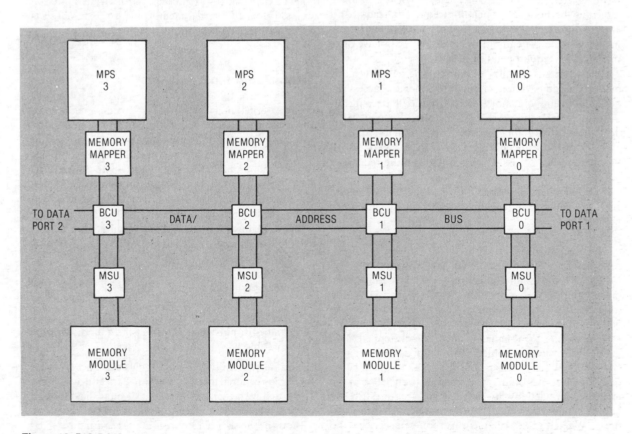

Figure 10. RIC RAM system.

March 1982

there is no interference between processors and memory. This bus allows a direct path from each MPS to its own memory module. If each MPS accesses its own memory module, four simultaneous memory accesses can occur. If MPSs access memory modules other than their own, they can cause memory interference. The bus is shared and multiple MPSs can access the same module; this can result in queued memory requests.

As shown in Figure 10, each memory module has a memory scheduling unit, or MSU and a bus control unit, or BCU. When an MPS accesses its own memory module, it is directly connected through its BCU to its MSU. The MSU indicates whether there are pending memory requests; if there are none, the access occurs immediately. If accesses are pending, the MSU queues a tag indicating which MPS requested memory service. An MSU queues MPS requests on a first-come-first-served basis. The MSU signals the MPS when its request reaches the head of the queue. The MPS then reissues its request, and memory is accessed immediately.

When an MPS accesses a memory module other than its own, the BCUs are configured to make the connecting bus a shared bus, as shown in Figure 11. The MPS first waits for access to the shared bus. The shared bus is scheduled according to a round robin by demand discipline; the first MPS or memory module gets access to the bus in round-robin order. After an MPS gains access to the bus, it sends the memory information and a destination tag indicating the destination memory module. The destination module sends the memory request's queue position on a separate bus. If 00 is sent, the memory request is processed immediately. Otherwise, the two-bit number indicates the number of pending requests ahead of the current one. Any memory module can have, at most, four memory requests pending, since an MPS can have only one memory request pending at a time.

Each MPS has circuitry to monitor the bus. When the memory module with a pending MPS memory access request completes a memory access, that MPS decrements the number of pending requests by one. When an MPS decrements this number to zero, it means that its request is at the head of the queue. When this MPS gains control of the shared bus, it reissues its request and the request is processed immediately.

The operations described above support the memory accesses made by independent, lockstepped, and pipelined MPSs.

The RIC memory system. The RIC memory system supports lockstepped and pipelined MPSs. Lockstepped MPSs can make simultanenous requests to their own memory modules. It is possible for lockstepped MPSs not to receive their requests for memory service simultaneously, since the queue length at one MPS memory module can differ from the queue length at another lockstepped MPS queue. To avoid this problem, lockstepped MPSs are synchronized. When lockstepped MPSs make a memory request, a wired-AND line, which connects all MPSs in the lockstep, is pulled low by each MPS. After an MPS has had its memory request serviced, it quits pulling this line down. When the last MPS memory request is serviced, the line rises to a logic one, indicating that the lockstep pro-

cess can continue. Lockstepped processors can also access memory modules other than their own. In such a case, each lockstepped MPS issues its request when it gets access to the bus. The lockstepped memory access is synchronized as above.

In addition to accessing memory, pipelined modes send data over the data bus to other MPSs in the pipe. Figure 12 shows the BCU configuration for pipelined data transfers. In this BCU configuration, the data bus is segmented to allow all adjacent MPSs in the pipe to transfer data in parallel; this includes transfers to MPSs on different RICs.

Each memory module is addressed with a 16-bit address. This allows for eventual growth of up to 64K bytes of directly addressable space for each of four MPSs. However, an MPS supports both 16- and 32-bit addresses. Sixteen-bit addresses are used to directly access an MPS's own memory module. Thirty-two bit addresses are used to access other memory modules or external memory. In accessing other memory modules, the most significant 14 bits comprise a tag indicating that the address is for an internal memory module. The next two significant bits select one of four memory modules. The remaining 16 bits point to an address in an internal memory module. If a 32-bit address does not point to an internal memory mdoule directly, it is either an external or a mapped address, depending upon MPS control. If the address is designated as external, it is sent to the external memory interface for processing. Otherwise, it is sent to the memory mapper. The memory mapper uses an associative search to determine whether the address is internal or external. If it is internal, the associated internal address is sent to internal memory. If it is external, it is sent to the external memory interface.

External interface

The external interface for the RIC is designed to support multiple RIC configurations, interchip communication, and data-path communication between system memory and system I/O. Two versions of pin assignment are planned, an 82-pin version with two 16-bit data/address ports and a 114-pin version. They are the same, except that the 114-pin version has two 32-bit data/address ports. The 82-pin RIC is discussed below.

Figure 13 illustrates the RIC pin assignment. There are five types of pin functions for the RIC:
- data/address,
- control
- interrupt,
- status, and
- power/clock.

The number of pins dedicated to each function group is listed in Table 1.

Data port. Each of the two 16-bit data/address ports has 16 bidirectional lines for carrying data and addresses. Associated with each port is a pair of handshake signals for gaining control of shared resources. The arbitration method for the shared resources is either round robin by demand, master-slave, or determined by external cir-

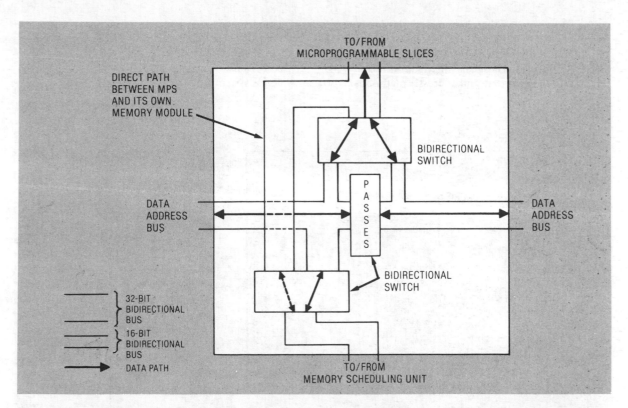

Figure 11. Bus control unit configuration for shared data/address mode. The upper bidirectional switch makes connections as shown for the MPS's BCU, which controls the data/address bus. All other corresponding switches in other BCUs block signals. The dashed connection in the lower bidirectional switch is made when an MPS accesses its own memory module.

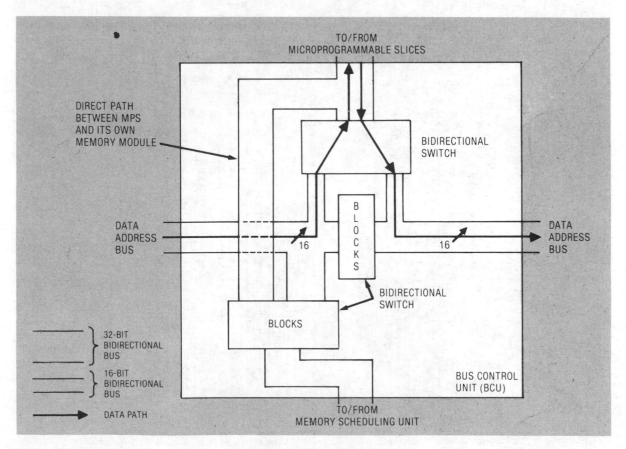

Figure 12. BCU configuration for the pipeline data transfer mode.

March 1982

cuitry. The round-robin and master-slave arbitration methods support a shared bus, while the external arbitration circuitry supports a network with a general topology. Each port has a pair of signals for synchronizing the sending and receiving of data and addresses on the bus. Also, each port has a bidirectional set of three signals to indicate bus status. These three signals indicate four types of read/write operations:

- access a user-specified RIC,
- access system RAM,

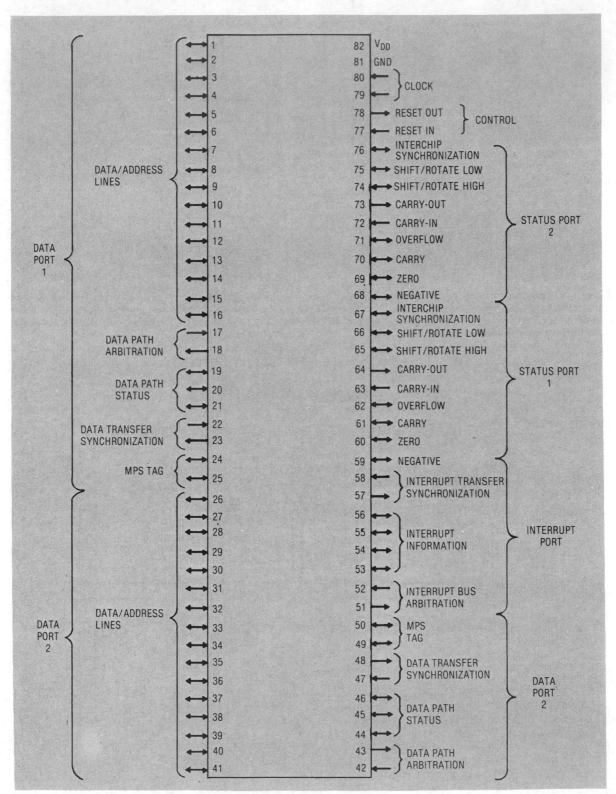

Figure 13. A restructurable IC pin assignment.

- access system I/O, and
- access the resource whose destination address is sent at the beginning of the access.

Finally, each port has a bidirectional pair of MPS tag identifiers. These are used to indicate the source MPS at the sender and/or the destination MPS at the receiver. The two data/address ports are independent. However, they can be combined into one port by internally performing the same operation to both ports concurrently and externally treating the two as one.

Status port. There are two identical status ports. Their main function is to provide the signal to lockstep two MPSs on different RICs. Status port 1 can be used to lockstep MPS 0, MPS 1, or to lockstep MPSs 1 and 0 to external MPSs. Status port 2 can be used to lockstep MPS 2, MPS 3, or internal locksteps, including MPS 2 and/or MPS 3 to external MPSs. In Figure 6, a 64-bit-wide external lockstep is formed with four MPSs on four RICs. There are four pin functions in each status port:
- ALU result status,
- carry linkage,
- shift/rotate linkage, and
- MPS synchronization.

There are four ALU result status pins:
- the negative result status, N;
- the zero result status, Z;
- the carry result status, C; and
- the overflow result status, V.

These four signals are connected to a bus via a wired-AND configuration. This bus connects all externally lockstepped status ports. The signals are encoded to indicate up to one of 16 ALU result outcomes. The carry linkage is a carry-in signal and a carry-out signal. The carry-out signal of one RIC is connected to the carry-in signal of the next most significant RIC. The shift/rotate linkage is used to perform shift operations between externally lockstepped MPSs. The shift/rotate high signal of a RIC is connected to the next most significant RIC's shift/rotate low signal. The shift/rotate high signal of the most significant RIC is connected to the shift/rotate low signal of the least significant RIC, to provide the shift/rotate linkage. The MPS synchronization pin ensures that externally lockstepped MPSs are executing the same instruction in phase. Without synchronization,

MPSs in an external lockstep can get out of phase because other MPSs on a RIC might be operating independently of the externally lockstepped MPSs. Thus, the time to fetch a microinstruction can vary among the RICs containing lockstepped MPSs. The MPS synchronization pin serves as a flag to indicate that each MPS has finished the previous instruction, fetched the next microinstruction, and is ready to execute it. The MPS synchronization pins are connected in a wire-AND configuration. When all externally lockstepped MPSs are ready to execute the next instruction, the MPS synchronization line is high. If one or more MPSs is not ready, the line is pulled low. When the MPS synchronization line is high, execution begins on the next clock cycle. (All RICs with MPSs in a common external lockstep must use the same system clock.) Shortly after execution begins, the MPS synchronization line is pulled low; it stays low until all MPSs are ready to execute the next instruction.

Interrupt port. The interrupt port serves two purposes. First, it receives and processes interrupts in a manner similar to that of conventional microcomputers and microprocessors. Second, it provides for interchip communication. The interrupt concept has been generalized to include the capability of sending interrupts to other receivers, thus providing for interchip communication. The purpose of interchip communication is to coordinate RICs to carry out a task, to initiate a task, and to transfer information. The interchip communication system is used to transmit commands and/or small amounts of data. The bulk data part of an information transfer is communicated between memories. For example, a disk read operation is initiated by using the interrupt port of a RIC to send commands to a disk controller. The data transfer between the disk system and the memory system is accomplished on a separate data path. The interrupt port contains pins for arbitration, information, and data transfer synchronization.

The interrupt port of the RIC has eight pins. Two are for arbitration of shared resources used during the sending of an interrupt. The same three arbitration modes used for the data ports apply to the interrupt port: round robin, master-slave, and general arbitration.

Four pins of the interrupt port are dedicated to data transfer. The data protocol has minimal specification and maximal user definition. In the interchip communication mode, the first information sent on these pins is an address. The length of the address is designated by the user. When an interrupt is sent, all chips on a common interrupt bus receive and store the address. The status signals indicate whether the information lines carry address or data. The receiver buffers the address portion as long as the status indicates address bits are being sent. After the destination address has been sent, each receiver uses the address to access a bit in the chip's RAM, to determine whether this chip is an intended receiver of the interrupt. In the conventional interrupt scheme, the first information sent is the interrupt level. The remaining two pins of the communication port are used for interrupt bus status. The four status values are:
- sending address,
- sending data,

Table 1.
Pin assignment by groups.

FUNCTION	NUMBER OF PINS
DATA/ADDRESS (2 PORTS)	50
CONTROL	2
INTERRUPT	8
STATUS (2 PORTS)	18
POWER/CLOCK	4
	82

- data/address nibble received, and
- interrupt information transfer completed.

After gaining control of the bus, the interrupt data is sent. The amount of data sent is determined by the user. The design permits an optional destination address of variable length, a variable-length data portion, and an optional source address of varying length to be components of the information sent during an interrupt. The interrupt information is buffered at the destination by the RIC's external interrupt manager. The external interrupt manager interrupts the destination MPS and passes to it the length of the message and a pointer to the interrupt message block.

Control lines. There are two control lines, reset in, or RI, and reset out, or RO. The RI and RO signals from all RICs are connected. The RI signal is active high. When it is raised to a one, the RICs begin to initialize themselves for operation. The RO signals are connected in a wired-AND configuration. When a RIC has completed the initialization operation, the RO signal, which has been pulled low, is allowed to float. When all RICs have completed initialization, the RO signal goes high to indicate this.

Power/clock. The RIC will use two power pins, +3 volts and ground, and an on-chip clock generator. This will allow a crystal to be placed across the two clock inputs; alternatively, an external clock can replace the crystal.

The major goal of the RIC project is to create a highly flexible part that can be used to form a wide variety of specific hardware designs through programmation. The designed-in flexibility of the RIC provides for this programmation. RIC's flexibility features include user-de-finable microlanguage and assembly language, user-programmable microcode, dynamic coordination of multiple internal processors, coordination of processors on multiple RICs, internal memory that can be used either as a cache or as an element of a virtual memory hierarchy, general topology for interchip communication and external data paths, and a user-definable interrupt mechanism.

A single RIC has been programmed to implement a VAX-11/780 instruction set processor. The implementation contains two pipelined stages. The memory unit stage fetches instructions and operands; the execution unit stage operates upon the operands. When implemented with one-micron technology, the RIC VAX approaches the performance of the VAX-11/780. A prototype one-micron NMOS IC has been developed in conjunction with the RIC specification. ■

Acknowledgment

This work is funded in part by the Defense Advanced Research Projects Agency under contract MDA 903-79-C-0433.

References

1. R. G. Arnold and E. W. Page, "A Hierarchical, Restructable Multi-Microprocessor Architecture," *Proc. 3rd Ann. Symp. Computer Architecture,* 1976, pp. 40-45.

2. S. I. Kartashev and S. P. Kartashev, "A Multicomputer System with Dynamic Architecture," *IEEE Trans. Computers,* Vol. C-28, No. 10, Oct. 1979, pp. 704-720.

3. *Bipolar Microcomputer Components Data Book,* Texas Instruments, Inc., Dallas, Tex., 1977.

4. *The Am2900 Family Data Book,* Advanced Micro Devices, Inc., Sunnyvale, Calif., 1979.

Section V-B
Signal Processing

VLSI Architectures for Digital Signal Processing

Robert E. Owen, Saratoga, CA

VLSI circuits for digital signal processing (DSP) abound today, and many more are promised. No VLSI conference is without its DSP session; the VLSI sessions at DSP conferences are standing-room-only; and the VLSI sessions at application conferences (such as those for telecommunications) are all about DSP. System designers face a confusing number of seemingly unrelated DSP circuits, many of which are only paper products of no practical interest. Here we shall survey some of the past and most of the recent commercially available chips from the standpoint of their architecture and how they fit into the systems where they are used.

The recent profusion of circuits is very closely related to today's levels of integration and speed improvement. Digital processing of signals is characterized by extensive processing and high speed. The greater need for processing occurs in at least three areas. There are many more multiplications and additions than in normal data processing. The precision is often higher: 16 bits is only marginal; in many cases, 24- or 32-bit precision is preferred. Moreover, data structures can be so complex that address computation is also substantial.

To meet the requirements of arithmetic-intensive processing and high speed, most programmable digital signal processors have taken the general form shown in Figure 1. Data and program memories are separate, to permit greater speed. The data memory itself may have multiple ports or else consist of two separate memories, in order to increase data rates. The arithmetic logic unit (ALU) is limited to operations on data, with separate adders or ALUs for data-address generation and program control. The data ALU has a parallel multiplier configured with an adder for rapid accumulation of products of two operands. Multiple buses may be used to bring both operands simultaneously from a double-ported data memory. Data input and output ports are placed on this same bus; this position reflects their major role. The approach is duplication or paralleling of functions to permit a higher processing rate than in conventional processors (in which functional elements are shared or multiplexed in time). The penalty that this approach must pay is in complexity and a larger silicon area. Another technique used to increase speed is *pipelining*, or subdividing functions into stages so that new data can start processing in an input stage while older data is still finishing in a stage that is nearer the output. The repetitive nature of most signal processing makes this technique practical.

This general framework for a digital signal processor or processing system can also include the quantitative measures shown in Figure 2. These measures help not so much for

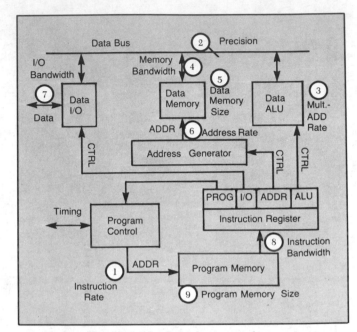

FIGURE 1. Basic functions and parameters of a digital signal processing system.

comparison between devices or systems as they do within one system to determine whether resources are matched for a particular task. The units have been chosen for consistency, for a convenient range of numbers, and for an increasing measure of performance. The basic precision of the system and the instruction execution rate are fundamental for all comparisons of speed and bandwidth. The data ALU is measured on the basis of the rate at which it does the basic multiplication and addition operation. Data memory has a cumulative bandwidth and maximum size. Addresses for the data memory that require an addition operation, a comparison operation, and a concatenation operation have a cumulative rate determined by the addresser. Data can be input to and output from the processor at a cumulative rate. All of the processing is governed by a varying pattern of bits from the program memory, whose byte rate is a measure of its flexibility or adaptability. The maximum size of the program memory measures how much flexibility can be exercised at one time.

VLSI circuits have generally taken three forms in satisfying the functional requirements of the system shown in Figure 1. The earliest forms were just the individual blocks themselves, configured internally to be as generally applica-

Reprinted with permission from *VLSI Design*, pp. 20, 21, 22, 24, 26, and 28, June 1984.
Copyright © 1984 by CMP Publications, Inc., 111 East Shore Road, Manhasset, NY 11030.

Signal Processing System Parameters

1 Instruction execution rate. Instructions/µsec.

2 Data precision. Basic system data bus and data memory width. Bytes.

3 Data processing rate. Rate of a multiply and an addition operation at system precision. Mult-Add/µsec.

4 Data memory bandwidth. Cumulative rate if multiple memories and buses. Bytes/µsec.

5 Data memory size. Address space in K bytes.

6 Address processing rate. Rate of an increment, compare and concatenate operation for a data memory address. Cumulative if multiple memories with independent addressers. Add-CMP-‖/µsec.

7 Data input/output bandwidth. Bytes/µsec.

8 Instruction bandwidth. Bytes/µsec.

9 Program memory size. Address space in K bytes.

FIGURE 2. Signal processing system parameters.

ble and flexible as possible. They are in no sense individually programmable, being instead controlled only by interconnections. We shall call these elements "building blocks." The well-known parallel multipliers are representative members of this group. The next circuits put all or most of the system functions (including program control) on one chip. These circuits are designed for general-purpose use, to accommodate as many applications as possible. These we'll call "programmable processors," because of their defined instruction set and control structure. The most recent form of this device is a variation of the programmable processor which is configured for one limited set of applications. These devices are called "single application circuits." A high-speed modem chip is a typical member of this third group.

Building Blocks

Some of the recent signal-processing ICs of the building-block type are shown in Table 1. Many of these chips are replications or minor variations on the popular TRW bipolar parallel multiplier MPY series and multiplier-accumulator MAC series; but they do represent all of the functional digital blocks shown in Figure 1. Distinctive speeds, packages, and fabrication technology are noted if known.

Architecturally, the progression began with the highly regular data-path multiplier and related functions. These devices offered high functional replacement value with little control complexity. The initial problems of arithmetic number systems (signed or unsigned, fractional or integer, sign-magnitude or two's-complement), rounding, overflow detection, and register clocking were solved; after that, the function changed little in six years. As Table 2 shows, during that same time, smaller geometries have steadily reduced the size (cost), power, and multiplication times of bipolar DSP devices. CMOS and recent improvements in algorithm designs have made an even more dramatic impact. Simple full-precision adders were missing until recently because of packaging-power limitations and low functional replacement value. But now nMOS and CMOS let full functions (such as barrel shifting, normalizing, and registers) be added to a full-precision ALU in a single package. The Analog Devices ADSP 1200 in Figure 3 is a recent example of one such circuit (Nuttall and Oxaal 1984). A 96-pin leadless chip carrier (LCC) and a pin-grid-array (PGA) package allow 36 bits of control and several full-size data ports without enlarging the small package made possible by CMOS technology.

Company	Recent DSP Building Blocks
TRW	TMC 2220 CMOS 4×32 bit correlator mask configurable TMC 216H CMOS MPY-16 145 nsec TMC 2008/9/10 CMOS MAC-8/12/16 100/135/165 nsec TDC 1030 Bipolar FIFO Memory 64 × 9 bits
AMD	29325 32b FP ALU & Mult Bipolar 144 pin 100 nsec 29516A/7A Bipolar MPY-16/µP mult. 45 nsec 29L516/7 Bipolar MPY-16/µP mult. 85 nsec 29C516/7 CMOS MPY-16/µP mult. 125 nsec 29509 Bipolar MAC-12 29-/L/C510 Bipolar & CMOS MAC-16 70/100/140 nsec 29540 Bipolar FFT Addresser
Analog Devices	ADSP 1080/12/16 CMOS MPY-8/12/16 100/130/170 nsec ADSP 1008/09/10 CMOS MAC-8/12/16 120/150/190 nsec ADSP 1024 CMOS MPY-24 84 pin 235, 275 nsec ADSP 1110 CMOS 16 bit MAC 28 pin 190, 240 nsec ADSP 1200 CMOS ALU shifter 96 pin 75 nsec
TI	74ALS1616 Bipolar 16 bit mult. 55 nsec 74AS888/90 Bipolar 8 bit ALU/Microcontroller 74AS897 Bipolar Address Generator 74AS898 Bipolar Register File
Rockwell	CMOS/SOS MAC-16 120 nsec
Synertek	SY 66016B/A/- nMOS MPY-16 90/150/200 nsec
IDT	IDT 7216/17 CMOS MPY-16/µP mult. 75, 90, 145 nsec
Matsushita	MK 9981 CMOS FP Multiplier 80 nsec, 28 pin
Toshiba	T 7429 CMOS/SOS MPY-16 27 nsec
NEC	CMOS 16 bit multi. 45 nsec
IMI	MPY-12IMI CMOS MPY-12 150 nsec gate array
Weitek	WTL 1016/A nMOS MPY-16 140/100 nsec WTL 1516/A nMOS µP mult. 140/90 nsec WTL 1010A/-/L nMOS MAC-16 90/120/140 nsec WTL 1032/33 nMOS 32 bit FP Mult/Adder 200 nsec/stage
Logic Devices, Inc.	LMU08/12/16 CMOS MPY-8/12/16 90/100/140 nsec LMU8U/13/17 CMOS µP mult. 8/12/16 bit 90/100/140 nsec LRF08 CMOS Multiple Port Register File 5 Ports

TABLE 1. Recent VLSI building blocks for digital signal processing.

The precision of data-path functions has steadily increased from 8- to 12-, 16-, and now to 24-bit integers as densities have increased. Two floating-point precision formats (22 and 32 bits) exist. The 32-bit precision will be more widely used because it conforms to the new IEEE standard; but 22-bit precision will prevail where speed remains critical. Because of the additional computational complexity of floating-point operations, the propagation delays are long relative to bus and I/O cycle times. Computation stages are often pipelined to increase the effective data rate for repetitive operations without increasing the latency time for a single operation.

Higher-precision parts have generally remained in standard 64-pin and smaller packages. The use of low pin-count packages was accomplished through time-multiplexing data ports or the most and least significant portions of each port. This technique lowers throughput rates and favors either

Device	TRW MPY016A	TRW MPY016H	TRW MPY016K	TRW TMC216H
Process	Bipolar	Bipolar	Bipolar	CMOS
Year Described	1977	1980	1983	1984
Minimum Feature Size	4μm	2μm	1μm	2μm
Area (sq. mils)	78K	48K	32.5K	29.9K
Number of Transistors	18K	18K	15K	6.5K
Pins	64	64	64	64
Power	5.0 W	4.4 W	4.0 W	0.50 W
Precision	16 × 16	16 × 16	16 × 16	16 × 16
Speed	230 nsec	145 nsec	45 nsec	145 nsec

TABLE 2. Technology trends for the simplest building block: the multiplier.

shared wide buses or multiple narrow buses. It also restricts chip functions because so few pins remain for control. The newest products are using the large-pin-count LCC and PGA packages to permit much more functional variation and higher throughput regardless of the system bus architecture. Even for integer building blocks, more pipelining will become common for higher precisions as the pin-sharing I/O bottlenecks are removed thanks to higher-pin-count packages, and propagation delays again will dominate. But higher-pin-count packages are expensive; where there is only one system data bus, the additional data pins offer no speed advantage. ADI's new ADSP 1110 integer multiplier-accumulator (Cox 1983) and Matsushita's MK 9981 floating-point multiplier (Uya *et al.* 1984) are for just that bus configuration. With only a single 16-bit data port, they can have additional functions like shifting, status, and overflow saturation values while still being housed in a small, inexpensive 28-pin dual-in-line package.

Program controllers for general-purpose data processors have been used in DSP systems, but now controllers are being designed for the special needs of real-time signal processing. These needs include faster interrupt servicing and context-switching structures other than the simple stack. Controllers are also adapting to the pipelined instruction flow that must match the pipelined data flow. Because general-purpose processors contain no data addressers, no devices (not even rudimentary ones) have been developed for this critical DSP functional block. Even bit-slice ALUs were difficult to configure in order to perform the repeated incrementing, shifting, masking, and concatenation that must often be done for each data address. One addresser device, the AMD 29540, exists already for the important but special case of the fast Fourier transform (FFT); but more general-purpose devices are now being designed. The often-avoided memory functions are now available in special configurations, such as multiple ports that increase DSP system speed, or first-in-first-out (FIFO) memory, which eases data synchronization.

Programmable Processors

The high density and lower power of MOS technologies have made it possible to fit all of the functional blocks of the programmable digital processing system of Figure 1 onto a single VLSI circuit. The advantages of small size and cost are clear; but high performance at lower power is another potential advantage, because all elements can be intimately connected without the time delays and power lost through going onto and off of the chip. The architectural challenge is to find configurations that are computationally efficient, small enough to be producible, and flexible enough to find wide application. Table 3 lists eight families of programmable processors from manufacturers which are trying to meet that challenge. These processors are at least potentially available for use by all system designers. Because they were developed over a period of at least six years, they reflect changing design methods and fabrication technologies as well as practical experience in useful applications. Two of the families have second-generation parts.

Table 3 also highlights the major technology-related facts for the DSP programmable processors, and lists the functional sizes of multipliers and program and data memories. Although a full comparative study of their architectures and performance parameters is beyond the scope of this article, some general trends are clear—both for the processors as a whole and for the individual functional blocks.

Early designs of data ALUs had small parallel or serial multipliers in order to conserve area. Current designs are completely parallel at full system-level precision. (This precision is limited more by companion data-memory area than by multiplier area.) Precisions cluster around 16 bits with larger accumulators. But as memory limitations are relaxed, precision will tend to increase to the 24- to 32-bit levels that are preferred for DSP. The success of Hitachi's low-precision floating-point processor is not clear at this time because of the limited availability of this part. All ALUs benefit from having a single data-number representation so that shifters, overflow levels, and multiplication alignments are consistent.

Data memory is distributed to increase speeds and minimize size; coefficient memory is often space-efficient ROM that is separately bused to the ALU. A few higher-precision accumulation registers are placed directly at the ALU output, and variable-storage RAM is small and has multiple ports. Early designs suffered from limited on-chip variable-data memory with no easy or fast way to increase it externally. Both second-generation processors tried to correct this shortcoming. But the cost was high, both in terms of area and in terms of the additional pins needed to address external memory.

As in the building blocks, address generation has suffered from neglect. Only simple counters and comparison functions generally exist. But the Fujitsu MB 8764 in Figure 4 represents a clear change, with two sets of address arithmetic (Kikuchi *et al.* 1983). Larger memories intensify this need, because they allow more complex applications than the highly recursive algorithms that were application targets for earlier processors.

Like data memory, program memory is smaller in ROM form on the chip. But even in erasable form (EPROM), making changes is time consuming, and program memory size cannot be increased. One of the major architectural features of the popular Texas Instruments TMS 320 family (see Figure 5) is that it is available with and without program ROM but always provides the possibility of more external

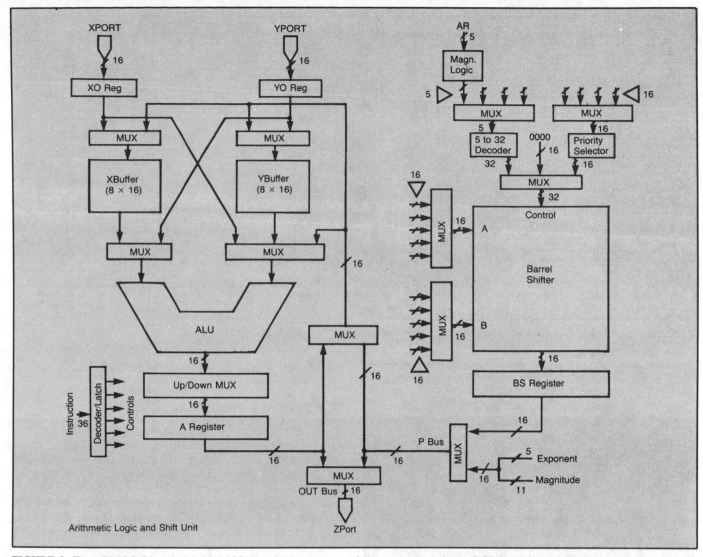

FIGURE 3. The CMOS 96-pin ADSP 1200 illustrates some of the new trends in DSP building blocks.

program memory (RAM, ROM, or EPROM). Although more flexible program control, such as high-speed subroutine context-switching, can reduce the size requirements of a DSP program, the need for alterable external program memory will remain. Again, second generation processors have expanded on-chip program memories and have allowed external additions. American Microsystems' S28211/2 family and the new Toshiba T6386/7 family (Sugai *et al.* 1984) both have two chips with large and small pin-out packages to accommodate the different program memory needs. The MB 8764 commits unshared pins to both data and address for off-chip program memory, as well as to ample wide-word memory on the chip.

Digital signal processing requires a great deal of computation. It is also data intensive; but the data is not all new data. Therefore, the input/output data rates of processors, although high by most standards, are often a small fraction of the arithmetic computation rates and of the internal data-transfer rates. For the single-chip processor with limited functional resources, this means that the required external I/O data rates are often less than 100 Kbytes per second. Most single-chip processors have taken advantage of this circumstance by using a narrow byte-wide data bus with relatively slow microprocessor bus-transfer protocols. Even processors with slow bit-serial ports have found application. But, as we have seen, the individual resources on a single chip proved inadequate and had to be expanded externally. Furthermore, the total processing resources on one chip have not been sufficient for some applications; as a result, multiple chips have had to be used. Voice-bandwidth vocoding is a good example of such a case. Solutions to both of these problems have called for much higher I/O data rates than those needed for only "new" data. Not all types of data transfers can be combined; but it is clear now that because of pin and power limitations, early designs provided too little aggregate I/O bandwidth.

Some early VLSI processors were designed to stand alone, without requiring another host or control microprocessor. The microprocessor was considered to be incompatible in speed, and was thought to be an unnecessary additional expense. Experience has shown that although DSP solutions are expensive, one of the ways they pay for themselves is their easy alterability or change of function. A control microprocessor is a very cost-effective way to change the DSP proces-

Device	AMI S2811	AMI 28211/2	Bell DSP-1	Fujitsu MB 8764	Hitachi 61810	Intel 2920/21	NEC 7720/P20	NEC 77220	TI 320	Toshiba T6386/7
Process	nMOS	nMOS	nMOS	CMOS	CMOS	EPROM/ nMOS	nMOS/ EPROM	CMOS	nMOS	CMOS
Minimum Feature Size	4.5 microns	3 microns	4.5 microns	2.3 microns	3 microns	6/4 microns	3 microns	2 microns	2.7 microns	2 microns
Year Described	1978	1983	1980	1983	1982	1978/81	1980	1984	1982	1983
Area (sq. mils)	41,000	—	106,000	145,000	79,000	47,000	44,000	—	70,000	76,000/ 74,000
Number of Transistors	30,000	—	45,000	91,000	55,000	20,000	40,000	—	—	66,000/ 48,000
Pins	28	28/64	40	88	40	28	28	—	40	28/64
Power	1 W	0.7 W	1.25 W	0.290 W	0.200 W	1 W	1 W	—	0.9 W	0.360 W
Precision	16	16	20	16	16 FP	25	16	24	16	16
Multiplier	12×12=16	12×12=16	4(4×20)=36	16×16=26	12×12=16	25×1=25	16×16=31	24×24=48	16×16=31	16×16=31
Speed	300 ns	300 ns	4(200) ns	100 ns	250 ns	600/400 ns	250 ns	100 ns	200 ns	250 ns
Program Memory	256×17 ROM	512×18 ROM/EXT	1K×16 ROM,EXT	1K×24 ROM,EXT	512×22 ROM	192×24 EPROM/ROM	512×23 ROM/EPROM	4096 ROM	4K×16 ROM,EXT	512×16 ROM/EXT
Data RAM	128×16	256×16	128×20	256×16	200×16	40×25	128×16	1024×24	144×16	128×16
Data ROM	128×16	128×16	In program	In program	128×16	16×4	512×13	1024	In program	512×16 ROM/EXT

TABLE 3. Technology trends for single-chip programmable digital signal processors.

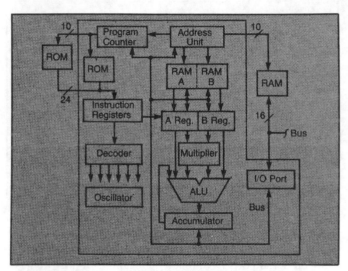

FIGURE 4. The CMOS 88-pin MB 8764 from Fujitsu illustrates some of the latest trends in programmable processors.

sor function and to provide control for low-bandwidth I/O and peripherals. A standard microprocessor interface is now seen as a necessary part of the data-I/O function. Using multiple DSP processors calls for not only high-bandwidth data exchanges as mentioned above, but also synchronization of the multiple processors. This synchronization must take place at the basic system clock rate so that tightly interconnected functions can communicate without long delays.

The most recent DSP programmable processors can incorporate most of the I/O enhancements mentioned above, thanks to CMOS's lower power, together with higher pin-count packaging. Outputting data from a chip consumes power on the chip. With bipolar and nMOS fabrication technologies, little was left in the power budget for I/O unless very large packages were used. With CMOS, more function and speed can be added on-chip, and more output pins can be powered for I/O; yet, power consumption is still less, thus allowing smaller packages.

DSP Devices for Single Applications

The availability of general-purpose programmable processors for DSP made it possible to try a digital solution in many existing applications. In some of them it was clear that cost and sometimes performance were compromised because the trade-offs in the general-purpose design were not the best for that application. In cases in which the potential sales volume has justified the design cost, programmable processors have been developed for a single application area. Most have been in the areas of data communication, image processing, and speech synthesis and recognition. A recent example is the NEC μPD 7764 speech-recognition chip shown in Figure 6 (Iwata et al. 1983). Note the six different types of memory, the five ALU-related functions, and two I/O controllers in a special configuration. This chip obviously performs functions that would be inefficient on a general-purpose DSP processor. The programming allows for flexibility (part of the program memory is RAM) in tasks and in algorithms, because speech recognition is a rapidly changing technology itself.

Trends in Design Methods

Of course, all early DSP VLSI circuits were custom designed. The complexity of these circuits made space-efficient circuit design critical. It is still important; DSP solutions are still expensive. But there are some interesting departures. At the simplest level, final-metal mask variations are being encouraged for architectural configuring, as in the TRW TMC 2220 correlator. Some building blocks, such as the MPY-12IMI from International Microcircuits, Inc. (Santa Clara, CA), are being designed with gate arrays. Other companies, such as Logic Devices (Sunnyvale, CA), are assembling their building-block families from higher-level design cells. First products have replicated older block functions with the design advantage of a shorter time to market, but products

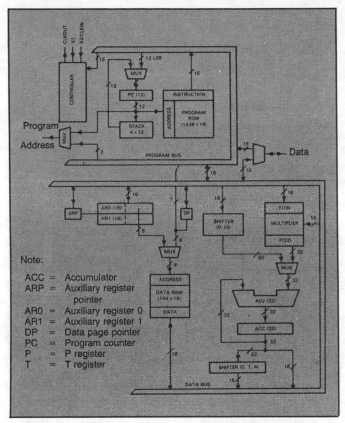

FIGURE 5. The comtemporary standard programmable digital signal processor: the TI TMS 320.

Note:

ACC = Accumulator
ARP = Auxiliary register pointer
AR0 = Auxiliary register 0
AR1 = Auxiliary register 1
DP = Data page pointer
PC = Program counter
P = P register
T = T register

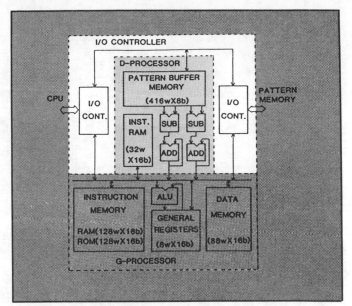

FIGURE 6. An application-specific digital signal processor: the programmable NEC μPD 7764 for speech recognition. (Photo courtesy of IEEE.)

with new functions will probably start appearing soon. Similarly, among the programmable processors, the Fujitsu MB 8764 is a manually laid out cell-based design. One result of Fusitsu's standard-cell approach is a chip with an extremely large area of 145 thousand square mils; another result is state-of-the-art architecture and performance in advance of that offered by other manufacturers. If market acceptance is good, the chip will surely be made smaller by a more area-efficient redesign. Few signal-processing markets are large enough yet to justify large design investments by manufacturers. The more CAD can be used to lower the design costs, the more standard VLSI circuits there will be for the DSP system designer. The expansion of single-application processors beyond a few large applications clearly depends on the development of higher-level DSP cell designs that can be reconfigured quickly.

Packaging Trends

One of the advantages of CMOS technology for DSP applications, aside from the reduction in power dissipation compared with bipolar technology, is the capability to use smaller IC packages. The development of leadless chip carriers and pin-grid-array packages has not only reduced the size of CMOS DSP chips but also made higher pin counts feasible. This greater pin count is needed for architectural flexibility and will ensure the use of these packages for all but the most cost-sensitive chips.

Summary

A functional architectural framework has been developed for digital signal processing systems. Then VLSI circuits designed for signal processing have been reviewed on the basis of their functions to determine where they fit within the framework. Common functions or groupings of functions have been reviewed for architectural trends that are the result of changing fabrication methods, design and packaging technologies, or experience with applications. □

References

Cox, J. 1983. "Building a DSP System Around CMOS Parts," *WESCON*, San Francisco, CA.

Iwata, T., H. Ishizuka, M. Watari, T. Hoshi, Y. Kawakami, and M. Mizuno. 1983. "A Speech Recognition Processor," *IEEE International Solid-State Circuits Conference*, New York, NY.

Kikuchi, H., T. Inaba, Y. Kubono, H. Hambe, and T. Ikesawa. 1983. "A 23K Gate CMOS DSP with 100 ns Multiplication," *IEEE International Solid-State Circuits Conference*, New York, NY.

Nuttall J., and J. Oxaal. 1984. "A 16-Bit Three Port Arithmetic Logic and Shift Unit," *IEEE International Conference on Acoustics, Speech, and Signal Processing*, San Diego, CA.

Sugai, M., M. Miyata, M. Shirakawa, M. Suzuki, E. Nishihara, K. Aoki, and S. Kawasaki. 1984. "2-Micron CMOS Digital Signal Processors," *IEEE Custom Integrated Circuits Conference*, Rochester, NY.

A Bit-Serial VLSI Architectural Methodology for Signal Processing

Richard F. Lyon

VLSI System Design Area, Xerox Palo Alto Research Center.
3333 Coyote Hill Road, Palo Alto, CA 94304 U.S.A.

INTRODUCTION

Applications of signal processing abound in the modern world of electronics. Telephones, stereos, radios, and televisions are the most common examples. In the general electronic communication and control market, there is a widespread demand for higher quality and lower cost signal processing components of all sorts. For many years, digital signal processing (DSP) techniques have been touted as "the way of the future," but have consistently failed to make a big impact on the market. We discuss here an *architectural methodology* designed to make digital signal processing "the way of the present."

An architectural methodology is a style, or school of design, that provides a basis for a wide range of architectures for different functions. The architectural methodology presented here is built on top of the logic, circuit, timing, and layout levels of VLSI system design methodology presented by Mead and Conway (1980). It includes a large component that is independent of the underlying technology.

Researchers working on DSP theory and applications have had a hard time implementing their ideas in cost-effective hardware, because the IC industry has not figured out how to support their needs. We hope to change that, based on the family of components described, in conjunction with the ability for designers to easily create their own more specialized system components from the silicon layout macros and composition rules that characterize our methodology. The standard-chip way of life has caused many inappropriate architectures to be proposed and tried in the past; the new freedom for non-specialists to easily design custom integrated systems will enable a wave of new applications and new architectures.

SIGNALS and SIGNAL PROCESSING

Signals are time-varying measurements or simulations of real-world phenomena; for example, an audio signal represents minute changes in air pressure as a function of time. In most familiar signal processing equipment, a signal is represented by an electrical analog, such as a continuously changing voltage; such analog signal representations can be directly processed through continuous-time components such as resistors, capacitors, inductors, diodes, amplifiers, etc. The modern alternative is to sample the signal at equally spaced instants of time, and to represent those sampled values either as discrete-time electrical analogs (e.g. amount of charge held on a capacitor) or as numbers (in some kind of a digital computing machine).

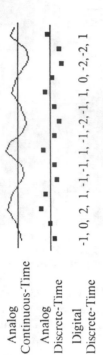

Analog Continuous-Time	
Analog Discrete-Time	
Digital Discrete-Time	-1, 0, 2, 1, -1, -1, 1, -1, -2, -1, 1, 0, -2, -2, 1

The theory of sampling and discrete-time signal processing is quite well developed, and applies to either analog or digital representations of signals. Switches, capacitors, and amplifiers are typical building blocks for analog discrete-time signal processing. For digital signal representations, the basic building blocks are memories, adders, and multipliers. Analog noise, analog drift, and digital roundoff effects of these components are also well understood.

Digital representations other than sequences of sample values are sometimes used; for example, delta-modulation and its variants use a fast sequence of one-bit values indicating whether the signal is above or below its predicted value at each time instant. These representations will not be considered here.

Using modern MOS VLSI technologies, large amounts of signal processing can be done with a small amount of silicon, compared to the more mature continuous-time analog technologies. Yet, signal processing devices remain relatively expensive and in various ways limited. We believe that difficulties in both design and usage of these devices is the reason. Therefore, we base our architectural decisions on our desire to jointly minimize design difficulty and usage (programming) difficulty, subject to maintaining the high performance promised by the technology. We also believe that it is not possible, for the range of applications we are considering, to get good enough performance from analog VLSI technologies; nor is it easy to design the required analog circuits. Accordingly, we have arrived at these basic decisions:

Question:	*Answer:*
Continuous-time or discrete-time? (so that MOS VLSI techniques, particularly digital techniques, can be used)	Discrete
Analog or digital sample representation? (to achieve highest quality and tractable design)	Digital
Bit-parallel or bit-serial number representation? (for high clock rate, and maximum flexibility and extensibility)	Bit-serial
Bus-oriented or dedicated signal paths? (for maximum extensibility, and efficiency)	Dedicated
General-purpose programmability or specialized? (for efficiency and ease of application)	Specialized
Fixed or variable filter parameters? (to cover the widest range of applications, including time-varying filters)	Variable
What number system? (for efficiency and ease of logic design)	Fixed point 2's comp. LSB first

Reprinted with permission from *Proc. 1st Int. Conf. Very Large Scale Integration*, J. P. Gray, Ed., Academic Press Inc., 1981, pp. 131–140.

DIGITAL FILTERS

Probably the most commonly needed signal processing component is a filter. It is simply a linear (usually time-invariant) system with memory (i.e. the output is a linear function of some history of the input). Most commonly, the input and output are single scalar signals, though this is not required; complex- and vector-valued signals will not be discussed, but are easily accommodated in this architecture. In a digital discrete-time architecture, a filter is a computation that operates on a sequence of numbers as input, and produces another sequence of numbers as output. The usual purpose of such computations is to pass some frequencies of signals, while attenuating others. See Moore (1978) for an introduction to the mathematics of digital filters and related signal processing components; for more detailed information, including four chapters on hardware implementations, see Rabiner and Gold (1975).

COMPONENTS AND HIERARCHY

The remainder of this paper explains how to implement signal processing components (operators and systems) in a style evolved from that presented by the Jackson, Kaiser, and McDonald (JK&M) in their 1968 paper "An Approach to the Implementation of Digital Filters." JK&M's notion of an *approach* is similar in some respects to our more developed notion of an architectural methodology, but lacks the notions of standardized interfaces and hierarchical composition of operators.

Our system-building strategy is to design top-down, decomposing blocks functionally into more detailed block diagrams, until we get down to low-level operators, such as adders and multipliers; then, to construct the system from the bottom up, by assembling operators into higher-level operators. To do this, we need conventions for the hierarchical definition and construction of operators. This methodology does not severely constrain the range of possible architectures, but unifies many architectures, allowing them to share components at many levels. The general features of the methodology and conventions are described in following sections.

An automatic gain control operator, composed of other operators.

The architectural methodology involves the use of heavily pipelined bit-serial arithmetic processors, all operating at a fixed throughput rate (words per second), and *multiplexed* over several channels or functions to match processing hardware rate to the signal sample rates. In this approach, it is important that the throughput of an adder be the same as that of a multiplier, which is in turn the same as that of a second-order filter section, or a delay element, or any other component (and the same as a wire, the elementary data path).

MULTIPLEXING

Digital time-division multiplexing is a technique developed by the Bell System to send several digital sampled data signals on a single wire pair (one-bit data path), by dividing time into periodically recurring time-slots, each of which could carry one signal sample. The number of signals that can be carried on one data path is then the word rate divided by the signal sample rate (the word rate is the bit rate divided by the number of bits per word, since a data path carries one bit at a time at a fixed rate). Bell's T1 carrier system is a good introductory example: it carries 24 voice-band signals, sampled at 8000 samples per second (8 ksps), with 8 bits per sample, on a wire pair with 1.544 Mbps total data rate (that's 24*8000*8 plus a few extra bits). These multiplexing concepts are easily extended from transmission of signals to processing of signals.

Bit-serial multiplexed signal representation

For many operators, multiplexing is as simple as time-interleaving the samples of several input signals, resulting in interleaved answers. But when the definition of an operator requires some *state* associated with a signal, multiplexing that operator requires that more state be saved, separately for each of the interleaved input signals. In digital signal processing, all state information is conventionally saved inside the unit-delay operator (called Z^{-1}, the inverse of the unit advance operator of Z-transform notation), which simply produces an output value equal to what the input value was one sample time earlier. The problem of saving state reduces to the problem of multiplexing the Z^{-1} operator. This operator must store one word for each multiplexed signal, so it is logically just a long shift register (if the signals being multiplexed do not all have the same sampling rate, things are more complicated, but still tractable).

CONVENTIONS

System-wide clocking, signalling, timing, and format conventions are needed in order to hierarchically design a system of potentially very high complexity. The basic approach we have adopted is to have system-wide synchronous clocks at the bit rate, and both centralized and distributed timing and control signal generation, as described below.

Clocking

The synchronous clocking scheme will include several versions of the bit clock, provided to accommodate the different types of technology from which the system will be constructed. These technologies and clock types are (1) LSTTL with positive edge-triggered clocking, and (2) high-performance NMOS with nonoverlapping two phase clocking. The relative clock phasing and data timing across technology boundaries will be fixed system-wide. In the NMOS part, the first clock gate encountered by a signal entering a subsystem will be designated Phi-2 (Phase-In), which will be high during the latter (low) part of the TTL clock cycle, when data bits are stable. Phi-1 (Phase-Out) will be the last clock gate encountered by a signal leaving an NMOS subsystem, and it will be timed such that output data, stable during Phi2, meets the TTL setup and hold requirements (see chapter 7 of Mead and Conway 1980).

Signalling

Signalling refers to the electrical representation of bits on wires. We adopted the electrical convention that a low voltage level would represent numerical digit 1, and a high level would represent numerical digit 0 (so that unconnected pulled-up inputs default to a zero signal). This is true within and between operators of any scale (subsystems, chips, or whatever). Bits are transmitted with a non-return-to-zero (NRZ) representation, which just means that the voltage changes to the voltage representing the new bit value after the appropriate clock edge, and remains throughout the clock period.

Timing signals are all active-high: the higher voltage state means logically true, asserting the named time state (e.g. the *LSBtime* signal discussed below is high during the least significant bit time of a word).

More electrical conventions for between-chip signalling are embodied in the I/O pad designs used in the NMOS parts. An informal statement of the conventions is that the NMOS parts be "compatible" with LSTTL (fanout of 10) in levels, noise tolerance, and transition times. The signalling conventions are about the only conventions that would need to be changed to accommodate new and different VLSI technologies as they become available. Some changes in clocking conventions may also be desirable at some point.

Timing

The centralized control signal generator is a time counter that keeps track of bits, words, etc. on a cycle equal to the slowest period of interest in the system. This control section may be thought of as a microprogram store with no conditional branching, a very wide horizontal control word output, and a long linear program (but the actual PLA encoding will be much more efficient).

The distributed control scheme is more widely used, and is more suitable for actually addressing the large-system issues, where design of a single controller for the many functions would be a task beyond the capabilities of a human designer, and would add too many interrelated constraints between parts of the system.

The basic distributed control strategy is that each data path be associated with a timing signal (which is used and generated locally) that represents time within a word; optionally, data paths may be associated also with other timing signals that represent time within a larger *frame*, as appropriate for a particular operator. The standard representation of time is simply for the signal to be true during the first time-slot of its frame, and false at all other times. The signal called *LSBtime* is true during the first (least significant) bit of each word, and the signal *Word0time* is true during the first word of each frame, for example.

Most arithmetic processing units need only clock, data, and one control input to identify the first bit of a new word. Thus, we adopt a convention to use an *LSBtime* signal in association with *every* operator input and output: operators accept *LSBtime* as a control input, and produce a delayed version as the *LSBtime* of their output(s). Thus data paths from one unit to another carry their own timing information, and units can be arbitrarily interconnected without concern about connection to a central controller. Still, the designer must take care to assure that merging signal paths (e.g. at an adder) and loops (e.g. in a recursive filter) have the correct total delay (instead, a special stretchable queue could be designed to automatically adjust delays, but this would greatly decrease the efficiency of the methodology at the chip level).

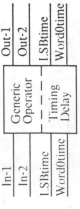

An operator with timing signals.

Format and numerical value restrictions

Some operators, such as multipliers, may take two input operands of different lengths. In such cases, the convention is that both operands are aligned to the same *LSBtime* signal (i.e. words of different lengths are right-adjusted). For short words, the bit slots between the sign bit of one word and the LSB of the next should be filled with sign extensions. Then values of words of any length can be described by the two's-complement binary integer interpretation of the entire 24-bit word (and can be written in binary, octal, or decimal as convenient).

Operators usually require that input values be limited to a range which is less than the total number of bits would allow. For example, the multiplier design for 24-bit words requires that the X input be in the range $[-2^{21}, 2^{21})$, while the Y input must be in the range $[-2^{2k-1}, 2^{2k-1})$, where k is the number of actual hardware stages in the multiplier (not over 12); both X and Y conform to the same format (24 bits per word, LSB first, two's complement), but have different value restrictions. Value restrictions can be enforced by either scalers or limiters, if necessary.

On consistent formats for signals and coefficients

We have deliberately chosen to have signals and coefficients conform to the same format, with their distinction being only in possibly different value restrictions. Many signal processing architectures do not have this property, thereby making it difficult to multiply (mix) two signals together, or to use a filter's output as another filter's coefficient input. These uses of multipliers are not the common time-invariant linear system uses, but are nevertheless widely needed for nonlinear and time-varying processing such as modulation, demodulation, automatic gain control, adaptive filtering, correlation detection, etc. Of particular importance to us is the application of a time-varying filter for speech synthesis; the coefficients of the synthesis filter are themselves the outputs of lowpass (interpolation) filters.

On interpretation of coefficients as fractional values

We have spoken of signal samples with integer values, but we often need to consider the representation of fractional values, especially for coefficients. The interpretation of operands as fractional values derives from the definition of the

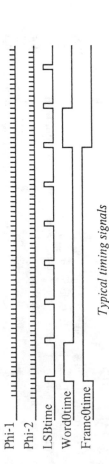

Typical timing signals

Six performance regimes

We are familiar with signal processing system designs that span six distinct regimes, in terms of the amount of multiplexing and/or functional parallelism that they incorporate. Several chips mentioned below have been designed by members of the author's team at Xerox, and will be described in forthcoming papers.

The low-performance regime (level 0) is characterized by a lack of multiplier hardware or other special signal processing features, and is exemplified by software systems running on simple general-purpose processors.

Bit-serial architectures begin to look interesting at level 1, characterized by the use of a single multiplier for different functions. This level is exemplified by Jim Cherry's "Synth" chip, a tenth-order lattice filter for speech synthesis; its block diagram has two multipliers per lattice stage and one overall gain control multiplier, while its implementation consists of one multiplier, one adder, one delay shift register, a little control logic, and a handful of data path switches.

When more total performance is required, level 2 is appropriate. This is basically the simple multiplexed filter approach described by JK&M, which uses one hardware multiplier for each multiplier in the block diagram of a section, but shares them over several sections by multiplexing. Lyon's 32-channel second-order "Filters" chip exemplifies this regime; it has almost no data path switching or control logic. Rich Pasco's "FOS" (first-order section) is another example, which uses scalars and switches instead of multipliers, for a range of specialized applications.

For signals of higher bandwidth, the multiplexing factor may be reduced to unity. We call this level 3, meaning one hardware filter per signal, with no sharing. Rich Pasco's "NCO" chip is a rather simplified example; it is a first-order integrator with no multiplexing, so the signal sample rate can be very high (500 ksps).

When the application requires sample rates higher than the word rate obtainable with the target technology, more parallelism is needed. At level 4, systems take several bits per clock cycle into a distributor, which de-interleaves and sends simple serial data to several operators in parallel; their results can later be combined into the net answer, in a partial-result combiner. Jim Cherry's "FIR" chip is intended to be used this way in the implementation of two-dimensional video filters for image understanding. The chip is one of the operators to be run in parallel; the associated distributor, control, combiner, and line memories are quite simple.

Finally, when the sample rate is higher than the achievable bit rate (level 5), another level of parallelism is needed, as in GE's "VFFT-10" system mentioned above. In this "very fast Fourier transform" system, blocks of signals are transmitted along many serial paths in parallel; signal samples are delivered on a collection of data paths by some faster signal source that can deal with the high bandwidths.

THE BUILDING BLOCKS — LOGIC DESCRIPTION

When designing the logic of many of the low-level and system-level operators, it is not important to know what technology will be used to implement those operators. Thus, this level of the design is relatively long-lived and independent of scale of integration, etc.

function performed by the multiplier. For example, the "5-level recoded" or "modified Booth's algorithm" pipelined multiplier design of Lyon (1976) computes $XY/2^{2k-2}$ for a k-section layout, with Y restricted to $[-2^{2k-1}, 2^{2k-1})$. So if we simply regard Y as a value in $[-2, 2)$, we can say that the multiplier computes XY (as long as we interpret X and XY consistently, either both as integers or both as fractions). Another way of saying this is that a multiplier which accepts n-bit coefficients regards the radix point as being n−2 places from the right. Multipliers can easily be designed with other scale factors, to give other range interpretations to the operands. The scaling described here is popular for signal processing.

Pipelining delay and composition of operators

Another important convention is that all operators have some constant positive integer number of clock cycles of delay (pipeline delay or transport delay) between the LSB of an input word and the LSB of the corresponding output word. For some operators, such as adders, each output bit depends only on the corresponding and lower-order input bits, so the delay can be very small--but not zero. We require that no output bit be combinationally related to any input bit, so at least one cycle of delay is required. This allows the system clock rate to be fast, independent of how many adders and other operators are cascaded (long ripple propagation delays are excluded, unlike typical bit-parallel systems).

The delay of an operator is required to be a constant, known at design time, rather than being dependent on control parameters. The designer must satisfy delay constraints, such as equal *LSBtime* for signals that merge in an adder, by inserting null operators with carefully chosen delay wherever needed (i.e. shift registers are added to make signals line up in time). Loop delay constraints are sometimes difficult or impossible to satisfy if operators in the loop have too much delay. Fortunately, all delay information is known at design time, and is easy to represent and design with, by using simple notations.

FUNCTIONAL PARALLELISM and MULTIPLEXING ALTERNATIVES

We have briefly discussed the use of multiplexing to accommodate many separate operations on relatively slow signals by using a small amount of fast hardware. In the other direction, *functional parallelism* can be used to perform operations on very fast signals, by using a larger number of hardware operators running in parallel, with appropriate techniques for combining their partial computations. Thus, it is possible to apply the serial-arithmetic architectural methodology to very wideband applications, such as radar and video processing.

There is really no limit on the bandwidth obtainable, as illustrated by the VFFT series of Fourier transform processors described by Powell and Irwin (1978), which use slow 3 Mbps (custom PMOS) serial chips to compute transforms of up to 10 Msps radio telescope data (i.e. the sample rate is actually faster than the bit rate!).

The techniques for use of functional parallelism in nonrecursive (finite impulse response) filters are straightforward. Similar techniques for recursive filters have been demonstrated by Moyer (1976). The point is that the bit-serial architectural techniques do not overconstrain the system-level architecture, and do not limit the applications to low-bandwidth areas.

For most operators, such as adders, multipliers, and filters, we design logic in the style popularized by synchronous edge-triggered TTL MSI families; that is, the exact nature of the clocking is suppressed, and all memory is in D-type flip-flops, sometimes connected as parallel registers or shift registers. Some of the operators, however, such as the larger memories, cannot be adequately represented by a logic-level design; most of the design work for those parts is in the technology-dependent circuit and layout level.

Given logic designs of this form, it is straightforward to apply a collection of techniques to translate them into circuits for the target technology. The techniques we use for NMOS are those described by Mead and Conway (1980), making heavy use of pass-transistors and dynamic storage to produce designs that are much more efficient than gate-based circuits. The logic designs therefore do not imply any particular gate structure or circuit design. Examples of CMOS implementations of multipliers and serial memories for digital filtering can be found in Ohwada, et al. (1979).

There is not room here to cover details of logic designs, so we just mention some the more useful operators that have been built. They are signal combiners (adder, multiplier, and variations and combinations), point operators (scaler, overflow detector/corrector or limiter, full-wave and half-wave rectifiers, square, and sine), serial memories (Z^{-1} block, random-access serial register file, and serial ROM), and filters (configurable biquadratic sections, parallel FIR blocks, lattice structures, and multipurpose first-order sections for zero-order hold, interpolation, differentiation, sum-and-dump, oscillators, etc.)

An interesting aspect of the memory-intensive operators is the serial-parallel-serial (SPS) memory organization for both read-write and read-only memories, random or sequential access, that simultaneously minimizes area and power while keeping a high input and output bandwidth. Since memory is always accessed for full words, the parallel internals of the memory can operate on a cycle that is much slower than the serial bit clock, allowing minimum-size dynamic memory cells and slow, low power addressing and bus logic. Only the input and output shift registers and a small control PLA need to run at the bit rate.

ON VLSI LAYOUT STYLES

When actually creating layouts for a system designed through our architectural methodology, it is important to use a simple layout style that facilitates placement and interconnection of components, at many levels of the hierarchy. For NMOS, we have chosen a standard grid style with fixed width (200λ) and variable height cells. A component may occupy any number of 200-lambda wide cells, subject to chip size constraints. The grid may be thought of as a distribution network for VDD, Ground, and two clock phases, into which components and wires are dropped (possibly automatically).

This style is comparable to the polycell layout style, except that the cells being placed and interconnected are considerably larger than gates and registers, and are already specialized to our architectural methodology. This allows considerably more efficient layouts than are possible with polycells, by leaving room for flexible optimizations of logic, circuits, and layout of low-level operators. System designers need never deal with this intra-component level of optimization, but there is always the opportunity for "wizards" to design efficient new low-level components to add to the available library.

Some operators are not designed on the 200-lambda grid, for sake of their layout efficiency. For example, the serial sequential-access memory used in the "Filters" chip is designed with a custom cell array style, appropriate for the large collection of memory cells and the controller PLA.

CONCLUDING REMARKS

As JK&M predicted in 1968, serial arithmetic is alive and well as a simple and efficient way to implement digital filters and other signal processing systems in silicon. By extending their approach into an architectural methodology, and by combining this with a VLSI design and implementation methodology, we suddenly enable designers everywhere to build their own novel low-cost signal processing systems. The methodologies provide a sensible context for the accumulation of a library of component and system designs, at logic and layout levels. So start now, and design the future accordingly.

REFERENCES

Jackson, L. B., Kaiser, J. F., and McDonald, H. S. (1968), An Approach to the Implementation of Digital Filters, *IEEE Trans. on Audio and Electroacoustics* AU-16, 413-421 (*reprinted in Rabiner and Rader, Digital Signal Processing,* IEEE Press).

Lyon, R. F. (1976), Two's Complement Pipeline Multipliers, *IEEE Trans. on Communications* COM-24, 418-425 (*reprinted in Salazar, Digital Signal Processing,* IEEE Press).

Mead, C. A. and Conway, L. A. (1980), *Introduction to VLSI Systems.* Addison-Wesley, Reading, Massachusetts.

Moore, F. R. (1978), An Introduction to the Mathematics of Digital Signal Processing, *Computer Music Journal* Vol. 2, No. 1 and 2.

Moyer, A. L. (1976), An Efficient Parallel Algorithm for IIR Filters, *IEEE International Conference of Acoustics, Speech, and Signal Processing.*

Ohwada, N., Kimura, T., and Doken, M. (1979), LSI's for Digital Signal Processing, *IEEE Journal of Solid-state Circuits* SC-14, 214-220.

Powell, N. R. and Erwin, J. M. (1978), Signal Processing with Bit-serial Word-parallel Architectures, *Real-Time Signal Processing,* Proc. SPIE Vol. 154, 98-104.

Rabiner, L. R., and Gold, B. (1975), *Theory and Application of Digital Signal Processing.* Prentice-Hall, Englewood Cliffs, New Jersey.

Bibliography

Note: Most of the recent work on digital VLSI system design, design automation, simulation and testing are mainly reported in the current issues of:

1. The IEEE Journal of Solid State Circuits

2. The IEEE Transactions on Computer Aided Design

3. The IEEE Spectrum

4. The IEEE Proceedings

5. The IEEE Transactions on Circuits and Systems

6. The IEEE Transactions on Computers

7. The IEEE Computer Magazine

8. The IEEE Micro Magazine

9. The Digests of Technical Papers of the IEEE Solid State Circuits Conferences

10. The Digests of the Custom Integrated Circuits Conferences

11. The Digests of the Design Automation Conferences

12. The Digests of the Canadian VLSI Conferences

13. The Proceedings of the European Solid State Circuits Conferences

14. The Proceedings of the International Conference on VLSI

15. The Proceedings of the International Conferences on Computer Aided Design

16. The Proceedings of the International Conferences on Computer Design

17. The IEE Proceedings on Electronic Circuits and Systems

18. The IEE Proceedings on Software and Microsystems

19. The IEE Computer Aided Engineering Journal

20. The IEE Proceedings on Computers and Digital Techniques

VLSI System Design

(1) A series of articles on Intel 432, IEEE J. Solid-State Circuits, Vol. SC-16, No. 5, pp. 514-537, October 1981.

(2) J. Abelson and P. Andreae: "Information transfer and area-time trade-offs for VLSI multiplication", Comm. ACM, Vol. 23, No. 1, pp. 20-23, 1980.

(3) A.V. Aho, et al.: The Design and Analysis of Computer Algorithms, Addison-Wesley, Reading Mass., 1974.

(4) A.V. Aho, et al.: Data Structures and Algorithms, Addison-Wesley, Reading Mass., 1983.

(5) A.V. Aho and J.D. Ullman: Principles of Compiler Design, Addison-Wesley, Reading Mass., 1977.

(6) A.V. Aho, et al.: "On notions of information transfer in VLSI circuits", Proc. Fifteenth Annual ACM Symposium on the Theory of Computing, pp. 133-139, 1983.

(7) M. Ajtai, et al.: "An O(n log n) sorting network", Proc. Fifteenth Annual ACM Symposium on the Theory of Computing, pp. 1-9, 1983.

(8) R. Aleliunas: "Randomized parallel computation", Proc. ACM Symp. on Principles of Distributed Computing, 1982.

(9) J. Allen: Conversion of algorithms to custom integrated circuits -- An MIT perspective, Proc. IEEE ICCC'80, October 1980.

(10) B. Aspvall and Y. Shiloach: "A polynomial time algorithm for solving systems of linear inequalities with two variables per inequality", Proc. Twentieth Annual IEEE Symposium on Foundations of Computer Science, pp. 205-217, 1979.

(11) M.J. Atallah and S.R. Kosaraju: "Graph problems on a mesh-connected processor array", Proc. Fourteenth Annual ACM Symposium on the Theory of Computing, pp. 345-353, 1982.

(12) D.E. Atkins, et al.: Overview of an arithmetic design system, Proc. 18th DA Conf., pp. 314-321, June 1981.

(13) A.J. Atrubin: "An interactive one-dimensional real time multipler", IEEE Trans. on Computers, EC-14, No. 3, pp. 394-399, 1965.

(14) R. Ayres: "Silicon compilation -- a hierarchical use of PLA's", Proc. Sixteenth Design Automation Conference, pp. 314-326, 1979.

(15) M.R. Barbacci: "Instruction set processor specifications (ISPS): The notation and its applications", IEEE Trans. Computers, Vol. C-30, pp. 24-40, January 1981.

(16) K. Batcher: "Sorting networks and their applications", Spring Joint Computer Conference, Vol. 32, pp. 307-314, AFIPS Press, Montvale, N.J., 1968.

(17) C.G. Bell, et al.: Designing Computers and Digital Systems Using Register-Transfer Module pdp/6, Maynard, MA, Digital Press, Digital Equipment Corporation, 1972.

(18) R. Bergman, et al.: "MOS array design: Universal array, APAR or custom", RCA Engineer, Vol. 21, No. 1, June/July 1975.

(19) J.W. Beyers, et al.: "A 32-bit VLSI chip", IEEE J. Solid-State Circuits, Vol. SC-16, No. 5, pp. 537-541, October 1981.

(20) S.N. Bhatt and C.E. Leiserson: "How to assemble tree machines", Proc. Fourteenth Annual ACM Symposium on the Theory of Computing, pp. 77-84, 1982.

(21) R. Bianchini and R. Bianchini Jr.: "Wireability of an ultracomputer", DOE/ER/03077-177, Courant Inst., New York, 1982.

(22) G. Bilardi, et al.: "A critique and appraisal of VLSI models of computation", in Kung, Sproull, and Steele, pp. 81-88, 1981.

(23) A. Borodin, et al.: "Fast parallel matrix and GCD computations", Proc. Twenty-Third Annual IEEE Symposium on Foundations of Computer Science, pp. 65-71, 1982.

(24) R.P. Brent and L.M. Goldschlager: "Some area time tradeoffs for VLSI", SIAM J. Computing, Vol. 11, No. 4, pp. 737-747, 1982.

(25) R.P. Brent and H.-T. Kung: "The chip complexity of binary arithmetic", Proc. Twelfth Annual ACM Symposium on the Theory of Computing, pp. 190-200, 1980a

(26) R.P. Brent and H.-T. Kung: "On the area of binary tree layouts", Information Processing Letters, Vol. 11, No. 1, pp. 44-46, 1980b.

(27) R.P. Brent and H.-T. Kung: "The area-time complexity of binary multiplication", J. ACM, Vol. 28, No. 3, pp. 521-534, 1981.

(28) S.A. Browning: "The tree machine: a highly concurrent computing environment", Ph.D. thesis, Dept. of Computer Science, CIT, Pasadena, CA., 1980.

(29) A.K. Chandra, et al.: "A complexity theory for unbounded fan-in parallelism", Proc. Twenty-Third Annual IEEE Symposium on Foundations of Computer Science, pp. 1-13, 1982.

(30) T.L. Chang and P.D. Fisher: "A programmable systolic array", IEEE COMPCON, pp. 48-53, 1982.

(31) B.M. Chazelle and L.M. Monier: "Model of computation for VLSI with related complexity results", Proc. Thirteenth Annual ACM Symposium on the Theory of Computing, pp. 318-325, 1981.

(32) F.Y. Chin, et al.: "Efficient parallel algorithms for some graph problems", Comm. ACM, Vol. 25, No. 9, pp. 659-665, 1982.

(33) W.A. Clark: "Macromodular computer systems", AFIPS Conference Proc., 1967.

(34) T. Edge and J. Gould: "Standard transistor array radix", Proc. Seventeenth Design Automation Conference, pp. 556-562, 1980.

(35) J.R. Egan and C.L. Liu: "Bipartite Folding and Partitioning of a PLA", IEEE Trans. Computer-Aided Design, Vol. CAD-3, No. 3, pp. 191-199, July 1984.

(36) A.L. Fisher: "Systolic algorithms for running order statistics in signal and image processing", in Kung, Sproull, and Steele, pp. 265-272, 1981.

(37) A.L. Fisher, et al.: "Design of the PSC: a programmable systolic chip", Proc. Third Caltech Conf. on VLSI, pp. 287-302, Computer Science Press, Rockville, Md., 1983.

(38) H. Fleisher and L.I. Maissel: "An introduction to array logic", IBM J. Research and Development, Vol. 19, pp. 98-109, March 1975.

(39) H. Fleisher: "Introduction to special section on programmable logic arrays", IEEE Trans. Computers, Vol. C-28, No. 9, p. 593, September 1979.

(40) S. Fortune and J. Wyllie: "Parallelism in random access machines", Proc. Tenth Annual ACM Symposium on the Theory of Computing, pp. 114-118, 1978.

(41) M.J. Foster and H.-T. Kung: "The design of special purpose VLSI chips", Computer, Vol. 13, No. 13, pp 26-40, 1980.

(42) Z. Galil and W.J. Paul: "An efficient general-purpose parallel computer", Proc. Thirteenth Annual ACM Symposium on the Theory of Computing, pp. 247-262, 1981.

(43) J.W. Greene and A. El Gamal: "Area and delay penalties in restructurable wafer-scale arrays", Proc. Third Caltech Conference on VLSI, pp. 165-184, Computer Science Press, Rockville, Md., 1983.

(44) D.L. Greer: "An associative logic matrix", IEEE JSSC, Vol. SC-11, October 1976.

(45) L.J. Guibas, et al.: "Direct VLSI implementation of combinatorial algorithms", Proc. Caltech Conf. on VLSI, pp. 509-525, 1979.

(46) L.J. Guibas and F.M. Liang: "Systolic stacks, queues, and counters", Proc. MIT Conf. on Advanced Research in VLSI, 1980.

(47) W.R. Heller and S. Triebwasser: "Design of digital systems in LSI", Proc. IEEE ICCC'80, pp. 90-92, Rye, N.Y., October 1980.

(48) F.J. Hill and G.R. Peterson: Digital Systems: Hardware Organization and Design, New York, Wiley, 1978.

(49) J.W. Jones: "Array logic macros", IBM J. Research and Development, pp. 98-109, March 1975.

(50) L. Junker and J.P. Roth: "User's guide to a program for multiple output logic array minimization", IBM Technical Report RC6894, IBM T.J.Watson Research Centre, Yorktown Heights, N.Y., 1978.

(51) A. Kaminker, et al.: "A 32-bit microprocessor with virtual memory support", IEEE J. Solid-State Circuits, Vol. SC-16, No 5, pp. 537-541, October 1981.

(52) E. Laporte, et al.: "Towards an integrated minicomputer based LSI design system", Proc. 14th ESSCIRC, pp. 163-165, September 1978.

(53) C.E. Leiserson, et al.: "Optimizing synchronous circuitry by retiming", Proc. Third Caltech Conference on VLSI, pp 87-116, Computer Science Press, Rockville, Md., 1983.

(54) C.E. Leiserson and J.B. Saxe: "Optimizing synchronous systems", Proc. Twenty-Second Annual IEEE Symposium on Foundations of Computer Science, pp. 23-36, 1981.

(55) T. Lengauer and K. Mehlhorn: "On the complexity of VLSI computations", in Kung, Sproull, and Steele, pp. 89-99, 1981.

(56) W.K. Luk: "A regular layout for parallel multiplier of $O(\log^2 n)$ time", in Kung, Sproull, and Steele, pp. 317-326, 1981.

(57) R. Lyon: "Simplified design rules for VLSI layout", Lambda, Vol. 2, No. 1, pp. 54-59, 1981.

(58) K.F. Mathews and L.P. Lee: "Bipolar chip design for a VLSI microprocessor", IBM J. Research and Development, Vol. 26, No. 4, pp. 464-474, July 1982.

(59) C.A. Mead and L.A. Conway: Introduction of VLSI Systems, Addison-Wesley, Reading Mass., 1980.

(60) D.I. Moldovan: "On the design of algorithms for VLSI systolic arrays", Proc. IEEE, Vol. 71, No. 1, pp. 113-120, January 1983.

(61) G. Moore: "VLSI: Some fundamental challenges", IEEE Spectrum, Vol. 16, pp. 30-37, April 1979.

(62) A. Mukhopadhyay: "Hardware algorithms for nonnumeric computation", IEEE Trans. on Computers, C-28, No 6, pp. 384-394, 1979.

(63) D. Nassimi and S. Sahni: "Finding connected components and connected ones on a mesh-connected parallel computer", SIAM J. Computing, Vol. 9, No. 4, pp. 744-757, 1980.

(64) R. Noto, et al.: "Automated universal array", IEEE Circuits Systems Magazine, Vol. 3, No. 3, pp. 14-18, 1981.

(65) C.H. Papadimitriou and M. Sipser: "Communication complexity", Proc. Fourteenth Annual ACM Symposium on the Theory of Computing, pp. 196-200, 1982.

(66) A.M. Patel and L.C. Cote: "Partitioning for VLSI placement problems", Proc. 18th DA Conference, pp. 411-415, June 1981.

(67) S.S. Patil and T.A. Welch: "A programmable logic approach for VLSI", IEEE Trans. Computers, Vol. C-28, No 9, pp. 594-601, September 1979.

(68) F.P. Preparata: "New parallel-sorting schemes", IEEE Trans. on Computers, C27, No. 7, pp. 669-673, 1978.

(69) F.P. Preparata: "A mesh-connected area-time optimal VLSI integer multiplier", in Kung, Sproull, and Steele, pp. 311-316, 1981.

(70) F.P. Preparata and J.E. Vuillemin: "Area-time optimal VLSI networks for multiplying matrices", Information Processing Letters, Vol. 11, No. 2, pp. 77-80, 1980.

(71) F.P. Preparata and J.E. Vuillmin: "Area-time optimal VLSI networks for computing integer multiplication and discrete Fourier transform", Proc. 8th International Colloquium on Automata, Languages, and Programming, pp. 29-40, Springer-Verlag, New York, 1981.

(72) J.H. Reif and L.G. Valiant: "A logarithmic time sort for linear size networks", Proc. Fifteenth Annual ACM Symposium on the Theory of Computing, pp. 10-16, 1983.

(73) J.P. Roth: Computer Logic, Testing and Verification. Potomac, MD: Computer Science Press, 1980.

(74) J. Roth: "Programmed logic array optimization", IEEE Trans. Computers, Vol. C-27, No. 2, pp. 174-176, 1978.

(75) C. Savage: "Parallel algorithms for graph-theoretic problems", R-784, Dept. of Computer Science, University of Illinois, Urbana, Ill., 1977.

(76) J.E. Savage: "Area-time tradeoffs for matrix multiplication and related problems in VLSI models", J. Computer and System Sciences, Vol. 20, No. 3, pp. 230-242, 1981b.

(77) M. Schmookler: "Design of Large ALUs using multiple PLA macros", IBM J. Reserach and Development, Vol. 24, No. 1, pp. 2-14, 1980.

(78) J.T. Schwartz: "Ultracomputers", ACM Trans. on Programming Languages and Systems, Vol. 2, No. 4, pp. 484-521, 1980.

(79) Y. Shiloach and U. Vishkin: "Finding the maximum, merging and sorting in a parallel computation model", J. Algorithms, Vol. 2, No. 1, pp. 88-102, 1981.

(80) M. Stefik, et al.: "The partitioning of concerns in digital system design", HPP-82-2, Dept. of Computer Science, Stanford University, Stanford, CA., 1982.

(81) H.S. Stone: "Parallel processing with the perfect shuffle", IEEE Trans. on Computers, C-20, No. 2, pp. 153-161, 1971.

(82) I.E. Sutherland and C.A. Mead: "Microelectronics and computer science", Scientific American, September 1977.

(83) A. Svoboda: "The concept of term exclusiveness and its effect on the theory of boolean functions", J. ACM Vol. 22, No. 3 pp. 425-440, June 1975.

(84) A. Svoboda and D.E. White: Advanced Logical Circuit Design Techniques. Garland Press, N.Y., 1979.

(85) E. Upfal: "Efficient schemes for parallel communications", Proc. ACM Symposium on Principles of Distributed Computing, 1982.

(86) L.G. Valiant: "Universality considerations in VLSI circuits", IEEE Trans. on Computers, C-30, No. 2. pp. 135-140, 1981.

(87) L.G. Valiant: "A scheme for fast parallel communication", SIAM J. Computing, Vol. 11, No. 2, pp. 350-361, 1982a.

(88) L.G. Valiant: "Fast parallel computation", TR-17-82, Aiken Computation Lab., Harvard University, Cambridge, Mass., 1982b.

(89) L.G. Valiant and G.J. Brebner: "Universal schemes for parallel communication", Proc. Thirteenth Annual ACM Symposium on the Theory of Computing, pp. 263-277, 1981.

(90) J.E. Vuillemin: "A combinatorial limit to the computing power of VLSI circuits", Proc. Twenty-First Annual IEEE Symposium on Foundations of Computer Science, pp. 294-300, 1980.

(91) A. Weinburger: "High speed programmable logic array adders", IBM J. Research and Development, Vol. 23, pp. 163-168, March 1979.

(92) A. Weinberger: "Large scale integration of MOS complex logic: A layout method", IEEE J. of Solid State Circuits, SC-2, No. 4, pp. 182-190, 1967. bi.R.A. Wood: "A high density programmable logic array chip", IEEE Trans. Computers, Vol. C-28, pp. 602-608, September 1979.

(93) A.C. Yao: "Some complexity questions related to distributive computing", Proc. Eleventh Annual ACM Symposium on the Theory of Computing, pp. 209-213, 1979.

Design Automation

(1) B. Ackland and N. Weste: "Functional verification in an interactive symbolic IC design environment", Proc. Second Caltech Conference on VLSI, Pasadena, CA., 1981.

(2) T. Adachi, et al.: "Hierarchical top-down layout design method for VLSI chip", presented at Nineteenth Design Automation Conference, 1982.

(3) S.B. Akers, et al.: "IC mask layout with a single conductor layer", Proc. Seventh Design Automation Conference, San Francisco, pp 7-16, June 1970.

(4) J. Allen and P. Penfield: "VLSI design automation activities at MIT", IEEE Trans. Circuits and Systems, p. 645, July 1981.

(5) R.L. Arndt and D.L. Dietmeyer: "DDLSIM -- A digital design language simulator", Proc. National Electronics Conference, Vol. 26, pp. 116-118, 1970.

(6) M.H. Arnold and J.K. Ousterhout: "Lyra: A new approach to geometric layout rule checking", Proc. Nineteenth Design Automation Conference, pp. 530-536, 1982.

(7) P.B. Arnold: "Complexity results for channel routing", TR-21-82, Aiken Computation Lab., Harvard University, Cambridge, MA., 1982a.

(8) P.B. Arnold: "Complexity results for single row routing", TR-22-82, Aiken Computation Lab., Harvard University, Cambridge, MA., 1982b.

(9) R. Auerbach: "FLOSS: Macrocell compaction system", Proc. Sixteenth Design Automation Conference, E. Lansing, Mich., 1979.

(10) J.P. Avenier: "Digitizing, layout, rule checking -- The everyday tasks of chip designers", Proc. IEEE, Vol. 71, No. 1, pp. 49-56, January 1983.

(11) R. Ayres: "ICL reference manual", SSP File #1364, Caltech, 1978.

(12) R. Ayres: "Silicon compilation -- A hierarchical use of PLA's", Sixteenth Design Automation Conference, p. 315, 1979.

(13) R. Ayres: "IC specification language", Proc. Sixteenth Design Automation Conference, pp. 307-309, June 1979.

(14) H.S. Baird and Y.E. Cho: "An artwork design verification system", Proc. Twelfth Design Automation Workshop, Boston, MA, pp. 414-420, June 1975.

(15) H. Baird: "Fast algorithms for LSI artwork analysis", Proc. Fourteenth Design Automation Conference, New Orleans, Louisiana, pp. 303-311, June 1977.

(16) C.M. Baker and C. Terman: "Tools for verifying integrated circuit design", Lambda, Vol. 1, No. 3, pp. 22-30, 1980.

(17) A.E. Baratz: "Algorithms for integrated circuit signal routing", Ph.D. thesis, Dept. of EECS, MIT, Cambridge, MA., 1981.

(18) M.R. Barbacci and A.W. Nagel: "An ISPS simulator", Dept. CSEE, CMU, Pittsburgh, PA, Technical Report, November 1977.

(19) M.R. Barbacci, et al.: "The ISPS computer description language", Technical Report, Computer Science Dept., Carnegie-Mellon University, Pittsburgh, PA., 1977.

(20) M.R. Barbacci: "Automated exploration of the design space for register transfer systems", Ph.D. thesis, Computer Science Dept., Carnegie-Mellon University, Pittsburgh, PA., 1974.

(21) M.R. Barbacci: "A comparison of register transfer languages for describing computers and digital systems", IEEE Trans. Computers, Vol. C-24, February 1975.

(22) J. Batali, et al.: "The DPL/Daedelus design environment", in VLSI-81 (J. P. Gray, ed.), Academic Press, New York, pp. 183-192, 1981.

(23) J. Batali and A. Hartheimer: "The design procedure language manual", VLSI Memo 80-31, MIT, Cambridge, MA., September 1980.

(24) G.M. Baudet: "On the area required by VLSI circuits", in Kung, Sproull, and Steele, pp. 100-107, 1981.

(25) W. Beeby: "The future of integrated CAD/CAM systems: The Boeing perspective", IEEE Computer Graphics and Applications, pp. 51-56, January 1982.

(26) H. Beke, et al.: "CAL-MP: An advanced computer aided layout program for MOS-LSI", Proc. ESSCIRC, 1980.

(27) H. Beke and W. Sansen: "CALMOS: A portable software system for the automatic interactive layout of MOS/LSI", Proc. Sixteenth Design Automation Conference, pp. 102-108, 1979.

(28) G. Bell and A. Newell: "The PMS and ISP descriptive system for computer systems", Proc. AFIPS SJCC, 1970.

(29) J.L. Bentley and T.A. Ottmann: "Algorithms for reporting and counting geometric intersections", IEEE Trans. on Computers, Vol. C-28, No. 9, pp. 643-647, 1979.

(30) J.L. Bentley and D. Wood: "An optimal worst-case algorithm for reporting intersections of rectangles", IEEE Trans. on Computers, Vol. C-29, No. 9, pp. 571-577, 1980.

(31) N. Bergmann: "A case study of the F.I.R.S.T silicon compiler", Proc. Third Caltech Conference on VLSI, pp. 413-430, Computer Science Press, Rockville, Md., 1983.

(32) F.C. Bergsten: "Computer-aided design, manufacturing, assembly and test (CADMAT)", Proc. Eighteenth Design Automation Conference, pp. 873-880, June 1981.

(33) L. Berman: "On logic comparison", Proc. Eighteenth Design Automation Conference, pp. 854-861, 1981.

(34) A. Bilgory: "Compilation of register transfer language specifications into silicon", UIUCDCS-R-82-1091, Dept. of Computer Science, University of Illinois, Urbana, Ill., 1982.

(35) K.M. Black and K.P. Hardage: "Advanced symbolic artwork preparation (ASAP)", Hewlett-Packard Journal, Vol. 32, No. 6, pp. 8-10, June 1981.

(36) W.R. Blood: "Computer aided design: The engineering interface for macrocell array circuits", Proc. Sixteenth Design Automation Conference, pp. 182-185, 1979.

(37) G. Bogo, et al.: "CASSANDRE and the computer aided logical system design", Proc. IFIP Congress, Ljubljana, Yugoslavia, p. 1056, 1971.

(38) A. Borodin and J.E. Hopcroft: "Routing, merging, and sorting on parallel models of computation", Proc. Fourteenth Annual ACM Symposium on the Theory of Computing, pp. 338-344, 1982.

(39) D. Borrione: "LASCAR: A language for simulation of computer architecture", International Symposium on Computer Hardware Description Languages and their Applications, New York, 1975.

(40) D. Borrione and J.F. Grabowiecki: "Informal introduction to LASSO: A language for asynchronous systems specification and simulation", Proc. Euro IFIP'79, London, September 1979.

(41) H.N. Brady and R.J. Smith: "Verification and optimization for LSI and PCB layout", Proc. Eighteenth Design Automation Conference, pp. 365-371, June 1981.

(42) R. Brayton, et al.: "A taxonomy of CAD for VLSI", Proc. ECTD, p. 34, August 1981.

(43) R.K. Brayton, et al.: "A comparison of logic minimization strategies using EXPRESSO: An APL program package for partitioned logic minimilization", Proc. IEEE International Conference on Circuits and Computers, 1982.

(44) M.A. Breuer: "Min-cut placement", J. Design Automation and Fault-Tolerant Computing, Vol. 1, No. 4, pp. 343-362, 1977.

(45) M.A. Breuer: "General survey of design automation of digital computers", Proc. IEEE, Vol. 54, No. 12, pp. 1708-1721, December 1966.

(46) K.D. Brinkmann and D.A. Mlynski: "Computer-aided chip minimization for IC layout", Proc. 1976 IEEE ISCAS, p. 314, 1976.

(47) Broster: "'CELLMOS': An automated LSI design technique", Proc. IEE CADMECCS, pp. 43-46, July 1981.

(48) D.J. Brown and R.L. Rivest: "New lower bounds for channel width", in Kung, Sproull, and Steele, pp. 178-185, 1981.

(49) D.W. Brown: "A state-machine synthesizer-SMS", Proc. Eighteenth Design Automation Conference, pp. 301-305, June 1980.

(50) H. Brown and M. Stefik: "Palladio: An expert assistant for integrated circuit design", HPP-82-5, Dept. of Computer Science, Stanford University, Stanford, CA., 1982.

(51) I. Buchanon: "Modelling and verification in structured integrated circuit design", Ph.D. thesis, University of Edinburgh, 1980.

(52) H.D. Caplanar and J.A. Janku: "Top-down approach to LSI system design", Computer Design, Vol. 13, No. 8, pp. 143-148, August 1974.

(53) J. Carey and W. Blood: "Macrocell arrays -- An alternative to custom LSI", Proc. Semicustom IC Technology Symposium, pp. 19-37, June 1981.

(54) P. Carmody, et al.: "An interactive graphics system for custom design", Proc. Seventeenth Design Automation Conference, pp. 430-434, June 1980.

(55) W.S. Chan: "A new channel routing algorithm", Proc. Third Caltech Conference on VLSI, pp. 117-140, Computer Science Press, Rockville, Md., 1983.

(56) B.M. Chazelle and D.P. Dobkin: "Detection is easier than computation", Proc. Twelfth Annual ACM Symposium on the Theory of Computing, pp. 146-153, 1980.

(57) T. Chiba, et al.: "SHARPS: A hierarchical layout system for VLSI", Proc. Eighteenth Design Automation Conference, pp. 820-827, 1980.

(58) F.Y. Chin and C.A. Wang: "Optimal algorithms for the intersection and the minimum distance problems between planar polygons", TR82-8, Dept. of Computer Science, University of Alberta, Edmonton, Alberta, 1982.

(59) Y.E. Cho, et al.: "FLOSS: An approach to automated layout for high-volume designs", Proc. Fourteenth Design Automation Conference, pp. 138-141, 1977.

(60) Y. Chu: "An ALGOL-like computer design language", Commun. ACM, Vol. 8, pp. 607-615, October 1965.

(61) S. Chuquillanqui and T. Perez-Segovia: "PAOLA: A tool for topological optimization of large PLA's", Proc. Nineteenth Design Automation Conference, pp. 300-306, 1982.

(62) R. Cleghorn: "PRIMEAIDS: An integrated electrical design environment", Proc. Eighteenth Design Automation Conference, pp. 632-638, June 1981.

(63) L. Conway, et al.: "MPC79: The large-scale demonstration of a new way to create systems in silicon", Lambda Magazine, Vol. 1, No. 2, 1980.

(64) L.V. Corbin: "Custom VLSI electrical rule checking in an intelligent terminal", Proc. Eighteenth Design Automation Conference, pp. 697-701, June 1981.

(65) W.E. Cory, et al.: "An introduction to the DDL-P language", Technical Report 163, Computer Science Lab., Stanford University, Stanford, CA., March 1979.

(66) W.E. Cory: "Symbolic simulation for functional verification with ADLIB and SABLE", Proc. Eighteenth Design Automation Conference, pp. 82-89, June 1981.

(67) A. Cranswick and K. Bather: "MICROCELL -- A user oriented custom design system for IC's", Proc. IEEE ICCC'80, Rye, N.Y., October 1980.

(68) J.D. Crawford: "A unified hardware description language for CAD programs", ERL Memo No. UCB/ERL M79/64, University of California Berkeley, May 1978.

(69) M.E. Daniel and C.W. Gwyn: "Hierarchical VLSI circuit design", Proc. 1980 IEEE ICCC'80, Rye, N.Y., p. 92, October 1980.

(70) M.E. Daniel and C.W. Gwyn: "CAD systems for IC design", IEEE Trans. CAD, Vol. CAD-1, No. 1, pp. 2-12, January 1982.

(71) J.A. Darringer: "The application of program verification techniques to hardware verification", Proc. Sixteenth Design Automation Conference, pp. 375-381, 1979.

(72) J.A. Darringer and W.H. Joyner: "A new look at logic synthesis", Proc. Seventeenth Design Automation Conference, p. 543-549, 1980.

(73) J.A. Darringer, et al.: "Experiments in logic synthesis", Proc. Proc. IEEE International Conference Circuits and Computers, Rye, N.Y., pp. 234-237, 1980.

(74) J.A. Darringer, et al.: "Logic synthesis through local transformations", IBM J. Research and Development, Vol. 25, No. 4, pp. 272-280, July 1981.

(75) H.W. Daseking, et al.: "VISTA: A VLSI CAD system", IEEE Trans. CAD, Vol. CAD-1, No. 1, pp. 36-52, January 1982.

(76) E. Dekel, et al.: "Parallel matrix and graph algorithms", TR 79-10, Dept. of Computer Science, University of Minnesota, Minneapolis, MI., 1979.

(77) D.N. Deutsch: "A dogleg channel router", Proc. Thirteenth Design Automation Conference, pp. 425-433, 1976.

(78) D.N. Deutsch and P. Glick: "An over-the-cell router", Proc. Seventeenth Design Automation Conference, pp. 32-39, 1980.

(79) R.C. deVries and A. Svoboda: "Multiple output minimization with mosaics of Boolean functions", IEEE Trans. Computers, Vol. C-24, No. 8, pp. 777-785, August 1975.

(80) D.L. Dietmeyer and J.R. Duley: "Register transfer languages and their translation", in Digital System Design Automation (M. Breuer, ed.), Woodland Hills, CA., Computer Science Press, pp. 117-218, 1975.

(81) D.L. Dietmeyer and M.H. Doshi: "Automated PLA synthesis of the combinational logic of a DDL description", Design Automation and Fault-Tolerant Computing, Vol. 3, pp. 241-257, April 1980.

(82) S. Director, et al.: "A design methodology and computer aids for digital VLSI systems", IEEE Trans. Circuits and Systems, Vol. CAS-28, No. 7, pp. 634-644, July 1981.

(83) D. Dolev, et al.: "Optimal wiring between rectangles", Proc. Thirteenth Annual ACM Symposium on the Theory of Computing, pp. 312-317, 1981.

(84) M.H. Doshi and D.L. Dietmeyer: "Automated PLA synthesis of the combinational logic of a DDL description", University of Wisconsin, Madison, Technical Report ECE-78-17, November 1978.

(85) K.A. Duke and K. Maling: "ALEX: A conversational hierarchical logic design system", Proc. Seventeenth Design Automation Conference, pp. 318-327, 1980.

(86) J.R. Duley and D.L. Dietmeyer: "A digital system design language (DDL)", IEEE Trans. Computers, Vol. C-17, p. 850-861, September 1968.

(87) J.R. Duley and D.L. Dietmeyer: "Translation of a DDL digital system specification to Boolean equations", IEEE Trans. Computers, Vol. C-18, pp. 305-313, April 1969.

(88) A.E. Dunlop: "SLIP: Symbolic layout of integrated circuits with compaction", Computer Aided Design, Vol. 10, No. 6, pp. 387-391, 1978.

(89) A.E. Dunlop: "SLIM -- The translation of symbolic layouts into mask data", Proc. Seventeenth Design Automation Conference, pp. 595-602, 1980.

(90) R. Dutton: "Stanford overview in VLSI research", IEEE Trans. CAS, p. 654-665, July 1981.

(91) T.H. Edmondson and R.M. Jennings: "A low cost hierarchical system for VLSI layout and verification", Proc. Eighteenth Design Automation Conference, pp. 505-509, June 1981.

(92) J.R. Egan and C.L. Liu: "Optimal bipartite folding of PLA", Proc. Nineteenth Design Automation Conference, pp. 141-146, 1982.

(93) R.A. Eustace and A. Mukhopadhyay: "A deterministic finite automaton approach to design rule checking for VLSI", Proc. Nineteenth Design Automation Conference, pp. 712-717, 1982.

(94) A. Feller: "Automatic layout of low-cost quick-turnaround random-logic custom LSI devices", Proc. Thirteenth Design Automation Conference, pp. 79-85, June 1976.

(95) A. Feller, et al.: "Standard cell approach for generating custom CMOS/SOS devices using a fully automatic layout program", IEEE Circuits and Systems Magazine, Vol. 3, No. 3, pp. 9-13, 1981.

(96) A. Feller and R. Noto: "A speed-oriented, fully automatic layout program for random logic VLSI devices", National Computer Conference Proc., pp. 303-311, 1978.

(97) C.M. Fidducia and R.M. Mattheyes: "A linear-time heuristic for improving network partitions", Proc. Nineteenth Design Automation Conference, pp. 175-181, 1982.

(98) M.J. Fischer and M.S. Paterson: "Optimal tree layout", Proc. Twelfth Annual ACM Symposium on the Theory of Computing, pp. 177-189, 1980.

(99) R.W. Floyd and J.D. Ullman: "The compilation of regular expressions into integrated circuits", J. ACM, Vol. 29, No. 2, pp. 603-622, 1982.

(100) M.J. Foster and H.-T. Kung: "Recognize regular languages with programmable building blocks", in VLSI-81 (J.P. Gray, ed.), Academic Press, New York, pp. 75-84, 1981.

(101) J.R. Fox: "The MacPitts silicon compiler: A view from the telecommunications industry", VLSI Design, pp. 30-37, May 1983.

(102) T.D. Friedman and S. Yang: "Methods used in an automatic logic design generator", IEEE Trans. Computers, Vol. C-18, No. 7, pp. 593-614, July 1969.

(103) D. Gibson and S. Nance: "SLIC -- Symbolic layout of integrated circuits", Proc. Thirteenth Design Automation Conference, pp. 434-440, June 1976.

(104) L.A. Glasser: "An interactive PLA generator as an archetype of a new VLSI methodology", Proc. IEEE ICCC'80 Conference, Rye, N.Y., pp. 608-611, October 1980.

(105) G.B. Goates and S.S. Patil: "ABLE: A LISP-based layout modeling language with user-definable procedural models for storage/logic array design", Proc. Eighteenth Design Automation Conference, pp. 322-329, 1981.

(106) G.B. Goates, et al.: "SLAs for VHSIC", Proc. Semicustom IC Technology Symposium, pp. 191-207, June 1981.

(107) L. Goldschlager: "A unified approach to models of synchronous parallel machines", Proc. Tenth Annual ACM Symposium on the Theory of Computing, pp. 89-94, 1978.

(108) T.F. Gonzalez and S.-L. Lee: "An O(n log n) algorithm for optimal routing around a rectangle", TR-116, Prog. in Math. Sciences, University of Texas, Dallas, TX., 1982.

(109) D.F. Gorman and J.P. Anderson: "A logic design translator", Proc. FJCC, pp. 251-261, 1962.

(110) S. Goto: "An efficient algorithm for two dimensional placement problem in electrical circuit layout", IEEE Trans. Circuits and Systems, Vol. CAS-28, No. 1, pp. 12-18, January 1980.

(111) S. Goto: "An automated layout design system for masterslice LSI chip", Proc. European Conference on Circuit Theory and Design, pp. 631-635, 1981.

(112) J. Gray, et al.: "Caltech silicon structures project", Proc. Sixteenth Design Automation Conference, pp. 305-326, 1979.

(113) J.P. Gray: "Introduction to silicon compilation", Proc. Sixteenth Design Automation Conference, pp. 305-306, June 1979.

(114) G.D. Hachtel, et al.: "Some results in optimal PLA folding", IEEE ICCC'80, Rye, N.Y., pp. 1023-1028, October 1980.

(115) G.D. Hachtel: "On the sparse tableau approach to optimal layout", Proc. IEEE ICCC'80, Rye, N.Y., pp. 1019-1023, October 1980.

(116) G.D. Hachtel, et al.: "Techniques for programmable logic array folding", Proc. Nineteenth Design Automation Conference, pp. 147-155, 1982.

(117) L.J. Hafer and A.C. Parker: "Register-transfer level automatic digital design: The allocation process", Proc. IEEE Design Automation Conference, Vol. 15, 1978.

(118) L.J. Hafer and A.C. Parker: "A formal method for the specification, analysis and design of register-transfer level digital logic", Proc. Eighteenth Design Automation Conference, pp 846-853, June 1981.

(119) L.J. Hafer and A.C. Parker: "Automated synthesis of digital hardware", IEEE Trans. Computers, Vol. C-31, No. 2, pp. 93-109, February 1982.

(120) K. Hardage: "ASAP: Advanced symbolic artwork preparation", Lambda, Vol. 1, No. 3, 1980.

(121) K. Hardage: "ASAP: Advanced symbolic artwork preparation", Conference on Advanced Research in Integrated Circuits, MIT, Cambridge, MA, January 1980.

(122) R.L. Harris, et al.: "The design and description of a waveform description and test programming language for CAD of digital systems", Proc. European Conference on Design Automation, pp. 203-207, Septemer 1981.

(123) A. Hashimoto and J. Stevens: "Wire routing by optimal channel assignment within large apertures", Proc. Eighth Design Automation Conference, pp. 155-169, 1971.

(124) C.L.F. Haynes: "The use of ADA as a design language for digital systems", Proc. European Conference Elec. Design Automation, pp. 139-143, September 1981.

(125) W.R. Heller: "Contrasts in physical design between LSI and VLSI", Proc. Eighteenth Design Automation Conference, pp. 676-683, June 1981.

(126) J.L. Hennessy: "SLIM: A simulation and implementation language for VLSI microcode", LAMBDA, pp. 20-28, April 1981.

(127) D.W. Hightower: "The interconnection problem: A tutorial", Computer, Vol. 7, No. 4, pp. 18-32, 1974.

(128) D.W. Hightower and R.L. Boyd: "A generalized channel router", Proc. Eighteenth Design Automation Conference, pp. 12-21, 1980.

(129) D.D. Hill: "ADLIB: A modular, strongly-typed computer design language", Proc. Fourth International Symposium on Computer Hardware Description Languages, Palo Alto, CA., pp. 75-81, 1979.

(130) K. Hirabayashi and M. Kawamura: "MACLOS -- mask checking logic simulator", IEEE J. Solid State Circuits, Vol. SC-15, No. 3, pp. 368-370, 1980.

(131) D.S. Hirschberg, et al.: "Computing connected components on parallel computers", Comm. ACM, Vol. 22, No. 8, pp. 461-464, 1979.

(132) R.B. Hitchcock: "Timing verification and the timing analysis program", Proc. Nineteenth Design Automation Conference, pp. 594-604, 1982.

(133) D. Hoey and C.E. Leiserson: "A layout for the shuffle-exchange network", Proc. IEEE International Conference on Parallel Processing, 1980.

(134) S.J. Hong and R. Nair: "Wire-routing machines -- New tools for VLSI physical design", Proc. IEEE, Vol. 71, No. 1, pp. 57-65, January 1983.

(135) S.J. Hong, et al.: "MINI: A heuristic approach for logic minimization", IBM J. Research and Development, Vol. 18, pp. 434-458, September 1974.

(136) J.E. Hopcroft and J.D. Ullman: Introduction to Automata Theory, Languages, and Computation, Addison-Wesley, Reading Mass., 1979.

(137) S. Horiguchi, et al.: "An automatic designed 32b CMOS VLSI processor", ISSCC'82, pp. 54-55, 1982.

(138) T. Hosaka, et al.: "A design automation system for electronic switching systems", Proc. Eighteenth Design Automation Conference, pp. 51-58, 1981.

(139) M.-Y. Hsueh and D.O. Pederson: "Computer-aided layout of LSI circuit building-blocks", Proc. 1979 International Symposium on Circuits and Systems, Tokyo, Japan, pp. 447-477, 1979.

(140) M.Y. Hsueh: "Cabbage -- Symbolic layout and compaction of integrated circuits", Ph.D. dissertation, Dept. EECS, University of California (Berkeley), 1979.

(141) M. Hsueh: "Symbolic layout compaction", in Computer Design Aids for VLSI Circuits (P. Antognetti, et al., eds.), Noordhoff, Holland, 1981.

(142) M. Hsueh: "Symbolic layout and compaction of integrated circuits", Memorandum No. UCB/ERL M79-80, Electronics Research Lab, University of California (Berkeley), December 1979.

(143) K.E. Iverson: A Programming Language, New York, Wiley, 1962.

(144) D. Johanssen: "Bristle blocks: A silicon compiler", Proc. Caltech Conference on VLSI, pp. 303-310, 1979. See also Proc. Sixteenth Design Automation Conference, pp. 310-313, 1979.

(145) S.C. Johnson: "Code generation for silicon", Proc. Tenth ACM Symposium on Principles of Programming Languages, 1983.

(146) S.C. Johnson, et al.: "Teachable station brings order to VLSI chip design", Electronics, p. 108, January 27, 1982.

(147) S.C. Johnson and S.A. Browning: "The LSI design language i", TM-1980-1273-10, Bell Laboratories, Murray Hill, NJ, 1980.

(148) R. Kamikawai, et al.: "Placement and routing program for masterslice LSIs", Proc. Thirteenth Design Automation Conference, pp. 245-250, 1976.

(149) S. Kang and W.M. vanCleemput: "Automatic PLA synthesis from a DDL-P description", Proc. Eighteenth Design Automation Conference, pp. 391-397, 1981.

(150) S. Kang: "Automated synthesis of PLA based systems", Ph.D. dissertation, Stanford University, 1981.

(151) K. Kani, et al.: "ROBIN: A building block LSI routing program", Proc. 1976 IEEE International Conference on Circuits and Systems, pp. 658-661, 1976.

(152) K. Kani, et al.: "CAD in the Japanese electronics industry", presented at the Symposium CAD of Digital Electronic Circuits and Systems, Brussels, Belgium, November 1978.

(153) A.R. Karlin, et al.: "Experience with a regular expression compiler", Proc. IEEE Conference on Computer Design/VLSI in Computers, 1983.

(154) K. Karplus: "CHISEL, an extension to the programming language C for VLSI layout", Ph.D. thesis, Dept. of Computer Science, Stanford University, Stanford, CA., 1982.

(155) H. Kato, et al.: "On automated wire routing for building-block LSI", Proc. 1974 IEEE International Conference on Circuits and Systems, pp. 309-313, 1974.

(156) T. Kawamoto and Y. Kajitani: "The minimum width routing of 2-row, 2-layer polycell layout", Proc. Sixteenth Design Automation Conference, pp. 290-296, 1979.

(157) H. Kawanishi, et al.: "A routing method of building block LSI", Proc. Seventh Asilomar Conference on Circuits, Systems and Computers, pp. 119-124, 1973.

(158) H. Kawanishi, et al.: "An algorithm on graphical manipulation of LSI mask pattern", Trans. Inst. Electron. Commun. Eng. Japan, Vol. J65-A, pp. 611-618, 1980 (Japanese).

(159) N. Kawato, et al.: "An interactive logic synthesis system based upon AI techniques", Proc. Nineteenth Design Automation Conference, 1982.

(160) N. Kawato, et al.: "Design and verification of large scale computers by using DDL", Proc. Sixteenth Design Automation Conference, pp. 360-366, 1979.

(161) G. Kedem and H. Watanabe: "Optimization techniques for IC layout and compaction", TR-117, Dept. of Computer Science, University of Rochester, Rochester, NY., 1982.

(162) Z. Kedem: "Optimal allocation of computational resources in VLSI", Proc. Twenty-Third Annual IEEE Symposium on Foundations of Computer Science, pp. 379-386, 1982.

(163) B.W. Kernigan, et al.: "An optimum channel-routing algorithm for polycell layouts of integrated circuits", Proc. Tenth Design Automation Conference, pp. 50-59, 1973.

(164) A. Kishimoto, et al.: "An interconnection check algorithm for mask pattern", Proc. IEEE International Conference on Circuits and Systems, pp. 669-672, 1978.

(165) H.-J. Knobloch: "Description and simulation of complex digital systems by means of the register transfer language TRS Ia", in Computer Design Aids for VLSI Circuits (P. Antognetti, et al., eds.), Noordhoff, Holland, p. 285, 1981.

(166) B.J. Korenjak: "PLOTS: A user-oriented language for CAD artwork", RCA Engineer, Vol. 20, No. 4, p. 20, December 1974.

(167) A.J. Korenjak and A.H. Teger, "An integrated CAD data base system", Proc. Twelfth Design Automation Workshop, Boston, MA, June 1975.

(168) T. Kozawa, et al.: "Block and track method for automated layout generation of MOS LSI array", ISSCC, pp. 62-63, 1972.

(169) T. Kozawa, et al.: "Advanced LILAC -- An automated layout generation system for MOS/LSIs", Proc. Eleventh Design Automation Workshop, pp. 26-46, 1974.

(170) T. Kozawa, et al.: "A concurrent pattern operation algorithm for VLSI mask data", Proc. Eighteenth Design Automation Conference. pp. 563-570, 1981.

(171) N. Kubo, et al.: "Algorithm for testing graph isomorphism", Proc. IEEE International Conference on Circuits and Systems, pp. 641-644, 1979.

(172) K. Kusunoki: "DEX-design automation system cooperation development", Proc. Tenth Design Automation Workshop, pp. 187-192, 1973.

(173) J.C. Latomb: Artificial Intelligence and Pattern Recognition in CAD, New York, North-Holland, 1978.

(174) W.W. Lattin, et al.: "A methodology for VLSI chip design", Second Quarter, Lambda, pp. 34-44, 1981.

(175) B. Lattin: "VLSI design methodology -- The problem of the 80's for microprocessor design", Proc. Sixteenth Design Automation Conference, pp. 548-549, 1979.

(176) U. Lauther: "A min-cut placement algorithm for general cell assemblies based on a graph representation", Proc. Sixteenth Design Automation Conference, pp. 1-10, June 1979.

(177) A. leBlond, et al.: "Automatic layout of symbolic MD-MOS circuits", Proc. IEEE ICCC'80, Rye, N.Y., pp. 772-774, October 1980.

(178) B.K. Lee and C. Jones: "CAD tools must change to meet the needs of VLSI", Electronics, November 17, 1981.

(179) C.M. Lee, et al.: "Automatic generation and characterization of CMOS polycells", Proc. Eighteenth Design Automation Conference, pp. 220-224, June 1981.

(180) C.Y. Lee: "An algorithm for path connections and its applications", IRE Trans. Electron. and Computers, Vol. EC-10, No. 3, pp. 346-364, 1961.

(181) G. Leive: "The design, implementation and analysis of an automated logic synthesis and module selection system", Ph.D. thesis, Dept. of Electrical Engineering, Carnegie-Mellon University, Pittsburgh, PA., January 1981.

(182) T. Lengauer: "The complexity of compacting hierarchically specified layouts of integrated circuits", Proc. Twenty-Third Annual IEEE Symposium on Foundations of Computer Science, pp. 358-368, 1982.

(183) G. Lev, et al.: "A fast parallel algorithm for routing in permutation networks", IEEE Trans. on Computers, Vol. C-30, No. 2, pp. 93-100, 1981.

(184) Y.-Z. Liao and C.K. Wong: "An algorithm to compact a VLSI symbolic layout with mixed constraints", IEEE Trans. on Computer Aided Design of Integrated Circuits and Systems, Vol. CAD-2, No. 2, pp. 62-69, 1983.

(185) K.J. Lieberherr and S.E. Knudsen: "Zeus: A hardware description language for VLSI", to appear in Proc. Twentieth Design Automation Conference, 1983

(186) B.W. Lindsay and B.T. Preas: "Design rule checking and analysis of IC mask designs", Proc. Thirteenth Design Automation Conference, June 1976.

(187) H.M. Lipp: "Methodical aspects of logic synthesis", Proc. IEEE, Vol. 71, No. 1, pp. 88-97, January 1983.

(188) R.J. Lipton, et al.: "ALI: A procedural language to describe VLSI layouts", Proc. Nineteenth Design Automation Conference, pp. 467-474, 1982.

(189) R.J. Lipton and R. Sedgewick: "Lower bounds for VLSI", Proc. Thirteenth Annual ACM Symposium on the Theory of Computing, pp. 300-307, 1981.

(190) R.J. Lipton, et al.: "Programming aspects of VLSI", Proc. Ninth ACM Symposium on Principles of Programming Languages, pp. 57-65, 1982.

(191) R.J. Lipton and J. Valdes: "Census functions: An approach to VLSI upper bounds", Proc. Twenty-First Annual IEEE Symposium on Foundations of Computer Science, pp. 13-22, 1981.

(192) J.C. Logue, et al.: "Techniques for improving engineering productivity of VLSI designs", IBM J. Research and Development, Vol. 25, No. 3, pp. 107-115, May 1981.

(193) K.G. Loosemore: "IC layout -- The automatic approach", Proc. European Solid State Circuit Conference, pp. 48-50, 1979.

(194) K.G. Loosemore: "Integrated CAD for LSI", Proc. Symposium on CAD, Brussels, pp. 13-21, 1978.

(195) K.G. Loosemore: "Gaelic -- an automatic solution to complex chip design", Proc. European Conference Elec. Design Automation, pp. 102-105, September 1981.

(196) P. Losleben: "Computer aided design for VLSI", in Very Large Scale Integration: Fundamentals and Applications (D.F. Barbe, ed.), New York, Springer-Verlag, pp. 89-127, 1980.

(197) M. Marshall: "VLSI pushes super-CAD techniques", Electronics, pp. 73-80, July 1980.

(198) G. Martin, et al.: "CELTIC -- Solving the problems of LSI design with an integrated polycell DA system", Proc. Eighteenth Design Automation Conference, pp. 804-811, June 1981.

(199) P. Marwedel: "The MIMOLA design system: Detailed description of the software system", Proc. Sixteenth Design Automation Conference, pp. 59-63, June 1979

(200) T. Matsuda, et al.: "LAMBDA: A quick, low cost layout design system for master-slice LSIs", presented at Nineteenth Design Automation Conference, 1982.

(201) R. Matthews, et al.: "A target language for silicon compilers", IEEE COMPCON, pp. 349-353, 1982.

(202) J.F. McCabe and A.Z. Muszynski: "A bipolar VLSI custom macro physical design verification strategy", IBM J. Research and Development, Vol. 26, No. 4, pp. 485-496, 1982.

(203) W.J. McCalla and D. Hoffman: "Symbolic representation and incremental DRC for interactive layout", Proc. IEEE ISCAS, Chicago, pp. 710-715, 1981.

(204) E.M. McCreight: "Efficient algorithms for enumerating intersecting intervals and rectangles", CSL-80-9, Xerox PARC, Palo Alto, CA., 1980.

(205) M. McFarland: "The value trace: A data base for automated digital design", M.Sc. thesis, Carnegie-Mellon University, Pittsburgh, PA., 1978.

(206) T.M. McWilliams: "Verification of timing constraints on large digital systems", Proc. Eighteenth Design Automation Conference, pp. 139-147, 1980.

(207) T.M. McWilliams and L.C. Widdoes: "SCALD: Structured computer-aided logic design", Proc. Fifteenth Design Automation Conference, Las Vegas, pp. 271-277, 1978.

(208) C.A. Mead and L.A. Conway: Introduction to VLSI Systems, Boston, MA., Addison-Wesley, 1980.

(209) K. Mikami and K. Tabuchi: "A computer program for optimal routing of printed circuit conductors", Proc. IFIP Congress'68, pp. 1475-1478, 1968.

(210) G.J. Milne: "A model for hardware description and verification", Proc. Twenty-First Design Automation Conference, pp. 251-257, 1984.

(211) T. Mitsuhashi, et al.: "An integrated mask artwork analysis system", Proc. Seventeenth Design Automation Conference, pp. 277-284, 1980.

(212) N. Miyahara, et al.: "A new CAD system for automatic logic interconnection verification", Proc. IEEE International Conference on Circuits and Systems, pp. 114-117, 1981.

(213) R. Mosteller: "REST -- Stick diagram editing system", M.Sc. thesis, Caltech, 1981.

(214) T. Moto-Oka, et al.: "Logic design system in Japan", Proc. Twelfth Design Automation Conference, pp. 241-250, June 1975.

(215) S. Murai, et al.: "The effects of initial placement techniques on the final placement results constructive vs. top-down techniques", Proc. 1980 IEEE International Conference on Circuits and Computers, pp. 80-82, 1980.

(216) M. Nagatani, et al.: "An automated layout system for LSI functional blocks: PLASMA", in Monograph of Technical Group on Design Technology of Electronics Equipment of Information Processing Society Japan, 41, pp. 1-10, 1980.

(217) D. Nash: "Topics in design automation databases", Proc. Fifteenth Design Automation Conference, pp. 463-474, June 1978.

(218) D. Nash and H. Willman: "Software engineering applied to computer-aided design (CAD) software development", Proc. Eighteenth Design Automation Conference, pp. 530-539, June 1981.

(219) R. Newton, et al.: "Design aids for VLSI: The Berkeley perspective", IEEE Trans. CAS, Vol. CAS-28, No. 7, p. 666-680, July 1981.

(220) R. Newton: "Computer-aided design of VLSI circuits", IEEE Proc., p. 1189, October 1981.

(221) C. Niessen: "Hierarchical design methodologies and tools for VLSI chips", Proc. IEEE, Vol 71, No. 1, pp. 66-75, January 1983.

(222) G. Odawara, et al.: "A symbolic functional description language", Proc. Twenty-First Design Automation Conference, pp. 73-80, 1984.

(223) Y. Ozawa, et al.: "Master slice LSI computer aided design system", Proc. Eleventh Design Automation Conference, pp. 19-25, 1974.

(224) J.F. Paillotin: "Optimization of the PLA area", Proc. Eighteenth Design Automation Conference, pp. 406-410, 1981.

(225) A. Parker, et al.: "The CMU design automation system: An example of automated data path design", Proc. Sixteenth Design Automation Conference, pp. 73-80, 1979.

(226) A.M. Patel and L.C. Cote: "Partitioning for VLSI placement problems", Proc. Eighteenth Design Automation Conference, pp. 411-415, June 1981.

(227) M.S. Paterson, et al.: "Bounds on minimax edge length for complete binary trees", Proc. Thirteenth Annual ACM Symposium on the Theory of Computing, pp. 293-299, 1981.

(228) D. Pederson: "Computer aids in integrated circuits design", in Computer Design Aids for VLSI Circuits (P. Antognetti, et al., eds.), Noordhoff, Holland, 1981.

(229) G. Persky, et al.: "LTX -- Minicomputer-based system for automatic LSI layout", J. Design Automation and Fault Tolerant Computing, No. 3, pp. 217-255, May 1977.

(230) G. Persky, et al.: "LTX -- A system for the directed automatic design of LSI circuits", Proc. Thirteenth Design Automation Conference, pp. 399-407, June 1976.

(231) R. Piloty, et al.: "CONLAN: A formal construction method for hardware description languages: Basic principles, language derivation, language application", 3 papers in National Computer Conference, Anaheim, CA., AFIPS Conference Proceedings, Vol. 49, 1980.

(232) R.Y. Pinter: "Optimal routing in rectilinear channels", in Kung, Sproull, and Steele, pp. 160-177, 1981.

(233) R.Y. Pinter: "On routing two-point nets across a channel", Proc. Nineteenth Design Automation Conference, pp. 894-902, 1982a.

(234) R.Y. Pinter: "The impact of layer assignment methods on layout algorithms for integrated circuits", Ph.D. thesis, MIT, Cambridge, Mass., 1982b.

(235) R.Y. Pinter: "River routing: Methodology and analysis", Proc. Third Caltech Conference on VLSI, pp. 141-164, Computer Science Press, Rockville, Md., 1983.

(236) B.T. Preas, et al.: "A hierarchical approach to VLSI layout", Dig. COMPCON, San Francisco, February 1980.

(237) B.T. Preas and W.M. vanCleemput: "Routing algorithms for hierarchical IC layout", Proc. IEEE ISCAS Conference, 1979.

(238) B.T. Preas: "Placement and routing algorithms for hierarchical integrated circuit layout", Ph.D. thesis, Stanford University, Stanford, CA., 1979.

(239) B.T. Preas and C.W. Gwyn: "General hierarchical automatic layout of custom VLSI circuit masks", J. Design Automation and Fault-Tolerant Computing, Vol. 3, No. 1, pp. 41-58, 1979.

(240) B.T. Preas and W.M. vanCleemput: "Placement algorithms for arbitrarily shaped blocks", Proc. Sixteenth Design Automation Conference, June 1979.

(241) B.T. Preas and C.W. Gwyn: "Methods for hierarchical layout of custom LSI circuit masks", Proc. Fifteenth Design Automation Conference, pp. 206-212, 1978.

(242) B.T. Preas, et al.: "Automatic circuit analysis based on mask information", Proc. Thirteenth Design Automation Conference, June 1976.

(243) J. Prioste: "Macrocell approach customizes fast VLSI", Electronic Design, Vol. 29, No. 11, pp. 159-166, June 7, 1980.

(244) R. Raghavan and S. Sahni: "Optimal single row router", Proc. Nineteenth Design Automation Conference, pp. 38-45, 1982.

(245) F.R. Ramsey: "A remote design station for customer uncommitted logic array designs", Proc. Eighteenth Design Automation Conference, pp. 498-504, 1981.

(246) S.M. Reddy, et al.: "A gate level model for CMOS combinational logic circuits with application to fault detection", Proc. Twenty-First Design Automation Conference, pp. 504-509, 1984.

(247) S.P. Reiss and J.E. Savage: "SLAP -- A silicon layout program", CS-82-17, Dept. of Computer Science, Brown University, Providence, R.I., 1982.

(248) D.G. Ressler: "Simple computer-aided artwork system that works", Proc. Eleventh Design Automation Workshop, Denver, CO, June 1974.

(249) R.L. Rivest: "The PI (placement and interconnect) system", Proc. Nineteenth Design Automation Conference, pp. 475-481, 1982.

(250) R.L. Rivest, et al.: "Provably good channel routing algorithms", in Kung, Sproull, and Steele, pp. 153-159, 1981.

(251) R.L. Rivest and C.M. Fiduccia: "A 'greedy' channel router", Proc. Nineteenth Design Automation Conference, pp. 418-424, 1982.

(252) L.M. Rosenberg: "The evolution of design automation toward VLSI", J. Digital Systems, Volume V, No. 4, pp. 301-318, 1981.

(253) L.M. Rosenberg: "The evolution of design automation to meet the challenges of VLSI", Proc. Seventeenth Design Automation Conference, Minneapolis, Minn., 1980.

(254) L. Rosenberg and C. Benbassat: "CRITIC: An integrated circuit design rule checking program", Proc. Eleventh Design Automation Workshop, Denver, CO, pp. 14-18, June 1974.

(255) J.A. Rowson: "Understanding hierarchical design", Ph.D. thesis, Caltech, 1980.

(256) W.L. Ruzzo and L. Snyder: "Minimum edge length planar embeddings of trees", in Kung, Sproull, and Steele, pp. 119-123, 1981.

(257) S. Sahni and A. Bhatt: "The complexity of design automation problems", Proc. Seventeenth Design Automation Conference, pp. 402-411, 1980.

(258) T. Saito, et al.: "A CAD system for logic design based on frames and demons", Proc. Eighteenth Design Automation Conference, pp. 451-456, 1981.

(259) T. Sasaki, et al.: "Hierarchical design verification for large digital systems", Proc. Eighteenth Design Automation Conference, June 1981.

(260) K. Sato, et al.: "MILD -- A cell based layout system for MOS LSI", Proc. Eighteenth Design Automation Conference, pp. 828-836, 1981.

(261) K. Sato, et al.: "A grid-free channel router", Proc. Seventeenth Design Automation Conference, pp. 22-31, 1980.

(262) K. Sato, et al.: "MIRAGE -- A simple model routing program for the hierarchical layout design of IC masks", Proc. Sixteenth Design Automation Conference, pp. 297-304, 1979.

(263) K. Sato and T. Nagai: "A method of specifying the relative locations between blocks in a routing program for building block LSI", Proc. 1979 IEEE International Conference on Circuits and Systems, pp. 673-676, 1979.

(264) C. Savage: "A systolic data structure chip for connectivity problems", in Kung, Sproull, and Steele, pp. 296-300, 1981.

(265) J.E. Savage: "Planar circuit complexity and the performance of VLSI algorithms", in Kung, Sproull, and Steele, pp. 61-67, 1981.

(266) J.E. Savage: "Three VLSI compilation techniques: PLA's, Weinberger arrays, and SLAP, a new silicon layout program", CS-82-24, Dept. of Computer Science, Brown University, Providence, R.I., 1982.

(267) H. Schorr: "Computer aided digital systems design and analysis using a register-transfer language", IEEE Trans. Computers, Vol. C-13, pp. 730-737, December 1964.

(268) R. Segal: "Structure, placement and modelling", M.Sc. thesis, Caltech, 1980.

(269) M.I. Shamos and D. Hoey: "Geometric intersection problems", Proc. Seventeenth Annual IEEE Symposium on Foundations of Computer Science, pp. 208-215, 1976.

(270) E. Shapiro, et al.: "A physical design engine for VLSI design", Proc. Edinborough Conference on VLSI Design (J. Gray, ed.), Edinborough, Scotland, August 1981.

(271) H. Shiraishi and F. Hirose: "Efficient placement and routing techniques for master slice LSI", Proc. Seventeenth Design Automation Conference, pp. 458-464, 1980.

(272) I. Shirakawa, et al.: "A layout system for the random logic portion of MOS LSI", Proc. Seventeenth Design Automation Conference, pp. 92-99, 1980.

(273) S.G. Shiva: "Computer hardware description languages -- A tutorial", Proc. IEEE, Vol. 67, No. 12, pp. 1605-1615, December 1979.

(274) S.G. Shiva: "Use of DDL in an automatic LSI design system", Proc. International Symposium on CHDLS and Their Applications, pp. 28-32, October 1979.

(275) S.G. Shiva and J. Covington: "Modular hardware synthesis/description/simulation using DDL", Proc. Nineteenth Design Automation Conference, pp. 321-329, 1982.

(276) S.G. Shiva: "Combinational logic synthesis from an HDL description", Proc. Seventeenth Design Automation Conference, pp. 550-555, 1980.

(277) S.G. Shiva: "Automatic hardware synthesis", Proc. IEEE, Vol. 71, No. 1, pp. 76–87, January 1983.

(278) H.E. Shrobe: "The data path generator", IEEE COMPCON, pp. 340–344, 1982.

(279) A. Siegel and D. Dolev: "The separation required for general single-layer wiring barriers", in Kung, Sproull, and Steele, pp. 143–152, 1981.

(280) H.J. Siegel: "Interconnection networks and masking schemes for single instruction stream -- multiple data stream machines", Ph.D. thesis, Princeton University, Princeton, N.J., 1977.

(281) H.J. Siegel: "Interconnection networks for SIMD machines", Computer, Vol. 12, No. 6, pp. 57–65, 1979.

(282) D.P. Siewiorek and M.R. Barbacci: "The CMU RT-CAD system -- An innovative approach to computer aided design", Proc. AFIPS Conference, Vol. 45, 1976.

(283) J.M. Siskind, et al.: "Generating custom high performance VLSI designs from succinct algorithmic descriptions", Proc MIT Conference on Advanced Research in VLSI, 1982.

(284) E.A. Snow: "Automation of module set independent register transfer design", Ph.D. thesis, Carnegie-Mellon University, Pittsburgh, PA., 1978.

(285) J. Soukup: "Circuit layout", Proc. IEEE, Vol. 69, No. 10, pp. 1281–1304, 1981.

(286) Special Issue on "Computer-aided design", Proc. IEEE, Vol. 69, No. 10, pp. 1185–1376, October 1981.

(287) Special Invited Papers on "VLSI design aids", IEEE Trans. Circuits and Systems, Vol. CAS-28, No. 7, pp. 617–680, July 1981.

(288) J.D. Stauffer: "SADIST (The SAndia Data Index STructure)", Sandia National Lab., Albuquerque, NM, Technical Report SAND80-1999, November 1980.

(289) M. Stefic, et al.: "The partitioning of concerns in digital system design", Proc. 1982 Conference on Advanced Research in VLSI, MIT, Cambridge, MA, 1982.

(290) J.A. Storer: "The node cost measure for embedding graphs on a planar grid", Proc. Twelfth Annual ACM Symposium on the Theory of Computing, pp. 201–210, 1980.

(291) T. Sudo, et al.: "CAD systems for VLSI in Japan", Proc. IEEE, Vol. 71, No. 1, pp. 129–143, January 1983.

(292) T. Sudo: "VLSI and CAD", Proc. IPSJ, Vol. 22, No. 8, 1981 (Japanese).

(293) T. Sudo: "Present status of CAD technologies in Japan", to be published.

(294) T. Sudo: "VLSI design methodology", Monograph of Technical Group on Design Technology of Electronics Equipment of Information Processing Society Japan, 1-1, 1979 (Japanese).

(295) T. Sudo and Y. Sugiyama: "Custom VLSI design system", Proc. Information Processing Society Japan, Vol. 22, No. 8, pp. 791–796, 1981 (Japanese).

(296) Y. Sugiyama, et al.: "A subnano second 12K gate bipolar 32 bit VLSI processor", Proc. of Custom Integrated Circuit Conference, pp. 73–77, 1981.

(297) N. Sugiyama, et al.: "An integrated circuit layout design program based on graph-theoretical approach", ISSCC, 1970

(298) Y. Sugiyama, et al.: "Routing program for multichip LSIs", Proc. USA-Japan Design Automation'75, pp. 87–94, August 1975.

(299) G.J. Sussman, et al.: "Design aids for digital integrated systems, an artificial intelligence approach", Proc. IEEE International Conference Circuits and Computers, pp. 612–615, 1980.

(300) I. Suwa and W.J. Kubitz: "A computer aided design system for segmented-folded PLA macro-cells", Proc. Eighteenth Design Automation Conference, pp. 398–405, June 1981.

(301) R.E. Swanson, et al.: "An AHPL compiler/simulator system", Proc. Sixth Texas Conference Computer Systems, pp. 1–11, 1977.

(302) S. Swerling: "Computer-aided engineering", IEEE Spectrum, pp. 37–41, November 1982.

(303) A.A. Szepieniec and R.H.J.M. Otten: "The top-down layout of a structured IC layout design", Proc. European Conference Elec. Design Automation, pp. 95–97, September 1981.

(304) C. Tanaka, et al.: "An integrated computer aided design system for gate array masterslices: Part 1. Logic reorganization system LORES-2", Proc. Eighteenth Design Automation Conference, pp. 59–65, 1981.

(305) C. Tanaka, et al.: "An integrated computer aided design system for gate array masterslices: Part 2. The layout design system for master-slice LSIs", presented at Nineteenth Design Automation Conference, 1982.

(306) D.E. Thomas: "The automatic synthesis of digital systems", Proc. IEEE, Vol. 69, No. 10, pp. 1200–1211, October 1981.

(307) D.E. Thomas and D.P. Siewiorek: "Measuring designer performance to verify design automation systems", IEEE Trans. Computers, Vol. C-30, No. 1, pp. 48–60, January 1981.

(308) D.E. Thomas and G. Lieve: "A technology relative design system", Proc. IEEE ICCC'80, Rye, N.Y., October 1980.

(309) D.E. Thomas: "The design and analysis of an automated design style selector", Ph.D. dissertation, Dept. Electrical Engineering, CMU, Pittsburgh, PA., June 1977.

(310) C.D. Thompson: "Area-time complexity for VLSI", Proc. Eleventh Annual ACM Symposium on the Theory of Computing, pp. 81–88, 1979.

(311) C.D. Thompson: "A complexity theory for VLSI", Ph.D. thesis, Carnegie-Mellon University, Pittsburgh, Pa., 1980.

(312) C.D. Thompson: "The VLSI complexity of sorting", in Kung, Sproull, and Steele, pp. 108–118, 1981.

(313) C.D. Thompson and H.-T. Kung: "Sorting on a mesh connected parallel computer", Comm. ACM, Vol. 20, No. 4, pp. 263–271, 1977.

(314) T. Tokuda, et al.: "A hierarchical standard cell approach for custom VLSI design", IEEE Trans. Computer-Aided Design, Vol. CAD-3, No. 3, pp. 172–177, July 1984.

(315) M. Tompa: "An optimal solution to a wire routing problem", Proc. Twelfth Annual ACM Symposium on the Theory of Computing, pp 161–176, 1980.

(316) H.W. Trickey: "Good layouts for pattern recognizers", IEEE Trans. on Computers Vol. C-31, No. 6, pp. 514–520, 1981.

(317) H.W. Trickey and J.D. Ullman: "A regular expression compiler", IEEE COMPCON, pp. 345–348, 1982.

(318) S. Trimberger, et al.: "A structured design methodology and associated design tools", IEEE Trans. CAS, p. 618–634, July 1981.

(319) S. Trimberger: "Combining graphics and layout language in a single interactive system", Proc. Eighteenth Design Automation Conference, 1981.

(320) C. Tseng and D.P. Siewiorek: "The modelling and synthesis of bus systems", Proc. Eighteenth Design Automation Conference, pp. 471–478, June 1981.

(321) K. Ueda: "Computer aided layout design for LSI: State of the art", Proc. Second USA-Japan Computer Conference, pp. 556–561, 1975.

(322) K. Ueda, et al.: "An automated layout system for masterslice LSI: MARC", IEEE J. Solid State Circuits, Vol. SC-13, No. 5, pp. 716–721, 1978.

(323) J.D. Ullman: "Combining state machines with regular expressions for automatic synthesis of VLSI circuits", STAN-CS-82-927, Computer Science Dept., Stanford University, Stanford, CA., 1982.

(324) W.M. vanCleemput: "A hierarchical language for the structured description of digital systems", Proc. Fourteenth Design Automation Conference, pp. 377–385, 1977.

(325) W.M. vanCleemput: "SDL: A structural design language for computer aided design of digital systems", Stanford CSL Technical Report 136, 1979.

(326) W.M. vanCleemput: Computer Aided Design Tools for Digital Systems, IEEE Press, 1979.

(327) B. Vergnieres: "Macro generation algorithms for LSI custom chip design", IBM J. Research and Development, Vol. 24, No. 5, pp. 612–621, 1980.

(328) M. Watanabe: "CAD tools for designing VLSI in Japan", Proc. ISSCC'79, pp. 242–243, February 1979.

(329) N. Weste and B. Ackland: "A pragmatic approach to topological symbolic IC design", Proc. VLSI'81, pp. 117–129, 1981.

(330) N. Weste: "MULGA -- An interactive symbolic layout system for the design of integrated circuits", BSTJ, p. 823, July–August 1981.

(331) N. Weste: "Virtual grid symbolic layout", Proc. Eighteenth Design Automation Conference, pp. 225–233, June 1981.

(332) G. Wiederhold, et al.: "A database approach to communication in VLSI design", IEEE Trans. CAD, Vol. CAD-1, No. 2, pp. 57–62, April 1982.

(333) J.D. Williams: "STICKS -- a graphical compiler for high level LSI design", Proc. 1978 National Computer Conference, pp. 289-295, AFIPS Press, Montvale, N.J., 1978.

(334) J.D. Williams: "STICKS -- A new approach to VLSI design", MSEE thesis, MIT, 1977.

(335) J.A. Wilmore: "Efficient Boolean operations on IC masks", Proc. Eighteenth Design Automation Conference, pp. 571-579.

(336) D.S. Wise: "Compact layouts of banyan/FFT networks", in Kung, Sproull, and Steele, pp. 186-195, 1981.

(337) W. Wolf, et al.: "Dumbo, a schematic-to-layout compiler", Proc. Third Caltech Conference on VLSI, pp. 379-393, Computer Science Press, Rockville, Md., 1983.

(338) A. Yamada: "Design automation status in Japan", Proc. Eighteenth Design Automation Conference, pp. 43-50, 1981.

(339) S. Yamada and T. Watanabe: "A mask pattern analysis system for LSI (PAS-1)", Proc. IEEE International Conference on Circuits and Systems, pp. 858-861, 1979.

(340) K. Yoshida, et al.: "A layout checking system for large scale integrated circuit", Proc. Fourteenth Design Automation Conference, pp. 322-330, 1977.

(341) J. Yoshida, et al.: "PANAMAP-B. A mask verification system for bipolar IC", Proc. Eighteenth Design Automation Conference, pp. 690-695, 1981.

(342) H. Yoshimura, et al.: "An algorithm for resistance calculation from IC mask pattern information", Proc. IEEE International Conference on Circuits and Systems, pp. 478-481, 1979.

(343) T. Yoshimura and E.S. Kuh: "Efficient algorithms for channel routing", IEEE Trans. on Computer Aided Design of Integrated Circuits and Systems, Vol. CAD-1, No. 1, pp. 25-35, 1982.

(344) H. Yoshizawa, et al.: "Automatic layout algorithms for master slice LSI", Proc. 1979 IEEE International Conference on Circuits and Systems, pp. 470-473, 1979.

(345) G. Zimmerman: "The MIMOLA design system: A computer aided digital processor design method", Proc. Sixteenth Design Automation Conference, pp. 53-58, 1979.

(346) G. Zimmerman: "Computer aided design of control structures for digital computers", Proc. IEEE International Conference Circuits and Computers, p. 103, October 1980.

(347) ----: "Quality of designs from an automatic logic generator", IBM Research Report RC 2068, April 1968.

(348) ----: "Translation of a DDL digital system specification to Boolean equations", IEEE Trans. Computers, Vol. C-18, pp. 305-313.

Simulation

(1) J. Abraham: "MINI -- A logic minimization program", Technical Report, University of Illinois, Urbana, Ill., 1981.

(2) J.M. Acken and J.D. Stauffer: "Logic circuit simulation", IEEE Circuits Systems Magazine, Vol. 1, No. 2, pp. 3-12, June 1979.

(3) T. Adachi, et al.: "Two-dimensional semiconductor analysis using finite-element method", IEEE Trans. Electronic Devices, Vol. ED-26, pp. 1026-1031, July 1979.

(4) V.D. Agrawal, et al.: "A mixed mode simulator", Proc. Seventeenth Design Automation Conference, Minneapolis, 1980.

(5) D.A. Antoniadis, et al.: "SUPREM -- A program for IC process modelling and simulation", Stanford Electronics Lab, Stanford University, Stanford, CA, Technical Report SEL78-020, 1978.

(6) G. Arnout and H. DeMan: "The use of threshold functions and Boolean-controlled network elements for macromodelling of LSI circuits", IEEE JSSC, Vol. SC-13, pp. 326-332, June 1978.

(7) R. Beaufoy and J.J. Sparkes: "The junction transistor as a charge-controlled device", Automat. Teleph. Electric Communication J., London, Vol. 4, p. 310, 1957.

(8) R.K. Brayton, et al.: "A new efficient algorithm for solving differential-algebraic systems using implicit backward differentiation formulas", Proc. IEEE, Vol. 60, pp. 98-108, January 1972.

(9) R.K. Brayton and R. Spence: Sensitivity and Optimization, New York, Elsevier Scientific Publications, 1980.

(10) R.K. Brayton, et al.: "Some results on sparse matrices", Math. Comput., Vol. 24, No. 112, 1970.

(11) R.K. Brayton, et al.: "A survey of optimization techniques for integrated-circuit design", Proc. IEEE, Vol. 69, No. 10, pp. 1334-1362, October 1981.

(12) M.A. Breuer: "Digital system design automation: Languages, simulation and data base", Book: Pitman, Woodland Hills, CA, 1975.

(13) R.E. Bryant: "An algorithm for MOS logic simulation", LAMBDA, Vol. I, No. 3, Fourth Quarter, pp. 46-53, October 1980.

(14) R.E. Bryant: "MOSSIM: A switch level simulator for MOS LSI", Proc. Eighteenth Design Automation Conference, pp. 786-790, 1981a.

(15) R.E. Bryant: "A switch-level simulation model for integrated logic circuits", MIT/LCS/TR-259, MIT, Cambridge, Mass., 1981b.

(16) R.E. Bryant: "A switch-level model and simulator for MOS digital systems", 5065:TR:83, California Institute of Technology, Pasadena, CA, 1983.

(17) E. Buturla, et al.: "Finite element modelling in 2 or 3 space dimensions and time", IBM J. Research and Development, July 1981.

(18) P.K. Chatterjee and J.E. Leiss: "An analytic charge-sharing predictor model for submicron MOSFETs", IEEE IEDM Digest of Technical Papers, Washington, D.C., pp. 28-33, 1980.

(19) B.R. Chawla, et al.: "MOTIS -- An MOS timing simulator", IEEE Trans. Circuits and Systems, Vol. CAS-22, pp. 901-909, December 1975.

(20) B.R. Chawla: "Circuit representation of the integral charge-control model of bipolar transistors", IEEE J. Solid-State Circuits, Vol. SC-6, p. 262, 1971.

(21) R.C. Chen and J.E. Coffman: "MULTI-SIM, a dynamic multilevel simulation", Proc. Fifteenth Design Automation Conference, 1978.

(22) L.K. Cheung and E.S. Kuh: "The bordered triangular matrix and minimum essential set of digraph", IEEE Trans. Circuits and Systems, Vol. CAS-21, pp. 633-639, 1974.

(23) J.D. Crawford, et al.: MOTIS-C User's Guide, Electronics Research Laboratory, University of California, Berkeley, June 1978.

(24) E.M. DaCosta and K.G. Nichols: "TWS -- timing and waveform simulator", Proc. IEE Conference on CADMECCS, pp. 189-193, July 1979.

(25) L.M. Dang: "A simple current model for short-channel IGFET and its application to circuit simulation", IEEE J. Solid-State Circuits, Vol. SC-14, pp. 358-367, 1979.

(26) R. Dang, et al.: "Complete process/device simulation system and its application to the design of sub-micron MOSFET", Proc. 1981 Symposium VLSI Technology, pp. 86-87, 1981.

(27) C.B. Davis, et al.: "Digital simulation as a design tool", RCA Engineer, Vol. 17, No. 4, pp. 34-39, December 1971.

(28) C.B. Davis and M.I. Payne: "R-CAP: An integrated circuit simulator", RCA Engineer, Vol. 21, No. 1, June/July 1975.

(29) H.C. De Graaff: "Review of models for bipolar transistors", Process and Device Modelling for Integrated Circuit Design (F. van de Wiele, et al., Eds.), Leyden, The Netherlands, Noordhoff, pp. 283-306.

(30) F.H. De la Moneda: "Threshold voltage from numerical solution of the two-dimensional MOS transistor", IEEE Trans. Circuit Theory, Vol. CT-20, pp. 666-673, 1973.

(31) H. DeMan, et al.: "Mixed mode circuit simulation techniques and their implementation in DIANA", Proc. NATO Advanced Study Institute, SOGESTA, Urbino, Italy, August 1980.

(32) H. DeMan: "Computer aided design for integrated circuits: Trying to bridge the gap", IEEE J. Solid-State Circuits, Vol. SC-14, pp. 613-621, June 1979.

(33) H. DeMan, et al.: "DIANA: Mixed mode simulator with a hardware description language for hierarchical design of PLAs", Proc. IEEE ICCC'80, Rye, N.Y., p. 356-360, October 1980.

(34) H. DeMan, et al.: "Mixed-mode simulation techniques and their implementation in DIANA", in Computer Design Aids for VLSI Circuits (P. Antognetti, et al., eds.), Noordhoff, Holland, 1981.

(35) G. De Micheli and A.L. Sangiovanni-Vincentelli: "Numerical algorithms for the timing analysis of VLSI circuits", Proc. 1981 European Conference on Circuit Theory and Design, August 1981.

(36) G. De Micheli, et al.: "New algorithms for timing analysis of large circuits", Proc. IEEE International Symposium Circuits and Systems, 1980.

(37) G. De Micheli and A. Sangiovanni-Vincentelli: "Numerical properties of algorithms for analysis of MOS VLSI circuits", Proc. European Conf. Circuit Theory and Systems, August 1981.

(38) R.W. Dutton, et al.: "Correlation of fabrication process and electrical device parameter variations", IEEE J. Solid-State Circuits", Vol. SC-12, pp. 349-355, August 1977.

(39) J.M. Early: "Effects of space-charge layer widening in junction transistors", Proc. IRE, Vol. 40, p. 1401, 1952.

(40) J.J. Ebers and J.L. Moll: "Large signal behaviour of junction transistors", Proc. IRE, Vol. 42, p. 1761, 1954.

(41) Y.A. El-Mansy and A.R. Boothroyd: "A simple two-dimensional model for IGFET operation in the saturation region", IEEE Trans. Electron Devices, Vol. ED-24, pp. 254-262, 1977.

(42) W.L. Engl, et al.: "Device modelling", Proc. IEEE, Vol. 71, No. 1, pp 10-33, January 1983.

(43) W.L. Engl, et al.: "MEDUSA -- A simulator for modular circuits", IEEE Trans. CAD, Vol. CAD-1, No. 2, pp. 85-93, April 1982.

(44) S.P. Fan, et al.: "MOTIS-C: A new circuit simulator for MOS LSI circuits", Proc. IEEE ISCAS, pp. 700-703, April 1977.

(45) R. Fletcher and M.J.D. Powell: "A rapid convergence descent method for minimization", Comput. J., Vol. 6, pp. 163-168, June 1963.

(46) Y. Fujinami, et al.: "A logic simulation system for MOS LSI circuits with bidirectional elements", Monograph of Technical Group on Circuits and Systems of Inst. Electronic Communication Eng. Japan, February 1982.

(47) T. Fujisawa and E.S. Kuh: "Piecewise-linear theory of nonlinear networks", SIAM J. Applied Math, Vol. 22, No. 2, March 1972.

(48) T. Fukazawa, et al.: "Logic simulator by data flow machine", Monograph of Technical Group on Design Technology of Electronics Equipment of Information Processing Society Japan, 6-3, October 1980 (Japanese).

(49) M. Fukuma and Y. Okuto: "Analysis of short channel MOSFETs in saturation", Monograph of Technical Group of IECE Japan, SSD78-6, 1978 (Japanese).

(50) R.I. Gardner and P.B. Weil: "Hierarchical modelling and simulation in VISTA", Proc. Sixteenth Design Conference, pp. 403-405, June 1979.

(51) C.W. Gear: "The automatic integration of stiff ODEs", Proc. IFIPS Congress, pp. A81-A85, 1968.

(52) J.A. George and J.W.H. Liu: "A quotient graph model for symmetric factorization", SIAM Sparse Matrix Proc. (I.S. Duff and G.W. Stewart, Eds.), pp. 154-175, 1978.

(53) J.A. George, et al.: User Guide for SPARSEPAK, Dept. of Computer Science, University of Waterloo, Waterloo, Ontario, Canada, June 1979.

(54) J.A. George and J.W.H. Liu: "A fast implementation of the minimum degree algorithm using quotient graphs", ACM TOMS, Vol. 6, No. 3, pp. 337-358, September 1980.

(55) J.A. George and F.G. Gustavson: "A new proof on permuting to block triangular form", IBM RC Report 8238, 10 pp., April 1980.

(56) J.A. George: "Nested dissection of a regular finite element mesh", SIAM J. Numerical Analysis, Vol. 11, pp. 345-363, 1974.

(57) J.A. George: "An automatic one-way nested dissection algorithm for irregular finite element problems", SIAM J. Numerical Analysis, Vol. 17, pp. 740-751, 1980.

(58) G. Guardabassi and A. Sangiovanni-Vincentelli: "A two-level algorithm by tearing", IEEE Trans. Circuits and Systems, Vol. CAS-23, pp. 783-791, 1976.

(59) G. Guardabassi: "A note on minimal essential sets", IEEE Trans. Circuit Theory, Vol. CT-18, pp. 557-560, 1971.

(60) H.K. Gummel and H.C. Poon: "An integral charge control model of bipolar transistors", Bell Systems Technology J., Vol. 49, pp. 827-852, May-June 1970.

(61) H.K. Gummel: "A self-consistent iterative scheme for one-dimensional steady-state transistor calculation", IEEE Trans. Electronic Devices, Vol. ED-11, pp. 455-465, October 1964.

(62) H.K. Gummel: "A charge control relation for bipolar transistors", Bell Systems Technical J., Vol. 49, p. 115, 1970.

(63) F.G. Gustavson: "Some basic techniques for solving sparse systems of linear equations", Sparse Matrices and Their Applications (D.J. Rose and R.A. Willoughby, Eds.), New York, Plenum Press, 1971.

(64) F.G. Gustavson: "Finding the lower triangular form of a sparse matrix", Sparse Matrix Computations (J. Bunch and D. Rose, Eds.), New York, Plenum Press, 1976.

(65) G.D. Hachtel, et al.: "Semiconductor analysis using finite elements. Part I. Computational aspects", IBM J. Research and Development, July 1981.

(66) G.D. Hachtel, et al.: "The sparse tableau approach to network analysis and design", IEEE Trans. Circuit Theory, Vol. CT-18, pp. 101-113, January 1971.

(67) G.D. Hachtel and A.L. Sangiovanni-Vincentelli: "A survey of third generation simulation techniques", Proc. IEEE, Vol. 69, No. 10, pp. 1264-1280, October 1981.

(68) I.N. Hajj: "Sparsity considerations in network solution by tearing", IEEE Trans. Circuits and Systems, Vol. CAS-27, pp. 357-366, May 1980.

(69) I.N. Hajj: "Solution of interconnected subsystems", Electronic Letter, Vol. 13, No. 3, pp. 78-79, February 1977.

(70) K. Hayamizu and T. Nanya: "An automatic design system for PLA", Proc. National Conference Information Processing Society Japan, 5G-4, 1978 (Japanese).

(71) E. Hellerman and D. Rarick: "Reinversion with the preassigned pivot procedure", Math. Programming, Vol. 1, pp. 195-216, 1971.

(72) E. Hellerman and D. Rarick: "The partitioned preassigned pivot procedure (P4)", Sparse Matrices and Their Applications (D.J. Rose and R.A. Willoughby, Eds.), New York, Plenum Press, 1972.

(73) J. Hennessy: "SLIM, a simulation and implementation language for VLSI microcode", Lambda, Vol. 2, No. 1, pp. 20-30, 1981.

(74) D.D. Hill and W.M. vanCleemput: "SABLE: A tool for generating structured, multi-level simulations", Proc. Sixteenth Design Automation Conference, pp. 272-279, June 1979.

(75) D.D. Hill and W.M. vanCleemput: "SABLE: Multi-level simulation for hierarchical design", Proc. IEEE ISCAS, Houston, pp. 431-434, April 1980.

(76) D.D. Hill: "Language and environment for multi-level simulation", Ph.D. dissertation, Stanford University, Stanford, CA, 1980.

(77) K. Hirabayashi and J. Watanabe: "MATIS -- Macromodel timing simulator for large scale integrated MOS circuits", Proc. Third USA-Japan Computer Conference, 26-1-1, 1978.

(78) S.S. Hirschorn, et al.: "Functional level simulation in FANSIM 3", Proc. European Conference Electronic Design Automation, pp. 193-197, September 1981.

(79) C. Ho, et al.: "The modified nodal approach to network analysis", IEEE Trans. Circuits and Systems, Vol. CAS-25, pp. 504-509, June 1975.

(80) B. Hoeneisen and C.A. Mead: "Current-voltage characteristics of small size MOS transistors", IEEE Trans. Electronic Devices (Correspondence), Vol. ED-19, pp. 382-383, 1972.

(81) B. Hofflinger, et al.: "Model and performance of hot-electron MOS transistors for VLSI", IEEE Trans. Electronic Devices, Vol. ED-26, pp. 513-520, 1979.

(82) A.J. Hoffman, et al.: "Complexity bounds for regular finite difference and finite element grids", SIAM J. Numerical Analysis, Vol. 10, pp. 364-369, 1973.

(83) M.Y. Hsueh, et al.: "The development of macromodels for MOS timing simulation", Proc. IEEE ISCAS, pp. 1-4, April 1978.

(84) M. Ishii, et al.: "Automatic input and interactive editing systems of logic circuit diagrams", Proc. Eighteenth Design Automation Conference, pp. 639-645, June 1981.

(85) M. Itoh, et al.: "A logic design automation system", Monograph of Technical Group on Design Technology of Electronics Equipment of Information Processing Society Japan, 4-3, March 1980 (Japanese).

(86) E.D. Johnson, et al.: "Transient radiation analysis by computer program (TRAC)", in User's Guide, Harry Diamond Labs, Autonetics Division, North American Rockwell, Anaheim, CA, Vol. 1, 1968.

(87) N.P. Jouppi: "TV: An NMOS timing analyzer", Proc. Third Caltech Conference on VLSI, pp. 71-86, Computer Science Press, Rockville, Md., 1983.

(88) R. Kasai, et al.: "Threshold voltage analysis of short and narrow channel MOSFETs by three-dimensional computer simulation", IEEE Trans. Electronic Devices, to be published.

(89) J. Katzenelson: "An algorithm for solving nonlinear resistive networks", Bell Systems Technology J., Vol. 44, pp. 1605-1620, 1965.

(90) K. Kawakita and T. Ohtsuki: "NECTAR-2 -- Circuit analysis program based on piecewise linear approach", Proc. IEEE International Conference Circuits and Systems, pp. 46-51, 1975.

(91) N. Kawato, et al.: "Design and verification of large-scale computers by using DDL", Proc. Sixteenth Design Automation Conference, pp. 360-366, June 1979.

(92) A.K. Kevorkian and J. Snoek: "Decomposition in large scale systems: Theory and applications in solving large sets of nonlinear simultaneous equations", Decomposition of Large Scale Problems (D.M. Himmemblau, Ed.), Amsterdam, The Netherlands, North Holland, 1973.

(93) A.K. Kevorkian: "A decompositional algorithm for the solution of large systems of linear algebraic equations", Proc. IEEE International Symposium on Circuits and Systems, pp. 116-120, 1975.

(94) F.M. Klaassen and C.W.J. de Groot: "Modeling of scaled-down MOS transistors", Solid-State Electronics, Vol. 23, pp. 237-242, 1980.

(95) T. Kojima and K. Watanabe: "CARD (computer assisted research and development) for electrical circuit", Fujitsu, Vol. 24, No. 7, pp. 175-189, 1973 (Japanese).

(96) M. Konaka, et al.: "Development of an MOS device simulator (MODEST) and some related problems in the two-dimensional numerical calculation", Inst. Electronic Commun. Eng. Japan, Technical Report SSD81-10, 1981 (Japanese).

(97) H.E. Krohn: "Vector coding techniques for high speed digital simulation", Proc. Eighteenth Design Automation Conference, pp. 525-529, June 1981.

(98) T. Kurobe, et al.: "Logic simulation system: LOGOS 2", Monograph of Technical Group on Design Automation of Information Processing Society Japan, DA31-2, 1977 (Japanese).

(99) E. Lawler: Combinatorial Optimization, Networks and Matroids, New York, Holt, Rinehart and Winston, 1976.

(100) C. leFaou and J.C. Renaud: "Functional simulation of complex electronic circuits", ESSCIRC'77, September 1977.

(101) E. Lelarasmee and A.L.M. Sangiovanni-Vincentelli: "RELAX: A new circuit simulator for large scale MOS integrated circuits", Proc. Nineteenth Design Automation Conference, pp. 682-690, 1982.

(102) Y.H. Levendel, et al.: "Accurate logic simulation models for TTL totempole and MOS gates and tristate devices", BSTJ, Vol. 60, No. 7, pp. 1271-1287, September 1981.

(103) F.A. Lindholm and D.J. Hamilton: "Incorporation of the early effect in the Ebers-Moll model", Proc. IRE, Vol. 59, p. 1377, 1971.

(104) J.G. Linvill: "Lumped models of transistors and diodes", Proc. IRE, Vol. 46, p. 1141, 1958.

(105) B. Magnahagen: "DIGSIM II, An in-circuit simulator", Proc. IEEE ICCC'80, Rye, N.Y., 1980.

(106) H. Masuda, et al.: "Characteristics and limitation of scaled-down MOSFETs due to two-dimensional field effect", IEEE Trans. Electron Devices, Vol. ED-26, pp. 980-986, 1979.

(107) J.A. Meijerink and H.A. van der Vorst: "An interactive solution method for linear systems of which the coefficient matrix is a symmetrical M-matrix", Math. Comput., Vol. 31, pp. 148-162, January 1977.

(108) G. Merckel: "A simple model of the threshold voltage of short and narrow channel MOSFETs", Solid-State Electronics, Vol. 23, pp. 1207-1213, 1980.

(109) J. Mermet, et al.: "Functional simulation", in Computer Design Aids for VLSI Circuits (P. Antognetti, et al., eds.), Noordhoff, Holland, 1981.

(110) N. Miyahara, et al.: "Timing simulator (LOTAS)", Monograph of Technical Group on Electronic Devices of Inst. Electronic Communication Eng. Japan, EDD82-13, pp. 61-69, 1982 (Japanese).

(111) H. Miyashita, et al.: "PLACAD: PLA design automation system", Proc. National Conference Inst. Electronic Communication Eng. Japan, 410, 1981 (Japanese).

(112) M.F. Moad: "A sequential method of network analysis", IEEE Trans. Circuit Theory, Vol. CT-17, pp. 99-104, February 1970.

(113) M.S. Mock: "A two-dimensional mathematical model of the insulated-gate field effect transistor", Solid-State Electronics, Vol. 16, pp. 601-609.

(114) J.L. Moll: "Large-signal transient response of a junction transistor", Proc. IRE, Vol. 42, p. 1773, 1954.

(115) M. Murakami, et al.: "Logic verification and test generation for LSI circuits", Proc. Test Conference, pp. 467-472, November 1980.

(116) L.W. Nagel: "SPICE2: A computer program to simulate semiconductor circuits", ERL Memo ERL-M520, University of California, Berkeley, May 1975.

(117) L.W. Nagel: "ADVICE for circuit simulation", Proc. ISCAS, Houston, Texas, April 1980.

(118) A.R. Neureuther: "IC Process Modeling and Topography Design", Proc. IEEE, Vol. 71, No. 1, pp. 121-128, January 1983.

(119) A.R. Newton: "Timing, logic and mixed mode simulation for large MOS integrated circuits", Proc. NATO Advanced Study Institute on Computer Design Aids for VLSI Circuits, Urbino, Italy, August 1980.

(120) A.R. Newton: "Timing, logic and mixed-mode simulation for large MOS integrated circuits", in Computer Design Aids for VLSI Circuits (P. Antognetti, et al., eds.), Noordhoff, Holland, 1981.

(121) A.R. Newton: "The simulation of large-scale integrated circuits", IEEE Trans. CAS, Vol. CAS-26, No. 9, pp. 741-749, September 1979.

(122) A.R. Newton: "Techniques for the simulation of large-scale integrated circuits", IEEE Trans. on Circuits and Systems, Vol. CAS-26, No. 9, pp. 741-749, 1979.

(123) H.N. Nham and A.K. Bose: "A multiple delay simulator for MOS LSI circuits", Proc. Eighteenth Design Automation Conference, pp. 610-617, 1980.

(124) Y. Ohno, et al.: "Logic verification system for very large computers using LSI's", Proc. Sixteenth Design Automation Conference, pp. 367-374, June 1979.

(125) Y. Ohno and Y. Okuto: "Computer aided Si-MOSFET process designing", Proc. Eleventh Conference Solid State Devices, Tokyo, Japan, Vol. 19, sup. 19-1, pp. 65-69, 1980.

(126) T. Ohtsuki, et al.: "Existence theorems and a solution algorithm for piecewise-linear resistor networks", SIAM J. Mathematical Analysis, Vol. 8, pp. 69-99, 1977.

(127) H. Oka, et al.: "Computer analysis of a short-channel BC MOSFET", IEEE Trans. Electronic Devices, Vol. ED-27, pp. 1514-1520, August 1980.

(128) Y. Okuto, et al.: "Device process designing aids for MOS LSIs", Proc. Second NASECODE Conference, 1981.

(129) J.M. Ortega and W.C. Rheinboldt: Iterative Solution of Nonlinear Equations in Several Variables, New York, Academic Press, 1970.

(130) J.K. Ousterhout: "Crystal: A timing analyzer for NMOS VLSI circuits", Proc. Third Caltech Conference on VLSI, pp. 57-70, Computer Science Press, Rockville, Md., 1983.

(131) E. Polak and Teodoru: "Newton derived methods for nonlinear equations and inequalities", in Nonlinear Programming, 2, Academic Press, 1975.

(132) E. Polak: "On the global convergent stabilization of locally convergent algorithms", Automatica, Vol. 12, pp. 337-342, 1976.

(133) H.C. Poon: "Modelling of bipolar transistor using integral charge control model with application to third-order distortion studies", IEEE Trans. Electronic Devices, Vol. ED-19, p. 719, 1972.

(134) H.C. Poon and J.C. Meckwood: "Modelling of avalanche effect in integral charge control model", IEEE Trans. Electronic Devices, Vol. ED-19, p. 90, 1972.

(135) H.C. Poon and H.K. Gummel: "Modelling of emitter capacitance", Proc. IEEE, Vol. 57, p. 2181, 1969.

(136) A. Popa: "An injection level dependent theory of the MOS transistor in saturation", IEEE Trans. Electronic Devices, Vol. ED-19, pp. 774-781, 1972.

(137) N.B. Rabbat, et al.: "A multilevel Newton algorithm with macromodelling and latency for analysis of large-scale nonlinear networks in the time domain", IEEE Trans. CAS, Vol. CAS-26, No. 9, September 1979.

(138) P.G. Raeth, et al.: "Functional modelling for logic simulation", Proc. Eighteenth Design Automation Conference, pp. 791-795, June 1981.

(139) A.E. Ruehli, et al.: "Time analysis of large scale circuits using one-way macromodels", Proc. IEEE International Symposium Circuits and Systems, pp. 766-770, 1980.

(140) A.E. Ruehli and G.S. Ditlow: "Circuit analysis, logic simulation, and design verification for VLSI", Proc. IEEE, Vol. 71, No. 1, January 1983.

(141) A.E. Ruehli: "Survey of analysis, simulation and modelling for large scale logic circuits", Proc. Eighteenth Design Automation Conference, pp. 124-129, June 1981.

(142) A. Ruehli, et al.: "Macromodelling: An approach for analyzing large-scale circuits", Computer Aided Design, Vol. 10, pp. 121-130, March 1978.

(143) A.R. Ruehli, et al.: "Time analysis of large-scale circuits containing one-way macromodels", IEEE Trans. CAS, Vol. CAS-29, No. 3, pp. 185-190, March 1982.

(144) K. Sakalla and S.W. Director: "An activity directed circuit simulation algorithm", Proc. IEEE ICCC'80, Rye, N.Y., pp. 1032-1035, October 1980.

(145) A. Sangiovanni-Vincentelli, et al.: "A new tearing approach Node-tearing nodal analysis", Proc. IEEE Symposium on Circuits and Systems, Phoenix, AZ, pp. 143-147, April 1977.

(146) A. Sangiovanni-Vincentelli and T.A. Bickart: "Bipartite graphs and an optimal bordered triangular form of a matrix", IEEE Trans. Circuits and Systems, Vol. CAS-26, No. 10, pp. 880-890, October 1979.

(147) A.L. Sangiovanni-Vincentelli: "Circuit simulation", NATO Advanced Institute on Computer Aids for VLSI Circuits, SOGESTA, Urbino, Italy, July 1980.

559

(148) A. Sangiovanni-Vincentelli and N.B.G. Rabbat: "Techniques for the time domain analysis of VLSI circuits", Proc. IEE, Part G, Vol. 127, pp. 292-301, December 1980.

(149) W. Sansen, et al.: "Computer design aids for VLSI circuits", NATO Advanced Institute on Computer Aids for VLSI Circuits", SOGESTA, Urbino, Italy, July 1980.

(150) T. Sasaki, et al.: "MIXS: A mixed level simulator for large digital system logic verification", Proc. Seventeenth Design Automation Conference, pp. 626-633, 1980.

(151) B. Schendel, et al.: "SAMSON -- Synchronous and asynchronous mixed simulator of digital networks", Proc. European Conference Electronic Design Automation, pp. 188-192, September 1981.

(152) S. Selberherr, et al.: "MINIMOS -- A program package to facilitate MOS design and analysis", Numerical Analysis of Semiconductor Devices (B.T. Brown and J.J.H. Miller, Eds.), Dublin, Ireland, Boole Press, 1979.

(153) J. Sherman and W.J. Morrison: "Adjustment of an inverse matrix corresponding to changes in the elements of a given column or row of the original matrix", Annual Math. Statistics, Vol. 20, p. 621, 1949.

(154) W. Sherwood: "A Hybrid scheduling technique for hierarchical logic simulators of 'close encounters of the simulated kind'", Proc. Sixteenth Design Automation Conference, pp. 249-254, San Diego, CA, 1979.

(155) W. Sherwood: "A MOS modeling technique for 4-state true-value hierarchical logic simulation", Proc. Eighteenth Design Automation Conference, pp. 775-785, 1981.

(156) T. Shikage and H. Sato: "A circuit simulation method based on macromodelling of circuit blocks", Proc. National Conference Semiconductors of Inst. Electronic Communication Eng. Japan, p. 62, 1979 (Japanese).

(157) W. Shockley: "The theory of p-n junction in semiconductors junction transistors", Bell Systems Technical J., Vol. 28, p. 435, 1949.

(158) H.L. Stone: "Iterative solution of implicit approximations of multidimensional partial differential equations", SIAM J. Numerical Analysis, Vol. 5, pp. 530-558, September 1968.

(159) V. Strassen: "Gaussian elimination is not optimal", Numerical Mathematics, Vol. 13, pp. 354-356, 1969.

(160) M. Sugimori: "DEMOS E circuit analysis program: ECSS", Proc. National Conference Inst. Electronic Communication Eng. Japan, p. 1784, 1974 (Japanese).

(161) S.A. Szygenda and E.W. Thompson: "Digital logic simulation in a time-based, table-driven environment. Part I. Design Verification", IEEE Trans. Computers, pp. 24-36, March 1975.

(162) S.A. Szygenda and E.W. Thompson: "Modelling and digital simulation for design verification and analysis", IEEE Trans. Computers, Vol. C-25, pp. 1242-1253, December 1976.

(163) N. Tanabe, et al.: "MOSTAP: An MOS circuit simulator for LSI circuits", Proc. IEEE International Conference on Circuits and Systems, pp. 1035-1038, 1980.

(164) N. Tanabe, et al.: "Timing simulation: A fast analysis method", Proc. National Conference Inst. Electronic Communication Eng. Japan, 450, p. 215, 1981 (Japanese).

(165) K. Taniguchi, et al.: "Two-dimensional computer simulation models for MOS LSI fabrication process", IEEE Trans. Electronic Devices, Vol. ED-28, May 1981.

(166) R.E. Tarjan: "Depth first search and linear graph algorithms", SIAM J. Comput., Vol. 1, pp. 146-160, 1972.

(167) G.W. Taylor: "The effects of two-dimensional charge sharing on the above-threshold characteristics of short-channel IGFETs", Solid-State Electronics, Vol. 22, pp. 701-717, 1979.

(168) M. Tokoro, et al.: "A module level simulation technique for systems composed of LSIs and MSIs", Proc. Fifteenth Design Automation Conference, pp. 418-427, June 1978.

(169) M. Tomizawa, et al.: "Two-dimensional device simulator for gate level characterization", Solid-State Electron.

(170) M. Tomizawa, et al.: "An accurate design method of bipolar device using a two-dimensional device simulator", IEEE Trans. Electronic Devices, Vol. ED-28, pp. 1148-1153, October 1981.

(171) T. Toyabe, et al.: "A two-dimensional analysis of I^2L with multistream function technique", Proc. NASECODE First Conference, pp. 290-292, 1979.

(172) T. Toyabe and S. Asai: "Analytical models of threshold voltage and breakdown voltage of short-channel MOSFETs derived from two-dimensional analysis", IEEE Trans. Electronic Devices, Vol. ED-26, pp. 453-461, 1979.

(173) T. Toyabe and S. Asai: "Analytical models of threshold voltage of short-channel MOSFETs derived from two-dimensional analysis", IEEE J. Solid-State Circuits, Vol. SC-14, pp. 375-383, April 1979.

(174) F.N. Trofimenkoff: "Field-dependent mobility analysis of the field-effect transistor", Proc. IEEE (Correspondence), Vol. 53, pp. 1765-1766, 1965.

(175) P.M. Trouborst and J.A.G. Jess: "Macromodelling by systematic code reduction", IEEE ICCC'80 Conference Proc., Rye, N.Y., pp. 337-340, October 1980.

(176) M. Tsuboya, et al.: "PLA design support system", Proc. National Conference Information Processing Society Japan, 5G-5, 1978 (Japanese).

(177) E.G. Ulrich: "Time sequenced logical simulation based on circuit delay and selective tracing of active network path", Proc. ACM Twentieth National Conference, pp. 437-448, 1965.

(178) E.G. Ulrich: "Exclusive simulation of activity in digital networks", Communication ACM, Vol. 12, No. 2, pp. 102-110, February 1969.

(179) W.M.G. vanBokhoven: "Macromodelling and simulation of mixed analog-digital networks by piecewise-linear system approach", Proc. IEEE ICCC'80, pp. 361-365, October 1980.

(180) A. Vladimirescu and D. Pederson: "A computer program for the simulation of large scale integrated circuits", Proc. IEEE ISCAS, pp. 111-113, April 1981.

(181) J. Watanabe: "Macromodel timing simulator -- MATIS", Proc. National Conference Semiconductors of Inst. Electronic Communication Eng. Japan, p. 63, 1979 (Japanese).

(182) J. Watanabe, et al.: "Seven value logic simulation for MOS LSI circuits", Proc. IEEE International Conference Circuits and Computers, pp. 941-944, October 1980.

(183) W.T. Weeks, et al.: "Algorithms for ASTAP -- A network analysis program", IEEE Trans. Circuit Theory, Vol. CT-20, pp. 628-634, November 1973.

(184) A. Westerberg: "La Scala -- A programming package for the solution of linear systems", Proc. SIAM National Meeting, Knoxville, TN, 1979.

(185) A. Westerberg and T. Berna: "Decomposition of very large-scale Newton-Raphson based flowsheeting problems", Computer Chem. Eng., Vol. 2, No. 1, pp. 61-63, 1978.

(186) C.L. Wilson and J.L. Blue: "Two-dimensional finite element charge-sheet model of a short-channel MOS transistor", Solid-State Electronics, Vol. 25, pp. 461-477, 1982.

(187) F.F. Wu: "Solution of large-scale networks by tearing", IEEE Trans. Circuits and Systems, Vol. CAS-23, pp. 706-713, December 1976.

(188) F.F. Wu: "Diakoptic network analysis", Proc. IEEE Power Industry Computer Applications (PICA), New Orleans, LA, pp. 364-371, June 1975.

(189) J. Yamada, et al.: "A large-scale circuit simulator (LOTAS)", Proc. National Conference Semiconductors of Inst. Electronic Communication Eng. Japan, p. 65, 1979 (Japanese).

(190) K. Yamaguchi, et al.: "Two-dimensional analysis of stability criteria of GaAs FET's", IEEE Trans. Electronic Devices, Vol. ED-23, pp. 1283-1290, December 1976.

(191) P. Yang, et al.: "SLATE: A circuit simulation program for the simulation of large scale integrated circuits", Proc. IEEE, ICCC'80, pp. 353-355, October 1980.

(192) L.D. Yau: "A simple theory to predict the threshold voltage of short-channel IGFETs", Solid-State Electronics, Vol. 17, pp. 1059-1063, 1974.

(193) K. Yokoyama, et al.: "Application of Fletcher-Powell's optimization method to process/device simulation of MOSFET characteristics", Solid-State Electronics, Vol. 25, pp. 201-203, March 1982.

(194) K. Yokoyama, et al.: "Threshold sensitivity minimization of short-channel MOSFETs by computer simulation", IEEE Trans. Electronic Devices, Vol. ED-27, pp. 1509-1514, August 1980.

(195) A. Yoshii, et al.: "A three-dimensional analysis of semiconductor devices", IEEE Trans. Electronic Devices, Vol. ED-29, pp. 184-189, February 1982.

(196) ----: "On block elimination for sparse linear systems", SIAM J. Numerical Analysis, pp. 585-603, 1974.

(197) ----: "An efficient heuristic cluster algorithm for tearing large-scale networks", IEEE Trans. Circuits and Systems, Vol. CAS-24, pp. 709-717, December 1977.

(198) ----: "On bordered triangular or lower N forms of an irreducible matrix", IEEE Trans. Circuits and Systems, Vol. CAS-23, pp. 621-624, 1976.

Testing

(1) J. Abadir and Y. Deswarbe: "Run-time program for self-checking single board computer", International Test Conference, Philadelphia, IEEE Cat.No. 82CH1808-5, pp. 205-213, 1982.

(2) M.I. Aboulhamid and E. Cerny: "A class of test generation for built-in testing", IEEE Trans. Computers, Vol. C-32, No. 10, pp. 957-959, October 1983.

(3) J.A. Abraham: "Design for testability", Digest of papers of the 1983 Custom Integrated Circuits Conference, IEEE Pub.83CH1859-8, pp. 278-283, 1983.

(4) J.A. Abraham and D.D. Gajski: "Design of testable structures described by simple loops", IEEE Trans. Computers, Vol. C-30, No. 11, pp. 875-884, November 1981.

(5) J.A. Abraham: "Functional level test generation for complex digital systems", Digest of Papers of the International Test Conference, pp. 461-462, October 1981.

(6) C. Acken: "A matrix of digital systems and its use in automatic test pattern generation", Sandia Report Sand82-0792, May 1982.

(7) V.D. Agrawal and M.R. Mercer: "Testability measures -- what do they tell us?", IEEE Proc. International Test Conference, pp. 391-396, November 1982.

(8) V.D. Agrawal and S.K. Jain: "Test generation for MOS circuits using D-algorithm", Proc. Twentieth Design Automation Conference, Miami Beach, pp. 64-70, June 1983.

(9) V.D. Agrawal and E. Cerny: "Store and generate built-in testing approach", Proc. Eleventh International Symposium on Fault-Tolerant Computing, pp. 35-40, June 1980.

(10) V.D. Agrawal, et al.: "Automation in design for testability", Proc. IEEE Custom IC Conference, Rochester, N.Y., IEEE Pub.84CH1987-7, pp. 159-163, May 1984.

(11) V.D. Agrawal: "Multiple fault testing of large circuits by single fault test sets", IEEE Trans. Computers, Vol. C-30, No. 11, pp. 855-865, November 1981.

(12) H. Ando: "Testing VLSI with random access scan", COMPCON'80, Digest of Papers, IEEE Pub. 80CH1491-OC, pp. 50-52, February 1980.

(13) P. Banerjee and J. Abraham: "Fault characterization of VLSI MOS circuits", Proc. International Conference on Circuits and Computers, New York, pp. 564-568, October 1982.

(14) P.H. Bardell and W.H. McAnney: "Self-testing of multi-chip modules", Proc. International Test Conference, Philadelphia, IEEE Cat.No. 82CH1808-5, pp. 200-204, 1982.

(15) Z. Barzilai, et al.: "The weighted syndrome sums approach to VLSI testing", IEEE Trans. Computers, Vol. C-30, No. 12, pp. 996-1000, December 1981.

(16) A. Bhattacharyya: "On a novel approach of fault detection in an easily testable sequential machine with extra inputs and extra outputs", IEEE Trans. Computers, Vol. C-32, No. 3, pp. 323-325, March 1983.

(17) D.D. Bhavsar and R.W. Heckelman: "Self-testing by polynomial division", Proc. International Test Conference, October 1981.

(18) A. Blum: "VLSI checking", IBM Technical Disclosure Bulletin, Vol. 23, No. 2, pp. 647-648, July 1980.

(19) A. Bose, et al.: "A fault simulator for MOS LSI circuits", Proc. Nineteenth Design Automation Conference, July 1982.

(20) S Bozorgui-Nesbat and E.J. McCluskey: "Structured design for testability to eliminate test pattern generation", Proc. International Fault-Tolerant Computing Symposium, pp. 158-163, October 1980.

(21) M. Brashler, et al.: "An integrated test development system", Proc. IEEE Custom IC Conference, Rochester, N.Y., IEEE Pub.1987-7, pp. 169-171, May 1984.

(22) M.A. Breuer (Ed.): Diagnosis and Reliable Design of Digital Systems, Computer Science Press, Rockville, 300p, 1976.

(23) F. Brglez: "Testability in VLSI", Technical Digest of the 1983 Canadian Conference on VLSI, Waterloo, Ontario, pp. 90-95, October 1983.

(24) R.E. Bryant and M.D. Schuster: "Fault simulation of MOS digital circuits", VLSI Design, Vol. IV, No. 6, pp. 24-30, October 1983.

(25) M.E. Buehler: "Comprehensive test patterns with modular test structures: The 2 by N probe pad array approach", Solid State Technology, Vol. 22, No. 10, pp. 89-94, October 1979.

(26) M.E. Buehler, et al.: "Role of test chips in coordinating logic and circuit design and layout aids of VLSI", Solid State Tech., Vol. 24, No. 9, pp. 68-74, September 1981.

(27) W.C. Carter: "Signature testing with guaranteed bounds for fault coverage", International Test Conference, Philadelphia, IEEE Cat.No. 82CH1808-5, pp. 75-82, 1982.

(28) H.Y. Chang, et al.: Fault Diagnosis of Digital Systems, Wiley-Interscience, New York, 1970.

(29) Y. Crozet and C. Landrault: "Design of self-checking MOS-LSI circuits: Applications to a four-bit microprocessor", IEEE Trans. Computers, Vol. C-29, No. 6, pp. 532-537, June 1980.

(30) W. Daehn and J. Mucha: "A hardware approach to self-testing of large programmable logic arrays", IEEE Trans. Computers, Vol. C-30, No. 11, pp. 829-833, November 1981.

(31) S Das Gupta, et al.: "A variation of LSSD and its implications on design and test pattern generation in VLSI", International Test Conference, Philadelphia, IEEE Cat.No. 82CH1808-5, pp. 63-66, 1982.

(32) R David: "Testing by feedback shift register", IEEE Trans. Computers, Vol. C-29, No. 9, pp. 668-673, July 1980.

(33) A.P. Duncan: "The impact of microprocessors on production-line testing: A problem which can provide its own solution", Microprocessors in Automation and Communications, Second International Conference on Microprocessors in Automation and Communication, London, England, pp. 253-263, January 1981.

(34) E.B. Eichelberger and E. Lindbloom: "Random-pattern coverage enhancement and diagnosis for LSSD logic self-test", IBM J. Research and Development, Vol. 27, No. 3, pp. 265-272, May 1983.

(35) E.B. Eichelberger and T.W. Williams: "A logic design structure for LSI testability", Proc. Fourteenth Design Automation Conference, pp. 464-468, June 1977.

(36) H. Eiki, et al.: "Autonomous testing and its applications to testable design of logic circuits", Proc. International Fault-Tolerant Computing Symposium, pp. 173-178, October 1980.

(37) I.I. Eldumiati and R.N. Gadenz: "Logic and fault simulations of the DSP, a VLSI digital signal processor", Proc. IEEE Conference on Circuits and Computers, Port Chester, N.Y., pp. 948-952, October 1980.

(38) Y. El-zig: "Classifying, testing and eliminating VLSI MOS failures", VLSI Design, Vol. IV, No. 5, pp. 30-35, September 1983.

(39) Y. El-zig and R. Cloutier: "Functional-level test generation for stuck-open faults in CMOS VLSI", Proc. IEEE Test Conference, Philadelphia, PA, pp. 536-546, October 1983.

(40) Y. El-zig: "Fault diagnosis of MOS combinational circuits", IEEE Trans. Computers, Vol. C-31, No. 2, pp. 129-139, February 1982.

(41) Y. El-zig: "Automatic test generation for stuck-open faults in CMOS VLSI", Proc. Eighteenth Design Automation Conference, Nashville, June 1981.

(42) W. Engl, et al.: "Self-testing ICs slash overall testing costs", Electronic Design, Vol. 31, No. 12, pp. 195-198, June 1983.

(43) E.H. Frank and R.F. Sproull: "Testing and debugging custom integrated circuits", ACM Computing Surveys, Vol. 13, pp. 425-451, December 1981.

(44) A.D. Friedman and P.R. Menon: Fault Detection in Digital Circuits, Prentice-Hall, Englewood Cliffs, 220p. 1971.

(45) H. Fujiwara and K. Kinoshita: "A design of programmable logic arrays with universal tests", IEEE Trans. Computers, Vol. C-30, No. 11, pp. 823-828, November 1981.

(46) H. Fujiwara and S. Toida: "The complexity of fault-detection problems for combinational circuits", IEEE Trans. Computers, Vol. C-31, No. 6, pp. 555-560, June 1982.

(47) H. Fujiwara and S. Toida: "The complexity of fault detection: An approach to design for testability", Proc. Twelfth Annual Fault-Tolerant Computing Symposium, pp. 101-108, June 1982.

(48) H. Fujiwara and K. Kinoshita: "Testing logic circuits with compressed data", J. Design Automation and Fault-Tolerant Computing, Vol. 3, pp. 211-225, Winter 1979.

(49) H. Fujiwara, et al.: "Universal test sets for programmable logic arrays", FTCS-10, Digest of Papers, pp. 131-136, 1980.

(50) S Funatsu, et al.: "Test generation system in Japan", Proc. Twelfth Design Automation Conference, pp. 112-114, June 1975.

(51) S Funatsu, et al.: "DESIGN: Designing digital circuits with easily testable consideration", 1978 Semiconductor Test Conference, pp. 98-102, 1978.

(52) J. Galiay, et al.: "Physical versus logical fault models MOS LSI circuits: Impact on their testability", IEEE Trans. Computers, Vol. C-29, No. 6, pp. 527-531, June 1980.

(53) M. Gianfagna: "A unified approach to test data analysis", Proc. Fifteenth Design Automation Conference, Las Vegas, Nevada, June 1978.

(54) P. Goel: "Test generation costs analysis and projections", Proc. Seventeenth Design Automation Conference, Minneapolis, pp. 77-84, June 1980.

(55) L.H. Goldstein: "Controllability/observability analysis of digital circuits", IEEE Trans. Circuits and Systems, Vol. CAS-25, No. 9, pp. 685-693, September 1979.

(56) S. Goshima, et al.: "Diagnostic system for large scale logic cards and LSIs", Proc. Eighteenth Design Automation Conference, pp. 256-259, June 1981.

(57) J. Grason and A.W. Nagle: "Digital test generation and design for testability", Proc. Seventeenth Design Automation Conference, Minneapolis, MN, pp. 175-189, June 1980.

(58) G. Grassl and H.-J. Pleiderer: "A function-independent self-test for large programmable logic arrays", Integration the VLSI Journal, North-Holland, Vol. 1, No. 1, pp. 71-80, April 1983.

(59) W.A. Groves: "Rapid digital fault isolation with FASTRACE", Hewlett-Packard Journal, pp. 8-13, March 1979.

(60) K. Gutfreund: "Integrating the approaches to structured design for testability", VLSI Design, Vol. IV, No. 6, pp. 34-42, October 1983.

(61) J.P. Hayes: "A fault simulation methodology for VLSI", Proc. Nineteenth Design Automation Conference, Las Vegas, June 1982.

(62) J.P. Hayes: "Transition count testing of combinational logic circuits", IEEE Trans. Computers, Vol. C-25, No. 6, pp. 613-620, June 1976.

(63) J.P. Hayes: "Testing memories for single-cell pattern-sensitive faults", IEEE Trans. Computers, Vol. C-29, No. 3, pp. 249-254, March 1980.

(64) H.I. Hellman: "TESTGEN: An interactive test generation and logic simulation program", RCA Engineer, Vol. 21, No. 1, June/July 1975.

(65) F.J. Henley: "Functional testing and failure analysis of VLSI using a laser probe", Proc. IEEE Custom IC Conference, Rochester, N.Y., IEEE Pub.84CH1987-7, pp. 181-186, May 1984.

(66) F.C. Hennie: Finite State Models for Logical Machines, Wiley, New York, 466p. 1968.

(67) R.D. Hess: "Testability analysis: An alternative to structured design for testability", VLSI Design, Vol. III, No. 2, pp. 22-29, March/April 1982.

(68) S.J. Hong and D.L. Ostapko: "FITPLA: A programmable logic array for function-independent testing", FTCS-10, Digest of Papers, pp. 131-136, 1980.

(69) T. Hoshino, et al.: "A test pattern generation system by using a block sensitizing method", Monograph of Technical Group on Design Technology of Electronics Equipment of Information Processing Society, Japan, 9-3, June 1981 (Japanese).

(70) C. Huang, et al.: "Latch-up analysis in CMOS static 4K RAMs using laser scanning", Proc. IEEE Custom ICC Conference, Rochester, N.Y., IEEE Pub 84CH1987-7, pp. 176-180, May 1984.

(71) O.H. Ibarra and S.K. Sahni: "Polynomially complete fault detection problems", IEEE Trans. Computers, Vol. C-24, No. 3, pp. 242-249, March 1975.

(72) S.K. Jain and V.D. Agrawal: "Test generation for MOS circuits using D-algorithm", Twentieth Design Automation Conference, Miami Beach, pp. 64-70, June 1983.

(73) S.K. Jain and A.K. Susskind: "Test strategy for microprocessors", ACM-IEEE Twentieth Design Automation Conference, Miami Beach, pp. 703-708, June 1983.

(74) W.A. Johnson: "Behavioural-level test development", Proc. Sixteenth Design Automation Conference, San Diego, CA, pp. 171-179, June 1979.

(75) M.G. Karpovsky: "Error detection in digital devices and computer programs with the aid of linear recurrent equations over finite commutative groups", IEEE Trans. Computers, Vol. C-26, No. 3, pp. 208-218, March 1977.

(76) R.H. Katz, et al.: "An experimental design frame for VLSI circuit prototyping", VLSI Design, Vol. III, No. 3, pp. 57-58, May/June 1982.

(77) M. Kawai and J.P. Hayes: "An experimental MOS fault simulation program CSASIM", Proc. Twenty-First Design Automation Conference, pp. 2-9, 1984.

(78) S.D. Kelley: "Imbedded memory test methods", IBM Technical Disclosure Bulletin, Vol. 21, No. 12, pp. 4911-4913, May 1979.

(79) J. Khakbaz and E.J. McCluskey: "Concurrent error detection and testing for large PLAs", IEEE Trans. Electronic Devices, Vol. ED-12, No. 4, pp. 756-764, April 1982.

(80) C.R. Kime and K.K. Saluja: "Test scheduling in testable VLSI circuits", Proc. Twelfth International Symposium on Fault-Tolerant Computing, pp. 406-412, June 1982.

(81) K. Kinoshita, et al.: "Test generation for combinational circuits by structure description functions", FTCS-10, pp. 152-154, October 1980.

(82) A. Kobayashi, et al.: "Flip-flop circuit with FLT capability", Proc. National Conf. Inst. Electronic Commun. Eng. Japan, No. 892, 1968.

(83) B. Koehler: "Designing a microcontroller 'supercell' for testability", VLSI Design, Vol. IV, No. 6, pp. 44-46, October 1983.

(84) B. Koenemann, et al.: "Computer aided test for VLSI", Proc. 1984 IEEE CICC, Rochester, N.Y., IEEE Pub.84CH1987-7, pp. 172-175, May 1984.

(85) B. Koenemann, et al.: "Built-in logic block observation techniques", Digest of Papers, International Test Conference, IEEE Pub. 79CH1509-9C, pp. 37-41, October 1979.

(86) D. Komonytski: "LSI self-test using level sensitive scan design and signature analysis", International Test Conference, Philadelphia, IEEE Cat.No. 82CH1808-5, pp. 414-424, 1982.

(87) B. Konemann, et al.: "Built-in test for complex digital integrated circuits", IEEE J. Solid-State Circuits, Vol. SC-15, No. 3, pp. 315-318, June 1980.

(88) P.G. Kovijanic: "Testability analysis", Digest of Papers, Test Conference IEEE Pub. 79CH1509-9C, pp. 310-316, October 1979

(89) H. Kubo: "A procedure for generating test sequence to detect sequential circuit failures", NEC Research Development, No 12, pp 69-78, 1968.

(90) J.D. Lesser and J.J. Shedletsky: "An experimental delay test generator for LSI logic", IEEE Trans. Computers, Vol. C-29, No. 3, pp. 235-248, March 1980.

(91) T. Li and L.A. Hollaar: "On the testability of the direct implementation of asynchronous circuits", Proc. 1984 Conference on Advanced Research in VLSI, MIT, Cambridge, pp. 117-122, January 1984.

(92) L. Longo: "VLSIIC test insert design and related implementation activities", Report No. IC83-8, VLSI Implementation Centre, Queen's University, September 1983.

(93) J. Losq: "Efficiency of random compact testing", IEEE Trans. Computers, Vol. C-27, No. 6, pp. 516-525, June 1978.

(94) D.J. Lu and E.J. McCluskey: "Recurrent test patterns", International Test Conference, Philadelphia, IEEE Cat.No. 83CH1933-1, pp. 76-82, 1983.

(95) A. Mahmood, et al.: "Concurrent fault detection using a watchdog processor and assertions", International Test Conference, Philadelphia, IEEE Cat.No. 83CH1933-1, pp. 622-627, 1983.

(96) T.E. Mangir and A. Avizienis: "Failure modes for VLSI and their effect on chip design", Proc. IEEE International Conference on Circuits and Computers, Port Chester, N.Y., pp. 685-688, October 1980.

(97) E.J. McCluskey and D.J. Lu: "Recurrent test patterns", Proc. 1983 International Test Conference, IEEE Pub 83CH1933-1, pp. 76-82, 1983

(98) E.J. McCluskey: "Built-in verification test", Proc. International Test Conference, IEEE Pub. 82CH1808-5, pp. 183-190, 1982.

(99) E.J. McCluskey and S. Bozorgui-Nesbat: "Design for autonomous test", IEEE Trans. Computers, Vol. C-30, No. 11, pp. 866-875, November 1981.

(100) K.L. Meinert: "Designing for testability with GUEST", Hewlett-Packard Journal, p. 28, March 1982.

(101) F. Motika, et al.: "An LSSD pseudo random pattern test system", International Test Conference, Philadelphia, IEEE Cat.No. 83Ch1933-1, pp. 283-288, 1983.

(102) J. Mucha: "Hardware techniques for testing VLSI circuits based on built-in test", Proc. COMPCON'81, pp. 366-369, February 1981.

(103) E.I. Muehldorf and A.D. Savkar: "LSI testing -- An overview", IEEE Trans. Computers, Vol. C-30, No. 1, pp. 1-17, January 1981.

(104) C. Mulder, et al.: "Layout and test design of synchronous LSI circuits", 1979 International Solid-State Circuits, pp. 248-249, 1979.

(105) A.F. Murray, et al.: "Self-testing in bit serial VLSI parts: High coverage at low cost", International Test Conference, Philadelphia, IEEE Cat.No. 83CH1933-1, pp. 260-268, 1983.

(106) J.C. Muzio and D.M. Miller: "Spectral techniques for fault detection", Twelfth International Symposium on Fault Tolerant Computing, Santa Monica, IEEE Cat.No. 82CH1760-8, pp. 297-302, 1982.

(107) M. Namjoo: "Techniques for concurrent testing of VLSI processor operation", International Test Conference, Philadelphia, IEEE Cat.No. 82CM1808-5, pp. 461-468, 1982.

(108) S. Nanda and S.M. Reddy: "Design of easily testable microprocessors -- A case study", Proc. International Test Conference, IEEE Pub. 82CH1808-5, pp. 480-483, 1982.

(109) T. Ogihara, et al.: "Test generation for scan design circuits with tri-state modules and bidirectional modules", IEEE Twentieth Design Automation Conference, Miami Beach, pp. 71-78, June 1983.

(110) K.P. Parker: "Software simulator speeds digital board test generation", Hewlett-Packard Journal, pp. 13-19, March 1979.

(111) R Parthasarathy and S.M. Reddy: "A testable design of iterative logic arrays", IEEE Trans. Computers, Vol. C-30, No. 11, pp. 833-841, November 1981.

(112) R.D. Patrie: "The macroassembler as a tool in LSI test pattern generation", Digest of Papers, International Test Conference, Philadelphia, pp. 457-466, November 1980.

(113) W.W. Peterson and E.J. Weldon, Jr.: Error-Correcting Codes, Cambridge, MIT Press, 1972.

(114) D.K. Pradhan and J.J. Stiffler: "Error detecting codes and self-checking circuits", Computer (IEEE), Vol. 13, No. 3, pp. 27-37, March 1980.

(115) D.K. Pradhan: "Sequential network design using extra inputs for fault detection", IEEE Trans. Computers, Vol. C-32, No. 3, pp. 319-323, March 1983.

(116) D. Radhadrishnan and G.K. Maki: "Test derivation for MOS switch logic networks", Proc. Twenty-First Allerton Conf. on Communication, Control and Computing, Illinois, pp. 786-795, October 1983.

(117) I.M. Ratui, et al.: "VICTOR -- A fast VLSI testability analysis program", IEEE Proc. International Test Conference, pp. 391-396, November 1982.

(118) D.R. Resnick: "Testability and maintainability with a new 64K gate array", VLSI Design, Vol. IV, No. 2, p. 34, March/April 1983.

(119) A.L. Rosenberg: "The Diogenes approach to testable fault-tolerant networks of processors", CS-1982-6.1, Dept. of Computer Science, Duke University, Durham, N.C., 1982.

(120) K.K. Saluja and R. Dandapani: "Comments on a data compression technique for VLSI testing", Proc. Twenty-First Allerton Conf. on Communication, Control and Computing, Illinois, pp. 796-805, October 1983.

(121) K.K. Saluja and C.S. Venkatraman: "Generalized data compression for VLSI testing", Proc. Twentieth Allerton Conf. on Communication, Control and Computing, pp. 229-230, October 1982.

(122) J. Savir: "Good controllability and observability do not guarantee good testability", IBM Research Report RC9432, June 1982.

(123) J. Savir, et al.: "Random pattern testability", IEEE Trans. Computers, Vol. C-33, No. 1, pp. 79-90, January 1984.

(124) J. Savir: "Syndrome -- Testing of 'syndrome-untestable' combinational circuits", IEEE Trans. Computers, Vol. C-30, pp. 606-608, August 1981.

(125) J. Savir: "Syndrome -- Testing design of combinational circuits", IEEE Trans. Computers, Vol. C-29, No. 6, pp. 442-451, June 1980 (+ corrections, pp. 1012-1013, November 1980).

(126) J. Savir, et al.: "Random pattern testability", Proc. Thirteenth International Fault-Tolerant Comp. Symposium, pp. 80-89, June 1983.

(127) J. Savir: "Syndrome-testable design of combinational circuits", IEEE Trans. Computers, Vol. C-29, No. 6, 442-451, June 1980.

(128) R.M. Sedmark: "Design for self-verification: An approach for dealing with testability problems in VLSI-based designs", Proc. IEEE Test Conference, pp. 112-124, October 1979.

(129) R. Seger: "The impact of testing on VLSI design methods", IEEE Journal of Solid State Circuits", Vol. SC-17, No. 3, pp. 481-486, June 1983.

(130) J.P. Shen and M.A. Schuette: "On-line self-monitoring using signatured instruction streams", International Test Conference, Philadelphia, IEEE Cat.No. 83CH1933-1, pp. 275-282, 1983.

(131) T. Shimono, et al.: "An efficient test generating algorithm based on testability measurements", Monograph of Technical Group on Design Technology of Electronics Equipment of Information Process Society Japan, 13-4, February 1982 (Japanese).

(132) M.W. Sievers and A. Avizienis: "Analysis of a class of totally self-testing functions implemented in a MOS LSI general logic structure", Proc. FTCS-10, pp 256-261, 1980.

(133) D.P. Siewiorek and L.K.-W. Lai: "Testing of digital systems", Proc. IEEE, Vol. 69, No. 10, pp. 1321-1333, October 1981.

(134) D.P. Siewiorek and R.S. Swarz: The Theory and Practice of Reliable System Design, Digital Press, Bedford, MA, 1982.

(135) K. Son and D.K. Pradhan: "Design of PLAs for testability", Digest of Papers, International Test Conference, Philadelphia, pp. 111-116, November 1980.

(136) T. Sridhar and J.P. Hayes: "A functional approach to testing bit-sliced microprocessors", IEEE Trans. Computers, Vol. C-30, No. 8, pp. 563-571, August 1981.

(137) T. Sridhar and J.P. Hayes: "Design of easily testable bit-sliced systems", IEEE Trans. Computers, Vol. C-30, No 11, pp. 842-854, November 1981.

(138) T. Sridhar, et al.: "Analysis and simulation of parallel signature analyzers", International Test Conference, Philadelphia, IEEE Cat.No. 82CH1808-5, pp. 656-661, 1982.

(139) N.R. Strader and T.J. Brosnan: "Error detection for serial processing elements in highly parallel VLSI processing architectures", Proc. 1984 Conference on Advanced Research in VLSI, Cambridge, MIT, pp. 184-193, January 1984.

(140) A.K. Susskind: "Testing by verifying Walsh coefficients", IEEE Trans. Computers, Vol. C-32, No. 2, pp. 198-201, February 1983.

(141) A.K. Susskind: "Survey of VLSI test strategies", Proc. IEEE Custom IC Conference, Rochester, N.Y., IEEE Pub.84CH1987-7, pp. 276-280, May 1984.

(142) A.K. Susskind: "Testing by verifying Walsh coefficients", Proc. Eleventh Annual Symposium on Fault-Tolerant Computing, Portland, MA, pp. 206-208, June 1981.

(143) S.M. Thatte, et al.: "An architecture for testable VLSI processors", Proc. International Test Conference, IEEE Pub. 82CH1808-5, pp 484-492, November 1982.

(144) S.M. Thatte and J.A. Abraham: "Test generation for microprocessors", IEEE Trans. Computers, Vol. C-29, No. 6, pp. 429-441, June 1980.

(145) U. Theus and H. Leutiger: "A self-testing ROM device", Digest of Technical Papers of the ISSCC'81, New York, pp. 176-177, February 1981.

(146) F.F. Tsui: "In-situ testability design (ISTD) -- A new approach for testing high-speed LSI/VLSI logic", Proc. IEEE, Vol. 70, No. 1, pp. 59-78, January 1982.

(147) P. Varma, et al.: "On-chip testing of embedded RAMs", Proc. 1984 CICC, Rochester, N.Y., IEEE Pub.84CH1987-7, pp. 286-290, 1984.

(148) J. Wakerly: Error-Detecting Codes, Self-Checking Circuits and Applications, New York, North-Holland, 1978.

(149) F.C. Wang and D.D. Bhavsar: "A bus-organized self-test processor architecture", Proc. IEEE Custom IC Conference, Rochester, N.Y., IEEE Pub.84CH1987-7, pp. 164-167, May 1984.

(150) I. Watson, et al.: "ICTES: A unified system for functional testing and simulation of digital ICs", Proc. Cherry Hill International Test Conference, Philadelphia, PA, 1982.

(151) D. Westcott: "The self-assist test approach to embedded arrays", Digest of Papers of the IEEE International Test Conference, pp. 203-207 1981.

(152) T.W. Williams: "Design for testability: What's the motivation?", VLSI Design, Vol, IV, No. 6, pp. 21-23, October 1983.

(153) T.W. Williams and K.P. Parker: "Design for testability -- A survey", IEEE Trans. Computers, Vol. C-31, No. 1, pp. 2-15, January 1982.

(154) T.W. Williams and K.P. Parker: "Design for testability -- A survey", Proc. IEEE, Vol. 71, No. 1, pp. 98-112, January 1983.

(155) T.W. Williams and K.P Parker: "Testing logic networks and design for testability", Computer (IEEE), Vol. 12, No. 10, pp. 9-21, October 1979.

(156) S. Yajima and T. Aramaki: "Autonomously testable programmable logic arrays", FTCS-11 Digest of Papers, pp. 41-43, 1981.

(157) T. Yamada, et al.: "TESTA-1: Test generation system using state diagram tracing approach", NEC Research and Development, No. 65, pp. 68-73, April 1982.

(158) A Yamada, et al.: "Automatic system level test generation and fault location for large digital systems", Proc. Fifteenth Design Automation Conference, pp. 347-352, June 1978.

(159) Y. You and J.P. Hayes: "A self-testing dynamic RAM chip", Proc. 1984 Conference on Advanced Research in VLSI, MIT, Cambridge, pp. 159-168, January 1984.

Author Index

Subject Index

Editor's Biography

Mohamed I. Elmasry (S'69–M'73–SM'79) received the B.Sc. degree from Cairo University, Cairo, Egypt, and the M.A.Sc. and Ph.D. degrees from the University of Ottawa, Ontario, Canada, all in electrical engineering.

He has worked in the area of digital integrated circuit and system design for the last 20 years. He has been with the Department of Electrical Engineering, University of Waterloo, Waterloo, Ontario, Canada, since 1974, where he is a Professor and founding Director of the VLSI Research Group. He has a cross appointment with the Department of Computer Science at the University of Waterloo. He has served as a consultant to research and development laboratories of companies in Canada and the U.S., including Bell Laboratories, GE, Ford Microelectronics, Linear Technology, Xerox, and BNR, in the area of LSI/VLSI digital circuit/subsystem design. During 1980–1981 he was at the Micro Components Organization, Burroughs Corporation, San Diego, CA, on leave from the University of Waterloo. He has authored and coauthored over 70 papers on integrated circuit design, including contributions to bipolar emitter-function-logic (EFL), I^2L, single-device-well (SDW) MOSFET's, and current mode NMOS. He has several patents to his credit. He is the Editor of the IEEE Press book *Digital MOS Integrated Circuits* (1981), and the author of the book *Digital Bipolar Integrated Circuits* (John Wiley, 1983).

Dr. Elmasry has served in many professional organizations in different positions and is currently the Chairman of the Technical Advisory Committee of the Canadian Microelectronics Corporation, a founding member of the Canadian VLSI Conference, and the founding President of PicoElectronics Inc. He is a member of the Association of Professional Engineers of Ontario.